Analytical Methods in CORROSION SCIENCE AND ENGINEERING

CORROSION TECHNOLOGY

Editor
Philip A. Schweitzer, P.E.
Consultant
York, Pennsylvania

Corrosion Protection Handbook: Second Edition, Revised and Expanded, edited by Philip A. Schweitzer

Corrosion Resistant Coatings Technology, Ichiro Suzuki

Corrosion Resistance of Elastomers, Philip A. Schweitzer

Corrosion Resistance Tables: Metals, Nonmetals, Coatings, Mortars, Plastics, Elastomers and Linings, and Fabrics: Third Edition, Revised and Expanded (Parts A and B), Philip A. Schweitzer

Corrosion-Resistant Piping Systems, Philip A. Schweitzer

Corrosion Resistance of Zinc and Zinc Alloys: Fundamentals and Applications, Frank Porter

Corrosion of Ceramics, Ronald A. McCauley

Corrosion Mechanisms in Theory and Practice, edited by P. Marcus and J. Oudar

Corrosion Resistance of Stainless Steels, C. P. Dillon

Corrosion Resistance Tables: Metals, Nonmetals, Coatings, Mortars, Plastics, Elastomers and Linings, and Fabrics: Fourth Edition, Revised and Expanded (Parts A, B, and C), Philip A. Schweitzer

Corrosion Engineering Handbook, edited by Philip A. Schweitzer

Atmospheric Degradation and Corrosion Control, Philip A. Schweitzer

Mechanical and Corrosion-Resistant Properties of Plastics and Elastomers, Philip A. Schweitzer

Environmental Degradation of Metals, U. K. Chatterjee, S. K. Bose, and S. K. Roy

Environmental Effects on Engineered Materials, edited by Russell H. Jones

Corrosion-Resistant Linings and Coatings, Philip A. Schweitzer

Corrosion Mechanisms in Theory and Practice: Second Edition, Revised and Expanded, edited by Philippe Marcus

Electrochemical Techniques in Corrosion Science and Engineering, Robert G. Kelly, John R. Scully, David W. Shoesmith, and Rudolph G. Buchheit

Metallic Materials: Physical, Mechanical, and Corrosion Properties, Philip A. Schweitzer

Corrosion Resistance Tables: Metals, Nonmetals, Coatings, Mortars, Plastics, Elastomers and Linings, and Fabrics: Fifth Edition, Philip A. Schweitzer

Corrosion of Ceramic and Composite Materials, Second Edition, Ronald A. McCauley

Analytical Methods in Corrosion Science and Engineering, Philippe Marcus and Florian Mansfeld

ADDITIONAL VOLUMES IN PREPARATION

Analytical Methods in CORROSION SCIENCE AND ENGINEERING

Edited by

Philippe Marcus
Florian Mansfeld

CRC Taylor & Francis
Taylor & Francis Group

Boca Raton London New York Singapore

A CRC title, part of the Taylor & Francis imprint, a member of the Taylor & Francis Group, the academic division of T&F Informa plc.

Published in 2006 by
CRC Press
Taylor & Francis Group
6000 Broken Sound Parkway NW, Suite 300
Boca Raton, FL 33487-2742

Printed in the United States of America on acid-free paper
10 9 8 7 6 5 4 3 2 1

International Standard Book Number-10: 0-8247-5952-4 (Hardcover)
International Standard Book Number-13: 978-0-8247-5952-0 (Hardcover)
Library of Congress Card Number 2005009633

Library of Congress Cataloging-in-Publication Data

Analytical methods in corrosion science and technology / Edited by Philippe Marcus and Florian Mansfeld.
p. cm.
Includes bibliographical references and index.
ISBN-13: 978-0-8247-5952-0 (alk. paper)
ISBN-10: 0-8247-5952-4 (alk. paper)
1. Corrosion and anti-corrosives. I. Marcus, P. (Philippe), 1953- II. Mansfeld, Florian.

TA418.74.A44 2005
620.1'1223--dc22 2005009633

T&F informa

Taylor & Francis Group
is the Academic Division of T&F Informa plc.

Visit the Taylor & Francis Web site at
http://www.taylorandfrancis.com

and the CRC Press Web site at
http://www.crcpress.com

Preface

In order to understand the mechanisms of corrosion, solve the corrosion problems encountered in service conditions, and improve the corrosion resistance of commonly used materials as well as materials of the future, the use of modern surface analytical and electrochemical techniques is essential.

Corrosion phenomena take place at the surface of materials exposed to chemically aggressive environments. Not only the initiation of corrosion, but also its propagation is essentially a surface chemical and electrochemical process. Therefore no significant progress can be achieved in this area without the use of modern surface chemical and structural characterization methods combined with electrochemical techniques.

Our aim is to provide the readers, for the first time, with a book in which both the major surface analytical techniques and the electrochemical techniques are presented, discussed, and illustrated by selected examples of applications to corrosion science and engineering.

In order to cover the large number of surface analytical and electrochemical techniques used for corrosion science and engineering, this book contains 19 chapters. The surface analytical techniques provide the necessary information on the surface chemical composition, structure, topography, defects, mechanical and electronic properties, in close relationship with the kinetic data on corrosion that are obtained by electrochemical techniques.

Chapter 1 to Chapter 8 deal with the presentation of the major surface analytical methods. The principles, the instrumentation, and the exact nature of the information derived from the measurements are presented. Electron spectroscopies (x-ray photoelectron spectroscopy and Auger electron spectroscopy), secondary ion mass spectrometry (SIMS), megaelectronvolt (MeV) ion beam analysis, and nanoprobe microscopies (STM and AFM) are described. A chapter is dedicated to the use of the synchrotron methods. Two chapters are focused on optical spectroscopies: infrared spectroscopy and glow discharge optical emission spectroscopy. Chapter 9 presents the recent developments in the application of radiotracer methods, and Chapter 10 describes and discusses the evaluation of mechanical properties of surface films by nanoindentation and nanoscratching.

Chapter 11 to Chapter 19 deal with topics that range from an introduction to electrochemical instrumentation to traditional dc techniques, well-established ac techniques such as electrochemical impedance spectroscopy, and newer techniques such as electrochemical noise analysis. Other chapters discuss local techniques such as the scanning electrode technique, the Kelvin probe technique, and various microelectrochemical techniques that can be used to study the details of anodic and cathodic processes that lead to corrosion phenomena. The chapter dealing with electrochemical instrumentation should be a welcome addition to the corrosion literature because it makes the user aware of the requirements for accurate and meaningful electrochemical measurements and illustrates the errors that can occur in such measurements and how they might be avoided. The other chapters give a thorough background of each experimental technique and illustrate the use of that technique for a number of different corrosion systems including practical applications in industry.

The contributors are from major universities and national research laboratories in Europe, the United States, Canada, and Japan. They are all internationally recognized experts in corrosion science, materials science, surface chemistry and electrochemistry, and surface analysis.

Philippe Marcus
Paris, France

Florian Mansfeld
Los Angeles, California

About the Editors

Philippe Marcus, Ph.D., is Director of Research at CNRS (Centre National de la Recherche Scientifique) and Director of the Laboratory of Physical Chemistry of Surfaces at Ecole Nationale Supérieure de Chimie de Paris, France. The author/co-author of over 300 professional papers, book chapters, and books in the areas of corrosion science, surface analysis, surface chemistry and electrochemistry, and materials science, he serves on the editorial boards of the *Corrosion Science Journal, Materials and Corrosion*, and *Corrosion Engineering, Science, and Technology*. He is the editor of *Corrosion Mechanisms in Theory and Practice*. Currently, he is the chairman of the European Federation of Corrosion Working Party on Surface Science and Mechanisms of Corrosion and Protection and the chairman of Division 4 of the International Society of Electrochemistry on Electrochemical Materials Science. Dr. Marcus earned his Ph.D. (1979) in chemistry from the Université Pierre et Marie Curie, Paris, France.

Florian Mansfeld, Ph.D., is Professor of Materials Science and Engineering and Director of the Corrosion and Environmental Effects Laboratory at the University of Southern California. He is the author of over 400 professional papers in the areas of corrosion science and technology, editor of eight books, and holds nine patents in the area of corrosion control. Dr. Mansfeld is a Fellow of The Electrochemical Society and of NACE. His honors and awards include the Whitney Award of NACE and the H.H. Uhlig Award of The Electrochemical Society. Dr. Mansfeld earned his Ph.D. (1968) in physical chemistry from the University of Munich, Germany.

Contributors

Arne Bengtson
Swedish Institute for Metals Research
Stockholm, Sweden

H. Böhni
Swiss Federal Institute of Technology
Zürich, Switzerland

James E. Castle
University of Surrey
School of Engineering
Guildford, United Kingdom

Francesco Di Quarto
Dipartimento di Ingegneria Chimica dei Processi e dei Materiali
Università di Palermo
Palermo, Italy

G.S. Frankel
Fontana Corrosion Center
The Ohio State University
Columbus, Ohio

Michael J. Graham
Institute for Microstructural Sciences
National Research Council Canada
Ottawa, Ontario, Canada

Dirk Lützenkirchen-Hecht
Institut für Materialwissenschaften und Fachbereich Physik
Bergische Universität Wuppertal
Wuppertal, Germany

Ewald Heitz
Karl Winnacker Institute of DECHEMA e.V.
Frankfurt, Germany

G. Horányi
Institute of Chemistry
Chemical Research Center
Hungarian Academy of Sciences
Budapest, Hungary

Francois Huet
Centre National de la Recherche Scientifique
Laboratoire Interfaces et Systèmes Electrochimiques
Université Pierre et Marie Curie
Paris, France

Magnus Johnson
Department of Materials Science and Engineering
Division of Corrosion Science
Royal Institute of Technology
Stockholm, Sweden

E. Kálmán
Institute of Chemistry
Chemical Research Center
Hungarian Academy of Sciences
Budapest, Hungary

D. Landolt
Materials Department
Swiss Federal Institute of Technology
Lausanne, Switzerland

P. Leblanc
Fontana Corrosion Center
The Ohio State University
Columbus, Ohio

Christofer Leygraf
Department of Materials Science and Engineering
Division of Corrosion Science
Royal Institute of Technology
Stockholm, Sweden

R. Scott Lillard
Material Science and Technology Division
Los Alamos National Laboratory
Los Alamos, New Mexico

Florian Mansfeld
Corrosion and Environmental Effects Labs
Dept. of Materials Science and Engineering
University of Southern California
Los Angeles, California

Philippe Marcus
Laboratoire de Physico-Chimie des Surfaces
Centre National de la Recherche Scientifique
Ecole Nationale Supérieure de Chimie de Paris
Paris, France

Vincent Maurice
Laboratoire de Physico-Chimie des Surfaces
Centre National de la Recherche Scientifique
Université Pierre et Marie Curie
Ecole Nationale Supérieure de Chimie de Paris
Paris, France

N. Stewart McIntyre
Surface Science Western and Department of Chemistry
Western Science Centre
The University of Western Ontario
London, Ontario, Canada

C.-O.A. Olsson
Materials Department
Swiss Federal Institute of Technology
Lausanne, Switzerland

M. Rohwerder
Max Planck Institut für Eisenforschung
Department of Interface Chemistry and Surface Engineering
Düsseldorf, Germany

Monica Santamaria
Dipartimento di Ingegneria Chimica dei Processi e dei Materiali
Università di Palermo
Palermo, Italy

Carl-Albrecht Schiller
Zahner-elektrik
Kronach, Germany

Didier Schmaus
Institut des Nanosciences de Paris
Université Deuis Diderot
Centre National de la Recherche Scientifique
Paris, France

Masahiro Seo
Laboratory of Interfacial Electrochemistry
Graduate School of Engineering
Hokkaido University
Sapporo, Japan

M. Stratmann
Max Planck Institut für Eisenforschung
Department of Interface Chemistry and Surface Engineering
Düsseldorf, Germany

Hans-Henning Strehblow
Institut für Physikalische Chemie
Heinrich Heine Universität
Düsseldorf, Germany

C. Sunseri
Dipartimento di Ingegneria Chimica dei Processi e dei Materiali
Università di Palermo
Palermo, Italy

Thomas Suter
Swiss Federal Institute of Technology
Zürich, Switzerland

Ian Vickridge
Institut des Nanosciences de Paris
Centre National de la Recherche Scientifique
Université Pierre et Marie Curie
Paris, France

Contents

1
X-Ray Photoelectron Spectroscopy in Corrosion Research

Hans-Henning Strehblow
Institut für Physikalische Chemie, Heinrich-Heine-Universität, Düsseldorf, Germany

Philippe Marcus
Laboratoire de Physico-Chimie des Surfaces, Centre National de la Recherche Scientifique, Ecole Nationale Supérieure de Chimie de Paris, Paris, France

Contents

1.1 INTRODUCTION

Corrosion is traditionally subdivided into two main topics, high-temperature corrosion with the attack of solid surfaces by hot and aggressive gases and aqueous corrosion with the degradation of materials in contact with aqueous electrolytes. Even in the case of atmospheric corrosion the surface of materials is exposed to thin electrolyte layers sometimes with successive wet and dry periods. Most important are metal surfaces although corrosion may also happen to nonmetallic materials such as semiconductors, insulators, and even polymers. The methods to study corrosion depend decisively on the systems. Aqueous corrosion was traditionally a domain of electrochemical research and various electrochemical methods have been applied to the study of corrosion of metals in electrolytes. These methods provide a good qualitative and quantitative insight. However an unambiguous interpretation of the chemistry and kinetics as well as the mechanisms of the observed processes requires a very detailed information on the surface of the materials and its changes. For these reasons there exists a strong demand for the application of surface analytical methods. X-ray photoelectron spectroscopy (XPS) is a very successful surface analytical tool for corrosion research but also for the investigation of corrosion failures in industry, related to various environments. Although XPS requires a sample transfer from the electrolyte to the ultrahigh vacuum (UHV) with a loss of the contact to the electrolyte and the control of the electrode potential, it provides reliable data on the chemical situation of the surface. This information is required to get a sound base for the interpretation of the mechanisms of corrosion processes and their kinetics.

XPS involves the ionization of atoms on solid surfaces (or species in the gas phase) by absorption of photons. The use of an x-ray source (AlKα or Mg Kα) permits the ejection of electrons from all electronic levels, especially core levels (within the energy range of the x-ray beam). The energy spectrum of the ejected electrons is characteristic for the elements involved in this process. In 1954 Kai Siegbahn and his group developed a high-resolution electron spectrometer, which permitted the precise determination of the energy of XPS signals, i.e. the energy of the core levels where the electrons come from. In 1958 the same Swedish group detected the chemical shift of XPS signals, i.e. the influence of the charge of atoms and their chemical environment on their energy. This discovery was a decisive step to develop this process to an analytical method for the study of the chemistry of solid surfaces and surface layers. Therefore the method was called electron spectroscopy for chemical analysis (ESCA), synonymous with XPS. In 1981 Kai Siegbahn received the Nobel Prize in physics for his pioneering work. Up to the mid-1970s XPS was a surface method applied to very fundamental studies of solids in contact with vacuum and low-pressure gas phase. When commercial spectrometers became available with an efficient and fast entry lock for specimens to the UHV, it developed to a widely applied method for fundamental and applied studies. Nowadays XPS is an essential method to study surfaces that have been exposed to gaseous and liquid environments as, for example, in catalysis, electrode kinetics, materials science, and corrosion. It is indispensable for many fields in research and industry including the optimization and control of products.

The chemical composition of a metal surface may be obtained by qualitative and quantitative XPS analysis. One thus need not rely on thermodynamic arguments, which are questionable if the kinetics are dominating the corrosion processes. If the active dissolution of an alloy is examined, for appropriate systems the amount of dissolved cations may be investigated qualitatively and quantitatively by the analysis of the electrolyte or with a rotating ring disc electrode with a sufficient time resolution down to ca. 1 s. However, a preferential dissolution of one or some of the components will cause the accumulation of the others at the metal surface, leading to a characteristic elemental distribution. The related distribution profiles are best investigated with XPS, often together with depth profiling by argon ion sputtering. If the metal surface forms an anodic surface film its composition and composition profile require surface analysis, which yields the fractions of the anions and cations even for multilayer structures. For a simple homogeneous anodic film forming on a binary alloy, one needs the determination of at least four fractions for two cations and two anions (oxide and hydroxide) for the surface layer and two atomic fractions for the metal surface underneath. The number of variables increases accordingly if the cations may be present in more than one oxidation state. Furthermore the surface film has often a bilayer or multilayer structure or a characteristic depth profile, which have to be evaluated to understand the corrosion properties of the system under study. This increases further the number of variables. All these problems cannot be solved with a simple application of electrochemical methods and they require the use of surface analytical methods like XPS to fully understand the corrosion system under investigation.

XPS has several advantages in comparison to other methods. It is a very soft method. Most specimens are not affected by the x-rays and they may be examined with quasi *in situ* conditions with no change of the composition of their surface and even with conservation of the oxidation state of the specimens. For this purpose an appropriate specimen preparation and transfer is an important aspect to avoid any uncontrolled changes due to the exposure of the specimen to the laboratory atmosphere, as discussed in Section 1.2.5.1. Other methods using electrons or ions as a probing beam like Auger electron spectroscopy (AES) or ion scattering spectroscopy (ISS) might be more harmful. An important feature is the chemical shift of XPS signals, which provides information on the oxidation state and the chemical binding of the species. The information depth of XPS is related to the escape depth of the photoelectrons, which is in the range of a few nm, thus sampling near surface layers and thin anodic films of this thickness. XPS has been widely used since the mid-1970s to solve various problems in research and industry. This chapter will give an introduction to XPS and its application to various corrosion problems.

1.2 METHOD AND EQUIPMENT

1.2.1 Principles of XPS

XPS involves the irradiation of a specimen with x-rays. The absorption of a high-energy x-ray photon causes the ionization of an electronic level. A part of the photon's energy is used for the ionization, while the excess appears as the kinetic energy of the photoelectron. Depending on the ionization energy of the electronic level the kinetic energy is large or small. The energy spectrum of the electrons is taken with an electrostatic energy analyzer and there appears a peak for each electronic level. Usually for the excitation the radiation of an Al- or Mg-coated x-ray anode is used. The Al Kα and the Mg Kα radiations with energies $h\nu = 1486.6$ and 1253.6 eV have a line width of 1.0 and 0.8 eV, respectively. This line width restricts in many cases the spectral resolution. It can be improved by an x-ray monochromator to 0.5 eV or less with a strong loss of the intensity of the exciting beam. This loss can be compensated for modern spectrometers by the use of magnetic lenses and of more sensitive detectors. Solid specimens of any kind may be measured like metals, semiconductors, and insulators. Gases can also be analyzed. The XPS signals of gases are usually very sharp due to the sharp electronic levels of free atoms and molecules. The core levels of species of a solid specimen are sharp as well whereas those of electrons from the outer levels are broadened due to an overlap of their orbitals and the formation of bands like the conduction and the valence bands. Figure 1.1 depicts the principle of the excitation of core levels with x-ray photons $h\nu$, leading to sharp peaks in the spectrum of the photoelectrons. If the excitation of electrons occurs from the valence band of a solid a broad signal reflects the broad energy distribution of their levels as also shown in Figure 1.1. All binding energies E_B of species of solid materials are given relative to the Fermi level E_F whereas for gases the energies are given relative to the vacuum level, i.e., ionization energies are used instead. Figure 1.2 depicts schematically the energetic relation of a specimen and the energy analyzer of the spectrometer with a simple model of two potential wells.

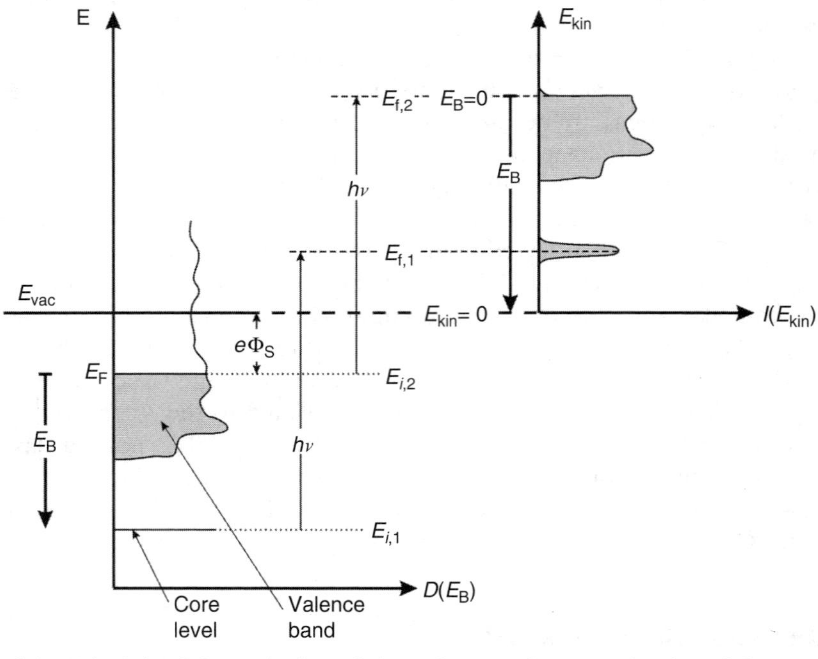

Figure 1.1 Principle of the excitation of photoelectrons from core levels and the conduction band of a metal. E_{vac}, vacuum level; E_F, Fermi level; E_i and E_f, initial and final states of electrons; $e\Phi_S$, work function of sample; E_B, binding energy, E_{kin}, kinetic energy of electrons, $D(E_B)$, density of states; $I(E_{kin})$, related XPS-intensity. The difference of the work functions of the sample and the electrostatic energy analyzer has been neglected in this figure for simplicity.

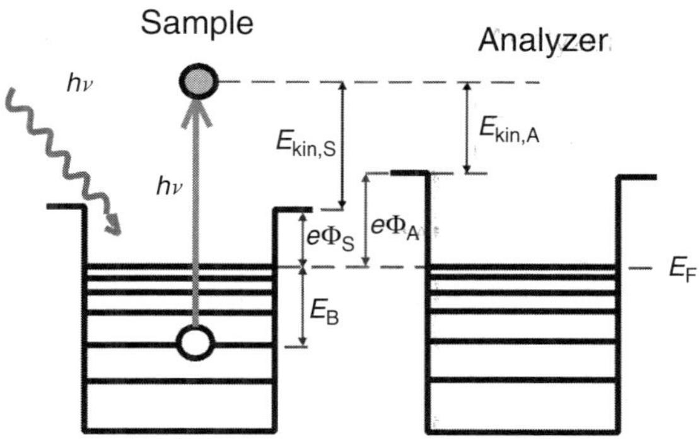

Figure 1.2 Energy diagram for experimental XPS, $e\Phi_S$ and $e\Phi_A$, work function of sample and analyzer; $E_{kin,S}$ and $E_{kin,A}$, kinetic energy with reference to the vacuum level of the sample and the analyzer.

The electron is ejected from its level by absorption of an x-ray photon to the vacuum with the kinetic energy $E_{kin,S}$ and $E_{kin,A}$ with respect to the sample and the analyzer. These different values are a direct consequence of the difference of the work functions $e\Phi_S$ and $e\Phi_A$ of the specimen and the analyzer, which causes a contact potential difference and thus two different vacuum levels. Equation (1.1) describes the energy balance for the photoelectrons which can be taken directly from Figure 1.2. This relation shows that with a known energy $h\nu$ of the radiation source the measured kinetic energy $E_{kin,A}$ and the work function $e\Phi_A$ one may calculate E_B. Usually the constant work function of the analyzer is calibrated by measurements of well-characterized standards and then compensated internally. Thus $e\Phi_A$ may be omitted which simplifies Equation (1.1) appropriately

$$h\nu = E_B + E_{kin,\,A} + e\Phi_A \tag{1.1}$$

According to the above principles, each occupied level of a species will generate a signal in the energy spectrum of the photoelectrons. The survey spectrum thus shows the various peaks of the different species located in a surface near range and in most cases several peaks for each species. Figure 1.3 gives an example for an Fe/Si alloy

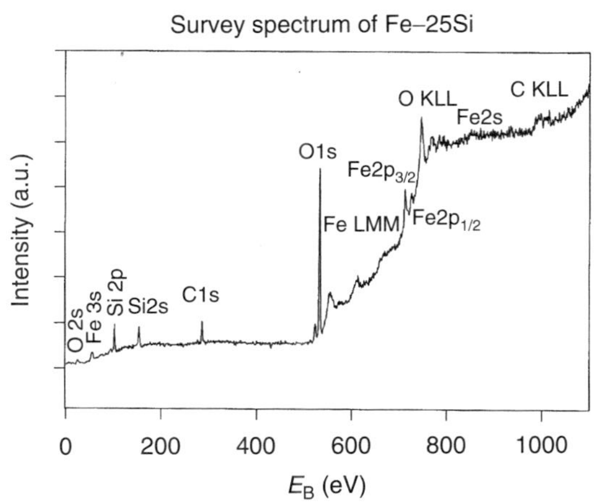

Figure 1.3 Example of XPS survey spectrum with XPS and Auger signals of Fe–25Si alloy passivated in acetate buffer solution at pH = 5.0 for 300 s at E = 0.9 V.

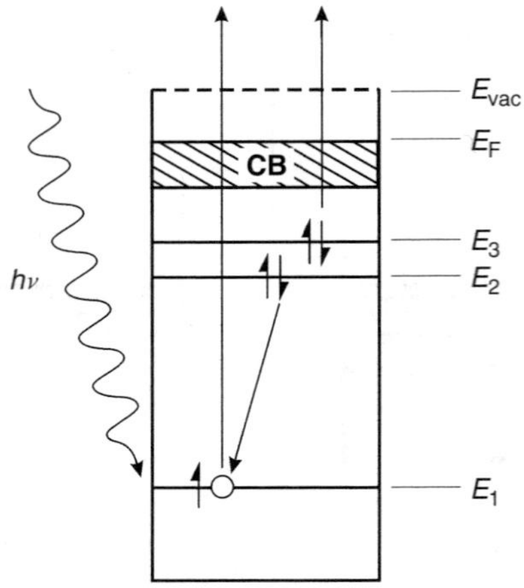

Figure 1.4 Principle of Auger electron emission with involved energy levels E_1 to E_3. CB, conduction band with Fermi level E_F.

containing the contributions of Fe, Si, and O of a thin oxide film at the surface and the metal underneath. This spectrum also contains Auger lines. When the absorption of an x-ray photon causes the ionization of a core level (E_1) an outer electron of energy E_2 will fall in this free deep level. The energy difference E_2-E_1 of these two levels may cause x-ray fluorescence or as a competing process the emission of a further electron from a sufficiently high level E_3 (Figure 1.4). The kinetic energy of the electron emitted from E_3 is $E_{1,2,3} = E_3-E_2-E_1$. For example, if the levels 1, 2, and 3 correspond to energy levels K, L, and L, respectively, the kinetic energy of the Auger electron emitted in the KLL transition will be denoted as E_{KLL}. As for electron-excited Auger spectroscopy three electron levels are involved, which partially explains the complicated and broad structure of the Auger lines. X-ray-induced AES provides additional information contained in XP-spectra, which may be used for the analysis of surface species. In some cases, it serves as an additional source for chemical information as will be discussed in Section 1.3.2.5.

The labeling of the XPS signals follows the description of the levels from where the ejected electrons come from. With exception of the s-levels, all levels are doublets due to the spin–orbit splitting. A missing electron of a fully occupied p-, d-, or f-level is equivalent to an orbital occupied by one electron only with an appropriate orbital quantum number. The spin–orbit interaction causes the energetic splitting to a doublet, i.e., for the Fe 2p signal to $2p_{3/2}$ and $2p_{1/2}$ (Figure 1.3) or for the Mo 3d signal to $3d_{5/2}$ and $3d_{3/2}$. The energy of the level with the higher total quantum number is higher with a smaller binding energy due to the parallel orientation of the orbital quantum number and the spin quantum number. Therefore $2p_{1/2}$ electrons have a smaller kinetic energy in comparison to the $2p_{3/2}$. The integrated intensity of the $2p_{1/2}$ signal is half of that of $2p_{3/2}$ due to the difference of the energetic degeneration of both terms ($2j+1$) (Figure 1.5a).

During the emission of a photoelectron one may get an additional excitation of a second electron. The related distribution of the energy $h\nu$ to both processes causes a shift of the photoelectron to larger binding energy, which yields a satellite peak. If the second electron is also ejected the satellite is called a shake-off peak, if it is only excited to a higher electronic level it is called a shake-up peak. Shake-up satellites are observed for divalent Cu ions (Figure 1.5b). They have only partially filled d-levels, which may accept an excited electron from anions of an appropriate compound.

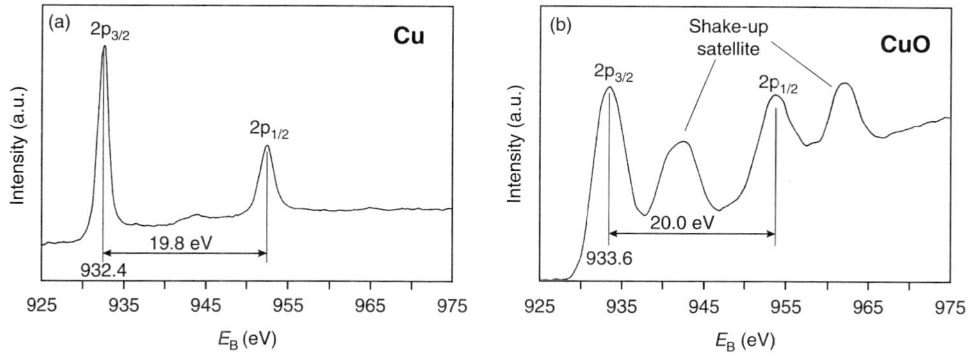

Figure 1.5 Cu 2p spectrum showing (a) the $2p_{3/2}$ and $2p_{1/2}$ doublet for Cu and (b) the $2p_{3/2}$ and $2p_{1/2}$ doublet and the shake-up satellites of CuO.

Cu-metal and Cu(I)-ions have a filled d-shell so that a shake-up to these levels is not possible. Therefore XPS spectra of CuO or $Cu(OH)_2$ layers exhibit a shake-up satellite of the Cu 2p signal due to additional excitation of electrons from O^{2-} or OH^- ions to the d-levels of Cu^{2+} (Figure 1.5b) whereas those of monovalent Cu-compounds and Cu-metal have no satellite (Figure 1.5a).

XPS peaks often show a pronounced tailing. This is a consequence of the excitation of electrons within the valence band during the emission of the photoelectron (Figure 1.6a). If the sample is an insulator or a semiconductor with a large bandgap ΔE_g the intensity of the XPS signal goes back to the level at the leading edge with a smaller binding energy for several eV and then increases (see, for example, Figure 1.6c). This feature mirrors the bandgap of the material (Figure 1.6b). For metals, electrons may be excited with any energy and no gap is observed. This effect is responsible for the asymmetry observed on the high binding energy side of core level spectra of metals. These features of the peak shape are important for the background subtraction that must be done for a quantitative evaluation of XPS data.

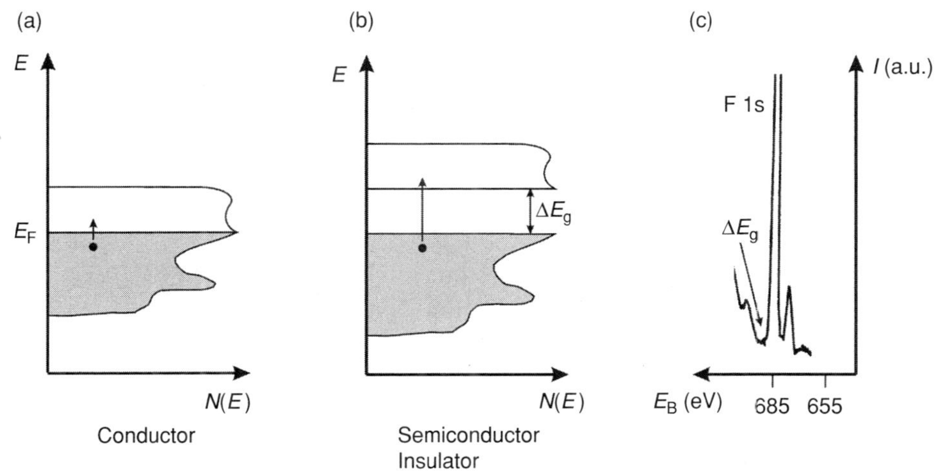

Figure 1.6 The effect of the band situation on the background. (a) Additional excitation of an electron within the conduction band of a conductor, (b) effect of the bandgap of a semiconductor or an insulator, and (c) example of an insulator (LiF).

1.2.2 Chemical Shift

A major property of XPS is the chemical shift, which opens up the possibility of obtaining detailed information on the chemical states of surface species. Due to the charge of the absorbing atom and the polarity of its bond the energy of the photoelectrons is submitted to a slight variation. A positive charge increases the binding energy and a negative charge lowers it up to several eV. One therefore observes a whole sequence of XPS energies associated with the different binding energies of the surface atoms. Figure 1.7 shows an example for an Fe specimen covered with an anodic oxide layer. Besides the metal signal at $E_B = 707\,eV$ one obtains an Fe(II) and an Fe(III) signal shifted to 709.5 and 710.5 eV, respectively [1]. Depending on the amount of species at the surface, the intensity of the related peaks is changing. The Fe(0) contribution decreases with increasing layer thickness due to the attenuation of the photoelectrons of the metal substrate by an increasing layer thickness on top. The chemical shift for the Fe-ions is relatively small with a large overlap of the signals. A quantitative evaluation requires a careful decomposition of the Fe $2p_{3/2}$ spectra by curve fitting (or peak synthesis) as described in Section 1.2.4.1. Another example with a much better separation of the peaks is given in Figure 1.8 for the C 1s signal of an organic compound [2]. Four types of carbon signals are obtained with F, O, and only H as ligands. With increasing electronegativity and number of the ligands the peaks are shifted by up to 8 eV [2]. The peaks are well separated and are of equal intensity due to the same number for each type of C atom in the compound. The peak energies of many organic and inorganic compounds are published in the literature so that one can decide on an empirical basis which oxidation number and binding situation is found for the species on the surface of a specimen [3,4].

This short introduction gives an insight into the qualitative interpretation of XP spectra and the determination of the chemical composition of surfaces and surface layers. The interpretation of corrosion phenomena requires also a quantitative

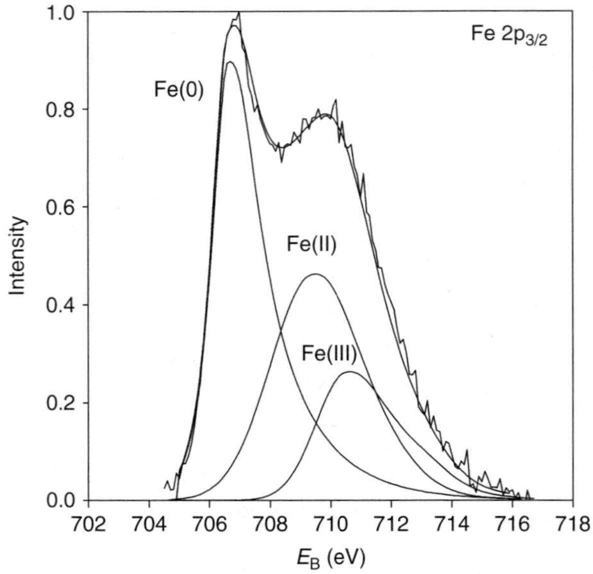

Figure 1.7 Chemical shift of the Fe $2p_{3/2}$ signal for Fe covered with a thin anodic layer containing Fe(II) and Fe(III) [1].

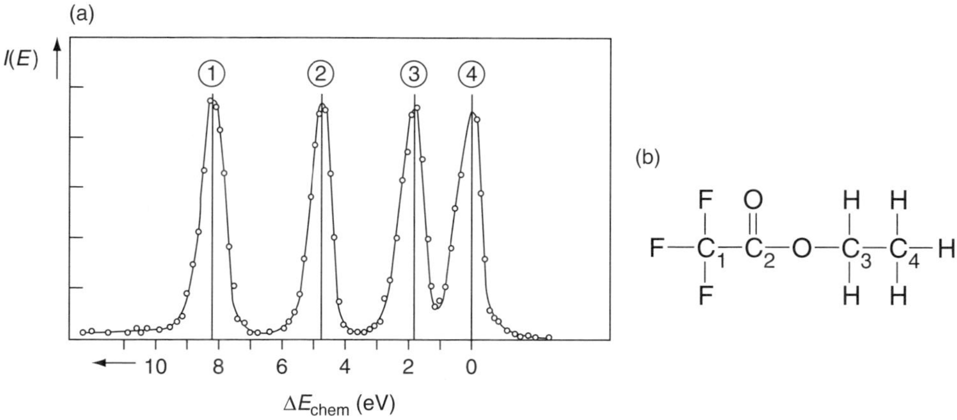

Figure 1.8 Chemical shift of C 1s (a) in ethylfluoroacetate (b) [2].

evaluation of the composition of a material surface, which will be discussed in Section 1.2.4.

1.2.3 UPS and Band Structure

Photoelectrons may be excited also by radiations in the UV range, which may be obtained by a UV source supplied with He or Ne gas. In most cases the HeI or HeII radiation is used with energies of 21.1 and 40.8 eV, respectively. These energies permit only the excitation of photoelectrons from the conduction band of metals or the valence band of semiconductors or insulators. The core levels of most elements may not be ionized due to the relatively small energy of the radiation. UV photoelectron spectroscopy (UPS) provides a detailed information on the density of states of the conduction or valence band. Figure 1.9a gives an example for a sputter-cleaned Cu specimen. The spectrum starts at the Fermi edge with $E_B = E_F = 0\,\text{eV}$ and ends at the cutoff edge of the secondary electrons. Electrons from the Fermi edge have the largest kinetic energy whereas those at the cutoff edge have lost all their kinetic energy due to inelastic processes with the electrons of the metal before they enter the vacuum and thus keep only the ionization energy, i.e., the work function $e\Phi_s$ of the metal specimen. The energy width ΔE of the whole UP-spectrum thus contains $e\Phi_W$ ($e\Phi_W = e\Phi_s$ for metals) according to Equation (1.2). An equivalent equation (Equation (1.3)) for semiconductors contains the threshold energy $e\Phi_{Th}$, i.e., the energy difference of the upper valence band edge and the vacuum level.

Since for semiconductors the Fermi level is in the bandgap, the photoelectrons with the highest kinetic energy are excited from the upper valence band edge. Therefore for semiconductors $e\Phi_{Th} \neq e\Phi_s$

$$\Delta E = h\nu - e\Phi_W \tag{1.2}$$

$$\Delta E = h\nu - e\Phi_{Th} \tag{1.3}$$

Equation (1.3) is very important for the determination of the energy of the conduction and valence bands within the scales related to the vacuum or the standard hydrogen electrode (SHE). These aspects will be discussed for the case of passive copper in Section 1.3.2.5. A correct application of UPS for the determination of work functions and threshold energies requires the application of a negative bias to

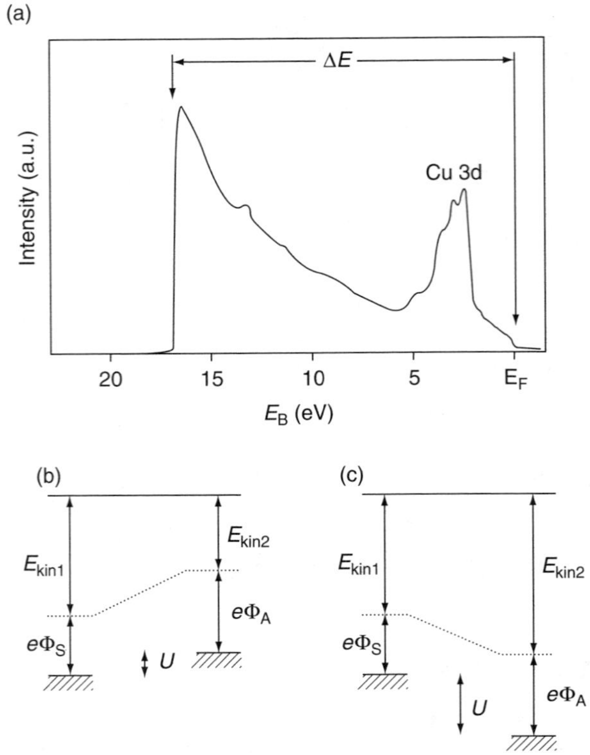

Figure 1.9 (a) UP-spectrum of a sputter-cleaned Cu surface. (b) and (c) Schematic energy diagrams for UPS with the kinetic energy E_{kin} of photoelectrons and the work functions of the sample and the analyzer, $e\Phi_S$ and $e\Phi_A$, with positive (b) and negative (c) bias U of the sample relative to the energy analyzer, explaining the necessary effect of a negative bias on the contact potential difference to get an undisturbed spectrum at the cutoff edge.

the specimen, usually ca. 5 V, to overcome the contact potential difference between the specimen and the analyzer (Figure 1.9b and c). The secondary electrons thus do not lose energy during their transfer in vacuum from the specimen surface to the analyzer. The work function of the analyzer may be larger than that of the specimen. Figure 1.9a also shows the well-pronounced structure of the d-band of Cu. An emersed Cu electrode contains additional structures that correspond to electronic states of adsorbed water or anodic oxide films (see Figure 1.22).

1.2.4 Quantitative XPS

1.2.4.1 Background Subtraction, Reference Spectra (Standards), and Peak Fitting

A quantitative evaluation of XPS signals requires an appropriate background subtraction and separation of the spectral components by curve fitting. Several methods of background subtraction have been proposed. A linear background sets a line between the base levels of the leading and falling edges of the peak. The method of Shirley [5] assumes that the background increases with the binding energy proportionally to the part of the signal immediately before a given energy. This idea assumes that the background increases by the additional excitation of electrons from the occupied to empty levels of the conduction band of a metal and from the valence to the conduction band of a semiconductor as described in Section 1.2.1. Procedures of background subtraction and the corresponding computer softwares have been improved with time. An additional method is that of Tougaard [6]. The method used depends usually on the specimen and the specific analysis to be performed. The Shirley background subtraction is the most widely used.

The subsequent separation into signal components by peak fitting requires the knowledge of the energy position and shape of the individual peaks. A good procedure involves the measurements of XP-spectra of well-characterized standards. For surface studies of electrochemical corrosion problems a reliable procedure starts with the preparation of the standards under well-controlled electrochemical conditions with subsequent appropriate specimen transfer to prevent any chemical changes. As an example the calibration of the peak of Fe(II) hydroxide requires a specimen preparation and transfer without contact to the laboratory atmosphere. Experimental details of the equipment and procedures for a specimen handling without air contact are described in Section 1.2.5. Usually $Fe(OH)_2$ is immediately oxidized to Fe(III)–hydroxide or –oxide when oxygen has access to the specimen surface. A previous Ar-ion sputtering of the Fe sample ensures the removal of any Fe(III)-or Fe(II)-residues in order to start with a clean sample surface before the growth of a Fe(II)–hydroxide layer in an alkaline solution such as 1 *M* NaOH. A reliable Fe(III)-free $Fe(OH)_2$-film as a Fe(II)-standard may be grown best by 5-min reduction of anodic oxide at $E = -1.0$ V in 1 *M* NaOH, which has been grown before at $E = 0.34$ V for 5 min. A 5% Cr content of the metal ensures a reduction to $Fe(OH)_2$ only whereas a film on pure Fe will be reduced to Fe-metal [7]. For the above-described Fe(II)-standard the contribution of the Fe-metal signal has to be subtracted because its XPS signal is only partially absorbed by the thin Fe(II)-layer of only a few nm. A sputtered Fe specimen serves as a metal standard which also permits a reliable correction of the Fe(II) standard. An appropriate Fe(III)-standard is formed by anodic oxidation of Fe in 1 *M* NaOH at $E = 0.34$ V where no Fe^{2+} contribution is observed. Controlled oxidation of metals under O_2 in the preparation chamber attached to the spectrometer is also performed to prepare well-defined reference materials to be used as standards for recording XPS reference spectra (e.g., NiO [8,9], Cr_2O_3 [10], or Al_2O_3 [11]).

The standard peaks are usually described by Gauss–Lorenz curves to which a tailing function is added. Figure 1.10a gives an example for the Fe-species and Figure 1.10b and c for Si and SiO_2 and the various oxygen species which may be present on a passivated Fe/Si alloy [12]. All peaks are sufficiently separated so that a quantitative evaluation of an actual passivated Fe/Si specimen is possible. For this procedure the shape and energy position of the standard peaks are preserved and only their intensity is changed to fit the measured data.

Another possibility is the use of commercial powder of those compounds whose presence at the surface is suggested. These often electronically insulating specimens have the disadvantage of a larger surface charging during XPS analysis with a related broadening of the signals. Other problems are the deviation of the chemical composition at the surface relative to the bulk and the surface contamination by carbonaceous species. The electrodeposition of compounds is another possibility. The electrodeposition of NiOOH on Pt or Au by oxidation from a weakly acidic Ni^{2+} solution is an example [9]. This Ni(III) standard is transformed to a $Ni(OH)_2$ film by its reduction in alkaline solutions. These few examples are only a small selection of possibilities which have been realized. The position and shape of standard signals may in some cases depend on the matrix. Thus it may be necessary to form an oxide

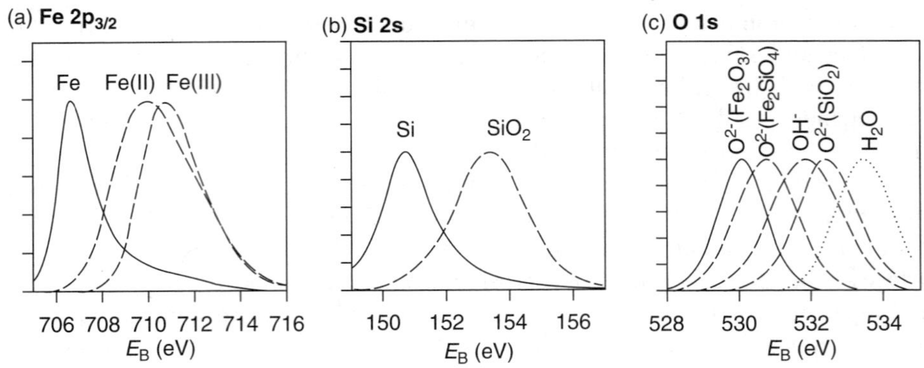

Figure 1.10 Reference spectra (standards) for (a) Fe $2p_{3/2}$ (Fe, Fe(II), and Fe(III)), (b) Si 2s (Si, Si(IV)), and (c) O 1s (species as indicated). Standards: sputter-cleaned Fe and Si, reduced anodic oxide ($Fe(OH)_2$, passive layer of Fe at positive potentials in 1 *M* NaOH), SiO_2 on Si wafer, anodic oxide on Fe50Si [12].

layer on the surface of the alloy under study instead of using pure metals as this includes the effect of the alloying elements.

As shown above, the evaluation of the XPS data of actual specimens is done in keeping constant the energy position and the shape of the standard peaks, and varying their intensity relative to each other to fit the measured spectrum. The integrated intensity of these signals is a measure of the amount of the related species.

1.2.4.2 Intensity Ratios and Layer Composition

The integrated intensity $I_{i,\infty}$ of a species present in a thick sample with homogeneous composition is given by Equation (1.4). k describes the special spectrometer settings including the x-ray flux, the analyzed area, and the yield of the detector. T is the transmission function of the energy analyzer, σ_i the photoionization cross-section, λ is the attenuation length of the electrons in the analyzed material, and D_i the density of the species i at the surface or within the surface layer. In order to become independent of k, intensity ratios are used, which are sufficient for the quantitative determination of the surface composition. The transmission function T and its dependence on the kinetic energy of the photoelectrons depend on the analyzer and its settings. Equation (1.5) gives the expression of T for the VG ESCALAB Mk2 spectrometer. The transmission functions of new spectrometers using a combination of magnetic and electrostatic lenses are more complex. Therefore they are provided in the software of commercial instruments, such as the Kratos Axis Ultra or the VG ESCALAB 250.

The photoionization cross-section σ_i is published in the tables of Scofield [13]. For the attenuation length of the electrons, λ_i, the empirical equation (1.6) proposed by Briggs and Seah is often used. It assumes a square root dependence on the kinetic energy of the electrons [14,15]. The influence of the matrix is considered approximately by a factor B with a different value for elements, inorganic compounds, and organic compounds. Theoretical and experimental values of attenuation lengths of electrons in pure elements and in selected compounds are also available [16,17]

$$I_{i,\infty} = kT\sigma_i\lambda_i D_i \tag{1.4}$$

$$T = 0.01\omega^2 E_p^{1.5} E_{kin}^{-0.5} (\text{VG ESCALAB Mk2}) \tag{1.5}$$

where E_p is the pass energy of the analyzer and ω is the angle of acceptance of electrons in the analyzer.

$$\lambda = B\sqrt{E_{kin}}. \tag{1.6}$$

where $B = 0.054\,\text{nm/eV}^{1/2}$ for elements, $B = 0.097\,\text{nm/eV}^{1/2}$ for inorganic compounds, and $B = 0.087\,\text{nm/eV}^{1/2}$ for organic compounds.

Intensity ratios of surface species have the advantage that specific spectrometer settings cancel. If the intensity ratios of two different elements are considered, the intensity ratio is given by Equation (1.7) (in which k has disappeared). $\sqrt{E_{kin}}$ from the transmission function T and λ will also cancel in the case of a VG ESCALAB spectrometer. Equation (1.7) still contains the photoionization cross-sections of both elements. In this way the quantitative composition of a surface may be obtained from XPS intensities:

$$\frac{I_A}{I_B} = \frac{\sigma_A T_A \lambda_A D_A}{\sigma_B T_B \lambda_B D_B} \tag{1.7}$$

If A and B correspond to different oxidation states of one element, k, T, λ, and σ in Equation (1.2) will cancel so that I_A/I_B is equal to the concentration ratio D_A/D_B.

1.2.4.3 Angular-Resolved XPS (ARXPS) and the Analysis of Thin Films

The contribution of the atoms to the XPS signal decreases with the depth of their position. The attenuation length of the photoelectrons λ is in the range of a few nm, with an exponential decay of the intensity with the length of the path the electrons have to travel within the layer to reach the vacuum. Therefore the outer part of a thin film contributes most to the signals. Equation (1.8) describes the signal I_{ox}^{Me} of a metal cation, within an oxide layer as a function of its thickness d with a self-attenuating factor (square brackets). The oxide layer is considered to be continuous (i.e., covering completely the surface) and its thickness is assumed to be the same on the whole surface. $I_{ox,max}^{Me}$ is the maximum intensity for a thick oxide film, i.e., with a thickness which exceeds ca. 3λ. The term $1/\cos\Theta$ reflects the increase of the path of the photoelectrons within the specimen with increasing emission angle Θ. Θ is the angle between the surface normal and the direction in which the photoelectrons are analyzed. It may be varied systematically by a tilt of the specimen relative to the direction of the entrance slit of the energy analyzer. For a thin film on top of the substrate or a surface layer the expression of the intensity of the substrate includes an exponential term that describes the attenuation of the signal by the surface layer. Equation (1.9) describes the attenuation of the signal of a metal substrate by a thin anodic oxide of thickness d:

$$I_{ox}^{Me}(\Theta) = I_{ox,\ max}^{Me}\left[1 - \exp\left(-d/\lambda_{ox}^{Me}\cos\Theta\right)\right] \tag{1.8}$$

$$I_{Me}^{Me}(\Theta) = I_{Me,\ max}^{Me}\exp\left(-d/\lambda_{ox}^{Me}\cos\Theta\right) \tag{1.9}$$

$$\frac{I_{ox}^{Me}(\Theta)}{I_{Me}^{Me}(\Theta)} = \frac{D_{ox}^{Me}}{D_{Me}^{Me}}\,\frac{\left[1 - \exp\left(-d/\lambda_{ox}^{Me}\cos\Theta\right)\right]}{\exp\left(-d/\lambda_{ox}^{Me}\cos\Theta\right)} \tag{1.10}$$

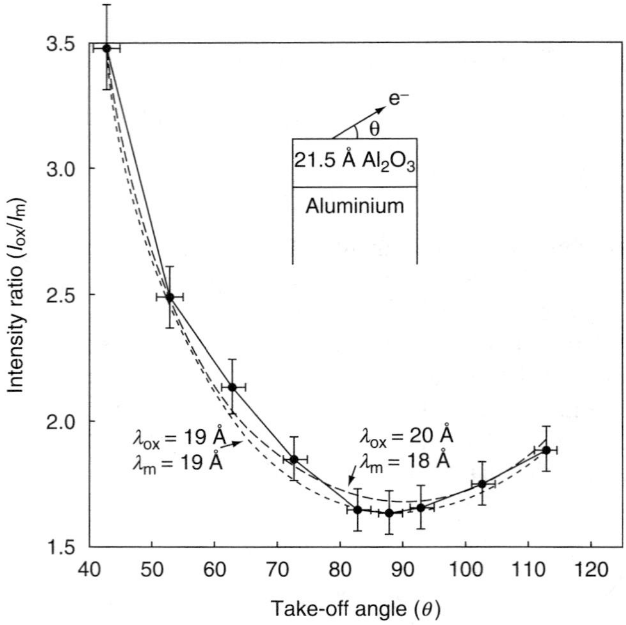

Figure 1.11 Intensity ratio I_{ox}/I_m versus take-off-angle of photoelectrons (angle between the surface and the direction in which the electrons are analyzed) for a thin layer of Al_2O_3 (21.5 Å) on Al [17]. The points are the experimental data, and the dotted lines are the theoretical curves obtained with different values of the attenuation lengths of the photoelectrons in the metal (λ_m) and in the oxide (λ_{ox}).

$$d = \lambda_{ox}^{Me} \cos\Theta \ln\left[\frac{I_{Ox}^{Me} D_{Me}^{Me}}{I_{Me}^{Me} D_{Ox}^{Me}} + 1\right] \tag{1.11}$$

As already mentioned, intensity ratios are convenient because they are independent of the special spectrometer settings. As an example the intensity ratio of the signals of the cation of an oxide layer I_{ox}^{Me} relative to that of the metal substrate I_{Me}^{Me} is given in Equation (1.10). This equation can be rearranged to Equation (1.11), which yields the layer thickness as a function of the intensity ratio and the related take-off angle. As an example, Figure 1.11 shows the intensity ratio of the Al 2p signals from Al^{3+} and Al (I_{ox}/I_m) versus the take-off angle Θ for thin Al_2O_3 layers (21.5 Å) on Al [17] (Note that here Θ, the take-off angle, is the angle between the surface and the direction in which the photoelectrons are analyzed.) The increase of the take-off angle Θ causes a decrease of the intensity ratio due to the position of the oxide layer on top of the substrate.

A more complex example is the intensity ratio of two species within two layers on top of each other. Figure 1.12 gives an example for Fe passivated in 1 *M* NaOH with a duplex film with an Fe(II)-containing inner and Fe(III)-containing outer part [7]. These studies require a specimen transfer via a noble gas atmosphere. Traces of oxygen would oxidize immediately Fe(II) to Fe(III), which would alter the results completely. The related equipment and procedures will be described in Section 2.5.1. The combination of two relations of the type of Equation (1.8) for the cations in the two layers with thicknesses d_1 and d_2, including an attenuation factor for the inner Fe(II) by the outer Fe(III)-containing layer yields Equation (1.12) which describes quantitatively the results of Figure 1.12:

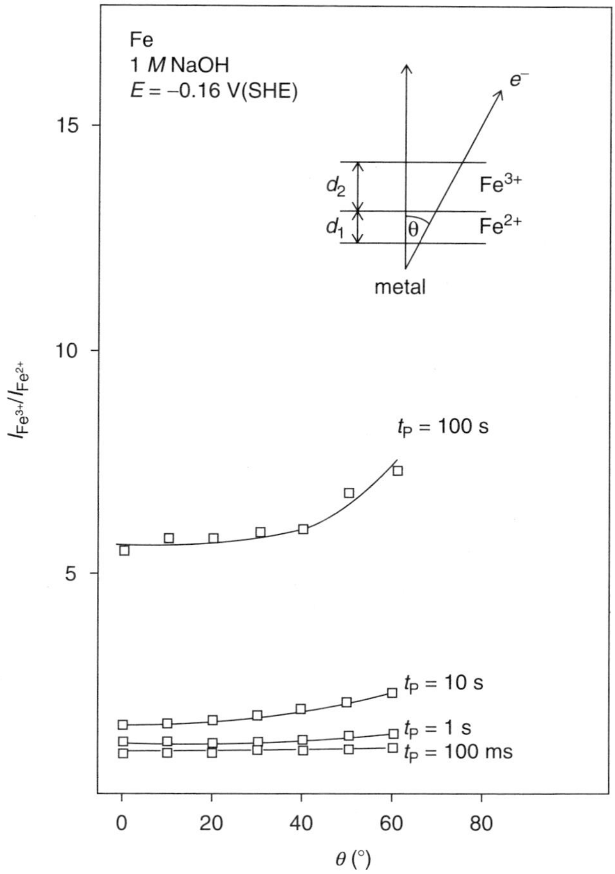

Figure 1.12 Intensity ratio $I_{Fe^{3+}}/I_{Fe^{2+}}$ versus emission angle Θ (angle between the surface normal and the direction in which the electrons are analyzed) for Fe passivated in 1 *M* NaOH for 100 ms up to 100 s as indicated, development of Fe(III) on top of Fe(II) with time of passivation [7].

$$\frac{I_{\mathrm{Fe(III)}}}{I_{\mathrm{Fe(II)}}} = \frac{D_{\mathrm{Fe(III)}}\left[\exp\left(\frac{d_2}{\lambda_{\mathrm{ox}}\cos\Theta}\right) - 1\right]}{D_{\mathrm{Fe(II)}}\left[1 - \exp\left(\frac{-d_1}{\lambda_{\mathrm{ox}}\cos\Theta}\right)\right]} \tag{1.12}$$

Short oxidation times of less than 1 s yield a very small Fe(III) to Fe(II) ratio. Apparently only Fe(II) has been formed, but no Fe(III) at this point. No change of the intensity ratio with Θ is observed. The intensity ratio is growing with time due to the formation of Fe(III), and a layered structure evolves for $t > 1$ s with the outer position of Fe(III) which is detected by the upward bending of the intensity ratio with Θ. This kind of ARXPS study is usually part of a detailed XPS investigation of the growth of passive layers. Such data lead to the interpretation in terms of a bilayer or multilayer structure of the film. Indeed passive layers on most pure metals and alloys show a multilayer structure. This has been also confirmed by XPS sputter depth profiles and in some cases by ISS. However, a bilayer or multilayer structure might be too simple. In various cases, a continuous distribution of the cations and anions is more appropriate. Valencies of layer components are difficult to follow by sputter depth profiles because of the reduction of the oxides due to preferential sputtering of oxygen (see next section). Therefore for very thin-films ARXPS is the appropriate method to detect a gradient of cations with different oxidation states.

1.2.4.4 Depth Profiling

The combination of sputtering of a surface by noble gas ions with surface analytical techniques such as XPS, AES, or ISS provides profiles of the elemental distribution within surfaces and surface layers. Usually a beam of argon ions with an energy in the range of 0.1 keV to several keV is used. With a homogeneous lateral distribution of the ions within the beam a homogeneous removal of the material may be achieved. Scanning of a focused beam on the specimen surface causes also a laterally homogeneous sputter process with time. XPS is then used to analyze the surface, at fixed intervals of sputtering time, to determine its elemental composition. As a result the elemental composition as a function of depth, called the sputter depth profile, is obtained. This method is often used to determine the elemental profiles in surface layers and thin films. The layer thickness is related to the sputter time. However, the absolute determination of the thickness requires a calibration of the sputter rate of the investigated material. Such a calibration requires the knowledge of the thickness of the standards. This is in most cases not available. Therefore the thickness scale of sputter depth profiles is often relative to a common standard, which may be easily tested in most laboratories. For this purpose anodically grown Ta_2O_5 on Ta is an accepted standard [18]. The profiles of Ta-oxide have the form of steps with sharp slopes at their beginning and end. Usually the layer thickness is defined as the distance of the two points of inflexion of the plateau in between. The measured depth profile of the standard and the related sputter rate correspond to the special settings of the ion source of the spectrometer. The same setting should be used for the calibration with the Ta_2O_5 standard and the profile of the actual specimen. In some cases the sputter profiles of well-known oxide standards have been measured relative to Ta-oxide and thus may serve to obtain an absolute depth scale with factors that are specific for the layer composition. This was done for e.g. Al_2O_3 [19].

As an example, a depth profile of Alloy 690 (Ni-30% Cr-8%Fe) with an oxide layer formed in high-temperature water is given in Figure 1.13 [20]. Here the thickness calibration was performed by nuclear reaction analysis (NRA). The profile clearly reveals an enrichment of chromium in the inner part of the oxide layer.

There are a few drawbacks to this method. The major one is the preferential sputtering of the light elements like oxygen in an oxide film, possibly causing a reduction of oxide and hydroxide layers, and thus an unwanted change of the valency of the cations. These possible changes of the oxidation state make the determination by ion sputtering of the in-depth distribution of cations with different oxidation states questionable. Preferential sputtering of components of a specimen surface will change the composition or their oxidation state. However, elements of similar mass are sputtered at the same rate so that a surface modification by preferential sputtering is not a problem for alloys containing elements of similar masses.

Other problems may affect depth profiles. The nonhomogeneous concentration profile within the ion beam causes a laterally varying removal of material, especially at the boundary of the sputtered area, and layers of different depths will contribute to the analysis. This crater effect may be avoided by sputtering a larger and analyzing a smaller area in the center, thus avoiding the contribution of the side walls. In this respect, sputter depth profiling by electron-excited AES is preferable, because the electron beam can be easily focussed on a much smaller area.

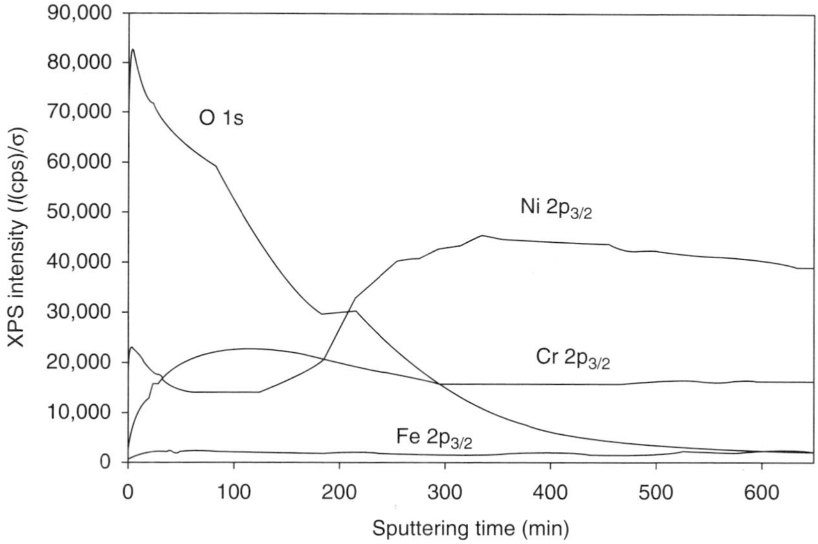

Figure 1.13 Sputter depth profile of the oxide layer formed on Alloy 690 (Ni–30Cr–8Fe) in high-temperature water (325°C, 50 h) showing the enrichment of chromium in the inner part of the oxide layer [20].

Ion mixing and recoil implantation are additional sources of problems in depth profiling experiments. Finally surface and interface roughening may occur. This may cause profiles with broad slopes at the end of plateaus even though the real profile might be sharp. For reactive metals, reoxidation of the metallic surface by the residual gas in the UHV chamber can occur and broaden the profile.

1.2.4.5 Ion Scattering Spectroscopy as a Complementary Method

The backscattering of noble gas ions from a solid surface may be used as an additional surface analytical method, which is extremely surface-sensitive. It allows to analyze qualitatively and quantitatively the outmost surface layer and thus is complementary to other methods like XPS and AES. It may be easily performed in an XPS spectrometer equipped with a focused scannable ion source with a well-defined primary energy of the noble gas ions (E_0). The energy spectrum of the backscattered ions is obtained with the electrostatic energy analyzer with inversed polarity of the hemispheres to allow the passage of the positively charged noble gas ions instead of electrons.

Backscattering of noble gas ions from a solid surface with a primary energy E_0 in the range of 1 to 3 keV is a binary collision process. This elastic collision causes an energy loss, which depends on the masses of the noble gas ions, m, and of the atoms in the target surface, M. A backscattering by $\Theta = 90°$ relative to the direction of the primary beam leads to a simple equation (the following equation) for the ratio E/E_0 of the energies of the noble gas ions after and before the collision:

$$\frac{E}{E_0} = \frac{M - m}{M + m} \tag{1.13}$$

For each target mass M a peak in the energy spectrum of the backscattered ions will appear. The special property of ISS is its unique surface sensitivity. Only those noble gas ions, which are backscattered from the outmost atomic layer, can keep their charge and thus may pass the electrostatic energy analyzer. All ions that are backscattered from deeper atomic layers are completely neutralized. Only target atoms with larger masses than the noble gas ions cause backscattering. He is used if oxygen is to be measured, Ne for heavier atoms, especially if their masses are close to each other and if their peaks have to be separated sufficiently as in the case of Fe and Cr or Cu and Ni. Examples for ISS depth profiles for these alloys are given in Sections 1.3.3.1 and 1.3.3.2. After calibration with well-characterized standards of known composition ISS is sufficiently quantitative to get the elemental composition of the outmost surface layer. Together with a soft sputter process, one obtains profiles of the composition of thin layers with a high depth resolution of one atomic layer.

1.2.5 Equipment

1.2.5.1 Specimen Preparation and Transfer

Surface analysis requires a careful specimen preparation to avoid artefacts and thus wrong conclusions on the corrosion process. For a specimen taken from its environment, this involves a careful cut to meet the special requirements of ca. 1 cm diameter without surface contamination. Most solid specimens may be introduced into the vacuum and analyzed without problems. There should not be large amounts of gaseous or easily vaporizable components at the surface, which could contaminate the UHV chambers. Small concentrations of degassing material of a sample may be removed by a sufficiently long storage in the fast entry lock, where the vacuum conditions are less critical. Soft ion sputtering often removes unwanted contaminations from the specimen surface to get information on the composition underneath. A fracture of the specimen in the UHV may provide an oxide-free metal surface with no contamination and a surface composition which is not affected by artefacts like preferential sputtering. Fracture stages with liquid nitrogen cooling are commercially available. They are often used for grain boundary segregation measurements after intergranular fracture. A large variety of specimens exposed to different environments has been analyzed in the past. Such analyses greatly helped to get reliable conclusions on problems in the production process or on the cause of environmental failure. Many surface analytical laboratories provide service analysis with the help and the expertise of their coworkers.

However, there are also special situations that require a special specimen treatment. For studies on corrosion mechanisms, it might be important to protect lower valent species from air oxidation, to prevent unwanted film growth after electrochemical specimen preparation, and to start electrochemical experiments with a well-characterized surface composition. Corrosion and film growth studies of reactive metals and alloys should start with an oxide-free surface with well-known composition. This may be best achieved by Ar-ion sputtering. XPS analysis is then used to control the surface conditions before further electrochemical measurements and preparation steps are performed. A specimen transfer within a closed system, which avoids any air contact, can then be used. The designs of several systems have been published in literature, which have been applied for electrochemical and corrosion studies [21–27] including commercial transfer vessels that may be mounted on the spectrometer and moved to a separate chamber such as a glove box for electrochemical experiments. Figure 1.14 depicts an equipment which has been applied for several studies in the laboratory of one of the authors [26,27]. It consists of a

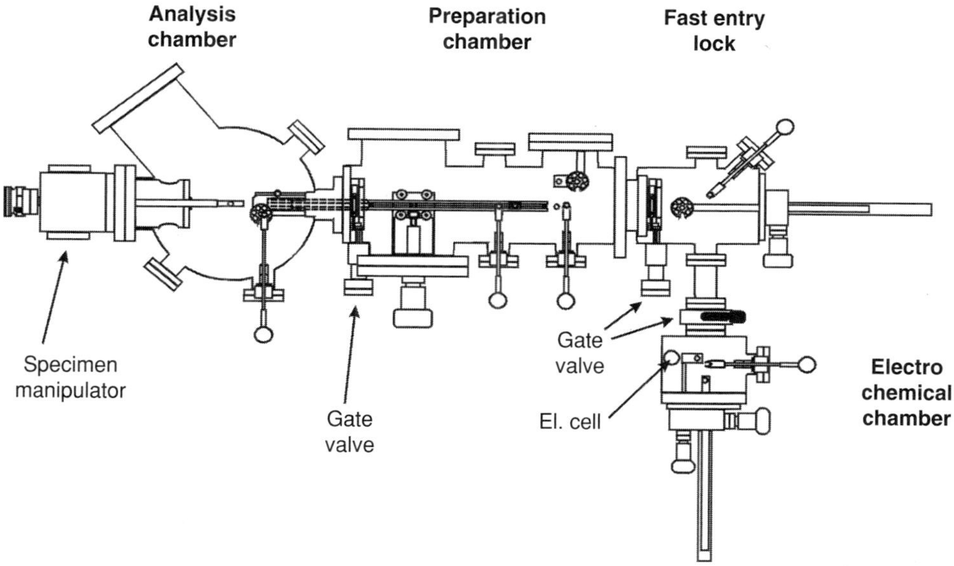

Figure 1.14 Schematics of an XPS instrument with an attached electrochemical chamber for transferring samples without air exposure [26, 27].

commercial three-chamber system, with the analysis and preparation chamber and the fast entry lock and an electrochemical chamber attached to it. The chambers are separated by valves. A rack and pinion drive system and wobble sticks at the ports permit an effective specimen transfer mounted on stubs between these chambers. The electrochemical chamber contains a small electrolyte vessel with a volume of ca. $2\,cm^3$ equipped with a Pt counter electrode and an electrolytic contact to a reference electrode outside of the vessel. Containers with the necessary electrolytes and pure water for cleaning and rinsing are connected *via* tubing and stopcocks. All electrolytes and water are purged with purified argon and smaller quantities may be taken to smaller containers and purged again before use.

A standard specimen preparation starts with an Ar-ion sputter cleaning within the preparation chamber, which may be controlled subsequently in the analysis chamber. The sample is then transferred *via* the fast entry lock to the electrochemical chamber, which is filled with a purified Ar atmosphere and fixed to a rotary drive. The sample surface is brought into contact with the electrolyte with a hanging meniscus geometry under potentiostatic conditions. Usually a negative potential is applied during immersion followed by a change to a potential of interest by pulse technique for a preset time in the range of milliseconds to hours. Finally the electrode is disconnected and emerged. The disconnection of the specimen from the potentiostatic circuit *via* a relay or stepping of the potential to a value with no electrochemical reaction permits passivation transients for very short times and negative potentials. Thus time-resolved oxidation studies become possible with a resolution of a millisecond. Also the reduction and further oxidation of anodically grown oxide films may be followed. Similarly active metal surfaces may be studied including their electrochemical double layer. It should however be mentioned that reactive metals will form a thin oxide film even if pretreated at very negative potentials due to water decomposition at open circuit potentials. Noble and seminoble metals like Cu and Ag may be kept free of any oxide during emersion and transfer via the Ar-atmosphere [26,28].

1.2.5.2 Energy Analyzers

Different electrostatic analyzers have been developed, the retarding mirror analyzer, the cylindrical mirror analyzer, and the spherical sector analyzer. The latter is used in modern XPS spectrometers together with magnetic and electrostatic lenses, which focus the photoexcited electrons to the entrance slit of the analyzer. Figure 1.15a depicts schematically the main parts of an XPS spectrometer. In addition to their imaging properties, the electrostatic lenses apply a negative bias. The electrons ejected by the photons of the x-ray source have to overcome this retarding voltage to get between the two hemispheres of the spherical sector analyzer. A preset voltage allows the passage of the electrons to the detector with a fixed pass energy. This mode of operation is called the constant pass energy mode. Usually a pass energy $E_{pass} =$ 100 eV is used for survey spectra and $E_{pass} = 20$ eV or less for obtaining a high-resolution spectrum over a narrower energy region. The retardation voltage is changed linearly with time to allow the photoelectrons with different kinetic energies to pass across the analyzer when a spectrum is recorded. The size of the entrance slit of the analyzer and the electrostatic lens may restrict the analyzed surface area of the specimen but the analyzed area can also be determined by the x-ray spot produced by a monochromator with focussing properties. The electron detector consists of a set of discrete channeltrons or a channel plate. This improves the signal and thus the detection limit of the species at the surface. The spectra recorded with the different channeltrons of an array or a plate have to be shifted properly and added to each other by a computer to obtain the XPS spectrum.

Imaging XPS. Modern spectrometers provide imaging facilities. A schematic representation of a spectrometer with a parallel imaging system is shown in Figure 1.15b. The same analyzer is used for spectroscopy (Figure 1.15a) and imaging. The photoelectrons are collected from the whole field of view simultaneously. A combination of lenses before and after the energy analyzer focus the photoelectrons on a channel plate detector. The hemispherical analyzer permits electrons with only a narrow band of energy to reach the detector. With such a system, images are produced without scanning the x-rays or the photoelectrons. In another design, a scanning system inside the analyzer input lens is used to raster the virtual image of the analyzed area. A spherical sector analyzer is added to the spectrometer to provide imaging facility. In a third design, the exciting electron beam in the x-ray source is scanned over the anode. The x-rays are then focussed by a monochromator onto the sample surface and, as a result, the focussed x-ray beam is scanned over the sample, and an image of the analyzed surface can be obtained. This design requires the use of an ellipsoidal (instead of the conventional spherical) diffracting crystal in the x-ray monochromator. The spatial resolution achieved with the new commercial instruments (VG ESCALAB 250, Kratos Ultra Axis, Physical Electronics Quantum 2000) is claimed to be from 3 μm upward.

X-Ray Sources. The x-ray source contains a filament as an electron source, which is run with ca. 20 mA and an acceleration voltage of ca. 15 keV. The x-ray cathode is a Cu cylinder, which is cooled inside with water and covered with Mg or Al. Usually a twin anode is used with an Al layer at one side and a Mg layer at the other side. Two filaments directly opposite to the related part of the anode permit the choice of the Al- or Mg-Kα x-ray source. The relatively small line width of these radiations, 1.0 and 0.8 eV, respectively, and their sufficiently large energies of $h\nu =$ 1486.6 and 1253.6 eV are appropriate for most applications in XPS. The energy resolution may be significantly improved by a monochromator (to ca. 0.5 eV),

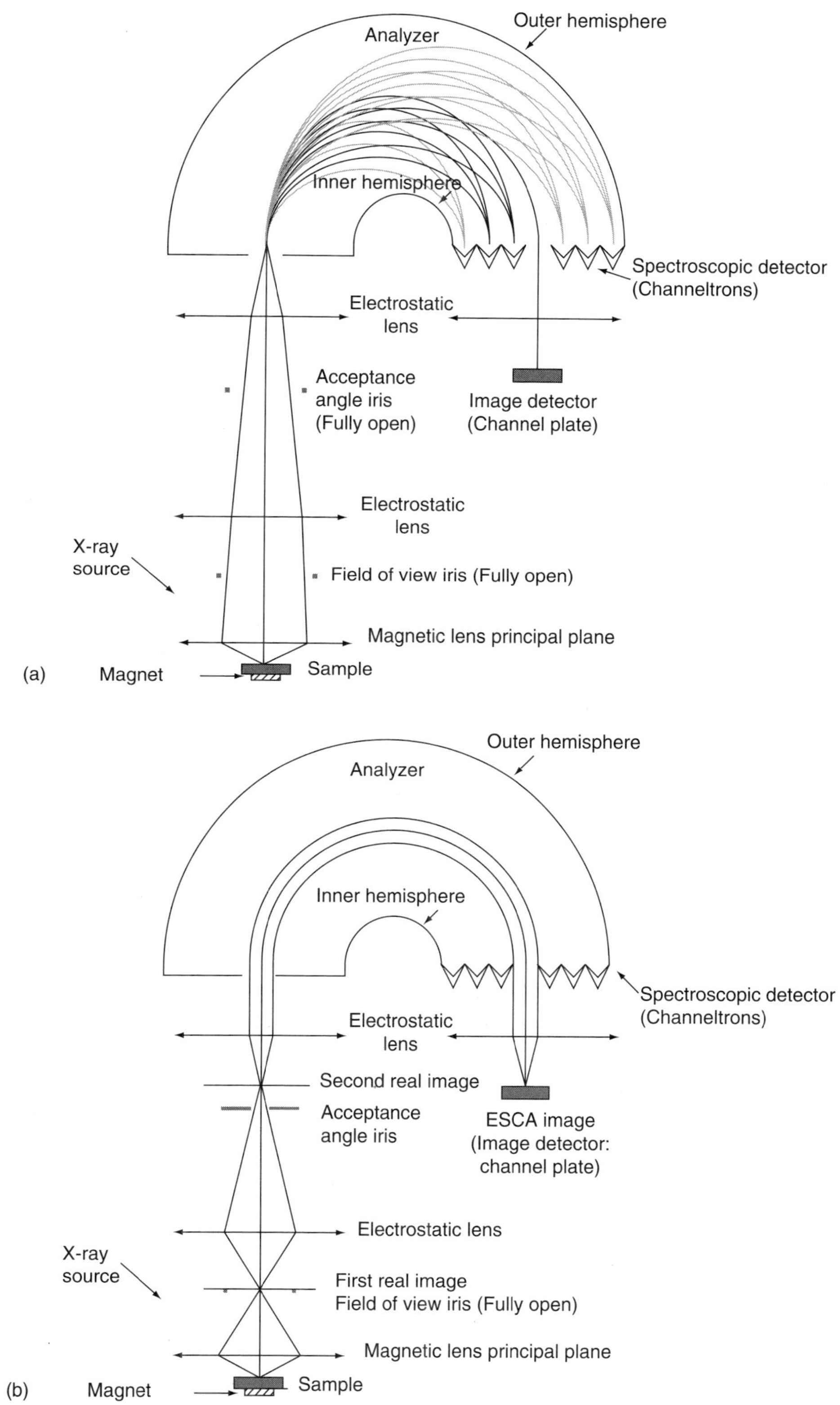

Figure 1.15 Schematics of an XPS spectrometer: (a) spectroscopic, and (b) imaging modes.

however with a loss in intensity. For special applications requiring a very high-energy resolution or variable wavelength of the excitation radiation, a synchrotron source can be used, but for most applications, the x-ray sources available on commercial equipments are sufficient.

1.2.5.3 Ion Sources

There are two kinds of ion sources. The penning source, with a broad ion profile with a large flat plateau, is used for cleaning of a specimen surface and sputter depth profiling of relatively thick anodic layers. The homogeneous sputtering may be achieved within the width of the plateau of the ion beam of 0.5 to 4 cm, depending on the acceleration voltage, which may be varied up to 5 keV, and the distance to the sample.

Another kind of source uses electrons from a heated filament to ionize noble gas ions, which are then accelerated to some keV. The ion beam is usually focussed to 20 to 100 μm and may be scanned by quadrupoles to raster the specimen surface. This ion source is used for a slow depth profiling of thin surface layers, or for ISS analysis.

1.3 APPLICATIONS TO CORROSION SCIENCE AND ENGINEERING

1.3.1 Active Dissolution, Nickel, and Nickel–Iron Alloys Containing Sulfur

Only a few metals may be examined with XPS in an active state, i.e., without a passivating oxide layer at the surface. Usually a metal surface held at potentials in the range of active dissolution will be covered with an oxide layer during emersion, with the exception of surfaces covered by species promoting the dissolution and preventing the oxidation, such as sulfur (see below). Noble or seminoble metal surfaces may be preserved in an active state, the latter when precautions are taken using protective noble gas atmosphere during sample transfer to the UHV.

During the dissolution of nickel and nickel–iron alloys containing sulfur, the sulfur remains and accumulates at the metal surface: this is the so-called anodic segregation phenomenon [29]. Sulfur is first adsorbed on the surface and then it forms a thin, nonprotective, sulfide layer, keeping the metal in the active state even at potentials above the active/passive transition of the sulfur-free metal. The surface has been analyzed by XPS and the spectra of the Ni $2p_{3/2}$ and S 2p regions are shown in Figure 1.16 [29]. In contrast to the passivated nickel surface (Figure 1.16a) the Ni $2p_{3/2}$ region of the S-doped nickel (Figure 1.16b and c) clearly reveals that the surface is not passivated (no contribution from Ni^{2+} in NiO at 854.4 eV and no signal from O^{2-} in NiO at 529.8 eV). A strong signal is observed in the S 2p region at 162 eV, a binding energy corresponding to nickel sulfide, formed during the anodic segregation process. There is no significant chemical shift between nickel metal and nickel sulfide and therefore the Ni 2p3/2 signal remains unchanged.

1.3.2 Passivity of Metals

In aqueous solution, reactive metals form a thin oxide layer that may be studied with XPS. The metal surface underneath may be examined as well. The XPS signals of the metal are generally sufficiently separated from those of the oxide due to the chemical

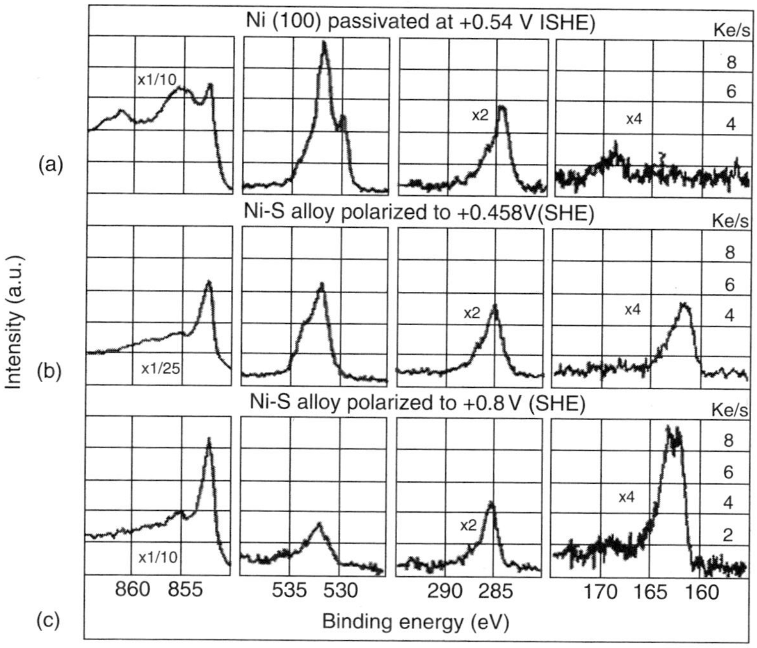

Figure 1.16 XPS spectra (Ni 2p3/2, O1 s, C1s, and S2p) showing the surface enrichment of S by anodic segregation on an S-doped nickel electrode [29].

shift. In this section, examples of XPS investigations of passive layers on pure metals are given. Furthermore the application of UPS to passive layers on Cu is added as an example for the determination of the energy of the upper valence band edge and the threshold energy. Examples of XPS studies of passive films formed on alloy surfaces will be reported in Section 1.3.3.

1.3.2.1 Nickel

Nickel is a reactive metal that passivates in acidic solutions due to the very slow dissolution of its anodic oxide. The first evidence of the bilayer structure of the passive film on nickel was provided by an XPS study of the passivation of Ni in 0.05 M H_2SO_4 [30]. The $Ni2p_{3/2}$ spectrum shown in Figure 1.17a exhibits a signal at 852.8 eV emitted by the metal under the passive film, which shows that the passive film is extremely thin, and thus the photoelectrons from the substrate are not totally attenuated. The peaks at higher binding energies are well fitted by two components associated with Ni^{2+} in NiO (854.4 eV) and Ni^{2+} in $Ni(OH)_2$ (856.6 eV). A slight ion etching removes essentially $Ni(OH)_2$, which shows that $Ni(OH)_2$ is in the outer part of the film and NiO in the inner part, in agreement with the model presented in the Figure 1.17b (the more detailed structure with alternating cation and anion planes in Figure 1.17b is derived from combined XPS and STM data). The O 1s region provides direct evidence of this, with an O^{2-} signal at 529.8 eV and an OH^- signal at 531.6 eV that strongly diminishes after a slight ion etching. Similar findings were also obtained by angle-resolved measurements. The quantitative analysis of the intensities, using the formalism described in the preceding part of this chapter, allowed the authors to calculate the thickness of the passive layer, which was found to be 1 nm under the passivation conditions that had been used.

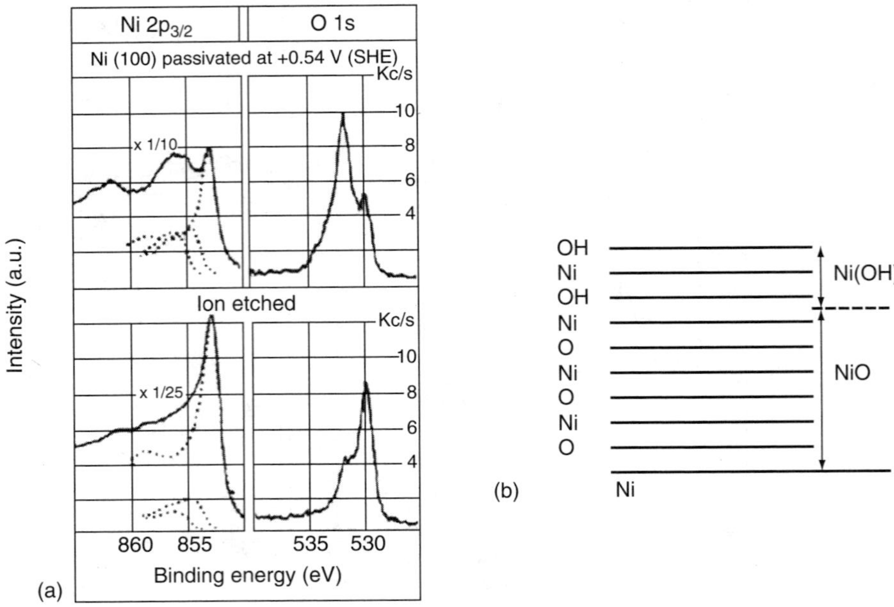

Figure 1.17 XPS spectra of passivated nickel: (a) Ni $2p_{3/2}$ and O 1 s regions after passivation in 0.05 *M* H_2SO_4, and after a short ion etching [30]. (b) Model of the bilayer structure of the passive film.

In alkaline solutions (1 *M* NaOH) the anodic oxide grows continuously up to 1.2 V where it achieves a total thickness of ca. 5 nm [9,31]. The oxide film grows linearly with the potential up to 0.94 V to ca. 3.5 nm. Ni(III) forms in the transpassive potential range due to the insolubility of NiOOH in 1 *M* NaOH. Its presence is shown by an additional peak in the Ni $2p_{3/2}$ spectrum for $E > 0.64$ V, which gets more pronounced in the potential range of vigorous oxygen evolution at $E = 1.64$ V where Ni(III) standards may be formed. Furthermore the O 1 s signal suggests at 1.64 V the presence of equal amounts of OH^- and O^{2-} ions. One therefore has to assume the oxidation of $Ni(OH)_2$ to NiOOH when the potential approaches the transpassive range [9,31]. Starting at $E = 0.5$ to 0.6 V all signals related to the anodic layer are shifted to higher binding energy by $\Delta E_B = 0.8$ eV. Similarly the work function of the anodic oxide is decreased by 0.8 eV. These changes have been explained by a 0.8 eV shift of the Fermi level to smaller values. This shift causes an increase of the work function and the same decrease of the binding energies of all core levels, which is a consequence of the relation of these values to the Fermi level [9,31]. The decrease of the Fermi level is caused by the formation of donor levels due to the oxidation of Ni(II) levels to Ni(III) when the oxidation of NiO starts close to the transpassive potential range.

1.3.2.2 Iron

Fe is a good example for the XPS analysis of a system in which the composition of the passive layer changes with potential. Figure 1.18 indicates a bilayer structure with an inner Fe(II) and an outer Fe(III) part when the film has been formed at potentials positive to the Flade potential E_{p2}. E_{p2} is the potential of passivation of Fe in acidic electrolytes, which shifts by 59 mV/pH according to the following equation:

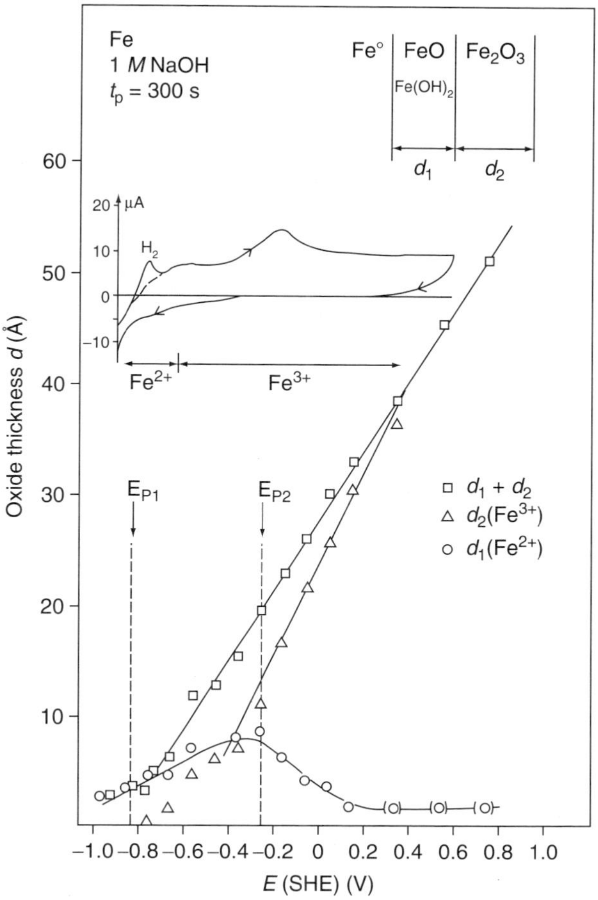

Figure 1.18 Partial layer thickness d_1 and d_2 of Fe(II) and Fe(III) in dependence of potential for Fe passivated for 300 s in 1 *M* NaOH deduced from XPS results. Inset: polarization curve of sputter-cleaned Fe with same potential scale, indicating soluble Fe ions as determined with an RRD electrode [32].

$$E_{p2} = 0.58 - 0.059\text{pH} \tag{1.14}$$

$$2Fe_3O_4 + H_2O \longrightarrow 3Fe_2O_3 + 2H^+ + 2e^- \tag{1.15}$$

Its value does not match the thermodynamic data for oxide formation from the metal for the known Fe-oxides and hydroxides which are ca. 0.56 V more negative. It has been suggested that E_{p2} corresponds to the oxidation of Fe_3O_4 to Fe_2O_3 according to Equation (1.15). The lower valent oxides of Fe dissolve too fast in acidic electrolytes to be protective. However, they dissolve sufficiently slowly in alkaline solutions where they are in dissolution equilibrium. Therefore the anodic oxides may be preserved even at potentials negative to E_{p2}. Figure 1.18 presents XPS data of Fe passivated in 1 *M* NaOH [7,32]. For comparison, a potentiodynamic polarization curve of a sputter-cleaned Fe electrode in 1 *M* NaOH is inserted with the same potential scale. The specimens have been prepared and transferred within the closed system of the spectrometer to avoid any oxidation by air. The data have been evaluated on the basis of a bilayer model with an inner Fe(II) and an outer Fe(III) part of thickness d_1 and d_2. Oxide formation starts at E_{p1} with a major contribution of Fe(II) but also some Fe(III). Above E_{p2} the amount of Fe(II) decreases. At $E > 0.3$ V the presence of Fe(II) cannot be confirmed by XPS due to the increasing thickness of the outer layer of Fe(III), which attenuates the contribution of the inner Fe(II) part. The decrease of Fe(II) matches the anodic peak of the polarization curve which corresponds to the oxidation of Fe(II) to Fe(III). An evaluation of the Fe $2p_{3/2}$ signals as an oxide with mixed valency at $E = -0.2$ V yields a composition Fe_3O_4, which supports the interpretation of the flade-potential E_{p2} by Equation (1.15) [7]. According to the O 1s signal anodic Fe(II) forms a hydroxide which turns to an oxide during its oxidation to Fe(III). These anodic reactions are reversible and the reduction of Fe(III) to Fe(II) and finally Fe-metal may be followed at sufficiently negative potentials. Parallel to the reduction, the oxide changes back to a hydroxide. These changes between the different oxidation states of Fe may be observed also for Fe/Cr alloys. A Cr content prevents the reduction of Fe(III) oxide to metal and it remains in the Fe(II) state.

1.3.2.3 Chromium

The passive film on chromium is composed of Cr^{3+} ions bonded to O^{2-} and OH^- ions and contains H_2O ligands. The XPS measurements evidence a stratification of the passive film (the OH^-/O^{2-} signal ratio is higher in the O 1s spectrum recorded with a take-off angle of 45°), interpreted by a bilayer model with a slightly hydrated inner part of Cr_2O_3 and a highly hydrated outer part of $Cr(OH)_3$. The development of the inner oxide part is strongly favored by an increase of the potential and by aging under anodic polarization. Typical XPS spectra are shown in Figure 1.19 [33].

The presence of higher valent cations (Cr^{6+}) would be revealed by an additional peak at higher binding energy in the Cr $2p_{3/2}$ spectrum. This occurs only at higher potential. The dissolution rate of passivated chromium is extremely small (less than 0.1 $\mu A/cm^2$). It is interesting to note that passive Cr does not show breakdown of passivity and localized corrosion. For these reasons, chromium is a major alloying element in corrosion-resistant alloys.

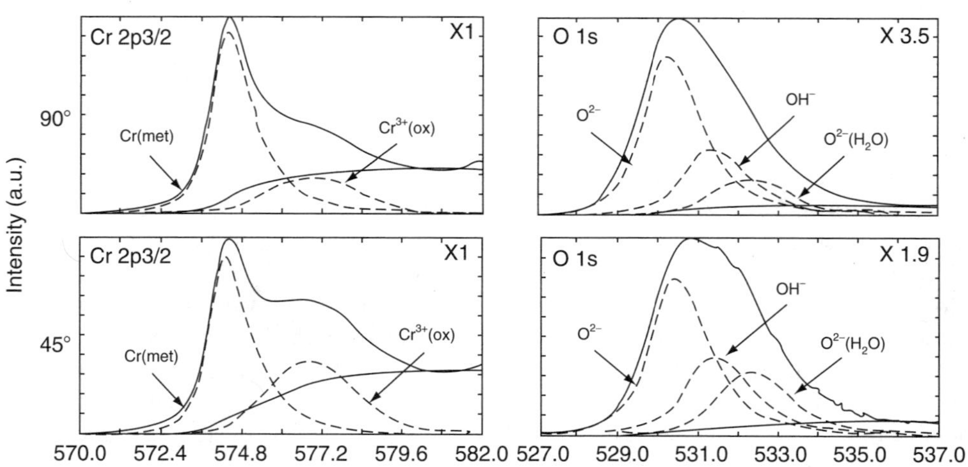

Figure 1.19 XPS spectra recorded at two different take-off angles (90° and 45°) for Cr passivated in 0.5 *M* H_2SO_4, showing the presence of an inner layer of Cr_2O_3 and an outer layer of hydrated $Cr(OH)_3$ [33].

1.3.2.4 Cobalt

Cobalt does not passivate in strongly acidic electrolytes. In alkaline solutions the polarization curve shows a potential range of primary and secondary passivity. In 0.1 *M* NaOH (pH 13), under conditions of primary passivity at $E = -0.4$ to 0.15 V the layer consists exclusively of Co(II) hydroxide whereas Co(III) oxide is dominating at $E > 0.3$ V (Figure 1.20a). The comparison of the O 1s signal with those of standards suggests the presence of Co(III) oxide Co_2O_3, oxy-hydroxide CoOOH, and Co_3O_4 oxide of the spinel type (Figure 1.20b) [34,35]. ARXPS proves the outer position of hydroxides, i.e., of CoOOH, and the inner location of oxides, i.e., of Co_2O_3 and the spinel. The layer composition changes also with time. In the secondary potential range $Co(OH)_2$ forms first which is changed to Co(III) species with time, i.e., the layer changes within some 100 s at $E = 0.55$ V to its final composition. This reaction gets faster with increasing electrode potential [35].

1.3.2.5 Copper

Copper is again a metal that passivates in weakly acidic and alkaline solution only. It follows the predictions of the Pourbaix diagram and one has to assume that passivity

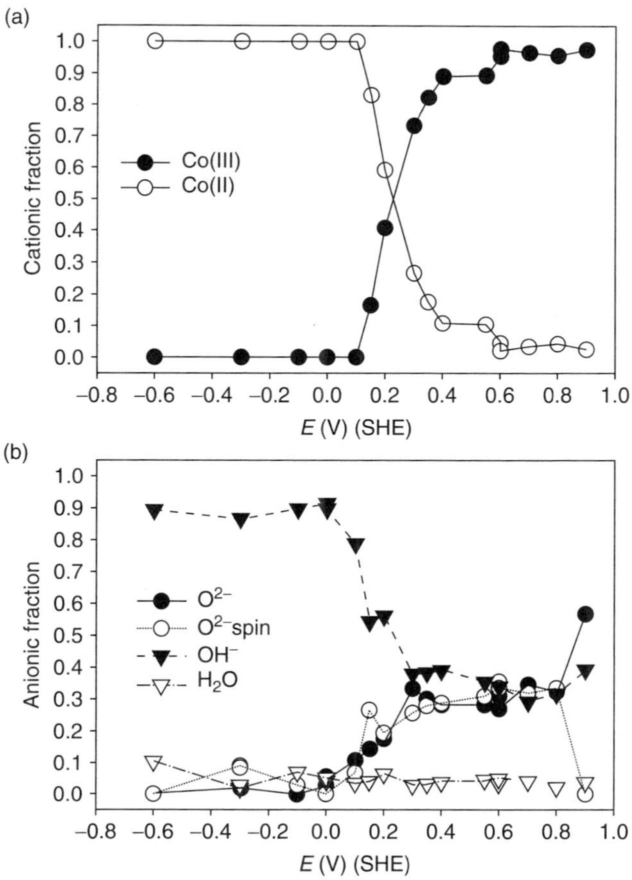

Figure 1.20 Quantitative analysis of the composition of the passive layer formed on Co after 300 s in 0.1 *M* NaOH as a function of the electrode potential, deduced from XPS data: (a) cationic fractions, and (b) anionic fractions [35].

is obtained only if the anodic layer is close to its dissolution equilibrium. Cu_2O forms at $E = 0.2$ V followed by a large anodic peak in the range of 0 to 0.4 V and the passive range at $E = 0.4$ to 0.9 V. Due to an almost negligible chemical shift of the Cu $2p_{3/2}$ level, Cu_2O cannot be distinguished from Cu metal. Therefore x-ray-induced Auger lines have been evaluated quantitatively to determine the amount of Cu_2O on the Cu-surface [36]. Figure 1.21 shows the peak fitting of an experimental x-ray-induced Auger line (Cu LMM) with the contributions of the Cu substrate and a Cu_2O film on the basis of the peaks of pure standards. The integrated peaks yield an oxide thickness in the range of some nm, which agrees well with the value obtained from the reduction charge of Cu_2O to Cu metal. Anodic oxidation in the passive range of 0.4 to 0.9 V yields a duplex film with an outer CuO, $Cu(OH)_2$, and an inner Cu_2O part. The outer Cu(II) layer may be easily detected by a chemical shift of the Cu $2p_{3/2}$ signal to $E_B = 933.6$ eV and the appearance of a satellite at $E_B = 942$ eV (see Figure 1.5). ISS depth profiles support the presence of this bilayer structure [37].

Additional information is obtained from UPS spectra of Cu covered with Cu oxides. As already discussed in Section 1.2.3 threshold energies of semiconducting or insulating layers can be measured similar to the determination of the work function of metals. According to Equation (1.3) the width of the spectrum yields $e\Phi_{Th} = 5.15$ and 4.6 eV for Cu_2O and CuO, respectively. With these measurements the upper valence band edge is determined relative to the vacuum level and thus also relative to the SHE.

The potential of SHE has been measured relative to the vacuum level to −4.6 V [28,38–41]. Together with the bandgap energy obtained from the wavelength dependence of electrochemical photocurrent measurements and the flat band potential from the potential dependence of the photocurrent, the band model of passive layers on Cu has been determined quantitatively [1,42]. Figure 1.22 gives an example of a UPS spectrum of a thin passive layer of Cu_2O on Cu. Besides the Cu 3d levels, the O 2p states are well pronounced. The valence band shifts by ca. 0.8 eV with respect to the UPS spectrum of pure Cu (see Figure 1.9a), which leads to a decrease δE of the width

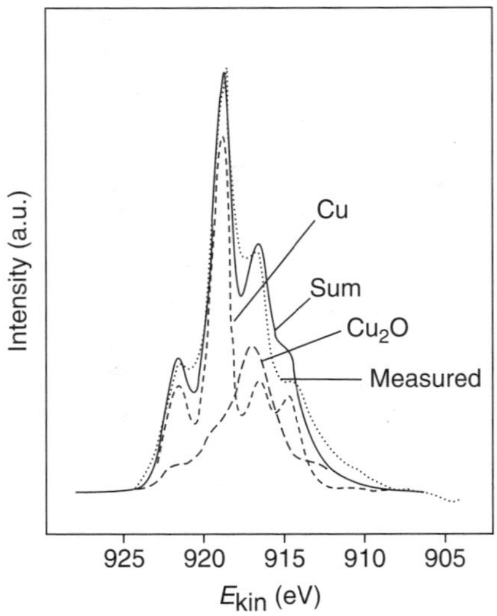

Figure 1.21 X-ray-induced Auger signal for anodically oxidized Cu (ca. 2 nm Cu_2O on Cu). Peak fitting is based on Cu and Cu_2O standards (dashed lines). The fitted signal is in full line and the experimental signal in dotted line [36].

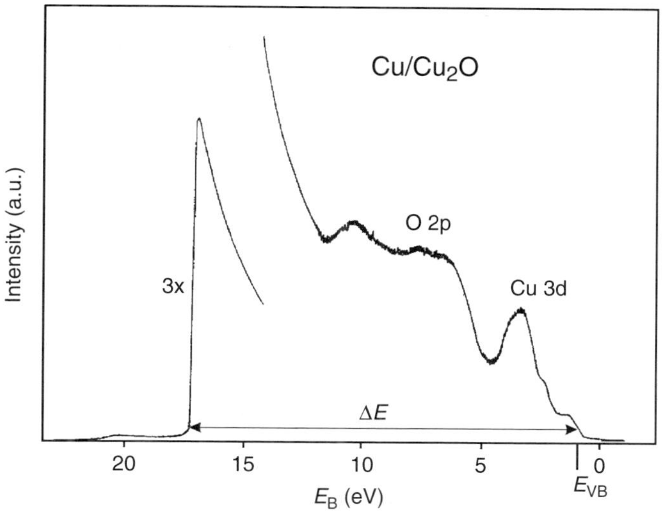

Figure 1.22 UP-spectrum of a thin passive layer of Cu_2O on Cu with Cu 3d and O 2p contributions, E_{VB} valence band edge of Cu_2O [42]. The width of the spectrum yields the threshold energy of the Cu_2O layer, $e\Phi_{Th}$ (see Equation (1.3)).

of the UPS spectrum ΔE and a shift of the valence band edge by this energy relative to the Fermi energy of Cu metal (see Equations (1.2) and (1.3)) [42].

1.3.3 Passivity of Alloys

1.3.3.1 Stainless Steels

Surface analysis, in particular with XPS, has been extensively used to analyze the passive films formed on both ferritic and austenitic stainless steels [43–57] and has provided the basis of our knowledge of the chromium enrichment in the passive films and of the duplex nature of the passive films on these alloys.

Indeed on stainless steels (and also on nickel-based stainless alloys [58]) the passive film can be described by a bilayer model (inner oxide and outer hydroxide layers). It has been shown by XPS that the concentration of Cr^{3+} in the inner oxide layer is much higher than the nominal chromium content of the alloy. As an example, a Cr concentration of 65% was measured for Fe–17Cr alloys passivated in acidic solution (at 25°C) [52], and 75% for Fe–20Cr [43], also in acidic solution. The thickness of the passive film on these alloys in acidic solution is in the range 1.5 to 2 nm (values also determined by XPS using the quantitative analysis described in the first part of this chapter). If the passive film is strongly enriched with Cr^{3+}, it is not, however, pure chromium oxide, and it also contains Fe^{2+} and Fe^{3+} cations.

Another important finding of the surface analysis by XPS is the fact that the enrichment of Cr^{3+} in the inner barrier oxide layer increases during aging of the passive film in aqueous solution. This is clearly seen for stainless steels (Fe–22Cr), on the XPS spectra of the Cr $2p_{3/2}$ and O 1s regions reported in Figure 1.23 [54]. The intensity of the signals corresponding to Cr^{3+}(ox) and O^{2-}, located at 576.7 and 530.2 eV, respectively, increases with polarization time, with respect to the intensities corresponding to the outer hydroxide layer (Cr^{3+} (hydroxide) at 577.2 eV and OH^- at 531.3 eV).

The XPS spectra recorded from an austenitic stainless, Fe18Cr14.3Ni2.5Mo (at.%) after passivation in 0.1 M HCl + 0.4 M NaCl at two differents potentials

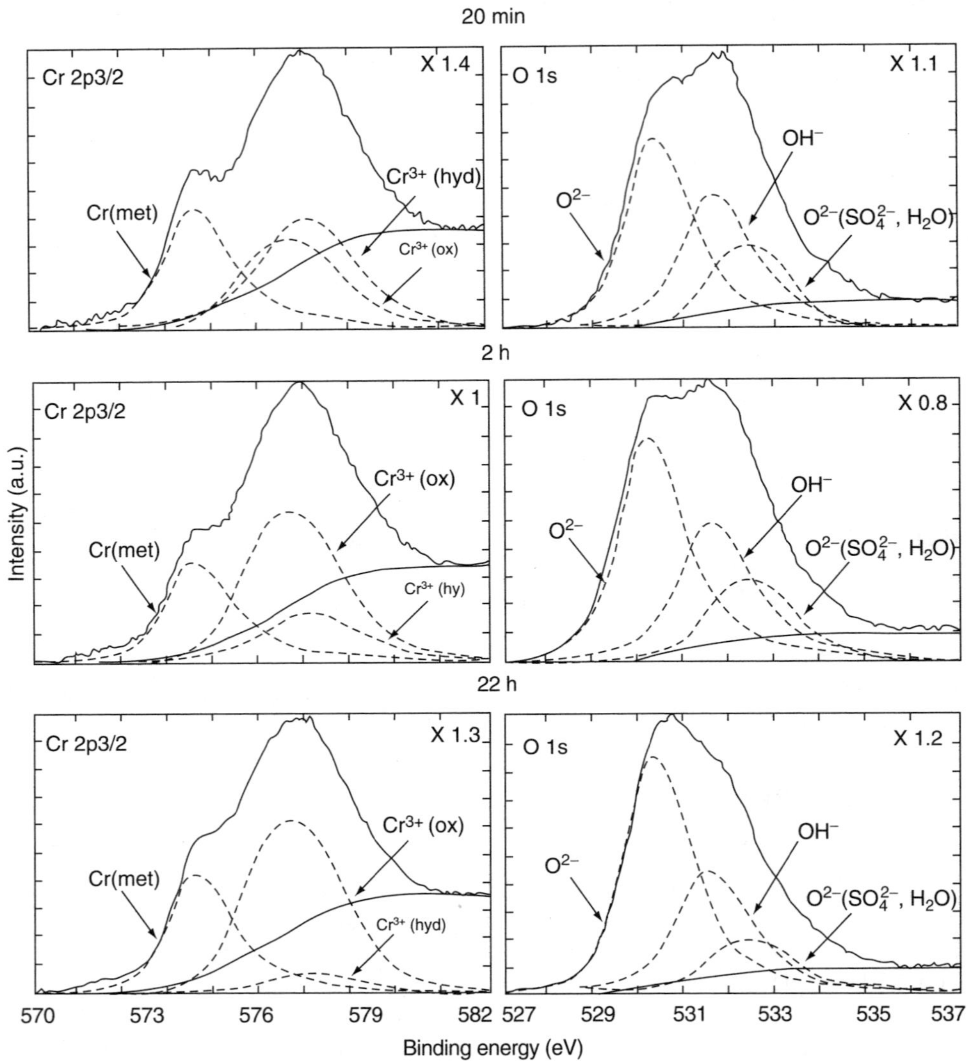

Figure 1.23 Aging of passive films on stainless steels studied by XPS [54]. The intensities of the signals corresponding to the inner layer of Cr_2O_3 (Cr^{3+}_{ox} in the Cr 2p3/2 spectrum and O^{2-} in the O 1s spectrum) increase with aging time.

are shown in Figure 1.24 [44]. The analyzed XPS regions are Ni $2p_{3/2}$, Fe $2p_{3/2}$, O 1s, C 1s, Mo $3d_{3/2,5/2}$, and Cl $2p_{1/2,3/2}$. The recording of such data was completed by measurements at different take-off angles and by depth profiles using argon ion sputtering [44]. In addition to the duplex nature of the film and the Cr^{3+} enrichment, these data revealed that molybdenum was present in the film as Mo^{4+} and Mo^{6+} being preferentially located in the outer hydrated layer. It is interesting to note here that Mo^{4+} (as MoO_2) was shown to be the main cation in the passive film formed on Mo in 0.1 *M* HCl [45]. It is also the XPS analysis of such alloys that revealed that Ni and Mo are enriched in the metallic state at the alloy surface underneath the passive film [57], and that nitrogen is anodically segregated on N-bearing stainless steels [57].

Fe–Cr alloys also provide a good example of the combination of XPS and ISS for a detailed study of the chromium enrichment in passive films [1,53]. Figure 1.25 shows

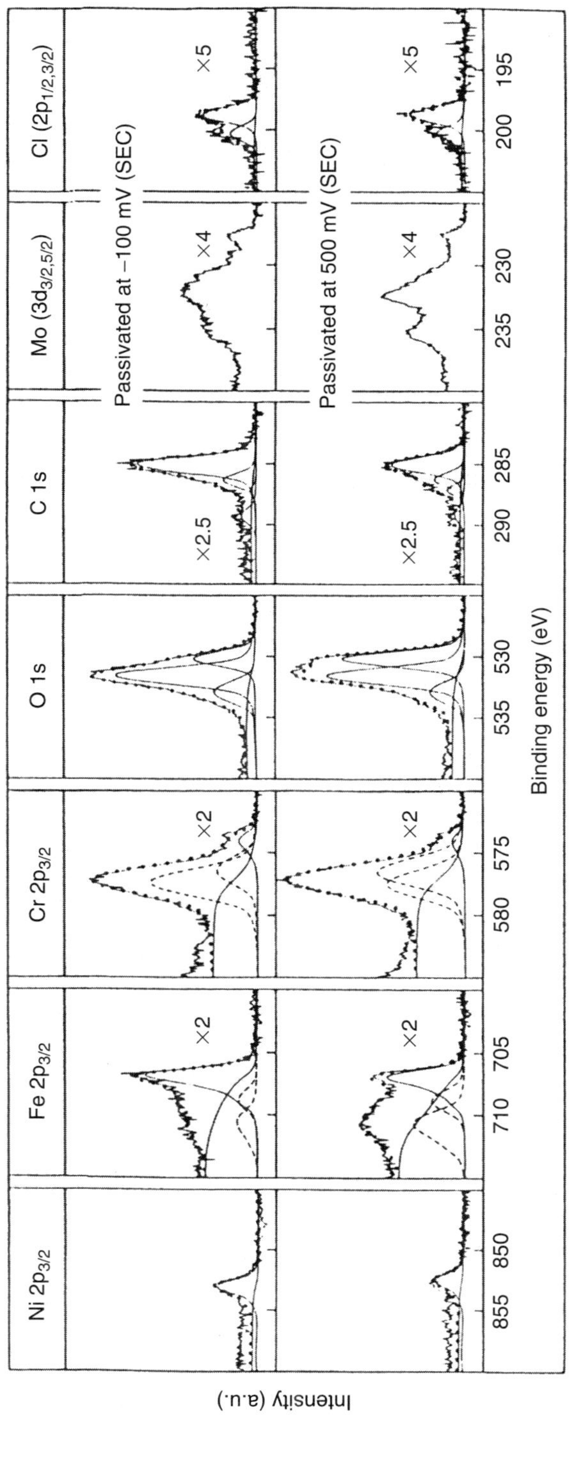

Figure 1.24 XPS spectra of an austenitic stainless steel (Fe18Cr14.3Ni2.5Mo) passivated in 0.1 *M* HCl + 0.4 *M* NaCl [44]

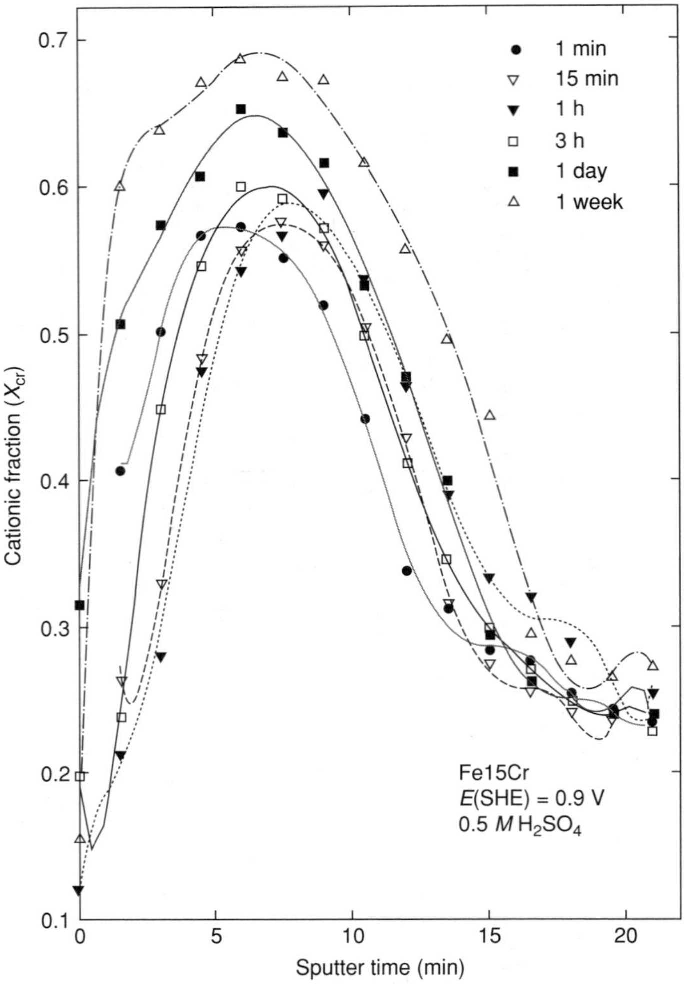

Figure 1.25 ISS profiles as a function of passivation time for Fe15Cr alloy passivated in 0.5 *M* H_2SO_4 at $E = 0.90$ V (SHE) [1, 53].

a series of ISS profiles as a function of passivation time (ranging from 1 min to 1 week). These data also show the progressive enrichment of chromium in the film with increased polarization time, and a broadening of the chromium distribution in the film.

In alkaline solutions, the dissolution of Fe(III) oxide is no longer possible and thus an Fe-rich oxide layer accumulates at the surface on top of the Cr-rich film [49].

1.3.3.2 CuNi Alloys

An interesting combination is the alloying of a reactive and a seminoble metal. This is the case for Cu/Ni alloys. These alloys show a poor passive behavior in strongly acidic electrolytes due to intense Cu dissolution at positive potentials. The current density increases with the Cu content of the alloy and decreasing pH, up to several mA/cm^2 [59]. However, in less acidic and in alkaline solutions, Ni enters preferentially the layer and forms a protecting film [59]. The passive film on Cu–20Ni has a pronounced multilayer structure, which has been deduced from ARXPS measure-

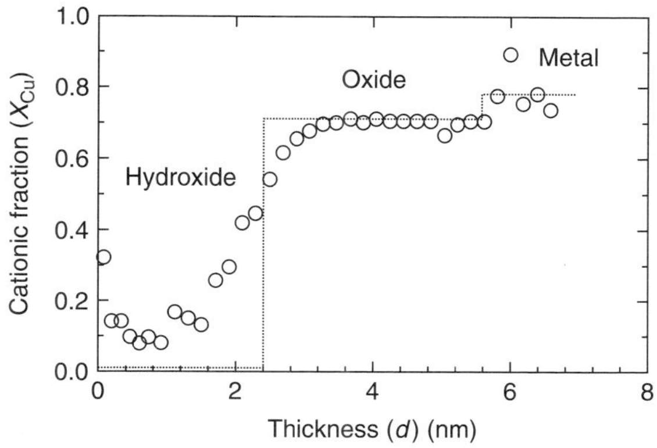

Figure 1.26 ISS profiles (points) and XPS profiles (dotted line) of Cu–20Ni passivated for 300 s in 1 *M* NaOH at 0.44 V (SHE) [1, 59].

ments. Ni is mainly located within the outer hydroxide layer whereas Cu enters only at $E > 0.50$ V. The inner oxide layer starts to form at $E > 0.0$ V and contains preferentially CuO (Figure 1.26). This layer structure has also been observed by ISS depth profiles (Figure 1.26). The results of both methods agree quantitatively [1,59]. Cu accumulates slightly at the metal surface due to preferential oxidation of the less noble Ni. This effect is more pronounced for alloys with higher Cu content and is less visible for the example of Figure 1.26.

1.3.3.3 FeSi Alloys

The addition of reactive elements such as Al and Si to Fe leads to their accumulation within the passive layer. Twenty percent of Si is required to improve passivity whereas less protection is obtained for less Si even when compared with pure Fe. A very pronounced accumulation of Si in a surface film is obtained in strongly acidic electrolytes [1,60]. Figure 1.27a and b compares XPS and ISS depth profiles of anodic oxide films formed on Fe–25Si at $E = 1.0$ V in phthalate buffer pH 5.0. ISS shows a large enrichment of SiO_2 within the layer due to its higher surface sensitivity and the related better depth resolution of this method. Almost pure SiO_2 is found in the center of the layer. XPS averages over a larger depth and thus smoothens the profile. A pure SiO_2 film grows in acidic electrolytes on high Si alloys preferentially at low potentials. Fe^{2+} is dissolved and leaves insoluble SiO_2 on the surface. With increasing potential, Fe(III) enters the layer. Fe silicides with a high Si content are frequently used in strongly acidic media due to their resistance against dissolution. The formation of dense SiO_2 films provides a good protection against pitting in chloride solution (chlorides often cause passivity breakdown and pitting of metals and alloys [61]).

1.3.3.4 FeAl Alloys

Passivation of FeAl alloys in weakly acidic and alkaline solutions yields similar results. Al(III) is accumulated in the center of the anodic layers [62,63]. XPS analysis yields a threefold accumulation, i.e., a maximum of 60 at%. A fourfold accumulation is found by ISS profiles. The evaluation of the XPS data yields a linear increase of the thickness of the oxide from 1 to 7 nm with the potential in the range of $E = -0.2$ to

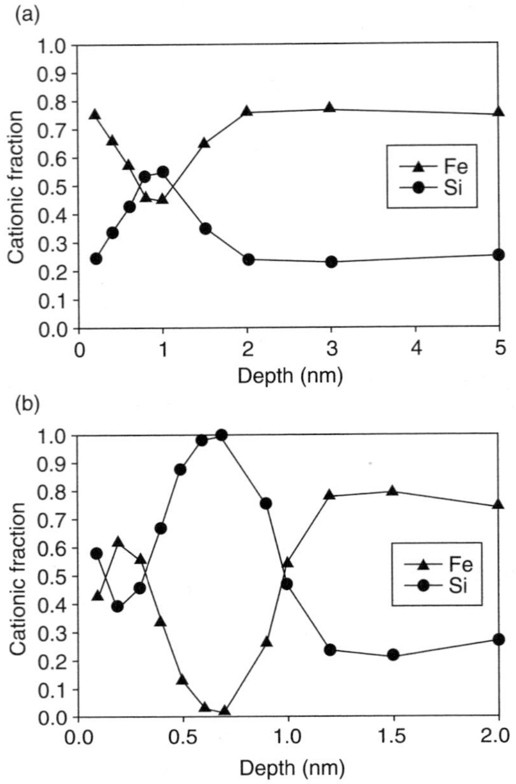

Figure 1.27 XPS (a) and ISS (b) sputter depth profiles for Fe–25Si passivated for 5 min at 1.0 V (SHE) in phthalate buffer pH 5.0 [1,60].

1.2 V. This result is in agreement with the high-field mechanism of oxide growth. The addition of Al to Fe decreases efficiently the active dissolution of the metal in weakly acidic solutions (pH 5.0). The formation of an Al-rich surface film apparently improves the protection of the metal surface [62,63].

1.3.4 Conversion Coatings

Chromate conversion coatings are very effective in protecting metals and alloys, for example, Al alloys, against localized corrosion, and promoting paint adhesion. The chemical composition of chromate conversion coatings has been investigated by XPS [64]. The data show that the layer is a mixture of chromium oxides, other components from the coating bath, and components from the substrate. Chromium is present as Cr(III) and Cr(VI), with Cr(VI) predominantly in the outer layer. Typical XPS spectra are shown in Figure 1.28 [64].

With the present need for replacing chromates by other nontoxic species, conversion coatings based on rare earth oxides have been developed. XPS has been used to analyze the composition of the new conversion coatings. This is illustrated by XPS spectra in Figure 1.29, obtained after treatment of magnesium in a cerium salt solution [65], in the course of a study of rare earth conversion coatings on Mg alloys. The Ce spectrum is quite complex, but the data indicate that the majority of Ce is in the form of Ce^{4+} [65].

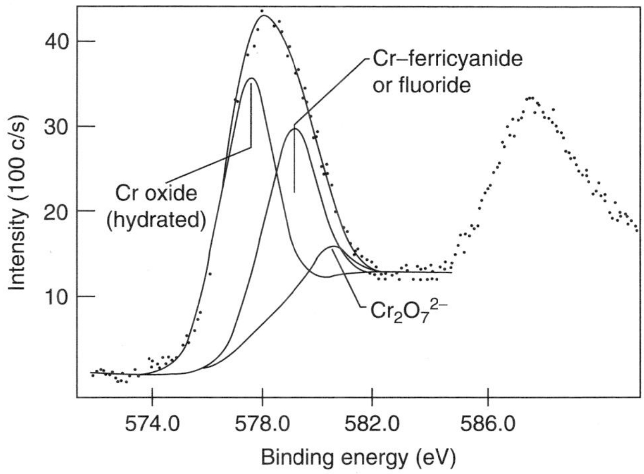

Figure 1.28 XPS spectra of chromate conversion coatings on Al [64].

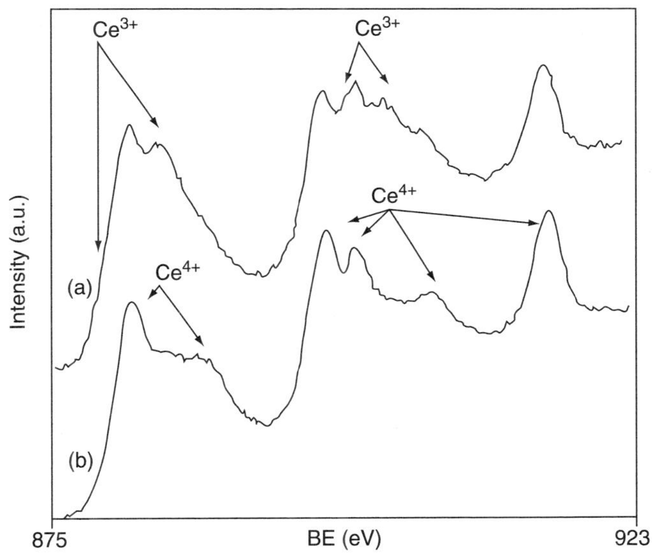

Figure 1.29 Ce 3d XPS spectra of a cerium-based conversion coating on magnesium after (a) 15 s and (b) 15 min of treatment in the cerium salt solution. The spectra show that the conversion coating contains essentially Ce(IV) [65].

1.4 CONCLUSION

The basic principles of XPS and the equipment for surface analysis have been reviewed. Applications to corrosion science and engineering have been discussed, with emphasis on passivity of metals and alloys, a field where decisive progress has been made by the use of quantitative XPS analysis. Other examples have been given on surface treatments, including chromate and chromate-free conversion coatings.

The qualitative and quantitative evaluation of XP-spectra gives detailed answers on the distribution of cations, including their oxidation states, and anions within passive layers and at the metal surface. The structure of surface films is usually very complex and multilayer structures or characteristic depth profiles are found which reflect the electrochemical properties of pure metals and the components of alloys. XPS permits to follow the changes of the composition and chemical structure of surface layers with potential, time, and the environmental and electrochemical conditions. It thus allows to follow changes during the growth or reduction of surface layers. XPS is a very valuable tool for a reliable interpretation of corrosion phenomena. In this regard ISS is a complementary method, which provides a detailed insight into the in-depth distribution of cations.

Besides fundamental studies XPS is a very valuable method for applied studies and may help to solve many problems in industry and questions related to the application of materials in various environments. Service analysis with XPS is now also widespread in failure analysis and problem solving in corrosion prevention.

REFERENCES

1. H.-H. Strehblow, in *Advances in Electrochemical Science and Engineering*, R.C. Alkire and D.M. Kolb, (eds.), Wiley-VCH, Weinheim, 2003, Vol. 8, p.270–374.
2. K. Siegbahn, C. Nordling, A. Fahlman, R. Nordberg, K. Hamrin, J. Hedman, G. Johansson, T. Bergmark, S.-E. Karlsson, I. Lindgren, and B. Lindberg, *ESCA: Atomic, Molecular and Solid State Structure, Studied by Means of Electron Spectroscopy*, Nova Acta Reg. Soc. Upsaliensis, Ser. IV, Vol. 20, Almqvist & Wiksells, Uppsala, 1967.
3. C.D. Wagner, W.M. Riggs, L.E. Davies, J.F. Moulder, and G.E. Mullenberg (eds.), *Handbook of X-Ray Photoelectron Spectroscopy*, Perkin Elmer Corporation, Eden Prairie, Minnesota, 1978.
4. C.D. Wagner, A.V. Naumkin, A. Kraut-Vass, J.W. Allison, C.J. Powell, and J.R. Rumble, NIST X-ray Photoelectron Spectroscopy Database, 2003.
5. D.A. Shirley, *Phys. Rev.* 135, 4709 (1972).
6. S. Tougaard, *Surf. Interf. Anal.* 11, 453 (1988).
7. S. Haupt, H.-H. Strehblow, *Langmuir* 3, 873 (1987).
8. P. Marcus, J. Oudar, and I. Olefjord, *J. Microsc. Spectrosc. Electron.* 4,63 (1979).
9. H.W. Hoppe and H.-H. Strehblow, *Surf. Interf. Anal.* 14, 121 (1989).
10. V. Maurice, S. Cadot, and P. Marcus, *Surf. Sci.* 458, 195 (2000).
11. I. Olefjord, H.J. Mathieu, and P. Marcus, *Surf. Interf. Anal.* 15, 681 (1990).
12. C. Schmidt and H.-H. Strehblow, *J. Electrochem Soc.* 145, 834 (1998).
13. H. Scofield, *J. Electron. Spectroc.* 8, 129 (1976).
14. M. P. Seah and W.A. Dench, *Surf. Interf. Anal.* 1, 2 (1979); M.P. Seah, *Surf. Interf. Anal.* 9, 85 (1986).
15. D. Briggs and M.P. Seah, *Practical Surface Analysis*, Wiley, 1983, p. 187.
16. S. Tanuma, C.J. Powell, and D.R. Penn, *Surf. Interf. Anal.* 17, 911 (1991) and 17, 927 (1991).
17. P. Marcus, C. Hinnen, and I. Olefjord, *Surf. Interf. Anal.* 20, 923 (1993).
18. S. Hofmann and J.M. Sanz, *Surf. Interf. Anal.* 5, 210 (1983).
19. I. Olefjord, P. Marcus, H.J. Mathieu, and S. Hofmann, in Proceedings of ECASIA'95, Montreux, Switzerland. H.J Mathieu, B. Reihl, and D. Briggs (eds.), Wiley, Chichester, 1996, p. 188.
20. A. Machet, A. Galtayries, and P. Marcus, *Surf. Interf. Anal.* 34, 197 (2002).
21. I. Olefjord and B.-O. Elfström, *Corrosion* (NACE) 38, 46 (1982).
22. P. Marcus, J. Oudar, and I. Olefjord, *Mater. Sci. Eng.* 42, 191 (1980).
23. E. Yeager, W.E. O'Grady, M.Y.C. Woo, and P. Hagans, *J. Electrochem. Soc.* 125, 348 (1978).
24. A.T. Hubbard, R.M. Ishikama, and J. Katekaru, *J. Electroanal. Chem.* 86, 274 (1978).
25. F.T. Wagner and P.N. Ross, *J. Electrochem. Soc.* 130, 1789 (1983).
26. S. Haupt, U. Collisi, H.D. Speckmann, and H.-H. Strehblow, *J. Electroanal. Chem.* 194, 179 (1985).

27. S. Haupt C. Calinski, H.W. Hoppe, H.D. Speckmann, and H.-H. Strehblow, *Surf. Interf. Anal.* 9, 357 (1986).
28. D. Lützenkirchen-Hecht and H.-H. Strehblow, *Electrochim. Acta* 43, 2957 (1998).
29. P. Marcus, J. Oudar, and I. Olefjord, *Mater. Sci. Eng.* 42, 191 (1980).
30. P. Marcus, J. Oudar, and I. Olefjord, *J. Micros. Spectrosc. Electron.* 4, 63 (1979).
31. H.W. Hoppe and H.-H. Strehblow, *Corr. Sci.* 31, 167 (1990).
32. H.-H. Strehblow, *Surf. Interf. Anal.* 12, 363 (1988).
33. V. Maurice, W.P. Yang, and P. Marcus, *J. Electrochem. Soc.* 141, 3016 (1994).
34. A. Foelske and H.-H. Strehblow, *Surf. Interf. Anal.* 29, 548 (2000).
35. A. Foelske and H.-H. Strehblow, *Surf. Interf. Anal.* 34, 125 (2002).
36. H.D. Speckmann, S. Haupt, and H.-H. Strehblow, *Surf. Interf. Anal.* 11, 148 (1988).
37. H.-H. Strehblow and B. Titze, *Electrochim. Acta* 25, 839 (1980).
38. S. Trasatti, *J. Electroanal. Chem.* 52, 313 (1974).
39. R. Gomer and S. Tryson, *J. Chem. Phys.* 66, 4413 (1977).
40. D.M. Kolb, *Z. Phys. Chem.* NF 154, 179 (1987).
41. E. R. Kötz, H. Neff, and K. Müller, *J. Electroanal. Chem.* 215, 331 (1986).
42. H.-H. Strehblow, U. Collisi, and P. Druska, *Control of Copper and Copper Alloys Oxidation*, Edition de la Revue Métallurgie, Paris, No. 6, 33 (1992).
43. I. Olefjord and B. Brox, in *Passivity of Metals and Semiconductors*, M. Froment (ed.), Elsevier, 1983, p. 561.
44. I. Olefjord, B. Brox, and U. Jelvestam, *J. Electrochem. Soc.* 132, 2854 (1985).
45. C.R. Clayton and Y.C. Lu, *J. Electrochem. Soc.* 133, 2465 (1986).
46. P. Marcus and I. Olefjord, *Corrosion Sci.* 28, 589 (1988).
47. S. Mishler, H.J. Mathieu, and D. Landolt, *Surf. Interf. Anal.* 11, 182 (1988).
48. R. Kirchheim, B. Heine, H. Fischmeister, S. Hofmann, H. Knote, and U. Stolz, *Corrosion Sci.* 29, 899, (1989).
49. C. Calinski and H.-H. Strehblow, *J. Electrochem. Soc.* 136, 1328 (1989).
50. J.E. Castle and J.H. Qiu, *Corrosion Sci.* 29, 591 (1989).
51. E. De Vito and P. Marcus, *Surf. Interf. Anal.* 19, 403 (1992).
52. W. Yang, D. Costa, and P. Marcus, *J. Electrochem. Soc.* 141, 2669 (1994).
53. S. Haupt and H.-H. Strehblow, *Corrosion Sci.* 37, 43 (1995).
54. V. Maurice, W. Yang, and P. Marcus, *J. Electrochem. Soc.* 143, 1182 (1996).
55. V. Maurice, W. Yang, and P. Marcus, *J. Electrochem. Soc.* 145, 909 (1998).
56. D. Hamm, C.-O. A. Olsson, and D. Landolt, *Corrosion Sci.* 44, 1009 (2002).
57. C.R. Clayton and I. Olefjord, in *Corrosion Mechanisms in Theory and Practice*, 2nd ed., P. Marcus (ed.), Marcel Dekker, New York, 2002, p. 217–241.
58. P. Marcus and J.-M. Grimal, *Corrosion Sci.* 33, 805 (1992).
59. P. Druska, H.-H. Strehblow, and S. Golledge, *Corrosion Sci.* 38, 835 (1996).
60. C. Schmidt and H.-H. Strehblow, *J. Electrochem. Soc.* 145, 834 (1998).
61. H.-H. Strehblow, in *Corrosion Mechanisms in Theory and Practice*, 2nd ed., P. Marcus (ed.), Marcel Dekker, New York, 2002, p. 243–285.
62. D. Schaepers and H.-H. Strehblow, *J. Electrochem. Soc.* 142, 2210 (1995).
63. D. Schaepers and H.-H. Strehblow, *Corrosion Sci.* 39, 2193 (1997).
64. A.E. Hughes, R.J. Taylor, and B.R.W. Hinton, *Surf. Interf. Anal.* 25, 223 (1997).
65. H. Ardelean, P. Marcus, and C. Fiaud, *Mater. Corrosion* 52, 889 (2001).

2
Auger Electron Spectroscopy

James E. Castle
University of Surrey, School of Engineering, Guildford, United Kingdom

Contents

2.1 INTRODUCTION

In probing a metallic surface with a view to understanding either its resistance to environmental degradation or, if corroded, to understanding the mechanism of attack, we will generally be interested in composition as a function of depth, and position. Auger electron spectroscopy (AES) can provide this information with a depth resolution of the order of nanometers and a spatial resolution of the order of 10 nm. In many cases information on the chemical or valence state of the surface species can also be obtained. The present review will enable full recognition of the range of information that AES might offer to the investigation of corrosion processes.

AES is an electron spectroscopic method and is closely related to x-ray photoelectron spectroscopy (XPS): each deriving its surface sensitivity from the limited

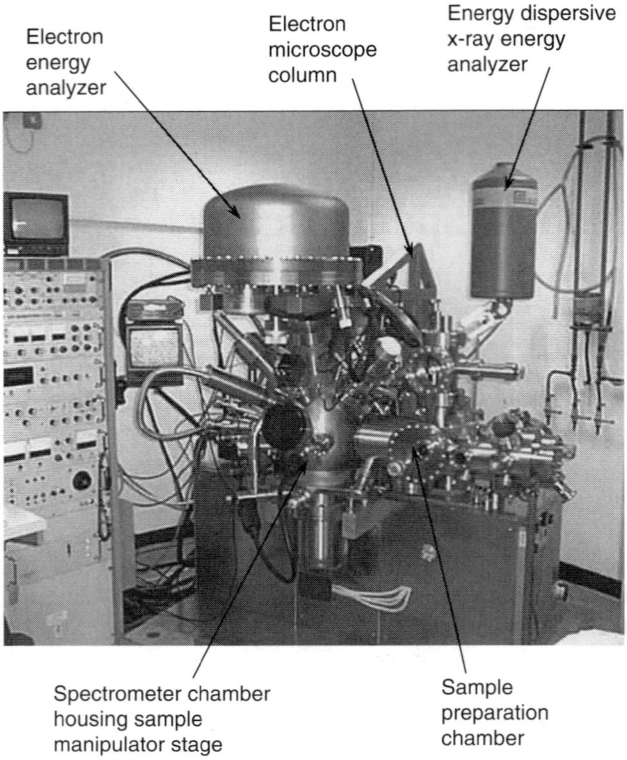

Figure 2.1 An Auger electron microscope in the author's laboratory. Ion guns and other ancillaries are mounted on the spectrometer chamber: the fracture stage and similar facilities are mounted on the preparation chamber.

mean free path, in a solid, of the emergent signal. AES is also a vacuum technique and cannot be used for *in situ* study of corrosion but is not entirely confined to *ex situ* examination of test pieces. A typical spectrometer (Figure 2.1) is often equipped with a sophisticated preparation chamber. Such a facility can enable high-temperature oxidation using a hot stage, will have a fracture stage, and may be fitted with a retractable electrochemical cell.

AES has been available in commercial form for more than 30 years, a similar history to XPS, and both have reached a mature stage of development. XPS and AES differ from each other in the nature of the energetic beam used to excite an analytical signal from the sample: an x-ray beam for XPS; a finely focused electron beam in the case of AES, with all the attendant facilities for microscopy. Researchers from the field of corrosion are likely to choose XPS for studies in which the surface is expected to be uniform in composition over large areas and for which the main need is information on surface composition and on the valence state of the elements present, for example, in passivated surfaces of alloy steels. By contrast, AES will be chosen when it is necessary to examine areas of highly localized corrosion, such as cracks and pits.

Although AES gives this possibility of excellent microscopy, the use of an electron beam as the excitation source brings with it all the problems of electrostatic charging, as found in the scanning electron microscope (SEM). Unlike the case of the SEM, however, the charging problem cannot be overcome by coating the surface:

this would simply mask the chemical information in the surface layer. AES is thus overwhelmingly a technique for the study of metallic surfaces and the thin layers of oxides and minerals that form on them.

2.2 AES: A BRIEF OVERVIEW

It is likely that the corrosion scientist intending to use AES will have an initial understanding of the sample surface, gained by use of SEM and energy dispersive x-ray analyses (EDS). Some idea of the contrast between the information provided by EDS with that arising from use of AES can be obtained from Figure 2.2. Figure 2.2(a) shows a typical inclusion group in a sample of a stainless steel with a composition similar to that of AISI316. The steel had been exposed to a solution of 0.5 *M* NaCl, 0.5 *M* Na_2SO_4, and 0.08% H_2O_2 for 10 sec in order to initiate corrosion at the

(a) (b) (c) (d) (e)

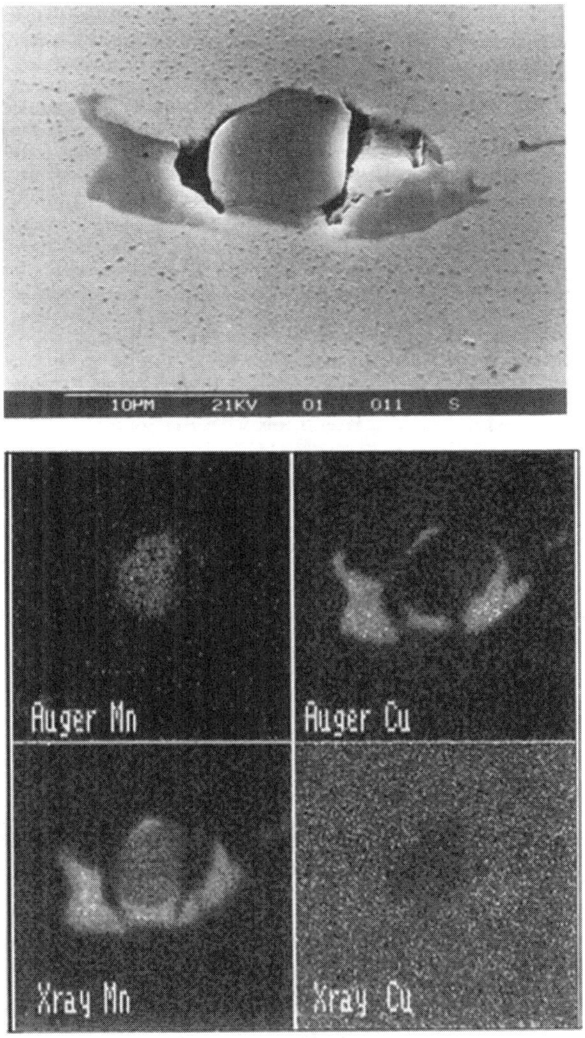

Figure 2.2 Secondary electron image, x-ray and scanning Auger micrographs of an inclusion group in stainless steel. (Courtesy M.A. Babes.)

inclusion. This inclusion is formed from manganese sulfide that has nucleated on the side of a central oxide particle, as is revealed clearly by the distribution of manganese in the x-ray image, Figure 2.2(d). The oxide particle contains some manganese in the left-hand half but the most intense signal comes from MnS. No copper is present in the inclusion but some is distributed throughout the steel, albeit at a very low level, as shown in Figure 2.2(e). Information of this nature is commonplace in the corrosion literature and the corrosion scientist might be forgiven for considering that this is all that might be obtained from this specimen. The peculiar value of AES to corrosion studies is revealed in the AES images of the same region (Figure 2.2(b) and (c)). In the surface layers, analyzed by the Auger electrons, MnS is replaced by CuS, which forms a protective layer, preventing further hydrolysis of the sulfide. The surface film of copper sulfide is found only on the MnS phase and the manganese present in the oxide inclusion remains unaltered. The AES images were obtained simultaneously with the x-ray images, using the same electron beam (15 kV) to excite x-ray and AES images and what is seen here is a comparison of surface information with that arising from the bulk phase. The SEM fitted with both x-ray detection and Auger electron detection becomes a microprobe analyzer operating in two depth zones, the micrometer and the nanometer, and doing so simultaneously.

The ability of AES to reveal surface chemical detail that eludes x-ray mapping is what makes the technique so valuable to the corrosion scientist. In many cases, it will lead to mechanistic interpretation that is different entirely from that which might have arisen from consideration of the x-ray images alone. The Auger spectrum is more complex in appearance than the x-ray spectrum and in many cases more can be gained from understanding its structure. However, images of the type shown in Figure 2.2 are obtained by setting "windows" around given peaks in the spectrum in exactly the manner that is familiar from EDS analysis.

The characteristic peaks found for a given element in the electron spectrum form the basis of the two techniques known as Auger depth profiling and scanning Auger microscopy (SAM). This chapter therefore opens with an explanation of the features found in the Auger electron spectrum. An understanding of the spectrum is not strictly necessary for this limited purpose and some readers might wish to jump directly to the sections dealing more directly with the use of AES in depth profiling and in mapping. The use of AES in the general field of corrosion research has been reviewed at regular intervals during its development (1–5). AES and its importance is such that it is included in the short list of suggested key words for corrosion techniques published by *Corrosion Science*. A description of its mode of operation and general principles of use can be found in encyclopedias (6,7) and it is treated as a spectroscopy in standard textbooks.

2.2.1 Auger Electron Spectroscopy

The Auger electron, which is central to the technique, is named after Pierre Auger (8), who discovered and explained the Auger effect, which he observed in experiments with cloud chambers in the mid-1920s. An Auger electron is generated by transitions within the electron orbitals of an atom, following excitation of an electron from one of the inner levels. This is described schematically in Figure 2.3. An incident primary electron beam of sufficient energy can ionize a target atom by ejecting a core electron. The excited atom relaxes by an electron, from an outer level, making a transition to a lower energy state and filling the vacant hole. The excess energy resulting from the change in binding energies of the transition electron is then released by the emission

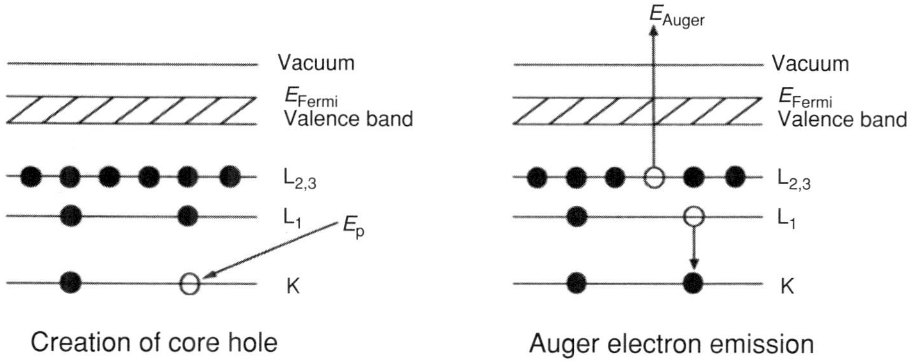

Figure 2.3 The energy levels involved in the creation of a KLL Auger electron.

of an Auger electron. The kinetic energy of the Auger electron is independent of the primary beam energy but is characteristic of the atom and electronic shells involved in its production. It is measurement of this kinetic energy in an electron spectrometer that gives the primary information as to the species present. Moreover, because these energies are low, ranging in values up to about 2000 eV, the information relates only to the top few atomic layers from which such low-energy electrons can escape.

The kinetic energy of the Auger $KL_1L_{2,3}$ transition shown in Figure 2.3 is given by

$$EKL_1L_{2,3} = EK - EL_1 - EL_{2,3} \tag{2.1}$$

where EK, EL_1, and $EL_{2,3}$ are the binding energies of the electronic shells involved in the process. The basic equation given above needs refinement as EL_1 and $EL_{2,3}$ are not energy levels in their normal ground state. The binding energy of the L_1 level is perturbed by the presence of a hole in the K shell and the hole in the L_1 shell similarly affects the binding energy of the $L_{2,3}$ shell electron. An empirical expression (9) that gives closer agreement with experimentally observed data is given by

$$EKL_1L_{2,3} = EK - (E_{(Z)}L_1 - E_{(Z+1)}L_1)/2 - (E_{(Z)}L_{2,3} - E_{(Z+1)}L_{2,3})/2 \tag{2.2}$$

that is, the binding energies of the outer shells are approximated by the mean value of the element concerned, $E_{(Z)}$, and that next higher in the periodic row, $E_{(Z+1)}$. In fact, however, peak positions are not usually identified by such first-principle calculations but by use of look-up tables embedded in the data system associated with the instrument.

As has been stated, the characteristic energy of the Auger transition is given by a combination of the binding energies of the three levels involved. Typically, in an LMM transition, the hole formed at the 2p level by electron impact is filled by a transition from the 3d level and the energy transferred to a second electron, which is emitted as the Auger electron.

Figure 2.4 shows the Auger electron spectrum for the LMM region of copper. The three principal peaks represent the three combinations resulting from the filling of the $2p_{3/2}$ orbital by either the 3d or the 3p electron. Binding energies are well known from XPS; thus, it is possible to test the energy relationship expressed in Equation (2.1), using, e.g., the most prominent peak of the set,

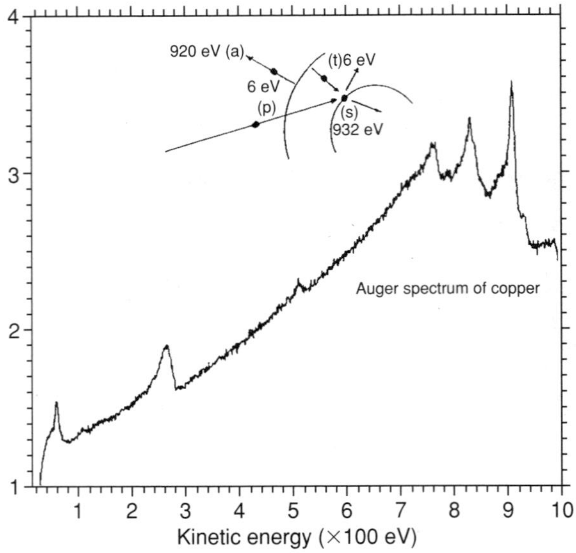

Figure 2.4 The Auger spectrum of copper. The inset shows the energies of the transitions according to the equivalent core model.

$$\mathrm{CuL_3M_5M_5 = Cu2p_{3/2} - Cu3d - Cu3d} \tag{2.3}$$

where each term represents the kinetic energy (left-hand side) or the binding energy (right-hand side). This gives (932–2–2) eV, yielding the value for $CuL_3M_5M_5$ of 928 eV compared with an observed value of 919 eV. A much closer agreement with the observed value is found by using the so-called "equivalent core model" (9) given in Equation (2.2). This yields (932–6–6) eV, that is, 920 eV.

The Auger electron spectrum of Figure 2.4 contains peaks for all the possible permutations by which the $Cu2p_{3/2}$ level can be filled. The three large peaks are the $CuL_3M_5M_5$, the $CuL_3M_3M_5$, and the $CuL_3M_3M_3$ transitions but there is a subsidiary set based on the filling of the $Cu2p_{1/2}$ level. Together these form the LMM Auger transitions for copper and they have an exact parallel with the $L\alpha$ and $L\beta$ series of x-ray lines for copper. (Description of a particular Auger transition is not commonly given using the orbital notation [1s, 2p, 3d, etc.] but the x-ray nomenclature [K, L_2, M_5, etc.] is preferred.) It is readily seen that there will be a series based on the filling of the 1s level, in parallel with the K series of x-ray lines.

The K, L, M, and N series of lines move across the energy spectrum, their kinetic energy increasing with atomic number. However, the probability of Auger transitions occurring in preference to the x-ray transition decreases sharply above a certain atom number for each series. Thus, in practice, each row of the periodic table has a favored set of peaks that are prominent within the energy range of the spectrometer (usually 0 to 2 keV). The KLL lines in the Auger spectrum are important for the light elements, Be to Ar, with the intense $KL_{2,3}L_{2,3}$ peak dominating the intensities: the KLL line for carbon or oxygen will usually be prominent in any Auger spectrum of a technological surface. The 3d transition metals have the characteristic LMM triplet of peaks seen in Figure 2.4 and the 4d-transition metals are readily identifiable by their MNN peak series and the well-resolved $M_4N_{4,5}N_{4,5}$ and $M_5N_{4,5}N_{4,5}$ doublet.

2.2.2 Chemical State Information

Binding energy is defined as the difference in total energy of the system before and after ionization of the atom or ion. Thus, because of the changing ionization state of the atom relative to its neighbors during these transitions, there will be polarization of the surrounding ions, locking up some energy and modifying the measured binding energies. These, so-called, final-state effects are sensitive to chemical and structural information and give rise to a shift in the kinetic energy of the Auger electron (10). The chemical shift is similar in magnitude to that observed in XPS and of the order of 2 to 4 eV. In cases where transitions occur between core-like orbitals that are well defined in energy, notably for elements either side of a transition series of the periodic table, the oxide and element can be distinguished *better* by use of AES compared with XPS because of the additional ionization that occurs on Auger emission. The chemical shift is widely used in studies of the corrosion of copper and zinc, and even more so in the cases of aluminum and silicon. An example of the use of the chemical shift for aluminum is given later in the chapter together with an illustration of chemically sensitive mapping.

As indicated above, transitions between core-like, i.e., filled, orbitals give very well resolved Auger lines. It often happens, however, that the key elements with which we are concerned in a given example of corrosion are transition elements; that is, the orbits are not full but are part of the valence band. In the case of most steel-forming elements, the 3d level is the degenerate valence band and the shape and breadth of the Auger peak result from the fact that it is a self-convolution of the electron density of states in the conduction band. The splitting of the 3d n levels leads to many transitions of similar energy that are often unresolvable, broadening the important peaks of the LMM triplet, which thus have a width at the half-maximum of 10 eV or even more. In these elements, the broad width is sufficient to mask the chemical shift and the metal and its oxide cannot easily be separated by a shift in position. Notwithstanding the lack of a distinctive shift, much progress has been made in separating chemical states by use of the change in peak shape. This is done using factor analysis, in which the characteristic peak shape becomes the target pattern in a series, such as obtained in a depth profile. Examples of the use of factor analysis and other data analysis techniques will be given in the section on depth profiling.

The lack of chemical state information in the steel-forming elements had a significant effect on the development of instrumentation of Auger spectrometers. High-energy resolution simply gave no advantage to those studying the cracking of steels in early nuclear power stations in the United States and thus energy resolution was discarded in favor of high sensitivity and fast acquisition. Spectra were normally obtained and displayed as the differential of the spectrum, as shown in Figure 2.5. Subsequently, however, instrumental developments have taken place, which mean that such difficult choices do not have to be made: modern Auger instruments can operate in high-energy resolution or in high spatial resolution modes, which would have low sensitivities, or in high-sensitivity modes, as required. Digital acquisition of signals means that the computer-based data systems can display spectra in the differential mode or in the normal mode, as required by the operator.

2.2.3 Quantification

Auger spectra can be quantified but with more difficulty than, for example, similar spectra obtained by XPS. There are two principal reasons why even relative methods

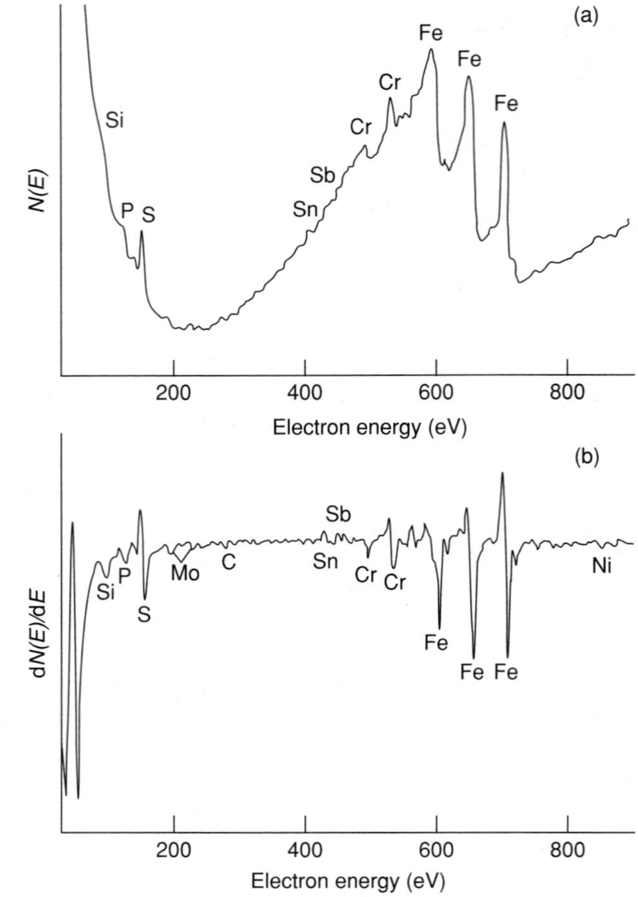

Figure 2.5 The Auger spectrum from the fractured surface of a chromium steel. (After Bishop, H.E., In: Walls, J.M. (ed.) *Methods of Surface Analysis*. Cambridge: Cambridge University Press, pp. 87–126 (1990).) The spectrum is shown in the normal mode and (lower) in the differentiated mode. Note the improvement in visibility of elements such as silicon, which lie on the steeply rising part of the spectrum.

are difficult to apply universally: first, many elements are not available in conducting form and thus a single scale of sensitivities relative to a single element, for example, fluorine or carbon, as in XPS, has never been produced; second, the initial core hole can be produced by backscattered electrons, other Auger electrons, or by locally produced x-rays. This latter effect gives rise to a medley of fluorescence and backscattering corrections that differ from sample to sample. A matrix correction term can be included in the quantification, which takes account of backscattered electron-induced Auger emission (11) and other matrix-dependent parameters (12). Alternatively, a local standard as similar as possible to the test piece is often employed to establish compositions. Hall et al. (13) established a very useful basis for quantification by use of a series of binary alloys containing the elements of the unknown material. Consider an alloy ABCD for which the intensity measured for element A is given by I_A. Then the concentration of A, in atomic %, is given by

$$n_A = 100I_A/(I_A + I_B/F_{B,A} + I_C/F_{C,A} + I_D/F_{D,A}) \tag{2.4}$$

where the factors F are obtained from the intensity ratios measured on the alloys of known composition, that is,

$$F_{B,A} = I'_B n'_A / I'_A n'_B \tag{2.5}$$

The primed values are those obtained from the reference material. Hall and Morabito (14) showed that the matrix factors were negligible for most commercial alloys and that the factors, F, were independent of composition across virtually the whole range. Once these are obtained for a given instrument then it is straightforward to obtain the composition of any material from the measured intensities. Ions such as chloride and oxide, in which the corrosion scientist is particularly interested, can be obtained from compounds such as silver chloride and magnetite that are both stable and sufficiently conducting for the Morabito method to be used.

2.3 AUGER ELECTRON DEPTH PROFILING

To produce an etch profile the electron gun is operated, either sequentially or continuously, with an ion gun, so that a series of analyses are made as material is removed by the etching action of the ion beam. A particular advantage of AES is that the analyses are obtained from a small region in the center of the etch pit so that the analysis has the benefit of the most uniform etch rate and is unhindered by redeposition of material that may occur at the fringe of the pit. A typical profile is shown in Figure 2.6 (15). The intensity of a given peak for each element is plotted against etch time and the profiles are used to contrast the difference between the oxide scale on a heavily worked surface and that in which all dislocations have been removed by chemical etching. While these qualitative profiles are suitable for comparative measurements, interinstrument and interlaboratory comparisons need to be made

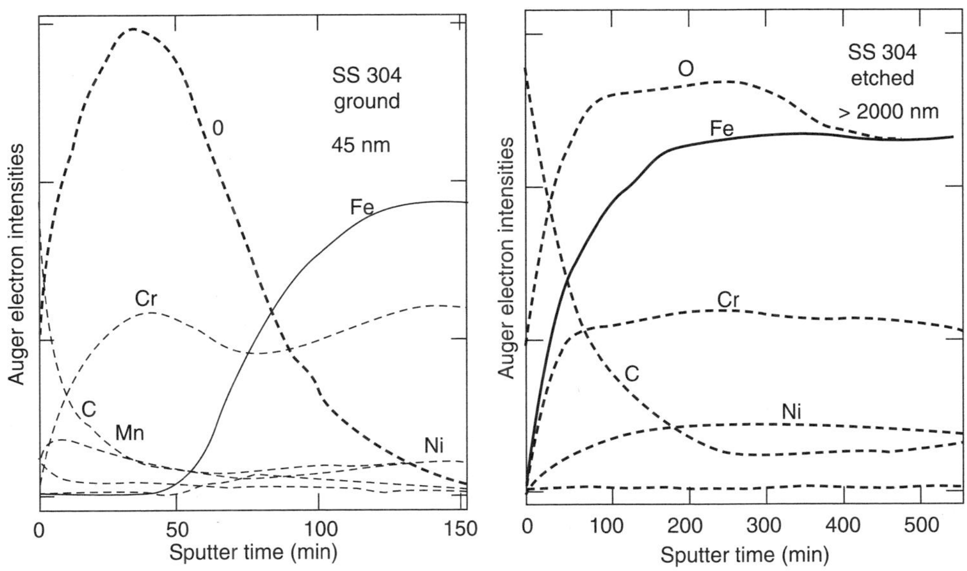

Figure 2.6 A typical set of depth profiles, taken from the corrosion product on a 9% Cr steel: left, on a surface prepared by grinding; right, on a chemically etched surface. (After Grabke, H.J., *Surface and Interface Analysis* 30, 112 (2000). With permission.)

on the basis of quantitative depth profiles. To enable this, the profile has to be transformed into a plot of composition against etch depth. The etch rate depends on the gun parameters, mainly the ion current density at the surface of the sample. It is usual to calibrate etch rate against known standards, such as anodic oxide films on tantalum, that are available as sets covering a range of thicknesses from national standards laboratories. Calibrations of etch rate for tantalum oxide against other oxides are available for a number of oxides (16). The peak intensity needs to be calibrated against local standards and normalized to give compositions in atomic %, by, for example, the methods suggested by Hall et al. Asterman et al. (17) have shown that local standards, for example, for oxygen, should include the use of mixed oxides, such as iron chromium spinels, which are often found in work on corrosion. The importance of this is seen in the depth profiles of Figure 2.7. The upper profile uses an oxygen signal calibrated against a simple oxide whilst the lower profile is calibrated using data from a series of mixed Fe–Cr oxides. The lower profile gives the expected value of 60 At% for the oxygen content.

Corrosion scientists are often interested in the oxide–metal interface. The interface region in a depth profile is broadened by virtue of the significant information depth carried by the Auger electron and because of the roughening that takes place during ion etching. Roughening can be minimized by sample rotation as pioneered by Zalar (18). The information depth, however, depends on the energy of the Auger electron and on the angle to the surface at which the electrons are collected. The probability of an electron escaping from the surface region

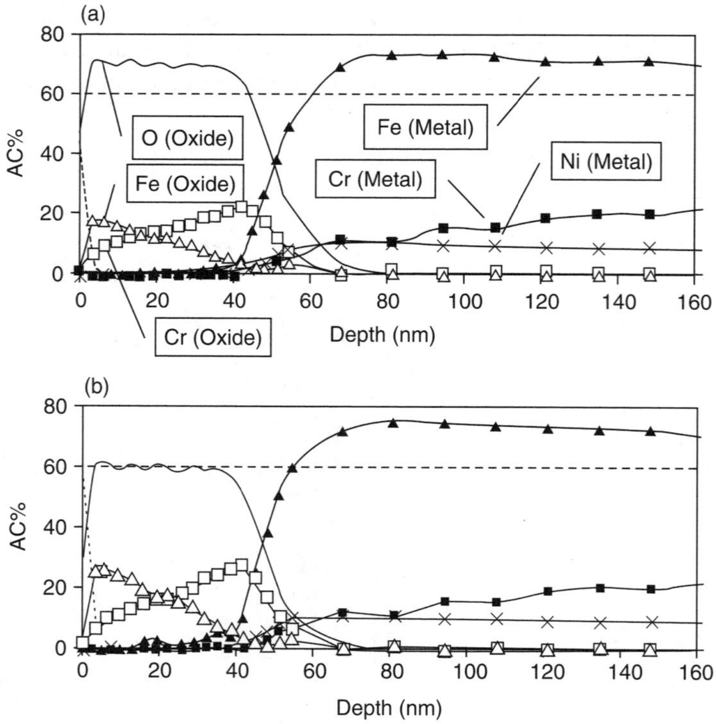

Figure 2.7 Depth profiles through an oxide on 304L stainless steel. The lower spectrum has been quantified by the method suggested by Astermann et al. (After Astermann, H., et al., *Surface and Interface Analysis* 34, 234 (2002). With permission.)

decreases exponentially with its depth of emission, d, according to the relationship (19)

$$I_d = I^\circ \exp(-d/\lambda \sin\theta) \tag{2.6}$$

where I_d is the measured intensity and I° the intensity from the same layer located at the surface. The characteristic length, λ, in this relation is the effective attenuation length of the electron (EAL), and θ is the angle subtended at the surface by the detector. In earlier publications, the parameter represented by λ is referred to as the inelastic mean free path (imfp). The imfp is a material property and depends on electron energy and some material properties (21). However, it is now realized that elastic scattering modifies the signal derived from a surface layer and that the depth resolution offered by AES will depend on instrumental parameters such as take-off angle. The parameter determined by use of overlayers of calibrated thicknesses is the EAL (22). This varies from peak to peak along the spectrum (23) but as an easily remembered rule of thumb is given in Ångstrom units, by half the square root of the kinetic energy. The range of analytical depth, say 1 to 5 nm, is around one-thousandth of the x-ray analytical depth, as used in EDS analysis in the electron microprobe analyzer.

A good discussion of depth profiling can be found in a review by Hofmann (24) and in recent work he has shown how broadening of the interface region can be minimized by use of a depth resolution function.

Many of the early studies on passive films were undertaken by AES: typical examples are seen in the publications of Mathieu and coworkers (25). Etch profiles based on electron spectroscopy have the disadvantage that depth resolution for such thin films (ca. 2 to 5 nm) is dominated by the EAL and thus the information depth is comparable to the film thickness. However, the oxygen profile does give a good indication of oxide thickness. Depth profiling is an excellent method of revealing the presence of minor elements, either derived from the electrolyte or from the substrate material, for example, sulfur and other trace elements have been revealed in the passive oxide by use of AES. The distribution of specific chemical states within the profile is much more problematic. Many materials rapidly degrade under the action of both the ion gun and the electron gun and to this can be added the difficulty that many chemical shifts are lost in the broad Auger peaks.

Although any discrete peak shifts may be lost within the broad peak, this does not means that there are no changes consequent on change of chemical state. Often there are quite clear changes in peak shape. The individual shapes can be picked out of the spectra obtained as ordered sets during profiling by means of principle component analysis (PCA). There are various mathematical forms of PCA but they all share the principle of determining a set of shapes that together add up to the overall shape of the spectrum. These shapes, or factors, do not even need to have a physical meaning but the expectation is that through careful choice of particular factors each will be representative of the contribution made to the overall spectrum by a given chemical state.

A common form of PCA is known as Factor Analysis. The complete set of spectra acquired in the depth profile is subjected to factor analysis and, when successful, yields (a) the chemical components of the spectra, with a signal-to-noise ratio that is much better than any individual spectrum and (b) the distribution in depth of these components. Gaarentroom (26) first pointed to the advantage of this technique and a good review with examples has been given by Solomon (27). Hofmann and Steffen (28) used the technique to interpret AES spectra of the early

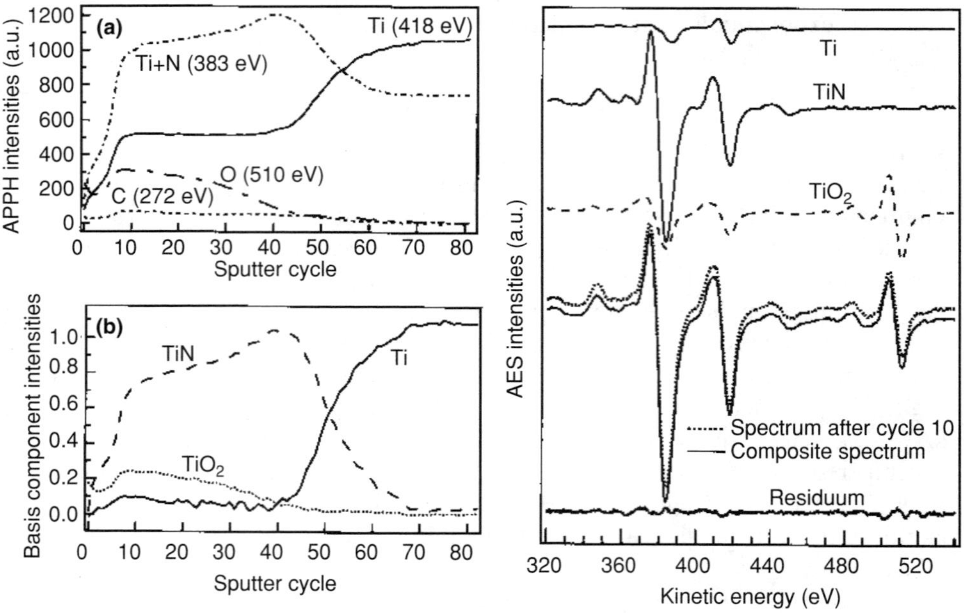

Figure 2.8 Decomposition of a depth profile into its component species using a least squares fitting of the reference spectra. (After Kovac, J., et al., *Surface and Interface Analysis* 30, 190 (2000). With permission.)

stages of the oxidation of a $NiCr_{21}Fe_{12}$ alloy and, in the case of the passive film on nickel, revealed a change of nickel valence state at the interface. Palacio and Mathieu have described the use of factor analysis in the oxidation of chromium (29). An alternative to factor analysis has been described by Kovac et al. (30) in which the depth profiles are decomposed into their chemical states by applying linear least squares fitting of spectra obtained from each of the pure components believed to be present. In contrast to factor analysis, in which principle components are blindly extracted from the overall shape, this technique, when it can be done, has the advantage that each of the principle components refers to a physical reality, which is the spectrum of a known chemical state. Figure 2.8 shows the results of this process applied to reveal the depth distribution of titanium oxide on the surface of titanium nitride. A typical spectrum, obtained after the 10th etch cycle, is shown to illustrate the close match that can be obtained by addition of component spectra of pure constituents in the appropriate amounts. The differing amounts used to fit each etch cycle are then plotted to give the resultant depth profile, as shown in Figure 2.8(a) and (b).

2.4 SCANNING AUGER MICROSCOPY

The Auger spectrum is the basis of SAM (6) in which an elemental map is created using the intensities of Auger peaks excited by the electron beam of an ultrahigh vacuum (UHV) SEM. In addition to the good depth resolution of AES, SAM enables the excellent spatial resolution to be fully exploited. Over the past 30 years, the spatial resolution has steadily improved, from the 20 to 50 μm obtainable on early instruments, to that of the modern Auger microscopes capable of acquiring spectra from spot sizes below 10 nm. The spot size controls the resolution in SAM, since, unlike x-ray spectroscopy, there is no spreading of the electron beam within the analytical depth of AES. Modern Auger microscopes have high-brightness sources delivering a higher electron flux for a given beam diameter and improved transfer lenses and multidetector analyzers for increased count rates. The development of electron sources in SAM has closely followed the progress made in SEM. At present, the two most popular methods for generating electron emission from a source are thermionic and Schottky emissions.

The source available has more impact than any other factor on the resolution attainable in the SAM image. Thermionic emission involves heating the source until electrons at the surface have enough energy to overcome the work function. The standard source is a hairpin-shaped tungsten wire. Tungsten filaments have the advantage of being cheap, simple to align, and not needing stringent vacuum requirements. For smaller spot sizes LaB_6 filaments are used. LaB_6 has a lower work function than tungsten and consequently has a higher brightness and also has a longer lifetime, up to 1000 h. To prevent poisoning of the filament, a vacuum of 10^{-6} mbar or lower is required and differential pumping is needed during ion beam profiling. Schottky emitters and other field emission sources have an even higher brightness, some 50 times greater than that of LaB_6, due to the small emission area and lens effect of the curved emitting area. As with LaB_6, a good vacuum is required for stable operation and the source needs differential pumping. The Schottky effect describes the effective decrease of the potential barrier faced by conduction band electrons when an electric field is applied to the surface. The most successful Schottky emitters, zirconium-coated tungsten, have enabled the problems with filament contamination of field emitters to be overcome. Schottky, zirconium-coated tungsten, tips run at about 1850 K; this high working temperature and the lower reactivity of Zr compared to W reduce adsorption of residual gases. Electron spot sizes of 15 nm can be achieved at 15 keV accelerating voltage. UHV is standard in Auger microscopes and these sources are fitted on most new dedicated Auger microscopes.

2.4.1 Image Acquisition

The acquisition of a set of images always requires choices to be made in order to make most efficient use of the available instrument time. This is particularly important if it is necessary to minimize the exposure time of the sample, because in-spectrometer degradation is a problem, or if the sample is undergoing dynamic changes, for example, when mounted on a hot stage to study surface changes during annealing. Such topics have been the subject of detailed study (31) but it is useful here to summarize some of the key issues. The choices to be made will involve: spatial resolution, that is, the number of pixels in the image for a given magnification; the energy resolution, in particular whether chemical state information is to be resolved; and the signal-to-noise ratio in the image, determined largely by the dwell time at each pixel. An image of resolution 128×128 pixels will contain more than 16,000 analyses per peak and, since the peak intensity must be defined by the intensity at, at least, two points on the spectrum this gives 32,000 analyses per element. A minimum description of the surface state in, for example, a study of chloride attack on a manganese sulfide inclusion in a 316 stainless steel will require analysis for nine elements: Fe, Cr, Ni, Mb, O, C, Mn, S, and Cl. This then gives a requirement for some 300,000 separate analyses. In order to obtain a reasonable signal-to-noise in the image a minimum acquisition time of 10 msec would be required, which, allowing for the computer overhead incurred in moving the beam, setting the analysis energy, and analyzer parameters, and recording the data, adds up to a minimum total acquisition time of 2 h. This time will be reduced if the instrument can be set up so that one detector counts on the peak position while another counts on the background intensity. However, it should be noted that, since the intensity relating to a given element is defined by only two sets of counts, peak top and peak bottom, it contains no chemical state information. Thus, energy resolution can be set to a minimum value. To obtain chemical state information the energy resolution must be increased so that the peak positions for the different states can be selected and this usually means that the total acquisition time is increased by a significant factor.

The acquisition times may seem surprisingly long compared with the time taken for mapping using EDS on a typical SEM. The difference stems from two factors. First, EDS is a parallel analyzer, that is, all energies are acquired simultaneously whereas the electron spectrometer is a serial analyzer in which the energies are scanned in sequence. Second, the volume analyzed in SAM is so much smaller than in x-ray analysis. In the latter, a volume of approximately 1 μm cube contributes to the signal, whereas in SAM it is of the order of 100 $nm^2 \times 10$ nm in depth, that is, smaller in volume by a factor of 10^4. These inherently long acquisition times have important consequences for the use of SAM in corrosion science. The engineer studying the structure of solid-state devices will know exactly where to direct attention in order to confirm a manufactured structure but the corrosion scientist has no certainty about where to examine a surface in order to understand what might have led to its degradation. To make efficient use of the expensive instrumentation the work carried out needs to be well planned, as described in the case histories later in this chapter.

2.4.2 Topographical Correction

Corrosion studies do not always involve flat surfaces. When they are rough there will be some regions facing the detector and giving a very strong signal while others may be facing away from the detector and thus give very little signal. This "topographic" variation in signal will often be much stronger than that arising from the compositional changes on the sample surface and must be removed for the results to be interpreted in any meaningful manner. Fortunately, the intensities of both the peak, P, and the background, B, signals vary in the same way. An Auger electron image, reflecting principally the chemical changes across the area, can be obtained by basing the map on the algorithm (P − B)/B. Other algorithms have also been suggested, the most widely used being (P − B)/(P + B) (32). Although these algorithms remove the primary cause of intensity variation in an Auger peak, small intensity variations might still arise because of artifacts associated with rough surfaces. For the corrosion scientist an important one arises with passive films that will appear thicker on surfaces that have glancing angle emission relative to the detector than on those that are normal to the detector. This is well understood and in XPS is the basis for nondestructive depth profiling.

2.4.3 Correlations in Maps

Auger maps can offer a limited form of chemical state information when those for different elements are examined in correlation with each other. Since each pixel of the image represents a complete analysis it is possible to produce correlation or anticorrelation maps of the surface (33). A typical example, a metallographic cross-section

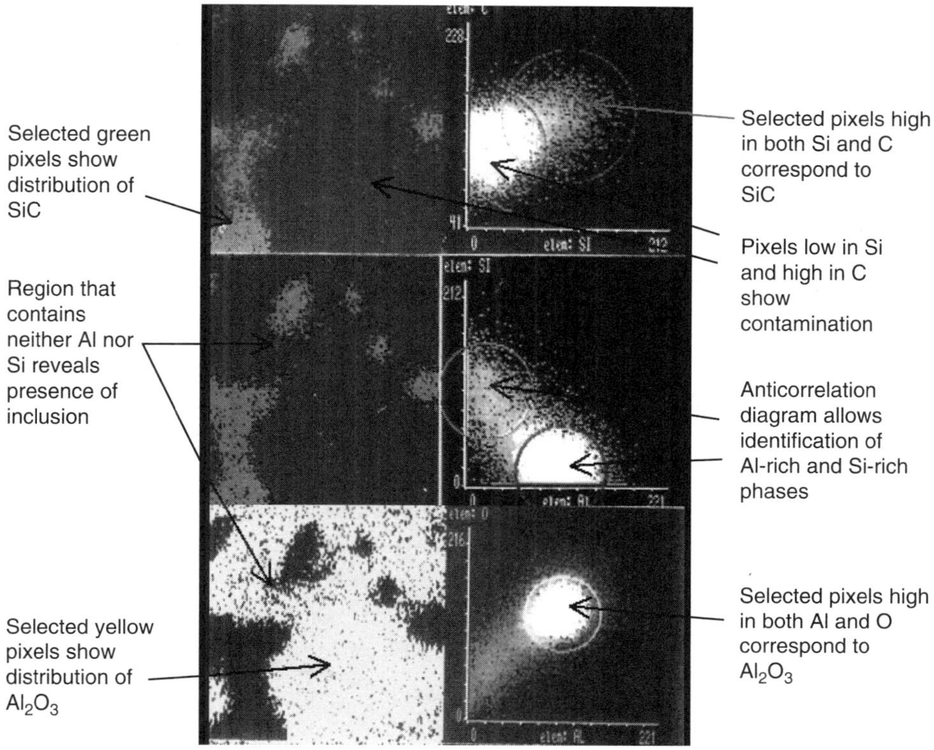

Figure 2.9 The pixels identified as being of significant interest are located by means of a scatter diagram (right images) and used to color-code the distribution of surface phases (left images).

though an aluminum alloy reinforced by silicon carbide particles, is shown in Figure 2.9. A scatter diagram is first produced in which the intensities of two given elements are plotted against each other for each pixel of the map. Phases containing both elements show up as clusters on the diagram. The mean ratio of the intensities constituting this cluster then can be interpreted as a *phase*, often of known composition, and the pixels of this phase can be highlighted in a map. Other regions, where there is no correlation between the elements, show up as two clusters, in which one or other of the elements has a low or zero concentration. Examples of correlated images can be seen in the plots for the elements found in a mapped region of the passivated surface of the silicon carbide/aluminum composite (34). By selecting the appropriate pixels in a scatter diagram, it is possible to plot the distribution of the phase Al_2O_3. Similarly, the phase SiC can be plotted. This has two advantages: first, only carbon associated with the carbide is selected for plotting, ignoring that associated with contamination of the surface; second, any constituents that are neither the surface film of aluminum oxide nor one of the reinforcing particles are shown up for separate analysis. The scatter diagram for aluminum and carbon shows anticorrelation, indicating that there is no phase containing these two elements.

EDS is an excellent adjunct facility for a scanning Auger microscope. When this technique is available on the microscope, the bulk and the surface maps can be generated simultaneously by the same scanning beam. In this case, because there is a perfect register between the pixels of each map, scatter diagrams can be made that include a single element, in surface (SAM) and bulk (EDS) materials. The method of data analysis by use of scatter diagrams was pioneered by Prutton and coworkers who have shown how many different types of images might be combined to give much additional confidence in the data quality (34).

2.5 APPLICATIONS IN CORROSION SCIENCE

2.5.1 Point Analyses

An analysis made on a fixed point on the surface is frequently taken as a precursor to more detailed investigations by depth profiling and by SAM. The usual purpose in this case will be to set up the required windows for each of the elements to be acquired. However, in other cases, the point analysis provides all that is required. An interesting example is provided by a set of analyses taken from a sample of Fe20Cr alloy as it was raised in temperature on a hot stage. The data are shown in Figure 2.10, taken from the work of Brooker (36). At temperatures up to 360°C, the

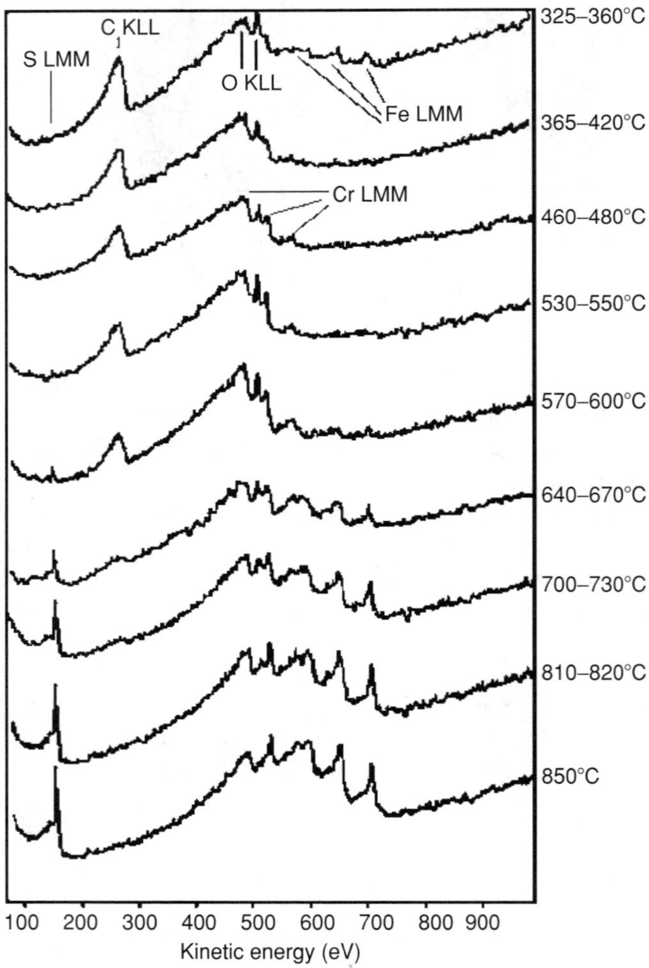

Figure 2.10 Illustration of the role of point analyses: the changing composition of the surface of a chromium steel as it is heated in vacuum. The changes are driven by thermodynamic free energy.

surface oxide contains both iron and chromium ions. As the temperature is increased beyond this, chromium oxide replaces iron completely as a result of a solid-state redox reaction. There are other significant changes: by 640°C carbon has decreased to zero as a result of reaction with the oxide and as the oxide is destroyed, so sulfur segregates to the surface, becoming very apparent as the temperature reaches 700°C. Moreover, as soon as the chromium oxide is lost then iron returns to the surface to give a final composition that reflects the composition of the bulk alloy. These spectra give an illustration of thermally driven changes in the surfaces of metals that will often involve the replacement of an oxide by another that is thermodynamically more stable. Manganese oxide and silicon oxide are two that frequently form the most stable surface on steels. There are two messages for the corrosion scientist from this type of study: first, vacuum annealing is not necessarily the best means of starting an experiment; second, similar changes can be induced by overlong exposure to the electron beam itself. The segregation of sulfur to the oxide-free surface is driven by the thermodynamic requirement to minimize the surface free energy and was shown by Hondros and Seah (37) to be a major cause of certain types of embrittlement and has been reviewed by Seah and Lea (38). Recently, Greef et al. (39,40) have examined the competition between sulfur segregation and oxygen adsorption as a function of temperature and oxygen exposure (in Langmuirs). Their papers give good examples of quantitative depth profiles and of surface composition as a function of experimental parameters using AES in the context of the oxidation of 9Cr1Mo steel used in the PWR-type of nuclear boilers.

Point analyses taken from recognizable features in an oxidized surface are frequently used as an aid to mechanistic interpretation. Examples may be found in the work of Grabke that again shows differences in the segregation of sulfur and other elements to the interfaces revealed by spalling (16).

A series of point analyses is often an effective method of obtaining a line scan across a feature or of assessing whether to carry out a mapping procedure for a given field of view. Flis et al. (41) have given a good example of the use of point analysis to demonstrate the accumulation of carbon beneath the nitride layer on a plasma-nitrided steel. The series of point analyses were made across the crater obtained by the use of a ball-cratering device. Figure 2.11 shows, diagrammatically, the spherical cut through the surface layers and the results of the Auger analyses. Ball cratering and other methods of producing a shallow cut though the surface layers are a useful alternative to ion etching, particularly when, as is often the case in corrosion films, there is the possibility of beam-induced reaction between reactive components.

2.5.2 Depth Profiles

Auger depth profiling is used for a number of purposes in corrosion studies and is the most frequent reason for the adoption of Auger spectroscopy in a corrosion investigation. Predominantly, it is used to measure the thickness of a thin (submicrometer) layer, most usually by determining the extent of the oxygen profile. Often, this is done by selecting an "end point" on the oxygen profile; for example, the thickness at which the oxygen intensity drops to a mid-point value between the maximum and the minimum values (42). The "Round Robin" reported by Olefjord and Marcus (42) showed the good reproducibility of this measurement made at a number of collaborating laboratories. A second reason for the use of AES depth profiling is to determine the distribution of the metallic cations derived from the alloy. Third, it is used to find the position, relative to the oxygen profile, of any

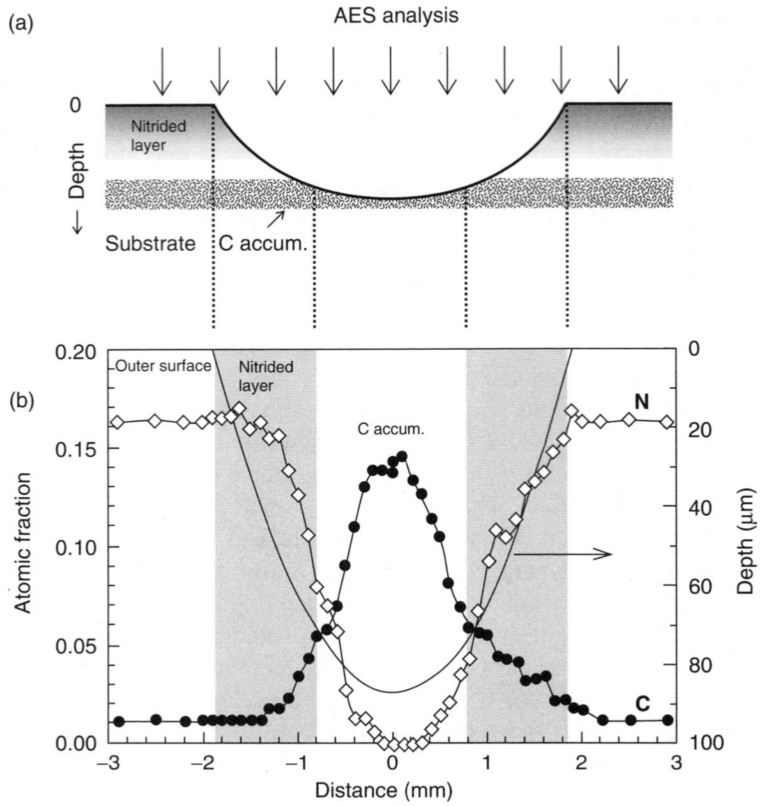

Figure 2.11 Point analyses taken as a line-scan across a cratered sample, in this case identifying carbon as an interface layer in a nitrided steel. (After Flis, J., et al., *Corrosion Science* 42, 313 (2000). With permission.)

adverse anions such as chloride (5, p. 1383) and sulfide or of any film enhancing ions such as phosphate or borate. Probably its most useful and longstanding role (43) has been to examine the enrichment or depletion of particular alloying components in the corrosion film. Many examples in the literature are concerned with the passivation of stainless steels in aqueous media (44,45) and depth profiling has confirmed the depletion of iron and nickel within the passive oxide. Metallic nickel is shown to enrich at the oxide–metal interface but there is no such enrichment of iron and its depletion results from preferential dissolution. Good agreement has been found in the oxide thickness on stainless steel measured by AES and XPS. Hakiki et al. (46) have used depth profiles in the form of the Cr/Fe ratio to show that the iron oxide segregates to an outer layer on 304 stainless steel and that the thickness of this layer increases steadily with oxidation temperature.

A vivid illustration of the precision of AES depth profiling is given by an oxidation study of W720 maraging steel in air over a temperature range of 300 to 600°C (Figure 2.12). This investigation, by Greyling et al., required the measurement of oxide thickness for a large number of samples, in each case by AES depth profiling: as can be seen the measurements all fall on smooth curves with remarkably little scatter (47). The same authors give a particularly interesting use of AES depth profiling in the context of mechanistic studies, using AES to find the location of a gold marker after intervening periods of oxidation. The method is illustrated in

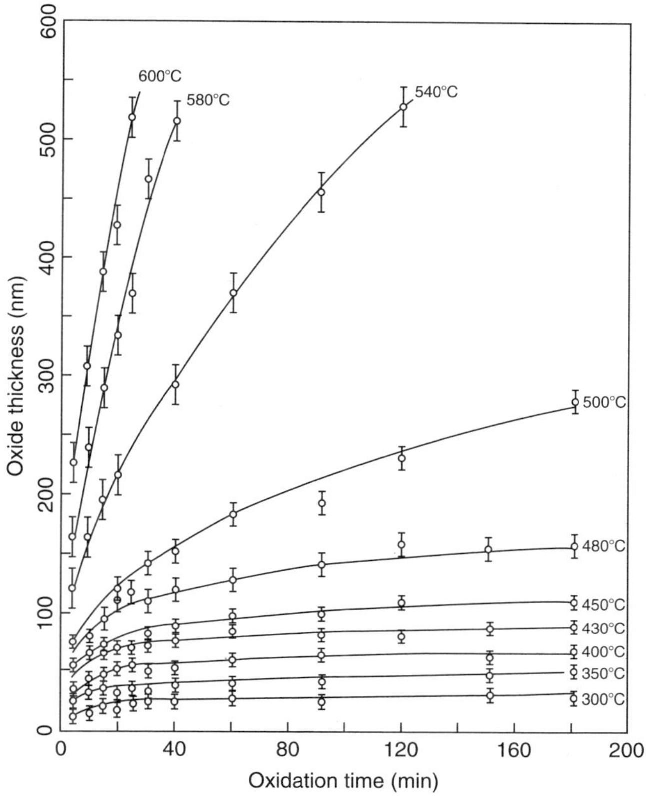

Figure 2.12 Oxidation kinetics determined using AES depth profiling to measure oxide thickness. (After Greyling, C.J., et al., *Surface and Interface Analysis* 16, 293 (1990). With permission.)

Figure 2.13. The diagram illustrates the manner in which a gold film, deposited after the first oxidation stage, becomes buried by the newly formed oxide, a clear indication that the film grows by cation diffusion. The use of inert markers is a classic method for the investigation of diffusion processes. The advantage of AES, as shown in this work, is that it can be applied to layers that are <100 nm in thickness, much less than is achievable by the usual techniques of metallography.

2.5.3 SAM in Aqueous Corrosion

An illustration of the use of maps to show the distribution of phases has already been given (Figure 2.9). Scanning Auger micrographs are particularly useful when their ability to locate deposits of nanometer thickness ranges is exploited. SAM is, however, a time-consuming technique and some form of systematic approach is required if cost-effective results are to be obtained. In the case of aqueous corrosion the systematic approach can be based on that required to identify sites and areas of electrochemical activity on the surface. Immersion of a metal in an electrolyte frequently leads to electrochemical activity and to the establishment of anodic and cathodic sites on the surface. The surface chemistry, that is, the ions adsorbed and the material deposited as a result of corrosion, rapidly changes to reflect the difference in electropotential and this can be detected in the SAM. The ions or material found on the surface need to be associated with either: (a) current flow through, or (b) a

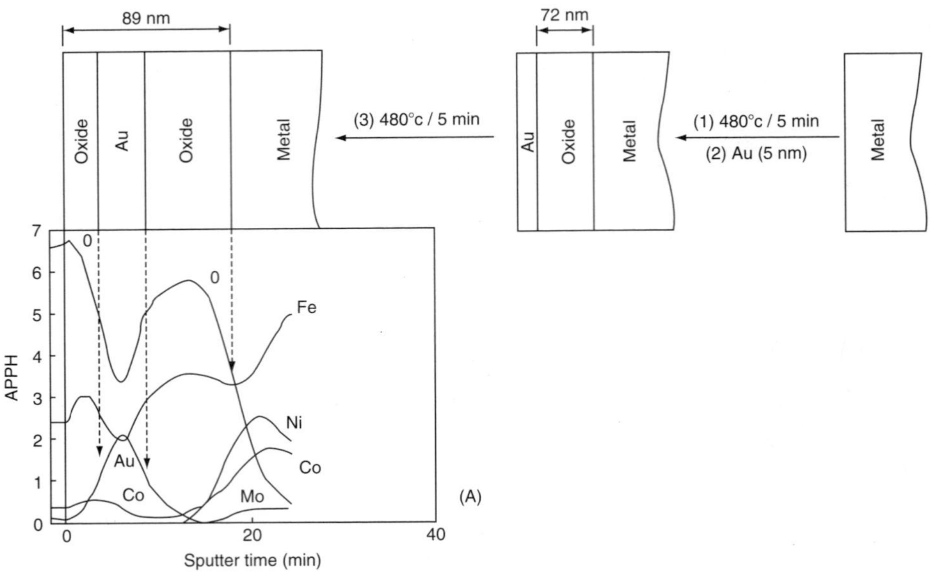

Figure 2.13 A marker experiment in which AES depth profiling is used to locate the position of the gold marker layer after a brief period of oxidation. (After Greyling, C.J., et al., *Surface and Interface Analysis* 16, 293 (1990). With permission.)

particular electropotential on that part of the surface. The current flow or potential difference then implies electrochemical activity and hence the sites that were active at the time of removal from the electrolyte can be established. The circumstances that lead to ions being segregated within the surface layer were discussed in a review by Castle (48). As an example, we can consider the change in the pH within the boundary layer as a result of the electrochemical current density at any local cathodes on the surface. The change in concentration of OH ions may lead to the solubility product of some metal hydroxides being exceeded, leading to the formation of a deposit, decorating the cathodic areas. This is such a useful device for the investigation of surface electropotentials, and the associated current flows, that it is often useful to add marker ions, such as Mg^{2+}, to the test solution.

Both of the common cathode reactions create a change in local pH. Generally, the local pH will rise until the rate of escape of hydroxyl ions by diffusion is equal to their rate of production. We can then write

$$i/A = (F/\delta)\{D_{OH^-}\Delta[OH] - D_{H^+}\Delta[H]\} \qquad (2.7)$$

where i/A is the current density at the cathode, F is the Faraday constant, δ is the boundary layer thickness, D_{OH^-} and D_{H^+} are the respective diffusion coefficients, and Δ[OH] and Δ[H] represent the concentration drop across the thickness of the boundary layer. Addition of magnesium ions to the electrolyte will, at a critical concentration, cause the precipitation of $Mg(OH)_2$, which is readily detected by SAM. The determination of the critical current density for precipitation through the cathode is possible by solution of the equation derived from (2.7) using the solubility product and the ionic product of water. Having bracketed, with SAM, the magnesium concentration and pH value at which a deposit appears, the corresponding current density at which the cathode was operating can be readily determined (34).

At anodes, deposition of a salt depends on the concentration of metal ions and thus on the *anodic* current flowing through the exposed surface. This will give rise, for example, to the chloride deposits that are readily observed by Auger spectroscopy and have been shown as SAM maps by Castle (49) and by Baker and Castle (50,51). Chlorides, however, are soluble and therefore most likely to be found in occluded cells and hidden from the limited depthwise view of the electron spectrometer. Moreover, we cannot probe the metal ions within the pit by varying the concentration of chloride ions in the electrolyte: the link with the concentration in the pit is too indirect. Complementary studies may be made using effects that depend on local potential rather than current density. In this case, the deposition of a marker on the surface needs to exploit a redox reaction. Typical of these is the reduction of copper ions to copper metal at the cathode. Nickel salts have also been found useful and many other cations might find use depending on the potential range of the source to be investigated. The ions chosen should neither be part of the metal system, nor should they stimulate corrosion.

The presence of potential differences on the surface can lead to species transport that can be recognized in the Auger map. Comparison of the microscale chemistry determined by SAM with that obtained on whole electrodes held at known potentials enabled Castle and Ke (52) to estimate the potential differences set up at the site of an incipient pit. Figure 2.14 from their work illustrates how sulfur generated by hydrolytic attack on a sulfide inclusion is captured by the cathodic annulus surrounding the anode associated with the inclusion.

When corrosion reaches the stage of visible pits, the surface selectivity of SAM is invaluable in establishing the presence of salts, derived from the electrolyte, within the cavity. High spatial resolution SAM has enabled investigations to be conducted very shortly after the pit has formed. After an immersion time of 10 sec, examination of pit nucleation and initial growth at oxide inclusions and evidence of pits propagating laterally under the passivating oxide films has been obtained in the combined SEM and Auger-EDS imaging study summarized by Baker (53).

2.5.4 Chemical State Imaging

In certain cases it is useful to exploit the differences in chemical state that are apparent in the spectra of the nontransition elements. The example taken here concerns the influence of an oxide film on the fracture toughness of a spray-formed A15.2Crl.4Zr alloy being developed for the aerospace industry. The surface was exposed by fracturing the sample using a UHV fracture stage (53). Aluminum and its alloys are ideal substrates to examine using AES and SAM because of the good chemical shift between the peaks for the metal and those for the oxide. Point spectra of the Al KLL region taken from two different areas on the fracture surface are shown in Figure 2.15, spectrum (a) and (b). The lower spectrum (a) is taken from the general fracture surface and shows a typical spectrum of Al metal with the main KLL peak at 1393.5 eV. The upper spectrum (b) is taken from the cratered region in the lower part of the secondary electron image (center) and shows an additional peak at 1386.0 eV due to a shift of the KLL peak, characteristic of Al_2O_3. In order to reveal the chemical state information separate images are made using each of these peaks. The spatial distribution of the oxide-rich areas is revealed by subtraction of the Al (metal) image from the Al (oxide) image. This image (Figure 2.15, right) (54) shows that in the forming process, the oxide on the surface of the metal powder is not broken up and uniformly distributed throughout the bulk as expected, but can remain localized, causing a reduction in the mechanical properties of the material.

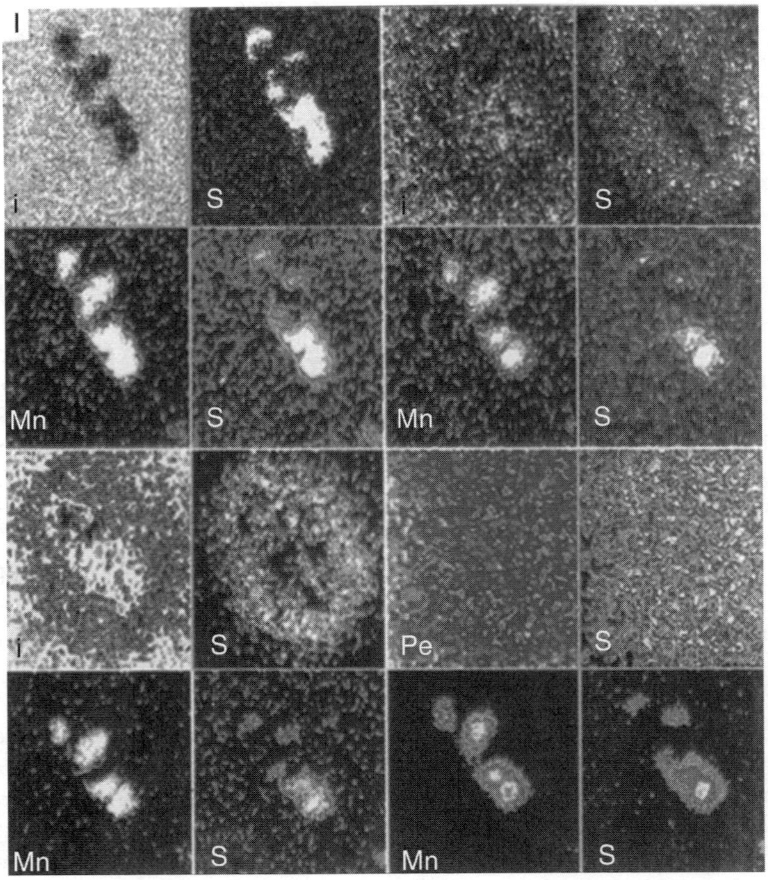

Fe A 0 sec	S A 0 sec	Fe A 900 sec	S A 900 sec
Mn X 0 sec	S X 0 sec	Mn X 900 sec	S X 900 sec
Fe A 2100 sec	S A 2100 sec	Fe A 4800 sec	S A 4800 sec
Mn X 2100 sec	S X 2100 sec	Mn X 4800 sec	S X 4800 sec

Figure 2.14 A set of EDX and SAM micrographs. As shown by reference to the key above, MnS hydrolyzes to release sulfur that forms an annulus around the inclusion (the overall length of the inclusion group is 4 μm).

2.6 CONCLUSION

This review has shown that Auger electron spectroscopy is a valuable tool throughout the entire range of corrosion science studies. As a spectroscopy, it is, in appropriate circumstances, able to reveal different chemical states. As a means of tracking different phases in thin films it is rivaled only by secondary ion spectroscopy but has the advantage of being essentially nondestructive. It has been shown that the test pieces

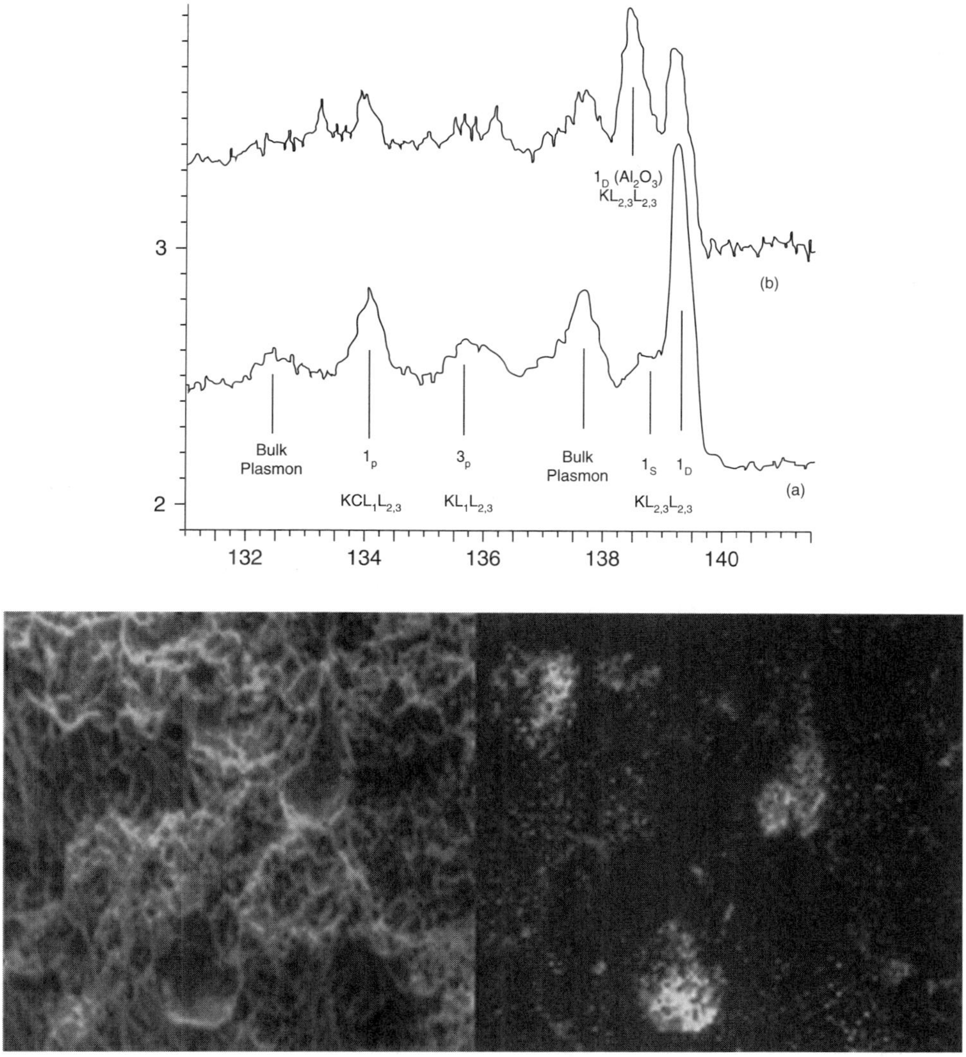

Figure 2.15 Fracture cross-sections of a spray cast aluminum alloy. The chemical shift of the oxide relative to aluminum metal enables the location of oxide in the fracture surface to be revealed. (After Baker, M.A. and Tsakiropoulus, P., *Surface and Interface Analysis* 20, 589 (1993). With permission.)

can be examined and returned for repeated exposure. It is much used for depth profiling through thin and moderately thick films and chemical states existing only at the interface can be detected. With the steady development of sensitive electron detection techniques and sources of higher brightness it is like the acquisition times for large maps, the major disadvantage of SAM will be much reduced.

REFERENCES

1. S. Briggs and J.T. Grant (eds.) *Surface Analysis by Auger and X-ray Photoelectron Spectroscopy*. Chichester, UK: Wiley (2003). See especially J.T. Grant, Chapter 3; M.P. Seah, Chapter 13; T. Wagner, J.Y. Wang and S. Hofman, Chapter 22.

2. J.M. Walls (ed.) *Methods of Surface Analysis*. Cambridge: Cambridge University Press (1990). See especially H.E. Bishop, pp. 87–126.
3. R.W. Cahn and E. Lifshin (eds.) *Concise Encyclopaedia of Materials Characterisation*. Oxford: Pergammon (1993), pp. 83–91. See especially: H.L. Marcus, pp. 28–34; S. Hofmann, pp. 83–91.
4. J.C. Vickerman (ed.) *Surface Analysis — The Principle Techniques*. New York: (1988). See especially H.-J. Mathieu, Chapter 4.
5. M.J. Graham, *Corrosion Science* **37**, 1377 (1995).
6. M.A. Baker and J.E. Castle, Scanning Auger microscopy, in: E. Lifshin (ed.) *Materials Science & Technology, Vol. 2B: Characterisation of Materials*, Chapter 13. Weinheim: VCH Publishers, pp. 219–239 (1993).
7. J.C. Rivière, *Ullmann's Encyclopaedia of Industrial Chemistry*, Vol. B6. New York: VCH Publishers, pp. 26–120 (1994).
8. P. Auger, *Surface Science* **48**, 1 (1975).
9. M.F. Chung and L.H. Jenkins, *Surface Science* **21**, 253 (1970).
10. G. Moretti, *Journal of Electron Spectroscopy and Related Phenomena* **95**, 95 (1998).
11. S. Ichimura, R. Schimizu, and J.P. Langeron, *Surface Science* **124**, 49 (1983).
12. M.P. Seah, in: D. Briggs and M.P. Seah (eds.) *Practical Surface Analysis,* Vol. 1. Chichester, UK: Wiley, pp. 201–255 (1990).
13. P.M. Hall, J.M. Morabito, and D.K. Conley, *Surface Science* **62**, 1 (1977).
14. P.M. Hall and J.M. Morabito, *Surface Science* **83**, 391 (1979).
15. H.J. Grabke, *Surface and Interface Analysis* **30**, 112 (2000).
16. H. Viefhaus, K. Hennesen, M. Lucas, E.M. Muller-Lorenz, and H.J. Grabke, *Surface and Interface Analysis* **14**, 59 (1994).
17. H. Asterman, R. Norling, J.-E. Svensson, A. Nylund, and L. Nyborg, *Surface and Interface Analysis* **34**, 234 (2002).
18. A. Zalar, B. Pracek, and P. Panjan, *Surface and Interface Analysis* **30**, 247 (2000).
19. C.R. Brundle, *Journal of Vacuum Science & Technology* **1**, 212 (1974).
20. A. Jablonski and C.J. Powell, *Surface Science Reports* **47**, 33 (2002).
21. M.P. Seah and W.A. Dench, *Surface and Interface Analysis* **1**, 2 (1979).
22. C.J. Powel, *Journal of Vacuum Science & Technology A* **21** (5), S42 (2003).
23. J.E. Castle, *Surface and Interface Analysis* **8**, 137 (1986).
24. S. Hofmann, *Surface and Interface Analysis* **30**, 228 (2000).
25. S. Mischler, H.J. Mathieu, and D. Landolt, *Corrosion Science* **37**, 467 (1995).
26. S.W. Gaarentroom, *Applied Surface Science* **26**, 561 (1986).
27. J.S. Solomon, *Surface and Interface Analysis* **10**, 75 (1999).
28. S. Hofmann and J. Steffen, *Surface and Interface Analysis* **14**, 59 (1989) (see also p. 1459).
29. C. Palacio and H.J. Mathieu, *Surface and Interface Analysis* **16**, 178 (1990).
30. J. Kovac, T. Bogataj, and A. Zalar, *Surface and Interface Analysis* **30**, 190 (2000).
31. A.D. Brooker and J.E. Castle, *Surface and Interface Analysis* **8**, 13 (1986).
32. M. Prutton, L.A. Larson, and H. Poppa, *Journal of Applied Physics* **54**, 374 (1983).
33. M. Prutton, M.M. El Gomati, Inst. Phys.Conf. Ser. No. 53 (5). Bristol, UK: Inst. Phys. (1988).
34. J.E. Castle, L. Sun, and H. Yan, *Corrosion Science* **21**, 1093 (1994).
35. M. Prutton, *Surface and Interface Analysis* **29**, 561 (2000).
36. A.D. Brooker, PhD Thesis, University of Surrey (1986).
37. E.D. Hondros and M.P. Seah, *International Metals Review* **22**, 262 (1977).
38. M.P. Seah and C. Lea, *Philosophical Magazine* **31**, 627 (1975).
39. A.P. Greeff, C.W. Louw, J.J. Terblans, and H.C. Swart, *Corrosion Science* **42**, 991 (2000).
40. A.P. Greeff, C.W. Louw, and H.C. Swart, *Corrosion Science* **42**, 1725 (2000).
41. J. Flis, J. Mankowski, and T. Zakroczmski, *Corrosion Science* **42**, 313 (2000).
42. I. Olefjord and P. Marcus, *Corrosion Science* **28**, 589 (1988).
43. J.B. Lumsden and R.W. Staehle, *Scripta Materialia* **6**, 1205 (1972).

44. G. Lorang, F. Basile, M. da Cunha Bela, and J.-P. Langeron, *Surface and Interface Analysis* **12**, 424 (1988).
45. G. Lorang, N. Jallerat, K. Vu Quang, and J.-P. Langeron, *Surface and Interface Analysis* **16**, 325 (1990).
46. N.E. Hakiki, M.F. Montemor, M.G.S. Ferreira, and M. da Cunha Bela, *Corrosion Science* **42**, 687 (2000).
47. C.J. Greyling, I.A. Kotze, and P.E. Viljoen, *Surface and Interface Analysis* **16**, 293 (1990).
48. J.E. Castle, *Proc. 12th Int. Corr. Congress*, Vol. 5B, NACE Int., pp. 3982–3390 (1993).
49. J.E. Castle, *Surface and Interface Analysis* **9**, 345 (1986) (see Figures 6 and 7).
50. M.A. Baker and J.E. Castle, *Corrosion Science* **33**, 1295 (1992).
51. M.A. Baker and J.E. Castle, *Corrosion Science* **34**, 667 (1993).
52. J.E. Castle and R. Ke, *Corrosion Science* **30**, 409 (1990).
53. M.A. Baker, *Surface and Interface Analysis* **20**, 535 (1993).
54. M.A. Baker and P. Tsakiropoulus, *Surface and Interface Analysis* **20**, 589 (1993).

3

Studies of Metal Corrosion and Oxidation Phenomenon Using Secondary Ion Mass Spectrometry

N. Stewart McIntyre
Surface Science Western and Department of Chemistry, Western Science Centre, The University of Western Ontario, London, Ontario, Canada

Michael J. Graham
Institute for Microstructural Sciences, National Research Council Canada, Ottawa, Ontario, Canada

Contents

3.1 INTRODUCTION

Corrosion processes begin on solid surfaces — often at microscopic locations on such surfaces. From there, it may propagate across the entire surface, down a grain boundary, or remain confined within pits or crevices. Thus, identification and analysis of the type and origin of the corrosion problem requires both surface and microscopic tools that can characterize the properties of internal structures as well as those of the outermost surfaces.

Surface analytical procedures have been used to study both corrosion processes that are carefully controlled electrochemically as well as those from industrial environments where multiple mechanisms may have been active during the exposure lifetime. The structural complexity of the corrosion products, particularly in the latter cases, frequently requires that several different techniques be used and that studies extend to depths in the material well below the outer surface. For such studies, analytical techniques that readily provide depth profiling as well as localized elemental distributions within the field of analysis have a strong advantage over surface techniques whose primary strength lies in determination of the composition of the original outermost surface.

Secondary ion mass spectrometry (SIMS) has acquired such strengths during its development over the past 20 years; the technique now seems well positioned to contribute increasingly to the corrosion literature and to the identification and correction of industry-based corrosion problems. On the other hand, techniques such as x-ray photoelectron spectroscopy (XPS) and Auger electron spectroscopy (AES), which have been more readily identified as part of the corrosion specialists "toolkit," will continue to provide specialized chemical information that is only slowly becoming available to the SIMS practitioner.

In this chapter, the principles and current instrumentation capabilities of SIMS will be reviewed, particularly with respect to the application of SIMS to corrosion specimens and problems. Then, issues associated with depth profiling and imaging will be reviewed, because both of these topics figure prominently in SIMS corrosion studies.

Finally, a comprehensive review will be made of past corrosion research that has been aided materially by SIMS, and some conclusions will be drawn about major roles that could be played by SIMS within the advancing field of corrosion science.

3.2 PRINCIPLES OF THE TECHNIQUE

SIMS is the mass spectrometric detection of ions emitted from a surface during its bombardment with a beam of particles. The bombarding particle beam (or primary beam) is composed either of positive or negative ions whose charge allows the beam to be focused to areas as small as 100 nm. The emitted (or secondary) particles are mostly neutral, but the $\sim$1% of positive or negative ions formed can be analyzed mass spectrometrically with great sensitivity after passing through an initial energy filtration. The SIMS process is represented schematically in Figure 3.1.

If the fluence of primary ions is sufficiently high, an appreciable number of surface atoms will be removed, with the effect that successive layers of the near surface region of a material can be analyzed sequentially by SIMS. This is a process called sputtering and the type of SIMS analysis that uses sputtering is called dynamic SIMS. Sputtering rates in corrosion studies may vary from 1 nm/min to 5 μm/h. If the primary beam fluence is less than 10^{13} atoms/cc, effectively no sputtering occurs, and the secondary ions emitted from the surface can be considered to come from the original surface and unaffected by possible damage by prior ion

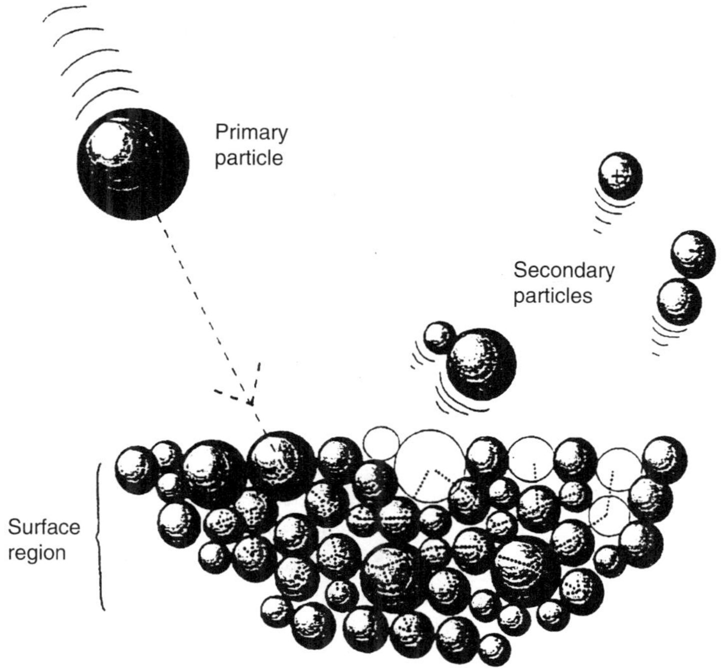

Figure 3.1 Schematic representation of the SIMS process. N. Lockyer, PhD, UMIST, 1996. (From Vickerman, J.C. (ed.), *Surface Analysis — The Principle Techniques*, John Wiley and Sons, New York, 1997. With permission.)

bombardment. This "static" form of SIMS is finding increasing use as a sensor of the chemistry of the outer surface.

By its nature, mass spectrometry has very high signal-to-background detection capabilities; this allows extremely high detection sensitivity of secondary ions with masses differing by less than 1 Da (one atomic mass unit). Because of the high mass resolution capability, it is often possible to identify highly complex molecular ions, on the basis of an accurate mass spectrometric measurement, frequently to better than one part in 10^5. The availability of high mass resolution in modern SIMS instruments has removed much of the ambiguity that earlier plagued spectral assignments.

The detection of a particular molecular ion has sometimes been used as an indicator of a certain chemical condition on the surface. In the case of dynamic SIMS, this practice is not encouraged, because sputtering could give rise to such species as part of recombination processes occurring at or above the surface. In a few cases shown below, some chemical inferences seem justified. In the case of static SIMS, the association of molecular ions with a chemical origin has been more generally adopted, but even here, some care must be exercised, because the emitting surface region has been subjected to a thermal shock during the ion production process.

Because corrosion and oxidation are often processes that are highly localized, there is much value in the ability of SIMS to recognize and display the distribution of localized oxidation products as a "map" of secondary ion distribution. In practice, this is accomplished by one of two different methods. The use of transfer ion optics in the secondary beam analysis column was previously found to produce ion images with a superior signal-to-noise ratio, as well as spatial resolution. More recently, the production of images by raster scanning a focused submicrometer primary beam has been found to produce images of higher resolution, with a comparable signal-to-noise ratio, and with no displacement between images of different mass due to acquisition at different times.

The dynamic SIMS process inevitably creates a record of the distribution of elements as a function of depth into the specimen surface, and, thus, it is possible to combine lateral and depth information into a three-dimensional (3D) image of the distribution of secondary ions produced by the specimen during the profile. Such information has proven to be extremely valuable to oxidation studies, because it provides a local picture of the nature of the oxide–metal interface. In addition, local grain boundary interface composition can be determined, not only at grain faces, but also at triple points. The presence of inclusions (such as oxides, hydrides, and carbides) can be readily identified, as can the effects of internal oxidation processes.

SIMS depth profiles have traditionally been associated with the analysis of semiconductor devices to discover the compositions of multilayer structures or dopant implants. In such cases, the effects of surface roughness, at least initially, do not have a major influence on the depth profile. By contrast, surfaces associated with corrosion processes are often quite rough, and even the process of mild oxidation is often studied on metal specimens whose surfaces are roughened by mechanical polishing. Thus, the depth information in SIMS profiles of corroded or oxidized surfaces is often affected by surface roughness. Some efforts have been made to compensate for roughness by measuring the surface topography before and after a depth profile, but this is not a procedure easily adapted to routine measurements.

As a technique that is sensitive to mass number, SIMS permits the use of isotopically labeled elements as a method for following the movement of specific elements during a particular period of an oxidation. Another virtue of the mass sensitivity is the ability to detect hydrogen, an element that is "invisible" when using surface analysis techniques such as XPS or AES. Further information on SIMS can be found in several published texts (1–3).

As is the case with all mass spectrometric techniques, SIMS requires the use of standards to produce quantitative responses. In the case of surfaces that usually have poorly defined compositions, standard addition protocols are the best approach to quantifying a particular element. Generally, this is best effected by using ion implantation to deposit a known reference quantity of a particular element into the region of interest. In most cases, corrosion-related elements (such as oxygen or chlorine) are introduced by implanting an isotope of minor natural abundance.

Three different types of mass analyzers are used for SIMS: the quadrupole, the magnetic sector, and the time-of-flight (ToF) mass spectrometers. An indication of the resolution and relative sensitivities of these mass spectrometers is presented in Table 3.1. All are used in corrosion studies. Quadrupoles, even with their low transmission, are commonly used for depth profiling of thin oxide films. ToF instruments have significant advantages for surface and near-surface analysis, but are now increasingly used for depth profiling. Magnetic sector instruments, typified by those produced by Cameca, provide higher ultimate detection sensitivity for most elements and are still the industry standard for use in the semiconductor community.

Each of these spectrometers can be equipped with one or more primary beam sources. Typical primary beams used in SIMS are cesium, oxygen, gallium, argon,

Table 3.1 Mass spectrometers for SIMS (from Physical Electronics)

Mass spectrometer	Resolution	Practical mass range	Transmission	Mass detection	Relative sensitivity
Quadrupole	10^2–10^3	$<10^3$	0.01–0.1	Sequential	1
Magnetic sector	10^4	$>10^4$	0.1–0.5	Sequential	10
Time-of-flight	$>10^4$	10^3–10^4	>0.5	Parallel	10^4

Table 3.2 Ion gun comparison (from Physical Electronics)

	Ar	O	Cs	Ga
Sputter rate	Slow	Good	Very good	Excellent
Beam size	200 μm	5 μm	5 μm	<20 nm
Maximum current	>4 μA	>3 μA	>1 μA	5 nA or lower
Ultimate sensitivity	Poor	Excellent	Excellent	Poor
Cost	Inexpensive	Expensive	Expensive	Very expensive

and sulfur hexafluoride. Oxygen enhances the yield of positive ions, and cesium the yield of negative ions. Sulfur hexafluoride imparts less damage to the surface, and is therefore used in studies of very shallow profiles and of complex organic thin films. Gallium can be focused to smaller dimensions than beams from any of the other commercially available sources; as such, it is used in acquiring most image data on ToF instruments. A comparison of the different ion guns is presented in Table 3.2.

In the next section, we describe in detail a number of studies involving corrosion and oxidation processes.

3.3 APPLICATIONS OF SIMS TO CORROSION AND OXIDATION STUDIES

3.3.1 Fundamental Studies

3.3.1.1 Oxidation and Corrosion of Iron and Iron–Chromium Alloys

The nature of electrochemically formed or "passive" oxide films on metals has been the subject of investigation since Schonbein and Faraday first developed the theory of a protective oxide skin over 200 years ago. It is now generally accepted that passivity is due to the presence of a thin oxide film 1 to 4 nm thick. However, there have been strong differences of opinion over the years as to whether the oxide is crystalline or amorphous, or whether it is anhydrous or contains hydroxyl ions or water. SIMS has proven to be a useful technique to characterize the nature of these oxides. For example, SIMS can be used to detect the presence of any H-containing species within the films, as well as studying oxide growth mechanisms and the air stability of oxide films.

In the case of iron, a number of SIMS studies have been performed in an attempt to determine whether hydrogen or hydroxyl ions are present in the passive film (4–6). Figure 3.2 shows an experimental SIMS depth profile of a passive film formed on iron in pH 8.4 borate buffer (6). The data show that the FeO^+ signal is much higher than that for $FeOH^+$, except on the outside of the oxide where a

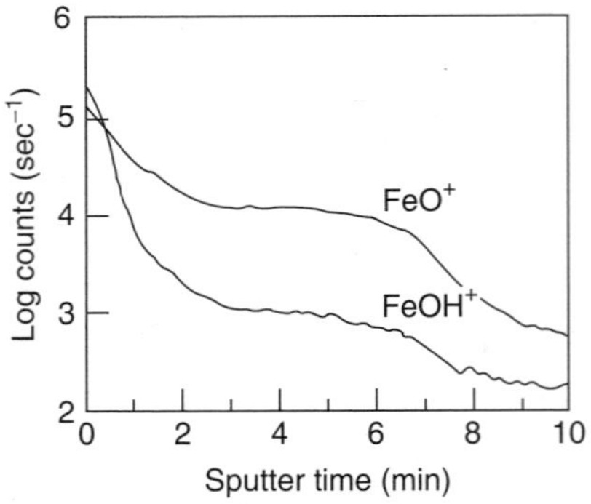

Figure 3.2 Positive polyatomic SIMS profile for a 3-nm-thick "passive" oxide film formed on iron in pH 8.4 borate buffer solution in the passive potential range. Sputtering was by 1 keV xenon. (From Mitchell, D.F. and Graham, M.J., *J. Electrochem. Soc.*, 133, 936, 1986. With permission.)

fraction of a monolayer of OH^- is adsorbed on the surface. The profile in Figure 3.2 is very similar to that for a "dry" Fe_2O_3 standard, until the oxide–metal interface is reached (after ~7 min of sputtering) when both the FeO^+ and $FeOH^+$ signals drop. Within the bulk of the oxide film, the hydroxyl content is essentially zero ($\pm$ 0.1%). Passive films formed on nickel in pH 3.0 sodium sulfate solution are also found to be hydroxyl free (6). Passive oxides on iron–chromium alloys contain both iron and chromium oxide species as shown in the SIMS profile in Figure 3.3. The chromium oxide concentration is higher in the inner part of the oxide (7). In Figure 3.3, the experimental CrO^+ and FeO^+ profiles are compared with calculated profiles using a layer-by-layer analysis model developed by Mitchell (8). The good agreement between the experimental and calculated SIMS profiles supports the view that the majority of the ions detected by SIMS were at the surface at the time of sputter removal. In order to obtain the results in Figure 3.3, it was necessary to sputter with heavy ions of low energy, i.e., 1 keV xenon ions. When similar sputtering experiments were performed with 4 keV argon, at near-normal incidence, as is frequently reported

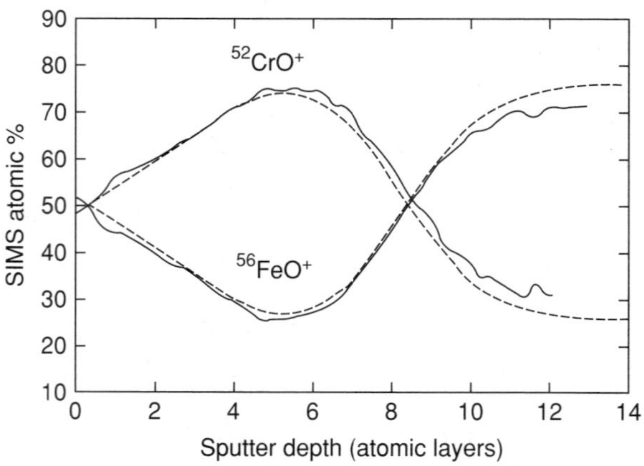

Figure 3.3 Comparison of experimental SIMS profile of a "passive" oxide film formed on Fe–26Cr (—) with the calculated profile (– – –) from the layer-by-layer analysis mode of Mitchell (8). Sputtering was by 1 keV xenon. (From Mitchell, D.F. and Graham, M.J., *Surf. Interface Anal.*, 10, 259, 1987. With permission.)

in the literature, no film composition model could be made to fit the experimental data because sputter mixing occurs. The films shown in Figure 3.3 were found to contain some hydroxyl ion (6): 0.6 $\pm$ 0.1% OH^- within the inner chromium-rich part of the film, and 1.1 $\pm$ 0.1% OH^- within the outer iron-rich part of the film. There is also a small amount of "extra" surface hydroxyl, as was found with passive films on iron.

Although it has been well known for many decades that chloride ion gives rise to local pitting corrosion of metals and alloys, the precise role of Cl^- in achieving pitting is not well understood. For example, it is still not clear whether Cl^- causes the initial local breakdown of the passive oxide film or simply interferes with the repassivation process after the film has broken down locally because of chemical dissolution. One of the favored models to explain the initiation of pitting corrosion involves the incorporation of Cl^- into the oxide lattice and its possible diffusion to the metal–oxide interface to initiate local breakdown events. To check this possibility, a considerable amount of research has been performed in an attempt to detect the presence of Cl^- in passive oxide films. SIMS is ideally suited to determine the presence of any incorporated Cl^-. Data in Figure 3.4 show that anodic films on nickel can contain up to ~4% Cl^- from the solution (9). This incorporated Cl^- makes the passive oxide film on nickel more defective, decreasing its resistance to open-circuit chemical dissolution (10). It might be expected that this incorporated Cl^- would also increase the susceptibility of nickel to pitting, but this is not the case. The resistance to pitting (based on the pitting potential) was actually much higher in

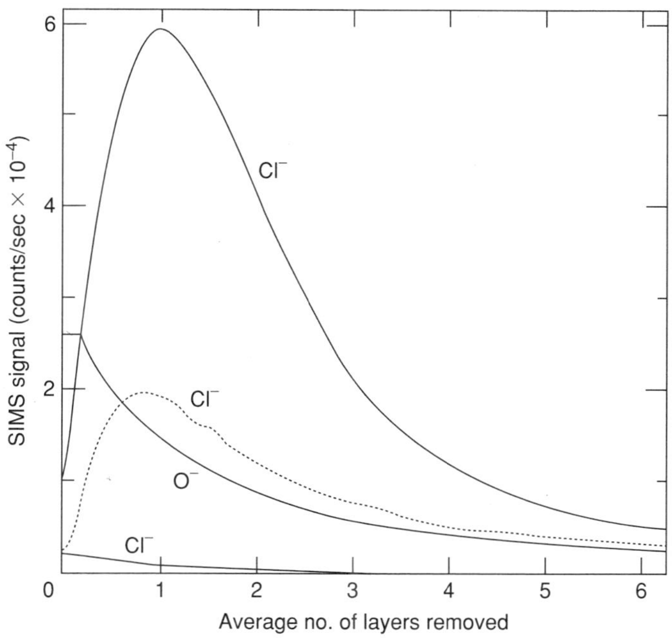

Figure 3.4 SIMS Cl^- (mass 35) sputter profiles of electropolished nickel, passivated either with or without cathodic reduction (CR) of the prior film, at 0.3 V for 15 sec in pH 4.0 Na_2SO_4 solutions with different [Cl^-]. (i) CR and passivation in a solution with [Cl^-] = 0 (- - -); (ii) no CR and [Cl^-] = 1.0 *M* (·); (iii) CR and [Cl^-] = 1.0 *M* (—). Also shown is a representative sputter profile for O^- (mass 16). Sputter rate using 0.5 keV xenon is 0.6 layer per min; 1 layer ~2 Å of NiO (each profile drawn through ~60 data points). (From MacDougall, B., Mitchell, D.F., Sproule, G.I., and Graham, M.J., *J. Electrochem. Soc.*, 130, 543, 1983. With permission.)

the case in which the passive film on nickel contained Cl^-. This suggests that Cl^- incorporation into the passive oxide film on nickel is not the reason for pit initiation. As found by SIMS, passive films on iron do not incorporate Cl^- (11), but Cl^- can be incorporated, to a level approaching 20%, into films on Fe–26Cr alloys after electropolishing in perchloric–acetic acid (12).

SIMS can also be used to study in detail the mechanism of removal by cathodic reduction of passive films formed on metals. A typical cathodic reduction profile of oxide formed on iron in borate buffer solution shows two arrests: the first is likely due to the reduction of γ-Fe_2O_3, and the second represents reduction of Fe_3O_4 (13). Oxide thickness data for iron in borate, determined by SIMS as a function of cathodic reduction (14), are presented in Figure 3.5. Experimentally, samples were anodized in 10% ^{18}O-enriched solution, and, after rinsing, were reduced to various extents in nonenriched solution. Oxide thicknesses were then determined from the ^{18}O concentration. The break point in Figure 3.5 corresponds to the sharp drop in potential observed in the reduction curve and provides direct evidence for two different reduction mechanisms corresponding to the two arrests in the cathodic reduction profile. Comparison of the slopes in Figure 3.5 with calculated charge efficiencies for various reactions indicates that the first reduction represents γ-$Fe_2O_3 \rightarrow Fe^{2+}$ in solution (conversion rate, 1.56 nm/mC/cm^2). The observed efficiency is not in agreement with reductions of γ-FeOOH, supporting the SIMS data discussed earlier indicating that the film contains no bound OH^-. The second reduction is most probably associated with $Fe_3O_3 \rightarrow$ Fe (metal), with a current efficiency of ~60%.

The air stability of passive films on iron and Fe–Cr alloys can be determined by SIMS from the analyzed ^{18}O level in the films compared with the known solution content (15). A lower level than the solution content indicates instability to air exposure. The decrease in ^{18}O concentration may be due to either film thickening upon air exposure, or to exchangeable oxygen in the film. For iron, where the prior

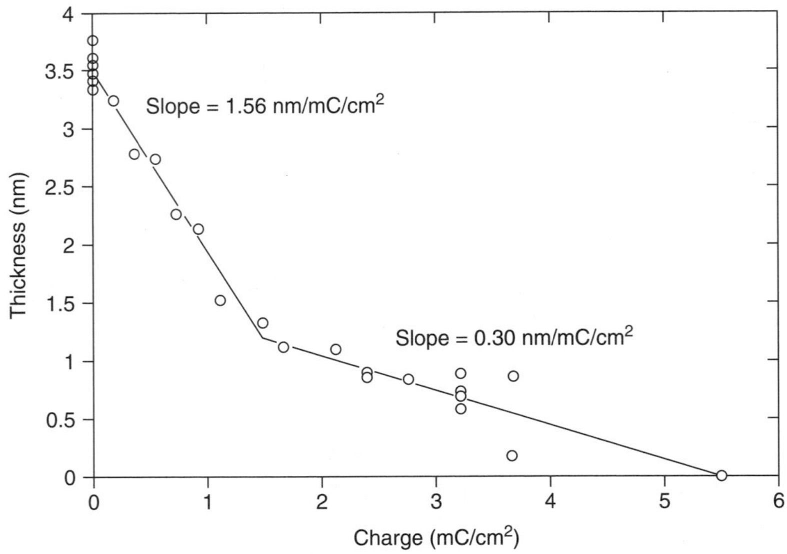

Figure 3.5 Oxide thickness as determined from ^{18}O/SIMS as a function of cathodic reduction of a passive film formed on iron in pH 8.4 borate buffer in the passive potential range; the slopes of the two lines are indicated. (From Bradwell, J.A., MacDougall, B., and Graham, M.J., *J. Electrochem. Soc.*, 135, 413, 1988. With permission.)

electropolished film can be completely cathodically reduced, ^{18}O/SIMS depth profiles of anodic films formed in ^{18}O-enriched borate buffer show an ^{18}O level equal to that of the solution over most of the passive potential range. (The data in Figure 3.5 were obtained for an air-stable film formed at 0.4 V.) Only at very low potentials does the enrichment level decrease, indicating that the films thicken (and are therefore diluted with ^{16}O) upon air exposure.

The results are summarized in Figure 3.6(a), where the total thickness (obtained from the SIMS interface using the Fe_3^+ signal) and the corrected thickness (calculated from the films' ^{18}O concentration, averaged over the total film thickness) are

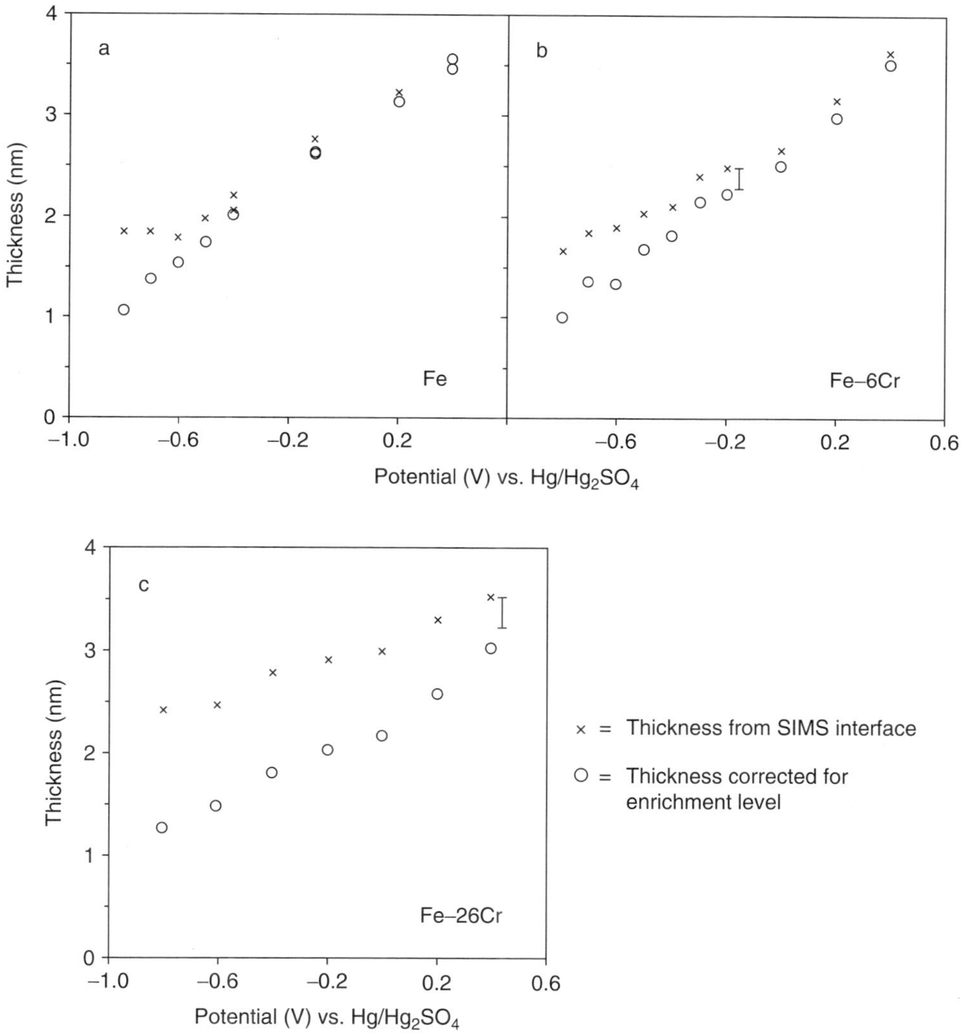

Figure 3.6 Thickness of the passive film on: (a) Fe, (b) Fe–6Cr, and (c) Fe–26Cr as determined from ^{18}O/SIMS. The samples were passivated for 1 h in ~8% ^{18}O-enriched borate buffer solution. The circles indicate the thickness obtained after correction for the overall ^{18}O enrichment level in the film. The "error bars" show a thickness of 0.2 nm in (b) and 0.3 nm in (c) for the amount of unreduced prior film. Sputtering was by 1 keV xenon. (From Graham, M.J., Bardwell. G.I., Sproule, D.F., Mitchell, D.F., and MacDougall, B.R., *Corros. Sci.*, 35, 13, 1993. With permission.)

plotted as a function of anodic potential (16). Clearly, the film is influenced by air exposure only at −0.5 V and below. In this case, it seems most likely that the films thicken upon air exposure because a constant thickness equal to the air formed thickness, ~1.7 nm, is reached upon air exposure at low potentials.

Thickness measurements from SIMS for Fe–6Cr alloys (15) are shown in Figure 3.6(b). The error bar represents a thickness of 0.2 nm, the amount of unreduced prior film. Points separated by a distance greater than the error bar indicate samples where a decrease in enrichment is observed beyond that which would be expected due to the irreducibility, thus indicating instability to air exposure. It is seen in Figure 3.6(b) that the passive film on Fe–6Cr is influenced by air exposure at −0.2 V and below. The film, therefore, is stable to air exposure over a smaller potential range than was the case for iron. Data for Fe–26Cr alloys (15) are shown in Figure 3.6(c). The error bar of 0.3 nm represents the upper limit for the amount of unreduced film. Clearly, the passive film on Fe–26Cr is not stable at any potential in the passive region. As seen in Figure 3.6(b) and (c) for the Fe–Cr alloys, the corrected and uncorrected thicknesses steadily diverge as the potential becomes less anodic. In contrast to iron, no constant thickness after air exposure is observed. The above results have important implications for *ex situ* analysis of surface films on Fe–Cr alloys. Clearly, caution should be exercised in the interpretation of the data, as the films may well have changed in the interval from removal from solution to installation in the ultrahigh vacuum systems. As seen in Figure 3.6, the results for the Fe–Cr alloys are quantitatively quite different from those for iron. In the case of iron, the unstable films thicken up to a constant thickness, whereas for the Fe–Cr alloys, the corrected and uncorrected thicknesses steadily diverge as the potential becomes less anodic.

The mechanism of oxide growth on iron can be studied by forming a film in nonenriched ^{18}O borate at −0.3 V, where the film is air stable, and then stepping the potential to say 0.1 V in ^{18}O-enriched solution for additional growth (17). SIMS profiles show that the additional oxide grows by inward oxygen transport, likely by both lattice and short-circuit diffusion paths.

SIMS analysis following sequential oxidation in $^{16}O_2$ and $^{18}O_2$ is an ideal way to study oxygen diffusion in high-temperature oxides, although interpretation of the data is complicated by problems associated with interface broadening due to nonuniform sputtering and ion beam mixing. Analyzing polyatomic species, for example, of the type $M^{16}O_2^-$, $M^{18}O_2^-$, and $M^{16}O^{18}O^-$, rather than the usual $^{16}O^-$ and $^{18}O^-$ ions, largely overcomes these difficulties (18). It is possible, using this polyatomic SIMS approach, to distinguish the grain boundary from lattice diffusion of oxygen, and to calculate diffusion coefficients. Imaging SIMS using liquid metal ion guns (Table 3.2), to achieve spatial resolutions better than 50 nm, is an ideal way to examine nonuniform oxide growth.

3.3.1.2 Oxidation and Corrosion of Chromium and Iron–Chromium Alloys: The "Reactive Element" Effect

Considering transport in α-Cr_2O_3 at high temperatures, Figure 3.7(a) shows a typical SIMS profile for (100) Cr oxidized at 825°C first in $^{18}O_2$ and then in $^{16}O_2$. The $Cr^{16}O_2^-$ and $Cr^{18}O_2^-$ profiles show that the ^{16}O-oxide exists as a separate layer on top of the ^{18}O-oxide, indicating that the major mass transport during oxidation is outward chromium diffusion. The mixed $Cr^{16}O^{18}O^-$ signal provides information on the extent of any inward oxygen diffusion; the profile in Figure 3.7(a) is

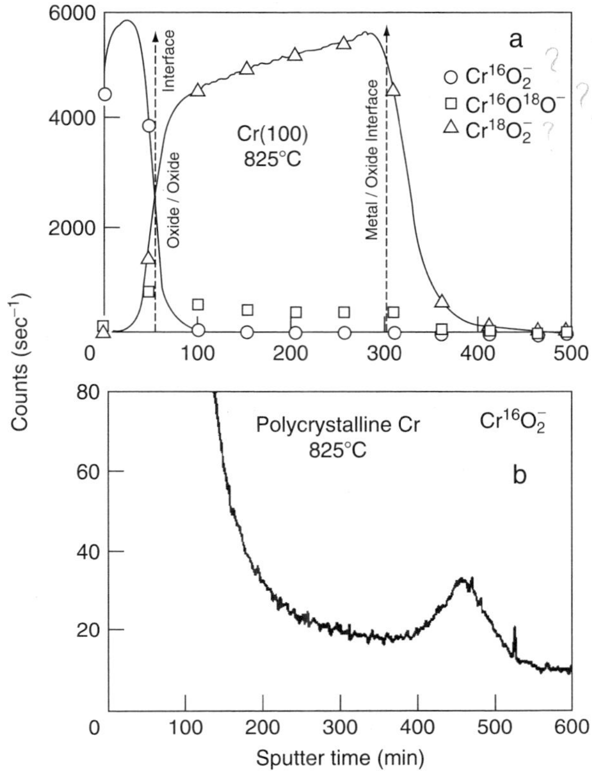

Figure 3.7 Negative polyatomic SIMS profiles: (a) for a 1.4-μm-thick chromium oxide (Cr_2O_3) film formed on a (100) chromium single crystal at 825°C (1.1 μm of oxide produced in $^{18}O_2$ followed by 0.3 μm produced in $^{16}O_2$); (b) expanded $Cr^{16}O_2^-$ profile for a Cr_2O_3 film formed on polycrystalline chromium at 825°C showing the location of the ^{16}O-oxide at the metal–oxide interface as a result of surface pretreatment. Sputtering by 4 keV xenon. (From Graham, M.J., *Corros. Sci.*, 37, 1377, 1995. With permission.)

asymmetric about the interface between the $^{16}O^-$- and $^{18}O^-$-oxide layers, indicating penetration into the inner oxide. Two independent oxygen diffusion processes are in effect. The first is isotropic oxygen self-diffusion occurring at the interface between the oxide layers. This isotropic diffusion has no net transport and would not contribute to film growth. The second process, associated with the tailing of the $Cr^{16}O^{18}O^-$ profile, is an anisotropic, localized inward oxygen diffusion assumed to be occurring down oxide grain boundaries. Oxygen diffusion coefficients can be calculated for the two processes from the areas of the $Cr^{16}O^{18}O^-$ profile (19). The diffusion coefficients for oxygen grain boundary diffusion are 2 to 3 orders of magnitude higher than for isotropic diffusion, but are still orders of magnitude too low to account for the observed oxidation rates. The major growth transport is by outward chromium diffusion; however, 1% of the oxide is created within the film as a result of inward oxygen diffusion. With time, this inward diffusion could give rise to sufficient compressive stress to cause detachment of the scale and loss of oxidation protection.

There is also a small rise in the $Cr^{16}O^{18}O^-$ and $Cr^{16}O_2^-$ signals in the vicinity of the metal–oxide interface that corresponds to the zone where the original 2-nm thick ^{16}O-oxide film formed during surface pretreatment is situated. Expansion of the

$Cr^{16}O_2^-$ profile confirms a small, but clearly distinct, peak in this region (Figure 3.7b) for polycrystalline chromium oxidized at 825°C (20). This result demonstrates that the ^{18}O/SIMS technique is sufficiently sensitive to allow a relatively thin layer of oxide to be detected underneath a very much thicker external layer. This thin prior oxide film in fact serves as an inert marker confirming oxide growth primarily by outward cation diffusion from the metal to the outer oxide surface, and formation of new oxide mainly on top of the existing scale.

The addition of small amounts of reactive elements such as cerium, yttrium, hafnium, thorium, lanthanum, or their oxide dispersions greatly increases the high-temperature oxidation resistance of Fe–Cr alloys under isothermal or cyclic conditions (21). Beneficial effects also result from ion implantation of the active element (22–25) or from surface-applied coatings (26–28). The ion implantation work of Bennett et al. (22–24) concerns the oxidation behavior of a 20Cr–25Ni–Nb stainless steel in CO_2 at temperatures in the range 900 to 1050°C. SIMS can be used to locate the position of the reactive element after oxidation. Figure 3.8 shows an SIMS profile (23) through cerium implanted 20Cr–25Ni–Nb stainless steel oxidized at 900°C for 216 h. The profile indicates that the variation of the steel constituents is similar to that in the scale, which develops at this temperature on the unimplanted steel. The cerium concentration increases from the outer surface to a maximum within the spinel layer near the interface with the underlying Cr_2O_3 layer. Silicon is enriched, presumably as silica, at the scale–steel interface. The broader peak does not necessarily indicate a thicker layer than the remainder of the scale, as it reflects the different sputtering rates of oxide and steel as the argon ion beam crosses that

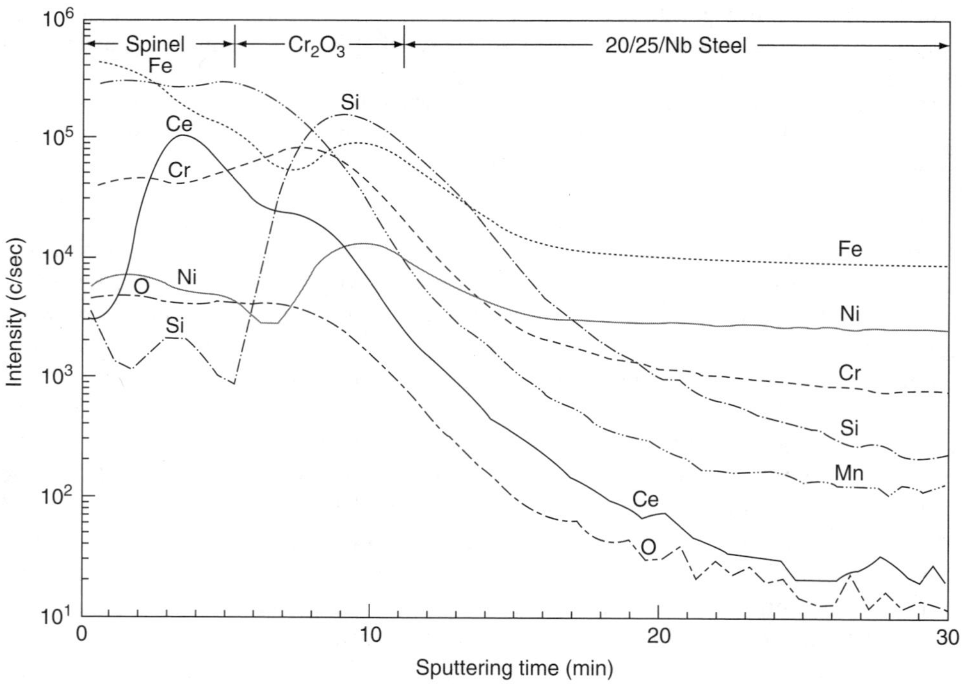

Figure 3.8 SIMS profile through the oxide scale formed on 20Cr–25Ni–Nb stainless steel implanted with 10^{17} yttrium ions per cm^2 after 216 h of oxidation in carbon dioxide at 900°C. Sputtering was by 10.5 keV argon. (From Bennett, M.J., Bishop, H.E., Chalker, P.R., and Tuson, A.T., *Mater. Sci. Eng.*, 90, 177, 1987. With permission.)

interface. The oxidation of the stainless steel changes between 900 and 1050°C from protective to nonprotective type behavior. Ion implantation of cerium, yttrium, and lanthanum provides continuing protection to the stainless steel to about 1000°C. Initial scale development is modified by the promotion of inward oxidant movement. The reactive element becomes incorporated as oxide grains and segregants along grain boundaries in the outer spinel layer of the scale adjacent to the inner Cr_2O_3 layer. Cation transport through these boundaries will be energetically less favorable, so that continued scale growth will be inhibited.

Regarding reactive element coatings on Fe–Cr alloys, very thin (4 nm) coatings can be applied by sputtering CeO_2 directly on to substrates, and SIMS used to determine the location of cerium within the oxide scale and to study its effect on transport processes (28–30). SIMS profiles for short-term oxidation at 900°C of coated Fe–26Cr alloy are shown in Figure 3.9(a). The scale is principally Cr_2O_3, with a small amount of iron in the outer part of the scale. The cerium in the oxide layer is detectable, and its maximum signal is observed to be away from

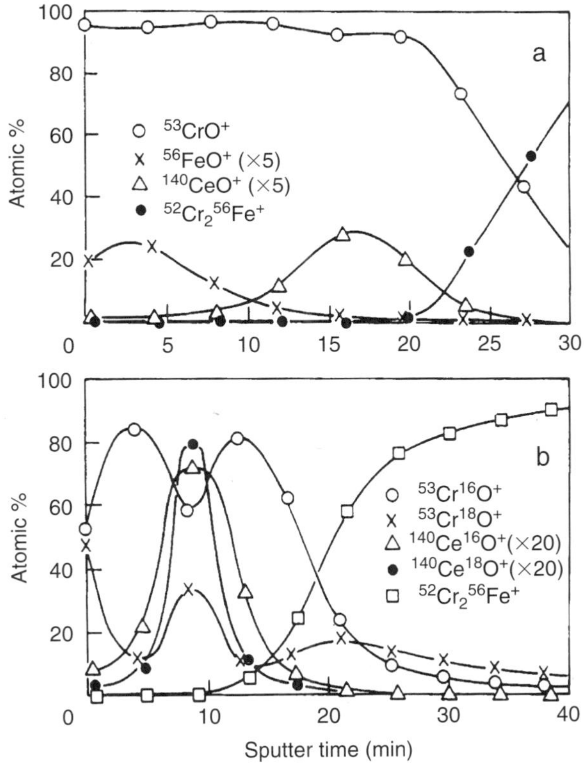

Figure 3.9 (a) Positive polyatomic SIMS profiles for a 76-nm-thick oxide formed on Fe–26Cr coated with 4 nm CeO_2 after 1 min oxidation at 900°C in 5×10^{-3} Torr oxygen. (b) Positive polyatomic SIMS profiles for a 0.36-μm-thick oxide film formed at 900°C in 5×10^{-3} Torr oxygen on Fe–26Cr coated with 4 nm CeO_2 after 0.7 h oxidation in $^{16}O_2$ followed by 5.9 h oxidation in $^{18}O_2$; oxide thicknesses were 0.17 and 0.19 μm, respectively. (The $^{52}Cr^{56}Fe^+$ profiles indicate the position of the metal–oxide interface.) Sputtering was by 4 keV xenon. (From Hussey, R.J., Papaiacovou, P., Shen, J., Mitchell, D.F., and Graham, M.J., in *Proc. Symp. on Corrosion and Particle Erosion at High Temp.*, Srinivasan, V. and Vedula, K., eds. TMS, 1989, p. 567. With permission.)

the alloy–oxide interface. Longer-term oxidations show that the location of the cerium maximum moves further toward the oxide–gas interface. If the CeO_2 coating can be considered to be a stationary marker, then its location toward the oxide–gas interface indicates that inward oxygen diffusion is becoming more predominant in governing oxide growth. $^{16}O_2/^{18}O_2$ experiments confirm that oxygen transport is occurring. Figure 3.9(b) shows data for a specimen oxidized at 900°C sequentially in $^{16}O_2$ and then $^{18}O_2$. In this longer-term experiment, the location of the cerium maximum is now in the middle of the scale. There are ^{18}O maxima ($^{53}Cr^{18}O^+$) at the gas–oxide interface, within the scale at the position of the cerium maxima, and at the alloy–oxide interface, showing that oxygen diffusion has occurred during oxidation. Analogous results have also been obtained on Y_2O_3-coated alloys. Examination of the SIMS data in Figure 3.9(b), in the form of the fraction of ^{18}O associated with the cerium- and chromium-bearing species through the scale, shows that a higher fraction is associated with the cerium-bearing species. If the cerium in the scale is located at oxide grain boundaries, the association of ^{18}O with cerium strongly suggests that transport of ^{18}O through the scale occurs along grain boundaries. SIMS lacks the lateral resolution to determine the precise location of the cerium, or its form and distribution within Cr_2O_3. However, comparison with the analytical TEM (transmission electron microscope) results of Cotell et al. (31), on the structure and location of yttrium within scales on chromium implanted with yttrium, suggests that cerium would be expected to be segregated as a Ce–Cr oxide phase at grain boundary regions. In summary, yttrium implantation or CeO_2 coatings causes a change in oxide growth mechanism of Cr_2O_3 from predominantly cation diffusion for the uncoated alloy to predominantly anion diffusion.

3.3.1.3 Nickel

For growth of NiO at 700 to 900°C, ^{18}O/SIMS experiments have confirmed that the rate controlling process is the outward diffusion of nickel through NiO (32–34). For example, data of Atkinson and Smart (33) show that the ^{18}O layer formed second after the first oxidation in $^{16}O_2$ is on the outside. Moon et al. (35) have also used SIMS in the sputter depth profiling and imaging modes to examine the penetration of ^{18}O into single and duplex NiO scales formed on nickel. It was found that a layer of $Ni^{18}O$ forms on top of a $Ni^{16}O$ layer, but also that ^{18}O has penetrated into the $Ni^{16}O$ layer, as also reported by Atkinson and Smart. It was proposed that gaseous O_2 penetrates the scale along pathways such as through-scale microchannells or closely spaced transient fissures.

The mechanism of oxidation of nickel can also be modified by cerium or yttrium implantation or CeO_2 coatings (36–39). NiO scale growth is inhibited by incorporated CeO_2 particles at temperatures below 1000°C, that is, in the short-circuit diffusion region, whereas the coating has little effect at higher temperatures, where scale growth is dominated by lattice diffusion. It was concluded that the reduced scale growth could be attributed to blocking of grain boundary diffusion (38). The location and form of implanted cerium and yttrium within the oxide scale can be established by SIMS (34). The results show that initially the peak reactive element concentration is located near the gas–oxide interface, but, with increasing time, the peak broadens, the maximum concentration gradually decreases, and the peak position is found at an increasing distance from the scale–gas interface. These results indicate that at least initially the reactive element ion implantation changes scale growth to predominantly inward oxidant movement. SIMS work by Czerwinski et al. (40,41) has confirmed that the mechanism of oxidation of nickel can be modified by cerium implantation or CeO_2 coatings, resulting in oxide growth occurring by a greater contribution from oxygen diffusion.

3.3.1.4 β-NiAl and FeCrAl Alloys

One of the main disadvantages of techniques such as SIMS is the relatively poor spatial resolution compared with TEM. However, developments continue to improve the resolution of surface-analytical techniques. As mentioned earlier, with a liquid metal ion gun, it is now possible to obtain SIMS images with a spatial resolution better than 50 nm. This is particularly useful for examining oxides on, for example, NiAl, where nonuniform Al_2O_3 growth occurs (42–44). A characteristic feature of the oxidation of β-NiAl is the growth of oxide ridges in the α-Al_2O_3 scale. The SEM micrograph in Figure 3.10(a) illustrates the ridge morphology observed at 1200°C. The oxide is much thinner between the ridges than in them. Figure 3.10(b) and (c) shows planar SIMS images of the ridged oxide. Figure 3.10(b) is an image in which the $^{18}O^-$ species has been detected in the surface of the oxide. The white lines indicate regions of high ^{18}O content, and they clearly correspond to the oxide ridges. The dark areas, between the white lines, indicate regions of scale with low ^{18}O content. Figure 3.10(c) is the complementary $^{16}O^-$ image for the same area of oxide. In this image, the areas between the ridges are brighter, which is indicative of higher ^{16}O levels.

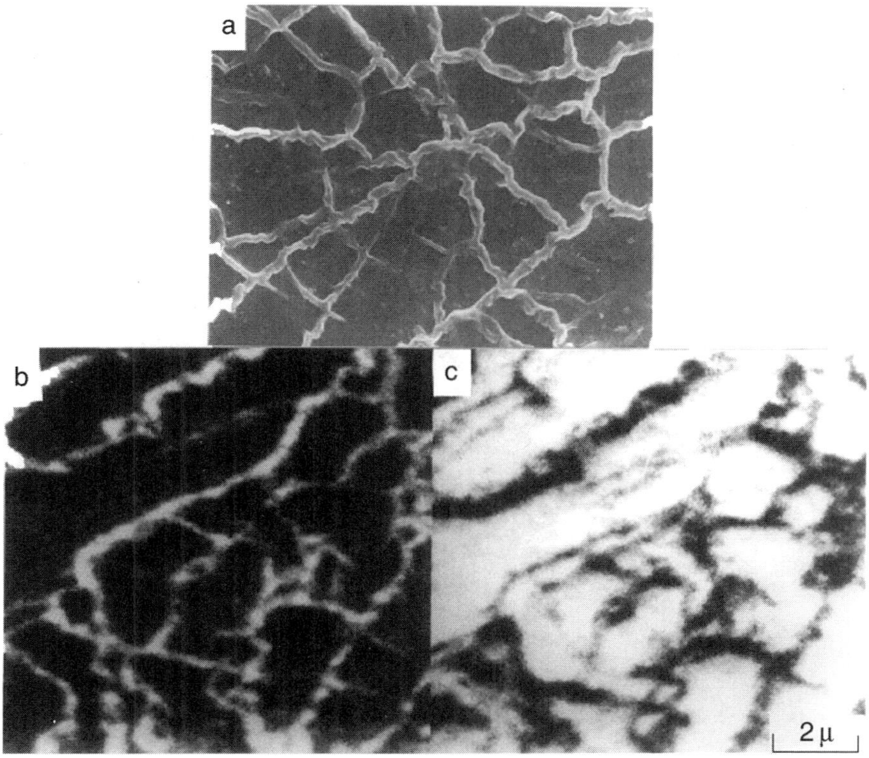

Figure 3.10 Oxide formed on β-NiAl at 1200°C after 4 h; 2 h in $^{16}O_2$ + 2 h in $^{18}O_2$. (a) SEM micrograph, (b) planar $^{18}O^-$ SIMS image, and (c) complementary $^{16}O^-$ image. (From Prescott, R., Mitchell, D.F., and Graham, M.J., in *Microscopy of Oxidation 2*, Newcomb, S.B. and Bennett, M.J., eds., Institute of Metals, London, 1993, p. 455. With permission.)

From these planar images of the outer surface of the oxide, it is clear that, after the formation of an α-Al_2O_3 layer, the bulk of the new surface oxide growth is confined to the ridges. In a series of planar images obtained from sputtering the oxide from the outer surface down to the substrate, it is possible to observe how the ^{18}O distribution changes with depth (43). The highest ^{18}O levels are found to be associated with the outer region of the oxide ridges, which shows that the ridges continue to develop throughout the period of oxidation in $^{18}O_2$. Because ^{18}O is detected on the outside of the α-Al_2O_3 ridges, it can be concluded that growth is occurring by outward diffusion of aluminum.

As the time of oxidation at 1200°C increases, the ridged structure coarsens. The ridges become higher and wider, and the flat areas between the ridges gradually disappear. Yttrium and zirconium alloy additions slow down the development of the ridges, and reduce, to some degree, the extent of outward aluminum diffusion. These thick scales can be analyzed as a taper or cross-section. For NiAl + 0.2% Zr after sequential oxidation for 100 h in $^{16}O_2$ followed by 100 h in $^{18}O_2$ (in Figure 3.11a), most of the ^{18}O is observed close to the scale–substrate interface or within channells or boundaries in the inner part of the oxide. The SIMS image clearly shows that inward diffusion of oxygen along short-circuit paths is now beginning to play a significant role in the growth process. This change in the growth mechanism appears to coincide with the coalescence or merging of the oxide ridges. The SIMS image in Figure 3.11(c) indicates the distribution of the zirconium; it is dispersed through the

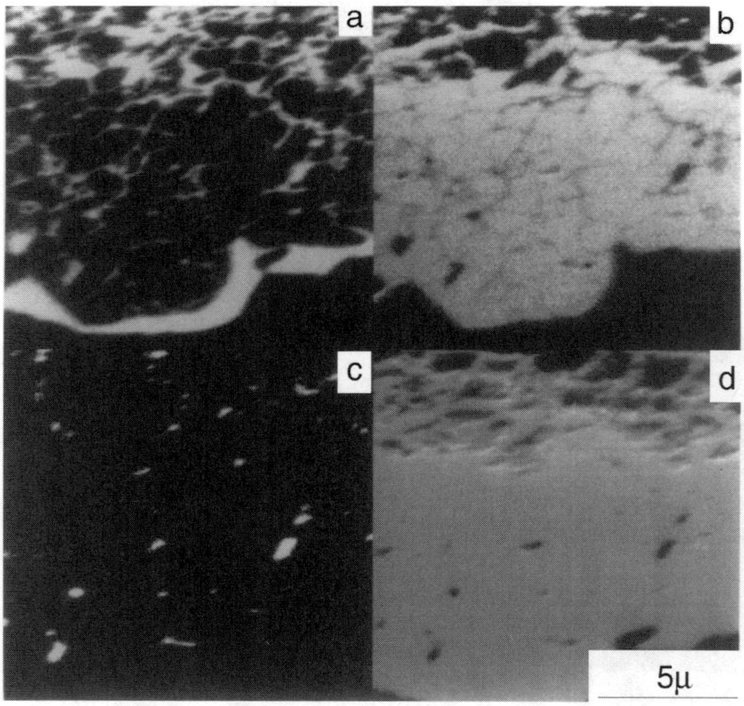

Figure 3.11 SIMS images of taper section (2.5×) of oxide formed on NiAl + 0.2%Zr after 200 h oxidation at 1200°C; 100 h in $^{16}O_2$ + 100 h in $^{18}O_2$: (a) $^{18}O^-$ image, (b) complementary $^{16}O^-$ image, (c) Zr signal, and (d) Al signal. Regions rich in the particular isotope or element appear white. (From Prescott, R., Mitchell, D.F., Graham, M.J., and Doychak, J., *Corros. Sci.*, 37, 1341, 1995. With permission.)

scale as ZrO_2 particles. The particles are finer in the outer portion of the scale and are coarser and more widely spaced close to the scale–substrate interface, probably reflecting the reduced rate of growth of the oxide as the film thickens. Confirmation of zirconium, yttrium, or sulfur, precisely at the oxide–substrate interface, is not possible by SIMS. Recent results (45) on samples previously analyzed by SIMS show that for NiAl + Zr no particles are present right at the oxide–substrate interface. Also, no interfacial voids are found, and sulfur is not detected at the oxide–substrate interface. Zirconium does segregate to this interface and to all α-Al_2O_3 grain boundaries, an observation in agreement with the results of Pint et al. (46). By contrast, no yttrium is detected at the oxide–substrate interface for NiAl + Y. While no sulfur is present at the interface, yttrium-rich particles containing sulfur are found in the alloy; presumably the yttrium has gettered the sulfur. These higher-resolution TEM studies therefore show that sulfur does not segregate to Al_2O_3–substrate interfaces where the oxide remains in good contact with the substrate.

α-Al_2O_3 scales on Fe–Cr–Al alloys, as seen in the ^{18}O/SIMS image in Figure 3.12(a), grow by both outward aluminum diffusion and inward oxygen diffusion (47), similar to alumina growth on β-NiAl described above. It is evident from the image that the second oxidant (^{18}O) is present as a band at the metal–oxide interface (left-hand side of the image), within boundaries in the scale, and on the outside of the oxide. An ^{18}O image of yttrium-containing alloy (Figure 3.12b) shows that ^{18}O is present in a wide band at the metal–oxide interface (left-hand side of the image). For the yttrium-containing alloy, the oxide grain structure changes from equiaxed to columnar, and STEM/EDS shows that yttrium has segregated to the α-Al_2O_3 grain

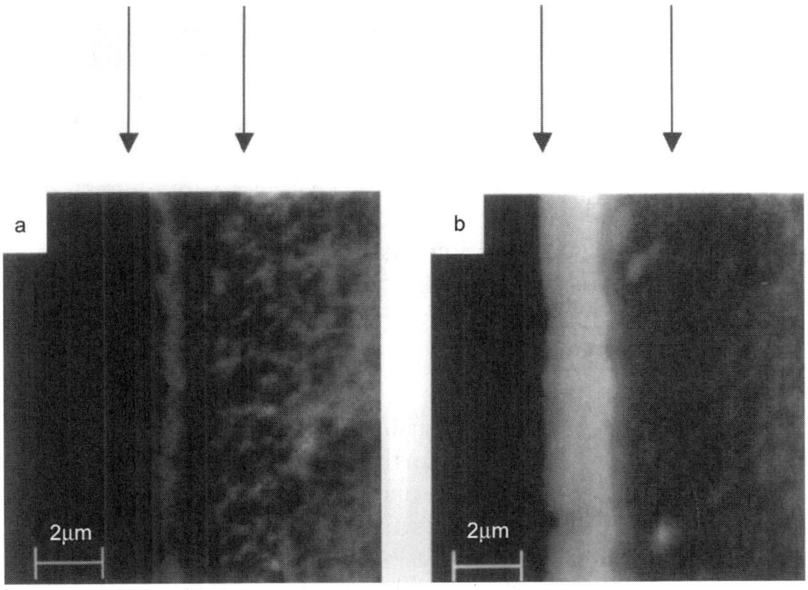

Figure 3.12 SIMS images of α-Al_2O_3 growth at 1200°C. (a) FeCrAl: $^{18}O^-$ SIMS image of oxide formed after 5 h:1.5 h in $^{16}O_2$ + 3.5 h in $^{18}O_2$; (b) FeCrAl + 0.1 wt.% Y; $^{18}O^-$ SIMS image of oxide formed after 5 h:1.5 h in $^{16}O_2$ + 3.5 h in $^{18}O_2$. Arrows in (a) and (b) indicate the position of the metal–oxide interface (left-hand side) and the oxide–gas interface (right-hand side). Regions rich in ^{18}O appear white. (From Mennicke, C., Schumann, E., Rühle, M., Hussey, R.J., Sproule, G.I., and Graham, M.J., *Oxid. Met.*, 49, 455, 1998. With permission.)

boundaries and to the alloy–oxide interface (47). The segregation of yttrium in the case of Fe–Cr–Al alloys has actually suppressed outward cation diffusion rather than reducing it as in the case of α-Al_2O_3 on β-NiAl or Cr_2O_3 on chromium and Fe–Cr alloys.

Similar SIMS results have been reported for Fe–20Cr–5Al (Kanthal) by Hultquist et al. (48). SIMS data for a foil oxidized at 1000°C in $^{16}O_2$ and then $^{18}O_2$ confirm that the alumina grows by oxygen anion transport. This material contained 2 ppm hydrogen. In contrast, alumina grows by predominantly outward cation transport if the material contains 40 ppm hydrogen, that is, hydrogen in the substrate induces increased cation transport, leading to increased oxide growth at the oxide–gas interface. This was also found to be the case for zirconium and copper, as well as for chromium (49). Hultquist et al. (48) also reported that Fe–20Cr–5Al alloy with 1% Y, charged with a certain amount of hydrogen, has the lowest oxidation rate and secondary neutral mass spectrometry (SNMS) data show that oxide growth takes place at both the metal–oxide and the oxide–gas interfaces.

Quadakkers et al. (50–52) have used both SIMS and SNMS to study the mechanism of alumina growth on Fe–20Cr–5Al and oxide-dispersion-strengthened (ODS) Incoloy alloy MA956 (Fe–20Cr–4.5Al–0.5Y_2O_3). Yttrium oxide dispersion was found to suppress outward scale growth. As shown in Figure 3.13, after a transient oxidation, the oxide on MA956 grows almost exclusively by oxygen grain boundary transport and so the thickening occurs by oxygen (^{18}O) grain boundary transport at the scale–metal interface by alumina formation. A "shoulder" in the ^{18}O profile (indicated by an arrow in Figure 3.13c) could represent a small amount of in-scale growth. The location of the shoulder could be correlated by SNMS with the relative maximum of the in-scale chromium concentration. According to Quadakkers et al. (51), apart from in-scale growth, the shoulder in the ^{18}O-concentration profile could be due to an enhanced isotope exchange between the oxygen, which is diffusing inward via grain boundaries with the oxygen in the grains. A possible reason for this effect would be a highly disturbed alumina lattice due to high amounts of dissolved chromium or the presence of a large number of interfaces due to coexistence of finely dispersed chromia particles with alumina grains. The shoulder in the ^{18}O profile was detected in the scales on both MA956 and Fe–20Cr–5Al; they only differed in their location relative to the oxide surface, because of the different contributions of cation diffusion to the overall oxide growth process on the two alloys.

3.3.1.5 Silicon

The ability of silicon to form a high-quality thermal oxide with a low density of interfacial states has been a major factor in the successful development of metal–oxide-semiconductor (MOS) technology and the microelectronics industry. The formation and growth of SiO_2 films have therefore been extensively researched, including SIMS studies of the extent of oxygen diffusion (53–55). Data at 1050°C (55) are shown in Figure 3.14 for Si(100) oxidized first in $^{16}O_2$ and then $^{18}O_2$ for different times. From the $Si_2^{18}O^-$ profiles, it is clear that oxide is growing at both the gas–oxide and SiO_2–Si interface, with the latter predominating. As seen in Figure 3.14, the amount of oxide formation at the SiO_2–Si interface increases with oxidation time. The SIMS data suggest that the inward diffusion of oxygen probably occurs along short-circuit paths, such as micropores extending to the SiO_2–Si interface.

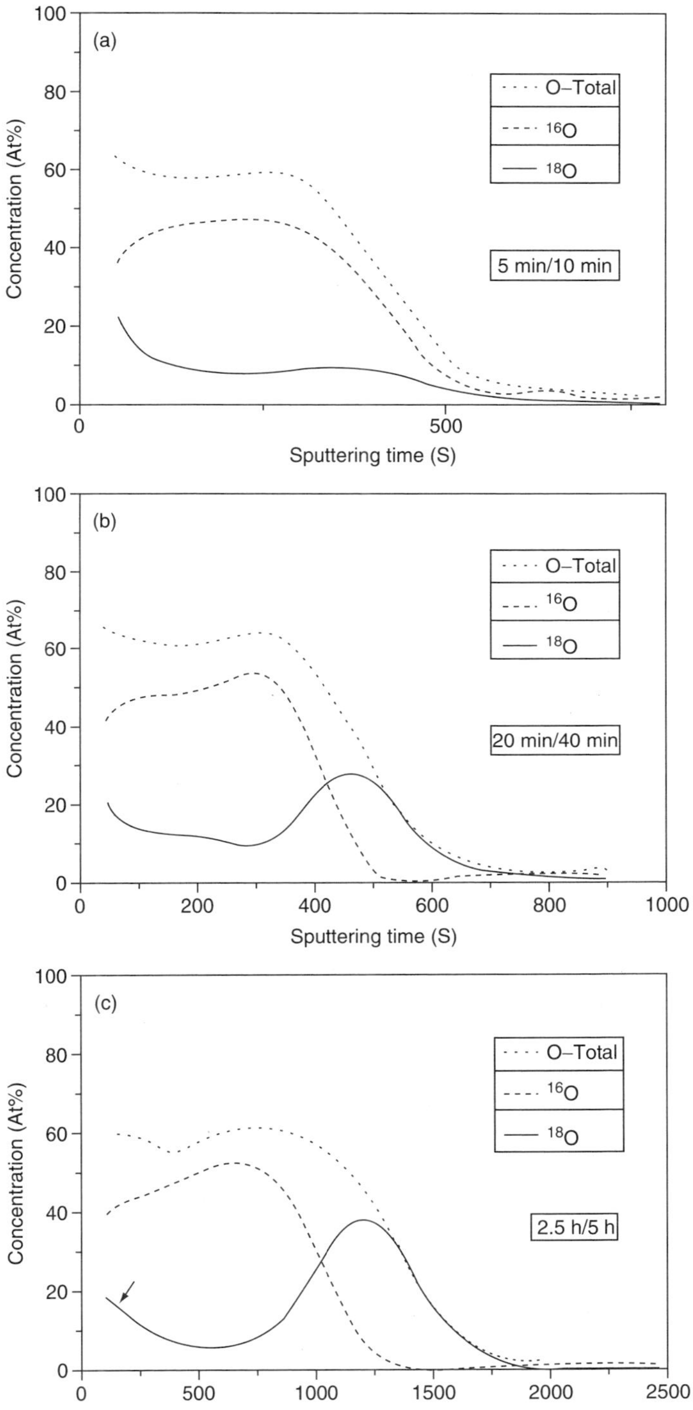

Figure 3.13 SNMS profiles: time dependence of oxygen isotope distribution (recalculated to 100% enrichment in ^{18}O) in oxide scales on MA 956 after two-stage oxidation in air and then ^{18}O-enriched air at 1000°C. Arrow indicates "shoulder" in the ^{18}O profile: (a) 5 min/10 min, (b) 20 min/40 min, and, (c) 2.5 h/5 h. Sputtering was by 0.8 keV argon. (From Quadakkers, W.J., Elschner, A., Spier, W., and Nickel, H., *Appl. Surf. Sci.*, 52, 271, 1991. With permission.)

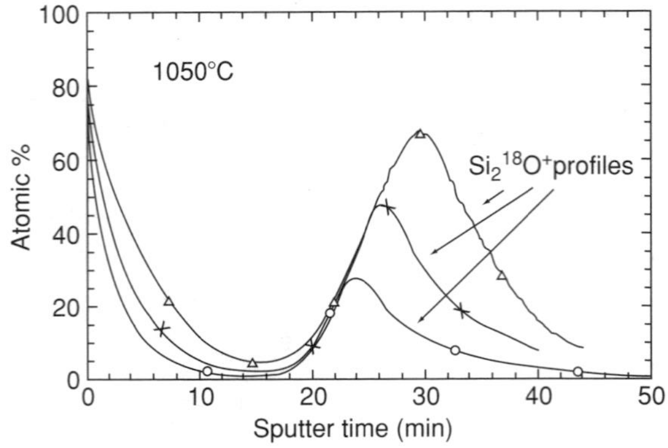

Figure 3.14 SIMS depth profiles for $Si_2^{18}O$ for oxide films formed on (100) Si at 1050°C in 10 Torr oxygen after 2 h in ^{16}O followed by 0.5 h (o), 1 h (+), and 2 h (Δ) in $^{18}O_2$. (From Hussey, R.J., Sproule, G.I., Mitchell, D.F., and Graham, M.J., in *Proc. 12th Int. Corrosion Congress*, Vol. 5B, Houston, TX, 1993, p. 3831. With permission.)

3.3.2 Applied Studies

3.3.2.1 Corrosion Studies of Steels and Ferrous Alloys

SIMS has been used to study the effectiveness of pickling methods that were intended to remove hydrogen from the near surface of high-strength steels (56). Such hydrogen embrittles the surface, leading to cracking and further degradation by corrosion. Polished samples were charged cathodically with deuterium from a D_2O solution because the spectrometer background for D is several orders of magnitude lower than for H. Samples were then pickled in a nitric–acetic–phosphoric (NAP) acid mixture, and the effects of this treatment on near-surface hydrogen were studied by SIMS. Quantitative measurements were calibrated using ion implants of deuterium.

In Figure 3.15, a quantitative depth profile of hydrogen concentration is shown for an as-charged surface of AISI 4140 and 1062 steel as well as following NAP pickling of the surface. In the first 2 μm of surface, the hydrogen concentration is about one order of magnitude lower following pickling. The pickling effect extends at least to a depth of 10 μm. It will be noted that no evidence of local hydride-containing structures can be seen. This is believed to be due to the high mobility of hydrogen in iron — in sharp contrast to the case for zirconium, as will be seen below.

Several studies have been carried out to determine crevice corrosion product distribution in cracks between mild (carbon) steel surfaces and Inconel 600 in steam generators (57). Of particular interest was the ability to obtain ion images of the distributions of elements such as boron, which is added as a corrosion inhibitor and is poorly detected by energy dispersive x-ray (EDX) and XPS analysis. By correlation of elemental distributions, it is possible to identify borates as being associated in crystallites with iron, magnesium, and manganese. Chlorides, which are known to be initiators of corrosion, appear to be excluded from the magnetite surface of the mild steel by the borate crystals. The value of such an analysis is that imaging allows unambiguous identification of the effectiveness of a particular inhibiting layer.

The corrosion of steel surfaces is frequently inhibited through the use of sacrificial coatings, such as zinc–iron compounds and tinplates. Pitting corrosion processes in the latter have been followed using imaging SIMS in a 3D depth profiling mode (58,59). Of particular interest were the mechanisms of pitting

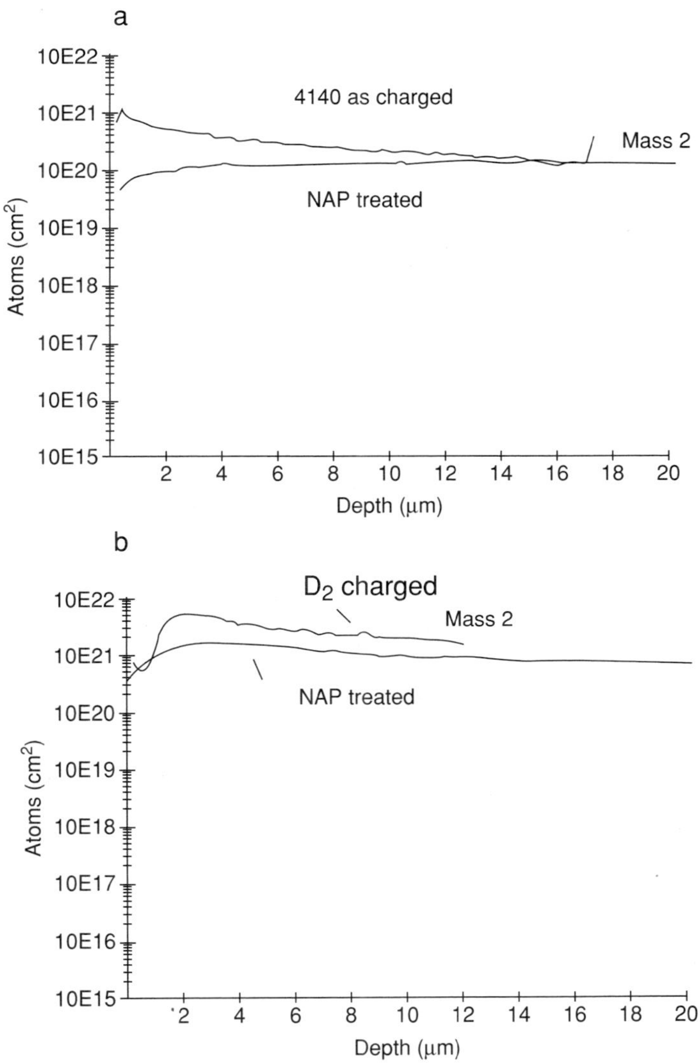

Figure 3.15 Deuterium concentration measured by SIMS as a function of depth in cathodically charged samples of: (a) AISI 4140 and (b) 1062 steels before and after pickling with NAP. Depth profiling was done with a Cs-positive primary beam and negative secondaries were collected. The depth scale was calibrated using a profilometer following sputtering. (From Sastri, V.S., Donepudi, V.S., McIntyre, N.S., Johnston, D., and Revie, R.W., *Metall. Trans.*, 19, A3071, 1988. With permission.)

produced by long-term exposures of the tinplate to fruit solutions under anaerobic conditions. The structure of tinplate on steel is complex; between the outer surface of pure tin (usually 0.15 to 2 μm thick), one or more iron–tin alloy phases form, which have their particular corrosion potentials. The most common is $FeSn_2$; SIMS calibration studies showed that this phase could be identified unambiguously in ion images using the $FeSn^-$ cluster ion. Three-dimensional SIMS profiling of specimens with different corrosion susceptibility was carried out, monitoring the near-surface distribution of this particular alloy phase as well as an FeO^- secondary ion to represent oxidized iron. The resultant composite image representations for two specimens with very different corrosion behavior are shown in Figure 3.16. These are produced from a "stack" of ion images, following their collection during a depth profile from the outermost tin-rich surface to the steel substrate. The sample shown in Figure 3.16(a) had exhibited a very low incidence of pitting corrosion. The cylindrical representations are, in fact, 2D views contrived to impart the essence of the 3D nature of the surface layers. The outer layer is pure tin, and no signal from the iron–tin alloy is registered. Its presence, however, can be detected as an interfacial layer of uneven depth between the pure tin and the steel substrate. The only oxide present is that detected within the steel immediately below the alloy–substrate interface. In Figure 3.16(b), the SIMS image representations are shown for the

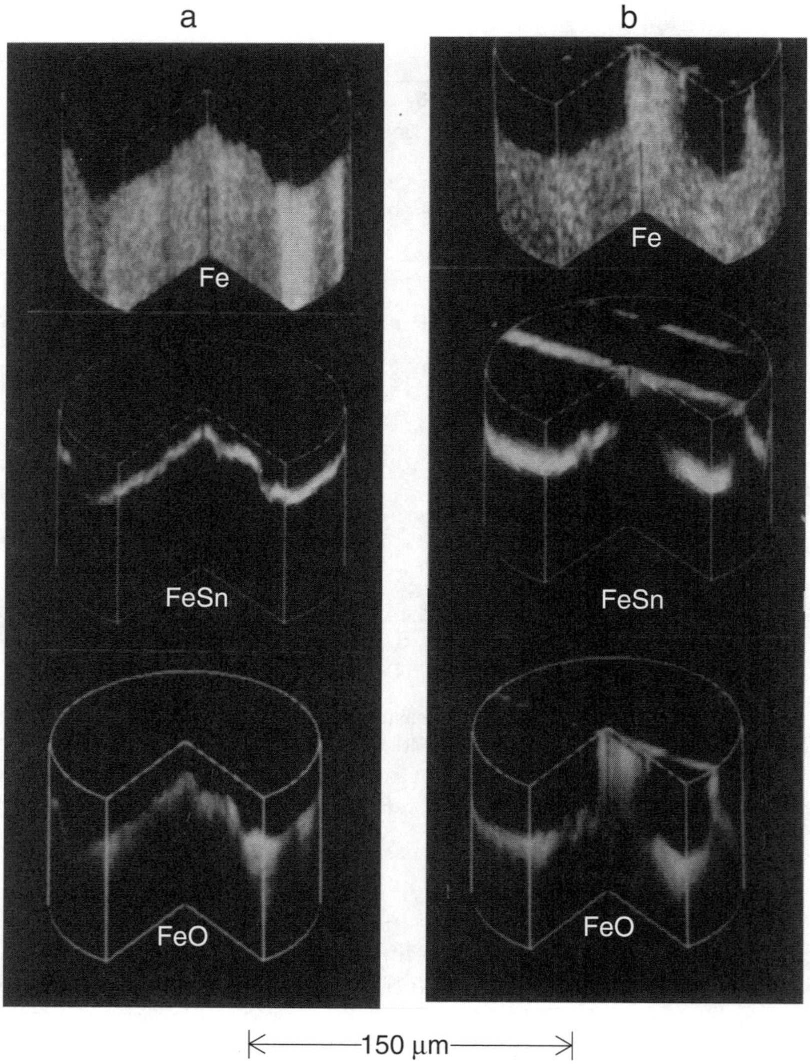

Figure 3.16 Representations of three-dimensional depth profiles of $FeSn^-$ (for $FeSn_2$ alloy) and FeO^- (for iron oxide) obtained for: (a) a tinplate surface with no visible pitting corrosion, and (b) a tinplate surface with 60% pitting incidence. The field diameter is 150 μm; the depth of each profile is 2 μm. (From Lu, S.F., Mount, G.R., McIntyre, N.S., and Fenster, A., *Surf. Interface Anal.*, 21, 177, 1994. With permission.)

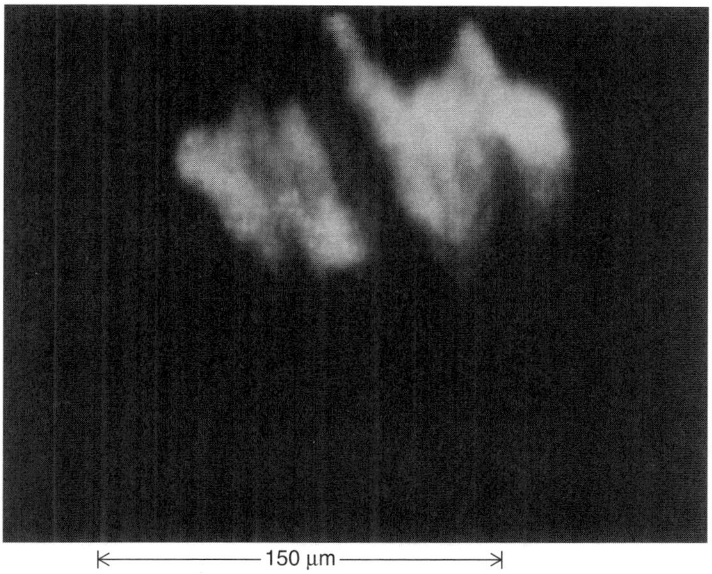

Figure 3.17 Volume-rendered three-dimensional image of the $FeSn_2$ interface produced from SIMS image depth profiling of $FeSn^-$. (From From Lu, S.F., Mount, G.R., McIntyre, N.S., and Fenster, A., *Surf. Interface Anal.*, 21, 177, 1994. With permission.)

same species for a sample that had exhibited significant pitting. The iron–tin alloy is detected on the outer surface, and the onset of corrosion is signaled by the partial oxidation of this alloy. The alloy layer was found to intersect the surface at regular intervals because rolling marks in the substrate brought portions of the alloy in contact with the liquid. This can be seen in the true 3D representation of the alloy layer (Figure 3.17) achieved by volume rendition processing software.

Aluminum–zinc corrosion protective coatings on steel have also been studied by SIMS depth profiling and imaging (60). An outer layer of Al–Zn alloy was detected, within which were small defect areas where iron and manganese were located. Beneath this layer was a quaternary alloy phase (zinc, silicon, iron, and aluminum) that was in contact with the iron substrate. The work, like that above, demonstrated the strength of the technique in tracing the diffusion of small concentrations of steel components through the protective layer.

Interfaces between galvanized steel and cathodic electrocoated paint systems have been characterized by SIMS in concert with XPS (61). Phosphated and non-phosphated e-coated panels were scribed mechanically and then exposed to a cyclic salt spray/drying/humidity test (GM scab corrosion test). SIMS spectra were obtained of areas where the coating had delaminated after the test, as well as of areas where coating adhesion was retained. For delaminated areas, SIMS and XPS spectra of each side of the interface were obtained. In Figure 3.18, SIMS spectra are shown for a delamination area comparing the metal and paint sides of interface. The spectrum of the metal side (a) shows peaks at 64, 81, 98, and 163 Da, representing oxides and hydroxides of zinc from the galvanized coating. Nitrogen-containing organic groups (urethanes?) can be inferred from the peaks at 28, 30, 42, and 44 Da, and aromatic fragments can be inferred from the peaks at 77, 91, and 115 Da. No evidence, however, of polyurethanes can be found; peaks should appear at 58, 59, and 135 Da. This latter set of peaks does occur on the paint side (b). It is

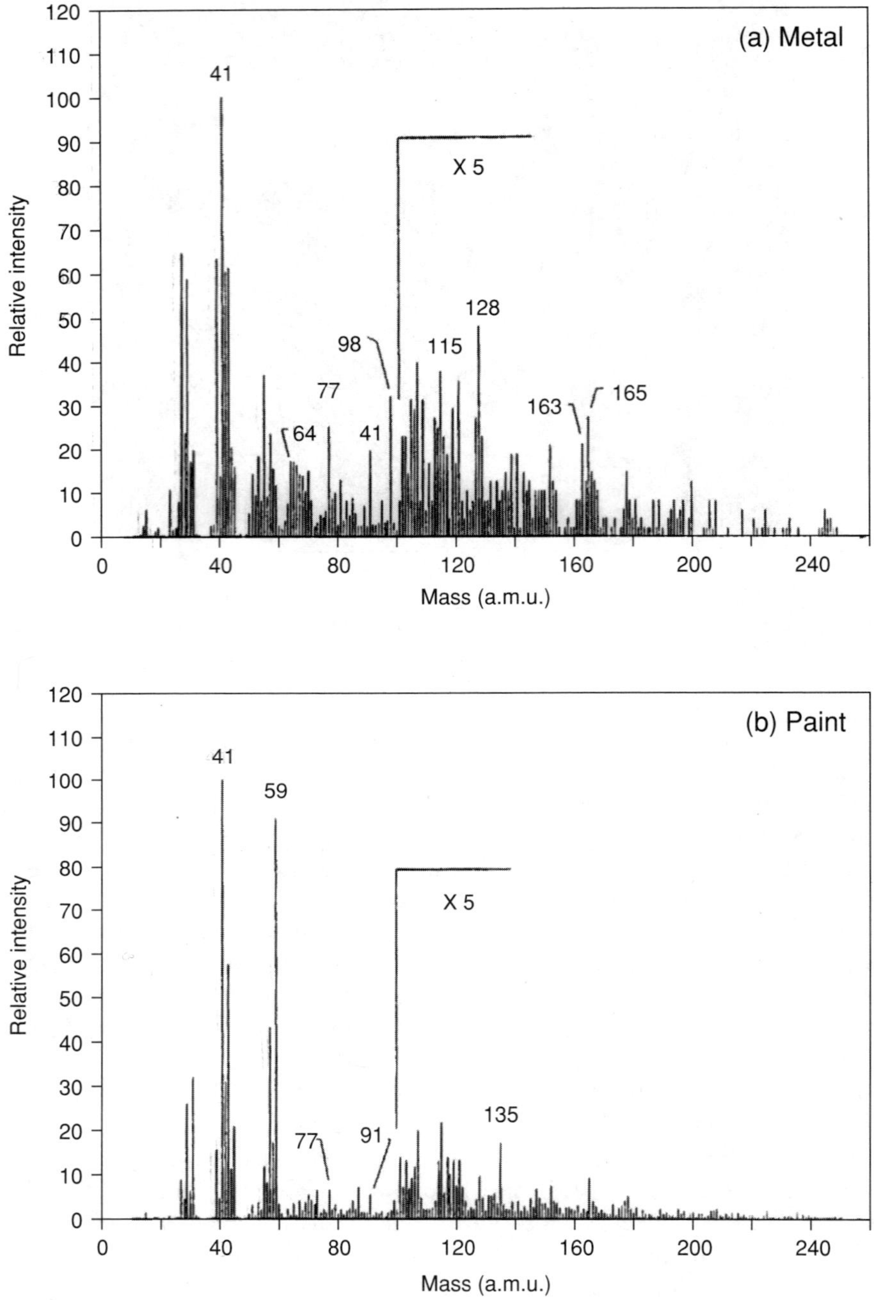

Figure 3.18 Positive SIMS spectrum of a delaminated paint interface from a nonphosphated galvanized mild steel panel that had been corrosion tested according to a GM scab corrosion test. Spectra from both sides of the interface are shown. (From van Ooij, W.J., Sabata, A., and Appelhans, A.D., *Surf. Interface Anal.*, 17, 403, 1991. With permission.)

thus inferred that the locus of failure occurs between the epoxy–urethane paint and a layer on the metal substrate that consisted of zinc hydroxides mixed with aromatic, nitrogen-containing paint components (perhaps an isocyanate). Effectively, no evidence of a corrosion-based origin was found for the failure, although it may exist.

By contrast, when a neighboring nondelaminated area was analyzed in the same fashion (see Figure 3.19), strong evidence for a metallic corrosion process was found. On the metal side (a), the spectrum shows a high intensity of positive sodium and sodium compounds (peaks at 23, 63 (NaOH), and 67 (COONa) Da); no Cl^- was

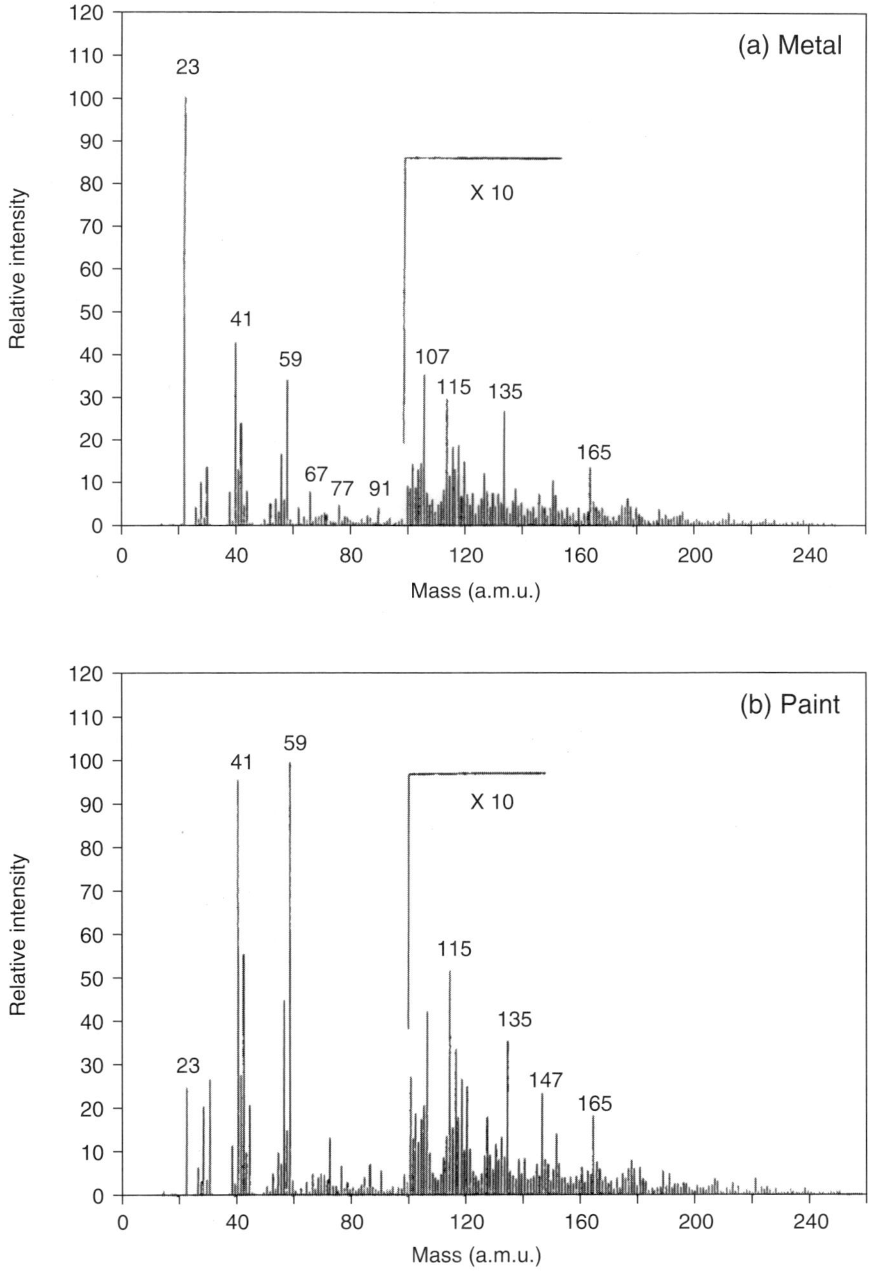

Figure 3.19 Positive SIMS spectrum of both sides of an interface between an e-coat paint layer and a galvanized coating on a mild steel panel. In contrast to the region described in Figure 3.18, the paint coating had continued to adhere to the substrate in this area. (From van Ooij, W.J., Sabata, A., and Appelhans, A.D., *Surf. Interface Anal.*, 17, 403, 1991. With permission.)

found in the negative ion spectrum. These two facts suggest that this region is the cathodic (high pH) side of an electrochemical corrosion cell. The anodic region was found to be at the contrived paint defect, where, in fact, a high Cl intensity was observed in the negative ion spectrum (not shown). Thus, a corrosion cell is established with a low pH anodic area at the defect and with a high pH cathodic area surrounding it approximately 1 mm away. The continued adhesion of paint to the cathodic areas is ascribed to reactions such as hydrolysis between the paint overlayer and the zinc hydroxides created in the cathodic region.

3.3.2.2 Stainless Steels

These alloys, as well as the nickel-based super alloys described later, have surfaces and grain interfaces that are often chromium-enriched. Thus, a major challenge for SIMS is to identify the distribution of this layer, as well as the other elements (if any) of which it is comprised. In addition, it is important to know how this layer reacts with typical corrosion initiators whose concentrations are low and whose distributions are localized.

A superb example of the power of SIMS to address such problems has been a study of the mechanisms of 316 stainless steel corrosion (62). Microscopic inclusions of MnS were identified on the surface of this alloy, both before and after its exposure to electrolyte. These inclusions are deposited in the bulk of the alloy during casting. Following corrosion, a finely focused Ga^+ beam was used to carry out a line profile of the chromium-rich surface adjacent to the inclusion. In composite Figure 3.20, the Cr/Fe ratio is plotted as a function of distance from an inclusion. The ratio changes regularly from a low of 0.5 at a distance of 100 nm from the inclusion, to a "normal" ratio of 1.0 found at a distance of 600 nm and greater. It is proposed that, during casting, the MnS inclusion reacts preferentially with the surrounding chromium, thus depleting this part of the alloy. Any surface that cuts through such an inclusion has a region adjacent to it where the oxide, which forms in air, is depleted in chromium.

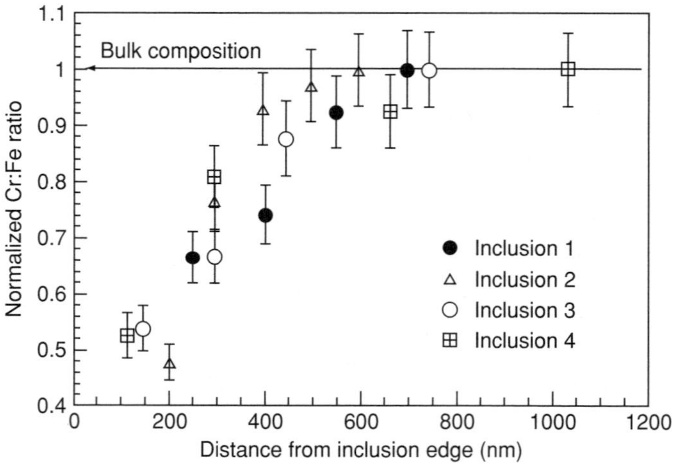

Figure 3.20 SIMS line profiles of iron chromium intensity ratios in the vicinity of a manganese sulfide inclusion on a 316 stainless steel surface. The plot shows the change in this ratio as a function of distance in nanometers from the edge of the inclusion. (From Ryan, M.P., Williams, D.E., Chater, R.J., and McPhail, D.S., *Nature*, 415, 771, 2002. With permission.)

Electrolyte brought in contact with this region would cause rapid dissolution of this Cr-depleted oxide and the formation of an active pit.

The enrichment of chromium within the MnS inclusions on austenitic stainless steel surfaces had previously been identified by SIMS imaging (63). No oxide film could be detected on the surface of such inclusions.

Corrosion of stainless steels by liquid alkali metals has been studied. Stainless steel alloys exposed in liquid Pb/Li alloy and pure Li were found to have been attacked strongly at their grain boundaries. Imaging SIMS was able to identify the penetration of Li into such boundaries (64).

3.3.2.3 Nickel, Nickel-Based Alloys, and Super Alloys

Nickel alloys are used in many demanding applications where the materials are stressed by very high temperatures or by hydrothermal corrosion. The alloys Monel 400 (28% to 34% Cu, 2.5% Fe, balance Ni), Inconel 600 (I600, 14% to 17% Cr, 6% to 10% Fe, balance Ni), Inconel 690 (I690, 27% to 31% Cr, 7% to 11% Fe, balance Ni), and Incoloy 800 (I800, 19% to 23% Cr, 30% to 35% Ni, balance Fe), are all in use as heat exchanger tubing in many of the world's nuclear generating stations. Of particular importance to such applications are those corrosion properties that could lead to pitting or cracking of each alloy, thus resulting in leaking of the radioactive primary circuit coolant into the secondary circuit.

The effects of a temporary swing to acidic coolant chemistry on the near-surface composition on each of the above alloys have been studied using XPS, AES, and SIMS (65). Alloy specimens were first exposed to normal conditions near pH 10 using hydrazine to maintain pH balance. Then, a chemical excursion was simulated with the injection of sufficient sulfuric acid (a typical "backwash" contaminant) to reduce the pH to levels as low as 5.

SIMS image depth profile analyses of selected microscopic regions of I600 and I800 exposed to such a simulated acid excursion are shown in Figure 3.21. These provide a cross-sectional view in much the same manner as in Figure 3.16. The intensities for each secondary ion studied (NiO^-, CrO^-, FeO^- [for I800], and S^-) are presented in a thermal color scale. The observed sulfur had been previously identified as a sulfide by XPS, a reaction product of sulfuric acid; sulfides have lower solubility in acid media compared to oxides and hydroxides. On I600, a single-phase surface oxide layer contains both chromium and nickel oxides as well as sulfur. Of particular note is the evidence for a localized corrosion attack of the underlying alloy from the appearance of a narrow filament of oxides and sulfur extending a fraction of a micrometer into the substrate. This may be the result of the extension of pitting corrosion or attack at a grain boundary. Also, within the substrate is seen evidence of some level of partial internal oxidation, particularly from the NiO^- image. The distribution of these oxides does not appear to be completely random; they may be present preferentially along grain boundaries. Sulfur inclusions are also observed.

For I800, a dual-oxide layer is observed with a compound chromium–nickel–iron oxide in the outer oxide layer, and an inner oxide that is primarily composed of chromium. Within the substrate, the chromium phase appears to have undergone more partial oxidation, in contrast to the case for I600.

SIMS has also been used to image the elemental composition in a through-wall crack in the inside diameter of an I600 boiler tube removed from service in a nuclear generating station (66). The observed intergranular cracking is a clear case of

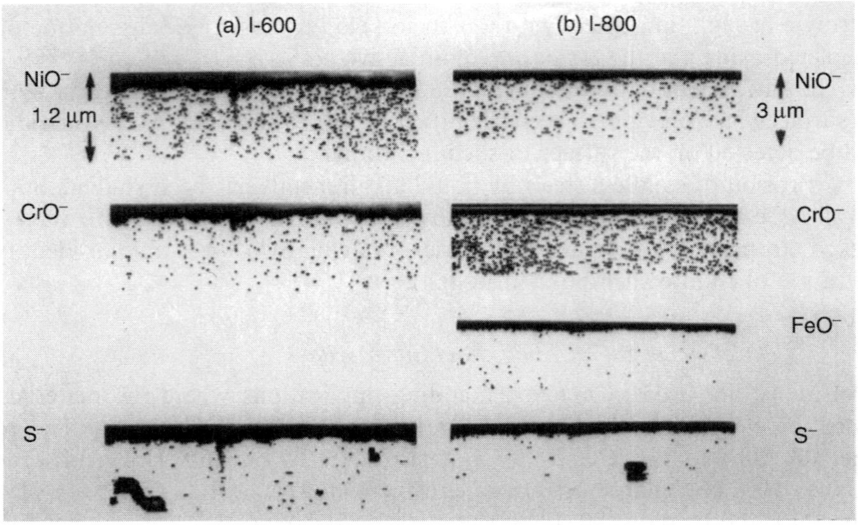

Figure 3.21 SIMS vertical cross-section images for oxides on Inconel 600 and Incoloy 800 under acid excursion conditions. The cross-sections are produced from the results of an imaging depth profile through the surface and near-surface regions. Images of NiO^-, CrO^-, FeO^-, and S^- are used to determine the different distributions of different oxide and sulfide phases. Some contributions from inclusions and precipitates in the alloy substrate can also be detected. The length of each section is 200 μm while the thicknesses are shown in the figure. The intensity of each ion is represented by a thermal color scale, with yellow as the highest intensity and black as the lowest. (From McIntyre, N.S., Davidson, R.D., Walzak, T.L., Brennenstühl, A.M., Gonzalez, F., and Corazza, S., *Corros. Sci.*, 37, 1059, 1995. With permission.)

stress corrosion cracking (SCC). Based on measurements of the spatial distributions of corrosion products in the crack, it was possible to draw some conclusions about the chemistry near the crack tip. The exact location of the crack tip was identified by SEM (image not shown). SIMS images are shown in Figure 3.22(a) to (f) for some negative secondary ion distributions near the crack tip. The entire crack length is outlined by the presence of oxygen in the crack wall (Figure 3.22a). In the region of the crack tip, strong signals from chlorine and sulfur are detected. By contrast, strong signals from corrosion products such as copper, hydrogen (as deuterium), and iron (not shown) are not found in the immediate region of the tip, but rather appear in a region 15 to 20 μm *behind* the crack tip. The exclusion of copper, zinc, and hydrogen from the crack tip region, and their appearance in a region somewhat behind the tip, suggests that cations of these elements are associated with a cathodic region, and that the tip itself is the anode couple, attracting stable anions such as chloride and sulfide. Thus, there seems to be no strong evidence of the involvement of hydrogen as hydrides in this particular cracking situation. Moreover, little evidence of oxygen in the grain boundaries ahead of the tip has been found, as has been suggested by the theory of "corrosion tunnels" (67).

Other specimens of I600, which have undergone SCC as a result of laboratory experiments, have been analyzed in more detail by imaging SIMS (68). The effects on cracking by the distributions of trace elements such as boron and carbon were explored using imaging cross-correlation techniques (69). Crack growth was found to be related to the amount of "free" boron that was detected in the boundaries without the presence of carbon. It is argued that the presence of carbon colocalized

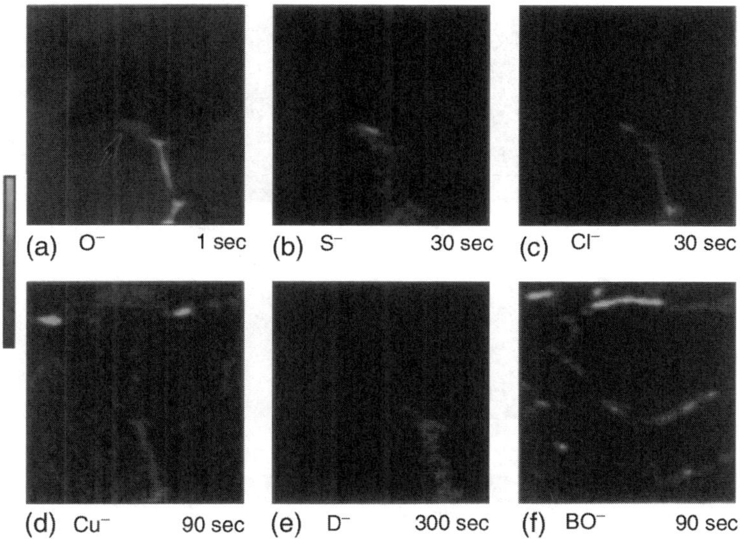

Figure 3.22 Secondary ion images of: (a) $^{16}O^-$, (b) $^{34}S^-$, (c) $^{35}Cl^-$, (d) $^{63}Cu^-$, (e) $^{2}D^-$, and (f) $^{27}BO^-$ intensity distributions. The field, in each case, is 150 μm. The thermal intensity scale represents an intensity range of 256 from black to yellow. (From McIntyre, N.S., Huctwith, C.M., Davidson, R.D., Brennenstühl, A.M., Lepik, O., and Clark, M., *Corros. Sci.*, 40, 1799, 1998. With permission.)

with boron locations effectively reduced the propensity of boron to assist the growth of the crack tip by reacting with it.

SIMS studies of the rate of intergranular oxidation for I600 specimens exposed to steam have shown that this alloy is susceptible to oxygen penetration through the grain boundaries, in contrast to specimens of I690 and I800 (70). These latter alloys are seen to have much thicker oxide layers on their outer surfaces, and it is surmised that the extent of internal oxidation is correspondingly lower. This oxygen penetration of the boundaries appears to occur via preferential oxidation of the chromium content of the boundaries. It is unclear from the experiments whether the rate of oxygen penetration is rapid enough to explain the crack growth.

The corrosion mechanisms of another Inconel alloy, Inconel 718, have been studied under typical primary circuit chemistry conditions (71). Among other techniques, SIMS was used to characterize the oxide growth rate, which was found to follow a parabolic rate law.

Monel 400 is a nickel–copper alloy used as heat exchanger tubing in Canadian nuclear reactors, as well as in marine applications. Imaging SIMS has recently been used (72) to image the distributions of nickel oxide and sulfur in a corrosion pit produced within a contrived crevice. Six sequential ion images are taken of different depths into the oxide and sulfide structures within and around the vicinity of a pit; these are presented in Figure 3.23(a) and (b) for NiO^- and S^-, respectively. The thin nickel oxide (or hydroxide) surface film in the pit is the likely result of a hydrolysis reaction during cooling down of the reactants; by contrast, the S^- image reveals sulfur concentrated outside the pit, implying cathodic deposition from sulfates present in the coolant.

Ni–Cr–Mo alloys are considered the most corrosion resistant of the nickel-based super alloys due to their high resistance to localized attack in chloride media. In nickel super alloys, it is important to optimize chromium and molybdenum

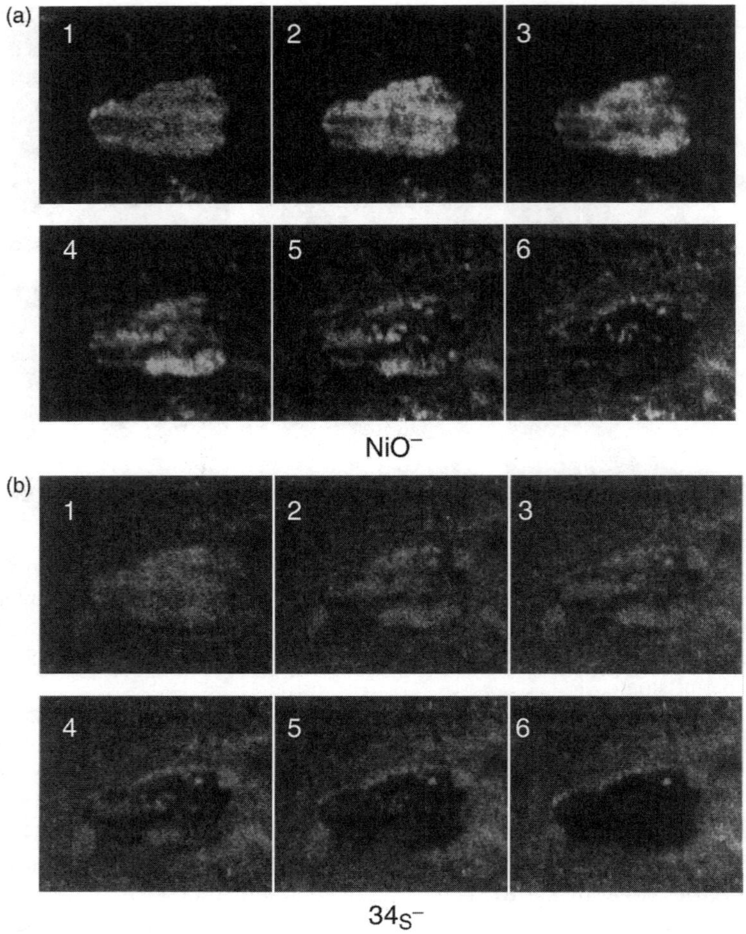

Figure 3.23 SIMS images of NiO^- and S^- distribution in the vicinity of a corrosion pit found in a contrived crevice consisting of a Monel 400 alloy. (From Francis, J.T., Brennenstühl, A.M., Ramamurthy, S., and McIntyre, N.S., *Surf. Interface Anal.*, 34, 2000. With permission.)

content in order to achieve this resistance to corrosion. The addition of chromium to nickel changes the electrochemical behavior of the alloy by lowering both the passivation potential and the passive dissolution current. Nickel-based alloys with higher chromium content, such as Alloy 22 (C-22–Cr > 20%), have the best corrosion resistance under aggressive testing conditions. Potentiostatic polarization experiments have been performed as a function of potential and temperature on C-22 and C-276 alloys (73). The objective was to compare the electrochemical behavior of the high chromium alloy (C-22) to the low chromium alloy (C-276–15% Cr). SIMS and XPS were used to determine the thickness and chemical composition of the films formed. The currents recorded were due to dissolution, with only minor increases in film thicknesses observed by these techniques. SIMS shallow depth profiling (profile depth <3 nm) was carried out using a low-energy cesium ion beam. The passive films on C-22 showed a distinct layered structure, consisting of an inner layer rich in chromium and nickel, and an outer layer enhanced in molybdenum (see Figure 3.24 and Figure 3.25). Oxide thickness increased with applied anodic potential. By contrast, the oxide film on C-276 did not show such a clear separation into layers, and the relative chromium content was much lower. The increase in oxide thickness

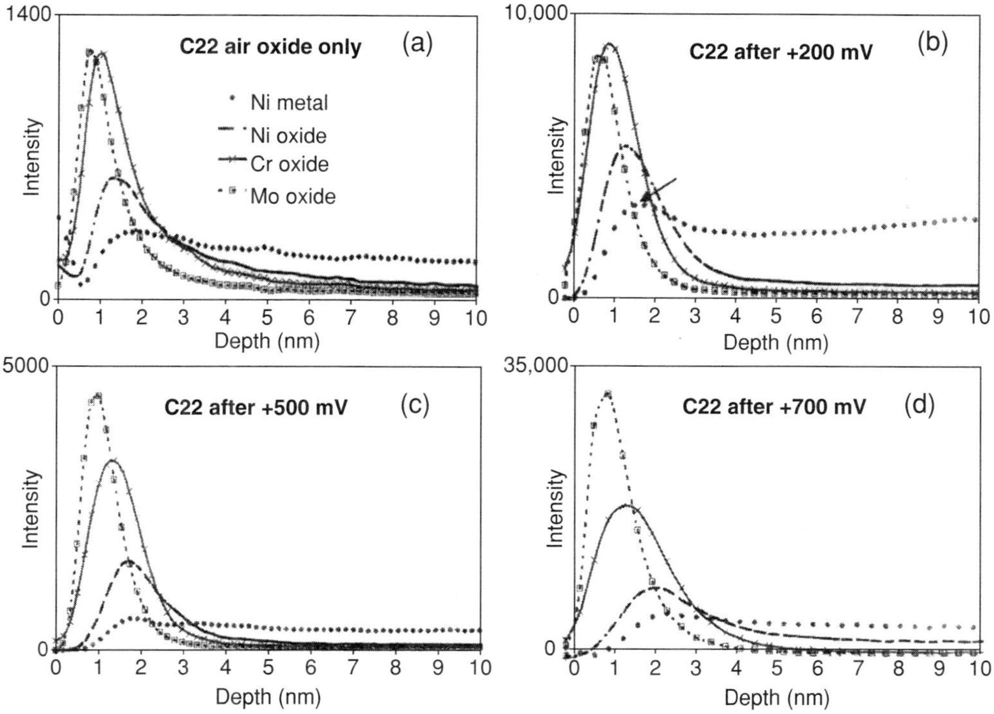

Figure 3.24 ToF-SIMS depth profiles for Ni–Cr–Mo alloy C-22 taken after: (a) 2 days air exposure, (b) polarization at 200 mV, (c) 500 mV, and (d) 700 mV. The film depth was calculated using a sputter rate of 3.2 nm/min. This sputter rate was determined by profilometry through the oxide and metal. As such, this sputter rate is a composite rate for both the oxide and metallic components. (From Lloyd, A.C., Shoesmith, D.W., McIntyre, N.S., and Nöel, J.J., *J. Electrochem. Soc.*, 150(4), B120–B130, 2003. With permission.)

with increasing anodic potential, and the low-temperature dependence of the passive current observed for C-22, are consistent with an oxide dissolution rate that is low compared to the rate of creation of oxygen vacancies. The absence of a dependence of film thickness on potential, and the higher temperature dependence of the passive currents on C-276 are consistent with overall control of the reaction at the solution–solid interface.

3.3.2.4 Aluminum

The very initial interactions between O_2 and an Al (100) surface at ambient temperatures were followed using a unique angle-resolved SIMS instrument that was capable of distinguishing between ordered and amorphous surface structures (74,75). Using this, it is possible to show that for oxygen doses of up to 200 Langmuirs, oxygen is only in a chemisorbed state on the metal surface; this follows from the observation that no changes are observed for the azimuthal angle distributions of Al^+ ions coming from the surface. The oxide that nucleates at higher exposures is shown to be amorphous, and evidence is introduced that oxide nucleation does not begin with island formation.

Quite recently, the oxide structures on clean polycrystalline aluminum and gold–aluminum alloy surfaces were studied by SIMS following ambient temperature exposure to deuterated water vapor using doses of 8000 Langmuirs (76). SIMS

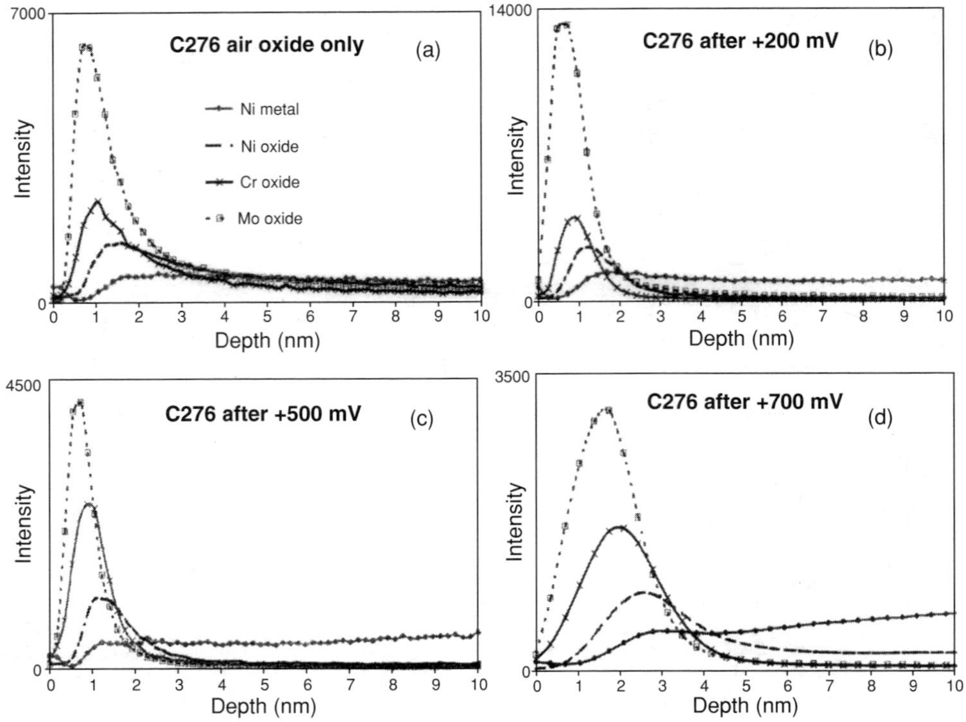

Figure 3.25 ToF-SIMS depth profiles for C-276 taken after: (a) 2 days air exposure, (b) polarization at 200 mV, (c) 500 mV, and (d) 700 mV. Film depth determined as described for Figure 3.24. (From Lloyd, A.C., Shoesmith, D.W., McIntyre, N.S., and Nöel, J.J., *J. Electrochem. Soc.*, 150(4), B120–B130, 2003. With permission.)

showed that the deuterium-containing layer was limited to the outermost 0.5 nm of the oxide layer, indicating that there is little hydration at this stage in the process. Of particular interest was the significant level of aluminum substrate oxidation that occurs even at ambient temperature. This contribution was quantified using isotopic oxygen implants and was found to amount to a concentration of several percent extending some distance into the substrate.

Clearly, SIMS has a valuable role to play in the understanding of internal oxidation of aluminum-based metals because it is possible to distinguish minor quantities of oxygen in the presence of a metallic phase.

The nitriding of aluminum in a radiofrequency discharge has been followed by ToF-SIMS, revealing different distributions of oxide and nitride species on the outer surface (77). Air oxidation of an AlN ceramic at high temperature (1100°C) has also been studied using depth profiling to follow the growth rate of the oxide phase (78).

3.3.2.5 Zirconium Alloys

The surface chemistry and metallurgy of zirconium and its alloys have been investigated extensively by SIMS. This comes, in part, because of the high technological importance of the alloys as critical components (fuel cladding and fuel channells) of the world's nuclear reactors. Much of this interest has to do with the performance of the alloys in contact with water, because metal hydrides can form under some conditions and these hydrides are very brittle. Thus, in some conditions, SCC is a particular concern.

An interesting early study involved an SIMS installation capable of handling highly radioactive fuel cladding (79). Radioiodine was found in areas of the cladding that had undergone SCC.

The surfaces of fuel channells removed from active service in a Canadian nuclear reactor were analyzed by SIMS for a number of minor and major elements, originating either as alloying elements or as impurities in the coolant (80). Of particular interest was hydrogen, which, as deuterium, was able to be quantified using nuclear reaction analysis of secondary standards. A correlation was found between the regions of the fuel channel that had been found to contain high concentrations of bulk hydrogen and the high concentrations of hydrogen throughout the depth profile of the surface oxide. This can be seen in Figure 3.26 and Figure 3.27,

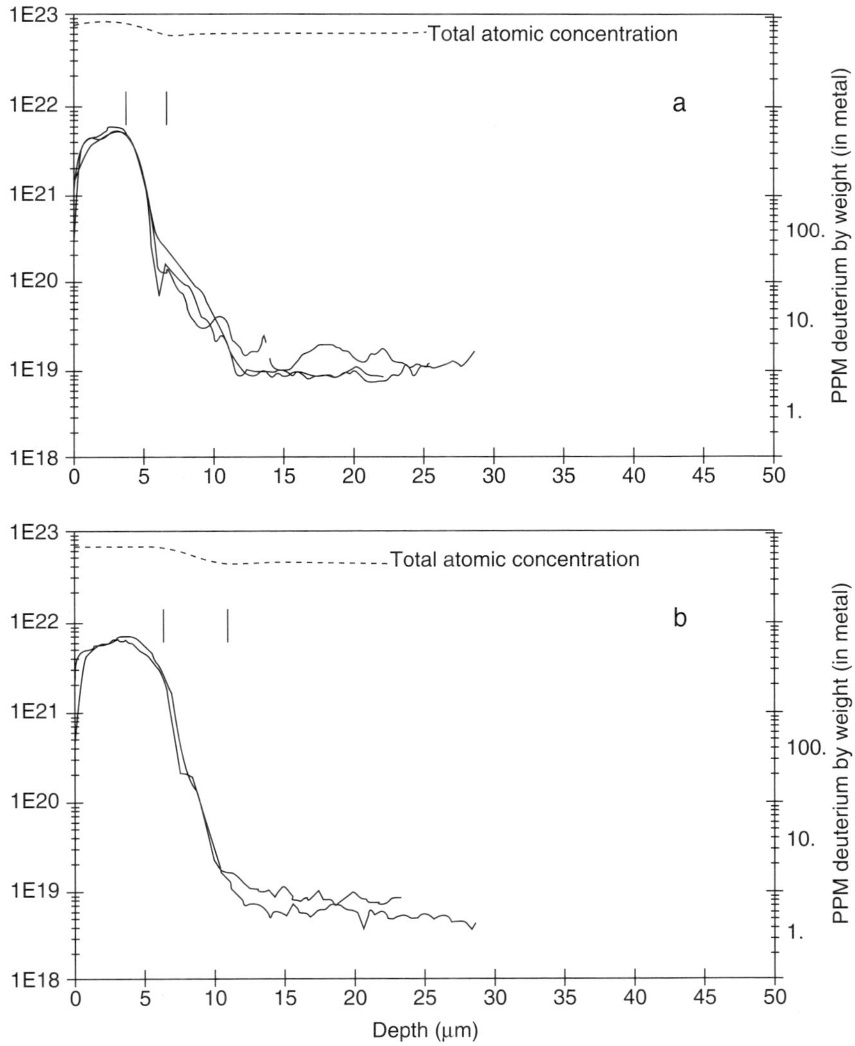

Figure 3.26 Quantitative SIMS deuterium profiles for the inside diameter: (a) inlet end and (b) outlet end of fuel channel J9. Three independent profiles are shown. The two vertical bars show the limits of the oxide–metal interface region. (From McIntyre, N.S., Weisener, C.G., Davidson, R.D., Brennenstühl, A.M., and Warr, B.D., *J. Nucl. Mater.*, 178, 80, 1991. With permission.)

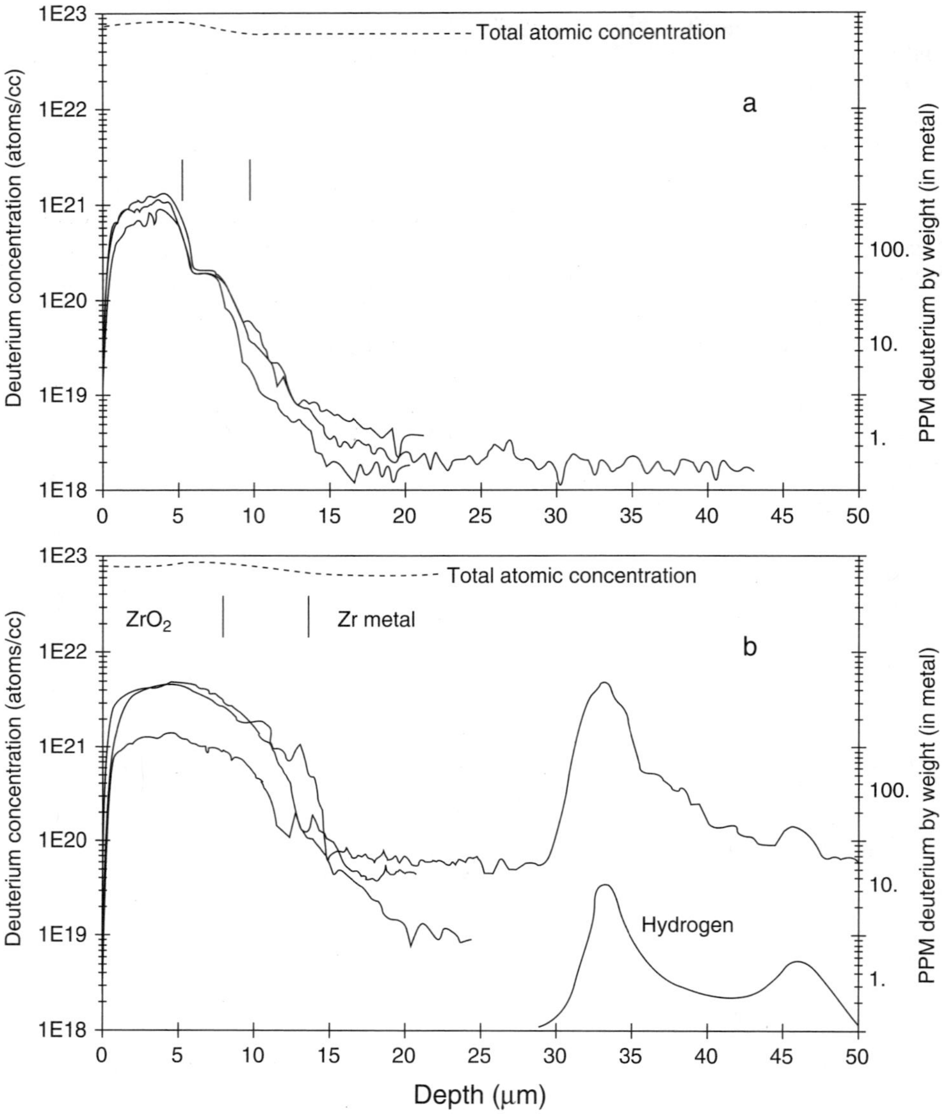

Figure 3.27 Quantitative SIMS deuterium profiles for fuel channel L9: (a) inlet end and (b) outlet end. (From McIntyre, N.S., Weisener, C.G., Davidson, R.D., Brennenstühl, A.M., and Warr, B.D., *J. Nucl. Mater.*, 178, 80, 1991. With permission.)

which show depth profiles for inlet and outlet ends of two different channells removed from the reactor: tube J9, which had exhibited low pickup of hydrogen according to bulk analysis, and tube L9, which had a particularly high hydrogen pickup. Thus, it appeared that movement of hydrogen across the oxide–metal interface is particularly apparent at the outlet end of L9. Later studies of cross-sections of this tube were able to identify regions of oxide breakdown on the basis of "tracks" of deuterium across the oxide layer that is nearest to the metal and which normally is the most protective oxide toward hydrogen ingress (81).

The diffusion of deuterium into and through oxide films on zirconium–niobium alloys has been investigated by analysis of SIMS images (82). Oxides grown on low

Zr/Nb alloys (typifying an alpha zirconium phase structure) were found to have higher permeability to water than oxide grown on higher Zr/Nb alloys where the beta niobium phase becomes more important. Using 2D cross-sections created from SIMS image depth profiling, at least two different surface oxide structures are identified: one is localized above grain boundaries and the other is more permeable to water ingress.

Another similar investigation (83) compared the oxidation rates of Zr-2 and 0.5%NbZr-2 before and after heat treatment to change the oxide structure from tetragonal to monoclinic. In the case of the Zr-2, it was found that the transition had no effect on the rate of oxygen uptake as determined by SIMS depth profiling. However, in the case of the 0.5%NbZr-2, the oxidation rate increased after the transition, and the concentrations of the minor elements iron, chromium, and nickel were found to increase at the columnar grain boundaries in the oxide.

Finally, SIMS has been one of the techniques used to investigate the role of massive amounts of hydride precipitate at the metal–oxide interface on the oxidation of Zr-2 alloy (84). No effect on oxidation rate was found, whether the hydrogen was present in solid solution, or distributed as fine precipitates or as a massive hydride. The effect of oxidation in each case was to cause the hydride to move further into the metal as the oxidation front advanced.

3.4 CONCLUSIONS

Because of its many advantages (see Table 3.3), over the past decade SIMS has become increasingly adopted as a major investigative tool for both fundamental and applied corrosion studies. Of particular note is the increasing capacity of the technique to identify and locate inorganic and organic corrosion precursors and modifiers of surface chemistry. This has been aided technologically by the development of high-precision mass measurements as well as improved image generation for surfaces that are nonconducting.

More accurate identification of surface chemistry requires knowledge that ties the observation of a polyatomic secondary ion to its molecular or ionic precursor on

Table 3.3 Advantages and disadvantages of SIMS

Advantages
Can detect H and D
Can distinguish isotopes
Chemical information obtained from the detection of sputtered molecular fragments
High sensitivity (ppm range)
High spatial resolution (20 nm)
Rapid imaging and mapping (a minute)
Good depth resolution
Rapid data collection
A developing user base and reasonable support data available
Disadvantages
Requires destruction of the sample
Wide sensitivity range for different elements
Difficult to quantify
Complex spectra

the surface. Greater confidence in this area is likely to come in the near future, as the use of soft and hard ionization sources is tried on different surfaces.

SIMS has an important role in defining the nature of the hydration sheath on surfaces and its role in corrosion processes. Careful experimentation with hydrogen and oxygen isotopes used under controlled atmospheres could provide the mechanistic definition of solution–surface exchange processes even though SIMS vacuum conditions preclude reproduction of real-world conditions. The next decade promises to be an exciting one for the application of SIMS to the understanding of corrosion.

REFERENCES

1. J.C. Vickerman and A.J. Swift, Secondary ion mass spectrometry — the surface mass spectrometry, in *Surface Analysis—the Principal Techniques*, J.C. Vickerman, ed., John Wiley and Sons, New York, 1997, pp. 135–214.
2. G.C. Smith, *Quantitative Surface Analysis for Materials Science*, The Institute of Metals, London, UK, 1991, pp. 29–36.
3. D. Briggs and M.P. Seah, *Practical Surface Analysis*, 2nd edn, Vol. 2, John Wiley and Sons, New York, 1990.
4. S.C. Tjong and E. Yeager, *J. Electrochem. Soc.*, **128**, 2251 (1981).
5. O.J. Murphy, T.E. Pou, J.O'M. Bockris, and L.L. Tongson, *J. Electrochem. Soc.*, **131**, 2785 (1984).
6. D.F. Mitchell and M.J. Graham, *J. Electrochem. Soc.*, **133**, 936 (1986).
7. D.F. Mitchell and M.J. Graham, *Surf. Interface Anal.*, **10**, 259 (1987).
8. D.F. Mitchell, *Appl. Surf. Sci.*, **9**, 131 (1981).
9. B. MacDougall, D.F. Mitchell, G.I. Sproule, and M.J. Graham, *J. Electrochem. Soc.*, **130**, 543 (1983).
10. B. MacDougall and M.J. Graham, *J. Electrochem. Soc.*, **131**, 727 (1984).
11. R. Goetz, B. MacDougall, and M.J. Graham, *Electrochim. Acta*, **31**, 1299 (1986).
12. M.J. Graham, in *Proc. 1st Int. Conf. on Microscopy of Oxidation*, University of Cambridge, 1990, The Institute of Metals, 1991, p.10.
13. M. Nagayama and M. Cohen, *J. Electrochem. Soc.*, **109**, 781 (1962).
14. J.A. Bardwell, B. MacDougall, and M.J. Graham, *J. Electrochem. Soc.*, **135**, 413 (1988).
15. J.A. Bardwell, G.I. Sproule, D.F. Mitchell, B. MacDougall, and M.J. Graham, *J. Chem. Soc. Faraday Trans.*, **87**, 1011 (1991).
16. M.J. Graham, J.A. Bardwell, G.I. Sproule, D.F. Mitchell, and B.R. MacDougall, *Corros. Sci.*, **35**, 13 (1993).
17. R. Goetz, D.F. Mitchell, B. MacDougall, and M.J. Graham, *J. Electrochem. Soc.*, **134**, 535 (1987).
18. D.F. Mitchell, R.J. Hussey, and M.J. Graham, *J. Vac. Sci. Technol. A*, **2**, 789 (1984).
19. R.J. Hussey, D.F. Mitchell, and M.J. Graham, *Werkstoffe und Korrosion*, **38**, 575 (1987).
20. M.J. Graham, *Corros. Sci.*, **37**, 1377 (1995).
21. J. Stringer and P.Y. Hou, in *Proc. Symp. on Corrosion and Particle Erosion at High Temperatures*, V. Srinivasan and K. Vedula, eds., TMS, 1989, p. 383.
22. M.J. Bennett, B.A. Bellamy, C.F. Knights, N. Meadows, and N.J. Eyre, *Mater. Sci. Eng.*, **69**, 359 (1985).
23. M.J. Bennett, H.E. Bishop, P.R. Chalker, and A.T. Tuson, *Mater. Sci. Eng.*, **90**, 177 (1987).
24. M.J. Bennett, J.A. Desport, M.R. Houlton, P.A. Labun, and J.M. Titchmarch, *Mater. Sci. Technol.*, **4**, 1107 (1988).
25. C.M. Cotell, G.J. Yurek, R.J. Hussey, D.F. Mitchell, and M.J. Graham, *Oxid. Met.*, **34**, 173 (1990).
26. D.P. Moon and M.J. Bennett, *Mater. Sci. Forum*, **43**, 269 (1989).

27. P.Y. Hou and J. Stringer, in *Proc. Symp. High Temp. Mat. Chem. III*, Vol. 86-2, Z.A. Munir and D. Cubicciotti, eds., The Electrochemical Society, 1986, p. 198.
28. R.J. Hussey, P. Papaiacovou, J. Shen, D.F. Mitchell, and M.J. Graham, in *Proc. Symp. on Corrosion and Particle Erosion at High Temp.*, V. Srinivasan and K. Vedula, eds., TMS, 1989, p. 567.
29. R.J. Hussey, P. Papaiacovou, J. Shen, D.F. Mitchell, and M.J. Graham, *Mater. Sci. Eng. A*, **120**, 147 (1989).
30. P. Papaiacovou, R.J. Hussey, D.F. Mitchell, and M.J. Graham, *Corros. Sci.*, **30**, 451 (1990).
31. C.M. Cotell, G.J. Yurek, R.J. Hussey, D.F. Mitchell, and M.J. Graham, *J. Electrochem. Soc.*, **134**, 1871 (1987).
32. H.E. Bishop, P.R. Chalker, and D.W. Smart, Harwell Laboratory Report AERE R-12771 (1987).
33. A. Atkinson and D.W. Smart, *J. Electrochem. Soc.*, **135**, 2886 (1988).
34. M.J. Graham and R.J. Hussey, *Oxid. Met.*, **44**, 339 (1995).
35. D.P. Moon, A.W. Harris, P.R. Chalker, and S. Mountfort, Harwell Laboratory Report AERE R-13097 (1988).
36. A.T. Chadwick and R.I. Taylor, *Solid State Ionics*, **12**, 343 (1984).
37. A.T. Chadwick and R.I. Taylor, *J. Microsc.*, **140**, 221 (1985).
38. D.P. Moon and M.J. Bennett, *Mater. Sci. Forum*, **43**, 269 (1989).
39. P.J. George, M.J. Bennett, H.E. Bishop, and G. Dearnaley, *Mater. Sci. Eng. A*, **116**, 111 (1989).
40. F. Czerwinski and W.W. Smeltzer, *J. Electrochem. Soc.*, **140**, 2606 (1993); **141**, L34 (1994).
41. F. Czerwinski, G.I. Sproule, M.J. Graham, and W.W. Smeltzer, *Corros. Sci.*, **37**, 541 (1995).
42. R. Prescott, D.F. Mitchell, and M.J. Graham, in *Microscopy of Oxidation 2*, S.B. Newcomb and M.J. Bennett, eds., Institute of Metals, London, 1993, p. 455.
43. R. Prescott, D.F. Mitchell, and M.J. Graham, *Corrosion*, **50**, 62 (1994).
44. R. Prescott, D.F. Mitchell, M.J. Graham, and J. Doychak, *Corros. Sci.*, **37**, 1341 (1995).
45. E. Schumann, J.C. Yang, M. Rühle, and M.J. Graham, *Oxid. Met.*, **46**, 37 (1996).
46. B.A. Pint, J.R. Martin, and L.W. Hobbs, *Oxid. Met.*, **39**, 167 (1993).
47. C. Mennicke, E. Schumann, M. Rühle, R.J. Hussey, G.I. Sproule, and M.J. Graham, *Oxid. Met.*, **49**, 455 (1998).
48. G. Hultquist, B. Tveten, E. Hörnlund, M. Limbäck, and R. Haugsrud, *Oxid. Met.*, **56**, 313 (2001).
49. B. Tveten, G. Hultquist, and T. Norby, *Oxid. Met.*, **51**, 221 (1999).
50. W.J. Quadakkers, H. Holzbrecher, K.G. Briefs, and H. Beske, *Oxid. Met.*, **32**, 67 (1989).
51. W.J. Quadakkers, A. Elschner, W. Spier, and H. Nickel, *Appl. Surf. Sci.*, **52**, 271 (1991).
52. W.J. Quadakkers, *Werkstoffe und Korrosion*, **41**, 659 (1990).
53. F. Rochet, B. Agius, and S. Rigo, *J. Electrochem. Soc.*, **131**, 914 (1984).
54. C.-J. Han and C.R. Helms, *J. Electrochem. Soc.*, **135**, 1824 (1988).
55. R.J. Hussey, G.I. Sproule, D.F. Mitchell, and M.J. Graham, in *Proc. 12th Int. Corrosion Congress*, Vol. 5B, Houston, TX, 1993, p. 3831.
56. V.S. Sastri, V.S. Donepudi, N.S. McIntyre, D. Johnston, and R.W. Revie, *Metall. Trans.*, **19**, A3071 (1988).
57. A. Pebler and G.C. Sweeney, Investigations of corrosion layers on mild steel with a direct imaging mass spectrometer, in *SIMS II, Proc. 2nd Int. Conf. on Secondary Ion Mass Spectrometry*, Stanford, CA, A. Benninghoven, C.A. Evans Jr., R.A. Powell, R. Shimizu, and H.A. Storms, eds., Springer-Verlag, Berlin, 1979, pp. 154–156.
58. S. Ramamurthy, T.L. Walzak, S.F. Lu, T.C. Lipson, and N.S. McIntyre, *Surf. Interface Anal.*, **17**, 834 (1991).
59. S.F. Lu, G.R. Mount, N.S. McIntyre, and A. Fenster, *Surf. Interface Anal.*, **21**, 177 (1994).

60. M. Van Craen, F. Adams, and G. Haemers, *Surf. Interface Anal.*, **4**, 56 (1982).
61. W.J. van Ooij, A Sabata, and A.D. Appelhans, *Surf. Interface Anal.*, **17**, 403 (1991).
62. M.P. Ryan, D.E. Williams, R.J. Chater, and D.S. McPhail, *Nature*, **415**, 771 (2002).
63. A. Rossi, B. Elsener, G. Hähner, M. Textor, and N.D. Spencer, *Surf. Interface Anal.*, **29**, 460 (2000).
64. M.G. Barker and I.E. Schreinlecher, *Surf. Interface Anal.*, **9**, 371 (1986).
65. N.S. McIntyre, R.D. Davidson, T.L. Walzak, A.M. Brennenstühl, F. Gonzalez, and S. Corazza, *Corros. Sci.*, **37**, 1059 (1995).
66. N.S. McIntyre, C.M. Huctwith, R.D. Davidson, A.M. Brennenstühl, O. Lepik, and M. Clark, *Corros. Sci.*, **40**, 1799 (1998).
67. N.A. Nielsen, in *2nd Int. Congr. on Metallic Corrosion*, London, 1962, pp. 200–210.
68. N.S. McIntyre, C.M. Huctwith, K.F. Taylor, E. Keating, N.O. Petersen, and A.M. Brennenstühl, *Surf. Interface Anal.*, **33**, 447 (2002).
69. M. Srivastava, N.O. Petersen, G.R. Mount, D.M. Kingston, and N.S. McIntyre, *Surf. Interface Anal.*, **26**, 186 (1998).
70. R.C. Newman, T.S. Gendron, and P.M. Scott, in *9th Int. Symp. on Environmental Degradation of Materials in Nuclear Power Systems — Water Reactors*, F.P. Ford, S.M. Bruemner, and G.S. Was, eds., The Minerals, Metals and Materials Society, Warrendale, PA, 1999, pp. 79–93.
71. O. Brucelle, J.G. Spilmont, J.M. Cloué, M. Foucault, and E. Andrieu, in *9th Int. Conf. on Environmental Degradation of Materials in Nuclear Power Systems — Water Reactors,* F.P. Ford, S.M. Bruemner, and G.S. Was, eds. The Minerals, Metals and Materials Society, Warrendale, PA, 1999, p. 197.
72. J.T. Francis, A.M. Brennenstühl, S. Ramamurthy, and N.S. McIntyre, *Surf. Interface Anal.*, **34**, 189 (2002).
73. A.C. Lloyd, D.W. Shoesmith, N.S. McIntyre, and J.J. Nöel, *J. Electrochem Soc.* **150**(4), B120 (2003).
74. L.L. Lauderback, A.J. Lynn, C.J. Waltmann, and S.A. Larson, *Surf. Sci.*, **243**, 229 (1991).
75. S.A. Larson and L.L. Lauderback, *Surf. Sci.*, **284**, 1 (1993).
76. H. Piao, M. Suominen Fuller, D. Miller, J.T. Francis, and N.S. McIntyre, *Appl. Surf. Sci.*, **187**, 266 (2002).
77. S. Gredelj, A.R. Gerson, S. Kumar, and N.S. McIntyre, *Appl. Surf. Sci.*, **199**, 234 (2002).
78. R. Yue, Y. Wang, Y. Wang, and C. Chen, *Appl. Surf. Sci.*, **148**, 73 (1999).
79. G. Bovalini, M. Cavallini, G. Grillo, D. Morani, and L. Ricciardi, *Energia Nucleare*, **27**, 157 (1980).
80. N.S. McIntyre, C.G. Weisener, R.D. Davidson, A.M. Brennenstühl, and B.D. Warr, *J. Nucl. Mater.*, **178**, 80 (1991).
81. M.C. Biesinger and N.S. McIntyre, in *Proc. 12th Int. Conf. on Secondary Ion Mass Spectrometry*, Brussels, Belgium, A. Benninghoven, P. Bertrand, H.-N. Migeon, and H.W. Werner, eds., Elsevier Science, Amsterdam, 2000, pp. 225–228.
82. A.H. Clarke and N.S. McIntyre, *Surf. Interface Anal.*, **25**, 948 (1997).
83. S. Nanikawa and Y. Yoshinori, *Nucl. Sci. Technol.*, **38**, 420 (2001).
84. M. Oskarsson, E. Ahlberg, U. Södervall, U. Andersson, and K. Pettersson, *J. Nucl. Mater.*, **289**, 315 (2001).

4

MeV Ion Beam Analytical Methods

Didier Schmaus and Ian Vickridge
Institut des Nanosciences de Paris, Université Deuis Diderot, Centre National de la Recherche Scientifique, Paris, France

Contents

4.1 INTRODUCTION

We present analytical methods that use swift (typically 0.1 to 1 MeV/u) ion beams to determine the composition and structure of solids, focussing on methods that may be applied to analyze the composition and crystallographic structure (including depth profiles) of films laterally homogeneous over a few square millimeters. These methods are rapid (typical spectrum acquisition times <10 min), quantitative (composition determined within a few percent), sensitive (down to 1/100 atomic monolayer in favorable cases), and offer depth resolution ranging from 1 to 100 nm, usually with negligible damage to the sample. It should be noted that ion beams can also be focused and scanned with micrometer resolution in the nuclear microprobe configuration for analysis of laterally inhomogeneous samples. The very high beam current densities needed for microbeam analysis often restrict use to those methods such as particle-induced x-ray emission (PIXE) having high cross-sections or samples able to withstand the damage incurred by the analysis.

This chapter is divided into two main parts. In the first (Sections 4.2 and 4.3), we present ion beam analytical methods that have found significant application in corrosion research: Rutherford backscattering spectrometry (RBS), elastic recoil detection (ERD), channelling, nuclear reaction analysis (NRA), and nuclear resonance profiling (NRP), together with the necessary underlying physical concepts. The physics is presented in order of decreasing probability of interaction between an incident ion and the atoms of the target: energy loss to electrons, Coulomb scattering from nuclei (and ion channelling effects), and, finally, nuclear reactions. In the second part (Section 4.4), a selection of recent illustrative examples of application of each of these methods is presented.

For the reader interested in exploring ion beam analysis more deeply, we particularly recommend reading Refs. (1–4).

4.2 BASIC PRINCIPLES

When a swift ion penetrates a solid, its energy, direction, and charge state are modified by interactions with the nuclei and electrons of the solid. We consider two interaction regimes that have quite different effects on the ion:

- Interactions with a high probability of occurrence. A given ion suffers many such interactions, but its characteristics are only slightly modified by each interaction. The influence of such interactions on the ion energy loss is described in Section 4.2.1.
- Interactions with a low probability of occurrence (close collisions between ion and target nuclei), which strongly modify the ion characteristics. A given ion generally suffers zero and occasionally one of these interactions (single collision regime). The analytical methods presented in Sections 4.2.2 to 4.2.4 are based on these infrequent interactions (large angle scattering, nuclear reaction).

4.2.1 Energy Loss — Stopping Power

An ion loses its energy (E) nearly continuously along its path x, through electromagnetic interactions with the solid, considered here as constituted of a collection of randomly distributed atoms. The average energy loss per unit path length is termed the "stopping power." Knowing this loss allows one to convert the ion energy scale into a depth scale. A semiempirical synthesis of experimental values for all ions in all elements is available (5).

From the measured stopping cross-section:

$$\varepsilon = -\frac{1}{N}\frac{dE}{dx}(\text{eV cm}^2) \tag{4.1}$$

one can deduce, knowing N as the atomic density of the solid, the "energy loss rate" $-dE/dx$ (eV/nm). The energy dependence of the stopping power exhibits a maximum whose value and energy position depends on the ion–solid system considered. As an example, the stopping power for ^{4}He ions in Ni (5) is presented in Figure 4.1: at about 1 MeV, it reaches its maximum value of $\varepsilon \approx 78\,\text{eV}/(10^{15}\ \text{atoms/cm}^2)$ (or $dE/dx \approx 710\,\text{eV/nm}$).

In the velocity regime considered, the stopping power is almost entirely due to interactions with target electrons, and is called the electronic stopping power to differentiate it from energy loss to nuclei, which becomes significant at much lower velocities.

Qualitatively, the maximum value observed at E_{max} can be understood as the result of two competing energy dependencies in the two energy domains separated by E_{max}.

When E is much greater than E_{max}, the ion is completely stripped inside the solid as its velocity, v, is much larger than the mean orbital velocity of its electrons. The corresponding stopping power is then correctly estimated by the nonrelativistic Bethe formula, derived from a quantum-mechanical treatment:

$$\frac{dE}{dx} = -N\frac{Z_1^2}{E}Z_2\frac{M_1}{m}2\pi\, e^4 \ln\left(\frac{2mv^2}{I}\right) \tag{4.2}$$

where Z_1 is the atomic number of the incident ion, Z_2 that of the solid, M_1 the atomic mass of the incident ion, m the electron mass, $e^2 = 1.44\,\text{eV nm}$, and I is a mean excitation potential, which can be approximated by $I \approx 10Z_2$ eV. In this "high-energy domain," the electronic stopping power is nearly inversely proportional to the ion energy.

In the "low-energy domain" ($E \ll E_{max}$), the electronic stopping power decreases as the ion energy decreases. This is because the ion is no longer fully stripped: the ion's electrons now screen the Coulomb field of its nucleus, reducing the magnitude of the ion–solid interactions, and the maximum energy transfer from the ion to the electrons in the solid is now of the same order as their binding energies, which

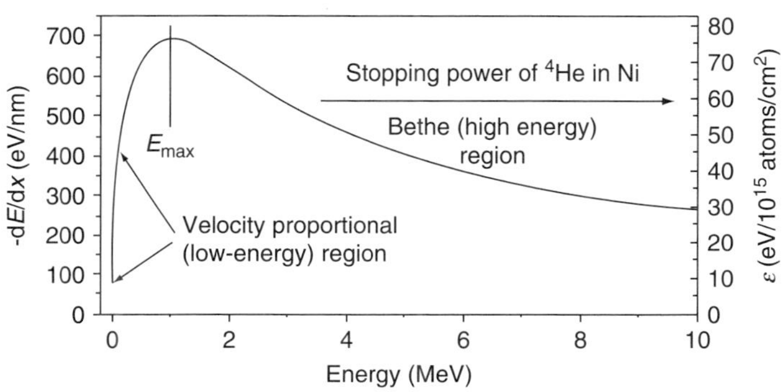

Figure 4.1 Stopping power for ^{4}He in Ni. The density of Ni is $= 9.13 \times 10^{22}$ atoms/cm^3.

prohibits many ionization and excitation processes. Theoretical estimation of the stopping power is rather involved in this velocity domain.

Reliable experimental values for compounds are generally not available but a reasonable estimation of the stopping cross-sections in any compound may be obtained from those tabulated for elements via Bragg's rule of linear additivity, which postulates that if the solid consists of atoms of type A, B, ... with stoichiometry $A_aB_b, \ldots$, and $a + b + \cdots = 1$, one has:

$$\varepsilon = a\varepsilon_A + b\varepsilon_B + \cdots \tag{4.3}$$

where ε_A, $\varepsilon_B, \ldots$ are the corresponding elemental stopping cross-sections. The corresponding dE/dx is given by Equation (4.1). In the energy domain considered here, this rule appears to be excellent for incident protons and in most cases leads to stopping powers accurate to within 10% for other light ions.

By integrating the stopping power along the ion trajectory, one obtains the range of the ion inside the target, tabulated in Ref. (5), which is typically of the order of 10 μm for ions and energies considered here.

Energy loss is a statistical process in which individual ion energy losses are distributed about the average energy loss incurred in penetrating to depth x. This broadening, first treated by Bohr (6), is called energy straggling, and may be represented by its full-width at half-maximum (FWHM) ΔE_{str}. For x greater than about 10 nm, $\Delta E_{str} \propto \sqrt{x}$.

4.2.2 Elastic Scattering

When an ion approaches a "target" atom of the solid, both particles experience a repulsive force due to the Coulomb interaction between their nuclei, screened by their electron clouds. This electrostatic interaction is the basis for RBS, ERD, and channelling. In the energy domain considered here, classical nonrelativistic concepts are valid: an incident ion can be considered as a particle, with a well-defined trajectory.

4.2.2.1 Rutherford Backscattering Spectrometry

In RBS, which exploits the infrequent very close elastic interactions between ions and target atoms, one can analyze the absolute number and the kinetic energy of the ions scattered at large angle by various target atoms into a detector. Target composition is deduced from the number of ions detected as a function of their energy — the "backscattering spectrum."

4.2.2.1.1 Mass Identification — Kinematic Factor. The minimum distance of approach between both particles in backscattering events is typically 10^{-3} Å, which is much smaller than the typical interatomic distance inside the solid, so that one can consider this event as an elastic collision between two isolated particles. Assuming that the target atom is free and initially at rest, and by applying conservation of kinetic energy (elastic) and momentum (isolated), we may obtain the kinetic energy E_1 of the ion after scattering as a function of its scattering angle in the laboratory frame θ (see Figure 4.2), its initial energy E_0, its atomic mass M_1, and the atomic mass of the target atom, M_2, by the following relation, written for $M_2 > M_1$:

$$E_1 = E_0 \left[\frac{(1 - (M_1/M_2)^2 \sin^2\theta)^{1/2} + (M_1/M_2)\cos\theta}{1 + (M_1/M_2)} \right]^2 = E_0 K(M_2, M_1, \theta) \tag{4.4}$$

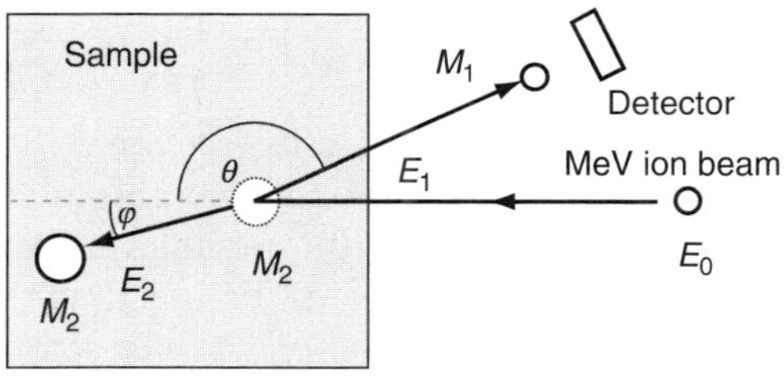

Figure 4.2 Typical backscattering geometry, in the laboratory frame, for an elastic collision between a target atom of mass M_2 initially at rest and an ion of mass M_1, kinetic energy E_0 before scattering and E_1 after scattering at angle θ.

This is fully valid for energy transfers much larger than the typical binding energy of the atom inside the solid (typically 20 eV).

The principle of mass identification in RBS lies in Equation (4.4): knowing E_1, one can deduce the corresponding mass M_2 of a target atom if one knows the detection angle θ, M_1, and E_0.

In a practical RBS experiment, one deduces M_2 by comparing the experimentally measured "kinematic factor," $K = E_1 / E_0$, to values of $K(M_2, M_1, \theta)$ calculated with Equation (4.4). Mass resolution better than 1 u is obtained up to about $M_2 = 40$ u under routine RBS measurement conditions ($^4He^+$, $E_0 = 2$ MeV, $\theta = 165°$).

4.2.2.1.2 Absolute Numbers of Target Atoms — Rutherford Cross-Section. The absolute quantitativity of RBS and its general applicability (for not too light target atoms) are related to the validity of the simple and well-known Rutherford cross-section for large angle scattering. Because the distance of closest approach of the ion to the target atom is sufficiently small in the kinetic-energy domain considered, the potential energy, E_P, is accurately represented by the Coulomb form:

$$E_P = \frac{Z_1 Z_2 e^2}{r} \tag{4.5}$$

where r is the internuclear distance and other symbols have been defined previously.

In this case, by equating the initial kinetic energy in the center-of-mass frame to the potential energy we obtain the collision diameter, or distance of closest approach, b_c, in a head-on collision

$$b_c = \frac{Z_1 Z_2 e^2}{E}\left(1 + \frac{M_1}{M_2}\right) \tag{4.6}$$

For 1 MeV ^{4}He incident on Ni, $b_c = 8.6 \times 10^{-4}$ Å. Since this is more than an order of magnitude smaller than the radius of the innermost electronic orbits of Ni, the screening of the Coulomb potential of the Ni nucleus by any of its electrons can be neglected. Furthermore, b_c is also more than an order of magnitude larger than the radius of the Ni nucleus ($\approx 5 \times 10^{-5}$ Å), so any nuclear contribution to the interaction can also be neglected: the interaction potential is purely Coulombic and the well-known Rutherford cross-section is valid. It can be written in the laboratory frame as

$$\frac{d\sigma}{d\Omega} = \frac{Z_1^2 Z_2^2 e^4}{4E^2} \frac{1}{\sin^4\theta} \frac{[(1 - (M_1/M_2)^2 \sin^2\theta)^{1/2} + \cos\theta]^2}{(1 - (M_1/M_2)^2 \sin^2\theta)^{1/2}} \tag{4.7}$$

Numerical values of this differential cross-section for each experimental case are readily calculated or determined from tables (1,2,4). They are generally expressed in barn/steradian (b/sr), with 1 barn $= 10^{-24}$ cm^2. For ^{4}He of 2 MeV scattered at 165° from Ni, $d\sigma/d\Omega = 1.04$ b/sr.

The simple scaling of the Rutherford cross-section with $d\sigma/d\Omega \propto Z_1^2\, Z_2^2/E^2$ makes RBS a powerful method for determining the absolute numbers of various target atoms, with much better sensitivity (10^{-2} monolayers or 10^{13} atoms/cm^2) for "heavy" atoms (large Z_2) than for "light" atoms. For very high precision RBS, small deviations from the Rutherford cross-section (7), due to screening effects, may be significant for the highest values of b_c, given by Equation (4.6), corresponding to low-energy ions on heavy target atoms.

4.2.2.1.3 Main Features of an RBS Spectrum for a Thin Compound Sample. The distinction between thin and thick samples for a given method (typically about 10 nm for standard RBS) is made on the basis of the corresponding depth resolution, which is discussed in Section 4.3.3. The experimental RBS spectrum obtained with a monoenergetic ion beam (initial energy E_0) incident on a compound thin sample will consist of several narrow peaks, of width equal to the energy resolution, detected at various energies E_1 (see Figure 4.3). Each of these peaks corresponds to target atoms of a particular mass M_2, which may be identified through the corresponding values of the kinematic factor $K = E_1/E_0$ (see Section 4.2.2.1.1).

From the integral of each peak (number N_D of detected ions around a given energy) one can deduce the absolute amount of the corresponding type of atoms inside the target. If this peak is due to elastic scattering of ions, by target atoms indexed by 2, at the detector angle θ, into the solid angle $\Delta\Omega$ that it subtends, N_D can be expressed as

$$N_{\mathrm{D}} = N_1 N_2 \Delta x \left(\frac{\mathrm{d}\sigma}{\mathrm{d}\Omega}\right)_{Z_1,Z_2,M_1,M_2,E,\theta} \Delta\Omega \qquad (4.8)$$

where N_1 is the total number of incident ions, N_2 is, as in Section 4.2.1, the atomic density of type 2 target atoms, Δx the target thickness, and $\mathrm{d}\sigma/\mathrm{d}\Omega$ the differential scattering cross-section given by Equation (4.7). From Equation (4.8), one can determine, if all other quantities are known, $N_2\Delta x$, that is, the number of type 2 target atoms per unit area. When repeated for each peak, one obtains the stoichiometry of the target and the absolute content of each type of atom.

In principle, one can obtain absolute quantities of any atom in the target by performing a single RBS experiment on this target: for that one needs to know very

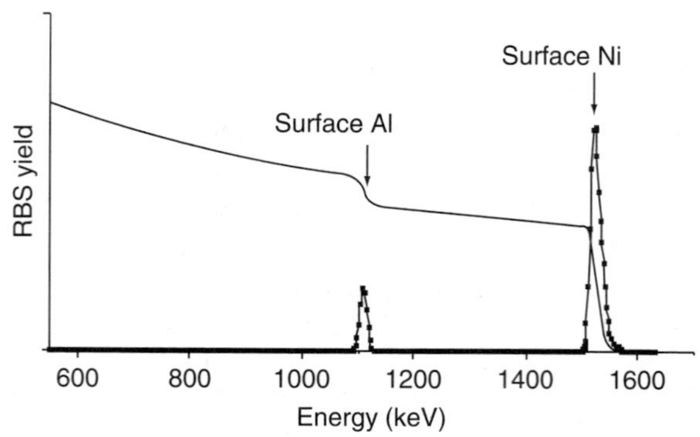

Figure 4.3 Idealized RBS spectra of a thick and a thin (1×10^{15} atoms/cm^2) NiAl film for 2-MeV $^4He^+$ detected at $\theta = 165°$. The spectrum from the thin film (squares) has been multiplied by 400 for visibility.

precisely the detection angle θ, solid angle $\Delta\Omega$, and the absolute number of incident ions N_1 (generally obtained by charge integration on the isolated sample, with care taken to cater for electron emission from the target).

Practically, one often proceeds by comparing two RBS spectra, registered in the same run under identical experimental conditions (same θ and $\Delta\Omega$): one for the sample to be analyzed and one for a standard sample, containing a known amount of a known species (8). This amount being known with a 2% precision, one can obtain by RBS, for sufficiently high counting statistics, an overall precision of about 2% for the absolute amount of various target atoms and about 1% for their relative amount (stoichiometry).

4.2.2.1.4 Main Features of an RBS Spectrum for a Thick Compound Sample. In a "thick" sample the energy loss due to the stopping power is larger than the energy resolution of the detection system, and one can depth profile target atoms by converting the detection energy scale into a depth scale for each type of atom.

In this conversion, for an ion having suffered a backscattering event on an atom of mass M_2 at depth x inside the target, one must consider the energy lost by the ion due to its stopping power along its way in ($\Delta E_{\text{in}}(x)$) and its way out ($\Delta E_{\text{out}}(x)$), and also the energy loss due to scattering, which is $(E_0 - \Delta E_{\text{in}}(x))(1 - K)$, E_0 being the energy of the incident beam and K the kinematic factor. For an incident beam normal to the target surface, one can calculate $\Delta E_{\text{in}}(x) = x|\overline{(\mathrm{d}E/\mathrm{d}x)}_{\text{in}}|$ and $\Delta E_{\text{out}}(x) = |(1/\cos\theta)\overline{(\mathrm{d}E/\mathrm{d}x)}_{\text{out}}|$ from mean values $\overline{(\mathrm{d}E/\mathrm{d}x)}_{\text{in}}$ and $\overline{(\mathrm{d}E/\mathrm{d}x)}_{\text{out}}$ of the rates of energy loss before and after backscattering. The energy at which such an ion is detected at an angle θ can then be written as

$$E_{\text{det}}(x) = KE_0 - x\left[K\left|\overline{\left(\frac{\mathrm{d}E}{\mathrm{d}x}\right)}_{\text{in}}\right| + \frac{1}{|\cos\theta|}\left|\overline{\left(\frac{\mathrm{d}E}{\mathrm{d}x}\right)}_{\text{out}}\right|\right] = KE_0 - x\overline{S} \tag{4.9}$$

with $K = K(M_2, M_1, \theta)$, given by Equation (4.4). The mean backscattering energy loss factor, $\overline{S} = \{K\Delta E_{\text{in}}(x) + \Delta E_{\text{out}}(x)\}/x$, allows, for each value of M_2, an energy–depth conversion (it depends on M_2 through K and $(\mathrm{d}E/\mathrm{d}x)_{\text{out}}$). From Equation (4.9), it is clear that an ion scattered by a target atom of mass M_2 situated at depth x inside the target is detected at the energy $E_{\text{det}}(x) = E_{\text{det}}(x=0) - x\overline{S}$, which is lower than if the atom was at the surface.

Thus, an RBS spectrum of a thick compound target consists of several superimposed contributions, each of them corresponding to target atoms of a given mass as shown in Figure 4.3. For each of these contributions one can determine:

- The corresponding mass M_2, from K deduced from its maximum energy (supposing that the atomic species is present at the surface of the sample): $E_{\text{det}}(x=0) = KE_0$.
- The depth x at which a backscattering event occurred on an atom of mass M_2, by: $x = \{E_{\text{det}}(x=0) - E_{\text{det}}(x)\}/\overline{S}$, which can be calculated in each experimental case from the tabulated values of K (1,2,4) and of $\mathrm{d}E/\mathrm{d}x$ (5). As an example, one can determine the thickness Δx of a self-supporting target or of a deposit on a lighter substrate by $\Delta x = \{E_{\text{det}}^{\max} - E_{\text{det}}^{\min}\}/\overline{S}$ in cases where not only the high-energy edge $E_{\text{det}}^{\max}$ but also the low-energy edge $E_{\text{det}}^{\min}$ ($\Delta x < 1\ \mu\text{m}$) appears on the RBS spectrum.

When depth profiling an atomic species i in a thick compound target one wants to determine, at a given depth x, the atomic density N_i of this type of atom. One obtains this quantity simply by using again Equation (4.8) in which Δx now represents the depth interval $\Delta x = \Delta E \ / \ \overline{S}$ associated with the energy width ΔE per spectrum channel and N_D the number of counts in the considered channel, corresponding to depth x. One must be careful for N_D to consider only those counts due to backscattering events on atoms of type i: one often needs to subtract an underlying contribution due to a heavier element. In particular, for a homogeneous compound, where contributions of different target atoms appear as plateaux of different heights (N_{D2}, $N_{D3}, \ldots$), one may obtain the stoichiometry by

$$\frac{N_2}{N_3} = \frac{N_{D2}}{N_{D3}} \frac{(d\sigma/d\Omega)_3}{(d\sigma/d\Omega)_2} \frac{\overline{S}_2}{\overline{S}_3} \tag{4.10}$$

For a rapid and rough evaluation of stoichiometry one can approximate the ratio of the differential cross-sections by $(Z_3/Z_2)^2$ and consider the ratio of the backscattering energy loss factors as equal to unity.

In practice, extracting detailed information from RBS spectra on thick compound targets with complicated depth profiles requires comparison of the experimental and simulated spectra. Two widely used simulation codes are RUMP (9) and SimNRA (10).

4.2.2.2 Elastic Recoil Detection

In this method, one can readily analyze the components of the sample for which $M_2 < M_1$ (in this sense it is complementary to RBS).

Instead of detecting, as in RBS, the scattered particle, one detects and analyzes in energy the target atoms themselves, which "recoil," after interaction, with a given angle ϕ, relative to the incident particle direction (see Figure 4.4). For an elastic collision, with the same assumptions as those used in deriving Equation (4.4), the kinetic energy of the recoil atom (E_2) can be simply expressed in the laboratory frame by

$$E_2 = E_0 - E_1 = E_0 \frac{4M_1/M_2}{(1 + M_1/M_2)^2} \cos^2\phi = E_2^{\max} \cos^2\phi \tag{4.11}$$

with the maximum energy transfer ($E_2 = E_2^{\max}$) corresponding to a head-on collision ($\phi = 0$). More generally, for a given (M_1, M_2) combination, large (easily measurable)

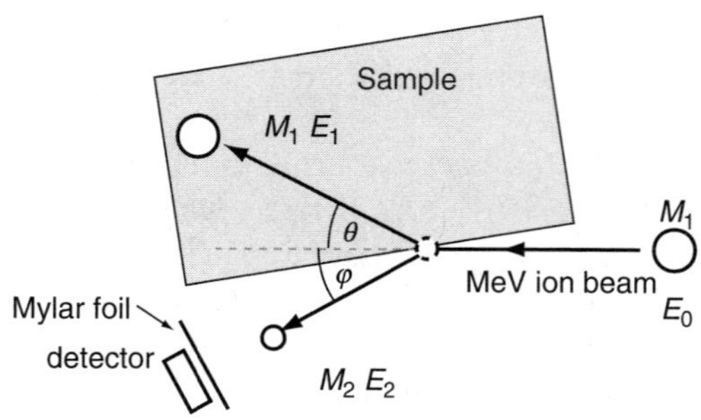

Figure 4.4 Typical ERD grazing incidence and detection geometry, in the laboratory frame, for an elastic collision between a target atom of mass M_2 initially at rest and recoiled with energy E_2 at angle φ by an ion of mass M_1, kinetic energy E_0 before scattering and E_1 after scattering at angle θ.

values of E_2 correspond to small values of ϕ (few 10°), and hence one needs to adopt a grazing incidence and detection geometry (see Figure 4.4).

ERD, employing He incident ions and using charged particle detectors readily available in a standard IBA laboratory (which we refer to as He-ERD), is particularly useful for profiling ^1H and ^2H. With higher energy and heavier incident ions, heavier target atoms can be analyzed; however, the larger number of recoiling species requires specialized particle identification systems based on measurements of total energy and at least one other parameter such as time-of-flight (velocity) or measurement of energy loss rate ($\Delta E - E$ detectors). The high flux of scattered primary ions may be kept from the detection system by use of a thin absorber foil of thickness chosen to let through the lighter recoiled ions (^1H, ^2H) whilst stopping the heavier scattered primary beam in the case of He-ERD. For heavy ion ERD this may be achieved by placing the detector at $\theta > \theta > \theta_{max} = \sin^{-1}(M_2/M_1)$, since for $M_1 > M_2$ scattering at angles greater than θ_{max} is impossible.

From the recorded ERD energy spectra one can deduce, in a similar way to RBS, the atomic masses of the various target atoms by measuring the upper energy edge of each contribution, and their depth profiles, by the shape of the corresponding contribution.

For mass identification, peaks appearing at the highest energy on the ERD energy spectrum correspond to the heaviest atom (for given values of M_1 and ϕ the recoil energy E_2 is maximum for $M_2 = M_1$, with $E_2^{max} = E_0$).

For a quantitative analysis, a cross-section derived from a purely Coulomb interaction is not valid for He-ERD, due to the particularly small value of the collision diameter (Equation (4.6)) for such a light ion–light target atom combination: there is now a nuclear contribution to their interaction (see Section 4.2.4.1). For absolute determination of ^1H (or ^2H) content by He-ERD, which can be achieved with a sensitivity better than 10^{15} atoms/cm^2, one needs a stable standard, with a known content of these atoms (11).

4.2.3 Channelling

When a well-collimated ion beam is incident on a single crystal along one of its major crystallographic axes or planes, one observes a strong reduction (by factors up to 50 for axial alignment) of interactions between ions and target atoms involving small (typically 0.1 Å or less) distances of approach, such as those encountered in RBS or NRA. These reductions and their angular dependencies are called channelling effects.

Axial channelling effects on RBS spectra are illustrated by the two RBS spectra in Figure 4.5: the "random" spectrum corresponds to a "random" direction (typically a few degrees between the crystal axis and the ion beam direction), and the "aligned" spectrum corresponds to the alignment of these two directions.

On the [1 1 1] aligned spectrum, we note both the appearance of a peak at the high-energy edge, called the surface peak (SP), corresponding to backscattering events very near the surface, and a strong reduction of the backscattering yield in the bulk of the crystal. One may experimentally characterize the amplitude of channelling effects, related to the crystal quality, by the ratio χ_{min} of the heights of the aligned and random spectra just behind the surface peak.

If they are ignored, channelling effects can lead to serious misinterpretations of the depth profiling of target atoms by RBS or NRA (12). Conversely, the sample may be oriented with a goniometer to take advantage of channelling effects for structural characterization of surfaces, interfaces, and thin layers (3,4,13).

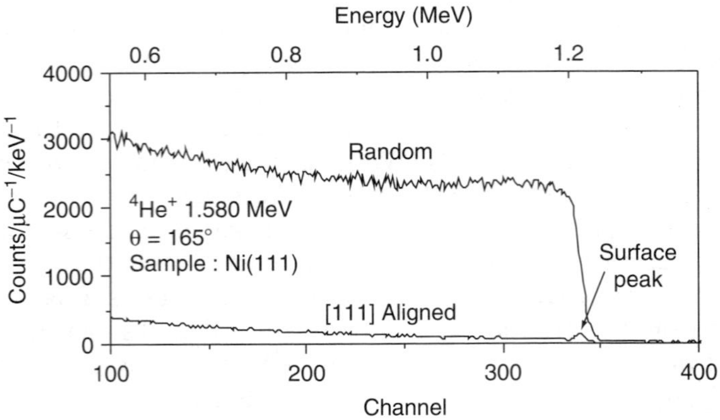

Figure 4.5 Experimental random and [1 1 1] aligned RBS spectra ($\chi_{min} = 2.9\%$) from a Ni(1 1 1) single crystal at $T = -105°C$. Obtained by the authors with an incident beam of 1580-keV $^4He^+$ and detection at $\theta = 165°$.

Channelling effects are due to the influence of the crystal structure on the ion trajectories. Under axial alignment conditions the successive ion–target atom interactions are strongly correlated, giving rise to "shadowing" and "steering" effects.

4.2.3.1 Shadowing Effect — Surface Peak: Surface Analysis

Consider the portion of a uniform flux of ions incident along a crystalline axis in the vicinity of an atomic row (see Figure 4.6). The first atomic plane is fully "seen" by the incident beam (it has a normal backscattering yield). The repulsive force (screened Coulomb potential) of this first atom deflects any incident ion away from the lattice position of the second and following atoms belonging to this row (they are in the shadow cone of the first atom). After a few atomic distances, the radius of the shadow cone becomes large with respect to the thermal vibration amplitude of the crystal atoms about their lattice sites. The sum of the backscattering yield of the first few atomic planes, where the shadowing is not yet total, can be obtained (14) from the integrated area under the SP in an aligned RBS spectrum. Since the SP area gives information on the first few angstroms of the crystal, remarkable depth resolution

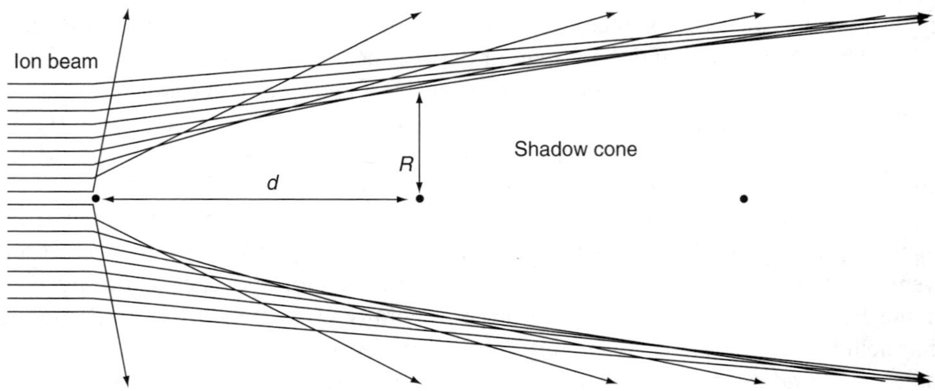

Figure 4.6 Schematic figure showing how a shadow cone is formed by elastic deflection of charged particles of a uniform beam incident along a major crystallographic axis; *d* represents the interatomic distance along the row and *R* the shadow cone at the level of the second atom of the row.

can be obtained with channelling, even when the depth resolution associated with the energy resolution of the detection system is much poorer.

The SP area can either be expressed in number of atoms per unit area or, more generally, in number of atoms per row ("seen" as random by RBS). For an ideal crystal, without any disorder or vibration, the SP area should correspond to one atom per row. In real crystals, this number is between one and ten atoms per row, depending on the ion, its energy, the crystal, its temperature, and the considered crystallographic direction. As an example, the SP area of the aligned RBS spectrum presented in Figure 4.5 corresponds to two atoms per row.

The value of the SP area and its temperature dependence gives information on the surface disorder and the surface thermal vibrations. The SP area dependence on the tilt and detection angle allows very precise determination of the surface relaxation or location of adsorbates. For these types of studies, one also often uses the blocking effect (shadowing effect by which the ions scattered by an atom in the direction of a row are deflected away from this direction by the other atoms of this row). Surface atom positions can be determined with a precision of the order of 3 pm by determining the tilt (or detection) angle at which the shadowing (or blocking) effect is maximum (13,15–17). It is worth noting that, in contrast to scanning tunneling microscopy, which is sensitive to the surface electronic density of states, shadowing and blocking experiments give direct access to the surface nuclei positions.

Quantitative analysis of experimental data requires theoretical evaluations of the SP, for which Monte Carlo simulations are performed for various crystal configurations (13,15). In these simulations, where small angle scattering and therefore distant interactions are concerned, electron screening of the Coulomb potential must be accounted for (see Section 4.2.3.2).

4.2.3.2 Steering Effect — Channelled Ion Trajectories: Subsurface Defect Analysis

We consider now the very large fraction of ions that has not suffered a large angle scattering event in the surface region. They will continue to travel in the crystal at a high velocity, nearly parallel to the atomic rows and not very close to them, and hence suffer correlated successive small angle scattering events. The trajectory of such ions may be thought of as being due to interaction with a continuous string (18). The interaction potential $U(r)$ with this string is obtained by averaging along the atomic row an appropriate screened Coulomb ion-atom potential of the Thomas–Fermi type with a screening radius typically between 10 and 20 pm.

The steering effect on the ion into the "channels" defined by the rows is due to this repulsive potential. For calculating ion trajectories, one needs to take into account several rows of the crystal. However, in what follows, we shall consider a single-row model, sufficient for determining the ion–row distance of closest approach. Let us project the ion motion onto the "transverse" plane (perpendicular to the row direction) and consider the ion "transverse energy" in this motion ($E_\perp$), which is the sum of its potential energy $U(r)$ and of its "kinetic transverse energy." It can be expressed, in the small angle approximation, as

$$E_\perp = U(r) + E\{\Psi(r)\}^2 \quad (4.12)$$

where E is, as usual, the kinetic ion energy and $\Psi(r)$ is the angle between the row direction and the ion velocity (three-dimensional) when the ion–row distance is r.

If one assumes a negligible energy loss then the value of the transverse energy of a given ion is conserved along its path and determined by its conditions of entrance inside the crystal ($U(r_{in})$, Ψ_{in}). The distance of closest approach r_{min}, for which $\Psi = 0$, is such that $E_\perp = U(r_{min})$.

This can be used for determining the critical angle for channelling (Ψ_C); that is, the value of the tilt angle above which the steering effect of the rows no longer holds and the concept of continuous row potential is no longer applicable. In the simplest approximation, this is obtained when the ion–row distance of closest approach is equal to the two-dimensional root mean square thermal vibration amplitude ρ (often calculated in the Debye model). This critical angle is such that one has $r_{min} = \rho$ even for a particle entering the crystal "very far from any row"; that is, with $U(r_{in}) = 0$. With $E_\perp = E\{\Psi_C\}^2 = U(\rho)$, one obtains for the critical angle for channelling (expressed in radians): $\Psi_C = [U(\rho) / E]^{1/2}$. As an example, one finds $\Psi_C \approx 0.85°$ for the experimental conditions corresponding to Figure 4.5.

Numerous applications of channelling for locating, identifying, and depth profiling defects in the bulk of crystals are described in detail in Ref. (3).

4.2.3.3 Thin Film Analysis, Interfaces, Epitaxial Layers

Channelling can be used to reduce the substrate signal in RBS so that the contribution of an overlying film appears more clearly on the RBS spectrum. This is particularly useful for thin films containing light elements, such as oxides. The background due to backscattering events on crystal atoms may be even further reduced with a grazing detection geometry. This modifies the correspondence between the detection energy scale and the depth scale. With a detection angle θ near 90° the factor $\overline{S}$ defined by Equation (4.9) increases markedly and so both the thickness $\Delta x = \Delta E / \overline{S}$ associated with the energy width (ΔE) per channel and the number of scattering atoms contained in this slab decrease.

Another feature of interest appearing on such an RBS spectrum is the SP of the underlying crystal, which depends on the crystalline quality of the interface. As discussed earlier the SP area depends on the amplitude of shadowing effects; hence, it increases if the presence of the film induces any disorder at the film–crystal interface. By measuring the SP area along various crystallographic directions and comparing it to the corresponding values on the bare crystal, one can observe interface disorder and analyze its nature (see Section 4.4.3).

Possible crystallinity of any film on a crystalline substrate can also be observed and analyzed by channelling effects occurring under alignment conditions (see Section 4.4.3). Channelling is particularly useful for studying the heteroepitaxial growth of thin films (19), and measuring relative structural distortion of the overlayer as small as 0.5%.

4.2.4 Nuclear Reaction Analysis

We designate a nuclear reaction A + b → c + D by A(b,c)D where the projectile b (M_1, Z_1) impinges on a target A (M_2, Z_2), producing a light reaction product c (M_3, Z_3) and a heavy product D (M_4, Z_4). For example, using obvious shorthand notation, RBS on ^{28}Si is represented by ^{28}Si(α,α)^{28}Si. Mass may not be conserved in nuclear reactions. The associated energy, $Q = (M_1 + M_2 - M_3 - M_4)c^2$, may be positive or negative (exo- and endothermic reactions, respectively) and must be taken into account for kinematics calculations.

Nuclear reactions are likely to occur when the incident ion approaches the target nucleus sufficiently closely that the attractive nuclear force overcomes the

Coulomb repulsion. The energy barrier that must be overcome by the incident particle before the nuclear force dominates is referred to as the Coulomb barrier B. Above B, an electrostatic description of the interaction no longer adequately explains the cross-section and kinematics: the Rutherford cross-section may not be valid and nuclear reactions can occur. A simplistic estimate of B may be obtained simply by substituting an estimate of the nuclear force range into the expression for the collision diameter b_c given in Equation (4.6). Assuming that the range of the nuclear force for a proton (atomic mass Mp) is about the diameter of the proton r_p and that the diameter of nuclei scales as the cube root of the nuclear mass number, we obtain:

$$B = \frac{Z_1 Z_2 e^2}{r_p} \frac{M_p^{1/3}}{(M_1^{1/3} + M_2^{1/3})} \left(1 + \frac{M_1}{M_2}\right) \tag{4.13}$$

Equation (4.13) is dominated by the Z dependence: nuclear reactions occur most easily for light ions on light targets; typically, hydrogen and helium isotopes of energies up to a few megaelectronvolts on targets up to about sulfur. For energies below B, penetration to the nucleus can still occur by quantum mechanical tunneling through the barrier.

4.2.4.1 Non-Rutherford Elastic Scattering

Deviations from the Rutherford cross-section may occur at energies below $B/2$ for elastic scattering, such as (p,p) and (α,α), and in fact the cross-section may even present sharp resonances that may be exploited for analytical purposes. It has been verified experimentally (1) that for elastic alpha scattering the cross-section is strictly Rutherford (deviation <2%) up to 2 MeV on elements down to oxygen (for which B from Equation (4.13) is 6.3 MeV). However, above this energy, the cross-section for oxygen shows strong variations. The first strong resonance, at 3.045 MeV, is often used to enhance the sensitivity to oxygen of RBS analysis (see Section 4.4.3).

4.2.4.2 Nuclear Reactions

It is beyond the scope of this work to describe in detail all of the various nuclear reaction mechanisms; however, it is worth noting that deuterium beams are particularly rich for NRA, since the deuterium nucleus consists of a weakly bound proton and neutron. The proton is repulsed by the positive charge of the target nucleus, whereas the uncharged neutron is not; therefore, it may become separated from the proton during the interaction giving rise to nuclear reactions since for the neutron there is no Coulomb barrier to overcome. Nuclear reaction cross-sections are generally much smaller than elastic scattering cross-sections: for example, the Rutherford cross-section for 1-MeV deuterium ions scattered at $\theta = 165^\circ$ on ^{16}O is 110 mb/sr, whereas typical nuclear reaction cross-sections for deuterium on ^{16}O are less than 10 mb/sr. As a consequence, much larger detection solid angles and beam currents are used than for RBS. As for ERD, an absorber foil of thickness typically between 3 and 30 μm is often used to eliminate the high flux of scattered primary beam. In contrast to RBS, where $Q = 0$ and $d\sigma/d\Omega$ are known analytically, in NRA the cross-sections and reaction energies can vary strongly with incident beam energy, detected product angle and nuclear reaction. This gives much scope for optimizing the analysis (incident particle energy, absorber foil thickness, detection angle), but also complicates interpretation, so that each analysis problem has to be considered individually. Theoretical extrapolations outside of measured ranges are of limited value and for practical use measured cross-section data are the most reliable. A very useful review of NRA for elements from hydrogen to neon is given in Ref. (20), and a substantial database of experimental nuclear reaction cross-sections for IBA is freely available from http://www-nds.iaea.org/ibandl/. A useful presentation may also be found in Ref. (21). In Figure 4.7 we show a selection of typical experimental parameters and cross-sections associated with some of the more commonly used nuclear reactions for NRA.

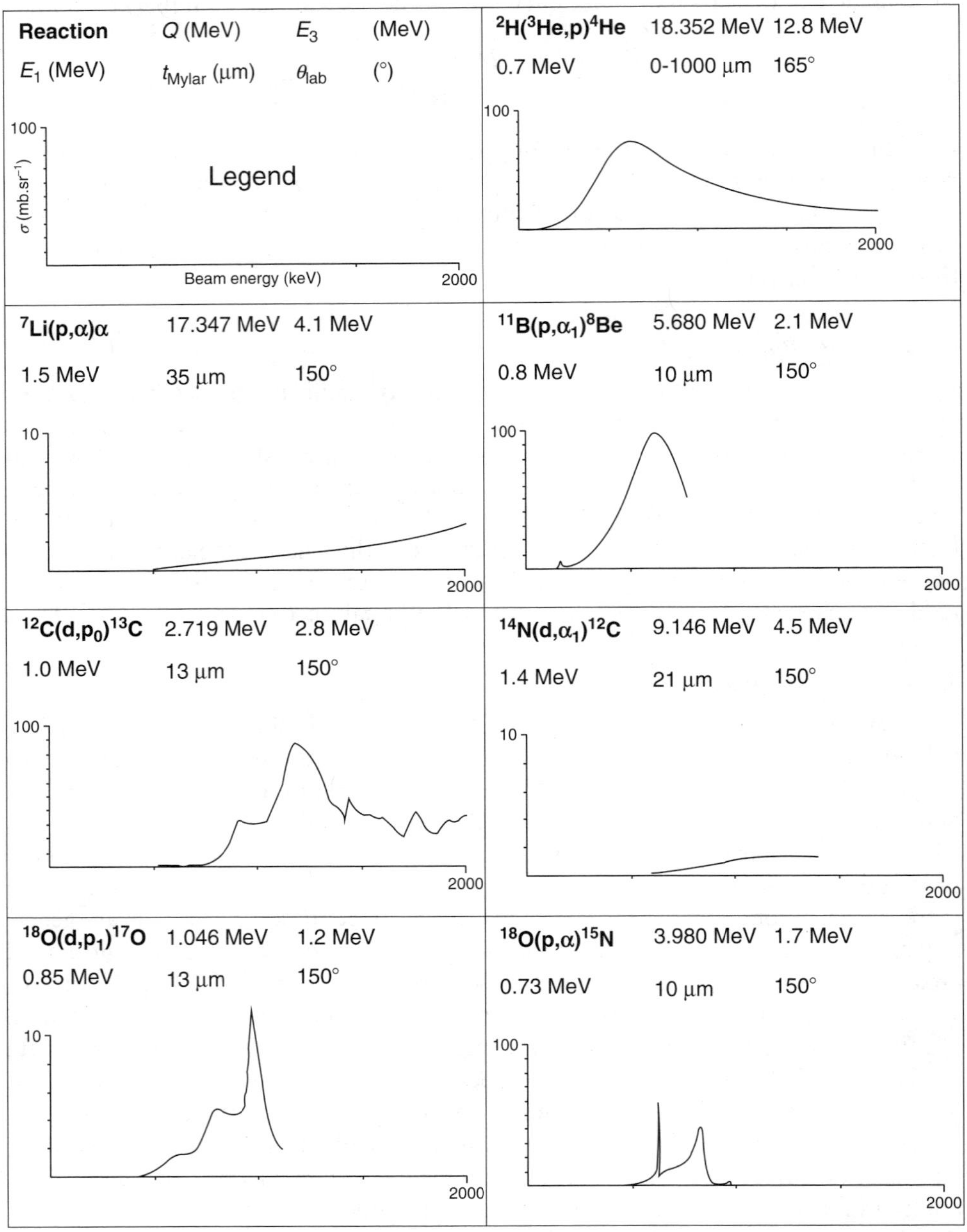

Figure 4.7 A selection of nuclear reactions and typical experimental conditions used for NRA of light elements; t_{mylar} is the thickness of a mylar absorber foil. The cross-sections have been taken from the sigmabase (http://www-nds.iaea.org/ibandl/), where they are fully referenced.

4.2.4.2.1 NRA of Thin Samples. NRA may be used in much the same way as RBS to determine absolute areal atomic densities in thin layers. Where possible, an incident energy is chosen such that the cross-section varies only slowly with energy, so that it is constant over the range of the energy lost by the primary beam in traversing the layer to be analyzed. In this case, Equation (4.8), where $d\sigma/d\Omega$ refers now to the differential cross-section of the nuclear reaction of interest, gives the number of detected particles. As for RBS, absolute values are obtained by comparing the yields from the sample to be analyzed with that from a reference sample; however, since the ratios of $d\sigma/d\Omega$ for different nuclear reactions are not known with very good accuracy, a separate reference sample is needed for each species analyzed. Accurate and stable thin reference samples for oxygen isotopes may readily be made by anodization of tantalum (22), and accurate cross-section ratio measurements on stoichiometric frozen gas targets (23) allow these references to be used for nitrogen and carbon determination as well. For other elements, reference targets are fabricated on a case-by-case basis.

4.2.4.2.2 NRA of Thick Samples. As for RBS, in NRA of thick samples the incident beam and exiting particles lose energy as they penetrate the sample, meaning that concentration depth profiling may be performed. Generally, the depth resolution is poorer, being limited by the larger detection solid angles and consequent kinematic spread of the reaction products (the reaction product energy varies appreciably over the angular spread of the detector), and by energy straggling in the absorber foil. NRA spectra may be quite complex, consisting of a mixture of protons and alpha particles of different energies produced by various reactions with different cross-sections. Quantitative interpretation invariably involves iterative comparison of experimental spectra with spectra calculated for various assumed target compositions, until satisfactory agreement between the experimental and simulated spectra is obtained. Two well-tested and widely used NRA simulation programs are SIMNRA (10) and SENRAS (24).

4.2.4.2.3 Nuclear Resonance Profiling. Some nuclear reaction cross-sections present isolated narrow peaks that may be exploited in NRP to obtain concentration–depth profiles in a different way to that used in NRA and RBS. Consider a beam of energy E greater than that E_r of a narrow resonance, incident on a target to be analyzed. The cross-section is zero at the incident beam energy, and only becomes appreciable in the vicinity of E_r, when the incident beam has lost energy $\Delta E = E - E_r$. As an illustration, consider Figure 4.8(a) where we show an incident proton beam of 154.5 keV on a sample of SiO_2. The beam traverses 34 nm of SiO_2 and loses 3.5 keV before reaching the resonance energy, which is 151 keV for the resonance in $^{18}O(p,\alpha)^{15}N$ taken as an example here. The yield of reaction products is proportional to the ^{18}O concentration at a depth of 34 nm. As the incident beam energy is increased, the resonance probes the concentration deeper in the sample. The yield of the resonance as a function of incident beam energy is an image of the concentration profile in the sample, and is referred to as an excitation curve.

In practice, the excitation curve is an imperfect image of the concentration profile. A number of effects reduce the precision with which we can know the depth in the sample from which the detected nuclear reaction products come. The resonance width itself provides the limit to the depth resolution: typical resonances may be as narrow as 100 eV. The incident beam has an energy distribution, which in the most favorable cases may be of the same order. Finally, the slowing down process itself introduces further beam energy spread due to the ion energy straggling. Whilst

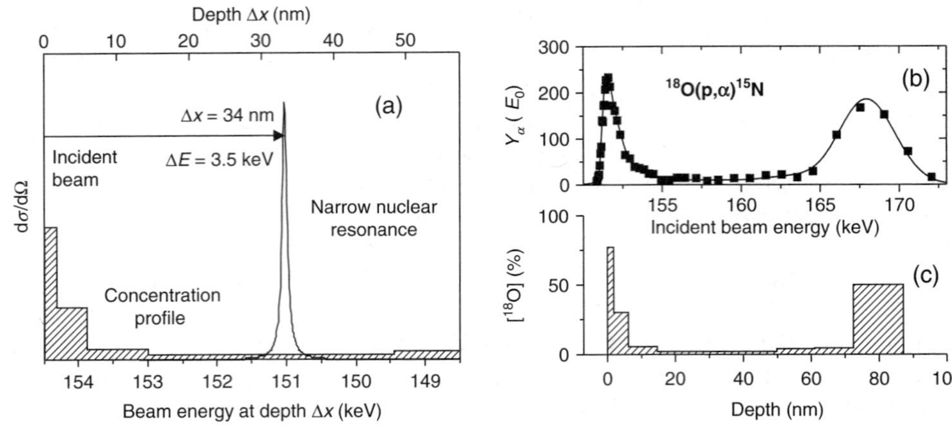

Figure 4.8 A schematic representation of the NRP method (a), and a typical measured excitation curve (b) obtained by the authors with the $^{18}O(p,\alpha)^{15}N$ resonance at 151 keV on an isotopically inhomogeneous thermal oxide of silicon (in inclined geometry to improve depth resolution). The solid line is the simulated curve corresponding to the concentration profile shown in (c), calculated with SPACES (25). In the energy-depth correspondence between (b) and (c) the sample tilt angle is taken into account.

energy straggling is only a secondary effect for RBS and NRA, in NRP it rapidly becomes the dominant contribution (see Section 4.3.3). For accurate interpretation of measured excitation curves it is necessary to take a detailed account of the energy loss in each individual ion–atom collision, which is of the order of the beam energy resolution. This is accomplished in the framework of a stochastic theory of energy loss, which has been embodied in a computer code that calculates very accurate excitation curves for arbitrary concentration profiles (25).

4.3 PRACTICAL ASPECTS OF IBA

Ion beam analysis was born from low-energy nuclear physics and much of the equipment used reflects this heritage. Figure 4.9 shows schematically the equipment needed: (i) a source of appropriate ions of suitable intensity, energy, and angular divergence; (ii) a beam transport system under vacuum to direct the beam onto the sample to be analyzed; (iii) a chamber in which samples and detectors may be mounted; and (iv) signal acquisition and treatment electronics.

4.3.1 Ion Beam Production

By far the most widespread sources of ions for IBA are small electrostatic particle accelerators. These consist of a system for generating and maintaining a positive potential, V, up to some MV above earth potential, and a means of producing positive ions — usually an RF gas discharge — which are then accelerated by the electrostatic field. The high potential may be produced by mechanical means, such as the Van de Graaff generator, which maintains a high potential on a metal dome, referred to as the terminal, by transporting charge to it via a rubberized canvas belt, or by solid-state voltage multiplier columns. The entire accelerator assembly needs to be insulated from earth potential. For voltages up to 500 kV this is feasible using air

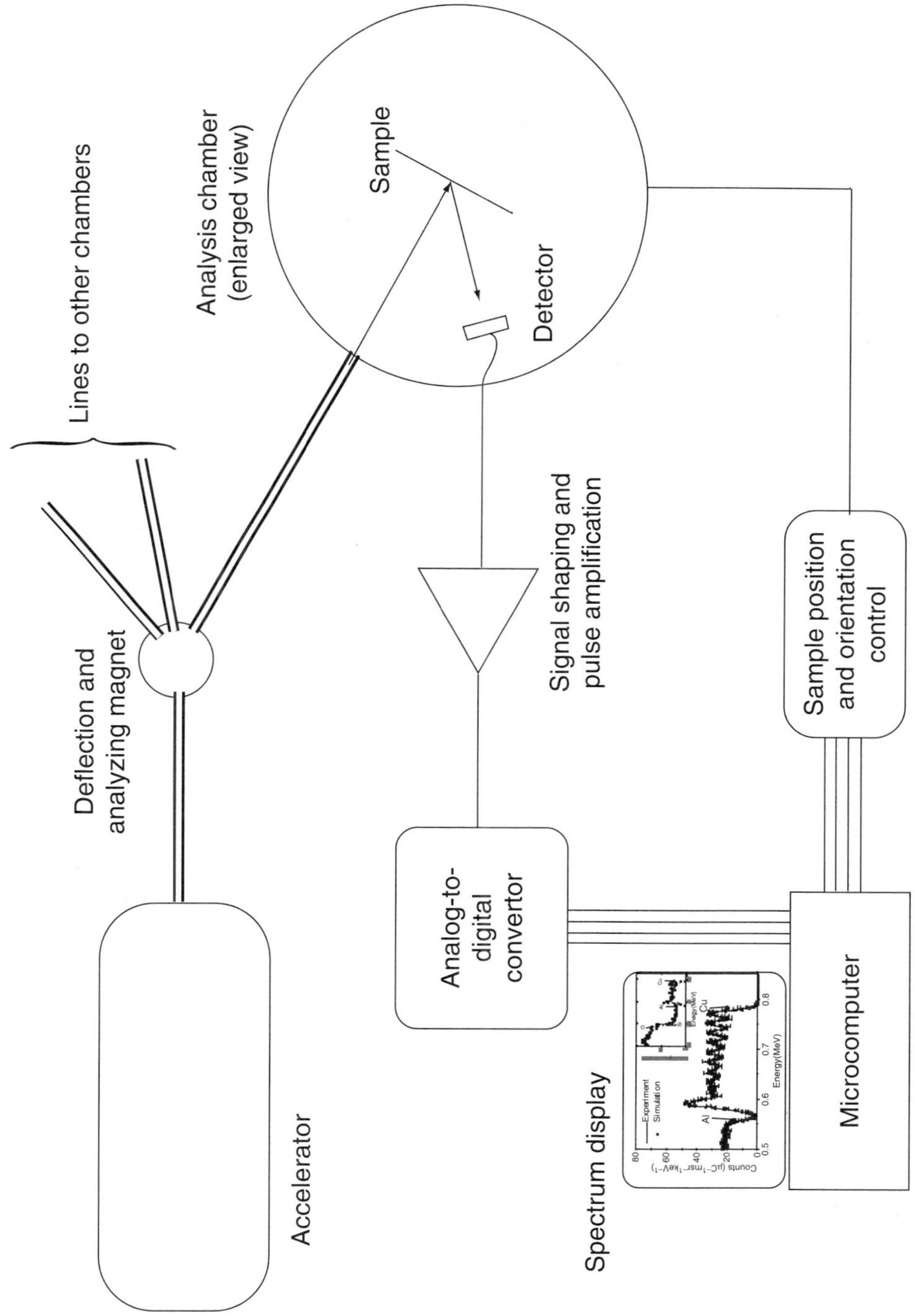

Figure 4.9 Schematic diagram of a typical IBA laboratory setup

insulation, but for higher voltages it is much more compact to enclose the accelerator in a pressure vessel containing a high-pressure insulating gas mixture.

The accelerated beam of particles of mass m, charge q, and energy $E=qV$ is focused and steered into the analyzing magnet where the magnetic field is set so that only particles of the desired energy and q/m ratio are transmitted further down the beamlines and into the target chamber. Typical beam currents range from 10^{-9} to 10^{-6} Å in a spot size ranging from fractions of a millimeter to a few millimeters for collimated beams. Submicrometer beam diameters can be obtained with careful focusing, in the "nuclear microprobe" configuration (26).

To obtain the energies required for high-energy ERD (Section 4.2.2.2), the potential of the terminal is used twice, in the "Tandem" accelerator. An ion source outside the pressure vessel generates negative ions. These are accelerated to the terminal, where n electrons are removed from an ion as it passes through a very thin gas or foil "stripper." The now positively charged ions are then further accelerated back down to earth potential, with a final energy equal to nV eV.

4.3.2 Charged Particle Detection

The development of IBA was in part sparked by the invention of the solid-state charged-particle detector. These devices are small, inexpensive, robust, and easy to use. A Schottky barrier is constructed on one face of a slab of semiconducting silicon. When a voltage is applied across the slab, electrons and holes are pumped from within the silicon until the voltage due to the internal electric field matches the externally imposed voltage. This forms a zone depleted in charge — the depletion region — in which there is an electric field, which extends from the surface to some depth in the detector. When a charged particle penetrates into this zone, it loses energy by ionization, producing on average one electron–hole pair for each 3.1 eV lost. If a particle that loses E eV in the depletion zone produced exactly $E/3.1$ charge pairs, the resolution with which the energy could be measured would depend only on how well we can count the number of charge pairs created. However, some energy is converted to phonons and Frenkel pair production in the detector, introducing fluctuations into the actual number of electron–hole pairs created for a given deposited energy, setting a fundamental limit of about 10 keV (FWHM) to the resolution achievable for typical RBS and NRA measurement conditions, no matter how well the number of charge pairs is measured. Detection solid angles are typically 1 msr for RBS and He-ERD and 50 msr for NRA. For minimizing the detector energy resolution ΔE_{det}, which sets the limit to depth resolution in RBS (see Section 4.3.3), efforts have been made to use detectors based on deflection by electric (27) or magnetic (28) fields. For particles of up to a few hundred kiloelectronvolts energy, ΔE_{det} well below 1 keV is achievable, at the expense of substantially increased experimental complexity. This variant of RBS is termed medium-energy ion scattering (MEIS) and may prove to be very useful for corrosion problems where films of nanometric thickness are formed or modified.

For particles fully stopped in the depletion zone of a solid-state detector the measured charge pulse is proportional to the particle energy, but for particles that have a range greater than the depletion zone, only that portion of the energy lost in the depletion zone contributes to the charge pulse. The thickness of the depletion zone may be varied by adjusting the applied voltage. This is useful for separating particles of the same energy but different ranges (29). For example, signals from protons, which have greater ranges than alpha particles of the same energy, may be moved to lower energies in a spectrum without altering the alpha particle spectrum in the same region by reducing the thickness of the depletion zone to less than the range of the protons.

The charge pulses are converted to voltage pulses by a charge-sensitive preamplifier, and, after further shaping and amplification, a pulse of around 1 μsec width is generated, whose height is proportional to the energy of the charged particle. This signal may then be sampled and converted by an analog-to-digital converter for storage and display in a computer.

4.3.3 Depth Resolution

The depth resolution, Δx, of an IBA method depends on the energy resolution through the conversion of an energy scale to a depth scale. Assuming that energy broadening effects are independent and of Gaussian form, we may write

$$\Delta x = \frac{\sqrt{\Delta E_{\text{beam}}^2 + \Delta E_{\text{str}}^2 + \Delta E_{\text{foil}}^2 + \Delta E_{\text{det}}^2}}{S} \quad (4.14)$$

where, for RBS, $S = \overline{S}$ as defined by Equation (4.9). For NRA, S may be defined analogously and for NRP $S = \mathrm{d}E/\mathrm{d}x$. In RBS, the dominant term is ΔE_{det}, which for $\Delta E_{\text{det}} = 10\,\text{keV}$ and $\theta = 165^\circ$ translates to $\Delta x \approx 8\,\text{nm}$ for 2 MeV $^4He^+$ in Ni, or 15 nm for 500 keV $^4He^+$ in Si. Depth resolution can be subnanometric for MEIS. For NRA, the straggling of the detected particles in the absorber foil, ΔE_{foil}, predominates and Δx is typically 100 nm. For NRP, ΔE_{det} may be taken to represent the resonance width Γ, and ΔE_{beam} needs to include thermal Doppler beam energy broadening (30). Generally, $1.5\,\text{nm} < \Delta x < 100\,\text{nm}$, depending on which of the terms dominates: Γ, ΔE_{beam}, or ΔE_{str} at the given depth in the target. Depth resolution may be improved by tilting the sample to give near-surface depth resolution of the order of 1–2 nm for RBS with extreme grazing geometry, although at very grazing incident or detection angles the lateral spread of the beam further degrades the depth resolution below the surface. A general and complete treatment of depth resolution in IBA is given in Ref. (31), and the associated DEPTH computer code for calculating Δx is available from the IBIS website (http://www.kfki.hu/~ionhp/doc/proidx.htm).

4.4 ILLUSTRATIVE APPLICATIONS OF IBA IN CORROSION RESEARCH

4.4.1 Application of RBS

RBS is easy to perform, qualitative information is readily obtained by simple inspection of the spectra, and reliable fitting codes are freely available, which is perhaps why RBS is the most widely used IBA method across almost all disciplines, including corrosion research. When applied to alloys of metals of similar masses (such as stainless steels for example) the overlapping signals from the alloy components interfere. On the other hand, when a light element is alloyed with a rather heavier one, the signal of the heavy element may be well separated from that of the light element, and even for relatively low concentration, the heavy component gives a substantial RBS signal due to the Z_2^2 dependence of the Rutherford cross-section in Equation (4.7). The group of G.E. Thompson, P. Skeldon, and colleagues from the University of Manchester (see, e.g. (32–35)) has extensively exploited this mass separation in basic studies of the anodic oxidation of binary alloys of aluminum with heavier metals. The basic approach is to grow thin alloy films on anodized aluminum substrates by simultaneous sputtering of pure metal targets in order to study the alloy behavior under different anodization conditions. The distribution of the heavier metal in the anodic film and remaining alloy are readily observed via RBS.

In Figure 4.10(a), we show data extracted from Ref. (34) representing RBS spectra from an as-deposited and an anodized binary Al–0.4% Au alloy. The Au and Al signals from the as-deposited alloy show a homogeneous alloy composition with depth. After anodization, the high-energy edge of the Au signal is shifted to lower energy and a peak appears. This indicates formation of a gold-free anodic alumina film, also visible in the Al signal, and an enrichment of gold in the alloy immediately below the anodic film. The solid lines are quantitative simulations made with RUMP (9) assuming formation of pure alumina, and conservation of the total number of gold atoms.

In Figure 4.10(b), we show data extracted from Ref. (35) representing RBS spectra from an anodized binary Al–27% Ce alloy. Here, in contrast to the example above, the high-energy region of the spectrum shows that Ce is incorporated into the growing oxide, with enrichment in the oxide surface. This enrichment is a consequence of the faster migration rate of cerium ions compared to that of Al^{3+} ions in the anodic oxide layer. One observes from the ratios of the corresponding plateau heights in the spectrum that in the bulk of the oxide film, Ce and Al are present in the same ratio as in the unoxidized alloy. This interpretation is confirmed quantitatively by the RUMP simulation, shown as the solid line.

4.4.2 Application of ERD

ERD has been applied for studies of barrier film formation in Al–Mg–Cu alloys (36) using 35 MeV Cl^{7+} ions, and for studies of the oxidation behavior of Zr and related alloys with 60 MeV ^{127}I (37) and 230 MeV ^{209}Bi (38). In the last study, Forster et al. (39–41) studied formation, during corrosion processes, of deuterides that embrittle the Zr–2.5% Nb alloy, which is used as a pressure-tube material in CANDU nuclear power reactors (where heavy water is used as both moderator and coolant). In particular, they considered the effects of a 10-min etch in 0.1% NH_4HF_2 in D_2O on the surface composition. In Figure 4.11 (data taken from Ref. (38)) we show the ERD spectra of various recoiled species obtained in a single measurement on an etched alloy film, using a gas-filled detector that measures the total energy and the rate of energy loss. The signals from each species, which overlap on a simple energy spectrum, are separated by their different energy loss rates (see Section 4.2.2.2). The gas-filled $\Delta E - E$ detector was unable to detect recoiled atoms lighter than boron here, and the hydrogen and deuterium profiles were obtained simultaneously with a semiconductor detector and absorber, as for He-ERD. For each spectrum, the higher energy corresponds to the surface region of the sample. The energy–depth correspondence must be considered individually for each spectrum since the energy lost on the outward path depends on the recoiled species. As with RBS, calculation of the relationship between signal height and composition is facilitated by the fact that recoil cross-section ratios follow the laws of Coulomb scattering. The fluorine contamination, deuterium uptake, and formation of an oxide layer are clearly and quantitatively observed here, demonstrating the power of the ERD method.

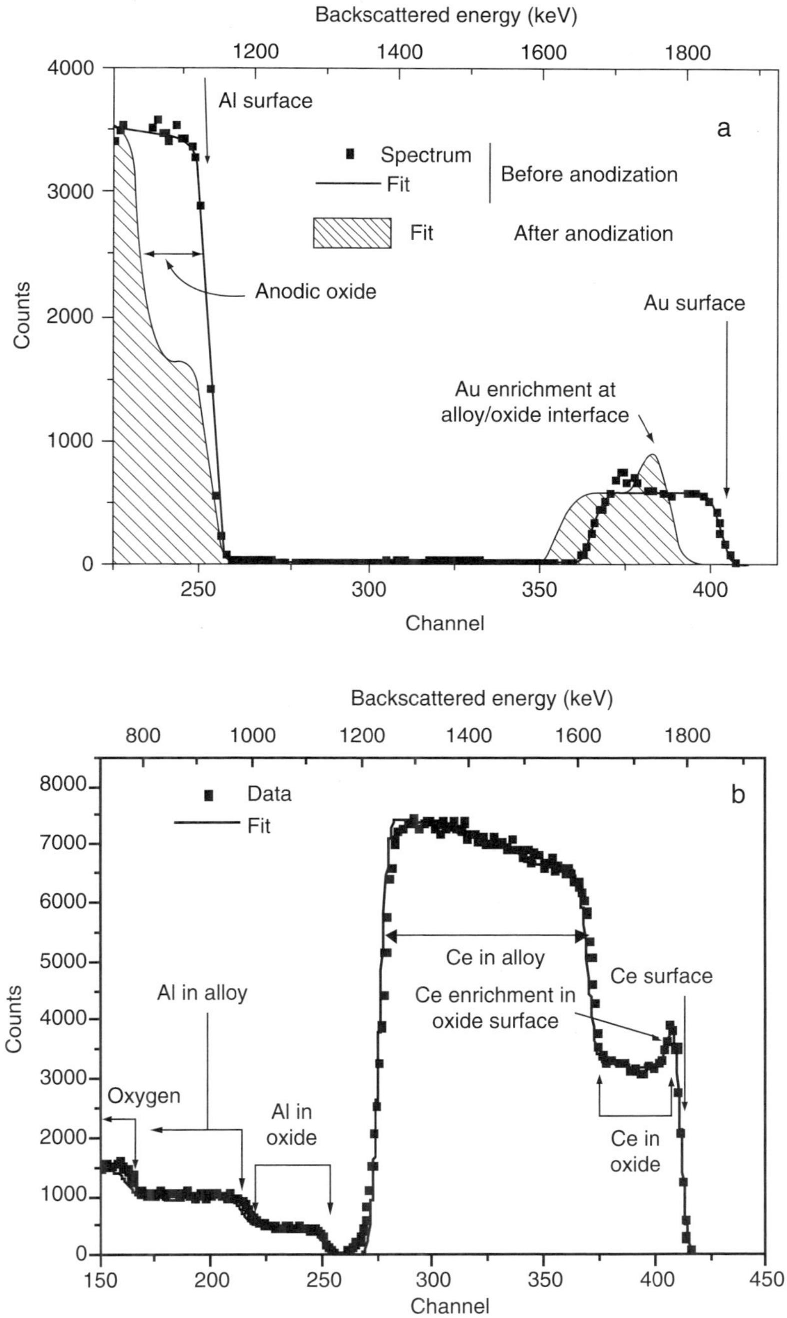

Figure 4.10 RBS spectra showing heavy element segregation and accumulation during anodic oxidation of a dilute (4.5% Au) binary Au–Al alloy (a) before (full line and representative data points) and after (shaded curve) oxidation and a binary Al–27% Ce alloy (b). The solid lines are theoretical spectra calculated with RUMP as an aid to interpretation.

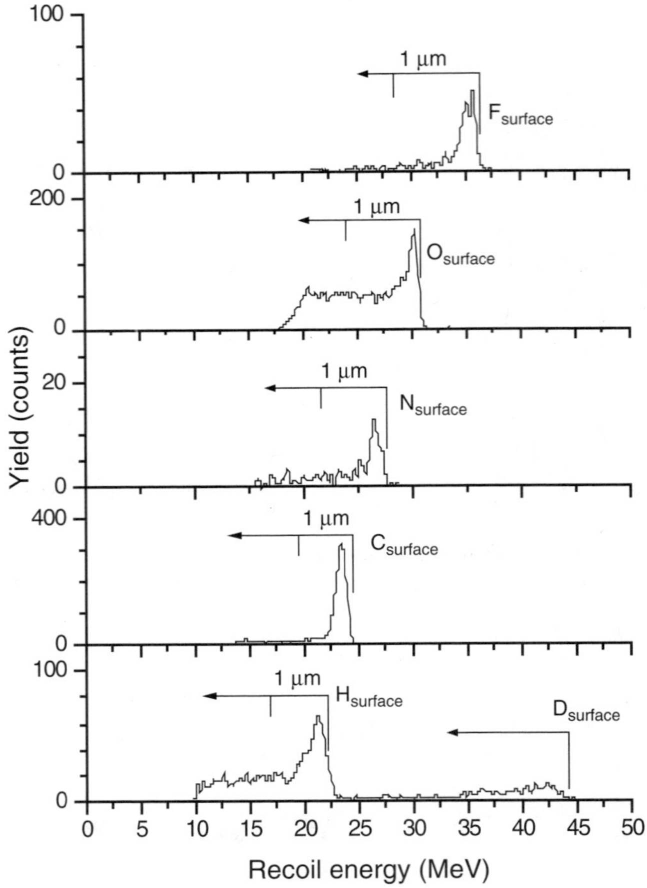

Figure 4.11 Energy spectra of various recoiled target atoms obtained by heavy ion ERD (230-MeV ^{209}Bi incident beam) of an etched Zr–2.5% Nb alloy. The hydrogen isotope energies have been multiplied by 10 to allow representation on the same scale as the other elements. (From J.S. Forster, R.L. Tapping, J.A. Davies, R. Siegele, and S.G. Wallace, *Surf. Coat. Technol.* 83, 201 (1996). With permission.)

4.4.3 Application of Channelling

We present two applications in corrosion research of channelling on single crystals, concerning in one case passive film growth on Ni–Mo alloys and in the second case surface modification by leaching of UO_2.

The first study is devoted to natural oxide and passive films formed on Ni–6% Mo alloys (42). The RBS spectra represented in Figure 4.12 have been obtained for a passive oxide film, covering a (1 0 0) Ni–Mo crystal, under random and [1 0 0] alignment incidence (both a Mo and a Ni SP are clearly seen on the latter spectrum).

Several conclusions can be obtained by analyzing these spectra:

1. By comparing the levels of the Mo and Ni plateaux in the random spectrum one confirms the 6% Mo concentration in the bulk crystal (see Section 4.2.2.1.4).
2. By measuring the ratio of the levels on the aligned and random spectra behind the SP ($\chi_{min} = 3\%$ both for the Mo and Ni levels) and comparing it

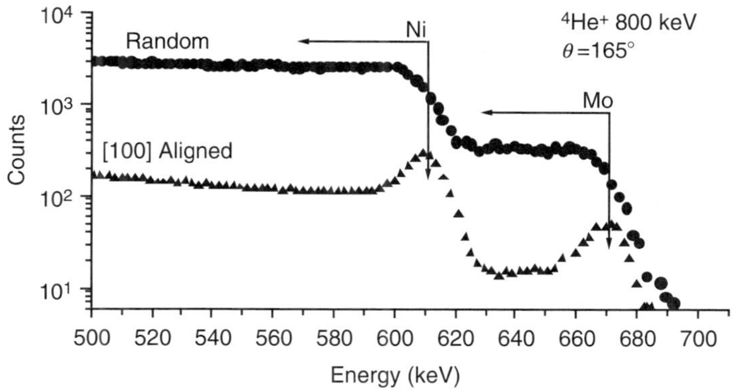

Figure 4.12 Random and [1 0 0] aligned RBS spectra on a Ni–6% Mo(1 0 0) alloy covered by a passive oxide film. Vertical arrows indicate ions scattered from Ni and Mo surface atoms. Spectra obtained with 800-keV $^4He^+$, $\theta = 165°$, 2.3 keV/channel. Note the logarithmic vertical scale.

to theoretical predictions one concludes that the bulk crystal is free of any structural distortion due to the presence of Mo.

3. By measuring the Mo and Ni SP areas one determines the absolute number of metallic atoms in the passive film, 5×10^{15} atoms/cm^2, which represents a film thickness of about 10 Å (assuming a density equivalent to that of NiO).
4. From the relative values of the Mo and Ni SP areas one determines a 12% Mo concentration and hence evidences a strong Mo segregation in the passive oxide film (in contrast with the case of natural oxide).
5. By comparing these SP areas to those obtained for a [1 1 0] alignment incidence one evidences an interface disorder characterized by random static displacements (with an rms value of 0.08 Å) of the atoms belonging to the first crystal plane. These displacements, which occur in a plane parallel to the surface, whatever is the crystal orientation (both the (1 0 0) and (1 1 0) orientations have been studied), can be attributed to strains or vacancies at the interface, induced by the passive films.

In addition to the data concerning the metallic-atom proportions and structure (obtained by RBS-channelling experiments), the absolute oxygen content of these films has been measured by NRA, using the $^{16}O(d,p)^{17}O$ nuclear reaction at 900 keV (see Section 4.2.4). The oxygen content of the passive film exceeds, by about 12×10^{15} atoms/cm^2 (only 1×10^{15} atoms/cm^2 in the case of the natural oxide), the number of metallic atoms. This excess can be attributed either to oxygen containing molecules or radicals remaining on the film after the passivation treatment (H_2O, OH^-, SO_4^{2-}) or to any further contamination.

The second study concerns the transformation of UO_2 (1 1 0) single crystals (fluorite cubic structure) upon leaching (43), which is of relevance for understanding aqueous corrosion of spent nuclear fuel in deep geological repositories. This study was performed by backscattering and channelling with a 3.07-MeV ^4He beam, taking advantage, in the analysis of the O sublattice, of the $^{16}O(\alpha,\alpha)^{16}O$ resonant (non-Rutherford) scattering occurring at 3045 keV (see Section 4.2.4.1). As a consequence, in the spectra, superimposed on a background almost entirely due to backscattering from U atoms, an oxygen resonance peak is clearly visible, which corresponds to backscattering from O atoms and measures the O content at a depth of $\approx$60 nm. By analyzing (using the RUMP code (9)) a random spectrum obtained from a UO_2 crystal modified by leaching (oxidation in demineralized water at 180°C for 24 h), these authors have determined the presence of a surface layer of composition $UO_{2.28}$ and of thickness $\approx$140 nm. Information on the structure of this layer has been obtained from channelling experiments. In Figure 4.13, reproduced from Ref. (43), are represented angular scans of backscattering yields, along the (0 0 1) plane, across three major axes of this modified UO_2 crystal. These yields are obtained by integration over an adequate detected-energy window on the spectra and are normalized to their random value.

In this figure, the open circles represent the yield for O atoms belonging to the transformed layer (O resonant peak area after subtraction of the U background), the open squares correspond to a yield for U atoms belonging to the transformed layer, whereas the full circles correspond to a yield for U atoms belonging to the bulk crystal, behind the transformed layer.

The following can be deduced from these scans:

1. By observing dips for both O and U *surface* yields, the crystallinity of the transformed layer is demonstrated.
2. This is confirmed by observing that the angular position of O dip always coincides with the *surface* U dip.

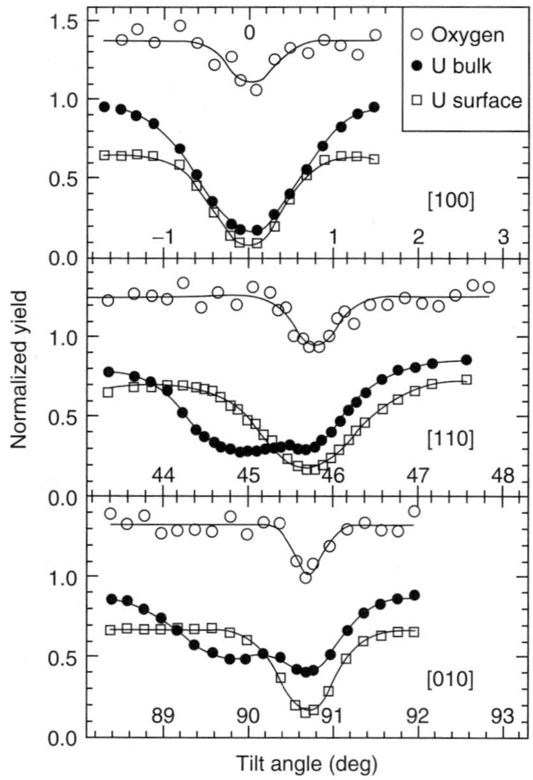

Figure 4.13 Angular scans (see text) across various axes of a UO_2 crystal, after leaching. The tilt angle measures, in the (0 0 1) plane, the angular deviation from the [1 0 0] direction. (From L. Nowicki, A. Turos, C. Choffel, F. Garrido, L. Thomé, J. Gaca, M. Wojcik, and Hj. Matzke, *Phys. Rev. B* 56, 534 (1997). With permission.)

3. By observing that the angular positions of the three dips for the [1 0 0] axis coincide, one can conclude that the direction of this axis in the transformed layer coincides with the one of the bulk crystal.
4. By observing that the angular position of the O and *surface* U dips for the [110] axis are not at 45° (where it is for an unleached UO_2 (1 1 0) crystal) but at 45.75° and
5. similarly that the angular positions of the O and *surface* U dips for the [0 1 0] axis are not at 90° (where it is for an unleached UO_2 (1 1 0) crystal) but at 90.75° one can conclude that the directions of the (1 1 0) and (0 1 0) axes in the transformed layers are shifted by 0.75° from their positions in the bulk cubic structure.

Hence, it appears that leaching of a UO_2 (1 1 0) crystal leads to a quasiepitaxial growth of a monoclinic phase of composition $UO_{2.28}$ with the peculiar feature that, in this transformation, the [1 0 0] conserved axis is not normal to the crystal surface but at 45° of this normal (it is monoclinization and not tetragonalization).

These measurements were complemented by x-ray diffraction experiments and comparison of the experimental scans to those obtained from Monte Carlo simulations, which reproduce the splitting in two dips of the *bulk* U scans across the [1 1 0] and [0 1 0] axes and show that in the transformed layer O and U atoms are somewhat displaced from their lattice sites.

4.4.4 Application of NRA

NRA is most useful for determination of amounts and profiles of light elements in the surface region of heavy substrates. Soon after its introduction, NRA was used extensively, together with stable isotopic tracing, in quantitative studies of anodic oxidation of tantalum and aluminum (44), duplex Al/Nb and Ta/Nb layers (45), and anodic oxidation of zirconium (46,47) amongst others. NRA is now routinely used for absolute determination of ^{16}O via $^{16}O(d,p_1)^{17}O$ in many oxidation studies including corrosion research, and also for numerous other light elements. We present here examples of ^{14}N and deuterium depth profiling.

Figure 4.14 shows a typical example of nitrogen depth profiling via $^{14}N(d,\alpha_1)^{12}C$ described in Ref. (48). Note that protons from $^{14}N(d,p)^{15}N$ reactions can interfere, but a thin barrier zone was used for the detector, as described in Section 4.3.2, so that only alpha particles are detected in this energy range. A convenient plateau in the nuclear reaction cross-section was exploited (see Figure 4.7) and the alpha-particle spectra give, to first order, a reasonable view of the nitrogen concentration depth profile, as shown by the indicative depth and concentration scales. Full quantitative interpretation still requires fitting of the spectra for assumed sample compositions. The solid lines are calculations for assumed composition profiles, which show that: (i) the main part of the nitride films are Ti_2N (x-ray diffraction confirmed this structure, as opposed to a mixture of Ti and TiN), (ii) there is a thin (<200 nm) surface layer of higher nitrogen concentration, and (iii) the compositional interface with the underlying metal is gradual, spread over a depth of the order of 1 μm.

As another illustration, we consider the work of Forster and colleagues who have studied deuterium ingress and egress during anodization of a Zr–2.5%Nb alloy, which is first cathodically charged with deuterium. Experiments in which the anodization step was carried out with an electrolyte prepared in $H_2^{16}O$ had shown loss of

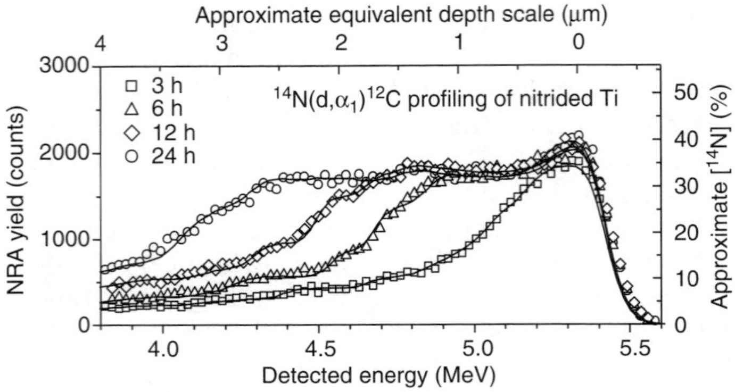

Figure 4.14 Alpha particle energy spectra measured at $\theta = 150°$, induced in $^{14}N(d,\alpha_1)^{12}C$, by a 1.4-MeV deuterium beam on commercially pure Ti nitrided in pure, ultra-dry N_2 at atmospheric pressure and 850°C. The depth and concentration scales are only indicative (see text).

deuterium during anodization, possibly due to exchange between the deuterium in the alloy and hydrogen from the electrolyte (39). To investigate this possibility, anodization was carried out with an electrolyte prepared in $D_2^{16}O$ (40) and deuterium depth profiled via the $^2H(^3He,p)^4He$ reaction (see Figure 4.7). In Figure 4.15, we show proton spectra extracted from Ref. (40) for Zr–2.5% Nb before and after anodization in the deuterated electrolyte. The shift to higher energy of the proton signal after anodization shows that the deuterium is deeper in the sample (the anodic film is essentially free of deuterium) since here, in contrast to RBS analysis, the detected particle energy increases with the depth of emission due to the reaction kinematics (49). Furthermore, the authors of this work fitted the proton spectra to extract depth profiles and were able to conclude that the total amount of deuterium in the alloy was reduced by the anodization process.

4.4.5 Application of NRP

As is the case for NRA, NRP is most useful for investigating concentration–depth profiles of lighter elements in heavier substrates. For example, Ligthart et al. (50) used the narrow ($\Gamma \sim 100$ eV) resonance at 992 keV in $^{27}Al(p,\gamma)^{28}Si$ to investigate distributions of Al implanted into pure copper and showed that corrosion resistance

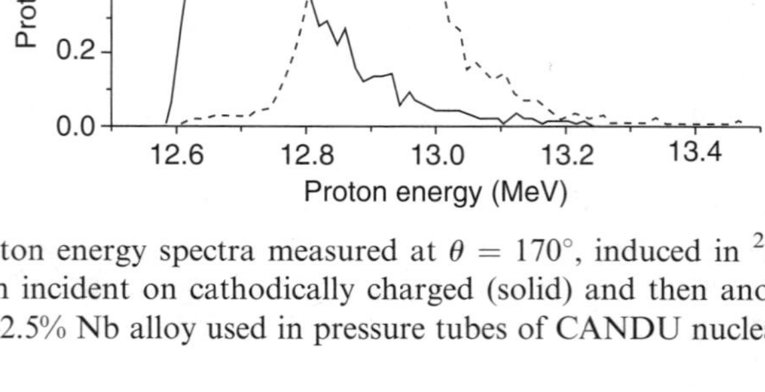

Figure 4.15 Proton energy spectra measured at $\theta = 170°$, induced in $^2H(^3He,p)^4He$, by a 800-keV ^3He beam incident on cathodically charged (solid) and then anodically oxidized in $D_2^{16}O$ (dotted) Zr–2.5% Nb alloy used in pressure tubes of CANDU nuclear reactors.

was improved via the formation of a protective Al_2O_3 layer on copper. Noli et al. (51) used this same resonance to investigate depth distributions of Al implanted in AISI 321 stainless steel for the same purpose. Koskinen et al. (52) used the narrow ($\Gamma \sim 120$ eV) resonance at 429 keV in $^{15}N(p,\alpha\gamma)^{12}C$ to determine profiles and concentration of nitrogen in AISI stainless steel coatings deposited by arc discharge in a nitrogen atmosphere of natural isotopic composition, and Nielsen et al. (53) used a resonance at 1005 keV in $^{52}Cr(p,\gamma)^{53}Mn$ to determine profiles of Cr implanted into an M50 bearing steel to improve its corrosion resistance.

Leslie et al. (54) used the resonance at 872 keV ($\Gamma = 4.5$ keV) in $^{19}F(p,\alpha\gamma)^{16}O$ for a study related to that on deuterium ingress in Zr–2.5% Nb alloy described in Sections 4.4.2 and 4.4.4. Since it had been observed that acidic fluoride etchants used for surface preparation could lead to substantial hydrogen and deuterium uptake from water, a link between this uptake and fluorine contamination was suspected. In a study of the etching of Zr, Zircalloy-2 and Zr–2.5% Nb alloy in 0.1% NH_4HF_2 in D_2O, fluorine profiles were determined and the total fluorine contamination compared to deuterium ingress. Figure 4.16, drawn from data in Ref. (54), shows excitation curves from a thin CaF_2 film (used as a reference) and a representative etched Zircalloy-2 sample. The corresponding solid lines are simulated curves for the thin CaF_2 film, where the form of the excitation curve is dominated by the resonance lineshape, and for a distribution of fluorine that is maximum at the surface and decreases exponentially with an average depth of 50 $\mu g/cm^2$ ($\sim$75 nm in Zr). This result was typical of the depth profiles from all the etched alloy samples, whereas deuterium, analyzed in these samples via the $^2H(^3He,p)^4He$ reaction, has a maximum concentration about 100 nm below the surface. Furthermore, the authors used the fact that the area under an excitation curve is directly proportional to the projected number of corresponding atoms to show that fluorine contamination and deuterium ingress are practically unrelated for a given alloy: deuterium ingress is much more closely related to alloy type. They conclude that although fluorine is necessary to activate hydrogen ingress, it need only be present above a certain threshold level to sufficiently weaken the oxide and allow deuterium ingress.

From the studies cited here, it appears to us that there remains considerable scope for application not only of NRP but also NRA in the study of corrosion phenomena, since corrosion invariably involves transport of atoms (atomic or ionic species) of light elements through surface films, for which these analytical methods

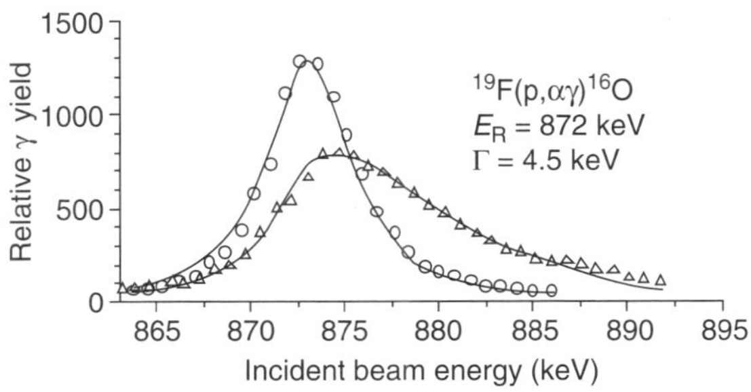

Figure 4.16 Excitation curves obtained via the $^{19}F(p,\alpha\gamma)^{16}O$ resonance at 872 keV from a thin CaF_2 sample (circles) and a Zircalloy-2 alloy etched with 0.1% NH_3HF_3 in D_2O solution at 300 K (triangles). The solid lines are excitation curves calculated for assumed concentration profiles (see text).

are especially well suited. In addition, it seems to us that the isotopic sensitivity of the nuclear-reaction methods, which was exploited in the 1970s and 1980s, deserves to be re-examined now that they have become mature analytical techniques, and reliable and simple programs for interpretation and simulation of NRA spectra and NRP excitation curves have been developed.

ACKNOWLEDGMENTS

We record here our appreciation for numerous colleagues and authors, and in particular P. Skeldon (University of Manchester Corrosion Centre, UK), F. Garrido (CSNSM, Orsay, France), and J.S. Forster (retired from AECL, Canada), who have generously given us their time in discussion, and permission to use their figures or data to illustrate aspects of IBA in corrosion research. The IBA analytical methods described here owe much to many members of our laboratory, whose contribution we are delighted to acknowledge here. We are also grateful to Dr G. Prévot for Figure 4.6, and warmly thank Dr C. Cohen for a careful reading of the methodological sections of the manuscript, and an anonymous reviewer for numerous helpful observations.

REFERENCES

1. J.R. Tesmer and M. Nastasi (eds.). *Handbook of Modern Ion Beam Materials Analysis*. Materials Research Society, Warrendale, PA (1995).
2. W.K. Chu, J.W. Mayer, and M.A. Nicolet, *Backscattering Spectrometry*. Academic Press, New York (1978).
3. L.C. Feldmann, J.W. Mayer, and S.T. Picraux, *Material Analysis by Ion Channelling*. Academic Press, New York (1982).
4. L.C. Feldmann and J.W. Mayer, *Fundamentals of Surface and Thin Film Analysis*. Elsevier Science, New York (1986).
5. J.F. Ziegler, J.P. Biersack, U. Littmark, *The Stopping and Ranged of Ions in Solids*, Pergamon Press, New York (1985).
6. N. Bohr, *Dan. Mat. Fys. Medd.* 18 (8), (1948).
7. J.L'Ecuyer, J.A. Davies, and N. Matsunami, *Rad. Effects* 47, 229 (1980).
8. C. Cohen, J.A. Davies, A.V. Drigo, and T. Jackmann, *Nucl. Instrum. Meth.* 218, 147 (1983).
9. L.R. Doolittle, *Nucl. Instrum. Meth. B* 15, 227 (1986).
10. M. Mayer, Report IPP 9/113, Max-Planck-Institut fur Plasmaphysik, Garching, Germany (1997).
11. V. Quillet, F. Abel, and M. Schott, *Nucl. Instrum. Meth. B* 83, 47 (1993).
12. A. L'Hoir, D. Schmaus, J. Cawley, and O. Jaoul, *Nucl. Instrum. Meth.* 191, 357 (1981).
13. J.F. Van der Veen, *Surf. Sci. Rep.* 5, 199 (1985).
14. I. Stensgaard, L.C. Feldman, and P.J. Silverman, *Surf. Sci.* 77, 513 (1978).
15. C. Cohen, A. L'Hoir, J. Moulin, D. Schmaus, M. Sotto, J.L. Domange, and J.C. Bouillard, *Surf. Sci.* 339, 41 (1995).
16. S. Robert, C. Cohen, A. L'Hoir, J. Moulin, D. Schmaus, and M.G. Barthes-Labrousse, *Surf. Sci.* 365, 285 (1996).
17. C. Cohen, H. Ellmer, J.M. Guigner, A. L'Hoir, G. Prévot, D. Schmaus, and M. Sotto, *Surf. Sci.* 490, 336 (2001).
18. J. Lindhard, *Mat. Fys. Medd. Dan Vid. Selsk.* 34 (14), 1 (1965).
19. Y. Kuk, L.C. Feldman, and P. Silverman, *Phys. Rev. Lett.* 50, 511 (1983).

20. *Proceedings of the Third International Conference on Chemical Analysis*, Namur, Belgium, 8–12 July1992. *Nucl. Instrum. Meth. B* 66 (1992).
21. B. Agius et al., *Surfaces Interfaces et Films minces*, Bordas, Paris (1990).
22. G. Amsel, J.P. Nadai, C. Ortega, S. Rigo, and J. Siejka, *Nucl. Instrum. Meth.* 149, 705 (1978).
23. J.A. Davies, T.E. Jackman, H. Plattner, and I. Bubb, *Nucl. Instrum. Meth.* 218, 141 (1983).
24. G. Vizkelethy, *Nucl. Instrum. Meth. B* 45, 1 (1990).
25. I.C. Vickridge and G. Amsel, *Nucl. Instrum. Meth. B* 45, 6 (1990).
26. F. Watt and G.W. Grimes (eds.). *Principles and Applications of High-Energy Ion Microbeams*. Adam Hilger, Bristol (1987).
27. R.M. Tromp, M. Copel, M.C. Reuter, M. Horn von Hoegen, J. Speidell, and R. Koudijs. *Rev. Sci. Instrum.* 62 (11), 2679 (1991).
28. B. Hjorvarsson, H. Enge, B. Anderberg, and W.A. Lanford. *Nucl. Instrum. Meth. B* 136–138, 1177 (1998).
29. G. Amsel, F. Paszti, E. Szilagyi, and J. Gyulai, *Nucl. Instrum. Meth. B* 63, 421 (1992).
30. B. Maurel, G. Amsel, and J.P. Nadai, *Nucl. Instrum. Meth.* 197, 1 (1982).
31. E. Szilágyi and F. Pászti, *Nucl. Instrum. Meth. B* 85, 616 (1994).
32. P. Skeldon, G.E. Thompson, G.C. Wood, X. Zhou, H. Habazaki, and K. Shimizu, *Corros. Sci.* 41, 291 (1999).
33. I. Felhosi, H. Habazaki, K. Shimizu, P. Skeldon, G.E. Thompson, G.C. Wood, and X. Zhou, *Corros. Sci.* 40, 2125 (1998).
34. H. Habazaki, K. Shimizu, P. Skeldon, G.E. Thompson, G.C. Wood, and X. Zhou, *J. Phys. D: Appl. Phys.* 30, 1833 (1997).
35. A.C. Crossland, G.E. Thompson, P. Skeldon, G.C. Wood, C.J.E. Smith, H. Habazaki, and K. Shimizu. *Corros. Sci.* 40 (6), 871 (1998).
36. G.E. Thompson, P. Skeldon, G.C. Wood, X. Zhou, U. Kressig, E. Weiser, H. Habazaki, and K. Shimizu, *Surf. Interface Anal.* 27, 57 (1999).
37. H.J. Whitlow, Y. Zhang, Y. Wang, T. Winzell, N. Simic, E. Ahlberg, M. Limbäck, and G. Wikmark. *Nucl. Instrum. Meth. B* 161, 584 (2000).
38. J.S. Forster, R.L. Tapping, J.A. Davies, R. Siegele, and S.G. Wallace, *Surf. Coat. Technol.* 83, 201 (1996).
39. J.S. Forster, D. Phillips, T.K. Alexander, R.L. Tapping, T. Laursen, and J.R. Leslie, *Nucl. Instrum. Meth. B* 48, 489 (1990).
40. J.S. Forster, T.K. Alexander, D. Phillips, R.L. Tapping, T. Laursen, and J.R. Leslie. *Nucl. Instrum. Meth. B* 56/57, 821 (1991).
41. J.S. Forster, R.L. Tapping, T. Laursen, and J.R. Leslie, *Nucl. Instrum. Meth. B* 89, 153 (1994).
42. P. Marcus, M. Moscatelli, C. Cohen, J. Gyulai, D. Schmaus, and M. Sotto, *J. Electrochem. Soc.* 135, 2706 (1988).
43. L. Nowicki, A. Turos, C. Choffel, F. Garrido, L. Thomé, J. Gaca, M. Wojcik, and Hj. Matzke, *Phys. Rev. B* 56, 534 (1997).
44. J. Siejka, J.P. Nadai, and G. Amsel, *J. Electrochem. Soc.* 118 (5), 727 (1971).
45. S. Rigo and J. Siejka, *Solid State Commun.* 15 (2), 255 (1974).
46. C. Ortega and J. Siejka, *J. Electrochem. Soc.* 129 (9), 1895 (1982).
47. C. Ortega and J. Siejka, *J. Electrochem. Soc.* 129 (9), 1905 (1982).
48. I.C. Vickridge, W.J. Trompetter, I.W.M. Brown, and J.E. Patterson, *Nucl. Instrum. Meth. B* 99, 454 (1995).
49. D. Dieumegard, D. Dubreuil, and G. Amsel, *Nucl. Instrum. Meth.* 166, 431 (1979).
50. H.J. Ligthart, E. Gerritson, P.J. van den Kerkhoff, S. Hoekstra, R.E. Van de Leest, and E. Keetels, *Nucl. Instrum. Meth. B* 19/20, 209 (1987).
51. F. Noli, P. Misaelides, H. Baumann, and A. Hatzidimitriou, *Corros. Sci.* 38 (12), 2235 (1996).

52. J. Koskinen, P. Torri, J.-P. Hirvonen, A. Mahiout, and A. Stanishevsky, *Surf. Coat. Technol.* 80, 57 (1996).
53. B.R. Nielsen, B. Torp, C.M. Rangel, M.H. Simplicio, A.C. Consiglieri, M.F. DaSilva, F. Paszti, J.C. Soares, A. Dodd, J. Kinder, M. Pitaval, P. Thevenard, and R.G. Wing, *Nucl. Instrum. Meth. B* 59/60, 772 (1991).
54. J.R. Leslie, T. Laursen, J.S. Forster, and R.L. Tapping, *J. Nucl. Mater*. 218, 314 (1995).

5

Scanning Tunneling Microscopy and Atomic Force Microscopy

Vincent Maurice and Philippe Marcus
Laboratoire de Physico-Chimie des Surfaces, Centre National de la Recherche Scientifique, Ecole Nationale Supérieure de Chimie de Paris, Paris, France

Contents

5.1 INTRODUCTION

Data on the chemistry and structure at the liquid/solid interface are necessary to understand and predict the corrosion behavior of metals and alloys in aqueous solutions and to improve the corrosion resistance of metallic materials in corrosive environments. Surface sensitive analysis techniques such as x-ray photoelectron spectroscopy (XPS), Auger electron spectroscopy (AES), secondary ion mass spectrometry (SIMS), nuclear reaction analysis (NRA), infrared reflection absorption spectroscopy (IRRAS), described in this book, are very useful to obtain data on the chemical composition of corroded surfaces. The panel of techniques that can be used to obtain structural data is more restricted. This is, in part, due to the difficulty of any structural analysis of very thin films on often rough substrates, using reflection high-energy electron diffraction (RHEED) and grazing incidence x-ray diffraction (GXRD), or low-energy electron diffraction (LEED), which is restricted to the *ex situ* ultrahigh vacuum (UHV) analysis of single-crystal surfaces. The advent of near-field microscopies (STM, scanning tunneling microscopy, and AFM, atomic force microscopy) in the 1980s has opened up new prospects in this field. Direct imaging of the surface structure on a scale ranging from a few micrometers down to atomic or molecular resolution can now be performed not only in *ex situ* (UHV or air) but most relevantly in *in situ* conditions. Moreover, time-resolved imaging allows the investigation of the dynamics of the structure modifications produced by corrosion processes.

The objective of this chapter is to give the reader a detailed description of the basics of the principle and operation of STM and AFM and their implementation to electrochemistry (electrochemical STM, ECSTM, and electrochemical AFM, ECAFM). The applications of ECSTM and ECAFM to corrosion science and corrosion engineering are then illustrated by selected examples on the topics of anodic dissolution, corrosion inhibition, passivation, passivity breakdown, localized corrosion, and tip-assisted nanoengineering of metals and alloys surfaces. Data obtained on the molecular scale are emphasized.

5.2 SCANNING TUNNELING MICROSCOPY

STM was invented in 1982 by G. Binnig and H. Rohrer, who measured electron tunneling between a surface and a tip separated by a vacuum gap of controlled width (1), the exponential decay of the tunneling current with the increasing width of the gap (2), and could image single atom steps (3) and the (7 × 7) reconstruction of the Si(111) surface (4). G. Binnig and H. Rohrer were awarded the Nobel Prize for this invention in 1986. The technique is unique in that it provides three-dimensional (3D) real-space images and it allows localized measurements of geometric and electronic structures.

5.2.1 Principle

The basic principle of STM is the tunneling of electrons between two electrodes separated by a potential barrier. In vacuum, the wave function ψ decays inside the barrier of width Z according to

$$\psi = \exp\left(-\frac{\sqrt{2m(V_{\mathrm{B}} - E)}}{\hbar} Z\right) \tag{5.1}$$

where m is the electron mass, $\hbar$ is Planck's constant, E is the energy of the state, and V_{B} is the potential in the barrier. For states at the Fermi level, the height of the tunneling barrier ($V_{\mathrm{B}} - E$) equals ϕ, the work function. The transmission probability, or the tunneling current, thus decays exponentially with barrier width according to

$$I \propto \exp(-2\kappa Z) \tag{5.2}$$

where κ is the decay constant for the wave function in the barrier

$$\kappa = \hbar^{-1}\sqrt{2m\phi} \tag{5.3}$$

Figure 5.1 illustrates the tunneling between two electrodes biased by a voltage V, which is by definition the difference between the two Fermi levels. Electrons within an electronvolt of the Fermi level tunnel from the occupied states of the negatively biased electrode to the unoccupied states of the positively biased electrode. Assuming a constant density of these electronic states, the tunneling current through the planar barrier increases linearly with the bias V according to

$$I \propto V \exp(-2\kappa Z) \tag{5.4}$$

Since most work functions are around 4 to 5 eV, $2\kappa \sim \text{Å}^{-1}$ and we find from Equations (5.3) and (5.4) that for a bias of 0.1 V, a tunneling current of ~1.5 nA can be observed for a separation between the two electrodes of ~0.9 nm. The tunneling current drops by nearly one order of magnitude for every increase of 0.1 nm of the barrier width between the two electrodes. A direct consequence of this exponential decay is that, if one electrode is a tip terminated by an atomic asperity protruding by ~0.1 nm, ~90% of the current is spatially limited on the electrode in regard to this atomic asperity at the apex of the tip.

5.2.2 Operation

The principle of operation of the instrument is shown in Figure 5.2. The tip is placed on a tripod formed by three independent piezoelectric tubes, P_x, P_y, P_z, respectively, controlled by the voltages V_x, V_y, V_z. This allows a precise control of the position of the tip with respect to the surface to be analyzed. Some instruments utilize a single piezoelectric tube partitioned in three sections P_x, P_y, P_z.

At a given bias voltage, the tip is scanned along the two lateral directions X and Y parallel to the surface, while a feedback circuit constantly adjusts the tip vertical position along the axis Z, so as to keep the current to a constant setpoint value

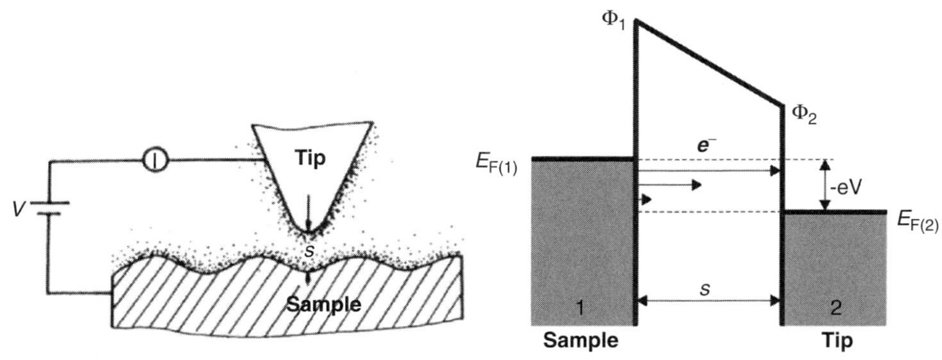

Figure 5.1 Principle of electron tunneling through a potential barrier between sample and tip electrodes separated by a distance s and polarized by a voltage V lower than the work functions, Φ_1 and Φ_2.

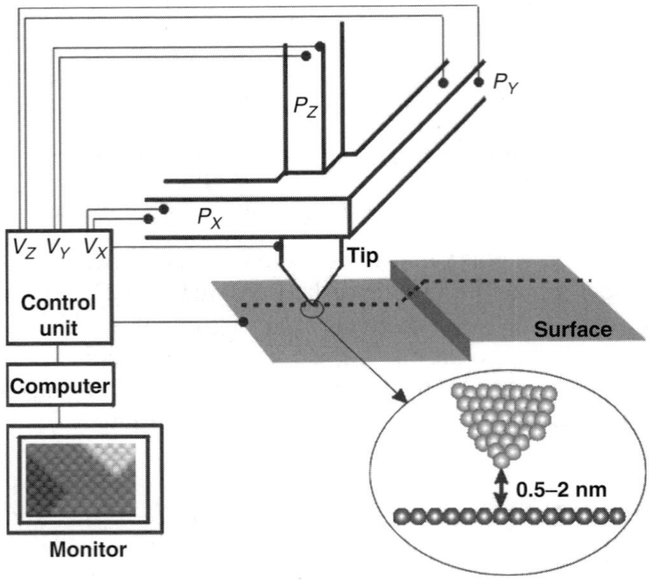

Figure 5.2 Schematics of the scanning tunneling microscope. A tripod of piezoelectric ceramics (P_X, P_Y, P_Z) allows, via the applied voltages V_X, V_Y, V_Z, to precisely control the position of the tip above the surface. During scanning along the X and Y axes, the feedback loop of the control unit adjusts the Z position of the tip in order to keep the distance (i.e., current) constant between surface and tip. The $V_Z = f(V_X, V_Y)$ values are used to construct the topographic image on the monitor.

chosen by the operator. This is the "constant current" mode where κZ is kept constant. If the local work function ϕ remains constant, a constant current corresponds to a constant Z, a constant distance between tip and surface. The shape of the surface is then reproduced by the path of the tip. The values of $V_z = f(V_x, V_y,)$ are used to produce a 3D image of the surface most often represented using a color or gray scale.

It is also possible to slow down or even cut off the feedback, so that the tip height remains constant over the surface. Small features are then reflected by current fluctuations rather than tip height adjustments. This "constant height" mode is not sensitive to topography variations and can only be used safely (i.e., without damaging the tip) on extremely flat surfaces such as surfaces obtained by cleavage (e.g., highly oriented pyrolitic graphite) or extremely well prepared surfaces of single crystals.

The setpoint current used can vary from a few tens of picoamperes for poorly conductive surfaces to a few nanoamperes for metal or doped semiconductor surfaces. Bias voltages can vary from a few volts to a few millivolts. High-quality tips can be prepared from tungsten or Pt–Ir wires by electrochemical etching (5). The lateral (spatial) resolution of the instrument depends on the quality of the tip preparation; it is ~0.1 nm. The vertical (depth) resolution depends on the quality of the damping system used to isolate the instrument from external vibrations and on the precision of the control of the tip–sample distance; it is ~0.001 nm at best.

In commercial STMs, the various parameters of operation of the instrument are controlled by the operator via a computer software that also permits the display of the STM image on monitors, the filtering of images, and the measurements of topography parameters after acquisition.

5.2.3 Interpretation of Images

STM images, revealing features on the nanometer scale and topography variations on a larger scale, can be interpreted as surface topographs somehow complicated by local variations of the work function (i.e., the tunneling barrier height). Known distances in the sample structure (e.g., step height on a single-crystal surface matching the reticular distance in the bulk) can be used to identify the atomic surface structure. Images with atomic lateral resolution also contain information on the local electronic structures of the surface and tip.

The theory of STM has been developed by Tersoff and Hammann (6); for a review see Ref. (7). It describes a situation of weak coupling between the electrons of the two electrodes with the first-order perturbation theory. At low temperatures the current is

$$I = \frac{2\pi e}{\hbar} \sum_{\mu,v} f(E_\mu)[1 - f(E_v + eV)]\left|M_{\mu v}\right|^2 \delta(E_\mu - E_v) \quad (5.5)$$

where $f(E)$ is the Fermi function, V is the voltage across the barrier, and $E_\nu(E_\mu)$ is the energy of the state $\nu(\mu)$ running over the surface (tip). The tunneling matrix element $M_{\mu\nu}$ has been approximated by Bardeen (8) as

$$M_{\mu v} = \frac{\hbar}{2m} \int \mathrm{d}S(\psi_\mu^* \nabla \psi_v - \psi_v \nabla \psi_\mu^*) \quad (5.6)$$

where the integral is over any surface lying entirely within the barrier region and ψ_ν (ψ_μ) is the wave function of the state $\nu(\mu)$.

The real difficulty to calculate the tunneling current and hence the STM images is to have explicitly the wave functions of the surface and tip, the actual atomic structure of the tip being not known. Assuming that the tip consists of a mathematical point source of current (6), the tunneling current at low temperature and for very small voltages (~10 mV) can be approximated to

$$I \propto \sum_v |\psi_v(r_t)|^2 \delta(E_v - E_F) \equiv \rho(r_t, E_F) \quad (5.7)$$

since any contributions to the matrix element are proportional to the amplitude of ψ_ν at the position of the tip r_t. Thus, this ideally simplified STM would simply measure $\rho(r_t, E_F)$, the local density of states at E_F, or charge density from states at the Fermi level, for the surface at the position of the tip.

This result has been generalized for larger voltages of a few volts ($eV < \kappa^2$) for which the tunneling current arises from a range of states lying within an electronvolt of the Fermi level. The tunneling current can be approximated using planar tunneling models as

$$I = \int_0^{eV} \rho_s(r, E)\rho_t(r, E - eV)T(E, eV)\, \mathrm{d}E \quad (5.8)$$

where $\rho_s(r, E)$ and $\rho_t(r, E)$ are, respectively, the density of states of the sample and tip at the location r and the energy E, measured with respect to their individual Fermi level; at negative sample bias, $eV < 0$, and at positive sample bias, $eV > 0$. The tunneling transmission probability $T(E, eV)$ for electrons with energy E and applied bias voltage V is given by

$$T(E, eV) = \exp\left(-2\hbar^{-1} Z\sqrt{2m}\sqrt{\phi + \frac{eV}{2} - E}\right) \quad (5.9)$$

Thus, at constant tunneling current, the contour followed by the tip and therefore the atomically resolved images are fairly complicated functions of the density of states of both sample and tip, together with the tunneling transmission probability.

5.2.4 Tunneling Spectroscopy

Tunneling spectroscopy, that is, the voltage dependence of the tunneling current, can be investigated with the STM (9). Voltage-dependent STM imaging is the simplest way of obtaining spectroscopic information, by acquiring conventional STM "topographic" images at different applied voltages and comparing the results.

More quantitative information regarding the symmetry properties and spatial localization of surface electronic states is obtained using modulation techniques to measure dI/dV at constant average tunneling current as a function of V. Such measurements can reveal structures in the surface density of states that can arise from critical points in the surface-projected bulk band structure or from true surface states associated with surface reconstructions or adsorbates. This technique is usually referred to as scanning tunneling spectroscopy (STS).

More complete information can be obtained over a wider energy range from complete I–V measurements measured at fixed tip position. The recording is then spatially localized at a single point of the surface. The recording can be performed at the atomic or even subatomic scale using atomically sharp tips or at nanometer scale using blunt tips. The measurements must be repeated at different locations in order to obtain a surface map of the I–V spectra.

The interest of the physical chemist in these techniques is the possibility to acquire laterally resolved chemical information on surfaces.

5.2.5 Electrochemical STM

STM was originally designed for operation in UHV. It was soon after realized that it should also work in solution (10–15). A major difference is the decrease of the tunneling barrier height in solution. Experimental values of ~1.5 eV have been reported (16,17), which is much smaller than the values of ~4 eV for UHV operation. It follows that for the same bias voltage and setpoint current (or tunneling gap resistance $R_t = V/I$), the microscope will operate at larger tip–surface distance in solution than in UHV. Assuming a bias of 0.1 V and a tunneling current of 2 nA ($R_t = 5 \times 10^7\ \Omega$), a tunneling distance of 0.89 nm is calculated using Equations (5.3) and (5.4) and a barrier height of 4 eV. It increases to 1.45 nm with a barrier height of 1.5 eV, in excellent agreement with values measured in solution (18).

5.2.5.1 Tip Insulation

Figure 5.3 illustrates schematically the configuration of an ECSTM. The tip is immersed in the electrolyte and acts as an electrode with electrochemical reaction taking place at the tip–electrolyte interface and generating a faradic current. The faradic current at the tip is superimposed to the tunneling current and affects the STM operation since the feedback circuit cannot discriminate between the two currents. It is therefore necessary to minimize the faradic current at the tip relative to the tunnel current.

The most effective minimization of the faradic current at the tip is obtained by reduction of the tip surface in contact with the electrolyte. This is done by covering

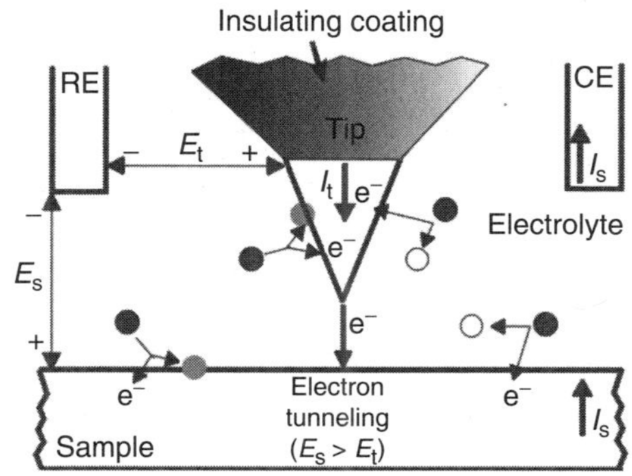

Figure 5.3 Schematic configuration of the electrochemical STM. RE and CE are the currentless reference electrode and counter electrode, respectively. E_S and E_t are the potentials of the sample and tip, respectively. I_S and I_t are the sample current and tip current, respectively. I_t is the sum of the faradic current resulting from electrochemical reaction at the tip surface and tunneling current between tip and sample.

most of the immersed portion of the tip by an insulating coating that can be made of glass (19), epoxy varnish (20), silicone polymer (21), Apiezon wax (22), nail polish (23), or electrophoric paint (24). Well-prepared coatings will leave only ~10 μm or less uncoated at the very end of the tip. Further minimization of the faradic current at the tip is achieved by polarizing the tip in the double-layer charging region (PtIr) or at the corrosion potential (W). This combined minimization allows to decrease the faradic current to values of 50 pA or less, extremely small relative to the usual set point tunneling current of 0.5 to 10 nA used in ECSTM, thus ensuring the absence of interference with the functioning of the STM.

5.2.5.2 *Instrumentation and Electrochemical STM Cell*

ECSTM is performed in a four-electrode cell (tip, substrate, auxiliary, and reference electrodes). This requires the use of a bipotentiostat that enables to control independently the potentials of the tip and substrate relative to a common reference electrode. The bias voltage for STM operation is then given by the difference of potentials between tip and substrate. The current resulting from the electrochemical reaction taking place at the sample–electrolyte interface is measured by the auxiliary electrode, as in a conventional electrochemical cell. In the electrochemical version of commercial STMs, the bipotentiostat is controlled together with the microscope by the computer software.

Figure 5.4 illustrates the design of an ECSTM cell. Other designs are described in Refs. (25,26). The substrate is used as the base of the electrochemical cell. A plastic (Kel-F or Teflon) piece forms the body of the cell, which is pressed against the substrate surface via an O-ring for sealing. The plastic cell is clamped to the cell holder to fix the whole assembly. The counter and quasi-reference electrodes are simply Pt wires brought in through and sealed in the walls of the cell. Pt quasi-reference electrode allows the stability (±15 mV) of potential control required in most corrosion experiments as long as their surface is not modified by corrosion products from the working electrode. The counter electrode is arranged symmetrically with respect to the substrate in order to provide a uniform potential drop between substrate and solution. This design allows a rapid replacement and electric contact of the substrate. It also permits the thorough cleaning in acid baths and

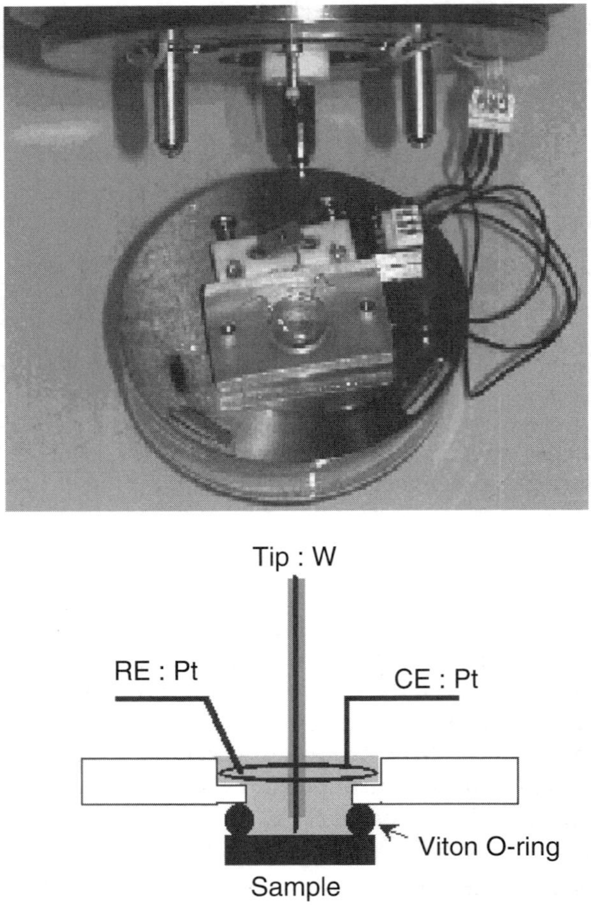

Figure 5.4 Modified electrochemical STM cell used with the Pico SPM (Molecular Imaging, Phoenix, AZ).

subsequent rinsing of the cell assembly, which are necessary to prevent the substrate surface from contamination originating from the cell itself. Where possible, the use of true rather than quasi-reference electrode is preferable using small reference electrodes in contact with the cell through small salt bridges. However, this can be detrimental to the cleanliness of the cell.

The tip approaches the cell from above. Since a long tip can lead to vibration problems, the amount of solution above the sample is minimized by limiting the height of the cell. Thus, most ECSTM cells have quite small volumes of 500 μl or less. Most conventional (non-STM) electrochemical experiments are performed in solutions deaerated by bubbling inert gases prior to and during measurements. This prevents oxygen, which can be reduced at electrodes, and impurities in the atmosphere from entering the solution. Deaeration is difficult in ECSTM cells because of the limited volume of the cell, the enhanced evaporation of the solution, and the movement of bubbles in the solution. Experiments are often performed in solutions exposed to the atmosphere. Some designs allow placing the cell in a sealed chamber that can be filled with inert gas, or alternatively the whole STM head can be placed in a glove box filled with inert gas. The degree of deaeration thus obtained is, however, lower than that reached by constantly bubbling an inert gas during measurements, which is incompatible with the absence of vibration required in precise STM measurements. When the electrochemical cell can be placed in a sealed chamber or glove box, the evaporation of the solution, which causes changes of concentration, can be minimized by saturating the chamber with water vapor.

5.2.5.3 Limitations of Electrochemical STM

As any method, ECSTM has its limitations and drawbacks. Relative to UHV-STM, the limitation resides in the difficulty to perform bias-dependent measurements. With the potential of the substrate fixed by the electrochemical process under study and with the potential of the tip adjusted to minimize the faradic current and optimize the STM measurement, the tunnel voltage can no longer be adjusted in a wide range. Thus, voltage-dependent imaging, STS, or local I–V measurements are often not possible in an ECSTM cell. However, the performance in tip isolation is such that faradic current at the tip can remain extremely small over potential regions of nearly 1 V, the limits being set by water decomposition or anodic oxidation of the tip material. This allows measurements over a moderate range of tunnel voltage.

Another limitation is related to the local nature of the measurements. The high spatial resolution capability of the instrument, which provides the basis for being able to image individual atoms or molecules, brings down the analyzed area to a tiny fraction of the total surface area (typically [1 μm × 1 μm] or less). A correlation between STM-derived and electrochemical information (integrated over areas of 1 cm^2 or less) is really meaningful if the area probed by STM is representative of the whole electrode surface. This is particularly true for molecular-scale STM studies performed on well-prepared single-crystal surfaces but also valid for lower-resolution studies.

Limitations can also arise from the possible interference of the tip with the electrochemical process at the working electrode surface. The close proximity of the tip can cause shielding effects for reaction at the sample–solution interface. An interference of sample and tip electric double layer is also possible (27).

Corrosion studies by ECSTM may also be limited by the rate of mass transport that can be high with respect to the acquisition time of the images, of the order of seconds to minutes with commercial microscopes. This limits the investigation to low-resolution imaging, which integrates larger surface areas or to low current corrosion processes for which the rate of mass transport is considerably reduced and compatible with high-resolution imaging. Finally, another limitation for corrosion studies results from the combined effect of the small volume of the ECSTM cell. It is indeed not possible to investigate the substrate modifications resulting from extensive corrosion processes since the observation of the substrate surface would be masked by the redeposition of corrosion products.

Despite these limitations, ECSTM investigations can provide invaluable information, in particular molecular-scale information, on the structural modifications related to corrosion processes such as active dissolution, adsorption, passivation, and initial stages of localized corrosion of metallic materials.

5.3 ATOMIC FORCE MICROSCOPY

The invention in 1986 of the force microscope (28) has been motivated by the limitation of STM to conducting samples. In force microscopy (FM), neither the sample nor the tip need to be conductive. The instrument, derived from STM, can measure and image local forces between a surface and a tip on a length scale of 10^{-11} to 10^{-7} m, including van der Walls, Born repulsion, electrostatic and magnetic forces, friction, and adhesion. The adaptability of the instrument is perhaps its greatest advantage. Topographic measurements (AFM) can be combined with imaging of electrostatic forces (EFM), magnetic forces (MFM), lateral forces, i.e., friction (LFM) and adhesion forces derived from force vs. distance measurements (29–32). We will restrict this presentation of AFM to topographic measurements derived from the measurement of local forces on a length scale of 10^{-10} to 10^{-8} m.

5.3.1 Principle and Instrumentation

The principle of AFM can be illustrated by the Lennard-Jones potential diagram shown in Figure 5.5, which describes the variation of the atomic interactions between two atoms or molecules separated by a few angströms. At very short distances, the interaction is dominated by the electrostatic repulsion felt by the electrons of each molecule as the orbitals start to overlap. The interaction is repulsive. At larger distances, the interaction is attractive and dominated by the dipole–dipole interaction between molecules, i.e., the van der Walls interaction. These interactions vary as inverse power laws of the distance, $1/d^{12}$ and $1/d^{6}$ for repulsive and attractive interactions, respectively. Summing the repulsive and attractive terms gives the total intermolecular potential U shown as a plain line in Figure 5.5. The potential energy minimum corresponds to the equilibrium separation. In AFM and derived FM, this is the force ($-\delta U/\delta d$) or negative force gradient ($-\delta^2 U/\delta d^2$) that is measured.

In AFM, the variations of the interactions between the atoms of the tip and those of the surface are probed while scanning the tip relative to the surface (or vice versa). The tip is positioned at the end of a cantilever spring. In most of the commercial instruments, the deflection of the cantilever is measured by an optical setup consisting in a laser beam reflected on the backside of the cantilever and impinging a partitioned photodiode. This is illustrated in Figure 5.6. The voltage difference between the upper and lower sectors of the photodiode is proportional to

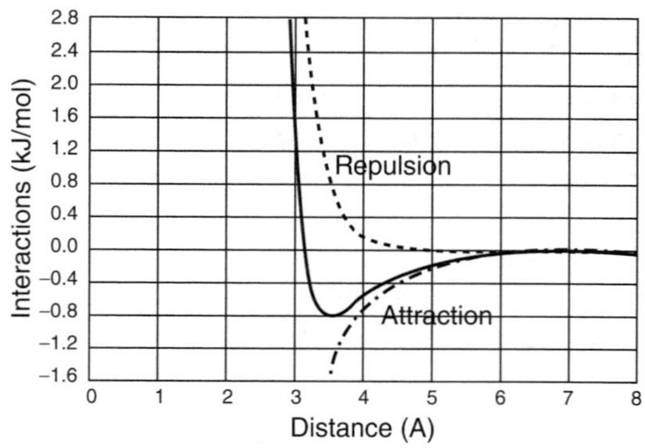

Figure 5.5 Lennard-Jones potential diagram representing the interaction between atoms at short distances. The repulsive part is dominated by electrostatic interactions (Born repulsion). The attractive part is dominated by dipole–dipole interactions (van der Waals interactions) and is less sensitive to the distance.

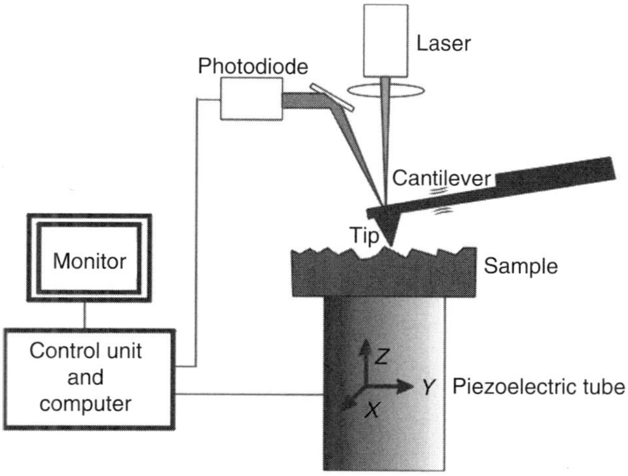

Figure 5.6 Schematics of the atomic force microscope. The lateral scanning along the *X* and *Y* axes causes deflections of the cantilever supporting the tip detected on the photodiode. The feedback loop of the control unit adjusts the *Z* position of the sample in order to maintain a constant deflection and associated force.

the deflection, z, of the cantilever. The force on the cantilever is simply $-k \cdot z$, k being the cantilever's spring constant.

As in STM, a piezoelectric setup allows the precise control of the tip position relative to the surface. A feedback circuit is generally used to adjust the tip–surface distance in order to maintain the deflection (force) of the cantilever at a setpoint value chosen by the operator. This is the "constant force" mode of imaging in which the tip reproduces the topographic profile of the surface. Imaging forces requires to cut off the feedback circuit in order to scan at constant height above the surface. The reconstitution of the images is identical to that used in STM.

The topographic interpretation of the features obtained on a nanometer scale or larger in AFM images is straightforward, provided the user is aware of the possible artifacts in the imaging (see below). The origin of the contrast in the atomic-scale images remains in many cases not fully understood (29,32). This will not be discussed in this chapter since most applications of AFM to corrosion science are limited to nanometer scale or lower resolution imaging.

5.3.2 Operation

There are several modes of operation of the AFM that can be discussed with the help of the typical force curve, that is, force on the cantilever tip vs. the tip–sample distance when approaching and leaving the surface, shown in Figure 5.7. The net forces are attractive (negative) and repulsive (positive) below and above the x-axis, respectively. Point A corresponds to the maximum attractive force or pull-on force as the tip–sample distance decreases. Point B corresponds to a zero-applied load for which attractive and repulsive forces cancel each other. Point C corresponds to a load applied to the sample by the cantilever, that is, a repulsive interaction, whose amplitude is chosen by the operator. At point C, the load is progressively removed by increasing the tip–sample distance. Point D corresponds to the maximum adhesive or pull-off force prior to return to the rest position at large tip–sample distance. The

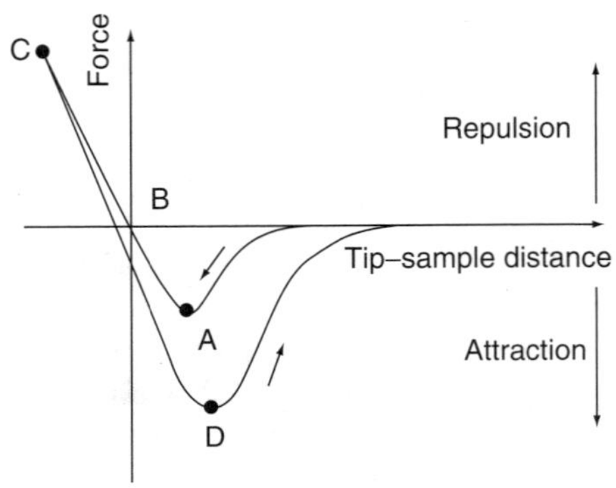

Figure 5.7 Typical force curve showing the force on the cantilever tip versus the tip–sample distance: (A) maximum attractive force; (B) zero-applied load; (C) maximum applied load; (D) maximum adhesive force.

maximum adhesive force measured after tip–sample contact is most often greater than the maximum attractive force measured in the absence of tip–sample contact.

The AFM user can choose to operate the microscope at any point along the force curve. Contact or noncontact operation modes depend on whether tip and sample are touching each other, which can be determined by modulating the tip–sample distance and noting when the modulation amplitude suddenly increases as the average tip–sample distance decreases. Contact occurs in the nonmodulated force curve shown in Figure 5.7 at the inflection point before A where the curvature changes from attractive to repulsive. Contact increases after this point as the tip–sample distance decreases. Contact-mode imaging permits high-resolution imaging, at the risk of a potentially destructive interaction that can damage the tip and the sample.

Typically, the adhesive force (point D in Figure 5.7) measured in air between a silicon nitride tip and a mica surface is 10^{-7} to 10^{-8} N (33). It results mostly from the presence of a condensed layer of water or other contaminants on the surface that forms a capillary bridge between tip and surface. Assuming that this force is exerted over a distance of 0.1 nm, an energy of 10^{-17} to 10^{-18} J ($\sim$100 to 10 eV) is dissipated. This is quite large with respect to the energy of chemical bonds at solid surfaces ($\leq$1 eV) and implies that the interaction will involve a large number of atoms at the tip apex and surface, which is detrimental for the spatial resolution of the imaging, and will modify the surface locally. Combined with the adhesive force, the dragging motion of the tip on the surface involves lateral shear force that can distort the measurement severely and induce lateral motion and tearing of surface features. The adhesive force is drastically reduced (typically $\leq 10^{-9}$ N) in liquid applications where the whole cantilever tip/surface assembly is immersed in solution, thus preventing the formation of a capillary bridge between tip and surface. However, the energy dissipated ($\leq$1 eV) remains of the order of the energy of chemical bonds and imaging in these conditions remains potentially destructive. The lateral spatial resolution of the instrument is $\sim$0.1 nm when the interaction can be spatially limited to the atomic asperities of the tip. The vertical (depth) resolution is $\sim$0.02 nm.

The contact mode of operation is not suitable for imaging of surfaces covered with weakly adsorbed layers or deposits, self-assembled monolayers of organic molecules, or surfaces of soft materials. These limitations can be overcome by alternately placing the tip in contact with the surface to provide high force resolution and then lifting the tip off the surface to avoid dragging the tip across the surface. In this resonant (or Tapping®) mode, the cantilever tip assembly oscillates at or near its resonance frequency with an amplitude of several nanometers when the tip is not in contact with the surface. The oscillating tip is then moved toward the surface until it lightly "touches." During scanning, the vertically oscillating tip alternately contacts the surface for force measurement and lifts off for implementation of the X, Y scan. When the tip passes over a bump in the surface, the cantilever has less room to oscillate and the amplitude of oscillation decreases. Conversely, when the tip passes over a depression, the cantilever has more room to oscillate and the amplitude increases. The tip–sample separation is adjusted via the feedback loop of the microscope to maintain a constant oscillation amplitude and force on the sample. This resonant mode inherently prevents the tip from sticking to the surface and causing damage during scanning. Unlike contact mode, it has sufficient oscillation amplitude to overcome the tip–sample stray adhesive forces, preventing the surface material to be moved sideways by shear forces during scanning since the applied force is always vertical.

The noncontact mode can also be used to image fragile samples. The tip and sample are not in contact before the curvature (i.e., derivative) of the force curve in Figure 5.7 changes from negative to positive. The measured forces are then only attractive van der Waals forces and there is no damage caused to either tip or sample. Unfortunately, the attractive van der Waals forces are substantially weaker than the forces used in contact mode and the force resolution is poor. Consequences are a poor spatial resolution and a problematic thermal drift. The force resolution can be improved by slightly oscillating the tip and using ac detection methods to detect the small forces between tip and sample.

This imaging mode can be used to image delicate samples with large features, provided that the sample–tip separation is properly controlled (thermal drift is minimized). Even then, during measurements, the probe can be frequently drawn to the sample surface by the surface tension of the adsorbed gases resulting in unusable data and sample damage similar to that caused by the contact technique. The noncontact mode is more appropriate for UHV operation where the sample and tip can be cleaned from any contaminants and where thermal drift is much less problematic. True atomic resolution can then be achieved when the tip–sample attractive force is such that the interaction is limited to a single atom at the tip apex.

5.3.3 Electrochemical AFM

The implementation of AFM to electrochemical measurement is easier than that of STM. This results from the fact that the material used for the integrated cantilever tip assembly that are commercially available is inert in most aqueous solutions. In addition, even in the case where electrochemical reaction would occur at the solution/tip interface, this has no direct effect, in contrast with STM, on the measurement of the signal that is used to probe the surface. In AFM, the cantilever tip assembly does not constitute a fourth electrode and a conventional potentiostat can be used to control the sample potential.

The same general design as for ECSTM cells can be used. The limitations of ECAFM regarding the correlation between the local information and integrated macroscopic information, the possible tip-shielding effects on the corrosion behavior, the kinetic of the mass transport, and the small volume of the electrochemical cell are the same as those described above for ECSTM.

Immersing the cantilever tip assembly and surface in solution has several consequences. It avoids the formation of a capillary bridge between tip and surface, which allows a better control of the adhesive force in contact mode (see above). In resonant contact mode, the fluid medium tends to damp the normal resonance frequency of the cantilever, which complicates this mode of operation. The noncontact mode becomes impractical because the van der Waals forces are even smaller, which can be a substantial limitation (e.g., for biological applications).

ECAFM is now a widely used technique for corrosion studies. It is often preferred to ECSTM despite the generally lower level of spatial resolution that can be achieved, because the obtained topographic information is independent the conductivity of the corrosion products. Moreover, the setting up of the experiment is easier than that of ECSTM.

Examples of applications in corrosion science and engineering are presented in the following part of this chapter.

5.4 APPLICATIONS TO CORROSION SCIENCE AND ENGINEERING

The first applications of STM and AFM to *Corrosion Science* studies appeared in the late 1980s (11,14,15). The objective of these early studies was to test the possibility of real-time observation of corrosion processes. The authors observed that corrosion processes occurred preferentially at existing defects of the electrode but did not attempt to make optimal use of the high-resolution capability of the instruments. Nowadays, the most part of corrosion studies performed by means of ECSTM or ECAFM imaging are still low-resolution studies not aiming at structural characterization of the corrosion processes at the molecular level. This limitation in spatial resolution persists in part due to the rate of the corrosion processes that can be too high with respect to the acquisition time of the images to be followed with molecular precision by STM or AFM. However, the appropriate control of the electrochemical conditions allows in some cases to minimize the rate of mass transport to a point compatible with STM or AFM imaging, as illustrated in the examples given below. The limitation in spatial resolution is also due to the fact that high-resolution studies require a high quality of surface preparation and control, that is, atomically flat surfaces of well-defined structure. This is achieved with single crystals of known orientation whose surface has been appropriately prepared to have terrace and step topography. Surface preparation will not be addressed in this chapter, and details are to be found in the experimental sections of the cited papers. Fundamental studies aiming at an improved understanding of the corrosion processes at the molecular level become then feasible and this *Surface Science* approach has brought about significant progress in the understanding of the role of the surface structure in corrosion processes.

In the following, we present selected examples of high-resolution data obtained on the topics of active dissolution of metals, inhibition of corrosion, growth and structure of passive films, and passivity breakdown. Examples of low-resolution studies are selected to illustrate the use of STM and AFM in corrosion science and engineering on the topics of corrosion inhibition, localized corrosion by pitting, and microstructure-related localized corrosion.

5.4.1 Active Dissolution of Metals and Alloys

The active dissolution of metals has been and still is extensively studied with ECSTM, which is conveniently used because the surface is free of passivating oxides and, for pure metals, the electrode surface remains smooth (i.e., terminated by an atomically flat terrace and step topography during the corrosion process). This is a very favorable situation for the atomic-scale imaging of dynamic processes at surfaces. Investigations can then focus on the structure modifications related to the adsorption of anions in the double-layer potential region preceding the onset of dissolution and on the dissolution of the metal atoms at specific sites. In the ECSTM experiments, sequences of consecutive images of a selected region are recorded while the potential is scanned or stepped to values of interest. The appropriate control of the potential allows controlling the etching rate so that the corrosion rate and the time resolution of the STM imaging (of the order of tens of seconds per image with commercial microscopes) are compatible. Studies have been performed on a number of metals: Cu (34–40), Ni (41,42), Ag (43,44), Co (45), Pd (46,47). A few alloys have been studied, mainly Cu–Au, a model system to investigate the fundamental aspects of selective dissolution and dealloying (48–5).

Copper has been the most investigated metal. The results obtained on the (001) face are chosen to illustrate this topic. A major result evidenced by ECSTM studies is that dissolution proceeds via a step-flow mechanism at moderate potentials (i.e., slow etching rate). This was first evidenced by Suggs and Bard on Cu(1 1 1) (34) and Cu(1 0 0) (35). This process is illustrated in Figure 5.8 for Cu(0 0 1) in H_2SO_4 and HCl (38). It shows the preferential etching of the surface at preexisting step edges. The selective etching of atoms at the step edges results from their lower coordination to nearest neighbor atoms than terrace atoms. In the absence of adsorbed anions, the etching process stabilizes the step edges oriented along the close-packed

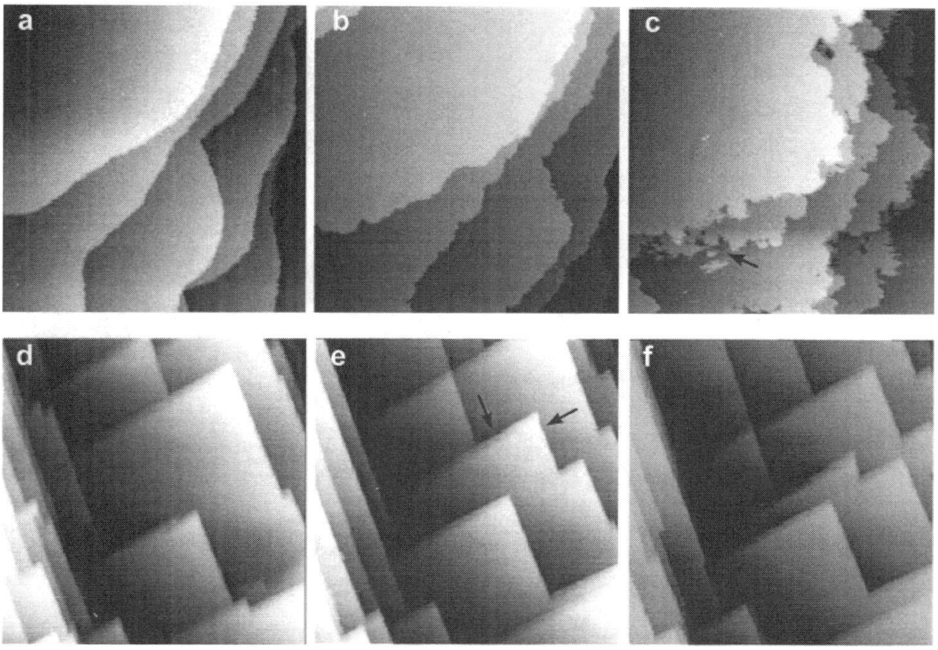

Figure 5.8 Series of ECSTM images of Cu(0 0 1) recorded (a), (b), (c) in 10 mM H_2SO_4 (180 nm × 180 nm) and (d), (e), (f) in 1 mM HCl (160 nm × 160 nm), showing the isotropic and anisotropic step flows caused by anodic dissolution in H_2SO_4 and HCl, respectively. (From Magnussen et al., *Appl. Phys. A* 66: S447, 1998, Springer Verlag. With permission.)

crystallographic directions of the crystal where the nearest neighbor coordination of the atoms is the highest, along the atomically smooth ⟨1 $\bar{1}$ 0⟩ directions on Cu(1 1 1) (34), Ni(1 1 1) (41), Ag(1 1 1) (43), and ⟨1 0 $\bar{1}$ 0⟩ on Co(0 0 0 1) (45).

The anisotropy of the etching is also dependent on the superstructure formed by adsorbed ions in the double-layer region. On Cu(0 0 1), a disordered and highly mobile sulfate adlayer is formed in sulfuric acid solution and the steps are etched isotropically without any preferential direction (37). In contrast, a highly ordered $c(2 \times 2)$ adlayer is formed in hydrochloric or hydrobromic acid solutions that stabilizes the step edges along the close-packed ⟨1 0 0⟩ directions of the superstructure (35–38,40). This difference of anisotropy of the etching is illustrated in Figure 5.8. Figure 5.9 shows the observed square lattice corresponding to the adsorbed Cl layer. The formation of this well-ordered superstructure, more corrugated than the metallic lattice, facilitates the high-resolution STM imaging of the surface structure. Atomic-scale ECSTM imaging has shown that the dissolution proceeds at structurally well-defined kinks of the [1 0 0]- and [0 1 0]-oriented steps via removal of the primitive unit cells of the Cl adlayer containing two Cu and one adsorbed Cl atoms. In the double-layer potential range, equilibrium fluctuations of these step edges have been observed that result from local removal/redeposition processes. At more anodic potentials, the dissolution processes prevail and a net removal is observed but the redeposition still occurs. These fluctuations are also illustrated by the sequence of images shown in Figure 5.9 (37,40).

Another effect of the $c(2 \times 2)$ adlayer evidenced by ECSTM imaging is an induced anisotropy of the dissolution process along the symmetrical ⟨1 0 0⟩ direc-

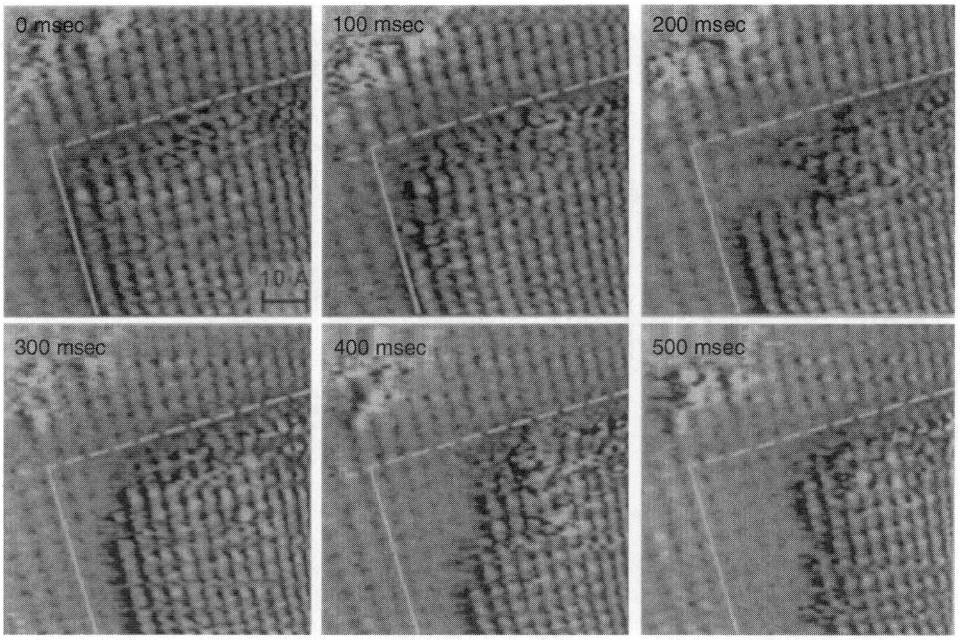

Figure 5.9 Series of video ECSTM images recorded on Cu(0 0 1) in 0.01 *M* HCl at −0.17 V/SCE showing the progressing dissolution of a Cu terrace starting at the outer terrace corner. The observed lattice is the $c(2 \times 2)$ ad-lattice formed by adsorbed chlorine. The markedly different dissolution behavior at the active and stable steps whose initial position is marked by the solid and dashed lines, respectively, is clearly visible. (From Magnussen et al., *Electrochim Acta* 46: 3725, 2001, Elsevier. With permission.)

tions of the step edges (37,38,40,47). This is also visible in Figure 5.9 where one orientation of the step edges dissolves more rapidly than the other. This is related to the structural anisotropy of the step edges induced by the $c(2 \times 2)$ adlayer, as illustrated by the model shown in Figure 5.10. This has been tentatively explained by the coordination of the outmost Cu atom forming the dissolving kink to the adjacent Cl adsorbates (40). At the reactive kink site, the Cu atom is closely coordinated to two adsorbed Cl atoms whereas at the stable kink site it is closely coordinated to one adsorbed Cl atom. Hence, the formation of the $CuCl_2^-$ dissolving complex is already initiated at the reactive kink site, generating a preferential dissolution along one of the symmetrical $\langle 1\ 0\ 0 \rangle$ directions. A similar anisotropic effect of the $c(2 \times 2)$ ordered adlayer on the dissolution of the (0 0 1)-oriented metal substrate has been reported for Br on Cu (38), S on Ni (42), I on Ag (44), and Pd (46).

The latest developments of the STM technique have also been applied to the study of the atomic-scale processes during anodic dissolution of Cu(0 0 1) in HCl (40). The direct observation of these rapid dynamic processes was studied with a high-speed electrochemical "video" STM capable of acquiring up to 25 atomic resolution images per second. It was observed at the atomic scale that the dissolution proceeds by a local fluctuating removal/addition process of the atoms at the kink sites at the steps on the crystal surface. The preferential nucleation of kinks at outer terrace corners rather than at straight step edges or inner terrace corners was demonstrated. The individual kinks condense into larger facets due to statistical fluctuations in the kink motion and mutual stabilization. This leads to the observation of an apparent collective dissolution of the atomic rows at the edges of the terraces, also illustrated in the images shown in Figure 5.9. Quantitative data on the kinks dynamics at the onset of Cu dissolution yield an average reaction rate at kinks of 10^3 to 10^5 atoms per second.

Dealloying by selective dissolution of Cu–Au alloys has also been studied by ECSTM (49,51). Real-time measurements were performed on the atomic scale on

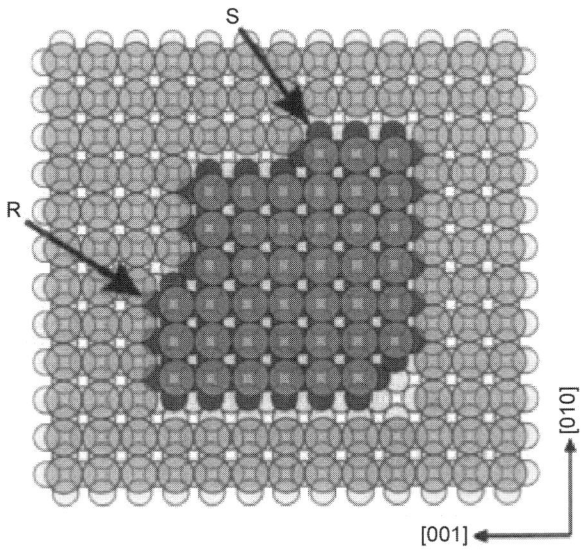

Figure 5.10 Model of the (1 0 0) surface of a face-centered cubic metal (e.g., Cu) covered by an ordered $c(2 \times 2)$ adlayer (e.g., Cl) showing the different structure of steps and kinks along the [0 1 0] and [0 0 1] directions. R and S mark the reactive and stable Cu atoms at the outer kinks, respectively. (From Magnussen et al., *Electrochim Acta* 46: 3725, 2001, Elsevier. With permission.)

(1 1 1) oriented Au–25Cu alloys (50) and Cu–25Au alloys (51) as a function of anodic polarization. Below the critical value of the potential, it is observed that Cu dissolution occurs by aggregation of vacancies in the first layer leading to a structure of voids that are one atom deep and grow two-dimensionally. No defects are created in the second layer during growth of these voids. This is attributed to an inhibited dissolution caused by the replacement of the Cu atoms of the second layer by the diffusing gold atoms of the selectively dissolved first layer. Above the critical value of the potential, porous surface structures are formed. This has been modeled by rapid stripping of the Cu atoms from the terraces and subsequent diffusion of the exposed gold atoms. The newly exposed Cu atoms dissolve to form deeper pits and eventually pores (52). It was observed experimentally that the formation of 3D pits is initiated well below the critical potential (51).

5.4.2 Corrosion Inhibition

The topic of corrosion inhibition of metallic surfaces is becoming increasingly addressed with scanning probe microscopies. Many of the reported studies are low-resolution AFM studies combined with electrochemical measurements and other surface analyses dedicated to the optimization of the efficiency of the inhibitor. Some examples will be presented in the second part of this section. In the first part, selected examples of high-resolution ECSTM studies applied to corrosion inhibition are reported.

Similarly to the studies of the anodic dissolution of metals, high-resolution ECSTM imaging has been applied on well-prepared single-crystal surfaces exhibiting a terrace and step topography to study the effect of corrosion inhibitors below and at or near the onset of dissolution. On copper, the effect of benzotriazole (BTAH, $C_6H_4N_3H$) has been studied in sulfuric, hydrochloric, and perchloric acid solutions on the Cu(0 0 1) surface (38). The effect of the surface structure has been, 53–55 addressed by a comparative study on Cu(1 1 1) in sulfuric acid solution (56) and on Cu(1 1 1) and Cu(1 1 0) in perchloric acid solution (55). The effect of strongly bonded self-assembled monolayers (alkanethiol), modeling corrosion inhibitors, has also been reported for Cu(0 0 1) in HCl solution (57).

A major outcome of these high-resolution ECSTM studies is the direct observation of the modifications of the mechanism of Cu dissolution. Instead of proceeding exclusively by a step flow process in the absence of the inhibitor as described above, the Cu dissolution in the presence of the inhibitor predominantly proceeds by the nucleation and growth of monoatomic pits resulting in the roughening of the surface. This modified mechanism is illustrated by the image sequence shown in Figure 5.11. An anodic shift of the onset of dissolution ranging from 0.05 to 0.3 V has been observed with ECSTM, depending on the concentration of BTAH in the electrolyte. The pits presumably form at the defects of the corrosion-inhibiting adlayer. The effect has been observed for submonolayer coverage of the inhibitor. It can be understood by the stabilization or “passivation” of the steps against dissolution that occurs even for submonolayer coverage, assuming that the inhibiting species adsorb preferentially at these defect sites.

Comparative ECSTM observations made in the double-layer potential region in sulfuric and hydrochloric acid solutions on Cu(0 0 1) shed some light on the reduced inhibiting efficiency of BTAH on the dissolution of Cu in the presence of chloride. In both solutions, commensurate ordered adlayers of BTAH have been observed for potentials less than $-0.6\ V_{SCE}$. They form a $(\sqrt{5} \times \sqrt{5})R26.6°$ structure assigned to a parallel stacking of BTAH molecules oriented perpendicular (or slightly tilted) to the

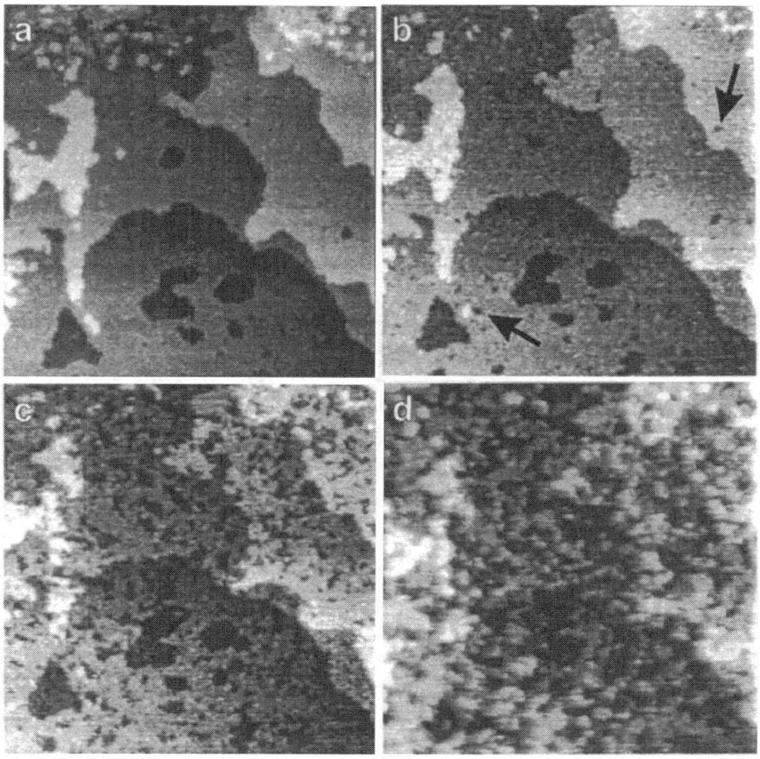

Figure 5.11 Series of ECSTM images (150 nm × 150 nm) showing the anodic dissolution of Cu(1 1 1) in 0.01 *M* H_2SO_4 + 10^{-4} *M* BTAH recorded at −0.33 V/SCE (a, b) and −0.06 V (c, d). The arrows in (b) point the formation of etch pits. (From Polewska et al., *J. Chem. Phys. B* 103: 10440, 1999, ACS. With permission.)

Cu(0 0 1) surface. At potentials greater than $-0.6\,V_{SCE}$ in HCl solutions, the structure and the dynamic behavior of the surface are identical to that found in inhibitor-free HCl solutions and described above. The displacement of the ordered structure of adsorbed BTAH species by adsorbed chloride ions forming the same *c*(2 × 2) structure at the same potential as in BTAH-free solutions is illustrated in Figure 5.12. In contrast, in sulfuric acid solutions the BTAH adlayer is observed in the entire underpotential region with formation of a chain-like structure with increasing potential assigned to the polymerization of the BTAH adlayer. These differences in the initial stages of formation of the inhibiting Cu(I)BTA film that forms at higher anodic potential in the Cu(I) oxidation region have been advanced to explain the lower inhibition efficiency in chloride-containing solution. In HCl solution, the direct conversion of the chemisorbed BTAH layer into the polymerized Cu(I)BTA phase is hindered by the formation of the *c*(2 × 2) adlayer of adsorbed chloride, which is not the case in H_2SO_4 solution. Consequently, in HCl solution BTAH or Cu(I)BTA is not present on the surface in the dissolution potential range and the dissolution proceeds by the same step flow mechanism. The dissolving species ($CuCl_2^-$) reacts in the solution with BTAH to form Cu(I)BTA, which subsequently precipitates on the Cu surface. However, this dissolution/redeposition mechanism produces an inhibiting layer less protective against corrosion than the inhibiting layer directly produced in the Cu(I) oxidation range by a surface reaction between adsorbed polymerized BTAH and Cu(I) ions in sulfuric acid solutions.

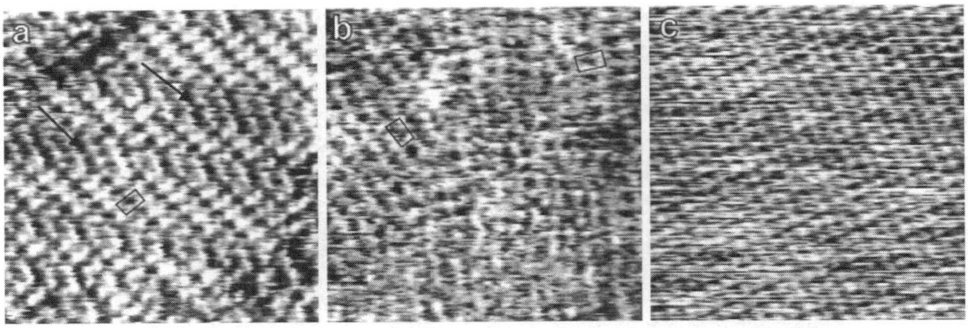

Figure 5.12 ECSTM images of Cu(0 0 1) in 0.1 *M* HCl + 0.075 *M* BTAH showing large (a) and smaller (b) ordered domains of the chemisorbed BTAH adlayer at −0.7 V/SCE (10 nm × 10 nm) and (c) the *c*(2 × 2) ordered adlayer at −0.45 V/SCE (7.5 nm × 7.5 nm). (From Vogt et al., *J. Chem. Chem. B* 102: 5859, 1998, ACS. With permission.)

ECAFM or ECSTM have also been applied in combination with other techniques like quartz-crystal microbalance (QCM), electrochemical impedance spectroscopy (EIS), XPS, AES, Fourier-Transform infrared, and/or IRRAS with the objective to optimize corrosion resistance. For example, the corrosion inhibition of copper by 2-mercaptobenzoxazole (MBO) has been studied by ECSTM (58). In the optimized conditions of inhibition, the topographs showed the rapid formation of a compact 3D inhibitor film on the surface containing cuprous ions and S and N with a modified chemical environment determined by XPS. It was concluded that the inhibitor reacts with Cu^+ species, resulting from corrosion, to produce a water-insoluble copper complex constituting the inhibiting film. Also on copper, ECAFM was used to optimize the corrosion protection by sodium heptanoate, a nontoxic "green" inhibitor (59). It was observed that in the optimal conditions of inhibition (pH 8, $[C_7H_{13}O_2Na] = 0.08$ *M*), a continuous metallic soap thin film, mainly constituted of copper heptanoate, was formed, acting as a blocking barrier whereas at pH 5.7 or 11, a discontinuous and therefore poorly protective film constituted of crystals of copper heptanoate or copper oxides was formed.

On iron, the corrosion inhibition in the presence of *N*,*N*-di(phosphonomethyl)-glicine (DPMG) and the synergistic effect of divalent cations (Ba^{2+}, Sr^{2+}, and Zn^{2+}) have been studied (60). At optimal concentration of the inhibitor (3×10^{-4} *M*) and most effective cation/inhibitor ratio (2:1), it is assumed that inhibition results from the formation of a DPMG/metal complex on the surface. Time-dependent measurement ECAFM topographs provided data on the corrosion protection at the micrometer scale and the development of pores for increasing corrosion rates. With Sr^{2+}, the protective surface layer becomes porous and less protective after 1 h. With Ba^{2+}, a smooth layer with very few pores and high corrosion resistance is formed. With Zn^{2+}, the protective surface layer becomes thicker with time and the surface becomes covered with bell-shaped insoluble zinc hydroxide particles identified by XPS. These ECSTM and ECAFM data evidence the relationship between the efficiency of the corrosion inhibition and the morphology of the protective surface layer formed in the presence of the inhibitor.

5.4.3 Passivation of Metals and Alloys

The growth and structure of passive films have been studied in detail with STM and AFM on pure metals, Cu (61–65), Ni (14,66–69), Fe (70–73), Cr (74,75), Al (76), Co

(45), and on ferritic and austenitic stainless steels (77–81). Prior to the STM studies, it was generally considered that the ultrathin passive films were amorphous, and that the absence of crystallinity was favorable to the corrosion resistance. This view has now changed, because STM studies have revealed that in many cases the passive films are crystalline.

Similarly to the topics of anodic dissolution and corrosion inhibition, the best resolved measurements have been obtained with ECSTM on well-prepared single-crystal surfaces of metals and alloys. Data obtained on copper and nickel are selected here to illustrate the use of STM in studying the growth and structure of passive films.

5.4.3.1 The Growth of Passive Films

The study of the growth of passive films requires the investigation of the structural modifications occurring both in the potential range of oxide formation and in the potential range preceding oxide formation, called underpotential range of oxidation. This has been done on Cu(1 1 1) in sodium hydroxide solution with ECSTM (62,63). Figure 5.13 shows the polarization curves characterizing the growth of the passive films for this system. A major result of the STM measurements is the evidence of the formation of an ordered adsorbed layer of hydroxide in the underpotential range of oxidation, which plays the role of a structural precursor for the growth of the Cu_2O(1 1 1) oxide layer in the Cu(I) oxidation range.

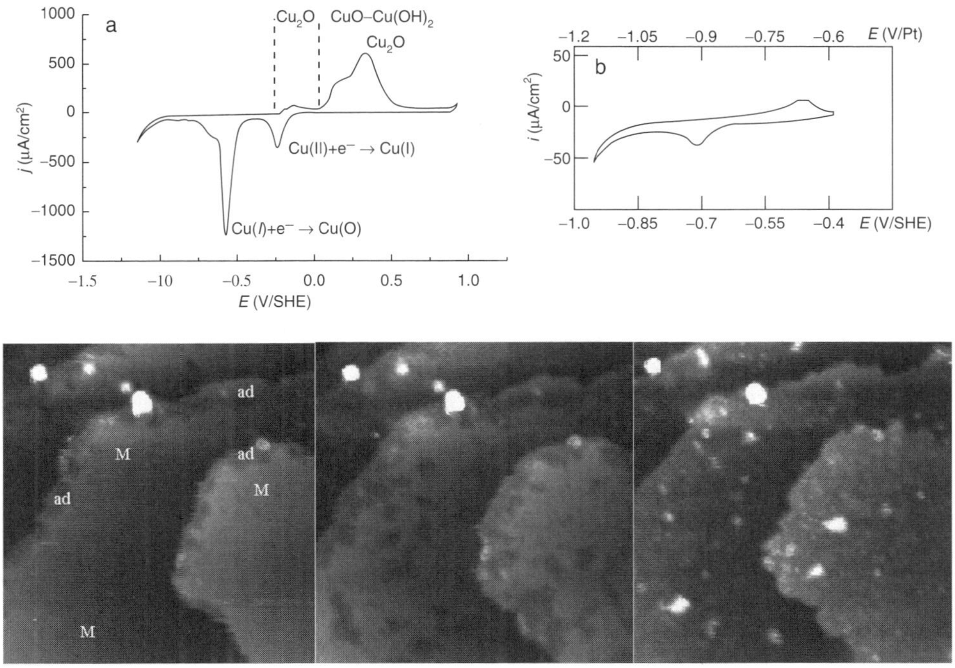

Figure 5.13 Voltamograms recorded for Cu(1 1 1) in 0.1 *M* NaOH between the hydrogen and oxygen evolution limits (a) and in the potential range below oxidation (b). The passive film consists of a single Cu(I) oxide layer and a duplex Cu(I)/Cu(II) oxide layer depending on the potential regions. The sequence of ECSTM images (40 nm × 40 nm, 38 sec/image) shows the growth of an adlayer of OH groups at $-0.6\ V_{SHE}$ at the onset of the anodic peak in (b). (From Maurice et al., *Surf. Sci.* 458: 185, 2000, Elsevier. With permission.)

The adsorption of the hydroxide layer is illustrated by the sequence of images shown in Figure 5.13, obtained after stepping anodically the potential from a region where no adsorption takes place to $-0.6\,V_{SHE}$ at the onset of an anodic peak observed at $-0.46\,V_{SHE}$ in the underpotential range of oxidation. This anodic peak is assigned to the adsorption of OH groups with a charge density transfer of $55\,\mu C/cm^2$ corresponding to a surface coverage of 0.19 OH per Cu atom of the Cu(1 1 1) surface, assuming the following one electron adsorption reaction: $Cu + OH^- \rightarrow Cu_{ads}{-}OH + e^-$. The atomically smooth terraces of the surface (marked M in Figure 5.13) become progressively covered by islands of darker appearance (marked ad) that grow laterally and coalesce to cover completely the terraces. The darker appearance of the ad-islands in STM images is a typical effect of adsorbed O species systematically observed on metal surfaces (82). It is an electronic effect resulting from an oxygen-induced decrease of the density of states near the Fermi level. Thus, the lowering of ~0.05 nm of the apparent height of the terraces in the OH ad-islands is not a topography variation. This illustrates the difficulty in discriminating topographic and electronic information in STM images at this scale.

The images in Figure 5.13 show that the adsorbed layer preferentially grows at the step edges, confirming the preferential reactivity of these defect sites of the surface. They also show that the terraces grow laterally by a displacement of the step edges, and that monoatomic ad-islands are formed at the end of the growth process. These two features are assigned to the reconstruction of the topmost Cu plane with a decrease of density induced by the adsorption of the OH groups. This causes the ejection of Cu atoms above the surface. The ejected atoms diffuse on the surface and are trapped at step edges, which causes the observed lateral displacement of the step edges. In the final stages of the adsorption process where most of the surface is already covered by the adlayer, the ejected atoms have a reduced mobility on the OH-covered terraces and aggregates to form the observed monoatomic ad-islands.

The reconstruction of the Cu topmost layer was confirmed by the atomically resolved image of the ordered adlayer shown in Figure 5.14. A hexagonal lattice with

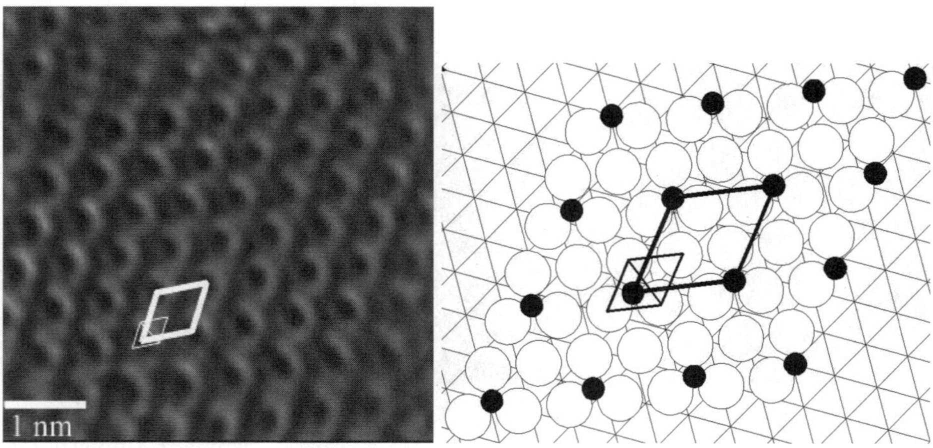

Figure 5.14 ECSTM image and model of the ordered structure of adsorbed OH groups formed on Cu(1 1 1) in 0.1 *M* NaOH in the potential region below oxidation. The large and small cells mark the lattice of adsorbed OH and of reconstructed copper, respectively. (From Kunze et al., *Electrochim Acta* 48: 1157, 2003, Elsevier. With permission.)

a parameter of 0.6 ± 0.02 nm is measured. Each unit cell contains one minimum and four maxima of intensity. By assigning the minima to the position of the adsorbed OH group, in agreement with the variation of apparent height caused by adsorbed OH species, a coverage of ~0.2 OH per Cu(1 1 1) atom is deduced, in excellent agreement with the coverage of 0.19 obtained from the electrochemical charge transfer measurement. It follows that the four maxima of intensity can be assigned to the surface Cu atoms. Their interatomic distance is ~0.3 nm, which is larger than the interatomic spacing of 0.256 nm in Cu(1 1 1) from bulk parameter values. This confirms the reconstruction of the topmost Cu plane into a lower density plane. In addition, the image in Figure 5.14 evidences that the adsorbed OH groups sit in threefold hollow position with respect to the reconstructed Cu plane and form a (2 × 2) structure with respect to the reconstructed Cu surface.

A striking feature of this adlayer structure is its similarity with the arrangement of the O and Cu planes in Cu_2O(1 1 1). In the [1 1 1] direction, cuprite consists of Cu planes embedded between two O planes. The hexagonal packing of the O planes has a (2 × 2) periodicity with respect to that of the Cu planes with parameters of 0.604 and 0.302 nm, respectively. The O atoms are in the threefold sites of the Cu planes. Thus, the adsorbed OH layer mimics the structure of the O and Cu planes in the (1 1 1)-oriented cuprite forming a structural precursor for the growth of the Cu(I) anodic oxide at more anodic potential.

In the potential range of Cu(I) oxidation, a Cu_2O(1 1 1) oxide film is formed with a faceted surface (61,63). The images shown in Figure 5.15 illustrate the influence of the oversaturation potential on the nucleation, growth, and crystallization of this Cu(I) oxide. At low oversaturation (see Figure 5.15a), poorly crystallized and one-monolayer-thick islands covering partially the substrate are formed after preferential nucleation at step edges. They are separated by islands of the ordered hydroxide adlayer. At higher oversaturation (see Figure 5.15b), well-crystallized and several-monolayer-thick films are formed, and the step edges are not preferential sites of nucleation. The equivalent thickness of the oxide layer can be deduced from subsequent measurements of the charge transfer during cathodic reduction scans. It was ~0.5 and 7 equivalent monolayers (ML, one ML corresponds to one (1 1 1)-oriented O^{2-}–Cu^{+}–O^{2-} slab) after growth at −0.25 and −0.2 V_{SHE}, respectively.

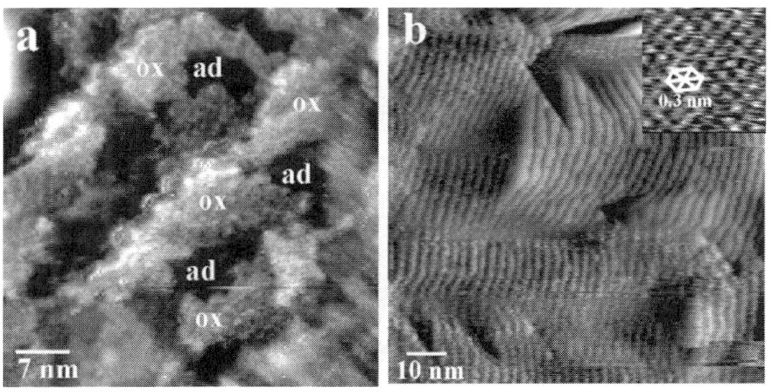

Figure 5.15 ECSTM images of the Cu(I) oxide grown on Cu(1 1 1) in 0.1 *M* NaOH at −0.25 V_{SHE} (a) and −0.20 V_{SHE} (b), showing the effect of the overstaturation potential on the growth of the passive layer. In (a), noncrystalline oxide islands (ox.) separated by the adsorbed OH layer (ad) cover partially the substrate. In (b), a crystalline oxide layer fully covers the substrate. Its atomic lattice shown in the inset corresponds to that of Cu_2O oriented (1 1 1). (From Kunze et al., *J. Phys. Chem. B* 102: 4263, 2001, ACS. With permission.)

The observed lattice of the oxide layer is hexagonal with a parameter of ~0.3 nm, consistent with the lattice of the Cu^+ planes in the (1 1 1)-oriented cuprite. It is thought that the surface of the oxide layer is hydroxylated in the aqueous solution and that the measured lattice corresponds to OH and/or OH^- groups forming a (1×1) layer on the Cu^+ planes of the (1 1 1)-oriented cuprite. In the potential range of Cu(II) oxidation, a crystalline $Cu_2O/CuO,Cu(OH)_2$ duplex film is formed (63). On Cu(0 0 1), the adsorption of OH groups also induces the reconstruction of the substrate to form a precursor to the growth of a (0 0 1)-oriented Cu_2O film in the Cu(I) oxidation range (61,65).

5.4.3.2 The Structure of Crystalline Passive Films

STM has also been applied to investigate *ex situ* and *in situ* the structure of the passive film formed on metals and alloys. Nickel in acid solution is the system for which the first atomic resolution images of passivated metal surfaces have been obtained (66,67), confirming by *ex situ* STM measurements the crystallinity of the passivated surface previously observed by RHEED (83). The *ex situ* results were later confirmed by *in situ* ECSTM measurements (41,68,69) and by GXRD measurements (84). In these studies, also performed on well-prepared single-crystal surfaces (Ni(1 1 1)), the metallic surface was cathodically reduced to remove the native oxide film formed during sample transfer in air to the electrochemical cell. Figure 5.16 illustrates the topography obtained prior to passivation with atomically flat oxide-free terraces separated by monoatomic steps oriented along well-defined directions of the metal lattice (69).

For the ECSTM experiments, it is necessary to step anodically the potential to the value of interest in the passive region rather to scan the potential. This is because extensive anodic dissolution occurs on nickel in acid solution prior to blocking by the passive film. Scanning the potential to the passive range increases anodic dissolution as shown by the intense active peak of the polarization curve shown in Figure 5.16, hence, producing a large amount of dissolved cations that precipitates on the sample surface because of the small cell volume and absence of agitation. The deposited corrosion products can form a relatively thick layer not characteristic of the passive film. Besides, its conductivity may be too low for STM imaging. The formation of such a film is likely to be the reason for the absence of stable tunneling conditions reported in the first ECSTM measurements of nickel passivated in sulfuric acid solution in which the potential was scanned rather than stepped to the passive region (14). The drawback of stepping the potential in the passive range to ensure a fast growth of the passive film and to minimize the surface modifications produced by anodic dissolution is that the growth of the passive film is too fast with respect to the time resolution of the imaging, thus precluding any precise time-resolved measurements of the growth process.

The ECSTM image of the passivated surface shown in Figure 5.16 evidences the crystallinity of the passive film formed on nickel in sulfuric acid solution (69). The passivated surface exhibits terraces and steps. The measured lattice is hexagonal with a parameter of 0.3 ± 0.02 nm assigned to NiO(1 1 1), NiO being the constituent of the inner part of the passive film. It must be pointed out that the (1 1 1) surface of NiO, which has the NaCl structure, is normally polar and unstable. It is, however, the surface that is obtained by passivation. The reason for this is that the surface is stabilized by adsorption of a monolayer of hydroxyl groups or by the presence of a monolayer of β-$Ni(OH)_2$ in parallel epitaxy with the NiO surface. Hydroxylated

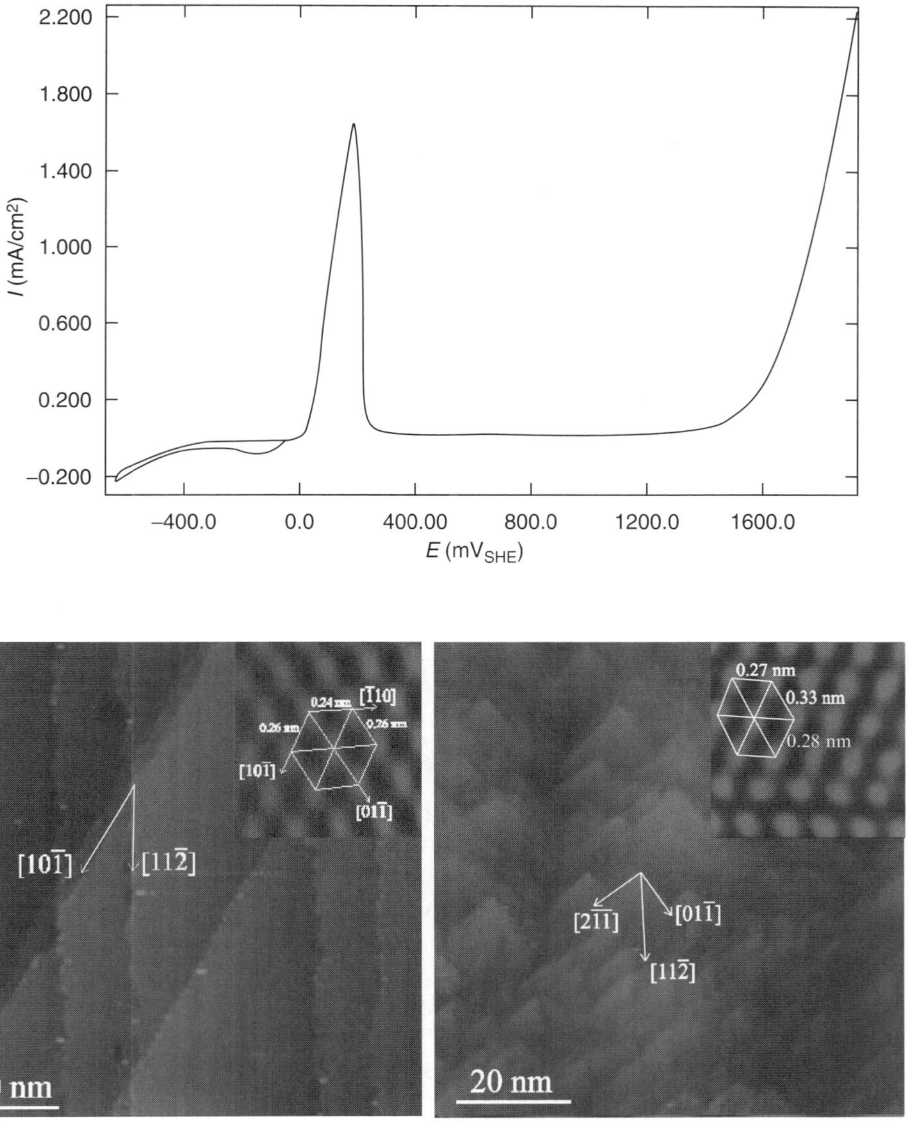

Figure 5.16 Polarization curve and ECSTM images of the Ni(1 1 1) surface recorded in 0.05 M $H_2SO_4 + 0.095$ M NaOH (pH ~2.9). Left: terrace topography of the metallic surface at −0.06 V/SHE below the passivation peak. Right: crystalline and facetted inner structure of the passivated surface at +0.95 V/SHE. The orientations of the substrate directions are marked. The insets show the atomic lattices. (From Zuili et al., *J. Electrochem. Soc.* 147: 1393, 2000, The Electrochemical Society. With permission.)

Ni(II) cations form the outer part of the passive film. These data show that the direction of growth of the oxide film is governed, at least, in part, by the minimization of the oxide surface energy by the hydroxyl/hydroxide groups. The presence of water is thus a major factor for the structural aspects of the growth mechanism. It has been suggested that the incorporation of water molecules causes the roughening and disordering of the hydroxide outer layer of the passive film (69).

The presence of terraces at the surface of the crystallized passive film is indicative of a slightly tilted epitaxy between the NiO(1 1 1) lattice and the Ni(1 1 1) substrate terraces. This tilt may partly relax the interfacial stress associated with the mismatch of 16% between the lattice parameters of the oxide and those of the metal substrate. This tilt has been confirmed by grazing x-ray measurements on Ni(1 1 1) (84). A similar surface faceting is observed for the Cu(I) oxide layer grown on copper for which the lattice misfit between Cu_2O and Cu lattices is similar (61,63).

On iron, the crystallinity of the passive film formed in borate buffer solution (pH 8.4) has also been evidenced by STM (71) and AFM (72). GXRD data have shown that the structure has the same sublattice of O anions as in γ-Fe_2O_3 and Fe_3O_4 but with a different site occupancy by cations (85).

The structure of passive films containing chromium oxide has also been investigated by STM on chromium (74,75) and on ferritic and austenitic stainless steels (77–81). A major finding of the STM data for these systems is that potential and aging under polarization are critical factors for the development of crystalline passive films with an atomic lattice consistent with α-Cr_2O_3(0 0 0 1) and for the dehydroxylation of the passive film as evidenced by combined XPS measurements. On chromium, the crystals of Cr_2O_3 can be very small at low potential where the oxide inner layer is not fully developed and where the passive film is highly hydrated and consists mainly of hydroxide (as shown by combined XPS measurements). This supports the view that the passive film on chromium can have a nanocrystalline structure. At high potential where the passive film is dehydrated and consists mostly of chromium oxide, larger crystals are formed. The nanocrystals and the larger crystals have a lattice consistent with α-Cr_2O_3(0 0 0 1). The basal plane of the oxide is parallel to Cr(1 1 0). A special feature of the passive film on chromium is that the oxide nanocrystals are cemented together by the chromium hydroxide outer layer. It has been suggested that the role of cement between grains played by chromium hydroxide, and of course the high stability of chromium oxide and chromium hydroxide makes this passive film extremely protective against corrosion.

Structural changes also occur during aging under anodic polarization of the passive layers formed on stainless steels in aqueous solution. The major modification is an increase of the crystallinity of the film and the coalescence of Cr_2O_3 islands in the inner oxide as observed on Fe–22Cr and Fe–18Cr–13Ni alloys studied over time periods of up to 65 h. This is illustrated by the images shown in Figure 5.17. For short polarization times ($\leq$2 h), the crystallinity of the passive films decreases with increasing Cr content of the alloy (77,78). The comparison of the rates of crystallization of Fe–22Cr(1 1 0) and Fe–18Cr–13Ni(1 0 0) revealed that the rate of crystallization is more rapid on the austenitic stainless steel than on the ferritic one (79,81). This is tentatively explained by a regulating effect of Ni on the supply of Cr on the alloy surface, a lower rate of Cr enrichment being in favor of a higher degree of crystallinity (86).

5.4.4 Passivity Breakdown

The high level of resolution achieved in STM measurements performed on well-defined single-crystal surfaces has been used (87) to show that the presence of crystalline defects at the surface of passive films can play a key role in the dissolution in the passive state. The detailed mechanisms must be elucidated to understand passivity breakdown and initiation of localized corrosion by pitting in the presence of chlorides. For these experiments, the Ni(1 1 1) surface was first passivated in a

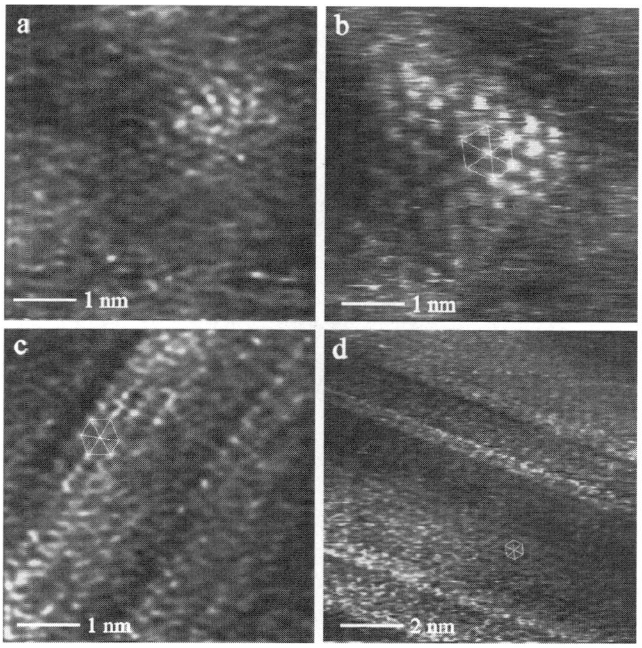

Figure 5.17 STM images of the Fe–22Cr(1 1 0) (left) and Fe–18Cr–13Ni(1 0 0) (right) surfaces recorded after passivation in 0.5 *M* H_2SO_4 at +0.5 V/SHE for 2 h (a), (b) and for 22 h (c), (d). The nearly hexagonal lattice is marked. The effect of aging under polarization is evidenced by the extension of the observed crystalline areas. (From Maurice et al., *J. Electrochem. Soc.* 143: 1182, 1996 for parts (a) and (c); Maurice et al., *J. Electrochem. Soc.* 145: 909, 1998 for parts (b) and (d). With permission.)

chloride-free sulfuric acid solution (pH 2.9) to produce the characteristic surface described above. Chlorides (0.05 *M* NaCl) were subsequently introduced in the electrolyte without changing the pH. Sequences of ECSTM images of a selected area were then recorded to follow the dynamics of the dissolution in the passive state. A sequence of three images is shown in Figure 5.18.

The observed dissolution is a two-dimensional (2D) process leading to a progressively decreasing size of the dissolving facets by a step flow process. The process is similar to that of the anodic dissolution at moderate potential with no pit forming. The 2D step flow process is dependent of the step orientation: the steps oriented along the closed-packed directions of the oxide lattice dissolve much less rapidly, due to the higher coordination of the surface atoms of the edges in these directions. At 0.05 V below the stable pitting potential, the process was found to be independent of the presence (0.05 *M*) or not of chloride in the electrolyte. The dissolution rate can be calculated from the decreasing areas of the dissolving facets. The average value of 0.44 nm^2/sec (~6 cations/sec) was obtained. Consistently with the absence of effect of chloride on the dissolution rate, the atomic lattice of the passive film formed in the absence of chloride was unmodified after the addition of chloride.

STM has also been applied to study passivity breakdown on well-defined single-crystal surfaces. For Ni(1 1 1) passivated in sulfuric acid (pH 2.9) and subsequently exposed to 0.05 *M* NaCl, *ex situ* images (88) have revealed a density of the order of 10^8 cm^{-2} of small metastable pits having a lateral dimension ranging from 40 to 150 nm and a depth of about 3 nm. An example is shown in Figure 5.19. The

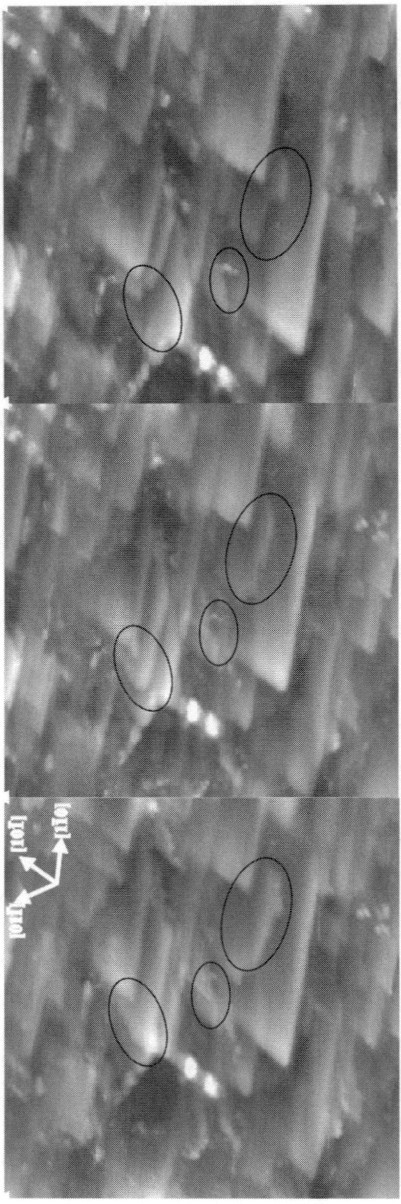

Figure 5.18 Sequence of ECSTM images (75 nm × 75 nm, 63 sec/image) showing the localized dissolution of the passivated Ni(1 1 1) surface at +0.85 V_{SHE} in 0.05 *M* H_2SO_4 + 0.095 *M* NaOH + 0.05 *M* NaCl (pH 2.9). The oxide crystallographic directions are indicated. The circles show the localized areas of dissolution. (From Maurice et al., *Surf. Interface Anal.* 34: 139, 2002. With permission.)

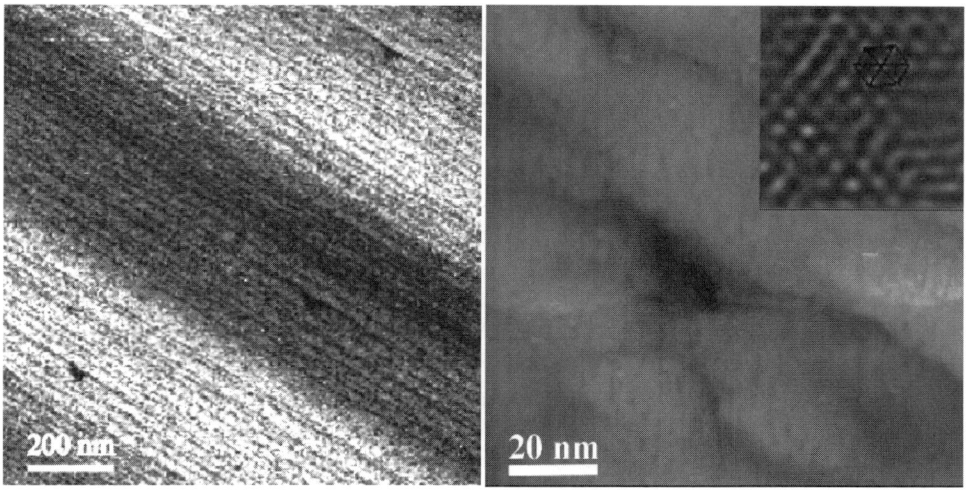

Figure 5.19 STM images of Ni(1 1 1) passivated in 0.05 *M* H_2SO_4 + 0.095 *M* NaOH (pH ~2.9) at 0.9 V/SHE for 30 min and subsequently exposed to 0.05 *M* NaCl for 90 min, showing metastable pits at two magnifications. The inset in the right images shows the crystalline lattice observed inside the metastable pit. (From Maurice et al., Critical Factors in Localized Corrosion III. The Electrochemical Society, 1999 (for left image); Maurice et al., *Electrochem. Solid-State Lett.* 4: 2001, The Electrochemical Society, 1999 (for right image). With permission.)

metastable character of the pit was deduced from the absence of a marked increase of current during the corrosion test. Assuming a two-electron reaction ($Ni(0) \rightarrow Ni(II) + 2e^-$), the amount of charge corresponding to the volume of a pit is of the order of 10^{-14} C. This is hardly detectable by electrochemical current transient measurements and shows the capability of the STM measurements to detect pits in their very initial stage of formation. The difficulty resides, however, in the localization of the pit with the instrument, given its limited scan range. This can be reasonably achieved provided the density of the pits is high enough.

These metastable pits were formed at and aligned along the substrate step edges (89), showing a direct relationship between the surface structure and preferential sites of breakdown of the passive film and pit initiation. The pits were also preferentially observed at the boundaries between the oxide crystals of the passive film (88), showing that grain boundaries of the passive film are preferential sites of passivity breakdown. The observed atomic structure inside these metastable pits (see Figure 5.19) is crystalline with lattice parameters similar to the lattice parameters measured on the passivated surface prior to exposure to chlorides, showing a similar structure of the surface repassivated in the presence of chlorides and confirming the metastable character of the pit.

5.4.5 Localized Corrosion

STM and AFM providing direct local topographic observations, their application to study localized corrosion processes such as pitting seems obvious. It is, however, limited by the necessary compromise between scanning range and resolution of the instrument. High-resolution measurements (mostly STM) are usually performed with scanning probes that can explore areas of (1 μm × 1 μm) or less. The scanning range may then not be large enough to localize pitting events that may occur with a low density, $<10^8$ cm^{-2}, and high-resolution measurements are thus limited to localized corrosion events that occur with a large enough density, such as the metastable pitting illustrated above. In contrast, lower-resolution (mostly AFM) measurements are often performed with scanning probes that can explore areas of $\sim$(100 μm × 100 μm). Combined with optical microscopic observations, they are well suited for the study of localized pitting events. However, the spatial resolution is in the nanometer range and decreases with increasing scan range. This does not allow the identification and investigation of the very initial stages of localized corrosion and low-resolution measurements are limited to more advanced stages of localized corrosion such as stable pit growth (pit propagation). The reported works on the pitting of passivated surfaces are low-resolution AFM studies performed *in situ* or *ex situ* on Al alloys or stainless steels (90–94). The use of ECAFM is preferred over ECSTM in these low-resolution studies since topographic images are obtained irrespective of the conductivity of the corrosion products.

Pitting corrosion at or near inclusions has been studied. The localization of pitting is done by previously localizing inclusions emerging at the sample surface. Localized corrosion near iron-rich particles was observed on the Al-6061-T6 alloy (91,93). The images in Figure 5.20 illustrate the development of a corrosion trench around an iron-rich inclusion. It was found that corrosion was initiated in regions where the ratio of inclusions to host matrix surface area was high. After formation of the trench, the dissolution of the matrix was uniform and proceeded radially from the inclusion to form a circular pit (93). While the extent of dissolution around the inclusion was at first independent of the inclusion size, it became distinctly dependent in later stages with larger cavities being formed around the larger inclusions. The formation of the trench was promoted by a -500 mV cathodic overpotential as shown in Figure 5.20. Based on comparative measurements in aerated and deaerated solutions, it has been proposed that the intermetallic inclusions act as cathodic site for the reduction of oxygen (93) causing the formation of an alkaline environment in the immediate vicinity of each inclusion, confirmed by local pH measurements. This local alkalization is thought to promote the local dissolution of the Al matrix.

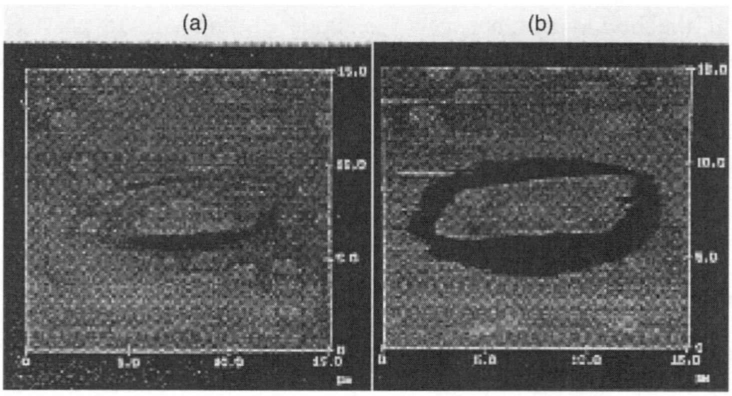

Figure 5.20 ECAFM images (15 μm × 15 μm) of corrosion around an Al_3Fe inclusion in Al-6061 alloy in 0.6 *M* NaCl at (a) open circuit potential, 3 h, and (b) -0.5 V cathodic overpotential, 3 min. (From Park et al., *J. Electrochem. Soc.* 146: 517, 1999, The Electrochemical Society. With permission.)

On stainless steels, ECAFM has also been used to observe the development at low resolution of pits near or at inclusions. On 304 SS, no pit initiation was observed around or at the MnS inclusions whose morphological changes were assigned to the buildup of corrosion products resulting from dissolution of the inclusion itself (91). As pits were initiated elsewhere on the surface, it was suggested that scanning with the AFM probe tip might affect the pit initiation. A more recent study (92) combining *ex situ* AFM with scanning electrochemical microscopy (SECM) and EDX analysis has shown that certain inclusions (sulfides) concentrate chloride, by electromigration, under a sulfur crust as a result of their dissolution evidenced by AFM. It has been proposed that the high local current density (evidenced by SECM), the electromigration of chloride, and the sulfur crust (evidenced by EDX) generate an, occluded extreme environment in which the stainless steel depassivates. In contrast, pitting was detected near titanium nitride or carbide inclusions in a Fe–17Cr ferritic alloy (90).

Apart from inclusions, localized corrosion of stainless steels can be initiated at defects such as dislocations or grain boundaries. Dislocations are similar to surface steps that have been observed to be preferential sites of initiation of pitting on pure nickel as described in Section 5.4.4. The dissolution of grain boundary carbide precipitates at early stage of intergranular corrosion (94) was recently reported for 304 SS in NaCl solution. Selective dissolution is another form of localized corrosion for duplex stainless steels, which have a two-phase microstructure. It was observed with low-resolution ECSTM for 2205-type duplex stainless steel having large separate volumes of ferrite and austenite in approximately equal fractions with a grain size of $\sim$5 μm (95). In 0.05 *M* H_2SO_4 + 1 *M* NaCl, no significant active dissolution was observed but some selective dissolution of the ferrite phase revealed the phase boundary region by forming a step for potentials higher than E_{corr} + 1 V. In 4 *M* H_2SO_4 + 1 *M* HCl, the selective dissolution of ferrite grains was observed at E_{corr} + 0.05 V. At E_{corr} + 0.15 V, the dissolution of the austenite phase was also observed by a stepwise process starting from the phase boundary region, previously revealed by selective dissolution of the ferrite grains.

5.4.6 Tip-Induced Localized Corrosion and Nanoengineering

Using the tip of an STM or AFM to locally induce the corrosion of the surface is one of the promising applications of the nanoprobes to corrosion science and to the nanostructuring of metallic surfaces. In ECSTM, the tip has been used to induce defects that can act as nucleation sites for metal deposition (96,97) and to generate metal clusters due to so-called jumps-to-contact (98–100). Recently, it has been shown that ECSTM can be used to induce metal dissolution spatially confined to a region underneath the tip at electrode potentials negative to the M/M^{z+} equilibrium potential, provided that the tip potential is positive to the equilibrium value (1 0 1). The result obtained on Cu(1 1 1) is shown in Figure 5.21. The tip-induced dissolution of the electrode was attributed to the role of counter-electrode played by the tip in the close vicinity of the electrode. The proposed mechanism involves electron transfer from the surface Cu atoms, not possible to the empty states of the electrode, to the empty states of the tip in regard. This requires the Fermi level of the tip to be lower than that of the electrode and below the M/M^{z+} equilibrium potential and the Fermi level of the electrode to be above the M/M^{z+} equilibrium potential. The phenomenon is inhibited for electrode potential far negative (<0.1 V) to the equilibrium potential. The process has been observed to lead to localized dissolution of the Cu surface in

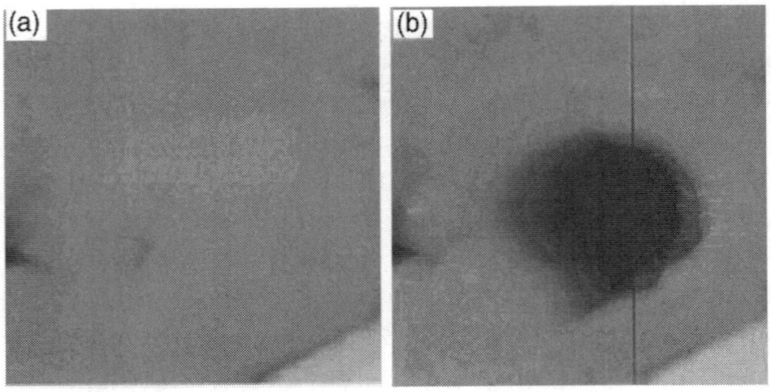

Figure 5.21 ECSTM images of a Cu(1 1 1) surface in a Cu^{2+} containing sulfuric acid solution at −0.05 V vs. Cu/Cu^{2+}, $E_{tip} = 0$ V vs. Cu/Cu^{2+}, $I = 2$ nA, prior to (a) and after (b) the tip has been scanned over a 25 nm^2 area at −0.05 V vs. Cu/Cu^{2+}, $E_{tip} = +0.02$ V vs. Cu/Cu^{2+}, $I = 2$ nA. (From Xie et al., *J. Electroanal. Chem.* 481: 177, 2000, Elsevier. With permission.)

acid electrolyte (1 0 1). It is expected to lead to localized oxide growth in neutral or basic solution in which copper oxides are stable. In both cases, it represents a potential means of nanostructuring of surfaces by locally induced corrosion.

ECAFM can also be used to induce localized corrosion. It was observed that the AFM tip–surface interaction can be used to locally increase the dissolution rate of Al (102,103). This requires a moderate corrosivity of the solution (0.1 *M* NaCl) since the effect was not observed in a noncorroding solution (demineralized water) and the thin film was corroding too fast to observe the effect of the scanning tip in 1 *M* NaCl. The enhanced dissolution of Al was attributed to the energy dissipation from frictional forces arising from scanning the tip over the surface (contact mode AFM). Local heating of the substrate may provide enough energy to overcome the activation energy barrier for the chemical reaction of the chloride ion with aluminum, thereby accelerating its dissolution in the aqueous solution.

ECAFM has also been used to perform high-resolution scratching experiments (104,105). It was found that the rastering with the AFM tip in contact mode in chloride solution resulted in accelerated dissolution of pure Al and alloy AA2024-T3. Rastering of pure Al resulted in enhanced uniform dissolution while on alloy AA2024-T3 it was dependent on the exact location in the microstructure that was scratched. The abrasion associated with AFM in contact mode resulted in the immediate dissolution of the Al–Cu–Mg particles, otherwise stable for hours. This behavior was attributed to the influence of unstable surface films that provided some protection when present but could be easily destabilized by the tip (104). In the presence of dichromate concentrations of 0.005 *M* or more, localized attack was not observed on pure Al perhaps owing to the formation of a harder film more resistant to AFM scratching. On AA2024-T3, the addition of 0.0005 *M* dichromate was sufficient to protect the Al–Cu–Mg particles from destabilization by the rastering tip. At higher applied tip forces, the Al–Cu–Mg particles could be destabilized by the rastering tip, but not the Al matrix (105).

These two examples illustrate the potential use of ECAFM to selectively corrode surfaces depending on their microstructure and to locally induce corrosion of the metallic substrate by destabilization of the protecting surface layers. Such effects could be positively used for nanostructuring metallic surfaces by controlled corrosion at the submicrometer scale.

5.5 CONCLUSION

In this chapter, the principles, instrumentation, and implementation to electrochemistry of STM and AFM have been presented. Both methods can provide 3D imaging of surfaces in the direct space during corrosion processes. ECSTM is better suited for high-resolution studies at the (sub)nanometer scale but restricted to conductive surfaces. Atomic scale studies require a high level of control to produce atomically smooth surfaces. ECAFM is well suited for (sub)micrometer scale studies of surfaces irrespective of their conductivity. Both methods can be used to locally induce corrosion, although the different mechanisms involved are complex. The selected examples of applications to corrosion science and engineering show the effect of the surface atomic structure on active dissolution and dealloying, correlate the structure of adsorbed inhibitors with the corrosion protection, show the role of the structure of adsorbed OH in the growth of passive films, emphasize the crystallinity of passive films on numerous substrates, the role of surface defects in passivity breakdown and the role of surface heterogeneities in pit initiation. The obtained data have greatly contributed to the advances in the understanding and control of corrosion of metals and alloys. In the future, ECSTM and ECAFM should also emerge as powerful tools, using corrosion in a constructive way to produce nanostructured surfaces.

REFERENCES

1. G. Binnig and H. Rohrer, *Helv. Phys. Acta* 55: 726, 1982.
2. G. Binnig, H. Rohrer, Ch. Gerber, and E. Weibel, *Appl. Phys. Lett.* 40: 178, 1982.
3. G. Binnig, H. Rohrer, Ch. Gerber, and E. Weibel, *Phys. Rev. Lett.* 49: 57, 1982.
4. G. Binnig, H. Rohrer, Ch. Gerber, and E. Weibel, *Phys. Rev. Lett.* 50: 120, 1983.
5. G. Rohrer. The preparation of tip and sample surfaces for STM experiments. In: D.A. Bonnell, ed. *Scanning Tunneling Microscopy and Spectroscopy. Theory, Techniques and Applications*. New York: VCH, 1993, pp. 155–187.
6. J. Tersoff and D.R. Hammann, *Phys. Rev. B* 31: 805, 1985.
7. J. Tersoff. Theory of scanning tunneling microscopy. In: D.A. Bonnell, ed. *Scanning Tunneling Microscopy and Spectroscopy. Theory, Techniques and Applications*. New York: VCH, 1993, pp. 31–50.
8. J. Bardeen, *Phys. Rev. Lett.* 6: 57, 1961.
9. R.J. Hamers. Methods of tunneling spectroscopy with the STM. In: D.A. Bonnell, ed. *Scanning Tunneling Microscopy and Spectroscopy. Theory, Techniques and Applications*. New York: VCH, 1993, pp. 51–103.
10. R. Sonnenfeld and P.K. Hansma, *Science* 232: 211, 1986.
11. B. Drake, R. Sonnenfeld, J. Schneir, and P.K. Hansma, *Surf. Sci.* 181: 92, 1987.
12. P. Lustenberger, H. Rohrer, R. Christoph, and H. Siegenthaler, *J. Electroanal. Chem.* 243: 225, 1988.
13. J. Wiechers, T. Twomey, D.M. Kolb, and R.J. Behm, *J. Electranal. Chem.* 248: 451, 1988.
14. O. Lev, F.-R. Fan, and A.J. Bard, *J. Electrochem. Soc.* 135: 783, 1988.
15. F.-R. Fan and A.J. Bard, *J. Electrochem. Soc.* 136: 166, 1989.
16. A. Vaught, T.W. Jing, and S.M. Lindsay, *Chem. Phys. Lett.* 236: 306, 1995.
17. G.E. Engelmann, J.C. Ziegler, and D.M. Kolb, *Surf. Sci.* 401: L420, 1998.
18. M. Binggeli, D. Carnal, R. Nyffenegger, H. Siegenthaler, R. Christoph, and H. Rohrer, *J. Vac. Sci. Technol. B* 9: 1985, 1991.
19. K. Itaya and E. Tomita, *Surf. Sci.* 201: L507, 1988.

20. A.A. Gewirth, D.H. Craston, and A.J. Bard, *J. Electranal. Chem.* 261: 477, 1989.
21. D.J. Trevor, C.E.D. Chisey, and D.N. Loiacono, *Phys. Rev. Lett.* 62: 929, 1989.
22. L.A. Nagahara, T. Thundat, and S.M. Lindasy, *Rev. Sci. Instrum.* 60: 3128, 1989.
23. S.-L. Yau, C.M. Vitus, and B.C. Schardt, *J. Am. Chem. Soc.* 112: 3577, 1990.
24. C.E. Bach, R.J. Nichols, W. Beckmann, H. Meyer, A. Schulte, J.O. Besenhard, and P.D. Jannakoudakis, *J. Electrochem. Soc.* 140: 1281, 1993.
25. A. Bard and F.-R.F. Fan, Applications in electrochemistry. In: DA Bonnell, ed. *Scanning Tunneling Microscopy and Spectroscopy. Theory, Techniques and Applications*. New York: VCH, 1993, pp. 287–333.
26. H. Siegenthaler. STM in electrochemistry. In: R. Wiesendanger and H.J. Guntherodt, eds. *Scanning Tunneling Microscopy II: Further Applications and Related Scanning Techniques*. Berlin: Springer, 1995, pp. 7–50.
27. Z.-X. Xie and D.M. Kolb, *J. Electroanal. Chem.* 481: 177, 2000.
28. G. Binnig, C.F. Quate, and Ch. Gerber, *Phys. Rev. Lett.* 56: 930, 1986.
29. N.A. Burnham and R.J. Colton, Force microscopy. In: D.A. Bonnell, ed. *Scanning Tunneling Microscopy and Spectroscopy. Theory, Techniques and Applications*. New York: VCH, 1993, pp. 191–249.
30. R. Wiesendanger and H.J. Guntherodt, eds. *Scanning Tunneling Microscopy II: Further Applications and Related Scanning Techniques*. Berlin: Springer, 1995.
31. R. Wiesendanger, H.J. Guntherodt, eds. *Scanning Tunneling Microscopy III: Theory and Related Scanning Probes Methods*. Berlin: Springer, 1996.
32. R.J. Hamers, *J. Phys. Chem.* 100: 13103, 1996.
33. A.L. Weisenhorn and P.K. Hansma, *Appl. Phys. Lett.* 54: 2651, 1989.
34. D.W. Suggs and A.J. Bard, *J. Am. Chem. Soc.* 116: 10725, 1994.
35. D.W. Suggs and A.J. Bard, *J. Phys. Chem.* 99: 8349, 1995.
36. T.P. Moffat, *Mater. Res. Soc. Symp. Proc.* 451: 75, 1997.
37. M.R. Vogt, A. Lachenwitzer, O.M. Magnussen, and R.J. Behm, *Surf. Sci.* 399: 49, 1998.
38. O.M. Magnussen, M.R. Vogt, J. Scherer, and R.J. Behm. *Appl. Phys. A* 66: S447, 1998.
39. P. Broekmann, M. Anastasescu, A. Spaenig, W. Lisowski, and K. Wandelt, *J. Electroanal. Chem.* 500: 241, 2001.
40. O.M. Magnussen, L. Zitzler, B. Gleich, M.R. Vogt, and R.J. Behm, *Electrochim. Acta* 46: 3725, 2001.
41. T. Suzuki, T. Yamada, and K. Itaya, *J. Phys. Chem.* 100: 8954, 1996.
42. S. Ando, T. Suzuki, and K. Itaya, *J. Electroanal. Chem.* 412: 139, 1996.
43. M. Dietterle, T. Will, and D.M. Kolb, *Surf. Sci.* 327: L495, 1995.
44. T. Teshima, K. Ogaki, and K. Itaya, *J. Phys. Chem. B* 101: 2046, 1997.
45. S. Ando, T. Suzuki, and K. Itaya, *J. Electroanal. Chem.* 431: 277, 1997.
46. K. Sashikata, Y. Matsui, K. Itaya, and M.P. Soriaga, *J. Phys. Chem.* 100: 20027, 1996.
47. K. Itaya, In: A. Wieckowski, ed. *Interfacial Electrochemistry — Theory, Experiments and Applications*. New York: Marcel Dekker, 1999, p. 187.
48. I.C. Oppenheim, D.J. Trevor, C.E.D. Chidsey, P.L. Trevor, and K. Sieradzki, *Science* 254: 687, 1991.
49. T.P. Moffat, F.-R. Fan, and A.J. Bard, *J. Electrochem. Soc.* 138: 3224, 1991.
50. S.J. Chen, F. Sanz, D.F. Ogletree, V.M. Hallmark, T.M. Devine, and M. Salmeron, *Surf. Sci.* 292: 289, 1993.
51. M. Stratmann and M. Rohwerder, *Nature* 410: 420, 2001.
52. J. Erlebacher, M.J. Aziz, A. Karma, N. Dimitrov, and K. Sieradzki, *Nature* 410: 450, 2001.
53. M.R. Vogt, W. Polewska, O.M. Magnussen, and R.J. Behm, *J. Electrochem. Soc.* 144: L113, 1997.
54. M.R. Vogt, R.J. Nichols, O.M. Magnussen, and R.J. Behm, *J. Phys. Chem. B* 102: 5859, 1998.
55. M. Sugimasa, L.-J. Wan, J. Inukai, and K. Itaya, *J. Electrochem. Soc.* 149: E367, 2002.

56. W. Polewska, M.R. Vogt, O.M. Magnussen, and R.J. Behm, *J. Phys. Chem. B* 103: 10440, 1999.
57. J. Scherer, M.R. Vogt, O.M. Magnussen, and R.J. Behm, *Langmuir* 13: 7045, 1997.
58. C.W. Yan, H.C. Lin, and C.N. Cao, *Electrochim. Acta* 45: 2815, 2000.
59. E. Rocca, G. Bertrand, C. Rapin, and J.C. Labrune, *J. Electroanal. Chem.* 503: 103, 2001.
60. J. Telegdi, M.M. Shaglouf, A. Shaban, F.H. Karman, I. Bertoti, M. Mohai, and E. Kalman, *Electrochim. Acta* 46: 3791, 2001.
61. N. Ikemiya, T. Kubo, and S. Hara, *Surf. Sci.* 323: 81, 1995.
62. V. Maurice, H.-H. Strehblow, and P. Marcus, *Surf. Sci.* 458: 185, 2000.
63. J. Kunze, V. Maurice, L.H. Klein, H.-H. Strehblow, and P. Marcus, *J. Phys. Chem. B* 105: 4263, 2001.
64. J. Kunze, V. Maurice, L.H. Klein, H.-H. Strehblow, and P. Marcus, *Electrochim. Acta* 48: 1157, 2003.
65. J. Kunze, V. Maurice, L.H. Klein, H.-H. Strehblow, and P. Marcus, *J. Electroanal. Chem.* 554–555: 113, 2003.
66. V. Maurice, H. Talah, and P. Marcus, *Surf. Sci.* 284: L431, 1993.
67. V. Maurice, H. Talah, and P. Marcus, *Surf. Sci.* 304: 98, 1994.
68. S.-L. Yau, F.-R. Fan, T.P. Moffat, and A.J. Bard, *J. Phys. Chem.* 98: 5493, 1994.
69. D. Zuili, V. Maurice, and P. Marcus, *J. Electrochem. Soc.* 147: 1393, 2000.
70. R.C. Bhardwaj, A. Gonzalez-Martin, J.O.'M. Bockris, *J. Electroanal. Chem.* 307: 195, 1991.
71. M.P. Ryan, R.C. Newman, and G.E. Thompson, *J. Electrochem. Soc.* 142: L177, 1995.
72. J. Li and D.J. Meier, *J. Electroanal. Chem.* 454: 53, 1998.
73. I. Diez-Pérez, P. Gorostiza, F. Sanz, and C. Müller, *J. Electrochem. Soc.* 148: B307, 2001.
74. V. Maurice, W. Yang, and P. Marcus, *J. Electrochem. Soc.* 141: 3016, 1994.
75. D. Zuili, V. Maurice, and P. Marcus, *J. Phys. Chem. B* 103: 7896, 1999.
76. R.C. Bhardwaj, A. Gonzalez-Martin, and J.O.'M. Bockris, *J. Electrochem. Soc.* 138: 1901, 1991.
77. M.P. Ryan, R.C. Newman, and G.E. Thompson, *Philos. Mag. B* 70: 241, 1994.
78. M.P. Ryan, R.C. Newman, and G.E. Thompson, *J. Electrochem. Soc.* 141: L164, 1994.
79. V. Maurice, W. Yang, and P. Marcus, *J. Electrochem. Soc.* 143: 1182, 1996.
80. H. Nanjo, R.C. Newman, and N. Sanada, *Appl. Surf. Sci.* 121: 253, 1997.
81. V. Maurice, W. Yang, and P. Marcus, *J. Electrochem. Soc.* 145: 909, 1998.
82. F. Besenbacher and J.K. Norskov, *Progr. Surf. Sci.* 44: 5, 1993.
83. J. Oudar and P. Marcus, *Appl. Surf. Sci.* 3: 48, 1979.
84. O.M. Magnussen, J. Scherer, B.M. Ocko, and R.J. Behm, *J. Phys. Chem. B* 104: 1222, 2000.
85. M.F. Toney, A.J. Davenport, L.J. Oblonsky, M.P. Ryan, and C.M. Vitus, *Phys. Rev. Lett.* 79: 4282, 1997.
86. P. Marcus, V. Maurice, Passivity and its breakdown, Joint ECS/ISE Meeting, Paris, France, 1997, eds. P. Natishan, H.S. Isaacs, M. Janik-Czachor, V.A. Macagno, P. Marcus, and M. Seo, In: The Electrochemical Society Proceedings Series, PV 97–26, Pennington, NJ, 1998, pp. 254–265.
87. V. Maurice, L. Klein, and P. Marcus, *Surf. Interf. Anal.* 34: 139, 2002.
88. V. Maurice, L. Klein, and P. Marcus, *Electrochem. Solid-State Lett.* 4: B1, 2001.
89. V Maurice, V Inard, and P Marcus, In: P.M. Natishan, R.G. Kelly, G.S. Frankel, and R.C. Newman, eds. *Critical Factors in Localized Corrosion III.* The Electrochemical Society Proceedings Series, PV 98–17, Pennington, NJ, 1999, pp. 552–562.
90. G. Gugler, J.D. Neuvecelle, P. Mettraux, E. Rosset, and D. Landolt, In: P. Marcus, B. Baroux, and M. Keddam, eds. *Modifications of Passive Films.* EFC Publications No. 12. The Institute of Materials, 1994, p. 274.
91. R.M. Rynders, C.-H. Paik, R. Ke, and R.C. Alkire, *J. Electrochem. Soc.* 141: 1439, 1994.

92. D.E. Williams, T.F. Mohiuddin, and Y.Y. Zhu, *J. Electrochem. Soc.* 145: 2664, 1998.
93. J.O. Park, C.-H. Paik, Y.H. Huang, R.C. Alkire. *J. Electrochem. Soc.* 146, 517, 1999.
94. R.E. Williford, C.F. Windisch Jr, and R.H. Jones, *Mater. Sci. Eng. A* 288: 54, 2000.
95. M. Femenia, J. Pan, C. Leygraf, and P. Luukkonen, *Corros. Sci.* 43: 1939, 2001.
96. W. Li, J.A. Virtanen, and R.M. Penner, *Appl. Phys. Lett.* 60: 1181, 1992.
97. W. Li, J.A. Virtanen, and R.M. Penner, *J. Phys. Chem.* 96: 6529, 1992.
98. R. Ullmann, T. Will, and D.M. Kolb, *Chem. Phys. Lett.* 209: 238, 1993.
99. D.M. Kolb, R. Ullmann, and T. Will, *Science* 275: 1097, 1997.
100. D.M. Kolb, R. Ullmann, and J.C. Ziegler, *Electrochim. Acta* 43: 2751, 1998.
101. Z.-X. Xie and D.M. Kolb, *J. Electroanal. Chem.* 481: 177, 2000.
102. L. Chen and D. Guay, *J. Electrochem. Soc.* 141: L43, 1994.
103. L. Roue, L. Chen, and D. Guay, *Langmuir* 12: 5818, 1996.
104. P. Schmutz and G.S. Frankel, *J. Electrochem. Soc.* 145: 2295, 1998.
105. P. Schmutz and G.S. Frankel, *J. Electrochem. Soc.* 146: 4461, 1999.

6

Synchrotron Methods for Corrosion Research

Dirk Lützenkirchen-Hecht
Institut für Materialwissenschaften und Fachbereich Physik, Bergische Universität Wuppertal, Wuppertal, Germany

Hans-Henning Strehblow
Institut für Physikalische Chemie, Heinrich-Heine-Universität, Düsseldorf, Germany

Contents

6.1 INTRODUCTION

Corrosion is an interdisciplinary field and the complicated environmental conditions and the interaction of several factors for corroding systems require as much information as possible to understand its leading mechanisms. Usually various surface methods are applied to learn about the chemical composition of surfaces and thin surface films that decisively influence the kinetics of corrosion processes. The application of photoelectron spectroscopy and ion spectroscopies together with *in situ* techniques like infrared (IR) spectroscopy provides a detailed insight into the chemistry of metal surfaces. However, the atomic structure of most surfaces and surface layers still remains unknown. An interesting activity started within the last 10 to 15 years when high-resolution scanning methods like scanning tunneling microscopy (STM) and scanning force microscopy (SFM) became available. These methods may be used as *in situ* techniques to study surfaces even with atomic resolution. Adsorbates and thin films have a pronounced influence on the corrosion properties of metals and their detailed structure plays a decisive role. Therefore, their structure is of much interest. However, the mentioned scanning methods are surface methods and thus sample the outmost atomic layer. Only in some cases does STM yield structural information on an outer and an inner part of a passive layer through a variation of the conditions for the tunnel current. Furthermore, the results of these tunnel methods refer to a small surface area only. The high lateral resolution permits the study of an extremely small fraction of the total surface. For these reasons laterally integrating methods are complimentary alternatives.

Modern synchrotron methods have been applied for the study of the structure of electrode surfaces and surface films. In this sense, the structure of adsorbates of metal ions on foreign substrates, the so-called under-potential deposits, has been studied. A more interesting problem for corrosion is the structure of anodic layers as passive films especially when grown on their own metal substrate. There are usually two kinds of methods that may be applied: x-ray diffraction (XRD), which requires a long-range order of the system that is studied, and x-ray absorption spectroscopy (XAS), which samples the near-range order of a central absorbing atom that is surrounded by more or less ordered coordination shells. XAS is thus a method that needs no well-ordered crystalline structure, which is the case for many thin anodic films that may not be examined by XRD. XAS works even for species dissolved within the electrolyte. At least the first coordination shell may be measured and thus one may sample its contributions to the near-range order. In principle, XAS and XRD are complimentary methods. XAS measures the near-range order like the radius of coordination shells, their coordination numbers, and their order or disorder. XRD samples the long-range order and thus gives the atomic order including the distances of a crystalline structure. A sufficiently high energy of x-rays for diffraction studies usually helps to apply the method in the presence of electrolyte and thus under potential control. This is also possible for XAS, but the energy of the absorption edge should be at least in the range of 8 keV in order to prevent total absorption of the beam within the solution. This can be achieved for most heavy metals like Cu, Ag, Mo, etc. With some restriction it may be still applied for a metal like Ni. An interesting feature is the possibility to see a structure in the vicinity of different kinds of absorber atoms including anions like bromide, which might be present in the structure of interest. Thus, the x-ray absorption fine structure (XAFS) from different kinds of atoms provides complementary results.

The investigation of the position and form of the absorption edge — the so-called near-edge x-ray absorption spectroscopy (NEXAS) — provides information on the chemical nature of the species like oxidation state and binding situation. Although these results may be obtained much more easily by XPS in the home laboratory it is nevertheless important as they provide information on an electrode surface for *in situ* conditions within the electrolyte under potential control. Furthermore, the size of the absorption edge is a measure for the amount of material and its change has been taken for the quantitative determination of its loss by dissolution in the electrolyte. In addition, since XAS is also applicable to liquid systems, the species dissolved in the electrolyte can be investigated.

This short introduction shows that synchrotron methods are valuable tools for the study of corrosion phenomena and especially for structure analysis. The only disadvantage is the requirement of intense x-ray beams, which are only available at a few synchrotron sources and require much experimental effort amid restricted beam time. These methods should be seen therefore as valuable tools for basic corrosion research and to a lesser extent as routine methods to solve actual corrosion problems. One should have applied already the usual methods like electrochemistry and surface analysis in the ultrahigh vacuum to design synchrotron experiments on the basis of a sound knowledge on the system of interest in order to be as successful as possible.

6.2 METHODS AND EQUIPMENT — WHY SYNCHROTRON RADIATION

In our review, we will focus on x-ray methods; accordingly, we will therefore put our major attention toward x-ray sources. Sealed x-ray tubes are the most widely used x-ray sources in the laboratory. The Bremsstrahlung continuum and the characteristic x-ray emission lines are excited by an electron beam that is accelerated toward a metal target within a vacuum tube. Depending on the details of the x-ray tube such as the type and the material of the anode, the focus size, and the cooling system, the maximum electrical input power of an x-ray tube is limited to about 2 to 25 kW. However, typically only fractions of a percent of this electric input power are transformed into x-ray radiation leaving the tube through x-ray windows (typically beryllium), so that the resulting x-ray intensities that impinge on the sample are relatively weak. This is a serious drawback for *in situ* investigations of electrodes and related corrosion processes, since the parasitic absorption by the electrolyte further reduces the intensity that is available for the experiments (see below). Having in mind that at least one x-ray photon per second should reach the detector even in the photon counting mode using sophisticated detector equipment, and assuming, e.g., a realistic rotating anode x-ray tube that provides about 10^6 monochromatic x-ray photons per second and mm^2, a Bragg reflectivity of a real sample (such as a passive film on a metal substrate) of about 10^{-3}, and a transmission of the electrochemical cell of about 10%, we end up with a single photon per second only. This means that we have to consider a very long time for real *in situ* experiments under electrochemical conditions if a sufficient signal-to-noise ratio, e.g., in a diffraction experiment, is desired. Nevertheless, *in situ* experiments with laboratory equipment have been reported in the literature (see, e.g., Refs. (1, 2)).

Synchrotron radiation (SR) facilities provide the most intense x-rays. Extended review articles covering the properties of SR and their application for numerous different scientific problems have appeared in the past (see, e.g., Refs. (3–6)). In this chapter, we will focus on some relevant points only. SR is generated when charged high-energy particles — usually electrons or positrons with kinetic energies of about 2 to 8 GeV — are moving on a curved path that is induced by a magnetic structure. In earlier times, SR was produced in an unwanted manner in particle storage rings. But with the recognition of its useful properties, especially with regard to materials science investigations, dedicated synchrotron laboratories were built all around the world. A list of built and planned sources can be found, e.g., in Ref. (7) or in the Internet (e.g., Ref. (8)). The new third-generation sources (European Synchrotron Radiation Facility (ESRF), Grenoble, France, Advanced Photon Source (APS), Argonne, U.S.A., and Spring 8, Harima, Japan) were constructed in order to optimize beam characteristics such as intensity, collimation, brilliance, and coherence. The increased availability of intense SR sources has its benefit for *in situ* investigations of corrosion processes, such as localized corrosion or passivation.

One big advantage of SR is that its continuous emission spectrum ranging from the IR to hard x-rays is well defined and can be calculated with the knowledge of the curvature and the energy of the stored electrons or positrons (5,9). The quality of the emitted radiation can be characterized by its brightness and brilliance (5), the latter parameter is defined by the number of photons that is emitted in one second from a source area of 1 mm^2 into a cone defined by 1 $mrad^2$ normalized to a spectral band width of 0.1%. In Figure 6.1, the brilliance of some synchrotron sources is illustrated as a function of the photon energy. For comparison, the radiation emitted by a rotating anode with a point focus is about 10^{10} photons/(s mm^2 $mrad^2$ 0.1%), which is out of the lower scale of Figure 6.1 by two orders of magnitude. In contrast to the discrete line spectrum, which is provided by the emission lines of an x-ray anode, the spectrum emitted by a bending magnet covers a large range of energy from the ultraviolet to hard x-rays. Thus, the energy (wavelength) for the experiments can be selected freely from this continuum and allows an optimization depending on the actual sample or problem under investigation. For the application of extended x-ray absorption spectroscopy (EXAFS, see below), the use of SR is mandatory, since the photon energy is scanned over about 1 keV in these experiments. Even for a simple transmission mode experiment the Bremstrahlung-continuum of x-ray tubes is too weak in intensity.

While second-generation synchrotron sources make use of the radiation emitted by bending magnets, sophisticated periodic magnet-structures (insertion devices, wigglers, and undulators) are installed at third-generation synchrotron sources. The periodic field inside these devices forces the passing electrons and positrons on a sinusoidal path. Depending on the number of antiparallel arranged magnets, the magnetic field strength, and the magnetic gap, the individual emission processes are without any phase relation (wiggler: strong field, small number of periods) or in phase (undulator: weak field, large number of undulator periods), improving the spectral brightness by at least two orders of magnitude compared to a bending magnet station (see Figure 6.1). In other words, SR is able to provide highly intensive monochromatic radiation with extremely small lateral and horizontal divergences. These properties are an indispensable prerequisite, e.g., for surface x-ray scattering or diffraction experiments (see, e.g., Refs. (10, 11)).

When considering the unique properties of SR, its time structure also has to be mentioned. The charge carriers in a storage ring are accumulated in small bunches, the size of which is measured in terms of the time that is necessary to pass the focal point of the experiment. While the length of an individual bunch is of the order of several picoseconds, the distance between two bunches amounts to several

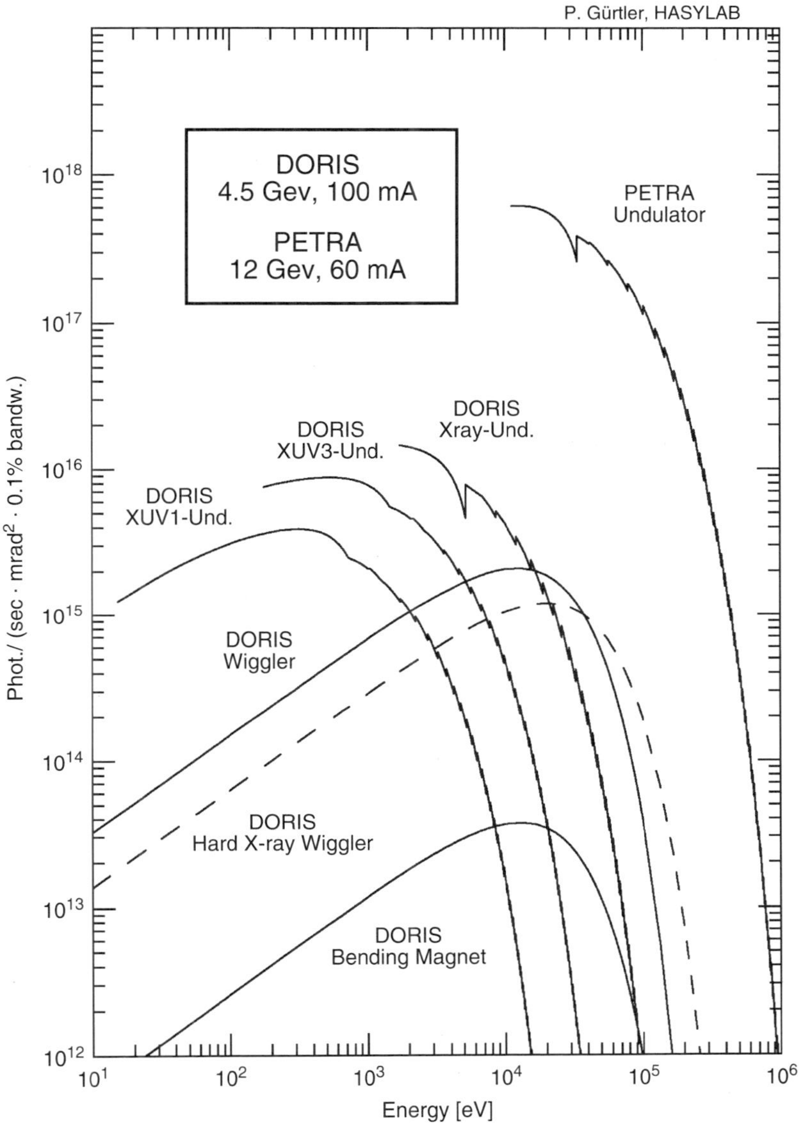

Figure 6.1 Average brilliance for different radiation sources at HASYLAB (DESY, Germany), BESSY (Berlin, Germany), and the ESRF (Grenoble, France). For comparison, the brilliance expected for future sources (Free electron lasers) is also depicted (courtesy of P. Gürtler, HASYLAB).

nanoseconds — therefore SR enables time-resolved experiments using this time structure of the storage ring.

Furthermore, due to the fact that the accelerated particles are kept in the orbit plane, the emitted synchrotron light is generally linearly polarized in this plane. This is important with regard to investigations of highly anisotropic systems such as, e.g., adsorbates on single-crystal surfaces. In addition, slightly below and above the orbit plane, SR light is circularly polarized, which gives unique experimental opportunities for the investigation of magnetic materials (see, e.g., Refs. (12, 13)). Newly constructed insertion devices provide circularly polarized light with extremely high brilliance even in the orbit plane (14, 15).

6.2.1 Synchrotron Radiation Sources: Beamlines and their Specifications

As we will focus on x-ray methods in this chapter, we will put more attention toward x-ray beamlines and beamline optics. A detailed review about beamline design and instrumentation can be found, e.g., in Refs. (4, 16). A typical design of an x-ray beamline is presented in Figure 6.2. Three different sections can be seen: the first section contains the radiation source—a bending magnet, a wiggler, or an undulator—beam position monitors, a slit system, and the main beam shutter, which are all located inside the concrete shielding of the storage ring. While the slit system defines the size of the beam, the beam position monitors are essential tools for the alignment of all the beamline components and a possible active feedback of the beamline to the electrons (or positrons) in the storage ring. If the main beam shutter is opened, the radiation from the source enters the second beamline section, which mainly consists of a monochromator and mirrors. The double-crystal monochromator selects a discrete energy from the continuous radiation emitted from the x-ray source by means of two Bragg reflections. Depending on the photon energy that is required for the experiment, several crystal materials and orientations with different *d*-spacings are used. Typical materials are Si(1 1 1) ($d = 3.136$ Å) for photon energies between 2.4 and about 15 keV, Si(3 1 1) ($d = 1.638$ Å) for energies from about 10 to 30 keV, and Si(5 1 1) ($d = 1.045$ Å) for energies above ca. 25 keV. For photon energies lower than about 2 keV, exotic monochromator crystal materials such as YB_{66} ($d = 5.88$ Å, (4 0 0)-reflection (17)) are used. However, the spectral range below about 5 keV can hardly be used for *in situ* corrosion experiments because of the extremely strong parasitic absorption of the electrolyte (see below). According to Bragg's law:

$$n\lambda = 2d \sin \Theta \tag{6.1}$$

where λ is the x-ray wavelength, d the lattice spacing of the monochromator crystal, and Θ the scattering angle between the impinging radiation and the lattice planes; not only the fundamental wave ($n = 1$) but also higher harmonics ($n > 1$) are transmitted

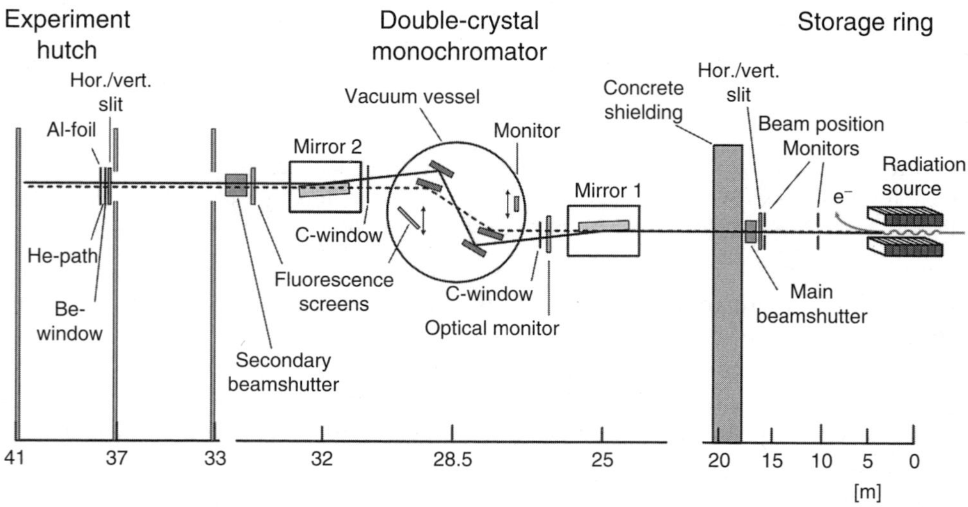

Figure 6.2 Schematic drawing of a typical x-ray beamline at a synchrotron radiation source, showing the unmirrored (- - - -) and the mirrored (——) beam. The main parts of the beamline and their functions are described in the text in more detail

by the monochromator. These higher harmonics may disturb the actual measurement, for example, additional Bragg peaks may occur in a diffraction experiment, and the absorption coefficients determined in an XAS study will be erroneous, so that a suppression of higher harmonics is highly desired. This may be achieved using different techniques such as (a) detuning of the monochromator crystals (18) or (b) the use of a mirrored beam. Detuning makes use of the fact that the Bragg angle width of higher harmonics is generally much smaller compared to the fundamental wave (19), so that a slight decline of both crystals with respect to each other effectively suppresses the transmitted intensity of higher harmonics.

The x-ray reflectivity of any surface generally decreases with the photon energy, thus a mirror can also be used for the harmonic rejection if the critical energy is smaller than the energy of the corresponding harmonic wave (16, 20). This way, an energy resolution of typically 1 eV at around 10 keV photon energy can be achieved.

In general, the heat load from the SR source affects the crystals of the monochromator, inducing a variety of surface slope errors, such as thermal bumps, thermal bowing, as well as lattice constant variations on monochromator crystals, which has severe implications on the monochromator layout, especially if a wiggler or an undulator at a third-generation source is used. While simple water cooling of the first crystal is sufficient for a monochromator at a second-generation source, more elaborated designs are required for monochromators at high heat insertion device beamlines. Either highly efficient cooling systems using, e.g., microchannell systems (21), direct water jet cooling (22), liquid gallium (23), cryogenic cooling (24), and mechanical systems for the correction of the thermally induced deformations (25) have been published. It should be mentioned that the heat load on the monochromator is reduced if a mirror is used in front of the monochromator. Furthermore, the mirrors can also be used for the focusing of the x-ray beam (26–29): depending on the surface of the mirror (flat, spherical, cylindrical) and its bending radius, the impinging radiation is either divergent or focused in one or two directions. Using an undulator as source, the point focus can be of a size of a few tenths of micrometers only, so that spectromicroscopy or microspectroscopic investigations are feasible, even under electrochemical conditions.

One major point that has to be mentioned with regard to SR measurements is that the experiments have to be located in a hermetically shielded hutch, which cannot be accessed when the beamshutter is opened for safety reasons. This implies that the electrochemical equipment has to be remote controlled from outside the hutch.

6.2.2 Detector Equipment

Ionization chambers are used for many purposes; a schematic representation of one such chamber is given in Figure 6.3. Basically, the charges (electrons and positively charged ions) created by the absorption of x-ray radiation are separated by an electric field between two metal plates, and the measured current is proportional to the intensity of x-rays. The main advantage of an ionization chamber is its easy and robust handling; however, for each photon energy a certain gas or a gas mixture has to be chosen. In addition, it is not possible to discriminate photons with different energies, such as elastically scattered and fluorescence photons in a diffraction experiment. While they are well suited for the measurement of very intense or high-energy x-ray beams — this is the reason why they are used, e.g., for XAS experiments and the measurement of the impinging x-ray intensity in any type of x-ray experiment — they

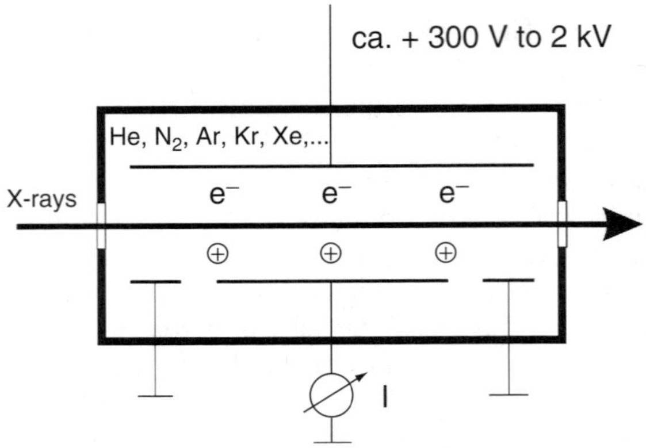

Figure 6.3 Schematic drawing of a gas-filled ionization chamber with two metal electrodes with a potential difference of several hundred volts and x-ray windows.

are not so good for detecting very small intensities. In this case, photon counting techniques have to be applied. These comprise scintillation counters such as Tl-doped NaI. Here, each absorbed photon causes a light pulse in the scintillator material, the intensity of which is proportional to the energy of the impinging photon. The light pulse is converted into a small current pulse via an optical contact and a photomultiplier. The pulses are sampled by simple counting electronics in order to obtain a measure of the x-ray intensity. Since the pulse height is proportional to the energy of the counted photon, these detectors also offer the opportunity of energy discrimination using a pulse height analysis (multichannell analyzer). This type of detector is typically used for x-ray fluorescence (XRF), x-ray reflectivity, or x-ray scattering measurements. The main disadvantage of a scintillation counter is the limited maximum count rate, which is mainly given by the time that is needed for the pulse formation process and its analysis. In order to avoid saturation of a scintillation counter, an absorber material with known attenuation has to be placed in front of the detector.

Based on semiconducting materials such as silicon, germanium, HgI_2, or GaAs, a large number of solid-state detectors (SSDs) with different operating principles has been developed in the last years. The basic idea of operating such a detector is to use it as a reversely biased diode (e.g., Ref. (30)). The electron–hole pairs created by the absorption of an x-ray photon, the number of which is proportional to the x-ray energy, are separated by the electric field in the device and the resulting current pulses are sorted with respect to their height and counted accordingly. A Si-pin-diode, where an intrinsic conduction layer of several micrometers thickness separates p- and n-conducting materials, is the simplest type of an SSD. In an Si(Li) detector, the pin-junction is made by diffusion of Li into p-Si, which means that this type of detector has to be cooled during operation, either by liquid nitrogen or a Peltier element, to avoid the damage or alteration of the structure. In addition, the noise level and the energy resolution of such a detector are limited by the capacitance of the junction, which is high for the above-mentioned SSDs. The limits can be overcome, for example, by using a silicon drift detector (see, e.g., Ref. (31); see also Figure 6.4). Silicon drift detectors with an integrated junction field effect transistor (JFET) are able to cope with high count rates (32). Incidence of an x-ray photon in a silicon drift detector creates a charge cloud in the n-type silicon bulk. If appropriate voltages are applied to the entrance window and the p^+ ring structure on the back-side of the

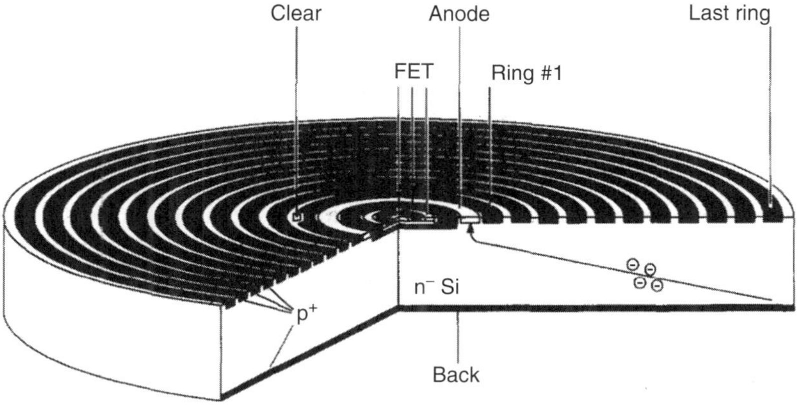

Figure 6.4 Schematic drawing of one element of a silicon drift detector with an integrated junction field effect transisitor, the gate of which is connected to the collecting anode. The entrance window for the x-rays is the nonstructured back-side of the device. Several of these elements may be arranged to form a multielement detector. (From P Lechner et al., *Nucl. Instrum. Meth. Phys. Res. A* 377: 346–351, 1996. With permission.)

detector, an inhomogeneous electric field is created within the detector, which forces the negative charges to drift to a very small readout anode. This concept was first proposed by Gatti and Rehak (33). State-of-the-art silicon technology allows the integration of a JFET monolithically in the center of the active silicon area (34), where the anode is connected to the gate contact of the JFET, leading to extremely small detector capacitance. In addition, the charge collecting time is only about 100 nsec, so that, besides a good energy resolution and low noise, a high count rate capability also results (35), which makes this detector type favorable for investigations with SR.

In recent times, two-dimensional (2D) area detectors have been increasingly used for single-crystal diffraction experiments. Charged coupled device (CCD) cameras with typically 10^6 individual pixels have been used frequently. A CCD camera for x-rays is built up by several components: basically, a single pixel in a CCD chip is composed of a metal–oxide–semiconductor structure. Photons impinging on the chip create electrons by means of photoabsorption in the intrinsically conducting Si layer. These electrons are stored in pixel cells and are read out by a shift register subsequently. Due to the high energy of x-rays, they need to be converted into visible light to be detected in the CCD chip in order to avoid damage to the chip — this can be done, e.g., by a phosphorous screen or any other scintillating material. Nevertheless, the chip needs to be cooled during operation. The camera head is coupled to the CCD chip by means of an optical contact, such as a fiber optic. Using such a device, a large number of Bragg peaks can be measured simultaneously, thus the time for a diffraction experiment can be minimized accordingly.

6.2.3 *In Situ* Cells for the Investigation of Corrosion-Related Processes

For the *in situ* investigation of corrosion processes, suitable cells in which the samples are kept under potential control during the experiments are an indispensable prerequisite. *In situ* studies of electrode surfaces with x-rays generally suffer from the parasitic absorption by the electrolyte. While this effect can be tolerated at high x-ray energies of about 15 keV or more, the absorption of the electrolyte increases dramatically with decreasing photon energies. For example, for energies around 25 keV, a 10 mm electrolyte layer has an x-ray transmission of about 65%, while at the energy of the Cu K_{α}-emission line (ca. 8.8 keV), the transmission of an electrolyte layer of only 3 mm decreases to about 12%. While for x-ray scattering or diffraction experiments this parasitic absorption by the electrolyte can be circumvented simply by using higher photon energies, they are a crucial drawback for XAS, since the electrode material with its characteristic absorption edges predefines the energy range that is necessary for the experiments. Depending on the scientific problem and the chosen experimental technique, different types of *in situ* cells have been applied, meeting both the experimental requirements for electrochemistry and the chosen analytical methods and techniques. Several cell geometries are presented schematically in Figure 6.5. A rather simple transmission type cell (Figure 6.5a) is suitable for the investigation of electrodes by XRD and x-ray reflectivity experiments and it is also suited for the study of corrosion products within the electrolyte. A number of studies of pitting corrosion applied this geometry for the investigation of artificial pits. A second type of electrochemical cells reduces the electrolyte to a thin film with a typical thickness well below 10 to 20 μm, which is trapped between the sample (working electrode) and an organic polymer film (polypropylene, Mylar, Kapton, etc.) of a few micrometers thickness (see, e.g., Refs. (36, 37)). Such a cell — presented schematically in Figure 6.5(b) — is well suited for fluorescence mode detection experiments and x-ray standing wave techniques (see, e.g., Refs. (38, 39)), since the path length of the detected fluorescence photons is minimized and independent of the incidence angle Θ. However, for surface-sensitive studies, the grazing incidence geometry with glancing angles of typically below 0.5° is mandatory, so that the path length of the x-rays in the electrolyte and the window material increase dramatically causing an enhanced parasitic absorption of the impinging radiation and also the emerging radiation accordingly. As a result, *in situ* studies of electrode surfaces are currently limited to x-ray energies of above ca. 6000 eV, even if state-of-the-art detector equipment such as multi-element SSDs (see, e.g., Refs. (37,

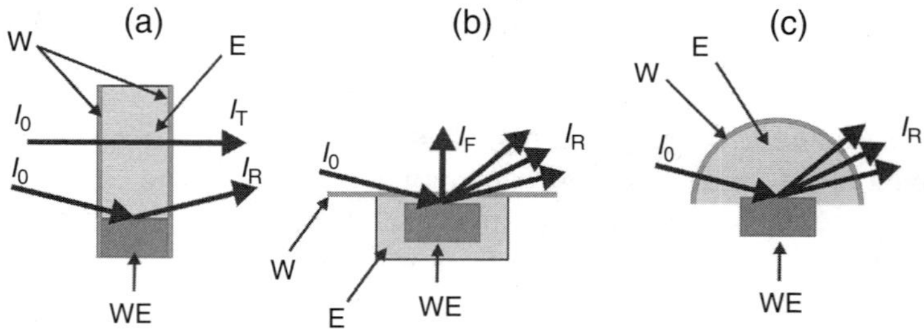

Figure 6.5 Schematic representation of different types of electrochemical cells for *in situ* investigations of corrosion processes. (a) A transmission-type cell can be used for the investigation of the electrode with x-ray diffraction, reflection mode spectroscopy as well as for the investigation of the electrolyte in front of the electrode, e.g., for the spectroscopic characterization of corrosion products. (b) A thin layer geometry is favorable for fluorescence detection or diffraction experiments of electrode surfaces. While the window is pressed on the surface of the electrode for the x-ray measurements with a remaining thin electrolyte layer in order to retain potential control, the window is inflated for the electrochemical preparation to allow for an unhindered mass transport from and to the electrode surface. (c) Cell geometry that is optimized for diffraction experiments (WE, working electrode; W, x-ray window; E, electrolyte).

40)) and high flux insertion device beamlines in third-generation storage rings are used (41). A third type of cell is optimized for XRD experiments (Figure 6.5c) (42). It mainly consists of a cylindrical body with a sample holder and a hemispherical x-ray window of fused silica (42). Due to its symmetry, all solid angles in the hemisphere above the sample surface are accessible for the incident and exiting beams. Since the x-ray path length in the electrolyte and the window material are angle independent, this cell design is ideal for diffraction, surface diffraction, as well as for x-ray reflectivity studies. Since cells with thin layer geometry possess a strongly restrained mass transfer from and to the working electrode, they cannot be used for the investigation of dynamical processes consuming species from the solution or generating significant amounts of soluble reaction products. For the investigation of corrosion reactions, therefore, cells without mass transfer limitations have to be applied or the electrode has to reach a stationary state with low reaction rates prior to the measurement in order to avoid erroneous results or misinterpretations.

6.3 X-RAY ABSORPTION FINE STRUCTURE

The basic process related to XAS is shown in Figure 6.6. An incident photon with sufficient energy is absorbed by an atom and excites a core electron to unoccupied levels (bands) or to the continuum. This excited atom can relax by filling the empty core level by an electron from a higher occupied core level. The excess energy can be released as a photon, which has an energy equivalent to the difference of the involved energy levels or an Auger electron. Both XRF and Auger-electron spectroscopy (AES) are element sensitive and can be used for the determination of the chemical composition of the samples. However, Auger electrons will not be considered in more detail, since electron-based techniques require a vacuum environment, while photon-based techniques can be applied for *in situ* electrochemical experiments.

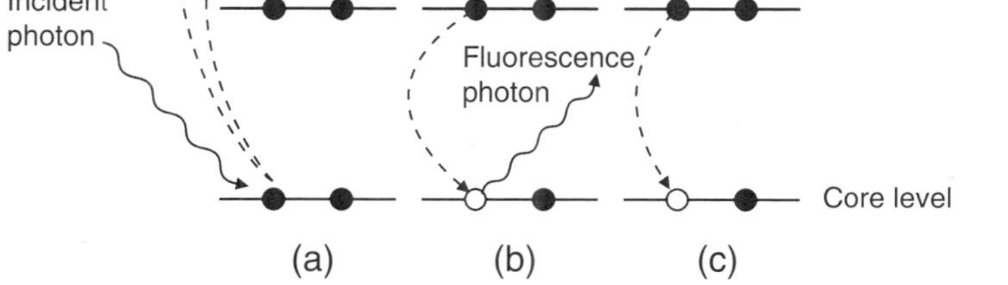

Figure 6.6 Schematic presentation of the x-ray absorption process: The energy levels of the atom and their occupation are shown before and after the photoabsorption took place (a). Channels for secondary decay processes (x-ray fluorescence (b) and Auger electrons (c)) are also displayed.

6.3.1 Basic Principles of XAFS

XAFS is related to the details of the photon absorption process as presented in Figure 6.6. For XAFS, the absorption of x-rays within the sample is measured as a function of the x-ray energy $E = h\nu$. More specifically, the x-ray absorption coefficient $\mu(E) = -\mathrm{d}\ \ln I/\mathrm{d}x$ is determined from the decay in the x-ray beam intensity I with distance x. Monochromatic radiation is used, and its energy is increased to the point at which core electrons can be excited to unoccupied states close to the continuum as shown in Figure 6.6. In Figure 6.7, an absorption spectrum of a silver metal foil is shown for an extended photon energy range. The experimental data show three general features: (1) an overall decrease in x-ray absorption with increasing energy; (2) the presence of sharp rises at certain energies called edges, which roughly resembles step-function increases in the absorption; and (3) above the edges, a series of wiggles or an oscillatory structure occurs which modulates the absorption typically by a few percent of the overall absorption cross-section as can be seen in the inset of Figure 6.7. It should be mentioned at this stage that the availability of intense new synchrotron x-ray sources also permits the investigation of highly diluted systems such as impurities or trace elements within an alloy or adsorbates and thin films on a surface of a foreign substrate. Especially for the investigation of flat electrode surfaces, a simple transmission mode experiment is not adequate; however, fluorescence mode detection or grazing incidence techniques (see below) are well suited even for samples in an electrochemical environment (see, e.g., Refs. (43–45)).

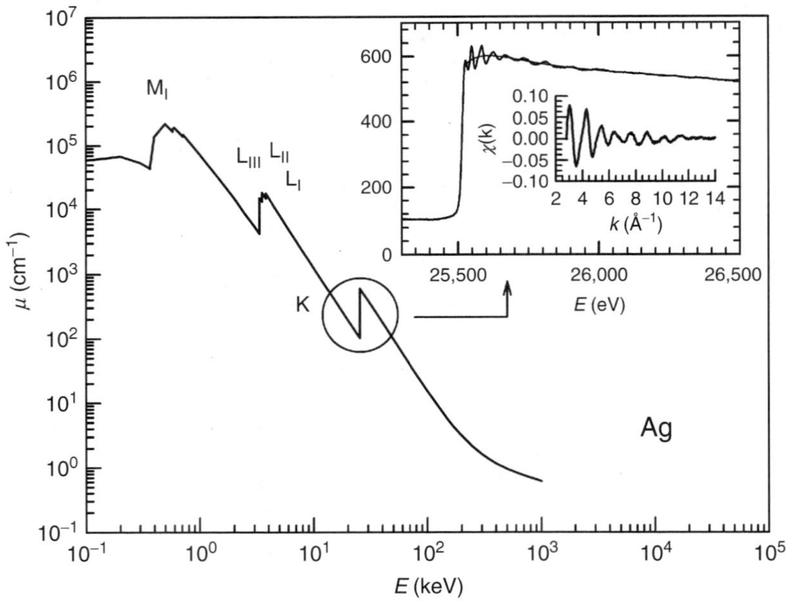

Figure 6.7 Schematic view of the absorption spectrum of an Ag foil in a wide energy range. Steep increases of the absorption coefficient are related to the K-, L-, and M-edges, respectively. The inset shows the absorption at the Ag K-edge on a magnified scale. The oscillatory behavior of the absorption coefficient $\mu(E)$ above the edge known as EXAFS is clearly resolved. The extracted fine structure oscillations $\chi(k)$ after the removal of the smooth background function $\mu_0(E)$ — which is indicated by a thin line — are also displayed as a function of the photoelectron wave vector scale.

The general decrease of μ with E can be related to the well-understood quantum mechanical phenomenon of x-ray absorption by atoms, which can be described, e.g., by Fermi's golden rule (see, e.g., Ref. (46)). Like the energy of the fluorescence photons, the energy of the absorption edge is characteristic for the absorbing material and hence the measured edges are signatures of the atomic species present in a material. For example, for the transition metals Fe, Co, Ni, and Cu, the K edges are located at 7112, 7709, 8333, and 8979 eV, respectively, i.e., they are separated by about 600 eV. Each edge corresponds to a quantum-mechanical transition that excites a particular atomic core-orbital electron to free or unoccupied levels. The nomenclature for the x-ray absorption edges reflects the origin of the core electron; i.e., K edges refer to transitions that excite the innermost 1 s electron, while L edges are related to those of 2 p and 2 s initial states. These transitions always occur to unoccupied states, i.e., to states with a photoelectron above the Fermi energy leaving behind a core hole. The resulting excited electron is referred to as a photoelectron. In a solid, it generally has enough kinetic energy to propagate freely through the material. Since the excited states are almost always extended states, i.e., quasifree states in molecules and conduction-band states in solids, a propagation of the photoelectron is possible even within insulators. From the change of the absorption coefficient at the edge, the concentration of the respective element can easily be calculated (see, e.g., Refs. (47, 48)). With regard to corrosion experiments, the element-specific absorption within the electrolyte can also be used for the analysis of soluble corrosion products in the active, passive, or transpassive state of a metal or its oxide (49, 50). In addition, experiments of this type enable the investigation of the selective dissolution of a single element in a multi-element specimen such as an alloy (51–54).

The exact energy of the absorption edge is a sensitive function of the valency of the excited atom as depicted in Figure 6.8 for some selected Ti compounds.

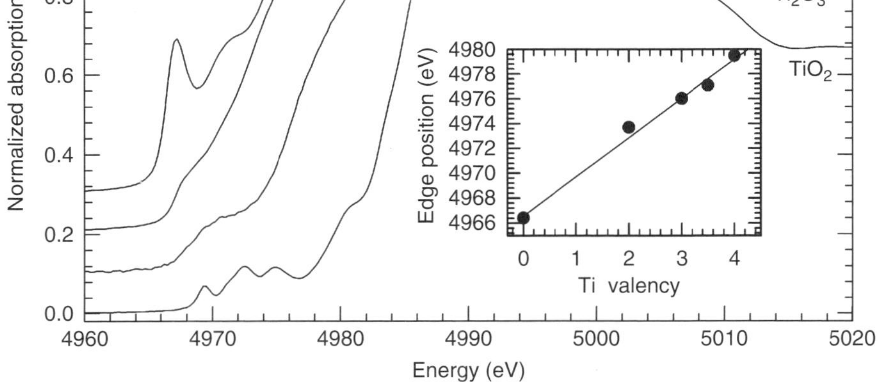

Figure 6.8 Near-edge x-ray absorption spectra at the Ti K-edge of selected Ti reference compounds with different oxidation numbers: Ti-metal (0), TiO (+2), Ti_2O_3 (+3), and anatase TiO_2 (+4). The approximately linear shift of the absorption edge towards higher energy with increasing Ti valency is depicted in the inset.

Obviously, all the spectra show very different shapes and features. Most obvious is a shift of the actual edge position as given by the first maximum of the derivative spectrum toward higher photon energies with increasing Ti valency. As shown in the inset, the edge energy shift is approximately linear with a slope of about 3.1 eV per valency unit in the case of the Ti oxycompounds. More generally, this effect provides an easy access for the determination of the valency of a selected element (see, e.g., Refs. (55–57)), even for trace elements in complex systems. In addition, as can already be anticipated from Figure 6.6, there are also features in the absorption spectrum below the edge. These so-called pre-edge peaks can be attributed to transitions from the excited photoelectron into unoccupied electronic levels of the sample. Therefore, an x-ray absorption spectrum probes the density of unoccupied states of the absorbing element (55, 58).

The detailed structure of the absorption coefficient within about 50 eV above the edge is referred to as near-edge x-ray absorption fine structure or x-ray absorption near-edge structure (XANES). Besides the above-mentioned transitions into unoccupied states, the near-edge structure also involves transitions of the photoelectron into the continuum, i.e., into states above the vacuum level. In the latter case, the photoelectron has a low kinetic energy and accordingly a large (inelastic) mean free path (see, e.g., Ref. (59)), so that it can undergo multiple scattering events (58, 60). The shape of the edge is thus highly sensitive to both the valency and the coordination of the absorbing atoms. Although the complexity of the involved processes makes a full theoretical treatment very difficult, it is currently possible to calculate XANES on an *ab initio* basis, i.e., on the assumption of a model structure (61, 62). This approach works quite well for molecules and complexes — which is interesting for the investigation of corrosion products within the electrolyte. However, for crystalline materials it is still a quite challenging task to fit an experimental spectrum using *ab initio* calculations since the theory of XANES is not as fully quantitative and requires several physical considerations (see, e.g., the review by Rehr and Albers (62)). Thus, for the practical analysis, the near-edge spectra are most commonly used as fingerprint techniques: edges are collected for a number of standard compounds with known valence and crystal structure and these are compared to the spectra of the actual samples. Certain compounds such as Cr^{6+} or Mo^{6+} give rise to distinct pre-edge peaks; thus, these species can be identified qualitatively and quantitatively. For the application of fingerprint techniques to electrochemical probes, see, e.g., Refs. (49, 52, 63–65).

For samples in which the absorber element is contained in more than one chemical form, however, the data interpretation is often not straightforward. For these problems, the application of sophisticated mathematical techniques such as principal component analysis (PCA) (66) is recommended. Compared to a simple fingerprint technique, PCA is able to specify whether or not a chosen reference compound contributes to the spectrum of the actual sample, i.e., the application of the PCA yields the number of reference compounds that are present within the sample (67, 68). Furthermore, in combination with a least-squares fit, PCA also provides the concentrations of all the identified phases (67–70).

The third feature in the absorption spectrum is the oscillation of the absorption coefficient that is visible up to about 1000 eV above the edge. The so-called EXAFS originates from scattering processes of the emitted photoelectrons with neighboring atoms as shown schematically in Figure 6.9. The photoelectron that is ejected from the absorbing atom can be described as a propagating electron wave with a wave vector k and a de-Broglie wavelength λ:

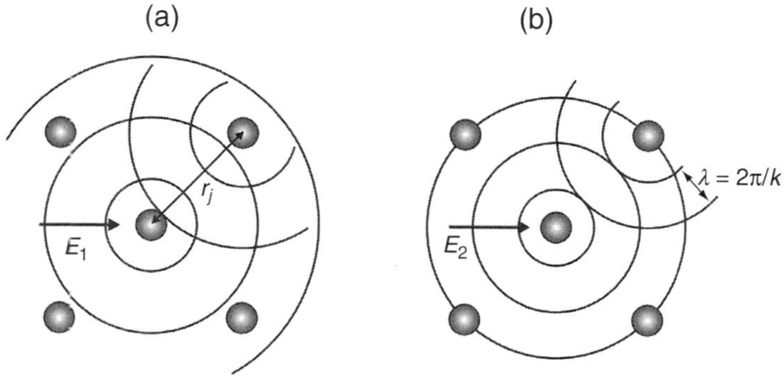

Figure 6.9 Scheme of the scattering processes related to the extended x-ray absorption fine structure (EXAFS). The photoelectron ejected from the central atom (absorber atom) propagates as a spherical wave and is backscattered from neighboring atoms in a radius r_j. Depending on the photon energy E (and thus on the wavelength λ of the photoelectron), the interference between the incoming photoelectron wave and the backscattered wave can be (a) destructive or (b) constructive, resulting in a reduced and an increased absorption for E_1 and E_2, respectively.

$$k = \frac{\sqrt{2m(E - E_0)}}{\hbar}, \qquad \lambda = 2\pi/k \tag{6.2}$$

where E is the actual photon energy, E_0 the edge energy, m the electron mass, $\hbar = h/2\pi$, and h is Planck's constant. This wave can be scattered by neighboring atoms, so that the outgoing wave interferes with the backscattered wave. Depending on the bond distance r_j related to the coordination shell of interest and the wavelength of the photoelectron, this interference can be constructive or destructive, resulting in an increase or a decrease of the absorption coefficient by a few percent. This phenomenon was first theoretically described by the short-range order theory developed by Sayers et al. (71). In contrast to the near-edge structures in the absorption coefficient, the theory of the extended XAFS is today well understood with acceptable tolerances from the experimental results (see, e.g., Refs. (72–76)). The EXAFS spectrum $\chi(E)$ is defined as the oscillatory part of the x-ray absorption above a given absorption edge, i.e.,

$$\chi(E) = \frac{\mu(E) - \mu_0(E)}{\mu_0(E)} \tag{6.3}$$

where $\mu(E)$ is the energy-dependent absorption coefficient and $\mu_0(E)$ is the smoothly varying, atomic-like background absorption (see Figure 6.7). $\mu_0(E)$ contains contributions from other edges and other elements within the sample. As recently shown, an oscillatory behavior can also be observed in the atomic background absorption, which has implications especially for the structures in the XANES region (77–79). Although many of these features can be associated with multielectron excitations (see, e.g., Refs. (80–82)), systematic changes of the atomic background were observed during the electrochemical charging of Pt/C electrodes (79,83), which indicates that information about the electrochemical state of the absorbing atom might also be extracted from these small features in the background. According to the plane-wave concept and the single-scattering approximation, the EXAFS oscillations $\chi(k)$ can be expressed by the following equation:

$$\chi(k) = \sum_j S_0^2 N_j \frac{|f_j(k)|}{kr_j^2} \sin(2kr_j + \delta_j(k)) e^{-2r_j/\lambda(k)} e^{-\sigma_j^2 k^2} \tag{6.4}$$

where N_j is the coordination number of the central atom in the jth shell, $f_j(k)$ is the backscattering amplitude, σ_j is the mean relative displacement of the atoms in the jth shell, and δ_j is a phase shift associated with the interactions of the photoelectron wave with the potential of the absorbing and the backscattering atoms. Since different backscattering atoms have their own characteristic backscattering phases and amplitudes, EXAFS enables the identification of the nature of the atoms in each coordination sphere around the absorbing atom. In contrast to XRD, EXAFS is very sensitive toward light backscattering atoms, which makes it an important technique for the investigation of, e.g., oxides and passive layers (43,84–86). The disorder σ^2 is partly due to thermal effects, which cause all of the atoms to oscillate around their equilibrium positions. Effects of structural disorder are similar, giving an additional contribution to σ^2. Both effects can be separated by means of temperature-dependent measurements. As a consequence of the $1/r_j^2$ dependence of χ(k) and the limited mean free path of the photoelectrons characterized by λ, EXAFS is useful in providing information about the local short-range order around the absorbing atom; i.e., the fine structure contains precise information about the local atomic structure around the atom that absorbed the x-rays. The structural information is typically obtained by Fourier filtering of the experimental $\chi(k)$ data into distance-space, giving a radial distribution function from which coordination numbers, interatomic distances, and the local disorder are determined (71). This turns out to be a unique signature of a given material, since it depends on both the detailed atomic structure and its vibrational properties. For this reason, EXAFS has become an important probe especially for disordered materials, where XRD can hardly be applied due to the absence of any long-range order correlations. An entire data analysis comprises the separation of the individual shells contributing to the radial distribution function and the fitting of these peaks with phases and amplitude functions (87).

An example for a simple data analysis is given in Figure 6.10 for a Mo-metal foil. In Figure 6.10(a), the radial distribution function, i.e., the magnitude of the Fourier transform (FT) of the k^2-weighted XAFS data, $\chi(k)*k^2$, is shown. The peaks in the FT($\chi(k)*k^2$) correspond to the first few coordination shells in the vicinity of the absorbing atom. Owing to the photoelectron phase shift arising from the scattering processes, all peaks in these FTs are generally shifted to lower distances compared to the crystallographic distances (see, e.g., Ref. (87)). Thus, the dominant peak doublet at ≈2.44 and 2.86 Å belongs to the first and second Mo–Mo coordinations at 2.73 and 3.15 Å, and the peaks in the FT between 4 and 6 Å can be attributed to Mo–Mo coordination spheres at 4.45, 5.22, 5.46, and 6.30 Å. For further data analysis, the data in the FT between ca. 1.6 and 3.2 Å were isolated by means of a filter function and back-transformed into k-space (Figure 6.10b), where a least-squares fit using a two-shell model was performed. Currently, phase and amplitude functions, which are used for such a fit, are calculated on an *ab initio* basis, and r_j, N_j, and σ_j are varied in order to obtain a best fit between the experiment and the simulation. The fit data were also transformed into r-space to show that both data sets also match quantitatively in r-space. In this context, it should be mentioned that the fit can, in principle, also be performed in r-space. A number of different but standardized data analysis packages are currently available for this purpose; a list can be found in the web (see, e.g., Ref. (88)). The near-range order information provided by the fit are the coordination distances, numbers (CN), and the disorder for each coordination shell. For the Mo foil in Figure 6.10, the results are $R_1 = 2.722$ Å, $N_1 = 8.02$, $\sigma_1^2 = 0.0038$ Å^2, $R_2 = 3.151$ Å, $N_2 = 5.05$, and $\sigma_2^2 = 0.0034$ Å^2. While the determined distances as well as the coordination number of the first shell agrees quantitatively with

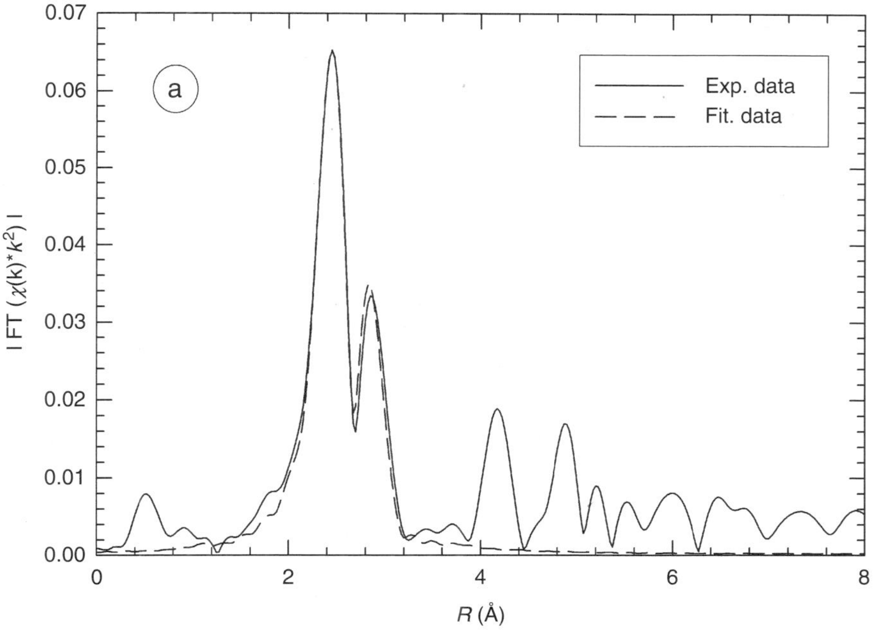

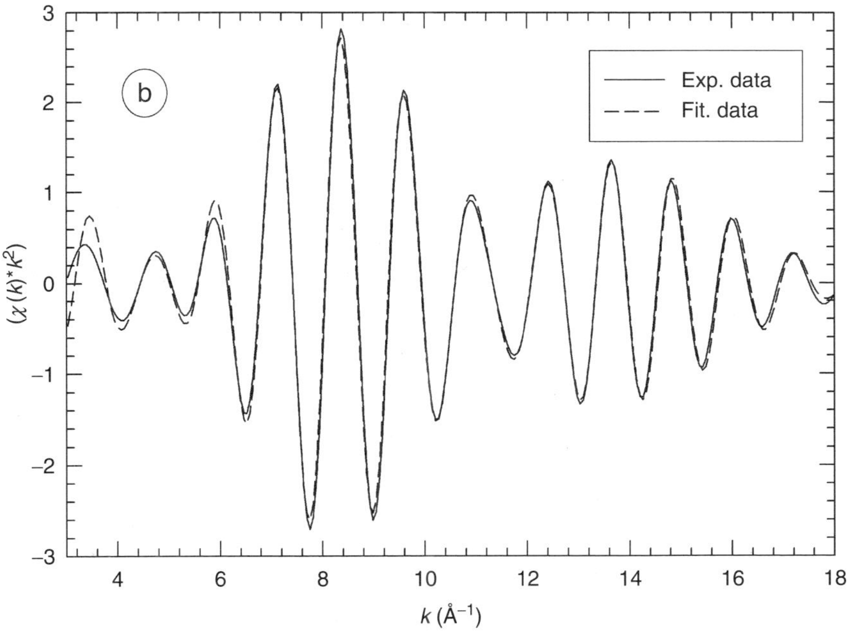

Figure 6.10 (a) Magnitude of the Fourier transform of the k^2-weighted absorption fine structure obtained from a Mo-metal foil at room temperature (k-range from 3.02 to 18.0 Å^{-1}, data not phase-shift corrected). Contributions of the first and the second coordination shells were isolated by means of a filter function and back-transformed into k-space. (b) Back-transformed experimental $\chi(k)*k^2$ data (solid line) and fit (dashed line) with a two-shell model. Fit results are $R_1 = 2.722$ Å, $N_1 = 8.02$, $\sigma_1{}^2 = 0.0038$ Å^2, $R_2 = 3.151$ Å, $N_2 = 5.05$, and $\sigma_2{}^2 = 0.0034$ Å^2. For comparison, the radial distribution function of the fit data is also shown in (a) as a dashed line.

literature data of polycrystalline Mo metal ($R_1 = 2.728$ Å, $N_1 = 8$, $R_2 = 3.150$ Å, $N_2 = 6$), the CN of the second shell is slightly reduced. This is due to influences of multiple scattering processes, which can however be accounted for in the current theory (61, 62).

More generalized models take into account the crystallographic structure of the sample under investigation, and scale all individual scattering paths accordingly using an expansion factor. Furthermore, different models for the disorder, such as the Debye or the Einstein model (see, e.g., Ref. (89)), can be used. Having in mind that at least four parameters are required for a single shell — in addition to R, N, and σ an individual E-shift is usually necessary for the fit — such models significantly reduce the number of individual fit parameters especially if a larger cluster of atoms is used for a fit.

6.3.2 Grazing Incidence XAFS

Due to the large penetration depth of hard x-rays in matter, XAFS is, in general, not surface sensitive; i.e., it is not suitable for the investigation of surface phenomena such as passive oxide film formation, adsorption, or corrosion, without any special precautions. However, the whole spectrum of x-ray analytical methods (XAS, XRD, XRF, x-ray topography, etc.) can be made surface sensitive using the grazing incidence geometry that is schematically shown in Figure 6.11. The energy-dependent index of refraction $n(E)$ in the hard x-ray regime is given by

$$n(E) = 1 - \delta(E) - i\beta(E) \tag{6.5}$$

where β and δ are small positive quantities. Depending on the material and the actual photon energy, typical values amount to about 10^{-4} to 10^{-8}. The absorptive correction β is proportional to the linear absorption coefficient μ, i.e., $\beta = \lambda\mu/4\pi$. Therefore, condensed matter is optically thinner than vacuum. Thus, x-rays entering into matter are refracted away from the surface normal, as indicated in Figure 6.11. According to Snell's law of refraction, which has the same form as in the visible spectral range,

$$n_1 \cos\phi_1 = n_2 \cos\phi_2 \tag{6.6}$$

there exists a glancing angle ϕ_c (critical angle) for which $\phi_2 = 0$. Assuming $n_1 = 1$ (vacuum), ϕ_c can be calculated from $\cos\phi_c = n_2$. If the absorption (β) can be neglected, a simple calculus leads to $\phi_c \approx (2\delta_2)^{1/2}$. Typical values for ϕ_c are 0.39° for Cu and 0.22° for Al for 8600 eV photon energy. For 10 keV photons, total reflection only occurs for grazing incidence below 0.5° (90). In this case, there is no x-ray wave propagation inside the material, and the electric field amplitude in the sample is damped exponentially with depth. The decay constant, i.e., the penetration depth z_0, amounts to only few nanometers below the critical angle (90). This is

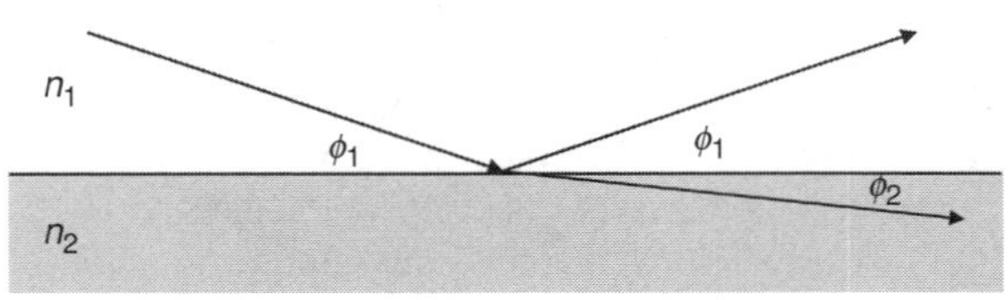

Figure 6.11 Schematic presentation of the grazing incidence geometry: the refracted beam is diffracted toward the surface.

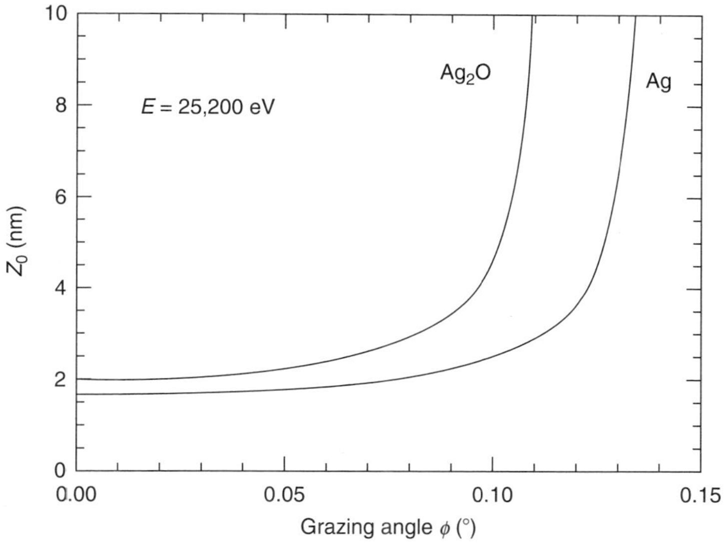

Figure 6.12 Angle dependence of the penetration depth z_0 for metallic silver and silver oxide Ag_2O at $E = 25.2$ keV. The critical angle amounts to $\approx 0.13°$ for Ag and $\approx 0.11°$ for Ag_2O.

illustrated in Figure 6.12 for Ag and Ag_2O, where a value of about 2 to 3 nm is obtained below the critical angle, while a steep increase is observed above ϕ_c. Therefore, the combination of EXAFS and the use of glancing angles in the vicinity of the critical angle of total reflection provide short-range order structural information about the surface region of the sample. However, in contrast to transmission EXAFS, both the real part and the imaginary part of the complex refractive index contribute to the reflectivity fine structure. Therefore, the reflectivity fine structure becomes highly angle dependent and cannot directly be compared to the conventional EXAFS theory (91–95). Several simple approximations have been given in the literature, e.g., for glancing angles well below the critical angle (93, 96). Often, the absorption coefficient $\mu(E)$ was extracted simply by $\mu(E) = (1 - R(E))/(1 + (R(E))$ (96). The applicability of such simple approaches on layered systems seems, however, very questionable. Due to their limited validity, more generalized approaches have been developed (94, 97, 98). In general, two situations have to be distinguished:

1. If the penetration depth of the x-rays is smaller than the film thickness of a surface layer (even an adsorbate might be considered as a thin film), and the sample might be treated as a homogeneous and isotropic phase, the EXAFS may be extracted from reflectivity fine structure data (94, 99–101).
2. If the film thickness is smaller than the penetration depth, a straightforward extraction of the EXAFS is not possible, especially if a thin oxide or hydroxide film is present on a substrate of the same element. In the latter case, model calculations for the reflectivity using, e.g., the Fresnel theory (90, 94, 102), have to be performed, and the measured data have to be compared to the calculations (97, 98, 100, 103).

Considering the first case, the normalized reflectivity fine structure $\chi_R(E)$ (Equation (6.7))

$$\chi_R(E) = \frac{R(E) - R_0(E)}{R_0(E)} = \frac{\Delta R(E)}{R_0(E)} \tag{6.7}$$

where $\Delta R(E)$ denotes the reflectivity fine structure superimposed on a structureless background $R_0(E)$, has to be analyzed in detail. Assuming that the experimental data can be fitted to a linear superposition of the oscillatory components $\Delta\beta$ and $\Delta\delta$ of the refractive index n; i.e., $\beta(E) = \beta_0(E) + \Delta\beta(E)$ and $\delta(E) = \delta_0(E) + \Delta\delta(E)$, and

$$\chi_R(E) = a\Delta\delta(E) + b\Delta\beta(E) \tag{6.8}$$

$\Delta\beta(E)$ and $\Delta\delta(E)$ can be extracted from the above equation by means of a Kramers–Kronig transform (98,99). Thus, the absorption coefficient can be determined from the reflectivity data of sufficiently thick surface films. It has been shown that this procedure works well for thin film analysis such as vacuum deposited films or anodic oxide layers of sufficient thickness, i.e., well above the penetration depth of the x-rays (101,104,105). An example of such a data analysis for a thin gold film (thickness $\approx$41 nm) on a glass support at the Au L_3 edge is given in Figure 6.13. In the inset, the absorption data $\chi(k)^*k^3$ extracted from the reflectivity spectrum are compared to those obtained from a gold foil measured in transmission. Obviously, both data sets are almost identical up to $k \approx 15\,\text{Å}^{-1}$; i.e., the atomic short-range order structure of both samples is identical. This has been proved by a detailed data analysis, i.e., the coordination distances and numbers as well as the disorder are identical within the error limits (101).

However, if the film thickness is smaller than the x-ray penetration depth, an extraction of the absorption coefficient is generally not possible. In this situation, reflectivity spectra of layered systems can be calculated using a procedure that is described in detail in Refs. (97,100); we therefore give only a brief summary of the calculation procedure. Similar to the calculation of a Fresnel reflectivity $R(\phi)$, the thickness of the surface layer as well as the energy-dependent indices of refraction of both the substrate and the top layer must be known. The latter quantities can be

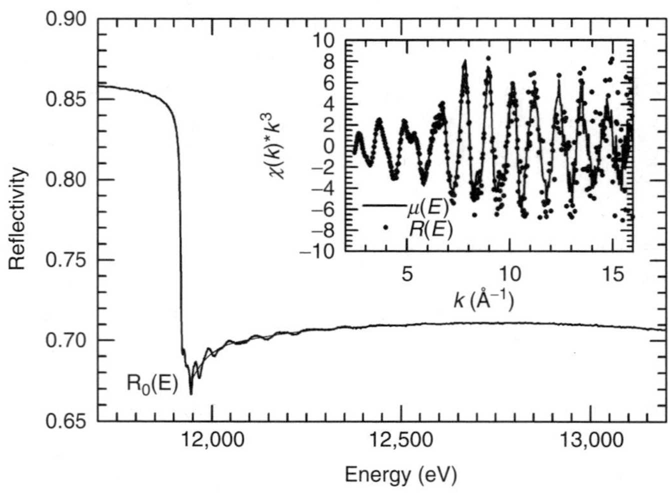

Figure 6.13 Reflectivity EXAFS scan of a gold film (thickness 41 nm) covering the Au L_3 edge for a glancing angle $\phi = 0.247°$. The smooth background function $R_0(E)$ is also indicated. In the inset, the k^3-weighted absorption fine structure data extracted from this reflectivity scan (•, $R(E)$) are compared to those obtained from a gold foil measured in transmission ($\mu(E)$, ——).

extracted from transmission EXAFS spectra of a suited reference compound by means of a Kramers–Kronig transform. For an oxidized metal surface such as a passivated metal electrode, the optical constants of the underlying metal surface as well as those of the oxides of which the passive layer consists have to be known. While $\beta(E)$ is correlated with the linear absorption coefficient $\mu(E)$ as already mentioned, $\delta(E)$ can be calculated from $\beta(E)$ by means of a Kramers–Kronig transform. With $n(E)$ (Equation (6.1)) from each sublayer, the reflectivity spectra were obtained as energy-dependent Fresnel reflectivities $R(E, \phi)$ (100). Due to the calculation procedure, these simulated reflectivity spectra $R(E, \phi)$ contain the near-range order structural information of the chosen model compounds. Alternatively, computer programs like FEFF (see, e.g., Refs. (73,74)) can also be used as sources for phases and amplitudes for the calculation of such reflectivity spectra (97,100) Therefore, experimental reflectivity data can be fitted to structure models, i.e., the determination of nearest neighbor distances, coordination numbers, and Debye–Waller factors of surface layers is enabled by adjusting the input parameters of the calculation so that the calculated and measured data show a close fit. A calculation scheme for these procedures is given in Figure 6.14 as a flow diagram for a layered system consisting of a thin Ag_2O layer on top of a silver metal surface. If the fit is not adequate, new model compounds or different thicknesses of the involved materials might be chosen. In addition, also a multilayered structure instead of a single surface layer has to be used for the simulations in order to obtain a close fit to the experiments. Furthermore, also the individual surface and interface roughnesses of the contributing phases contribute to the signal and have to be considered accordingly (100). It should be stressed at this point that a variation of the glancing angle allows depth-resolved experiments according to the changes of the penetration depth with glancing angle (see Figure 6.12). Thus, in contrast to scanning probe techniques, grazing incidence x-ray methods offer the opportunity to probe buried interfaces also.

6.3.3 Time-Resolved X-Ray Absorption Spectroscopy

In a conventional EXAFS experiment, a double-crystal monochromator is moved in small energy steps from lower to higher energies in order to obtain a spectrum. Such a measurement scanning over $\approx$1000 eV typically takes about 15 to 20 min for bulk specimen. The investigation of diluted samples or the detection of reflectivity spectra often require acquisition times of more than 1 h, i.e., those EXAFS experiments cannot be used to monitor time-dependent phenomena. There are two possibilities to speed up the time that is required for the measurement of an x-ray absorption spectrum:

1. In the quick-scanning-EXAFS (QEXAFS) mode, the double-crystal monochromator is slewed continuously at a high speed by means of microstepper motors and each data point is integrated for typically 0.01 to 0.2 sec, so that a complete spectrum is measured on the fly (106,107). Depending on the details of the monochromator and the sample under investigation, a spectrum can be obtained in less than 1 min or even a couple of seconds only (106,107).
2. The dispersive EXAFS method (DEXAFS) employs a curved polychromator crystal and the sample is placed in the focal point of the polychromator fan. Due to the fixed energy–position correlation in the diffracted energy

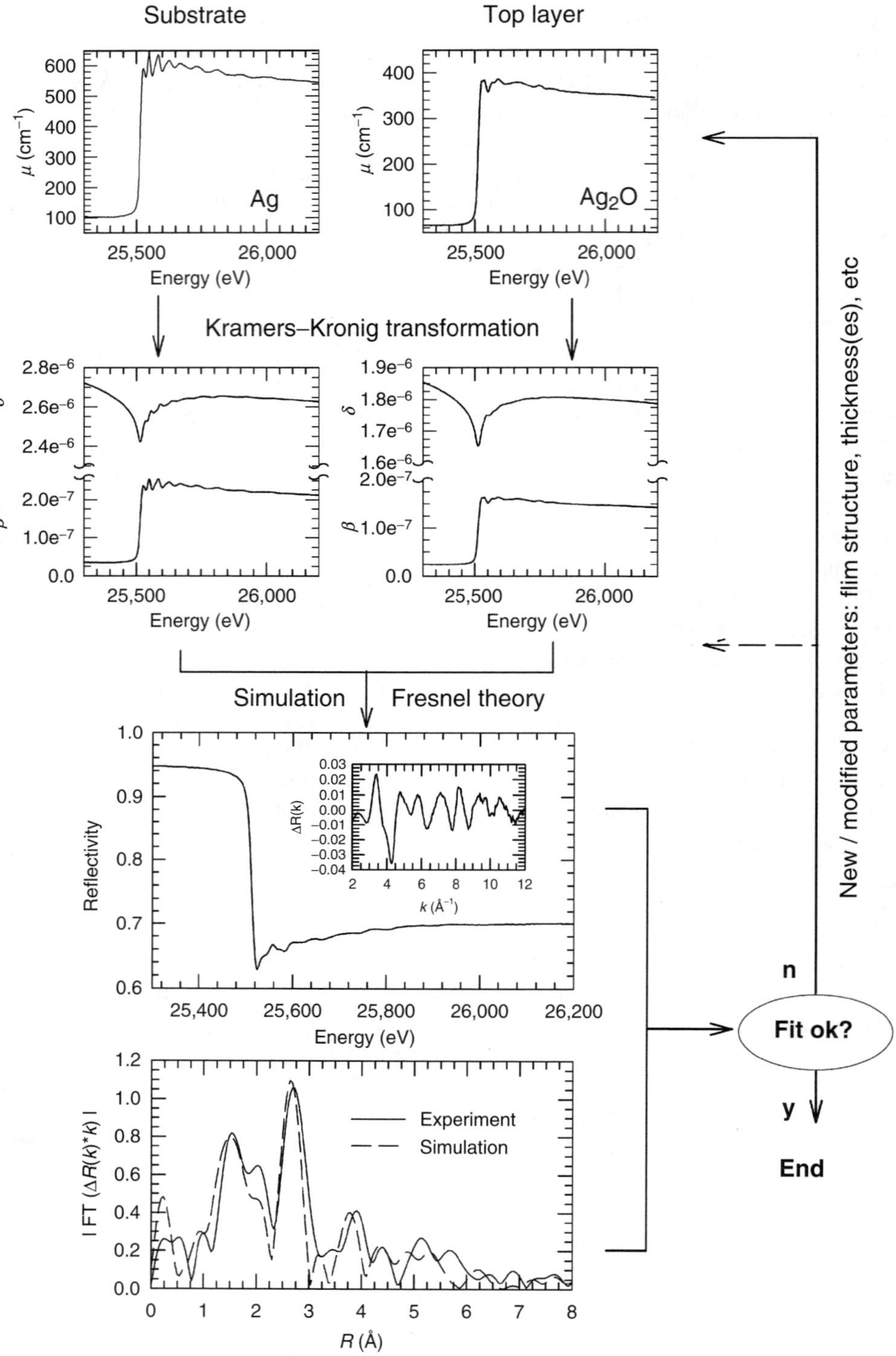

Figure 6.14 Flow diagram for the fit of a reflection mode XAFS spectrum of a layered system. For details, see text.

band, a complete absorption spectrum can be obtained by parallel acquisition of the intensity distribution by a position-sensitive detector (108–114). A measuring time in the millisecond range for an absorption spectrum is feasible at most currently available dispersive spectrometers.

Both approaches have successfully been adopted for *in situ* investigations of electrodes under working conditions (see, e.g., Refs. (115,116)). Due to the fact that QEXAFS measurements use conventional monochromator equipment, they can be easily combined with, e.g., fluorescence or electron-yield detection and reflection-mode EXAFS experiments. However, DEXAFS investigations are restricted to transmission experiments, and cannot make use of fluorescence detection in an efficient way, which is, on the other hand, essential for the investigation of diluted specimen such as biological systems or those samples that cannot be penetrated by x-rays (e.g., thick samples).

6.3.4 Experimental Procedures

A typical sketch of an x-ray absorption experiment showing the most important instrumentation is given in Figure 6.15. In general, a slit system is placed in the beam path downstream the monochromator in order to define the size of the beam to the size of the sample. For a simple transmission mode experiment, ionization chambers filled with noble gases are used as detectors for the incoming (I_0) and the transmitted (I_1) intensity from which the absorption of the sample may be calculated by $\ln(I_0/I_1)$. It has to be mentioned in this context that any parasitic absorption of the x-rays caused, e.g., by an air path in the beam, x-ray windows, or the electrolyte, which is essential for *in situ* experiments, has to be carefully subtracted from this absorption in order to derive the true absorption coefficient $\mu(E)$. A third chamber (I_2) is used to measure a reference sample simultaneously with each sample in order to precisely calibrate the energy scale for each spectrum. Parallel fluorescence data acquisition (I_F) can be done using a pin-photodiode or an SSD. The same detector equipment can also be used for grazing incidence studies. However, the vertical beam size has to be adopted for the finite length of the sample and the actual grazing angle for the experiment, resulting in typical values for the vertical beam size of about 10 to 50 μm only. In addition, a diffractometer or a reflectometer with the possibility to adjust the sample and detector positions with high accuracy is required.

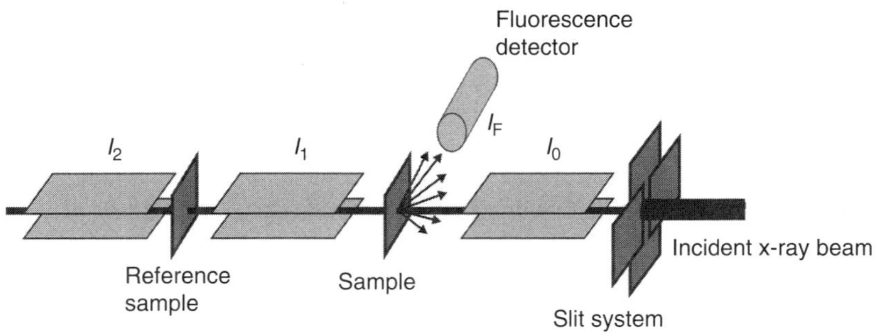

Figure 6.15 Schematic view of an x-ray absorption experiment. For grazing incidence studies, the sample and the detector for the reflected beam have to be mounted on rotatable stages.

6.4 X-RAY DIFFRACTION

6.4.1 Basic Principles of X-Ray Diffraction

When considering XRD, an adequate mathematical description of the crystal lattice is required. In real space, each point on the lattice can be described as a vector **R**

$$\mathbf{R} = n_1\mathbf{a}_1 + n_2\mathbf{a}_2 + n_3\mathbf{a}_3 \tag{6.9}$$

where $\mathbf{a_1}$, $\mathbf{a_2}$, and $\mathbf{a_3}$ are the basis of the elementary cell unit, and n_1 to n_3 are integer numbers. However, this notation is not well suited for the description of diffraction phenomena. Since diffraction is directly related to the lattice planes inside any crystal lattice, a description that bases on theses planes is favorable. Such a representation of the crystal is given by the reciprocal lattice, which is given by

$$\mathbf{Q} = h_1\mathbf{b}_1^* + h_2\mathbf{b}_2^* + h_3\mathbf{b}_3^* \tag{6.10}$$

Again, h_1 to h_3 are integer numbers and the reciprocal lattice is connected to the real space representation by $\mathbf{a}_i \cdot \mathbf{b}_j{}^* = 2\pi\delta_{ij}$ (for $i = 1, 2, 3$; $\delta_{ij} = 1$ for $i = j$, $\delta_{ij} = 0$ for $i \neq j$). The well-known Miller indices (h, k, l) — which are characteristic for a selected diffraction plane — are obtained by dividing h_1, h_2, and h_3 by their largest common devisor. It follows directly that the reciprocal lattice vector lies perpendicular to the corresponding plane in real space. Thus (see Figure 6.16), a Bragg-reflected intensity, i.e., a diffraction peak, corresponds to the situation that the difference between the wave vector $\mathbf{K}_i$ of the incoming x-ray wave and that of the reflected wave $\mathbf{K}_f$ is equal to a reciprocal lattice vector **Q**, i.e., $\mathbf{Q} = \mathbf{K}_f - \mathbf{K}_i$ (Laue condition, see, e.g., Ref. (89)). The Bragg angle Θ_B is then identical to the angle between the impinging radiation and the lattice plane $(h\,k\,l)$. Using $|\mathbf{K}_f| = |\mathbf{K}_i| = 2\pi/\lambda$, introducing the lattice spacing d_{hkl}:

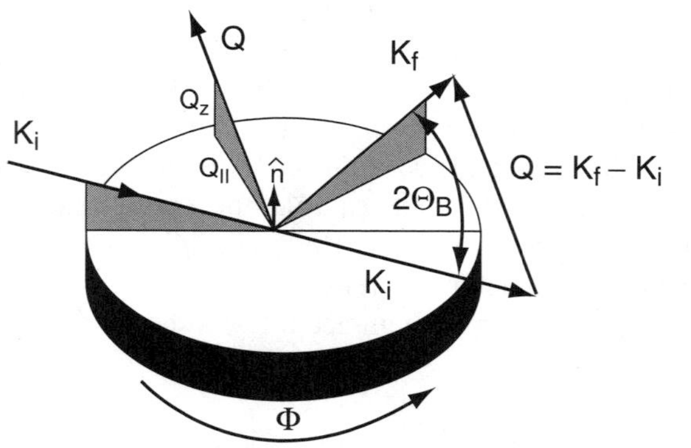

Figure 6.16 Schematic presentation of the scattering geometry in an x-ray diffraction experiment (Laue–case): a Bragg peak is observed, if the difference between the wave vector of the Bragg-reflected beam $\mathbf{K}_f$ and the impinging radiation $\mathbf{K}_i$ is identical to a reciprocal lattice vector **Q**. The Bragg angle Θ_B is identical to the angle between the diffracted wave and the $(h\,k\,l)$ lattice plane, which is not necessarily parallel to the surface of the crystal under investigation. Three different types of scans are indicated: (1) Out-of-plane scan ($\mathbf{Q}z$ or $\mathbf{Q}_\perp$-scan): the rotation (Φ) of the crystal is fixed, and data are measured by varying the component of **Q** normal to the surface (which is named $\mathbf{Q}_z$ or $\mathbf{Q}_\perp$) while the component parallel to the surface ($\mathbf{Q}_\parallel$) is fixed. (2) In-plane radial scan ($\mathbf{Q}_\parallel$-scan): the component of **Q** along the surface is varied. The direction of **Q** is retained, but its magnitude is varied by changing the incident and diffracted angles (Θ and 2Θ) symmetrically. (3) Phi-scan: the crystal rotates about its surface normal with the positions of the incident and diffracted beams fixed. The magnitude of **Q** is fixed for this scan while the direction of **Q** rotates.

$$|\mathbf{K}_f| \sin \Theta_B = \frac{1}{2}|\mathbf{Q}| \tag{6.11}$$

$$d_{hkl} = \frac{2\pi}{|h\mathbf{b}_1^* + k\mathbf{b}_2^* + l\mathbf{b}_3^*|} \tag{6.12}$$

and keeping in mind that

$$|\mathbf{Q}| = n|h\mathbf{b}_1^* + k\mathbf{b}_2^* + l\mathbf{b}_3^*| = \frac{2\pi n}{d_{hkl}} \tag{6.13}$$

the well-known form of the Bragg equation (Equation (6.1)) follows immediately. The interpretation of Equation (6.1) is illustrated in Figure 6.17: constructive interference only occurs if the optical path difference between the x-rays, which are diffracted from neighboring lattice planes, is equal to an integer multiple of the x-ray wavelength λ. Each crystal structure is characterized by the distribution of its lattice atoms in the unit cell; thus, the resulting diffraction pattern is representative for the crystal structure. In more detail, structure factors F_{hkl}, which include the positions of the atoms in the crystal structure under investigation as well as the occupation of these positions by different types of atoms, can be calculated according to the following equation:

$$F_{hkl} = \sum_n f_n \exp(2\pi i \mathbf{Q} \cdot \mathbf{R}_n) \tag{6.14}$$

where $\mathbf{R}_n$ is the lattice vector of the nth atom and $\mathbf{Q}$ the reciprocal lattice vector that represents scattering processes at the hkl-lattice planes (89). The summation includes all the atoms in the unit cell, and f_n is the atom form factor, which represents the scattering power of an individual atom. Basically, f_n is determined by the number and the distribution of the electrons belonging to a single atom (89). Values for f_n are tabulated, e.g., in Ref. (117). The diffracted intensity then is given by

$$I \propto |F_{hkl}|^2 \tag{6.15}$$

thus not only the positions of the diffraction peaks are representative for the crystal structure, but also their intensities. This can be illustrated in a simple example: KCl and KBr grow in the same crystal structure, namely that of NaCl. However, the

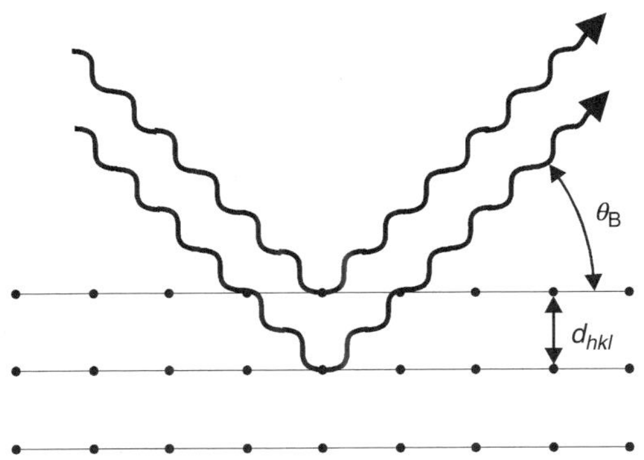

Figure 6.17 Basic principle of x-ray diffraction: for constructive interference, i.e., the observation of a Bragg peak, the path length difference between the interfering x-rays has to be an integer multiple of the x-ray wavelength λ.

observed diffraction patterns are not identical: while strong (1 1 1), (2 0 0), (2 2 0), (3 1 1), (3 3 1), and homologous reflections are observed for KBr, only (2 0 0), (2 2 0), (2 2 2), and homologous peaks are detectable for KCl. Thus, the observed differences must be due to the occupation of the identical atomic sites with different types of atoms. For the NaCl crystal structure, F_{hkl} can be calculated to as

$$F_{hkl} = \begin{cases} 4f_{\mathrm{Cl^-}} + 4f_{\mathrm{Na^+}} & \text{for } h, k, l \text{ even} \\ 4f_{\mathrm{Cl^-}} - 4f_{\mathrm{Na^+}} & \text{for } h, k, l \text{ odd} \end{cases} \tag{6.16}$$

that is, two sets of diffraction peaks are expected. Since both K^+ and Cl^- have the same number of electrons, their atomic form factors are almost identical: $f(K^+) \approx f(Cl^-)$. Thus, the diffraction peak intensities for (h k l) peaks with odd numbers for (h k l) have vanishing intensities according to Equation (6.16). In contrast, for KBr, where $f(K^+) \neq f(Br^-)$, a Bragg intensity results for both subsets of diffraction peaks. This can also be understood if the NaCl structure is investigated in more detail: assuming that identical atoms are present on both Na^+ and Cl^- sites, a simple cubic structure with only 50% of the original unit cell length follows. For this type of unit cell, only diffraction peaks with even values for (h k l) are allowed, leading finally to the diffraction pattern that is observed for KCl. One might therefore conclude that a representative diffraction pattern results for each individual crystal structure, which can be used for its identification, for example, in a mixture of different materials.

However, the analysis of diffraction patterns of real samples might be more complicated for different reasons as follows. For example, the crystallites (grains) inside a polycrystalline sample might have a preferred orientation (texture). As a consequence, the Bragg reflexes of all other orientations are extremely suppressed in their intensity compared to those expected from calculated structure factors. Such a behavior is well known, e.g., for vapor-deposited thin films (see, e.g., Refs. (118,119)). Furthermore, a diffraction pattern is also modified by the presence of defects in the crystal lattice. For example, the presence of stacking faults in the sample results in systematic shifts and a characteristic broadening of the diffraction peaks (120,121), i.e., some of the peaks remain stable in position and half-width, while others systematically move toward larger or smaller Bragg angles in conjunction with a broadening, and even peak splitting is observed for special defect structures. Stress and strain also cause systematic Bragg peak shifts (122). Thus, the detailed analysis of measured peak positions, their widths, and intensities can be used for the identification of existing defects.

When finite size effects are included, the peaks are found to be inversely related to the dimension of the diffracting region of the crystal; thus, the line broadenings provide information about the particle size D of small crystallites, for example, by application of the Scherrer formula (Equation (6.18)):

$$D = \frac{K\lambda}{\Delta\Theta \cos\Theta} \tag{6.17}$$

where K is the Scherrer constant, which depends on the shape of the crystallites under investigation and in most cases is close to 0.9, and $\Delta\Theta$ is the line width at half-maximum (in radians) after correction for instrumental broadening (see, e.g., Ref. (123)). Information about the particle size distribution can be obtained from a more complete line shape Fourier analysis (123–126). However, it is not trivial to separate the different contributions (123,124,126). In this context, the use of SR may simplify the data interpretation because, compared to a laboratory experiment, the experimental line width is drastically reduced with SR due to the extremely small divergence of the beam.

Limitations of XRD, even when using SR, are particles or domain sizes of the investigated samples of at least 2 to 3 nm to yield a measurable diffraction pattern in the sense of the Bragg equation. Thus, particles or domains with size smaller than about 2 to 3 nm will appear to be x-ray amorphous in an XRD experiment and will not reveal any sharp Bragg reflexes. In conclusion, XRD gives the opportunity of very detailed crystal structure determinations, although, in practice, the interpretation of the measured diffraction pattern may be complicated.

6.4.2 Surface-Sensitive X-Ray Diffraction

Due to the weak interaction of x-rays with matter, the intensities diffracted from crystalline thin films or electrode surfaces are generally weak as already mentioned. Therefore, surface-sensitive XRD tools are highly desired in order to amplify these contributions. In principle, there are two ways to perform surface-sensitive diffraction experiments: first, by using the small penetration depth that results for incidence angles below the critical angle of total reflection as already described in Section 6.3.2, and second by measuring Bragg peaks that result from the broken symmetry at the surface of an electrode (surface x-ray scattering, SXS). For both situations, the geometry of the diffraction experiment is highly asymmetric compared to the situation outlined in Figure 6.17. While the use of the grazing incidence geometry is applicable for all kinds of crystalline electrodes, SXS experiments require a working electrode with a well-defined structure, i.e., these studies are restricted to single-crystal electrodes.

In the grazing incidence geometry, the incident x-rays impinge on the surface of the sample at small angles ϕ and the detector is scanned along 2Θ in the same plane as the incident beam and the surface normal in order to record the x-rays diffracted from crystal planes inclined to the surface of the sample. As already seen in Section 6.3.2 and Figure 6.12, the penetration depth z_0 rapidly increases for glancing angles above the critical angle ϕ_c. Since the diffracted beams originate from this region of variable depth, the diffraction pattern for multilayered materials with depth-dependent structures, such as a metal electrode with a passive layer with a possible substructure on top, will depend on the incidence angle. Assuming that a phase is located in a depth between $z = a$ and $z = b$ from the surface, it follows that the diffracted intensity I is proportional to (127,128)

$$I \propto z_0(\phi)[e^{-a/z_0(\phi)} - e^{-b/z_0(\phi)}] \tag{6.18}$$

Thus, for a surface oxide film with $a = 0$ and $b =$ thickness t, one obtains an intensity proportional to

$$I \propto z_0(\phi)[1 - e^{-t/z_0(\phi)}] \tag{6.19}$$

For polycrystalline samples, each coherently diffracting region diffracts with an intensity given by Equation (6.19), and the total intensity measured by the detector is the sum of the contributions from all these regions. If they are not uniform in depth but have a distribution of distances a and b given by $f(a,b)$, then the total intensity diffracted from the considered phase is given by the convolution of $f(a,b)$ with Equation (6.19) (127,128). Thus, depth-dependent concentrations can be determined by assuming a model for the sample with given distribution functions $f_j(a, b)$ for the species j of interest, a calculation of the diffracted intensities as described above and the comparison to the experimental data (127,128). At this point, however, it should be mentioned that the peak positions of the measured diffraction peaks differ from the corresponding Bragg-angle values due to the refraction of the incident x-rays as they penetrate the air (or liquid)/solid interface (129).

When investigating the diffraction of a surface, it is reasonable to consider a 2D crystal of finite thickness. The basic cell vector $\mathbf{a}_3$ (see Equation (6.9)) perpendicular to the surface is chosen to equal this thickness. Diffraction is then still sharply peaked in both directions parallel to the surface; however, the Laue condition for the direction perpendicular to the surface is relaxed, and the intensity is continuous in the out-of-plane direction. This is due to the fact that the reciprocal lattice is made of rods perpendicular to the surface plane as schematically shown in Figure 6.18. If we define $\mathbf{Q} \cdot \mathbf{a}_3 = 2\pi l$, l is now taken as a continuous variable since intensity is present also for noninteger values of l. Intensity variations along the rod, i.e., as a function of Q_z or l, are solely contained in the structure factor; thus, they are related to the positions (z-coordinates) of the atoms within the unit cell of the 2D crystal. Generally, the rod modulation period gives the thickness of the layer and its amplitude is related to normal atomic displacements of the atoms within the layer. This is the case, e.g., for a reconstructed surface for which rods are found for fractional order values of h and k, i.e., well outside scattering from the bulk material (130–132). In reality, the surface layer is always present on top of a truncated crystal — such a crystal can be modeled by the product of a step function describing the electron density variation as a function of the coordinate perpendicular to the surface, and an infinite lattice. The resulting diffraction pattern is then the convolution of the 3D reciprocal lattice with the FT of the step function (133). An infinite number of Fourier components are necessary to build a step function and, as a consequence, nonzero intensity can be found in between the Bragg-peaks thus connecting them as a function of l (130,133) as schematically indicated in Figure 6.18. Similar to the alteration of specularly reflected intensity, also the intensity of the CTR is modified by the presence of surface roughness: If the surface is rough on an atomic level, the step function of the perfect crystal surface can be approximated by fewer Fourier components, so that

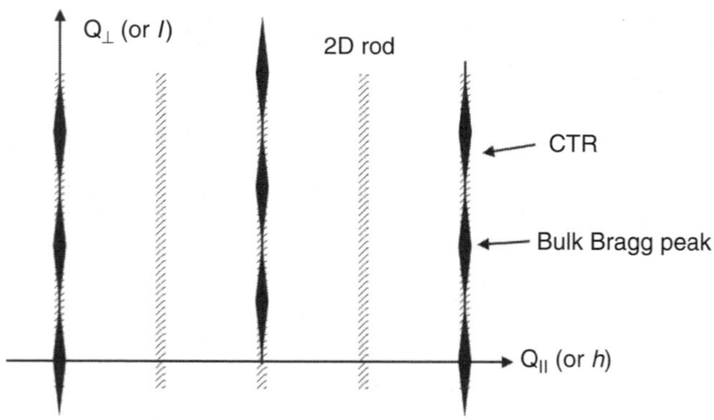

Figure 6.18 Schematic view of the reciprocal lattice of a 2D crystal consisting of continuous rods (shaded areas) and a 3D-crystal truncated by the surface, giving rise to the so-called crystal truncation rods (CTR). The latter have intensity maxima at the positions of the bulk Bragg positions and a quickly varying intensity in between.

the intensity between the Bragg peaks is generally smaller for a rough surface compared to a perfectly sharp interface (131,133).

6.4.3 Procedures and Data Evaluation

If it is desired to measure the intensity in the vicinity of a point in the reciprocal space, then the related motor positions of the diffractometer have to be known accordingly, and if the motors of the diffractometer are moved, it is essential to know the associated positions in reciprocal space. The required conversion mathematics is described in the literature (134,135). As can directly be anticipated from Figure 6.16 the net planes are not necessarily parallel to the surface of the sample, and thus the careful orientation and alignment of the sample and the used diffractometer is mandatory for proper XRD experiments. As indicated in Figure 6.16, **Q** can be separated into a component parallel ($\mathbf{Q}_{\|}$) and perpendicular to the surface ($\mathbf{Q}_z$ or $\mathbf{Q}_{\perp}$). The absolute value of $\mathbf{Q}_{\perp}$ is a simple function of the incidence and exit angles ϕ_i and ϕ_f as follows by geometrical considerations: $|\mathbf{Q}_{\perp}| = k(\sin\phi_i + \sin\phi_f)$. If ϕ_i and ϕ_f are very small, i.e., $\mathbf{Q} \approx \mathbf{Q}_{\|}$, the scattering plane is nearly parallel to the surface, and the diffracting net planes are perpendicular to it. Thus, in this scattering geometry, one has only to rotate the sample about its surface normal to bring these net planes to diffracting conditions, which occurs when they form an angle Θ with respect to both the incident beam and the scattered beam. This way, the long-range periodicity of the sample can easily be probed. In addition, it is often very useful to measure the scattered intensity as a function of $\mathbf{Q}_{\perp}$, which is obtained, e.g., by increasing ϕ_f while keeping grazing incidence. In general, for each diffraction peak under investigation, the measured intensities have to be corrected for the primary beam profile, the acceptance of the detector, and other experimental conditions such as the polarization of the used SR, which are usually not identical for all diffraction peaks (see, e.g., Ref. (130)). Thus, all measured Bragg intensities have to be corrected accordingly in order to be comparable to the corresponding calculated structure factors. Instructions for different types of diffractometers are given in Refs. (136,137). Since the intensity is proportional to F_{hkl}^2, the difference between calculated and measured structure factors has to be minimized for a number of different diffraction peaks taking into account also the uncertainties in the measurement of each diffraction peak, thus weak Bragg peaks are weighted less than strong ones. For the structure determination, a model has to be assumed and the atomic positions and their occupancies have to be varied in order to minimize the deviation between experiment and calculation by least-squares fit routines.

6.5 APPLICATIONS

6.5.1 Thin Oxide Films

Many metals — among those the technologically important transition metals Fe, Ni, and Cr — are protected against corrosion in aqueous environments by ultrathin oxide/hydroxide films with a thickness of typically 3 nm. Others like the valve metals Al, V, Ti, Ta, or Zr may form layers of thickness of several tenths of nanometers. The formation of these so-called passive layers stabilizes the reactive metal surface against corrosion. The enormous technological importance of passivity has led to intensive studies of the structure and chemistry of passive layers and the closely related protective properties. Especially the atomic defect structure of the passive films is of fundamental importance with regard to the corrosion resistance of the underlying metal. For example, the reductive dissolution of the passive film — which leads to a breakdown of passivity — has been associated with valence changes within the film in the case of iron as follows. In acidic solutions the outer part of the passive layer on iron mainly consists of Fe_2O_3. This Fe(III) oxide will be reduced at potentials negative to the Flade potential, i.e., negative to E_{Fl} = 0.59 to 0.059 pH to an Fe(II)-rich oxide layer that dissolves and leaves an unprotected Fe surface. In alkaline solutions, the Fe(II)-rich layer is more stable due to its low solubility at high pH and thus is still protective. Therefore, the change of Fe(III) to Fe(II) ions within the layer may be followed with XPS on immersed electrodes when the potential is decreased below E_{Fl} (138,139). In addition, the oxide turns to a hydroxide during reduction. At sufficiently negative potentials, it can be reduced even to metallic Fe, a process that is inhibited in the presence of Cr, i.e., for passive layers on FeCr alloys (139). Sudden changes of the potential will lead to chemical changes within the passive film, which then is submitted to stress with a related film breaking and may be one mechanism for the nucleation of corrosion pits in the presence of chloride within the electrolyte (140,141). Electrostriction may be another reason for film breaking during sudden potential changes and a serious increase of film breaking and pit nucleation. In this section, results of some recent investigations of passive films on iron, nickel, and copper will be presented.

6.5.1.1 Passive Layers on Iron

Passive iron has been studied since the 18th century for a comprehensive summary of the history; see, e.g., Refs. (138,142,143). However, although passive films on iron have been studied extensively for more than 250 years, there are contradictionary results about the atomic structure of these thin oxide layers. Electrochemical experiments and electron diffraction studies suggest that the film consists of an outer layer of γ-Fe_2O_3 and an inner layer of Fe_3O_4 (139,144–146). However, electron diffraction experiments must be performed *ex situ*, i.e., in a vacuum system after the removal of the electrode from the electrolyte, due to the incompatibility of electron beams and the aqueous environment. This, however, implies that the film structure may have altered, e.g., by the loss of OH, water, or a recrystallization procedure after the loss of potential control and the transport to the vacuum. Thus, *in situ* studies, among those applying EXAFS (63–65,147–149), surface enhanced Raman spectroscopy (SERS) (150,151), and Mössbauer spectroscopy (152,153), have been performed, leading to the conclusion that the passive film on iron is amorphous or has a very small grain size, and has a structure similar to iron hydroxides or oxyhydroxides. Recent *in situ* x-ray absorption near-edge data gave no evidence for octahedral Fe–O bonds such as in α-Fe_2O_3, α-FeOOH, or γ-FeOOH, but are compatible with a disordered structure with distorted coordination polyhedra (63). However, these x-ray absorption data can also be interpreted in terms of a spinell structure similar to γ-Fe_2O_3 or Fe_3O_4 (63). The latter interpretation is consistent with *in situ* STM experiments (154). Further complications arise from the preparation of the passive layer, since it is well known that significant amounts of Fe might dissolve during the oxidation of iron, especially in unbuffered solutions of sulfate or perchlorate (155). The products of any dissolution processes are ferrous ions, which can subsequently be oxidized to give a deposit like γ-FeOOH on the Fe-electrode surface (see, e.g., Refs. (156,157)). Therefore, particular care must be taken to avoid such artifacts during the preparation of passive layers. It has been shown that using potential steps to the passive region and a slightly alkaline borate buffer solution, any detectable iron dissolution can be prevented and the oxide film can grow by a solid-state reaction (63,139,158).

Therefore, XRD experiments with SR were performed on single-crystal Fe electrodes using well-defined electrochemical conditions in order to clarify the passive film structure in detail (159,160). A focused x-ray beam with energy of about 10 keV (≈1.24 Å x-ray wave length) was used. Solid-state Ge or Si(Li) detectors (discrimination of the Fe fluorescence) equipped with soller slits were used to analyze the diffracted beam. In the cited study (159,160), borate buffer (pH 8.4) was used. The crystals were immersed at −1.4 V (vs. mercurous sulfate electrode) and thus oxide formed in the laboratory air was removed by prolonged cathodic reduction. The iron oxidation was performed using potential steps to +0.4 V, i.e., well in the passive potential region, anodic to the active dissolution peak. While *in situ* data were collected with the potential continuously held at +0.4 V, the electrode was removed from the electrolyte after 60 min of oxidation and placed into a high-purity He atmosphere for the *ex situ* experiments. As already mentioned, the *ex situ* data do not suffer from background scattering from the electrolyte and thus enable the measurement of diffraction peaks with very low intensities above the background. *In situ* data were measured in an electrochemical cell (161,162) for comparison. This is illustrated in Figure 6.19 for the 4 0 $\bar{4}$ reflection on the Fe(1 1 0) electrode. The appearance of Bragg peaks first of all clearly demonstrates that at least a fraction of the passive layer is crystalline, with a well-developed crystallographic structure. While the peak position, intensity, and shape are identical for the *in situ* and the *ex situ* experiments, the pronounced increase of the background level is obvious. A series of different peaks was measured *in situ* and *ex situ* to ensure that the removal of the electrode did not change the structure of the passive layer. As a stringent test of the stability of the passive layer after the removal from the electrolyte and the loss

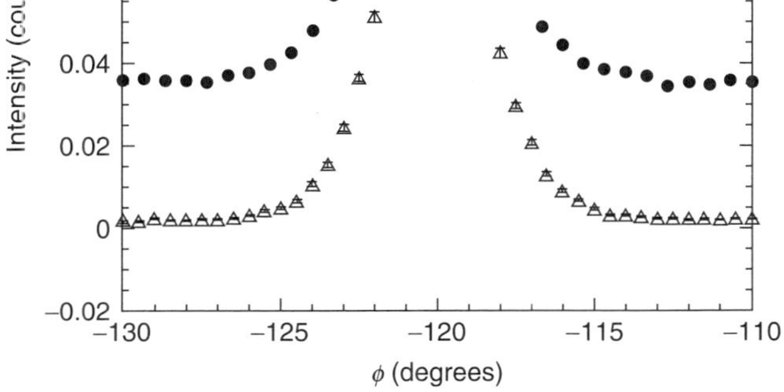

Figure 6.19 Comparison of the (4 0 $\bar{4}$)-diffraction peaks (ϕ-scans) of the passive iron oxide layer on Fe(110) measured *in situ* and *ex situ*. Note the increased background scattering level for the *in situ* experiment. (Taken from AJ Davenport et al., *J. Electrochem. Soc.* 147: 2162–2173, 2000. With permission.)

of potential control, structure factors were determined for the *in situ* and *ex situ* measurements. These are shown in Figure 6.20. Obviously, the *ex situ* structure factors are all within the error bars of the measured *in situ* peaks. Thus, it can be concluded that the immersed passive layer on iron is stable and not altered for the chosen conditions. A typical diffraction peak pattern is shown in Figure 6.21. The high quality of such *ex situ* data sets was then used for the crystallographic analysis and the refinement of the iron oxide layer structure as will be discussed below. From this diffraction pattern, the epitaxy of the passive film is apparent. For both iron surfaces, the [1 $\bar{1}$ 0] direction of the oxide layer is parallel to the [1 0 0] direction of the iron substrate. However, the (001) planes of the oxide layer on Fe(0 0 1) are parallel to the iron surface, while for Fe(1 1 0), the (1 1 1) plane of the oxide is parallel to the iron substrate surface (159,160). This orientational relation seems to be pinned by the similarity between the oxide ($\bar{2}$20) and the metal (100) spacings that amount to 2.96 and 2.89 Å, respectively. From these diffractograms, the lattice parameters of the oxide layer can be inferred to be about 8.39 $\pm$ 0.01 Å parallel to the iron surface and 8.3 $\pm$ 0.1 Å in the perpendicular direction. For comparison, the lattice constant of Fe_3O_4 is 8.394 Å (163), while that of γ-Fe_2O_3 is 8.3396 and 8.3221 Å (164). It should be mentioned that the lattice parameters from other bulk Fe oxides or Fe oxyhydroxides differ significantly from those measured for the passive film.

A closer inspection of the measured intensities of the diffraction peaks yields a more detailed insight into the structure of the passive iron oxide layer. Based on the measured symmetry, the lattice constants, and the detected Bragg peak intensities, the passive layer data are consistent with Fe_3O_4 and γ-Fe_2O_3, i.e., oxides of the spinell type. All other crystalline iron oxides, hydroxides, or oxyhydroxides are not compatible to the measured diffractograms. However, from Figure 6.22, it can be concluded that the passive layer structure is nevertheless different from Fe_3O_4 and γ-Fe_2O_3, as can be seen from the differences in the structure factors of the passive iron oxide layer and the spinell oxides. Even linear combinations of Fe_3O_4 and γ-Fe_2O_3 cannot adequately describe the experimental data. The detailed processing of the diffractograms comprised the analysis of 68 symmetry inequivalent peaks. It should be noted that the packing density of the oxygen anions in spinell oxides determines the lattice constant (165). From the similarity of the lattice constants of the passive layer and those of Fe_3O_4 and γ-Fe_2O_3, it can be derived that all the oxygen sites are fully occupied. Since the free energy of formation of vacancies and interstitials is small in spinells (166), models with randomly distributed cation vacancies and

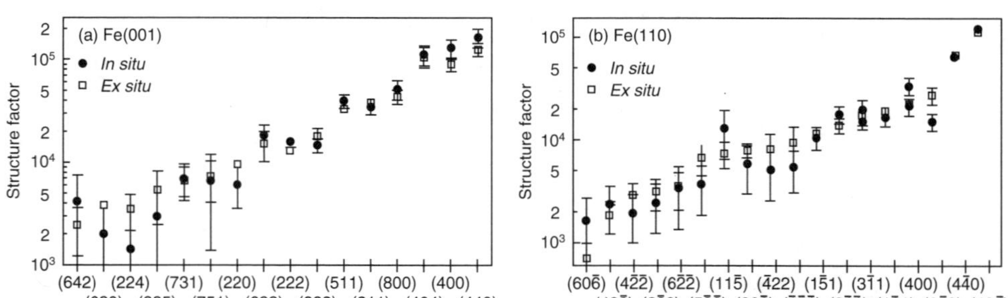

Figure 6.20 Comparison of the structure factors (F^2) determined *in situ* (filled circles) and *ex situ* (open squares) for the passive iron oxide layer on (a) Fe(0 0 1) and (b) Fe(1 1 0). (Taken from AJ Davenport et al., *J. Electrochem. Soc.* 147: 2162–2173, 2000. With permission.)

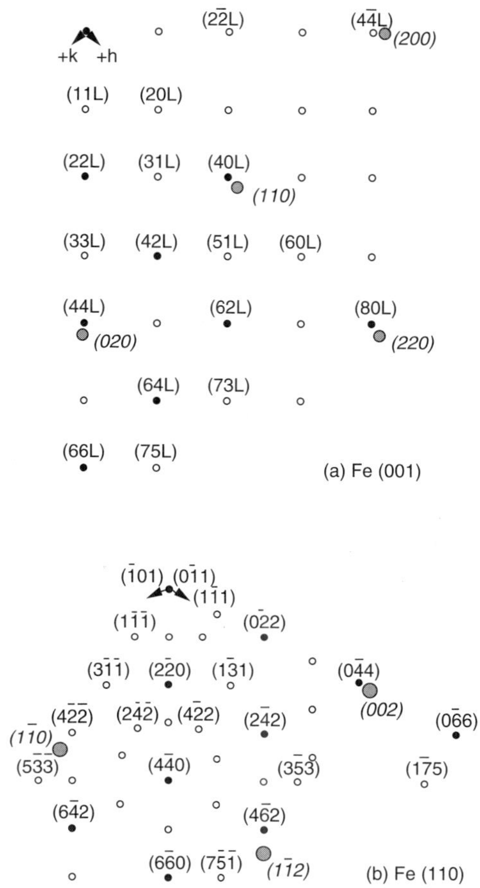

Figure 6.21 A single quadrant of the diffraction pattern for the passive iron oxide layer on (a) Fe(0 0 1) and (b) Fe(1 1 0). While the diffraction peaks of the iron metal substrate are represented by large gray filled circles, the peaks related to the passive layer are depicted as small circles (open and filled for pure in-plane peaks and nearly in-plane peaks, respectively). Only part of the out-of-plane peaks (labeled with an L in (a)) is shown for clarity. (Taken from MF Toney et al., *Phys. Rev. Lett.* 79: 4282–4285, 1997. With permission.)

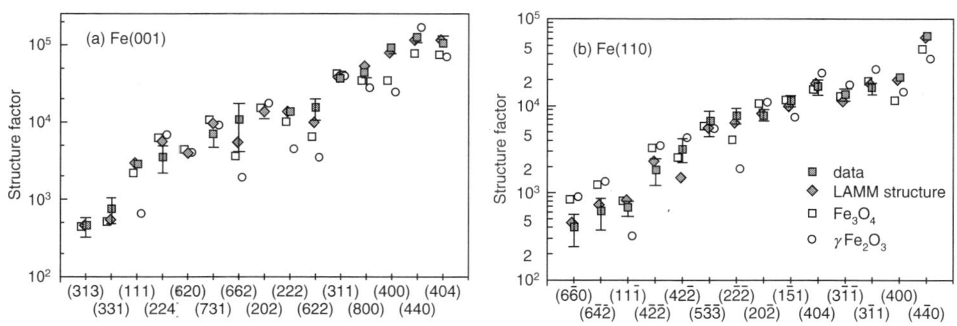

Figure 6.22 Comparison of the experimental structure factors (filled squares) with those of selected Fe-oxide model compounds for the passive layer on (a) Fe(0 0 1) and (b) Fe(1 1 0): open squares for Fe_3O_4, open circles for γ-Fe_2O_3, and filled diamonds for the "LAMM"-phase. (Taken from AJ Davenport et al., *J. Electrochem. Soc.* 147: 2162–2173, 2000. With permission.)

interstitials (called LAMM structure) have to be taken into account. Therefore, four additional fit parameters are needed: the occupancies of the octahedral and tetrahedral Fe sites, and the concentration of the octahedral and tetrahedral interstitial sites. The iron and oxygen atoms were fixed in the position for an ideal spinell with the lattice parameters given above. Besides, the lateral Debye–Waller factor for the Fe cations was varied in the range $\sigma_{\|}(\mathrm{Fe}) = 0.1$ to 0.15 Å, while the other Debye–Waller factors were fixed to the following values: $\sigma_{\perp}(\mathrm{Fe}) = 0.28$ Å, $\sigma_{\|}(\mathrm{O}) = 0.1$ Å, and $\sigma_{\perp}(\mathrm{O}) = 0.28$ Å (159,160). Compared to Fe_3O_4 and γ-Fe_2O_3, such a model fits the experimental data very well, as can be seen in Figure 6.22. Most of the structure factors calculated for the LAMM structure lie within the error bars of the experiment, and the goodness-of-fit parameter (reduced χ-squared) is significantly smaller. A detailed view of the LAMM structure is given in Figure 6.23. It should be mentioned that the spinell unit cell, which contains 32 oxygen anions, 16 octahedral and 8 tetrahedral cation sites (165), is fully occupied in Fe_3O_4, while for γ-Fe_2O_3, a quarter of the octahedral sites have only 33% occupancy, the remaining octahedral and tetrahedral sites being fully occupied. Both spinell reference compounds do not contain any interstitial ions. The authors found an octahedral site occupancy of 80% $\pm$ 10% and a tetrahedral site occupancy of 66% $\pm$ 10% for the passive film in the LAMM structure. While no evidence for tetrahedral interstitials was found, there are cations occupying 12% $\pm$ 4% of the available octahedral interstitial sites (159). The large Debye–Waller factors (0.1 up to 0.28 Å), which are needed to fit the experimental data, are indication of the static disorder in the film, which results from the bond length variations induced by the interstitials and vacancies.

From the geometric situation in the LAMM structure it can be anticipated that there is a correlation between tetrahedral vacancies and octahedral interstitials, since the interstitial cations have a high probability of occupying sites close to an unoccupied tetrahedral site. In addition, since the appearance of iron in a higher oxidation

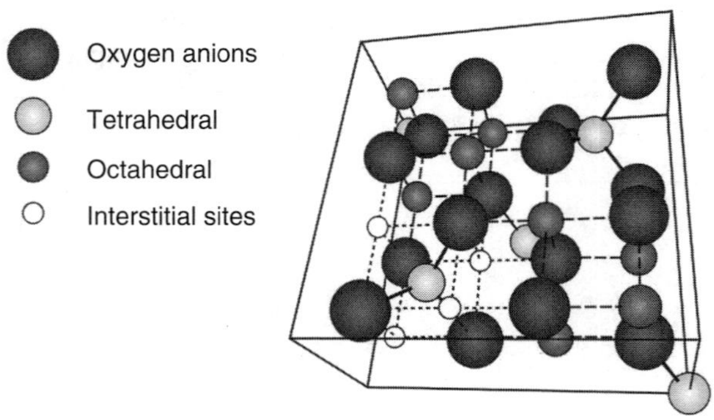

Figure 6.23 Schematic illustration of the passive film structure (LAMM structure). The solid black lines are the borders of the bottom half of the unit cell (big spheres: oxygen anions, fully occupied; smaller light gray spheres: tetrahedral cation sites, 66% occupancy; dark gray spheres: octahedral cation sites, 80% occupancy). Four of the eight octahedral interstitial sites are indicated by small white spheres, four additional sites are located in the upper right hand section, which is not shown. Note that the interstitials are shown in well-defined positions for clarity, while in the LAMM structure used for the fit of the diffractograms, they are distributed randomly in the structure. Bonds are also indicated for clarity: Bold, dashed, and dotted lines are for the tetrahedral, octahedral, and octahedral interstitial bonds, respectively. (Adapted from MF Toney et al., *Phys. Rev. Lett.* 79: 4282–4285, 1997. With permission.)

state is more likely at the oxide/solution interface than at the inner metal/oxide interface, it seems to be likely that interstitials are located at the inner interface with a higher probability, while vacancies are expected at the surface toward the solution accordingly. Such a nonuniform distribution of interstitial and vacancy sites is however not included in the data analysis so far and might be subject for future investigations. Nevertheless, it has to be stressed that the proposed LAMM structure is different from the duplex layer structure with an inner Fe_3O_4 and an outer γ-Fe_2O_3 sublayer as proposed by some authors (144–146) for the following reasons: first, the octahedral vacancies are randomly distributed in the passive film and not located at specific positions as in γ-Fe_2O_3 or absent as in Fe_3O_4. Moreover, in γ-Fe_2O_3 as well as in Fe_3O_4, the tetrahedral sites are fully occupied and there are no octahedral interstitials (159,160). In addition, the stoichiometry $Fe_{1.9\pm0.2}O_3$ determined from the best-fit structure suggests that most of the iron in the passive film is in the Fe^{3+} state, in agreement with other investigations (63,139,158).

Besides the crystallographic structure of the passive layer, which is given by the peak positions of the detected Bragg peaks, more information is contained in the width of these peaks. First, the lateral crystallite size can be calculated to about 60 Å for the passive film on Fe(0 0 1) and ca. 45 Å for that on Fe(1 1 0) (159,160). Therefore, the passive film is best described as a nanocrystalline material. In addition, the peak width also indicates that the oxide has numerous additional planar defects, namely stacking fault and antiphase boundaries. Without going too far into the details, a summary of the defect information derived from the diffractograms is given in Figure 6.24. As the related data analysis is far beyond the scope of this article, we therefore refer to the original literature (160).

6.5.1.2 The Passive Layer on Ni(1 1 1) in Sulfuric Acid

The excellent corrosion behavior of nickel was previously related to a densely packed oxide film that was investigated with various different experimental techniques in the past (167–172). It was found that the passive layers formed in sulfuric acid mainly consist of NiO and $Ni(OH)_2$ with a thickness of only a few monolayers (169–171). Especially the results of depth-sensitive surface analytical experiments suggest a duplex structure of the film, with an inner NiO layer and an outer $Ni(OH)_2$ layer (170). Furthermore, the inner, nearly stoichiometric oxide layer was assumed to be crystalline, while the outer hydroxide was attributed to be amorphous (172,173). *Ex situ* STM of passivated Ni electrodes with atomic resolution revealed hexagonal structures with lattice constants close to bulk NiO and a characteristic arrangement of steps (174,175). The latter was interpreted to an oxide phase that is tilted with regard to the underlying Ni metal electrode, as confirmed by *in situ* STM investigations (171,176,177). However, although STM enables the direct visualization of single atoms in real space, the lattice constants determined by STM are afflicted with relatively high uncertainties. For the NiO/$Ni(OH)_2$ system, this is a major limitation, since the lattice constants of both Ni compounds (4.177 Å (163) and about 4.166 to 4.178 Å (178)) are only slightly different. In addition, due to the fact that STM only allows a local characterization of the passive layers at their surfaces, no insight into the internal passive film structure is possible and a statistical assessment of the measured data is rather difficult and has not been reported in the literature so far. However, all these limitations can be overcome by *in situ* XRD experiments, which allow the accurate determination of the crystalline structure of the passive layer on Ni electrodes as will be described below (179).

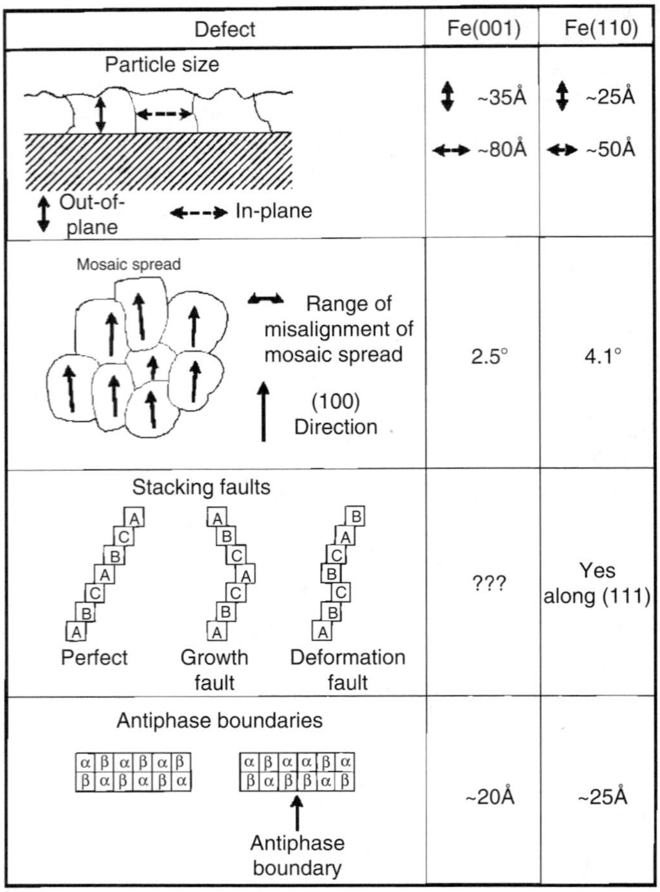

Figure 6.24 Summary of the defect information on the passive layers on Fe(0 0 1) and Fe(1 1 0) determined from x-ray diffraction experiments. The schematic pictures are sideviews representing the corresponding defect structure. (Taken from AJ Davenport et al., *J. Electrochem. Soc.* 147: 2162–2173, 2000. With permission.)

Polished and electropolished single-crystal Ni(1 1 1) electrodes that were subsequently annealed under a reductive atmosphere in order to obtain atomically smooth surfaces were used as working electrodes. The experiments were performed in 0.05 *M* H_2SO_4 (pH 1.0) in an *in situ* cell with a thin window, which was deflated for the electrochemical reductions and oxidations. The x-ray measurements were performed with an x-ray wavelength of 1.20 Å (corresponding to an energy of about 10,300 eV) and a liquid nitrogen cooled Ge detector (179). The hexagonal coordinate system of the Ni(1 1 1) substrate (180) served as a basis for the diffraction experiments, thus all the detected Bragg peaks are transformed to the following scattering wave vector units with $\mathbf{Q} = (a_{Ni}^*, b_{Ni}^*, c_{Ni}^*) \cdot (H, K, L)$ (179). The reciprocal lattice vectors parallel and perpendicular to the surface are $a_{Ni}^* = b_{Ni}^* = 8\pi/\sqrt{6}a_{Ni} = 2.912\,\text{Å}^{-1}$ and $c_{Ni}^* = 2\pi/\sqrt{3}a_{Ni} = 1.029\,\text{Å}^{-1}$, respectively, with $a_{Ni} = 3.5238$ Å. A saturated Ag/AgCl electrode, to which all the potentials given are quoted, served as reference electrode. Prior to the formation of a passive layer on the Ni electrodes, the oxide layer formed in the laboratory air was removed by electrochemical reduction at −0.60 V for typically 15 min. The complete reduction was proved by the absence of any oxide-related diffraction peaks in the diffractogram. Subsequently, the anodic formation of the passive layer was initiated by a potential step into the passive range. As indicated by x-ray reflectivity measurements and additional STM investigations, the passivation is accompanied by a significant increase of the surface roughness, which might be related to a possible Ni dissolution.

Peak profiles measured for three low-order diffraction peaks are shown in Figure 6.25, where the intensities measured for the air-formed oxide are compared to those of the passive layer. The diagram shows that very similar structures are formed in both cases. A more detailed analysis of the detected peak positions and intensities further indicates the absence of a crystalline $Ni(OH)_2$ phase that is in quantitative agreement with a previous investigation of the passive layer on Ni, which indicated that the outer $Ni(OH)_2$ part of the passive layer is x-ray amorphous (171). The integrated NiO peak intensities of the passive film are in reasonable agreement with the structure factors calculated for crystalline NiO. However, there are some substantial differences in both diffractograms: while the air-formed oxide shows a diffraction peak doublet for the (0, 0.836, *L*) and the (0.836, 0, *L*) scans, only a single peak was found in the related diffractograms of the electrochemically formed passive layer (179). More generally, while the air-formed oxide obeys peaks for $(a_{NiO}^* \cdot h', a_{NiO}^* \cdot k', c_{NiO}^* \cdot (h' + 2k' + 3l'))$ and at $(a_{NiO}^* \cdot h', a_{NiO}^* \cdot k', c_{NiO}^* \cdot$

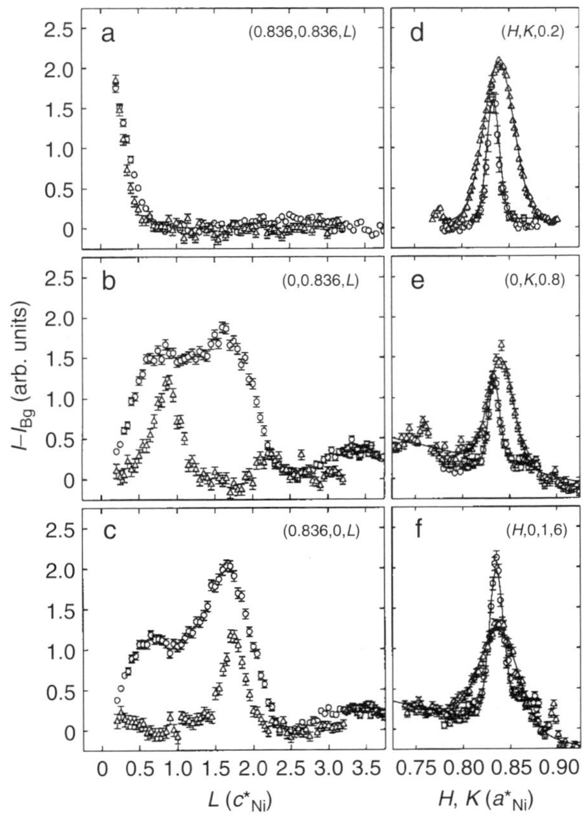

Figure 6.25 Diffraction peak profiles measured for the air-formed oxide (circles) and the passive film (triangles) formed at +0.50 V in 0.05 *M* H_2SO_4 (pH 1.0) on Ni(1 1 1) for different scan directions: (a) to (c) radial scans, (d) to (f) transverse scans. The background scattering (I_{BG}) by the electrolyte and the x-ray windows was already subtracted from the *in situ* data. (Taken from OM Magnussen et al., *J. Phys. Chem. B* 104: 1222–1226, 2000. With permission.)

$(2h' + k' + 3l'))$ where $h', k', l' = 0, \pm 1, \pm 2, \ldots$, the passive layer only shows the latter series of peaks. The diffraction pattern of the air-formed oxide can be related to two twin-related, (111)-oriented NiO phases with the in-plane orientation parallel or antiparallel to the underlying Ni metal substrate. Therefore, the ABC stacking sequence of the Ni-fcc-lattice is followed by an ABC- or a BAC-sequence in the NiO film. This implies that the electrochemically grown passive layer only shows one orientation, namely the antiparallel orientation, while the parallel oriented NiO phase is completely absent (179). It should be mentioned that this preferred antiparallel orientation of the NiO passive layer can also be obtained if the reduction procedure of the air-formed oxide layer is omitted; thus, the structure can be reverted from a film with nearly equal amounts of parallel and antiparallel aligned domains to a preferred antiparallel oriented one.

One might argue that the parallel oriented fractions of the air-formed film are simply converted to an antiparallel orientation under electrochemical conditions, implying that the latter orientation is more stable in the electrolyte. However, the real situation is much more complex because the air-formed oxide is dissolved in a first step and reformed in the antiparallel orientation in a second stage (181). According to the presented results, the crystalline part of the oxide structure is directly affected by the environment, which again demonstrates the importance of *in situ* measurements.

It should be mentioned that the peak widths of the measured diffraction peaks (σ_R and σ_T for the radial and transverse widths) are in quantitative agreement with the assumption of a perfectly (anti-)parallel lattice orientation of the air-formed Ni oxide film with respect to the underlying Ni(1 1 1) substrate, i.e., independent their actual L-position, the same σ_R is found for all peaks belonging to the same orientation. Thus, the (1 1 1)-oriented planes of the air-formed Ni oxide are perfectly parallel to those of the substrate. However, for the antiparallel orientation, the measured peak widths ($\approx 0.034\,\text{Å}^{-1}$) are nearly twice as large as those corresponding to the parallel orientation ($\approx 0.020\,\text{Å}^{-1}$). If the domain size is calculated using the Scherrer formula, these peak widths correspond to crystallites of about 70 to 120 Å (179). All transverse peaks are generally well described by gaussians as can be seen in Figure 6.25(d)–(f), the widths of these peaks are only slightly broader compared to the radial direction.

Obviously, the situation is much more complex for the passive films formed in sulfuric acid. In this case, both σ_R and σ_T depend strongly on the L-position of the diffraction peaks. For example, for the $(0, a_{NiO}^*, c_{NiO}^*)$, $(a_{Ni}^*, 0, 2c_{NiO}^*)$, and $(0, 0, 3c_{NiO}^*)$ reflections, the average values determined for σ_R from six independent measurements are 0.042, 0.062, and $0.110\,\text{Å}^{-1}$, respectively. A similar L-dependence was also found for the transverse widths of these diffraction peaks as well as for higher-order peaks. It was shown that this behavior can be related to a slight tilt of the passive layer lattice with regard to the underlying Ni electrode (179) as schematically shown in Figure 6.26, in agreement with *ex situ* STM results by Maurice and coworkers (174,175). They observed a highly stepped film morphology, which was attributed to a tilted growth of the passive layer lattice on the Ni surface. The tilt angle α of about $(8 \pm 5)^\circ$ was calculated from the mean terrace width L and the step height d ((1 1 1) lattice spacing) by $\sin \alpha = d/L$ as schematically indicated in Figure 6.26 (174,176). The authors claimed that the epitaxial mismatch of the tilted oxide lattice with respect to the (1 1 1) Ni surface is minimized. In addition, also for anodic Cu oxide layers formed in 0.1 M NaOH, a crystalline oxide phase tilted with respect to the underlying Cu metal was found (182). A schematic view of the diffraction

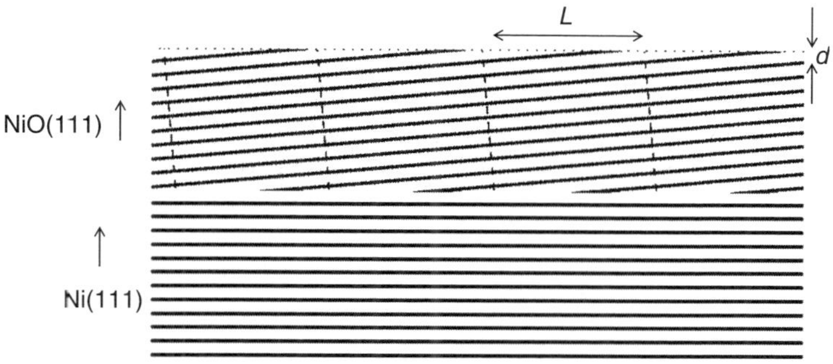

Figure 6.26 Schematic representation of a section view through the Ni(1 1 1) electrode covered with an crystalline NiO layer. The orientation of the (1 1 1) layer of the oxide phase is slightly tilted against that of the metal (1 1 1) orientation as indicated. The resulting periodic array of monoatomic steps (length L and step height d corresponding to the (1 1 1) lattice spacing of the oxide) at the surface toward the electrolyte, which was observed by several *in situ* and *ex situ* STM investigations (171,174,177), can easily be seen. Please note that the lattice spacings of the oxide and the metal are not drawn to scale to their real crystallographic values.

pattern, which is expected for a tilted NiO phase on the (1 1 1) Ni surface, is depicted in Figure 6.27. As can be seen, a tilt of the NiO lattice by an angle α would result in a shift of the (0, 0, $3c_{NiO}{}^*$) reflection away from the specular axis. Similarly, the ($a_{NiO}{}^*$, 0, $2c_{NiO}{}^*$) and the (0, $a_{NiO}{}^*$, $c_{NiO}{}^*$) peaks shift to higher L and lower H, and lower L and lower K, respectively. Due to the threefold lattice symmetry and the twofold tilt symmetry, six different domains should be present in the diffractograms, as schematically depicted in the inset in Figure 6.27, and all of these domains should have the same probability for symmetry reasons if all domains are tilted by the same angle α. However, peaks related to such a NiO phase with a well-defined lattice tilt were not observed by the authors (179). Instead, a detailed mapping of the reciprocal space around the ($a_{NiO}{}^*$, 0, $2c_{NiO}{}^*$) and the (0, $a_{NiO}{}^*$, $c_{NiO}{}^*$) diffraction peaks shows that there are no discrete satellite spots as shown in Figure 6.28. Moreover, the space

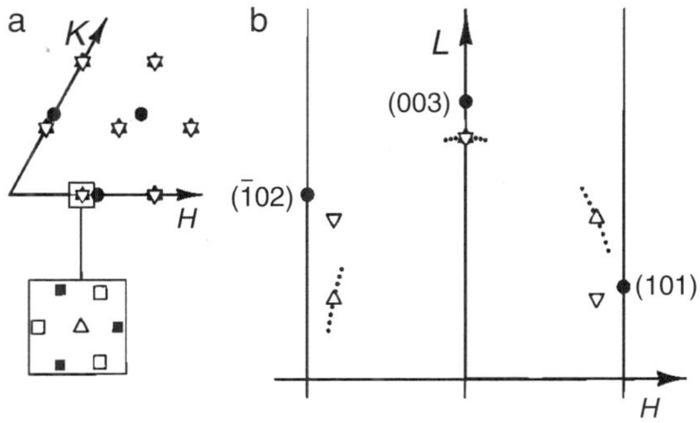

Figure 6.27 Reciprocal space pattern showing (a) the in-plane and (b) the out-of-plane positions of the Ni substrate peaks (filled circles) as well as the crystal truncation rods (solid lines). Peaks of the NiO lattice with parallel (◁) and antiparallel (△) orientations are indicated as well as a shift of these peaks induced by a tilt of the oxide lattice with regard to the metal substrate by $\alpha = \pm 2, \pm 4, \pm 6$, and $\pm 8°$ (small dots). In the inset of (a), the effect of a tilt by $\alpha = +8°$ (open squares) and $\alpha = -8°$ (filled squares) on the position of the ($a_{NiO}{}^*$, 0, $2c_{NiO}{}^*$) diffraction peak are indicated. (Taken from OM Magnussen et al., *J. Phys. Chem. B* 104: 1222–1226, 2000. With permission.)

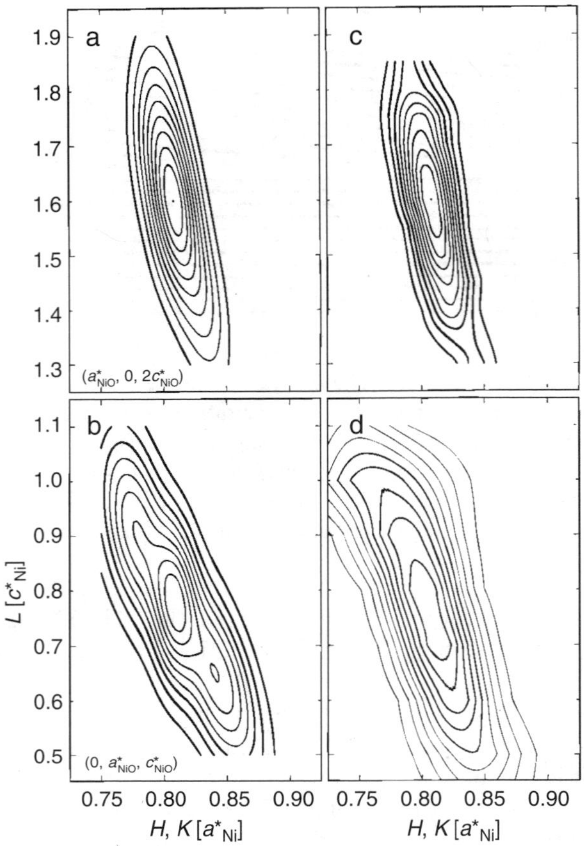

Figure 6.28 Experimental ((a), (b)) and calculated ((c), (d)) contour plots around the (a_{NiO}*, 0, 2c_{NiO}*) and the (0, a_{NiO}*, c_{NiO}*) diffraction peaks of the electrochemically formed passive layer on Ni(1 1 1) in 0.05 *M* H_2SO_4. For the calculations, continuous distributions of the tilted phases were assumed (for details, see text). (Taken from OM Magnussen et al., *J. Phys. Chem. B* 104: 1222–1226, 2000. With permission.)

map can neither be described as an ellipsoid elongated along *L*, which one may expect for an untilted phase. The curved shape that is observed can be described by a distribution of oxide phases with small, slightly different tilt angles. The tilt distribution was modeled by a distribution function $f(\alpha)$ and the scattering from each of these domains was treated as a 3D gaussian with widths σ_L and $\sigma_H = \sigma_K$. For the fitting procedure, the width, position, and amplitude of the distribution function $f(\alpha)$ was varied. Several different types of functions were tested, the best result was obtained for a Gaussian distribution with a peak position $\alpha_0 = \pm 3.3°$ and a width $\sigma_a = 1.8°$ (179). It has to be mentioned that this result is in agreement with recent *in situ* STM examinations of the passive layer on Ni(1 1 1), which showed a local tilt between 0.7° and 3.4° with regard to the (1 1 1) axis (171,177). Besides the tilt distribution, the fit procedure also provides values for the in-plane and out-of-plane lattice spacings, which are $a_{\mathrm{NiO}} = 2.959 \pm 0.004$ Å and $c_{\mathrm{NiO}} = 7.50 \pm 0.01$ Å (179). While the in-plane spacing of the passive film is identical to the corresponding value of the bulk material ($a_{\mathrm{NiO}} = 2.9549$ Å at 20°C), along the surface normal, a lattice expanded by about 3.7% relative to the bulk value ($c_{\mathrm{NiO}} = 7.227$ Å at 20°C) was found. In contrast, the air-formed oxide is expanded in both directions by

1.1% (in-plane) and 2.5% (out-of-plane). Furthermore, an averaged in-plane domain size of about 80 Å and an oxide thickness of about 24 Å were calculated from the fits for the electrochemically formed passive layer. These values agree with previous *ex situ* investigations by XPS (169,170) and are considerably larger compared to the air-formed oxide layer. A more detailed analysis of the NiO film structure like that for the Fe oxides, which we discussed in the previous section, is not possible in the case of the Ni passive-oxide since the number of Bragg reflections that are accessible in an *in situ* experiment is too small for a sophisticated refinement of the film structure. The low intensity of higher-order reflections together with the broad distribution of tilt angles further complicates the interpretation of the diffraction patterns. In addition, changes of the film structure after the removal from the electrolyte and the loss of potential control prevent investigations in a quasi *in situ* manner as described for the passive layer on iron in the previous section.

However, the results of the presented x-ray scattering study have important consequences for the corrosion stability of Ni electrodes. The coexistence of parallel and antiparallel domains in the air-formed oxide implies the presence of twin boundaries between the individual domains. These boundaries reveal a rather open structure in the direction normal to the surface, which provide pathways for an easy diffusional mass transport from or to the underlying Ni metal electrode (183). In the case of the electrochemically formed passive film, the density of these twin boundaries is at least significantly reduced compared to the air-formed film. In addition, the occurrence of small-angle boundaries between the individual domains still allows a dense packing of the structure. Thus, the passive layer does not show up these channells with facilitated mass transport, which might — together with the higher thickness — explain the better corrosion resistance of the passive film compared to the air-formed oxide film.

6.5.1.3 *The Passive Layer on Polycrystalline Cu*

Anodic oxide layers and passive layers on Cu have been intensively examined with electrochemical and optical techniques as well as x-ray and UV photoelectron spectroscopy (XPS and UPS), and x-ray reflectivity measurements (184–194). There is general agreement that the oxidation of Cu in alkaline media proceeds in two well-separated steps: starting from a reduced metal surface, first a thin Cu(I) layer, mainly consisting of Cu_2O, is formed, while at higher potentials, a duplex-type layer with a Cu(II)-oxide/hydroxide on top of the above mentioned Cu_2O layer is formed (188,191). A typical cyclic voltammogram is given in Figure 6.29. In this figure, the reduction of the air-formed Cu oxide layer, which mainly consists of a 2 to 4 nm thick Cu_2O layer (100,195), is also indicated by a dashed line. In addition, a schematic representation of the Cu phases that are detected at certain potentials is given. According to surface analytical studies, the thickness of the Cu(I) layer amounts to about 1 to 2 nm, while the total thicknesses of the duplex layer are in the range from 2 to 5 nm (188). While the atomic structure of thermally formed, thick (about 100 nm thickness) oxide layers on copper has been examined with XRD (196) and XAS (197) in the past, only little is known about the structure of thin passive films on Cu surfaces, which are formed during the anodic oxidation of copper in aqueous media. An *in situ* study has recently shown that the anodic oxidation and reduction of Cu can be monitored using x-ray reflectivity measurements (194): the presence of a thin surface oxide modulates the measured reflectivities as a function of the grazing angle. Furthermore, the data evaluation reveals that the surface roughness of the thin

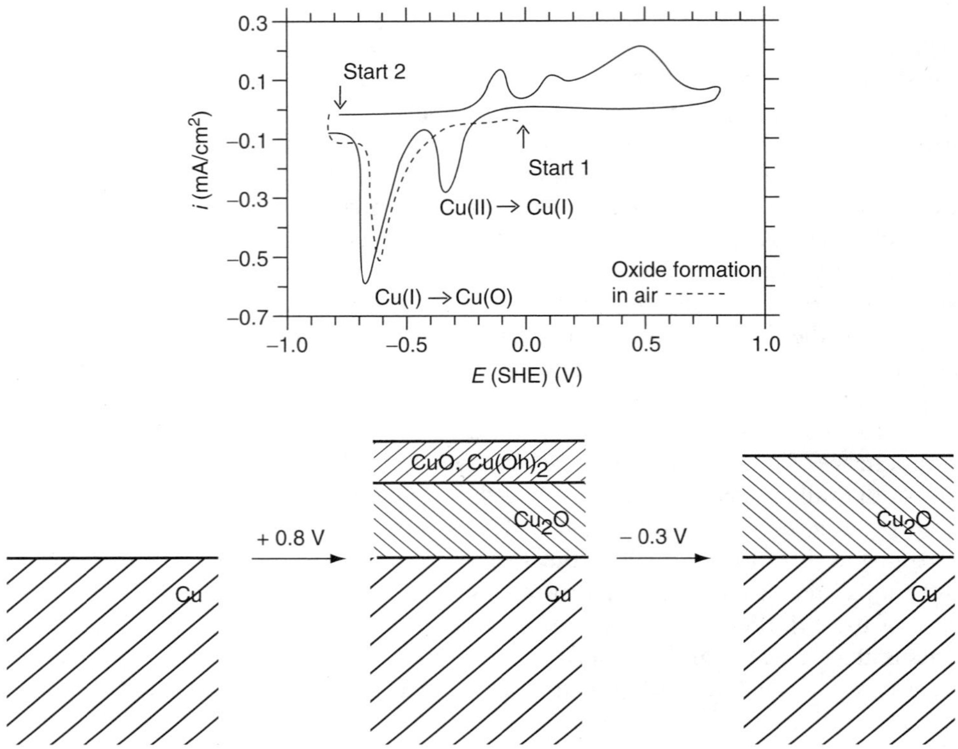

Figure 6.29 Cyclic voltammogram of Cu in 0.1 *M* NaOH (solid line) together with a schematic representation of the potential-dependent composition of the passive layer. The dashed line indicates the reduction of the air-formed oxide prior to the cyclic voltammogram. (From P Borthen. Grundlagen und Anwendungen der Röntgen-Absorptionsspektroskopie bei streifendem Einfall. PhD-dissertation, Heinrich-Heine-Universität Düsseldorf, 1996 (in German). With permission.)

film copper electrode in alkaline borate solution increases significantly from about 11 to about 15 Å after a complete oxidation/reduction cycle (194). In addition, the thickness of the metallic copper layer has decreased from 28.5 to about 27.5 nm after such a cycle, indicating that copper dissolution occurs to a measurable amount (194), as confirmed by earlier studies with a rotating ring disc electrode (198). In the cited study, however, the anodic oxide layer on Cu was treated as a homogeneous medium, which is however not true as revealed by *ex situ* XPS and ISS measurements (184,188,198,199). Thus, more elaborated experiments are necessary to clarify this situation in more detail. In Figure 6.30(a), near-edge x-ray absorption data measured in the vicinity of the Cu K-edge of a Cu electrode in 0.1 *M* NaOH are presented for different potentials as indicated (100). Significant changes were observed as a function of the oxidation potential. Calculations based on the application of the Fresnel theory using the optical constants of polycrystalline Cu, Cu_2O (cuprite), and CuO (tenorite) are given in Figure 6.30(b). By comparing the experiments and the calculations, it is obvious that the schematic presentation of the passive film structure given in Figure 6.29 is able to describe the measured data. For potentials up to 0.0 V, only Cu_2O signals were found. The growth of a Cu(II) oxide starts for potentials anodic to 0.0 V. The presence of this Cu(II) layer dramatically influences the shape of the measured spectra at +0.2 and +0.8 V. Having in mind that the amplitude of the x-ray field

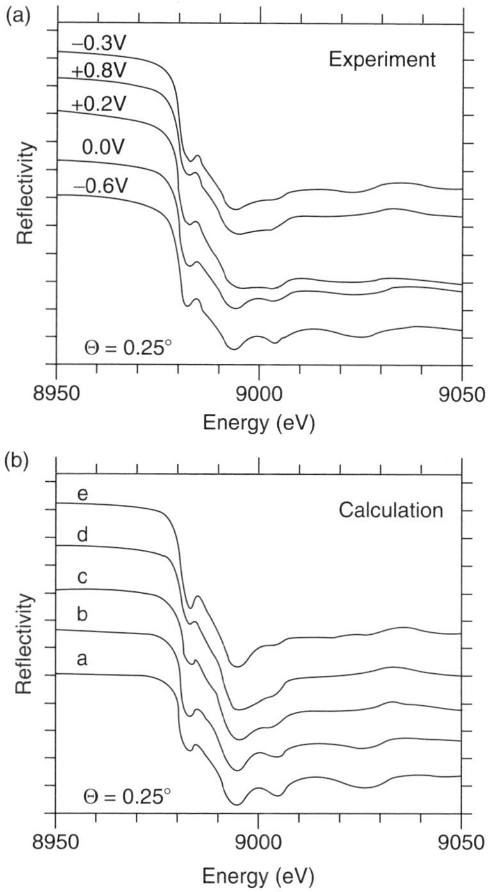

Figure 6.30 (a) Reflection-mode x-ray absorption spectra (glancing angle $\phi = 0.25^\circ$) of a copper electrode in 0.1 *M* NaOH for different potentials as indicated. (b) Calculated reflectivity spectra for a Cu electrode covered with a duplex oxide layer consisting of two sublayers with different individual thicknesses (t_1 = outer CuO sublayer, t_2 = inner Cu_2O sublayer) as follows: (a) oxide-free Cu surface ($t_1 = 0$, $t_2 = 0$), (b) $t_1 = 0$, $t_2 = 1$ nm, (c) $t_1 = 1$ nm, $t_2 = 1$ nm, (d) $t_1 = 1$ nm, $t_2 = 1.5$ nm, (e) $t_1 = 0$, $t_2 = 3$ nm. (From P Borthen. Grundlagen und Anwendungen der Röntgen-Absorptionsspektroskopie bei streifendem Einfall. PhD-dissertation, Heinrich-Heine-Universität Düsseldorf, 1996 (in German). With permission.)

decreases exponentially with depth, it is clear that the spectra are governed by those parts of the electrode that are in contact with the electrolyte. Thus, the experiments are able to prove the model of the passive layer that is given in Figure 6.29. In addition, the results suggest that the CuO layer is crystalline, in agreement with recent STM investigations with atomic resolution. In these studies, the cuprite structure is found for the Cu_2O layer in the passive range of anodic peak at about −0.1 V (SHE) of the polarization curve (Figure 6.29) (182,200). For the duplex films at potentials positive to ca. +0.3 V (SHE) the outer CuO part has the structure of tenorite (201). Both oxide films have an epitaxial relation to the orientation of the substrate crystals. The reduction of the CuO layer at −0.3 V leads to significant changes of the near-edge spectrum that are in quantitative agreement with the presence of a homogeneous Cu_2O passive layer.

Results of *in situ* EXAFS measurements of a passivated Cu electrode in 0.1 *M* NaOH are presented in Figure 6.31 and Figure 6.32. For these experiments, the Cu electrode was oxidized at +0.8 V for 5 min and, subsequently, the CuO sublayer was reduced at −0.3 V. This procedure was applied in order to obtain a passive layer that only consists of Cu(I) oxide (100,184,191). Obviously, the presence of the passive layer is damping the intensity of the fine structure oscillations of the metal; this can easily be seen by comparing the spectrum of the reduced metal surface (spectrum (a) in Figure 6.31) with that of the oxidized surface at the same grazing angle (spectrum (c)). In Figure 6.32, the magnitude of the FTs is presented for the oxide-free surface (a) and the oxide-covered surface (b). The dashed lines are fits to the experimental data. The agreement between measured and calculated data is good for both situations, i.e., the passive layer structure on Cu is very similar to cuprite for the present situation. The thickness of this Cu_2O layer is 3.0 nm according to these calculations, which agrees quantitatively with results of surface analytical studies (188,191).

Another interesting experiment on the passive layer on Cu and its corrosive attack was performed using specular and nonspecular x-ray reflectivity (202). The passive layer of about 4 nm thickness was prepared anodically in a sodium bicarbonate solution, which is known to show pitting of the Cu electrode for sufficiently high anodic potentials. Off-specular reflectivity measurements show that the lateral correlation length of the oxidized surface (without pits) is less than about 10 nm, indicating the presence of small oxide crystallites with a roughness of 2.2 nm, compared to about 1.8 nm for the oxide-free surface (202). The application of nonspecular reflectivity measurements now enables to study the development and the evolution of pits as function of time and potential *in situ*, since the presence of pits modifies the diffuse scattering profiles in a characteristic manner. For these experiments, the detector is fixed in a certain 2Θ- or q_z-position, while the sample angle is slightly varied from the specular conditions. This is shown in Figure 6.33 for a potential of +0.5 V (vs. SCE) and different holding times. The side lobes that develop on both sides of the broadened specular peak (transverse momentum transfer $q_x = 0$) are features associated with scattering from the pits (202). A detailed analysis and

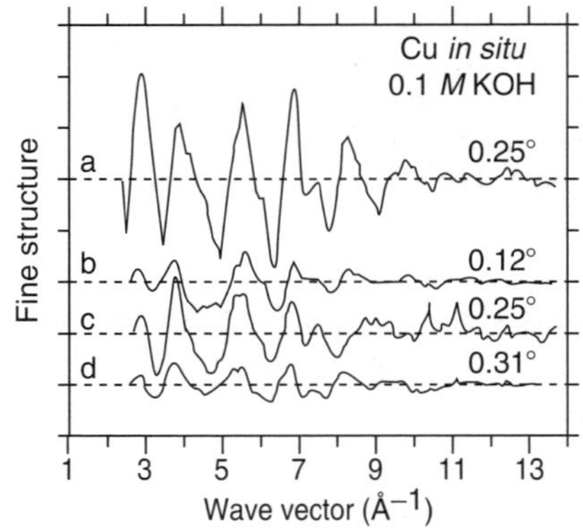

Figure 6.31 Fine structure oscillations determined from reflection-mode EXAFS investigations of a Cu electrode in 0.1 *M* NaOH for different potentials: (a) reduced Cu electrode at −0.6 V. (b) to (d) Oxidation of the electrode at +0.8 V for 5 min and subsequent reduction of the CuO sublayer at −0.3 V. Note the reduction of the fine structure oscillations in spectrum (c) compared to (a), which is caused by the presence of the passive layer. (From P Borthen. Grundlagen und Anwendungen der Röntgen-Absorptionsspektroskopie bei streifendem Einfall. PhD-dissertation, Heinrich-Heine-Universität Düsseldorf, 1996 (in German). With permission.)

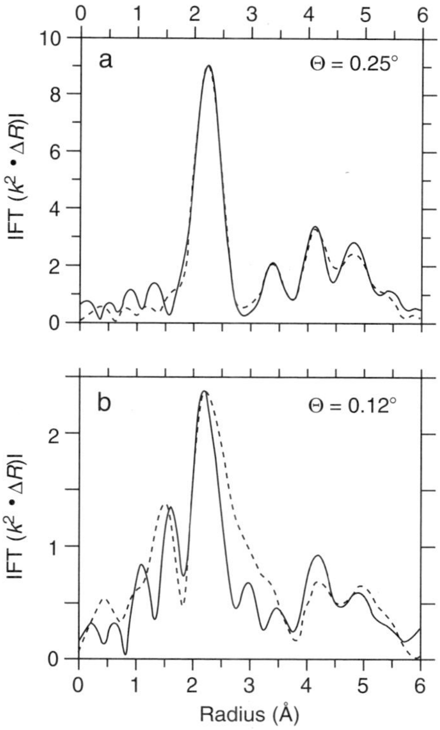

Figure 6.32 Fourier transforms of the fine structure oscillations determined from reflection-mode EXAFS investigations of a Cu electrode in 0.1 *M* NaOH for different potentials: (a) reduced Cu electrode at −0.6 V ($\phi = 0.25°$). (b) Oxidation of the electrode at +0.8 V for 5 min and subsequent reduction of the CuO sublayer at −0.3 V ($\phi = 0.12°$). Full lines: experimental data; dashed lines: fit lines assuming a pure, polycrystalline Cu electrode in (a). In (b), this Cu electrode is covered with a Cu_2O (cuprite) layer of 3 nm thickness. (From P Borthen. Grundlagen und Anwendungen der Röntgen-Absorptionsspektroskopie bei streifendem Einfall. PhD-dissertation, Heinrich-Heine-Universität Düsseldorf, 1996 (in German). With permission.)

modeling of the angular position, height, and width of these features yields the average size, the overall density, and the nearest neighbor distance of the pits. In addition, it could be evidenced that the pits are not distributed randomly on the surface of the oxide-covered Cu electrode, i.e., there are clusters that contain a certain number of individual pits. Within such a cluster, a mean distance of ca. 7.5 to 8.5 μm was found, and a mean size of the individual pits of ca. 4 μm. Furthermore, longitudinal off-specular scans (measurements of the side lobes as a function of q_z, i.e., the incidence angle) were performed in order to determine the depth distribution of the pits. The results show that the average pit depth has a bimodal distribution, with the pits being preferentially distributed near the surface and at ca. 4 nm below the surface (202). The detailed analysis of these scans indicates that the pits below the surface become deeper with time. This behavior is consistent with the pitting mechanism of film breaking, where pits are created at the interface of the oxide and the bulk metal, by a transport of Cu through the oxide film, and the formation of voids and eventually film collapse at a later stage (203,204). In the case of Cu in $NaHCO_3$, the presented results indicate that there are two coexisting kinds of pits in the early stages of corrosion: one having the

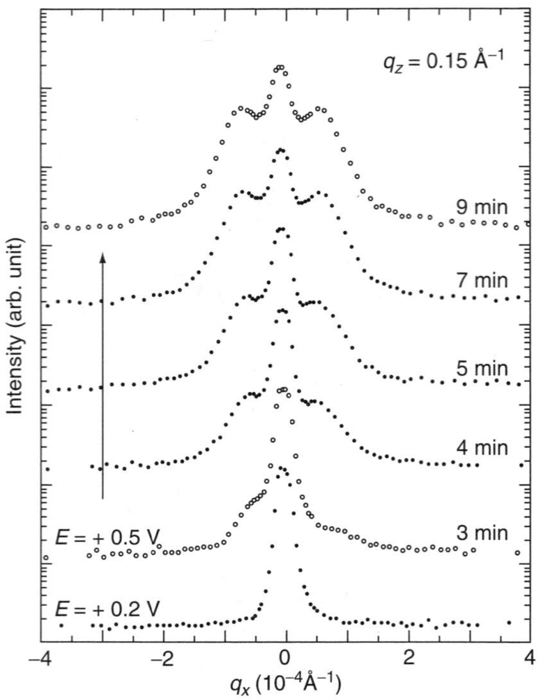

Figure 6.33 Transverse off-specular x-ray reflectivities (x-ray wavelength = 1.125 Å) of a Cu electrode in 0.01 *M* $NaHCO_3$: while the electrode is covered with an oxide film at +0.2 V, pitting corrosion occurs at a potential of +0.5 V. The sidepeaks in the specular peak at $q_x = 0$ that develop with pit formation time can be associated to scattering from the pits. (Taken from YP Feng et al., *Physica B* 221: 251–256, 1996. With permission.)

oxide film completely removed and being exposed to the top surface, and a second, which is buried near the Cu/Cu-oxide interface (202).

6.5.2 EXAFS of Corrosion Products

The products formed during the corrosion of a metal usually appear in front of the corroded electrode. Depending on the electrode material, the electrolyte, the applied potential, and other details of the system under investigation, the concentration of the corrosion product may vary from very low values up to saturated solutions. Part of the corrosion products in acidic media is incorporated into the salt layers that are formed during the intense electrochemical dissolution of metal electrodes. These salt layers play an important role, e.g., in passivation (205–208) and localized corrosion (207–209), electropolishing (210), and electromachining (211). For example, some metals require salt films prior to the formation of a passivating oxide film at the inner interface between the salt layer and the metal (205–207). In addition, localized corrosion (crevice corrosion, pitting corrosion, etc.) is a major cause of corrosion failure of stainless steels in halide-containing media (208,209,212–215). Synchrotron techniques may be used for a detailed analysis of the salt layer and the corrosion products that are transferred into the electrolyte. Especially the high brilliance of synchrotron sources, which enables focused beams of less than 10 μm in size providing x-rays with high intensities at the same time, can effectively be used, e.g., for *in situ*

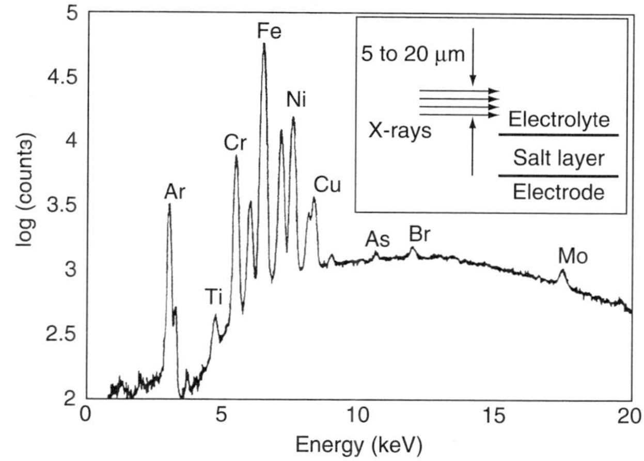

Figure 6.34 Energy-dispersive x-ray fluorescence spectrum from the solution 50 μm above a dissolving stainless steel electrode as depicted in the insert. By varying the distance of the x-ray beam relative to the surface of the corroded metal surface, the chemical composition of the salt layer and the electrolyte in front of the electrode can be monitored. (Adapted from HS Isaacs et al., *J. Electrochem. Soc.* 142: 1111–1118, 1995. With permission)

x-ray microprobe analysis in connection with an electrochemical environment (216–218). In Figure 6.34, an example for an x-ray microprobe analysis of the electrolyte in front of a corroded steel electrode is presented (218). This spectrum was obtained by using a polychromatic x-ray beam that was collimated by a pinhole of 20 μm size as schematically shown in the inset of Figure 6.34. By a variation of the distance from the beam center to the electrode surface, the composition of the salt layer as well as the species that have been dissolved into the solution may be studied in detail by measuring the element-specific fluorescence radiation. For this purpose, a solid-state lithium-doped Si detector was used in the cited study (218). The beam was positioned about 50 μm above the surface of a commercial austenitic stainless steel with a composition as follows: C: 0.11, Mn: 1.7, P: 0.01, S: 0.002, Si: 0.54, Cr: 17.73, Ni: 12.81, Cu: 0.12, Mo: 0.28, Fe: 66.70%. A solution of 0.5 M HCl + 0.5 M NaCl was used as electrolyte and the steel was anodically dissolved at fixed potentials. Obviously, contributions from the main metal constituents of the working electrode are detected in the solution. Using a better-focused beam, a detailed analysis of the thickness and the composition of the salt layer is possible. The experiments have shown that the salt layers formed on Ni–Cr steels are generally enriched in Fe, but contain little Ni and only very small amounts of Cr. The salt layer thickness increased approximately linearly with the corrosion potential reaching values of typically 1 to −10 μm. In addition, the concentration of the involved species decreases with the distance from the corroded metal surface, which suggests that the salt layer has a rather porous structure, in agreement with its properties as a barrier for the electromigration and diffusion of anions (218,219) as has been already concluded from galvanostatic pulse measurements (213).

Besides the detection of the different constituents of the salt layer and their changes, e.g., with corrosion potential and time, also a detailed *in situ* structural analysis of these species is possible using XAS. First approaches have appeared in the literature just very recently (219). An example is presented in Figure 6.35 for the salt layer formed on an iron-based stainless steel (Fe–16.8% Cr–12.0% Ni–2.0% Mo) in a

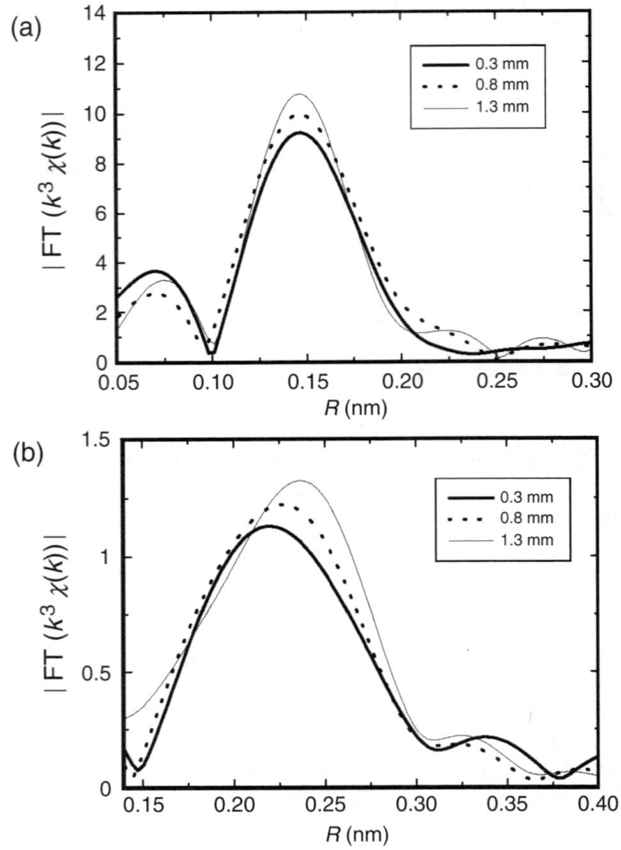

Figure 6.35 Fourier transforms of the extended x-ray absorption fine structure for the species detected in the salt layer on a stainless steel electrode during its corrosion in 1 *M* LiBr: (a) Cr–K-edge, (b) Br–K-edge for different distances above the electrode. (Taken from M Kimura et al., *J. Synchrotron Rad.* 8: 487–489, 2001. With permission.)

1 *M* LiBr solution (219). For the potentiostatic conditions chosen, the resulting corrosion current density amounts to typically 100 mA/cm^2. Extended x-ray absorption spectra were measured for different positions of the x-ray beam with respect to the corroded metal surface. In Figure 6.35, FTs of these x-ray absorption spectra are presented for the experiments at the Cr–K edge (a) and the Br–K edge (b). While the FTs calculated from the Cr–K spectra remain unchanged with respect to the position of the beam, the FTs obtained from the Br-edge data clearly change with the distance of the x-ray beam from the electrode surface as can be seen in Figure 6.35(b). The comparison with reference compounds (solid and dissolved $CrBr_3$, $Cr(OH)_3$ solid, dissolved $CrCl_3$, solid and dissolved LiBr) shows that the peaks in the FTs can be associated mainly with Cr–Br and Cr–O interactions (219). The results obtained by a detailed fit suggest that the Cr^{3+} ions in the salt layer are coordinated by OH and H_2O in the pit, and that some of the ligands are exchanged by Br especially close to the interface, which may explain the more obvious changes that are detected at the Br-edge. More specifically, the Br spectra detected near the interface to the metal resemble those of $CrBr_3$ salt, however, with strong contributions of oxygen (219). However, quantitative values for the coordination numbers and the bond distances have not been published so far. This is, on the one hand, related to existing

limitations of the beam size and the fact that the interface between the salt layer and the metal is not ideally smooth and flat. In consequence, the information obtained from an absorption spectrum is the convolution of spectra belonging to different regions of the salt layer, which might have different structures and compositions. For future investigation of the salt layer and its interfaces to the metal electrode and the electrolyte, it is very promising to use focusing optical elements such as Kirkpatrick–Baez mirror systems (26) or x-ray lenses (220,221) instead of a simple pinhole for the beam collimation. The new microfocus beamlines installed at third-generation synchrotron sources, e.g., at the ESRF (http://www.esrf.fr), the APS (http://www.aps.anl.gov), or the Swiss Light Source (SLS, http://www.psi.ch/index_e_sls.shtml) provide a high photon flux and at the same time a beam size well below 10 μm.

The corrosion products that have dissolved into the electrolyte can be detected qualitatively and quantitatively with high accuracy and time resolution using electrochemical techniques such as a rotating ring disk electrode (see, e.g., Refs. (222,223)). Here, the potential-dependent oxidation and reduction of soluble corrosion products is applied. However, electrochemical methods cannot provide structural information. XRD is a powerful tool providing detailed structural data of dissolved species within electrolytes (see, e.g., Ref. (224)). The information obtained by XRD is the sum of all atomic pair correlation functions in the solution. However, usually the concentration of corrosion products is very low compared to the components of a given bulk electrolyte. Therefore, it is very difficult to separate all the detected correlations and identify those of the species of interest from the dominating background. The situation is further complicated since long-range order correlations of the electrolyte introduced, e.g., by water hydrogen bridges further reduce contributions of the species of interest. XRD can be made element specific by the application of the anomalous scattering technique (225); thus, all contributions of different atomic pair correlation functions can be separated more easily. However, these measurements are highly time consuming and so cannot be used for the structural characterization of time-dependent processes such as corrosion phenomena with a change of the concentration c of the corroded species with time. Furthermore, the corrosion products are inhomogeneously distributed in the liquid phase as a consequence of their formation at the electrode surface and their diffusion into the electrolyte, which further complicates the situation.

The lack of any long-range order correlations of dissolved corrosion products makes EXAFS, with its inherent ability to specify the local short-range order around a selected element, an ideal method for the investigation of the atomic structure and the chemical state of the corrosion products in the liquid phase (47,48,226). The use of state-of-the-art detectors and stable monochromators also enables the investigation of highly diluted solutions. Furthermore, time-resolved investigations are easily possible as will be shown below. As an example, we will report about the corrosion product formed during the active dissolution of Mo in alkaline KOH solutions (47,48). In Figure 6.36, the cyclic voltammogram of a Mo metal electrode in 1 M KOH (pH 13.8) is presented. The increase of the current density above ca. −0.4 V can be associated to the anodic dissolution of the Mo electrode. Potentiostatic polarization experiments at potentials positive to −0.35 V yield corrosion currents that are constant with time, suggesting the active dissolution of the electrode (48). The insets of Figure 3.36 depicts the XANES region of Mo K-edge absorption spectra recorded at different times after potential steps from −1.0 V to −0.2 V (a) and −0.3 V (b), respectively. The increasing magnitude of the Mo absorption edge clearly indicates

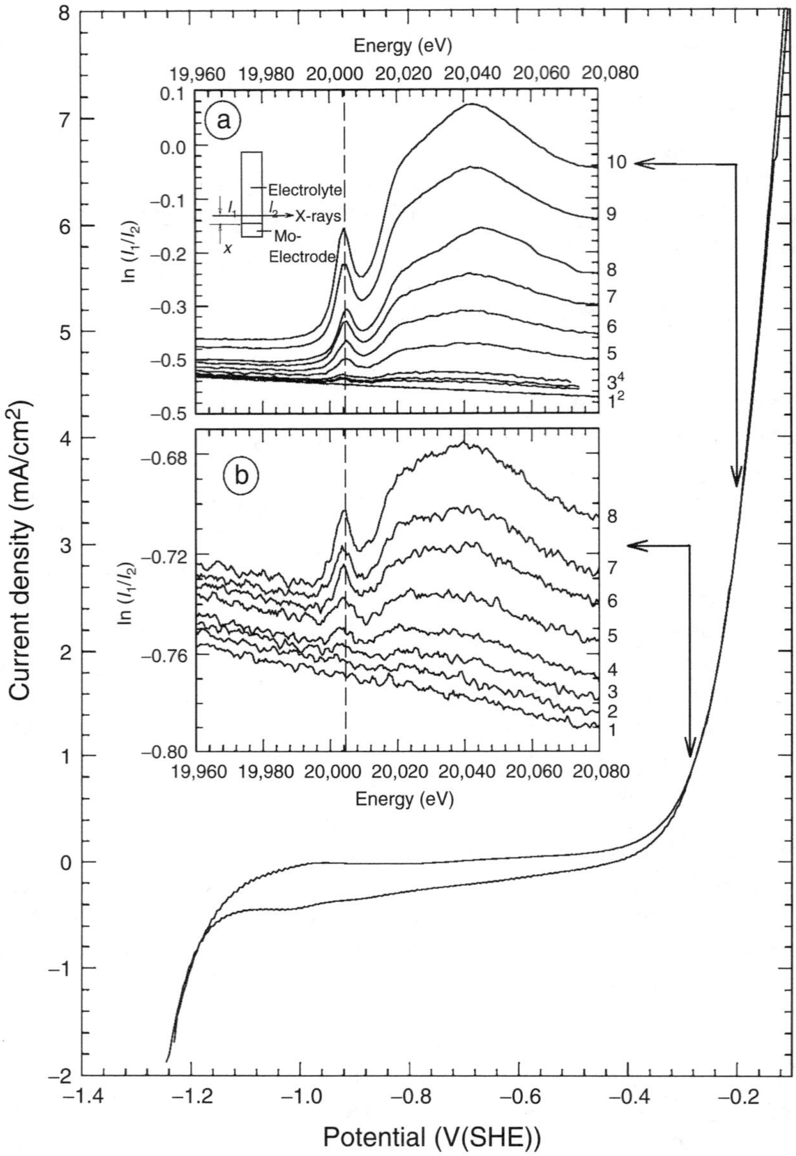

Figure 6.36 Potentiodynamic polarization curve (scan rate 100 mV/sec) for a Mo electrode in 1 *M* KOH (pH 13.8). In the insets, XANES spectra obtained during corrosion of the electrode after potential steps from −1.0 V to (a) −0.3 V and (b) −0.2 V are presented. The x-ray beam passes the electrolyte parallel to the surface of the Mo electrode in a distance x as schematically shown in the insert of (a). (a) *In situ* spectra 1 to 8 denote 0, 1000, 2000, 3000, 4030, 6100, 8060, and 10,040 sec oxidation times, $x = 0.57$ cm. The spectra were recorded with an integration time of 0.1 sec/data point resulting in a total scan time of 70 sec from 19,950 to 20,300 eV. (b) *In situ* spectra 1 to 10 denote 0, 400, 650, 920, 1950, 2500, 3400, 4300, 5300, and 6300 sec of oxidation time, $x = 0.35$ cm. Spectra 1 to 4 were recorded in 8.8 sec (0.08 sec/data point), spectra 5 to 10 in 12.1 sec/spectrum (0.1 sec/data point).

the increasing concentration of Mo in the solution with time for both experiments. Similar to the fluorescence experiments presented above, the concentration of the corroded metal in the solution can be determined as a function of the corrosion time and the potential (48). During the above-mentioned corrosion of stainless steel in chloride-containing electrolytes, the metal cations are accumulated in the solution above the salt layer to concentrations higher than 1 *M*. In the case of Mo dissolution, only small metal concentrations of typically 1 mmol/l were found (48). In addition, the time dependence of the absorption spectra was used for the determination of the diffusion coefficients of the dissolving species (48).

Besides the time dependence of the concentration, the XANES spectra presented in Figure 3.36 provide additional electronic and structural information of the dissolved Mo species. The most conspicuous features in the measured spectra are the intense pre-edge peaks at about 20005 eV. XANES data of some molybdenum reference compounds MoO_2 (Mo^{4+}), MoO_3 (Mo^{6+}), and metallic Mo are depicted in Figure 6.37 for comparison. As expected, the edge position shifts to higher binding energy with higher oxidation state. No prepeak can be observed for metallic Mo and MoO_2, while MoO_3 reveals a structure at ca. 20,007 eV. It is well known that the electronic structure of a given atom — and therefore also its near-edge x-ray absorption spectrum — depends on the geometry of its ligands in its direct vicinity (see, e.g., Refs. (227,228)). A pseudo-atomic model leads to the conclusion that the observed pre-edge structures may be due to 1s → 4d, 1s → 5s, and 1s → 5p transitions, i.e., to unoccupied electronic levels (229). The first of these transitions is partially allowed in a tetrahedral oxygen sphere, but it is strictly forbidden for an octahedral oxygen arrangement (229). Therefore, the interpretation of the near-edge spectra of the Mo oxides is straightforward. MoO_2 crystallizes in the rutile structure, which consists of Mo–O octahedra (230); thus, MoO_2 does not reveal a prepeak (cf. Ref. (231)). In contrast, the oxygen coordination around Mo in MoO_3 has a more complex structure with four short Mo–O distances between 1.67 and 1.94 Å with a structure close to a

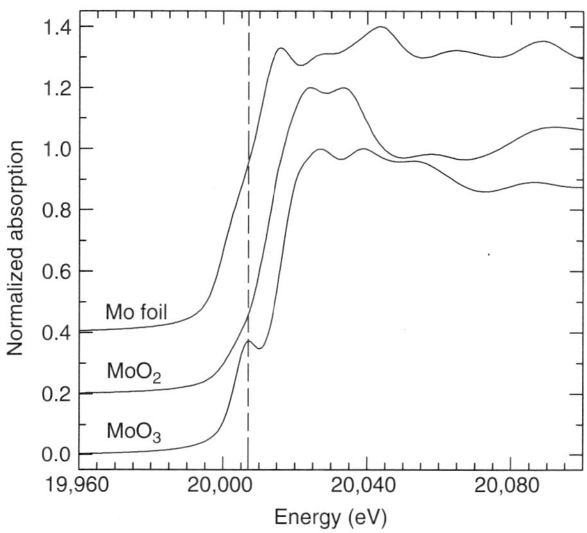

Figure 6.37 Mo K-edge XANES spectra of some selected Mo reference compounds: Mo metal foil, MoO_2 and MoO_3 powders. The dashed vertical line indicates the position of the MoO_3 pre-edge peak.

tetrahedron, and two additional long Mo–O distances of 2.25 and 2.33 Å, which complete a strongly distorted octrahedral coordination (232); thus, the existence of a pre-edge absorption maximum can easily be explained for MoO_3. It is well known that the accurate energy and intensity of such a prepeak depends strongly on the local environment of the x-ray absorbing atom, i.e., on the exact positions of the neighboring atoms. For example, in Na_2MoO_4, Ba_3TiMoO_4, and $PbMoO_4$, which are all composed by nearly ideal MoO_4 tetraeders, the prepeak is more intense and shifted toward a lower photon energy of about 20,005 eV compared to MoO_3, while ammonium heptamolybdate $(NH_4)_6Mo_7O_{24} \cdot 4H_2O$, which reveals a distorted octahedral symmetry similar to MoO_3, reflects the spectral features of MoO_3 (231). From the analysis of the Mo pre-edge data and the position of the absorption edge one can easily conclude that the Mo corrosion product in KOH is a Mo^{6+} species, very likely in a tetrahedral bond geometry. This is further supported by *ab initio* calculations using the FEFF code (61), a result of such a calculation is presented in Figure 6.38 together with an experimental near-edge spectrum of the corrosion products. For these calculations, a charged MoO_4^{2-} ion with four oxygens in 1.75 Å radial distance (see below) was assumed. As can be seen, the positions and the intensities of all relevant features in the absorption spectrum can be reproduced by these *ab initio* calculations. This example clearly illustrates the applicability and the relevance of *ab initio* calculations for the *in situ* analysis of species that are involved in corrosion processes.

The findings of the XANES data analysis are in agreement with those of extended x-ray absorption data. In Figure 6.39, an extended XAFS spectrum recorded after 5300 sec of anodic dissolution of a Mo electrode at −0.2 V is presented. For comparison, the spectrum recorded at −1.0 V, i.e., before the initiation of the corrosion of the Mo metal at $t = 0$ sec is also shown. Each data point was integrated for 0.15 sec resulting in a total time of ≈150 sec for a complete spectrum

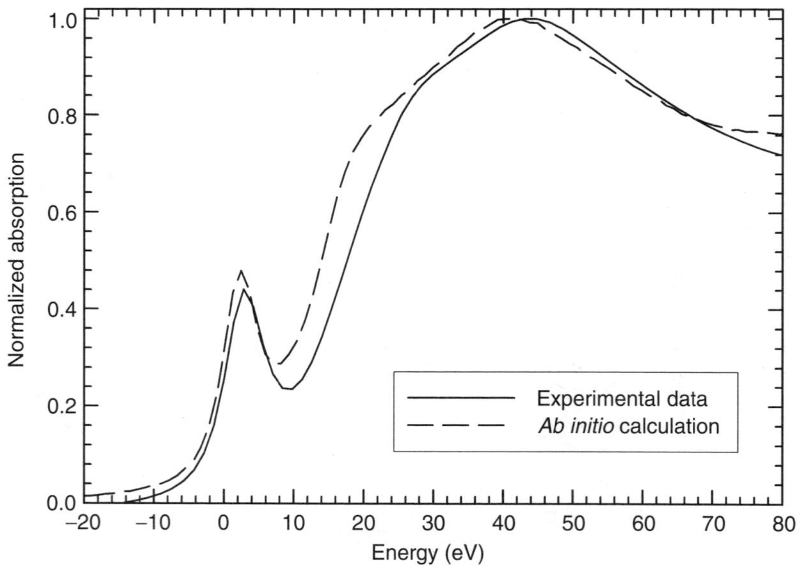

Figure 6.38 Comparison of the near-edge spectrum of the Mo species detected in front of the corroded Mo electrode and *ab initio* calculation using the FEFF 8 code (61) for a MoO_4^{2-} ion with four oxygen ions in coordination with the absorbing Mo ion in an ideal tetrahedron with 1.75 Å Mo–O bond distance.

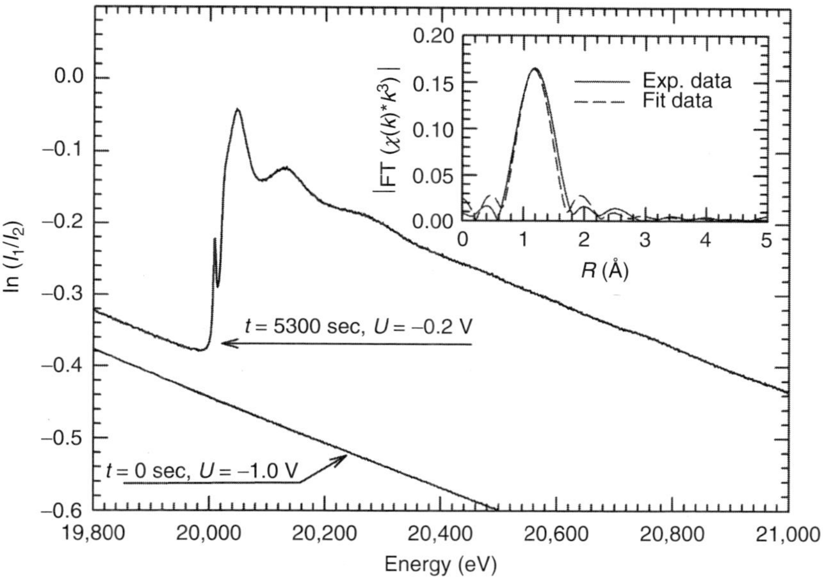

Figure 6.39 Transmission EXAFS spectrum recorded *in situ* after oxidation of a Mo electrode for 5300 sec at −0.2 V. For comparison, the spectrum recorded at −1.0 V is also presented. In the inset, the magnitude of the Fourier transform of the k^3-weighted fine structure oscillations is shown as a full line (k-range for the FT: $2\,Å^{-1} \leq k \leq 10\,Å^{-1}$, data not phase shift corrected). The dashed line presents the fit of the experimental data with a single Mo–O coordination shell with $R_1 = 1.75(4)$ Å, $N_1 = 3.96(9)$, and $\sigma_1 = 0.067$ Å.

ranging from 19,800 to 21,000 eV. Besides the pre-edge features that are discussed above, EXAFS modulations are clearly visible in this absorption spectrum, although the concentration determined from the edge jump was only 0.025 *M* (48). In the FT, which is depicted in the inset of Figure 6.39, only one single coordination shell belonging to the nearest neighbor bond is visible at about 1.2 Å radial distance; i.e., there is only one single coordination shell of backscattering atoms detectable. For further data evaluation, the contributions to this coordination shell between $R =$ 0.5 Å and $R =$ 1.8 Å were separated by means of a filter function, back-transformed into k-space, and fitted with phases and amplitudes calculated with FEFF (73) in the k-range from 2.2 to $10.0\,Å^{-1}$. In the inset of Figure 6.36, the FT of the fit is also shown for comparison. Keeping in mind that the concentration of Mo in the solution is very small and that the absorption data were collected in a couple of seconds only, the fit quality is quite reasonable. The averaged values determined for the nearest neighbor distance R_1, the coordination number N_1, and the mean square displacement σ_1 are $R_1 = 1.75 \pm 0.02$ Å, $N_1 = 4.06 \pm 0.11$, and $\sigma_1 = 0.061 \pm 0.005$ Å. From the determined coordination number of about $N = 4$, the absence of further coordination shells in the FTs of our EXAFS data and the presence of a Mo^{6+} species, it can be concluded that MoO_4^{2-} ions are formed during corrosion of Mo in alkaline media in agreement with the XANES data analysis. In should be mentioned that the small Mo–O bond distance of about 1.76 Å is very typical for molybdates, i.e., Mo in a tetrahedral arrangement with four surrounding oxygen atoms (see, e.g., Ref. (233)). From the presented data we can therefore give strong evidence for the presence of MoO_4^{2-} ions during the electrochemical corrosion of Mo metal in alkaline solution. Further investigations of this type may give additional information about the processes that occur during the active dissolution of a metal electrode and may help to prove the presence of characteristic species, and thus support existing models for these phenomena.

The last example in this section deals with the corrosion of lead-containing brass in drinking water (234). Dissolved lead in potable water constitutes a significant threat to exposed populations. Since lead compounds are often amorphous and mixed with cations from other metals, their identification with conventional techniques such as XRD is difficult. On the other hand, the nature of the formed lead compounds has to be known to predict a possible lead release, especially with regard to the corrosion of lead–tin solder and lead brasses (235). Surface-sensitive EXAFS experiments enable the detailed investigation of the thin layer deposits that contain Cu and Pb. The samples were obtained from a potable water distribution system for which lead problems have been documented. After exposure to potable water, a thin layer of a corrosion deposit was found on the surface of the lead brass (234). According to electron and XRD, this layer is amorphous. A detailed data analysis of the x-ray absorption spectra lead to the conclusion that malachite ($Cu_2(OH)_2CO_3$) and hydrocerussite ($Pb_3(OH)_2(CO_3)_2$) are the predominant phases that constitute the deposit.

6.6 CONCLUSIONS AND FUTURE OUTLOOK

The aim of this review was to give examples of the information that has been gained so far by using SR for the study of corrosion phenomena. Various methods and instruments were described in the initial sections and these are now well established for *in situ* investigations of corrosion processes. Many diffractometers with high accuracy and dedicated high flux x-ray absorption beamlines have been installed at different SR sources all around the world to perform new experiments. The strength and the many possibilities of all the techniques have been highlighted to some detail in many case studies such as the investigation of the detailed atomic structure of passive layers of thickness of a few nanometers with XRD or EXAFS studies of corroded metals. However, it is essential to point out that all the techniques have certain limitations. For example, one limitation for the grazing incidence surface XRD experiments is the need for single-crystal surfaces of very high quality, which are often difficult to obtain and handle, especially under electrochemical conditions. It has to be mentioned that structurally more complicated surfaces than the low indexed (1 0 0), (1 1 0), or (1 1 1) surfaces or polycrystalline samples are generally very difficult to study. However, making use of the very intense and brilliant beams of undulator sources installed at third-generation synchrotron sources such as the ESRF, Spring 8, or the APS should allow even this. Furthermore, a complete characterization of the samples using synchrotron methods alone certainly does not make too much sense, and complementary experiments using laboratory equipment are highly recommended before conducting an experiment at a synchrotron. In this context, besides pure electrochemical techniques, especially surface analytical methods such as XPS, AES, ISS (LEIS), or LEED have proven their applicability for corrosion phenomena.

Further aspects for future investigations include time-resolved experiments with an improved time resolution. This is necessary, e.g., for studying the growth dynamics of passive films in more detail, especially with regard to the initial stages of film formation. Furthermore, also competing side reactions such as thinning of passive layers and layer breakdown by pitting can be assessed. The recently proposed Piezo-QEXAFS technique seems to have unique properties for such investigations. Piezo-QEXAFS is based on the quick-scanning data acquisition technique as described in Section 6.3; however, the time for a single spectrum is reduced to a few milliseconds by using piezo-driven tilt tables on which the individual monochromator crystals are installed (236). Due to the limited travel in the Bragg angle that can be achieved using piezo translators, this technique is restricted to near-edge X-ray absorption investigations. An example of the technique is given in Figure 6.40, where reflection mode spectra ($\phi = 0.2°$) obtained from a gold electrode using the Piezo-QEXAFS technique in about 50 msec are compared to a stepscan measured in about 10 min. Obviously, all structures that are representative for the gold electrode are reproduced by the fast measurements, so that this technique is well suited to follow fast changes of electrode structures or valencies as described above in more detail. Latest developments of the fast QEXAFS techniques include measurements of extended energy regions around a given absorption edge by employing a cam driven monochromator (237). This technique seems to be well suited to follow, e.g., the changes that occur at the surface of electrochemically oscillating systems such as the

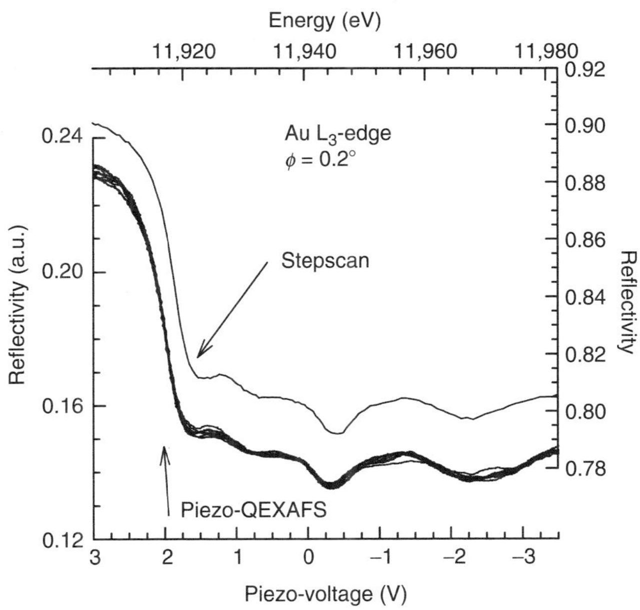

Figure 6.40 Reflection-mode EXAFS investigations ($\phi = 0.2°$) of a gold electrode at the L_3-edge (11,919 eV). A measurement performed in the stepscan mode in about 10 min is compared to Piezo-QEXAFS investigations (10 Hz oscillation frequency corresponding to about 50 msec for a single spectrum; 10 subsequently measured spectra are shown).

dissolution and repassivation of cobalt electrodes in alkaline solutions (238): polarization of cobalt electrodes in 0.1 *M* NaOH leads to an oscillatory behavior of the current measured under potentiostatic conditions in a certain regime in the passive region. As shown in Figure 6.41(a), the time scale of the observed current oscillations amounts to roughly a second, as can be seen more clearly in the Fourier-transformed data in Figure 6.41(b). It should be possible to follow the state of the Co electrode with a time resolution of well below 1 sec by combining fast transmission mode experiments

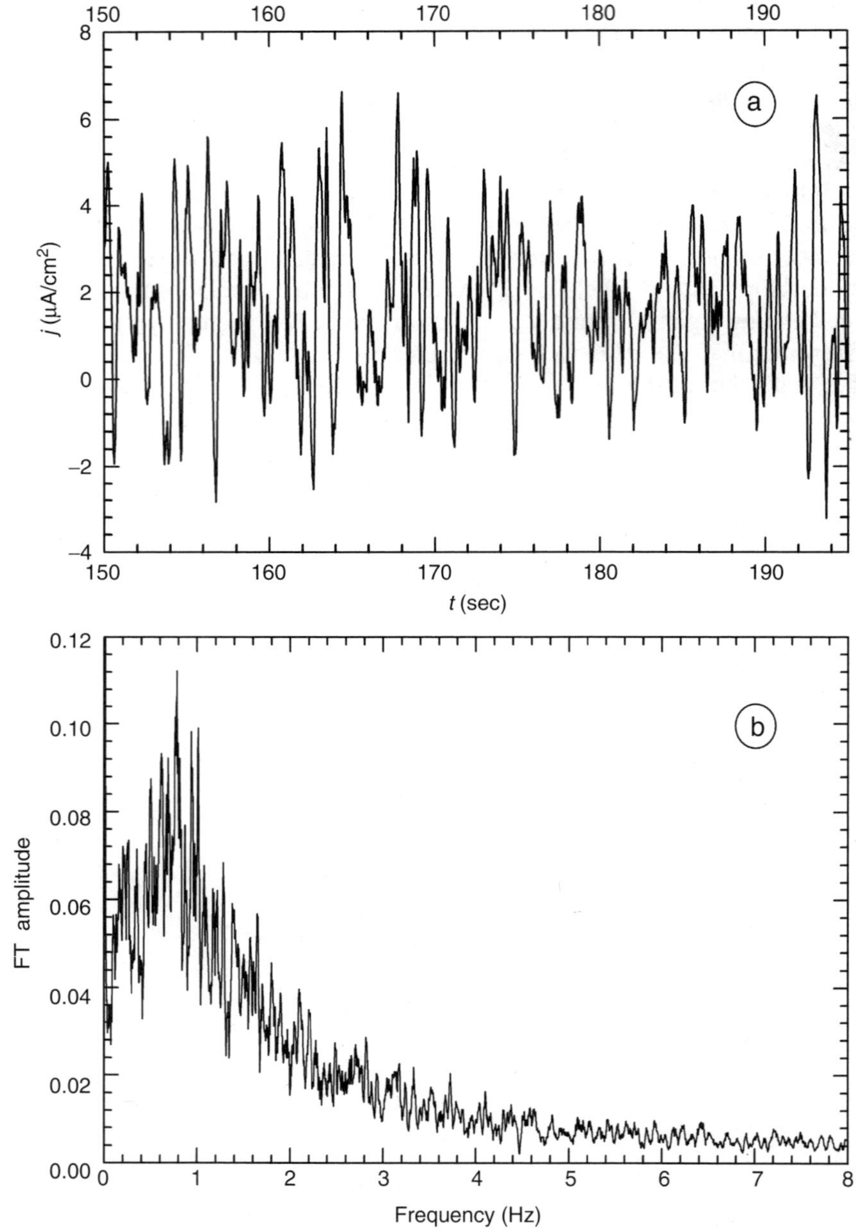

Figure 6.41 (a) Current oscillations observed for a Co electrode in 0.1 *M* NaOH at $E = +0.35\,V(SHE)$. A time scale of about 1 sec results, which can be more clearly seen in (b), where the Fourier-transformed data are shown. (From A Foelske, unpublished results. With permission.)

with a focused x-ray beam parallel to the electrode surface and reflection mode experiments using grazing angles well below the critical angle. This way, all the species dissolved in the electrolyte as well as those remaining on the electrodes surfaces and their changes with time can be monitored, and proposed time and pH-dependent electrochemical reactions such as the oxidation of Co_3O_4 or the reduction of CoOOH should be proved. It has to be stressed that keeping track of these species cannot be performed using, e.g., x-ray photoelectron spectroscopy, where, however, detailed information about passive layers on Co has been gained (239,240), but XPS cannot reach the time domain of the observed electrochemical oscillations.

Furthermore, the x-ray techniques discussed above can also be applied under active dissolution conditions, i.e., the combination of transmission and reflection mode experiments should allow for a full characterization, even during active dissolution (241). In summary, the application of modern synchrotron methods really permits completely new experiments. It is very likely that understanding all these features will provide further insight into the fundamental processes of corrosion, passivation, and of transport of species through passive films and perhaps even for corrosion protection and the beneficial effects of chromium.

ACKNOWLEDGMENTS

The authors like to thank A.J. Davenport, O.M. Magnussen, P. Borthen, T. Suzuki, and their coworkers for the provision of their data for publication in this article.

REFERENCES

1. M Fleischmann, A Oliver, J Robinson. *In situ* x-ray diffraction studies of electrode solution interfaces. *Electrochim. Acta* 31: 899–906, 1986.
2. ED Chabala, T Rayment. Time resolved in situ and real time study of an electrode process by differential x-ray diffraction: Ag underpotential depostition on Au(111). *J. Electroanal. Chem.* 401: 257–261, 1996.
3. C Kunz. *Synchrotron Radiation, Techniques and Application.* Topics in Current Physics, Vol. 10. Heidelberg: Springer, 1979.
4. EE Koch. *Handbook of Synchrotron Radiation, Vol. 1, Characteristics, Instrumentation and Principles of Research Applications.* Amsterdam: North-Holland, 1983.
5. G Margeritondo. *Introduction to Synchrotron Radiation Research.* Oxford: Oxford University Press, 1988.
6. P Coppens, D Cox, E Vlieg, IK Robinson. *Synchrotron Radiation Crystallography.* New York: Academic Press, 1992.
7. H Kobayakawa, K Huke. Worldwide census of synchrotron radiation facilities. *Rev. Sci. Instrum.* 60: 2548–2561, 1989.
8. http://www.esrf.fr/navigate/synchrotrons.html.
9. J Schwinger. Electron radiation in high energy accelerators. *Phys. Rev.* 70: 798–799, 1946.
10. H Dosch. *Critical Phenomena at Surfaces and Interfaces.* Berlin: Springer Verlag, 1992.
11. M Tolan. *X-Ray Scattering From Soft-Matter Thin Films.* Berlin: Springer Verlag, 1999.
12. G Schütz, W Wagner, W Wilhelm, P Kienle, R Frahm, G Materlik. Absorption of circularly polarized x-rays in iron. *Phys. Rev. Lett.* 58: 737–740, 1987.
13. G Schütz, R Frahm, P Mautner, R Wienke, W Wagner, W Wilhelm, P Kienle. Spin-dependent extended x-ray-absorption fine structure: probing magnetic short-range order. *Phys. Rev. Lett.* 62: 2620–2623, 1989.

14. J Goulon, P Elleaume, D Raoux. Special multipole wiggler design producing circularly polarized synchrotron radiation. *Nucl. Instrum. Meth. A* 254: 192–201, 1987.
15. XM Marechal, T Tanaka, H Kitamura. An elliptical wiggler for SPring-8. *Rev. Sci. Instrum.* 66: 1937–1939, 1995.
16. BW Batterman, DH Bilderback. X-ray monochromators and mirrors. In: *Handbook of Synchrotron Radiation, Vol. 3: X-Ray Scattering Techniques and Condensed Matter Research*, G. Brown and D. Moncton (eds.). Amsterdam: North Holland, 1991, pp. 105–120.
17. J Wong, G Shimkaveg, W Goldstein, M Eckart, T Tanaka, ZU Rek, H Tompkins. YB66: a new soft-x-ray monochromator for synchrotron radiation. *Nucl. Instrum. Meth. A* 291: 243–249, 1990.
18. A Krolzig, G Materlik, J Zegenhagen. A dynamic control and measuring system for x-ray rocking curves. *Nucl. Instrum. Meth.* 208: 613–619, 1983.
19. G Materlik, VO Kostroun. Monolithic crystal monochromators for synchrotron radiation with order sorting and polarizing properties. *Rev. Sci. Instrum.* 51: 86–94, 1980.
20. P Sainctavit, J Petiau, A Manceau, R Rivallant, M Belakhovsky, G Renaud. Two-mirror device for harmonic rejection. *Rev. Sci. Instrum.* 60: 2027–2029, 1989.
21. J Arthur, WH Tompkins, C Troxel Jr, RJ Contolini, E Schmitt, DH Bilderback, C Henderson, J White, T Settersten. Microchannell water cooling of silicon x-ray monochromator crystals. *Rev. Sci. Instrum.* 63: 433–436, 1992.
22. JP Quintana, M Hart, DH Bilderback, C Henderson, D Richter, T Setterston, J White, D Hausermann, M Krumrey, H Schulte-Schrepping. Adaptive silicon monochromators for high-power insertion devices. Tests at CHESS, ESRF and HASYLAB. *J. Synchrotron Rad.* 2: 1–5, 1995.
23. RK Smither, GA Forster, DH Bilderback, M Bedzyk, K Finkelstein, C Henderson, J White, LE Berman, P Stefan, T Oversluizen. Liquid gallium cooling of silicon crystals in high intensity photon beams. *Rev. Sci. Instrum.* 60: 1486–1492, 1989.
24. CS Rogers, DM Mills, W-K Lee, PB Fernandez, T Graber. Experimental results with cryogenically cooled, thin, silicon crystal x-ray monochromators on high-heat-flux beamlines. *Proc. SPIE* 2855: 170–179, 1996.
25. H Schulte-Schrepping, J Heuer, B Hukelmann. Adaptive indirectly cooled monochromator crystals at HASYLAB. *J. Synchrotron Rad.* 5: 682–684, 1998.
26. P Kirkpatrick, AV Baez. Formation of optical images by x-rays. *J. Opt. Soc. Am.* 38: 766–774, 1948.
27. JA Howell, P Horowitz. Ellipsoidal and bent cylindrical condensing mirrors for synchrotron radiation. *Nucl. Instrum. Meth.* 125: 225–230, 1975.
28. SM Heald, JB Hastings. Grazing incidence optics for synchrotron radiation x-ray beamlines. *Nucl. Instrum. Meth.* 187: 553–561, 1981.
29. D Pauschinger, K Becker, R Ludewig. Side-cooled 1200 mm silicon x-ray mirrors with pneumatic benders. *Rev. Sci. Instrum.* 66: 2177–2179, 1995.
30. GF Knoll. *Radiation Detection and Measurement*. New York: Wiley, 1979.
31. P Kappen, L Tröger, G Materlik, C Reckleben, K Hansen, J-D Grundwaldt, BS Clausen. Silicon drift detectors as a tool for time-resolved fluorescence XAFS on low concentrated samples in catalysis. *J. Synchrotron Rad.* 9: 246–253, 2002.
32. L Strüder, C Fiorini, E Gatti, R Hartmann, P Holl, N Krause, P Lechner, A Longoni, G Lutz, J Kemmer, N Meidinger, M Popp, H Soltau, U Weber, C von Zanthier. High-resolution high-count-rate x-ray spectroscopy with state-of-the-art silicon detectors. *J. Synchrotron Rad.* 5: 268–274, 1998.
33. E Gatti, P Rehak. Semiconductor drift chamber—an application of a novel charge transport scheme. *Nucl. Instrum. Meth. Phys. Res. A* 225: 608–614, 1984.
34. P Lechner, S Eckbauer, R Hartmann, S Krisch, D Hauff, R Richter, H Soltau, L Strüder, C Fiorini, E Gatti, A Longoni, M Sampietro. Silicon drift detectors for high resolution room temperature x-ray spectroscopy. *Nucl. Instrum. Meth. Phys. Res. A* 377: 346–351, 1996.

35. L Strüder. High-resolution imaging x-ray spectrometers. *Nucl. Instrum. Meth. Phys. Res. A* 454: 73–113, 2000.
36. L Blum, HD Abruna, J White, JG Gordon, GL Borges, MG Samant, O Melroy. Study of underpotentially deposited copper on gold by fluorescence detected surface EXAFS. *J. Chem. Phys.* 85: 6732–6738, 1986.
37. Z Shi, S Wu, J Lipkowski. Coadsorption of metal atoms and anions: Cu UPD in the presence of SO_4^{2-}, Cl^- and Br^-. *Electrochim. Acta* 40: 9–15, 1995.
38. G Scherb, A Kazimirow, J Zegenhagen, TL Lee, MJ Bedzyk, H Noguchi, K Uosaki. *In situ* x-ray standing wave analysis of electrodeposited Cu monolayers on GaAs(001). *Phys. Rev. B* 58: 10800–10805, 1998.
39. J Zegenhagen. Surface structure determination with x-ray standing waves. *Surf. Sci. Rep.* 18: 200–271, 1993.
40. AJ Davenport, M Sansone, JA Bardwell, AJ Aldykiewicz Jr, M Taube, CM Vitus. *In situ* multielement XANES study of formation and reduction of the oxide film on stainless steel. *J. Electrochem. Soc.* 141: L6–L8, 1994.
41. H Oyanagi, R Shioda, Y Kuwahara, K Haga. Surface sensitive XAFS in the hard x-ray region with sub-monolayer sensitivity. *J. Synchrotron Rad.* 2: 99–105, 1995.
42. G Scherb, A Kazimirov, J Zegenhagen. A novel thick-layer electrochemical cell for *in situ* x-ray diffraction. *Rev. Sci. Instrum.* 69: 512–516, 1998.
43. M Hüppauff, B Lengeler. Valency and structure of iridium in anodic iridium oxide films. *J. Electrochem. Soc.* 140: 598–602, 1993.
44. A Tadjeddine, A Lahrichi, G Tourillon. *In-situ* x-ray absorption near-edge structure: a probe of the oxidation state of underpotentially absorbed metals. *J. Electroanal. Chem.* 360: 261–270, 1993.
45. A Cognigni, I Ascone, S Zamponi, R Marassi. A quasi-solid state electrochemical cell for in situ EXAFS measurements on biological samples. *J. Synchrotron Rad.* 8: 987–989, 2001.
46. A Messiah. *Quantum Mechanics*, Vol. 1. New York: Wiley, 1976.
47. D Lützenkirchen-Hecht, R Frahm. Time resolved XAFS investigations of the anodic dissolution of Mo. *J. Synchrotron Rad.* 6: 591–593, 1999.
48. D Lützenkirchen-Hecht, R Frahm. The corrosion of Mo in KOH: time resolved XAFS investigations. *J. Phys. Chem. B* 105: 9988–9993, 2001.
49. P Schmuki, S Virtanen, AJ Davenport, CM Vitus. Transpassive dissolution of Cr and sputter deposited Cr oxides studied by *in situ* x-ray near edge spectroscopy. *J. Electrochem. Soc.* 143: 3997–4005, 1996.
50. S Virtanen, P Schmuki, M Büchler, HS Isaacs. Electrochemical behaviour of Fe in phosphate solutions studied by in situ x-ray absorption near edge structure. *J. Electrochem. Soc.* 146: 4087–4094, 1999.
51. AJ Davenport, HS Isaacs, JA Bardwell, B MacDougall, GS Frankel, AG Schrott. *In situ* studies of passive film chemistry using x-ray absorption spectroscopy. *Corros. Sci.* 35: 19–25, 1993.
52. AJ Davenport, M Sansone, JA Bardwell, AJ Aldykiewicz Jr, M Taube, CM Vitus. *In situ* multielement XANES study of formation and reduction of the oxide film on stainless steel. *J. Electrochem. Soc.* 141: L6–L8, 1994.
53. AJ Davenport, RC Newman, LJ Oblonsky, MP Ryan. XANES study of reactivation of Fe–Cr alloys near the 13% threshold. In: *Surface Oxide Films*, JA Bardwell (ed.), Electrochem. Soc. Proc. PV 96-18, Pennington, 1996.
54. P Schmuki, S Virtanen, HS Isaacs, MP Ryan, AJ Davenport, H Böhni, T Stenberg. Electrochemical behaviour of Cr_2O_3/Fe_2O_3 artificial passive films studied by in situ XANES. *J. Electrochem. Soc.* 145: 791–801, 1998.
55. B Lengeler. X-ray absorption and reflection in materials science. *Adv. Solid State Phys.* 29: 53–73, 1989.
56. JG Chen, CM Kim, B Frühberger, BD DeVries, MS Touvelle. A NEXAFS determination of the oxidation state of vanadium carbide on V(110): observation of a charge transfer from vanadium to carbon. *Surf. Sci.* 321: 145–155, 1994.

57. T Ressler, J Wong, J Roos. Manganese speciation in exhaust particulates of automobiles using MMT-containing gasoline. *J. Synchrotron Rad.* 6: 656–658, 1999.
58. JC Fuggle, JE Inglesfield. *Unoccupied Electronic States: Fundamentals for XANES, EELS, IPS and BIS*. Topics in Applied Physics, Vol. 69. Berlin: Springer, 1992.
59. MP Seah, WA Dench. Quantitative electron spectroscopy of surfaces: a standard data base for electron inelastic mean free paths in solids. *Surf. Interface Anal.* 1: 2–11, 1979.
60. J Petiau, G Calas, P Sainctavit. Recent developments in the experimental studies of XANES. *J. Phys. (Paris) Colloq.* C9: 1085–1096, 1987.
61. AL Ankudinov, B Ravel, JJ Rehr, SD Conradson. Real-space multiple-scattering calculation and interpretation of x-ray-absorption near-edge structure. *Phys. Rev. B* 58: 7565–7576, 1998.
62. JJ Rehr, RC Albers. Theoretical approaches to x-ray absorption fine structure. *Rev. Mod. Phys.* 72: 621–654, 2000.
63. AJ Davenport, M. Sansone. High resolution in situ XANES investigation of the nature of the passive film on iron in a pH 8.4 borate buffer. *J. Electrochem. Soc.* 142: 725–730, 1995.
64. AJ Davenport, JA Bardwell, CM Vitus. *In situ* XANES study of galvanostatic reduction of the passive film on iron. *J. Electrochem. Soc.* 142: 721-724, 1995.
65. P Schmuki, S Virtanen, AJ Davenport, CM Vitus. *In situ* x-ray absorption near-edge spectroscopic study of the cathodic reduction of artificial iron oxide passive films. *J. Electrochem. Soc.* 143: 574–582, 1996.
66. ER Malinowski, DG Howery. *Factor Analysis in Chemistry*. New York: John Wiley and Sons, 1980.
67. SR Wasserman. The analysis of mixtures: application of principal component analysis to XAS spectra. *J. Phys. IV (France)* 7: C2-203–C2-205, 1997.
68. T Ressler, J Wong, J Roos, IL Smith. Speciation of manganese particles in exhaust fumes of cars utilizing MMT. *Environ. Sci. Technol.* 34: 950–958, 2000.
69. T Ressler, J Wienold, RE Jentoft. Formation of bronzes during temperature-programmed reduction of MoO_3 with hydrogen — an *in-situ* XRD and XAFS study. *Solid State Ionics* 141–142: 243–252, 2001.
70. J-D Grunwaldt, D Lützenkirchen-Hecht, M Richwin, S Grundmann, BS Clausen, R Frahm. Piezo x-ray absorption spectroscopy for the investigation of solid-state transformations in the millisecond range. *J. Phys. Chem. B* 105: 5161–5168, 2001.
71. DE Sayers, EA Stern, FW Lytle. New technique for investigating noncrystalline structures: Fourier analysis of the extended x-ray absorption fine structure. *Phys. Rev. Lett.* 27: 1204–1207, 1971.
72. N Binsted, SL Cook, J Evans, GN Greaves, RJ Price. EXAFS and near-edge structure in the cobalt K-edge absorption spectra of metal carbonyl complexes. *J. Am. Chem. Soc.* 109: 3669–3676, 1987.
73. JJ Rehr, RC Albers, SI Zabinski. High-order multiple-scattering calculations of x-ray-absorption fine structure. *Phys. Rev. Lett.* 69: 3397–3400, 1992.
74. SI Zabinski, JJ Rehr, A Ankudinov, RC Albers, MJ Eller. Multiple-scattering calculations of x-ray-absorption spectra. *Phys. Rev. B* 52: 2995–3009, 1995.
75. A Filipponi, A Di Cicco, CR Natoli. X-ray-absorption spectroscopy and *n*-body distribution functions in condensed matter. I. Theory. *Phys. Rev. B* 52: 15122–15134, 1995.
76. A Filipponi, A Di Cicco. X-ray-absorption spectroscopy and *n*-body distribution functions in condensed matter. II. Data analysis and applications. *Phys. Rev. B* 52: 15135–15149, 1995.
77. JJ Rehr, SI Zabinsky, A Ankudinov, RC Albers. Atomic XAFS and XANES. *Physica B* 208 & 209: 23–26, 1995.
78. JJ Rehr, CH Booth, F Bridges, SI Zabinsky. X-ray absorption fine structure in embedded atoms. *Phys. Rev. B* 49: 12347–12350, 1994.
79. WE O'Grady, DE Ramaker. Atomic x-ray absorption fine structure: a new tool for examining electrochemical interfaces. *Electrochim. Acta* 44: 1283–1287, 1998.

80. A Kodre, I Arcon, R Frahm. Exact atomic absorption background for Rb K-edge EXAFS. *J. Phys. IV (France)* 7: C2-195–C2-197, 1997.
81. A Filipponi. Double excitation effects above inner shell x-ray absorption edges. *Physica B* 208 & 209: 29–32, 1995.
82. JP Gomilsek, A Kodre, I Arcon, A-M Loreau-Lozac'h, S Bénazeth. Atomic background in 4p elements. *J. Synchrotron Rad.* 6: 304–305, 1999.
83. WE O'Grady, X Qian, DE Ramaker. Systematic chemical effects observed in "atomic" x-ray absorption fine structure. *J. Phys. Chem. B* 101: 5624–5626, 1997.
84. L Bosio, R Cortes, A Defrain, M Froment. EXAFS from measurements of x-ray reflectivity on passivated electrodes. *J. Electroanal. Chem.* 180: 265–271, 1984.
85. J Kruger, GG Long, Z Zhang, DK Tanaka. The use of x-ray absorption spectroscopic techniques to study the influence of alloying elements on passive films. *Corros. Sci.* 31: 111–120, 1990.
86. P Druska, H-H Strehblow. *In-situ* examination of electrochemically formed Cu_2O layers by EXAFS in transmission. *J. Electroanal. Chem.* 335: 55–65, 1992.
87. D Koningsberger, R Prins. *X-ray Absorption: Principles, Applications, Techniques of EXAFS, SEXAFS and XANES*. New York: John Wiley and Sons, 1988.
88. http://ixs. csrri.iit.edu/database/.
89. C. Kittel. *Introduction to Solid State Physics*, 7th edn. New York: John Wiley and Sons, 1996.
90. LG Parratt. Surface studies of solids by total reflection of x-rays. *Phys. Rev.* 95: 359–369, 1954.
91. G Martens, P Rabe. The extended x-ray absorption fine structure in the reflectivity at the K edge of Cu. *J. Phys. C: Solid State Phys.* 14: 1523–1534, 1981.
92. B Poumellec, R Cortes, F Lagnel, G Tourillon. A new method to extract the x-ray absorption fine structure from the reflectivity spectra: application to the study of (Ti, Nb)O_2 amorphous solid solutions. *Physica B* 158: 282–283, 1989.
93. SM Heald, H Chen, JM Tranquada. Glancing-angle extended x-ray-absorption fine structure and reflectivity studies of interfacial regions. *Phys. Rev. B* 38: 1016–1026, 1988.
94. P Borthen, H-H Strehblow. X-ray reflectivity fine structure from homogeneous materials in the hard-energy range. *J. Phys.: Condens. Matter* 7: 3779–3787, 1995.
95. DT Jiang, ED Crozier. Anomalous dispersion effects in glancing angle XANS of ultrathin films. *J. Phys. IV (France)* 7: C2-249–C2-250, 1997.
96. F d'Acapito, S Mobilio, P Cikmas, V Merlo, I Davoli. Temperature modification of the Nb oxidation at the Nb/Al interface studied by REFLEXAFS. *Surf. Sci.* 468: 77–84, 2000.
97. P Borthen, H-H Strehblow. Near surface structure determination using x-ray reflection absorption spectroscopy. *Physica B* 208 & 209: 421–422, 1995.
98. P Borthen, H-H Strehblow. Grazing incidence x-ray absorption spectroscopy of near surface regions: possibilities and limitations. *J. Phys. IV (France)* 7: C2-187–C2-189, 1997.
99. P Borthen, H-H Strehblow. X-ray reflectivity fine structure and EXAFS. *Phys. Rev. B* 52: 3017–3019, 1995.
100. P Borthen. Grundlagen und Anwendungen der Röntgen-Absorptionsspektroskopie bei streifendem Einfall. PhD-dissertation, Heinrich-Heine-Universität Düsseldorf, 1996 (in German).
101. D Lützenkirchen-Hecht, R Frahm. Structural investigations of sputter deposited thin films: reflection mode EXAFS, specular and non-specular X-ray scattering. *Physica B* 283: 108–113, 2000.
102. B Lengeler. X-ray absorption and reflection in the hard X-ray range. In: *Photoemission and Absorption of Solids and Interfaces with Synchrotron Radiation*, M Campagna, R Rosei (eds.). Amsterdam: North Holland, 1990, pp. 157–202.
103. D Hecht, P Borthen, H-H Strehblow. An x-ray absorption fine structure study of the initial stages of the anodic oxidation of silver. *Surf. Sci.* 365: 263–277, 1996.

104. D Hecht, P Borthen, H-H Strehblow. Near range order structural information of anodic silver oxide layers by means of reflectivity fine structure data evaluation. *J. Phys. IV (France)* 7: C2-707–C2-708, 1997.
105. D Lützenkirchen-Hecht, R Frahm. Reflection mode XAFS investigations of reactively sputtered thin films. *J. Synchrotron Rad.* 8: 478–480, 2001.
106. R Frahm. New method for time dependent x-ray absorption studies. *Rev. Sci. Instrum.* 60: 2515–2518, 1989.
107. R Frahm. QEXAFS in milliseconds. *Synchrotron Rad. News* 8 Vol. 2: 38, 1995.
108. T Matsushita, RP Phizackerley. A fast x-ray absorption spectrometer for use with synchrotron radiation. *Jpn. J. Appl. Phys.* 20: 2223–2228, 1981.
109. RP Phizackerley, ZU Rek, GB Stephenson, SD Conradson, KO Hodgson, T Matsushita, H Oyanagi. An energy-dispersive spectrometer for the rapid measurement of x-ray absorption spectra using synchrotron radiation. *J. Appl. Crystallogr.* 16: 220–232, 1983.
110. M Hagelstein, S Cunis, R Frahm, W Niemann, P Rabe. The energy dispersive x-ray absorption spectrometer DEXAFS at HASYLAB. *Physica B* 158: 324–328, 1989.
111. H Blank, B Neff, S Steil, J Hormes. A new energy dispersive x-ray monochromator for soft x-ray applications. *Rev. Sci. Instrum.* 63: 1334–1337, 1992.
112. PL Lee, MA Beno, G Jennings, M Ramanathan, GS Knapp, K Huang, J Bai, PA Montano. An energy dispersive x-ray absorption spectroscopy beamline, X6A, at NSLS. *Rev. Sci. Instrum.* 65: 1–6, 1995.
113. M Hagelstein, C Ferrero, M Sanchez del Rio, U Hatje, T Ressler, W Metz. XAFS with an energy dispersive Laue monochromator. *Physica B* 208 & 209: 223–224, 1995.
114. M Hagelstein, A San Miguel, T Ressler, A Fontaine, J Goulon. The beamline ID24 at the ESRF for energy dispersive x-ray absorption spectroscopy. *J. Phys. IV (France)* 7: C2-303–C2-308, 1997.
115. D Hecht, R Frahm, H-H Strehblow. Quick-scanning EXAFS in the reflection mode as a probe for structural information of electrode surfaces with time resolution: an *in situ* study of anodic silver formation. *J. Phys. Chem. B* 100: 10831–10833, 1996.
116. A Dent, J Evans, M Newton, J Corker, A Russell, MB Abdul Rahman, S Fiddy, R Mathew, R Farrow, G Salvini, P Atkinson. High quality enrgy dispersive XAFS on the 1 s timescale applied to electrochemical and catalyst systems. *J. Synchrotron Rad.* 6: 381–383, 1999.
117. *International Tables for X-ray Crystallography, Vol III: Mathematical Tables* (Published for the International Union of Cryst.). Birmingham: Kynoch Press, 1962.
118. D Schumacher. Surface scattering experiments with conduction electrons. *Springer Tracts Mod. Phys.* 128: 1–97, 1993.
119. W Schlemminger, D Stark. The influence of deposition temperature on the crystalline and electrical properties of thin silver films. *Thin Solid Films* 137: 49–57, 1986.
120. MS Paterson. X-ray diffraction by face centered cubic crystals with deformation faults. *J. Appl. Phys.* 23: 805–811, 1952.
121. H Neff. *Grundlagen und Anwendungen der Röntgenfeinstrukturanalyse*. München: Oldenbourg, 1962 (in German).
122. RW Vook, F Witt. Quantitative, *in situ* x-ray diffractometer investigations of evaporated metal films. In: *Basic Problems in Thin Film Physics*. Proc. Int. Symp. Clausthal–Göttingen. Göttingen: Vandenhoek und Ruprecht, 1965, pp. 188–193.
123. HP Klug, LE Alexander. *X-ray Diffraction Procedures*. New York: Wiley, 1974.
124. BE Warren, BL Averbach. The effect of cold worked distortions on x-ray patterns. *J. Appl. Phys.* 21: 595–599, 1950.
125. BE Warren, BL Averbach. The separation of cold-work distortion and particle size broadening in x-ray patterns. *J. Appl. Phys.* 23: 497, 1952.
126. R Hosemann, A Preisinger, W Vogel. Über den α-Fe- $FeAl_2O_4$-Mischkristall in aktivierten Ammoniakkatalysatoren. *Ber. Bunsenges. Phys. Chem.* 70: 797–802, 1966.
127. MF Toney, TC Huang, S Brennan, Z Rek. X-ray depth profiling of iron oxide thin films. *J. Mater. Res.* 3: 351–356, 1988.

128. MF Toney, S Brennan. Structural depth profiling of iron oxide thin films using grazing incidence asymmetric Bragg x-ray diffraction. *J. Appl. Phys.* 65: 4763–4768, 1989.
129. MF Toney, S Brennan. Observation of the effect of refraction on x-rays diffracted in a grazing-incidence asymmetric Bragg geometry. *Phys. Rev. B* 39: 7963–7966, 1989.
130. R Feidenhans'l. Surface structure determination by x-ray diffraction. *Surf. Sci. Rep.* 10: 105–188, 1989.
131. IK Robison, DJ Tweet. Surface x-ray diffraction. *Rep. Prog. Phys.* 55: 599–651, 1992.
132. G Renaud. Oxide surfaces and metal/oxide interfaces studied by grazing incidence x-ray scattering. *Surf. Sci. Rep.* 32: 1–90, 1998.
133. IK Robinson. Crystal truncation rods and surface roughness. *Phys. Rev. B* 33: 3830–3836, 1986.
134. JM Bloch. Angle and index calculations for "*z*-axis" x-ray diffractometer. *J. Appl. Crystallogr.* 18: 33–36, 1985.
135. M Lohmeier, E Vlieg. Angle calculations for a six-circle surface x-ray diffractometer. *J. Appl. Crystallogr.* 26: 706–716, 1993.
136. C Schamper, HL Meyerheim, W Moritz. Resolution correction for surface x-ray diffraction at high beam exit angles. *J. Appl. Crystallogr.* 26: 687–696, 1993.
137. E Vlieg. Integrated intensities using a six-circle surface x-ray diffractometer. *J. Appl. Crystallogr.* 30: 532–543, 1997.
138. H-H Strehblow. Passivity of metals. In: *Advances in Electrochemical Sciences and Engineering*, R Alkire, DM Kolb (eds.). Weinheim: Wiley-VCH, 2003, pp. 271–378.
139. S Haupt, H-H Strehblow. Corrosion, layer formation and oxide reduction of passive iron in alkaline solution, a combined electrochemical and surface analytical study. *Langmuir* 3: 873–885, 1987.
140. H-H Strehblow. Mechanisms of pitting corrosion. In: *Corrosion Mechanims in Theory and Practice*, P Marcus (ed.). New York: Marcel Dekker, 2002, pp. 243–285.
141. H-H Strehblow. Pitting corrosion. In: *Ecyclopedia of Electrochemistry, Vol. 4 Corrosion and Oxide Films*, AJ Bard, M Stratmann, GS Frankel (eds.). Weinheim: Wiley-VCH, 2003, pp. 308–343.
142. HH Uhlig. History of passivity, experiments and theories. In: *Passivity of Metals*, RP Frankenthal, J Kruger (eds.). Princeton: The Electrochemical Society, 1978, pp. 1–28.
143. MB Ives, JL Luo, JR Rodda. Active–passive dissolution of stainless steel in hot concentrated sulphuric acid. In: *Passivity of Metals*, MB Ives, BR MacDougall, JA Bardwell (eds.). Pennington, NJ: The Electrochemical Society, 1999, pp. 915–920.
144. M Nagayama, M Cohen. The anodic oxidation of iron in a neutral solution. *J. Electrochem. Soc.* 109: 781–790, 1962.
145. CL Foley, J Kruger, CJ Bechtoldt. Electron diffraction studies of active, passive and transpassive oxide films formed on iron. *J. Electrochem. Soc.* 114: 994–1001, 1967.
146. K Kuroda, BD Cahan, G Nazri, E Yeager, TE Mitchell. Electron diffraction study of the passive film on iron. *J. Electrochem. Soc.* 129: 2163–2168, 1982.
147. M Kerkar, J Robinson, AJ Forty. *In situ* structural studies of the passive film of iron and iron/chromium alloys using x-ray absorption spectroscopy. *Faraday Discuss. Chem. Soc.* 89: 31–40,1990.
148. GG Long, J Kruger, DR Black, M Kuriyama. Structure of passive films on iron using a new surface-EXAFS technique. *J. Electroanal. Chem.* 150: 603–610, 1983.
149. RW Hoffman. Study of passivity of iron by *in situ* methods: Moessbauer and EXAFS. In: *Passivity of Metals and Semiconductors*, M. Froment (ed.). Amsterdam: Elsevier Science, 1983, pp. 147–162.
150. JC Rubim, J Dünnwald. Enhanced Raman scattering from passive films on silver-coated iron electrodes. *J. Electroanal. Chem.* 258: 327–344, 1989.
151. J Gui, TM Devine. *In situ* vibrational spectra of the passive film on iron in buffered borate solution. *Corros. Sci.* 32: 1105–1124, 1991.

152. WE O'Grady. Moessbauer study of the passive oxide film on iron. *J. Electrochem. Soc.* 127: 555–563, 1980.
153. J Eldridge, RW Hoffman. A Mössbauer spectrocopy study of the potential dependence of passivated iron films. *J. Electrochem. Soc.* 136: 955–960, 1989.
154. MP Ryan, RC Newman, GE Thompson. An STM study of the passive film formed on iron in borate buffer solution. *J. Electrochem. Soc.* 142: L177–L179, 1995.
155. B MacDougall, JA Bardwell. Passivation of iron in sulfate, perchlorate, and borate solutions: role of borate in the passivation process. *J. Electrochem. Soc.* 135: 2437–2441, 1988.
156. M Cohen, K Hashimoto. The cationic reduction of gamma-Fe_2O_3 and oxide films on iron. *J. Electrochem. Soc.* 121: 42–45, 1974.
157. M Büchler, P Schmuki, H Böhni. Iron passivity in borate buffer. Formation of a deposit layer and its influence on the semiconducting properties. *J. Electrochem. Soc.* 145: 609–614, 1998.
158. LJ Oblonsky, AJ Davenport, MP Ryan, HS Isaacs, RC Newman. *In situ* x-ray absorption near edge structure study of the potential dependence of the formation of the passive film on iron in borate buffer. *J. Electrochem. Soc.* 144: 2398–2404, 1997.
159. MF Toney, AJ Davenport, LJ Oblonsky, MP Ryan, CM Vitus. Atomic structure of the passive oxide film formed on iron. *Phys. Rev. Lett.* 79: 4282–4285, 1997.
160. AJ Davenport, LJ Oblonsky, MP Ryan, MF Toney. The structure of the passive film that forms on iron in aqueous environments. *J. Electrochem. Soc.* 147: 2162–2173, 2000.
161. MG Samant, MF Toney, GL Borges, L Blum, OR Melroy. Grazing incidence x-ray diffraction of lead monolayers at a silver (111) and gold (111) electrode/electrolyte interface. *J. Phys. Chem.* 92: 220–225, 1988.
162. MF Toney, JG Gordon, MG Samant, GL Borges, OR Melroy, LS Kau, DG Wiesler, D Yee, LB Sorensen. Surface x-ray-scattering measurements of the substrate-induced spatial modulation of an incommensurate adsorbed monolayer. *Phys. Rev. B* 42: 5594–5603, 1990.
163. RC Weast (ed.). *Handbook of Chemistry and Physics*, 61st edn. Boca Raton, FL: CRC Press, 1981.
164. C Greaves. A powder neutron diffraction investigation of vacancy ordering and covalence in γ-Fe_2O_3. *J. Solid State Chem.* 49: 325–333, 1983.
165. RW Wyckoff. *Crystal Structures*, Vol. 3. New York: Interscience Publishers, 1965.
166. R Dieckmann. Point defects and transport properties of binary and ternary oxides. *Solid State Ionics* 12: 1–22, 1984.
167. N Sato, G Okamoto. Electrochemical passivation of metals. In: *Electrochemical Materials Science*, JOM Bockris, BE Conway, E Yeager, RE White (eds.). New York: Plenum Press, 1981, pp. 193–245.
168. B MacDougall, MJ Graham. Growth and stability of passive films. In: *Corrosion Mechanisms in Theory and Practice*, P Marcus, J Oudar (eds.). New York: Marcel Dekker, 1995, pp. 143–174.
169. P Marcus, J Oudar, I Olefjord. XPS-study of the passive film on nickel. *J. Microsc. Spectrosc. Electron.* 4: 63–72, 1979.
170. H-W Hoppe, H-H Strehblow. XPS and UPS examinations of the formation of passive layers on Ni in 1 M sodium hydroxide and 0.5 M sulfuric acid. *Surf. Interface Anal.* 14: 121–131, 1989.
171. D Zuili, V Maurice, P Marcus. *In situ* investigation by ECSTM of the structure of the passive film formed on Ni(111) single-crystal surfaces. In: *Proceedings of the Symposium on Passivity and its Breakdown*, P Natishan, HS Isaacs, M Janik-Czachor, VA Macagno, P Marcus, M Seo (eds.). Pennington: The Electrochemical Society, 1998, pp. 1013–1024.
172. B MacDougall, M Cohen. Anodic oxide films on nickel in acid solutions. *J. Electrochem. Soc.* 123: 191–197, 1976.
173. J Oudar, P Marcus. Role of adsorbed sulphur in the dissolution and passivation of nickel and nickel–sulfur alloys. *Appl. Surf. Sci.* 3: 48–67, 1979.

174. V Maurice, H Talah, P Marcus. *Ex situ* STM imaging with atomic resolution of Ni(111) electrodes passivated in sulfuric acid. *Surf. Sci.* 284: L431–L436, 1993.
175. V Maurice, H Talah, P Marcus. A scanning tunneling microscopy study of the structure of thin oxide films grown on Ni(111) single crystal surfaces by anodic polarization in acid electrolyte. *Surf. Sci.* 304: 98–108, 1994.
176. T Suzuki, T Yamada, K Itaya. *In situ* electrochemical scanning tunneling microscopy of Ni(111), Ni(100), and sulfur-modified Ni(100) in acidic solution. *J. Phys. Chem.* 100: 8954–8961, 1996.
177. D Zuili, V Maurice, P Marcus. Surface structure of nickel in acid solution studied by *in situ* scanning tunneling microscopy. *J. Electrochem. Soc.* 147: 1393–1400, 2000.
178. RS McEwen. Crystallographic studies on nickel hydroxide and the higher nickel oxides. *J. Phys. Chem.* 75: 1782–1789, 1971.
179. OM Magnussen, J Scherer, BM Ocko, RJ Behm. *In situ* x-ray scattering study of the passive film on Ni(111) in sulfuric acid solution. *J. Phys. Chem. B* 104: 1222–1226, 2000.
180. X Wang, B Ocko, AJ Davenport, HS Isaacs. *In situ* x-ray-diffraction and -reflectivity studies of the Au(111)/electrolyte interface: reconstruction and anion adsorption. *Phys. Rev. B* 46: 10321–10338, 1992.
181. B MacDougall, DF Mitchell, MJ Graham. Changes in oxide films on nickel during long-term passivation. *J. Electrochem. Soc.* 132: 2895–2898, 1985.
182. J Kunze, V Maurice, LH Klein, H-H Strehlow, P Marcus. *In situ* scanning tunneling microscopy study of the anodic oxidation of Cu(111) in 0.1 M NaOH. *J. Phys. Chem. B* 105: 4263–4269, 2001.
183. JV Cathcart, GF Petersen, CJ Sparks. The structure of thin oxide films formed on nickel crystals. *J. Electrochem. Soc.* 116: 664–674, 1969.
184. H-H Strehblow, B Titze. The investigation of the passive behaviour of copper in weakly acid and alkaline solutions and the examination of the passive film by ESCA ans ISS. *Electrochim. Acta* 25: 839–850, 1980.
185. DW Shoesmith, S Sunder, MG Bailey, GJ Wallace, FW Stanchel. Anodic oxidation of copper in alkaline solution. Part IV, nature of passivating film. *J. Electroanal. Chem.* 143: 153–165, 1983.
186. C-H Pyun, SM Park. *In situ* spectroelectrochemical studies on anodic oxidation of copper in alkaline solution. *J. Electrochem. Soc.* 133: 2024–2030, 1986.
187. MM Lohrengel, JW Schultze, H-D Speckmann, H-H Strehblow. Growth, corrosion and capacity of copper oxide films investigated by pulse techniques. *Electrochim. Acta* 32: 733–742, 1987.
188. H-D Speckmann, S Haupt, H-H Strehblow. A quantitative surface analytical study of electrochemically formed copper oxides by XPS and x-ray induced auger spectroscopy. *Surf. Interface Anal.* 11: 148–155, 1988.
189. L Grasjo, M Seo, N Sato. *In-situ* gravimetry of passivation of copper by means of quartz crystal microbalance. *Corros. Sci.* 31: 299–304, 1990.
190. M Shirkhanzadeh, GE Thomson, V Ashworth. A study of the initial stages in oxitation of copper in alkaline solutions. *Corros. Sci.* 31: 293–298, 1990.
191. U Collisi, H-H Strehblow. The formation of Cu_2O layers on Cu and their electrochemical and photoelectrochemical properties. *J. Electroanal. Chem.* 284: 385–401, 1990.
192. S de Sanchez, EAL Berlouis, DJ Schiffrin. Difference reflectance spectroscopy of anodic films on copper and copper base alloys. *J. Electroanal. Chem.* 307: 73–86, 1991.
193. DT Schwartz, RH Müller. Oxidation films on copper in alkaline media: intensity modulated photoelectrochemical and Raman spectroscopy studies. *Surf. Sci.* 248: 349–358, 1991.
194. H You, CA Melendrews, Z Nagy, VA Maroni, W Yun, RM Yonco. X-ray-reflectivity study of the copper-water interface in a transmission geometry under in situ electrochemical control. *Phys. Rev. B* 45: 11288–11298, 1992.
195. TL Barr. *Modern ESCA*. Boca Raton, FL: CRC Press, 1994.

196. B Borie. A diffraction measurement of the structure of Cu_2O films grown on copper. *Acta Crystallogr.* 13: 542–545, 1960.
197. DA Fisher, GG Cohen, NJ Shevchik. EXAFS investigation of oxidation processes in metallic Cu. *J. Phys. F* 10: L139–L142, 1980.
198. H-H Strehblow, H-D Speckmann. Corrosion and layer formation of passive copper in alkaline solutions. *Werkst. Korr.* 35: 512–519, 1984.
199. H-D Speckmann, MM Lohrengel, JW Schultze, H-H Strehblow. The growth and reduction of duplex oxide films on copper. *Ber. Bunsen Ges. Phys. Chem.* 89: 392–402, 1985.
200. J Kunze, V Maurice, LH Klein, H-H Strehblow, P Marcus. *In situ* STM study of the anodic oxidation of Cu(001) in 0.1 M NaOH. *J. Electroanal. Chem.*, 554–555: 113–125, 2003.
201. J Kunze, V Maurice, LH Klein, H-H Strehblow, P Marcus. *In situ* STM study of the duplex passive films formed on Cu(001) and Cu(111) in 0.1 M NaOH. *Corros. Sci.* 46: 245–264, 2004.
202. YP Feng, SK Sinha, CA Melendres, DD Lee. X-ray off-specular reflectivity studies of electrochemical pitting of Cu surfaces in sodium bicarbonate solution. *Physica B* 221: 251–256, 1996.
203. H Böhni. Breakdown of passivity and localized corrosion processes. *Langmuir* 3: 924–930, 1987.
204. DD Macdonald. The point defect model for the passive state. *J. Electrochem. Soc.* 139: 3434–3449, 1992.
205. TP Hoar. The anodic behaviour of metals. In: *Modern Aspects of Electrochemistry*, Vol. 2, JO'M Bockris, BE Conway (eds.). London: Butterworths, 1963, pp. 262–342.
206. RR Beck. Formation of salt films during passivation of iron. *J. Electrochem. Soc.* 129: 2412–2418, 1982.
207. RC Alkire, D Ernberger, TR Beck. Occurrence of salt films during repassivation of newly generated metal surfaces. *J. Electrochem. Soc.* 129: 1382–1388, 1978.
208. KJ Vetter, H-H Strehblow. Entstehung und Gestalt von Korrosionslöchern bei Lochfraa an Eisen und theoretische Folgerungen zur Lochfraakorrosion. *Ber. Bunsenges. Phys. Chem.* 74: 1024–1031, 1972.
209. H-H Strehblow. Mechanisms of pitting corrosion. In: *Corrosion Mechanisms in Theory and Practice*, P Marcus (ed.). New York: Marcel Dekker, 2002, pp. 243–285.
210. D Landolt. Fundamental aspects of electropolishing. *Electrochim. Acta* 32: 1–11, 1987.
211. M Datta, D Landolt. On the role of mass transport in high rate dissolution of iron and nickel in ECM electrolytes: I. Chloride solutions. *Electrochim. Acta* 25: 1255–1262, 1980.
212. H-H Strehblow, J Wenners. Determination of the growth of corrosion pits on iron and nickel in an early stage of development and its relation to metal dissolution in concentrated chloride media. *Z. Phys. Chem. NF* 98: 199–206, 1975.
213. H-H Strehblow, J Wenners. Investigation of the processes on iron and nickel electrodes at high corrosion current densities in solutions of high chloride content. *Electrochim. Acta* 22: 421–427, 1977.
214. HS Icaacs. The behavior of resistive layers in the localized corrosion of stainless steel. *J. Electrochem. Soc.* 120: 1456–1462, 1973.
215. RD Grimm, AC West, D Landolt. AC impedance study of anodically formed salt films on iron in chloride solution. *J. Electrochem. Soc.* 139: 1622–1630, 1992.
216. Y Wu, AC Thompson, JH Underwood, RD Gaiuque, K Chapman, ML Rivers, KW Jones. A tunable x-ray microprobe using synchrotron radiation. *Nucl. Instrum. Meth. A* 291: 146–151, 1990.
217. ML Rivers, S.R. Sutton, K.W. Jones. Synchrotron x-ray fluorescence microscopy. *Synchrotron Rad. News* 4(2): 23–26, 1991.
218. HS Isaacs, J-H Cho, ML Rivers, SR Sutton. *In situ* x-ray microprobe study of salt layers during anodic dissolution of stainless steel in chloride solution. *J. Electrochem. Soc.* 142: 1111–1118, 1995.
219. M Kimura, M Kaneko, T Suzuki. *In situ* observation of pitting of stainless steel by XAFS. *J. Synchrotron Rad.* 8: 487–489, 2001.

220. B Lengeler, C Schroer, J Tümmler, B Benner, M Richwin, A Snigirev, I Snigireva, M Drakopoulos. Imaging by parabolic refractive lenses in the hard x-ray range. *J. Synchrotron Rad.* 6: 1153–1167, 1999.
221. A Snigirev, V Kohn, I Snigireva, B Lengeler. A compound refractive lens for focusing high-energy x-rays. *Nature* 384: 49–51, 1996.
222. WJ Albery, ML Hitchman. *Ring Disk Electrodes.* Oxford: Clarendon Press, 1971.
223. G Engelhardt, D Schaepers, H-H Strehblow. The hydrodynamically square-wave modulated rotating ring disk electrode: theoretical basis and its experimental verification. *J. Electrochem. Soc.* 139: 2170–2175, 1992.
224. G Palinkas, E Kalman. Diffraction investigations of aqueous electrolyte solutions. In: *Diffraction Studies on Non-Crystalline Substances,* I Hargittai, WJ Orville-Thomas (eds.). Amsterdam: Elsevier Science, 1981, pp. 293–373.
225. J Kortright, W Warburton, A Bienenstock. Anomalous x-ray scattering and its relationship to EXAFS. In: *EXAFS and Near Edge Structure,* A Bianconi, L Incoccia, S Stipcich (eds.). Berlin: Springer Verlag, 1983, pp. 362–372.
226. D Lützenkirchen-Hecht, CU Waligura, H-H Strehblow. An *in situ* EXAFS study of corrosion products. *Corros. Sci.* 40: 1037–1042, 1998.
227. SP Cramer, TC Eccles, FW Kutzler, KE Hodgson, LE Mortenson. Molybdenum x-ray absorption edge spectra. The chemical state of molybdenum in nitrogenase. *J. Am. Chem. Soc.* 98: 1287–1288, 1976.
228. FW Kutzler, CR Natoli, DK Misemer, S Doniach, KO Hodgsen. Use of one-electron theory for the interpretation of near edge structure in K-shell x-ray absorption spectra of transition metal complexes. *J. Chem. Phys.* 73: 3274–3288, 1980.
229. NS Chiu, SH Bauer, MFL Johnson. Cobalt/molybdenum/aluminium oxide catalysts structure determination by EXAFS. I. Mo K edge in the oxidized state. *J. Catal.* 89: 226–243, 1984.
230. BG Brandt, AC Stapski. Refinement of the crystal structure of MoO_2. *Acta Chim. Scand.* 21: 661–672, 1967.
231. LMJ von Hippel, F Hilbrig, H Schmelz, B Lengeler, H Knözinger. XANES investigation of molybdate catalysts supported on titania. *Collect. Czech. Chem. Commun.* 57: 2465–2474, 1992.
232. L Gmelin, *Handbuch der Anorganischen Chemie, 8. Aufl., Molbydän, Ergänzungsband B2.* Berlin: Springer Verlag, 1986.
233. SC Abrahams, JM Reddy. Crystal structure of the transition-metal molybdates. I. Paramagnetic alpha-$MnMoO_4$. *J. Chem. Phys.* 43: 2533–2543, 1965.
234. AI Frenkel, GV Korshin. EXAFS studies of the chemical state of lead and copper in corrosion products formed on the brass surface in potable water. *J. Synchrotron Rad.* 6: 653–655, 1999.
235. MR Schock. Understanding corrosion control strategies for lead. *J. Am. Water Works Assoc.* 81(7): 88–100, 1989.
236. H Bornebusch, BS Clausen, G Steffensen, D Lützenkirchen-Hecht, R Frahm. A new approach for QEXAFS data acquisition. *J. Synchrotron Rad.* 6: 209–211, 1999.
237. M Richwin, R Zaeper, D Lützenkirchen-Hecht, R Frahm. Piezo-XAFS — time resolved x-ray absorption spectroscopy. *Rev. Sci. Instrum.* 73: 1668–1670, 2002.
238. A Foelske, unpublished results.
239. A Foelske, H-H Strehblow. Passivity of cobalt in borate buffer at pH 9.3 studied by x-ray photoelectron spectroscopy. *Surf. Interface Anal.* 29: 548–555, 2000.
240. A Foelske, H-H Strehblow. Structure and composition of electrochemically prepared oxide layers on Co in alkaline solutions studied by XPS. *Surf. Interface Anal.* 34: 125–129, 2002.
241. D Lützenkirchen-Hecht, unpublished results.

7 Infrared Spectroscopy

Christofer Leygraf and Magnus Johnson
Department of Materials Science and Engineering, Division of Corrosion Science, Royal Institute of Technology, Stockholm, Sweden

Contents

7.1 INTRODUCTION

The purpose of this chapter is to give a general background of infrared (IR) spectroscopy, and to show the capability of IR spectroscopy in corrosion science. It starts with a section covering the underlying theory, continues with sections on the instrumentation used and the most important techniques available, and ends with a selection of papers, in which IR spectroscopy has been applied to understand corrosion-related phenomena. The selection is based on an extensive literature search that resulted in more than 300 papers published over the last 20 years or so. The literature search shows a significant increase in the number of papers published, from approximately 15 each year around 1990 to almost 40 each year 10 years later. Most studies use IR spectroscopy in the reflection–absorption mode (IRAS) under *ex situ* conditions, and IRAS is thus the technique that will be most extensively treated here. The IRAS technique frequently does not stand alone, but is used in combination with other analytical techniques, such as x-ray photoelectron spectroscopy and x-ray diffraction, to provide complementary analytical information of corrosion products formed. As opposed to many other analytical techniques, IR spectroscopy can provide surface-related information obtained under *in situ* conditions. Therefore, the emphasis in this chapter is on papers in which IR spectroscopy has been used as the main analytical technique, and in particular on studies that have taken advantage of the inherent *in situ* capability of IRAS. In general, no lateral resolution can be achieved, which excludes studies of pitting corrosion, crevice corrosion, stress corrosion cracking, and other forms of more localized corrosion. The papers, briefly presented in Section 7.5, have been divided into the following corrosion-related areas: adsorption in electrochemical environments, corrosion inhibition, surface treatments, corrosion product formation in aqueous environments, corrosion product formation in atmospheric environments, and microbial corrosion.

7.2 THEORY

7.2.1 Infrared Radiation

IR radiation, like all electromagnetic radiation, is a sinusoidal electromagnetic wave consisting of an electric field and a magnetic field, which always are perpendicular to each other and in phase. It is the electric field that causes interaction with molecules. The radiation travels in a direction perpendicular to the two fields, and can be characterized by its wavelength λ (m), which is the distance between two consecutive maxima, its frequency ν (Hz), which is the number of waves per unit time, and its velocity c (m/sec), which is the speed of light in vacuum (=2.997925×10^8 m/sec). These quantities are related according to the following equation:

$$c = \nu\lambda \tag{7.1}$$

When traveling through another medium than vacuum, the velocity ν_m (m/sec) is given by $\nu_m = c/n$, where n is the refractive index of the medium. The intensity I (W/m^2) of the radiation is proportional to the square of the amplitude A (V/m) of the field. In spectroscopy, the wavenumber $\bar{\nu} = 1/\lambda$ (cm^{-1}) is a commonly used unit, since it is proportional to the frequency and the energy E (J) of the radiation, given by the following equation:

$$E = h\nu = hc/\lambda \tag{7.2}$$

where h is Planck's constant (=6.6261×10^{-34} J sec). Since IR radiation encompasses a long range of wavelengths, it is convenient to introduce subregions (as given in Table 7.1). The division is based on how the radiation is generated and detected in the different regions.

Table 7.1 The three infrared regions and corresponding parameters

Infrared region	Wavelength λ (μm)	Frequency ν (Hz)	Wavenumber $\bar{\nu}$ (cm^{-1})
Near	0.78–2.5	3.8×10^{14}–1.2×10^{14}	12,800–4000
Middle	2.5–50	1.2×10^{14}–6×10^{12}	4000–200
Far	50–1000	6×10^{12}–3×10^{11}	200–10

Because the wavelength of the IR radiation is much longer than the length of a molecule, it is in order to assume that the whole molecule experiences the same electric field. IR spectroscopy is mostly performed in the mid-IR region (1–4), in which qualitative analyses are common, but this region also enables fruitful quantitative studies. Due to the versatility of this region, this chapter will be mostly focused on mid-IR spectroscopy. When performing a qualitative analysis of a spectrum, there are two important subregions in the mid-IR region, the *group frequency region* spanning the range 1200 to 3600 cm^{-1}, and the *fingerprint region* covering the range 600 to 1200 cm^{-1}. In the group frequency region, it is possible to determine what functional groups are present in the molecule, such as C–H, C=O, and O–H. Affected by the surroundings, the group frequencies will not be found at exactly the same wavenumber for all different compounds, but instead in a certain wavenumber range characteristic of that bond. The characteristic frequencies of some important vibrations in the mid-IR region are shown in Table 7.2.

An extensive treatment of IR characteristic frequencies is found in Refs. (5,6). The fingerprint region is used for the determination of the whole structure of the molecule by comparison to known spectra, since the overall structure gives rise to very specific spectral features in this region. The near-IR region (7) is mostly used for quantitative measurements of, for example, alcohols, proteins, and moisture measurements in the atmosphere. The far-IR (8) region is among other fields suitable for studying hydrogen bonds, vibrations of ligands bonded to metal atoms, and rotations in gases.

7.2.2 Molecular Vibrations

IR radiation has an energy high enough to cause rotational and vibrational excitations in molecules, but the energy is not energetic enough to cause electronic excitations. For the purpose of this chapter, the vibrational excitations are the most important forms of transitions, so rotations and electronic excitations will not be treated any further. A molecule in the classical model is assumed to consist of points (the nuclei) connected by massless springs (the bonds or electrons). The nuclei can move in different ways to each other, and it can be shown (9) that a nonlinear molecule has $3N - 6$, and a linear molecule $3N - 5$ different vibrations, where N is the number of atoms in the molecule. These vibrations are called *normal modes* of the molecule, and can sometimes be degenerate. For each normal mode, the atoms of the vibrating molecule pass their equilibrium positions at the same time, and vibrate with the same frequency. During a vibration, the bond length or angle changes, and the various kinds of vibrations are shown in Figure 7.1.

The normal modes are not independent of each other, resulting in *vibrational coupling*. This means that different vibrations affect each other, so the absorption frequencies of a functional group change slightly depending on the surroundings, as

Table 7.2 Characteristic absorption bands of various compounds

Bond	Type of vibration	Type of compound	Wavenumber range (cm^{-1})	Intensity[a]
C–H	Stretch	Alkanes	2850–2970	S
C–H	Deformation	Alkanes	1340–1470	S
C–C	Stretch	Alkanes, skeletal	700–1260	W
O–H	Stretch	Alcohols, phenols (dilute)	3590–3650	V, sh
		Alcohols, phenols (H-bonded)	3200–3550	S, br
		Carboxylic acids (H-bonded)	2500–3000	V, br
C–O	Stretch	Alcohols, carboxylic acids	1050–1300	S
C=O	Stretch	Aldehydes	1690–1740	S
		Carboxylic acids	1710–1780	S
		Amides (dilute solutions)	1630–1690	S
N–H	Stretch	Amines, amides	3300–3500	M
C≡N	Stretch	Nitriles	2210–2280	S
$-NO_2$	Stretch (asym.)	Nitro compounds	1500–1570	S
	Stretch (sym.)	Nitro compounds	1300–1370	S
S–H	Stretch	Mercaptans, aliphatic thiols, thiophenols (free)	2550–2600	W
$>SO_2$	Stretch (asym.)	Organic sulfones (dilute solution)	1290–1360	S
	Stretch (sym.)		1120–1170	S
SO_4^{2-}	Stretch	Sulfate ion	1080–1130	S
	Several bands	Sulfate ion	580–680	M
CO_3^{2-}	Stretch	Carbonate ion	1410–1450	S, br
NO_3^-	Stretch	Nitrate ion	1340–1410	S, br
M=O	Stretch	Metal oxides	825–1100 (in general)	S
M–H	Stretch	Metal hydrides	1700–2270 (in general)	M, sh

[a]S, Strong; M, medium; W, weak; V, variable; br, broad; sh, sharp; M, Metal.

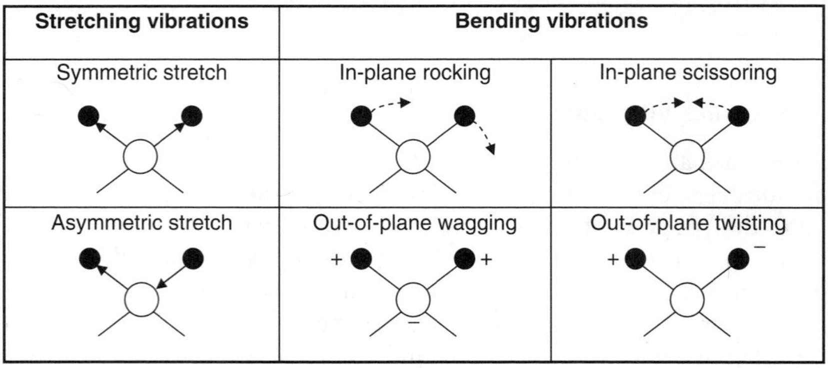

Figure 7.1 Stretching and bending vibrations. "+" indicates that the atom moves toward the reader and "–" indicates that the atom moves away from the reader.

mentioned above. As an example, the C–O stretching absorption band in methanol is found at $1034\,cm^{-1}$, but at $1053\,cm^{-1}$ in ethanol. In order to absorb IR radiation, the dipole moment of the molecule has to change as a result of the vibration, described by Equation (7.3). The frequency of the IR radiation has also to match the frequency of the vibrating dipole:

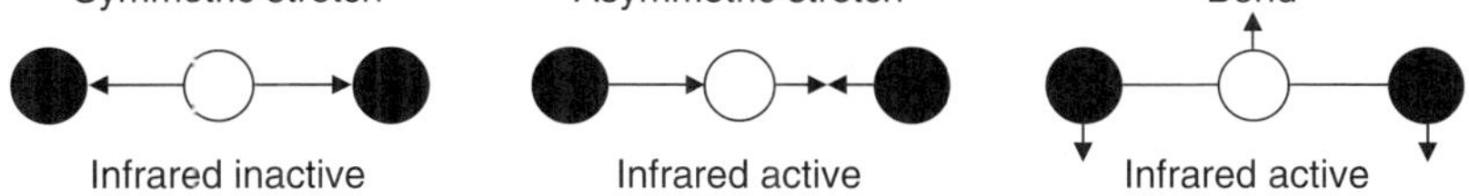

Figure 7.2 Normal modes of carbon dioxide.

$$\left(\frac{\partial \mu}{\partial Q}\right) \neq 0 \tag{7.3}$$

where μ is the dipole moment ($\mu = qr$, where q is the charge of the atom and r is the distance between the charges), and Q is the normal coordinate, describing the "amplitude" of the normal mode. The intensity of an IR band is proportional to the square of the change in dipole moment, $I \propto (\partial \mu / \partial Q)^2$. During an absorption, the amplitude of the molecular vibration changes, but the frequency remains constant. Most molecules will absorb IR radiation and therefore be *IR active*, with exceptions for compounds such as diatomic homonuclear molecules and single atoms. In carbon dioxide, for example, the asymmetric stretch and the bend are IR active, but the symmetric stretch is IR inactive since the dipole moment remains constant during that vibration, according to Figure 7.2.

A comparison can be made to *Raman spectroscopy*, where a change in the *polarizability* of the molecule is necessary for the molecule to be Raman active. In carbon dioxide, the symmetric stretch is Raman active, but the asymmetric stretch and the bend are Raman inactive. A consequence of these *selection rules* is that a molecule with a *center of inversion* cannot have modes that are both IR and Raman active, which is true for CO_2 molecule as shown above. A molecule is said to possess a center of inversion if it is possible to draw a straight line from any atom through a common point (the center of inversion), and at the same distance on the other side of the point find an identical atom.

Imagine a diatomic molecule that is assumed to obey Hooke's law, meaning that the restoring force is proportional to the deviation from the equilibrium bond length:

$$F = -k(r - r_e) \tag{7.4}$$

where F (N) is the restoring force, k (N/m) is the force constant (spring constant), $r - r_e$ is the deviation from the equilibrium distance r_e, and the minus sign indicates that the force is restoring. The force constant k determines the strength of the bond, and a large k is indicative of a strong bond. The calculation of k from an IR spectrum is an important application of vibrational spectroscopy. The above description of the force being directly proportional to the displacement is an approximation where the molecule is said to be a *harmonic oscillator*. The corresponding potential energy is given by the following equation, which is the formula of a parabola:

$$V(r) = \frac{1}{2}k(r - r_e)^2 \tag{7.5}$$

Equation (7.6) displays the amplitude independent *natural frequency* v_n of a vibration, each normal mode having its own natural frequency:

$$v_n = \frac{1}{2\pi}\sqrt{\frac{k}{u}} \tag{7.6}$$

where u is the *reduced mass*, given in the following equation:

$$u = \frac{m_1 m_2}{m_1 + m_2} \tag{7.7}$$

where m_1 and m_2 are the masses of the atoms. Equations (7.5) and (7.6) are derived in Ref. (9, pp. 383–385). To improve the description of a molecular vibration, it is necessary to treat it quantum mechanically, which is a result of the wavelike properties of a particle. The classical and the quantum models differ in the sense that the vibrational energies a molecule can possess are discrete in quantum mechanics, but continuous in the classical model. The following equation, derived in Ref. (10), displays the vibrational energy in the quantum model:

$$E_{\mathrm{vib}} = \left(v + \frac{1}{2}\right) h\nu_{\mathrm{n}} \tag{7.8}$$

where v is the *vibrational quantum number* ($v = 0, 1, 2, \ldots$). The difference in energy between two energy levels in the harmonic oscillator is $h\nu_{\mathrm{n}}$ for all consecutive levels, which equals the energy of the light required to excite the vibrational mode. When $v = 0$, the energy is $\frac{1}{2}h\nu_{\mathrm{n}}$, called the *zero point energy*. As seen in Equation (7.8), the vibrational energy can never attain a value of zero, which is a result of Heisenberg's uncertainty principle. The quantum mechanical selection rule for a harmonic oscillator states that the vibrational quantum number v has to change by ± 1 for an IR absorption to occur, which is similar to the classical picture where the IR radiation and the molecular dipole must have the same frequency.

In the real world there are no perfect harmonic oscillators. Instead all molecular vibrations are more or less *anharmonic*, meaning that the harmonic approximation is no longer valid. Figure 7.3 shows the harmonic and anharmonic oscillators, and it is clearly seen that the harmonic approximation is good only at small deviations from the equilibrium position.

The origin of anharmonicity and the ensuing breakdown of the harmonic approximation are consequences of the coulombic repulsion arising when two

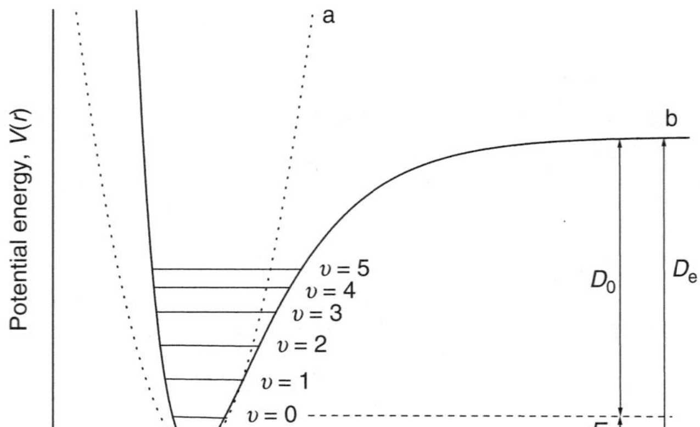

Figure 7.3 Potential energy diagrams for the harmonic (a) and the anharmonic oscillator (b). E_0 is the zero point energy, D_0 is the dissociation energy of the bond, and D_e is the sum of the zero point energy and the dissociation energy.

atoms come close together, and the decrease in restoring force when the two atoms start to dissociate. When anharmonicity prevails, the potential energy cannot be described on the quadratic form shown in Equation (7.5). Instead, the Morse function (11), given in Equation (7.9), is a commonly used function that better describes the vibrational energy:

$$V(r) = D_e(1 - e^{-a(r-r_e)})^2 \tag{7.9}$$

where D_e is defined in Figure 7.3, and $a = (\frac{1}{2}kD_e)^{1/2}$ is a constant determining the curvature of the curve. At small deviations from the equilibrium position, the Morse function reduces to the harmonic function (Equation (7.5)) as expected (4, p. 18). In an anharmonic potential energy curve, the spacing between two consecutive energy levels decreases with increasing vibrational number (see Figure 7.3). Transitions that were forbidden in the harmonic case are allowed in the anharmonic case, and therefore *overtones*, *combination bands*, and *hot bands* can be seen. At room temperature, most molecules will be in their vibrational ground state $v = 0$, according to the Boltzmann distribution, and therefore the transition $v = 1 \leftarrow v = 0$ is called the *fundamental transition*. Overtones occur when light with frequencies corresponding to approximately integer multiples of the fundamental transition ($\Delta v = \pm 2, \pm 3, \ldots$) is absorbed by dipoles performing anharmonic vibrations at those frequencies. Since the spacing between the energy levels decreases with increasing vibrational number for an anharmonic oscillator, overtones are found at frequencies slightly lower than multiples of the fundamental transition frequency. Combination bands are found when two (or more) different fundamental vibrational modes are excited by a single photon, and hot bands occur when a molecule not in the ground state is excited, for example, a transition from $v = 1$ to $v = 2$. Overtones and combination bands can be used for studies in the near-IR region, as described in Section 7.5.1. Usually, overtones, combination bands, and hot bands are much weaker than the fundamental transition. However, there are situations when overtones and combination bands can show much stronger intensity than expected, referred to as *Fermi resonance* (4, pp. 31–33). It occurs when a fundamental transition is close in frequency to a combination band or overtone of another vibrational mode.

7.3 INSTRUMENTATION

7.3.1 Sources of Infrared Radiation

There are many kinds of IR sources, most of them being solids electrically heated to a certain temperature. Commonly working in the temperature range between 1600 and 2400 K, these sources emit IR radiation with an energy distribution resembling a blackbody, according to Planck's law. The Nernst Glower is a rod mainly composed of rare earth oxides like yttrium, thorium, and zirconium, with wires connected to the ends enabling electrical heating. A Globar rod is made of silicon carbide and is also electrically heated. Usually, the Globar operates at temperatures lower than the Nernst Glower. A nichrome wire can be used as an IR source by shaping it into a coil and letting a current pass through it. Then the resistance of the wire produces IR radiation, which is lower in intensity than the radiation given off by the Nernst Glower and the Globar. For longer wavelengths ($\lambda > 50\ \mu m$), the high-pressure mercury arc is suitable. It consists of mercury vapor sealed within a quartz tube, and when passing a current through the vapor, IR radiation is generated. Tunable IR lasers can also be used as IR sources (12), with the advantage of producing radiation of much higher intensities than blackbody emitters do. A very powerful source for IR radiation is the synchrotron, which, with its higher stability, brightness, and absolute output, is a valuable tool especially when performing experiments in the low-frequency range.

7.3.2 Detectors

There are two main types of detectors: *thermal* and *semiconductor*. Thermal detectors, including thermocouples, bolometers, and pyroelectric detectors, work by measuring the increase in temperature on absorption of IR radiation. To reduce the noise level, the detector can be housed in a vacuum chamber, and the IR beam can be chopped in order to distinguish it from noise. Thermocouples consist of two different metals such as antimony and bismuth. These metals make up a junction, and the temperature change is measured as a change in potential. Bolometers work by detecting the change in resistance when IR radiation strikes the detector, and common materials used are platinum and nickel. Pyroelectric detectors are composed of thin layers of single crystals of pyroelectric materials placed between two electrodes. Other dielectric materials lose their polarization when an applied electric field is removed, but pyroelectric materials retain a polarization even afterwards. Since the retained polarization of the pyroelectric material is temperature dependent, the system will work as a capacitor, where the capacitance changes by temperature. A common pyroelectric detector is the DTGS (deuterated triglycine sulfate) detector. Semiconductor detectors can be made of mercury cadmium telluride (MCT), indium antimonide, and lead sulfide for example. When IR radiation is absorbed by the detector, electrons are excited from the valence band to the conduction band, thereby inducing a current that is a measure of the IR intensity. Advantages of semiconductor detectors are that they provide a higher sensitivity and have a faster response than pyroelectric detectors. Therefore, they are suitable in Fourier transform infrared (FTIR) spectrometers (Section 7.3.3). However, pyroelectric detectors can also be used in FTIR spectrometers. A drawback of semiconductor detectors is that radiation with a longer wavelength than the bandgap cannot be detected, since the energy then is too low to cause an excitation.

7.3.3 Spectrometers

There are three kinds of IR spectrometers, *dispersive*, *nondispersive*, and *Fourier transform* (FT). Earlier, dispersive spectrometers were most common, but now FT spectrometers have taken over. Dispersive spectrometers are mainly used for qualitative measurements, and require the use of a monochromator in order to separate a single wavelength from the incoming polychromatic IR radiation. A spectrum is obtained by stepping through the wavelength interval of interest, and recording the intensity at the detector as a function of wavelength.

Nondispersive spectrometers are used for quantitative analysis and are simpler than dispersive spectrometers. In contrast to dispersive spectrometers they do not employ a monochromator, but some of them use different filters, whereas some do not separate the wavelengths at all.

An FT spectrometer usually involves a *Michelson interferometer* (13) instead of a monochromator. In FTIR spectroscopy (14), an *interferogram* is obtained, in which destructive and constructive interference of the light gives rise to an intensity variation as a function of optical path difference between two beams in the spectrometer. The IR spectrum is subsequently obtained by taking the Fourier transform of the interferogram (1, p. 149). An FT spectrometer has many advantages over a dispersive spectrometer. The most important advantage is the much larger signal-to-noise ratio (S/N), which is especially important when the signal is weak. The origin behind the large S/N ratio is that in a dispersive instrument only a fraction of the incoming radiation reaches the detector, since most of the radiation is blocked by the exit slit of the monochromator, whereas in an FTIR spectrometer all radiation reaches the detector. Another advantage is that the time required to generate a spectrum is much shorter for an FT spectrometer, since all the polychromatic radiation is analyzed at the same time, and no scanning of wavelength is necessary as in dispersive instruments. An FT spectrum can nowadays be generated in less than a second, enabling a large number of scans without being too time consuming, thereby improving the S/N ratio even more (S/N $\propto \sqrt{n}$, where n is the number of scans). Other important advantages are the high resolution, which can be 0.1 cm^{-1}, and the very high wavelength accuracy.

7.4 EXPERIMENTAL TECHNIQUES

There are two main classes of experimental techniques in IR spectroscopy, *transmission* and *reflection* techniques. In mid-IR spectroscopy, both techniques are valuable tools for qualitative and quantitative analyses. A comprehensive discussion on IR spectroscopy techniques is found in Ref. (3, pp. 87–137). *Sum frequency generation* (SFG), a technique related to IR spectroscopy, is also treated.

7.4.1 Transmission Spectroscopy

Transmission experiments are, in general, probably the most common way of performing IR spectroscopy, but in corrosion science, reflectance techniques are more widely used. The technique of transmission can be used for examining nearly all kinds of samples — solids, liquids, and gases. In this technique, the amount of IR light transmitted through a sample is measured, as seen in the following equation:

$$T = \frac{I}{I_0} \tag{7.10}$$

where T is the transmittance, I_0 is the incident intensity, and I is the intensity of the transmitted light. The spectrum displays the percentage transmitted light as a function of wavelength. Another parameter often used is the *absorbance* A, defined in the following equation:

$$A = -\log_{10} T \tag{7.11}$$

The absorbance can also be determined through Beer's law (the following equation), which describes the absorption of light during transmission:

$$A = \varepsilon l c \tag{7.12}$$

where ε (l/mol/cm) is the molar absorptivity, l (cm) is the pathlength of the cell, and c (mol/l) is the concentration of the sample. A spectrum can be plotted in either transmittance or absorbance units, but when performing quantitative analyses, the absorbance must be used, since it is linearly proportional to concentration. An advantage of transmission spectroscopy is the high S/N ratio. A disadvantage may be a time-consuming sample preparation in some cases, and a limited sample thickness range. If the sample is too thin, the absorbance will be too weak to give a good signal, and if the sample is too thick, all light will be absorbed. One way of studying solids qualitatively and quantitatively in transmission is to compress a pellet of a sample ground into a fine powder and diluted in an inert matrix such as KBr. The pellet is then placed in the IR beam path. Mulls (mineral oils) can also be used as dispersing agents for solids. It is also possible to dissolve solids in inert liquids and place the solution in a cell. CS_2, CCl_4, $CDCl_3$, and $CDBr_3$ are suitable solvents at wavenumbers higher than $1200\,cm^{-1}$, since they do not have any absorption bands there. Water is difficult to use as a solvent, since it is a very strong absorber of IR radiation, and therefore path lengths of at most 50 μm are possible to use. For further information on solvents and their application ranges, see Ref. (15). Gases are rarely studied in transmission in corrosion science. However, transmission experiments on gases are, in general, performed using a cell consisting of a cylinder with IR transparent windows like KBr at the ends.

7.4.2 Reflectance Spectroscopy

There are various types of reflectance techniques, such as specular, diffuse (DRIFTS), attenuated total reflection (ATR), and infrared reflection–absorption spectroopy (IRAS), but they all have in common that IR light impinges on a surface, and subsequently the reflected light is detected. Further information on reflectance spectroscopy can be found in Refs. (16,17). Reflectance spectra differ from transmission spectra, but the two different techniques provide the same information, and there are ways of transforming the reflectance spectra so that comparisons to transmission spectra can be made. However, it should be emphasized that reflectance spectra give an estimate of the composition of the near-surface region, since the IR radiation usually only penetrates between 1 and 10 μm into the sample. This is to be compared to transmission spectra, which describe the bulk composition of the sample. An advantage of reflectance spectroscopy is that the spectrum is independent of the thickness of the sample, so sample preparation is facilitated compared to transmission spectroscopy. Another advantage is that reflectance techniques are nondestructive. However, reflectance techniques are not as general as transmission spectroscopy, since gases cannot be analyzed. Other disadvantages are that the S/N ratio is lower as a result of the difficulty of capturing all the reflected light, and quantitative measurements may be hard to perform, since the penetration depth of the IR radiation is unknown and dependent on the sample material and the angle of incidence. How the various reflectance techniques are performed practically is shown in Figure 7.4.

7.4.2.1 Specular Reflection

Specular reflection is the reflection from mirror-like surfaces, where the angle of incidence equals the angle of reflection, as shown in Figure 7.4(a). The surfaces giving rise to the best specular reflection spectra are smooth, level, lustrous, and have a large detection area. Metals are therefore good specular reflectors, but dielectric samples have a low specular reflection. Organic samples, for example, reflect only 5% to 10% of the IR radiation (18). A specular reflection spectrum differs substantially from a transmission spectrum because of the derivative shaped bands (Reststrahlen), a result

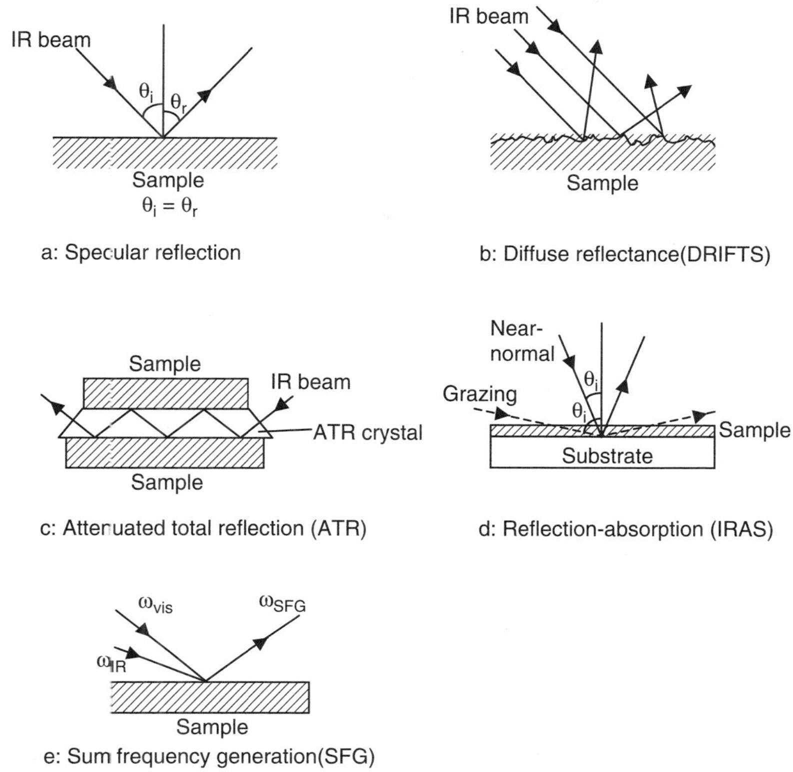

Figure 7.4 The experimental set-ups for specular reflection, DRIFTS, ATR, IRAS, and SFG.

of the specular reflection spectrum being a combination of the real and imaginary part of the refractive index, as shown in the following equation:

$$N = n + \mathrm{i}\kappa \tag{7.13}$$

where N is the complex refractive index, n is the real refractive index, i is the imaginary unit, and κ is the extinction coefficient, which describes the absorption (κ is related to the often encountered absorption coefficient k, by $k = 4\pi\kappa/\lambda_0$, where λ_0 is the wavelength in vacuum). It is possible to split the specular reflection spectrum into its components, the real refractive index spectrum and the extinction coefficient spectrum, by the use of the mathematical formulas known as Kramers–Kronig (KK) transformations (19). The resulting extinction coefficient spectrum corresponds to the absorbance spectrum obtained by transmission spectroscopy. The real part of the refractive index changes dramatically at wavelengths where an absorption occurs (20), which is seen in Figure 7.5.

7.4.2.2 Diffuse Reflectance Infrared Fourier Transform Spectroscopy

The technique of diffuse reflection (DRIFTS) is used to analyze dull samples as powders. In contrast to specular reflection, where the angle of incidence equals the angle of reflection for the whole beam, the IR radiation is reflected at all angles in the diffuse reflection process (21), shown in Figure 7.4(b). The specular reflection of dielectric materials follows the Fresnel law (17, p. 16), whereas diffuse reflection is described by the Lambert cosine law (17, p. 47). Usually, the sample powder is mixed

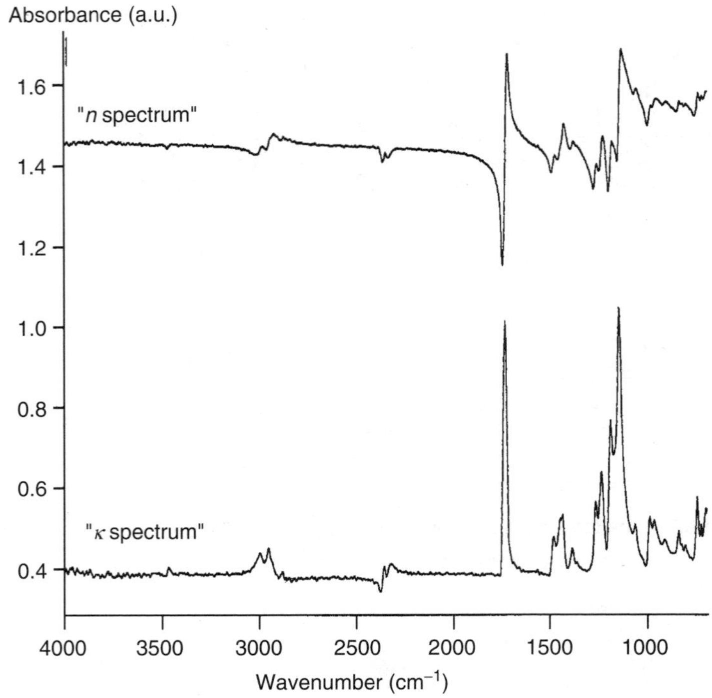

Figure 7.5 Real refractive index (n, top) and extinction (κ, bottom) spectra of Plexiglass calculated from the specular reflection spectrum by the use of Kramers–Kronig transformations. (Reprinted with permission from BC Smith. *Fundamentals of Fourier Transform Infrared Spectroscopy*. Boca Raton: CRC Press LLC, 1996, p. 129. Copyright Lewis Publishers, an Imprint of CRC Press. Boca Raton, Florida.)

into a reflecting, inert matrix, like KBr, but the mixture is not compressed into pellets. Instead, the IR beam is focused on a cup filled with the mixture, and therefore the sample preparation for qualitative experiments is easy. Quantitative analyses require very careful preparations, because of the difficulty of obtaining reproducible values of the scattering coefficient s (Equation (7.14)), which depends on the particle shape, size, and packing density. However, it is manageable if careful preparations are undertaken. Solids dissolved in liquids can be analyzed by dripping the solution onto the matrix and letting the solvent evaporate. In order to analyze large samples it is possible to rub a piece of silicon carbide paper against the sample material, and then perform the measurements on the piece of paper. Since diffuse reflectance is a weak process (4% to 10% reflectance, (18, p. 58)), large samples are beneficial to use. In a diffuse reflectance spectrum, the peak positions will be the same as in a transmission spectrum, but the peak intensities will become distorted, and low-intensity peaks will appear higher than in transmission. In order to perform quantitative analyses, the spectrum has to be transformed by the Kubelka–Munk equation (22), shown in the following equation:

$$f(R_\infty) = \frac{(1 - R_\infty)^2}{2R_\infty} = \frac{k}{s} \tag{7.14}$$

where $f(R_\infty)$ is the intensity in Kubelka–Munk units, R_∞ is the absolute reflectance of an infinitely thick sample, usually given as the intensity ratio of the sample mixture to a nonabsorbing reference such as KBr, the absorption coefficient $k = 2.303\varepsilon c$, and s is the scattering coefficient. From this follows that the Kubelka–Munk spectrum has peaks with intensities that increase linearly by concentration, and thereby is suitable for quantitative analyses.

7.4.2.3 *Attenuated Total Reflection*

ATR enables qualitative and quantitative experiments, and the experimental set-up is shown in Figure 7.4(c), depicting the sample and the ATR crystal. ATR relies on the process of total internal reflection, meaning that there is no wave transmitted into the sample from the ATR crystal. This process occurs when light is incident from a medium with higher refractive index to a medium with lower refractive index, and the angle of incidence θ_i is higher than the critical angle θ_c, defined by the following equation:

$$\theta_c = \sin^{-1}\left(\frac{n_1}{n_2}\right) \tag{7.15}$$

where n_1 is the refractive index of the sample, n_2 is the refractive index of the ATR crystal, and $n_2 > n_1$. Equation (7.15) is a result of Snell's law. Since the ATR crystal has a higher refractive index than the sample, the IR beam will undergo total internal reflection inside the crystal. Common crystals are ZnSe ($n = 2.42$), Ge ($n = 4.0$), and KRS-5 (58% TlI and 42% TlBr) ($n = 2.35$). When the IR light undergoes total internal reflection in the crystal, a small portion of the radiation will penetrate a short distance of around 0.1 to 5 μm into the sample in what is called an evanescent wave (13, p. 124, 23). This wave decays exponentially with distance into the sample, and the penetration depth, defined as the depth at which the intensity has decreased by a factor *e*, is dependent on the wavelength of the radiation, the refractive indices of the sample and the ATR crystal, and the angle of incidence, as shown in the following equation:

$$\mathrm{DP} = \frac{\lambda}{2\pi n_2(\sin^2\theta_i - (n_1/n_2)^2)^{1/2}} \tag{7.16}$$

where DP is the penetration depth. Where the sample has an absorption band, the evanescent wave gets absorbed, and the light reflected back into the ATR crystal becomes attenuated. However, the ATR spectra will differ in peak intensities from transmission spectra because the penetration depth in ATR is a function of wavelength, according to Equation (7.16). The short penetration depth of ATR makes it a more surface-sensitive technique than transmission spectroscopy. This is illustrated in Figure 7.6, showing a spectrum of a thin film of fluorinated polyethylene.

The transmission spectrum carries information from the whole sample and just shows a small C–F peak at around 1200 cm^{-1}, whereas the ATR spectrum carrying information from only the near-surface region shows a much more intense C–F peak. From these results, the conclusion can be drawn that the film has been fluorinated at the surface. An advantage of ATR is that the spectrum is invariant of the thickness of the sample, unless the sample is thinner than the penetration depth of the evanescent wave. This makes it a good technique for the studies of solids, liquids, polymers, and thin films, if the sample is well adhering to the crystal. It is also possible to analyze semisolids like asphalt, lipstick, and peanut butter, which are too absorbing to perform transmission experiments on. Since water strongly absorbs IR radiation, ATR is a suitable technique for studying aqueous solutions, as the light only penetrates a short distance into the sample, and many ATR crystals do not react with water. Very low solute concentrations cannot be detected, however, as a consequence of the short penetration depth. It is possible to improve the S/N ratio by utilizing multiple reflections as shown in Figure 7.4(c), instead of using a single reflection.

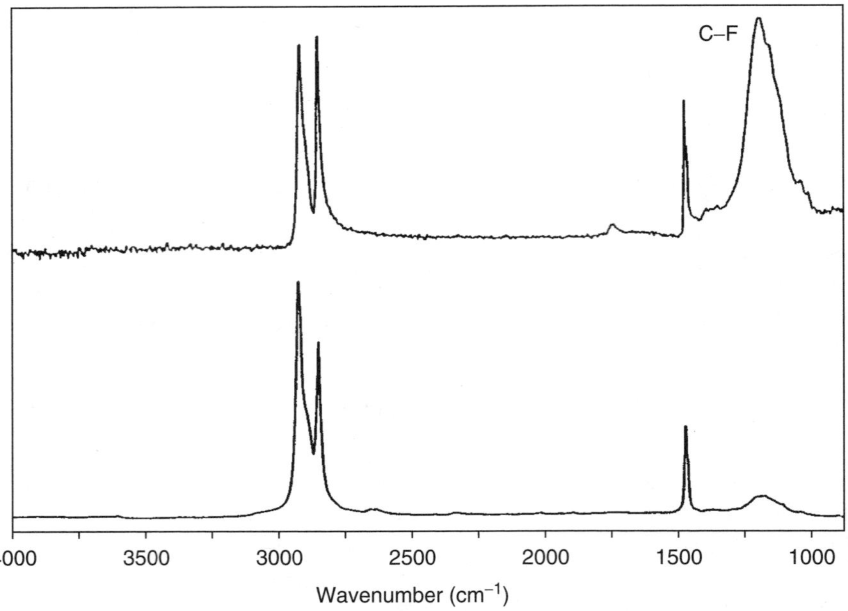

Figure 7.6 ATR (top) and transmission (bottom) spectra of a fluorinated film of polyethylene. (Reprinted from HJ Humecki. *Practical Guide to Infrared Microspectroscopy*. New York: Marcel Dekker, 1995, p. 61, by courtesy of Marcel Dekker. Inc. With permission.)

Polarized IR radiation can be used to determine the orientation of molecules, and if the orientation is known, it can be used for vibrational band assignments (24). These determinations are based on the fact that molecules oriented on a solid surface only absorb the component of the IR radiation that is polarized parallel to the direction of the transition moment of the vibration. Light polarized perpendicular to the transition moment is not absorbed at all. The orientation determinations are performed by comparing the intensities of bands in spectra taken with different polarizations. Polarization modulation can also be used in transmission and IRAS experiments.

7.4.2.4 Infrared Reflection–Absorption Spectroscopy

IRAS (also denoted IRRAS) studies are of great importance in corrosion science, both experiments performed *ex situ* and *in situ*, a field that is extensively explored in Section 7.5. The technique can be visualized as a double transmission process (25), where the IR beam passes through the thin sample placed on a substrate twice, according to Figure 7.4(d). To avoid losses, it is important that the substrate is highly reflective, and therefore metals are appropriately used as substrates. Interference fringes in the spectrum are avoided by making sure that the sample is in good contact with the substrate. Depending on the angle of incidence θ_i (measured from the surface normal), the technique is either classified as near-normal incidence when θ_i is small, or grazing angle incidence when θ_i is large. Grazing angle incidence is often used in the studies of thin films. When IR light is incident on a thin film on a metal substrate, the interaction will depend on the polarization of the light. Unpolarized IR radiation consists of two perpendicular electrical vectors, s-polarized light, which is perpendicular to the plane of incidence, and p-polarized light, which is parallel to the plane of incidence (the plane of incidence is the plane containing the surface normal and the incident IR ray). s-polarized light undergoes a phase shift close to 180° upon reflection for all angles of incidence. This leads to destructive interference between the incident and the reflected rays at the surface, and a cancellation of the electric field, thereby disabling any interaction with dipoles in the surface film. The phase shift of p-polarized light depends on the angle of incidence. At high angles of incidence, the incident and the reflected waves will be in phase at the surface, resulting in an enhancement of the amplitude of the electric field in the direction normal to the surface. The optimum angle of incidence, which is slightly dependent on the kind of metal used and the wavenumber, is reported to be around 88° (25). As a consequence of the different behavior of s- and p-polarized light, only adsorbates on the metal surface having a vibrating dipole component along the surface normal will absorb IR radiation. This is referred to as the *surface selection rule*. Along with the characteristics of s- and p-polarized light, the ensuing surface selection rule makes it possible to determine the orientation of adsorbates on metal surfaces (26). By modulating the polarization, the signal from molecules adsorbed on a metal surface can be distinguished from the same molecules in the surrounding gas or liquid, since the molecules adsorbed on the metal surface only will absorb p-polarized radiation, whereas molecules in the gas or liquid will absorb both s- and p-polarized IR radiation (27). Thus, the background contribution from the gas or liquid can efficiently be removed. The great fortification of the electric field at the surface for grazing angles, and the fact that a beam incident in a grazing angle will cover more molecules than at a near-normal angle of incidence, make grazing angle IRAS a valuable tool for the studies of thin films on metal substrates. Figure 7.7 shows an example of the enhanced surface sensitivity of grazing angle IRAS

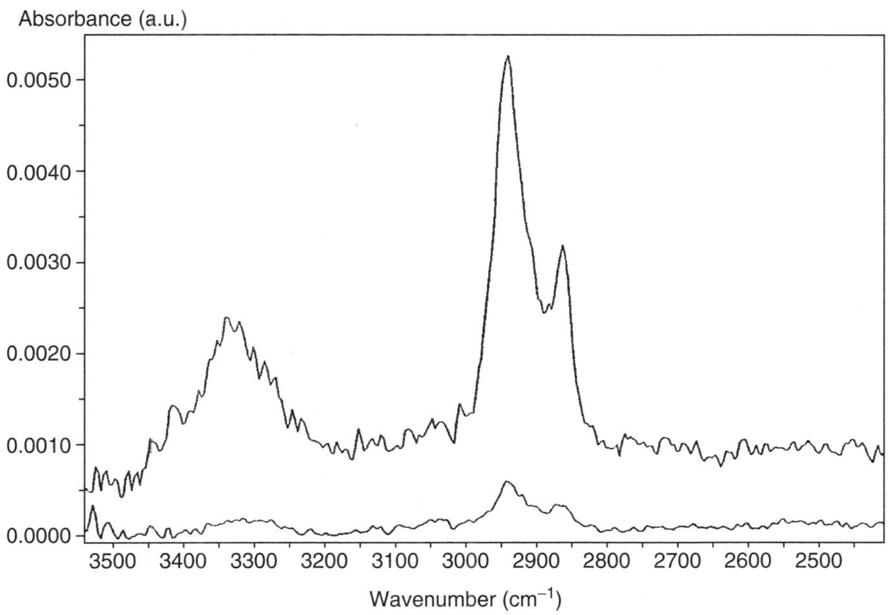

Figure 7.7 Grazing angle (top) and near-normal angle (bottom) incidence IRAS spectra of a thin film of polyurethane on gold. (Reprinted from HJ Humecki. *Practical Guide to Infrared Microspectroscopy*. New York: Marcel Dekker, 1995, p. 66, by courtesy of Marcel Dekker. Inc. With permission.)

compared to near-normal IRAS, resulting in a higher intensity of the bands in the grazing angle spectrum.

When comparing transmission and IRAS spectra, frequency shifts may be seen as a consequence of optical effects arising in the IRAS spectra. These shifts are due to the significant change in refractive index occurring when scanning over an absorption band, according to Figure 7.5. As seen in Figure 7.5, the refractive index is higher at the low wavenumber side than at the high wavenumber side of the absorption peak, resulting in a higher degree of reflection at both surfaces of the surface film for light at lower wavenumbers. As a consequence, a shift toward higher wavenumbers can be seen in the IRAS spectrum of a given band compared to the corresponding transmission spectrum. The shifts are more pronounced for strong and broad bands than for narrow and sharp bands.

IRAS is sometimes encountered in applications in which IR spectroscopy is combined with electrochemistry (Section 7.5.1), where the beam has to pass through transparent windows. In this case, it is of utmost importance that the windows do not react with the liquid, in order to avoid impurities from the windows into the liquid, and deterioration of the windows. A list of window and cell materials used in IR spectroscopy is shown in Table 7.3.

7.4.2.5 Infrared Microspectroscopy

As was pointed out in Section 7.1, IR spectroscopy in general possesses no lateral resolution. However, IR microspectroscopy can be obtained by integrating IR microscopy with light microscopy (28). For reasons given in Section 7.2.1, we limit the description to radiation in the mid-IR region, although IR microspectroscopy is used in the near-IR region as well.

Typically, IR microspectroscopy is obtained by allowing the IR beam to leave the interferometer through a hole in the spectrometer and direct it into the microscope. The IR beam is focused onto the sample through a condenser. In many cases the condenser is based on so-called Cassegrainian optic, a mirror arrangement in which the IR beam is first reflected through a small convex mirror, then collected

Table 7.3 Infrared transmission limits for some important cell and window materials used in reflection and transmission experiments. The upper transmitting limit is for all materials at least 16,000 cm^{-1}, which is in the visible region. A high refractive index results in higher reflection losses at the interfaces

Material	Lower transmission limit (cm^{-1})	Refractive index at 2000 cm^{-1}	Comments
Glass	3000	1.49 (at 5000 cm^{-1})	Water insoluble, resistant to acids
CaF_2	1100	1.40	Nearly water insoluble
BaF_2	870	1.45	Slightly water soluble, soluble in acids
NaCl	625	1.52	Soluble in water, methanol, and acids
ZnSe	550	2.4	Water insoluble, soluble in acids
KBr	400	1.54	Soluble in water and methanol, hygroscopic
KRS-5	275	2.38	Nearly water insoluble, soluble in acids, poisonous
CsI	200	1.74	Soluble in water and alcohols

through a large concave mirror before the focused beam impinges on the surface. The beam interacts with the sample either in the transmission mode or in the reflection mode and the sample can be placed on an X, Y translation stage allowing for different parts of the sample to be analyzed. After interaction, the beam is collected through an objective, which may be based on Cassegrainian optic similar to the condenser, but reversed. The beam then passes through an aperture that determines the sample area to be detected. Finally, the focused IR beam is analyzed by means of a high-sensitive detector, such as an MCT detector (see Section 7.3.2).

Spatial resolution of mid-IR microspectroscopy is limited to 10 to 20/μm (28) and the generation of spectral maps may take several hours because of the time required to monitor each individual spectrum. Data are usually obtained through line-imaging, in which IR spectra are recorded along a line of the sample, or through two-dimensional imaging, in which the images are acquired by selecting a single wavenumber or integrating over a wavenumber interval and plot the result versus sample position. Recent developments include the use of multichannel solid-state detectors, which will enable new imaging possibilities.

IR microspectroscopy has found numerous applications in forensic medicine, polymer technology, and biology. However, only a very limited number of papers have so far reported its use for corrosion-related applications (29).

7.4.2.6 Sum Frequency Generation

SFG (30,31) is a nonlinear, vibrational laser spectroscopy technique that is uniquely suitable for surface studies. An SFG process can only occur where the symmetry is broken. Therefore, randomly oriented molecules in bulk media will not give rise to any SFG signal. Only at interfaces where the symmetry is broken will the SFG process occur. A consequence of this selection rule is that it is possible, for example, to distinguish the surface molecules of a liquid from the vast excess of the same molecules in the bulk. This inherent surface sensitivity is an advantage when using SFG for surface studies compared to IR spectroscopy. All interfaces accessible by light can be studied by SFG, including, for example, the solid/gas, liquid/gas, liquid/liquid, and solid/liquid interfaces. The SFG process can be viewed as a combination of an IR and a Raman process, where two pulsed laser beams, one fixed visible and one tunable IR, are combined in time and space on a surface (Figure 7.4e). A third beam with the sum frequency of the two incoming beams is thus generated, described by the following equation:

$$\omega_{\mathrm{SFG}} = \omega_{\mathrm{vis}} + \omega_{\mathrm{IR}} \tag{7.17}$$

where ω_{SFG} is the frequency of the SF beam, ω_{vis} is the frequency of the visible beam, and ω_{IR} is the frequency of the IR beam. When ω_{IR} is in resonance with a vibrational frequency of a molecule, the intensity of the emitted sum frequency beam is greatly enhanced. The direction of the sum frequency beam is determined by the law of conservation of momentum. By scanning the IR wavelength, SFG makes it possible to obtain an IR spectrum of the molecules at the interface. Changing the polarization of the beams enables studies of vibrations in different directions, and thereby it is possible to perform orientational measurements. There are many advantages of SFG. Besides the ones mentioned above, SFG has submonolayer sensitivity, it is *in situ* and nondestructive, high pressure studies are possible, and a subpicosecond time resolution is possible. Disadvantages are that usually the signal is low, and the limited frequency range compared to IR spectroscopy. Related to SFG is the simpler technique of second harmonic generation (SHG) (32), which also is surface specific, but cannot probe surface vibrations.

7.5 APPLICATIONS OF INFRARED SPECTROSCOPY TO CORROSION-RELATED PHENOMENA

7.5.1 Adsorption in Electrochemical Environments

Characterization of the metal/liquid interface has traditionally been performed mainly through electrochemical techniques. By integrating IRAS into electrochemical measurements, it became possible to obtain atomic and molecular information of adsorbates or films of reaction products on electrode surfaces in electrochemical environments. A major constraint when detecting submonolayer quantities of adsorbates by IR spectroscopy is the absorption of IR radiation by the electrolyte, usually a water solution. Figure 7.8 displays schematically an electrochemical cell adapted to IRAS analysis of the electrode surface.

Minimization of IR absorption is achieved by forming a thin (usually in the range 1 to 20 μm) electrolyte layer between the electrode and the IR transparent window (see Table 7.3). However, such a narrow gap may cause a depletion of electroactive species during electrochemical polarization. This may be overcome by the use of a thin-layer flow cell with continuous replenishment of the thin layer, or by applying the polarization at a much wider gap (of the order of millimeters), and subsequently moving the electrode surface to the window for IR measurements. The detection of submonolayers of adsorbates also necessitates the modulation of either the electrode potential or the light polarization state. In this way the absorbance of the adsorbate can be differentiated from the absorbance of other species within the path of the IR beam (see Ref. (33) for further details).

One important development of IRAS in this context is "*in situ* electrochemical surface science" (34), in which studies performed on well-defined monocrystalline

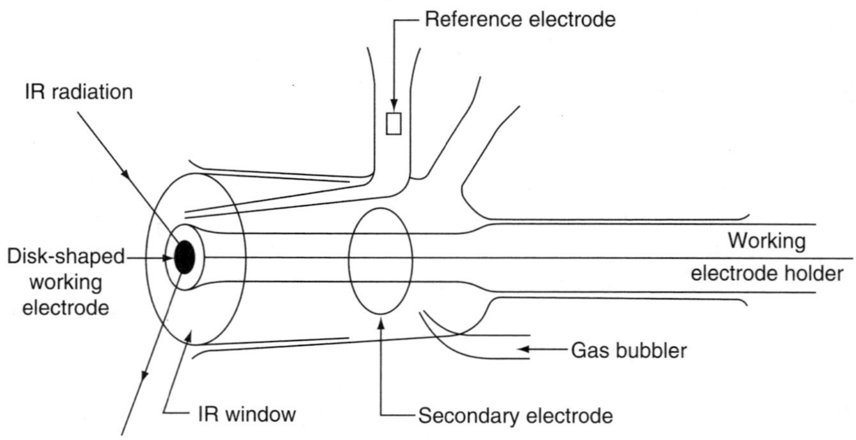

Figure 7.8 Electrochemical cell adapted to *in situ* measurements by IRAS. (Reprinted from *J. Electroanal. Chem.*, Vol. 160, A Bewick, K Kunimatsu, SB Pons, JW Russell, Electrochemically modulated infrared spectroscopy (EMIRS); Experimental details, pp. 47–61, Copyright 1984, with permission from Elsevier Science.)

substrates in an electrochemical environment overlap with studies of the same adsorbate/substrate system in an ultrahigh vacuum environment. A large body of studies concerns the adsorption of carbon monoxide on monocrystalline Pt-group transition metals. By comparing IR spectra of the chemisorbates obtained in electrochemical and ultrahigh vacuum environments, respectively, it has been possible to deduce the influence of, for example, double-layer effects or of solvation on adsorbate properties. Such studies have the potential to provide insight into the bonding between adsorbate and substrate as a function of applied electrode potential.

Of fundamental importance in corrosion science is the adsorption of hydrogen on metal surfaces, regarded as a prerequisite in proton-induced dissolution processes (35). From the large body of previously published electrochemical measurements of platinum in acid solutions, it was concluded that hydrogen adsorbs in two different states in the potential region prior to hydrogen evolution (36). By modulating the electrode potential in the region of appearance of the different adsorption states and recording the potential-modulated IR spectra, support for two existing adsorption states was obtained. A more "strongly bound" hydrogen occurred at high coordination sites and resulted in IR bands (between $\approx$4500 and 5500 cm^{-1}) characterized by a higher reflectivity. The strongly bound hydrogen, therefore, seems to behave more metallic with association to the conduction band of the metal. The more "weakly bound" hydrogen resulted in a combination band at around 5250 cm^{-1}, and was associated with the inner Hemlholtz plane where it bonds to water molecules through lone pair molecular orbitals, see Figure 7.9.

Self-assembled monolayers (SAM) provide an interesting way of modifying surfaces in order to obtain desired surface properties, including increased corrosion inhibition. The molecules involved in SAM anchor through chemisorption that results in strong adhesion to the substrate. The films are usually densely packed and their properties can be tailored by altering the chemical composition. IR spectroscopy has been used in several studies for analyzing, for example, the bonding mechanisms of SAM layers to the substrate, and the degree of crystallinity of the SAM layers. In what follows, a few selected studies are briefly discussed in which SAMs were prepared in solutions onto metal substrates, dried, transferred, and analyzed by IRAS with p-polarized light at grazing incidence.

The effect of chain length of *n*-alkanethiols on their corrosion protection ability of copper was investigated by IR spectroscopy in combination with electrochemical impedance spectroscopy (37). By varying the chain length ($CH_3(CH_2)_{n-1}SH$; $n = 8$, 12, 16, 18, 20, 22, 29), the crystallinity of SAM films was investigated through grazing

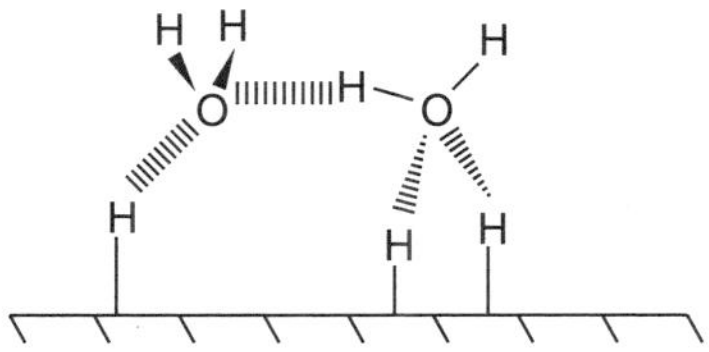

Figure 7.9 The structure and orientation of water showing how weakly bound hydrogen bonds water molecules to a platinum surface, while maintaining hydrogen bonding to bulk water. (Reprinted from *J. Electroanal. Chem.*, Vol. 132, A Bewick, JJ Russell, Structural investigation by infra-red spectroscopy of adsorbed hydrogen on platinum, pp. 329–344, Copyright 1982, with permission from Elsevier Science.)

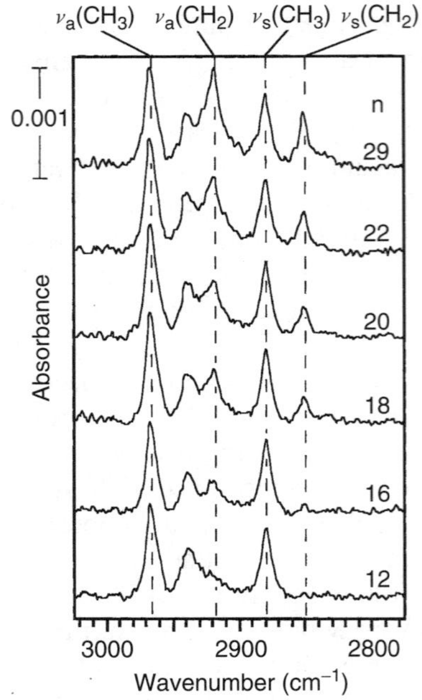

Figure 7.10 The transition from less ordered ($n = 12$) to more ordered ($n = 16$ and higher) alkanethiols on copper, as depicted from grazing incidence reflection–absorption infrared spectra of the C–H stretching region. The four dashed lines indicate the positions of primary modes representative of well-ordered chain formation. (Reprinted with permission from GK Jennings, JC Munro, TH Yong, PE Laibnis. Effect of chain length on the protection of copper by *n*-alkanethiols. *Langmuir* 14: 6130–6139, Copyright 1998 American Chemical Society.)

incidence IR spectra of the C–H stretching region (2850 to 2965 cm^{-1}), see Figure 7.10. Upon chemisorption to copper, the hydrocarbon chains anchor with the thiol group, whereby the chains are orientated almost normal to the surface. With $n = 16$ or more, the chains became crystalline and densely packed, whereby a much higher corrosion resistance was obtained compared to shorter chains. Thicker films also possessed greater intermolecular interactions that resulted in better structural stability when the SAM-covered copper surfaces were exposed to oxidizing conditions.

Structural rearrangement of SAMs upon exposure to corrosive substances is another important issue that has been explored by IR spectroscopy. *n*-Alkanoic acids ($C_nH_{2n+1}COOH$) form SAMs anchored by $-CO_2^-$ ligand species on native oxide-covered silver, copper, and aluminum (38). Upon exposure to hydrogen chloride (HCl) vapor for a few minutes, SAMs on aluminum exhibited no changes. Exposure for a few seconds of SAMs on silver and copper, on the other hand, resulted in protonation of the carboxylate head group, rapid dewetting, and rearrangement into clusters of alkanoic acids. The clusters exhibited disordered forms for acids with $n = 13$ or less, and ordered forms for acids with $n = 14$ or more. Ambient storage caused a partial rearrangement back to the original SAM structures. The IR vibrational modes of interest for these studies include the C–H stretching region (2850 to 2965 cm^{-1}), the symmetric $-CO_2^-$ stretching mode ($\approx$1400 cm^{-1}), and the antisymmetric $-CO_2^-$ stretching mode ($\approx$1510 cm^{-1}). Different behavior of SAMs on oxide-covered aluminum, silver, and copper surfaces was suggested to be due to a combination of differences in intermolecular interactions, $-CO_2^-$/metal substrate interaction, and reactivity of each metal toward the corrosive hydrogen chloride vapor. In a similar experimental approach, the rearrangement of SAMs on silver toward corrosive hydrogen sulfide (H_2S) vapor was studied (39). Rapid sulfidation of silver upon hydrogen sulfide exposure caused a destabilization of the $-CO_2^-$/silver interface, protonation of the carboxylate head group, and significant and rapid perturbation of the SAMs.

Other IR-based studies of SAMs on metal surfaces include S-alkyl thiosulfates [$CH_3(CH_2)_{n-1}S_2O_3^-Na^+$] on copper (40), *n*-alkanethiols [$CH_3(CH_2)_{n-1}SH$] on gold (41), terminated alkyl-phosphonic and phosphoric acid monoalkyl esters on aluminum (42), and chemically modified alkanethiols on copper (43). The cited references are all corrosion related. In addition, numerous studies have been reported in which SAMs adsorbed on gold have been investigated by IR-based techniques. However, these studies have not been included here.

In order to explore possible interactions between protein molecules and metal surfaces, IRAS, in combination with other surface analytical techniques, has been used to characterize adsorption characteristics of bovine serum albumin on pure chromium and AISI 304 or 316 stainless steel (44,45). Such studies are vital for the understanding of adhesion of microorganisms in areas such as microbial corrosion (Section 7.5.6), or degradation of biomaterials and food canisters.

SFG has been used in various electrochemical experiments to explore the electrode/electrolyte interface (46–48). The technique enables time-, potential-evolved, and time-resolved studies of the electrochemical interface. In Ref. (46), the interface between a Pt(1 1 1) electrode and mixtures of acetonitrile and water was studied in order to determine the orientation of the acetonitrile molecules at the interface. The SFG spectra reveal that the orientation of acetonitrile on the platinum surface is dependent upon the electrode potential, and that the acetonitrile molecules for some potentials is oriented with the methyl group directed toward the electrode and for other potentials with the nitrogen pointing toward the electrode. Other information possible to extract from the spectra is that water disrupts the orientation of the acetonitrile molecules at the platinum surface. The adsorption of hydrogen on platinum electrodes has been studied by SFG in the under- and overpotential ranges, and some basic characterization of the double-layer structure has also been done (49). In these experiments, potential-evolved SFG spectra of H–Pt vibrations in sulfuric acid have been recorded, revealing information about the adsorption sites of hydrogen on the platinum surface, and also about the bonding between adsorbed hydrogen and water molecules in the electrolyte. Potential-evolved SFG spectra of a platinum electrode in methanol/perchloric acid mixtures are also treated in Ref. (49), in order to identify the adsorbed species produced when methanol dissociatively adsorbs on the platinum surface. SAMs have been analyzed by SFG in various environments (50), where a long chain trichlorosilane, OTS [$CH_3(CH_2)_{17}SiCl_3$], was covalently adsorbed onto silica to form a stable monolayer. The results obtained showed, for example, that the SFG spectra of the SAMs taken in air, hexadecane, and carbon tetrachloride were similar, indicating that the nature of the monolayer is little influenced by the surroundings.

7.5.2 Corrosion Inhibition

Closely related to adsorption studies of molecules by IR spectroscopy is the investigation of mechanisms of corrosion inhibitors. IR-spectroscopy-based techniques in combination with electrochemistry have been used in a large number of studies to help elucidate the actual role of molecules that can act as corrosion inhibitors.

IRAS was used to study the potential-dependent bonding of benzotriazole (BTAH) to copper, a metal for which BTAH is a frequently used corrosion inhibitor (51). In agreement with other studies under similar conditions, for example (52), dissolved cuprous ions coordinate with BTA^- ions to form a polymeric complex, Cu(I)BTA. This multilayer thick film decomposes when changing the potential in the negative direction. The formation process was dependent on the anion in the supporting electrolyte: in chloride-containing solution the Cu(I)BTA film was more readily formed than in sulfuric acid solution. This difference was attributed to the coordination behavior between the anions (Cl^- and HSO_4^-, respectively) and Cu(I) in the Cu(I)BTA film. A derivative of BTA, 1-(methoxymethyl)-1*H*-benzotriazole (MeOMe-BTA), was also studied in a similar way. Due to blockage of one of the nitrogens in the triazole ring, this molecule was not able to form any complex with copper.

In all, this study shows that the structural information obtained by IR spectroscopy in combination with electrochemistry reveals surface molecular information that can explain the resulting corrosion-inhibiting ability. Similar studies, driven by finding less toxic and/or more efficient corrosion inhibiting molecules, have been performed *in situ* or *ex situ*. They include 1,7-diphosphono-heptane on monocrystalline Fe(1 1 0) (53), propargyl alcohol on iron (54), alkylamines, *N*,*N*-dimethylalkylamines and alkanethiols on iron (55), electropolymerized polyaniline on iron (56), conducting polymers on type 304 stainless steel (57), 1-phenyl-5-mercaptotetrazole on copper (58), undecylimidazole on copper (59), silane-modified polyvinylimidazole on copper (60), benzotriazole and totyltriazole on copper (61), and tetrazole derivatives on copper (62).

7.5.3 Surface Treatments for Corrosion Protection

Several methods have been explored that aim at improving the adhesion between a traditional corrosion protective polymeric coating and the metal substrate. In these methods an ultrathin film (<10 nm) of some adhesion promoter is applied, usually a polymer film anchored to the metal substrate through some functional group able to form strong bonds to the metal oxide. The development of these films is based on a mechanistic understanding of surface phenomena, including chemical bond formation and delamination processes. It takes advantage of surface-sensitive techniques able to reveal molecular information of the adhesion promoters.

In a series of papers, IRAS in combination with quartz crystal microbalance (QCM) was used to characterize thin polymer films on iron oxide surfaces (63,64). Figure 7.11 is a schematic display of the experimental set-up, which was used not only for comprehensive studies of formation of the polymer films, but also for characterizing the oxide covering the iron substrate (65).

The techniques involved act complementary. As an example, the QCM could record a significant weight decrease as a result of argon plasma treatment, which removed organic contaminants from the iron oxide surface. It could also record a weight increase as a result of subsequent oxygen plasma treatments, causing an increase of the existing oxide thickness of around 2 nm. The corresponding chemical changes could be followed by simultaneous IRAS at grazing incidence, in which the spectra after plasma treatments were compared with a reference spectrum recorded before treatment. Negative peaks at around 2900 cm^{-1} (corresponding to stretching

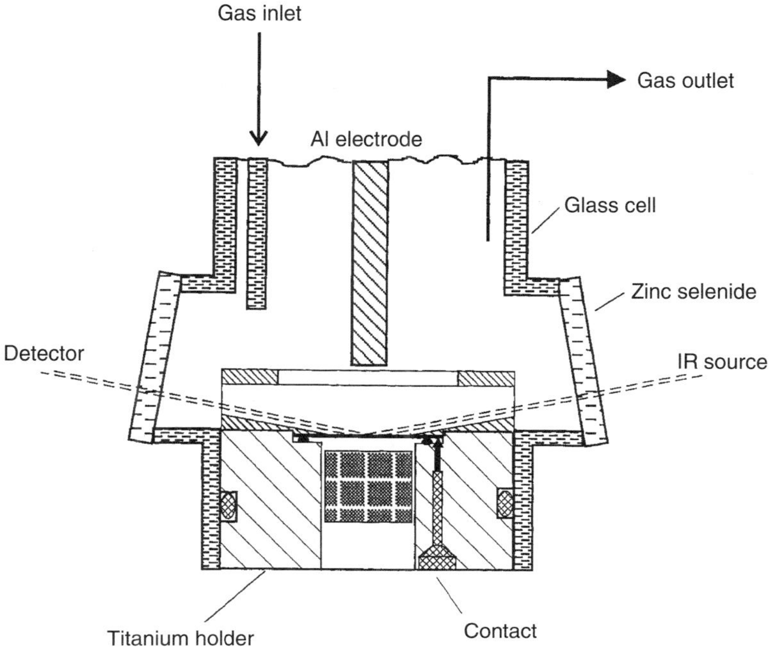

Figure 7.11 Experimental set-up for plasma modification of iron oxide covered surfaces and for deposition of ultrathin plasma polymers, and subsequent *in situ* analysis by IRAS in combination with QCM. (Reproduced with permission from G Grundmeier, E Matheisen, M Stratmann. Formation and stability of ultrathin organosilane polymers on iron. *J. Adhesion Sci. Technol.* 10: 573–588. Copyright 1996 International Science Publishers.)

vibrations of methylene and methyl groups) and at 1430 cm^{-1} (deformation vibrations of methylene groups) correspond to removal of organic contaminants. Positive peaks at 700 cm^{-1} (assigned to γ-Fe_2O_3), and at 900 and 1300 cm^{-1} (assigned to FeOOH) reflect the increase in oxide thickness. To conclude, both techniques can be used to follow chemical and mass changes upon plasma treatments with submonolayer sensitivity.

Related IR spectroscopy studies include silane-based pretreatments of iron and steel (66–68), silane-based treatments of copper (69), and carboxyl-based treatments of copper (70).

7.5.4 Corrosion Product Formation in Aqueous Environments

The formation of films of corrosion products in aqueous solutions, with or without electrochemical control, has been extensively studied with IR-spectroscopy-based measurements, both *in situ* and *ex situ*. In what follows, a few examples of *in situ* studies are briefly presented.

An *in situ* IRAS study was performed to follow the electrochemically controlled oxidation or reduction of lead in sulfuric acid (71). This system is of vital technical importance for the lead acid battery system, and has been extensively studied by means of electrochemical methods. A characteristic vibrational band at 631 cm^{-1} (the antisymmetric deformation vibration of the sulfate ion) was taken as evidence of lead sulfate ($PbSO_4$) formation. By applying galvanostatic conditions, the growth and reduction of lead sulfate was followed quantitatively in two ways: by integrating the area of the 631 cm^{-1} peak and by calculating the consumed charge. In the potential region where lead sulfate forms, a linear dependence between consumed charge and integrated peak area was found. At more anodic potentials, lead dioxide (PbO_2) formation also occurred, evidenced by an increased absorption band at around 5200 cm^{-1}. The high reflectivity of this compound impeded any quantitative evaluation of lead sulfate in the potential region in which both lead dioxide and lead sulfate form. In all, this study has shown that lead sulfate formation and reduction could be followed quantitatively with a detection limit thickness of 10 to 15 nm.

Other applications of IRAS to anodic film formation on various metals have been reported, albeit also under *ex situ* conditions. A multianalytical study of polycrystalline copper in alkaline electrolyte at pH 12 indicated the formation of Cu_2O in the potential range of Cu(I) (72). Evidence for this was strong Cu–O stretching bands at 880 and 1160 cm^{-1}, respectively. In the potential region of Cu(II), two strong CuO–H stretching bands appeared at 3300 and 3570 cm^{-1}, indicating the formation of $Cu(OH)_2$. The simultaneous use of *in situ* scanning force microscopy, electrochemical QCM, and grazing incidence x-ray diffractometry provided detailed insight into the solid-state growth mechanism of phases detected.

ATR spectroscopy was used to follow the photo- and potential-controlled roughening process of n-Si(1 1 1) in acidic fluoride containing solutions under *in situ* conditions (73). In the potential region around 0.3 V_{NHE}, photoassisted roughening occurs accompanied by a H-terminated silicon surface. This process could be monitored by the appearance of peaks originating from SiH (2080 cm^{-1}) and SiH_2 (2110 cm^{-1}). At more anodic potentials ($\geq$0.4 V_{NHE}), the formation of SiO_2 starts, as evidenced by peaks representative of SiO_2 at 1050 and 1230 cm^{-1}, respectively. The detection limit under present *in situ* analysis conditions appeared to be less than one monolayer of SiO_2. The photoassisted roughening occurs only in a narrow potential region and without any detection of SiO_2, which implies that the photoprocess in this potential region involves oxidation of Si to some soluble SiF compound.

Other film systems formed in aqueous solutions with or without electrochemical control and studied by IR spectroscopy methods include *in situ* monitoring of anodic film formation of iron in neutral phosphate solution (74), rust formation on steel in aqueous solutions (75), *in situ* studies of the reduction of aryldiazonium salts on iron or mild steel (76), cathodic film formation on aluminum (77), oxide and hydroxide film formation on tin in aqueous solutions (78), anodic oxide formation on niobium (79), TiO_2 layers produced on pure Ti and on Ti6Al4V through anodic oxidation and by sol–gel coating (80), corrosion of glass in aqueous solutions (81), transformation of bioactive glass granules during *in vitro* immersion (82), and apatite formation on a bioactive glass in simulated physiological solution (83).

7.5.5 Corrosion Product Formation in Atmospheric Environments

Atmospheric corrosion can proceed in the presence of a very thin aqueous film, from a few monolayers of water and thicker. This implies that strong water/IR radiation interaction, faced during *in situ* IR spectroscopy studies in bulk water solutions, is easier to overcome in atmospheric corrosion. Hence, IRAS has been developed for *in situ* studies of the influence of gaseous corrosion stimulators, such as SO_2 and NO_2, on the atmospheric corrosion rate of metals (84). The experimental set-up was later integrated with a QCM that enables simultaneous changes of surface chemistry (IRAS) and mass (QCM) to be monitored *in situ* upon introduction of corrosive atmospheric constituents (85,86). In this way, data can be obtained based on the simultaneous use of IRAS and QCM. The amplitudes of phases identified by IRAS during altering exposure conditions or as a function of exposure time can be compared with the simultaneous mass increase as obtained by QCM. In this way each phase can be identified and the corresponding mass of each phase determined. Figure 7.12 displays combined *in situ* results obtained during initial atmospheric corrosion, in which a copper surface was exposed to altering exposure conditions involving humidity, SO_2, and O_3. The figure shows that three phases can be detected (a copper sulfate, $CuSO_4 \cdot H_2O$, physisorbed H_2O, and a copper oxide, Cu_2O) and their corresponding masses (in $\mu g/cm^2$) estimated as a function of exposure conditions and time. This combination of *in situ* techniques has enabled to deduce a molecular understanding of the initial SO_2-induced atmospheric corrosion of copper.

All studies described so far in this section have used p-polarized light at grazing incidence, and have proven that reaction products can be detected with submonolayer sensitivity. A further improvement in sensitivity and versatility was obtained by introducing polarization modulation during *in situ* studies of atmospheric corrosion (87).

Both IRAS and QCM possess poor lateral resolution, and therefore a third technique has been introduced, *atomic force microscopy*, in order to provide complementary *in situ* information with submicrometer lateral resolution (88–90). With the combination of three independent techniques, the growth of micrometer-sized corrosion products could be monitored and their composition

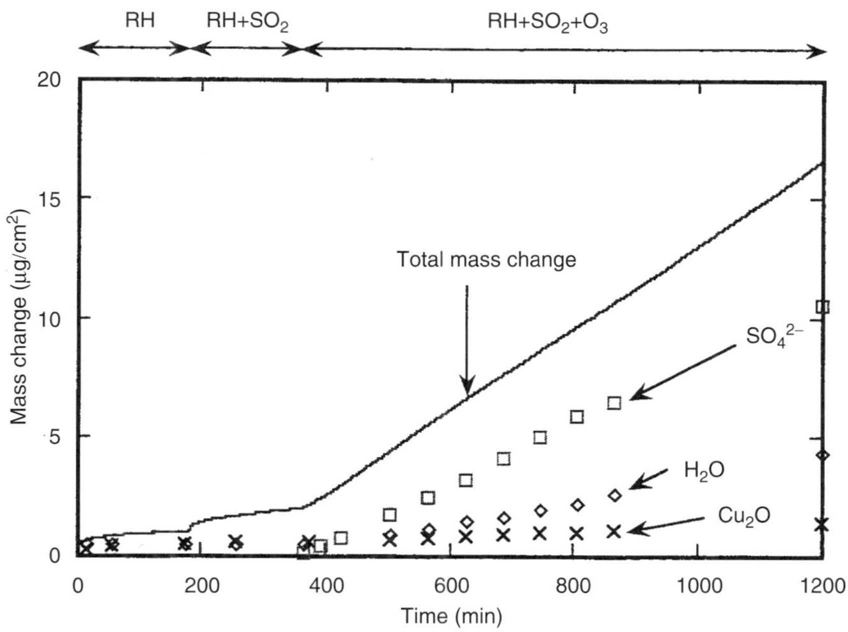

Figure 7.12 The atmospheric corrosion of copper during introduction of first 80% relative humidity, followed by introduction of 200 parts per billion of SO_2 in the humidified air, and completed by subsequent introduction of O_3. The figure displays the mass increase of the three phases detected (Cu_2O, H_2O, and $CuSO_4$), as deduced from IRAS combined with the QCM, and the total mass increase. (Reproduced with permission of The Electrochemical Society, Inc, T Aastrup, M Wadsak, C Leygraf, M Schreiber. *In situ* studies of the initial atmospheric corrosion of copper. Influence of humidity, sulfur dioxide, ozone, and nitrogen dioxide. *J. Electrochem. Soc.* 147: 2543–2551, 2000.)

investigated *in situ*. Another complementary technique to provide *in situ* information with improved lateral resolution (of the order of 50 μm) is the Kelvin probe, which enables Volta potential measurements to be performed simultaneously with IRAS (91).

The list of IR studies related to atmospheric corrosion is long, in fact longer than that of any other corrosion-related areas discussed in this chapter. In order to give an impression of the wide spectrum of IR spectroscopy applications, a list of strongly selected examples is given next. It includes *in situ* IRAS studies of corrosion product formation in confined surfaces (92), IR images to obtain lateral distribution of corrosion products (93), corrosion product formation in indoor atmospheres (94,95), degradation of fibers in ancient manuscripts (96,97), degradation of enamels from crafts and art objects (98), degradation of ancient glasses (99), *in situ* quantitative measurements of water at the interface between organic coatings and the substrate (100), atmospheric corrosion of aluminum caused by sulfuric acid (101), identification of compounds in copper patina (102,103), in nickel patina (104), and in corrosion products formed on ancient iron (105,106). One of very few examples of IR microspectroscopy applied to corrosion science includes a study of filiform corrosion processes on coated aluminum alloys (107).

7.5.6 Microbial Corrosion

Microbial corrosion is induced by the presence of microorganisms in aqueous environments at or near a metal surface. The result is an acceleration or inhibition of corrosion processes, caused by chemical, spatial, or other changes in the proximity of the corroding surface (108). The microorganisms attach to the surface, bacteria grow, and extracellular polymeric substances are produced, resulting in a biofilm at the metal surface. This film may cause local changes in, for example, pH or O_2 concentration. Corrosion-stimulating products, such as H_2S, may also be released from sulfate-reducing bacteria. The chemical complexity of the biofilm has frequently hampered a detailed understanding of the mechanisms of microbial corrosion.

FTIR spectroscopy is one of several techniques that has been successfully applied to improve the understanding of microbial corrosion through identification of biofilms and their impact on the metal surface (109,110). This capability can be demonstrated through *in situ* studies of the aqueous/copper interface by ATR (111,112). Extracellular acidic polysaccharides possess a corrosive behavior on copper, and this effect was investigated in the presence and absence of a protein-conditioning film on the copper substrate. The copper corrosion under the biofilm was monitored by coating a copper film of thickness around 10 nm onto a germanium internal reflection element. The evanescent wave (see Section 7.4.2.3) of IR radiation penetrates into the solution from the copper-covered germanium surface under total internal reflection, and organic molecules adsorbed onto the copper surface absorb radiation, which results in a spectrum. A C–O stretching band at 1037 cm^{-1} formed evidence of strong chemical interaction between dextran (one of the investigated polysaccharides) and the copper surface, whereas amide bands at around 1650 and 1550 cm^{-1} formed evidence of protein films on the copper surface. By exposing a copper-covered cylindrical germanium ATR crystal to a flow of different aqueous solutions, the corrosive action of dextran and other polysaccharides could be monitored, also the eventual shielding effect of a protein conditioning film against corrosive polysaccharides.

Other applications of ATR spectroscopy include the nondestructive monitoring of biofilm growth (113,114), and the interaction of bacteria with thin copper films evaporated onto a germanium internal reflection element (115).

In all, this and other examples have demonstrated the capability of FTIR spectroscopy for *in situ* studies of chemical changes of biofilms attached to metal surfaces. Inherent limitations of the technique are the strong absorption of IR light by water (Section 7.4.1), and the fact that living cells cannot be distinguished from dead cells (109).

7.6 CONCLUDING REMARKS

This chapter has focused on FTIR spectroscopy, because of its inherent advantages over dispersive and nondispersive IR spectroscopy. Since the first FTIR instruments became available in the late 1960s, FTIR has found numerous applications on, for example, metals, minerals, polymers, and biological specimens in such diverse fields as failure analysis and contaminant identification in the electronics industry, phase-identification in earth science, forensic paint examinations, analytical microscopy of pharmaceuticals, and the disclosure of mechanisms in corrosion science. Like many other analytical techniques, FTIR is constantly being pushed toward higher sensitivity (e.g., by combining gas chromatography and FTIR), increased lateral space resolution (through introduction of FTIR microspectroscopy), increased time resolution (e.g., by means of dynamic optical alignment), increased availability (e.g., through development of portable FTIR instruments), and lower cost. The summary of important applications of FTIR in corrosion science bears clear evidence of the versatility of FTIR-based techniques, and shows that most corrosion-related applications have been based on reflection–absorption spectroscopy, followed by ATR and diffuse reflectance spectroscopy. An experimental challenge is to use FTIR for revealing corrosion-related physicochemical processes under *in situ* conditions. The potential capability of an IR-related and extremely surface specific *in situ* technique, SFG, promises to generate new exciting research directions in corrosion science.

REFERENCES

1. M Diem. *Introduction to Modern Vibrational Spectroscopy*. New York: John Wiley & Sons, 1993.
2. A Fadini, FM Schnepel. *Vibrational Spectroscopy Methods and Applications*. Chichester: Ellis Horwood, 1989.
3. BC Smith. *Fundamentals of Fourier Transform Infrared Spectroscopy*. Boca Raton: CRC Press, 1996.
4. NB Colthup, LH Daly, SE Wiberly. *Introduction to Infrared and Raman Spectroscopy*, 3rd edn. San Diego: Academic Press, 1990.
5. G Socrates. *Infrared Characteristic Group Frequencies*, 2nd edn. Chichester: John Wiley & Sons, 1994.
6. RA Nyquist, RO Kagel. *Infrared Spectra of Inorganic Compounds (3800–45 cm^{-1})*. New York: Academic Press, 1971.
7. DR Burns, EW Ciurzak. *Handbook of Near-Infrared Analysis*. New York: Marcel-Dekker, 1992.
8. A Finch, PN Gates, K Radcliffe, FN Dickson, EF Bentley. *Chemical Applications of Far Infrared Spectroscopy*. New York: Academic Press, 1970.

9. DA Skoog, FJ Holler, TA Nieman. *Principles of Instrumental Analysis*, 5th edn. Philadelphia: Saunders College Publishing, 1998, p. 387.
10. IN Levine. *Quantum Chemistry*, 4th edn. New Jersey: Prentice-Hall, 1991, pp. 66–69.
11. PM Morse. Diatomic molecules according to the wave mechanics. II. Vibrational levels. *Phys. Rev*. 34: 57–64, 1929.
12. W Demtröder. *Laser Spectroscopy Basic Concepts and Instrumentation*, 2nd edn. Berlin: Springer-Verlag, 1996, pp. 297–334.
13. E Hecht. *Optics*, 3rd edn. Reading: Addison Wesley Longman, 1998, p. 124.
14. PR Griffiths, JA Haseth. *Fourier Transform Infrared Spectrometry*. New York: John Wiley & Sons, 1986.
15. R Silverstein, G Bassler, T Morill. *Spectrometric Identification of Organic Compounds*. New York: John Wiley & Sons, 1981.
16. G Kortüm. *Reflectance Spectroscopy*. Berlin: Springer-Verlag, 1969.
17. WWM Wendlandt, HG Hecht. *Reflectance Spectroscopy*. New York: Interscience Publishers, 1966.
18. HJ Humecki. *Practical Guide to Infrared Microspectroscopy*. New York: Marcel Dekker, 1995, p. 53.
19. R de L Kronig, HA Kramers. Atti Congr Intern Fisici, 1927, Como-Pavia-Roma (Published by Nicola Zanichelli, Bologna), 2: 545, 1928.
20. M O'Keefe. Infrared optical properties of cuprous oxide. *J. Chem. Phys*. 39: 1789–1793, 1963.
21. MP Fuller, PR Griffiths. Diffuse reflectance measurements by infrared Fourier transform spectrometry. *Anal. Chem*. 50: 1906–1910, 1978.
22. P Kubelka, F Munk. *Z. Tech. Phys*. 12: 593, 1931.
23. J Fahrenfort. Attenuated total reflection a new principle for the production of useful infra-red reflection spectra of organic compounds. *Spectrochim. Acta* 17: 698–709, 1961.
24. RJ Francel. Polarized infrared radiation in the study of the molecular structure of substituted nitrobenzenes. *J. Chem. Phys*. 74: 1265–1268, 1952.
25. RG Greenler. Reflection method for obtaining the infrared spectrum of a thin layer on a metal surface. *J. Chem. Phys*. 50: 1963–1968, 1969.
26. DL Allara, JD Swalen. An infrared reflection spectroscopy study of oriented cadmium arachidate monolayer films on evaporated silver. *J. Phys. Chem*. 86: 2700–2704, 1982.
27. WG Golden, DD Saperstein. Fourier transform infrared reflection–absorption spectroscopy of surface species. *J. Electron Spectrosc. Relat. Phenom*. 30: 43–50, 1983.
28. PJ Treado, MD Morris. Infrared and Raman spectroscopic imaging. In: MD Morris, ed. *Microscopic and Spectroscopic Imaging of the Chemical State*. New York: Marcel Dekker, 1993, pp. 71–108.
29. N Le Bozec, D Persson, A Nazarov, D Thierry. Investigation of filiform corrosion on coated aluminum alloys by FTIR microspectroscopy and scanning Kelvin probe. *J. Electrochem. Soc*. 149: B403–B408, 2002.
30. PB Miranda, YR Shen. Liquid interfaces: a study by sum-frequency vibrational spectroscopy. *J. Phys. Chem. B* 103: 3292–3307, 1999.
31. CD Bain. Sum-frequency vibrational spectroscopy of the solid/liquid interface. *J. Chem. Soc. Faraday Trans*. 91: 1281–1296, 1995.
32. YR Shen. Surface properties probed by second-harmonic and sum-frequency generation. *Nature* 337: 519–525, 1989.
33. R Nichols. IR spectroscopy of molecules at the solid-solution interface. In: J Lipkowski, PN Ross, eds. *Adsorption of Molecules at Metals Electrodes*. New York: VCH Publishers, 1992, pp. 347–389.
34. MJ Weaver, N Kizhakevariam, X Jiang, I Villegas, C Stuhlmann, A Tolia, X Gao. Vibrational spectroscopy at metal–solution interfaces: some perspectives and prospects for electrochemical surface science. *J. Electron Spectrosc. Relat. Phenom*. 64/65: 351–362, 1993.

35. W Stumm, G Furrer. The dissolution of oxides and aluminum silicates; examples of surface-coordination-controlled kinetics. In: W Stumm, ed. *Aquatic Surface Chemistry*. New York: Wiley, 1987, pp. 197–219.
36. BE Conway and L Bai. Vibrational spectroscopy at metal–solution interfaces: some perspectives and prospects for electrochemical surface science. *J. Electroanal. Chem.* 198: 149–175, 1986.
37. GK Jennings, JC Munro, TH Yong, PE Laibnis. Effect of chain length on the protection of copper by *n*-alkanethiols. *Langmuir* 14: 6130–6139, 1998.
38. YT Tao, GD Hietpas, DL Allara. HCl vapor-induced structural rearrangements of *n*-alkanoate self-assembled monolayers on ambient silver, copper, and aluminum surfaces. *J. Am. Chem. Soc.* 118: 6724–6735, 1996.
39. YT Tao W-L Lin, GD Hietpas, DA Allara. Infrared spectroscopy study of chemically induced dewetting in liquid crystalline types of self-assembled monolayers. *J. Phys. Chem. B* 101: 9732–9740, 1997.
40. AT Lusk, GK Jennings. Characterization of self-assembled monolayers formed from sodium S-alkyl thiosulfates on copper. *Langmuir* 17: 7830–7836, 2001.
41. RD Weinstein, D Yan, GK Jennings. Self-assembled monolayer films from liquid and supercritical carbon dioxide. *Ind. Eng. Chem. Res.* 40: 2046–2053, 2001.
42. I Maege, E Jaehne, A Henke, HJP Adler, C Bram, C Jung, M Stratmann. Self-assembling adhesion promoters for corrosion resistant metal polymer interfaces. *Progr. Org. Coat.* 34: 1–12, 1998.
43. M Itoh, H Nishihara, K Aramaki. Preparation and evaluation of 2-dimensional polymer-films by chemical modification of an alkanethiol self-assembled monolayer for protection of copper against corrosion. *J. Electrochem. Soc.* 142: 3696–3704, 1996.
44. C Rubio, D Costa, MN Bellon-Fontaine, P Relkin, CM Pradier, P Marcus. Characterization of bovine serum albumin adsorption on chromium and AISI 304 stainless steel, consequences for the *Pseudomonas fragi* K1 adhesion. *Colloids Surf B: Biointerfaces* 24: 193–205, 2002.
45. CM Pradier, D Costa, C Rubio, C Compère, P Marcus. Role of salts on BSA adsorption on stainless steel in aqueous solutions. I. FT-IRRAS and XPS characterization. *Surf. Interface Anal.* 34: 50–54, 2002.
46. S Baldelli, G Mailhot, P Ross, YR Shen, GA Somorjai. Potential dependent orientation of acetonitrile on platinum (1 1 1) electrode surface studied by sum frequency generation. *J. Phys. Chem. B* 105: 654–662, 2001.
47. P Guyot-Sionnest, A Tadjeddine. Spectroscopic investigations of adsorbates at the metal–electrolyte interface using sum frequency generation. *Chem. Phys. Lett.* 172: 341–345, 1990.
48. F Dederichs, A Petukhova, W Daum. Adsorption of CN at the Pt(1 1 1)/liquid interface by dissociation of acetonitrile and the potentiality of CO contamination: a sum-frequency generation study. *J. Phys. Chem. B* 105: 5210–5216, 2001.
49. A Tadjeddine, A Peremans. Vibrational spectroscopy of the electrochemical interface by visible infrared sum-frequency generation. *Surf. Sci.* 368: 377–383, 1996.
50. P Guyot-Sionnest, R Superfine, JH Hunt, YR Shen. Vibrational spectroscopy of a silane monolayer at air/solid and liquid/solid interfaces using sum-frequency generation. *Chem. Phys. Lett.* 144: 1–5, 1988.
51. ME Biggin, AA Gewirth. Infrared studies of benzotriazole on copper electrode surfaces. Role of chloride in promoting reversibility. *J. Electrochem. Soc.* 148: C339–C347, 2001.
52. C Törnkvist, D Thierry, J Bergman, B Liedberg, C Leygraf. Methyl substitution in benzotriazole and its influence on surface structure and corrosion inhibition. *J. Electrochem. Soc.* 136: 58–64, 1989.
53. P Poczik, I Felhosi, J Telegdi, M Kalaji, E Kalman. Layer formation by 1,7-diphosphono-heptane. *J. Serbian Chem. Soc.* 66: 859–870, 2001.

54. K Aramaki, E Fujioka. Spectroscopic investigations on the inhibition mechanism of propargyl alcohol for iron corrosion in hydrochloric acid at elevated temperatures. *Corrosion* 53: 319–326, 1997.
55. T Kawai, H Nishihara, K. Aramaki. Inhibition effects of amines and thiols on corrosion in anhydrous methanol solution containing $FeCl_3$. *J. Electrochem. Soc.* 143: 3866–3873, 1996.
56. MC Bernard, S Joiret, A Hugot-Le Goff, PD Long. Protection of iron against corrosion using a polyaniline layer: II. Spectroscopic analysis of the layer grown in phosphoric/ethanolic solution. *J. Electrochem. Soc.* 148: B299–B303, 2001.
57. LF D'Elia, RL Ortíz, OP Márquez, Y Martinéz. Electrochemical deposition of poly (o-phenylenediamine) films on type 304 stainless steel. *J. Electrochem. Soc.* 148: C297–C300, 2001.
58. XR Ye, XQ Xin, JJ Zhu, ZL Xue. Coordination compound films of 1-phenyl-5-mercaptotetrazole on copper surface. *Appl. Surf. Sci.* 135: 307–317, 1998.
59. S Yoshida, H Ishida. An investigation of the thermal-stability of undecylimidazole on copper by FT-IR reflection–absorption spectroscopy. *Appl. Surf. Sci.* 89: 39–47, 1995.
60. J Jang, H Ishida. Silane-modified polyvinylimidazole for corrosion protection on copper at elevated-temperatures. *J. Appl. Polym. Sci.* 49: 1957–1962, 1993.
61. D Sockalingum, M Fleischmann, MM Musiani. Near-infrared Fourier-transform surface-enhanced Raman scattering of azole copper corrosion-inhibitors in aqueous chloride media. *Spetcrochim. Acta A* 47: 1475–1485, 1991.
62. F Zucchia, G Trabanelli, M Fonsatia. Tetrazole derivatives as corrosion inhibitors for copper in chloride solutions. *Corros. Sci.* 38: 2019–2029, 1996.
63. G Grundmeier, E Matheisen, M Stratmann. Formation and stability of ultrathin organosilane polymers on iron. *J. Adhes. Sci. Technol.* 10: 573–588, 1996.
64. G Grundmeier, M Stratmann. Plasma polymerization — a new and promising way for the corrosion protection of steel. *Werkst und Korr* 49: 150–160, 1998.
65. G Grundmeier, M Stratmann. Influence of oxygen and argon plasma treatments on the chemical structure and redox state of oxide covered iron. *Appl. Surf. Sci.* 141: 43–56, 1999.
66. V Subramanian, WJ van Ooij. Effect of the amine functional group on corrosion rate of iron coated with films of organofunctional silanes. *Corrosion* 54: 204–215, 1998.
67. GP Sundararajan, WJ van Ooij. Silane based pretreatments for automotive steels. *Surf. Eng.* 16: 315–320, 2000.
68. TJ Lin, JA Antonelli, DJ Yang, HK Yasuda, FT Wang. Plasma treatment of automotive steel for corrosion protection — a dry energetic process for coatings. *Progr. Org. Coat.* 31: 351–361, 1997.
69. JS Jang, IK Jang, HC Kim. Adhesion promotion of the polyimide–copper interface using silane-modified polyvinylimidazoles. *J. Appl. Polym. Sci.* 68: 1343–1351, 1998.
70. H Yamabe. Stabilization of the polymer-metal interface. *Progr. Org. Coat.* 28: 9–15, 1996.
71. GLJ Trettenhahn, GE Nauer, A Neckel. *In situ* external reflection absorption FTIR spectroscopy on lead electrodes in sulfuric acid. *Electrochim. Acta.* 41: 1435–1441, 1996.
72. W Kautek, M Geua, M Sahre, P Zhao, S Mirwald. Multi-method analysis of the metal/electrolyte interface: scanning force microscopy (SFM), quartz microbalance measurements (QMB), Fourier transform infrared spectroscopy (FTIR) and grazing incidence x-ray diffractometry (GIXD) at a polycrystalline copper electrode. *Surf. Interface Anal.* 25: 548–560, 1997.
73. J Rappich, HJ Lewerenz. Photo- and potential-controlled nanoporous silicon formation on n-Si(111): an *in-situ* FTIR investigation. *Thin Solid Films* 276: 25–28, 1996.
74. CA Borras, R Romagnoli, RO Lezna. *In-situ* spectroelectrochemistry (UV-visible and infrared) of anodic films on iron in neutral phosphate solutions. *Electrochim. Acta* 45: 1717–1725, 2000.

75. S Music, M Gotic, S Popovic. X-ray diffraction and Fourier-transform infrared-analysis of the rust formed by corrosion of steel in aqueous-solutions. *J. Mater. Sci.* 28: 5744–5752, 1993.
76. A Adenier, MC Bernard, MM Chehimi, E Cabet-Deliry, B. Desbat, O Fagebaume, J. Pinson, F. Podvorica. Covalent modification of iron surfaces by electrochemical reduction of aryldiazonium salt. *J. Am. Chem. Soc.* 123: 4541–4549, 2001.
77. CF Lin, MD Porter, KR Hebert. Surface-films produced by cathodic polarization of aluminum. *J. Electrochem. Soc.* 141: 96–104, 1994.
78. BX Huang, P Tornatore, Y-S Li. IR and Raman spectroelectrochemical studies of corrosion films on tin. *Electrochim. Acta* 46: 671–679, 2000.
79. BX Huang, K Wang, JS Church, Y-S Li. Characterization of oxides on niobium by Raman and infrared spectroscopy. *Electrochim. Acta* 44: 2571–2577, 1999.
80. D Velten, V Biehl, F Aubertin, B Valeske, W Possart, J Breme. Preparation of TiO_2 layers on cp-Ti and Ti6Al4V by thermal and anodic oxidation and by sol–gel coating techniques and their characterization. *J. Biomed. Mater. Res.* 59: 18–28, 2002.
81. AA Salem, R Kellner, M Grasserbauer. Study of corrosion processes in glass by a multitechnique approach. Part 2. Infrared spectroscopy. *Glass Technol.* 35: 135–140, 1994.
82. S Radin, P Ducheyne, S Falaize, A Hammond. *In vitro* transformation of bioactive glass granules into CaP shells. *J. Biomed. Mater. Res.* 49: 264–272, 2000.
83. P Li, Q Yang, F Zhang, T Kokubo. The effect of residual glassy phase in a bioactive glass-ceramic on the formation of its surface apatite layer *in vitro*. *J. Mater. Sci. — Mater. Med.* 3: 452–456, 1992.
84. D Persson, C Leygraf. Initial interaction of sulfur-dioxide with water covered metal-surfaces. An *in-situ* IRAS study. *J. Electrochem. Soc.* 142: 1459–1468, 1995.
85. T Aastrup, C Leygraf. Simultaneous infrared reflection absorption spectroscopy and quartz crystal microbalance measurements for *in situ* studies of the metal/atmosphere interface. *J. Electrochem. Soc.* 144 (9): 2986–2990, 1997.
86. J Itoh, T Sasaki, M Seo, T Ishikawa. *In situ* simultaneous measurements with IR-RAS and QCM for investigation of corrosion of copper in a gaseous environment. *Corros. Sci.* 39: 193–197, 1997.
87. PW Faguy, WN Richmond, RS Jackson, SC Weibel, G Ball, J Payer. Real-time polarization modulation *in situ* infrared spectroscopy applied to study of atmospheric corrosion. *Appl. Spectrosc.* 52: 557–564, 1998.
88. T Aastrup, M Wadsak, C Leygraf, M Schreiber. *In situ* studies of the initial atmospheric corrosion of copper. Influence of humidity, sulfur dioxide, ozone, and nitrogen dioxide. *J. Electrochem. Soc.* 147: 2543–2551, 2000.
89. T Aastrup, M Wadsak, M Schreiner, C Leygraf. Experimental *in situ* studies of copper exposed to humidified air. *Corros. Sci.* 42: 957–967, 2000.
90. M Wadsak, T Aastrup, I Odnevall Wallinder, C Leygraf, M Schreiner. Multianalytical *in situ* investigation of the initial atmospheric corrosion of bronze. *Corros. Sci.* 44: 791–802, 2002.
91. D Persson, S Axelsen, F Zou, D Thierry. Simultaneous *in situ* infrared reflection absorption spectroscopy and Kelvin probe measurements during atmospheric corrosion. *Electrochem. Solid-State Lett.* 4: B7–B10, 2001.
92. F Zhu, X Zhang, D Persson, D Thierry. *In situ* infrared reflection absorption spectroscopy studies of confined zinc surfaces exposed under periodic wet–dry conditions. *Electrochem. Solid-State Lett.* 4: B19–B122, 2001.
93. B Lefez, S Jouen, J Kasperek, B Hannoyer. FT-IR microscopic base imaging system: applications for chemical analysis of Zn and Ni atmospheric corrosion. *Appl. Spectrosc.* 55: 935–938, 2001.
94. D Persson, C Leygraf. Metal carboxylate formation during indoor atmospheric corrosion of Cu, Zn and Ni. *J. Electrochem. Soc.* 142: 1468–1477, 1995.

95. AR Mendoza, F Corvo. Outdoor and indoor atmospheric corrosion of carbon steel. *Corros. Sci.* 41: 75–86, 1999.
96. R Hatada, K Baba. Preparation of hydrophobic diamond like carbon films by plasma source ion implantation. *Nucl. Instrum. Meth. B Phys. Res.* 148: 655–658, 1999.
97. M Carme Sistach, N Ferrer, MT Romero. Fourier transform infrared spectroscopy applied to the analysis of ancient manuscripts. *Restaurator* 19: 173–186, 1998.
98. C Troll, M Pilz. Simulation of degradation phenomena on model enamels with historical compositions. *Glastechnische Berichte-Glass Science Techn* 71: 307–314, 1998.
99. E Greiner-Wronowa, L Stoch. Influence of environment on surface of the ancient glasses. *J. Non-Cryst. Solids* 196: 118–127, 1996.
100. T Nguyen, E Byrd, D Bentz, CJ Lin. *In situ* measurements of water at the organic coating substrate interface. *Progr. Org. Coat.* 27: 181–193, 1996.
101. Q Dai, A Freedman, GN Robinson. Sulfuric acid-induced corrosion of aluminum surfaces. *J. Electrochem. Soc.* 142: 4063–4069, 1995.
102. JY Malvaulty, J Lopitaux, D Delahaye, M Lenglet. Cathodic reduction and infrared reflectance spectroscopy of basic copper(II) salts on copper substrate. *J. Appl. Electrochem.* 25: 841–845, 1995.
103. H Strandberg, LG Johansson. The formation of black patina on copper in humid air containing traces of SO_2. *J. Electrochem. Soc.* 144: 81–89, 1997.
104. D Persson, C Leygraf. Analysis of atmospheric corrosion products of field exposed nickel. *J. Electrochem. Soc.* 139: 2243–2245, 1992.
105. AVR Kumar, R Balasubramaniam. Corrosion product analysis of corrosion resistant ancient Indian iron. *Corros. Sci.* 40: 1169–1178, 1998.
106. R Balasubramanian, AVR Kumar. Characterization of Dehli iron pillar rust by x-ray diffraction, Fourier transform infrared spectroscopy and Mössbauer spectroscopy. *Corros. Sci.* 42: 2085–2101, 2000.
107. N Le Bozec, D Persson, A Nazarov, D Thierry. Investigation of filiform corrosion on coated aluminum alloys by FTIR microspectroscopy and scanning Kelvin probe. *J. Electrochem. Soc.* 149: B403–B408, 2002.
108. D Thierry, W Sand. Microbially influenced corrosion. In: P Marcus, J Oudar, eds. *Corrosion Mechanisms in Theory and Practice*. New York: Marcel Dekker, 1995, pp. 457–500.
109. G Chen, RJ Palmer, DC White. Instrumental analysis of microbiologically influenced corrosion. *Biodegradation* 8: 189–200, 1997.
110. GG Gesey, PJ Bremer. Applications of Fourier transform infrared spectrometry to studies of copper corrosion under bacterial biofilms. *Mar. Technol. Soc. J.* 24: 36–43, 1990.
111. KP Ishida, PR Griffiths. Theoretical and experimental investigation of internal reflection at thin copper films exposed to aqueous solutions. *Anal. Chem.* 66: 522–530, 1994.
112. KP Ishida, PR Griffiths. Investigation of polysaccharide adsorption on protein conditioning films by attenuated total reflection infrared spectrometry. *J. Colloid Interface Sci.* 213: 513–524, 1999.
113. DC White, AA Arrage, DE Nivens, RJ Palmer, RF Rice, GS Sayler. Biofilm ecology: on-line methods bring new insights into MIC and microbial biofouling. *Biofouling* 10: 3, 1996.
114. DE Nivens, JQ Chambers, TR Anderson, A Tunlid, J Smit, DC White. Monitoring microbial adhesion and biofilm formation by attenuated total reflection Fourier-transform infrared-spectroscopy. *J. Microbiol. Meth.* 17: 199–213, 1993.
115. PJ Bremer, GG Geesey. Laboratory-based model of microbiologically induced corrosion of copper. *Appl. Environ. Microbiol.* 57: 1956–1962, 1991.

8

Glow Discharge Optical Emission Spectroscopy

Arne Bengtson
Swedish Institute for Metals Research, Stockholm, Sweden

Contents

8.1 GLOW DISCHARGE DEVICES — AN INTRODUCTION

Glow discharge (GD) is a general term for low-power electrical discharges in reduced-pressure systems. A GD device consists of a vacuum vessel, with two physically separated surfaces inside that form a cathode and an anode. The geometrical shape of the vessel and electrodes can be varied almost infinitely. The vacuum vessel is evacuated and filled with a discharge gas to a pressure ranging from about 10 to 1000 Pa. Noble gases, most commonly argon, are used in analytical GDs in order to avoid chemical reactions inside the device. The current is low to moderate, varying between approximately 0.1 mA to a few amperes. The GD plasma is physically divided into more or less distinct regions. In the *cathode dark space*, which typically extends about a millimeter from the cathode, the electrons lose most of their kinetic energy through collisions with gas atoms, and practically the entire potential drop takes place here. This region is comparatively dark due to the fact that the cross-section for electron excitation of the atoms is low at high electron energies. The cross-section for ionization is also low at high electron energies, and relatively little ionization occurs in the dark space. Beyond this region, one finds the intensely luminous *negative glow*, which varies from several millimeters to a few centimeters in length depending on the design of the device. Here, the average electron energy is typically 20 to −30 eV, where the cross-sections for ionization and excitation are at a maximum. The actual degree of ionization varies, but very seldom exceeds a few tenths of a percent. If the distance between the electrodes is sufficiently long (as in, e.g., a neon tube), the negative glow is followed by a slightly less luminous region known as the *positive column*, which extends all the way up to the anode. In this entire region, the electrons have essentially thermal energies and the plasma is nearly on anode potential.

Depending on the voltage–current characteristics, a GD is classified as *normal* or *abnormal*. A normal GD is characterized by a constant voltage independent of the current, which is the case when the current density is very low. As the current is increased, the discharge gradually takes on an ohmic voltage–current characteristic; that is, the voltage is proportional to the current. The discharge is then classified as abnormal. As the current is further increased, the voltage suddenly drops to a very low value. At this point, the discharge changes characteristics from a GD to what is known as an *Arc discharge*.

In recent years, GD devices that are energized by a radio frequency (RF) alternating electric field instead of a direct current have become more common as sources for analytical spectrometry. The major advantage with these devices is their ability to sputter nonconducting materials, which extends the range of applications for Glow Discharge Optical Emission Spectroscopy (GD-OES) considerably. For more detailed information about the fundamentals of GDs, including RF varieties, the interested reader is referred to, e.g., Refs. (1,2).

8.1.1 Sputtering in Glow Discharges

As early as 1849, Grove (3) discovered that the cathode in an electrical discharge is eroded, or etched. Before the advent of modern atomic theory, this effect could not be properly understood. We now know that this erosion is caused by bombardment of ions and fast neutral species from the plasma, and the proper term for this effect is *cathodic sputtering*. The sputtering of the cathode is the very basis for the usefulness of GDs as tools for surface and depth profile analysis. By designing devices where the sample to be analyzed forms the cathode, the continuous erosion of the sample surface by the discharge is utilized in order to do elemental depth profile analysis. In a GD, the bombarding particles do not have well-defined energy, as is the case when an ion beam in high vacuum is used for sputtering. This is due to the fact that the mean free path for atoms and ions is only of the order of 0.01 to 0.1 mm, which means that the bombarding particles lose kinetic energy due to collisions before reaching the cathode. The short mean free path also has the consequence that a substantial fraction of the sputtered atoms returns to the cathode surface, where they can be redeposited or cause "self-sputtering." As the discharge is switched on, a "dynamic equilibrium" between sputtering and redeposition processes is rapidly established, and the sample is eroded at an "effective" sputtering rate, also referred to as the mass loss rate.

In spite of the complexity of the sputtering processes involved, the effective mass loss rate in several types of GD devices can be accurately described by the following simple empirical expression (4):

$$q_b = C_{qb} i (U - U_{0b}) \tag{8.1}$$

where q_b is the mass loss rate of sample b, i is the current, U is the voltage, and C_{qb}, U_{0b} are *sputtering rate constants* characteristic of the sample b. C_{qb} is related to the *sputtering yield* (probability of a sputtering event per ion impact) of the material; U_{0b} is a threshold voltage, below which the sputtering is reduced to zero. Equation (8.1) shows that the mass loss rate increases linearly with current and voltage. As is the case when sputtering with high-energy ion beams, the mass loss rate varies considerably with the material. The relative differences in sputtering rates observed between different materials follow closely those observed in sputtering with high-energy ion beams.

8.1.2 Optical Emission in Glow Discharges

GDs emit light, or *optical emission*, originating from atoms, ions, and, to a lesser extent, molecules in the plasma, all of these species being energetically excited by means of inelastic collision processes. More strictly, optical emission is defined here as emitted electromagnetic radiation in the wavelength range 100 to 1000 nm, comprising the ultraviolet, the visible, and the near-infrared wavelength ranges.

Most of the excitation processes giving rise to the optical emission results from collisions with electrons. However, a large number of other types of processes are also of considerable importance, such as charge and energy transfers in ion–atom, ion–ion, and atom–atom collisions. Since sputtered atoms from the cathode diffuse into the plasma, and there take part in the excitation–emission processes, the elemental composition of the cathode can be determined by means of spectral analysis of the optical emission. Thus, the combination of a GD source and an optical spectrometer is a tool for elemental depth profile analysis. The resulting analytical technique is called glow discharge optical emission spectroscopy.

8.2 GD-OES INSTRUMENTATION

In the majority of publications on depth profile analysis by GD-OES, the GD sources used are based on the design of Grimm (5). The principal layout of this source is provided in Figure 8.1. The lamp body is normally at ground potential, and the cooled, isolated front plate is at negative potential during operation. The tubular-shaped anode fits tightly into the central opening of the front plate, and extends to approximately 0.2 mm distance from the front surface. The inner diameter of the anode is typically 4 mm, but anodes of 8, 7, 5, 2, and 1 mm diameter are also used. The flat sample, which forms the cathode, is placed on the front plate, and sealed from the atmosphere by an O-ring of relatively soft material. When operating the source, the lamp house is first evacuated for a few seconds by a vacuum pump (pumpdown), after which time an argon flush valve is opened. After a period

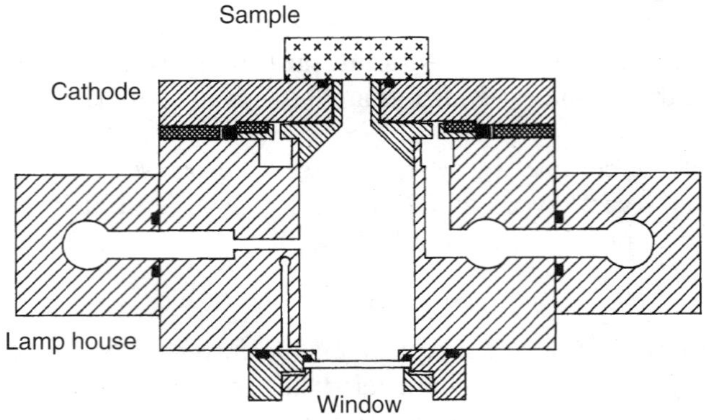

Figure 8.1 Principal layout of the Grimm-type glow discharge source. The sample is in contact with the cathode plate in front, and therefore forms the cathode of the discharge. The hollow anode is separated from the sample by a very short distance (approximately 0.2 mm). When in operation, the lamp body is evacuated by a pump while a constant flow of argon maintains a pressure of typically 5–10 mbar.

typically a few seconds to a minute (“preflush”) to stabilize the pressure at approximately 100 Pa, the discharge is initiated by applying a voltage of typically 500 to 1000 V. With a 4-mm anode, the operating current is normally in the range 10 to 50 mA. The geometry of the source constricts the discharge physically to the inside of the anode tube (obstructed discharge). The cathodic sputtering process during operation continuously erodes the sample surface, which forms the cathode of the device. Before being adsorbed on a cold surface inside the source, a substantial fraction of the sputtered atoms diffuse into the negative glow region of the plasma, where they are excited and emit element-characteristic optical emission.

With a relatively high current density of 50 to 500 mA/cm^2, the penetration rate in a Grimm-type GD is typically in the range 1 to 10 μm/min. The homogeneous electric field distribution in the active region ensures that the sample surface is sputtered rather evenly, resulting in a crater with a nearly flat bottom (Figure 8.2). Thus, by recording the optical emission signals as a function of sputtering time, an elemental depth profile is obtained. In order to make full use of the vast amount of analytical information provided from the GD, the spectrometer must be able to

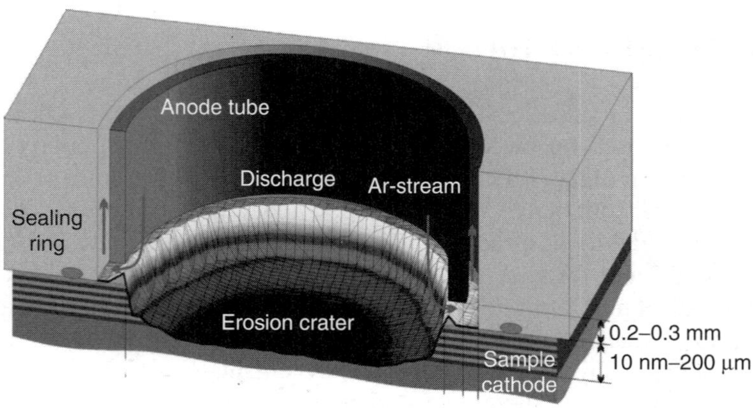

Figure 8.2 The crater shape after sputtering in a Grimm source.

sample the emission spectrum at a rate of at least 100 times per second. As a consequence, so-called multichannell spectrometers are most commonly used. A spectrometer of this type has a fixed grating, and is equipped with individual photomultiplier detectors for up to 60 analytical spectral lines (wavelengths), determined by exit slits in fixed positions. Depending on the analytical applications, one or occasionally two spectral lines for each element of interest are selected for the fixed analytical program. Recently, systems equipped with CCD (charge coupled device) arrays instead of photomultipliers have also become available; such systems give access to a complete spectral range. For the important light elements (C, S, P, N, O, H), the most sensitive analytical lines are found in the vacuum ultraviolet spectral region below 200 nm. Therefore, the spectrometer optical system must be transparent at these short wavelengths. This is usually accomplished by keeping the optics in vacuum, but using a purged nitrogen atmosphere is also very common. For increased flexibility, a complete spectrometer system often incorporates some scanning device; for example, an additional monochromator with fixed detector and a movable grating.

In commercially available GD-OES systems, the source is normally fixed to the spectrometer to form one integrated unit. A schematic diagram of a GD-OES spectrometer system, based on a Grimm-type source, is shown in Figure 8.3.

8.2.1 Instrumental Settings

When performing depth profile analysis, there are a number of practical considerations governing the instrumental settings. In most depth profile applications, the samples can only sustain a limited amount of heat, restricting the setting of the discharge parameters (power, voltage, and current). On the other hand, a high

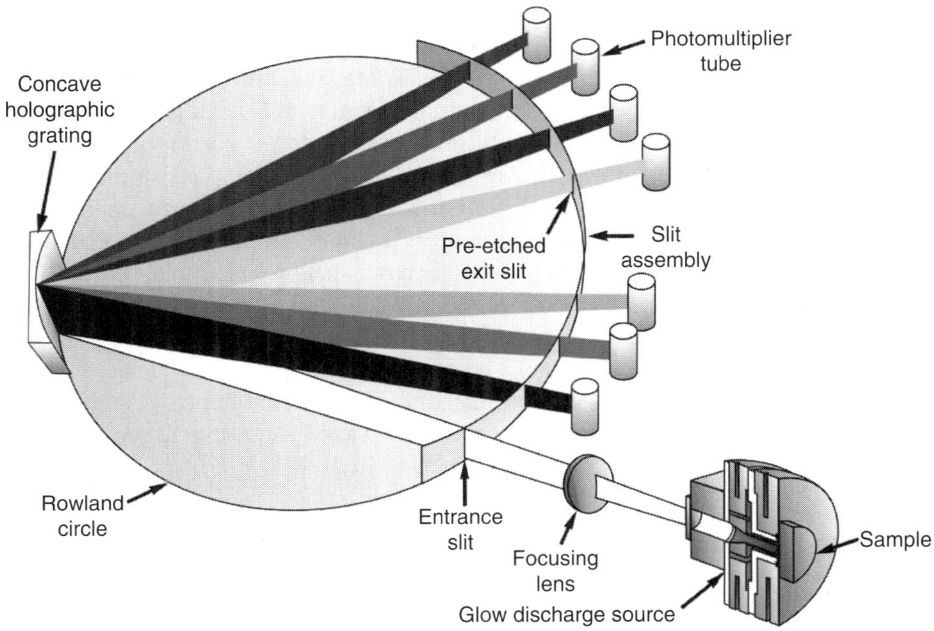

Figure 8.3 Principal layout of a glow discharge optical emission spectrometer, based on a polychromator with curved grating and fixed exit slits.

depth penetration rate is often desired in the interest of sample throughput and signal intensity. Optimizing the discharge parameters are therefore to a large extent the search for maximum sustainable input power. Special attention must be paid to RF systems as opposed to DC systems. In a DC source, the applied power is identical to the plasma power, and optimized settings for one source type can readily be transferred to another of a different manufacturer. This is not the case for RF sources, where a substantial part of the input power is lost in cables, connectors, etc. These losses vary to a large extent between sources of different design, and are typically in the range 10 to 60% of the applied power. Consequently, determination of maximum sustainable applied power in the analytical plasma must be carried out individually for each RF source type. In terms of analytical figures of merit, the depth resolution is normally of primary importance. It is well known that the shape of the sputtering crater is affected by the discharge parameters. General recommendations valid for several applications are not available, since the optimum conditions are highly material dependent. For each application, it is recommended to test at different voltage–current (or power–pressure) combinations to determine the best settings for a flat crater bottom.

In addition to suitable discharge parameters, data acquisition parameters must be selected. While it was stated previously that a speed of at least 100 measurements per second per channell was required for a good system, it is seldom meaningful to use such high speeds throughout the depth profile, since that leads to excessively large data files with largely redundant information. In several applications, a high speed during the first few seconds in order to see rapid transients in the top atomic layers is useful, then the speed can be progressively slowed down to the order of one measurement cycle per second at depths exceeding a few micrometers.

8.3 PRESENTATION AND QUANTIFICATION OF DATA

The primary data from a GD-OES depth profile analysis are in the form of intensities from the elemental detectors as a function of sputtering time. This information is normally presented in diagrams on the computer screen and/or on paper, which in a qualitative way represent elemental depth profiles through the corresponding surface layers (Figure 8.4). The word *qualitative* is used in order to emphasize the fact that the data presented in this way are not quantified into concentration vs. depth, which normally is the desired information. However, for many applications, a qualitative depth profile is quite sufficient. This can be the case when, for example, the analytical problem is simply to compare "good" samples with "bad" for trouble shooting in a production process. In order to extract full analytical information from the data however, quantification is necessary. The quantification problem can be separated into two parts: (i) elemental concentrations, and (ii) sputtered depth. As we shall see in the following sections, there is a connection between these two aspects of depth profile quantification that can be effectively exploited.

8.3.1 The Emission Yield Concept as a Basis for Quantification

Experimentally, it is easy to show that the emission intensities of analytical lines in GD-OES are not just proportional to the concentration of the corresponding element, but also to the sputtering rate of the sample. Intuitively, this observation is easy to accept; the emission intensity should be proportional to the sample atom density

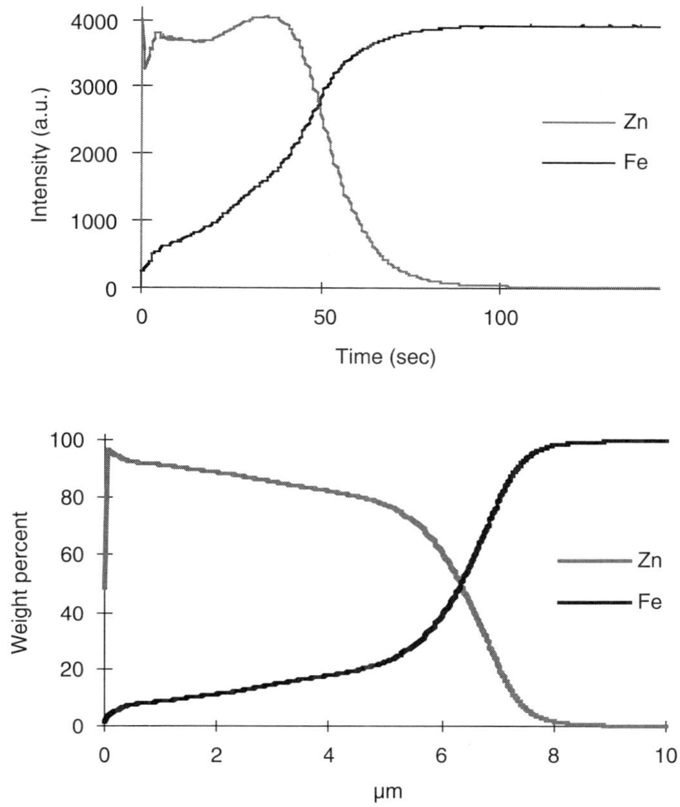

Figure 8.4 Qualitative and quantitative depth profiles through a ZnFe (Galvanneal) coating on steel.

in the plasma, which in turn must be proportional to the sputtering rate. If we allow the voltage, current, and pressure to vary, the situation becomes more complex. However, as long as these parameters remain at least nearly constant, the sputtering rate-intensity proportionality provides for an elegant solution to the quantification problem based on the concept of the *emission yield* (6,7), which can be defined as the emitted light of a spectral line per unit sputtered mass of the corresponding element. The emission yield is an element- and instrument-dependent quantity, which must be determined independently for each spectral line and instrument. The assumption that forms the basis for this quantification technique is that the "integrated signal intensity from one element (and spectral line) is proportional to just the sputtered mass of that element," which implies that the emission yield is independent of the sample matrix. This has been investigated by several authors (8,9), and is by now widely accepted to be valid, at least to a first approximation. Mathematically, the relation described above can be expressed:

$$R_{nm} = I_{nm}\delta t/\delta w_n \tag{8.2}$$

where δw_n is the sputtered mass of element n during time increment δt, I_{nm} is the emission intensity of spectral line m of element n, R_{nm} is the *emission yield* of spectral line m of element n, which is an atomic- and instrument-dependent quantity.

Equation (8.2) is equivalent to

$$I_{nm} = c_n q_b R_{nm} \tag{8.3}$$

where c_n is the mass fraction ("concentration") of element n in sample segment b, q_b ($=\delta w_n/\delta t$) is the sputtering (mass loss) rate in sample segment b.

Calibration is performed by determining the emission yields by means of calibration samples. These samples may be of bulk type with known concentrations, in which case it is necessary to determine the sputtering rate of each calibration sample. Alternatively, samples with coatings of known composition and thickness may be used. Due to the matrix-independence of emission yields, the calibration samples do not have to be of similar composition to the unknown samples that are to be measured. This fact increases the practical applicability of the emission yield quantification method over methods requiring matrix-matched calibration samples enormously. As an example, a set of calibration samples consisting of low alloy steels, high alloy steels, cast irons, a few different aluminum alloys, a few different brasses, and a high-purity copper will cover a wide range of depth profile applications. For analysis of oxides, nitrides, etc., this "basic" set of calibration samples needs to be supplemented with "high" samples for O, N, and H.

In practical analytical work, the most commonly used calibration function is the following (Equation (8.3) rewritten):

$$c_n q_b = k^*_{nm} I_{nm} - b_{nm} \tag{8.4}$$

where k_{nm} is a calibration constant (equal to the inverse of the emission yield) and b_{nm} is the background signal expressed in the same units as the mass loss rate.

Equation (8.4) is often augmented by a second-order term to handle any nonlinearity of the calibration curves. For several sensitive spectral lines, nonlinearity occurs as a result of a process known as *self-absorption*. Not described here, the calibration function normally also includes so-called line interference corrections to compensate for spectral line overlaps from other elements than the analyte.

When measuring an unknown sample, the calibration function gives the *sputtered mass* of each element per time increment. The *total* sputtered mass of the sample is obtained as the sum over all elements, and the concentrations are easily calculated as fractions of the sum.

8.3.2 Determination of Sputtered Depth

From the calibration, the effective sputtering rate is obtained in units of mass loss, that is, micrograms per second. The density must also be known in order to convert this information to depth. The density can only be estimated from the elemental concentrations, since no other information is available from the analysis. The most commonly used algorithm is based on summing over the fractional volumes of each element, using tabulated densities of the pure elements. This method gives very accurate results for all types of metallic alloys. Compounds containing light and gaseous elements as majors (oxides, nitrides, carbides, etc.) are a bit more uncertain. The densities assigned to these elements have to be taken as averages based on measured densities of several materials. The calculated densities for oxides, nitrides, etc., are generally accurate to within 10%, but deviations up to 30% have been observed in a few cases. An inaccurate density will translate into a corresponding systematic error in the determined sputtered depth.

8.3.3 Some Artifacts to Be Considered

A qualitative depth profile (intensity vs. time) is generally more or less distorted with respect to the true profile, in basically two respects. First, variations in the sputtering rate have the effect that the depth is not linear with the time scale. Second, since the emission intensity is proportional to the sputtering rate, these variations also cause a distortion in the apparent concentrations. These types of distortions disappear with a correct quantification. However, there are certain additional artifacts from the sputtering process that must be considered when interpreting a depth profile. This is in fact true of any depth profiling technique based on sputtering. First, the sputtering process fundamentally limits the depth resolution. Starting with a minimum "information depth" of the order of 1 nm on the top surface, the relative depth resolution (apparent interface width/sputtered depth) is rather constant and typically 10 to 15% (2). This means that at, for example, 10 μm depth, an "infinitely sharp" interface will appear as approximately 1-μm thick. This is typical of any depth profiling technique based on sputtering, and it must be kept in mind when, for example, concentration levels in interface regions are estimated. A very narrow peak of high concentration will be smeared out, with the effect that the determined peak concentration may appear considerably lower than the true value. However, the integrated elemental mass under such a peak, as expressed in, for example, grams per square meters, remains correct and is not affected by depth resolution. A more difficult form of artifact to deal with in depth profile analysis is what is known as "preferential sputtering." This occurs if the sample has a coarse microstructure with relatively large grains of different phases. An example is aluminum oxide particles on a zinc surface. Due to the very large difference in sputtering rates of these materials, the oxide appears to penetrate deeper into the zinc layer than it actually does.

In recent years, it has been discovered that another type of "matrix" effect, linked to the element hydrogen, can also affect the quantification of GD-OES depth profiles (10). It has been shown that the emission yields of spectral lines from other elements can be rather dramatically affected even by very minute concentrations of hydrogen in the plasma. The hydrogen can originate from the sample itself, contamination due to adsorbed water and pumping oil inside the source, or small vacuum leaks. Both enhancement and suppression effects can be observed for one element depending on the emission line used, showing that it is the excitation probability of excited levels that are affected rather than "plasma chemical" reactions at work. While source contamination and leaks should be minimized by improved source design, the large number of applications where hydrogen is found in the sample itself has made it necessary to introduce matrix correction algorithms to deal with these effects. At present, such algorithms have been introduced into software from all major manufacturers of GD-OES instruments.

8.4 APPLICATIONS OF GD-OES DEPTH PROFILE ANALYSIS FOR CORROSION RESEARCH

There are a large number of GD-OES depth profile applications relevant for corrosion research in mainly two respects. First, GD-OES can be used to investigate and characterize properties of coatings and other material surfaces designed to increase corrosion resistance. Second, GD-OES can sometimes be used to analyze actual corrosion products on surfaces, mainly oxides and hydroxides.

8.4.1 Galvanized Steels

Due to their excellent corrosion resistance, various types of galvanized steel sheets are used extensively in the automotive, home appliances, and construction industries. The platings typically consist of pure Zn or some Zn-base alloy, with coating thicknesses in the range 5 to −20 μm. These types of materials have become technically and economically the most important application of GD-OES depth profile analysis to date. One reason for this is the very high effective sputtering rate of GD-OES (up to 10 μm/min), enabling complete depth profiling of these rather "thick" coatings within minutes. Compared to any other type of equipment for depth profile analysis, this is exceptionally fast, giving GD-OES the capability for quality control in an industrial laboratory. Another factor contributing to the suitability of the technique for this type of industrial application is that the requirements on sample preparation are kept to a minimum.

8.4.1.1 ZnNi Electroplated Steels

This type of material, which has found widespread use primarily in the automotive industry, is illustrated in Figure 8.5. The major analytical problem is to determine the thickness, or coating weight, and the Ni content of the coating. Surface contamination (mainly hydrocarbons) may also be of interest, in particular, contamination on the original steel surface, appearing as enrichments in the Zn/Fe interface. Such contamination is relevant to the adherence of the coating and hence also the corrosion resistance.

8.4.1.2 Galvanneal

This type of material is produced by electroplating steel with pure zinc, followed by annealing at a temperature that allows all of the zinc to form ZnFe intermetallic phases. This is an example of a *conversion coating*, which is claimed to have superior wear and corrosion resistance compared to the original zinc coating. The major analytical

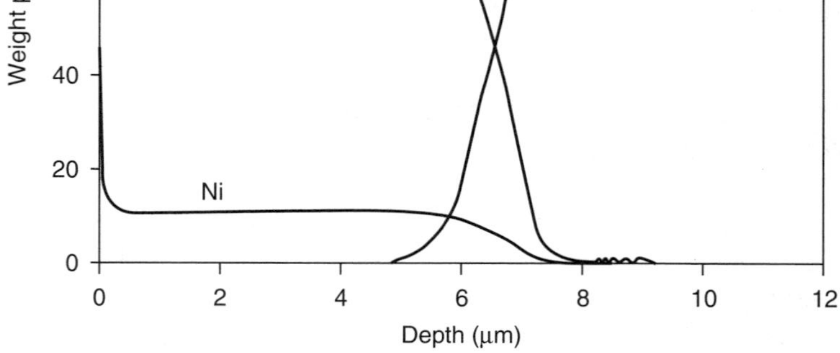

Figure 8.5 Quantitative depth profile of an electroplated ZnNi coating on steel.

problems are to determine the coating thickness and the variation of Fe content with depth. Qualitative and quantitative depth profiles are shown in Figure 8.4.

8.4.2 Hot Rolled Steels

Analysis of the oxide scale on hot rolled steels was among the first applications of GD-OES for depth profile analysis in the steel industry. The oxide thickness and stoichiometry as well as elemental segregation in the oxide/steel interface are analytical information of interest. The qualitative depth profile in Figure 8.6 shows that the oxygen content in the oxide scale decreases with depth, the outer part corresponding to Fe_3O_4 and the inner part to FeO. A quantified profile gives the same information; the intensity vs. time profile emphasizes the fact that GD-OES sputters through this approximately 5-μm thick oxide rather quickly. The conclusion that two oxide layers are present is supported by x-ray diffraction studies on the same material. Si, Cu, and Cr have segregated to the interface due to selective oxidation. If the pickling process just removes the oxide scale, the metal surface will have a composition different from the bulk material. Particularly for high alloy steels, such segregation effects can seriously impair the corrosion resistance and GD-OES is a fast, effective analytical tool for control of the efficiency of the pickling process.

8.4.3 TiN and Other Nonmetallic Coatings on Steels and Hard Materials

Physical vapor deposition (PVD) and chemical vapor deposition type coatings are becoming increasingly important applications for quantitative GD-OES analysis. TiN, TiCN, CrN, and other nitride compounds are examples of such "high-tech" surface coating materials used in order to improve both the wear and corrosion resistance of cutting tools. In order to control the production of such coatings, there is a need for analytical techniques capable of rapid and quantitative depth profile analysis. GD-OES has proven to be a very powerful tool for this purpose, and it is often used in R&D aimed at the development of new types of coatings.

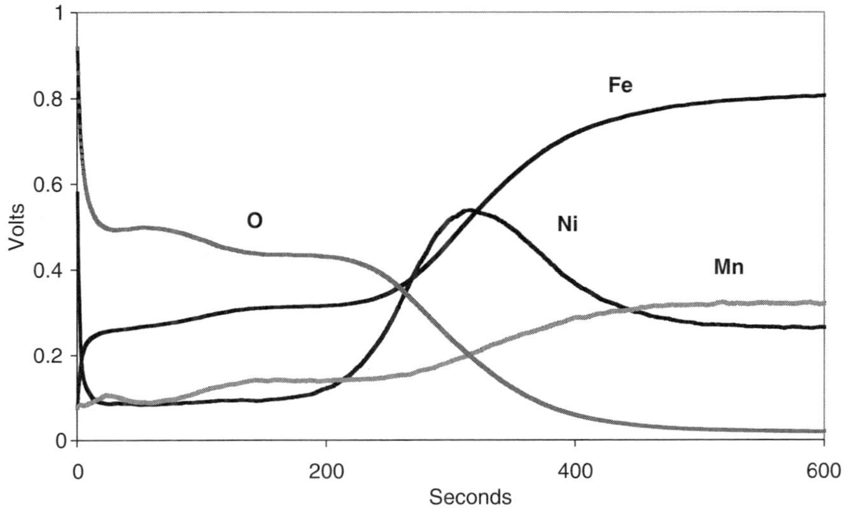

Figure 8.6 Qualitative depth profile through the oxide scale on a hot rolled steel.

Multilayer PVD structures are produced partly in order to improve the corrosion resistance of the material. Microscopic studies of PVD films often reveal small pores running through the entire depth of the film. In certain applications, this could mean that the substrate is subjected to corrosion attack from, for example, water vapor seeping though the pores. By producing multilayer films of alternating composition, the pores are "staggered" in depth, thereby reducing the probability of finding pores that connect the surface to the substrate. An example of such a structure with 15 sublayers of alternating TiN and TiAlN on steel is shown in Figure 8.7. This example also illustrates the fact that the practical depth resolution that can be attained by GD-OES is quite impressive, considering that the analyzed area has a diameter of several millimeters.

8.4.4 Passivation Layers on Stainless Steels

While it has been a widely held opinion that GD-OES is not suitable for near-surface analysis of extremely thin layers, recent development has shown that the minimum information depth is less than 1 nm. Using the most advanced quantification techniques, including compensation for the effect of hydrogen, fully quantitative depth profiles of passivation layers on stainless steels can be obtained (Figure 8.8). It has recently been shown that there is a strong correlation between such GD-OES depth profiles and the critical pitting temperature (CPT) of stainless steel surfaces, provided other parameters influencing the CPT are kept constant.

8.4.5 Polymer Coatings

Some of the most common types of surface coatings for corrosion protection are polymer coatings (paint). With the introduction of RF powered GD-OES sources, these and other nonconductive coatings can be sputtered and profiled. Figure 8.9

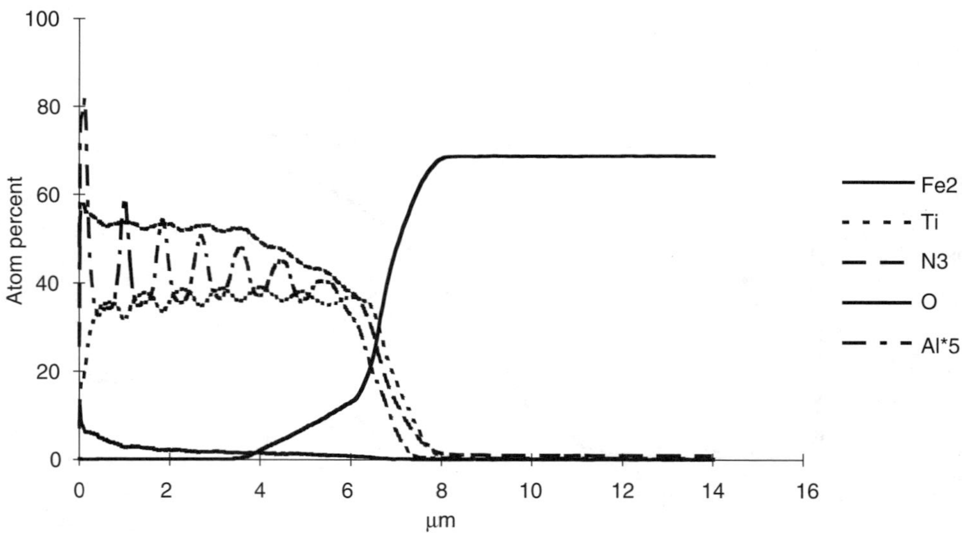

Figure 8.7 Quantitative depth profile of a coating with alternating TiN and TiAlN layers on steel.

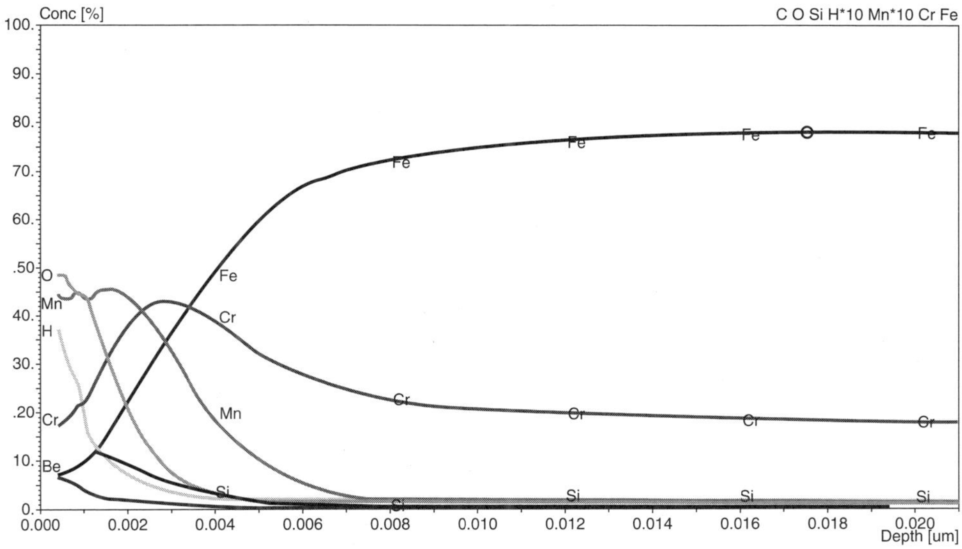

Figure 8.8 Quantitative depth profile of the passivation layer on a stainless steel.

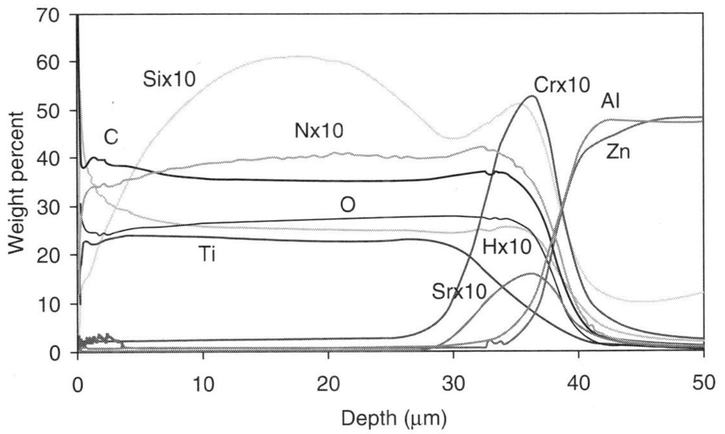

Figure 8.9 Quantitative depth profile of a polyester and primer paint coat on galvanized steel.

shows an example of a depth profile of a polyester topcoat on a primer. It should be emphasized, however, that calibration for polymer coatings in the frame of RF GD-OES analysis is considerably more difficult than calibration for inorganic coatings in DC mode. One obvious difficulty is the lack of calibration standards for certain elements frequently found in paint pigments; e.g., Sr and Ba. In the example of Figure 8.9, a wet chemical analysis of the coating itself was used for calibration of Sr. Another difficulty is the determination of the actual discharge parameters in the analytical plasma (power, voltage, current). As was mentioned briefly in Section 8.3.1, variations in these parameters affect the emission yields and hence the accuracy of the calibration and the analytical results. Furthermore, the "hydrogen" effect mentioned in Section 8.3.3 must be considered when quantifying data from polymer coatings, since the concentration of hydrogen in these materials is significant.

REFERENCES

1. R.K. Marcus (ed.). *Glow Discharge Spectroscopies*. New York: Plenum Press, 1993.
2. R. Payling, D. Jones, A. Bengtson. *Glow Discharge Optical Emission Spectrometry*. Chichester: John Wiley & Sons, 1997.
3. W.R. Grove. *Philos. Trans. R. Soc.* 142: 87, 1852.
4. P.W.J.M. Boumans. Analytical studies of sputtering in a glow discharge for spectrochemical analysis. *Anal. Chem.* 44: 1219–1228, 1972.
5. W. Grimm. Eine neue Glimentladungslampe für die Optische Emissionsspektralanalyse. *Spectrochim. Acta B* 23: 443–454, 1968.
6. J. Pons-Corbeau. Study of Emission and Sputtering Yields in Some Alloys and Oxides by Glow Discharge Optical Spectrometry: Quantification of Analysis (in French). IRSID Report RE 1141: 1–29, 1984.
7. K. Takimoto, K. Suzuki, K. Nishizaka, T. Ohtsubo. Quantitative Analysis of Zinc–Iron Alloy Galvanised Coatings by Glow Discharge Spectrometry and Secondary Ion Mass Spectrometry. Nippon Steel Technical Report 33: 28–35, 1987.
8. A. Naoumidis, D. Guntur, M. Mazurkiewicz, H. Nickel, W. Fischer. Beitrag zur Quantifizierung von Tiefenprofilen bei der Analyse von Korrosionsschichten an Ni-baslegierungen. *Proceedings of Analytische Glimmentladungsspektroskopie*, Julich, 1990, pp. 138–153.
9. A. Bengtson. A contribution to the solution to the problem of quantification in surface analysis work using glow discharge atomic emission spectrometry. *Spectrochim. Acta B* 40: 631–639, 1985.
10. A. Bengtson, S. Hänström. The influence of hydrogen on emission intensities in GD-OES — consequences for quantitative depth profile analysis. *Proceedings of the Fifth International Conference on Progress in the Steel and Metal Industries*, Luxemburg, 1998, pp. 47–54.

9

Recent Developments in the Application of Radiotracer Methods in Corrosion Studies

G. Horányi and E. Kálmán
Institute of Chemistry, Chemical Research Center, Hungarian Academy of Sciences, Budapest, Hungary

Contents

9.1 INTRODUCTION

Joliot's fundamental work, especially the publication in 1930 of the report on his method, can be considered a landmark in the application of radiochemical methods in electrochemistry. In the 1950s and 1960s, development in this field was marked by the simultaneous formation of working groups in Moscow (in Frumkin's Institute, Balashova and Kazarinov; in the Karpov Institute, Kolotyrkin and Losev), Leipzig and Dresden (Schwabe and coworkers), and in Pennsylvania (Bockris and coworkers) (see reviews (1–10)).

During the last 40 years, additional centers were formed in Hungary (Horányi and coworkers in Budapest (11–21); Varga, Kolics, and coworkers in Veszprém (22–29)), Poland (Sobkowski, Zelenay, and coworkers in Warsaw (30–35)), the United States (Wieckowski and coworkers in Urbana (36–44)), and Lithuania (Poskus and coworkers (45–49)). Together with the new groups, the application of radiotracer methods has continued in Bockris' laboratory (50–54) and the late Professor Kazarinov's laboratory (55–58). Results obtained by a French group (Marcus et al.) should also be mentioned (59).

The electrochemical phenomena studied by radiotracer techniques in these laboratories range from equilibrium adsorption of ions and neutral species to corrosion processes involving formation of surface layers (films), deposition and dissolution of metals, etc. The variety of fields involved in these studies and the development of the experimental technique have been well demonstrated in the reviews published during the last 20 years (2–10,60).

The first steps toward the application of radiotracer methods in corrosion studies were taken in the investigation of dissolution processes of metals and alloys. These studies are surveyed in an excellent monograph by Kazarinov and Andreev (3).

The principle of these methods is based on the labeling of a component of the metal phase by one of its radioactive isotopes and calculating the dissolution rate of the metal specimen by measuring either the increase in radiation coming from the solution phase, or the decrease in radiation coming from the solid phase.

The main steps characterizing radioactive tracer methods used for investigation of dissolution processes of metals and alloys can be summarized as follows:

1. Introduction of the radioisotope into the specimen. This task can be achieved at least by three methods: (a) through melting; (b) by electrolytic deposition of the radioactive metal; and (c) by subjecting the metal specimen to neutron irradiation in a nuclear reactor.
2. Measurement of the changes in radiation intensity caused by the dissolution process.

As mentioned above, this step can be performed in two ways:

1. Determination of the radioisotope concentration in electrolyte solution either through "sampling" or continuous measurement of the radiation intensity of the solution phase.
2. Determination of radiation intensity coming from the metal sample. This method was used in the case of amalgams. The dissolution process of solid electrodes can be studied by any radiotracer method used for the investigation of adsorption phenomena. (The principle of these methods will be discussed later.)

The application of radiotracer methods in corrosion research covers the following main topics:

1. Clarification of the role of the adsorption of various ions (HSO_4^-, Cl^-, $S_2O_3^{2-}$, etc.) in corrosion processes (25, 51–54,61–69). Attempts are made to make distinction between adsorption phenomena occurring on the metal surfaces and those on corrosion products (70–75).
2. Investigation of the adsorption of corrosion inhibitors in order to contribute to the clarification of the mechanism of inhibitor action (76, 77).
3. Study of radioactive contamination and decontamination problems of the nuclear industry. Study of the accumulation of radioactive components in corrosion products of cooling systems of reactors and corrosion problems connected with nuclear waste treatment and disposal (see, for instance, Refs. (78–93) and references cited therein).

The present survey is devoted mainly to questions related to the above problems.

9.2 EXPERIMENTAL METHODS

9.2.1 *In Situ* Radiotracer Techniques for Adsorption Studies

9.2.1.1 Cell Types

Various methods have been developed for *in situ* radiotracer adsorption studies depending on the requirements of the problems to be studied. The methods applied and the main factors leading to the choice of the most appropriate method were discussed in detail in review articles (2–6,60,94,95). In this chapter, only the main features of the most important methods will be considered.

In the case of the *in situ* studies the central problem is how to separate the signal (radiation) to be measured from the background radiation, and how to attain the optimal ratio of these quantities.

From this point of view methods can be divided into two main groups:

1. Radiation of the solution background is governed and minimized by self-absorption of the radiation, that is, by the attenuation of the radiation intensity by the radioactive medium itself (thin-foil method).
2. Background radiation intensity is minimized by mechanical means (thin-gap method).

9.2.1.1.1 The Thin-Foil Method. In the case of the thin-foil method, the detector "sees" simultaneously both components of the radiation coming through a thin foil (metal or metal-plated [gold] plastic film) forming the bottom of the cell. The adsorbent is either the foil itself or a thin layer deposited on the bottom of the cell serving as a mechanical support and electric conductor (if the foil is metal plated).

The situation can be visualized by the scheme shown in Figure 9.1.

The radiation measured (I_T) is

$$I_T = I_s + I_a \tag{9.1}$$

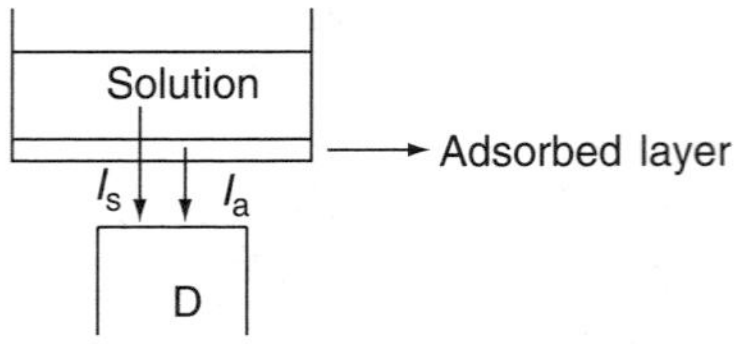

Figure 9.1 Scheme representing the principle of the thin-foil method.

where I_s and I_a are the intensities of the radiation coming from the solution phase and from the adsorbed layer, respectively. Using isotopes emitting soft β^--radiation (radiations characterized by high mass absorption coefficient (μ) (^{14}C, ^{35}S, and ^{36}Cl)), the self-absorption of the radiation in the solution phase is so high that the thickness of the solution layer effective in the measured solution background radiation is very low. (Similar phenomena could be observed for isotopes emitting low-energy x-rays.)

In these cases, on the basis of a simplified model, the solution background radiation is equal to

$$I_s = \alpha I_0 q \int_0^\infty c \, \exp[-\mu x \rho] \, dx = \alpha I_0 q \frac{c}{\mu \rho} \tag{9.2}$$

while the radiation coming from the adsorbed species is

$$I_a = \alpha I_0 q \gamma \Gamma \tag{9.3}$$

where α is a proportionality factor, I_0 is the specific activity of the labeled species present in concentration c in the solution phase, q is the geometric surface area of the electrode, γ is its roughness factor, μ is the mass absorption coefficient of the radiation, ρ is the density of the solution phase, Γ is the surface concentration of adsorbed species, and x is the coordinate measuring the distance from the electrode surface in the solution phase. The integration goes from 0 to the infinity. However, the "infinity," for instance, in the case of the soft β^--radiation emitted by ^{14}C is less than 10^{-2} cm, as the radiation is completely absorbed in a solution layer of this thickness.

On the basis of Equations (9.1) to (9.3), Γ can be determined easily:

$$\Gamma = \frac{I_a}{I_s} \frac{c}{\mu \rho \gamma} = \frac{I_T - I_s}{I_s} \frac{c}{\mu \rho \gamma} = \left(\frac{I_T}{I_s} - 1 \right) \frac{c}{\mu \rho \gamma} \tag{9.4}$$

It follows from the preceding simple equations that the solution background radiation is proportional to the concentration of the labeled species and reliable results can be expected only if $I_T/I_s \gg 1$ or $I_a/I_s > 1$. I_a is proportional to the roughness factor (Equation (9.3)). To give some orientation to the requirements for the application of the thin-foil method, consider the following simple example. In the case of Γ-values of about 10^{-10} mol cm^{-2} at smooth surfaces ($\gamma \approx 1$ to 2) using ^{14}C or ^{35}S for labeling ($\mu \approx 3 \times 10^2$ cm^2 g^{-1}). I_s becomes commensurable with I_a above concentrations about 10^{-4} mol dm^{-3} where $I_a/I_s \approx 1$. However, by increasing the roughness factor, for instance, by 100–200 I_a will increase by the same factor; thus, I_a/I_s reaches unity about 10^{-2} mol dm^{-3}.

Various technical versions of the foil method have been elaborated. For details of the technical description, we refer to the literature (2–5). A typical representative

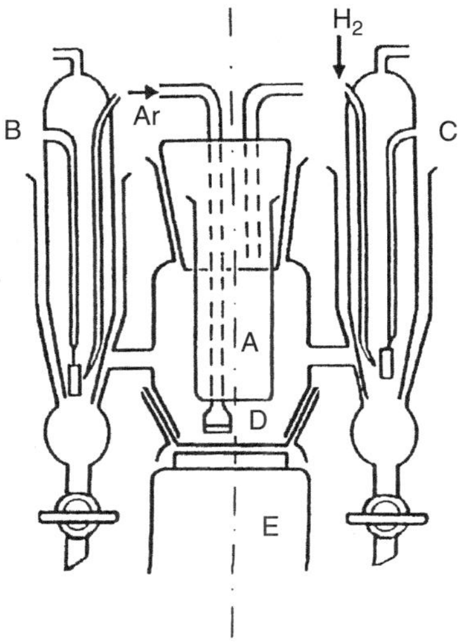

Figure 9.2 Representation of the cells characteristic for thin-foil method. A, central compartment; B, auxiliary electrode; C, reference electrode; D, main electrode, E, scintillation counter.

of this type of cells is shown in Figure 9.2 (2,16). These versions found their applications mostly in the study of adsorption phenomena on rough surfaces with high real surface area ($\gamma > 10$) using isotopes emitting β^--radiation.

As mentioned, Equations (9.2) to (9.3) were derived for ideal conditions, neglecting, for instance, the role of secondary radiation effects, the problems connected with the physical meaning of the roughness factor, etc. Therefore, an important question is how to get reliable surface concentration data from radiation intensity measurements under conditions that are far from the ideal. The procedure depends on the accuracy required. Detailed information and analysis concerning these problems can be found in the review papers cited (3–8). Some new development has been reported in Refs. (13, 18, 96–100).

9.2.1.1.2 The Thin-Layer or Gap Method. In cases where radiation coming from the solution background would be too high in comparison with that originating from the adsorbed layer (smooth surfaces, γ-radiation), some kind of mechanical means should be applied to reduce the role of the background radiation. The terms used to denote these techniques have undergone some change over the last two decades (see literature (1–6)). In the following discussion, we try to use the terms recently suggested.

Mechanical control of the background radiation can be achieved in two different ways: fixed rigid reduction of the solution layer thickness (classical thin-layer or gap method), and flexible mechanical or temporary reduction of the thickness of the solution layer (electrode-lowering technique). In the first case, specially designed cells are used where only a very thin solution layer flows between the electrode and the detector, as shown in Figure 9.3.

This method is not often used, as it has several drawbacks, such as complexity of the equipment (a pump for solution circulation), the problem of determining exact electrochemical parameters owing to the presence of a thin solution layer separating the auxiliary and reference electrodes from the main electrode, etc.

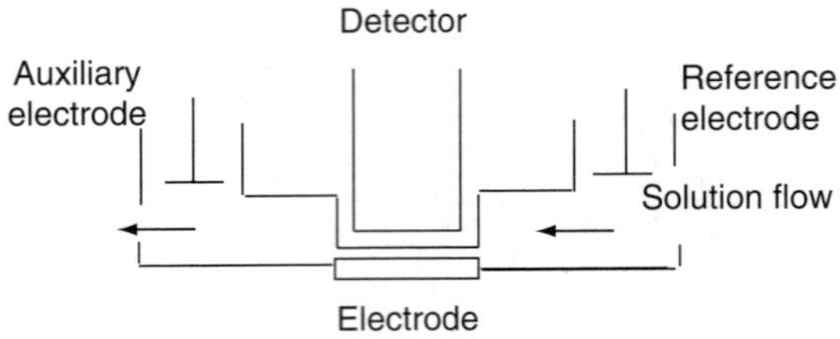

Figure 9.3 Classical version of thin layer or gap method.

These problems are avoided by the application of a flexible version, the so-called electrode-lowering technique. In this technique, the solution gap between the electrode and detector is minimized temporarily only during the time of measurement of the adsorption. The electrode is positioned in two positions, as shown by the scheme in Figure 9.4.

In the "lifted-up" position, attainment of the adsorption and electrochemical equilibrium (or steady state) proceeds without disturbance; the detector measures only the solution background radiation. In the pressed-down position, the intensity measured comes from the species adsorbed on the electrode surface and from the solution layer present in the gap between the bottom of the cell and the electrode. The thickness of this gap depends on the mechanical state of the electrode surface, the stability of the bottom of the cell, etc.

In the first version of this method, developed by Kazarinov and Andreev (3), the bottom of the cell was a plastic foil, that is, a material with not very much mechanical stability. This drawback was eliminated by Wieckowski and co-workers (100) using a glass scintillator embedded in a ceramic disk to form the bottom of the cell. Owing to this construction, the gap thickness attained in the electrode's pressed-down position could be about 1 to 2 $\times$ 10^{-4} cm. This thickness is one order of magnitude lower than the limiting thickness value determined by the self-absorption of β^--radiation of ^{14}C in the case of the foil method. The cell developed by Wieckowski and co-workers (100) is shown in Figure 9.5. Detailed analyses of the calculation of surface concentrations from radiation intensities are available in Refs. (4–6). Similar reduction of the gap thickness was reported by Poškus and Agafonovas (45). It is evident that Equations (9.2) to (9.4), even under ideal conditions, cannot be applied to these systems. First, the thickness of the gap and the radiation back-scattered by the electrode should be taken into consideration.

In the case of isotopes emitting β^--radiation instead of Equation (9.4), in the first approximation, the following relationship could be used for the calculation of Γ:

$$\Gamma = \frac{I_p}{I_b} \frac{c}{\mu\rho\gamma f_b(\exp - \mu\rho d)} \tag{9.5}$$

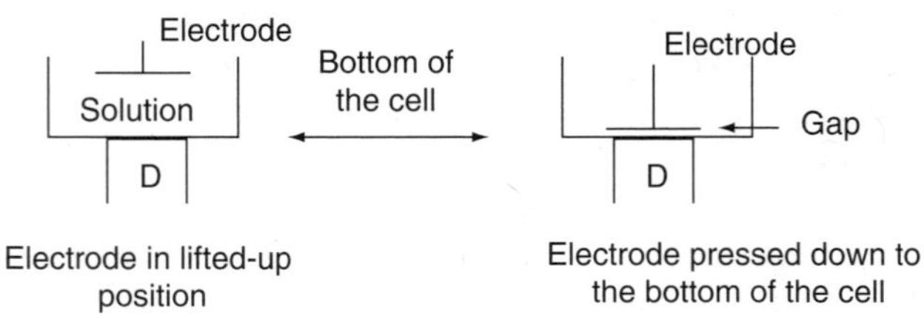

Figure 9.4 Schematic representation of the principle of the electrode-lowering technique.

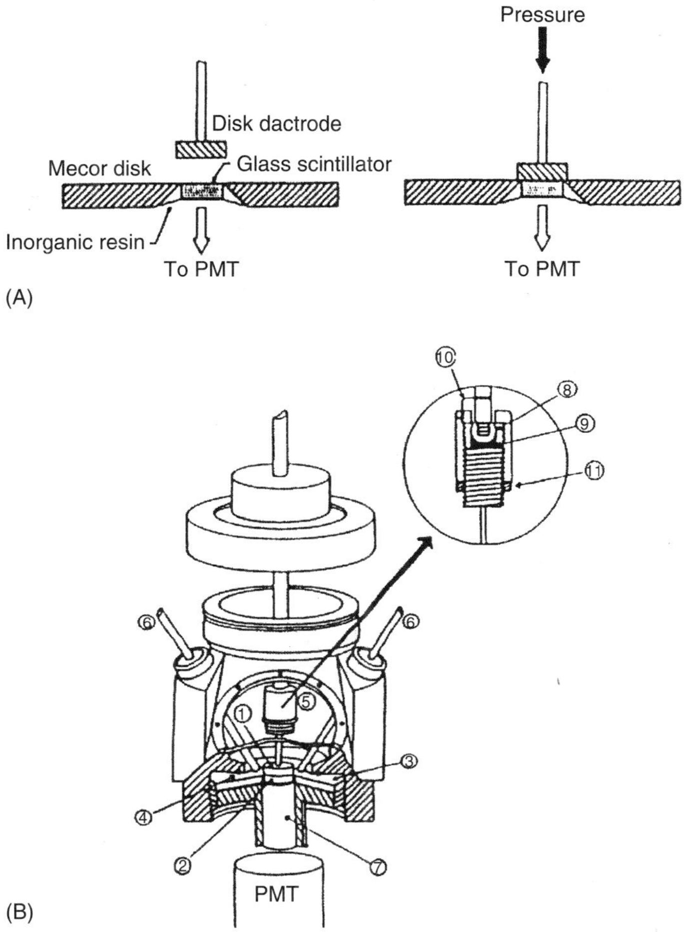

Figure 9.5 (A) Diagram of the cell elaborated by Krauskopf et al. Diagram of the electrode in the raised position for adsorption (left) and the "squeezed" position for adsorbate measurement (right). (B) Diagram of the Teflon cell: 1, platinum electrode; 2, glass scintillator; 3, Macor ceramic disk cell bottom; 4, Teflon O-ring; 5, flexible elbow (see insert); 6, cell ports (six); 7, light pipe; 8, stainless steel sphere; 9, concave Teflon spacer; 10, platinum wire for electrical contact; 11, lock nut.

where I_p is the counting rate when the electrode is pressed down against the detector surface, I_b is the counting rate when the electrode is far from the detector, d is the thickness of the solution layer between the electrode and the detector, and f_b is the backscattering factor (for the material of the electrode).

The problem of the reliability of the data obtained by thin-gap "electrode-lowering" technique has been revisited and discussed over the last few years in a number of publications (28, 44, 49). It is hoped that these discussions will lead to the development of sophisticated techniques to provide highly accurate measurements of adsorption.

9.2.1.2 Direct and Indirect Methods

As to the role of the labeled species in the radiotracer study of adsorption phenomena, two different versions of the method may be distinguished. In the first, the direct method, the species to be studied is labeled and the radiation measured gives direct information on the adsorption of this species. However, this method cannot be used in several cases owing to technical restrictions related to the very nature of the radiotracer method (the available concentration range is limited; no distinction can be made between the adsorption of the labeled compound studied and that of a product formed from it; the number of commercially available labeled compounds is restricted).

Considering all these problems, the use of the so-called indirect radiotracer methods was suggested. Instead of labeling the species to be studied, another adequately chosen labeled species (indicator species) is added to the system, and the adsorption of this component is followed by the usual radiotracer measuring technique. Evidently, the sorption of the indicator species should be in relation with that of the species to be studied. The nature of this link could be different in different systems. For instance, in some cases, competitive adsorption with the labeled species, while in other cases, induced adsorption of the labeled species may provide information on the adsorption behavior of a given molecule. The principle of the study in the former case can be demonstrated by the scheme presented in Figure 9.6.

9.2.1.3 State of the Specimen Studied

The nature (metals, alloys, oxides, modified electrodes, etc.) and the structure (powdered or compact, polycrystalline or single crystal materials, smooth or rough surfaces) of the specimen to be studied are important factors determining the approach to the sorption phenomena and the radiotracer methods used for their study.

The main groups are as follows:

9.2.1.3.1 Compact Metals and Electrodeposited Metal Layers. For a long period of time, the adsorption phenomena had been studied only on polycrystalline surfaces. Both smooth and rough surfaces with high real surface area were considered, using preferably the electrode-lowering technique in the former and the foil method in the latter cases.

For many years only platinized platinum electrode, as representative of electrodes with high real surface area, has been in the foreground of studies. During the last decades investigations have been extended to other metals as well, using copperized (101–103), silverized (104, 105), nickelized (106), rhodized (12, 107, 108), and aurized (102, 109) electrodes with high real surface area (roughness factor values above 20). The behavior of electrodeposited Tc and Re layers was studied in detail (110, 111). This extension of adsorption studies involved the elaboration of methods for the preparation of the electrodes and measurement of the roughness factors.

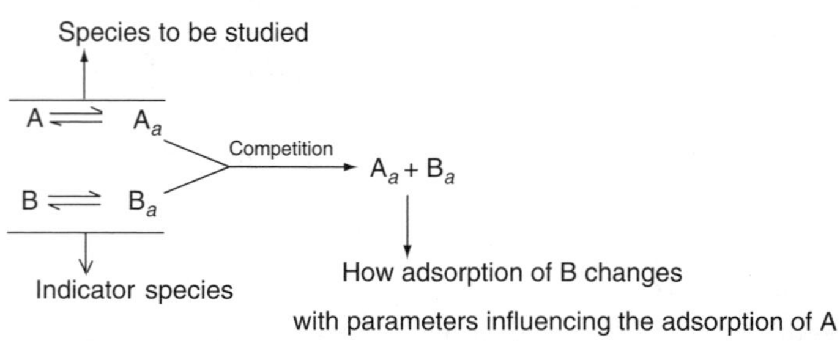

Figure 9.6 Scheme representing the principle of the indirect radiotracer technique.

The investigation of electrosorption phenomena at morphologically uniform and ordered single-crystal, i.e., well-defined, electrodes is a new development since the end of the 1980s (4–6,100). The version of electrode-lowering technique elaborated by Wieckowski and colleagues was created to this end (4–6,100) in conjunction with other methods for the study of single-crystal surfaces. At any rate, the method gives excellent results for polycrystalline smooth electrodes as well.

9.2.1.3.2 Powdered Adsorbents. The principle of the radiotracer method for the *in situ* study of adsorption phenomena on powdered metal samples and the experimental technique were reported more than 30 years ago. The use of powdered samples ensures higher reliability of tracer adsorption measurements as compared to those carried out with smooth surfaces of compact metal samples. It is, however, a drawback of the use of powdered materials that very often their structure cannot be unambiguously defined and determined.

It was demonstrated that in some cases a special version of the foil method, the sprinkling of the powdered metal on the bottom of the cell, could be a useful technique (61). A series of papers (61–64) were published presenting results obtained with stainless steel samples used as construction material in nuclear reactors. Later, the method was extended to other metals (75, 112) and various oxides as well (70–74).

The very principle of the method is the measurement of radiation intensity originating from species adsorbed on a powdered layer sprinkled on a thin gold-plated plastic foil that serves simultaneously as the window for radiation measurement. The scheme of the experimental setup is shown in Figure 9.7.

In the case of thin layers using soft β^--emitter for labeling, the radiation intensity measured for a given specific surface area of the powder should be proportional to the amount of the powder sprinkled on the bottom of the measuring cell.

However, with increasing thickness the radiation intensity tends to a limiting value in consequence of self-absorption of the soft β-radiation as shown in Figure 9.8 for the case of the adsorption of ^{35}S-labeled sulfuric acid on hematite.

It may be seen from this figure that above 10 mg cm^{-2} layer thickness ($\sim 5 \times 10^{-3}$ cm) we attain the limiting value of the radiation intensity.

The determination of Γ values from radiation intensity data requires three independent measurements (60):

1. Measurement of radiation intensity (I_1) coming from the solution phase in the presence of the labeled species but in the absence of powder.
2. Measurement of the total limiting radiation intensity originating from the powder layer following the adsorption of the labeled species (I_2). In this case, owing to the thickness of the powder layer the detector "does not see"

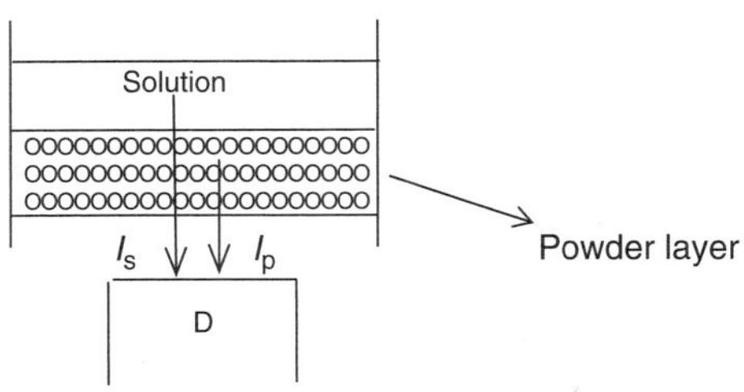

Figure 9.7 Scheme of the experimental setup in the case of powdered adsorbent: D, detector; I_s, radiation coming from the solution phase; I_p, radiation coming from the powdered oxide layer.

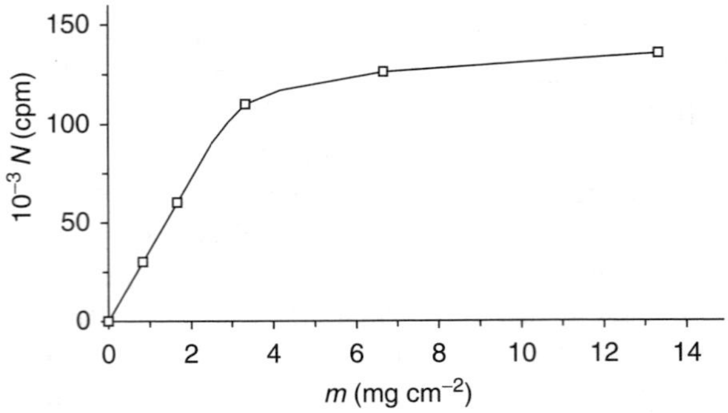

Figure 9.8 Count rate vs. surfacic density of hematite powder (referred to the geometric surface area) relationship. Labeled H_2SO_4: 2×10^{-4} mol dm^{-3} in 1×10^{-1} mol dm^{-3} $HClO_4$. (Radiation intensity is given in count per minute = cpm, m = mass of powder referred to the geometric surface area.)

I_1, that is, the radiation coming from the bulk solution phase is completely absorbed by the powder layer.

3. Measurement of the radiation intensity (I_3) without the adsorption of the labeled species on the powder surface or by addition of nonlabeled species to the solution phase, that is, by decreasing the specific activity to a very low value. In the case of the latter procedure, the addition of nonlabeled species, the decrease in specific activity does not result in a change in the radiation intensity coming from the solution phase as it follows from Equation (9.6). (A tenfold increase of c results in tenfold decrease in I_0.) In contrast, at a fixed Γ value (for instance, at saturation coverage) the radiation coming from the adsorbed species should decrease together with I_0 (see Equations (9.3) and (9.7)). This radiation corresponds to the radiation coming from the solution in the pores of the layer without the radiation coming from the bulk solution phase.

The schemes of the experimental setup in the case of the measurement of I_1, I_2, and I_3 are shown in Figure 9.9.

In accordance with Ref. (60), the following equations can be given:

$$I_1 = \alpha I_0 qc/\mu\bar{\rho}_0 \tag{9.6}$$

$$I_2 = (\alpha I_0 q/\mu\bar{\rho})[\varepsilon c + (1-\varepsilon)\rho_m a_s \Gamma] \tag{9.7}$$

$$I_3 = \alpha I_0 qc\varepsilon/\mu\bar{\rho} \tag{9.8}$$

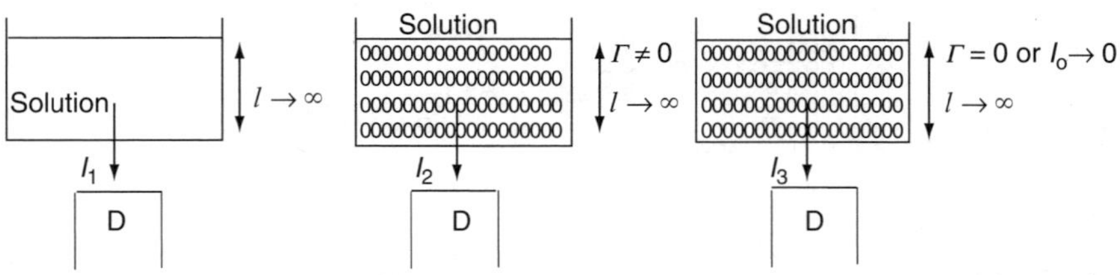

Figure 9.9 Schemes of the experimental setup in the case of the measurement of I_1, I_2, and I_3 (Equations (9.6), (9.7), and (9.8), respectively).

where α is the proportionality factor, I_0 is the specific activity of the adsorbate, μ is the mass absorption coefficient, c is the concentration of labeled species, ρ_0 is the density of the solution phase, ρ_m is the density of the dispersed material (in compact form), $\bar{\rho}$ is the average density of the layer ($\varepsilon\rho_0 + (1-\varepsilon)\rho_m$), ε is the porosity of the layer, a_s is the specific surface area of the powder, Γ is the extent of the adsorption, and q is the surface area of the bottom of the cell where the powder is dispersed.

The intensity originating from the adsorption (I_4) is

$$I_4 = I_2 - I_3 = [\alpha q I_0 (1-\varepsilon)\rho_m/\mu\bar{\rho}]a_s\Gamma \tag{9.9}$$

The porosity of the layer can be determined from the ratio

$$\frac{I_1}{I_3} = \frac{\bar{\rho}}{\varepsilon\rho_0} = \frac{\varepsilon\rho_0 + (1-\varepsilon)\rho_m}{\varepsilon\rho_0} \tag{9.10}$$

Denoting I_1/I_3 by A we obtain

$$\varepsilon = \frac{\rho_m}{(A-1)\rho_0 + \rho_m} \tag{9.11}$$

On the other hand,

$$\frac{I_4}{I_3} = \frac{1-\varepsilon}{\varepsilon}\frac{\rho_m a_s \Gamma}{c} \tag{9.12}$$

Denoting I_4/I_3 by B we obtain

$$B = (A-1)\rho_0 \frac{a_s\Gamma}{c} \tag{9.13}$$

$$\Gamma = \frac{B}{(A-1)\rho_0}\frac{c}{a_s} \tag{9.14}$$

In the case of layer thickness corresponding to the limiting radiation intensity, a very high reproducibility of the measurements can be realized (if adsorption equilibrium is attained). Despite an intensive stirring up of the solution resulting in the dispersion of the oxide layer in the liquid phase, after a short deposition period, the same radiation intensity can be measured as shown in Figure 9.10.

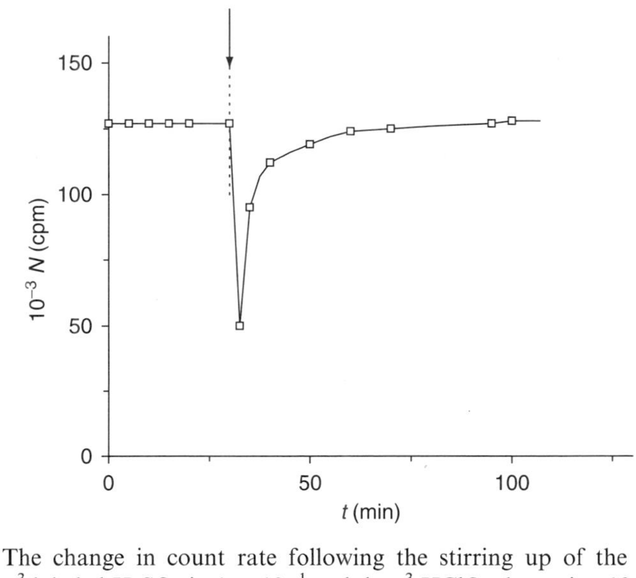

Figure 9.10 The change in count rate following the stirring up of the oxide layer; 1.5 $\times$ 10^{-4} mol dm^{-3} labeled H_2SO_4 in 1 $\times$ 10^{-1} mol dm^{-3} $HClO_4$; hematite: 10 mg cm^{-2}.

It may be seen from this figure that as a result of stirring the powder particles with adsorbed labeled species are lifted (up) from the bottom of the cell and the measured count rate decreases as the detector does not "see" these particles. Following the slow deposition of the powdered adsorbent the count rate increases and attains its original value.

A very important advantage of the powder method is that the slow dissolution of the powder (for instance, metals or oxides in acidic medium) does not exert any influence on the count rate if the amount of the powder is in the region of the limiting radiation intensity.

All these results verify that the method can furnish reliable results (for the determination of the character of pH and concentration dependence of the adsorption on various nonconducting or conducting powders).

However, in the case of a specimen available in the form of compact metal, a serious drawback is that the application of the method should be preceded by the preparation of a powder from the compact metal.

9.2.1.3.3 Polymer Films. A great variety of methods has been used for the investigation of polymer films on electrodes and one of these is the radiotracer method (55, 113–121), which enables us to obtain information on the ionic charge transport in electroactive polymer films labeling co- and/or counter ions as well as on the formation and destruction of the films formed from labeled monomers. The main problems were (119) embedding of electrolytes during the formation of polymer films, study of the motion of counter- and co-ions in the course of electrochemical measurements, and study of the exchange processes. For a more detailed discussion of the above problems, we refer the reader to Ref. (122).

The results of radiotracer polymer film studies could be of importance in corrosion science as well if we take into consideration that conducting polymers can form a corrosion protective coating for metals, especially for steel or anodized aluminum.

9.2.2 β-Backscattering: A Radiochemical Tool for the *In Situ* Study of the Electrochemical Formation and Dissolution of Thin Layers

The backscattering of β-radiation is a well known and widely studied phenomenon in the field of nuclear physics and serves as a basis for various analytical methods. This phenomenon is often used to measure the thickness of thin layers formed on various surfaces. In some cases changes in layer composition can be determined by measuring the intensity of the backscattered radiation.

The phenomenon of backscattering follows from the very nature of processes occurring in the case of interaction of β-particles with a material. A significant part of the β-radiation is absorbed; however, a great number of β-particles are scattered in various directions. Those particles that are scattered in a direction about 180° (in comparison to the direction of the incident radiation) are the so-called backscattered particles forming the "backscattered radiation." The ratio of the intensity of the backscattered radiation to the incident radiation depends strongly on the thickness, density, and atomic number of the scattering material.

Surprisingly, for a long period, β-backscattering has not been applied to the *in situ* study of electrode processes leading to the formation or dissolution of layers on a support and it has not been used to study transformations resulting in changes in the composition of layers covering electrode surfaces.

In Refs. (123, 124), it was shown that, under appropriately chosen experimental conditions, β-backscattering could be used to follow the electrochemical formation or dissolution of thin metal layers in a cell originally developed for radiotracer adsorption studies. ^{99}Tc was used as the β-radiation source ($E = 0.29$ MeV).

Layers formed by electrodeposition from solutions containing Bi^{+}, Ag^{+}, Cu^{2+}, and ReO ReO_4^- ions were studied. The supporting electrolytes were 1 mol dm^{-3} H_2SO_4 and $HClO_4$ solutions.

9.2.2.1 Principle of the Method

Only a brief survey can be given here of some of the main features of β-backscattering.

Two arrangements of the β-source can be considered: (i) internal, and (ii) external.

In the case of an internal source, the electrode surface or a layer deposited on it (support) is the source of β-radiation and the backscattering by a layer formed on the support (scatterer) is measured. A schematic representation of this arrangement is shown in Figure 9.11.

In this case the measured radiation intensity (I_T) consists of the intensity I of the direct radiation and the intensity I_b of the backscattered radiation:

$$I_T = I + I_b \tag{9.15}$$

In the case of an external source, the situation is quite different and there are several options depending on the position of the source. Two possible arrangements of the β-source are shown in Figure 9.12.

In this case the measured radiation intensity consists of the intensity I_b^s of the radiation backscattered by the support and the contribution ΔI_b to the intensity of backscattered radiation caused by the layer formed on the support:

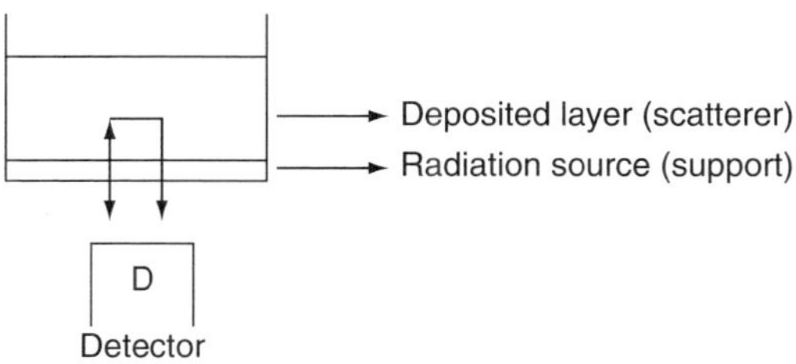

Figure 9.11 β-Backscattering: the schematic representation of the arrangement in the case of an internal source (support). (From *J. Electroanal. Chem.* G. Horányi, Elsevier, 1994. With permission.)

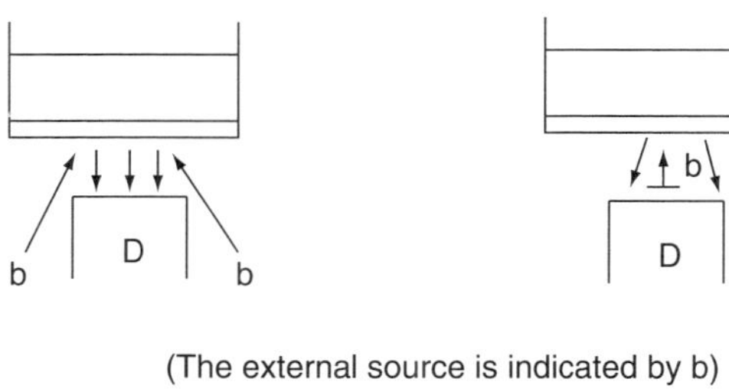

Figure 9.12 β-Backscattering: the scheme of arrangements in the case of external radiation source. (From *J. Electroanal. Chem.* G. Horányi, Elsevier, 1994. With permission.)

$$I_T = I_b^s + \Delta I_b \tag{9.16}$$

9.2.2.2 An Illustrative Example

In the present discussion, the characteristic features of backscattering will be considered in connection with the application of an internal β-source using a ^{99}Tc layer to this end.

The first step is the preparation of the radiation source by the electrodeposition of a ^{99}Tc layer on the gold-plated plastic foil forming the bottom of the cell.

The formation of this layer, the radiation source, can be monitored by measuring the increase in radiation intensity during the electrodeposition of the layer. In the case of thin layers ($d < 10^{-4}$ g cm^{-2}), the radiation intensity is proportional to the thickness *d* of the layer, which can be calculated by the usual method. Figure 9.13 shows the relationship between the amount deposited (layer thickness) and the radiation intensity measured. It follows from these results that a relatively high count rate can be attained at low thickness values (10^{-5} g cm^{-2}).

Since the electrodeposited ^{99}Tc layer serves as a support for the electrochemical deposition of other species, it is important to explore its stability range. Figure 9.14 shows cyclic voltammetric curves obtained with a ^{99}Tc layer and the count rate vs. potential curve in the potential range from 0 to 600 mV for a 1 mol dm^{-3} H_2SO_4 supporting electrolyte.

It can be seen from this figure that the ^{99}Tc layer may be considered as an inert support in the potential range studied. However, at potentials above 600 mV dissolution of the ^{99}Tc layer takes place as can be seen in Figure 9.15, where a positive sweep and the corresponding radiometric curve are shown. It follows from these results that the β-radiation source can be removed from the gold surface and the renewed surface can be used for other studies without any complications.

Electrodeposition from solutions of Cu^{2+}, Ag^{+}, ReO_4^-, and Bi^{3+} ions was studied. It follows from the discussion of the fundamental phenomena (123) that the higher the atomic number of the scatterer the higher will be the sensitivity of the method.

Curves a, b, and c in Figure 9.16 show how the intensity of the backscattered radiation changes with the thickness of the deposited layer (calculated from the charge used in the deposition process).

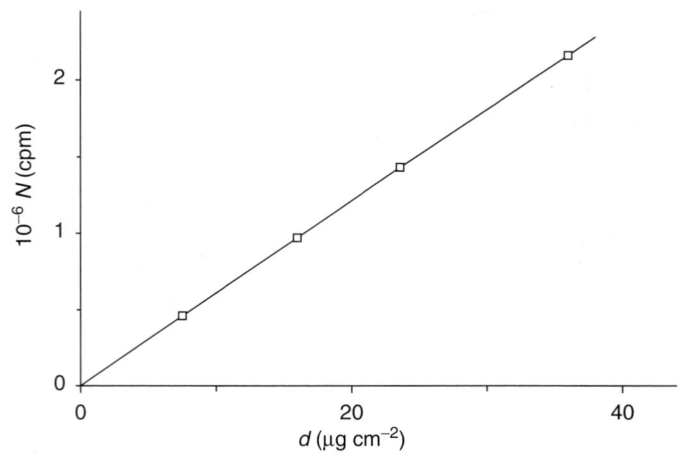

Figure 9.13 The relationship between the intensity of radiation coming from a ^{99}Tc layer and the amount (thickness) of material deposited. (From *J. Electroanal. Chem.* G. Horányi, Elsevier, 1994. With permission.)

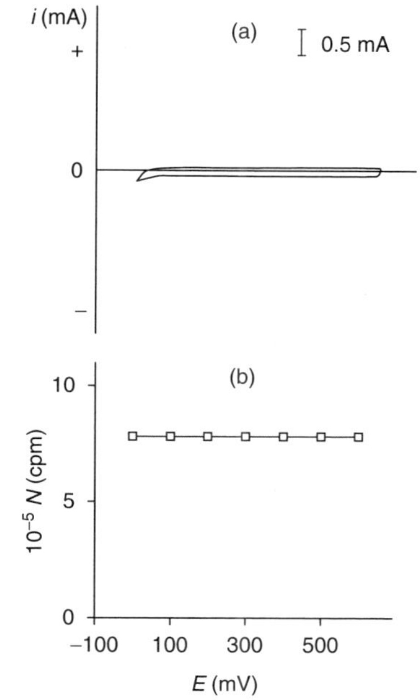

Figure 9.14 (a) Voltammetric curve (sweep rate, 5 mV sec^{-1}); (b) count rate vs. potential curve at a ^{99}Tc layer in the potential range 0 to 600 mV (on RHE scale) in 1 mol dm^{-3} H_2SO_4 supporting electrolyte. (From *J. Electroanal. Chem.* G. Horányi, Elsevier, 1994. With permission.)

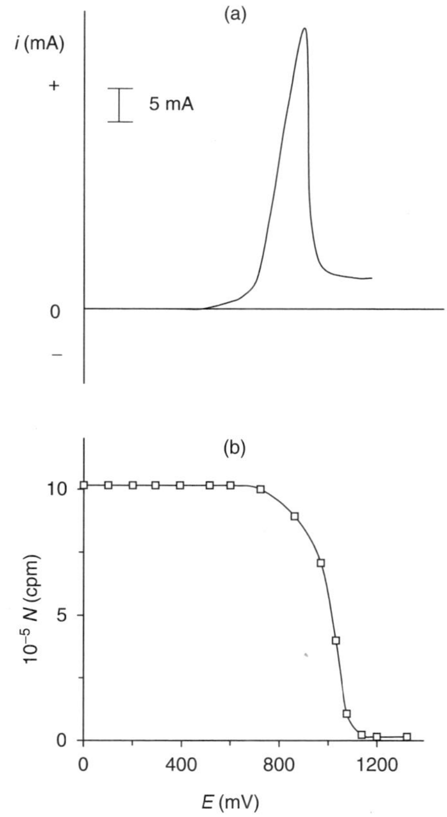

Figure 9.15 The dissolution of a ^{99}Tc layer during an anodic sweep (2.5 mV sec^{-1}): (a) voltammetric curve; (b) count rate vs. potential (RHE scale). (From *J. Electroanal. Chem.* G. Horányi, Elsevier, 1994. With permission.)

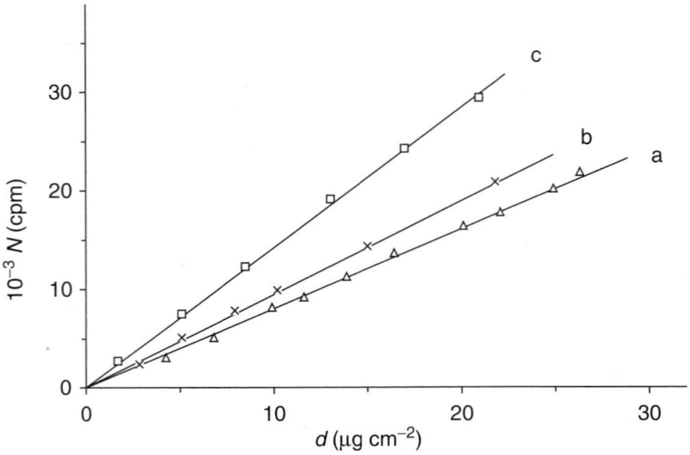

Figure 9.16 The dependence of the intensity of the backscattered radiation on the thickness *d* of the deposited layer: (a) Cu; (b) Ag; (c) Bi. (From *J. Electroanal. Chem.* G. Horányi, Elsevier, 1994. With permission.)

The intensity ΔN of the backscattered radiation at a given thickness is equal to the difference between the actual count rate N and the initial count rate N_0, that is, $\Delta N = N - N_0$. As the statistical error of the measurements is $\sigma = \sqrt{N} + \sqrt{N_0}$ in the case of initial intensities of 10^6 counts min^{-1} a very high sensitivity can be attained when the radiation intensity for a given thickness is measured for 5 to 10 min. It follows from a comparison of curves a, b, and c in Figure 9.16 that the higher the atomic number of the scatterer the higher is the slope of the ΔN–d relationship.

It is unlikely that the electrodeposited layers will have a strictly uniform thickness. Thus, the "thickness" values reported here should be considered as average values.

The results obtained with Re species cannot be expressed in terms of the amount deposited as the composition and valency state of the deposited species are unknown. Nevertheless, a continuous increase in the intensity of the backscattered radiation can be observed during the electroreduction of ReO_4^- species, which is shown in Figure 9.17.

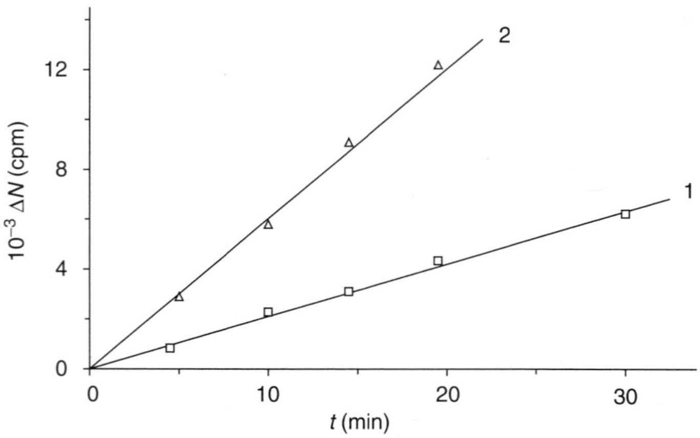

Figure 9.17 Increase in the intensity of the backscattered radiation during the electroreduction of ReO_4^- ions (1×10^{-3} mol dm^{-3}) at (1) 0 and (2) −100 mV (RHE scale) in 1 mol dm^{-3} $HClO_4$ solution. (From *J. Electroanal. Chem.* G. Horányi, Elsevier, 1994. With permission.)

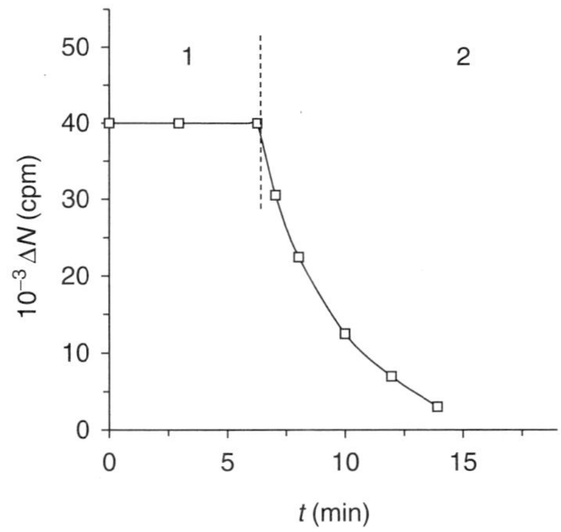

Figure 9.18 Change in the intensity of the backscattered radiation during the dissolution of an electrodeposited Re species following a potential switch from (1) 0 mV to (2) 500 mV. (From *J. Electroanal. Chem.* G. Horányi, Elsevier, 1994. With permission.)

It can be shown that not only the electrodeposition but also the dissolution of a layer can easily be monitored by measuring the changes in the radiation intensity. Figure 9.18 shows the results obtained during the dissolution of a layer of Re species. Simultaneous voltammetric and radiometric (voltradiometric) studies provide important information on the charge involved in the anodic dissolution process.

The results obtained from a voltradiometric study of the dissolution of a Bi layer are shown in Figure 9.19.

The periodic deposition and dissolution processes occurring during cyclic voltammetric measurements in the presence of Bi^{3+} ions can be demonstrated clearly by the changes in the backscattered radiation as shown in Figure 9.20.

Figure 9.20 clearly shows that in both directions above 200 mV a decrease in the count rate can be observed owing to the dissolution of the electrodeposited Bi while under 200 mV the increase is connected with electrodeposition of Bi.

The potentialities of the method using external sources with the application of five different β-emitters (^{36}Cl, ^{60}Co, ^{99}Tc, ^{134}Cs, and ^{204}Tl) were also investigated (124).

The experimental results reported in Ref. (124) demonstrated that β-backscattering, using an external β-source, can be a useful tool for the *in situ* study of electrodeposition and dissolution of thin layers. The most important advantage of this method over other methods using internal β-sources is that only minimal handling of radioactive solutions is required. The sealed sources can be used several times and the treatment of radioactive waste can be avoided. The other advantage of the external radiation source over the internal source is that a larger number of substances can be used for electrochemical studies. Thus, it offers a significant extension of the applicability of this method. Following appropriate calibration the backscattered radiation intensity values can be unambiguously transformed into layer thickness values.

9.2.3 Some Other Nuclear Methods

For a long period neutron irradiation was the most important tool to produce radiotracers for study of wear and corrosion of various parts of a machine without the necessity of stopping and disassembling it.

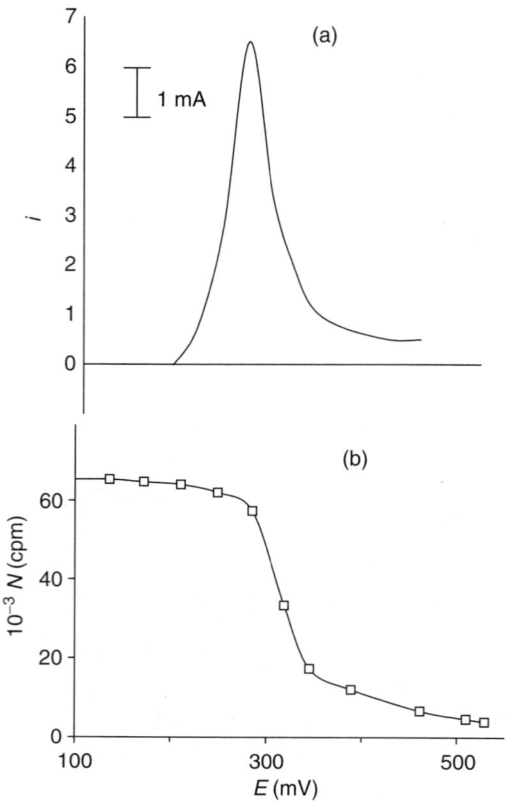

Figure 9.19 Dissolution of a Bi layer during an anodic sweep (sweep rate, 0.6 mV sec^{-1}): (a) current vs. potential relationship (on RHE scale) and (b) the simultaneous change in the intensity of the backscattered radiation. (From *J. Electroanal. Chem.* G. Horányi, Elsevier, 1994. With permission.)

In the last years, the most used methods have been thin layer activation (TLA) and ultrathin layer activation (UTLA) techniques (25, 126).

TLA is an ion beam technique. This method consists of an accelerated ion bombardment of the surface of interest of a machine part. Radioactive tracers are created by nuclear reactions in a well-defined volume of material. Loss of material owing to wear, corrosion, or abrasion phenomenon is characterized by monitoring the resulting changes in radioactivity.

There are two basic methods for measuring the material loss by TLA technique. One of them is based on remnant radioactivity measurements using a previously obtained calibration curve. The second is based on measuring the increasing radioactivity in the surrounding liquid phase.

The UTLA method is based upon the principle of recoil implantation by recoiling applied to radioactive heavy ions generated by a beam of light mass particles (p, d, ^{3}He, ^{4}He). These charged particles are easy to produce by an isocronous variable energy cyclotron. The most commonly used energies do not exceed 40 MeV.

A thin target (a few micrometers) of elementary composition **A** is bombarded by the primary beam and is activated following the nuclear reaction **A(a, b)B**. Some generated radioactive heavy ions **B** acquire sufficient kinetic energy (maximum

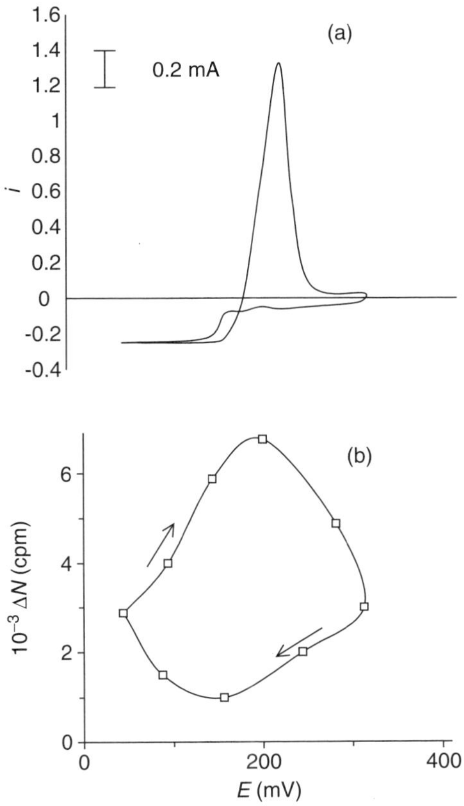

Figure 9.20 (a) Cyclic voltammetric and (b) voltradiometric curves obtained in the presence of 2×10^{-4} mol dm^{-3} Bi^{3+} in 1 mol dm^{-3} H_2SO_4 (sweep rate, 0.25 mV sec^{-1}) (potential on RHE scale). (From *J. Electroanal. Chem.* G. Horányi, Elsevier, 1994. With permission.)

energies of a few hundred kiloelectronvolts to a few megaelectronvolts) to recoil out of the target and be implanted in the material. Some of the of suitable nuclear reactions for the application of the UTLA method are as follows: ^{65}Cu(p, n)^{65}Zn, ^{56}Fe(p, n)^{56}Co, ^{48}Ti(p, n)^{48}V, ^{59}Co(p, pn)^{58}Co, and ^{55}Mn(α, n)^{58}Co.

The UTLA method presents numerous advantages compared to direct activation methods (126):

1. The activation being independent of the activated material composition and the deposition methods permitting to deposit a wide range of materials, the UTLA method may be applied to all kinds of materials. Moreover, the potentially damaging effects are far less than those induced by direct activation techniques using light or heavy ions.
2. The wide range of radioisotopes that can be generated and implanted allows carrying out investigations corresponding to various experimental requirements (depth to activate, chemical nature of the radioelement in the case of selective corrosion, etc.).
3. The generated activities are very low (a few kilobecquerels). Radioprotection precautions are therefore considerably reduced.
4. The use of light particle beams (p, d, ^{3}He, α) is common to a vast majority of accelerators.

It is, however, a drawback that the implantation takes place in a vacuum chamber; the application fields of this method are limited to reduced-dimensions samples.

An interesting application of the classical neutron activation method for the study of protective ability of paint coatings could be mentioned (127).

Investigations of the partial rates of corrosion and selective transfer of the metal substrate components and impurities into a given coating and medium are of extreme importance when developing corrosion proof steels and alloys as well as selecting constructive materials and coatings for equipment for the food and high-purity material industries. Such data are crucial insofar as the corrosion mechanism and the effectiveness of the organic coating are concerned.

The experimental procedure involves the following steps:

1. The production of radionuclides by bombardment of the metal specimens with neutron flux in a reactor.
2. Application of coating on the radioactive specimens in accordance with common application techniques.
3. Corrosion-electrochemical tests of the coated and uncoated specimens in different media with regular sampling for the presence of labeled corrosion products.
4. Gamma-spectrometric analysis of corrosion products to obtain data on physicochemical characteristics of the processes.

9.3 DETECTION OF INTERMEDIATE STATES FORMED IN THE COURSE OF DISSOLUTION OF METALS THROUGH THE RADIOTRACER STUDY OF ANION ADSORPTION

Study of the adsorption of radiolabeled anions in the course of dissolution of metals may contribute to the mechanistic interpretation of the phenomena observed.

Two illustrative examples will be presented in this chapter.

9.3.1 Radiotracer Evidences Proving the Role of Cu^+ Ions in the Behavior of the Cu^{2+}–Cu System (112)

In a recent work (112), combined EQCM and voltammetric studies in conjunction with radiotracer adsorption investigations were carried out in order to obtain direct information on the interfacial behavior of the Cu^{2+}–Cu system equilibrium and the dynamic conditions with the aim of clarifying the role and participation of Cu^+ ions formed by the interaction of cupric ions and copper metal. The appearance and accumulation of Cu^+ ions in the solution phase is clearly demonstrated through radiotracer experiments using ^{36}Cl labeled Cl^- ions.

The investigation of the specific adsorption of anions on a copper electrode by the radiotracer technique has been the subject of several studies (60). With the use of radiolabeled chloride ions the formation of CuCl on the surface of copper can be easily followed. The technique elaborated for powdered metals was used. In this case direct *in situ* observations can be carried out by measuring only the intensity of the radiation emitted by the isotope used. The main tendencies characteristic of the processes occurring at solid–liquid interface can be easily demonstrated without

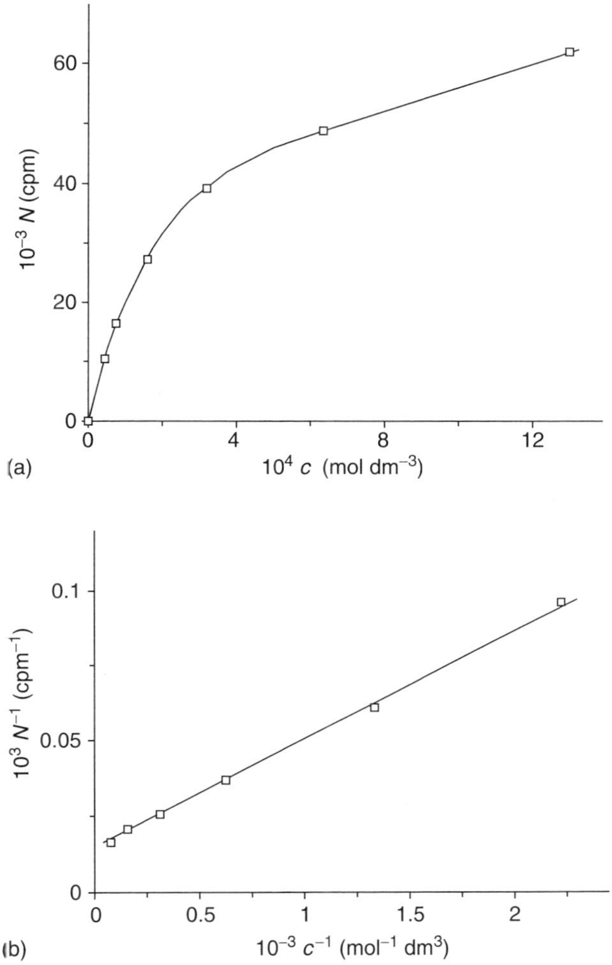

Figure 9.21 (a) Concentration dependence of the adsorption of labeled chloride ions expressed in radiation intensity: 50 mg cm^{-2} copper powder, 1 mol dm^{-3} $HClO_4$ + 1 × 10^{-4} mol dm^{-3} $Cu(ClO_4)_2$; (b) $1/I$ vs. $1/c$ plot of the data. (From *J. Electroanal. Chem.* G. Lang et al., Elsevier, 1994. With permission.)

detailed quantitative knowledge of the extent of adsorption and the specific surface area of the powder studied.

Figure 9.21(a) shows the concentration dependence of the adsorption of Cl^- ions in the presence of 1 *M* perchloric acid. Changes in the extent of adsorption are reflected by the corresponding changes in radiation intensities. As the data presented in Figure 9.21(b) give a linear relationship in the $1/I$ vs. $1/c$ plot, the concentration dependence of the adsorption can be interpreted in terms of a Langmuir isotherm.

It follows from these figures that the system studied behaves "normally"; no indication of processes differing from regular adsorption can be observed.

However, further experiments carried out at higher chloride and/or copper ion concentrations furnished results that cannot be reconciled with the picture based exclusively on adsorption phenomena.

The study of the exchange of adsorbed labeled species with nonlabeled species added to the solution phase is a widely used method in the investigation of the

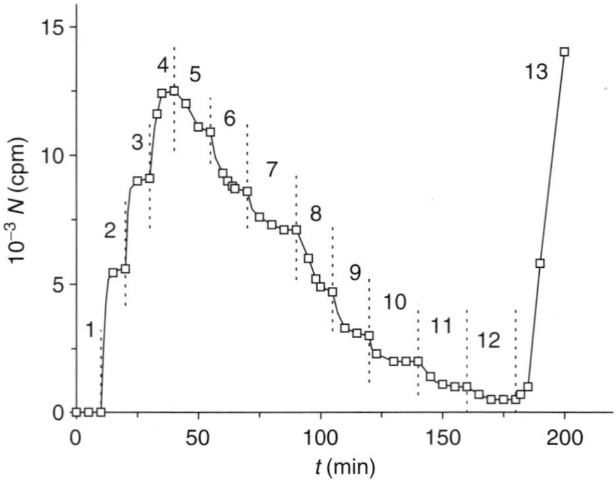

Figure 9.22 Count rate vs. time curve following the addition of labeled and nonlabeled chloride ions. Supporting electrolyte: 1 mol dm^{-3} $HClO_4$ + 10^{-4} mol dm^{-3} $Cu(ClO_4)_2$. Addition of labeled HCl: (1) 0; (2) 1; (3) 2; (4) 4 × 10^{-5} mol dm^{-3}. Addition of nonlabeled NaCl: (5) 0.8; (6) 1.6; (7) 3.2; (8) 6.4; (9) 13; (10) 26; (11) 52; (12) 104; (13) 208 × 10^{-4} mol dm^{-3}. (From *J. Electroanal. Chem.* G. Lang et al., Elsevier, 1994. With permission.)

mobility of adsorbed species and, consequently, for the demonstration of the simultaneous occurrence of adsorption and desorption processes determining the adsorption equilibrium that allows the treatment of the data in terms of an adsorption isotherm.

Figure 9.22 demonstrates that in the case of the adsorption of chloride ions, the gradual isotope dilution (i.e., the gradual addition of nonlabeled chloride to an equilibrium system containing radiolabeled species) at low concentration leads to phenomena expected on the basis of the isotherm presented in Figure 9.21(a). The decrease in the count rate measured follows from the very fact that the increase in the adsorption with increasing total chloride concentrations is overcompensated by the decrease of the specific activity. At concentrations where the coverage with respect to the adsorbed species is not far from its limiting value the dilution should lead to very low, practically zero, intensity values. This reflected by sections 10, 11, and 12 in Figure 9.22. However, above a certain concentration value a dramatic increase in the radiation intensity can be observed as is shown by section 13 in Figure 9.22. The amount of the chloride ions corresponding to this increase is by several orders of magnitude higher than that corresponding to a monolayer coverage. This means that a chloride-containing thick layer is formed, which should be a cuprous chloride layer.

Figure 9.23 shows that the addition of copper ions to an equilibrium adsorption system also leads to a significant increase in the radiation intensity, which cannot be ascribed to simple adsorption phenomena. The section following the addition of nonlabeled Cl^- (arrow 1, Figure 9.23) similarly to sections 5 to 12 in Figure 9.22 shows the effect of isotope dilution on the count rate in the absence of copper ions. Addition of copper ions to the solution phase (arrow 2) at a concentration commensurate with that of chloride ions results in a rapid increase in the radiation intensity indicating the formation of a Cl^--containing insoluble product (CuCl) on the metal surface. In agreement with the observations made in connection with the results presented in Figure 9.22, further increase in the concentration of chloride ions is

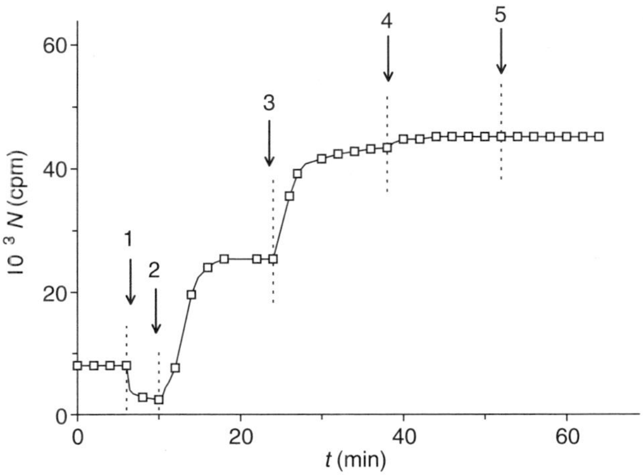

Figure 9.23 Effect of addition of Cu^{2+} and Cl^- ions on the count rate: initial state 2×10^{-4} mol dm^{-3} labeled HCl in 1 mol dm^{-3} $HClO_4$; (1) addition of 1×10^{-3} mol dm^{-3} NaCl (nonlabeled); (2) addition of 4×10^{-2} mol dm^{-3} $Cu(ClO_4)_2$; (3) addition of 2×10^{-2} mol dm^{-3} NaCl (nonlabeled); (4) addition of 6×10^{-2} mol dm^{-3} $Cu(ClO_4)_2$; (5) addition of 2.5×10^{-1} mol dm^{-3} HCl (nonlabeled). Additions were achieved at moments indicated by the arrows. (From *J. Electroanal. Chem.* G. Lang et al., Elsevier, 1994. With permission.)

connected with the further increase in the radiation intensity as well. This means that the thickness of the layer formed on the surface increases significantly. The addition of copper ions, however, exerts no further influence on the count rate. The last section was obtained following the addition of a significant amount of nonlabeled HCl (arrow 5). As the radiation intensity remains unchanged it can be stated that the mobility of the labeled species in the layer formed is very low, and no further thickening of the layer takes place.

The surface layer can be eliminated by repeated washing with NaCl solution or with a concentrated HCl solution as shown in Figure 9.24.

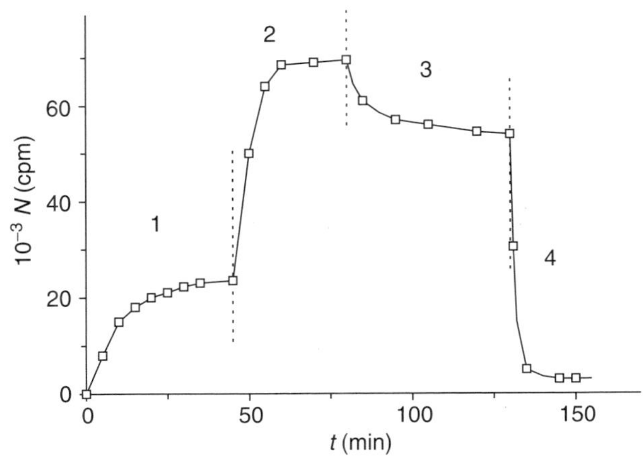

Figure 9.24 Formation and elimination of labeled CuCl layer: (1) addition of 10^{-2} mol dm^{-3} $Cu(ClO_4)_2$ to the system 1 mol dm^{-3} $HClO_4 + 2 \times 10^{-2}$ mol dm^{-3} NaCl; (2) further addition + 4×10^{-2} mol dm^{-3} $Cu(ClO_4)_2$; (3) washing of the Cu powder with water; (4) addition of concentrated HCl. Final concentration 1.5 mol dm^{-3}. (From *J. Electroanal. Chem.* G. Lang et al., Elsevier, 1994. With permission.)

Taking into consideration the results obtained from combined voltammetric and EQCM studies the following mechanistic conclusions were drawn. The interaction of Cu^{2+} ions with copper surfaces leading to the formation of Cu^{+} species involves adsorbed anions. This is well demonstrated by radiotracer experiments with Cl^{-} ions. Depending on the concentration of chloride ions, some surface sites on copper are occupied by adsorbed chloride species. The interaction of adsorbed chloride ions with cupric ions can be given by the following equation: Cu^{2+} (solution) $+ 2\,Cl^{-}$(ads.) $+ e^{-} =$ CuCl(surface) $+ Cl^{-}$(solution) while cuprous ion should appear in accordance with CuCl(surface) $= Cu^{+}$(solution) $+ Cl^{-}$(solution) and CuCl(surface) $+ Cl^{-}$(solution) $= CuCl_2^{-}$ (solution). (The assumption of the reaction Cu^{2+}(sol) $+ Cl^{-}$(ads) $+ e^{-} =$ CuCl(surface) should be rejected as the transfer of Cu^{2+} from the solution phase to the surface should be connected with the arrival of two electrons to the interface.)

The formation of CuCl(surface) is completely reversible under conditions where no formation of bulk CuCl occurs. This is reflected by the apparent adsorption isotherm of Cl^{-} ions shown in Figure 9.21(a) and by the mobility of adsorbed species shown in Figure 9.22.

The appearance and accumulation of Cu^{2+} ions in the solution phase are clearly demonstrated by the experiments presented in Figure 9.23 and Figure 9.24. Both the increase in Cu^{2+} and Cl^{-} concentrations results in an increase of radiation intensity that can be explained only by the massive deposition of CuCl formed from cuprous ions present in the solution phase as a result of interaction of Cu^{2+} ions with the copper surface.

9.3.2 Specific Adsorption of Radiolabeled Anions Indicating the State of the Electrode Surface in the Course of Deposition and Dissolution of Cd

The specific adsorption of ^{36}Cl labeled Cl^{-} ions and ^{35}S labeled HSO_4^{-} ions was studied in 1 mol dm^{-3} $HClO_4$ supporting electrolyte in the presence of Cd^{2+} ions at a gold substrate in a wide potential range corresponding to the electrodeposition, alloy formation, upd of Cd species, and the existence of an adatom free surface.

The problem with Cd electrodeposition in acid medium is connected with the very fact that it should be accompanied by simultaneous hydrogen evolution owing to the low standard potential of the Cd^{2+}/Cd system (−0.403 V). Fortunately, the upd of cadmium ions on gold results in a significant increase in the hydrogen overvoltage; thus, the deposition of Cd^{2+} ions can be carried out in strong acid media. A further problem is that the electrodeposited Cd layer is not stable under open circuit condition in acid medium. Its more or less rapid corrosion takes place according to the coupled reactions:

$$Cd \longrightarrow Cd^{2+} + 2e^{-}$$
$$2H^{+} + 2e^{-} \longrightarrow H_2$$

On the other hand, this very fact could be of interest as it can be considered as a model system for sorption studies in the course of rapid corrosion.

Figure 9.25 shows the potential dependence of Cl^{-} ion adsorption in the presence of Cd^{2+} in a wide potential range from −0.5 to +0.7 V. The curve, obtained by steady-state measurements under potentiostatic control starting from the most negative potentials where electrodeposition of Cd takes place, is composed of four

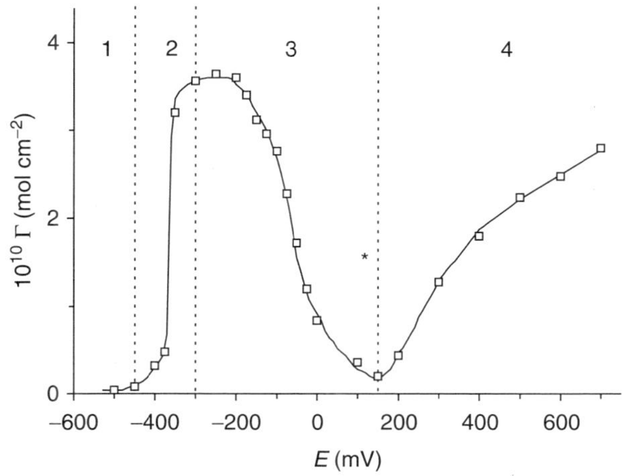

Figure 9.25 Potential dependence (RHE scale) of the adsorption of Cl^- ions ($c = 2 \times 10^{-5}$ mol dm^{-3}) in the presence of Cd^{2+} ions (8×10^{-3} mol dm^{-3}) in 1 mol dm^{-3} $HClO_4$ supporting electrolyte. (From *J. Solid State Electrochem.* G. Horányi, Springer-Verlag, 2002. With permission.)

distinct sections. In section 1, massive bulk deposition of Cd takes place and the low adsorption values indicate that no specific adsorption of anions occurs on the electrodeposited Cd layer. In section 2, a significant increase in anion adsorption can be observed by shifting the potential to more positive values. The decrease in anion adsorption with increasing potentials shown by section 3 is connected with the decrease of the coverage with respect to Cd adatoms that induce anion adsorption. The induced adsorption of anions by Cd adatoms is a well-known fact from the literature (2, 4, 60). Section 4 corresponds to the adsorption of chloride ions on pure gold surface.

Similar behavior was found in the case of labeled sulfate species.

A study of the open circuit behavior of the system (i.e., the spontaneous corrosion of the electrodeposited layer) gives an insight into the factors determining the observations presented above.

Figure 9.26 shows the potential versus time curve obtained following an electrodeposition at −600 mV, and thereafter the opening of the electric circuit.

Three well-defined sections can be distinguished in the curve presented in Figure 9.26. The first section corresponds to the dissolution of the bulk metal deposit and indicates the mixed potential of the two processes involved in the corrosion. The almost constant potential value characterizing this section could not be very far from the equilibrium Cd/Cd^{2+} potential owing to the high hydrogen overpotential on Cd. The second section, as it will be shown and discussed in detail, reflects the behavior of an Au–Cd alloy formed in the course of the deposition of Cd. In the potential range corresponding to the third section Cd should be present in form of adatoms.

The polarization behavior of the system in the potential ranges discussed above is shown in Figure 9.27. Evidently, no real steady-state polarization curves can be taken in the first two potential sections owing to the dissolution of bulk deposit (section 1) and alloy (section 2). Nevertheless, for short periods both anodic and cathodic branches can be observed. In the anodic branches the negative current corresponding to the discharge of protons is overcompensated by the dissolution of

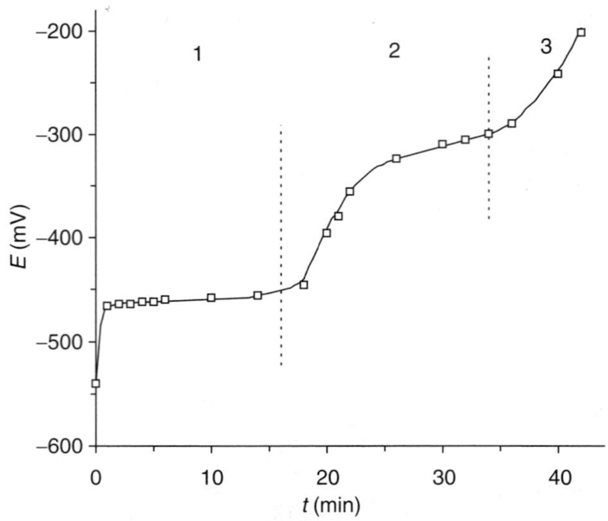

Figure 9.26 The potential vs. time curve following the electrodeposition of Cd at −600 mV and thereafter opening the electric circuit (data as in Figure 9.25). (From *J. Solid-State Electrochem.* G. Horányi, Springer-Verlag, 2002. With permission.)

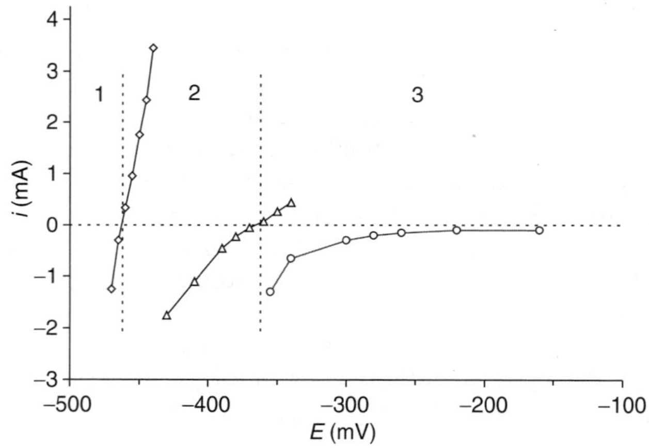

Figure 9.27 Polarization behavior of the system Au/Cd in the bulk electrodeposition region (1); the alloy formation region (2); and upd region (3) (potential on RHE scale). (From *J. Solid-State Electrochem.* G. Horányi, Springer-Verlag, 2002. With permission.)

Cd. In the region of upd (section 3) only the discharge of protons (the formation of H_2) could be the predominant process at negative potentials. Section 1 was obtained after bulk deposition of Cd at −600 mV while section 2 was determined following the dissolution of the main part of electrodeposited layer.

No adsorption of anions takes place on bulk Cd deposit and following a switch from a potential value where bulk deposition of Cd^{2+} takes place to the open circuit potential (i.e., interruption of the electric circuit) no change in the low anion adsorption value can be observed. Polarizing the electrode in small potential steps starting from the open circuit potential the results reflected by Figure 9.28 were obtained. In Figure 9.28, the current and radiation intensity values are plotted against time following potential switches in the presence of labeled sulfate ions. It may be seen

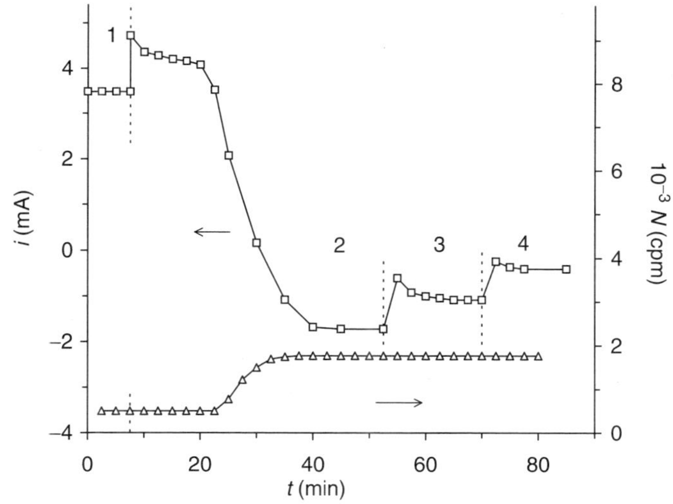

Figure 9.28 Simultaneous measurement of current and radiation intensity following potential (RHE scale) switches in the presence of labeled sulfate ions in the potential range where the dissolution of bulk Cd deposit takes place ($c_{H_2SO_4} = 2 \times 10^{-4}$, $c_{Cd} = 1 \times 10^{-2}$ mol dm^{-3} in 1 mol dm^{-3} $HClO_4$. (1) −440; (2) −430; (3) −410; (4) −390 mV. (From *J. Solid-State Electrochem.* G. Horányi, Springer-Verlag, 2002. With permission.)

from this figure that despite the anodic dissolution of the Cd layer no significant change in anion adsorption takes place. The extent of anion adsorption remains at a constant low level in a wide potential range as shown in by Figure 9.28. This potential range corresponds to that considered as "alloy formation region." Under potentiostatic control negative current flows, there is no bulk Cd on the surface.

At a certain potential value the appearance of a transient positive current and a continuous increase in the anion adsorption can be observed for a long period of time as shown in Figure 9.29. (The apparent Γ values are given owing to possible changes in the real surface area [see below]. Γ was calculated by the usual way, measuring the background radiation and using Equations (9.2) and (9.3).)

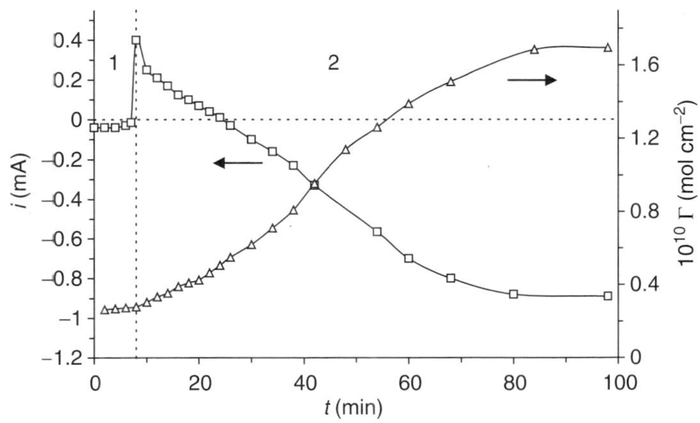

Figure 9.29 Current vs. time and adsorption (*Γ*) vs. time curves in the course of dissolution of Cd/Au alloy following a potential (RHE scale) switch from −365 to −360 mV (data as in Figure 9.28). (From *J. Solid-State Electrochem.* G. Horányi, Springer-Verlag, 2002. With permission.)

The data presented in Figure 9.29 clearly demonstrate a slow transformation indicating the slow oxidation of some kind of species. We have no other choice as to assume the oxidation of Cd dissolved in gold; that is, the dissolution of the Cd component of Cd/Au alloy formed during the electrodeposition of Cd^{2+} ions. The attainment of the steady state requires very long time (more than 1 h in the present case). This very fact seems to be in agreement with the assumption of alloy formation as in this case the slow transport of Cd atoms from the bulk alloy phase to the surface of the electrode could be the rate-determining step in the overall oxidation process. If dissolution of the alloy is completed there is no change in the extent of anion adsorption shifting the potentials by 100 to 150 mV to more positive values. This situation corresponds to maximum coverage with respect to Cd adatoms. A decrease in anion adsorption takes place only at potentials where the upd of Cd follows well-known patterns; that is, the coverage with respect to Cd adatoms decreases with increasing potentials and thus the extent of anion adsorption induced by the adatoms should decrease simultaneously (see the relevant literature on the adsorption of anions induced by adatoms (2, 4, 60)).

Alloy formation and thereafter its dissolution lead to a significant roughening of the gold surface reflected by an increase in the count rate corresponding to the anion adsorption. Figure 9.30 shows the increase in the count rate in the case of labeled chloride ions at potentials where the formation of adatoms can be neglected. The roughening was clearly demonstrated by the changes in the oxide layer reduction charge data.

9.4 STUDY OF THE ADSORPTION OF CORROSION INHIBITORS

9.4.1 Inorganic Inhibitors

9.4.1.1 Phosphate and Pertechnetate

A radiotracer study of the deposition of ^{32}P labeled phosphate and ^{99}Tc labeled pertechnetate (TcO_4^-) ions was reported in Ref. (128) on a 100-μm Al 1100 foil

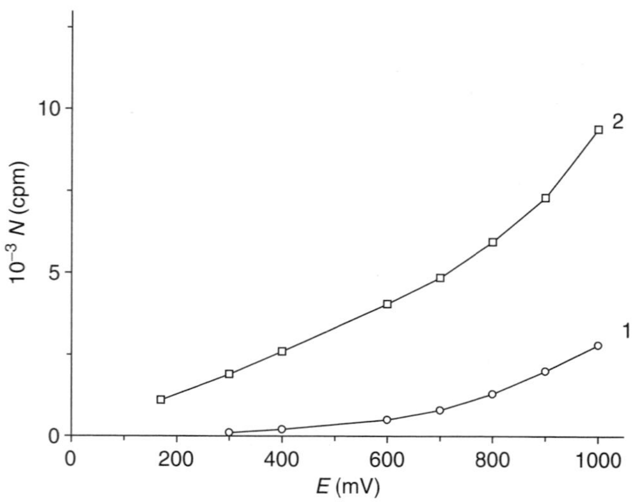

Figure 9.30 Radiation intensity measured on the gold surface before (1) and after (2) the alloy formation in the presence of labeled Cl^- species (data as in Figure 9.25). (From *J. Solid-State Electrochem.* G. Horányi, Springer-Verlag, 2002. With permission.)

positioned at the bottom of the measuring cell. The behavior of two corrosion-inhibiting oxy anions was compared as function of time and electrode potential.

In accordance with previous observations of the same group (129), it is assumed that the deposition of phosphate is initiated by the anodic dissolution of aluminum. The formation of technetium films was shown to proceed preferentially via cathodic processes and resulted in a greater extent of adsorption than that observed for phosphate.

It is assumed that the phosphate deposition is due to the result of adsorption, incorporation, and precipitation of insoluble phosphate compounds on the oxide film or corrosion products, which means that phosphate deposition is indirectly related to electrochemical reactions, that is, aluminum dissolution. The statement concerning phosphate deposition is in agreement with the results obtained with γ-Al_2O_3 (73) and is in correspondence with the reported pH dependence of phosphate adsorption on corroding Al and γ-Al_2O_3 (130).

On the other hand, the technetium deposition can be mainly ascribed to direct electrochemical reactions since pertechnetate adsorption is weak on aluminum oxide. The technetium deposit appears, in part, to be a film of $Tc(OH)_4$ formed by the reduction of pertechnetate. It can be assumed that a stepwise reduction of Tc(VII) species occurs (presumably in one-electron steps) and the consecutive charge transfer processes are accompanied by (a) complexation reactions, (b) disproportionation, (c) precipitation or deposition processes, and (d) transport phenomena.

A very rough approximation can be given by the following scheme (the oxidation state of the Tc species is marked in the simplest way as Tc(VII), Tc(VII), etc.):

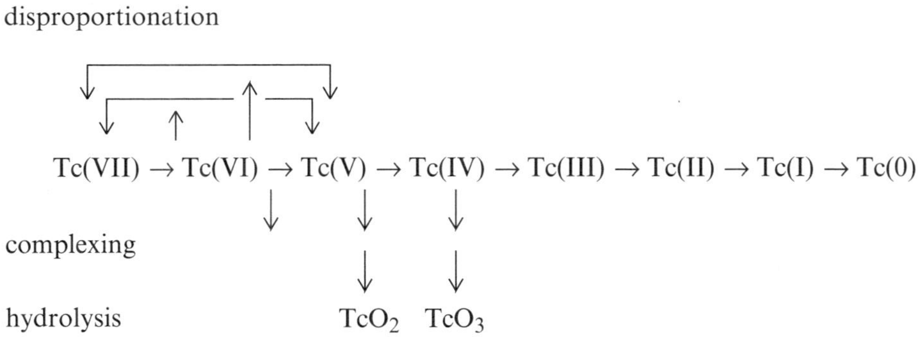

For instance, the existence of two parallel pathways in the reduction of Tc(IV) species in H_2SO_4 supporting electrolyte unambiguously follows from this scheme. As the stability of the Tc(IV) complex formed is relatively high in acid media, no further cathodic reduction of this anion can be expected. However, the ratio of the amount of complex Tc(IV) species in the solution phase to that of deposited Tc depends on the ratio of the rate w_1 of complexation to the rate w_2 of further reduction of Tc(IV) ions formed at the electrode surface:

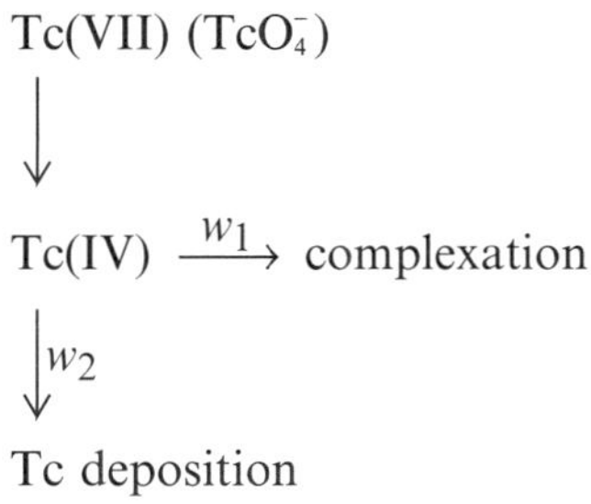

A parallel process to the complexation could be the hydrolysis of the Tc(IV) species:

$$Tc^{4+} + 2H_2O \longleftrightarrow TcO_2 + 4H^+$$

The occurrence of this reaction explains the formation of $TcO_2{\cdot}xH_2O$ deposit on the surface.

9.4.1.2 Adsorption of Chromate

As is well known, chromate conversion coating is used as a pretreatment for aluminum alloys; for instance, Alclad 2024 is used in aircraft applications. Although the composition and structure of these coatings are very complex, it may be assumed that chromate adsorption and absorption into Al_2O_3 formed on the surface and the adsorption processes on reduced chromate species, various Cr(III) oxides and hydroxides, could play an important role in the behavior of the coating. Thus, in first approximation, the study of chromate adsorption on Al_2O_3 and the investigation of anion adsorption on Cr_2O_3 may be suggested as models for adsorption studies.

The influence of solution pH on chromium(VI) deposition from 0.1 *M* NaCl + 0.1 *mM* Na_2CrO_4 solution onto Al 1100 alloy was studied under open circuit potential (OCP) condition using ^{51}Cr-labeled chromate species (131). The data obtained from the *in situ* radiochemical analysis were complemented by x-ray photoelectron spectroscopy (XPS) and electrochemistry. The results of these analyses indicated that both the kinetics and the extent of deposition are strongly pH dependent. Chromate deposition exhibited maximum surface coverage at pH 2. Chromate, even in 0.1 *mM* concentration, was found to effectively reduce the chloride content of the oxide film. In addition, the presence of chromate in the solution phase was found to alter the hydration and thickness of the passive film. XPS analyses revealed that chromate not only adsorbs but also electrochemically interacts with aluminum, creating a Cr(III,VI)-rich layer on the surface.

In order to gain a deeper insight into the role of aluminum and chromium oxides in the overall adsorption process an indirect radiotracer study of the adsorption of chromate ions on γ-Al_2O_3 and a direct study of the adsorption of sulfate ions on Cr_2O_3 using ^{35}S labeled H_2SO_4 were carried out by Horányi and Gáncs (74).

The results obtained reflect the specific adsorption of sulfate ions as ^{35}S-labeled sulfuric acid was present in low concentrations ($c < 10^{-3}$ mol dm^{-3}) in comparison to the large excess of perchlorate supporting electrolyte (0.25 to 1.0 mol dm^{-3}).

It has been found that the extent of adsorption is determined by the protonation of the surface sites, similar to the behavior of other oxides studied previously. A comparison of Cr_2O_3 and Al_2O_3 in this respect shows that the protonation of the former takes place at significantly lower pH values than that of the latter. In Figure 9.31 the pH dependence of sulfate adsorption on the two oxides is compared on a relative scale. These curves reflect the protonation equilibrium of the oxides. It follows from the comparison of the two curves that the protonation on Cr_2O_3 takes place mainly in the pH range where Al_2O_3 attains its limiting value. Therefore, it could be expected that in the case of the mixed presence of the two oxides the overall sorption behavior at pH values above 4 would be determined by that of Al_2O_3.

The indirect radiotracer study of the adsorption of chromate on Al_2O_3 was carried out using labeled sulfate ions as indicator species.

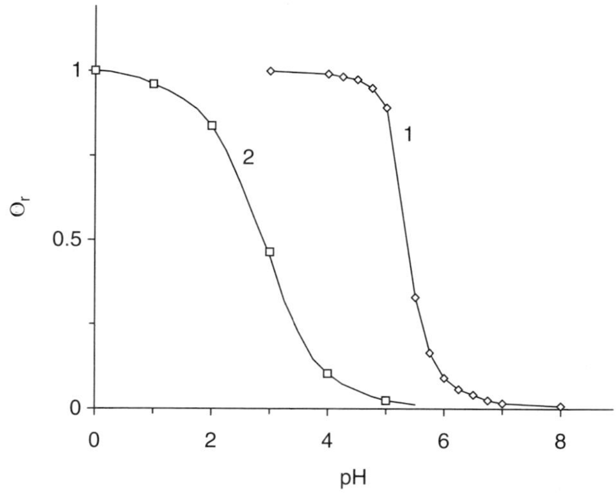

Figure 9.31 Comparison of the pH dependence of the adsorption of sulfate ions (relative coverage); the measured count-rate values were divided by the highest count rate obtained at lowest pH in the case of γ-Al_2O_3 (1) and Cr_2O_3 (2). (From *J. Solid State Electrochem.* G. Horányi, Springer-Verlag, 2002. With permission.)

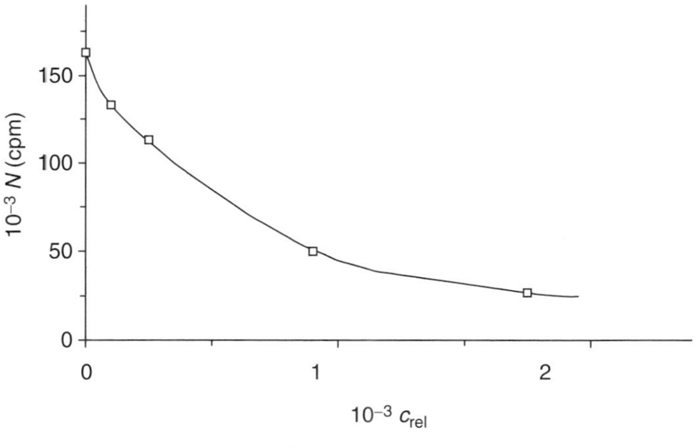

Figure 9.32 Effect of addition of CrO_4^{2-} species on the adsorption of labeled sulfate ions present in a concentration of 4×10^{-5} mol dm^{-3} in 0.25 mol dm^{-3} $HClO_4$ supporting electrolyte; 30 mg cm^{-2} Al_2O_3 (pH $\sim$ 0.7). Count-rate data are plotted against $c_{rel} = c_{CrO_4^-}/c_{SO_4^-}$.

The first step in the investigation was the study of the influence of chromate species on the adsorption of labeled sulfate ions. The result of this study is shown in Figure 9.32.

The data reported in Figure 9.32 were analyzed on the basis of theoretical considerations (60, 75) with the assumption that the adsorption behavior of both species, sulfate and chromate, can be described in terms of Langmuir-like behavior. The isotherm obtained for chromate is presented in Figure 9.33.

It follows from the results presented above that the adsorption strength of chromate species is lower than that of sulfate ions as a substantial displacement of adsorbed sulfate species occurs only at high CrO_4^- / SO_4^{2-} ratios. (It can be seen in Figure 9.32 that 50% of the adsorbed sulfate ions is displaced at a ratio of about 500.)

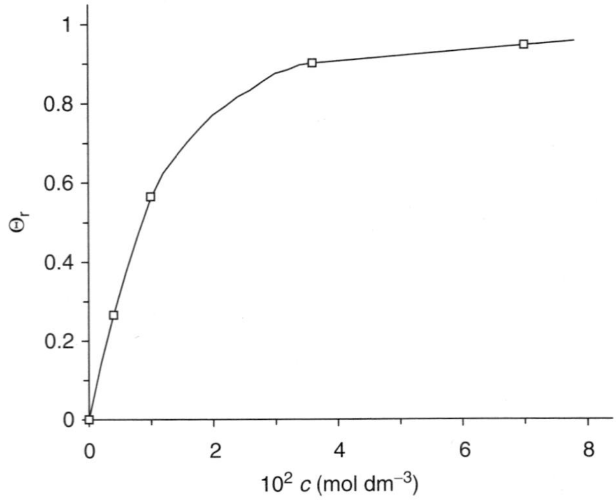

Figure 9.33 The calculated adsorption isotherm of CrO_4^{2-} species expressed in relative coverage (see caption of Figure 9.31). (From *J. Solid State Electrochem.* G. Horányi, Springer-Verlag, 2002. With permission.)

This means that in the presence of sulfate ions even at their low concentration no remarkable adsorption of chromate ions takes place at least under the condition studied here, as it is well known that the state of chromate species strongly depends on the pH.

The results obtained show that the adsorption strength of chromate species is low in comparison to sulfate ions. It follows from this that the results of the treatment of Al_2O_3-covered Al surfaces with chromate could be influenced by the presence of sulfate ions.

9.4.2 Organic Inhibitors

9.4.2.1 Study of the Adsorption of Organophosphono Compounds

Nowadays, the discharge of chemicals into rivers, lakes, or water reservoirs is controlled by strict environmental regulations. Owing to eutrophization, environmental regulations have also been proposed to decrease the phosphorous content of natural water, which means that even the application of nontoxic inorganic phosphates should be reduced. One way to achieve this is the application of organic phosphono compounds instead of inorganic phosphorous materials. The reason for this is that C–P bonds are much more resistant to hydrolysis in comparison to P–O bonds (132). A widely used model of organic phosphonic acid inhibitor is 1-hydroxyethane-1,1-diphosphonic acid (HEDP) (133, 134):

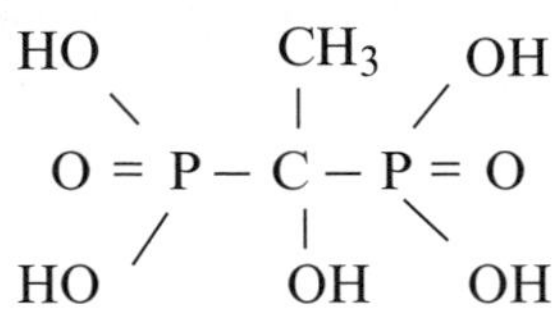

The research on environmentally friendly inhibitor compounds is the center of interest worldwide. One way is a possible decrease of the phosphorous content in anticorrosive molecules for water treatment while simultaneously retaining or increasing their inhibition efficiency. Such investigations involve the preparation of new molecules with lower phosphorous content. Phosphonate compounds containing the $-CH_2PO_3H_2$ fragment instead of, or together with, a carboxyl group have a number of features that are due to their distinctive stereochemistry, their greater electronegativity, and the greater potential of PO_4^{2-} than of COO^- for dentate formation. An important feature of phosphonates compared to aminocarboxyl chelates is their ability to form a quite stable protonated MeHL complex (134).

A systematic study of the corrosion behavior of organic compounds was made in order to prove the importance of phosphonic and carboxyl groups in inhibitory processes and to optimize inhibitor concentration with minimum corrosion rate and maximum scale inhibiting activity (135). As a result of these studies, a new class of environmentally friendly compounds has been developed based on amino acids (136). *N,N*-Di(phosphonometyl)glycine (DPG) was found to be an excellent corrosion inhibitor (137). Self-assembling molecules (SAMs) have been used recently as corrosion inhibitors (138).

The adsorption studies of ^{14}C labeled HEDP (139) on iron electrodes have shown that HEDP forms a loosely bound adsorption layer on iron oxide surface. The presence of zinc (^{65}Zn) and calcium (^{45}Ca) ions in aqueous 0.5 mol cm^{-3} $NaClO_4$ solutions leads to an increase in corrosion inhibition due to the formation of different weakly soluble complex compounds (140).

Coupled application of the *in situ* radiotracer "foil" method and voltammetry gives preliminary information on the time, potential, concentration, and pH dependence of HEDP adsorption on a polycrystalline gold electrode (141). It has been stated that the adsorption of HEDP on polycrystalline gold is highly dependent on the electrode potential, as well as on the composition, concentration, and the solution pH. The next step in the adsorption studies was the investigation of sorption phenomena taking place on the corrosion products, that is, on oxide surfaces.

The adsorption of DPG and HEDP was studied on a hematite surface by indirect radiotracer technique using ^{35}S-labeled H_2SO_4 as indicator species. As the adsorption of sulfate ions at hematite was studied in detail in Ref. (71) and a reversible adsorption was found, ^{35}S-labeled sulfate ions could be used as an ideal indicator species.

The adsorption of sulfate ions can be described by a Langmuir isotherm. Thus, taking into consideration the scheme in Figure 9.6, the following equations can be given:

$$\theta_A = \frac{b_A c_A}{1 + b_A c_A + b_B c_B} \tag{9.17}$$

"A" denotes the species studied

$$\theta_B = \frac{b_B c_B}{1 + b_A c_A + b_B c_B} \tag{9.18}$$

Transformation of Equation (9.18) leads to the expression

$$\frac{1}{\theta_B} = 1 + \frac{1}{b_B c_B} + \frac{b_A c_A}{b_B c_B} \tag{9.19}$$

At a fixed concentration of the indicator species (B) and taking into account that in accordance with Equation (9.19):

$$\frac{\Gamma_B}{\Gamma_B(\max)} = \theta_B = \frac{I}{I(\max)}$$

θ_B should be proportional to the radiation intensity measured (I)

$$\theta_B = QI \tag{9.20}$$

where $Q = 1/I(\max)$ is a proportionality factor characteristic for the system studied; we thus obtain

$$\frac{1}{I} = D + Fc_A \tag{9.21}$$

where

$$F = \frac{b_A Q}{b_B c_B}; \quad D = Q\left(1 + \frac{1}{b_B c_B}\right)$$

From Equations (9.17) and (9.18):

$$\frac{\theta_A}{\theta_B} = \frac{b_A}{b_B}\frac{c_A}{c_B} \tag{9.22}$$

then

$$\theta_A = F'Ic_A \tag{9.23}$$

Curve 1 in Figure 9.34(a) shows the effect of the addition of DPG on the radiation intensity originating from adsorbed labeled sulfate species present at a concentration of 2×10^{-4} mol dm^{-3} in 0.5 mol dm^{-3} $NaClO_4$ supporting electrolyte at pH 2.5. It may be seen from this figure that a significant decrease in the adsorption of sulfate ions takes place at DPG concentrations commensurable with that of sulfate species.

For the sake of comparison, the displacement of the adsorbed sulfate species by phosphoric acid is shown by curve 3 in Figure 9.34(a). Figure 9.34(a) clearly demonstrates that the adsorbability of DPG is higher than that of phosphoric acid. On plotting the $1/I$ values against the concentration of DPG, a linear relationship is obtained as expected on the basis of Equation (9.21) (Figure 9.34b).

By using Equation (9.23) for the calculation of the θ values the adsorption isotherm of DPG (relative scale) can be obtained as shown in Figure 9.35.

The displacement of sulfate species by DPG and HEDP is compared in Figure 9.34(a). It may be seen from this figure that the adsorbability of the two inhibitors is almost the same.

9.5 COMPARATIVE STUDY OF ADSORPTION PHENOMENA OCCURRING ON CORRODING METALS AND THE CORRESPONDING OXIDES/HYDROXIDES

The role of various anions in the corrosion of metals is of great interest. Relatively few direct *in situ* information is available on the adsorption of anions in the course of corrosion processes.

As mentioned in Section 9.2, the problems caused by continuous transformation and dissolution of the metal can be avoided by the application of isotopes emitting soft β-radiation and by using powdered (or less finely dispersed) metals as adsorbents.

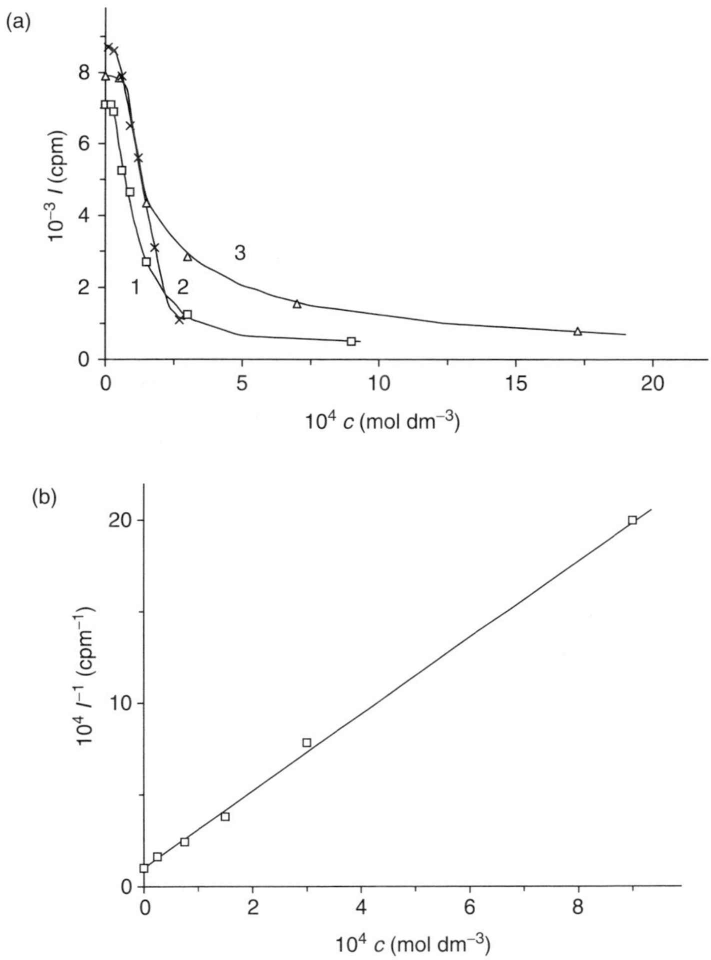

Figure 9.34 (a) The displacement of adsorbed labeled sulfate species by DPG (1), HEDP (2), and phosphoric acid (3); 2×10^{-4} mol dm^{-3} labeled H_2SO_4 in 0.5 mol dm^{-3} $NaClO_4$; pH 2.5 (2 mg Fe_2O_3/cm^2 referred to the geometric surface area of the bottom of the measuring cell). (b) $1/I$ vs. DPG concentration representation of data reported in (a). (From *J. Solid State Electrochem*. E. Kalmen and G. Horányi, Springer-Verlag, 2002. With permission.)

In this case the dissolution of the metal powder covered with adsorbed labeled species in acidic medium does not exert any influence on the radiation intensity measured if layer thickness remains to be higher than self-absorption thickness, that is, the detector "senses" the same thickness throughout the process in spite of the fact that part of the powder has been consumed. The same refers to the dissolving oxides.

A study of Zn, Al system was carried out in Ref. (75). Figure 9.36 shows how the extent of the specific adsorption of labeled sulfate ions on Zn and Al changes with the pH of the solution phase. (Owing to the great excess of $NaClO_4$ the adsorption observed should be considered as specific adsorption, without influence of electrostatic double-layer effects.) It may be seen from this figure that the extent of the adsorption goes through maximum. This behavior allows us to assume that anion adsorption takes place on the protonated surface sites of the corresponding oxide/hydroxide system.

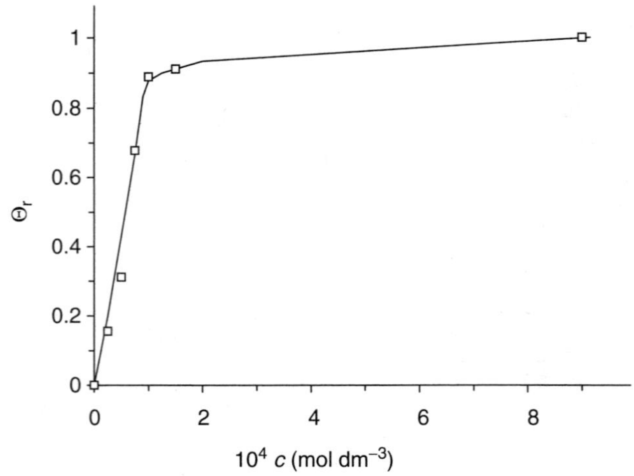

Figure 9.35 Adsorption isotherm of DPG in θ units calculated by Equation (9.23). (From *J. Solid State Electrochem.* E. Kalmen and G. Horányi, Springer-Verlag, 2002. With permission.)

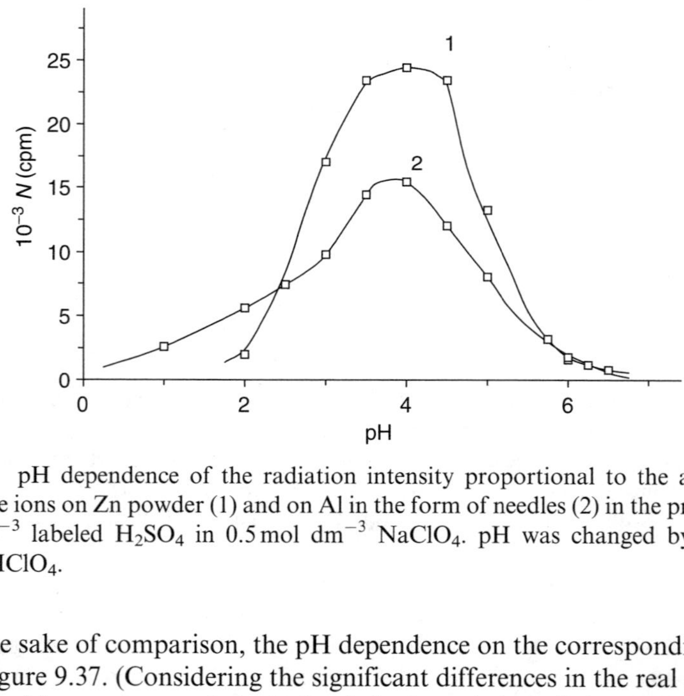

Figure 9.36 pH dependence of the radiation intensity proportional to the adsorption of labeled sulfate ions on Zn powder (1) and on Al in the form of needles (2) in the presence of 2×10^{-4} mol dm^{-3} labeled H_2SO_4 in 0.5 mol dm^{-3} $NaClO_4$. pH was changed by addition of NaOH and $HClO_4$.

For the sake of comparison, the pH dependence on the corresponding oxides is shown in Figure 9.37. (Considering the significant differences in the real surface area of the two oxide powders and the lack of the knowledge of the magnitude of the surface areas the comparison could be carried out only on a relative scale.)

Two competitive processes should be considered in acid medium:

1. Formation of surface oxides/hydroxides.
2. Dissolution of the surface species. Elementary reactions can be given by the following equations:

a. $Me \longrightarrow Me^{n+} + ne^-$
b. $2H^+ + 2e^- \longrightarrow H_2$

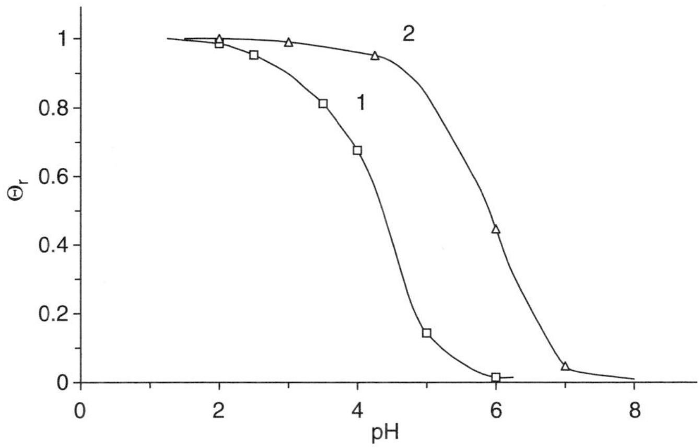

Figure 9.37 pH dependence of the adsorption of sulfate ions on ZnO (1) and on AlOOH (2) on relative scale. pH was changed by addition of NaOH and $HClO_4$.

c. $H_2O = H^+ + OH^-$
d. $Me^{n+} + nOH^- = Me(OH)_n$

The protonation of the hydroxide/oxide formed on the surface induces the specific adsorption of sulfate ions, as demonstrated in Refs. (70–72). With decreasing pH the surface hydroxide on the metal should dissolve and this leads to the decrease in the number of adsorption sites for anions.

Figure 9.38 shows starting from a perchlorate solution with pH 2 how the pH value and radiation intensity shift simultaneously to higher values.

The pH stabilizes about 4.5–5.0; however, owing to the accumulation of corrosion products on the metal surface radiation intensity increases. The addition of protons results in an increase in sulfate adsorption indicating that the protonation of the surface oxide/hydroxide species induces specific adsorption of labeled sulfate species as shown in Figure 9.39. There is proportionality between the count rate and

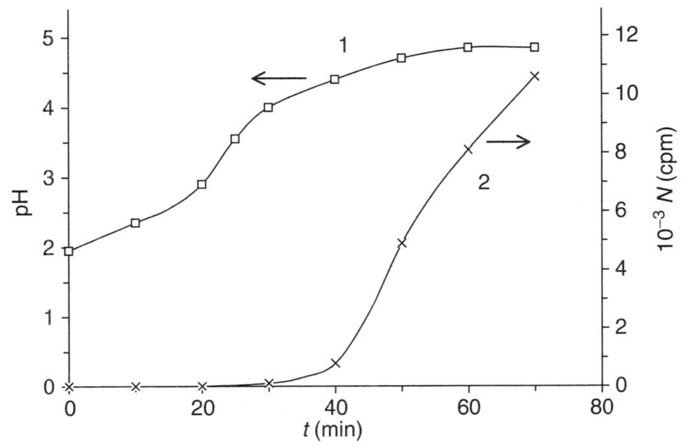

Figure 9.38 The changes in pH value (1) and count rate originating from labeled sulfate (2) in the course of the corrosion of Zn powder in the presence of 0.5 mol dm^{-3} $NaClO_4$ with initial pH 2. (From *Corros. Sci.* G. Horányi and E. Kalman, Pergamon, 2002. With permission.)

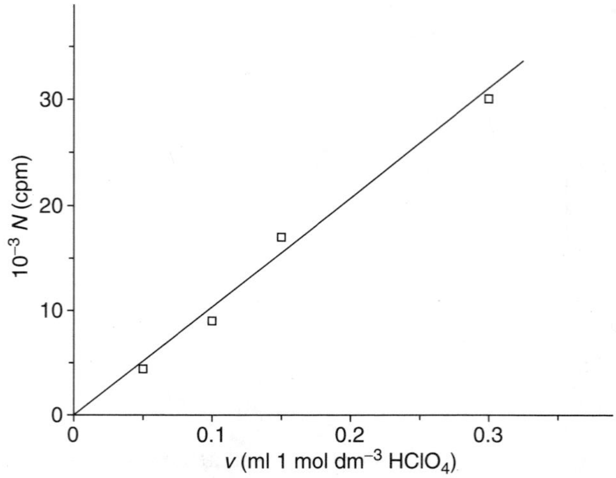

Figure 9.39 Effect of the addition of $HClO_4$ in small portions to 25 ml 0.5 mol dm^{-3} $NaClO_4$ solution in contact with 50 mg cm^{-2} Zn powder and remaining in the pH range between 4 and 5 (3×10^{-4} mol dm^{-3} labeled H_2SO_4). (From *Corros. Sci.* G. Horányi and E. Kalman, Pergamon, 2002. With permission.)

the amount of $HClO_4$ added in small portions within the pH range from 5 to 4.5. In this pH range, the proton sorption capacity of Zn oxides/hydroxides is very significant.

The protonation–dissolution sequence of the surface oxide is well reflected in the behavior of powdered ZnO alone in the presence of an excess of acid. The radiation intensity originating from adsorbed labeled sulfate species goes through a maximum as shown in Figure 9.40. This type of behavior was discussed in Ref. (70). The increase in the count rate should be in connection with the adsorption of labeled molecules induced by protonation. Owing to the presence of a great excess of acid the oxide dissolves continuously, the thickness of the oxide layer tends to zero,

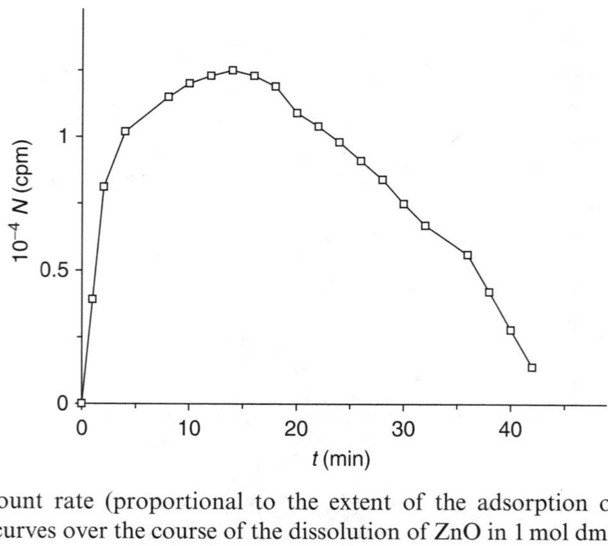

Figure 9.40 Count rate (proportional to the extent of the adsorption of labeled sulfate species) vs. time curves over the course of the dissolution of ZnO in 1 mol dm^{-3} $HClO_4$. Initial amount of the oxide powders referred to the geometric surface area is 30 mg cm^{-2}. (From *Corros. Sci.* G. Horányi and E. Kalman, Pergamon, 2002. With permission.)

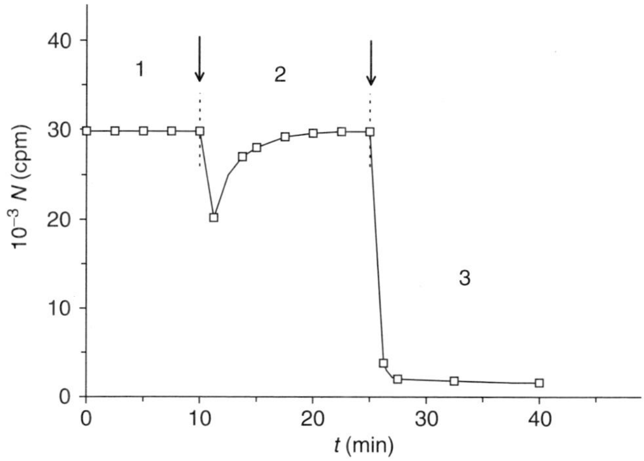

Figure 9.41 Effect of addition of Cl^- ions and nonlabeled H_2SO_4 on count rate in the presence of 4×10^{-4} mol dm^{-3} labeled H_2SO_4 in 0.5 mol dm^{-3} $NaClO_4$ at pH 3 in contact with Zn powder (1). Addition of 4×10^{-2} mol dm^{-3} NaCl (2). Addition of 4×10^{-2} mol dm^{-3} nonlabeled Na_2SO_4 (3) (the moment of addition is indicated by arrow). (From *Corros. Sci.* G. Horányi and E. Kalman, Pergamon, 2002. With permission.)

and the complete dissolution is indicated by the absence of any radiation above that of the background (i.e., the radiation coming from the solution phase).

Chloride ions do not exert significant effect on the adsorption of sulfate ions as shown by section 2 in Figure 9.41. The addition of Cl^- ions in a concentration of several orders of magnitude over that of labeled sulfate ions does not result in the displacement of the latter. However, the addition of nonlabeled SO_4^- results in a dramatic decrease in the count rate indicating the rapid exchange of adsorbed species with the molecules in the solution phase. This means that the labeled molecules are not strongly bound to the surface. Similar behavior was found for the corresponding oxides.

Similar mobility of the adsorbed species was found for Al, as shown by Figure 9.42.

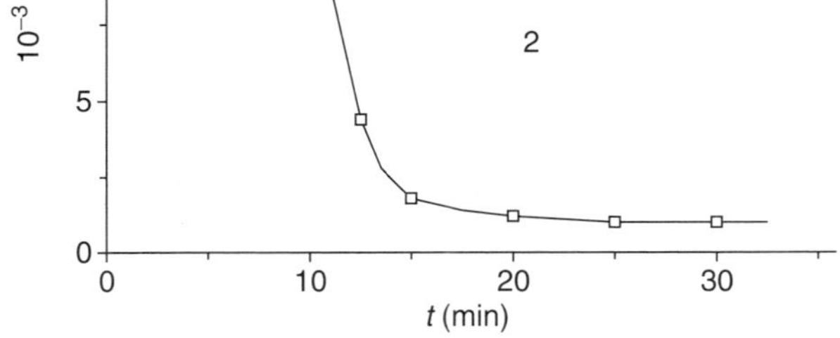

Figure 9.42 Exchange of labeled sulfate species (4×10^{-4} mol dm^{-3}) adsorbed on granulated Al with nonlabeled H_2SO_4 (4×10^{-2}) mol dm^{-3} (the moment of addition is indicated by arrow). (From *Corros. Sci.* G. Horányi and E. Kalman, Pergamon, 2002. With permission.)

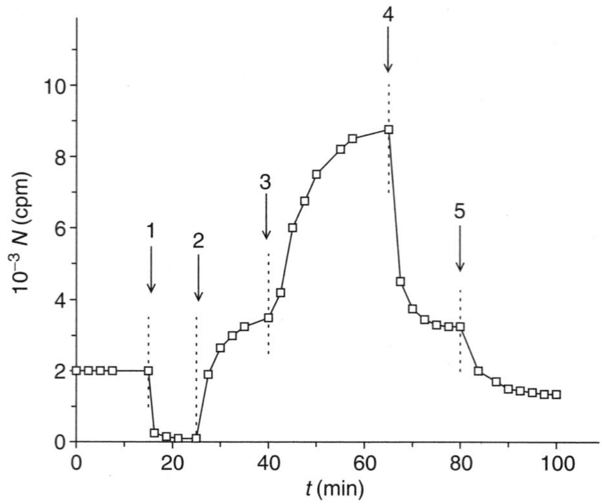

Figure 9.43 The effect of pH changes on the radiation intensity. Initial state Al granulated in 0.5 mol dm^{-3} $NaClO_4$; + labeled H_2SO_4: 4×10^{-4} mol dm^{-3} pH 5; (1) shift to pH 9; (2) shift to pH 5.5; (3) shift to pH 5; (4) shift to pH 4; (5) shift to pH 2.5 (pH was changed by addition of NaOH and $HClO_4$; the moment of addition is indicated by arrows). (From *Corros. Sci.* G. Horányi and E. Kalman, Pergamon, 2002. With permission.)

In the case of Al a shift of the pH to a slightly alkaline value (pH 9) could result in an increase in the sorption capacity of the surface after returning again to a lower pH value (see Figure 9.43). It is demonstrated in Figure 9.43 that starting from pH 5 a shift to pH 9 results in a decrease in the adsorption; however, returning again to pH 5 the adsorption value observed is significantly higher than it was at the beginning. It may be assumed that the alkaline shift modifies the state and thickness of the surface layer.

The experimental results presented demonstrate that the characteristic features of the overall sorption of sulfate ions observed in the course of the corrosion of Al and Zn are similar to those of the corresponding oxides/hydroxides. The extent of sorption at a fixed pH value is determined by the amount of the corrosion products accumulated on the surface and owing to this situation there is no possibility of obtaining information concerning the adsorption on the "pure" metal sites, if any.

The very fact that the adsorption of sulfate ions on Al and Zn goes through maximum in the pH range studied can be interpreted in terms of factors determining the sorption capacity of the surface oxide layers formed. The specific adsorption of sulfate molecules is determined by the extent of protonation of the oxides/hydroxides. As is known, the protonation of AlOOH or Al_2O_3 tends to a limiting value (Γ_{max}) with decreasing pH and the most characteristic changes occur in the pH range from 6 to 4. This pH range depends on the protonation equilibrium constant, K, which can be determined from the pH dependence of the anion adsorption on protonated sites. At the pH value where $\Gamma = \Gamma_{max}/2$ or $I = I_{max}/2$: pH = lg K (71, 72).

At very low pH values, the steady-state coverage onto corrosion products tends to zero owing to their dissolution.

On the other hand, at higher pH values, above pH 6, no significant protonation occurs; thus, the induced anion adsorption should be again very low.

9.6 A SPECIAL FIELD OF STUDIES: NUCLEAR INDUSTRY

In the nuclear industry (nuclear power stations and institutions processing radionuclides), the problems of radioactive contamination and decontamination play a very important role.

Without the knowledge of the mechanism of the contamination there is no possibility of finding effective methods for decontamination and, in addition, there would be a gap in the understanding of processes occurring in the course of the operation of nuclear reactors. The clarification of the corrosion and contamination processes occurring during the operation of the reactors is a very important, but not the only, problem in the nuclear industry.

The treatment of the radioactive waste and the prevention of possible corrosion of the containers, in order to avoid the possible leakage of radioactive materials into the soil, are important environmental problems.

Another fundamental question is how to achieve the dismantling of shutdown reactors without the danger of contamination of the environment.

The effect of the radiation on the behavior of structural materials should also be considered. It is an important question how their corrosion resistance is changing under continuous irradiation.

All these problems are, for instance, discussed in Refs. (78–93). Some of the main tendencies will be presented.

In Ref. (93) and the literature cited therein, the problems of contamination and corrosion of the cooling circuit of water-cooled nuclear reactors are discussed in detail in the light of results obtained from the study of adsorption of various ions on constructional materials (austenitic stainless steels) used in PWR (pressurized water reactor) systems.

These results contribute to the better understanding of corrosion and re-deposition processes causing the radioactive contamination of out-of-reactor-core surfaces.

The accumulation of radioactive cations such as $^{60}Co^{2+}$, $^{110m}Ag^{+}$, $^{51}Cr^{3+}$, and $^{54}Mn^{2+}$ from the reactor coolant onto the surface layer of corrosion products is a regular phenomenon in cooling systems; therefore, the study of sorption of these ions on constructional materials is a fundamental requirement.

In Ref. (93), the study of the accumulation of ^{110m}Ag on austenitic stainless steel is presented as a model for such studies starting from the problem that the contamination caused by ^{110m}Ag activated in the core appears to be a significant problem in some Soviet-made PWRs, and may lead to trouble in the continuous measurement of other contaminants in the primary side coolant. Under suitable model conditions, with the use of “foil” and “thin-gap” methods, the accumulation of ^{110m}Ag-labeled Ag^{+} from a model solution of the primary side coolant of PWRs onto an austenitic stainless steel was studied.

The main conclusions drawn from this study by the authors are as follows:

1. Accumulation of ^{110m}Ag on austenitic stainless steel is most likely due to the cementation (underpotential deposition) of radioactive silver ions.
2. Both the deposition (contamination) and dissolution of deposited silver (decontamination and surface prevention) can be affected by the electrochemical parameters of the system studied.

The decontamination of some parts of a nuclear reactor is required time to time in order to keep the radioactive contamination of these parts (coolant circuit surfaces, coolant pumps) (80, 142) under safe level.

To remove the radioactivity from a metal surface, chemical cleaning reagents have been widely applied. Such a method is the AP-CITROX decontamination procedure (AP — alkaline permanganate; CITROX — citric and oxalic acids). The composition of the oxide grown and that of the oxide deposited in PWR and boiling water reactor (BWR) systems play an important role in deciding whether or not the reagents can provide effective decontamination for coolant circuit surfaces. The main risk associated with decontamination is corrosion. Therefore, the investigation of the corrosion behavior of structural materials following chemical treatment is an important task. As to the details see, for instance, Refs. (80, 142).

A different approach to the contamination and decontamination problem is the minimizing of radiation build-up by some kind of preventive action. For instance, zinc oxide ceramics (natural zinc) are used to reduce radiation build-up in BWRs by incorporating zinc ions in the corrosion film of stainless steel pipes in the cooling system of the nuclear reactor.

It is demonstrated in Ref. (143) that soluble zinc at concentration levels of 5 to 20 ppb in the BWR reactor cooling water minimizes the radiation build-up by inhibiting the corrosion of the stainless steel recirculating pipework.

Attempt is made for the application of corrosion inhibitors (144). In Ref. (144) the properties of the transient species formed by reactions of e^-_{aq}, H atom, and OH radicals with a number of corrosion inhibitors have been determined using pulse radiolysis technique. These properties have been correlated with the corrosion rates of carbon steel samples in EDTA/ascorbic acid/citric acid (EAC) medium in the presence of these corrosion inhibitors. The importance of these results in the context of the use of these compounds in chemical decontaminant formulations for nuclear reactor applications is discussed.

The nuclear waste management, the storage of low- (LLW), medium- (MLW) and high-level (HLW) radioactive waste require the study of the corrosion behavior of containers for disposal of spent nuclear fuel and radioactive waste (78, 79, 81, 83, 85, 86, 90–92).

The purpose of a disposal is to isolate the radioactive waste from man and the environment. If the isolation is broken the leakage and transport of radioactive substances must be retarded. The package is one of several barriers, used to achieve these two main functions. According to the rules of the Swedish waste management (92) for short-lived, low- and intermediate-level waste standard containers of steel and concrete are used. Spent fuel is placed in a canister consisting of a pressure-bearing insert of cast nodular iron and an outer corrosion barrier of copper before it is deposited in a deep geological repository. A fundamental question is external copper corrosion. Under currently known conditions at the deep repository level, the canister is expected to remain intact for a very long time, much longer than the 100,000 years stipulated in the design premises. If water enters the canister, the cast iron insert will corrode anaerobically, since the groundwater is oxygen-free, with hydrogen generation and magnetite formation. A magnetite layer will build up on the iron surface and the corrosion rate is expected to be around 0.1 μm year^{-1}.

Corrosion investigations have been performed on the austenitic structural steel AISI 304L, in comparison with the structural steel AISI 316L, in an aerated solution and a deaerated solution, which was leached from low- and medium-level radioactive waste (90). On the basis of measured potentiodynamic anodic polarization curves and the results of cyclic polarization tests, it was found that both types of steel, as well as the corresponding welds, had a high pitting potential and high protective potential, which means that they have a strong tendency to form a compact and corrosion-resistant passive film. The repassivation capability of both types of steel prevents the occurrence of stress corrosion cracking at the level of concentration of chloride ions, which corresponds to the described type of waste, whereas absorbed atomic hydrogen does not reduce toughness or cause hydrogen embrittlement. The results of the research work confirmed that it is possible to use AISI 304L structural steel for the construction of containers for the temporary, 30-year storage of low- and medium-level radioactive waste.

In the case of low- and medium-level radioactive waste, γ-radiation does not have an effect on corrosion processes. In contrast, in the case of HLW containers the effect of γ-radiation on the corrosion behavior could be important (81).

In Ref. (81) the effect of γ-radiation on the passive layers of Ti and TiO·2Pd container materials for high-level waste disposal was studied. From the various types of radiation relevant for HLW only γ-radiation can penetrate the container wall and affect the passive layers. Two major effects should be considered: radiolysis of the surrounding salt brine environment with subsequent formation of radiolytic products such as H_2 and strong oxidizing agents like H_2O_2, ClO^-, etc., and the so-called γ-photoeffect that occurs directly within the protecting passive layer. The γ-photoeffect is caused by the hot electrons that dissipate their energy not only thermally but also by generation of electron/hole pairs in the semiconducting passive layers.

Pitting in hot brine as well as the effect of γ-radiation on Ti-passive layers were investigated on single Ti-grains. The γ-photoeffect was simulated by means of *in situ* UV laser illumination with high power density at high local resolution. Without illumination pitting occurred only at potentials much higher than the free corrosion potentials, that is, no pitting on any grain orientation has to be expected under conditions relevant for HLW disposal.

Furthermore, UV laser illumination of the sample surface in q-brine did not result in photoinduced pitting but a laser-induced oxide growth took place. In q-brine this reaction dominated in the entire potential range. It is caused by the potential change within the oxide and the Helmhotz layer due to accumulation of photoholes at the oxide/electrolyte interface resulting in an enhanced charge transfer through this interface.

On the basis of the results presented in Ref. (81), the authors arrived at the conclusion that Ti and TiO·2Pd are promising container materials for HLW disposal.

An interesting example for the investigation of corrosion processes in radioactive medium is the study of the corrosion of steels in highly concentrated tritiated water used in tritium reprocessing plants (78, 79).

Tritium decays with the emission of a β^--particle. The energy released in this decay is 5.7 keV per tritium tom and is high enough to decompose locally the molecules of water along the path of the β^--particle with the formation of radiolytic species.

It is essential in highly concentrated tritiated water processing facilities that there can be no leaks from the processing circuit to the environment. Organic polymers, seal or pump oils are decomposed by reactions induced by the energy released from β^--particles and cannot be used. To avoid corrosion, using very highly radioactive aqueous solutions, the equipment must be completely constructed of stainless steel.

In Refs. (78, 79), a detailed analysis can be found on how the storage conditions influence the corrosion of stainless steel containers.

The investigations of the corrosion of fuels and fuel cladding tubes constitute an important contribution to the solution of the long-term waste storage (mostly deep storage, geological disposal) (see, e.g., Refs. (85, 86)).

Zirconium alloys such as Zircaloy-4 (1.70 wt% Sn; 0.24 wt% Fe; 0.13 wt% Cr) are used in water pressurized reactors as fuel cladding tubes. During the reactor processing, the inner side of cladding tubes is oxidized and implanted by recoil fission products. Most of the fission products are located within the first 2 μm. Some of these fission fragments are long-living isotopes and pose a problem in case of long-term waste storage. In Ref. (85), the corrosion of zirconia was studied using europium surface inplantation as a marker with the aim to clarify the problems connected with long-term waste storage.

In Ref. (86), dissolution and corrosion of $(Th,U)O_2$ fuel were investigated at room temperature and near 100°C in near neutral and acidic water (pH 3) to evaluate the suitability of irradiated UO_2-doped thoria as a waste form for direct geological disposal. XPS and x-ray diffraction were used to study oxidation of $(Th,U)O_2$ fuel. The uranium in the surface of $(Th,U)O_2$ fuel undergoes oxidation similar to that observed in UO_2 fuel under similar conditions. Nevertheless, the dissolution rate of uranium from $(Th,U)O_2$ fuel in aerated solutions is much lower than that from UO_2 fuel under similar conditions. These studies suggest that a properly prepared $(Th,U)O_2$ fuel can be an acceptable waste form for permanent geological disposal.

The examples presented in this section are only illustrative ones from the wealth of corrosion studies in the field of nuclear industry.

ACKNOWLEDGMENT

Financial support from the Hungarian Scientific Research Fund is acknowledged (OTKA Grant T 031703, T037643, TO45888).

REFERENCES

1. NA Balashova, VN Kazarinov. Use of radioactive-tracer method for the investigation of electric double-layer structure. In: AJ Bard, ed. *Electroanalytical Chemistry*, Vol. 3. New York: Marcel Dekker, 1969, pp. 135–197.
2. G Horányi. Recent developments in the application of the radiotracer method to the investigation of adsorption and electrocatalytic phenomena. *Electrochim. Acta* 25:45–57, 1980.
3. VE Kazarinov, VN Andreev. Tracer methods in electrochemical studies. In: E Yeager, JO'M Bockris, BE Conway, eds. *Comprehensive Treatise on Electrochemistry*, Vol. 9. New York: Plenum Press, 1984, pp. 393–443.
4. A Wieckowski. *In situ* surface electrochemistry: radioactive labeling. In: JO'M Bockris, BE Conway, RE White, eds. *Modern Aspects of Electrochemistry*, Vol. 21. New York: Plenum Press, 1990, pp. 21, 65–119.
5. P Zelenay, A Wieckowski. Radioactive labeling: toward characterization of well-defined electrodes. In: HD Abruna, ed. *Electrochemical Interfaces: Modern Techniques for In Situ Surface Characterization*. New York: VCH Publishers, 1991, pp. 479–527.
6. EK Krauskopf, A Wieckowski. Radiochemical methods to measure adsorption at smooth polycrystalline and single crystal surfaces. In: J Lipkowski, PN Ross, eds.

Adsorption of Molecules at Metal Electrodes. Frontiers of Electrochemistry. New York: VCH Publishers, 1992, pp. 119–169.
7. G Horányi. New trends in the radiotracer study of electrosorption phenomena. B. *Electrochemistry* 5:235–241, 1989.
8. YE Sung, A Thomas, M Gamboa-Aldeco, K Franaszczuk, A Wieckowski. Adsorption characteristics by a radiochemical method on smooth electrode surfaces. *J. Electroanal. Chem.* 378:131–142, 1994.
9. G Horányi. Radiochemical studies of processes at electrodes. *Rev. Anal. Chem.* 14:1–58, 1995.
10. G Horányi. Electrosorption studies in electrocatalysis. In: JJ Spivey, ed. *Catalysis*. Specialist Periodical Report, Vol. 12. Cambridge: The Royal Society of Chemistry, 1996, pp. 254–301.
11. G Horányi, M Wasberg. Comparative radiotracer study of the adsorption of the adsorption of Cl^-, $HSO_4^-(SO_4^{2-})$ and $H_2PO_4^-$ anions on rhodized electrodes. *J. Electroanal. Chem.* 404:291–298, 1996.
12. G Horányi, M Wasberg. Study of the equilibrium underpotential deposition of copper and the accompanying anion adsorption on rhodized electrodes in the course of cyclic voltammetric measurements. *J. Electroanal. Chem.* 413:161–164, 1996.
13. G Horányi. A novel approach to platinized platinum electrodes: specific adsorption as a partition between the solution phase and the electrodeposited platinum black layer. *J. Electroanal. Chem.* 417:185–188, 1996.
14. G Horányi, A Aramata. Specific anion adsorption in the course of upd of Zn^{2+} ions on platinum. *J. Electroanal. Chem.* 434:201–207, 1997.
15. G Horányi. Radiotracer study of interactions of anions and adatoms on metal surfaces with high real surface area. *ACH Models Chem.* 134:33–47, 1997.
16. G Horányi, J Solt, F Nagy. Investigation of adsorption phenomena on platinized platinum electrodes by tracer methods I. Experimental procedure. *J. Electroanal. Chem.* 31:87–93, 1971.
17. G Horányi. Radiotracer study of the specific adsorption of HCO_3^- ions on a platinum electrode. *J. Electroanal. Chem.* 453:113–119, 1998.
18. G Horányi. Gap between cell and detector in certain versions of the thin foil method for radiotracer electrosorption studies. *ACH Models Chem.* 135:219–224, 1998.
19. G Horányi. Influence of the specific adsorption of anions on the "double layer region" of the cyclic voltammetric curves obtained with platinized platinum electrodes. *J. Solid State Electrochem.* 2:237–241, 1998.
20. G Horányi. Anion adsorption in the course of dichromate reduction on a smooth gold electrode in aqueous acid solutions. *J. Solid State Electrochem.* 4:153–158, 2000.
21. G Horányi. Specific adsorption of anions in the course of underpotential and bulk depositions and alloy formation of cadmium on a smooth gold support in acid medium. *J. Solid State Electrochem.* 4:177–180, 2000.
22. A Kolics, K Varga. Coupled *in situ* radiochemical and electrochemical study of cobalt accumulation on polycrystalline gold. *Electrochim. Acta* 40:1835–1844, 1995.
23. Z Németh, L Erdei, A Kolics. A new approach to the study of ion transport processes by an *in situ* radiotracer method based on measuring intensity changes and energy spectrum alterations of β-radiation. *J. Radioanal. Nucl. Chem. Lett.* 199:265–275, 1995.
24. Z Németh, L Erdei, A Kolics. A new approach to studying ion transport in corrosion protective coatings using an *in situ* radiotracer method. *Corros. Sci.* 37:1163–1166, 1995.
25. K Varga, P Baradlai, WP Barnard, G Myburg, P Halmos, JH Potgieter. Comparative study of surface properties of austenitic stainless steels in sulfuric and hydrochloric acid solutions. *Electrochim. Acta* 42:25–35, 1997.
26. K Varga, P Baradlai, A Vértes. *In-situ* radiotracer studies of sorption processes in solutions containing (bi)sulfite ions I. Polycrystalline gold. *Electrochim. Acta* 42:1143–1155, 1997.

27. K Varga, P Baradlai, D Hanzel, W Meisel, A Vértes. *In-situ* radiotracer studies of sorption processes in solutions containing (bi)sulfite ions II. Low carbon steel. *Electrochim. Acta* 42:1157–1167, 1997.
28. G Hirschberg, Z Németh, K Varga. A detailed study of the reliability of some crucial parameters used in the *in-situ* radiotracer sorption studies. *J. Electroanal. Chem.* 456:171–191, 1998.
29. G Hirschberg, P Baradlai, K Varga, G Myburg, J Schunk, P Tilky, P Stoddart. Accumulation of radioactive corrosion products on steel surfaces of VVER type nuclear reactors. I. ^{110m}Ag. *J. Nucl. Mater.* 265:273–284, 1999.
30. P Waszczuk, P Zelenay, J Sobkowski, Surface interaction of benzoic acid with a copper electrode. *Electrochim. Acta* 40:1717–1721, 1995.
31. M Szklarczyk, NN Hoa, P Zelenay. Interaction of thiourea with silver electrodes. *J. Electroanal. Chem.* 405:111–116, 1996.
32. S Smoliński, P Zelenay, J Sobkowski. Effect of surface order on adsorption of sulfate ions on silver electrodes. *J. Electroanal. Chem.* 442:41–47, 1998.
33. P Waszczuk, P Zelenay, J Sobkowski. Radiometric and voltammetric study of benzoic acid adsorption on a polycrystalline silver electrode. *Electrochim. Acta* 43:1963–1968, 1998.
34. S Smoliński, J Sobkowski. Adsorption of sulfate ions on monocrystalline copper electrodes: the structural effects. *J. Electroanal. Chem.* 463:1–8, 1999.
35. P Waszczuk, A Wnuk, J Sobkowski. Radiometric study of underpotential deposition of thallium on single crystal silver electrodes from perchloric acid solutions. *Electrochim. Acta* 44:1789–1795, 1999.
36. A Wieckowski. Comments on the paper entitled "In Situ Studies of Adsorption of Organic Compounds on Platinum Electrodes", by JO'M Bockris and K.T. Jeng. *J. Electroanal. Chem.* 352:313–320, 1993.
37. W Savich, S-G Sun, J Lipkowski, A Wieckowski. Determination of the sum of Gibbs excesses of sulfate and bisulfate adsorbed at the Pt(111) electrode surface using chronocoulometry and thermodynamics of the perfectly polarized electrode. *J. Electroanal. Chem.* 388:233–237, 1995.
38. E Herrero, JM Feliu, A Wieckowski, J Clavilier. The unusual adsorption states of Pt(111) electrodes studied by an iodine displacement method: comparison with Au(111) electrodes. *Surf. Sci.* 325:131–138, 1995.
39. A Kolics, AE Tomas, A Wieckowski. ^{36}Cl labelling and electrochemical study of chloride adsorption on a gold electrode from perchloric acid media. *J. Chem. Soc. Faraday Trans.* 92:3828–3736, 1996.
40. A Kolics, JC Polkinghorne, A Wieckowski. Adsorption of sulfate and chloride ions on aluminum. *Electrochim. Acta* 43:2605–2618, 1998.
41. A Kolics, JC Polkinghorne, AE Thomas, A Wieckowski. Sorption of sulfate and chloride anions on a well-characterized Al 2024 electrode. *Chem. Mater.* 10:812–824, 1998.
42. A Wieckowski, A Kolics, JC Polkinghorne, ED Eliadis, RC Alkire. Radioactive labeling, electrochemical, and ultrahigh vacuum study of sulfate adsorption on type 316 stainless steel. *Corros. Sci.* 54:800–813, 1998.
43. A Wieckowski, A Kolics. Radioactive Labeling Investigations of Ordered Interfaces: (Bi)Sulfate Adsorption on a Pt(111) Electrode. Meeting Abstract. The 1999 Joint International Meeting; 196th Meeting of The Electrochemical Society, Hawaii, October 17–22, 1999; The Electrochemical Society; Honolulu, Hawaii, 1999; No 2126.
44. A Wieckowski, A Kolics. Comments on "On the calculation of surface concentration from the measurement by the radiotracer 'electrode lowering' technique." *J. Electroanal. Chem.* 464:118–122, 1999.
45. D Poškus, G Agafonovas. Adsorption of cyanide-containing species from potassium cyanide solutions on gold electrodeposits. *J. Electroanal. Chem.* 393:105–112, 1995.

46. D Poškus, G Agafonovas, A Žebrauskas. A radiometric study of thallium adsorption on gold electrode. *Chemija* 71–76, 1996.
47. D Poškus, G Agafonovas, A Žebrauskas. Surface concentration of thallium on gold electrode in cyanide and dicyanoaurate alkaline solutions. *Chemija* 104–108, 1996.
48. D Poškus, G Agafonovas, I Jurgaitiene. Effect of thallium ions on the adsorption of cyanide-containing species from cyanide and dicyanoaurate solutions on a polycrystalline gold electrode. *J. Electroanal. Chem.* 425:17–34, 1997.
49. D Poškus. On the calculation of surface concentration from measurements by the radiotracer "electrode lowering" technique. *J. Electroanal. Chem.* 442:5–7, 1998.
50. JO'M Bockris, KT Jeng. *In situ* studies of adsorption of organic compounds on platinum electrodes. *J. Electroanal. Chem.* 330:541–581, 1992.
51. J Jovancicevic, JO'M Bockris, JL Carbajal, P Zelenay, T Mizuno. Adsorption and absorption of chloride ions on passive iron systems. *J. Electrochem. Soc.* 133:2219–2226, 1986.
52. P Zelenay, MA Habib, JO'M Bockris. Adsorption from solution on platinum an *in situ* FTIR and radiotracer study. *Langmuir* 2:393–405, 1986.
53. JO'M Bockris, M Gamboa-Aldeco, M Szklarczyk. Ionic adsorption at the solid-solution interphase using three *in situ* methods. *J. Electroanal. Chem.* 339:355–400, 1992.
54. JO'M Bockris, Y Kang. The protectivity of aluminum and its alloys with transition metals. *J. Solid State Electrochem.* 1:17–35, 1997.
55. VE Kazarinov, VN Andreev, MA Spytsin, AV Shlepakov. Role of anions in the electrochemical transformation processes of polyaniline. *Electrochim. Acta* 35:899–904, 1990.
56. VE Kazarinov, VN Andreev, MA Spitsyn. Study of the polyaniline film degradation process by radiotracer and electrochemical methods. *Electrochim. Acta* 41:1757–1760, 1996.
57. VN Andreev. Electrochemical synthesis and properties of polyaniline films on various substrates. *Russian J. Electrochem.* 35:735–739, 1999.
58. VN Andreev. Mechanism of the phenol action on the synthesis and properties of polyaniline films. *Russian J. Electrochem.* 35:740–743, 1999.
59. J-M Herbelin, N Barabouth, P Marcus. A novel method for the study of the effects of Cl^- ions on passivation using the Cl^{36} radioisotope. Application to the passivation of nickel. *J. Electrochem. Soc.* 137:3410–3414, 1990.
60. G Horányi. Radiotracer studies of adsorption/sorption phenomena at electrode surfaces. In: A Wieckowski, ed. *Interfacial Electrochemistry. Theory, Experiment, and Applications*. New York: Marcel Dekker, 1999, pp. 477–491.
61. K Varga, E Maleczki, G Horányi. Radiotracer study of the adsorption of SO_4^{2-} (HSO_4^-) and Cl^- ions on an austenitic stainless steel. *Electrochim. Acta* 33:25–31, 1988.
62. K Varga, E Maleczki, E Házi, G Horányi. Towards a multipurpose radiotracer method for the investigation of sorption phenomena on constructional material samples. *Electrochim. Acta* 35:817–823, 1990.
63. K Varga, E Maleczki, G Horányi. Radiotracer study of the simultaneous adsorption of SO_4^{2-} and HSO_4^- ions on an austenitic stainless steel type 08x18H10T. *Electrochim. Acta* 33:1167–1169, 1988.
64. K Varga, E Maleczki, G Horányi. On the simultaneous role of Cr in the passivity of stainless steel and in the sorption of HSO_4^- and Cl^- ions in the passive film. *Electrochim. Acta* 33:1775–1778, 1988.
65. P Marcus, J-M Herbelin. The entry of chloride ions into passive films on nickel studied by spectroscopic (ESCA) and nuclear (^{36}Cl radiotracer) methods. *Corros. Sci.* 34:1123–1145, 1993.
66. A Wieckowski, E Ghali, M Szklarczyk, J Sobkowski. The behaviour of iron electrode in CO_2-saturated neutral electrolyte II. Radiotracer study and corrosion considerations. *Electrochim. Acta* 28:1627–1633, 1983.

67. AE Thomas, Y-E Sung, M Gamboa-Aldeco, K Franaszczuk, A Wieckowski. Investigation of thiosulfate adsorption on 316 stainless steel in neutral solutions by radioactive labeling, electrochemistry, and Auger electron spectroscopy. *J. Electrochem. Soc.* 142: 476–484 1995.
68. JH Potgieter, WO Barnard, G Myburg, K Varga, P Baradlai, L Tomcsányi. Corrosion behaviour of duplex stainless steels containing minor ruthenium additions in reducing acid media. *J. Appl. Electrochem.* 26:1103–1110, 1996.
69. A Kolics, AE Thomas, A Wieckowski. Radiochemical, electrochemical and UHV study of anionic accumulation on Al and Al 2024 alloy. Extended Abstracts of 47th Meeting of International Society of Electrochemistry, Veszprém & Balatonfüred, 1996, P6a-12.
70. P Joó, G Horányi. Radiotracer study of the specific adsorption of anions on metal oxides in acid media: an experimental approach. *J. Colloid Interface Sci.* 223:308–310, 2000.
71. G Horányi, P Joó. *In situ* study of the specific adsorption of HSO_4^-/SO_4^{2-} ions on hematite by radiotracer technique. *J. Colloid Interface Sci.* 227:206–211, 2000.
72. G Horányi, P Joó. Application of the radiotracer technique for the study of the specific anion adsorption on Al_2O_3 in acidic medium. *J. Colloid Interface Sci.* 231:373–378, 2000.
73. G Horányi, P Joó. Some peculiarities in the specific adsorption of phosphate ions on hematite and γ-Al_2O_3 as reflected by radiotracer studies. *J. Colloid Interface Sci.* 247:12–17, 2002.
74. G Horányi, L Gáncs. Indirect radiotracer study of the adsorption of chromate ions on γ-Al_2O_3 and direct study of the adsorption of sulfate ions on Cr_2O_3 using ^{35}S-labelled H_2SO_4. *J. Solid State Electrochem.* 6:485–489, 2002.
75. G Horányi, E Kálmán. Novel application of radiotracer techniques for corrosion studies of zinc and aluminium in acidic medium. *Corros. Sci.* 44:899–907, 2002.
76. E Kálmán, G Horányi. Indirect radiotracer study of the adsorption of corrosion inhibitors on hematite. *J. Solid State Electrochem.* 6:253–258, 2002.
77. G Horányi, E Kálmán. Radiotracer study of the adsorption of corrosion inhibitors on oxide surfaces modeling the products of metal corrosion. Proceedings of International Symposium on Corrosion and Corrosion Protection, San Francisco, 2001, pp. 379–385.
78. G Bellanger, JJ Rameau. Corrosion of nickel–chromium deposit on AISI 316L stainless steel in radioactive water with and without fluoride at pH 4. *J. Nucl. Mater.* 226:104–119, 1995.
79. G Bellanger, JJ Rameau. Effect of slightly acid pH with or without chloride in radioactive water on the corrosion of maraging steel. *J. Nucl. Mater.* 228:24–37, 1996.
80. I-J Yang, M-Y Teng, W-I Huan, Y-L Sun. Decontamination of the reactor coolant pump in Maanshan nuclear power plant. *Nucl. Eng. Dis.* 167:91–97, 1996.
81. A Michaelis, S Kudelka, JW Schultze. Effect of γ-radiation on the passive layers of Ti and TiO.2Pd container-materials for high-level waste disposal. *Electrochim. Acta* 43:119–130, 1998.
82. A Bahadur. Development and evaluation of a low chromate corrosion inhibitor for cooling water systems. *Can. Met. Q.* 37:459–468, 1998.
83. J Engelhardt, G Marx. Contact corrosion measurements on the pair UO_{2+x} and carbon steel 1.0330 in brines and bentonite porewater with respect to direct waste disposal. *J. Nucl. Mater.* 264:161–168, 1999.
84. VA Kashparov, VP Protsak, N Ahamdach, D Stammose, JM Peres, VI Yoschenko, SI Zvarich. Dissolution kinetics of particles of irradiated Chernobyl nuclear fuel: influence of pH and oxidation state on the release of radionuclides in the contaminated soil of Chernobyl. *J. Nucl. Mater.* 279:225–233, 2000.
85. K Poulard, A Chevarier, JC Duclot, N Moncoffre, D Crusset. Use of RBS to validate europium surface implantation as a marker for the corrosion study of zirconia. *Nucl. Inst. Met. Phys. Res. B* 161–163:668–672, 2000.
86. S Sunder, NH Miller. XPS and XRD studies of $(Th,U)O_2$ fuel corrosion in water. *J. Nucl. Mater.* 279:118–126, 2000.

87. K Varga, P Baradlai, G Hirschberg, Z Németh, D Oravetz, J Schunk, P Tilky. Corrosion behaviour of stainless steel surfaces formed upon chemical decontamination. *Electrochim. Acta* 46:3783–3790, 2001.
88. K Varga, G Hirschberg, Z Németh, G Myburg, J Schunk, P Tilky. Accumulation of radioactive corrosion products on steel surfaces of VVER-type nuclear reactors. II. ^{60}Co. *J. Nucl. Mater.* 298:231–238, 2001.
89. M Kelin, J Dadoumont, Y Demeulemeester, V Massaut. Experience in decommissioning activities at the BR3 site. *Fusion Eng. Des.* 54:443–449, 2001.
90. L Vehovar, M Tandler. Stainless steel containers for the storage of low and medium level radioactive waste. *Nucl. Eng. Des.* 206:21–33, 2001.
91. B Kienzler, B Luckscheiter, S Wilhelm. Waste form corrosion modeling: comparison with experimental results. *Waste Management* 21:741–752, 2001.
92. T Hedman, A Nyström, C Thegerström. Du combustible nucléaire aux déchets: recherches actuelles. From nuclear fuels to waste: current research. *C. R. Physique* 3:903–913, 2002.
93. K Varga, G Hirschberg, P Baradlai, M Nagy. Combined application of radiochemical and electrochemical methods for the investigation of solid/liquid interfaces In: E Matijević, ed. *Surface and Colloid Science*, Vol. 16. New York: Kluwer Academic/Plenum Publishers, 2001, pp. 341–393.
94. ME Gamboa-Aldeco, K Franaszczuk, A Wieckowski. Radiotracer study of electrode surfaces. In: AT Hubbard, ed. *The Handbook of Surface Imaging and Visualization*. New York: CRC Press, 1995, p. 635.
95. G Horányi. Radiotracer studies of adsorption at electrode surfaces. In: AT Hubbard, ed. *Encyclopedia of Surface and Colloid Science*. New York: Marcel Dekker, 2002, pp. 4423–4437.
96. A Kolics, G Horányi. The role of secondary effects in the radiochemical study of interfacial phenomena at metal electrodes. Potentialities of methods based on the study of x-radiation-bremsstrahlung-induced by β^- particles. *J. Electroanal. Chem.* 376:167–177, 1994.
97. A Kolics, G Horányi. On the application of the ^{125}I isotope for radiotracer adsorption studies (radiochemical considerations). *J. Electroanal. Chem.* 372:261–263, 1994.
98. A Kolics, G Horányi. Some considerations about the applicability of the *in situ* radiotracer "foil" method for the study of accumulation processes on measurement of β-radiation. I. *Electrochim. Acta* 41:791–802, 1996.
99. A Kolics, G Horányi. Potentialities of a version of the radiotracer "foil" (Joliot) method for sorption studies based on the measurements of x-radiation. *Appl. Radiat. Isot.* 47:551–561, 1996.
100. EL Krauskopf, K Chan, A Wieckowski. *In situ* radiochemical characterization of adsorbates at smooth electrode surfaces. *J. Phys. Chem.* 91:2327–2332, 1987.
101. G Horányi, EM Rizmayer, P Joó. Radiotracer study of adsorption phenomena on copper electrodes. Study of the adsorption of chloride ions and thiourea on "copperized" electrodes. *J. Electroanal. Chem.* 149:221–235, 1983.
102. G Horányi, EM Rizmayer. Study of the adsorption of phosphoric acid on copper and gold electrodes by a radiotracer method. *J. Electroanal. Chem.* 176:349–352, 1984.
103. G Horányi, EM Rizmayer, P Joó. Radiotracer study of the adsorption of HSO_4^- ions on a "copperized" electrode and on underpotential deposited cadmium on copper in acidic medium. *J. Electroanal. Chem.* 154:281–286, 1983.
104. G Horányi, EM Rizmayer, J Kónya. Radiotracer study of anion adsorption at silver electrodes in acidic medium. *J. Electroanal. Chem.* 176:339–348, 1984.
105. G Horányi, G Vértes. Radiotracer study of the underpotential formation of a silver sulfide monolayer on silver electrodes in alkaline medium. *Electrochim. Acta* 31:1663–1665, 1986.
106. G Horányi EM Rizmayer. Radiotracer study of adsorption phenomena at nickel electrodes in acidic medium. *J. Electroanal. Chem.* 180:97–108, 1984.

107. G Horányi, M Wasberg. Comparative radiotracer study of the adsorption of Cl^-, HSO_4^- (SO_4^{2-}) and $H_2PO_4^-$ anions on rhodized electrodes. *J. Electroanal. Chem.* 404:291–298, 1996.
108. G Horányi, EM Rizmayer. Radiotracer study of the adsorption of sulphuric acid on rhodized electrode. *J. Electroanal. Chem.* 201:187–198, 1986.
109. G Horányi, EM Rizmayer, P Joó. Radiotracer study of the adsorption of Cl^- and HSO_4^- ions on a porous gold electrode and on underpotential deposited metals on gold. *J. Electroanal. Chem.* 152:211–222, 1983.
110. G Horányi, I Bakos. Coupled radiometric and electrochemical study of the reduction of TcO_4^- ions at a gold electrode in acidic medium (mechanistic aspects of Tc depositon). *J. Electroanal. Chem.* 370:213–218, 1994.
111. G Horányi, I Bakos. Investigation of the electrodeposition and behaviour of Re layers by coupled radiochemical and electrochemical methods. *J. Electroanal. Chem.* 378:143, 1994.
112. GG Láng, M Ujvári, G Horányi. New EQCM, voltammetric and radiotracer evidences proving the role of Cu^+ ions in the behavior of the Cu^{2+}–Cu system. *J. Electroanal. Chem.* 522:179–188, 2002.
113. G Inzelt, G Horányi. Combined cyclic voltammetric and radiometric study of polymer film electrodes. *J. Electroanal. Chem.* 200:405–410, 1986.
114. G Inzelt, G Horányi, JQ Chambers, EW Day. Combined electrochemical and radiotracer (cyclic voltradiometric) study of the motion of counter-ions in tetracyanoquinodimethane modified electrodes. *J. Electroanal. Chem.* 218:297–306, 1987.
115. G Inzelt, G Horányi, JQ Chambers. Radiotracer study of the sorption of counter- and co-ions in tetracyanoquinodimethame and poly(vinyl ferrocene) modified electrodes. *Electrochim. Acta* 32:757–763, 1987.
116. G Inzelt, G Horányi. Combined electrochemical and radiotracer study of anion sorption from aqueous solutions into polypyrrole films. *J. Electroanal. Chem.* 230:257–265, 1987.
117. G Horányi, G Inzelt. Anion-involvement in electrochemical transformations of polyaniline. A radiotracer study. *Electrochim. Acta* 33:947–952, 1988.
118. G Horányi, G Inzelt. Application of radiotracer methods to the study of the formation and behaviour of polymer film electrodes. Investigation of the formation and overoxidation of labeled polyaniline films. *J. Electroanal. Chem.* 257:311–317, 1988.
119. G Inzelt, G Horányi. Combined electrochemical and radiotracer study on the ionic charge transport coupled to electron transfer and ionic equilibria in electroactive polymer films on electrodes. *J. Electrochem. Soc.* 136:1747–1752, 1989.
120. G Horányi, G Inzelt. *In situ* radiotracer study of the formation and behaviour of polyaniline film electrodes using labelled aniline. *J. Electroanal. Chem.* 264:259–272, 1989.
121. AV Shlepakov, G Horányi, G Inzelt, VN Andreev. Izuchenie sorbtsii ionov khlora plenkami polianilina. *Elektrokhimiya* 25:1280–1283, 1989.
122. G Inzelt. Mechanism of charge transport in polymer-modified electrodes. In: A Bard, ed. *Electroanalytical Chemistry: A Series of Advances*, Vol. 18. New York: Marcel Dekker, 1994, pp. 89–241.
123. G Horányi. β Backscattering: a radiochemical tool for the *in-situ* study of the electrochemical formation and dissolution of thin layers. *J. Electroanal. Chem.* 370:67–72, 1997.
124. A Kolics, G Horányi. β-Backscattering: a radiochemical tool for the *in-situ* study of the electrochemical formation and dissolution of thin layers. II. Application of external β-sources. *J. Electroanal. Chem.* 374:101–107, 1994.
125. O Lacroix, T Sauvage, G Blondiaux, PM Racolta, L Popa-Simil, B Alexandreanu. Metrology conditions for thin layer activation in wear and corrosion studies. *Nucl. Instrum. Meth. Phys. Res. A* 369:427–430, 1996.
126. O Lacroix, T Sauvage, G Blondiaux, L Guinard. Ultra thin layer activation by recoil implantation of radioactive heavy ions: applicability in wear and corrosion studies. *Nucl. Instrum. Meth. Phys. Res. B* 122:262–268, 1997.

127. VG Lambrev, NN Rodin, VA Koftyuk. Application of the method of neutron activation for studying corrosion processes and protective ability of paint coatings. *Progr. Org. Coat.* 30:1–8, 1997.
128. AS Besing, P Waszczuk, A Kolics, A Wieckowski. Deposition of inhibiting anions on aluminum. Proceedings of International Symposium, Corrosion and Corrosion Protection, San Francisco, 2001, pp. 393–401.
129. A Kolics, P Waszczuk, L Gáncs, Z Németh, A Wieckowski. Enhancement and inhibition of phosphate deposition on aluminum. *Electrochem. Solid-State Lett.* 3:369–373, 2000.
130. Z Németh, L Gáncs, G Gémes, A Kolics. pH dependence of phosphate sorption on aluminum. *Corros. Sci.* 40:2023–2027, 1998.
131. L Gancs, AS Besing, R Bujak, A Kolics, Z Nemeth, A Wieckowski. Interaction of chromate with aluminum in NaCl solutions. *Electrochem. Solid-State Lett.* 5:B16–B19, 2002.
132. E Kálmán. Routes to the development of low loxicity corrosion inhibitors for use in neutral solutions. European Federation of Corrosion Publications, No. 11. A Working Party Report on Corrosion Inhibitors, Institute of Materials, 1994, p. 12.
133. E Kálmán, B Várhegyi, I Bakó, I Felhősi, FH Kármán, A Shaban. Corrosion inhibition by 1-hydroxy-ethane-1,1-diphosphonic acid — an electrochemical impedance spectroscopy study. *J. Electrochem. Soc.* 141:3357–3360, 1994.
134. I Felhősi, Z Keresztes, FH Kármán, M Mohai, I Bertóti, E Kálmán. Effects of bivalent cations on corrosion inhibition of steel by 1-hydroxyethane-1,1-diphosphonic acid. *J. Electrochem. Soc.* 146:961–969, 1999.
135. J Telegdi, E Kálmán, FH Kármán. Corrosion and scale inhibitor with systematically changed structure. *Corros. Sci.* 33:1099–1103, 1992.
136. E Kálmán, I Felhősi, FH Kármán, I Lukovits, J Telegdi. Corrosion and environmental degradation. In: M Schütze, ed. *Materials Science and Technology*, Vol. 19. Weinheim: Wiley-VCH, 2000, pp. 471–537.
137. J Telegdi, FH Kármán, M Shaglouf, E Kálmán. Influence of additives on the inhibition visualized by AFM. Proceedings of the 9th European Symposium on Corrosion Inhibitors (9SEIC), Ann Univ Ferrara NS Sez V Suppl N 11:249–259, 2000.
138. I Felhősi, J Telegdi, G Pálinkás, E Kálmán, Kinetics of self-assembled layer formation on iron. *Electrochim. Acta* 47:2335–2340, 2002.
139. L Várallyai, J Kónya, FH Kármán, E Kálmán, J Telegdi. A study of inhibitor adsorption by radiotracer method. *Electrochim. Acta* 36:981–984, 1991.
140. FH Kárman, E Kálmán, L Várallyai, J Kónya. The effect of bivalent cations in the adsorption of phosphonic acids on iron electrodes studied by the radiotracer method. *Z Naturforsch* 46a:183–186, 1991.
141. I Felhősi, R Ékes, P Baradlai, G Pálinkás, K Varga, E Kálmán. Coupled radiotracer and voltammetric study of the adsorption of 1-hydroxy-ethane-1,1-diphosphonic acid on polycrystalline gold. *J. Electroanal. Chem.* 480:199–208, 2000.
142. Z Homonnay, A Vértes, E Kuzmann, K Varga, P Baradlai, G Hirschberg, J Schunk, P Tilky. Effects of AP-CITROX decontamination procedure on the surface oxide layer composition of stainless steel originating from the primary circuit of a VVER-type nuclear rector. *J. Radioanal. Nucl. Chem.* 246:131–136, 2000.
143. R Puyané. Effectiveness of isotope depleted ZnO to minimize radiation build-up in boiling water nuclear reactors. *J. Mater. Proc. Technol.* 56:863–872, 1996.
144. GR Dey, DB Naik, K Kishore, CK Vinayakumar, B Yuvaraju, G Venkateswaran, PN Moorthy. Correlation between corrosion inhibition and radiation chemical properties of some organic corrosion inhibitors. *Radiat. Phys. Chem.* 51:171–174, 1998.

10

Nanoindentation and Nanoscratching Techniques for Evaluation of Mechanical Properties of Surface Films

Masahiro Seo
Laboratory of Interfacial Electrochemistry, Graduate School of Engineering, Hokkaido University, Sapporo, Japan

Contents

10.1 INTRODUCTION

Corrosion of metals depends on the corrosion products. If very thin and compact oxide films, i.e., passive films, are formed on the metals, the corrosion rate rapidly drops to almost zero, i.e., the metals are subjected to the passive state. Many studies on physicochemical properties of passive films such as composition, structure, thickness have been performed in relation to the corrosion resistances (1,2). The passive film is locally broken by attacks of aggressive ions such as chloride ions or stresses generated in the film. If the breakdown sites are not repaired the metals are eventually subjected to localized corrosion such as pitting corrosion. In spite of that, the breakdown of passive film is directly related to the mechanical properties of the film, and there are few studies on mechanical properties of passive films or passive metal surfaces (3).

Nanoindentation technique (4,5) was developed to evaluate the mechanical properties of surface films with thickness in the range of nanometers. A combination of the nanoindentation technique with scanning probe microscopy, such as atomic force microscopy (AFM), would be powerful to evaluate the mechanical properties of thin surface films such as passive films because the shape of the indent after the indentation tests can be confirmed by AFM (6).

The effects of mechanical deformation on chemical reactions or changes in the mechanical properties of a solid surface caused by chemical reaction are often observed (7), which come from mutual transformations of mechanical and chemical energies. The scientific field dealing with mutual transformations of mechanical and chemical energies is called "mechanochemistry" (7). Mechanochemistry is very important to understand more deeply the local breakdown of a passive film or the initial stage of localized corrosion such as pitting and stress corrosion cracking.

In this chapter, the instrumentation and method of nanoindentation are described, followed by a description of the results of its application to surface thin films such as passive or anodic oxide films. Furthermore, the results of *in situ* nanoindentation under electrochemically controlled conditions are described and discussed to explore the possibility for nano-mechano-electrochemistry.

10.2 NANOINDENTATION TECHNIQUE

The nanoindentation apparatus consists mainly of a force/displacement transducer that allows the simultaneous determination of displacement and load during indentation tests. During indentation, the indentation depth and the load are recorded simultaneously versus time. Hardness and elastic modulus can be determined from the load–depth curve and the contact area of indenter versus the contact depth function. Nanoindentation is performed with a pyramidal (Berkovich or cube corner) or conical diamond tip attached to the transducer. Figure 10.1 shows the geometry of typical indenters. The Berkovich pyramidal diamond tip (Figure 10.1a) has a total

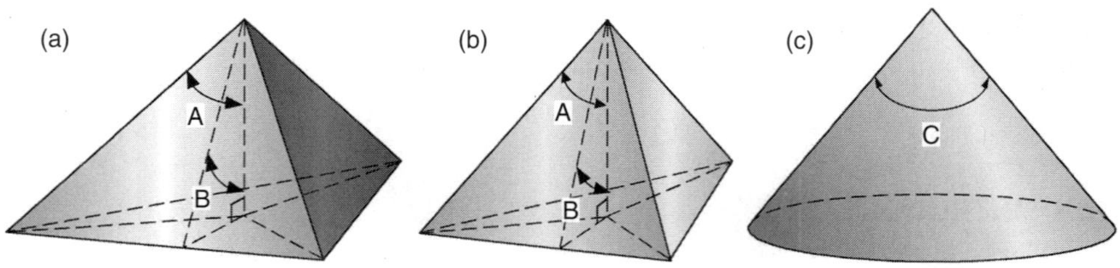

Figure 10.1 Geometries of typical indenters: (a) Berkovich tip ($A = 76.9°$ and $B = 65.3°$); (b) cube corner tip ($A = 54.7°$ and $B = 35.5°$); (c) conical tip ($C = 60°$ or $90°$).

included angle of 142.3° ($A = 76.9°$ and $B = 65.3°$) and a tip radius less than 0.1 μm. The cube corner pyramidal diamond tip (Figure 10.1b) has a total included angle of 90° ($A = 54.7°$ and $B = 35.5°$) and a tip radius less than 0.1 μm. The conical diamond tip (Figure 10.1c) has a conical angle of 90 or 60° and a tip radius of 1 to −100 μm. The Berkovich or cube corner tip is used for nanoindentation, while the conical tip can be used for both nanoscratching and nanoindentation. The cube corner tip or the more acute tip, which enables small projected contact area, is recommended for nanoindentation tests of ultrathin hard films on a soft substrate (8). In the case where indentation is performed in liquids, these indenters are attached to tungsten rods in order to avoid contact of the transducer with the solution.

10.3 DETERMINATION OF MECHANICAL PROPERTIES

10.3.1 Hardness

Mechanical properties of a solid surface such as hardness and elastic modulus can be determined from the measured load–depth curve (Figure 10.2). The hardness, H, is defined as the maximum load, L_{max}, divided by the projected contact area, A, of the indenter at L_{max} (5):

$$H = \frac{L_{max}}{A} \tag{10.1}$$

If the indenter has an ideal geometry, the projected contact area, A, can be obtained from the contact depth, h_c, at maximum load:

$$A = c_0 h_c^2 \tag{10.2}$$

where $c_0 = 24.5$ for a Berkovich pyramidal diamond tip and $c_0 = 2.598$ for a cube corner pyramidal diamond tip. The contact depth, h_c, differs from the indentation depth, h_{max}, at the maximum load in the measured load–depth curve since the outside

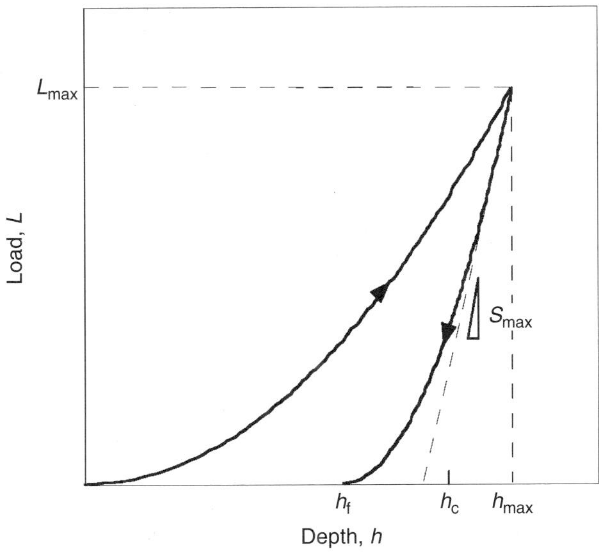

Figure 10.2 Schematic load–depth curve for determination of hardness, H, and elastic modulus, E_s.

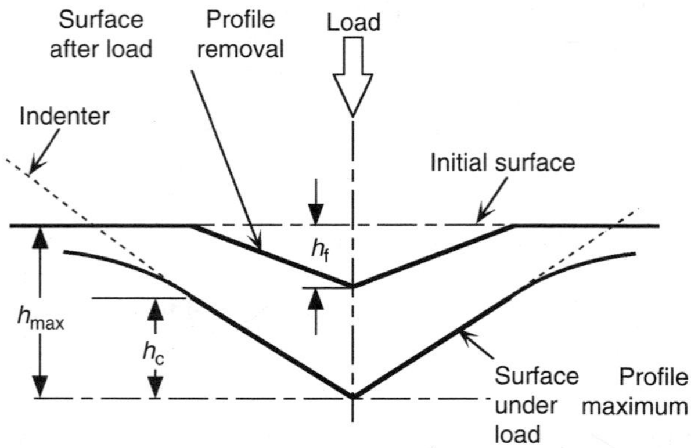

Figure 10.3 Schematic illustration of the surface in nanoindentation at a maximum load. (From W.G. Oliver, G.M. Pharr. *J. Mater. Res.* 7:1564–1583, 1992. Courtesy of Materials Research Society. With permission.)

area near the indenter is elastically deformed (shown schematically in Figure 10.3) (4). The following correction can be made to obtain h_c from h_{max}:

$$h_c = h_{max} - \frac{0.75L_{max}}{S_{max}} \tag{10.3}$$

where S_{max} is the stiffness, which is equal to the slope of unloading curve at the maximum load in the measured load–depth curve. The coefficient 0.75 in Equation (10.3) is characteristic to the pyramidal indenter.

Furthermore, the calibration of the indenter shape is necessary to exactly determine the hardness, because the geometry of the indenter is usually not ideal. For the calibration of the indenter shape, a fused quartz is used as standard material because the mechanical properties of fused quartz are known. The load–depth curves of the fused quartz at various maximum loads are measured to obtain the stiffness, S_{max}. Assuming that the elastic modulus of the fused quartz is constant, independent of indentation depth, the projected contact area, A, can be obtained as a function of stiffness, S_{max}, from (5):

$$A = \left(\frac{\pi}{4}\right)\left(\frac{S_{max}}{E_r}\right)^2 \tag{10.4}$$

$$\frac{1}{E_r} = \frac{(1-\nu_s^2)}{E_s} + \frac{(1-\nu_i^2)}{E_i} \tag{10.5}$$

where E_r is the reduced modulus, E_s the elastic modulus of the fused quartz, E_i the elastic modulus of the indenter, ν_s is Poisson's ratio of the fused quartz, and ν_i is Poisson's ratio of the indenter.

The projected contact area, A, computed from Equation (10.4) can be plotted versus the contact depth, h_c. The function $A(h_c)$ is called the area function, which is fitted to a fifth-order polynomial of the form:

$$A = c_0h_c^2 + c_1h_c + c_2h_c^{1/2} + c_3h_c^{1/4} + c_4h_c^{1/6} + c_5h_c^{1/8} \tag{10.6}$$

where c_i ($i = 0, 1, 2, 3, 4, 5$) are constants, independent of h_c and can be determined by curve fitting of the measured area function, $A(h_c)$. A typical example of the area function measured by using the cube corner diamond tip is shown in Figure 10.4.

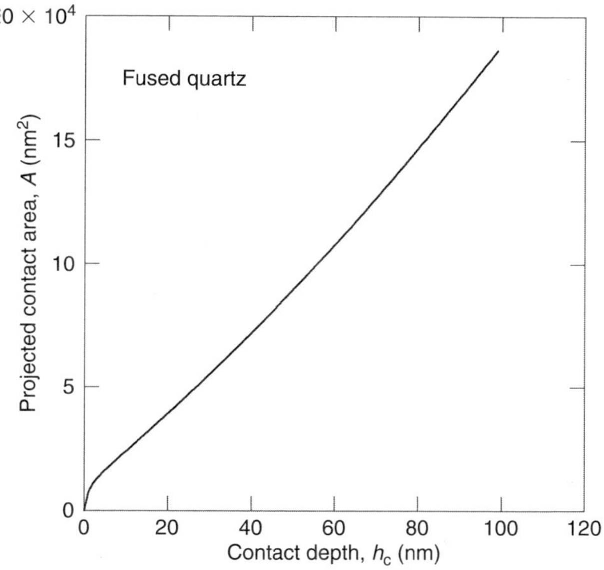

Figure 10.4 Relation between projected contact area, A, and contact depth, h_c, obtained for fused quartz by using the cube corner diamond tip.

Once the values of c_i ($i = 0, 1, 2, 3, 4, 5$) are determined, the projected contact area, A, at any h_c can be obtained from Equation (10.6), and thus the hardness, H, can be determined as a function of h_c. Equation (10.4) can be rewritten as follows:

$$S_{max}^{-1} = \frac{\pi^{1/2}}{2E_r} A^{-1/2} \tag{10.7}$$

10.3.2 Elastic Modulus

The elastic modulus of the specimen surface, E_s, can be determined by using Equations (10.7) and (10.5) from the load–depth curves measured at various maximum loads after calibration of the indenter shape. First, S_{max}^{-1} is plotted versus $A^{-1/2}$ and then the reduced elastic modulus, E_r, is determined from the slope of the linear relation between S_{max}^{-1} and $A^{-1/2}$. The elastic modulus of the specimen surface, E_s, therefore, is eventually determined by using Equation (10.5) if Poisson's ratio of the specimen, ν_s, is known.

10.4 MECHANICAL PROPERTIES OF SURFACE OXIDE FILMS EVALUATED BY NANOINDENTATION

10.4.1 Single Crystal Magnetite (100) Surfaces

Figure 10.5 shows the load–depth curves of the single crystal magnetite (1 0 0) surface measured at various maximum loads in air by using the cube corner pyramidal diamond tip. The single crystal magnetite (1 0 0) was prepared by cutting a natural octahedron magnetite crystal and then mechanically polished. The hardness of the single crystal magnetite (1 0 0) surface determined by the load–depth curves in Figure 10.5 is plotted versus contact depth, h_c, in Figure 10.6. A constant value of $H = 10.2$ GPa, independent of h_c, is noticed for $h_c > 20$ nm. The microhardness of

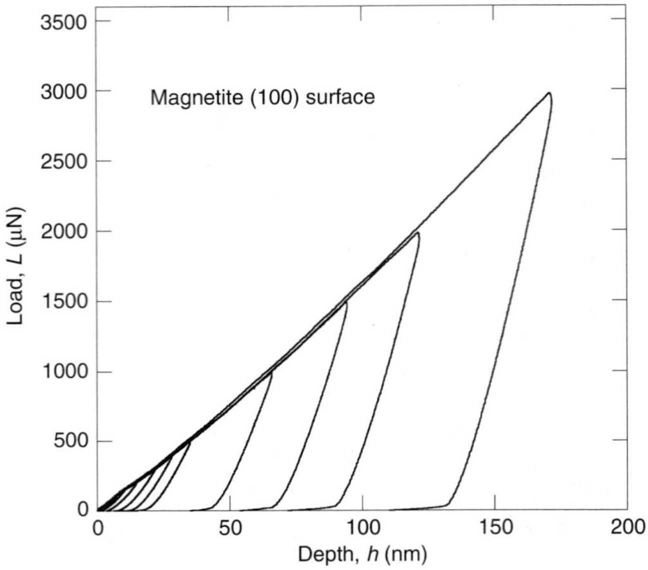

Figure 10.5 Averaged load–depth curves at various maximum loads for (100) magnetite surface measured in air. (From M. Seo, M. Chiba, *Electrochim. Acta* 47:319–325, 2001. Courtesy of Elsevier Science Ltd. With permission.)

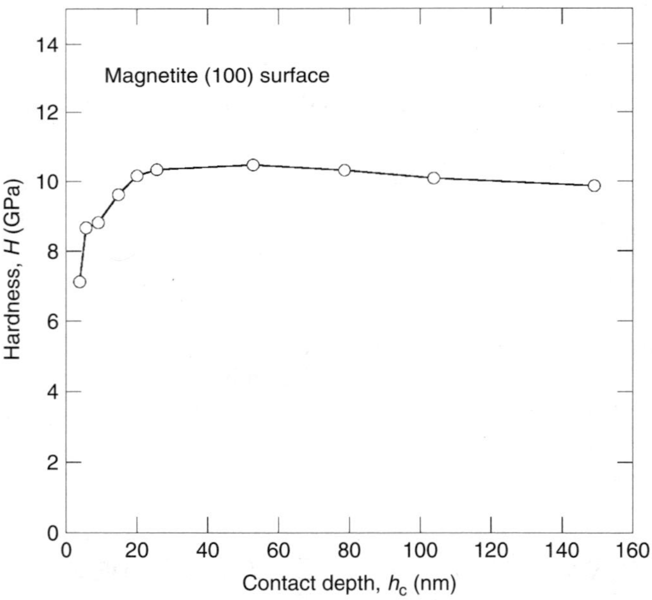

Figure 10.6 Relation between hardness, H, and contact depth, h_c, for (1 0 0) magnetite surface.

magnetite was reported to be $H = 4.7$ to 7.9 GPa at a maximum load of $L_{max} = 0.5$ N (9). The hardness, however, increases with h_c in the range of $h_c < 20$ nm. The dependence of hardness on h_c in the shallow indentation depth is called the indentation size effect (ISE) (10). The ISE is mainly caused by the difference in hardness between the surface layer and the substrate. The hardness of the surface layer of

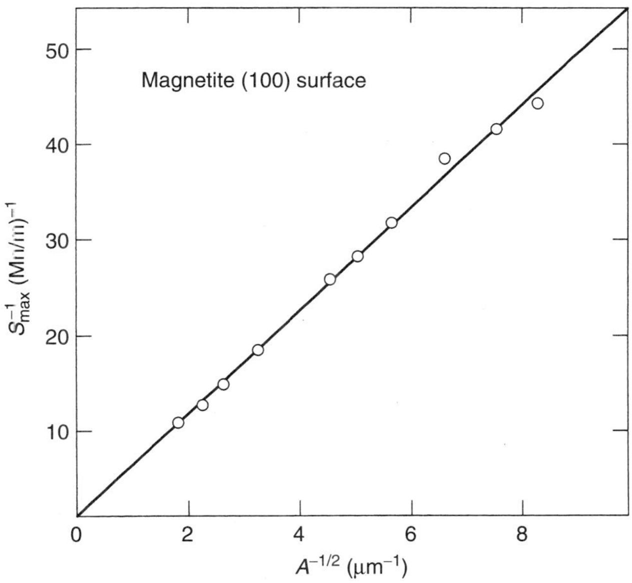

Figure 10.7 Relation between S_{max}^{-1} and $A^{-1/2}$ for (1 0 0) magnetite surface.

magnetite (1 0 0) may be less than that of the bulk probably due to depth-distribution of the chemical composition or impurity level.

Figure 10.7 shows the relationship between S_{max}^{-1} and $A^{-1/2}$ for the magnetite (1 0 0) surface. The elastic modulus of the magnetite (1 0 0) surface, $E_s = 174$ GPa, is determined from the slope of the linear relation in Figure 10.7 by assuming Poisson's ratio of the magnetite $\nu_s = 0.31$. The elastic modulus of the magnetite has not been reported so far, although the elastic modulus hematite, Fe_2O_3, was reported to be $E_s =$ 275 GPa (11).

10.4.2 ANODIC OXIDE FILMS ON TANTALUM

Anodic oxide films on valve metals such as aluminum or tantalum are widely used as electrolytic capacitors. However, there are no reliable data of mechanical properties of anodic oxide films. In the case where the indentation depth is larger than the film thickness, the measured hardness is a composite of the film and the substrate metal. An anodic oxide film thicker than the indentation depth has to be prepared in order to evaluate the mechanical properties of the oxide film itself. An anodic oxide film on tantalum was chosen as the typical case where the film thickness is larger than the indentation depth. Polycrystalline tantalum (purity: above 99.97%) surface was anodically oxidized up to a cell voltage of 200 V at a constant current density of 157 μA/cm^2 in a pH 8.4 borate solution after electropolishing in a solution mixture of HF and H_2SO_4. The thickness of the anodic oxide film formed on tantalum under the above conditions was about 310 nm as determined by ellipsometry.

The hardness, H, of the tantalum surface anodically oxidized up to 200 V is plotted versus contact depth, h_c, in Figure 10.8. The hardness of the electropolished tantalum surface is also plotted in Figure 10.8 for comparison. It is noted that the contact depth, h_c, corresponds to the actual indentation depth at a maximum load, L_{max}. The hardness of the anodized tantalum surface up to 200 V is in the range of 7 to −9 GPa, while the hardness of the electropolished tantalum surface is in the range

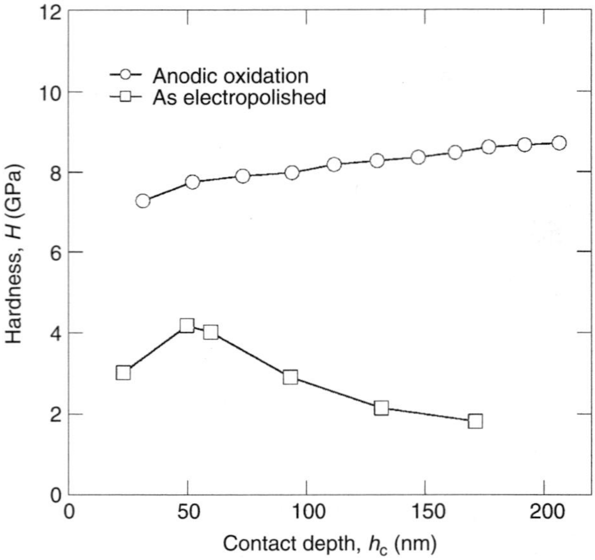

Figure 10.8 Relation between hardness, H, and contact depth, h_c, for polycrystalline tantalum surfaces anodized at a constant current density of 157 μA/cm^2 up to a cell voltage of 200 V in pH 8.4 borate solution.

of 2 to −4 GPa, indicating that the oxide film on anodized tantalum is harder than the tantalum substrate. Oxide films on anodized tantalum were found to be amorphous by x-ray diffraction. The hardness of anodized tantalum surface increases as the contact depth, h_c, increases, that is, the actual indentation depth approaches the film/substrate tantalum interface in spite of the fact that the hardness of the anodic oxide film seems harder than that of the substrate tantalum. It has been reported that boron species are incorporated in the outer part of the anodic oxide film when tantalum is anodized in ammonium pentaborate solution (12). The hardness of the outer part of the film containing boron species may be different from that of the inner part of the film without boron species. The outer part of the film containing boron species, however, was not beyond one-third of the whole film thickness (12). Bolshakov and Pharr (13, 14) pointed out that the pile-up of materials produced by nanoindentation could apparently increase the hardness. The increase of hardness in the inner layer of the anodic oxide film may result from the increase in pile-up with increasing contact depth. The hardness of 7 to −8 GPa at $h_c < 60$ nm for the anodized tantalum surface corresponds to that of the anodic oxide film itself since the ratio of h_c to the film thickness is less than 0.2 and the effect of the substrate may be negligibly small.

The electropolished tantalum surface is covered with an air-formed film with a thickness less than 2 nm. Although the reason why the hardness peak appeared in the range of $h_c = 50$ to −60 nm for the electropolished tantalum surface is not clear, the hardness value reaches a steady state of $H = 2$ GPa at $h_c > 150$ nm, which corresponds to the hardness of the tantalum substrate. The elastic modulus, E_s, of the anodic oxide film on tantalum evaluated from the linear relation between S_{max}^{-1} and $A^{-1/2}$ was 132 GPa, while that of the tantalum substrate was 341 GPa. Alcalá et al. (15) have reported that the hardness at $h_c = 50$ nm and the elastic modulus for the anodic oxide film (480 nm) prepared with anodic oxidation of sputter-deposited tantalum in 0.01 M ammonium pentaborate solution are $H = 5.3$ GPa and $E_s = 140$ GPa, respectively. It is seen that the hardness of the anodic oxide film on polycrystalline tantalum is higher than that of the anodic oxide film on sputter-deposited tantalum, while the elastic modulus is almost the same between them. Alcalá et al. (15) have also reported that the hardness at $h_c = 50$ nm and the elastic modulus for the anodic oxide film (490 nm) formed on aluminum foil in 0.01 M ammonium pentaborate solution are $H = 7$ GPa and $E_s = 122$ GPa, which are factors of 1.3–3.7 times lower than those of crystalline aluminas such as γ-Al_2O_3 and sapphire.

10.4.3 Anodic Oxide Films on Titanium

The ISE is important for the evaluation of mechanical properties of thin surface films with a thickness less than the indentation depth. Here, the anodic oxide film on titanium was chosen as the typical case where the thickness of anodic oxide film is less than the contact depth. Titanium thin films with a thickness of 250 nm were deposited on glass plates (5 mm × 40 mm × 210 μm) by magnetron sputtering and then anodically oxidized at 5.7 V (RHE) for 1 h in 0.1 M HCl. The thickness of the anodic oxide film formed on titanium is known to be 18 nm from ellipsometry (16). The nanoindentation hardness for the titanium thin films anodically oxidized is plotted versus contact depth, h_c, in Figure 10.8. The hardness of the titanium surfaces decreases with increasing contact depth, h_c. This ISE is common for a hard surface film on a soft substrate.

10.4.4 Separation of Hardness of Film from the Composite Hardness

If the effects of elastic compression and surface pile-up are ignored, the composite hardness, H, may be determined by a volume law of mixture:

$$H = H_f \frac{V_f}{V_t} + H_s \frac{V_s}{V_t} \tag{10.8}$$

where H_f is the hardness of surface film, H_s the hardness of substrate, V_f the plastically deformed volume of the film, V_s the plastically deformed volume of the substrate, and $V_t = V_f + V_s$. In the case of materials (e.g., stainless steel) with low σ/E values where σ is the uniaxial compressive yield stress and E Young's modulus, the plastically deformed volume can be calculated by Tabor's model (17). On the other hand, in the case of materials (e.g., high strength steels and ceramics) with high σ/E values, the expanding spherical cavity model (18) can be used for the calculation of the deformed volume. In Equation (10.8), it is assumed that the film/substrate interface is frictionless. In reality, however, some adhesion between film and substrate takes place that will result in shear stresses transmitted across the interface and thereby influencing the shape of the plastic zone, depending on H_f/H_s. It is necessary to incorporate a further weighing factor, χ, for obtaining a reasonable fit to experimental data. Equation (10.8) is modified for a soft film on a hard substrate,

$$H = H_f \chi^3 \frac{V_f}{V_t} + H_s \frac{V_s}{V_t} \tag{10.9a}$$

and for a hard film on a soft substrate,

$$H = H_f \frac{V_f}{V_t} + H_s \chi^3 \frac{V_s}{V_t} \tag{10.9b}$$

where χ is given by

$$\chi = \left(\frac{E_f H_s'}{E_s H_f'} \right)^n \tag{10.9c}$$

In Equation (10.9c), E_f and E_s are Young's moduli, H_f' and H_s' are characteristic hardness values of the film and substrate at some contact depth, respectively, and n, determined empirically, has a value ranging from 1/2 to 1/3.

Alternatively, Bhattacharya and Nix (19, 20) made a finite element simulation of the indentation process to investigate the elastic–plastic response of materials and calculated elastic and plastic deformations during submicrometer indentation of thin films on substrates by a conical indenter. The effect of elastic and plastic properties of both film and substrate on the composite hardness of the film/substrate system was studied by determining the average pressure under the indenter as a function of the indentation depth. They derived eventually empirical equations to predict the hardness variation with contact depth for two different cases. For the case of a hard film on a softer substrate, the composite hardness, H, can be described as

$$\frac{H}{H_s} = 1 + \left(\frac{H_f}{H_s} - 1 \right) \exp\left[-\frac{(H_f/H_s)}{(\sigma_f/\sigma_s)(E_f/E_s)^{1/2}} (h_c/d) \right] \tag{10.10a}$$

where σ_f and σ_s are the yield strengths of the film and substrate, respectively, and d is the film thickness. For the case of a soft film on a harder substrate, the composite hardness, H, can be given as

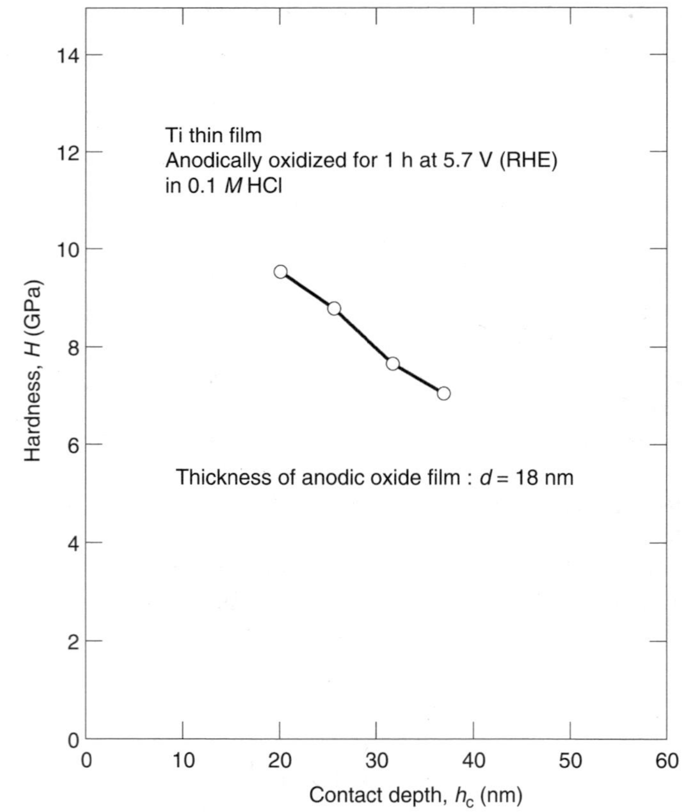

Figure 10.9 Relation between hardness, H, and contact depth, h_c, for titanium thin film anodized for 1 h at 5.7 V (RHE) in 0.1 M HCl.

$$\frac{H}{H_s} = 1 + \left(\frac{H_f}{H_s} - 1\right) \exp\left[-\frac{(\sigma_f/\sigma_s)}{(E_f/E_s)}(h_c/d)^2\right] \tag{10.10b}$$

Equation (10.10a) may be applied to the anodic oxide film/titanium substrate system since the anodic oxide film is harder than the titanium substrate as seen from the dependence of composite hardness on contact depth in Figure 10.9. The following deformation of Equation (10.10a) is useful for separation of H_f and H_s from H:

$$\ln\left(\frac{H}{H_s} - 1\right) = \ln\left(\frac{H_f}{H_s} - 1\right) - \left\{\frac{(H_f/H_s)}{K}\left(\frac{h_c}{d}\right)\right\} \tag{10.11a}$$

$$K = \left[\left(\frac{\sigma_f}{\sigma_s}\right)\left(\frac{E_f}{E_s}\right)^{1/2}\right] \tag{10.11b}$$

Since K, H_s, and H_f are constant, a linear relation should hold between $\ln((H/H_s) - 1)$ and (h_c/d). The slope of the linear line corresponds to $[(H_f/H_s)/K]$, while the intersection point of the linear line with the ordinate corresponds to $\ln((H_f/H_s) - 1)$. If the value of H_s is known, the value of H_f can be obtained from the linear relation between $\ln((H/H_s) - 1)$ and (h_c/d). The rearrangement of H vs. h_c curves in Figure 10.9 gave the good linear relation between $\ln((H/H_s) - 1)$ and (h_c/d) as shown in Figure 10.10. In Figure 10.11, the constant value of $H = 5.0$ GPa obtained in large

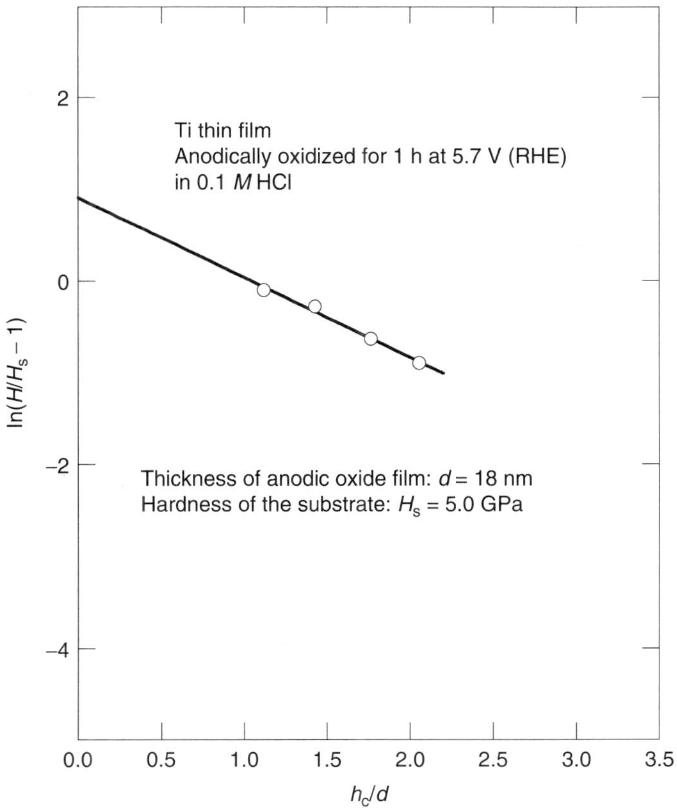

Figure 10.10 Relation between $\ln(H/H_s - 1)$ and h_c/d for titanium thin film anodized for 1 h at 5.7 V (RHE) in 0.1 M HCl.

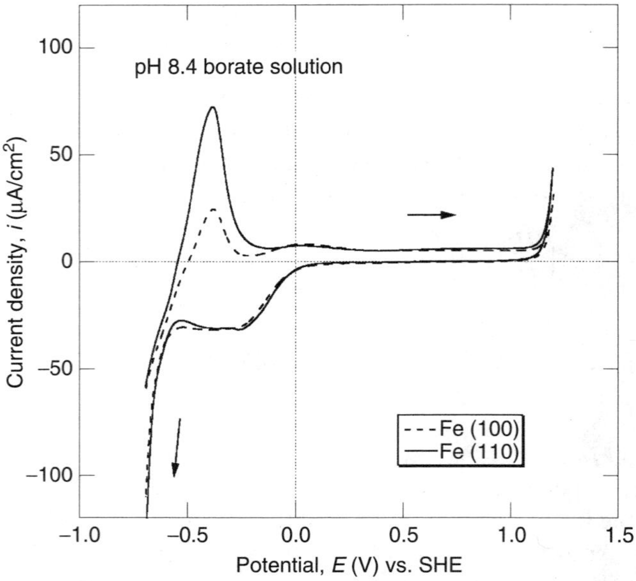

Figure 10.11 Potentiodynamic polarization curves of (1 0 0) and (1 1 0) iron surfaces measured at a potential sweep rate of 100 mV/min in pH 8.4 borate solution. (From M. Seo, M. Chiba. *Electrochim. Acta* 47: 319–325, 2001. Courtesy of Elsevier Science Ltd. With permission.)

contact depth for the titanium thin film as received was employed as H_s because the effect of air-formed film on the hardness can be neglected. As a result, the values of $H_f = 16.2 \pm 0.5$ GPa and $K = 3.7 \pm 0.8$ were determined for the titanium thin films anodically oxidized at 5.7 V (RHE) for 1 h in 0.1 *M* HCl, 0.1 *M* H_2SO_4, pH 6.9 phosphate, and pH 8.4 borate solutions. The values of $H_f = 16.2 \pm 0.5$ GPa are very close to that ($H = 15$ GPa) of Nb (0.05 wt%) doped rutile type of single crystal TiO_2 (0 0 1) surface determined by nanoindentation tests. The value of $H_s = 5.0$ GPa for the titanium thin film is fairly larger than that ($H_s = 2$ GPa) for the polycrystalline titanium thick plate. The titanium thin film prepared on a glass plate with magnetron sputtering has usually a columnar structure normal to the glass plate, which seems to harden the surface.

10.5 *IN SITU* NANOINDENTATION OF PASSIVE METAL SURFACES ELECTROCHEMICALLY CONTROLLED IN SOLUTIONS

Nanoindentation tests on metal surfaces are usually performed in air, that is, *ex situ* after the oxide film formation. Passive films on metals such as iron, nickel, chromium are so thin (1 to −5 nm) that the indentation depth is larger than the film thickness. In the case where the indentation depth is significantly larger than the film thickness, the passive film is compressed or elongated underneath or at the very near edge of the indentation tip during nanoindentation. If the indentation load exceeds the fracture strength of the film and the film is broken at the very near edge of the indentation tip, the bare metal surface emerges at the rupture site and may oxidize with oxygen in air.

Nanoindentation tests performed in solution (*in situ*) on passive metal surfaces may be useful for the evaluation of the mechanical properties of metal surfaces electrochemically polarized in the passive state. In the case where the passive film under electrochemically controlled conditions is broken at the very near edge of the indentation tip during *in situ* nanoindentation, an active dissolution of metal from the bare surface at the rupture site and followed by a repassivation may take place because the metal is kept electrochemically in the passive state. It is expected that a series of mechano-electrochemical reactions on passive metal surfaces, such as film breakdown, active dissolution, and repassivation, influences the load–depth curves measured by *in situ* nanoindentation. *In situ* nanoindentation tests on iron and titanium surfaces kept electrochemically in the passive state have been recently made from the above viewpoint (21–23). For that purpose, a miniature electrochemical cell was designed for *in situ* nanoindentation (23). The Berkovich or cube corner diamond tip was attached to a tungsten rod to avoid contact of the transducer with the solution during *in situ* nanoindentation.

10.5.1 (1 0 0) and (1 1 0) Single Crystal Iron Surfaces Electrochemically Polarized in the Passive State

The *in situ* nanoindentation test was applied to (1 0 0) and (1 1 0) single crystal iron surfaces electrochemically controlled in the passive state in a pH 8.4 borate solution. The cube corner diamond tip was used in accordance with recommendations for nanoindentation of ultrathin hard films on a soft substrate (8). The single crystal iron (1 0 0) and (1 1 0) specimens (disk-shaped plates with a diameter of 10 mm and a thickness of 1.0 or 1.4 mm) were mechanically polished with alumina abrasives and finally electropolished in a mixture of 70% $HClO_4$ and glacial CH_3COOH (1:20) at 17°C to remove the worked layer. Figure 10.10 shows the potentiodynamic polarization curves of the iron (1 0 0) and (1 1 0) surfaces measured at a potential sweep rate of 100 mV/min in pH 8.4 borate solution (22). In the active dissolution region, the anodic current peak of the iron (1 1 0) surface is three times as much as that of the iron (1 0 0) surface in spite of the fact that the (1 1 0) surface is energetically more stable than the (1 0 0) surface. On the other hand, there are no significant differences in passivity-maintaining current between iron (1 0 0) and (1 1 0) surfaces. The high anodic current density of the iron (1 1 0) surface, although not clear, may result from its high density of steps by which the anodic dissolution is faster than on the terrace.

Figure 10.12 shows the averaged load–depth curves measured at various maximum loads for (1 0 0) and (1 1 0) iron surfaces kept at 0.25 V (SHE) in the passive region for 1 h in a pH 8.4 borate solution (22). These load–depth curves were obtained by averaging the load–depth curves that were measured 20 times at each maximum load because of fluctuations in the load–depth curves. Similar load–depth curves at various maximum loads were obtained for iron surfaces kept at other potentials in the passive region. The hardness (2.9 to 3.3 GPa) of the passive (1 0 0) and (1 1 0) iron surfaces determined from load–depth curves at a maximum load of $L_{max} = 400\ \mu N$ in Figure 10.12 is shown as a function of the potential at which the passive film was formed in Figure 10.13. The hardness of (1 1 0) iron surfaces is larger by about 10% than that of (1 0 0) iron surfaces. Furthermore, the hardness increases slightly with increasing film formation potential. The slope of the linear relation between hardness and film formation potential is about 0.2 GPa/V or less. The indentation depth is in the range of 65 to 90 nm. An ellipsometry study (24) showed that the average thickness of passive films formed potentiostatically on a polycrystalline iron for 1 h in pH 8.4 borate solution increases linearly with the film formation potential and is in the range of 2 to 4.5 nm. The ratio of film thickness to indentation

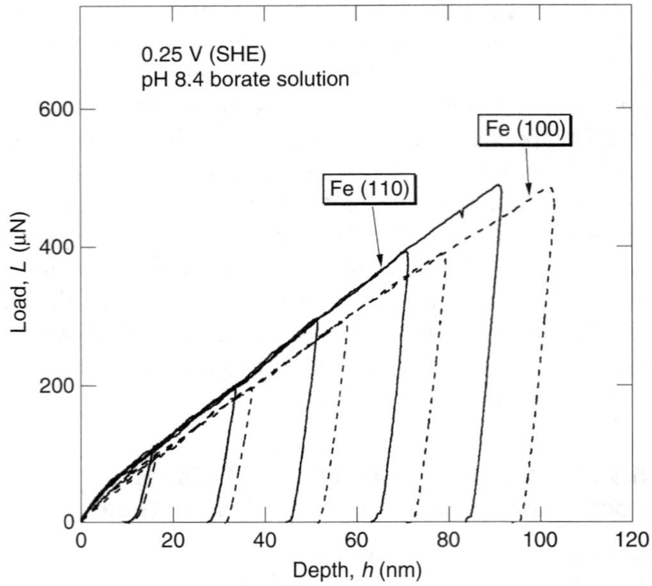

Figure 10.12 Averaged load–depth curves measured at various maximum loads for (1 0 0) and (1 1 0) iron surfaces kept at 0.25 V (SHE) in the passive region for 1 h in pH 8.4 borate solution. The load–depth curves were measured 20 times at each maximum load. (From M. Seo, M. Chiba. *Electrochim. Acta* 47:319–325, 2001. Courtesy of Elsevier Science Ltd. With permission.)

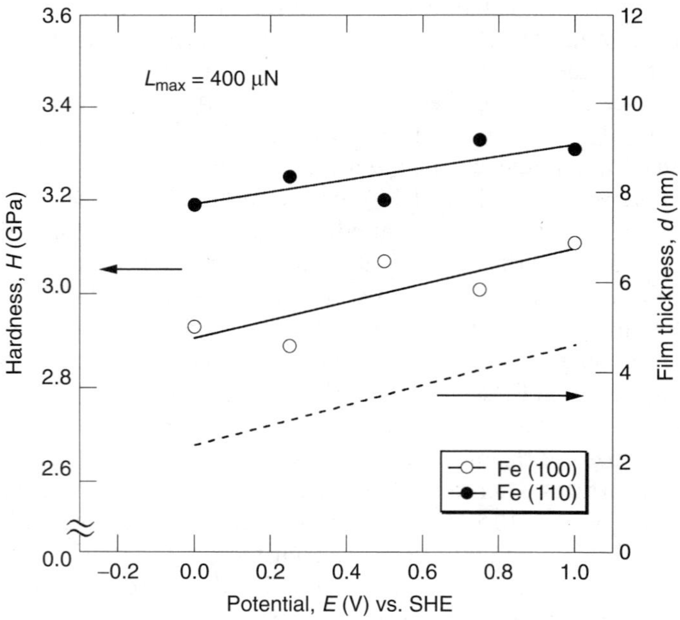

Figure 10.13 Hardness, H, at a maximum load of $L_{max} = 400$ μN as a function of formation potential of passive film, E, for (1 0 0) and (1 1 0) iron surfaces kept in the passive region for 1 h in pH 8.4 borate solution. The dashed line shows the thickness of passive films on iron obtained ellipsometrically by Sato et al. (24). (From M. Seo, M. Chiba. *Electrochim. Acta* 47: 319–325, 2001. Courtesy of Elsevier Science Ltd. With permission.)

depth is only from 0.03 to 0.05. The hardness (10.2 GPa) of magnetite surface (see Figure 10.8) is 3.4 times as much as that of passive iron surfaces.

It is known (25, 26) that the structure of passive films on iron single crystals (0 0 1) and (1 1 0) is similar to that of magnetite. Therefore, the slight increase in hardness with electrode potential in Figure 10.13 may be ascribed to the contribution of a passive film in spite of a ratio of film thickness to the indentation depth that is very small. The hardness of bare (1 0 0) and (1 1 0) iron surfaces, estimated from the potential dependencies of hardness and film thickness in Figure 10.13, is 2.7 and 3.0 GPa, respectively. Metals with a body center cubic structure (bcc) such as α-Fe tend to slip easily to the ⟨1 1 0⟩ direction. The nanoindentation performed normal to the (1 0 0) surface with a cube corner diamond tip has a vector component of the load to the ⟨1 1 0⟩ direction by which slip is induced easily leading to the decrease in hardness. In contrast, nanoindentation normal to the (1 1 0) surface has no vector component to the ⟨1 1 0⟩ direction. The difference in hardness between the bare (1 0 0) and (1 1 0) iron surfaces may result from the difference in easiness of slip between them.

10.5.2 Effect of Dichromate Treatment on Hardness of Passive Iron Surfaces

The addition of a passivator as chromate or dichromate to solutions permits to improve the corrosion resistance of metals. It is important to explore how a chromate or dichromate treatment influences the mechanical properties of passive iron surface in relation to its corrosion resistance. Figure 10.14 shows the load–depth curves for

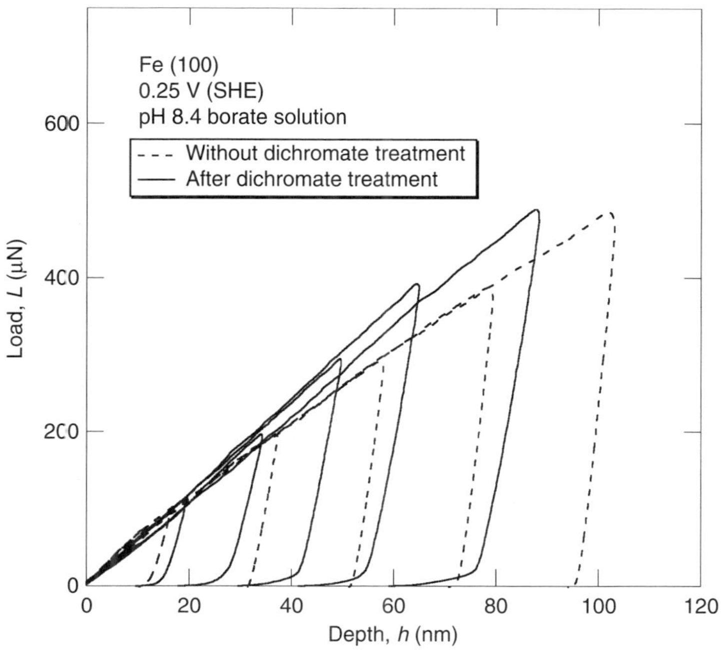

Figure 10.14 Averaged load–depth curves measured at various maximum loads for (1 0 0) iron surface kept at 0.25 V (SHE) in the passive region for 1 h in pH 8.4 borate solution without and after immersion in 5 × 10^{-2} *M* $K_2Cr_2O_7$ solution at room temperature for 24 h. (From M. Seo, M. Chiba. *Electrochim. Acta* 47: 319–325, 2001. Courtesy of Elsevier Science Ltd. With permission.)

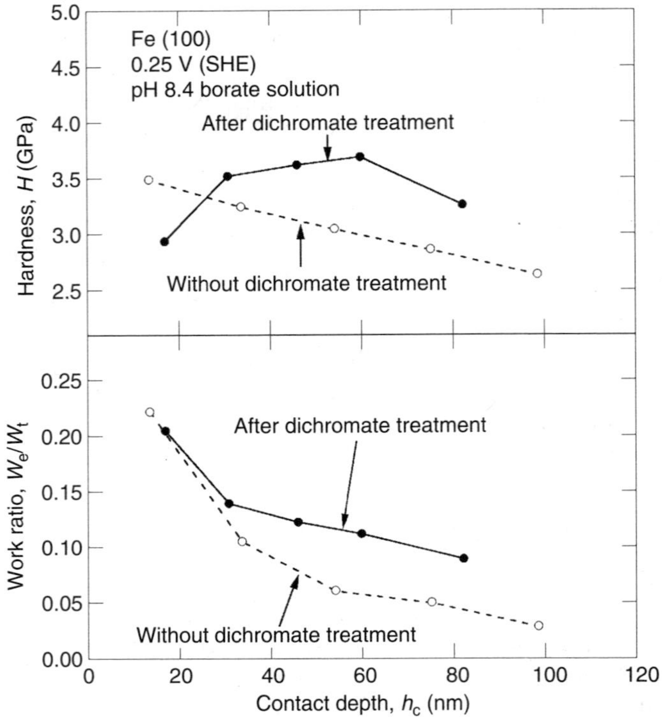

Figure 10.15 Hardness, H, and the ratio of elastic work to total work, W_e/W_t, as a function of h_c for (1 0 0) iron surfaces kept at 0.25 V (SHE) in the passive region for 1 h in pH 8.4 borate solution without and after a dichromate treatment. (From M. Chiba, M. Seo. *Corros. Sci.* 44: 2379–2391, 2002. Courtesy of Elsevier Science Ltd. With permission.)

the iron (1 0 0) surfaces kept at 0.25 V (SHE) in the passive region for 1 h in pH 8.4 borate solution without and after immersion in 5×10^{-2} M $K_2Cr_2O_7$ solution at room temperature for 24 h (22). The dichromate treatment tends to shift the load–depth curves to higher load and lower depth. Figure 10.15 shows the hardness, H, and the ratio of elastic work to total work, W_e/W_t, as a function of h_c for passive (1 0 0) iron surfaces kept at 0.25 V (SHE) without and after a dichromate treatment (23). The hardness, H, of the passive (1 0 0) iron surfaces increases with the dichromate treatment except for $h_c < 30$ nm.

The plastic work on the specimen surface during nanoindentation is also important for evaluation of the mechanical properties since hardness is associated with plastic deformation. An increase in hardness means an increased resistance for plastic deformation. The plastic work achieved during nanoindentation decreases with increasing hardness. The graphical integration between $h = 0$ and h_{max} of the loading curve shown schematically in Figure 10.16 gives the total work, W_t, during nanoindentation, while the graphical integration between h_f and h_{max} gives the work, W_e, related to elastic deformation during nanoindentation (23). The work, W_p, related to plastic deformation, can be obtained by subtracting W_e from W_t. The work ratio, W_e/W_t, as well as hardness, H, increase with the dichromate treatment at $h_c > 30$ nm (Figure 10.15). The increase in W_e/W_t corresponds to a decease in W_p/W_t, that is, the increase in hardness. It should be noted that the indentation depth, $h_c =$ 30 nm, is more than 10 times as large as the thickness of the passive film on iron. It is

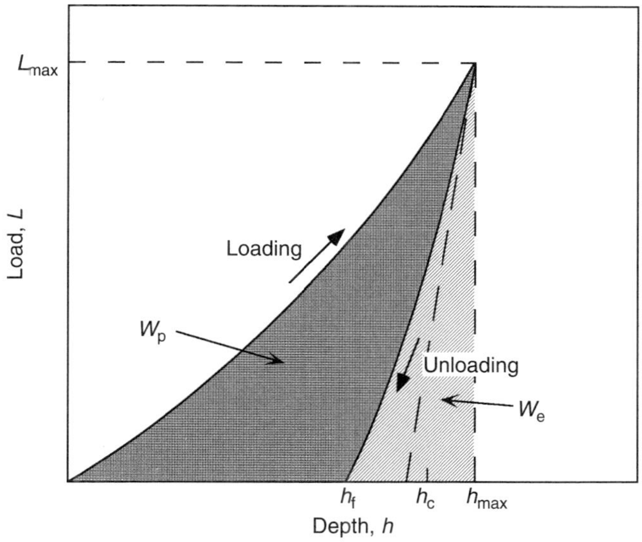

Figure 10.16 Schematic load–depth curve for calculation of the work ratio, W_e/W_t. W_e: elastic work done during nanoindentation; W_p: plastic work done during nanoindentation; W_t: total work done during nanoindentation ($=W_e + W_p$). (From M. Chiba, M. Seo. *Corros. Sci.* 44:2379–2391, 2002. Courtesy of Elsevier Science Ltd. With permission.)

known that the ellipsometrical thickness of passive film formed on polycrystalline iron at 0.25 V (SHE) for 1 h in pH 8.4 borate solution is about 3 nm (24).

The passive film at the near edge of the contact surface can be partly ruptured as the indentation depth increases during loading. The ruptured parts, however, will be repaired, depending on the repassivation rate. The rupture and repair, therefore, will be repeated during loading. The passive film, even if very thin, would act as a resistance against indentation to some extent. The film rupture promotes the indentation, while the film repair impedes the indentation. The nano-mechano-electrochemical properties of passive film, therefore, affect the load–depth curves. It is known that a chromium (mainly Cr^{3+})-enriched passive film is formed on iron by immersion in a chromate or dichromate solution (27, 28). The high corrosion resistance of chromium-enriched film results from not only the high barrier property of the film itself against ion transport but also from the high ability of self-healing at the defective or ruptured sites (29, 30). The self-healing of chromate conversion coatings has been confirmed by scratch tests (29). Although the passive film on iron after the dichromate treatment is significantly thinner than the chromate conversion coatings, the same mechanism should be operative in the passive film. The self-healing behavior means a high repassivation rate at ruptured sites. The hexavalent chromium ions slightly existing in the passive film would suppress the active dissolution from the ruptured sites to promote repassivation. At present, the promotion of repassivation at the ruptured sites is the most probable explanation for the increase in H and W_e/W_t due to the dichromate treatment.

Figure 10.17 shows the difference in hardness, H, and work ratio, W_e/W_t, between *in situ* and *ex situ* nanoindentation tests on passive (1 0 0) iron surfaces after a dichromate treatment. H and W_e/W_t obtained with *in situ* nanoindentation are larger than those obtained with *ex situ* nanoindentation, suggesting that the repassivation rate at the ruptured sites of the passive film during *in situ*

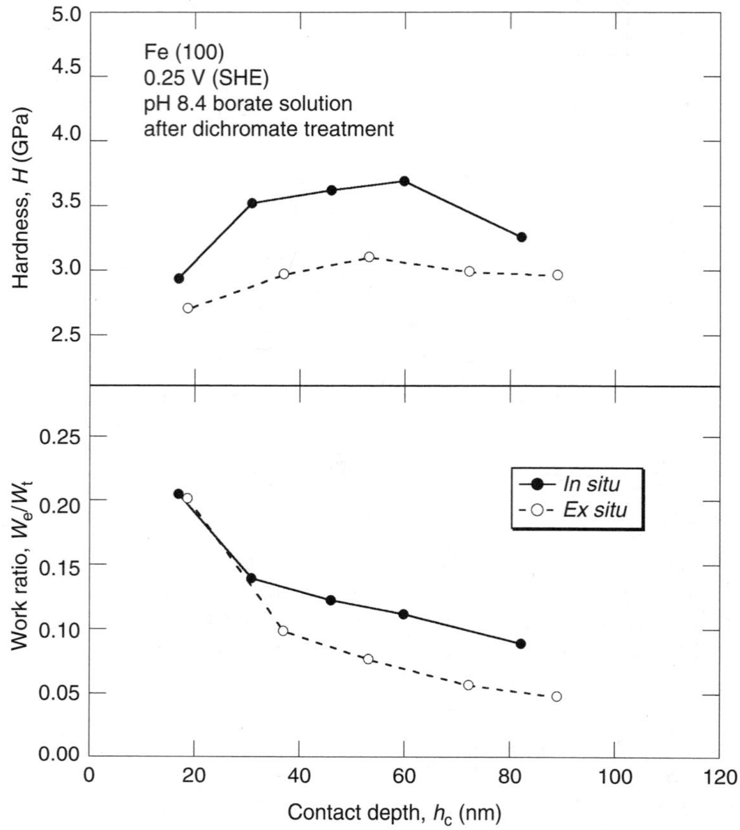

Figure 10.17 Hardness, H, and work ratio, W_e/W_t, as a function of contact depth, h_c, obtained with *in situ* and *ex situ* nanoindentation tests for passive (1 0 0) iron surfaces after a dichromate treatment. (From M. Chiba, M. Seo. *Corros. Sci.* 44:2379–2391, 2002. Courtesy of Elsevier Science Ltd. With permission.)

nanoindentation is higher than during *ex situ* nanoindentation. Figure 10.18 shows the differences in H and W_e/W_t between *in situ* and *ex situ* nanoindentation tests on (100) iron surfaces passivated without the dichromate treatment. In contrast to Figure 10.17, it is seen from Figure 10.18 that there are no significant differences in H and W_e/W_t between *in situ* and *ex situ* nanoindentation, suggesting that the repassivation rate at the ruptured sites of the passive film during *in situ* nanoindentation is almost the same as during *ex situ* nanoindentation. Consequently, it is deduced that repassivation due to the dichromate treatment is more significant in solution than in air.

10.5.3 Polycrystalline Titanium Surfaces Electrochemically Controlled in the Passive State

If passive or anodic oxide films are thicker, the difference in load–depth curves between *in situ* and *ex situ* nanoindentations would be more significant. Figure 10.19 shows the load–depth curves obtained in *in situ* (kept at 5 or 0 V (RHE) in solution) and *ex situ* (in air) nanoindentation tests at a maximum load of L_{max} = 300 μN for polycrystalline titanium surfaces anodized at 5 V (RHE) for 1 h in pH 8.4

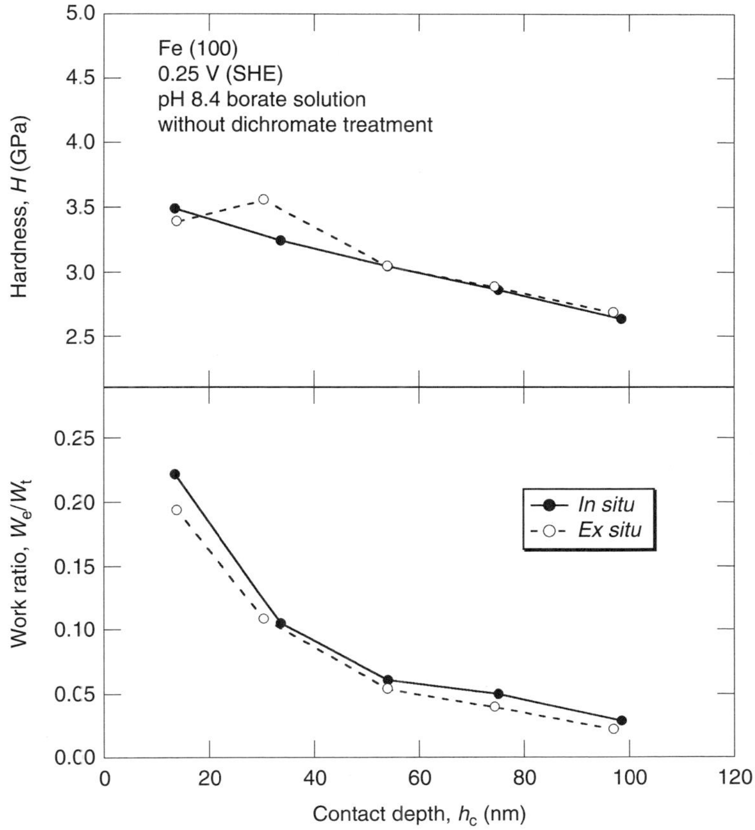

Figure 10.18 Hardness, H, and work ratio, W_e/W_t, as a function of contact depth, h_c, obtained with *in situ* and *ex situ* nanoindentation tests for passive (1 0 0) iron surfaces without the dichromate treatment. (From M. Chiba, M. Seo. *Corros. Sci.* 44:2379–2391, 2002. Courtesy of Elsevier Science Ltd. With permission.)

borate solution. A higher load at indentation depth larger than 7 nm is required for *in situ* nanoindentation (Figure 10.19). The ellipsometrical thickness of anodic oxide film formed on titanium at 5 V (RHE) for 1 h in pH 8.4 borate solution is about 15 nm (16), two times as large as the indentation depth above which the *in situ* load–depth curve deviates upward as compared with the *ex situ* load–depth curve. The rupture of the anodic oxide film at the near edge of the contact surface may start when the indentation depth exceeds the half thickness of the film. In the case of *in situ* nanoindentation, the repair of the ruptured sites would be promoted by the potential difference between titanium substrate and solution. The difference in load–depth curves between titanium surfaces kept at 5 and 0 V (RHE), respectively after anodization at 5 V (RHE) for 1 h may indicate that the repassivation rate is higher as the potential difference between titanium substrate and solution is larger. On the other hand, in the case of *ex situ* nanoindentation, the repair of the ruptured sites would be slow because the oxidation of the ruptured sites will be brought only in contact with oxygen in air. The large difference in load–depth curves between *in situ* and *ex situ* nanoindentations for the anodized titanium surfaces, therefore, is ascribed to the high repassivation rate at the large potential difference, that is, high electric field between titanium substrate and solution.

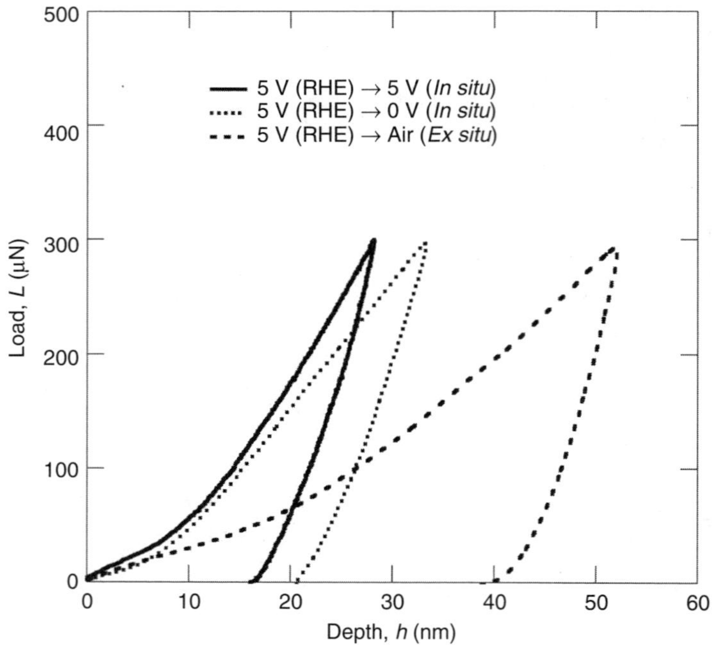

Figure 10.19 Load–depth curves obtained with *in situ* and *ex situ* nanoindentation tests at a maximum load of $L_{max} = 300\ \mu N$ for polycrystalline titanium surfaces anodized at 5 V (RHE) for 1 h in pH 8.4 borate solution. (From M. Seo, Y. Kurata. *Electrochem. Acta* 49:3221–3228, 2003. Courtesy of Elsevier Science Ltd. With permission.)

10.6 *IN SITU* NANOSCRATCHING ON PASSIVE METAL SURFACES IN SOLUTION

Nanoscratching tests are also useful for the evaluation of properties of passive metal surfaces such as the coefficient of friction and are performed by measuring a lateral force as a function of a scratching distance under a constant normal force. The coefficient of friction, μ', is defined as the ratio of lateral force, F_L, to normal force, F_N:

$$\mu' = \frac{F_L}{F_N} \tag{10.12}$$

Figure 10.20 shows the AFM image and height profile after *in situ* nanoscratching at a rate of 0.2 μm/sec along a lateral distance of 2 μm at a normal force of $F_N = 500\ \mu N$ on (1 0 0) iron surface kept at 0.25 V (SHE) in the passive region for 1 h in a pH 8.4 borate solution. The groove produced by the scratching and the resulting protrude can be observed from Figure 10.20. In Figure 10.21, the displacement in normal direction, D_N, and coefficient of friction, μ', are plotted as a function of displacement in lateral direction, D_L, during *in situ* nanoscratching at a normal force of $F_N = 500\ \mu N$ on (1 0 0) iron surface kept at 0.25 V (SHE) in the passive region. D_N is not constant and varies with scratching length, that is, D_L, while μ' increases with D_L and attains a steady state. The steady-state values of μ' obtained with *in situ* and

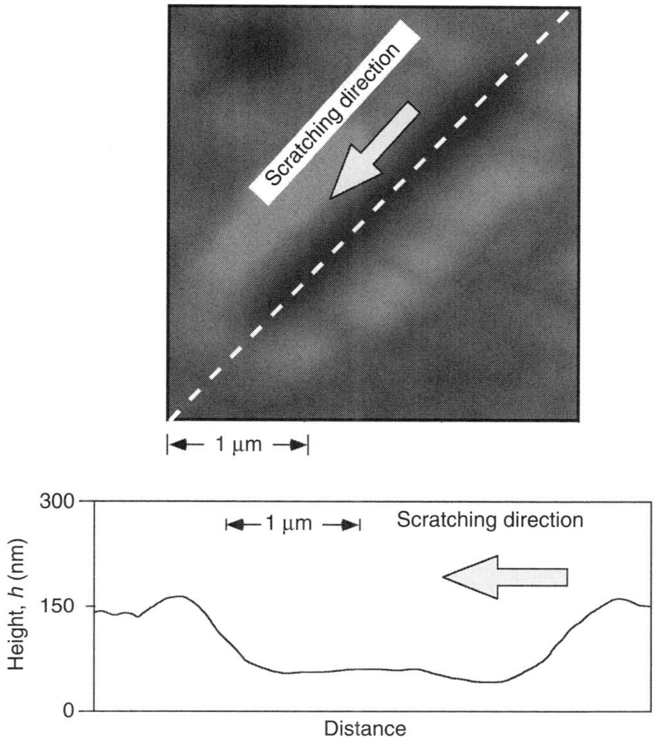

Figure 10.20 AFM image and depth–profile after nanoscratching at a rate of 0.2 μm/sec along a lateral distance of 2 μm at a normal force of $F_N = 500\,\mu N$ for (1 0 0) iron surface kept at 0.25 V (SHE) in the passive region for 1 h in pH 8.4 borate solution.

ex situ nanoscratching tests on passive (1 0 0) and (1 1 0) iron surfaces are plotted versus normal force, F_N, in Figure 10.22 (31).

According to Amontons–Coulomb's law of friction (32, 33), F_L is proportional to F_N and thus the values of μ' are constant, independent of F_N. However, the results of Figure 10.22 indicate that μ' increases with increasing F_N. Moreover, μ' obtained from *in situ* nanoscratching tests are always higher than those obtained from *ex situ* nanoscratching tests on both (1 0 0) and (1 1 0) iron surfaces. It is also remarkable that μ' for (1 0 0) iron surface is higher than for (1 1 0) iron surface, irrespective of *in situ* or *ex situ* nanoscratchings. Similar results were obtained for polycrystalline titanium surfaces anodized at 5 V (RHE) for 1 h in pH 8.4 borate solution as shown in Figure 10.23 (34). The values of μ' obtained with *in situ* nanoscratching tests are surprisingly higher than those obtained with *ex situ* nanoscratching tests.

The friction force, F_L, consists of an adhesion term, F_a, and a ploughing term, F_p (32, 33):

$$F_L = F_a + F_p \tag{10.13}$$

The adhesion term, F_a, is given by

$$F_a = \tau\left(\frac{F_N}{H}\right) \tag{10.14}$$

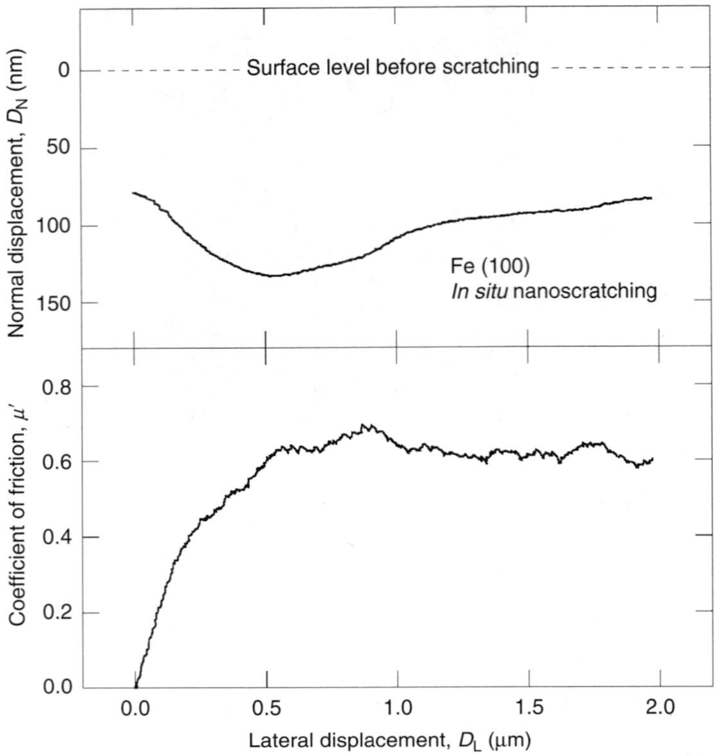

Figure 10.21 Normal displacement, D_N, and coefficient of friction, μ', as a function of lateral displacement, D_L, at a normal force of $F_N = 500\,\mu N$ for (1 0 0) iron surface kept at 0.25 V (SHE) in the passive region for 1 h in pH 8.4 borate solution.

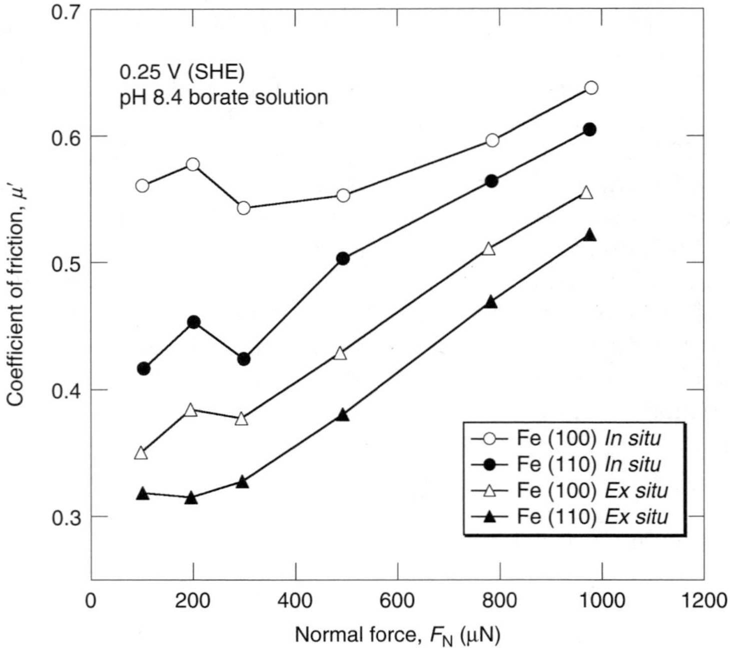

Figure 10.22 Relation between coefficient of friction, μ', and normal force, F_N, obtained with *in situ* and *ex situ* nanoscratching tests for passive (1 0 0) and (1 1 0) iron surfaces.

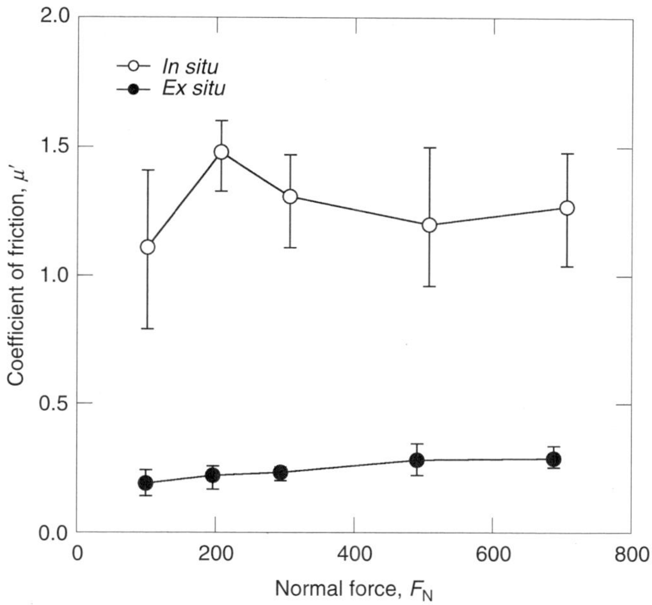

Figure 10.23 Coefficient of friction, μ', as a function of normal force, F_N, obtained with *in situ* and *ex situ* nanoscratching tests for polycrystalline titanium surfaces anodized at 5 V (RHE) for 1 h in pH 8.4 borate solution. (From M. Seo, Y. Kurata. *Electrochem. Acta* 49: 3221–3228, 2003. Courtesy of Elsevier Science Ltd. With permission.)

where τ is the shear stress, i.e., shear strength, and H is the hardness at the contact interface. The ploughing term, F_p, may be represented by

$$F_p = \rho_f \left(\frac{F_N}{H}\right)\left(\frac{A_v}{A_h}\right) \tag{10.15}$$

where ρ_f is the plastic flow pressure of the specimen against scratching, that is, the impedance against the movement of indenter tip to the lateral direction, A_v and A_h are the vertical and horizontal cross-sectional areas of the indenter tip in contact with the specimen surface, respectively. The ratio of A_v to A_h is constant, depending on the geometry of a conical indenter. If substituting Equations (10.14) and (10.15) into Equation (10.13), μ' is represented by

$$\mu' = \frac{\tau}{H} + r\left(\frac{\rho_f}{H}\right) \tag{10.16}$$

where r is A_v/A_h. The values of H for many metals are in the range of three to five times as much as those of τ. The contribution of the adhesion term to μ', therefore, is in the range of 0.2 to 0.33. The large values of μ' obtained with *in situ* nanoscratching tests as compared to *ex situ* nanoscratching tests on iron and titanium surfaces suggest that the contribution of the ploughing term to μ' is predominant. A series of mechano-electrochemical reactions at the passive metal surface such as film rupture, active dissolution, and repassivation may be repeated at the front side of the indenter tip during *in situ* nanoscratching to provide the high impedance against the movement of indenter tip. It seems that the predominant contribution of the ploughing term originates from the series of mechano-electrochemical reactions taking place at the front side of the indenter tip during *in situ* nanoscratching. The ploughing term may also be related to tribo-corrosion and wear (35, 36). Further experiments, however, are needed to analyze the content of the ploughing term and to find a clear connection with the mechano-electrochemical reactions of the passive metal surfaces including tribo-corrosion and wear.

10.7 SUMMARY

Nanoindentation and nanoscratching are useful techniques for evaluating the mechanical properties (hardness, elastic modulus, and coefficient of friction) of surface oxide films formed on metals. In the case of thick oxide films, such as anodic oxide films on tantalum, the mechanical properties of the oxide films such as hardness and elastic modulus can be evaluated from the load–depth curves measured by nanoindentation tests. In the case of thin oxide films, the mechanical properties measured by nanoindentation tests are composite properties of the surface oxide film and the substrate metal. The separation of composite hardness has been examined for the anodic oxide film/titanium system by using a model applicable to a hard film on a soft substrate. The hardness of titanium anodic oxide film thus separated was close to that of rutile type of TiO_2.

In situ nanoindentation tests under electrochemical control in solution as well as *ex situ* nanoindentation tests in air were performed on (1 0 0) and (1 1 0) single crystal iron surfaces, and on polycrystalline titanium surfaces. The differences in hardness values obtained between *in situ* and *ex situ* nanoindentation tests were not significant for both passive (1 0 0) and (1 1 0) iron surfaces. The hardness of the passive (1 1 0) iron surface was greater by 10% than that of the passive (1 0 0) iron surface. On the other hand, for the passive (1 0 0) iron surface after a dichromate treatment, the hardness obtained by *in situ* nanoindentation was fairly greater than that obtained by *ex situ* nanoindentation. The increase in hardness due to the dichromate treatment was ascribed to the high repassivation rate at the film-ruptured sites near the contact edge of the indenter during *in situ* nanoindentation as compared with *ex situ* nanoindentation in air. Moreover, for the anodized titanium surface, the hardness obtained by *in situ* nanoindentation was significantly larger than that obtained with *ex situ* nanoindentation, indicating that the repassivation rate at the film-rupture sites is promoted due to a high electric field under the potential control in solution as compared to air. The results obtained with *in situ* nanoindentation, therefore, may reflect the mechano-electrochemical properties of a passive film such as film rupture and repair.

The friction of the passive metal surfaces can be evaluated by nanoscratching tests. A high coefficient of friction on passive iron and titanium surfaces was obtained with *in situ* nanoscratching as compared to *ex situ* nanoscratching. Particularly, the difference in the coefficient of friction obtained by *in situ* and *ex situ* nanoscratching tests was significantly large for the anodized titanium surface. It has been pointed out that the friction of the passive metal surface is composed of two terms, namely adhesion and ploughing. The high coefficient of friction obtained with *in situ* nanoscratching indicates that the ploughing term is predominant and is influenced by a series of mechano-electrochemical reactions such as film rupture, active dissolution, and repassivation, which is repeated at the front side of indenter tip during *in situ* nanoscratching. The *in situ* nanoscratching as well as *in situ* nanoindentation are useful for the evaluation of the mechano-electrochemical properties of passive metal surfaces.

REFERENCES

1. N. Sato. Toward a more fundamental understanding of corrosion processes. *Corrosion* 45:354–368, 1989.
2. B.D. Craig. *Fundamental Aspects of Corrosion Films in Corrosion Science*. New York: Plenum Press, 1991.
3. G. Bombara. The role of oxide films in stress corrosion cracking initiation. *Corros. Sci.* 19:991–999, 1979.
4. W.C. Oliver, G.M. Pharr. An improved technique for determining hardness and elastic modulus using load and displacement sensing indentation experiments. *J. Mater. Res.* 7:1564–1583, 1992.
5. B. Bhushan. *Handbook of Micro/Nano Tribology*. New York: CRC Press, 1995, pp. 321–396.
6. E.T. Lilleodden, W. Bonin, J. Nelson, J.T. Wyrobek, W.W. Gerberich. *In situ* imaging of μN load indents into GaAs. *J. Mater. Res.* 10: 2162–2165, 1995.
7. E.M. Gutman. *Mechanochemistry of Materials*. Cambridge: Cambridge International Science Publishing, 1998.
8. Y. Yamada-Takamura, E. Shimono, T. Yoshida. Nanoindentation characterization of cBN films deposited from vapor phase. Proceedings of the 14th International Symposium on Plasma Chemisry, Vol. III, 1999, pp. 1629–1634.
9. G.V. Samsonov. *The Oxide Handbook*. New York: IFI/Plenum, 1973, p. 259.
10. P.M. Sargent. Use of the indentation size effect on microhardness for materials characterization. In: *Microindentation Techniques in Materials Science and Engineering*, P.J. Blau, B.R. Lawn, eds. ASTM Special Technical Publication 889. Philadelphia, PA: American Society for Testing and Materials, 1985, pp. 160–174.
11. S.K. Venkataraman, D.L. Kohlstedt, W.W. Gerberich. Continuous microindentation of passivating surfaces. *J. Mater. Res.* 8: 685–688, 1993.
12. K. Shimizu, K. Kobayashi, G.E. Thompson, P. Skeldon, G.C. Wood, A novel maker approach for the determination of transport numbers during the growth of anodic oxide films on tantalum. *Philos. Mag. B* 73:461–485, 1996.
13. G.M. Pharr. Measurement of mechanical properties by ultra-low load indentation. *J. Mater. Sci. Eng. A* 253:151–159, 1998.
14. A. Bolshakov, G.M. Pharr. Influences of pileup on the measurement of mechanical properties by load and depth sensing indentation techniques. *J. Mater. Res.* 13:1049–1058, 1998.
15. G. Alcalá, P. Skeldon, G.E. Thompson, A.B. Mann, H. Habazaki, K. Shimizu. Mechanical properties of amorphous anodic alumina and tantala films using nanoindentation. *Nanotechnology* 13:451–455, 2002.
16. T. Ohtsuka, M. Masuda, N. Sato. Ellipsometric study of anodic oxide films on titanium in hydrochloric acid, sulfuric acid, and phosphate solution. *J. Electrochem. Soc.* 132:787–792, 1985.
17. D. Tabor. *The Hardness of Metals*. Oxford: Clarendon Press, 1951.
18. K.L. Johnson. The correlation of indentation experiments. *J. Mech. Phys. Solids* 18:115–116, 1970.
19. A.K. Bhattacharya, W.D. Nix. Finite element simulation of indentation experiments. *Int. J. Solids Structures* 24:881–891, 1988.
20. A.K. Bhattacharya, W.D. Nix. Analysis of elastic and plastic deformation associated with indentation testing of thin films on substrate. *Int. J. Solids Structures* 24:1287–1298, 1988.
21. M. Seo, M. Chiba, K. Suzuki. Nano-mechano-electrochemistry of iron (100) surface in solution. *J. Electroanal. Chem.* 473:49–53, 1999.
22. M. Seo, M Chiba. Nano-mechano-electrochemistry of passive metal surfaces. *Electrochim. Acta* 47:319–325, 2001.
23. M. Chiba, M. Seo. Effects of dichromate treatment on mechanical properties of passivated single crystal iron (100) and (110) surfaces. *Corros. Sci.* 44:2379–2391, 2002.

24. N. Sato, K. Kudo, T. Noda. The anodic oxide film on iron in neutral solution. *Electrochim. Acta* 16:1909–1921, 1971.
25. M.F. Toney, A.J. Davenport, L.J. Oblonsky, M.P. Ryan, C.M. Vitus. Atomic structure of the passive oxide film on iron. *Phys. Rev. Lett.* 24:4282–4285, 1997.
26. A.J. Davenport, L.J. Oblonsky, M.P. Ryan, M.F. Toney. The structure of the passive film that forms on iron in aqueous environments. *J. Electrochem. Soc.* 147:2162–2173, 2000.
27. J.B. Lumsden, Z. Szklarska-Smialowska. The properties of films formed on iron exposed to inhibitive solutions. *Corrosion* 34:169–176, 1978.
28. M. Seo, N. Sato. Inhibition in the Context of Passivation. Corrosion-89. New Orleans: NACE, 1989, Paper No. 138, pp. 1–18.
29. J. Zhao, G. Frankel, R.L. McCreery. Corrosion protection of untreated AA-2024-T3 in chloride solution by a chromate conversion coating monitored with Raman spectroscopy. *J. Electrochem. Soc.* 145:2258–2264, 1998.
30. R.G. Kelly, J.R. Scully, D.W. Shoesmith, R.G. Buchheit. *Electrochemical Techniques in Corrosion Science and Engineering*. New York: Marcel Dekker, 2003, pp. 285–288.
31. M. Chiba, M. Seo. Mechano-electrochemical properties of passive iron surfaces evaluated by an *in situ* nano-scratching test. *J. Electrochem. Soc.* 150:B525–B529, 2003.
32. F.P. Bowden, D. Tabor. *The Friction and Lubrication of Solids I*. Oxford: Clarendon Press, 1954.
33. D.F. Moore. *Principle and Application of Tribology*. London: Pergamon Press, 1975.
34. M. Seo, Y. Kurata. Nano-mechano-electrochemical properties of passive titanium surfaces evaluated by *in situ* nano-indentation and nano-scratching. *Electrochim. Acta.* 49: 3221–3228, 2003.
35. S. Mischler, A. Spiegel, D. Landolt. The role of passive oxide films on the degradation of steel in tribocorrosion systems. *Wear* 225–229:1078–1087, 1999.
36. D. Landolt, S. Mischler, M. Stemp. Electrochemical methods in tribocorrosion: a critical appraisal. *Electrochim. Acta* 46:3913–3929, 2001.

11
Introduction to Electrochemical Instrumentation

Carl-Albrecht Schiller
Zahner-elektrik, Kronach, Germany

Contents

11.1 INTRODUCTION

Electrochemistry has developed into a key science in many areas currently under development. To take just a few of the most recent examples, we might mention hydrogen technology, microelectronics, nanotechnologies, and sensor elements.

Suitable professionally manufactured instruments are available nowadays for a wide spectrum of electrochemical methods of investigation — from simple, static measurements of potential through to the analysis of electrochemical noise. Modern devices allow high-precision measurements to be made conveniently. It has now become the exception to find home-made, customized electronics being used for these purposes in laboratories. This is because an in-house construction capable of the precision and reliability of professional instruments calls for high level of detailed understanding along with a high level of commitment for its manufacture, and this resource may well be lost from the project itself.

However, the brave new world of experimentation also has its downside. The notorious "black box" stands between the object of interest and the experimental result. Just what happens on the way from the electrical contacts at the measuring cell to the solid-looking digital display of a numerical value? Does this number, or this pretty graph in the printed document, truly represent the property that was looked for, or is it just an answer to a question that was never actually asked?

The fact is that modern instrumentation distances the scientist from the measurement process itself. A certain measure of understanding of how the measured data are acquired and what goes on within the black box is necessary if the capacity to be critical of the results is to be recreated.

It is for this reason that the present chapter takes on the task of reminding the reader of the fundamentals of electrochemical instrumentation. The principles underlying measurement set-ups for the most popular electrochemical methods are then further discussed. Significant emphasis is placed there on the recognition of parasitic effects. This should help us to avoid the artifacts of the measurement process, which would dramatically decrease the great meaning of correctly measured electrochemical information.

11.2 A REFRESHER: NECESSARY BASIC PRINCIPLES OF ELECTRICAL THEORY

Anyone who feels comfortable with terms such as "electrical resistance," "complex impedance," "parallel circuit," "series circuit," "capacitance," and "inductance" can confidently skip the following section, and go straight on to Section 11.3. Those who read on will be introduced briefly to those fundamental principles of electrical theory that are necessary for understanding electrochemical measurement procedures.

Electromagnetism is one of the fundamental forces of nature, along with gravity and the nuclear forces. Its components, electricity and magnetism, are linked to one another through Maxwell's equations (1, chapter 18–1 p. 18.1), while its force is transmitted by photons. When considering electrochemical processes it is possible to leave the relationship between electromagnetism and the other natural forces safely to one side.

The time-scale of electrochemical processes is usually long, and rapidly changing electrical and magnetic fields rarely occur. For the same reason, wave aspects are negligible. It is usually therefore sufficient, for electrochemical purposes, to focus on electricity in isolation.

The elementary force associated with electricity is electric potential. For different locations in space, r, it assigns different potential energies, $E(r)$, adopted by bodies within that field, depending on their charge, Q.

The potential defines the "electrical tension," V, between two positions in space with the dimension "Volt" [V]

$$V = \Delta E/Q[\mathrm{V}] \tag{11.1}$$

When a charge, Q, moves between two such spatial elements within a time, t, this process is referred to as an "electrical current"

$$I = Q/t\,[\mathrm{A}] \tag{11.2}$$

having the unit "Ampere" [A]. If the spatial elements differ in voltage by ΔV, then the energy

$$\Delta E = \Delta \mathrm{V} \cdot I \cdot t\,[\mathrm{V\ A\ sec}] \tag{11.3}$$

is released. The above processes can be more easily grasped if we restrict ourselves to their effects on precisely two spatial positions, the "poles," in a closed system. A full electrical description of such a "two-pole" comprises the voltage, $V(t)$, between the poles and the current, $I(t)$, that flows between the two poles.

Some special cases of $V(t)$ and $I(t)$ characterize particular electrical material classes. In particular, $I = 0$ at constant $V \neq 0$ applies to insulators, $V = 0$ at constant $I \neq 0$ describes superconductors, while $\mathrm{V} \propto I$[1] applies to Ohmic conductors.

If two two-poles are connected together at one end of each (series connection, Figure 11.1, left), then the voltages add to form the total voltage at the open pole ends. The same current flows through both two-poles. If the two two-poles are connected at both ends (parallel circuit, Figure 11.1, right), then those two two-poles have the same voltage. The total current through the joint pole is the sum of the partial currents in the individual two-poles

$$Z_s = Z_1 + Z_2 \tag{11.4}$$

[1]In the following the symbol $\propto$ is used to indicate proportionality between two magnitudes.

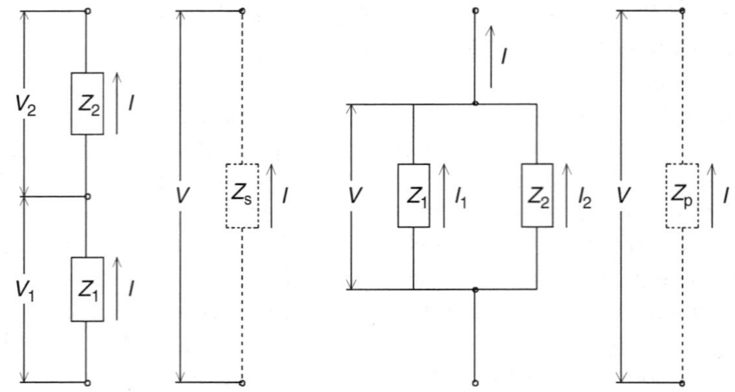

Figure 11.1 Serial and parallel connections of two-poles Z_i with corresponding voltage V and current I.

$$1/Z_p = 1/Z_1 + 1/Z_2 \tag{11.5}$$

The diagram moves straight on to illustrate the rules for the property of "impedance," $Z = V/I$, for two-poles (Equations (11.4) and (11.5)). These rules for serial and parallel connections can be extended to describe combinations of a number of two-poles to form a "network" (Figure 11.2, Equations (11.6) and (11.7)) (1, chapter 22–3, p. 22.7). Kirchhoff's laws state that the sum of all potentials along a closed path within a network is zero, just like all the differences of altitude occurring during a walk in the mountains add up to zero if you return to your starting point. They state further that the sum of all the currents at a node is also zero. This is quite easy to understand, if we think of the node as a point with zero dimensions having one-dimensional inflows and outflows: what flows in must also flow out

$$\sum_{i=1}^{n} V_i = 0 \tag{11.6}$$

$$\sum_{i=1}^{n} I_i = 0 \tag{11.7}$$

Two-poles such as Ohmic conductors whose behavior is independent of time can be completely electrically described by curves relating voltage and current, either of the

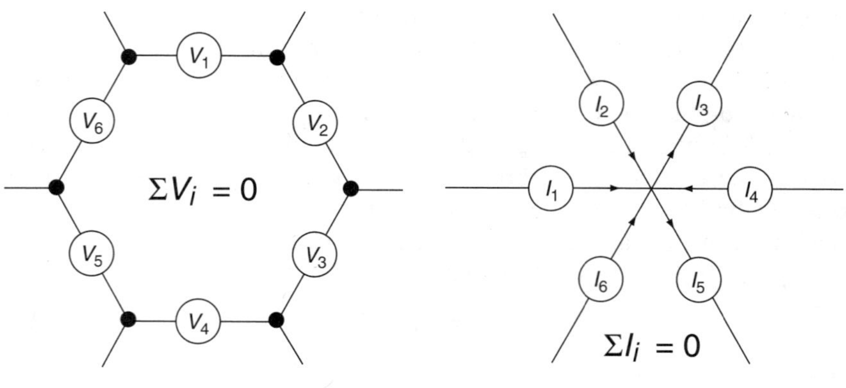

Figure 11.2 Kirchhoff's law: voltage and current in network meshes and nodes.

form $V = f(I)$, or expressed as $I = g(V)$. The most elementary representatives of this class are Ohmic conductors themselves, in which the function $f(I)$ is simply $(k \cdot I)$ where k is a constant; in other words, the relationship is one of direct proportionality.

The proportionality factor is the electrical resistance, R, and its unit is the Ohm [Ω]. The relationship $V \propto I$ is "Ohm's law":

$$V = R \cdot I\,[\mathrm{V}] \tag{11.8}$$

Ohmic relationships occur frequently in electrochemistry. Metallic and ionic conductors are the most frequent representatives of this group.

Current–voltage curves (Figure 11.3a) for Ohmic resistors are straight lines through the origin at $V = 0$ and $I = 0$. Two-poles, such as resistors, which exhibit zero current at constant $V = 0$ are referred to as "passive." An "active" two-pole results from connecting a passive two-pole in series with a voltage source. The voltage source symbolizes a special two-pole with a constant potential difference that is independent of the current flowing. An example of the current–voltage curve for this sort of object is shown in Figure 11.3(b). It is found to be useful to generalize the concept of resistance to nonlinear, active two-poles. Such a combination provides, for instance, a usable electrical model for the behavior of fuel cells or batteries when operating close to their nominal voltage. A nonlinear differential resistance is therefore defined

$$R_{\mathrm{diff}} = (\mathrm{d}V/\mathrm{d}I)_{\mathrm{V}} \tag{11.9}$$

for selected sections of a current–voltage curve at constant V. Both linear and nonlinear conductivities occur widely. Something approaching linearity (where R_{diff} is constant) is most often found when the potential between the poles under consideration affects the velocity with which charge is transported but does not have a significant effect on the number of charge carriers available for that transport. This is most often the case when the total potential difference is distributed more or less evenly over the space between the poles. If, on the other hand, high field strengths occur locally, as can happen when the conductivity mechanisms change at phase boundaries, this can form an active charge carrier. The activation processes lead to an exponential dependence of conductivity on the potential difference. The resulting current–voltage curve often has the form

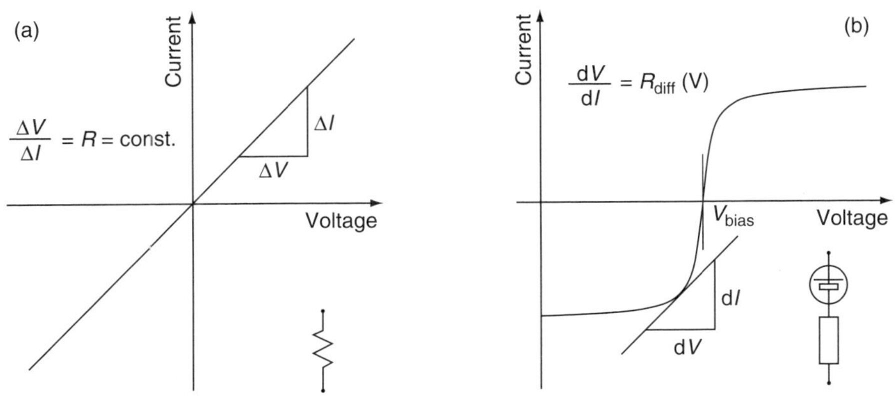

Figure 11.3 Symbol and current–voltage curve of an Ohmic resistance (a) and a nonlinear two-pole including a potential source (b).

$$I = I_1 \cdot \left(e^{(V-V_1)/b_1 \cdot V_T} - 1\right) + I_2 \cdot \left(e^{(V-V_2)/b_2 \cdot V_T} - 1\right) \cdots$$

$$I = \sum_{i=1}^{n} I_i \cdot \left(e^{(V-V_i)/b_i \cdot V_T} - 1\right) \tag{11.10}$$

where b_i represents the characteristic constants, I_i the current, V_i the activation potential, and V_T is the "temperature voltage." The magnitude V_T has a value of approx. 26 mV at room temperature, and is composed only of physical constants and the temperature

$$V_T = R \cdot T/F \tag{11.11}$$

Here R represents the general gas constant, F the Faraday constant, and T the absolute temperature. Common examples of current–voltage curves described by Equation (11.10) are the characteristic curves of semiconductor barrier layers (C. kittel, 1996) ("diode curves") and the electrochemical transfer reaction (the "Butler–Volmer" curve). Differentiating Equation (11.10) shows that the differential resistance of such a current–voltage curve in the simplest case of $n = 1$ takes the value

$$R_{\text{diff}} = b \cdot V_T/(I + I_1) \tag{11.12}$$

We can now gain a glimpse of the usefulness of the concept of differential resistance:

- The coefficients, b, can easily be determined from the curve of R_{diff} against current. Generally, they are around $b \approx z$, where z is the effective number of the charge carriers responsible for the current flow. In this way they reveal something about the mechanism of the current-carrying processes.
- At the point where $I = 0$ it is possible to determine the "exchange currents," I_i, leading us, for example, to reaction rates.
- The temperature dependency of R_{diff} allows us to determine band gaps in semiconductors and activation energies $E_i = V_i \cdot Q_e$ (where Q_e is the elementary electron charge) for electrochemical processes.

11.2.1 Voltage, Current, and Time: The Role of Time

So far we have restricted ourselves to cases in which the effect of time can be ignored, and in which the two-pole can be characterized by an unambiguous current–voltage curve. In fact, nature seldom approximates to this limiting case. The simple idea that "everything changes" is only one reason for this. Even in the absence of aging effects, in order to observe time-independent laws it is necessary first of all for temporal changes to entirely subside. The following example will illustrate this.

Metals are typical Ohmic conductors. However, the resistance of metals is, to a first approximation, proportional to temperature. The current–voltage curve of the tungsten filament of a conventional light bulb exhibits Ohmic resistance at small currents. If the curve is, however, traced at higher currents, the electrical power dissipated

$$N = V \cdot I \tag{11.13}$$

leads over time to a noticeable rise in temperature through electrical heating. A very rapid recording of the curve does not give the filament time to heat up, and illustrates the linear current–voltage relationship of the "cold resistance." If, on the other hand, the measurement is taken sufficiently slowly for the electrical power and

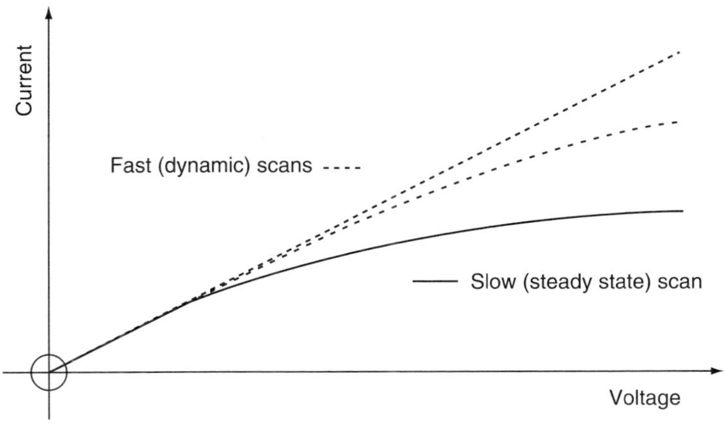

Figure 11.4 Fast "dynamic" (dotted lines) and slow "steady state" (solid line) current voltage curves of a light bulb.

the radiated heat to reach approximate equilibrium, then a significantly bent "steady-state" current–voltage curve (Figure 11.4) is obtained.

Two-poles such as incandescent bulbs therefore exhibit an "electrical memory." Both voltage and the current have an effect on the two-pole, so that the current no longer depends purely on the instantaneous value of the voltage at time t_1, but also on what has previously happened during $t < t_1$ (Equations (11.14) and (11.15)). Similar considerations apply to the dependence of the voltage on current

$$I(t_1) = f[V(t \leq t_1)] \tag{11.14}$$

$$V(t_1) = f[I(t \leq t_1)] \tag{11.15}$$

Objects with electrical memory can be described by time-dependent differential equations such as the following equations:

$$I(t) = a_0 \cdot V + a_1 \cdot \mathrm{d}V/\mathrm{d}t + a_2 \cdot \mathrm{d}^2V/\mathrm{d}t^2 + \cdots \tag{11.16}$$

$$V(t) = b_0 \cdot I + b_1 \cdot \mathrm{d}I/\mathrm{d}t + b_2 \cdot \mathrm{d}^2I/\mathrm{d}t^2 + \cdots \tag{11.17}$$

The most elementary representatives of such two-poles are, on the one hand, the capacitor and, on the other hand, as an example of objects that behave inductively, the coil. Capacitors are electrical components in which the voltage, V, between the terminals, is proportional to the electrical charge, Q, that has passed through the terminals (Equation (11.18)). The quotient, C, is known as the capacitance

$$V = Q/C \tag{11.18}$$

The current, I, which is the charge, Q, flowing in unit time, leads to the time law:

$$I = \mathrm{d}Q/\mathrm{d}t \tag{11.19}$$

for the capacitor in differential form, where the symbol for the time derivative of X, $\mathrm{d}X/\mathrm{d}t$ is $\dot{X}$:

$$I = C \cdot \frac{\mathrm{d}V}{\mathrm{d}t} = C \cdot \dot{V} \tag{11.20}$$

If Equation (11.20) is solved for V, then the electrical memory (Equation (11.15)) can be immediately seen at the limits of the integral:

$$V(t_1) = \frac{1}{C} \cdot \int_{t=0}^{t=t_1} I \, \mathrm{d}t \tag{11.21}$$

The electrical properties of a capacitor can be illustrated by interpreting Equations (11.20) and (11.21): When the voltage at the terminals is constant, no current flows. If the voltage changes, the current at the terminals is proportional to the rate of change in the voltage, $\mathrm{d}V/\mathrm{d}t$, and to the capacitance, C. The voltage, V, across the capacitor is proportional to the current that has flown over time, and therefore to the charge, Q, and inversely proportional to the capacitance. Let us avoid one widespread misconception by making it clear that a capacitor does not "store" any net charge. Just as in the case of resistors or of any other two-pole, the same amount of charge that flows into one pole emerges from the other — see Kirchhoff's law. The big difference between a capacitor and a resistor is the energetic reversibility. The electrical energy, $V \cdot I \cdot t$, which is dissipated to the environment by a resistor, is stored reversibly in a capacitor.

From the physical point of view, capacitive behavior is as elementary as is resistive behavior. The most common form of capacitance is the plate capacitor (Figure 11.5). The energy is stored by holding the difference in charge through an electrical field on two facing conductive surfaces such as two sheets of metal foil. The capacitance of a plate capacitor, C_{pc} (Equation (11.22)) is proportional to the area of the plates, A, and to the dielectric constant, ε, of the material between the plates, known as the dielectric, E. It also increases as the distance between the plates, d, is reduced

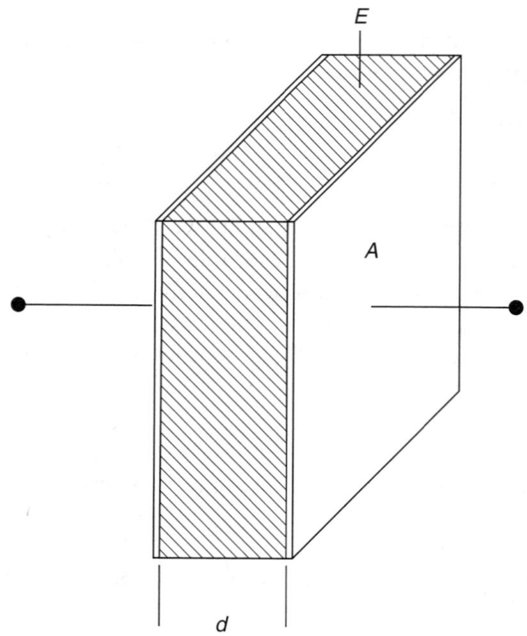

Figure 11.5 Plate capacitor with metal plates of area A and distance d, filled with a dielectric medium E.

$$C_{\mathrm{pc}} = \frac{\varepsilon \cdot \varepsilon_0 \cdot A}{d} \tag{11.22}$$

The dielectric constant ε_0 of vacuum is about 8.85×10^{-12} [A sec/V m]. Capacitances are often seen from their negative side in electrochemical applications: it is not just plate capacitors, but any other conductor extending through space that has a capacitance. The electrical leads to an object being measured are exclusively intended to transfer information between the object and the measuring instrument — but this information can, unfortunately, be corrupted by the "stray capacitance."

Structures analogous to plate capacitors also play a useful role in electrochemistry (3). Coating systems or other layers on the surfaces of metals form, together with the electrolytes, plate capacitor structures. Equation (11.22) shows that if the magnitudes C, ε, and A are known, it is possible to determine the layer thickness, d, which is useful in the study of coatings.

The capacitance of the electrochemical double layer (Figure 11.6) is also created by a "plate capacitor" consisting of electrolyte and electrode. The role of the dielectric is performed here by the — more or less — insulating energy barrier at the location of the transition between ion conduction and electron conduction (the "electrochemical double layer"). In contrast to a dielectric, in coating layers the tiny "thickness" of the double layer easily causes high field strengths to develop — with the consequential dependency of physical behavior on potential.

Capacitance of adsorption and crystallization and the capacitance of the blocked finite diffusion are typical examples of another type of capacitance that will be discussed in more detail in Chapter 13. Charge transport is linked here to the transport of material. It is not the effect of the quantity of charge on the electrical field that leads to capacitive behavior in accordance with Equation (11.21), but rather the electrochemical activity associated with the material concentration. In accordance with the Nernst equation (Equation (11.23)) it is possible to assign different potentials to charged species with different activities, a_i, in different regions of space (phases), so assigning one voltage, V, to two phases. The factors V_{T} and b have the significance explained above

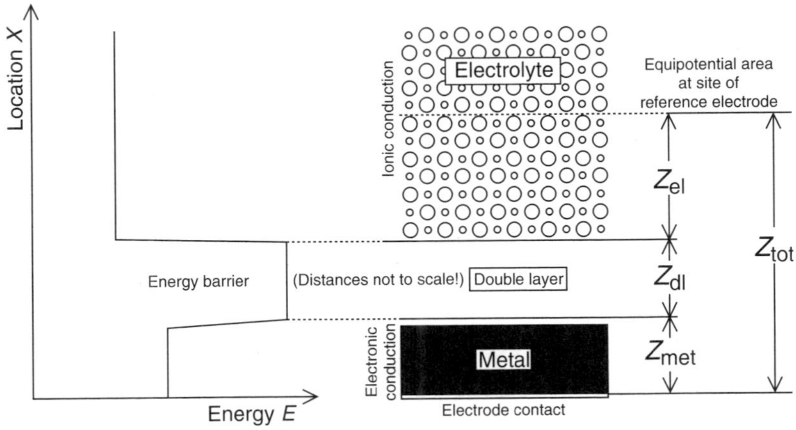

Figure 11.6 The electrochemical double layer acts as a plate capacitor. The Ohmic contribution of the metal Z_{met} and the electrolyte Z_{el} add up with the double layer impedance Z_{dl} to the total impedance Z_{tot} of the electrode.

$$V_{2,1} = b \cdot V_{\mathrm{T}} \ln\left(\frac{a_2}{a_1}\right) \qquad (11.23)$$

Equation (11.23) illustrates one of the great strengths of electrical methods applied to chemistry: if the activities, a_i, represent the concentrations, c_i, then concentration relationships can easily be determined from V. This provides the key to a whole range of electroanalytic methods.

Let us now discuss a second form of electrical memory: inductive behavior is displayed by electrical components in which the voltage, V, between the terminals is proportional to the rate of change of the current flowing in it, $\mathrm{d}I/\mathrm{d}t$. The physical properties follow the time-dependency law for inductance, L:

$$V = L\frac{\mathrm{d}I}{\mathrm{d}t} = L \cdot \dot{I} \qquad (11.24)$$

If Equation (11.24) is solved for I (Equation (11.25)), then the electrical memory (Equation (11.14)) of the inductance can be seen again in the limits of the integral, as it was in the case of capacitance

$$I(t_1) = \frac{1}{L} \int_{t=0}^{t=t_1} V \mathrm{d}t \qquad (11.25)$$

The behavior of inductance is complementary to that of capacitance: Equations (11.20) and (11.21) can be converted into Equations (11.24) and (11.25) by swapping the roles of voltage and current. An interpretation of Equations (11.24) and (11.25) shows that the voltage is zero when the current is constant in an inductive two-pole. If the current changes, the voltage at the terminals is proportional to the rate of change of current, $\mathrm{d}I/\mathrm{d}t$, and to the inductance, L. Like those associated with capacitance, the electrical processes in an inductance are reversible — inductors store electrical energy. Inductance is most typically found in an inductor constructed as a coil. The energy here, $V \cdot I \cdot t$ is stored as the difference in the "electrical impulse" $\Delta V \cdot t$ in the associated magnetic field. Inductances most usually show an unpleasant aspect in electrochemical applications: every conductor that extends through space shows the same kind of inductive behavior as a coil. Like the stray capacitance of electrical leads, their "stray inductance" can also lead to errors in the information provided about the measured object. In the section on constructional tips for electrical cells, the consequences of stray effects will be examined in more detail.

Similarly to the capacitor-like behavior, inductive behavior is not linked to the appearance of magnetic fields. Equation (11.25) shows that for such a system the current depends on the history of the potential. If a flow of mass is required, so that an electrochemical system can follow the potential curve by adopting a new current ("relaxation"), time-dependent laws similar to Equation (11.25) may apply. The same is true if the conductivity of a system rises significantly with the temperature, but the current corresponding to the electrical heating settles with some delay due to the system's thermal inertia.

Relaxation inductances can be seen in a system in which activation and repassivation processes take place corresponding to the changing potential — as is often the case in corrosion processes or in the CO poisoning of fuel cell anodes (Figure 11.7). Inductance due to thermal relaxation occurs as soon as electrical heating plays a noticeable part in electrolytes — fuel cells and electrolysis cells are examples of this.

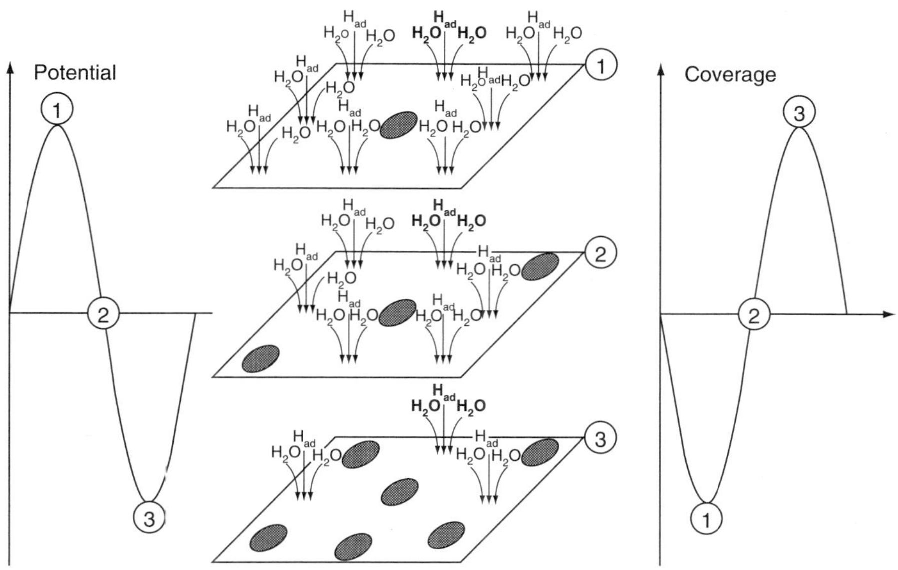

Figure 11.7 The potential dependency of the share of electrochemical active surface may lead to inductive appearance. In this example of a PEM fuel cell, carbon monoxide blocks the active surface for the main electrochemical reaction, the hydrogen oxidation. Three states of coverage are sketched, beginning with the most anodic state. The coverage will follow the potential with a certain delay in time that will cause "relaxation inductance."

11.2.2 From Differential Resistance to Impedance

We have seen that two-poles are generally described electrically by time-dependent differential equations for voltage and current. In the case of the elementary two-poles, capacitance, inductance, and, particularly, resistance, these time-dependent laws are manageable and easily understood. If, however, we imagine a network composed of these components it is soon found that a mathematical treatment of the resulting system of differential equations easily becomes difficult. If, in addition to these fundamental elements, more complicated, nonlinear descriptions of electrochemical processes are also found, analysis in the "time domain" becomes nothing but an exercise in mathematics.

There is, however, another approach available. As a first step, the signal (preferably the voltage) is divided into a stationary component and a dynamic component. In the stationary case, that is, the time-independent case, the system of differential equations simplifies itself to expressions similar to Equation (11.10). They determine the "working point" (bias) on the stationary current–voltage curve. The dynamic component is investigated in a second step. The key here is to — provisionally — restrict the analysis to periodic sinusoidal time signals of voltage and current. Sinusoidal signals have the convenient property that all the functions of the signals in linear systems are, in turn, sinusoidal and have the same period — regardless of how complicated the time-dependent system of differential equations is. This is, of course, due to the fact that all derivatives of sinusoidal signals and their linear combinations are also sinusoidal.

Finally, a simple trick makes it possible to model nonlinear systems: in cases where the signal $V(t)$ consists of a constant, stationary component V_{bias} onto which a very small dynamic component, $V_{\sim}$ (Figure 11.8; Equation (11.26)) is superimposed,

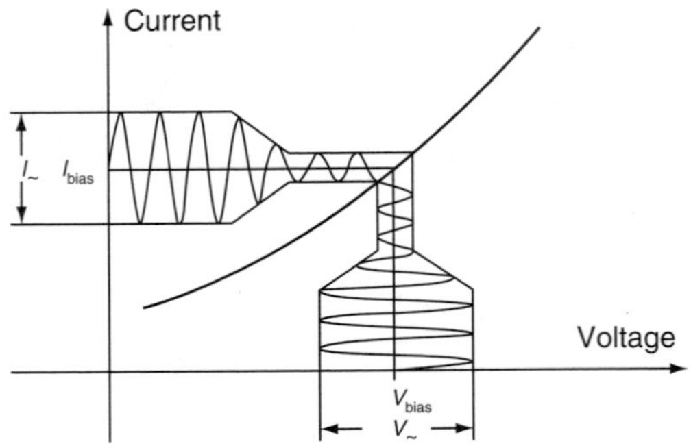

Figure 11.8 Modulation of the steady-state current–voltage curve at a certain bias by a small sinusoidal voltage results in a sinusoidal current. AC voltage and current exhibit the same periodicity with a constant phase shift.

nonlinear time-dependent laws can be developed as a power series, and truncated after the linear term

$$V(t) = V_{\text{bias}} + V_\sim \tag{11.26}$$

Like the voltage, the resulting current can be split into a stationary component, I_{bias}, and a dynamic component, $I_\sim$:

$$I(t) = I_{\text{bias}} + I_\sim \tag{11.27}$$

Sinusoidal time signals, $V_\sim(t)$, are usefully expressed using Euler's relation (4), Equation (11.28), in complex notation. In this way, cosine and sine components can conveniently be combined into one "periodicity factor," $e^{j\omega t}$, and separated from the peak value, $V_\wedge$, otherwise known as the amplitude, as in Equation (11.29), where j is the imaginary unit and $\omega = 2\cdot\pi\cdot f$ represents the angular frequency

$$e^{jx} = \cos x + j \sin x \tag{11.28}$$

$$V_\sim = V_\wedge \cdot e^{j\omega t} \tag{11.29}$$

$$I_\sim = I_\wedge \cdot e^{j(\omega t+\varphi)} = I_\wedge \cdot e^{j\omega t} \cdot e^{j\varphi} \tag{11.30}$$

Making the assumption, mentioned above, of a small value of $V_\sim$, the current associated with Equation (11.29) follows from Equation (11.30). The periodicity factor for $I_\sim$ is extended with a "phase shift" between the voltage and the current characteristic for the two-pole system with a value $-\pi < \varphi \leq +\pi$. The particular advantage of this procedure is that an "alternating current" analog to the differential resistance, R_{diff}, can be found. If the small components $V_\sim$ and $I_\sim$ are divided, analogously to the derivation of R_{diff}, the periodicity factors reduce to the component that represents the phase shift. We obtain the complex analog to R_{diff}, namely the AC "impedance" Z of the two-pole

$$Z(f) = \frac{V_\sim}{I_\sim} = \frac{V_\wedge}{I_\wedge} e^{-j\varphi} \tag{11.31}$$

Like the magnitudes $V_\sim$ and $I_\sim$, Z generally depends on the frequency, f, in a way characteristic for the two-pole. It is now possible to understand where the strength of the impedance concept lies as compared with stationary methods: if a number of Ohmic resistances are combined into any kind of network with two-poles (Figure 11.9), the values of the individual resistances cannot be determined from

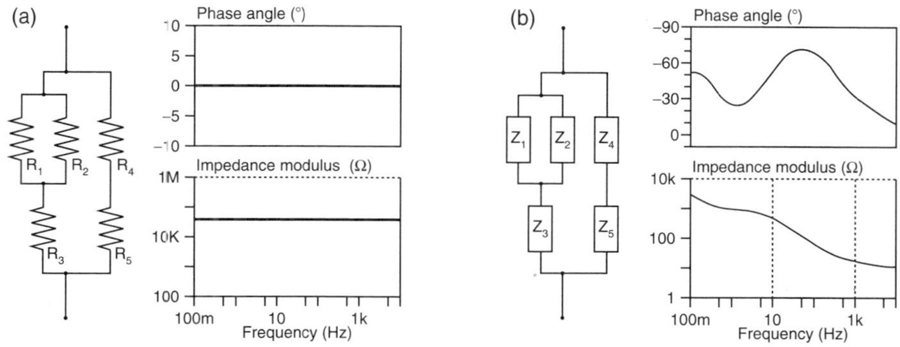

Figure 11.9 Two-poles built from resistance networks (a) and impedance networks (b) with typical impedance spectra (Bode diagrams, see Equation (11.37)).

the total resistance of the two-pole. The case of frequency-dependent impedances is, however, different: if the frequency response of the individual components is significantly different, it is possible in favorable cases to determine the full set of parameters if the network is examined at a sufficiently large number of different frequencies.

As has already been explained, characteristic impedances can be assigned to electrochemical processes. In electrochemical impedance spectroscopy (EIS), the impedance of an electrochemical cell is investigated over a very wide range of frequencies. An impedance spectrum, or an impedance transfer function, is measured. The network data decoded from this can be used to derive chemical and kinetic parameters, mechanisms, and electrode topologies.

So far, the route from the general time-dependent law (Equations (11.16) and (11.17)) to the impedance transfer function (Equation (11.31)) has only been presented in a brief and abstract form. The principles of the method for the simplest cases can be understood in the following derivation of the impedance for capacitance and inductance.

We begin with the time-dependent law for the capacitor in differential form (Equation (11.20))

$$I = C\frac{\mathrm{d}V}{\mathrm{d}t} = C \cdot \dot{V}$$

and expand the voltage function for the small, sinusoidal test function (Equation (11.29))

$$V_\sim = V_\wedge \cdot \mathrm{e}^{\mathrm{j}\omega t}$$
$$\mathrm{d}V/\mathrm{d}t = \dot{V} = V_\wedge \cdot \mathrm{e}^{\mathrm{j}\omega t} \cdot \mathrm{j}\omega = V_\sim \cdot \mathrm{j}\omega \tag{11.32}$$

Differentiating Equation (11.29), we obtain the expression in Equation (11.32) for the time derivative of the sinusoidal voltage, $\mathrm{d}V/\mathrm{d}t$.

The impedance Z_C is then easily obtained by inserting the result into Equation (11.20):

$$I_\sim = C \cdot \dot{V}_- = C \cdot V_\sim \cdot \mathrm{j}\omega \tag{11.33}$$

$$Z_C(f) = \frac{V_\sim}{I_\sim} = \frac{1}{2\pi f C}\frac{1}{\mathrm{j}}[\Omega] \tag{11.34}$$

The inductance behaves in a manner complementary to the capacitance — if the places of voltage and current are swapped in the above derivation, the impedance Z_L^* of the inductance (Equation (11.35)) is obtained in a manner similar to that of Equation (11.34)

$$Z_L(f) = \frac{V_\sim}{I_\sim} = 2\pi f L \mathrm{j}[\Omega] \quad [\Omega] \tag{11.35}$$

In principle, we have inserted the operator $e^{j\,\omega t}$ into the time law of the capacitor, which is a function of the three variables voltage, current, and time, converting it into a function of frequency, Z. Its origin as a function of several variables is indicated by the complex character of Z: instead of the original function of three variables, we now have obtained two functions of one variable, namely the functions for the real and imaginary components of Z. Real functions of one variable can be meaningfully represented in one plane. Generally speaking, two methods of graphical representation have become accepted for the complex function Z. In the "Nyquist diagram" or the "complex plane plot," the real and imaginary parts are plotted on linear axes against one another in one plane (Figure 11.10; Equation (11.36)). The frequency variable does not appear directly. If the logarithm of the complex impedance is formed, then the components of the complex logarithm (Equation (11.37)) can be plotted on a plane diagram of the two components against the logarithm of frequency (a Bode diagram)

$$x = \mathrm{Re}\{Z(f)\}, \quad y = \mathrm{lm}\{Z(f)\} \tag{11.36}$$

$$x = \ln f/\mathrm{Hz}, \quad y_{|z|} = \mathrm{Re}\{\ln(Z(f)\}, \quad y_\varphi = \{\ln(Z(f)\} \tag{11.37}$$

A wide range of methods for representing the frequency dependency of magnitudes directly derived from the impedance is in common use. Capacitance, inductance, the complex dielectric constant, the dielectric modulus, complex refractive index, and so forth may be displayed. What they all have in common is that, in contrast to the elementary approach above, specific models are hypothesized as to how the observed electrical impedance arises.

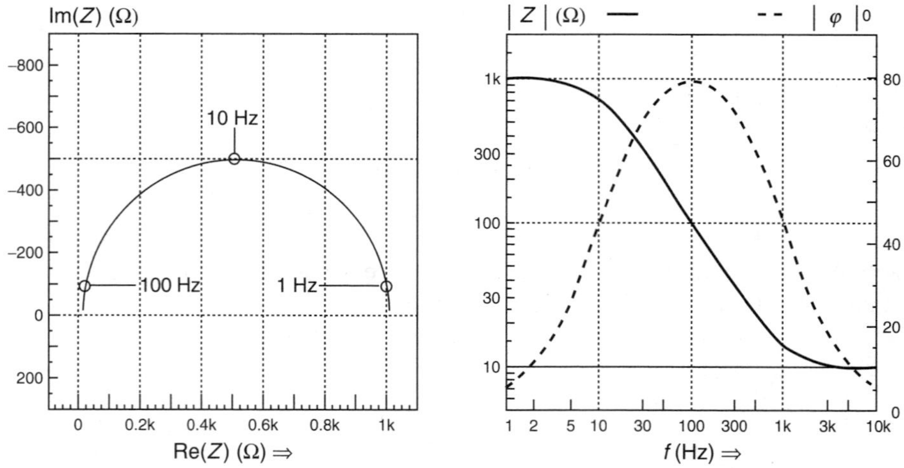

Figure 11.10 Typical plot of an impedance transfer function: complex plane plot (a) and Bode diagram (b).

We do not need to pursue these considerations any further here, but keep in mind the principle of AC impedance — many aspects of electrochemical instrumentation can be more easily understood with their aid.

This excursion into fundamental electrical principles has prepared us sufficiently to follow those who are more advanced into the next chapter.

11.3 FUNDAMENTALS OF ELECTROCHEMICAL INSTRUMENTATION: MEASUREMENT OF POTENTIAL AND CURRENT IN ELECTROCHEMICAL CELLS

Electrochemistry is the chemistry of redox processes. When the anodic part of the reaction (Equation (11.38), oxidation of a reduction agent) takes place in a separate site from the cathodic part of the reaction (Equation (11.39), reduction of an oxidation agent), then an electron exchange current is not immediately possible between the reaction partners

$$\mathrm{Red} \longrightarrow \mathrm{Red}^{m+} + m \cdot \mathrm{e}^- \tag{11.38}$$

$$\mathrm{Ox} + n \cdot \mathrm{e}^- \longrightarrow \mathrm{Ox}^{n-} \tag{11.39}$$

In order to maintain charge neutrality, the electrons must take the route through an additional electron conductor that electrically joins the anode and cathode sites. If the connection is accessible outside the electrochemical system, the exchange currents can be monitored in the external part of the electrical circuit. The sequence of electron conductor and ion conductor, and again of ion conductor and electron conductor, thus represents a full cell (two-electrode cells, Figure 11.11; arrangements with additional potential probes, and cells with three or four electrodes are discussed further below).

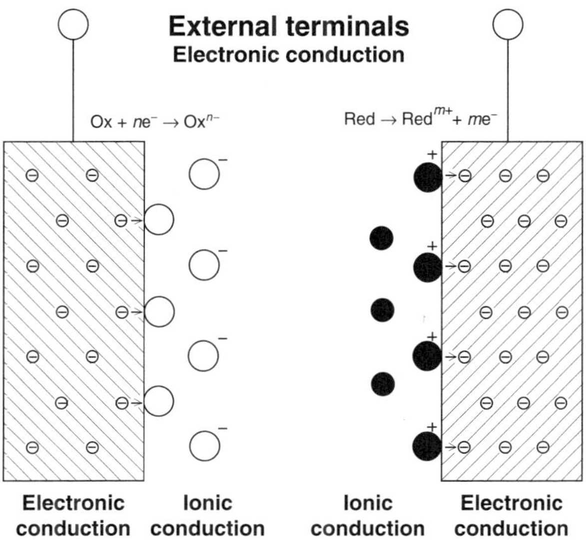

Figure 11.11 Two electrodes, connected by ionic conducting media, build an electrochemical "full cell" arrangement. The indicated reactions will take place if current flow between the external terminals is enabled.

Two elementary limiting cases must be distinguished in the two-electrode experiment:

1. The external electrical circuit is completely interrupted at a certain point (Figure 11.12a). At the phase boundaries between ionic conductors and electron conductors potential differences develop, and these precisely oppose the thermodynamic driving force of the partial reaction. Both anodic and cathodic forward and reverse reactions reach equilibrium. The net electrochemical reaction is thus inhibited, and no external current flows. As in any other electrical circuit, the sum of all the partial potentials in the loop is zero (Kirchhoff's law):

$$\sum_{i=1}^{k} V_i = 0 \tag{11.40}$$

   In the simplest case, shown in the sketch, only the partial potentials of the anode and the cathode need to be considered. This means that, apart from the arithmetic sign, the potential found at the terminals of the circuit interruption is equal to the sum of the anode and cathode potentials. The observable electrical magnitude is a voltage (the full cell's open circuit potential, OCP), the current being zero.
2. The external electrical circuit is completed by a load at open ends. In the example shown, this is an Ohmic resistance (Figure 11.12b). If the potential relationships from limiting case 1 are transferred to limiting case 2, a current given by

$$I_r = \frac{\text{OCP}}{R} \tag{11.41}$$

   should flow through the resistance (Equation (11.41)), in accordance with Ohm's law.

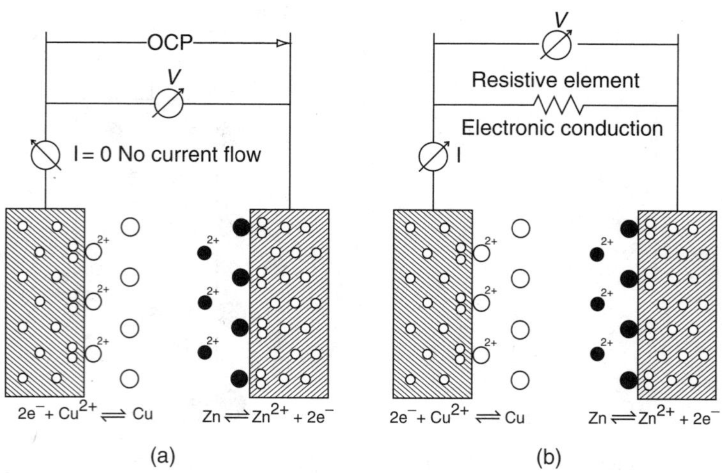

Figure 11.12 Two limiting cases for an electrochemical full cell: the electrochemical net reaction is inhibited due to an inhibited external current flow (a). The OCP appears at the terminals. In the case of an enabled external current flow (b), an electrochemical net reaction takes place according to the difference in the free energy.

Both anodic and cathodic forward and reverse reactions lose their equilibrium; the net electrochemical reaction begins. The electrical current loading and the chemical reactions, however, both develop additional inhibiting potential differences at the phase boundaries (e.g., charge transfer polarization and diffusion over-potential), the nature of which is described extensively in other chapters.

Consequently the current, I_r, cannot flow continuously. Instead, a time-dependent current $|I(t)| < |I_r|$, which can become stable after a certain time, arises. The observable physical magnitudes are the voltage at the resistor and the current in the electrical circuit.

11.3.1 The Ideal Measurement

Fundamentally, any measurement has an effect on the measured object — this is in fact a necessary consequence of the quantum nature of the world. In the classical, limiting case, it is still nevertheless possible to come close to the ideal of a measurement without influence.

In other words, an ideal instrument behaves as if it were not actually part of the system. Putting it more precisely, if undesirable side effects are to be avoided, the instrument must not remove any energy from the system under investigation ("zero-power measurement").

In the case of measuring potential as shown in Figure 11.12(a), it is easy to see that an ideal measuring instrument connected between the terminals of the interruption in the circuit must not cause any current to flow. Otherwise, the instrument itself has an effect similar to the load resistor in the arrangement in Figure 11.12(b), causing the potential, $V(t)$, to differ from the OCP.

11.3.2 Measuring Potential — An Easy Job?

Technical solutions for potential measuring instruments are not perfect. Figure 11.13 expresses the nonideal properties relevant to this case for a voltmeter in the form of an equivalent circuit diagram. Components are connected in parallel with the ideal virtual instrument here, resulting in the flow of parasitic current.

The symbol for the resistor R_{in} stands for the unwanted component in the current that is proportional to a constant measured potential in accordance with Ohm's law. The current source I_b, on the other hand, represents a component in the current that is independent of the measured potential (bias — or leakage current).

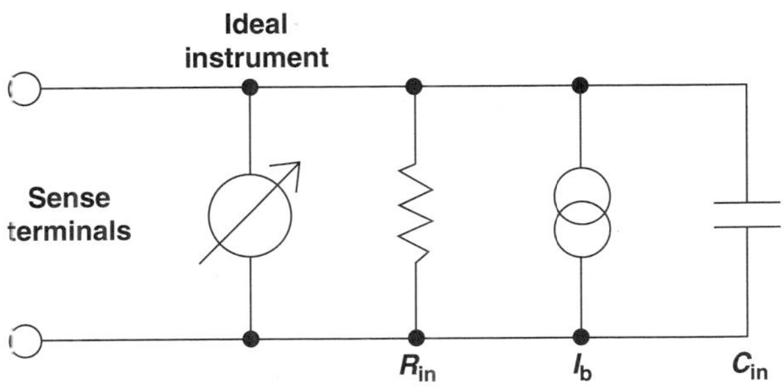

Figure 11.13 Input resistance, input capacitance, and bias current characterize a technical voltmeter.

In addition to the most important characteristics of the input resistance and leakage current, the input capacitance C_{in} is also significant. A reminder: capacitors are electrical components in which the voltage, V, between the terminals is proportional to the stored electrical charge, Q (see also Section 11.2; Equations (11.18) to (11.20)).

On the other hand, the current through the capacitor, I, is proportional to the rate of change of charge, dQ/dt. From this we can conclude that the input capacitance of a voltmeter is of no concern to us provided the voltage being measured is constant. We can assume that in the case of typical "DC" method[2] only the instrument input resistance and its leakage current need to be considered.

The following example, however, illustrates how complex an apparently simple measurement of equilibrium potential can be. The equilibrium potential of a metal plate (with an area of about 20 cm^2) protected from corrosion by an organic polymer (with a thickness of about 100 μm) is to be measured. The counter electrode is to be a low-resistance reference electrode, such as a silver mesh coated with AgCl in a chloride electrolyte, whose contribution to the total potential (V_{ref}) is assumed to be constant and known (Figure 11.14).

The metal–polymer–electrolyte phase sequence itself behaves in a manner similar to a capacitor; the contributions from reactions expressed in Equations (11.38) and (11.39) are assumed to be negligible. In order to describe the object being measured in a simplified way we make use of an equivalent circuit diagram, similar to the way in which we characterized a voltmeter. A description that includes a voltage source and a capacitor is sufficient for the sketched case.

We assume that there is no charge on the capacitor, C_c, before the measuring instrument is connected. If we could measure without side effects, the cell voltage would be given by $V = V_{ref}$. If, however, our instrument has a small input capaci-

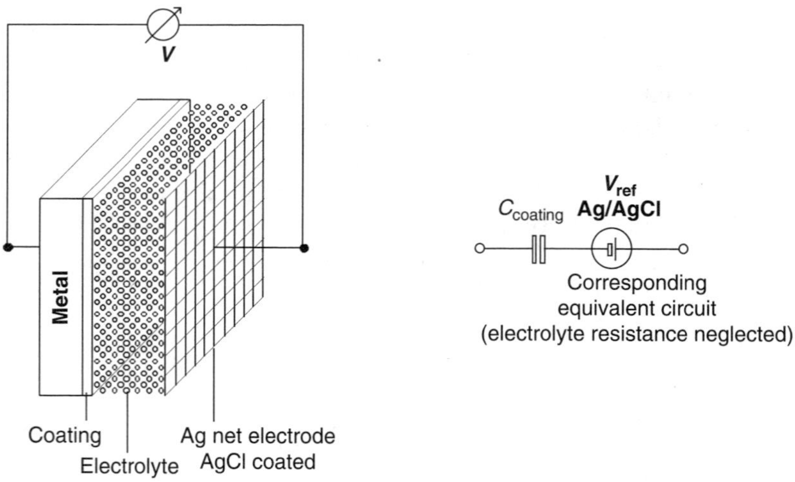

Figure 11.14 Potential measurement at a coated metal electrode (left). At first approximation, the electrode appears as the circuit sketched at the right-hand side.

[2]In what follows, "DC" methods refer to standard electrochemical methods in which the measured magnitudes only change slowly in comparison with the reaction times of the system under investigation ("stationary" or "quasistationary" methods).

tance, C_{in}, not negligible in comparison with the capacitance of the electrode, then after connection of the voltmeter a charge of

$$Q = V_{ref} . C_{ser} \quad \text{where } C_{ser} = (1/C_{in} + 1/C_c)^{-1}$$

is distributed over the series network, C_{ser}, consisting of the capacitances of the electrode and the measuring instrument. Only a fraction $V_{ref} = C_c/(C_{in} + C_c)$ is then measured. If the input capacitance of the measuring instrument is about 50 pF (a typical input capacitance for a broadband voltmeter between the working and reference electrode terminals), the measurement of the equilibrium potential of the film electrode (with an electrode capacitance of approximately 1 nF) is already distorted by about 5%. If, in the expectation of a better absolute precision, one uses a commercial $6\frac{1}{2}$-digit multimeter (with an input capacitance of approximately 1 nF), then only half the value of V_{ref} is measured!

The above example does not just demonstrate pitfalls of apparently simple measurement tasks, it was also selected in order to illustrate the problems associated with the concept of equilibrium potential: because the measurement of potential with no side effects is in fact an illusion, the determination of equilibrium potential is only meaningful when redox reactions following Equations (11.38) and (11.39) are proceeding at the boundaries between the phases. Their equilibrium may only be lightly disturbed by the unavoidable injection of charge involved in the measurement. Material reserves of the reagents in Equations (11.38) and (11.39) must therefore exist for buffering. If processes that can stabilize the potential are entirely absent, such as would be the case on an electrode with a perfect layer of insulation, then measurement of the equilibrium potential is not possible.

11.3.3 Requirements of an Instrument for Measuring Potential

We can now summarize our experience with the measurement of potential:

- A good voltmeter should not cause any additional current flow as a result of its presence in the measured circuit.
- Input resistance, leakage current, and input capacitance are the most important characteristics of a voltmeter.
- The input resistance should be large in comparison with the internal resistance of the measured object.
- The leakage current should be negligibly small in comparison with the electrochemical exchange current in the object.
- The dynamic current component resulting from the input capacitance when the potential changes over time must also be negligibly small in comparison with the exchange current. For this reason, the input capacitance must be particularly small when fast measurements are being taken.
- If these rules are not followed, such as would be the case when no potential stabilizing electrochemical reaction is taking place, the measurement of potential is irrelevant.

The suitability of the instrument depends heavily, of course, on the system being investigated. The parasitic properties of a simple hand multimeter ($\approx$10 MΩ input resistance, $\approx$1 nF input capacitance, leakage current $<$100 pA) do not make any noticeable contribution to the measuring error when measuring the cell voltage of batteries or fuel cells. However, the same instrument will yield grossly incorrect

Table 11.1 Typical Source Impedance, R_{src}, of the Electrochemical System, Acceptable Measuring Instrument Input Resistance R_{in}, with Input Capacitance, and Acceptable Leakage Current I_b

Application of potential measurement	R_{src}	R_{in}, C_{in}	I_b
Cell voltage of batteries and fuel cells	10^{-4}–$10\,\Omega$	$10^{4}\,\Omega$, 10^{-9} F	μA–mA
Single electrode potentials in batteries and fuel cells, corrosion in autopassive systems	10^{2}–$10^{6}\,\Omega$	$10^{8}\,\Omega$, 10^{-9} F	nA–μA
Corrosion under paints and coatings, potential measurement on microelectrodes and electrochemical sensors	10^{4}–$10^{9}\,\Omega$	$10^{12}\,\Omega$, 10^{-10} F	fA–pA
SECM and other nanoelectrodes	10^{4}–$10^{9}\,\Omega$	$10^{12}\,\Omega$, 10^{-12} F	fA–pA
Electrochemical potential noise	10^{2}–$10^{9}\,\Omega$	$10^{12}\,\Omega$, 10^{-10} F	fA–pA

measurements when measuring a pH value using a glass electrode. Table 11.1 provides a general overview of the requirements of voltmeters for particular applications.

11.3.4 Measuring the Cell Under Load: Measuring Current

So far we have only been concerned with measuring potential. Now let us consider the measurement of current. We recall limiting case 2, sketched above, in which the electrical circuit is completed by a load connected between the open terminals.

The current that flows is our second most important source of information about the system being investigated. Whereas the potential provides us in the first place with information about the thermodynamic driving forces of the reactions involved, the current is associated through Faraday's law directly with the net progress of the reactions. It thus also provides us with kinetic information.

If current is to be measured, an appropriate instrument must be inserted into the circuit. The requirements for suitable instruments are complementary to the requirements for measuring potential. An additional partial potential across the current measuring instrument is included in the zero sum of the potentials in the circuit. This therefore changes the partial potentials at the other locations. This means that the current flowing in the circuit is different from that which flows when the current is not being measured. In order to minimize the effect of the measurement on the state of the system, the partial potential across the instrument should be a minimum. An "ideal" instrument for measuring current therefore is effectively a short circuit and, analogously to the ideal measurement of potential, absorbs no power from the system.

Figure 11.15 indicates the relevant nonideal properties of a measuring instrument for the current measuring case in the form of an equivalent circuit diagram. Components are connected in series with the ideal virtual instrument here, resulting in the development of parasitic potentials. The series resistor represents the direct current internal resistance developing an error voltage proportional to the current. The voltage source represents an error voltage (offset voltage) independent of the measured current. In addition to the most important characteristics of series resistance and offset voltage, the series inductance may also be significant if the currents change in time at a significant rate. A reminder: an inductor is an electrical component in which the voltage, V, between the terminals is proportional to the rate of

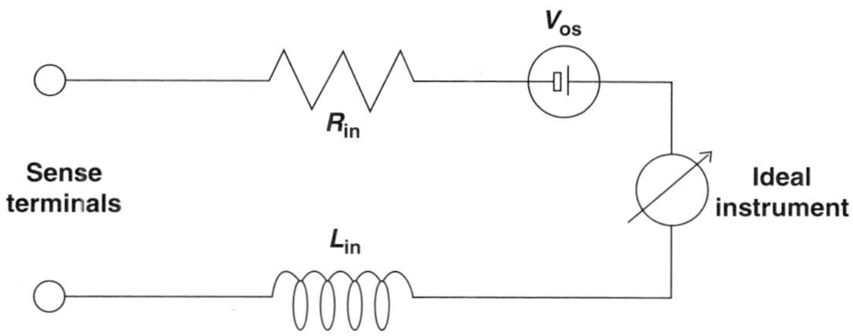

Figure 11.15 Series input resistance, series input inductance, and offset voltage mainly characterize a real world current meter.

change of the current flowing in it, dI/dt (see also Section 11.2, Equation (11.24)). It follows that this inductance is of no significance for DC measurements. It can, however, make a significant contribution to errors in dynamic measurements.

Table 11.2 provides a general overview of the requirements of current measuring instruments for particular applications.

Voltmeters are available that approach perfection, at least for slow measurements. Near-perfect current measuring instruments, however, are a great deal more difficult to implement, even for direct current. The reason for this is that the most accurate measurement principle is based on the generation and acquisition of a voltage signal proportional to the current. Accurate current measurement, therefore, is usually the measurement of the potential difference across a measuring resistor inserted into the circuit (the "load voltage" across a "shunt"). It is not difficult to guess that this kind of current measurement can be quite problematic in a number of application fields.

The following example demonstrates this in the case of measuring the current in a battery (Figure 11.16). A battery is intended to supply a current of 1 A to a defined load. The resulting terminal voltage at the battery is to be 1 V. An additional shunt resistance of 1 mΩ in the circuit whose current is being measured will create a

Table 11.2 Typical Source Potential V, Typical Currents I, Acceptable Series Impedance, Z_{ser}, in the Measuring Instrument and Acceptable Offset Voltage V_{offs} in Current Measurements

Current measurement application	Typical source potential	Typical currents	Acceptable measuring instrument series impedance	Acceptable offset voltage
Load current in batteries and fuel cells	1 V	10^{-1}–10^{3} A	10 μΩ–1 nH	1 mV
Corrosion in autopassive systems	0.1 V–1 V	10^{-10}–10^{-6} A	1 kΩ–10 μH	1 mV
Corrosion under paints and coatings, current measurement on microelectrodes and electrochemical sensors	0.1 V–1 V	10^{-13}–10^{-8} A	10 kΩ–100 μH	1 mV
SECM and other nanoelectrodes	1 mV–1 V	10^{-13}–10^{-9} A	10 kΩ–1 μH	0.1 mV
Electrochemical current noise	1 mV–0.1 V	10^{-14}–10^{-6} A	1 Ω–1 μH	0.1 mV

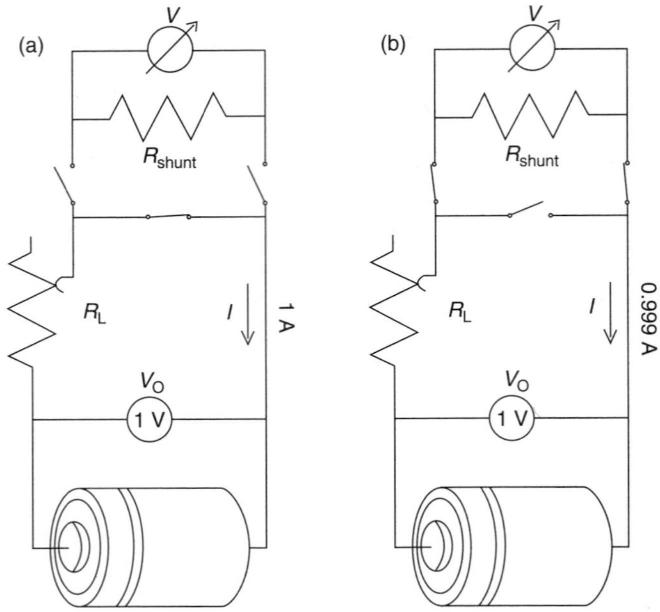

Figure 11.16 Introducing a shunt-type current meter influences the measuring object: the resulting current (b) is less than without current measurement (a).

potential between its terminals of approximately 1 mV. This will lower the current in the circuit by something of the order of 0.1%. This generates a corresponding systematic measurement error. In addition to this there is a stochastic measurement error with a similar order of magnitude created, for instance, by thermal error voltages present when measuring the already small load voltage of approximately 1 mV. Simply reducing the magnitude of the shunt resistance does not, therefore, result in an improvement of the overall precision.

The solution to the problem is found to lie in the application of active instrumentation amplifiers operating according to the current–voltage converter principle (Zero resistant amplifier, ZRA). The way these devices work will be considered in detail later. For the moment, let it just be said that they are not the magic bullet. It calls for an electronic component (an operational amplifier) to handle the entire measured current, and to pass it via a diversion from the first measurement terminal to the second. This is just about practicable for measured currents less than 1 A. ZRAs that are suitable for batteries or fuel cells are, however, not commercially standard.

Conventional ammeters, such as those found in multimeters, are mostly based on the shunt principle. When used for electrochemical measurements, it is essential that their load voltage (terminal voltage), which is liable to be between 0.1 and 1 V (!!!) when the measuring scale is being fully exploited, is taken into account: it is better to use a less sensitive measurement range to minimize the resulting error.

11.3.5 "Clear Relationships" — Active Control of the State of the Cell System

Describing the state of an electrochemical system on the basis of measurements obtained passively is often inadequate. It is often desirable to actively create certain conditions in the system, and in some cases to vary them.

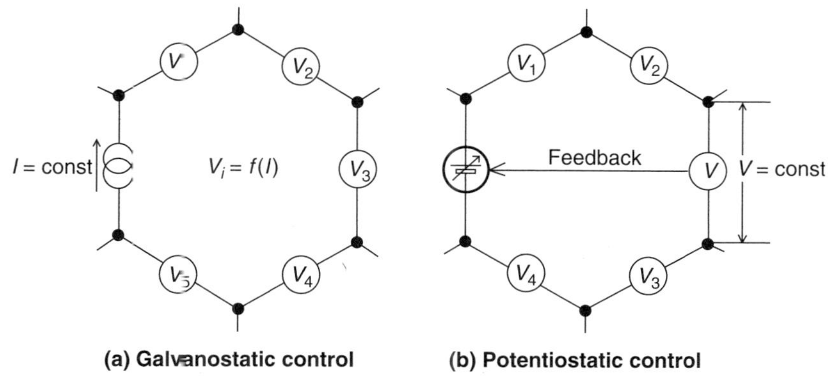

Figure 11.17 Controlling the conditions in an electric circuit. Current control (a) and voltage control (b).

In the example shown in Figure 11.16, it is possible to adjust the current roughly to around 1 A by carefully choosing the load, and then to check it with an ammeter. Instead of now using a time-consuming process of trial and error to repeatedly vary the load in order to approach the exact desired value of 1 A it is much more attractive to delegate the cycle of adjustment, comparison, and further adjustment to an automatic system. An automated system that adjusts the desired current in a circuit including the system under investigation is known in electrochemistry as a "galvanostat" (Figure 11.17a). The effect of the galvanostat on the circuit is the same as a constant current source. In a system under galvanostatic control, the current is the command variable, and the potentials are the response variables.

Instead of forcing a specific current in a system, it is also possible, with an appropriate regulation device, to specify a fixed potential between two points in the circuit (Figure 11.17b). This kind of "potentiostat" has the effect of a constant voltage source. In a system under potentiostatic control, the potential is the command variable, and the current is the response variable.

There are, furthermore, particularly elegant solutions for potentiostats and galvanostats to the problem of the load voltage when measuring current.

Before we consider these techniques and the associated methods of electrochemical cell construction more closely, it will be worthwhile recapitulating the principles of operational amplifiers.

11.4 ELECTRONIC AIDS FOR THE ELECTROCHEMICAL ENGINEER

A considerable number of active aids, including potentiostats, galvanostats, operational amplifiers, etc., have already been mentioned. Sometimes it is not enough simply to be aware of the function and the purpose of our electronic helpers. If we want accurate measurements, and if we want to be able to interpret them reliably, it is also important to understand the way in which these devices work.

11.4.1 The Principles of Operational Amplifiers

Generally speaking, amplifiers are used to create large effects out of small causes. In instrumentation technology they can help us to come closer to the ideal of measurements with no side-effects extracting power. Many amplifiers are designed as closed systems, in which the user can only indirectly, if at all, have any effect on their basic properties. Parameters such as the "amplification factor" or the "frequency range" are determined within the closed system. One example of this would be the amplification of the radio signals received by a mobile telephone as they pass from the antenna to become outputs for the ear-piece and for the display.

In many scientific applications, however, it is useful if the experimenter himself has a large measure of control over the function and properties of amplifiers. For this reason operational amplifiers (OPAs) were developed in the 1960s, stimulated by applications in analog computing. OPAs are amplifiers for electrical voltage with properties that come close to the ideal. The nature of the amplification can be specified over an enormous range through the connection of additional, external circuitry (5).

Through appropriately selected circuitry, a variety of basic functions can be implemented such as voltage followers, inverting and noninverting amplifiers, adders, and integrators. It is then possible to use these basic functions to construct devices useful for electrochemistry, such as instrumentation amplifiers, potentiostats, and galvanostats.

Figure 11.18 illustrates the circuit symbol for an OPA, together with the input and output signals, the symbols for power supply (the voltage supplies V_{EE} and V_{PP}), and the most important parasitic magnitudes. The power supply and the parasitic magnitudes will in most cases not be included in the sketches below for simplicity. The reader should nevertheless not forget that there are deviations from the ideal. The current that comes out of the amplifier, moreover, must ultimately derive from the power supply

$$V_{out} = (V_{in+} - V_{in-}) \cdot A_{ol} \tag{11.42}$$

The "magic formula," responsible for all the practical things that can be done with OPAs (Equation (11.42)), is quick to explain:

- The OPA's output voltage, V_{out}, is proportional to the difference between the input voltages V_{in+} at the noninverting input and V_{in-} at the inverting input.
- The proportionality factor is the OPA's open-circuit amplification, A_{ol}, and can be assumed to be close to ∞, independently of the output current, I_{out}, resulting from the load, Z_L.

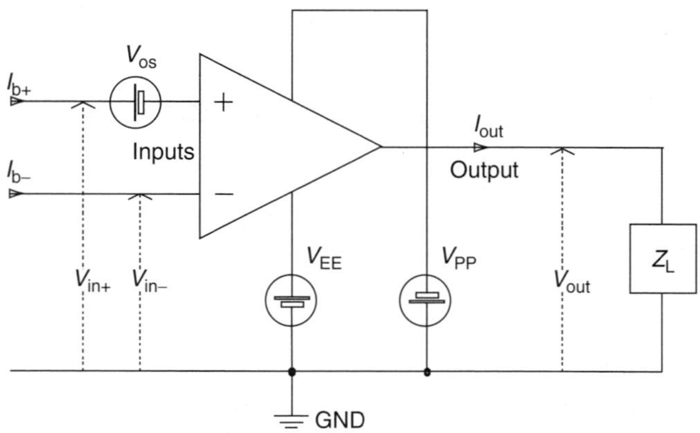

Figure 11.18 Operational amplifier symbol with the characteristic input and output voltages V_{in} and V_{out}, the supply voltages V_{EE} and V_{PP}, the output load current flowing through the load Z_L, and the parasitic input bias currents I_{b+} and I_{b-}.

- The parasitic magnitudes of the input leakage current I_{b+}, I_{b-} and the input offset voltage, V_{os}, are assumed to be negligibly small.

11.4.2 A Useful Helper for Potential Measurements: The Voltage Follower

The simplest basic configuration of an OPA is the voltage follower (Figure 11.19).

Equation (11.42) shows that even a small difference in the inputs ($V_{in+} - V_{in-}$) is sufficient to cause a large output voltage, V_{out}. If the output voltage is fed back unmodified to the inverting input, the output potential becomes $V_{out} \approx V_{in+} \cdot (1 - 1/A_{ol})$. In other words, the output voltage "follows" the input voltage, apart from the tiny difference V_{in+}/A_{ol}. V_{out} is an image of the input voltage regardless the load Z_L and the output current I_{out}. The output terminal of the OPA can be connected to subsequent signal processing electronics without having any loading effect on the signal source.

If the load Z_L is our multipurpose measuring instrument with an input resistance of $10\,M\Omega$, it can be seen that the combination of voltage follower and multimeter is quite suitable for acquiring the voltage of a glass electrode — provided that the absolute value of the OPA's input leakage current, I_{b+}, and its input offset voltage, V_{os}, are small enough.

The capability to decouple the current in a load circuit from a potential source also suggests the main application of a voltage follower: the acquisition and buffering of potentials from high-impedance sources (a buffer amplifier) at the input of the operational amplifier. A buffer is often required as an element in analog signal conditioning, for instance, when voltage dividers, filters, and so forth, cause significant source resistance, but the potential that develops must be further processed with as little modification as possible.

11.4.3 "Active Shielding" with a Voltage Follower — A Patent Solution for Interference?

One variation of the buffer amplifier helps to defuse a general problem that arises when taking measurements on high-impedance test pieces. The potential at high-impedance sources is not just disturbed by the measurement process, but is also

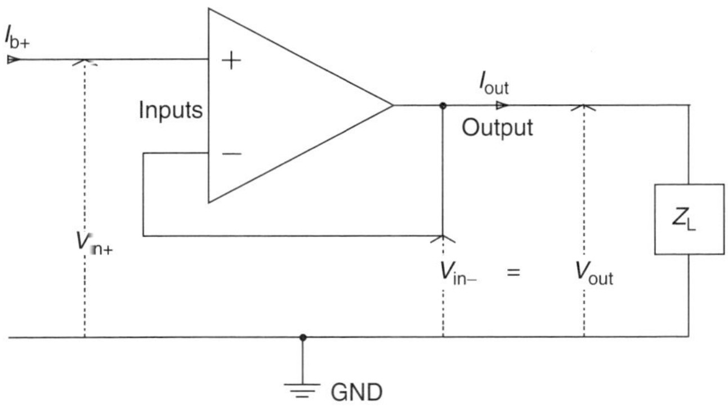

Figure 11.19 An OPA configured as a voltage follower. The output voltage V_{out} is near the input voltage V_{in+}, in spite of the load current I_{out} flowing through Z_L. A good voltage follower exhibits low input leakage current I_{b+}.

affected by interfering currents arising from electric fields in the neighborhood of the source. Stray capacitances lead to dynamic potential errors, while static errors can also be caused by incomplete insulation. The usual counter measure, electrostatic screening, will be discussed in more detail later. At this point it is necessary to explain that any type of screening itself leads to an increase in the stray capacitance, unless the potential of the screen is simultaneously raised to the signal potential. This is exactly what a buffer amplifier can do, if the screen is connected to the amplifier output. Figure 11.20 shows schematically how stray currents created by external fields are absorbed by the active screen and diverted to the output of the OPA. Provided the stray currents are small, the potential differences they will cause in the decoupling network, DN, between the signal lead and the screen will remain negligibly small. The field strength is, therefore, zero in the neighborhood of the signal lead, and the signal remains unaffected.

An examination of the circuit in Figure 11.20 may raise the question of why we do not omit the network DN when constructing an "ideal" active screen, since it does restrict the effectiveness of the circuit. Unfortunately, the nonideal dynamic properties of OPAs prevent the arrangement from working without the DN (although a simple resistor may be all that is necessary). The imperfections of OPAs in the AC domain play an important role in the practical aspects of using potentiostats and galvanostats. This will have to be considered in more detail later.

11.4.4 "Inverting Amplifiers" — Help with the Measurement of Current

So far we have considered OPA circuits in which the noninverting input (abbreviated to In_+ below) is used as the signal input. Signals can also be fed to the inverting input (abbreviated below to In_-). A number of useful circuits can be implemented in this way.

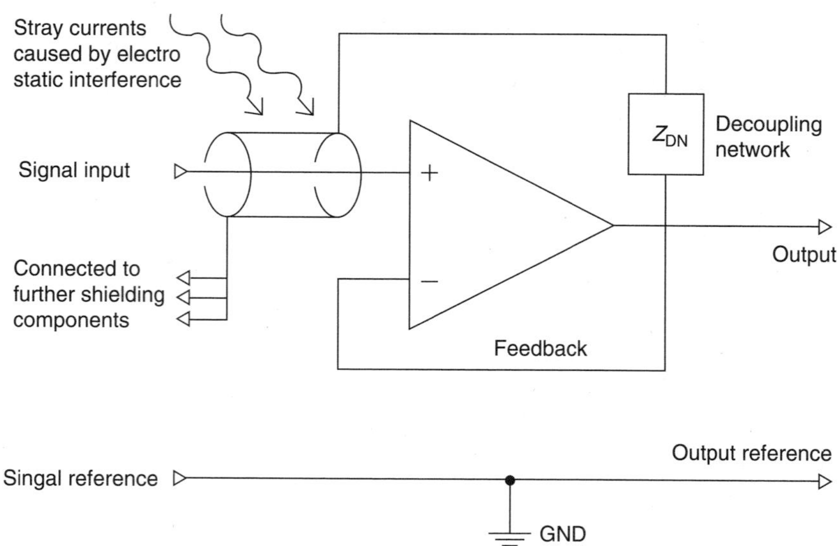

Figure 11.20 The output of a voltage follower can drive active shielding. Almost no additional load appears at the signal input line, due to a minimal voltage difference between signal and shielding.

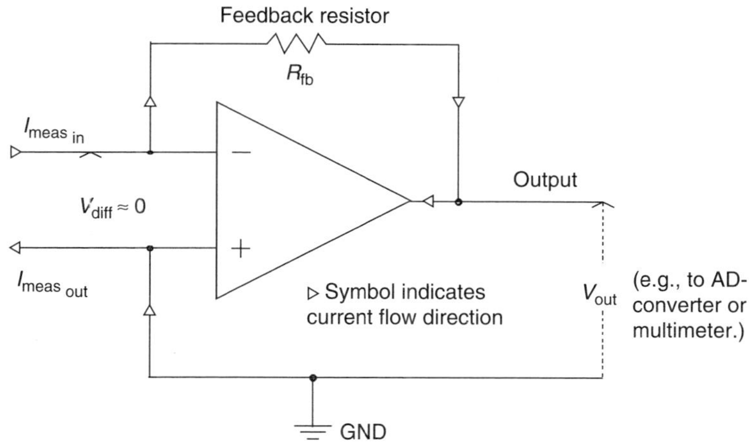

Figure 11.21 When a current I_{meas} is fed into the inverting input of an OPA, a current proportional output voltage V_{out} can be measured. This type of current-to-voltage converter is called zero-resistance amplifier due to diminishing voltage drop V_{diff} between the current inlet and outlet.

The principle can be demonstrated through a current–voltage converter or ZRA (Figure 11.21). In this case In_+ is connected to the reference potential. In_- is connected to the output via a resistor, R_{fb}. When idle (input current zero) the reference potential also appears at the output, so that the output voltage is zero. If a current I_{meas} is fed into the input (e.g., the current to be measured when the application is that of a current/voltage converter) then a potential given by $V_{out} = -I_{meas} \cdot R_{fb}$ develops at the OPA's output. The full input current flows through the resistor, R_{fb}, to the output of the OPA, and the difference in the input voltages is once again minimal. Because a significant voltage does not develop between the input and the output of the current (which is why the configuration is called a ZRA), the current–voltage converter can be used as an almost ideal method of measuring currents.

The inverting amplifier, or IA (Figure 11.22), is only one step beyond the ZRA. The ZRA at the current input is combined with the simplest form of voltage-to-current converter, that is, a resistor (R_{in}): R_{in} generates the current $I_{meas} = V_{in}/R_{in}$ at the input. This yields the IA's transfer function, g, where the output voltage is given by $V_{out} = -I_{meas} \cdot R_{fb}$

$$g = \frac{V_{out}}{V_{in}} = -\frac{R_{fb}}{R_{in}} \tag{11.43}$$

Example: To build an IA with an amplification factor of 10 ($g = -10$), R_{fb} must be 10 times larger than R_{in}; so that, for instance, $R_{in} = 100\,\Omega$ could be used with $R_{fb} = 1000\,\Omega$.

11.4.5 A Little Secret: Component Dimensioning for OPA Circuits

It is not the intention of this chapter to enter the details of OPA dimensioning. A few hints as to why in the example above $1000\,\Omega$ was recommended for R_{fb} rather than $1\,M\Omega$ or $1\,\Omega$ (with the corresponding choice for R_{in}) will, however, help us later to manage some of the difficulties of dynamic electrochemical measurements. The

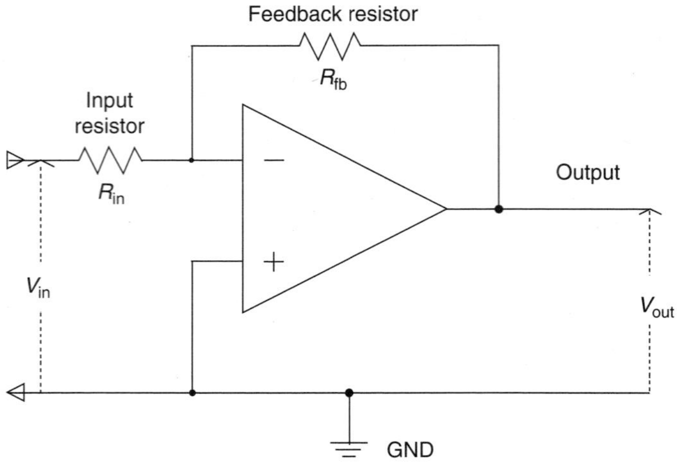

Figure 11.22 An inverting amplifier is built from a ZRA plus an input resistor as voltage-to-current converter.

dimensions are, of course, chosen so that the circuit satisfies the user's requirements as best as possible. A low-resistance design offers a number of advantages here over high-resistance dimensioning:

1 Leakage currents are less significant. If, for example, an OPA has an input bias current, I_b, of $\pm 100\,\mathrm{nA}$ (typical for precision OPAs with medium bandwidth), I_b causes an additional voltage error $V_e = R_{in} \cdot I_b$.

   With the dimensioning above this yields just $10\,\mu\mathrm{V}$ — this is in the same range as other small potential measurement errors such as V_{os}, and much smaller than, for instance, the potential tolerance of a commercially available reference electrode.
2. The energy distribution of the conductive electrons in resistors results in a noise voltage (observed as standard deviation of the nominal voltage), characteristic for the absolute temperature, T, the value of the resistance, R, and the frequency bandwidth, Δf:

$$V_r = \sqrt{4kTR\Delta f} \tag{11.44}$$

3. The noise contribution of the OPA's input leakage current ΔI_b adds to the total input bias current. It causes a voltage noise that increases with the resistor value. Low-resistance dimensioning reduces the effects of these unwanted potential variations on the signal being measured.
4. The electrostatic and magnetic properties of space and of any material that fills it cause pervasive, but usually undesirable, capacitive and inductive elements in the behavior of electronic components and circuits. Strictly speaking, a resistor only behaves as an Ohmic conductor when the voltage is constant. As the frequency of AC signals rises, its behavior is determined more and more by its stray capacitance and stray inductance. Down to a value of a few $100\,\Omega$, this is primarily noticeable, in the usual physical structures, as the stray capacitance, which, together with the stray capacitances of the OPA and of the circuit itself, typically amounts to a value of around 10 pF. At a frequency of 1 kHz a capacitance of this size has the

same admittance as a 15-MΩ resistor. The total impedance of the effective feedback network is dominated almost completely by the 1 kΩ resistor, suggested for R_{fb}. A nearly vanishing 0.007% error appears at this frequency.

All these considerations favor low-resistance dimensioning. There are, on the other hand, at least two arguments that represent a lower limit for the choice of resistance:

1. The stray inductance becomes significant when AC signals are present in resistors with less than 10 Ω.
2. Low values of resistance require high currents. A typical precision OPA can supply about 10 mA of output current. This limits the minimum value for R_{fb} to about 1 kΩ, if we assume that we plan to operate in a range of ±10 V, appropriate to electrochemistry. Generally speaking, low-resistance dimensioning has a disadvantageous effect on the current requirement of a circuit, and this must be considered particularly for battery operated constructions.

11.4.6 The Noninverting Amplifier

The step from the buffer amplifier to the noninverting amplifier (NIA) is as easy as the step from the ZRA to the IA. Not the entire output signal is fed back to the input of the OPA as was the case with the buffer. Instead of this, a voltage divider consisting of Z_1 and Z_2 is used to generate a fraction, b, of the output voltage V_{out}. This fraction b is used for feedback (Figure 11.23). In order to minimize the voltage difference at its inputs, the OPA then adopts a voltage of $V_{out} \approx V_{in} \cdot (1 + 1/b)$ at its output.

In order for the voltage divider ratio, b, to be real and to be nearly independent of frequency, the partial impedances Z_1 and Z_2 must have the same frequency characteristic. This is most easily achieved with two resistors:

$$b = R_2/R_1 \tag{11.45}$$

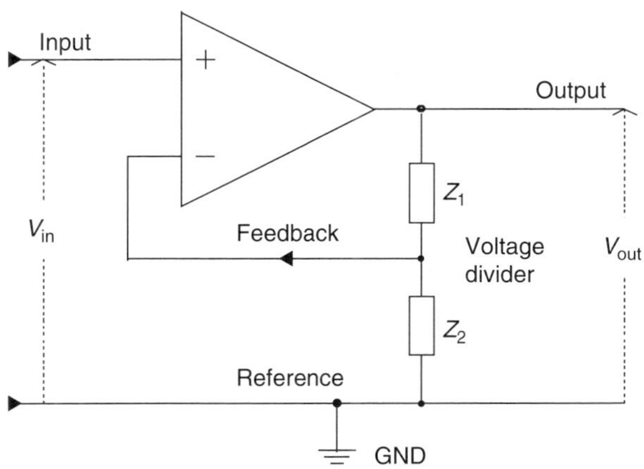

Figure 11.23 The noninverting amplifier (NIA) uses a voltage divider for the feedback signal.

This yields the simple relationship of Equation (11.46) for the amplification of the NIA:

$$g \equiv V_{out}/V_{in} = 1 + R_1/R_2 \tag{11.46}$$

Like the IA, the NIA is most often used for adapting levels, for example, to adapt an input signal to the useful range of an A/D converter. In contrast to the IA its amplification is positive, $g > 1$, and it has an almost infinite input resistance (at In_+). The IA and NIA play complementary roles in analog signal processing.

11.4.7 The Instrumentation Amplifier — A "Jack-of-All-Trades"

The properties of both amplifier types are combined in the "instrumentation amplifier" (INA). The disadvantage of the most simple form of construction, shown as a "difference amplifier" (DA) in Figure 11.24, is the low input resistance resulting from the resistors R_1 to R_4. Preceding it by two buffer amplifiers allows the DA to be extended to form an INA with a nearly infinite input resistance.

INAs are versatile modules, because voltage differences can be acquired and converted to relate to a new reference point such as GND. This makes it possible, for instance, to measure partial potentials in multiphase systems, or to use shunt measuring techniques to determine currents at any location within a circuit. Figure 11.25 shows how the potential difference between the two sides of a membrane in an electrolyte through which current is flowing can be measured with the aid of an INA.

Figure 11.25 also illustrates the special feature of a good INA: the output voltage, V_{out}, should depend only on the difference in voltages between the two

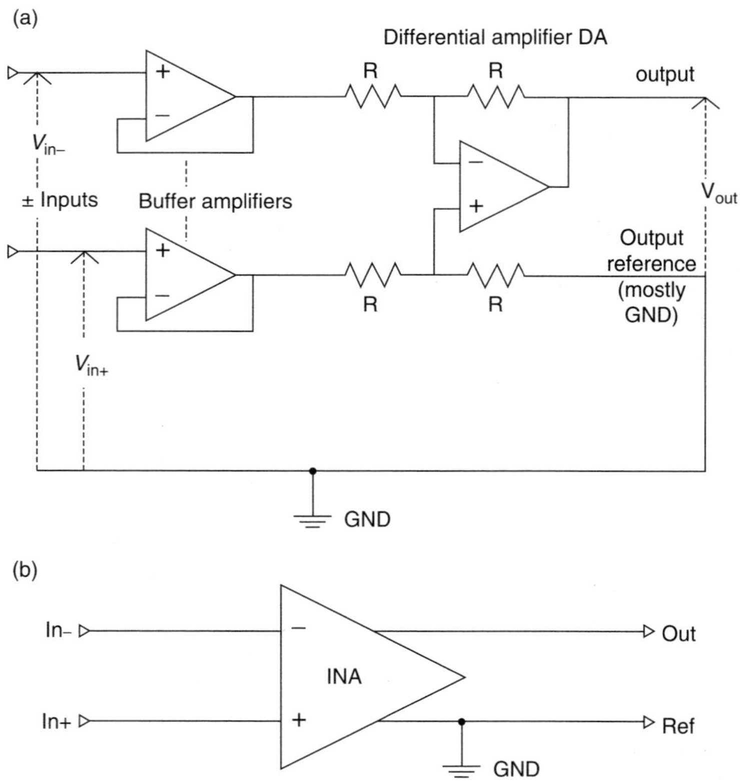

Figure 11.24 Instrumentation amplifier (INA): (a) detailed circuit; (b) INA symbol. An OPA-based INA contains a differential amplifier (DA) structure.

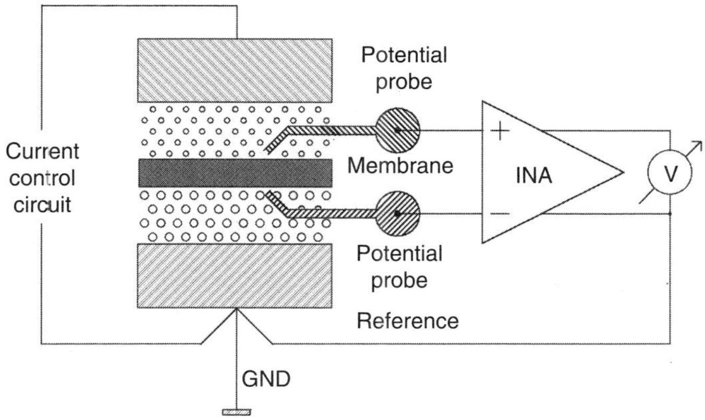

Figure 11.25 Measuring potentials floating to GND by means of an INA.

potential probes (the differential signal), and should be as little influenced as possible by their mean potential in comparison with the GND reference, the common mode signal $V_{cm} = (V_{in+} + V_{in-})/2$.

The parameter that serves as a figure of merit for his feature is the common mode rejection ratio, CMRR (Equation (11.47)). It indicates the factor by which the desired amplification, g, of the meaningful signal at the output is greater than the unwanted amplification, G, of the common-mode signal V_{cm}

$$\mathrm{CMRR} = \frac{g}{G}, \quad \text{where } g = \frac{V_{out}}{V_{in+} - V_{in-}} \quad \text{and} \quad G = \frac{2V_{out}}{V_{in+} + V_{in-}} \tag{11.47}$$

Figure 11.26 explains the role of the common mode rejection in the context of the task of measuring the single-cell potentials in a battery of cells (e.g., a stack of fuel cells).

11.4.8 The Three-Electrode Cell and the Potentiostat

The most important task of OPA technology in electrochemistry has to be the implementation of potentiostats for the operation of three-electrode cells. It is clearly

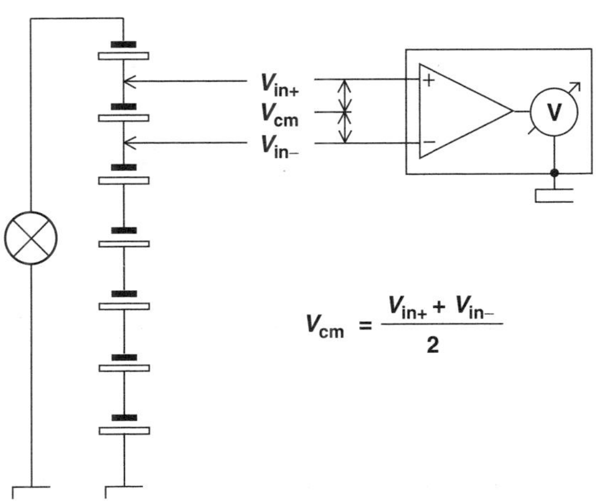

Figure 11.26 The nature of the common mode interference illustrated by the potential measurement in a fuel cell stack.

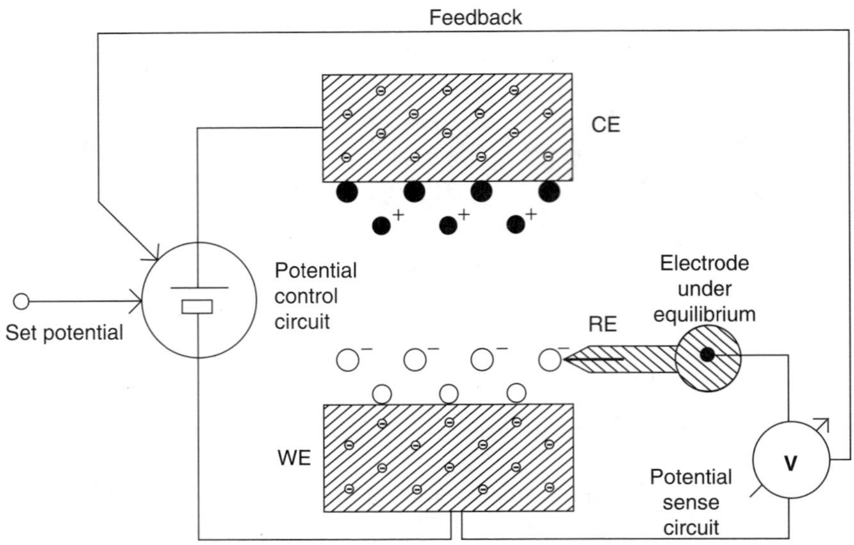

Figure 11.27 Potentiostatic control of an electrochemical cell. The voltage measured at the equipotential plane by means of a reference electrode (RE) in front of the working electrode (WE) is used as a feedback signal to control the counter electrode (CE).

very useful if, rather than merely passively measuring the potential at any one time in an electrode–electrolyte system (a half-cell), it is possible to actively control it. The potential measurement of the half-cell can indeed easily be implemented by means of a second electrode serving as a reference electrode (RE). In order to set up conditions other than those associated with the equilibrium position at the electrode under investigation (the working electrode, WE), it is, however, necessary to use another electrode (the counter electrode, CE). This allows the potential and the current in the working electrode to be controlled, while the reference electrode continues not to draw current. This makes the contribution of the reference electrode in the measurement of the total potential constant; changes in the potential can be assigned entirely to the working electrode. Figure 11.27 shows schematically the structure of a three-electrode cell with the characteristic measurement and control parameters.

The information obtained, without drawing current, about the actual potential difference between the working electrode and the reference electrode by using the potential measurement circuit serves as "negative feedback" for the control loop: the difference between the desired potential and the measured potential is highly amplified, and is used to adjust the potential and the current in the circuit between the working electrode and the counter electrode until the stage is reached where this difference is minimized.

11.4.9 The Simplest Possible Potentiostat: A Modified Noninverting Amplifier

We can see that this task can easily be performed by an NIA (Figure 11.23 and Figure 11.28): the potential, GND, on which all voltages are referred, is connected to the working electrode, the OPA In_+ input to the set potential, In_- to the reference electrode, and the OPA output to the counter electrode. The tap between the impedances Z_1 and Z_2 in Figure 11.23 is formed by the equipotential plane at the

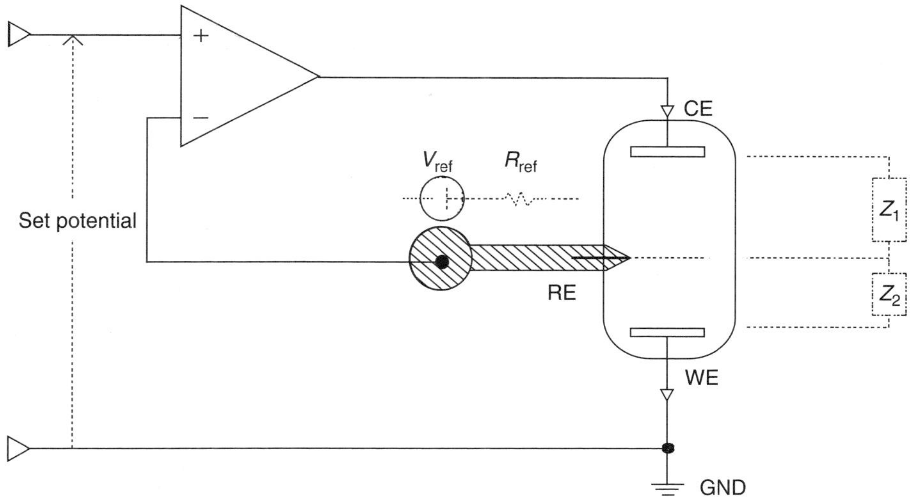

Figure 11.28 An NIA working as a basic potentiostat circuit.

opening of the reference electrode capillaries. Other than in Figure 11.23 the lead to In_- is seen to contain the potential offset, V_{ref}, caused by the reference electrode. Impedance Z_1 is formed by the region of material between the counter electrode and the equipotential plane, while impedance Z_2 is formed by the region between the equipotential plane and the working electrode.

The (not directly accessible) potential, V_{equ}, of the equipotential surface in front of the electrode under investigation adopts the value given by the following equation:

$$V_{eqn} \approx (V_+ - V_{ref})\left(1 - \frac{Z_1 + Z_2}{Z_2 A_{ol}}\right)$$
$$V_{eqn} \approx V_+ - V_{ref} \tag{11.48}$$

Equation (11.48) shows that an OPA does indeed work as a potentiostat, and that, apart from the offset potential of the reference electrode, fixes at In_- the set potential desired for the working electrode (Equation (11.48)). More precisely, the term $(Z_1 + Z_2)/(Z_2 A_{ol})$ in Equation (11.48) must remain negligibly small. This is only the case when the open circuit gain factor, A_{ol}, is large in comparison with the ratio of Z_1/Z_2.

Let us consider the very important practical consequences of this:

- The ratio of the impedance of the counter electrode, Z_1, and of the working electrode, Z_2, determines whether the potentiostat can perform its function!

When constructing three-electrode cells, many electrochemical engineers attempt to place the opening of the reference electrode capillary as close as possible in front of the working electrode. This is intended to minimize the influence of the column of electrolyte located above the electrode (the "Ohmic drop" $R_{el} \cdot I$).

If this is overdone, however, the result, instead of being more precise, may get unpredictable. Due to the double layer capacity the impedance at high frequencies of an electrode is dominated by R_{el}. As the total impedance of the counter electrode, Z_1, even rises when minimizing R_{el}, the impedance of the working electrode, Z_2, becomes very small, and the ratio Z_1/Z_2 then becomes dangerously large — the potentiostat no longer operates properly! These kinds of errors have particularly dramatic effects on "fast" measurements such as those required for impedance spectroscopy. Because of its fundamental significance, this aspect of the construction of measurements cells will be considered in more detail later. In the meantime, bearing the following rules of thumb in mind can give us general protection from unpleasant surprises of the above type:

- The position of the capillary opening should not be closer to the working electrode than one-tenth of the distance between the working electrode and the counter electrode.
- The area of the counter electrode should not be significantly smaller than the area of the working electrode.

A further aspect of the Z_1/Z_2 ratio should also be considered: if, as might happen with polarization measurements, we want to set a potential of $V_{equ} = \pm 1$ V at the equipotential surface in front of the working electrode, the potentiostat must provide a voltage of $V_{out} \approx V_{equ} \cdot Z_1/Z_2$ at the terminal of the counter electrode. If we stay within the above rules, this can just about be managed by a standard precision OPA operating from ± 15 V supply rails. Failing to observe this problem, for example, when the ratio $Z_1/Z_2 \approx 20$, will mean that the desired range for V_{equ} of ± 1 V can no longer be covered.

One obvious possibility would be to seek a solution through raising the available output voltage range (the "compliance voltage") of the potentiostat. It is indeed true that there are some tasks required in electrochemistry, such as the measurement of cell stacks, which can only be handled with an extended voltage range. Apart from the unavoidable loss in precision of the electronics used[3] in high-voltage potentiostats, high values of the ratio Z_1/Z_2 cause the errors resulting from leakage currents flowing between the counter electrode to the reference electrode and ground to become large. This is particularly noticeable in AC measurements on high-resistance electrolytes. Measurements involving high-resistance electrolytes do not necessarily require a higher output voltage, V_{out}, because as the resistance of the electrolyte columns increases, Z_1 and Z_2 are affected equally. The corresponding rise in the resistance from the capillary to the reference electrode is a bigger problem. The high source resistance of the potential probe (R_{ref} in Figure 11.28) turns out to be particularly problematic.

Ideally, no current should flow in the lead from the reference electrode to In_-. In that case, R_{ref} would not impair the precision of the measurements. In fact, leakage currents from this lead to the counter electrode and GND (significant for bulky reference electrodes) and in In_- can play a fatal role. The effects can, in particular, be dramatic in AC measurements when the ratio Z_1/Z_2 is unfavorable. The following points should therefore be taken to heart:

[3]Electrical components cannot be optimized equally well in respect of every important parameter. In particular, very high voltage or current requirements in amplifiers are incompatible with simultaneous requirements for low noise, low error voltages, small leakage currents, or fast response. Precision amplifiers for DC work best with ± 15 V power supplies and with output currents of up to ± 10 mA. When maximum speed (slew rate, settling time, and bandwidth) are important, supply voltages of ± 5 V are optimum, and output currents of up to ± 100 mA are possible. Because the parameters that are most important for electrochemical applications are scarcely satisfied by components available on the market, the manufacturers of potentiostats generally have to develop their own constructions.

- Reduction of the potential probe's source resistance and leakage current. Reference electrodes should be small, while electrolyte paths within the probe should be short and have large cross-section.
- If a high source resistance is unavoidable, then a buffer amplifier (Figure 11.20) should be inserted into the lead from the potential probe to In_- as close as possible to the location of the electrode.

11.4.10 The "Catch" in the Three-Electrode Arrangement

In our "minimal potentiostat," one side of the power supply in the potentiostatic control circuit and one side in the potential measurement circuit have a joint connection: the working electrode is connected to the joint GND reference for both circuits. This can lead to difficulties. A membrane cell, such as that sketched in Figure 11.25, or other similar arrangements of solid bodies fitted with entirely separate potential sensing circuits, cannot be operated successfully with the minimal potentiostat. There are also strict limits on measurements in conventional three-electrode cells: they only function with sufficient precision if the influence of the common lead to the working electrode is negligibly small. From its electrical role, the material of which the metallic lead is made belongs to the working electrode. If current in the control circuit leads to significant voltage differences along the lead, the actual potential will differ from what is wanted (Figure 11.29).

A cure is to entirely separate the current control circuit and the potential measuring circuit, representing a change to the four-electrode arrangement, also known as the Kelvin scheme, as shown in Figure 11.25. The potentiostat must, of course, also be capable of handling the two circuits separately. This is achieved by inserting an INA into the minimal potentiostat's potential measuring circuit in order to shift the reference of the measured potential to GND (Figure 11.30).

There are only a few kinds of electrochemical measurement that can be taken with a three-electrode arrangement at the potentiostat without any further thought. In typical DC methods, and when the currents in the control circuit do not exceed a few milliamperes, the errors resulting from the small potential differences in the common working electrode lead are negligible. In the presence of rapidly changing

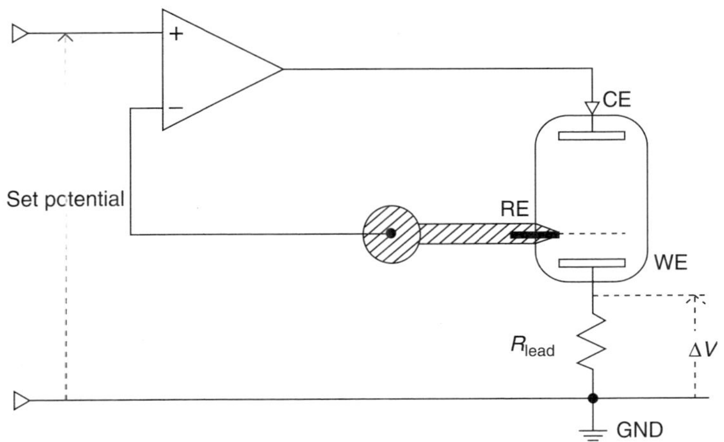

Figure 11.29 A potentiostatic three-electrode arrangement cannot avoid a difference ΔV between set voltage and actual voltage at the equipotential plane of the cell. It is due to the Ohmic drop in the working electrode to GND connection line with the resistance R_{lead}.

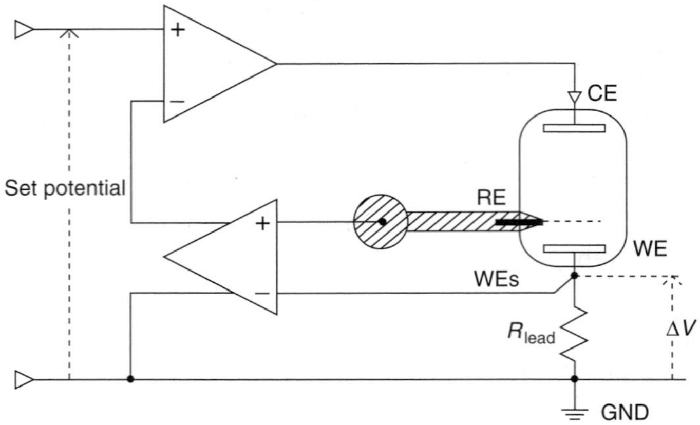

Figure 11.30 A potentiostatic four-electrode set-up adjusts the set potential precisely in spite of Ohmic drop ΔV across the resistance R_{lead} of the working electrode (WE) feeding line by means of an additional sensing line WEs.

signals and high frequencies, this lead may acquire an easily noticed, even dominating, influence over the potential of the working electrode, and it is essential that the following is borne in mind:

- Whenever possible, only use potentiostats that fully support the four-electrode arrangement.
- Even three-electrode cells should be connected according to the four-electrode scheme by providing separate leads for the working electrode to the current control circuit input and the potential measurement circuit input.

11.4.11 The Galvanostat — The Complement to the Potentiostat

Like potentiostats, galvanostats are systems with feedback. The information used by the control circuit for regulation purposes, however, does not originate from a potential measured at a particular location, but from the current flowing in the control circuit itself (Figure 11.31).

An NIA can help again when it is planned to make a specified current flow in a control circuit. Figure 11.32 illustrates how a minimal galvanostat can be constructed using a single OPA in the NIA configuration.

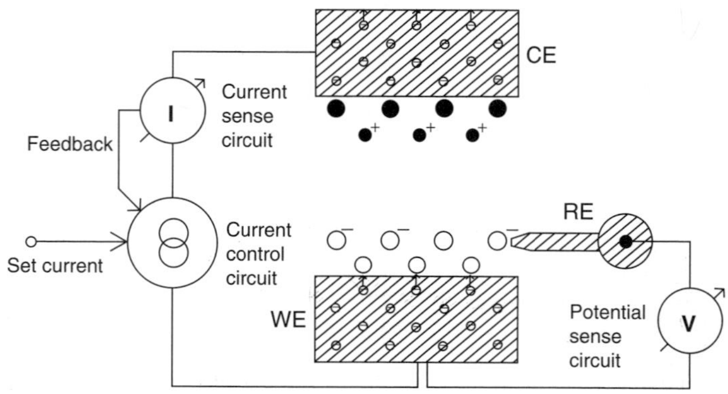

Figure 11.31 In a galvanostatic arrangement a current control unit feeds in a set current into an electrochemical full cell circuit. The potential sense circuit is optional.

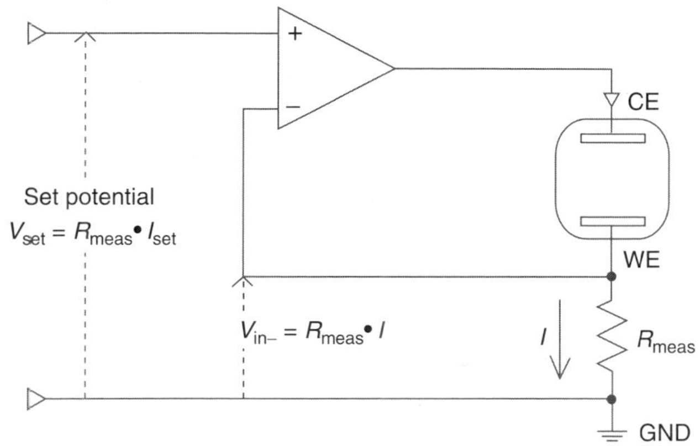

Figure 11.32 A simple galvanostat circuit based on an OPA in NIA configuration. The set current is adjusted by V_{set} and R_{meas}.

It functions according to a simple principle: because the NIA generates almost exactly the potential V_{in+} specified at In_+ at the common node of Z_1 and Z_2, it is only necessary to substitute an Ohmic resistor, R_{meas}, for Z_2. The current in the circuit between the OPA output through the series circuit represented by Z_1 and R_{meas} to GND is then given by $I \propto V_{in+}/R_{meas}$. The electrochemical cell is inserted in the place of Z_1. It is, therefore, clear that with a little skill, a potentiostat can also be used as a galvanostat, even if the device manufacturer had not planned this.

Simply connect the working electrode at the input for the reference electrode, and connect this point through a measuring resistor, R_{meas} (e.g., from a resistor decade box) to the potentiostat's working electrode output. The set value, once again, is a voltage. In order to create a specific current, I_{set}, specify a potential of $V_{set} = I_{set} \times R_{meas}$.

The problem of three- or four-electrode connection does not apply to a galvanostat, because the current is inserted through just two connections. Regulating to the set value does not require any potential to be measured. In order, however, for the experiment to sense the resulting potentials precisely, the same point applies as was explained above for the potentiostat: current-carrying and potential-measuring leads should always be separate.

A large number of electrochemical workstations offer the facility of choosing between the function of a potentiostat and of a galvanostat. The two methods of operation are equally fundamental for a large number of electrochemical techniques.

11.4.12 The "Best" Is Just Good Enough — But What Is the "Best?"

Just what is a "good" potentiostat or galvanostat, and what is a "bad" one? There is no simple answer to this question.

There is, of course, a whole string of trivial criteria: the device should, for instance, be electrically safe, easy to operate, reliable, have a long service life, and be economical. Most of these properties are not very easy to judge solely from the manufacturer's brochure, and the question of value for money can only be answered when the other criteria, together with the performance data, are considered along with the price.

For the technical data there are no binding standards governing which specifications are to be provided or how they are to be presented. Moreover, it is often the application that determines whether "more" or "less" of a particular property means that the device is “better” or “worse”: what might, for instance, appear indispensable when cycling rechargeable batteries may turn out to be a problem for sensitive methods such as the measurement of electrochemical noise.

As in the chapter on the measurement of voltage and current, some transparency to frequently cited general properties and technical data relevant to potentiostats, galvanostats, and other electrochemical instruments will be provided. This will also involve a discussion of aspects of the coupling of instruments to computers (A/D–D/A conversion, interfaces, etc.) that are so common nowadays.

After that a brief explanation of the principles of frequently applied electrochemical techniques will be provided, without making too many inroads into the comprehensive articles in other chapters of this book. The key point is to outline the method-specific requirements of the relevant instrumentation technology.

11.4.13 The Role of the Earth Potential

When we consider a particular experiment we tend to estimate our chances of success or failure on the basis, on the one hand, of the properties of the object being measured and, on the other hand, on the quality of the measuring instrument. This would indeed be enough, if only the experiment took place in a closed mini-universe. Unfortunately, however, the main cause of potential problems, namely the environment in which the measurements are taken, is then ignored. A few key words — screening, grounding, galvanic isolation, antenna effect — are enough to suggest that a small book could be written on the topic of “avoiding external influences” alone. If we wish to make at least a realistic estimate of the effect of sources of error, then we must also consider the environment of the experimental set-up, and we must form a usable model of it. The term “ground” (GND) has often been used above as the notional reference point, used as the zero in a measuring set-up. It is helpful to specify an electrical zero level to which all the voltages in the system are related.

The potential of the earth’s surface is an obvious reference to take for this purpose when modeling the environment in which measurements are taken. It can also be considered to be constant enough over a wide surrounding area. All conductors that are joined to this “ground potential,” such as water pipes, heating systems, the housing of earthed electrical devices, or protective earth contacts in the power supply itself are assigned a potential of zero in this simple model. These conductors are all included under the label “GND” in Figure 11.33. All other conductors not belonging to the item being measured, such as mains power conductors, lamp housings, the screen, and the observer of the experiment, have potentials unequal to zero. Commonly, these are also subject to variations over time: mains-operated devices demonstrate systematic oscillations at mains frequency, data display devices at the screen refresh frequency, while gas discharge lamps adopt a variety of often chaotic frequencies.

The pervasive potential variations at mains frequency are also transferred to the human body. Moving around in the area of an experiment, a human being will often also become electrostatically charged, undergoing a slow, but large change in potential. All these conductors are indicated in Figure 11.33, for the sake of simplicity, as “transmitting antennae,” because they fill the space around the experiment with disturbances to the electrical field. Then there is the experiment itself, including the leads to the measuring instrument: the conductors used here function as “receiving antennae” for the polluting field.

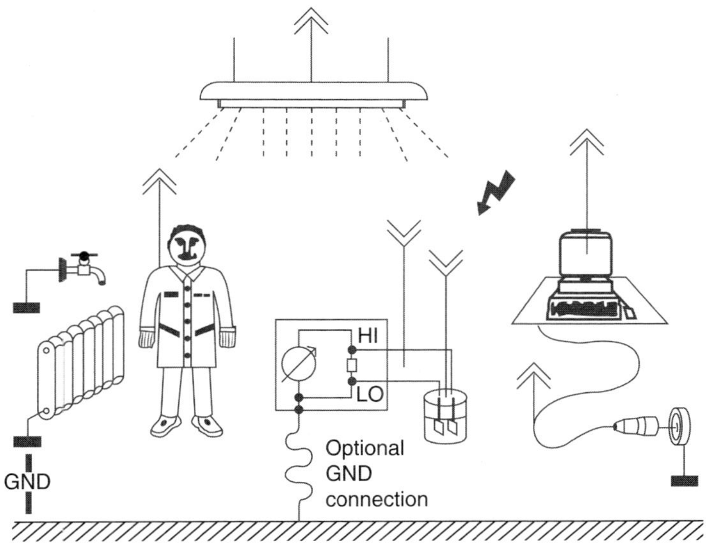

Figure 11.33 The role of the electric environment within an electrochemical experiment. The interference symbol is the flash. Interference sources and receivers are symbolized by antennas accordingly. Grounded objects are indicated by a GND symbol.

The possible potential difference between the reference potential of the measuring set-up (the "LO" reference)[4] and the ground or earth potential, GND, does not just have an effect on the influence of the different sources of interference, but is also used to classify various types of measuring set-ups. If there is a conductive connection (the GND path in Figure 11.33) between LO and GND, then we speak of a "grounded" or of a "nonfloating" arrangement. If there is no earth conductor, we speak of electrical isolation (a "floating" arrangement).

Misunderstandings of the significance of electrical isolation and grounding when constructing experimental set-ups are often causes of experimenters' headaches. Before we can list a number of clear pointers for construction, we need to understand the causes of interference from the measuring environment rather better.

As is usually the case with interference, the problem lies in the competition between the wanted and the unwanted electrical effects (signals and interference). For the important case of measuring small currents, Figure 11.33 illustrates that the flow of energy from the "transmitting" to the "receiving" antennae leads to unwanted interfering currents that compete with the signal current. It does not help much to attempt to interrupt the circuit of the interfering current by removing the earth conductor. The measuring set-up itself will act as an antenna, causing interference currents to flow as a result of radiation even in the absence of a ground conductor. The most effective method of countering interference arising from electric fields is to capture the currents that are created by means of "screens" in front of the measuring circuit's "receiving antennae," and to divert them to the ground potential before they affect the signals (Figure 11.34).

[4]Frequently, the structure of the signal inputs to an instrumentation circuit for voltage or current is, other than with an ideal INA, asymmetric in respect of the circuit's reference potential. The input that is "closer" to the reference potential (i.e., is not different from that of the reference, or has only a small difference) is referred to as the LO input, or the "cold terminal." The other input is referred to as the HI input, or the "hot terminal."

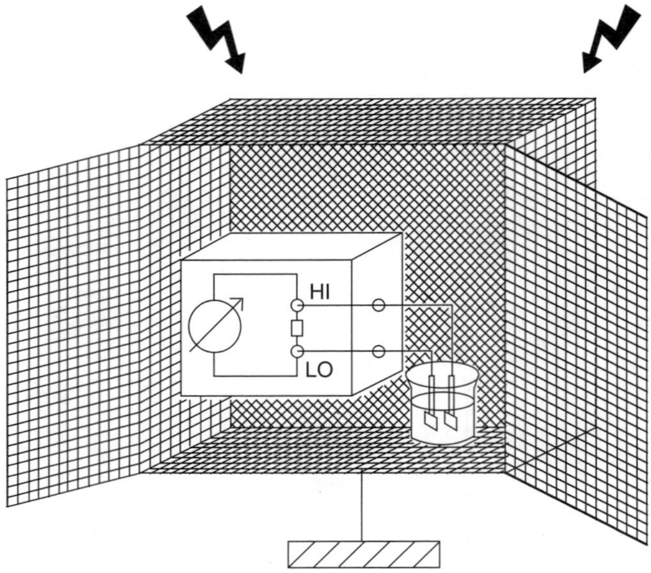

Figure 11.34 A shield of conductive material acts as a bypass for the disturbing currents caused by an electrostatic noisy environment.

If a ground connection is present it is easier to omit additional screening, since none of the conductors connected to ground have an interfering effect. When a measuring set-up is grounded, an observer can easily be changed from an "interferer" to a "suppressor" simply by touching something at ground potential such as the earthed housing of the measuring instrument.

One thing is clear now: an earth connection is "good." But are more connections better?

It can, in practice, often happen that there is already an earth connection in an electrochemical system being investigated. Picture, for instance, corrosion investigations on a pipeline or in the cooling circuit of a power plant.

Figure 11.35 illustrates what happens when an earthed instrument is used in such a system to measure the potential between a water pipe and a reference electrode

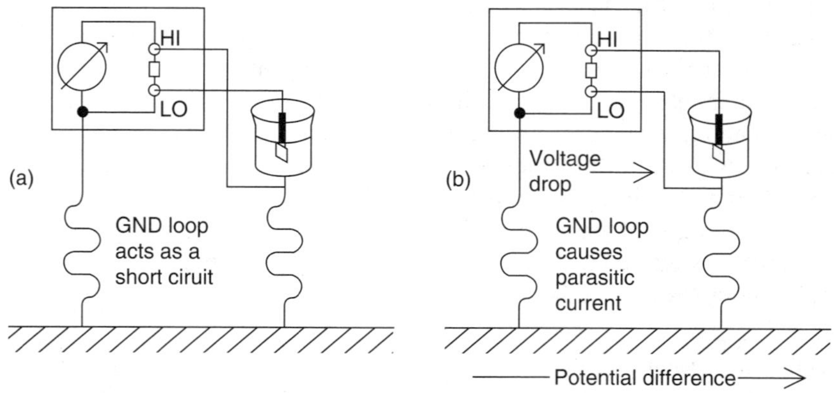

Figure 11.35 Two ways to connect a grounded meter to a grounded object: the bad one (a) and still not so nice (b). The metal beaker with the electrode and electrolyte inside symbolizes a segment of a water pipe.

that has been inserted. In the worst case, shown in Figure 11.35(a), the grounded LO input of the instrument is connected to the reference electrode, while the HI input is connected to the pipe. It is easy to see that this causes the potential of the reference electrode to be short-circuited via the ground connection of the LO input. It is obvious that this cannot work.

But the inverse case in Figure 11.35(b) still has its problems. In this case, the ground connections of the LO input on the one side and of the pipe on the other form a closed conductive path (an "earth loop"). If even slight differences in potential arise between the two earths, such as might be caused by the electrochemical potential differences of different materials, the low resistance of the loop can easily cause large currents to flow. This, in turn, means that noticeable potential differences arise in the connecting path between the LO input and the pipe, leading again to measurement errors.

In order to take precise measurements in such a system, the instrument itself must not have its own earth connection — in other words it must be electrically isolated.

At this point, we can formulate a few rules, to which more comments will be added below:

- For sensitive electrochemical measurements, the reference potential of the measurement set-up should be connected to earth at just one point.
- If the object being measured does not have its own earth connection, the simplest technique is to use an earthed instrument. If an electrically isolated instrument is nevertheless used, an additional earth connection should be made.
- On the other hand, if the object being measured is connected to earth at some point, then an electrically isolated instrument *must* be used.
- If possible, whichever part of the object being measured is likely to function most powerfully as an antenna should be connected to the instrument's LO input.
- If extensive screening is necessary, the instrument's LO input is connected to the screening, such as the Faraday cage.
- If an earthed instrument is used for the measurements, all the conductors at earth potential in the region around the measuring location contribute to the avoidance of interference. Conductors that are not earthed, however, function as transmitters of interference. Workstation supports, stands, screening cages, and any observers should therefore also be grounded.

These tips are first of all oriented to the frequently occurring situation in which electrostatic interference limits the accuracy of experimental measurements. This is typical for electrochemical measurements on coating layers, microelectrodes, or when pH electrodes are used, as well as generally when the typical currents are less than the order of magnitude of a microampere.

They should also help to avoid incorrect measurements of potential resulting from ground loops, which unfortunately also have effects when the typical currents are in the higher ranges, which otherwise are not problematic.

If the above rules are taken seriously, then it is usually possible to avoid the inconvenience of using a Faraday cage. As an example, when measuring the corrosion potential of the coated inner wall of a beverage can, it is necessary for the measuring arrangement to be protected very effectively from electrostatic interference. A

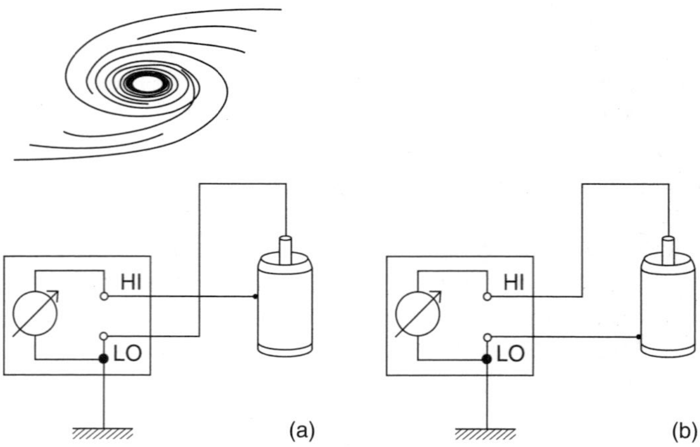

Figure 11.36 Two ways to measure the corrosion potential of the protective layer inside a beverage can: in arrangement (a) the potential is disturbed by the electric fields of the environment. If one connects the can to ground (b) much less interference will arise.

reference electrode is placed for this purpose inside the can. If, following typical electrochemical practice, the reference electrode is connected to the LO input of the earthed measuring instrument, the body of the can act as an antenna for interference (Figure 11.36a). If the body of the can is connected as the working electrode to the LO input, the can acts as a perfect screen for the reference electrode inside, connected to the HI input (Figure 11.36b).

Deciding on the optimum design for the measurement set-up is not always as easy as it is for a simple measurement of potential. In potentiostatic experiments we begin by thinking about four leads instead of just two. If there are additional signals (e.g., in disk-ring experiments) or more sophisticated methods (such as fast pulse and impedance measurements) it becomes even more complicated. The points made above remain just as important. Optimum design of the measuring set-up, however, sometimes requires additional knowledge about the instruments being used and how they are connected or isolated from ground.

11.4.14 The Computer as a Measurement Slave

At the information output of most of the circuits processing analog signals discussed above one or more symbols for measuring instruments have generally been drawn.

In our considerations so far, the function and method of operation of the modules has been our primary focus. We have, therefore, ignored the fact that in modern laboratories a large part of the information has, for a long time, not been passed directly from a display instrument to the observer's eye (Figure 11.37).

The signals that have been created are instead first digitized, and then further processed by a microprocessor. Although a PC is at the end of the line of measurement processing, an entire hierarchy of processors can sometimes be involved. Required signals such as the value and the modifications over time of the set potential and set current of the potentiostat or galvanostat are no longer simply controlled by manually adjustable voltage sources and function generators. The set magnitudes are instead specified by microprocessors as digital values, and are fed through digital/analog converters to provide analog electrical signals for controlling the equipment.

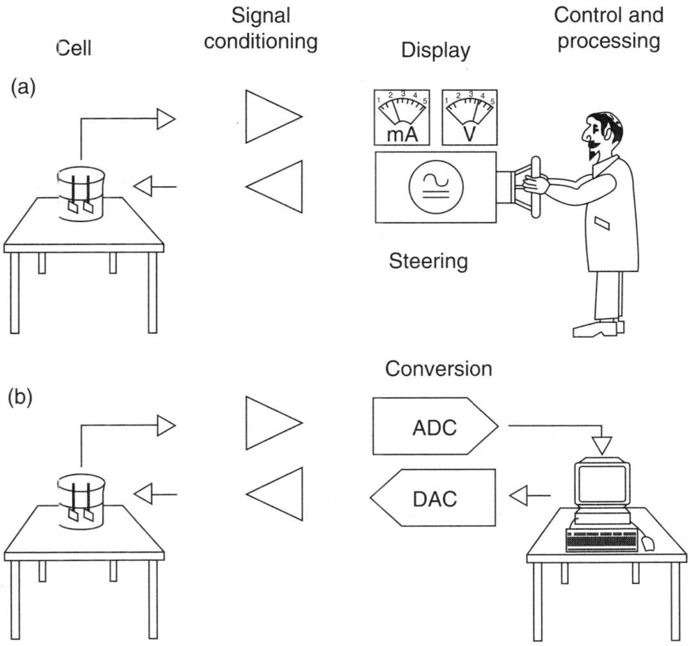

Figure 11.37 Traditional (a) and computerized (b) electrochemical experiment.

The crucial steps here are the conversion of the analog measurements into digital values — the analog/digital conversion (A/D) — and the inverse process of digital/analog conversion (D/A) (6). Analog measurements represent physical magnitudes. As such, and leaving aside quantum effects, they can be "resolved" to any degree, just like real numbers. In digital calculators, on the other hand, data have only a limited resolution. The resolution is usually indicated by the data's word length. This is the number of independent binary information units, or bits, in one data word. Eight such bits, each of which can adopt precisely two states, are known as one byte, a data word that therefore can adopt $2^8 = 256$ discrete states. One straightforward way of representing these states is to number them with integers from 0 to 255.

It is now easy to picture the digitization process. What is necessary is to distribute the expected range[5] of a measured physical magnitude in an appropriate way over the number of discrete states in a data word. One byte could be used to represent voltages from 0 to 255 V in steps of 1 V. If the zero point and the range of values are adapted rather more appropriately for typical electrochemical values, the voltage range from -1.28 to $+1.27$ V could be represented with a resolution of 10 mV. Smaller or larger values than the chosen interval can, of course, not be represented, and lead to an underflow or overflow situation. The expected range represents a segment of the series of real numbers, while the range of the digital states is a segment from the series of integers. It is, therefore, also possible to formally

[5]The literature often refers to the expected range of a physical magnitude as its dynamic range. Although an attempt is made in this chapter to restrict the use of the adjective "dynamic" to processes that vary in time, the term "dynamic range" will be used below in its accepted meaning of the expected range.

understand the digitization process as a linear mapping of an interval within the series of real numbers onto an interval within the series of integers (see also Figure 11.38).

This is a severe loss of information — or so it might appear. Lengthening the data word brings an improvement. If we represent the range of measurements discussed above, from −1.28 to 1.27 V, by means of two bytes instead of one, then we require a 16-bit A/D converter. This gives us the digital result with a resolution of $2.56\,V/2^{16}$. The resulting steps are then about 40 μV wide — it is hardly possible to determine an electrochemical potential more precisely than this.

The problem of D/A conversion is quite similar. In this case, the desired output range for the specified magnitude is represented by the interval of the discrete states of the data word that describes it. An 8-bit D/A converter scaled as above can, therefore, provide an output for the set value of a potentiostat over the range from −1.28 to +1.27 V with a resolution of 10 mV, while steps of about 40 μV are provided by a 16-bit D/A converter.

So — do we want more resolution, according to the slogan "a lot helps a lot?" No other isolated parameter related to digital measurement acquisition is so misleading as the number of bits in A/D and D/A converters, since the word length only indicates how many states are, in principle, available — how accurately occupied they are in fact remains quite open. The user can get a much clearer representation of the precision of an instrument from the traditional error specification. The manufacturer here quotes an absolute uncertainty along with a value-dependent relative uncertainty, and whichever is the larger will apply. The controlled output voltage of the potentiostat mentioned above, could, for instance, be specified as follows:

Output voltage: resolution 16 bit, uncertainty ±0.5 mV absolute,
±0.05% of the set value.

This would be quite a respectable specification for an electrochemical device — it is seldom for the output voltage of a potentiostat to really be more precise. The permissible error here is, as is usual when long word lengths are quoted, about an order of magnitude greater than suggested by the bit-resolution.

11.4.15 A/D and D/A Converter Families, Functional Principles, Intended Applications, and Conversion Parameters

If we pay attention to the traditional methods of quoting errors, then we are quite well protected from nasty surprises. If the reader finds this enough, he can skip the section that follows.

A more detailed consideration of the properties of A/D and D/A converters (in the following short ADC and DAC) is necessary if we want to orient ourselves reliably in the jungle of computer-controlled instrument specifications. The most important static specifications (i.e., relevant in the time average) of absolute and relative errors, as explained above, can easily be understood through the diagram of the representation process (Figure 11.38). On one side is a segment from the series of real numbers corresponding to the range of set or expected values. In the above example of the 8-bit DAC this extends from −1.280000... to +1.270000.... On the other side is the range of integers that indicates the available states of the digital word.

If the converter operates perfectly, the two ranges fit together precisely (Figure 11.38a). If the two segments are displaced with respect to one another, then the difference from the ideal value is a fixed amount, the offset error (Figure 11.38b). In

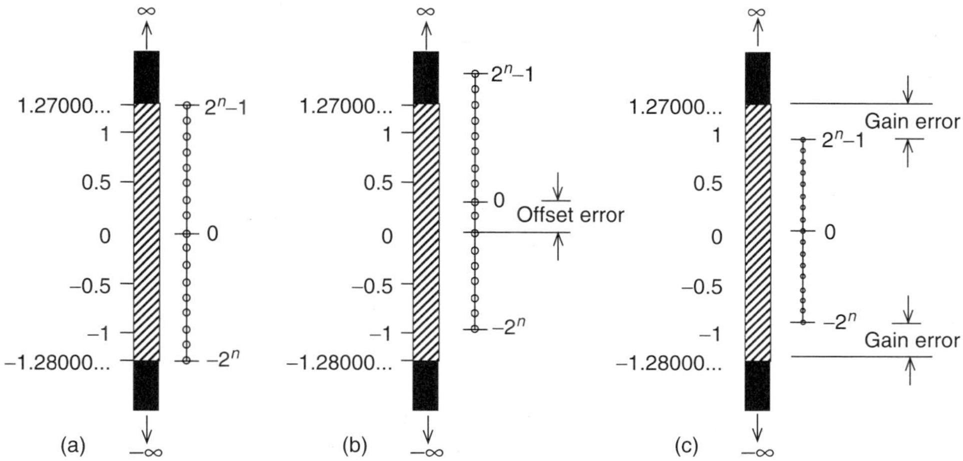

Figure 11.38 Proper conversion (a), conversion affected by offset error (b), and gain error (c).

addition to this, the length of the analog segment can differ from the intended value, leading to an error proportional to the analog value, the gain error (Figure 11.38c).

Both the offset error and the gain error make significant contributions to the total error. With word lengths of 13 bits or more, they usually dominate any digitization error associated with the resolution. The static precision of a converter can also be impaired by errors in the linearity of the mapping. Figures on linearity, however, appear to be the poor relation of manufacturers' specifications.

Nevertheless, they are enormously important for digitally generated control signals in the majority of dynamic measurements.

Figure 11.39 differs from the principle of Figure 11.38 primarily in that the association between an analog value and a digital code is represented on X and Y coordinates. An ideal converter is illustrated by a straight staircase function. It passes through the scaling points defined by the initial and final values of the intervals. Converters with offset or gain errors miss these points (Figure 11.39a). Figure

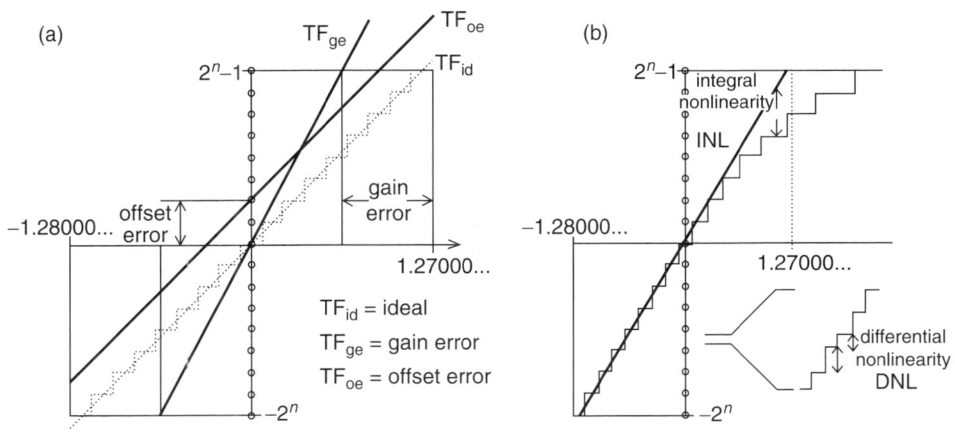

Figure 11.39 Converter transfer functions. Precise transfer function (TF) and affected ones by gain and offset error (a). TF affected by linearity error (b).

11.39(b) shows the characteristic curve of a converter with linearity errors. A bend in the staircase results in "integral nonlinearity, INL," while deviations from the constant step height result in "differential nonlinearity, DNL."

If the DNL does not exceed the average step height (the bit step for the least significant bit, LSB), then the contribution to the total static error can usually be neglected. If, however, the DNL is larger than one bit step, the transfer function is, at that location, not monotonic. Conversely, if a converter can be classified as monotonic, this means the same as DNL < 1 LSB. In systems involving feedback, such as digital controllers, nonmonotonic ADCs can cause greater problems than might be expected from their contribution to the static error. This point aside, ADCs with moderate DNL are a lot less problematic than corresponding DACs. This is because if DACs are used for a dynamic drive, the DNL contributes a strong noise to the signal output magnitude. There are a large number of electrochemical methods, such as cyclic voltammetry, where this leads to a drastic deterioration in the measuring precision. Figure 11.40 shows the cyclic voltammetry curve (CV) of a capacitor when the controlling DAC has significant INL and DNL errors.

The result of replacing analog ramps by digitally controlled staircase functions during voltammetry measurements will be discussed in more detail in the tips for the associated measuring methods. We can also see in the example illustrated in Figure 11.40 that the dynamic behavior of the converter is also significant, on top of the static error specification. Because their dynamic behavior is primarily determined by the way in which they function, the properties of the most popular A/D and D/A converter architectures are discussed below.

Table 11.3 provides a summary of the name, typical word lengths, conversion rates (in measurements/second), popular applications, and fundamental properties of ADCs.

A/D converters that use the principle of successive approximation (SAR-ADC) find a digital value representing the value of the input voltage for a short period as it changes over time. The input voltage must not change during the "conversion time" — that is the time between taking the sample and determination of the digital value. This is usually ensured by using an integrated sample-and-hold circuit (S/H) that acquires the available signal within a very short time span and then supplies it, as a constant value, for the converter to measure during the conversion time.

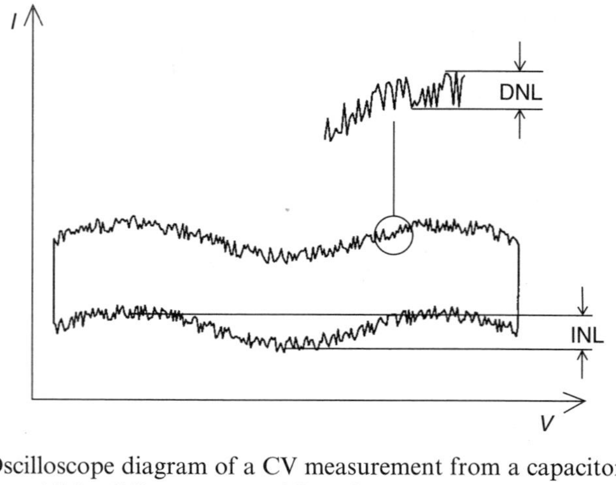

Figure 11.40 Oscilloscope diagram of a CV measurement from a capacitor, controlled by a DA converter that exhibits different types of linearity error.

Table 11.3

Functional principle/name	Word length	Conversion rate	Application	Comment
Successive approximation	8–16	100 kHz–300 MHz	Universal, multichannel measurement acquisition	Sampling principle
				S/H circuitry required
Sigma/delta	16–24	10 Hz–100 kHz, 192 kHz	Low-frequency, audio	Partially integrating
Ramp	Decimal 4–8 digits	0.1 Hz–10 Hz	Precision DC measurement, multimeters	Continuously integrating
Parallel	6–10	MHz–GHz	High-frequency, video, radar	Continuous

SAR-ADCs can be used very broadly, and their behavior is easy to understand. They are less suitable for the highly accurate measurement of slow processes or for DC. They are, however, most suitable as "workhorses" in electrochemistry for applications such as fast voltammetry, pulsed processes, and EIS.

SAR-ADCs can be nonmonotonic, but the INL is nevertheless sufficient for many applications. Economical 16-bit SAR-ADCs have a typical INL of ± 4 LSB and a DNL of ± 1 LSB at conversion rates of 100 kHz. In microprocessor-controlled applications the gain and offset errors can be calibrated out, so that a typical static precision of 14 bits is achieved. The precision of individual measurements does not drop dramatically even when acquiring signals that are changing rapidly.

The short signal acquisition time also means that measurements can be taken at signal frequencies that are much higher than the maximum conversion rate. It is, therefore, necessary to remain aware of the Nyquist criterion, which states that: The highest frequency, f_g, present in a mixed signal may not be higher than half the conversion rate, f_s:

$$f_s \geq 2f_g, \quad \text{The Nyquist criterion} \tag{11.49}$$

The Nyquist criterion quantifies a straightforward statement: if a signal changes so fast that a sequence of A/D conversions cannot keep up, then the digital values that are found do not represent the signal.

In practice, preceding low-pass filters are used in an attempt to prevent this problem of "aliasing." Such filters suppress frequencies in the signal mixture that are above a specified "limit frequency," and in this way reduce the speed with which the signal changes. If the limit frequency is set to a value significantly below half the conversion rate, as required by the Nyquist criterion, aliasing errors are avoided.

The unpleasant consequences of aliasing are discussed in the subsection describing EIS measurements.

Sigma-delta ADCs operate more slowly than SAR-ADCs, and the measurement represents the signal level over almost the full conversion time, or even a multiple of that (integrating measurement by approximation). If the measuring rate approaches the maximum conversion rate of the converter, the Nyquist criterion is satisfied approximately, even if higher frequencies are present.

Sigma–delta ADCs are generally also more precise. A monotone transfer characteristic and a low DNL are automatically ensured by the principle of operation. Integrated microcontrollers often compensate offset and gain errors by an automatic calibration. The INL, however, is usually significantly worse than the resolution might suggest. Systematic noise at certain frequency ranges also impairs the achievable precision. Because it is usually essential in electrochemical measuring systems for the electronic circuits described above for signal acquisition and preparation (buffers, amplifiers, INAs, filters) to precede the ADCs, and because these are not included in the autocalibration process, the attractive specifications of such converters are, in practice, of low use.

Sigma–delta ADCs are not very suitable for sampling a large number of channels, because they either require long conversion times or a number of conversions in order to settle to a stable final value. Sigma–delta ADCs for audio applications are not suitable for instrumentation purposes.

Still more time consuming are ramp ADCs. They are the best option when high static precision, stability, and low noise are important, and when time plays a subsidiary role. The conversion processes are self-calibrating and integrating. Most ramp ADCs are used in digital multimeters, for which reason they often supply their output in decimal rather than binary form, so that a display can be driven directly. This is not very suitable for the acquisition of measurements by microprocessor. The electrochemical engineer should nevertheless surround himself with a few ramp ADCs in the form of accurate multimeters. This allows him to keep close tabs on his instruments.

Parallel ADCs are specialists in acquiring rapidly changing signals. Rather than using a sequential procedure to determine the individual bits, as in the other processes, they are determined in parallel at the same time. Parallel ADCs allow achieving high sampling rates at the expense of low static precision, high price, and high power consumption.

As the cyclic voltammetry graph of Figure 11.40 demonstrates, the flow of data from the computer to the object being measured also calls for our attention.

Only the first two types of DAC listed in Table 11.4 are suitable for instrumentation purposes. Whereas the principle of construction of DACs based on

Table 11.4 Summary of the Name, Typical Word Lengths, Conversion Rates, Popular Applications and Fundamental Properties of Common D/A Converter Types

Functional principle/name	Word length	Conversion rate (DC)	Application	Comment
CMOS-R2R	8–16	1 MHz	Universal voltage divider	Multiplying
Switched current sources	6–16	100 MHz	Signal synthesis Video	Low noise High power consumption
Switched voltage divider	12–16	100 kHz	Digitally operated control loops	Construction-dependent monotonic
Single-bit converter	16–24	100 Hz, 192 kHz	Voltage sources, audio	Construction-dependent linear and monotonic

switched voltage dividers ensures that they are monotonic, the INL nevertheless leaves a lot to be desired. This is due to the fact that it is constructed internally of a number of segments, which is why it is alternatively known as a segmented DAC.

In this respect, the performance of the single bit converters, primarily familiar from audio technology, is much better. Their function is similar to an inverse of the sigma–delta A/D converters.

Like these, however, they also suffer from systematic noise at certain frequency ranges. This is not a problem in audio applications. The noise and the poor static precision of single bit converters mean, however, that they are of little use for electrochemical purposes in spite of their outstanding linearity.

CMOS-R2R DACs are also used as digitally programmable voltage dividers. They have the disadvantage for these applications that higher frequency interference does not become proportionally reduced when very low output voltages are set.

High-resolution converters of this type are nevertheless the first choice for control sources in electrochemistry. Because perfect linearity is difficult to achieve with these devices, only the high-price component batches with the best specifications should be used.

It must be recognized that the specification of DACs must be seen as more problematic than is the case with ADCs. It is therefore better, when attempting to choose devices, to first look at the specification for the output voltage and the output current of potentiostats and galvanostats. An impressive low uncertainty of a few microvolts when measuring the potential may look good in the brochure, but what use is this to the electrochemical engineer if the set voltage can be wrong by a full millivolt?

In the end, the method applied also determines whether a statement of resolution or of precision has any real relevance. This can be seen in an example from EIS: the stability, resolution, and precision of the stationary component of the set voltage in the potentiostat in use are just as important here as it is in most other electrochemical procedures. The principle of EIS, moreover, is based on the superimposition of a small excitation on top of the stationary magnitudes. The question, however, of whether the excitation amplitude here is 1 or 2 mV does not have any systematic effect on the result, provided only that it is sufficiently small. Under these circumstances, it is irrelevant whether the setting range of the excitation amplitude is set with a high resolution of 0.01% with the aid of D/A converters, or just in large steps of 1–2–5 using a long chain of electromagnetic relays. There is one property that usually does not appear in specifications, but is more important than the resolution: the setting procedure should ensure that the noise and interference voltage components also shrink proportionally at small excitation amplitudes. Otherwise it is almost impossible to take precise impedance measurements in markedly nonlinear systems.

It is also important in a computer-controlled electrochemical workstation for the A/D and D/A conversions to be perfectly synchronized. This implies that the appropriate option of integrating or sampling A/D conversion is used for the particular method.

11.5 THE STANDARD ELECTROCHEMICAL METHODS FROM THE POINT OF VIEW OF INSTRUMENTATION TECHNOLOGY

The electrochemical engineer would be likely to classify the various working techniques as observational, chemical-analytic, and preparative methods. The first group is the largest, with prominent representatives such as Tafel methods, EIS, cyclic voltammetry and electrochemical noise analysis, or ENA. The second group includes polarography and related methods (7). Preparative methods generally exploit electrolysis.

Such a division in accordance with the purpose of the application is unfortunately not very helpful in terms of the instrumentation techniques employed. A more suitable approach would be to make the classification according to the way time is handled, and whether the electrochemical system under test is actively controlled:

1. *Passive*: These methods are restricted to observation of the electrochemical system under investigation in an open electrical circuit. Examples of this would include measurements of equilibrium potential over time, or measurements of the electrochemical potential noise.
2. *Static*: Here the circuit including the electrochemical cell being investigated is closed under specific external electrical conditions. Polarization measurements, such as current against time in potentiostatic operation, the galvanostatic complement of this, or measurements of electrochemical current noise are typical examples of this group.
3. *Variable steady state*: As in the static case, the system being investigated is set into a specified state in a closed circuit, and the observable parameters are measured. The measurements are repeated a number of times as the conditions are systematically varied. It is expected that the system's response to the modified conditions has settled by then to a new, steady-state condition. Examples of this include the recording of current–voltage curves for Tafel evaluation, or measurements of the effect of loading on fuel cells.
4. *Dynamic*: As in the steady-state case, dynamic measurements are taken while the system is being actively controlled, while systematic changes are made over time. The difference is that it is, however, not expected that a steady state has been achieved with respect to the control magnitude. Instead, we are also, or even mainly, interested in the time response of the system. EIS, CV, linear and pulse voltammetry, polarography and current interruption methods are the most familiar representatives of dynamic procedures.

11.5.1 Passive Measurements of Equilibrium Potentials

The passive methods present the fewest headaches. The general requirements for an instrument to measure potential, such as high input resistance, good static precision, and CMRR, have already been extensively discussed. They are particularly significant for potentiometry in analysis and sensor technology.

When investigating electrochemical potential noise, however, the static precision is of less importance than a high dynamic range. In ENA, we are mostly interested in small fluctuations in the potential over a period of seconds. The electrodes being investigated often, in addition, have a much greater drift in the mean potential over long time periods (Figure 11.41). In order for the relevant small component of the signal to be recorded with sufficient precision, and for it to be separated through computation from the larger long-term trend, analog signal conditioning with the lowest possible intrinsic noise is first necessary. Further, the subsequent A/D converter requires a high resolution, and must be monotonic, although the INL is, in this exceptional case, of lower significance. This means that

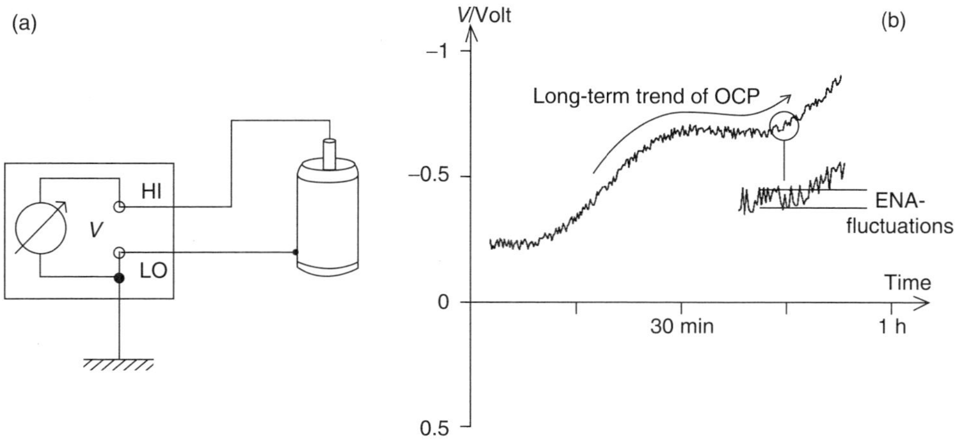

Figure 11.41 The electric arrangement for the ENA of the OCP is common with the usual passive potential measurement methods (a). In contrast to these methods, ENA is more interested in small fluctuations in short time scales that are superimposed to the long-term trend of the potential (b).

sigma–delta A/D converters with medium resolution (around 20 bits) and with a sufficiently high conversion rate (100 Hz or more) can usefully be employed.

11.5.2 Static System Control

The static methods generally require, in addition to the measuring instrument, a potentiostat or galvanostat to control the state of the system when the electrical circuit is closed. In order to measure electrochemical current noise it is sometimes sufficient to set the potential to zero. Using a ZRA to provide a virtual short-circuit for the current measurement is an adequate approach. The same considerations that apply to ENA of the potential apply, in this form of ENA, to the ZRA and the A/D converter.

If potentiostats like those of Figure 11.30 are used to generate a fixed potential for the measurement of current over time (chronoamperometry) it is of crucial importance that the generation of the set potential and the power stage are stable and that they produce little noise. Because the dynamic behavior of electrochemical cells over a wide range of times is dominated by the capacitive components of the double layer and transport processes, small fluctuations in the controlled voltage are powerfully amplified in the current response — the faster the fluctuations are, the higher is the amplification.

It is, therefore, important to pay attention to the specification of the frequency spectrum of the potentiostatic noise. If the application does not require high output voltages and currents then, as so often, it may be a case of "less is more": providing the facility for high power levels makes it more difficult to construct a low-interference potentiostat — less powerful devices usually create less noise and are more stable.

How much interference in the flow of the data itself finally makes its unwanted way through to the evaluation depends on the filtering and on the type of A/D conversion. As has already been mentioned when discussing the Nyquist criterion

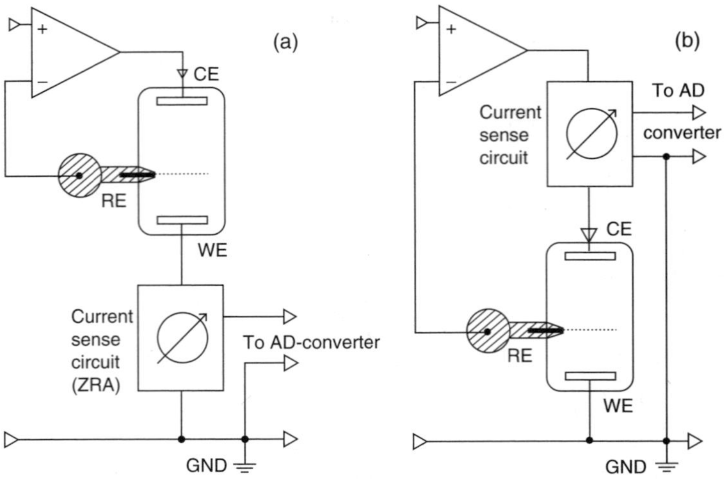

Figure 11.42 Two alternatives for the placement of the current sensing path within a potentiostatic arrangement. The solution (a) easily allows the application of high output voltage and current to the cell, due to low CMRR requirements. The more expensive version (b) has the advantage that the working electrode is insensitive to interference, due to their direct connection to the LO reference.

(Equation (11.49)), low-pass filtering or integrating A/D conversion processes can minimize the effects of high-frequency interference.

There is, furthermore, a question to be answered regarding the most favorable position for the current measuring set-up within the circuit. If the manufacturer inserts the ammeter in the current feed to the working electrode (Figure 11.42a), the requirements on the CMRR of the current sensing path are light, and do not depend on the voltage of the counter electrodes. It is, therefore, easy to achieve a large range of output voltages for the counter electrode in this arrangement.

Alternatively, Figure 11.42(b) shows an ammeter inserted in the circuit for the counter electrode. Because the instrument in this case "floats" at the potential of the counter electrode, a good CMRR is very important. Specially constructed INAs with wide frequency ranges are, therefore, used in fast potentiostats, although their range of DC potentials is restricted to less than ± 15 V. Potentiostats with the structure of Figure 11.42(b), therefore, have a restricted range of output voltages. The advantage of this arrangement is that the working electrode is connected directly to the arrangement's reference potential. Whereas the arrangement of Figure 11.42(a) means that all the electrodes in the cell are sensitive to electrostatic interference, it is even possible in the arrangement in Figure 11.42(b), like that of Figure 11.36(b), to touch the working electrode without fear of interference.

Galvanostatically controlled chronopotentiometry and ENA are less critical than their potentiostatic equivalents — as we shall see below, however, this does not apply to their dynamic cousins.

11.5.3 Measurements with Varying but Steady-State Potential and Current

First, it is, of course, true that in the physical world strictly-steady state parameters cannot be achieved — in fact, the term "quasi-steady state" would be more appropriate. The intention is to change the electrical conditions of the electrochemical system being investigated so slowly that the contributions of its "electrical memory" to the system response are negligible. In the case of the acquisition of a current–voltage curve this has the effect that noticeably slower or faster recording does not have any effect on the resulting measurements.

It should also be obvious that a steady state, according to the definition above, cannot be achieved in systems that are dominated by capacitance or inductance.

Having made these basic points, we can now turn to the instrumentation technology involved in steady-state methods. This does not differ greatly from that used for static system control.

The requirements for the dynamic current measuring range, however, do become sharper when the potential is being varied. Let us imagine a typical "state-of-the-art" A/D converter intended to measure the current. It has a resolution of 24 bits, and a precision of a full 18 to 19 bits — limited by the INL and by noise. If we begin recording our current–voltage curve with around 1 mA to exploit the converter's full range, we achieve a fantastic precision, better than $\pm 10^{-5}$. Nevertheless, measurements close to the equilibrium potential, displaying a few nanoamperes, can be incorrect by a full 100%! In other words, it is not possible to rely on the dynamics of the A/D conversion process when measuring current, any more than it is for the magnitudes that depend on it, such as impedance. A good electrochemical workstation has a dynamic range for current measurement of about 10^{12}, which is about a million times better than the best A/D converter. If currents over a range from 1 pA up to 1 A are to be measured accurately, the range of measurements must be adapted to the current flowing at any one time by switching the circuitry.

The requirements for rapid response in quasi-steady-state methods are relatively relaxed. For this reason, range switching during the measurements can be automatic.

When assessing the capacities of an electrochemical workstation it is therefore important to consider how many current measuring ranges are available and how great are the steps between them. It should be possible, when using slow methods, for the equipment to search for the optimum current measuring range itself.

A little skepticism is particularly appropriate when thinking about small currents. If the most sensitive current measuring range is quoted as having a full-scale value of ± 1 nA, and if a high-resolution A/D converter, as described above, is available, then it is not possible to estimate the sensitivity simply by combining the full-scale deflection with the resolution of the A/D converter. The result would be unrealistic. A universal electrochemical workstation cannot, of course, count individual electrons like peas. Measurements in the femtoampere range and below can only be successfully carried out using special probes in the immediate neighborhood of the object being measured, and in the presence of perfect screening. Painstakingly accurate control of temperature, air humidity, air ionization levels, and all sources of piezo-electric effects are essential here. A simple experiment can make this clear. Obtain a small 1 GΩ test resistor. Applying 10 mV should result in a current of 10 pA. In order to get a feel for the problem, do a small breath on the resistor. This will change the current dramatically.

Steady-state methods usually operate according to a fixed time scheme: this starts with setting up the new system conditions — this is followed by waiting until a quasi-steady state is achieved (settling phase). The steady-state measurements are then taken. The loop is then closed by starting the next cycle. Under these circumstances, continuously integrating measurements have no value — they would mean

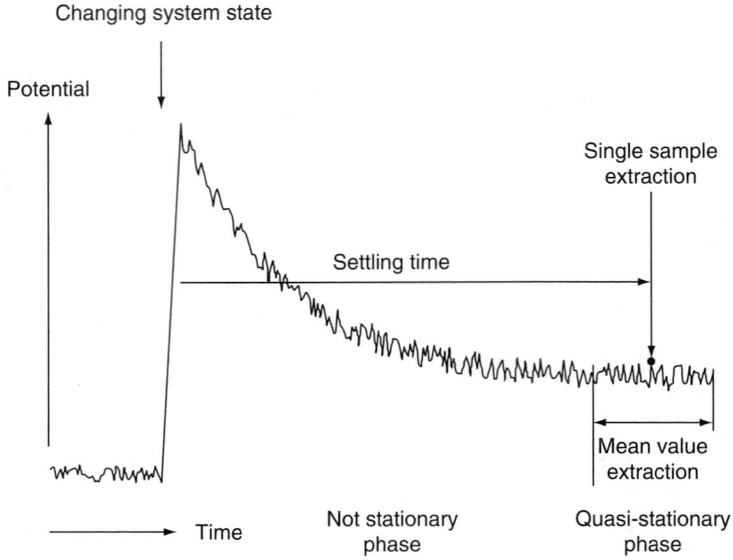

Figure 11.43 Applying steady-state methods means that one has to wait for a certain settling time after changing the system state. Instead of extracting just one sample which is representative for a very short time, one should take the mean value of a certain interval.

that the measurement was affected by information from the settling phase in addition to the wanted information relating to the measurement. Instead of this, a sample must be taken for measurement at the end of the settling phase (Figure 11.43). This does not allow aliasing effects to be completely avoided.

But instead of using a fast A/D converter to take precisely one measurement in each cycle, it is a better idea to form the mean over a rather large measuring interval. As has been explained above, the mains frequency is a principal component in the mixture of interfering signals. If the measuring system uses a multiple of the mains period as the measuring interval, the worst interference can be suppressed.

11.5.4 Dynamic Methods

It is a particularly demanding task for a universal electrochemical measuring system to master the broad field of dynamic measuring methods.

11.5.5 Dynamic Pulse Methods

We begin with dynamically sampled pulse methods, because the requirements here are at least similar to those of the quasi-steady-state methods. In contrast, however, the requirements for time response are not at all simple — on the contrary.

The quality of the measurements depends crucially on the precision of the timing, which often must be synchronized with external events. As a representative for a range of methods, most of which are applied to electro-analytic purposes, differential pulsed polarography (DPP) will be used.

It is well known that the attraction of mercury as an electrode material for polarography is due to two particular properties: one is that mercury has the highest hydrogen over-voltage of all metals. The other is that by periodically generating drops of liquid metal it is convenient to continuously provide a fresh electrode surface.

If the DPP is to yield reproducible relationships from drop to drop in relation to the electro-active surface while measurements are being taken, it is necessary for the acquisition of samples for the measurement accurately to be synchronized with the growth of the drops.

In DPP, it is assumed that the component of the current that cannot be assigned to the electrode reactions is approximately independent of the potential.

For this reason the stable, yet still unsaturated, electrode surface is used, so that by applying a potential step it is possible to separate this component of the current from the component of the Faraday current that is independent of potential, through computing the differences. Figure 11.44 illustrates the principle of DPP.

The time intervals for the individual phases of DPP lie in the range from seconds down to milliseconds, the droplet growth phase being the longest.

Even small time uncertainties have noise effects with an immediate influence on the measurement. A PC running a multitasking operating system cannot provide the necessary precision for time control. An effective instrument for polarography should, therefore, be controlled by a powerful microprocessor system of its own.

The high level of information yielded by sensitive electro-analytic methods depends on the accurate measurement of extremely small electrode currents. In addition to the need for accurate time control and current sensitivity in the instrument, the process is also challenging to the user. It is essential that he takes the guidance for avoiding electrostatic interference seriously.

11.5.6 Linear Voltammetry

The methods of linear voltammetry, or LSV, are typical examples of dynamic techniques. A potential changing at a constant rate — a ramp — is applied to the electrode under test. Cyclic voltammetry is the most widely applied example of these techniques. In this approach, ramps with rising and falling potential alternate periodically.

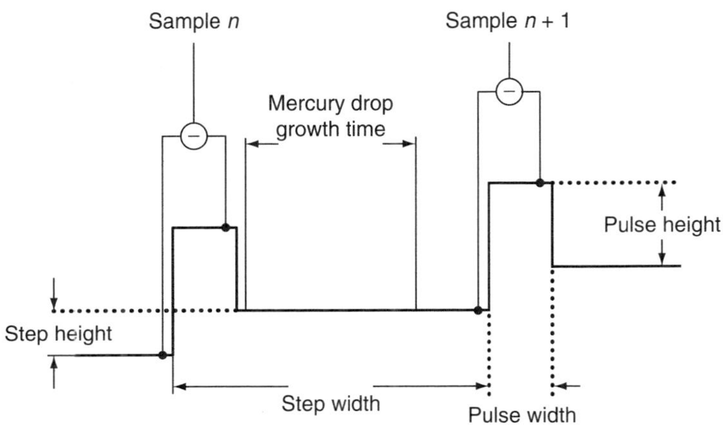

Figure 11.44 The principle of DPP. Potential steps are applied, clocked with the controlled growth of the mercury drops. The current is sampled at times just before and after a certain delay behind the steps. The current difference will be processed.

The ramp's rate of potential rise, $\Delta V/\Delta t$, has a characteristic effect on the dynamic component of the various current mechanisms in the electrode. Current due to capacitive effects rises in proportion to $\Delta V/\Delta t$. If diffusion processes are also responsible for limiting the charge transfer current, the current rises approximately in proportion to the square root of the rate of rise of potential. If the curve along the $\sqrt{t}$ axis is integrated, the effect of diffusion can be largely suppressed, and a quasi-stationary curve is obtained. Under ideal circumstances it is also possible to determine kinetic information such as the order of reactions, rate, valency, and degree of reversibility of the transfer reactions involved by carefully analyzing the potential level and the magnitude of anodic and cathodic peak currents and their dependency on $\Delta V/\Delta t$.

CV does sometimes demonstrate a certain weakness. The theories applied to interpret the measurements are based on the assumption that there is a constant rate of rise of the immediate electrode potential — a potentiostat, however, is only able to set the sum of the electrode potential and the voltage drop in the column of electrolyte above the electrode, U_Ω (Ohmic component). A significant Ohmic component impairs the analysis of the data. The traditional approach to this problem has been through modifications to the apparatus, by applying a variety of techniques to compensate for U_Ω. This will be considered again in the section covering current interruption methods. The "queen" of the dynamic methods, EIS, is not heavily disturbed by the Ohmic component — in the past, this has caused CV to rather trail behind EIS. However, the use of microelectrodes allows the Ohmic component to be largely avoided in the simplest way imaginable: a comparable value of electrolyte conductivity as is found with a traditional large-area electrode is here associated with only a small fraction of electro-active surface. This is equivalent to a strong reduction of the electrolyte resistance share on the total potential drop. With the increasing popularity of micro-electrodes and ultra-microelectrodes, and with the increase in the sensitivity of instruments that has occurred at the same time, CV has also regained some lost territory.

In the past, the linear voltage rise needed as the set value for the potentiostat has been generated with special analog OPA circuits.[6] Even if the circuitry is constructed carefully, the precision that can be achieved in this way is limited, especially when the rate of rise of potential is low.

As digital techniques have advanced, it has become natural to apply them to generate the linear rise using a digital controller and a D/A converter instead of using analog methods. In this way it is easy to generate any desired voltage waveform or slew-rate with a stable behavior, even down to zero voltage rise.

This method of controlling electrochemical experiments with a microprocessor has now become popular, but has presented CV with a new problem. The output voltage of a D/A converter fed with successively increasing numerical values is, when examined closely, not quite like a ramp but rather a staircase (Figure 11.39). Increasing the resolution makes little difference: instead of the smooth ramp, the steps of the staircase are still there, even though they are smaller.

Considering capacitive objects will help us to focus on the important difference between a ramp and a staircase. If one applies a voltage signal $V(t)$ to a capacitor, the current $I(t) = C \cdot dV/dt$ will flow. A constant slew rate results in a constant current. In

[6]If an IA is constructed with a capacitor feedback, C_{fb}, instead of a feedback resistor, R_{fb}, an integrator is obtained: the output voltage, V_{out}, is proportional here to the negative time integral of the input voltage. A constant input voltage, therefore, generates a linear voltage ramp at the output. If the input voltage is reversed when a particular maximum or minimum potential is reached, then the triangular voltage required for CV can easily be generated.

contrast, discrete steps will cause current pulses with δ-pulse shape — which means in theory infinite height for an infinite short time interval, but with a well-defined integral charge of $Q = C\Delta V$. In practice, the δ-pulse will be distorted to a short pulse of high amplitude accompanied by some ringing. Its shape is determined by the combined pulse response of the potentiostat and the electrochemical cell. If the measurement technique samples the response signal after a certain time delay relative to the step, the result must be wrong! The only way to get a correct current result is to measure the charge by integrating the total step interval and to calculate the mean current. This calculated current value is identical to the one measured under continuous analog sweep conditions. The most exact way to perform integrating measurement technique under these conditions is to use fast AD-conversion combined with low pass filtering — with a bandwidth still high enough to avoid significant distortion of the potential waveform — but with a high over-sampling AD-conversion rate.

Test your CV set-up. If your CV software offers you a facility for setting delay times, you should be warned. You will soon discover the wrong measurement strategy if you attempt to take a cyclic voltammetric graph on a capacitor: a 10 μF capacitor subject to a rate of rise of ±100 mV/sec should yield a rectangular cyclic voltammetry graph, with a current amplitude of ±1 μA within the limits of component tolerance.

Switching current ranges over many decades can only be managed using electromechanical relays if the requirements for accuracy relevant to electrochemistry are to be satisfied. The switching time of relays results in gaps, typically a few milliseconds long, in the continuous current being measured. For this reason, the rates of rise encountered in CV, which can be high, mean that switching the measuring range for the rapidly changing currents cannot be done automatically without itself disturbing the experiment. It is, therefore, better to carry out continuous CV measurement within a fixed current measuring range. Before the final measurement,

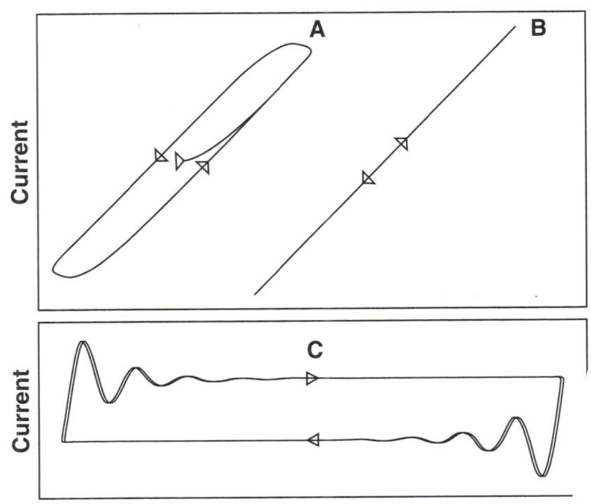

Figure 11.45 Fast CV test runs can help to detect high-frequency di curves exhibit what one has to expect testing a high Ohmic resistor. D the proper response curve B are due to an excess signal delay in the l (curve A). In curve C, the effect of a reference electrode with high sour when testing a capacitor (refer also to Figure 11.55).

therefore, manual or automatic tests should be carried out in order to estimate the most favorable measuring range.

High scan rates associated with the measurement of small currents present a serious technical challenge. Under such conditions, CV must be looked on as a fast method. The consequences in terms of parasitic effects are similar to those occurring with EIS (Figure 11.45). Constructional tricks described later for arrangements of cells that will function effectively at high frequency also help fast CV.

11.5.7 Current Interruption Methods

The electrical behavior of any electrode is described by a network that always contains two elementary components (Figure 11.6). These are the capacitance of the electrochemical double layer and the Ohmic resistance of the electrolyte. They form a series circuit. If the current flowing through the electrode is switched off suddenly the partial potentials on the capacitance and the resistance behave quite differently: whereas the partial potential across the double layer changes relatively slowly, the voltage drop across the electrolytic resistance, R_{el}, collapses with (almost) no delay. A potential jump can therefore be seen at the moment when the current is interrupted in the graph displaying the potential against time (Figure 11.46a). This corresponds to the product of the series resistance, R_{el}, and the current, I, before the interruption, which is the Ohmic share, U_{Ω}, of the total potential. It results, as was explained above in relation to CV, in an uncertainty in the determination of the immediate electrode potential. While it is possible to eliminate U_{Ω} after the measurement when using slow methods, it is worth striving in linear voltammetry to compensate for it within the measurement process. The traditional method for compensating U_{Ω} combines the voltammetry with phases of interrupted current. The value of U_{Ω} that has just been found is automatically added to the set voltage value. This method may be popular, but it has problems. One reason for this is that switching circuits on or off, particularly in systems with feedback such as the potentiostatic loop, are always associated with the injection of charge. In other

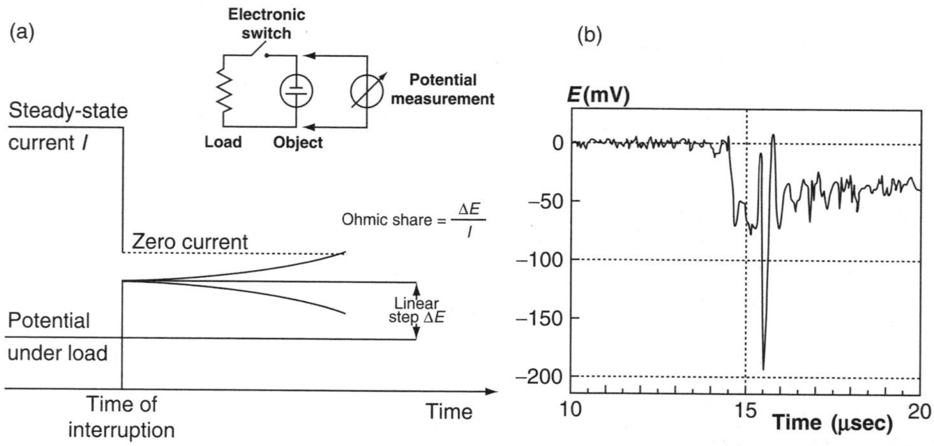

Figure 11.46 The principle of the current interruption method for determining the Ohmic component from the potential step response is shown in an idealized form in graph (a). (b) shows a real step response of a 300 cm^2 PEM fuel-cell when a current of 80 A is off.

words, the electrical system being investigated is continuously disturbed to an unknown extent by the injection of charge.

Anybody who has done much work with EIS knows another reason: determining the voltage jump, U_Ω, in the time domain corresponds to observing the high-frequency behavior of the electrode in the frequency domain. It can easily happen that the relevant frequency range is subject to heavy parasitic interference effects, so that equating the step response to U_Ω can lead to results with little significance (Figure 11.46b).

The strategy of determining the resistance of the electrolyte as precisely as possible beforehand, such as by measuring the impedance, is subject to fewer problems. The measuring system calculates the correction component from the current flowing at the time, and adds it to the set voltage while the voltammetry is being carried out (digital feedback). It is alternatively possible to feed back a signal directly via an adder to the potentiostat (analog feedback), which is proportional to R_{el} and the current being measured.

Both forms of coupling, however, do impair the electrical stability of the arrangement of cells. If we attempt to eliminate U_Ω perfectly, the potentiostat tends to oscillate. As a rule of thumb we should content ourselves with compensating for no more than 90% of the Ohmic component.

The Ohmic components of electrodes, or of complete cells, not only interfere with electrochemical measurements, but are also important characteristic parameters when assessing the efficiency of batteries, accumulators, fuel cells, and in electrolysis. As is also explained below, precise determination of the Ohmic component using EIS can fail, particularly in highly conductive systems, as a result of mutual inductance. It is, nevertheless, possible to obtain useful results under such circumstances from the high-current interruption method. It is, however, necessary to carry out an extensive mathematical analysis of the step response, as can be guessed from Figure 11.46.

11.5.8 Electrochemical Impedance Measurement

Getting involved with EIS (8,9) for the first time one may find it intuitively difficult to grasp the processes around. Not having an innate sense of alternating current is one reason for the problem. Another lies in the enormous dynamic range encountered in these methods: the conductivity of electrochemical cells in common use extends over a range of about 12 orders of magnitude — it extends from the behavior of an almost perfect conductor to almost total insulator. The highly conductive end of the range is marked by large batteries and fuel cells, while barrier coating systems are items that are close to being insulators. A 12 orders of magnitude are also necessary to cover the range of currents involved in stationary electrochemistry — but in EIS, frequency is added as a further dimension: with the electrochemically relevant range of frequencies starting at less than 10^{-5} Hz and extending to around 10^{7} Hz, EIS covers a quite enormous range. This is doubled again, if the complex nature of impedance is taken into account.

With so much room, it is easy to overlook the fact that physics places drastic restrictions on this freedom to use high frequencies: theory might like to consider impedance two-poles as objects with no dimensions and two connections — in practice, EIS provides impedances only as electrochemical cells, more or less extended in space. In addition to this, the connecting leads from a cell to the measuring instrument significantly add to the effective dimensions of the cell.

As a result, as cells are investigated at higher and higher frequencies, we cannot help but reach frequency ranges where the electrical behavior is no longer determined by the contributions of the two-pole impedance, but by undesired parasitic effects. In other words, every cell has a characteristic limiting frequency, above which EIS makes no sense, as illustrated in Figure 11.47. At lower frequencies, almost the whole dynamic range is available to the experimenter with sufficient measurement precision. At high frequencies, two fundamental effects lead to increasingly tight restrictions on the relevant dynamic range: the curve for high impedances is affected by the unavoidable stray capacitances that appear to be connected in parallel with the impedance under investigation. Their origins are primarily to be found in cable capacitances and in the input capacitances of measuring instruments. Contrary to what we might guess, the series inductance of the cables (complementary to the stray capacitance) bears less responsibility for the curves for small impedances with well-designed connection technology. This is because the effect of this inductance can be very effectively minimized, as described below. In contrast, the effects of electromagnetic coupling between the current-carrying leads on the one hand and the leads for measuring the potential on the other hand are quite dramatic. This mutual inductance is usually given very little consideration, but it is the main reason why impedance spectroscopy on fuel cells, batteries, and other low-resistance objects at high frequencies can be very difficult.

A general limitation for extremely high impedances results from the thermal noise in accordance with Equation (11.44): a resistance of 1 T Ω at room temperature, with a signal bandwidth of 1 kHz generates an effective noise voltage of about 4 mV — quite comparable to the small signal amplitudes that are preferred for EIS for the sake of maintaining linearity. About 12 orders of magnitude are, however, also the approximate limit of what can be covered by the best instruments in a fixed configuration of apparatus, because greater current sensitivity for even higher impedances prevents an instrument from being used to measure very small impedances — and vice versa.

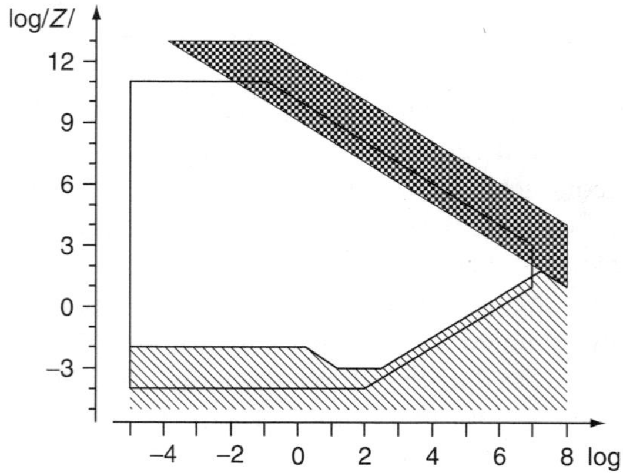

Figure 11.47 Estimated dynamic range for successful EIS measurements by means of a state-of-the-art instrument. The limitations at high frequencies are mainly caused by stray capacitance (high impedance limit) and mutual induction (low impedance limit). The shaded areas at the top of the diagram mark the typical impedance range of films and barrier coatings. The ones at the bottom sketch the impedance range of fuel cells. EIS measurements in both ranges are not easy to carry out.

We need to consider briefly the way in which EIS equipment functions if we are to design the experimental set-up to be used for successful EIS measurements.

It is well known that EIS is based on the principle of impressing a small excitation — one that allows to treat the investigated system as linear — onto a possibly nonlinear object in a stationary condition (Figure 11.8). In modern EIS measuring equipment, an integrated potentiostat holds the electrochemical cell in a stationary state as well as superimposing the test excitation. It is driven by a DC power source for stationary control, and by a frequency generator that is responsible for the dynamic component. The impedance cannot be measured directly, but is calculated from the measurable magnitudes of voltage and current. Figure 11.48 illustrates the principles according to which such equipment is constructed.

Although it might at first appear otherwise, the frequency response of the potentiostat is not, in this arrangement, critical for the achievable precision at high frequencies. The incomplete reproduction of the set signal by the potentiostat at the location of the cell has an equal effect both on the voltage measured and on the current. If the amplitude of the exciting signal is still sufficient in comparison with interfering magnitudes, the potentiostat's errors are fully cancelled when measuring impedance. The symmetry of the parallel signal processing routes for voltage and current is nevertheless crucial if a precise result is to be obtained.

If a ZRA is used to obtain information about the current, then the symmetry is violated before we start. A perfectly symmetrical arrangement results, on the other hand, if we use the shunt principle shown in the Figure 11.48 to measure the current. If, as the frequency curve is being registered, the shunt reference resistor used for measuring the current is adjusted to the measured impedance automatically by switching ranges, the amplitudes in the two signal routes remain of comparable size, and the error resulting from crosstalk is minimum. This is an effective way of coping with the possibly extreme dynamic range of EIS object. This route is, however, only open to impedance measurement carried out under potentiostatic control.

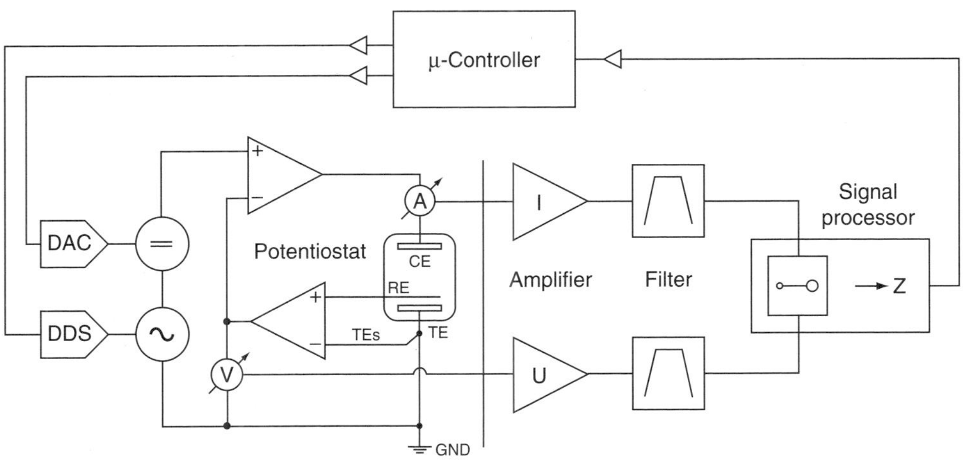

Figure 11.48 Functional scheme of a modern EIS equipment. The cell is controlled stationary as well as dynamically by a potentiostat. It is supplied by a DA-converter for the stationary set magnitude and a DDS signal generator for the dynamic excitation. The dynamic shares of the measured voltage and current are separated from the steady-state signals and processed in parallel. The impedance is calculated from the complex quotient of both signals.

If, instead of a small AC voltage, AC current is supplied as excitation, the current measuring range is determined. Instead of the many orders of magnitude that can be covered by switching ranges when using a potentiostat, only the internal dynamic range of the voltage measuring channel is available in galvanostatic mode. A fixed current amplitude also means that the resulting voltage amplitude when passing through the spectrum can not only become unwantedly small, but also can reach unacceptably high values in relation to the nonlinear properties of the electrodes being investigated. It is, therefore, generally a good idea to prefer potentiostatic operation, and to use galvanostatically controlled EIS only after careful thought, and to restrict it to objects with small impedance variation over frequency such as batteries.

Good measurement software can do one last thing to alleviate the situation. The desire for current control does not really spring from the need to actually specify the excitation amplitude in the form of a current. Voltage and current are, after all, of equal status in the realm of impedance. The point is rather that the flow of DC current is more often representative of the stationary state of a cell than a fixed cell voltage. We might consider in this respect the task of recording the impedance spectrum of a battery as it discharges. Small variations in process values like temperature or fuel gas pressure, which are changing the cell voltage, mean that similar considerations apply to fuel cells. Their operating conditions are also better characterized by a specific load current than by a stable potential.

In this situation, the measuring software in a potentiostatic set-up can exploit the short pauses between the impedance measuring samples to modify the potential, in order to regulate the DC current back to a constant value, while the impedance measurement then follows without difficulty under potentiostatic control ("pseudo-galvanostatic" EIS).

A similar strategy is followed by the "pseudo-equilibrium potential" measuring mode when EIS is used for corrosion research: pauses in the measurement are used here to give the system being investigated an opportunity, with the potentiostat switched off, of settling to the actual, free corrosion potential. Following this, the next impedance value is found at this potential, under potentiostatic control.

As explained in the section discussing the role of the measuring environment, the competition between wanted (signal) and unwanted (interference) components in the measurement plays a crucial role in determining the achievable precision. Let us "transform" this consideration to the consideration of impedance. It can be seen that the proportion of signals at interfering frequencies, $f_s \neq f$, that make unwanted contributions to the measurement is crucial for the precision of the impedance, $Z(f)$, at the desired signal frequency, f. There are clear advantages in using sine waves as the excitation signal.

Unfortunately, high spectral purity (low spectral bandwidth Δf) of a sine signal is directly associated with its oscillation duration Δt (Equation (11.50)) and in turn for the measurement time

$$\Delta f \geq \Delta t \qquad (11.50)$$

If sine waves are used, the interfering frequency components other than the signal frequency can be suppressed by means of electronic band pass filters.

In order to determine the impedance spectrum with sufficient frequency resolution in this way, the signal frequency is changed step by step. In the low-frequency part of the spectrum, this becomes a game of patience: after all, a single period at 10^{-5} Hz takes more than a whole day.

In terms of measuring time, it is not just the patience of the electrochemical engineer that is a problem, but the stability of the system under test, as will be described later.

Excitation functions and filtering can be designed in such a way that, instead of a mono-spectral sine wave, other harmonic frequency components[7] are exploited at the same time. The intriguing consequence is that rather than merely measuring one point, a number of points in the spectrum can be determined at the same time in a single measuring cycle. At low frequencies, however, the time that can be saved by using multispectral methods in comparison with sine wave methods is not very large (1, p. 18–1). The time advantage also comes at a high price: in an electrochemical cell, can, like in any nonlinear system, mono-spectral excitation result in interfering harmonics in addition to the regular system response. If we work with multispectral signals, the interfering components enter into the evaluation along with the desired higher-frequency system response. In contrast to sine wave measurements, where the formation of harmonics as a result of nonlinearity is easily detected as interference, when using multispectral methods they inevitably lead to measurement errors. If comparable information density and precision are required, the time required using multispectral methods can even be longer than what is needed for sine wave methods.

As a result, sine waves have become generally accepted for use in EIS. Multispectral methods can, however, claim to have a certain importance in applications where the time advantage is of significantly greater importance than the loss in precision.

Because the discrete Fourier transform (DFT)[8] is a precondition for the application of multispectral methods, it is often associated with the disadvantages of multispectral methods. Using the DFT here in combination with sine wave EIS can be very helpful: if fast A/D converters are used to sample the sine wave of the measuring signal synchronously with the measuring frequency at a large number of points, over-sampling can create a digital image of the signal with a high time resolution in the memory of a signal processor (Figure 11.49). If, additionally, the mean course of a large number of sequential periods is calculated in the memory (averaging), then interference that is not correlated with the measuring frequency is effectively eliminated. DFT can then be applied to decompose the stored and averaged image from a large number of periods into its frequency components. The DFT has the effect here of a powerful digital filter. The fundamental wave corresponds to the measuring frequency. The fundamental components of the voltage and current signals can be used to calculate the desired impedance without error, even if the potentiostat can only partially follow the provided sine wave. If, in addition to the fundamental wave, close harmonics are found in the system response, this is also an indication of nonlinearity. The ratio of the power of the fundamental wave and the

---

[7]If we restrict ourselves, as is indeed essential if digital filter algorithms are to be used, to periodic signal forms, then all the other frequency components are integral multiples (harmonics) of the fundamental frequency, which is itself the inverse of the period.

[8]The DFT is the discrete image in the digital domain for the familiar continuous Fourier transform. The DFT is most often applied in the form of the fast Fourier transform (FFT). If the digital image of a signal curve is available as a sequence of 2^n values with equal intervals between them, the FFT can very rapidly determine the frequency components of which it is composed.

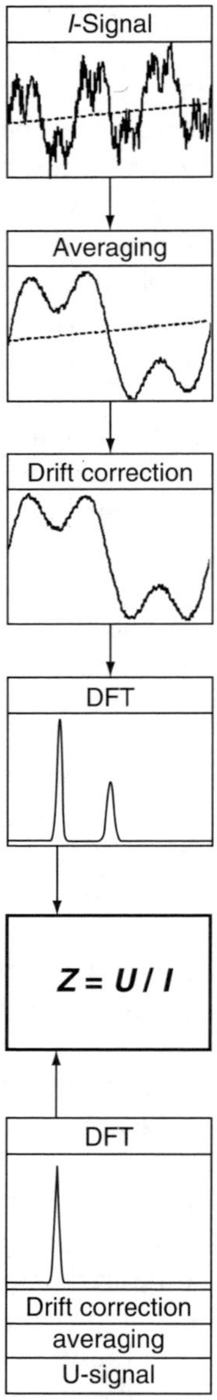

Figure 11.49 The principle of the high over-sampling impedance measurement technique. After the analog preprocessing the primarily acquired excitation and response signals are sampled synchronously with high rate. Several periods are averaged to one mean period course by adding up the samples of equal phase in consecutive cycles. Contributions of steady-state signal drift may be compensated optionally. The impedance is calculated as the quotient of the DFT fundamental share of both signals. An uncertainty estimation can be obtained from the weighted integral of the lower harmonics share.

weighted,[9] summed power of the harmonics (a kind of signal/noise ratio) can be used to estimate the uncertainty of measurement.

In summary, it can be stated that DFT sine wave methods involving a large number of samples (over-sampling methods) are, due to their insensitivity to interference and harmonics,[10] superior to those of their simpler cousins that must content themselves with satisfying just the Nyquist criterion (Nyquist methods). Modern devices, therefore, no longer usually work with Nyquist methods, as represented by the traditional digital correlation or analog lock-in techniques.

Let us return again to the problem of measuring time. Strictly speaking, electrical impedance is only defined for stationary systems. Nevertheless, EIS has also to be applied to systems that are not perfectly stable. It is, therefore, not impossible for the drift to be large enough for a noticeable change in the stationary component of the system response to be caused while one point on the spectrum is being measured. It is possible — assuming that periodic methods are being used — for this drift to be measured while the spectral measurement is proceeding, and to be compensated for thorough calculation. There are, however, strict limits to this kind of "online drift compensation" (Figure 11.49).

Even with a relatively weak drift it can still happen that the state of the object being investigated differs noticeably at the end of one complete measurement from its state at the beginning.

Impedance spectra from this kind of changing system are difficult to interpret. Time effects can, however, very easily be detected in EIS data. A normal electrochemical impedance spectrum corresponds, as a rule, to the spectrum of an ideal two-pole, and does not have any discontinuities, strong resonances, or jumps. For the Bode diagram (Figure 11.9 and Figure 11.10b; Equation (11.37)) of such spectra, the logarithmic Hilbert transformation for impedance two-poles, Z-HIT, predicts that the impedance modulus course, H, can be determined to a good approximation from the course of the phase angle, φ (Equation (11.51)) (10, 11)

$$\ln|H(\omega_0)| \approx \text{const.} + \frac{2}{\pi}\int_{\omega_s}^{\omega_0} \varphi(\omega)\,\mathrm{d}\ln\omega + \gamma\frac{\mathrm{d}\varphi(\omega_0)}{\mathrm{d}\ln\omega}, \quad \text{Z-HIT} \tag{11.51}$$

The integral of φ over the logarithm of the frequency, ω, is formed for this purpose. The additional correction provided by a fraction, γ, of the derivative of φ provides a high precision in the approximation.

If the impedance modulus course calculated in this way does not agree well with the impedance curve determined experimentally, differences in the low-frequency region of the spectrum suggest temporal drift, whereas those in the high-frequency regions indicate interference resulting from mutual inductance.

There is a whole family of methods based on the so-called Kramers–Kronig relations (12) with the intention to detect distortions in EIS data. Z-HIT is distinguished by simplicity: even without any mathematical treatment — simply by taking a close look at the Bode diagram — one is able to detect distortions of the proper

[9]As well as being caused by nonlinearity, lower-order harmonics can also indicate components of interfering frequencies close to the measuring frequency that can partially pass through the digital filter. In contrast to higher order harmonics, they should be taken seriously as an indication of measurement errors.

[10]Harmonics are also often generated when a potentiostat is required to operate simultaneously at high frequencies and high currents ("nonlinear distortion factor").

shape of an impedance spectrum by comparing sign and magnitude of the phase angle with the gradient of the impedance modulus course.

Looking at the correctly measured Bode-diagram in Figure 11.10(b) one observes that a phase-angle tending to zero for frequencies around 1 Hz and 10 kHz, corresponds to a gradient, tending to zero at these frequencies. The negative maximum of the phase angle in turn corresponds to the negative maximum gradient of the impedance at the 100 Hz inflection point — very close to the relationship between a mathematical function and its first derivative.

Differing from that, the impedance modulus course of a nonstationary (corroding) electrode in Figure 11.50 exhibits a divergence at low frequencies: while the average phase angle is near zero, the gradient of the measured impedance (rhomb symbols) is clearly negative. The solid line in the diagram, calculated from the phase angle after Equation (11.51), indicates the (correct) impedance course to be expected in the absence of any interference.

We can now summarize what has been said about the EIS measuring technique:

- Time must be invested if meaningful results are to be generated with EIS. Anyone who ignores the fundamental laws of nature will immediately be punished by loss of information.
- Time should also be taken when studying the specifications of EIS measuring equipment. With such complex structures, it is easy to get lost in the jungle of impenetrable technical data. The frequency and amplitude resolution of the excitation, isolated data regarding converter resolution, and the frequency range of the potentiostats are, like so much else, more camouflage than information.
- Of particular importance, in contrast, is the specification of the achievable accuracy of measurement in dependency on frequency, impedance value, and excitation amplitude. A clear table or graph, such as Figure 11.47, should not be missing. It is necessary to ensure here that this specification includes the entire measuring loop, including the potentiostat.

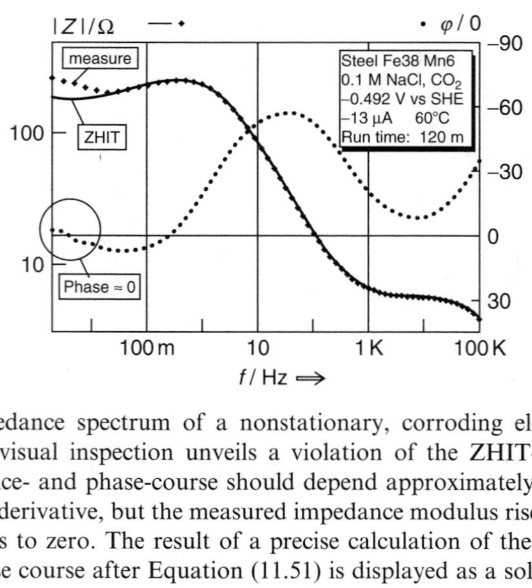

Figure 11.50 Impedance spectrum of a nonstationary, corroding electrode in the Bode representation. The visual inspection unveils a violation of the ZHIT-rule moving to low frequencies: impedance- and phase-course should depend approximately like a mathematical function and its first derivative, but the measured impedance modulus rises significantly, while the phase angle tends to zero. The result of a precise calculation of the expected impedance course from the phase course after Equation (11.51) is displayed as a solid curve.

- Devices that work with oversampling are to be preferred. The weaknesses of devices that operate on Nyquist limit cannot be discovered either through the data sheet or through measurements on a dummy cell. One will only get a clear picture through realistic electrochemical measurements under the most difficult conditions possible.
- Reliable data on electrochemical reaction mechanisms, kinetics, the electrical properties, and topology of covering layers and electrodes are some of the strengths of EIS. One of its differences from other dynamic methods such as CV is that it does not supply any direct thermodynamic information. If we want to deal with the problem that there are always a few more unknowns in a system than can be determined by EIS, it must be combined with other methods. A factor to include in the assessment of measuring equipment is the extent to which other electrochemical methods are also supported. The facility of combining EIS with the measurement of the other internal and external physical magnitudes such as potential, time, and temperature, and control of and by external events, are further important criteria.
- EIS measurements should be examined for signs of time effects — most easily by visually inspection based on Z-HIT. These effects, along with parasitic effects associated with the cell being measured and the cables are the main causes of incorrect measurements. Intelligent software can minimize the coarsest effects of drift as the measurements are taking place, and it should support subsequent correction by means of the interpolation of spectral series vs. time.

11.5.9 The Construction and Electrical Connection of Electrochemical Cells

The construction of the electrochemical cell and its connection to the measurement instrument play a key role in ensuring good-quality measurements. This has already been seen in the section describing the effect of the electrical environment. This role is not, however, limited to the antenna effect of cables and cells.

The researcher does not generally have the freedom to optimize cells unilaterally from the point of view of electrical instrumentation. Electrochemistry must be carried out also in autoclaves, cryostats, or furnaces, or it may be necessary to be able to move, observe, or irradiate the electrodes during the experiment. The intended application can also dictate the form of the cell from the beginning — impedance measurements on batteries, or corrosion tests on beverage cans, fall into this category.

However, the quality of the measurements — particularly those obtained from dynamic techniques — depends very sensitively on apparently insignificant details. For this reason it is vital to exploit the remaining possibilities for optimization.

If current flows, the geometry of the cell is of critical importance, which is easy to understand. Any cell is, necessarily, a three-dimensional structure, and the current density distributes itself through this body in accordance with the electrical field and the conductors distributed across space. But as far as the homogeneity of the electrode material permits it, we attempt to simplify the model in such a way that only one dimension, z, parallel to the direction of the electrical field, has to be observed. Lateral movements in the x,y-plane (the plane of the electrode being examined) should not find any differing relationships (Figure 11.51).

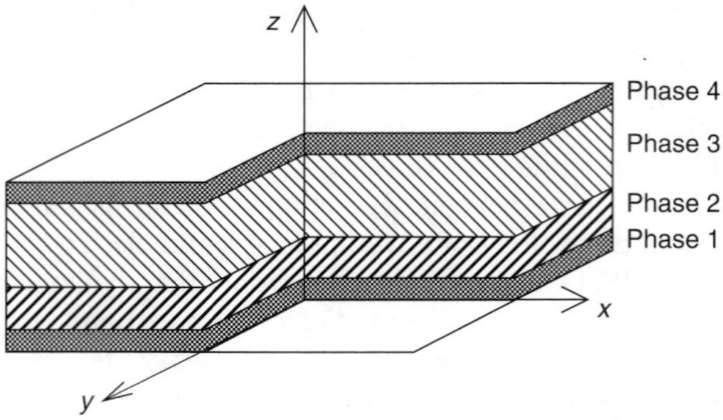

Figure 11.51 One-dimensional effective electrode model. In direction z, which is also the direction of the electric field vector, different phases of matter may be stacked. Along the x or y direction one will find no differences at different sites.

Figure 11.52 shows in two dimensions what must be observed with a planar arrangement of electrodes. Any possible effect of a reference electrode on the homogeneity of the field is ignored in the diagram.

A quasi-one-dimensional geometry is, of course, not restricted to the planar arrangement of electrodes. Figure 11.53 illustrates the important features of cylindrical electrodes. Similarly, when measuring with mercury droplets the counter electrode is best treated as an approximately spherical mesh electrode. It soon becomes clear that it is usually necessary to make compromises. An annular counter electrode with the droplet location in the center is certainly a better solution here than an electrode plate located on one side.

Unfavorable electrode geometry results in uneven current density over an electrode surface of uniform material. Moving away from one-dimensional geometry also makes it more difficult to separate the individual components of the total impedance into individual concentrated elements, such as the double layer capacitance and the electrolyte resistance. This is not such a severe problem for EIS. Methods based on the best possible compensation of the electrolyte resistance suffer more heavily. Because electrolyte columns of different heights are effectively in operation over different parts of the electrode surface, the compensation will fail.

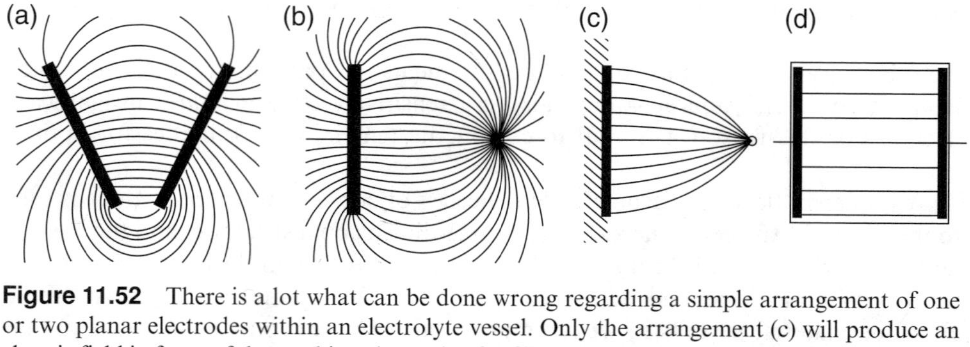

Figure 11.52 There is a lot what can be done wrong regarding a simple arrangement of one or two planar electrodes within an electrolyte vessel. Only the arrangement (c) will produce an electric field in front of the working electrode of sufficient homogeneity. The ideal arrangement (d) has identical shape of both electrodes, which is also the shape of all intersections along the electrolyte column.

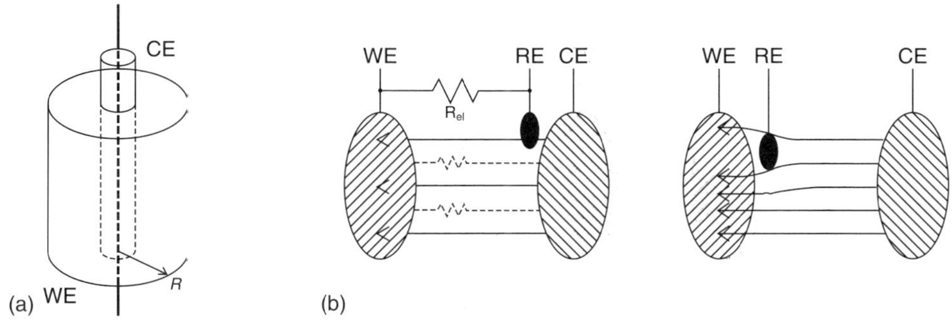

Figure 11.53 (a) A cylindrical working electrode should have a cylindrical counter electrode accordingly in a symmetric arrangement. (b) The position of the reference electrode obeys two contradicting rules: a site nearer to the counter electrode will increase the electrolyte resistance in an unnecessary manner. A position very close to the working electrode surface will distort the electric field nearby.

Figure 11.53 also suggests how the position of the reference electrode must be taken into account. Because no current flows in it, we might first think that it does not have any effect on the distribution of the field lines. In fact, both the position and the precise implementation of the reference electrode call for close attention. This has already been explained in the section about the role of the reference electrode position in the discussion of the potentiostatic arrangement (following Figure 11.28 and Equation (11.48)).

The reference electrode disturbs the shape of the field lines in front of the working electrode as soon as its size in the region of its opening (capillary tube) is comparable with the distance to the electrode. The potential determined by the capillary is then no longer representative of the undisturbed region of the working electrode. If we approach the working electrode even more closely, the region in which the potential is being measured is strongly shielded from the field of the counter electrode. Figure 11.54 illustrates this situation. The electrical model makes it clear that, instead of the undisturbed electrode potential, only a fraction of it is measured, which is delayed in time and phase-shifted by the distributed network R_2–Z_2.

The impedance spectrum from this kind of arrangement shows an unrealistically large negative phase angle, $\varphi < -90°$, in the frequency range that would normally be dominated by the double layer capacitance. This can be wrongly interpreted as a sign of a negative real component in the total impedance. The effects of this error on the impedance spectrum are still relatively easy to comprehend — in contrast to the situation with other dynamic methods.

If the electrode distance is made a little smaller again, the situation even becomes dangerous — at least for the cell, if a powerful potentiostat is connected. The function of the potentiostat turns into its opposite. Instead of stabilizing the potential at a fixed value, the potentiostat falls into powerful electrical oscillation. Voltages and currents reaching the limits of the power capacity have been known to destroy a carefully prepared electrode. It can, for that matter, also become dangerous if a broken electrical circuit or a gas bubble separates the potentiostat from the actual potential information of the reference electrode.

When discussing the potentiostatic arrangement, it was also mentioned that the part played by the reference electrode in the quality of measurements goes beyond

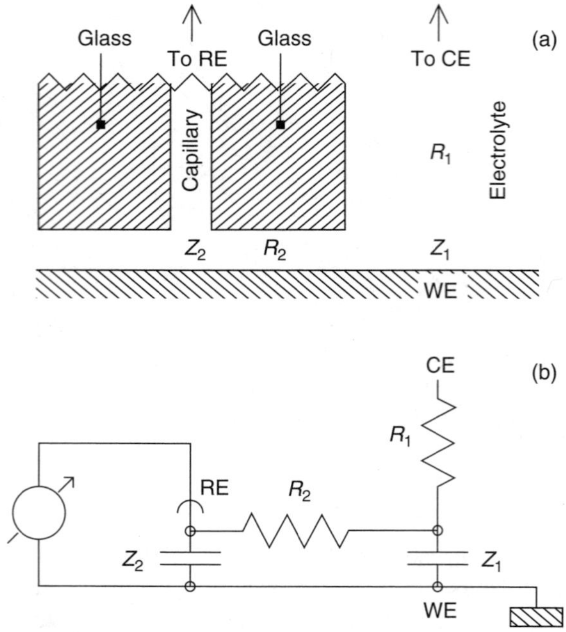

Figure 11.54 (a) A highly enlarged section around the RE capillary opening in front of the working electrode illustrates the role of the relevant impedance elements when determining the WE potential. (b) The corresponding (simplified) equivalent circuit shows that the measured potential at RE is no longer representative for the undisturbed WE potential at the site of Z_1, if the capillary is brought close to the WE. R_2 increases if the distance capillary–WE is decreased.

the effect of its geometry (13, 14). The main reason for this is the source resistance of the potential probe (R_{ref} in Figure 11.28). The combined effect of the unavoidable stray capacitances in the electrical structure distorts the potential information in a characteristic manner. These effects are most easily understood if we think in terms of EIS. Figure 11.55 illustrates an electrical model of the measuring loop in which the parasitic components of the reference electrode circuit are included. The working

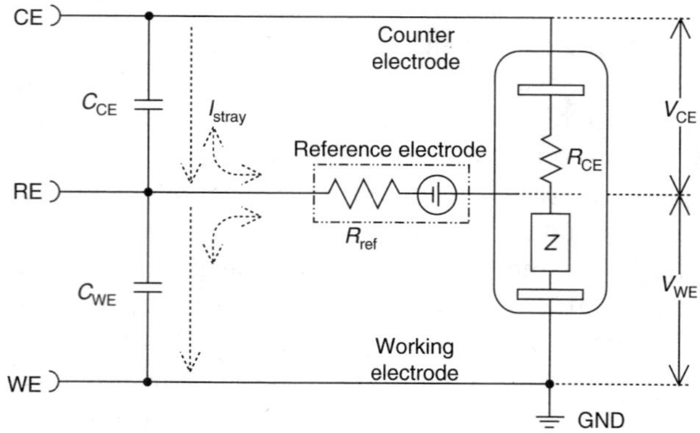

Figure 11.55 The bridge coupling effect describes the parasitic behavior of the reference electrode circuit at high frequencies. To calculate the impedance one is interested in the potential V_{WE} in front of the working electrode. But the only accessible voltage V_{RE} is affected by the current through the stray capacities C_{WE} and C_{CE}. High source resistance R_{ref} and high counter electrode resistance R_{CE} will favor strong high-frequency distortions.

electrode impedance being investigated, Z, together with the dominating electrolyte resistance of the counter electrode, R_{CE}, the source resistance of the reference electrode, R_{ref}, and its stray capacitances, C_{WE} and C_{CE}, form a voltage divider bridge for the measured voltage V_{RE} at the reference electrode terminal, across the test and counter electrodes.

It can be seen that, at high frequencies, V_{RE} depends almost exclusively on the voltage divider formed by C_{WE} and C_{CE}, and hardly at all on the working electrode potential, V_{WE}.

The voltage drop, V_{CE}, across R_{CE} is usually many times greater than V_{WE}. Therefore, as soon as the applied frequency is raised, leaving the interference-free low-frequency region, the flow of AC current from the counter electrode to the reference electrode through the stray capacitance C_{CE} is dominant. In this transition region, the error caused in the impedance spectrum can be approximately described as an inductance in series to Z.

We often see in the impedance spectrum of a classical three-electrode arrangement that, instead of the constant impedance that corresponds to the electrolyte resistance, inductive behavior is seen at high frequencies. This is not terribly surprising, because significant inductance in the feeds to the working electrode causes this to be expected. Sometimes, however, it is seen that the observed inductance is many times greater than the estimated lead inductance. It is a good idea not to ascribe this to some mysterious electrochemical process. It is more valuable to consider optimization of the reference electrode construction, because the cause usually lies in the bridge coupling effect described above.

Once the effect has been understood, it is easy to think of effective improvements: both lowering of the stray capacitance and a low resistance construction, in particular of the reference electrode, allow the disturbance to be moved to less interesting, higher frequency, ranges.

Reducing the size of the reference electrode and the length of capillary are very effective, as both of these have a favorable effect on lowering the stray capacitance and the internal resistance.

In addition to the reference electrode itself, its connecting leads make a significant contribution to the stray capacitance, and therefore a buffer amplifier close to the cell can often bring improvements, as has already been suggested in the section describing the voltage follower (Figure 11.19 and Figure 11.20).

The conductance of the reference electrode, $1/R_{ref}$, competes with the stray capacitances C_{WE} and C_{CE}. Lowering R_{ref}, therefore, improves the situation, even if only for AC. A little trick can be used to create a kind of AC bypass, in parallel with the DC path to the reference electrode, which does this job. Figure 11.56 illustrates how a low-resistance metallic auxiliary electrode (e.g., a Pt ring) is installed in the immediate neighborhood of the capillary opening. The potential measured there is taken to the outside via an insulated wire. It can be connected via a capacitance, C, which should be significantly larger than the stray capacitances, to the standard output of the reference electrode.

There are a few lessons that we should draw from what has been said about the role of cell geometry and reference electrode construction:

- Provided the electrochemical system being investigated permits it, it is better not to turn a one-dimensional problem into a multidimensional one. In planar systems one should strive for geometrical similarity of the working and counter electrodes and of the electrolyte column. If the electrolyte

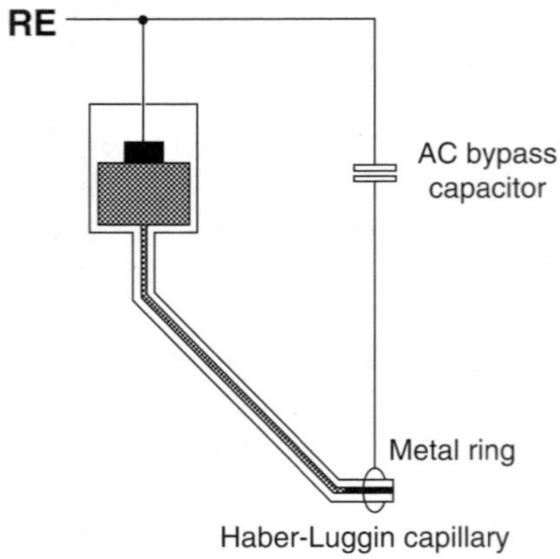

Figure 11.56 An AC bypass enhances the high-frequency performance of a usual reference electrode.

column is not constrained to the electrode surface, at least seal off the rear of the working electrode from the electrolyte. It is not difficult to approximate a quasi-one-dimensional arrangement in cylindrical, or sometimes even in spherical form.

- Do not attempt to minimize electrolyte resistance at any price — the cost can be too high. Estimates of favorable reference electrode positions have already been recommended in Equation (11.48) and the section following this equation. The unfortunate reinforcement of the bridge-coupling effect resulting from raising the ratio of V_{CE}/V_{WE} provides a further argument in this direction. The reference electrode should never be brought so close to the working electrode that it hinders the passage of the field lines from the counter electrode in the region of the capillary opening.
- If one wants to carry out EIS successfully using three- or four-electrode techniques in the medium- and high-frequency ranges too, or if one is interested in clean dynamic measurements, even at high rates of rise and with short pulse durations, the construction of the reference electrode should be at the center of your attention. Avoid large dimensions, long electrolyte paths, bridges of low cross-sectional area, and thick frits. The most reliable way to a good result can often be to construct a reference electrode or to fit the electrode with an AC bypass for the potential.
- Sometimes it is not possible to achieve the goal of a sufficiently low resistance reference electrode, perhaps because it is necessary to work in electrolyte of low conductivity. The best thing to do then is to decouple the reference electrode signal from the capacitance of the connecting leads and the input to the measuring instrument by means of a buffer amplifier. As shown in Figure 11.20, the buffer can be used at the same time to provide screening against electrostatic interference.

A large part of the previous discussion of ways of ensuring good measurement quality has been concerned with the situation that applies to high resistance systems. This is no surprise. If the measured currents are systematically small, the competition with parasitic currents is of great relevance to interference. However, there are specific problems that occur when taking measurements on particularly low-resistance systems. Once again it is valuable to understand the causes, if opportunities for good measurement quality are not to be wasted. It has already been mentioned when discussing the limits of measurement using EIS equipment (Figure 11.48) that mutual inductance is the main source of error there. This statement can be generalized to cover all the fast dynamic methods, with the exception of the current interruption technique. In order to take accurate electrical measurements on low-resistance objects, it is always necessary to employ four-conductor measurement techniques, in which there is a strict separation of the leads that carry currents from those that measure potential. This has already been mentioned when discussing the principle of the four-electrode potentiostat (Figure 11.30). This method of connection is also known as the Kelvin technique (Figure 11.57a) (15).

The Kelvin technique alone, however, only avoids static measurement errors on low-resistance objects. Mutual inductance must also be considered when taking dynamic measurements (Figure 11.57b). Mutual inductance occurs when the magnetic fields from the current-carrying conductors in a Kelvin arrangement induce interfering voltages in the potential-measuring conductors.

Because mutual inductance is based on magnetic coupling, the usual defense mechanisms, instinctive to the electrochemical engineer, fail: screened wires, Faraday cages, earthing, and similar techniques bring no improvement.

It is necessary to prevent as many magnetic field lines as possible from emerging out of a surface enclosing the current-carrying cables. One proven technique is to twist the pair of conductors together. The magnetic fields created by the current flowing to and from the object cancel each other almost perfectly because of their opposing directions. However, the finite distance between the pair of conductors means that a small residual field still emerges. If the pair of conductors used for the potential measurement is also twisted, the residual field causes almost identical interference in the two cables. Because the electronics that measures the potential only detects the potential difference between the two leads, the interference from the residual field has almost no effect.

Figure 11.58 illustrates the principle of cancellation. It should, however, also be emphasized that perfect cancellation can never be achieved in practice. A structure that is absolutely free from mutual inductance is only possible in a few special cases having unusual symmetry. The object being measured also has a finite size, so that there is always an opportunity for mutual inductance to arise, at least in the immediate neighborhood of the connections.

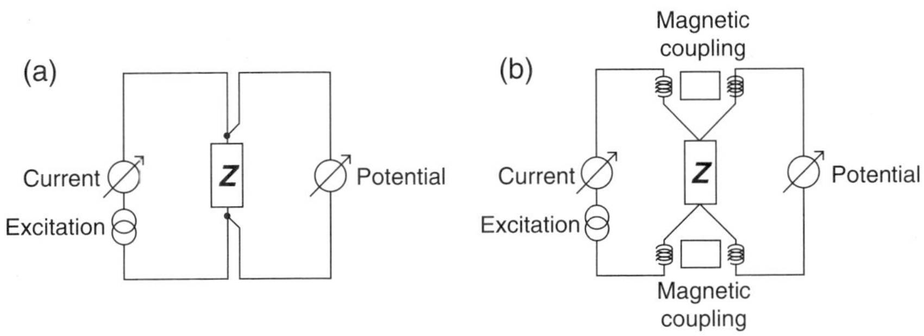

Figure 11.57 The principle of the four-electrode Kelvin connection scheme (a) and its high-frequency equivalent circuit (b) explaining the mutual induction problem. High-frequency mutual induction errors occur from the transformer like magnetic coupling between the current feeding and the potential sensing lines.

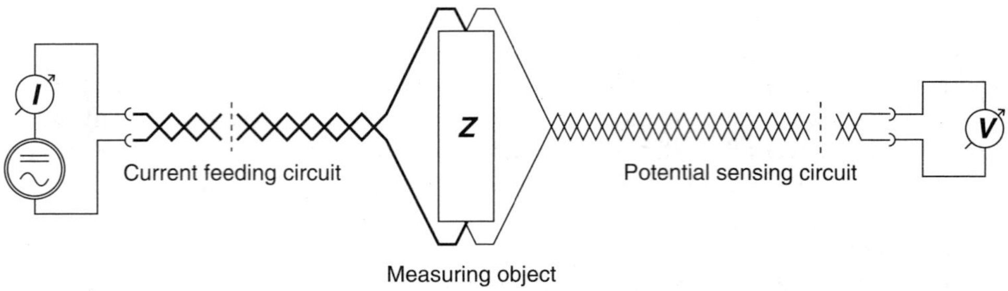

Figure 11.58 Using two pairs of twisted cables for the current feeding as well as the current sensing circuit suppresses the mutual induction interference best. However, mutual induction in the immediate neighborhood of the measuring object is nearly unavoidable.

As explained, this represents the boundary for all dynamic measurement methods based on the simultaneous acquisition of voltage and current. For EIS, these effects can be summarized into a rule of thumb giving the upper frequency limit, f_{max}, for meaningful impedance measurements on typical objects, according to the expected impedance value, Z

$$f_{max} \approx 10^6 \text{ Hz} \cdot Z/\Omega \tag{11.52}$$

REFERENCES

1. R.P. Feynman, R.B. Leighton, M. Sands, in *The Feynman Lectures on Physics,* Commemorative Issue, Pearson Addison Wesley, New York, 1989.
2. C. Kittel, in *Introduction in Solid State Physics*, 7th edition, John Wiley & Sons, New York, 1996.
3. R. Woods, in *Electroanalytical Chemistry*, vol. 9. ed. A.J. Bard, Marcel Dekker, New York, 1976.
4. M. Abramowitz, I.A. Stegun, in *Handbook of Mathematical Functions with Formulas Graphs and Mathematical Table,* Dover Publications, New York, 1968.
5. J.K. Roberge, in *Operational Amplifiers: Theory and Practice*, John Wiley & Sons, New York, 1975.
6. The Engineering Staff of Analog Devices, in *Analog–Digital Conversion Handbook, 3rd* edition, ed. D. Sheingold, Prentice-Hall, Englewood Cliffs, NJ, 1986.
7. A.J.Bard, L.R. Faulkner, in *Electrochemical Methods*, John Wiley & Sons, New York, 1980.
8. J.R. Macdonald, in *Impedance Spectroscopy*, John Wiley & Sons, New York, 1987.
9. M. Sluyters-Rehbach, J.H. Sluyters, in *Electroanalytical Chemistry*, vol. 4, ed. A.J. Bard, Marcel Dekker, New York, 1970.
10. W. Ehm, H. Göhr, R. Kaus, B. Röseler, C.-A. Schiller, *Acta Chimica Hungarica*, **137**, 2–3, 2000, 145.
11. W. Ehm, R. Kaus, C.-A. Schiller, W. Strunz, in *Z-HIT — A Simple Relation Between Impedance Modulus and Phase Angle, Providing a New Way to the Validation of Electrochemical Impedance Spectra*, eds. F. Mansfeld, F. Huet, O.R. Mattos, Electrochemical Society, Pennington, NJ, 2001, vol. 2000-24, p. 1.
12. P. Agarwal, M.E. Orazem, *J. Electrochem. Soc.*, **142**, 1995, 4159.
13. H. Göhr, M. Mirnik, C.-A. Schiller, *J. Electroanal. Chem.*, **180**, 1984, 273.
14. G. Hsieh, S.J. Ford, T.O. Mason, L.R. Pedersen, *Solid State Ionics*, **91**, 1996, 191.
15. E.W. Golding, F.G. Widdis, in *Electrical Measurements and Measuring Instruments*, 5th edition, Pitman, London, 1963.

12

DC Electrochemical Methods

Ewald Heitz
Karl–Winnacker–Institute of DECHEMA e. V., Frankfurt, Germany

Contents

12.1 INTRODUCTION

When a corrosion process proceeds by an electrochemical mechanism, electrochemical methods can be applied in addition to chemical, surface analytical, and other methods. Electrochemical corrosion processes consist of anodic and cathodic partial reactions that are coupled with an electronic current in the metal and an ionic current in the electrolyte. The partial processes and with them the partial current densities are uniformly distributed in time and location over the surface, forming a homogenous mixed electrode resulting in uniform corrosion (1). In the case where the anodic and cathodic partial current densities are spatially separated, anodic and cathodic regions are formed. Such electrodes are designated heterogenous mixed electrodes. Many localized corrosion phenomena such as, for example, pitting, crevice, galvanic, erosion and stress corrosion cracking (SCC), and corrosion fatigue are the result of the formation of heterogenous mixed electrodes.

For the investigation of electrochemical reactions potential measurements, current measurements, and polarization measurements with externally applied potential or current are used. These methods are further divided into stationary and nonstationary methods. DC electrochemical test methods belong to the type of stationary methods, although there exist some limiting cases to nonstationary methods that are also included in this chapter.

The aims of electrochemical measurements in general and of DC measurements in specific are numerous and can be characterized as follows:

- Determination of electrochemical corrosion rates.
- Assessment of the potential dependence of corrosion reactions.
- Determination of critical potentials.
- Evaluation of corrosion mechanisms.

12.2 TYPES OF EXPERIMENTS

Corrosion as well as electrochemical experiments can be conveniently divided into corrosion testing and corrosion investigations (2). In corrosion testing, the medium is often artificial and is chosen solely to produce a specific type of corrosion under standardized conditions. On the other hand, in corrosion investigations, the medium is comparable to that used in practice and the actual corrosion system is investigated in this case.

In corrosion testing the performance and evaluation of experiments are specified by standards, inspection sheets, schedules, etc. To this end, test media, size of test sample, test conditions, and duration of the test are particularly relevant. Corrosion testing serves essentially as a form of quality control of materials after production and fabrication.

In corrosion investigations of the general kind the size of the test sample, the test medium, the conditions and length of the test are not specified, but they must be relevant to the particular problem. Often, the cost factors must be carefully analyzed, particularly with regard to the length of time and the scale of the investigation.

Based on the time factor the following distinctions should be made:

- Long-term corrosion experiments
- Short-term corrosion experiments
- Accelerated corrosion experiments

In long-term corrosion experiments the exposure time should be similar to the service life required. This gives the safest prediction of the behavior of the corrosion system in service. In short-term corrosion experiments the exposure times are shortened by intensifying the corrosion conditions. Short-term corrosion experiments are mainly used for explaining corrosion reactions and corrosion mechanisms since they lead to applicable recommendations within a reasonable experimental time. However, it is important that in accelerating the corrosion conditions there is no change in the corrosion mechanism. If the corrosion mechanism is different from that occurring in practice, difficulties can arise in extrapolating the results to longer times (scaling up in time). In accelerated corrosion experiments, the exposure time is drastically shortened by appropriate choice of medium. Accelerated corrosion experiments are suitable only for comparison of certain material properties, for example, the resistance to intergranular corrosion or SCC.

According to the size of the test sample and the experimental setup the following distinctions can be made:

- Laboratory experiments
- Pilot scale experiments
- Plant and field tests

In laboratory experiments, an attempt is made to simulate the conditions experienced in practice, but using small specimens (model experiments). Thus, an appropriate relation between size of test specimen and volume of medium must be chosen according to the corrosion system. In cases of uniform corrosion, failure can occur due to edge effects and under constant corrosion conditions the specimen size has no effect on corrosion rate. The size of the test sample is, however, significant in galvanic corrosion, pitting corrosion, crevice corrosion, and in the mechanico-chemical types of corrosion such as SCC, corrosion fatigue, and flow-dependent corrosion. Changes of scale can lead to changes in corrosion phenomenology and, thereby, corrosion rate. Problems of geometric scaling up and scaling down have been discussed elsewhere (3,4).

Pilot scale experiments are carried out between actual plant scale and laboratory scale tests and serve particularly to mirror actual service conditions. They find particular application when the experimental parameters cannot be realized on the laboratory scale, for example, where high flow velocities, high pressures, or complex media such as process gases at high temperatures are involved. Pilot scale experiments are planned as a rule as model experiments in which the corrosion conditions correspond to the service conditions in a plant, machine, or other equipment. The newest developments are experiments in so-called miniplants that simulate the various steps of a process and can be used successfully for corrosion experiments (5). Although they are considerably more cost intensive than laboratory experiments, the practical value of the results obtained from pilot scale experiments is significant.

Plant and field tests are carried out directly in the operating plant or in natural surroundings. They are therefore necessary when the corrosion conditions cannot be reproduced either in the laboratory or in pilot scale. An example of a plant test is the exposure of test coupons or test samples stressed by bending in oil or gas extraction plants. Examples of field tests are on-site tests in water (e.g., seawater), aggressive atmospheres, or soils.

Specific information on corrosion investigations and corrosion testing is contained in technical regulations, standards, and guidelines published by institutions such as ASTM, NACE, ISO, DIN, BSI, etc. Some of these publications are of an older date, but are still useful and applicable.

12.3 ELECTRODES AND CELL CONSTRUCTION

Electrical circuits and instrumentation for electrochemical testing have been discussed in Chapter 11. For electrode and cell construction, a number of useful monographs and guidelines can be consulted (2,6–9,10). In the following section, a summary of the most important information for carrying out electrochemical tests will be presented.

12.3.1 Pretreatment of Test Samples

Since corrosion processes involve interfacial reactions, they can be very dependent on the state of the surface. For this reason, the preparation of the test samples used for carrying out the individual experiments is very important. The pretreatment of the test specimen must be appropriate to the particular investigation, from which it necessarily follows that there is no universally valid standardized surface finish. Pretreatment of the test specimen comprises heat treatment and surface preparation.

In principle any pretreatment should produce a surface as clean and as close to the service condition as possible. For example, in performing electrochemical experiments this can be achieved by:

- Wet polishing (water) with emery powder or paper of a given grain size (e.g., 150 or finer)
- Final degreasing in an organic solvent or in an aqueous degreasing bath

The mechanical cleaning process of polishing has the disadvantage that a very reactive metal surface is produced that often behaves differently in the "fresh" state than in the "aged" state. Differences between fresh and aged surfaces particularly become apparent in corrosion systems with relatively low corrosion rates, whereas in rapidly corroding systems (for example acid corrosion) the stationary state is reached in a shorter time. This is particularly significant in short-term experiments.

In preparing the specimen (cutting, polishing, etc.) residual stresses can be introduced into the surface. If the experiment depends on stress-free surfaces there are two possibilities:

1. A further heat treatment after producing the specimen followed by etching to remove the oxide film.
2. Removal of the cold-worked surface layer by electropolishing or severe etching.

When the purpose of the experiment is to simulate certain practical conditions in the laboratory, it is imperative to adhere as closely as possible to the practical case. Thus, it can be important to test the material in the state in which it is to be used, for example, in the as-received state. In certain cases it might be important to include joints, for example, welds, in the test.

12.3.2 Size of Electrodes

In order to study kinetics and mechanisms small plates, foils, spheres, disks, rods, or wires are used as electrodes. They permit high current densities with minimum ohmic heating or find use in cases where high-current sources are not available. If localized corrosion has to be studied, a certain minimum area must be guaranteed. The size has to be chosen so that the corrosion effect occurs with high probability on the electrode surface. This is especially valid when measuring the potential dependence of pitting, SCC, etc.

Another reason for designing electrodes of a certain size is the combination of weight loss with electrochemical measurements. Thus, very small electrodes should not be used when gravimetric measurements are required. On the other hand, the use of microelectrodes has attracted much interest in the study of corrosion effects on a microscopic scale. Microelectrochemical techniques are described in Chapter 17 of this book.

12.3.3 Current and Potential Distribution

In principle, current measurements give only integral values as current per total exposed electrode surface. On real electrode surfaces, the current distribution is more or less nonuniform. This limitation applies particularly for cases of nonuniform corrosion attack.

Since measurement of the current distribution is only possible with considerable effort (using, e.g., segmented electrodes), some general rules for minimizing nonuniformity will be presented (for theoretical background, see Refs. [10,11]).

If a working electrode of length L with a parallel counter electrode (distance h), as in Figure 12.1, is considered, the current distribution is the more uniform the larger the ratio L/h.

From this the following recommendations can be given:

- A design should be chosen with working and counterelectrodes as close as possible (taking into account the possible interference of anodic and cathodic reaction products).
- The smallest electrodes possible should be selected (without interfering with statistical effects, such as during pitting).

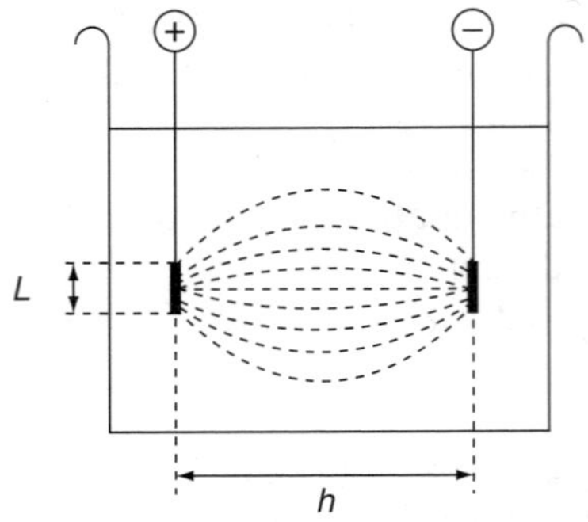

Figure 12.1 Current distribution on parallel plate electrodes of length L and distance h.

Based on the concept of the dimensionless Wagner Number (10), the following further conclusions are derived:

- If possible, low conductivity solutions should be avoided.
- Polarization resistances should be as high as possible.

This last statement suggests that electrodes producing higher currents during polarization show more nonuniform current distribution, since high currents are correlated with low polarization resistances.

The rules cited give some useful working/counter electrode designs that are shown in Figure 12.2. Best suited for minimizing ohmic drop in media of low conductivity is the cylindrical geometry.

When using plate electrodes it is necessary to consider whether the reverse side should be insulated. This is not necessary in systems with high conductivity and with sufficient distance between the electrode and the cell wall. Insulated plate electrodes produce problems as a result of crevice formation at the edge of the insulation. The best solution is, however, to provide a noninsulated plate electrode with two counter-electrodes situated on both sides of the plate.

12.3.4 Effect of Mass Transfer

Mass transfer influenced corrosion reactions can only be correctly investigated on electrodes that are exposed under well-defined hydrodynamic conditions. For basic

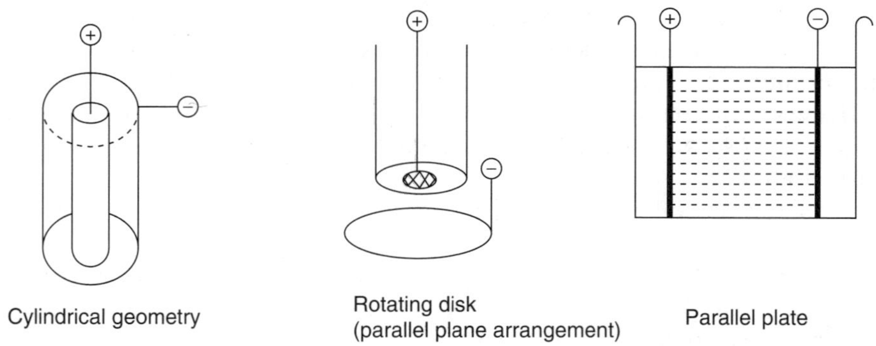

Figure 12.2 Electrode design with various geometries to minimize nonuniformity of current distribution.

studies the most suitable type is the rotating disk which is either placed in the end face of a rotating cylinder or used as free rotating system with an integral shaft (2,10). Further flow models are the rotating cylinder, free or coaxial, and pipe and channel flow (12). For kinetic studies the ring-disk and the double-ring electrode have proven successful (13). Pipe and channel flow are hydrodynamic systems that give valuable information on technical problems during high flow rates or multiphase flow (12).

12.3.5 Electrodes in Sensor Techniques

There are a number of examples for the use of electrodes in sensor techniques for monitoring corrosion that are thoroughly reviewed in Ref. (14). A new development is the use of multisensors, which allows for a higher degree of corrosion prediction reliability. Various kinds of DC sensors for polarization resistance, current–potential characteristics, current and potential measuring devices can be combined with classical sensors for pH, oxygen concentration, conductivity, etc. to give a multisensor surveillance system. By application of appropriate signal processing methods information on the probability of localized corrosion such as crevice corrosion and pitting can be obtained. A number of examples of the application of such multisensors in pilot plants of the process industry have been reported (15, 16).

Sensors for current and limiting current measurements belong to new investigation methods in microbiologically influenced corrosion. Selected application examples are described in Section 12.7.4.

12.3.6 Considerations Concerning the Electrolyte

During the measurement the concentration of the components of the electrolyte should not change significantly. However, changes due to anodic or cathodic reactions cannot be avoided. Therefore, if longer experiments are carried out, the electrolyte has to be exchanged regularly. Another possibility of solving this problem is the use of divided cells with separate anode and cathode compartments. Separating systems are ion exchange membranes or mechanical diaphragms based on polymer or ceramic materials. Further information on experimental handling of separation systems is discussed elsewhere (10).

12.4 POTENTIAL MEASUREMENTS

The open-circuit potential (OCP) of a corroding electrode is measured as voltage between this electrode and a stable reference electrode using a high-input resistance voltmeter (for a description of the electrical circuit see Chapter 11).

12.4.1 Reference Electrodes

The potential of all reference electrodes is based on that of the standard hydrogen electrode (SHE). Due to its great sensitivity to impurities and its impracticability, secondary standard metal/metal ion electrodes are usually employed in practice (6,17). The potential of these electrodes should always be checked against a suitable standard electrode, for example the SHE. The reproducibility of secondary electrodes increases with the anion concentration of the low-solubility metal salt.

With homogenous mixed electrodes the reference electrode can just be dipped in the electrolyte. With heterogenous mixed electrodes a narrow capillary filled with electrolyte is brought up to the surface. The design and position of these capillaries pose current and potential distribution problems. In order to minimize ohmic drop they are placed as close as possible to the electrode surface. However, if the distance is too small, they act as a current shield and nonuniform current distribution arises. In practice the tip of the Haber-Luggin probe should be at a distance of about 2*d* from the working electrode, where *d* is the external diameter of the capillary. With such an arrangement local potentials of a surface can be measured (Figure 12.3).

The choice of reference electrode depends on the electrolyte and the experimental conditions. If contamination of the electrolyte with chloride ions has to be avoided, then, for example, the Hg/Hg_2SO_4 electrode can be used. For alkaline solutions the Hg/HgO electrode is most suitable. The $Cu/CuSO_4$ electrode is simple to prepare in the laboratory and is robust for field experiments. A number of reference electrodes together with their field of application are presented in Table 12.1.

If resistance to high temperatures and pressures is required, special electrodes with their special methods of preparation have to be considered, for example, thallamide or silver electrodes. If a suitable reference electrode that can be inserted directly into the system cannot be found for technical reasons, a reference electrode in a separate vessel is used, which is connected to the measurement cell by an electrolyte bridge. With such an arrangement contamination of the electrolyte in both systems can be avoided. If the measurements are to be carried out in a nonaqueous electrolyte, for example, organic or inorganic solvents or molten salts, in general the reference electrodes described above are not applicable. In such cases it is necessary to refer to the specialized literature (18) or to develop a suitable reference electrode for the purpose. The behavior of methanol and ethanol solutions is more amenable, as in most of these an aqueous calomel or silver/silver chloride electrode can be used. It is of course important that the interdiffusion of water and alcohol is kept within certain limits, for example, by use of an additional electrolyte bridge.

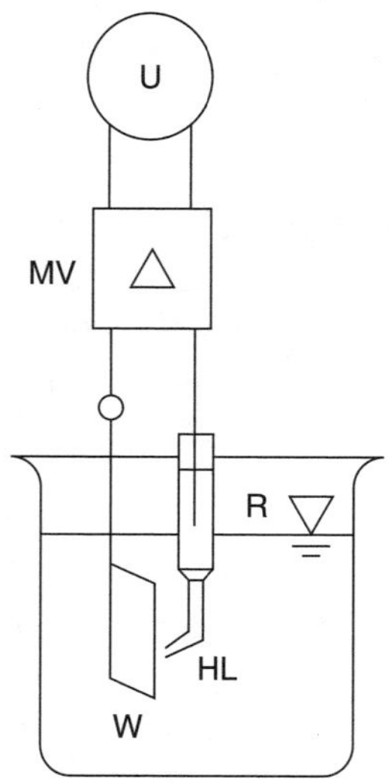

Figure 12.3 Principle of potential measurements: W = working electrode, R = reference electrode, HL = Haber-Luggin capillary, MV = amplifier, U = voltmeter.

Table 12.1 Potentials and Applications of Reference Electrodes in Use

Half cell	Electrolyte	U_{ref} vs. standard hydrogen electrode at 25°C (mV)	Temperature range (°C)	Temperature coefficient (mV/°C)	Application
$Hg/Hg_2Cl_2/Cl^-$ (calomel)	sat. KCl	+242	0 to 70	0.65	General in laboratory technology
$Hg/Hg_2SO_4/SO_4^{2-}$	sat. K_2SO_4	+710	0 to 70	—	Solutions containing sulfate
$Hg/HgO/OH^-$	1 *M* NaOH	+140			Alkaline solution
$Ag/AgCl/Cl^-$	3 *M* KCl	+207	−10 to +80	1.00	Water
$Ag/AgCl/Cl^-$ (e.g., silver)	3 *M* KCl	+207	0 to 130	1.00	Water
Hg Tl/TlCl (e.g., thallamide)	3 *M* KCl	−507	0 to 150	0.1	Hot media
$Cu/CuSO_4$	sat. $CuSO_4$	+320		0.97	Soils, water

12.4.2 Conversion of Measured Potentials to the Standard Hydrogen Scale

Conversion of the potential of the working electrode $E_{meas.}$ measured with a reference electrode with a potential E_{ref} to that of the SHE E_H is achieved by the equation

$$E_H = E_{meas.} + E_{ref} \tag{12.1}$$

where E_{ref} is the potential of the reference electrode on the SHE scale.

12.4.3 Fields of Application

The most important potential measurements are the determination of the OCP or the free corrosion potential E_{corr} of the particular metal/medium system. The OCP is quite generally the potential of an electrode that is not subjected to an external current, while the free corrosion potential is the potential of a corroding electrode without external current.

In detail, such measurements provide

- A method for determining whether the corrosion system is in the active or the passive state.
- Determination of the potential distribution on the corroding surface (heterogenous mixed electrodes).
- Data for corrosion monitoring in a plant or in the field.
- Information on the free corrosion potential as the starting point for the application of electrochemical protection methods.
- Data for control of the protection potential during electrochemical protection.

12.5 CURRENT MEASUREMENTS

12.5.1 Conversion of Current Densities into Corrosion Rates

In all corrosion processes that proceed by an electrochemical mechanism the consumption of material in the particular partial reaction is related to the electric current I (A) by Faraday's Law:

$$m = MIt/zF \tag{12.2}$$

where m is the amount of material consumed electrochemically (g), M is the molar weight (g mol^{-1}), F is the Faraday number (96487 A s mol^{-1}), t is the time (s), and z is the charge number. Differentiating with respect to t gives

$$\mathrm{d}m/\mathrm{d}t = MI/zF \tag{12.3}$$

The rate of chemical consumption of material, dm/dt, is proportional to I, and if Equation (12.3) is divided by the area, it follows that the corrosion rate corresponds to a corrosion current density. An important objective of corrosion investigations is the determination of such corrosion current densities.

12.5.2 Current Measurement in Galvanic Corrosion

The direct measurement of corrosion currents is difficult, because a corroding metal surface consists of many short-circuited corrosion cells with corresponding currents. This is immediately understandable for homogenous mixed electrodes, since the corrosion cells are uniformly distributed in time and space. However, with some heterogenous mixed electrodes there are some experimental arrangements where a cell current is measurable. Such a setup is possible, for example, when two different metals are in contact and anode and cathode reactions are spatially separated. By including in the circuit a low-resistance ammeter or standard resistance the cell current can be measured, as illustrated in Figure 12.4.

When carrying out such measurements, the resistance of the current measuring device must be small compared with the sum of all the resistances of the corrosion cell (polarization resistance, ohmic resistance). The zero resistance ammeter (ZRA) is a special case and is further described in Chapter 11.

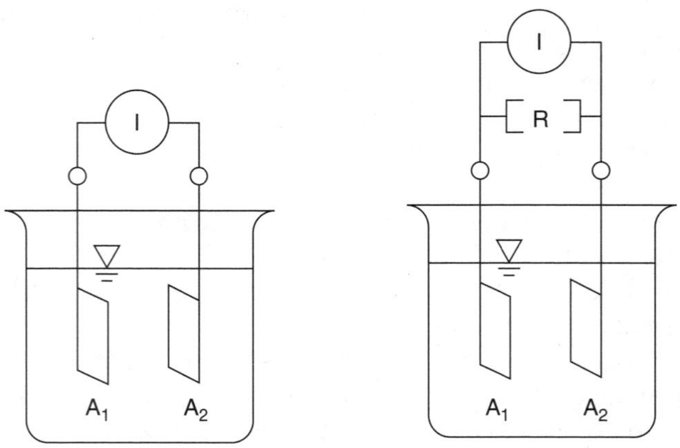

Figure 12.4 Principle of current measurement during galvanic corrosion: A_1, A_2 = working electrodes, I = ammeter, U = voltmeter, R = resistor.

The cell current measured in this way is equal to the corrosion current only if the anode and cathode reactions are completely separated, that is, if there are no currents flowing on the electrodes additional to those flowing in the connection. The main application of current measurements in corrosion reactions is the determination of the cell current in galvanic corrosion. The measurements can be performed either in a simplified manner, as shown in Figure 12.4, or by using a circuit with a current integrator (Chapter 11). It is important that the cell current measurements are combined with weight loss investigations in order to establish the equivalence between cell current and corrosion rate. If this is not done, then it is only possible to give a semiquantitative assessment of the corrosion system. Further details of the investigation of galvanic corrosion are given in several standards (e.g. [19]).

12.5.3 Current Measurements in the Electrolyte

Current measurements are also possible in the electrolyte. Since the currents in the metal phase and the electrolyte phase are equal, the determination of the current in the electrolyte gives the same information as the direct measurement in the metal. The current density i is obtained from the ohmic potential drop determined by two reference electrodes with Haber-Luggin capillaries and the conductivity of the electrolyte according to

$$i = \Delta E \kappa / a \tag{12.4}$$

where ΔE (V) is the potential difference between two reference electrodes, a (cm) is the distance of the tips of the Haber-Luggin capillaries, and κ is the specific conductivity of the electrolyte (Ω^{-1} cm^{-1}).

Measurements in the electrolyte are helpful if separation of anode and cathode in galvanic corrosion is not possible (e.g., in combinations of weld metal and base metal). Potential path measurements in the region of the contact can then lead to the cell current.

Current measurements in the electrolyte are of importance if large corrosion cells play a major role. An important example of this kind is the determination of corrosion currents of steel in reinforced concrete exposed to seawater or in localized microbial corrosion (current mapping on corroding surfaces). Further information on this method is given in Section 12.7.

12.6 POLARIZATION MEASUREMENTS

12.6.1 Measuring Principles and Fields of Application

When the potential of a free corroding metal surface is changed by connecting it with an external power source, the metal will be polarized. As a consequence of the large diversity of corrosion systems, there exist many correlations between current and potential designated as current–potential curves (polarization curves). For polarization measurements the metal to be investigated is connected as the working electrode together with a counterelectrode, a reference electrode, and the electrolyte. Such an electrochemical cell is shown in Figure 12.5.

There are two principal groups of electrochemical methods according to the position of the working potential to the corrosion potential:

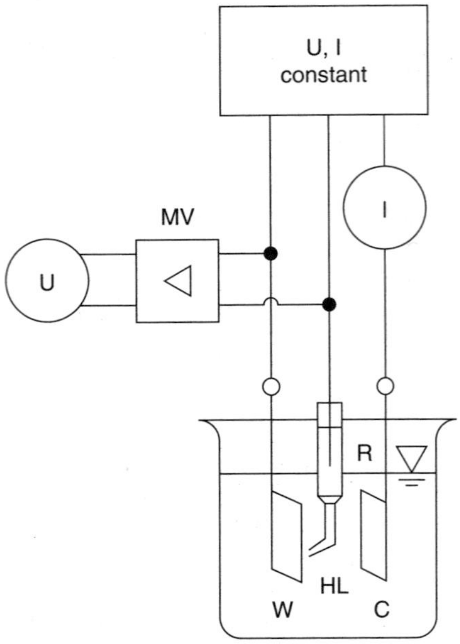

Figure 12.5 Electrochemical cell for polarization measurements: W = working electrode, C = counter electrode, R = reference electrode, HL = Haber-Luggin capillary, MV = amplifier, U = voltmeter, I = ammeter.

- Measurements in the vicinity of the corrosion potential.
- Measurements far away from the corrosion potential (>100 mV).

The first group comprises polarization measurements for the determination of the polarization resistance, potential and current measurements on corrosion cells, and various advanced methods presented below. Examples of measurements far from the corrosion potential are the determination of total polarization curves for electrochemical characterization of corrosion systems, the determination of pitting potentials, passivation and repassivation potentials, and of critical potentials that are connected with a number of local corrosion processes such as crevice corrosion, SCC, corrosion fatigue, and intergranular corrosion.

In general, the polarization curve near the corrosion potential has the form shown in Figure 12.6. This curve is the result of the superposition of several potential dependent reactions. In the simplest case these are the anodic dissolution of the metal (curve 1) and the cathodic reduction of an oxidant (curve 2, neglecting, for the time being, back reactions).

According to the theory of the additivity of partial reactions to give the total reaction, which is the basis of all electrolytic corrosion processes (1), the total polarization curve (3) is the sum of the partial polarization curves (1) and (2):

$$I_{\text{sum}} = I_{\text{anod.}} + I_{\text{cath.}} \tag{12.5}$$

From this equation, it can be concluded that at the corrosion potential the values of anodic and cathodic partial currents in Figure 12.6 are equal. If the potential is shifted in the anodic or cathodic direction, the total polarization curve approaches the anodic or cathodic partial polarization curve, respectively. For the anodic side this means that the measured current of a corrosion cell equals the anodic partial

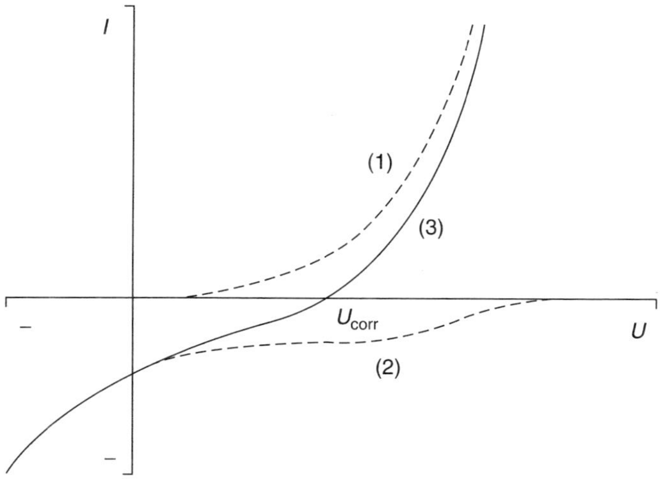

Figure 12.6 Potential dependence of the total and partial currents in corrosion reactions: U_{corr} = corrosion potential; (1) anodic partial current–potential curve; (2) cathodic partial current–potential curve; (3) total current–potential curve (sum curve).

current and therefore the metal dissolution rate. However, the anodic partial current can only be obtained by weight loss measurements at a fixed potential. From this it follows that conversion of anodic currents into corrosion rates (anodic partial currents) must be verified by weight loss measurements.

Measurements far from the corrosion potential give polarization curves of the types shown in Figure 12.7 and Figure 12.8, which represent actively corroding

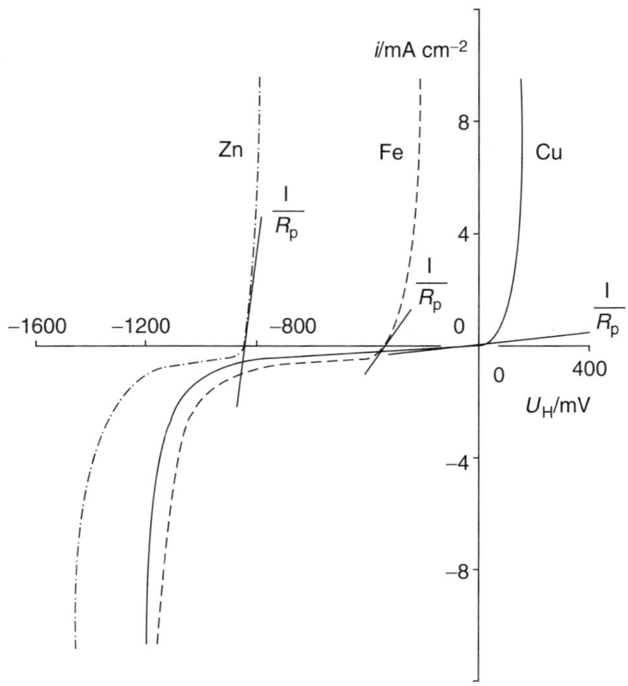

Figure 12.7 Current density–potential curves of actively corroding metals Zn, Fe, Cu, and $1/R_p$ in 3% NaCl solution, pH = 5.5.

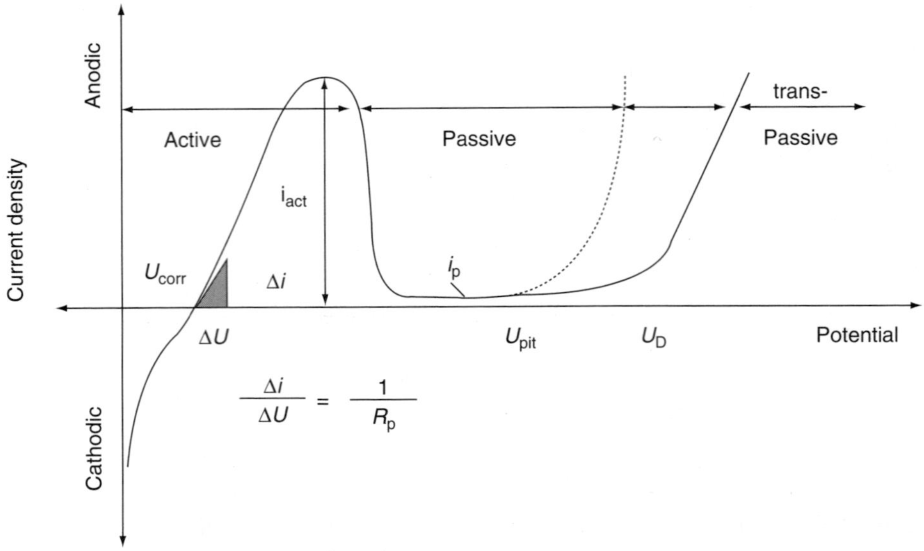

Figure 12.8 Current–potential curve of passive corrosion systems and determination of the polarization resistance R_p: i_{act} = maximum active current density, i_p = passive current density, U_D = breakthrough potential, U_{pit} = pitting potential.

metals and metals with a passivity region, respectively. The anodic branches of the curves in Figure 12.7 show an exponential rise, while the cathodic branches exhibit first a limiting region followed by exponential behavior. The limiting region is due to the diffusion controlled cathodic oxygen reduction. Exponential behavior is due to the charge transfer controlled anodic metal dissolution and cathodic hydrogen evolution reactions. Further analysis of such polarization curves is possible as discussed in the literature on electrode kinetics (20–24).

Important information can be taken from systems in which passivation occurs, as shown in Figure 12.8. An active, passive, and transpassive region can be distinguished: these depend greatly on the nature and composition of the corrosion system. In general, such curves cannot be evaluated quantitatively for the reasons explained above, but they give valuable information on critical potentials, such as

- Active region with peak current density i_{act}.
- Passive region with passive current density i_p.
- Transpassive region.
- Pitting potential E_{pit}.
- Slope of the polarization curve at the corrosion potential E_{corr}.

12.6.2 Measuring Techniques

In practice, measurements of polarization curves can be accomplished by the use of a potentiostat to maintain the working electrode at a constant electrode potential (Figure 12.5). This is achieved by maintaining a constant predetermined potential between the working electrode and the reference electrode that results in a current flow. The current source, amplifiers, devices for voltage, and current measurements are combined in the potentiostat. The potentiostat is able, as an electronic regulating device, to produce any required potential–time function. From a large number of possibilities the following classical techniques may be chosen:

- Potential hold experiments in which the potential is held constant for a given time of the experiment.
- "Potentiostatic" experiments for imposing quasistationary states in which the applied potential is maintained until the current has become constant. The applied potential is then changed to the next value.
- Potentiodynamic (potentiokinetic) experiments with a linear change of potential with time.
- Potential on, potential off, potential reversal, potential step experiments with very rapid changes in potential, as well as any other technique with a given potential–time program.

Among these methods only the potential hold measurements and potentiostatic measurements applied for given time intervals belong to the group of DC measurements.

A further important measuring arrangement for corrosion investigations is the galvanostatic polarization circuit in which the applied current is held constant and the potential is recorded as a function of time. The principle of this technique is also shown in Figure 12.5.

A particularly elegant galvanostatic circuit is possible with the help of a potentiostat. According to this principle the potentiostat controls a constant voltage across a resistor and thereby applies a constant current. The desired constant current flows through the measuring circuit that is connected in series with the resistor.

Other versions of the potentiostatic polarization circuit can also provide various current–time functions when operated galvanostatically. In addition to galvanostatic holding experiments and galvanostatic experiments with stepwise current changes, galvanostatic switching on, switching off, and reversal experiments have been used.

It is not possible to access the passive region from galvanostatic current potential curves. This is demonstrated by comparing the galvanostatic and potentiostatic polarization curves for 13% Cr steel in acidic and neutral solutions (Figure 12.9).

In contrast to curves (a) (acidic medium, potentiostatic), no passive region is obtained in curve (c) (acidic medium, galvanostatic). If parallel lines to the potential axis are drawn through the curves (a) in the anodic region, it can be seen that for every

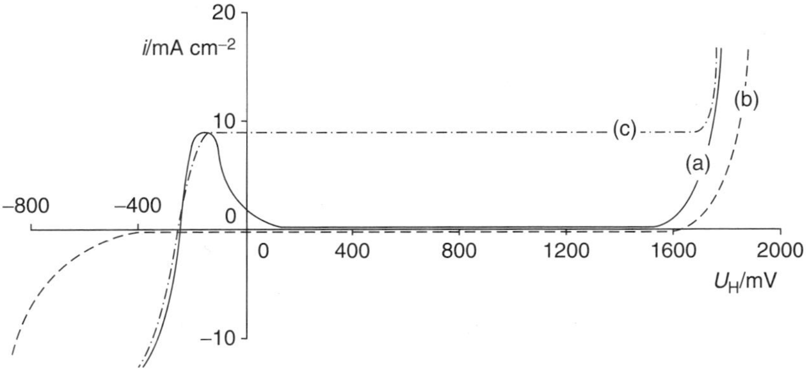

Figure 12.9 Current–density potential curves for a 13% Cr steel: (a) 0.2 *M* H_2SO_4, potentiostatic; (b) 0.2 *M* Na_2SO_4, potentiostatic; (c) 0.2 *M* H_2SO_4, galvanostatic.

current density there are three values of the potential. In potentiostatic measurements these values can be determined. For galvanostatic measurements with increasing current the polarization curve follows the left branch of the active region. When the current at the maximum in the active region is slightly increased, a current of this magnitude exists only in the transpassive region and therefore the electrode potential is displaced very rapidly by almost 2 V in the anodic direction in curve (c). Therefore, regions with decreasing current density–potential characteristics and minima in the current density cannot be recorded with the galvanostatic circuit and a potentiostatic circuit is required. The results in Figure 12.9(b) explain that for a 13% Cr steel in neutral solution an active region does not exist, which demonstrates that under these particular test conditions this material is not susceptible to corrosion.

12.6.3 Stationary Methods

Galvanostatic and potentiostatic measuring methods comprise the stationary and quasistationary methods. Their main applications are:

- Characterization of the electrochemical systems taking part in the corrosion process (metal electrode and redox systems, systems with active and passive behavior, electrode kinetics).
- Determination of the dependence of dissolution rates on potential, in particular the estimation of critical potentials such as the pitting potential.
- Determination of the potential dependence of rupture time and crack propagation rate in SCC and corrosion fatigue.
- Potential dependence of selective corrosion of structural components in intergranular corrosion.
- Potential dependence of dissolution rates during cathodic and anodic protection.
- Estimation of corrosion rates solely from current–potential curves.

Application of electrochemical methods requires careful consideration of their limitations. Since they are indirect methods that rely on Faraday's Law, a measured current gives no information concerning the chemical reactions involved with the current or the spatial distribution of corrosion phenomena (current distribution). On the other hand, current and voltage measurements can be made with great accuracy and speed and also they lend themselves to automation.

A disadvantage of electrochemical methods is the influence of the current and potential changes on the corrosion processes. They can cause more or less irreversible changes in important parameters of the system such as in surface microprofile and surface coverage or by products formed on the working or counterelectrode that can influence the quantities being measured. Based on the length of time and the magnitude of the impressed current and potential changes, such disturbances can be large or small.

12.6.4 Corrosion Rates from Polarization Resistance Measurements

The polarization resistance R_p is defined as the slope of the current density–potential curve at the corrosion potential E_{corr} as indicated in Figure 12.7 and Figure 12.8:

$$R_p = (dE/di)_{E \to E_{corr}} \tag{12.6}$$

where R_p is related to the corrosion current density i_{corr} by

$$i_{corr} = B/R_p \tag{12.7}$$

Thus, the corrosion rate of a material in open-circuit corrosion can be calculated from electrochemical data. The relationship is known as the Stern–Geary equation, which was published in 1957 (25) and is based on the Wagner–Traud mixed potential theory published in 1938 (1). The proportionality constant B is dependent on the system and must be determined experimentally. A summary of experimental B values has been compiled (26).

The measurement of the polarization resistance is of great practical importance and is also known as linear polarization resistance (LPR) measurements assuming a linear potential–current density relationship around E_{corr}. For application of the polarization resistance technique, the following limitations must be considered:

- The corrosive attack must be uniform.
- The OCP must not change during the period of measurement.
- The corrosion mechanism must not change during the measurement.
- No other electrochemical reactions must occur except the corrosion reaction.
- Ohmic resistances of the system must be small compared with the polarization resistance.

The theory, application, and limitations of polarization resistance measurements have been dealt with in a number of publications (27, 28). Ohmic drop evaluation and compensation, which is another important experimental problem since the experimental R_p value is the sum of the true R_p and the uncompensated solution resistance, are treated in Ref. (29).

Various commercial devices making use of polarization resistance measurements are available. They use two or three electrode probes and apply, for example, 10-mV perturbations. It is conventional to alternate the polarization direction and to use the average of the currents obtained in both directions. From measured R_p values and fixed B values corrosion rates can be estimated. Such devices are largely used for corrosion monitoring in the field and plants. Their significance is less in giving quantitative data of corrosion rates than in indicating changes with time and thus providing an alarm for the surveillance personnel when corrosion rates have reached unacceptable levels. Examples for R_p measurements are shown in Figure 12.7 and Figure 12.8.

12.6.5 Extrapolation of Stationary Current Density–Potential Curves

If it can be assumed that the anodic and cathodic partial reactions are completely independent of each other and if the measurements are made far from the corrosion potential, extrapolation of the anodic and cathodic branches of the polarization curve to E_{corr}, where the applied current is zero, provides the corrosion current. For extrapolation, the polarization curves are plotted in a log i/E graph ("Tafel lines") since most of the curves represent exponential functions. This method — usually called the Tafel extrapolation method — has the disadvantage, described at the beginning, of possibly significant changes of the electrode surface during the measurement due to the high polarization levels.

12.7 SELECTED EXAMPLES OF THE APPLICATION OF DC TECHNIQUES

12.7.1 Potential Distribution along a Partially Galvanized Steel Rod

For simulation of underground corrosion and corrosion protection the following laboratory experiment can be performed. A half-galvanized carbon steel rod is placed in a poorly conducting electrolyte (soil solution), and the potential is measured along the length of the specimen by means of a reference electrode with Haber-Luggin capillary (2). The potential distribution is shown in Figure 12.10. In the solution with the lower resistivity of 2000 Ω cm the potential distribution is more uniform than in the 10,000-Ω cm solution due to the unfavorable current distribution. This has the consequence that the protected region in the 10,000-Ω cm solution is smaller (5 mm) than in the 2000-Ω cm solution (45 mm).

From such an experiment general conclusions for cathodic protection under field conditions can be made. If Wagner's similarity law is applied some rules for spatial scaling up can be deduced along the lines shown in Section 12.3.3 (4,10).

12.7.2 Potential Mapping in Field Applications

Carbon steel in reinforced concrete generally is in the passive state. When a reinforced concrete structure is penetrated by aggressive chloride containing water, the steel becomes active. This results in a potential shift of some hundreds of millivolts in the negative direction. Figure 12.11 exhibits the results of a large potential mapping program on a street bridge and a street tunnel. It can be seen that the potential distribution of the two structures differs significantly. While the steel of the bridge is in the passive state, in the street tunnel it is in the active state.

It is evident that potential mapping of this kind (30) provides an excellent possibility for monitoring the corrosion state of constructions and apparatus in the field and in plants as well as the effectiveness of cathodic protection of such structures. Further information is given elsewhere (31).

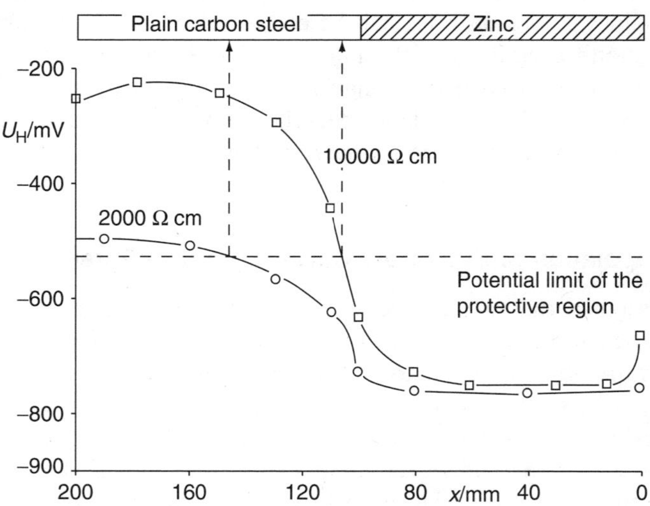

Figure 12.10 Potential distribution along a half-galvanized steel rod in two electrolytes of different resistivities simulating soil (2).

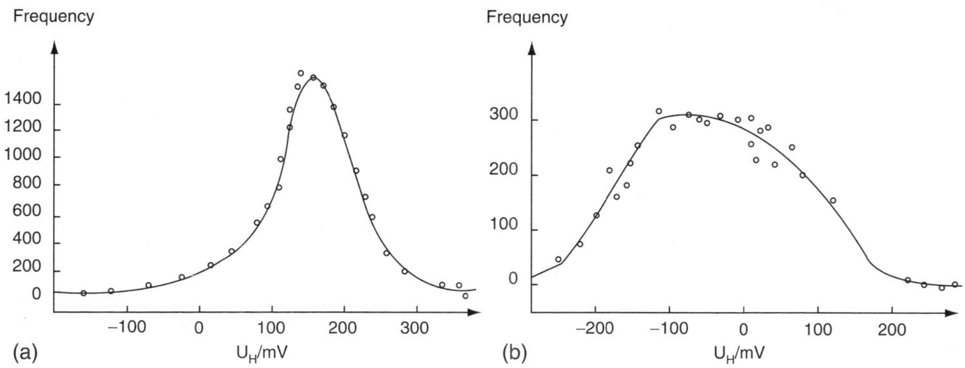

Figure 12.11 Frequency distribution of steel potentials in reinforced concrete structures (30). (a) Passive potentials of the reinforcement of a street bridge. (b) Active potentials of the reinforcement of a chloride contaminated street tunnel.

12.7.3 Potential Monitoring in Nuclear Power Stations

Potential measurements in high-temperature waters of nuclear power stations are classical examples of demonstrating the influence of water chemistry on intergranular stress corrosion cracking (IGSCC) of sensitized stainless steel (32). Figure 12.12 shows the potential dependence of stainless steel AISI 304 on the oxygen content and temperature under conditions of the steam generator of a nuclear power plant (32, 33). Potential values are raised by increasing the oxygen concentration to more than 600 mV, shifting the potential of the material into the critical region of IGSCC. Limits of allowed oxygen levels are 100 μg kg^{-1} (0.1 ppm) with chloride contents less than 3 ppm. Reduction of the potential is achieved by appropriate hydrogen/water chemistry.

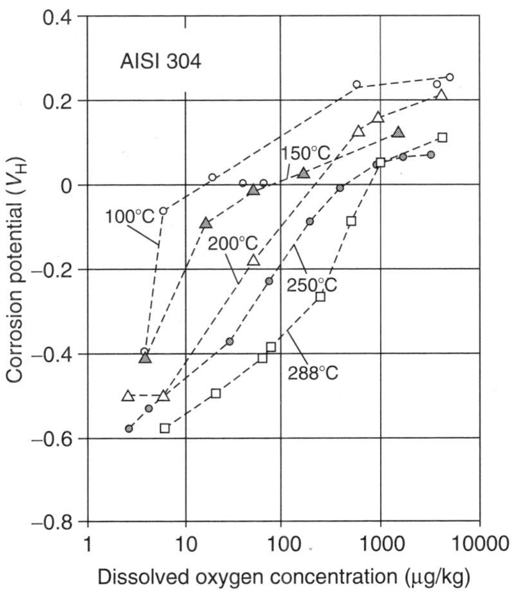

Figure 12.12 Corrosion potential of austenitic stainless steel AISI 304 as a function of oxygen content and temperature under conditions of a steam generator (21,32).

In Ref. (32) the design of an on-line potential monitoring device installed in a U-tube-type nuclear steam generator is described.

12.7.4 Current Measurements during Microbially Enhanced Crevice Corrosion of Stainless Steel and during Corrosion of Carbon Steel by Sulfate-Reducing Bacteria

An interesting application of current measurements concerns the influence of microbial slime layers on the corrosion properties of stainless steel in seawater (34). This application is of historical interest since it is one of the first experiments concerning the ennoblement of stainless steel in seawater due to the action of a biofilm (designated at that time as microbial slime layer). The aim of this work was the determination of the crevice corrosion susceptibility of stainless steel in natural and artificial seawater. Different cathode-to-anode area ratios were used and the crevice corrosion rate was determined by current measurements. The experimental results are plotted in Figure 12.13.

From these findings the following conclusions can be drawn:

- Natural seawater induces increasing crevice corrosion rates after an incubation time of a few days (formation of the biofilm).
- Higher cathode/anode ratios give larger corrosion rates.
- Artificial seawater does not induce crevice corrosion (absence of a biofilm).

In a new development sensors based on the principle of concentric disk-ring electrodes have been designed for the application of various stationary and nonstationary electrochemical methods (35, 36). Such sensors are successfully used for investigations of the corrosion of carbon steel by sulfate-reducing bacteria (SRB) (35, 36). By combining a small center electrode with a large ring electrode, localized biocorrosion can be initiated electrochemically or mechanically. When the arrangement is short-circuited, the biocorrosion process can be followed by measuring the current with a ZRA. Thus, parameter studies such as investigation of the effect of different biocides can be performed. Further work succeeded in corrosion current

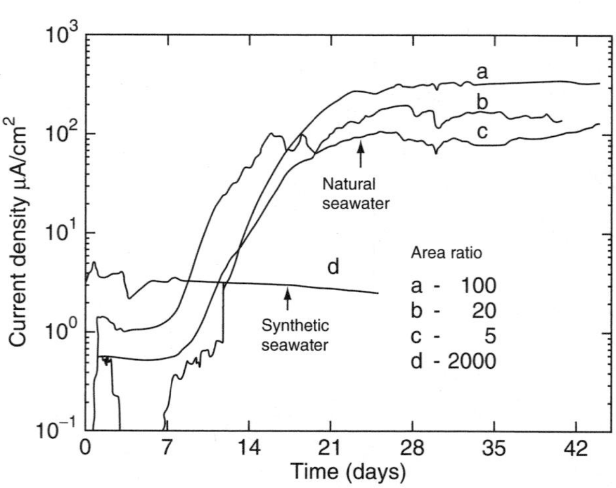

Figure 12.13 Crevice corrosion rate of stainless steel AISI 316 L in natural and synthetic seawater at different cathode to anode ratios (34).

mapping of steel surfaces attacked by SRB (36). By the use of microelectrode probes according to the principle of current measurements in the electrolyte (Section 12.5.3), it was possible to evaluate the influence of mechanical damage, biocide action, and time on the kinetics of biofilm growth. Moreover, this type of electrode combination has been found very suitable for electrochemical impedance spectroscopy and electrochemical noise analysis.

12.7.5 Limiting Current Measurement for Monitoring Biofilm Growth

Biofilms on metallic surfaces can be considered as porous layers that act as transport barriers in mass transfer controlled electrode reactions. This property is used to determine the thickness of such layers by limiting current measurements (37, 38). A new type of sensor on the basis of an impinging jet flow has been developed (37) and applied to monitoring of biofilm growth. The liquid jet hits the surface to be investigated, which is cathodically polarized to measure the limiting current of oxygen reduction in the test solution or of ferricyanide added as tracer compound. The current can be converted into biofilm thickness, which in turn is dependent on the flow rate and growth time. Typical results are shown in Figure 12.14, which shows that the biofilm thickness decreases with increasing flow rate and increases with growth time. Such limiting current measurements are very effective in following on-line and in real-time biofilm formation and prevention by biocides.

12.7.6 Corrosion Currents in Reinforced Concrete Constructions in Seawater

Offshore constructions of reinforced concrete with large surfaces in great depth suffer from corrosive attack if the steel is exposed to seawater in cracks (39). Consequently, such structures have to be examined for cracks and it is possible to detect the corrosive activity in cracks by electrochemical means. Figure 12.15 shows the current distribution in the vicinity of cracks with the components parallel and normal to the concrete surface. The currents are obtained by measuring the potential distribution by means of two reference electrodes as described in Section 12.5.3.

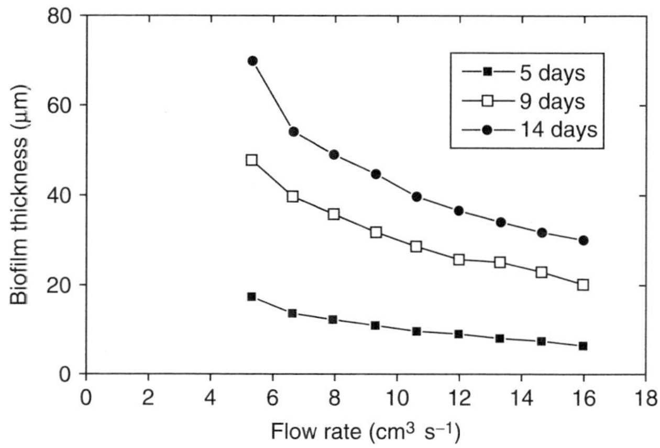

Figure 12.14 Variation of biofilm thickness in seawater vs. flow rate (volume flow of a jet) and growth time on a gold surface (38).

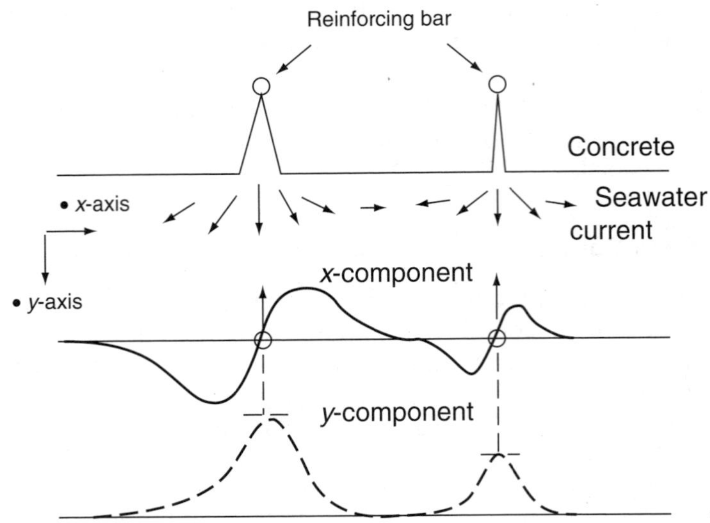

Figure 12.15 Components of the corrosion current vectors at different cracks in reinforced concrete (39).

In order to suppress fluctuations of potential signals due to noise and drift, a pair of electrodes is rotated. The corrosion signal can then be separated electronically from the interferences of other sources.

Due to the high conductivity of seawater the potential signals are in the range of microvolts to millivolts and the corresponding current densities in the front of the crack approach 6 μA cm^{-2}.

This method can also be used to check cathodic protection levels, for instance, for coated steel pipes in seawater.

12.7.7 Pitting Potential of Differently Tempered Stainless Steel

The determination of pitting potentials of passivating metals and alloys by means of polarization curves belongs to the most frequently applied electrochemical test methods. Pitting corrosion is a complex process with a number of stages, which differ in mechanism, as shown in a review paper (40). Especially for the pitting of stainless steel in chloride containing solutions many mechanistic problems are posed.

The following application example shows the influence of tempering conditions on the pitting potential of stainless steel AISI 304. Specimens have been tempered at two temperature levels of different lengths and then cooled down, all steps in air. The pitting test solution was an aerated, neutral 0.3 *M* chloride solution. Anodic polarization curves with a slow potential scan rate of $dE/dt = 100$ mV h^{-1} have been taken. With such small polarization rates quasistationary states are obtained.

From the results shown in Figure 12.16 it can be seen that the pitting potential (onset of the steep part of the anodic polarization curve) varies according to the heat treatment. Whereas the pitting potential of the nontreated material has a value of $E_p = 450$ mV, it is lowered to about 100 mV after tempering for 20 h at 500°C. This finding is in agreement with the practical experience that tarnish layers from welding are sites of preferred pitting. It is to be assumed that a passive film breaking mechanism preferentially occurs at the thicker oxide layers after the tempering procedure.

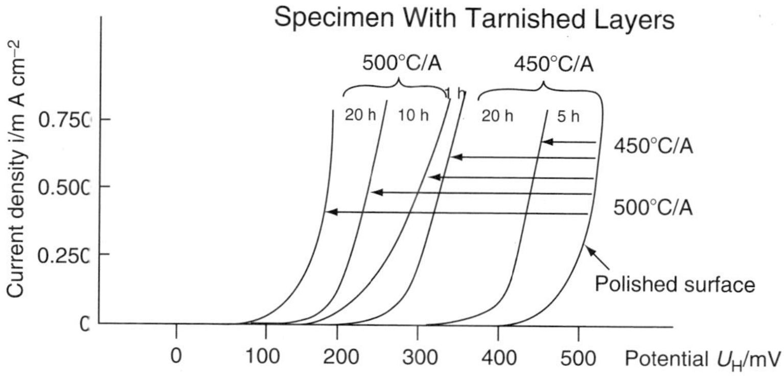

Figure 12.16 Anodic polarization curves of differently tempered AISI 304 stainless steel in neutral 0.3 *M* chloride solution.

12.7.8 Material Selection by Means of Polarization Curves and Polarization Resistance Measurements under Conditions of Erosion Corrosion

Material selection for hydraulic equipment under condition of erosion corrosion in particle-containing liquids is a good example of the application of electrochemical methods. Tests were made in a pilot plant for flow-induced corrosion. The medium was a quencher solution as it is used in flue gas desulfurization plants with flow velocities up to 24 m sec^{-1} (12, 41). Materials chosen were ferritic, ferritic austenitic, and austenitic stainless steels and cast alloys as they are used in hydraulic equipment.

Polarization curves: In order to obtain an overview of the electrochemical corrosion behavior, polarization curves were taken and plotted in Figure 12.17.

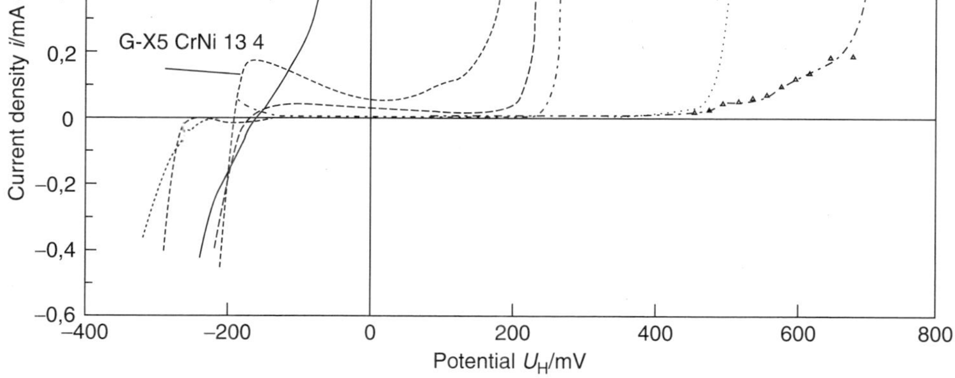

Figure 12.17 Polarization curves of various stainless cast alloys and steels in high flow rate quencher fluid of a flue gas desulfurization plant (30 g l^{-1} chlorides, pH = 3.5, 10% weight gypsum; 24 m sec^{-1}) (10, 41).

The polarization behavior of a number of alloys allows for the following conclusions:

- Cast alloys G-X5 CrNi 13 4 and G-X180 CrMo 25 2 exhibit active corrosion.
- Stainless steel X5 CrNi 13 4 has a lower passive current density as well as a smaller peak active current density than the corresponding cast alloy.
- The breakthrough potential of erosion corrosion increases in the order: G-X40 CrNiMo 27 5 < G-X3 CrNiMoCu 24 6 < X3 CrNiMoN 17 13 5.

From the materials tested, Alloy X3 CrNiMoN 17 13 5 shows the best corrosion behavior. These conclusions allow a first choice of materials, which has to be followed by further quantitative corrosion tests and parameter studies (41).

Polarization resistance (R_p): Measurements were performed for a parameter study of the pH dependence of erosion corrosion under conditions of flue gas desulfurization. Typical materials for hydraulic equipment were chosen and tested under high-flow load conditions as described in the preceding example (12). The R_p values were determined potentiostatically with $+10\,mV$ and $-10\,mV$ step polarization starting from the free corrosion potential. They were used as such and not converted to corrosion rates, because B values were not available for the given experimental conditions.

Plots of $1/R_p$ vs. pH are shown in Figure 12.18. It can be seen that below a certain pH value the reciprocal R_p (proportional to corrosion rate) increases sharply. These critical pH values vary with the composition of the alloy and exhibit the best erosion corrosion resistance for the material X3 CrNiMo 17 13 5. This finding is in agreement with its polarization behavior (Figure 12.17).

12.7.9 Anodic Polarization and Slow Strain Rate Testing in Stress Corrosion of Carbon Steels

Predictive approaches to SCC of C-Mn steels in alkaline, carbonate, and nitrate environments by means of potentiodynamic polarization curves have been successfully applied in a classical study (42). When slow strain rate testing is performed at constant potentials in various environments and if the ratio of t_{fs}/t_{fo} (time to fracture in solution to time to fracture in oil) is plotted as shown in Figure 12.19, curves with minima indicating SCC are observed. The minima indicate SCC at various potentials depending on the SCC-inducing medium with nitrate solution being the most effective. The sensitive potential regions can be correlated with anodic current densities in a polarization diagram that is the difference curve of polarization curves obtained at fast ($1000\,mV\,min^{-1}$) and slow ($10\,mV\,min^{-1}$) scan rates.

Minima of the ratio t_{fs}/t_{fo} correspond to maxima of anodic current densities indicating SCC susceptibility. Thus, potentiodynamic tests can predict the occurrence of SCC in various environments if critical stress conditions exist.

12.8 CONCLUSIONS

- A prerequisite for useful application of electrochemical methods is the consideration of the time factor and size of the experiments. According to the time factor a distinction in terms of long-term, short-term and

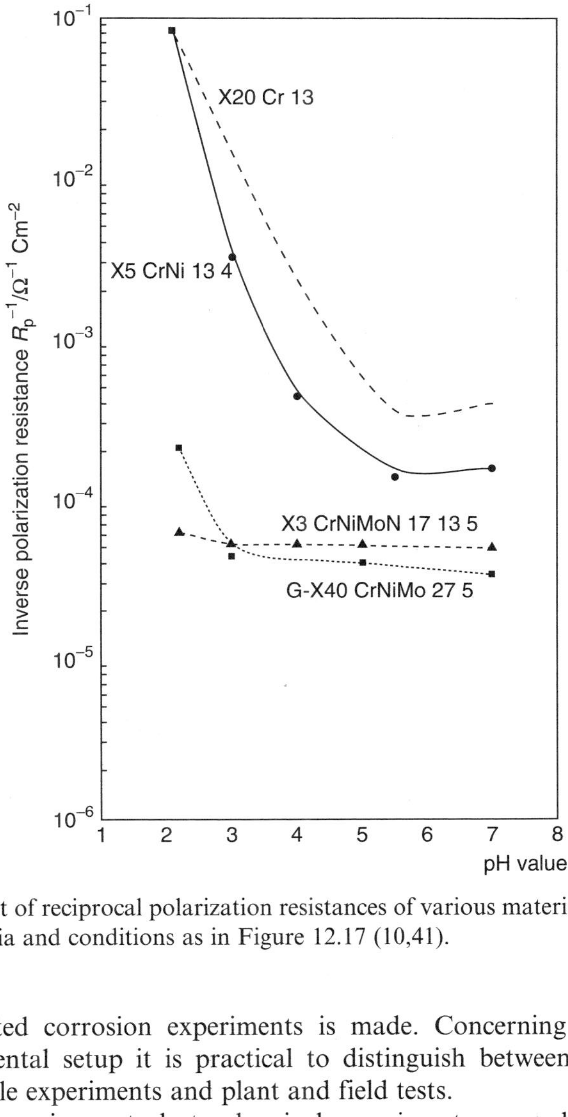

Figure 12.18 Plot of reciprocal polarization resistances of various materials as a function of the pH value. Media and conditions as in Figure 12.17 (10,41).

accelerated corrosion experiments is made. Concerning the size of the experimental setup it is practical to distinguish between laboratory and pilot scale experiments and plant and field tests.

- Before carrying out electrochemical experiments a number of important parameters have to be clarified: pretreatment and size of electrodes, effect of mass transfer of species to and from the electrode surface, and current distribution on the electrode surface.
- DC electrochemical methods comprise potential, current, and polarization measurements. They aim at the determination of corrosion rates, the assessment of the potential dependence of corrosion reactions, the determination of critical potentials, and the evaluation of corrosion mechanisms. Careful attention to limitations of each method is required. Electrochemical measurements can be made with great accuracy and speed and lend themselves to automation.
- Potential measurements are easy to perform. A prerequisite is the choice of a suitable reference electrode. Besides standard reference electrodes, which are commercially available, it might be necessary to use special reference electrodes, which may be developed for the particular application.

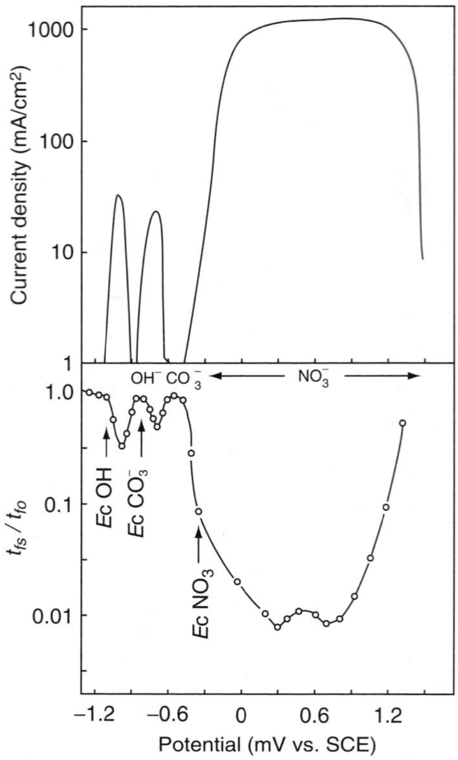

Figure 12.19 Anodic polarization curves and times to failure t_s vs. potential in nitrate, hydroxide, and carbonate/bicarbonate solutions for testing of SCC (42). E_c = corrosion potential; t_{fs} = time to failure in solution; t_{fo} = time to failure in oil (reference).

- Current measurements are generally applied for the investigation of galvanic corrosion. The chief area is the determination of the cell current. During such measurements it is important to establish the equivalence between cell current and corrosion rate with weight loss tests.
- Polarization measurements belong to the key methods in DC electrochemical investigations. Potential hold experiments and potentiostatic measurements are finding widespread application. A special case is the determination of the corrosion rate from the polarization resistance. The LPR method is of great practical importance. For reliable application of these methods, a number of limitations must be observed.
- The application examples presented here demonstrate the large diversity of corrosion problems that can be solved using DC electrochemical methods. They cover all industrial areas starting from classical potential measurements in nuclear power plants to modern approaches of microbially influenced corrosion in water systems as well as electrochemically based material selection for extreme conditions in practice. DC electrochemical methods belong to the basic tools for the evaluation of corrosion mechanisms.

REFERENCES

1. C. Wagner and W. Traud, *Z. Elektrochem.* 44: 391–402, 1938.
2. E. Heitz, R. Henkhaus, and A. Rahmel, "*Corrosion Science, an Experimental Approach*," Ellis Horwood Lim., Chichester, UK, 1992.
3. E. Heitz, *Werkstoffe Korros.* 36: 400–406, 1985.
4. L. Clerbois, E. Heitz, F. P. Ijsseling, J. C. Rowlands, and J. P. Simpson, *Br. Corros. J.* 20: 107–115, 1985.
5. U. Eul, H. von Rège, E. Heitz, and W. Sand, "Simulation and control of MIC by the miniplant technique," in "*Microbially Influenced Corrosion of Materials*," E. Heitz, H. C. Flemming, and W. Sand, eds., pp. 187–203, Springer, Berlin, 1996.
6. European Working Party Report, "Guidelines on Electrochemical Corrosion Measurements," No. 4, The Institute of Metals, London, 1990.
7. A. T. Kuhn, "*Techniques in Electrochemistry, Corrosion and Metal Finishing*," John Wiley & Sons, New York, 1987.
8. P. T. Kissinger and W. R. Heinemann, eds., "*Laboratory Techniques in Electroanalytical Chemistry*," 2nd ed., Marcel Dekker, New York, 1996.
9. Southampton Electrochemistry Group, "*Instrumental Methods in Electrochemistry*," Ellis Horwood Lim., Chichester, UK, 1985.
10. E. Heitz and G. Kreysa, "*Principles of Electrochemical Engineering*," VCH Verlag, Weinheim, 1986.
11. N. Ibl, "Current distribution," in "*Comprehensive Treatise of Electrochemistry*," E. Yeager, J. O. Bockris, B. Conway, and S. Sarangapani, eds., Vol. 6, pp. 239–315, Plenum Press, New York, 1983.
12. E. Heitz, "Chemo-mechanical effects of flow on corrosion," in "*Flow-Induced Corrosion: Fundamental Studies and Industry Experience*," K. J. Kenelly, R. H. Hausler, and D. C. Silverman, eds., pp. 1–1 to 1–29, NACE, Houston, 1996.
13. U. Stöckgen and K. E. Heusler, *Electrochim. Acta* 44: 2765–2770, 1999.
14. V. S. Agarwala and S. Ahmad, "Corrosion detection and monitoring — a review," CORROSION/2000, Paper 217, NACE, Houston, 2000.
15. J. Zosel, W. Fichtner, H. Kaden, and B. Rau, *Mater. Corros.* 47: 299–306, 1996.
16. U. Eul, E. Heitz, J. Zosel, and H. Kaden, *Chem. Techn.* 52: 144–151, 2000.
17. Bard, R. Parsons, and J Jordan, "*Standard Potentials in Aqueous Solutions*," Marcel Dekker, New York, 1985.
18. E. Heitz, "Corrosion of metals in organic solvents," in "*Advances in Corrosion Science and Technology*," M. G. Fontana and R. W. Staehle, eds., Vol. 4, pp. 149–243, Plenum Press, New York, 1974.
19. ASTM G71-1998, "*Standard Guide for Conducting and Evaluating Galvanic Corrosion Tests in Electrolytes*," ASTM, 1998.
20. A. Hamnett, C. W. Hamann, and W.Vielstich, "*Electrochemistry*," VCH-Wiley Weinheim, Germany, 1997.
21. A. J. Bard and L. R. Faulkner, "*Electrochemical Methods*," 2nd ed., Wiley & Sons, New York, 2001.
22. E. Gileadi, "*Electrode Kinetics*," VCH Publishers, New York, 1993.
23. H. Kaesche, "*Metallic Corrosion*," NACE, Houston, 1985.
24. M. Kowaka, "*Metal Corrosion and Protection Technology*," Allerton Press, New York, 1990.
25. M. Stern and A. L. Geary, *J. Electrochem. Soc.* 104: 56–59, 1957.
26. R. Grauer, P. R. Moreland, and G. Pini, "A Literature Review of Polarization Resistance Constant (B) Values for the Measurement of Corrosion Rate," Publication No. 52405, NACE, Houston, 1982.
27. F. Mansfeld, "The polarization resistance technique for measuring corrosion currents," in "*Advances in Corrosion Science and Technology*," M. G. Fontana and R. W. Staehle, eds., pp. 163–262, Plenum Press, New York, 1976.

28. E. Heitz and W. Schwenk, *Br. Corros. J.* 11: 74–77, 1976.
29. L. Clerbois and F. P. Ijsseling, "Evaluation and compensation of ohmic drop," in Working Party Report, "Guidelines on Electrochemical Corrosion Measurements," No. 4, pp. 40–54, The Institute of Metals, London, 1990.
30. B. Elsener and H. Böhni, "Methods for the inspection of reinforcement corrosion in concrete Structures — field experience," in "*Electrochemical Methods in Corrosion Research*," *Mater. Sci. Forum* 111–112: 635–646, 1992.
31. J. A. Gonzales, M. Benito, and S. Feliu, *Corrosion* 51: 145–152, 1995.
32. Stellwag, W. Beyer, and N. Wieling, "Electrode potential monitoring in hot water systems — a method to identify critical corrosion conditions," in "*Electrochemical Corrosion Testing*," E. Heitz, J. C. Rowlands, and F. Mansfeld, eds., Monograph Vol. 101, pp. 17–26, DECHEMA, Frankfurt, Germany, 1986.
33. R .W. Staehle, "Corrosion and Corrosion Cracking of Materials for Water-Cooled Reactors," EPRI-Report, NP-1741, 1981.
34. R. Holthe, P. O. Gartland, and E. Bardal, "Influence of the microbial slime layer on the electrochemical properties of stainless steel in sea water," in Eurocorr 87, pp. 617–623, DECHEMA, Frankfurt, Germany, 1987.
35. N. Montfort Moros, B. Tribollet, D. Festy, and R Marchal, "Parametric study of localized corrosion artificially initiated: application to a carbon steel biocorrosion sensor," CORROSION/2001, Paper No. 01262, NACE, Houston, 2001.
36. M. Keddam, N. Montfort Moros, B. Tribollet, and D. Festy, "Detection and mapping of srb influenced corrosion of C-steel," CORROSION/2002, Paper No. 02447, NACE, Houston, 2002.
37. D. Herbert-Guillou, B. Tribollet, and D. Festy, *Bioelectrochemistry* 53: 119–125, 2000.
38. D. Festy, D. Herbert-Guillon, and B. Tribollet, *Electrochem. Acta* 44: 1067–1075, 1999.
39. R. Polder, "Detection of local corrosion currents," in "*Electrochemical Corrosion Testing*," E. Heitz, J. C. Rowlands, and F. Mansfeld, eds., DECHEMA Monograph, Vol. 101, p. 125, DECHEMA, Frankfurt, Germany, 1986.
40. H.-H. Strehblow, "Mechanisms of pitting corrosion," in "*Corrosion Mechanisms in Theory and Practice*," P. Marcus and J. Oudar, eds., Marcel Dekker, New York, 1995.
41. I. Ehmann, E. Heitz, and A. Schnitzler, *Werkstoffe Korros.* 42: 520–527, 1991.
42. R. N. Parkins, *Corros. Sci.* 20: 147–166, 1980.

13

Electrochemical Impedance Spectroscopy

Florian Mansfeld
Corrosion and Environmental Effects Labs, Dept. of Materials Science and Engineering, University of Southern California, Los Angeles, California

Contents

13.1 INTRODUCTION

The electrochemical impedance spectroscopy (EIS) technique differ from all other techniques described in this book insofar as the properties of an electrode/electrolyte system are evaluated as a function of the frequency of a small ac signal that is applied at a fixed working point (E,I). In most studies of corrosion systems $E = E_{corr}$ and $I = 0$. Modern equipment allows collection of impedance spectra in a wide frequency range covering about 10 orders of magnitude. Evaluation of the experimental spectra as a function of the frequency f of the applied ac signal in Bode plots, which are the preferred display of EIS data (1), reveals one or more time constants in different frequency regions. In Bode plots the phase angle Φ and the logarithm of the impedance modulus $|Z|$ are plotted vs. log f. The minima and maxima of Φ and their positions at certain frequencies are sensitive indicators of system properties and their changes with changing experimental conditions.

Historically, modern dynamic system analysis was introduced in the 1960s by Epelboin and coworkers in Paris, France (2), who applied what was then called the ac impedance technique to a wide number of electrochemical fields including corrosion and its inhibition. Mansfeld and Kendig (3) were the first to use the term "electrochemical impedance spectroscopy" in a paper entitled "Electrochemical Impedance Spectroscopy of Protective Coatings" that appeared in Werkstoffe und Korrosion in 1985. J.R. Macdonald edited a book entitled *Impedance Spectroscopy*, which appeared in 1987 and contains a chapter dealing with the application of the technique in the study of corrosion of materials (4). The progress made in the understanding and application of EIS in corrosion science and technology can be followed by studying the proceedings of the International Symposia on Electrochemical Impedance Spectroscopy, which have been held in Bombannes, France (1989) (5), Santa Barbara, USA (1992) (6), Ysermonde, Belgium (1995) (7), Angra dos Reis, Brazil (1998) (8), and Marilleva, Italy (2001) (9). The proceedings of an ASTM Symposium on "Electrochemical Impedance: Analysis and Interpretation," held in San Diego, USA, in 1991 contains papers dealing with modeling and corrosion processes, applications of the Kramers–Kronig transformations, corrosion and inhibition, corrosion of aluminum, corrosion of steel in soil and concrete, as well as protective coatings on metals (10). Mansfeld and Lorenz (11) have provided a review of the application of EIS in corrosion science and technology. The application of the EIS method in corrosion research has also been discussed in great detail in the proceedings of the Symposia on Electrochemical Methods in Corrosion Research (EMCR), held in Manchester, U.K. (1982), Toulouse, France (1985), Zurich, Switzerland (1988) (12), Espoo, Finland (1991) (13), Sesimbra, Portugal (1994) (14), Trento, Italy (1997) (15), and Budapest, Hungary (2000) (16). Recently, the EIS technique was discussed in the symposium on "New Trends in Electrochemical Impedance Spectroscopy (EIS) and Electrochemical Noise Analysis (ENA)" (17). Little et al. (18) and Cottis and Turgoose (19) have described examples of the application of EIS in the series *Corrosion Testing Made Easy*, published by NACE.

The equipment for the analysis of the dynamic behavior of an electrochemical system consists of a frequency analyzer (FRA), spectrum analyzer, or network analyzer combined with a digital regulation device to interface the electrochemical cell and a computer system for collection and analysis of the experimental impedance data. Great progress has been made in recent years in hardware and software for collection and analysis of impedance data. As a consequence the EIS technique has become a very valuable tool for the evaluation of a wide variety of corrosion systems as will be discussed below.

13.2 EXPERIMENTAL APPROACH FOR THE COLLECTION OF EIS DATA

The experimental approach for collecting EIS data has been discussed in detail in the chapter by Schiller et al. In general, a three-electrode cell is being used for the determination of impedance spectra; however, two-electrode cells with two electrodes of the same material and a reference electrode are also used since this arrangement has become a common method for the simultaneous collection of electrochemical potential and current noise data (20, 21). As discussed in the chapter on the electrochemical noise technique by Huet, it is possible to determine the noise impedance from the analysis of potential and current noise data in the frequency domain and compare this result with EIS data (22, 23).

A common arrangement for EIS studies is shown in Figure 13.1, where the potentiostat contains the FRA and the computer contains the software for data collection. The electrochemical cell shown in Figure 13.1 is commonly used with flat plate electrodes, especially polymer coated metals or anodized aluminum samples.

Modern equipment allows data collection at very high (up to 1 MHz) and very low (1 mHz or less) frequencies. In collecting EIS data at very high frequencies care has to be taken to eliminate the phase shift caused by the reference electrode. This can be done by using a pseudoreference electrode that consists of an inert metal wire that is capacitatively coupled to the reference electrode (24). At very low frequencies a compromise has to be made between reduction of scatter of the EIS data and the time necessary for data collection. The magnitude of the ac signal has to be low enough to ensure linearity of the system. Usually, ac signals of about 10 mV are applied. For high-impedance systems such as polymer coated metals or anodized aluminum, larger signals of about 50 mV or more can be used.

13.3 DISPLAY AND ANALYSIS OF EIS DATA

The experimentally determined complex impedance $Z(j\omega) = Z' + jZ''$ is usually displayed in complex plane plots (Z' vs. Z'') or in Bode plots ($\log|Z|$ vs. $\log \omega$ and phase angle ϕ vs. $\log \omega$, with ($|Z| = (Z'^2 + Z''^2)^{1/2}$, $\phi = \arctan(Z''/Z')$, and $\omega = 2\pi f$).

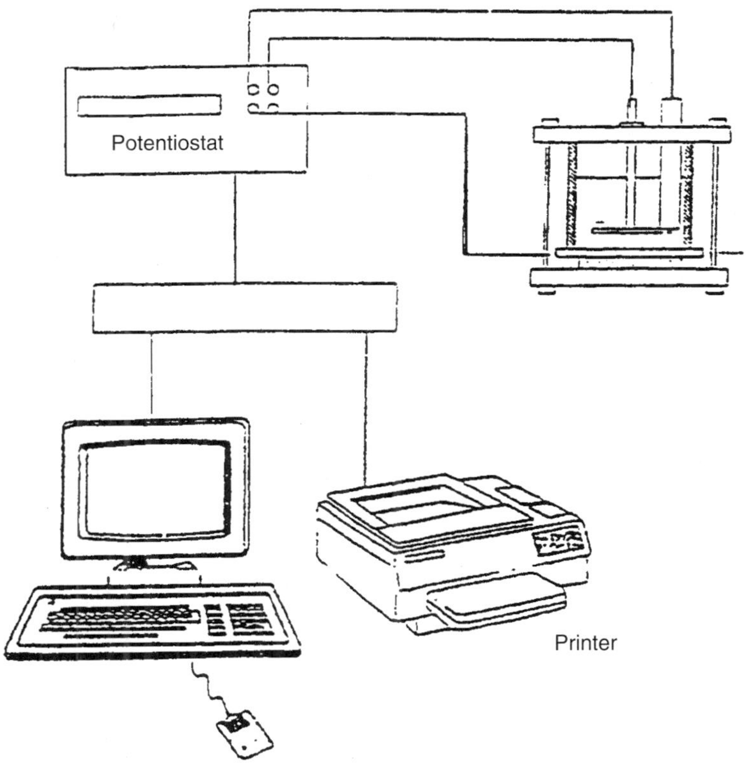

Figure 13.1 Experimental arrangement for the collection of EIS data.

Other possibilities for displaying experimental impedance data have been discussed elsewhere and will be described briefly below (25, 26).

Simple corrosion systems (strictly transfer-controlled corrosion reactions, uniform corrosion on homogenous surfaces) can be described by the simple equivalent circuit (EC) in Figure 13.2, which allows the establishment of correlations between electrochemical system parameters and characteristic impedance elements (25–27). All ohmic resistances in the measured system are contained in R_s. The corrosion current density (c.d.) i_{corr} is correlated with the polarization *n* resistance R_p by (28)

$$i_{corr} = \frac{B}{R_p} \tag{13.1}$$

where

$$B = \frac{b_a b_c}{2.3(b_a + b_c)}$$

In Equation (13.1), b_a and b_c are the Tafel slopes for the anodic and cathodic corrosion reaction, respectively. For very simple cases the capacitance C corresponds to the double-layer capacitance. The impedance for the EC in Figure 13.2 is given by

$$Z(j\omega) = R_s + \frac{R_p}{1 + j\omega C R_p} \tag{13.2}$$

with

$$Z' = R_s + \frac{R_p}{1 + (\omega C R_p)^2} \quad \text{and} \quad Z'' = -\frac{\omega C R_p^2}{1 + (\omega C R_p)^2}$$

Elimination of the frequency ω leads to

$$\left(Z' - R_s - \frac{R_p}{2}\right)^2 + (Z'')^2 = \left(\frac{R_p}{2}\right)^2 \tag{13.3}$$

which is the equation of a semicircle in the complex plane plot with R_p as diameter (Figure 13.3). The capacitance C can be determined from the resonance frequency ω_m at the maximum value of Z'':

$$C = \frac{1}{\omega_m R_p} \tag{13.4}$$

It is possible to use linear plots to determine the elements in Figure 13.2 (25, 26). A plot of Z' against $\omega Z''$ according to

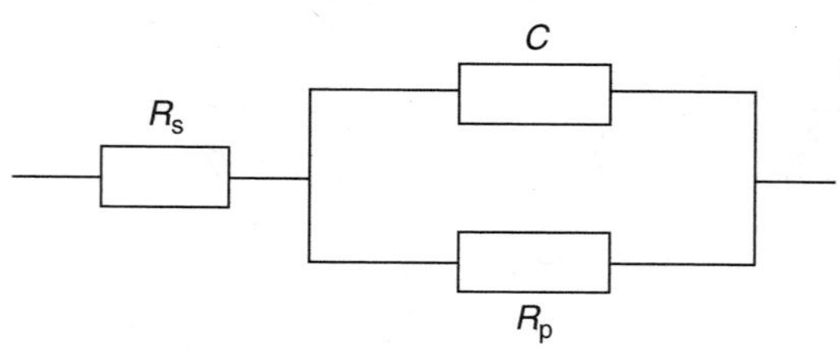

Figure 13.2 EC for simple corrosion systems.

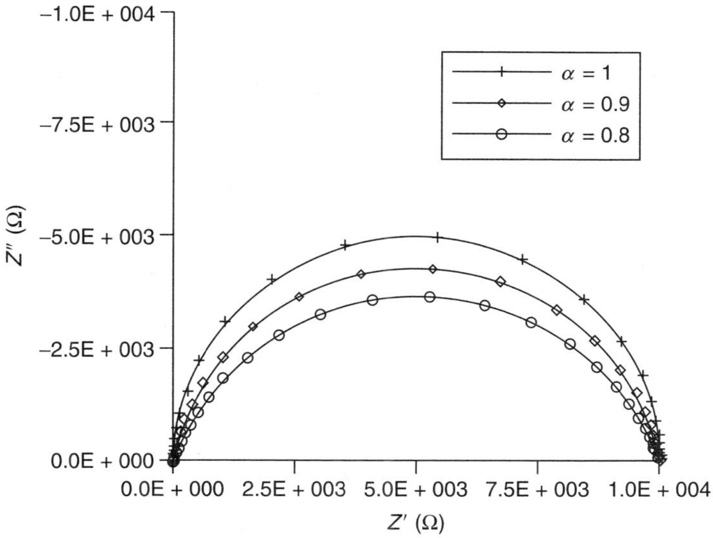

Figure 13.3 Complex plane plots for the EC in Figure 13.1; $R_s = 1\,\Omega$, $R_p = 10^4\,\Omega$, $C = 10^{-5}\,\mathrm{F}$, and $\alpha = 1.0$, 0.9, and 0.8.

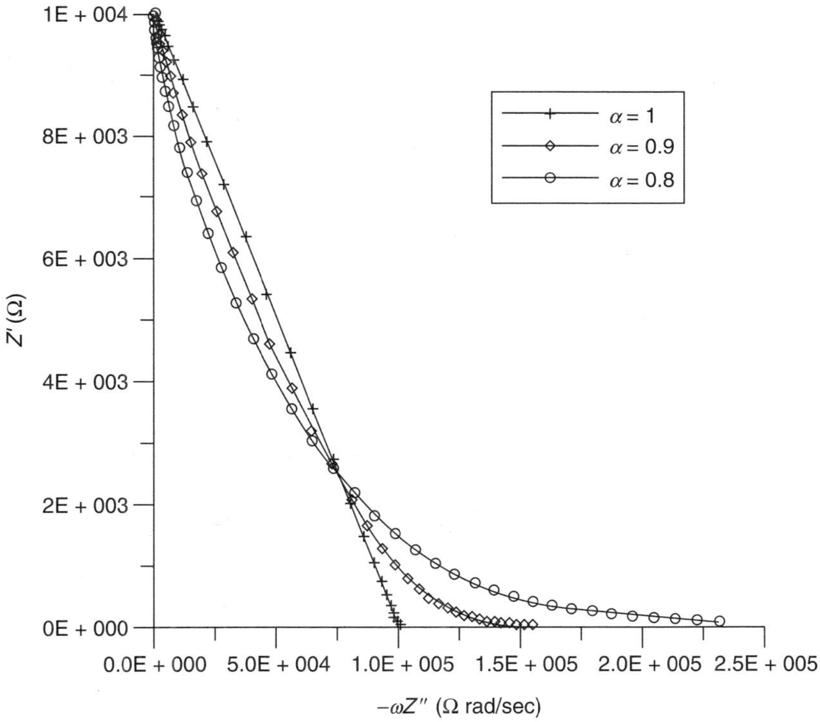

Figure 13.4 Determination of R_s, R_p, and C from linear plots; values as in Figure 13.2.

$$Z' = R_s + R_p - R_p \omega C |Z''| \tag{13.5}$$

leads to a straight line with a slope of $-R_pC$ and an intercept of $R_s + R_p$ for $\omega \rightarrow 0$ (Figure 13.4). If Z' is plotted against Z''/ω according to

$$Z' = R_s + \frac{|Z|''}{\omega R_p C} \tag{13.6}$$

R_s can be obtained from the intercept at $Z''/\omega = 0$. The capacitance C can be determined from the slope of the straight lines according to Equations (13.5) and (13.6) or from a $Z' - \omega Z''$ plot at $Z' = R_s$, since at this point $\omega Z'' = C^{-1}$. Similarly, for a $Z' - Z''/\omega$ plot at $Z'' = R_s + R_p$ one finds $Z''/\omega = R_p^2 C$.

For determination of the characteristic values R_s, R_p, and C it is useful to apply numerical fitting procedures such as the computer program CIRFIT (29) for the evaluation of complex plane plots. This is especially advantageous if the measured frequency range is not sufficient to produce a complete semicircle, if experimental scatter is to be eliminated, or when deviations from an ideal semicircle occur. Often experimental impedance plots obtained at E_{corr} have the shape of a depressed semicircle with the center of the circle below the real axis, as shown in Figure 13.3. Mansfeld and Kendig (30, 31) have accounted for such deviations from the ideal behavior by introducing an exponent α in Equation (13.2):

$$Z(j\omega) = R_s + \frac{R_p}{1 + (j\omega C R_p)^{\alpha}} \tag{13.7}$$

The use of the exponent α in Equation (13.7) is only a formal description of the experimental data, and the physical meaning of α is often not entirely clear. It will be noted that analysis of experimental impedance data according to Equation (13.7) usually leads to large values of the capacitance C. This situation suggests that most of the observed impedance data can be described better by the model of partially blocked surfaces of inhomogenous surfaces (32).

An integration method for the evaluation of experimental impedance data has been described by Mansfeld and Kendig (3). In this case the imaginary part of the impedance Z'' is plotted against log ω (Figure 13.5). By integration between $\omega \to \infty$ and $\omega = \omega_m$ (Equation (13.4)), R_p is determined according to the Kramers–Kronig relationship, assuming a symmetric function of Z'' with respect to ω_m:

$$R_p = \frac{4}{\pi} 2.3 \int_{\omega_m}^{\infty} Z''(\omega) d \log \omega \tag{13.8}$$

An advantage of this method is the more accurate determination of the capacitance $C = (1/\omega_m R_p)$, since ω_m can be determined more exactly by interpolation of the measured impedance data. The value of α can be calculated from R_p and Z''_m:

$$\alpha = \frac{4}{\pi} \arctan \frac{2|Z''_m|}{R_p} \tag{13.9}$$

Figure 13.5 shows the frequency dependence of Z''_m for $\alpha = 1.0$, 0.9, and 0.8 and the values of R_s, R_p, and C in Figure 13.3. An additional advantage of the integration method lies in the fact that only high-frequency data above $\omega = \omega_m$ are needed, which reduces the time required to carry out the measurement considerably (31, 33). However, this is possible only if it has been determined by preliminary tests that the corroding system can be described by the simple EC in Figure 13.2. In the case where two time constants occur, such as for polymer coated metals or certain corrosion processes in neutral aerated media (see below), the high-frequency time constant can be separated by a deconvolution method (26) and integration of the second time constant can be carried out separately. In the case where a maximum of

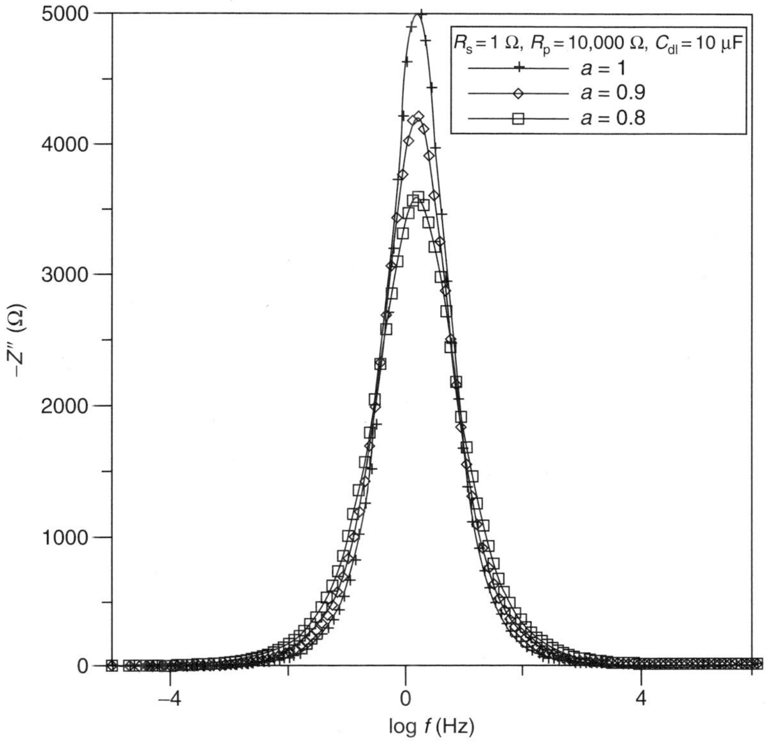

Figure 13.5 Illustration of the integration method (Equation (13.13)) for the EC in Figure 13.2 and the numerical values in Figure 13.3.

Z'' is not found in the measured frequency range, which could be due to very high values of R_p, or C, the integration method cannot be used. Mansfeld and Kendig (34) have proposed an extrapolation method in which a nonlinear least-squares analysis of the inverse of Equation (13.7) is performed to obtain the optimum least-square values of R_s, R_p, and C and α.

The integration method is especially suited for low-conductivity media since the frequency dependence of Z'' and therefore ω_m is independent of solution conductivity (R_s). This is shown in Figure 13.6(a) for an EC similar to that of Figure 13.2 for varying values of R_s. The corresponding Bode plots are shown in Figure 13.6(b). The Z'' vs. log ω plots show a sharp maximum of Z'' at ω_m, even for $R_s > R_p$. Experimentally, the evaluation of the data becomes more difficult with increasing R_s since the scatter of the data can increase. Mansfeld et al. (30, 34–36) have carried out impedance measurements in deionized water with or without inhibitors. John et al. (37) have demonstrated the use of impedance techniques for steel in concrete under immersed conditions.

Plots of $-Y''/\omega$ vs. Y'/ω as shown in Figure 13.7 are useful for accurate determination of the capacitance C, which is the diameter of the semicircle on the $-Y''/\omega$ axis. Y'' and Y' are the imaginary and real parts of the admittance, respectively. As for the linear plots of Z' vs. $-\omega Z''$ in Figure 13.4, even relatively small deviations from ideal capacitive behavior (i.e., $\alpha < 1$) lead to significant deviations from the straight line in Figure 13.4 or the semicircle in Figure 13.7, which makes accurate analysis of the experimental impedance data quite difficult.

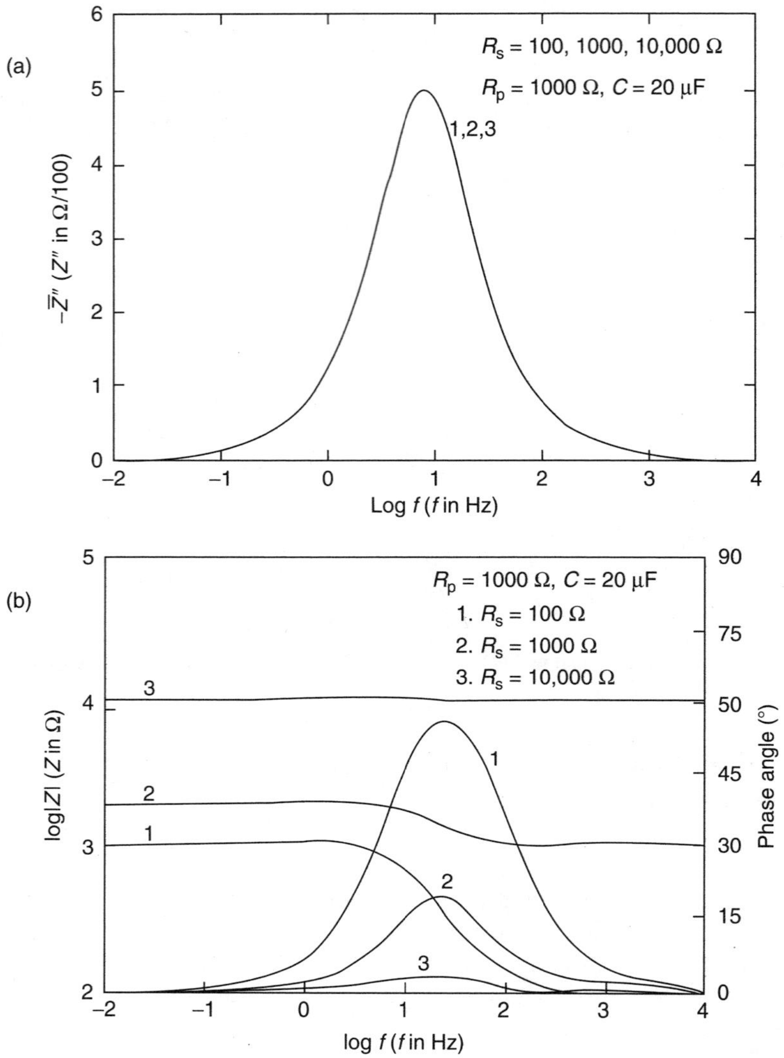

Figure 13.6 Z'' vs. $\log f$ as a function of R_s (a) and Bode plots for the same numerical values of R_s, R_p, and C (b).

The Bode plots corresponding to the EC in Figure 13.2 are shown in Figure 13.8 for $\alpha = 1$, 0.9, and 0.8, respectively. As for a complex plane plot, R_s and R_p can be determined from

$$R_s = \lim_{\omega \to \infty} |Z| \tag{13.10}$$

and

$$R_s + R_p = \lim_{\omega \to 0} |Z| \tag{13.11}$$

For $\alpha = 1$ the capacitance C can be determined from the impedance at $\omega = 1$ rad/sec, since at this frequency $\log|Z| = -\log C$.

For $\alpha = 1$ the breakpoint frequencies f_b^1 and f_b^2, for which $\phi = -45°$ (Figure 13.8), can be calculated as

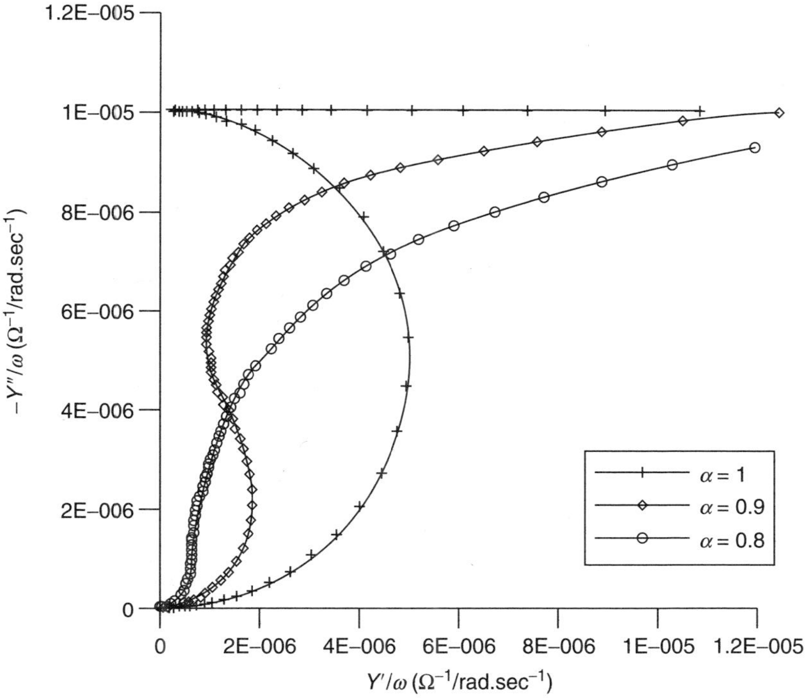

Figure 13.7 Plots of $-Y''/\omega$ vs. Y'/ω for the EC in Figure 13.2 and the numerical values in Figure 13.3.

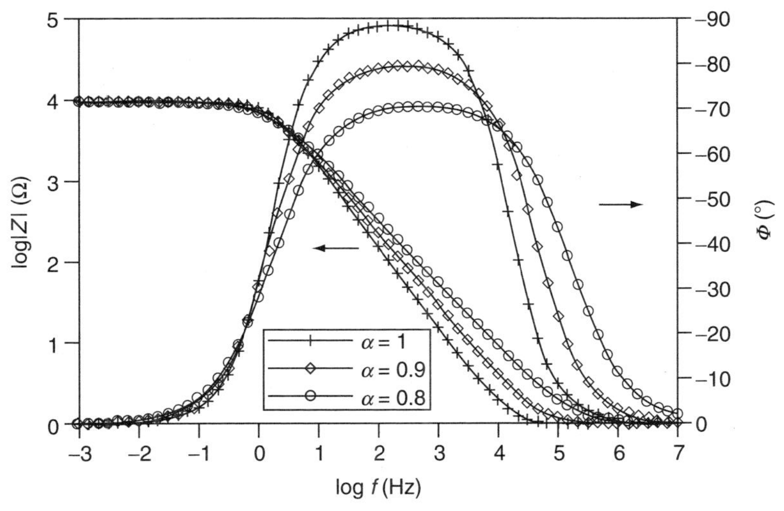

Figure 13.8 Bode plots for the EC in Figure 13.2 and the numerical values in Figure 13.3.

$$f_{\mathrm{b}}^{1,2} = \frac{1}{4\pi R_{\mathrm{p}} C}\left(1 \pm \sqrt{1 - 4\frac{R_{\mathrm{s}}}{R_{\mathrm{p}}} - 4\frac{R_{\mathrm{s}}^2}{R_{\mathrm{p}}^2}}\right) \tag{13.12}$$

The maximum of the absolute value of the phase angle $\phi_{\max}$ and the corresponding frequency $f_{\max}$ are given by

$$\phi_{max} = \arctan \frac{R_p}{2[R_s(R_s + R_p)]^{1/2}} \tag{13.13}$$

$$f_{max} = \frac{1}{2\pi C R_p}\sqrt{1 + \frac{R_p}{R_s}} \tag{13.14}$$

According to Equations (13.12) and (13.14), an increase of C at constants R_s and R_p moves both f_b^1 and f_b^2 as well as f_{max} to lower frequencies. A decrease of R_p increases f_b^2, while an increase of R_s decreases f_b^1 if the other two parameters do not change. In low-conductivity media (high values of R_s), ϕ_{max} can become less than 45°, which makes determination of f_b^1 and f_b^2 difficult, if not impossible.

In cases where mass transport becomes important, the Warburg impedance Z_w is placed in series with R_p, resulting in the Randles circuit (Figure 13.9a). Z_w is expressed as

$$Z_w = \sigma(1 - j)\omega^{1/2} \tag{13.15}$$

where σ is the Warburg coefficient. According to Equation (13.15), σ can be determined from the slope of the straight line in a Randles plot, which is a plot of Z' vs. $\omega^{-1/2}$. The open boundary finite element length diffusion (OFLD) element models the diffusion impedance when the diffusion layer has finite dimensions (Figure 13.9b). Using Boukamp's notation (38) the mass transport impedance is given by (39, 40):

$$Z_{OFLD} = \{\tan h(B(j\omega)^{1/2})\} / Y_0(j\omega)^{1/2} \tag{13.16}$$

where $B = l/(D)^{1/2}$ is the characteristic diffusion parameter, l is the diffusion length, D is the diffusion coefficient, and $Y_0 = (\sigma(2)^{1/2})^{-1}$. For infinite values of l the OFLD model becomes the Randles circuit. Mansfeld et al. (21) have used the OFLD model in the analysis of EIS data for polymer coated steel exposed to seawater.

Figure 13.9(c) shows a Bode plot, a complex plane plot, and a Randles plot for the parameters listed in the figure caption. The value of σ can be determined from the straight lines in the complex plane and Randles plots.

Constant phase elements (CPEs) (Figure 13.9b) have become common in fitting procedures of electrochemical impedance spectra. The impedance of a CPE is given by

$$Z(\text{CPE}) = Y_0^{-1}(j\omega)^{-n} \tag{13.17}$$

The term "constant phase element" was apparently first used by Brug et al. (41) in 1984, stating that "the CPE is generally believed to originate from a distribution in the current density along the electrode surface as a result of surface inhomogeneity."

While the use of CPEs usually increases the "goodness of the fit," the significance of the parameters Y_0 and n often is not discussed. Obviously, almost any spectrum can be fit to an EC containing a sufficiently large number of CPEs. However, this type of approach usually does not produce information concerning the physical properties of the system under investigation.

In studies of corroding systems, CPEs are most often used to describe the frequency dependence of nonideal capacitive behavior. The dimension of Y_0 is sec^n/Ω, while that of a capacitance is sec/Ω or F. Despite this difference Y_0 is often used as if it were the capacitance of the corroding system. Conversion of Y_0 data into C is very important when experimental capacitance data are to be used to determine quantitatively system parameters such as the thickness of anodized layers on Al alloys

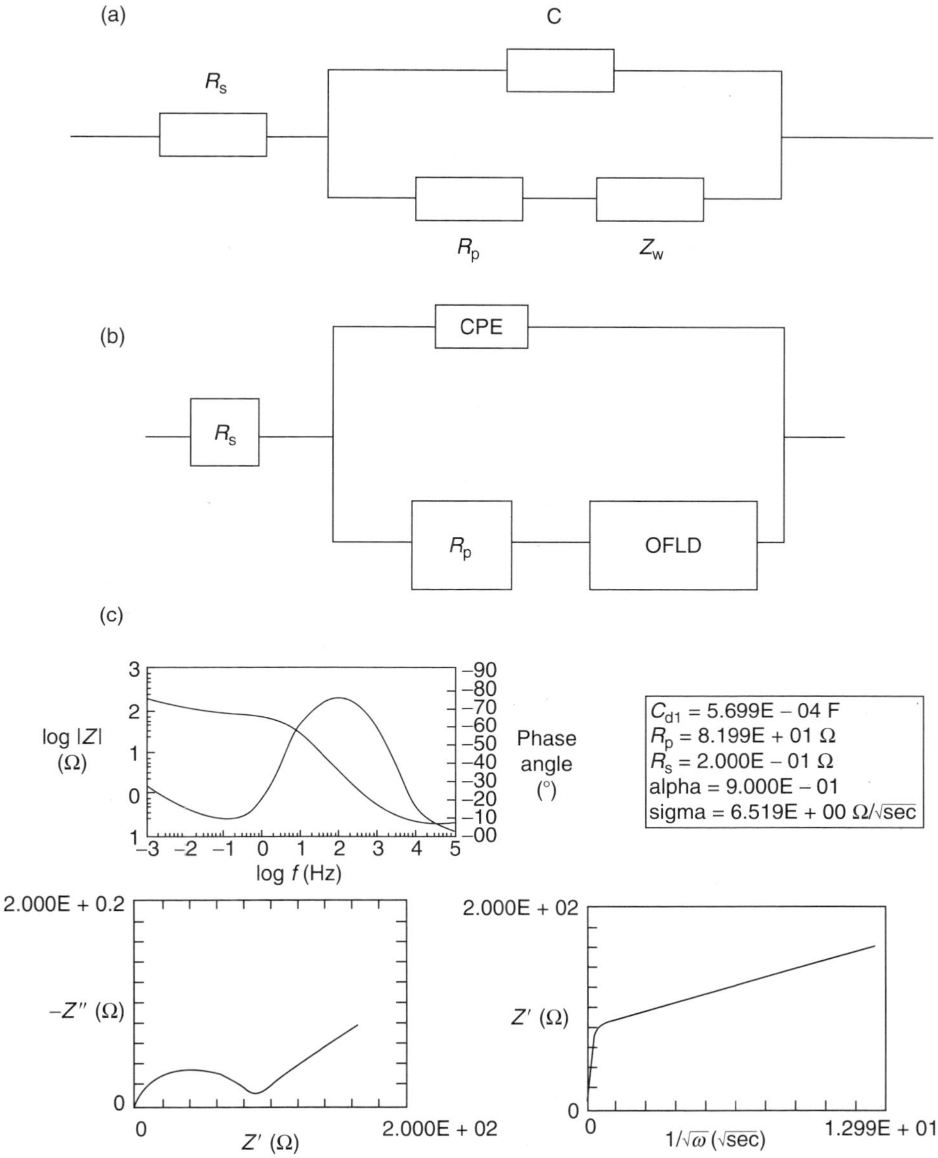

Figure 13.9 Randles circuit (a) and OFLD model (b) for systems with charge transfer and mass transport contributions. Bode plot, complex plane plot, and Randles plot for $R_s = 20\,\Omega$, $R_p = 83\,\Omega$, $C_{dl} = 570\,\mu F$, $\alpha = 0.90$, and $\sigma = 6.5\,\Omega/\text{sec}^{1/2}$ (c).

or the dielectric constant k of low-k materials. Calculations using Y_0 values instead of C will result in thickness values that are too low and k values that are too high.

A procedure for conversion of Y_0 into C was first described in the Ph.D. thesis of van Westing (42), who determined the water uptake of polymer coatings based on the time dependence of C. He suggested the following relationship:

$$C = Y_0(\omega)^{n-1}/\sin(n\pi/2) \tag{13.18}$$

The radial frequency ω (rad/sec) in Equation (13.18) was considered to be in the center of the logarithmic frequency range of the applied ac signal, that is, for a measurement between 1 and 10 kHz, f was given by $10^{3.5}$ Hz (42). Mertens et al. (43) suggested that the frequency to be used in Equation (13.18) should be the frequency at which the phase angle Φ is at a maximum. Most recently, Ilevbare and Scully (44) repeated this suggestion and proposed the use of Equation (13.18). However, detailed analysis has shown that Equation (13.18) is incorrect. As shown by Hsu and Mansfeld (45), the correct equation to be used to convert Y_o into C is

$$C = Y_o(\omega''_m)^{n-1} \tag{13.19}$$

In Equation (13.19) ω''_m is the frequency at which Z'' is at a maximum. Figure 13.10 shows the frequency dependence of Z'' and Φ. At ω''_m, which for a simple parallel R–C combination has the value $(RC)^{-1}$, $\Phi = -45°$ for $n = 1$, that is, ω''_m is also the breakpoint frequency for $n = 1$. It will be noted that ω''_m is independent of n, while the maximum of Φ changes with n. Obviously, C can be directly calculated from the experimental values of f''_m and R as

$$C = (2\pi f''_m R)^{-1} \tag{13.20}$$

Equation (13.19) is derived based on the fact that at ω''_m the real part of the impedance Z' is independent of n as shown in Figure 13.11. Therefore, the impedance of an ideal capacitance, $Z_C = (\omega C)^{-1}$, equals Z (CPE) Equation (13.17) at f''_m. For the parameters used for the simulations in Figure 13.10 and Figure 13.11, $Z'(f''_m) = R_p/2 = 5\,\text{k}\Omega$ independent of n, while for $n = 1$, $Z''(f''_m) = -Z'(f''_m)$ (45).

Table 13.1 lists simulation parameters (R_s, R_p, and C) and fit parameters (Y_o and n) for $R_s = 1\,\Omega$, $R_p = 10^4\,\Omega$, and $C = 10^{-5}$ F for different values of n. Conversion of Y_o into C using Equation (13.19) resulted in the correct value of C, while the use of Equation (13.18) produced some error (Table 13.1).

Software commonly used for the analysis of EIS data includes that provided by Boukamp (38). Mansfeld et al. have developed the ANALEIS software with modules for the analysis of one-time-constant (OTC) spectra (BASICS) as well as impedance spectra for polymer coated metals (COATFIT), anodized aluminum alloys

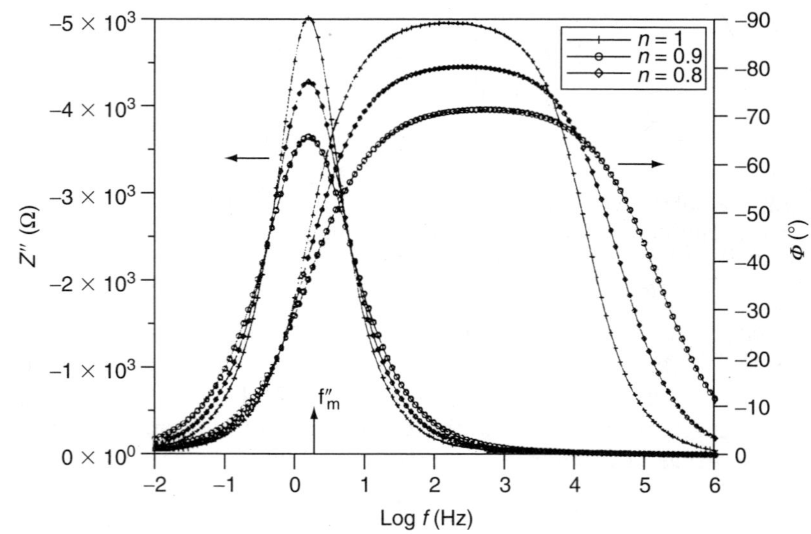

Figure 13.10 Frequency dependence of Z'' and Φ for different n values and the numerical values in Figure 13.5.

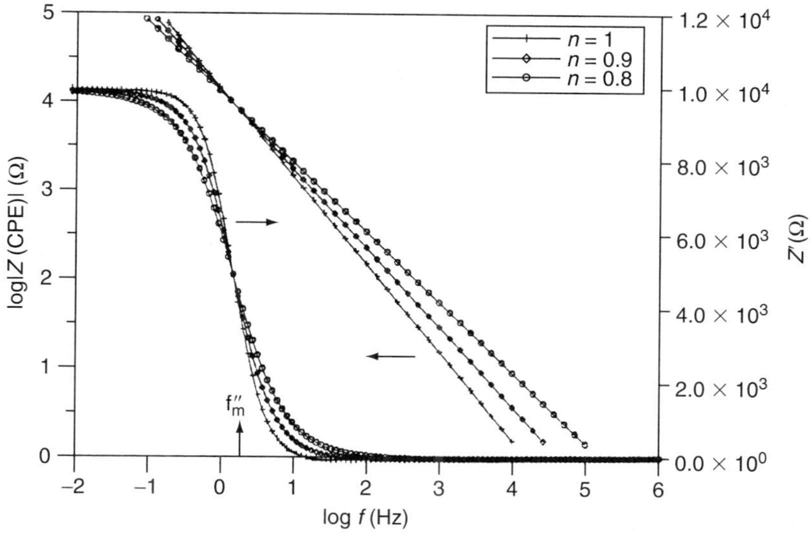

Figure 13.11 Frequency dependence of Z' and Z(CPE) for different n values and the numerical values in Figure 13.3.

Table 13.1 Calculation of C from Y

	R_s (Ω)	R_p (Ω)	C (F) or Y_0 (sec^n/Ω)	n	$C = Y_o\,(\omega''_m)^{n-1}$	$C = Y_o\,(\omega''_m)^{n-1}/\sin(n\pi/2)$
Simulation parameters	1	1×10^4	1×10^{-5}	1	—	—
Fit parameters	1	1×10^4	1×10^{-5}	1	—	—
Simulation parameters	1	1×10^4	1×10^{-5}	0.9	—	—
Fit parameters	1	1×10^4	1.2589×10^{-5}	0.9	1.0004×10^{-5}	1.0130×10^{-5}
Simulation parameters	1	1×10^4	1×10^{-5}	0.8	—	—
Fit parameters	1	1×10^4	1.5849×10^{-5}	0.8	1.0008×10^{-5}	1.0599×10^{-5}

(ANODAL), and pitting of aluminum alloys (PITFIT) (46–48). The manufacturers of hardware for EIS studies often also provide software that allows the users to design an appropriate EC and use it for data analysis.

13.4 APPLICATIONS OF EIS

The EIS technique has been applied successfully in many areas of corrosion research including determination of R_p, evaluation of the mechanism and efficiency of inhibitors, studies of the properties of protective polymer coatings and anodized layers and their changes with exposure to corrosive media, localized corrosion, as well as monitoring of microbiologically influenced corrosion (MIC) and corrosion inhibition by bacteria. Examples of the application of EIS in these areas will be given in the following, including a discussion of the models that have been used to describe the metal/electrolyte system in each case and to analyze the experimental impedance spectra. The proceedings of the EIS (5–9) and EMCR symposia (12–16) provide evidence of many successful applications of EIS.

13.4.1 Determination of the Polarization Resistance

For a simple OTC system, for which the EC in Figure 13.2 applies, R_p can be determined as the dc limit of the impedance according to Equation (13.11). Epelboin et al. (49) summarized the status of the use of the ac impedance technique for corrosion studies and corrosion rate determination in an ASTM symposium that was held in 1979. The classic system Fe/H_2SO_4 + propargylic alcohol was discussed in detail. In this system two inductive loops appear at the lowest frequencies, which makes the accurate determination of R_p difficult (Figure 13.12a). Epelboin et al. suggested that the transfer resistance R_t rather than R_p should be used to determine corrosion rates and concluded that R_t provided better agreement with weight loss data than R_p, as shown in Figure 13.12(b), which contains values of the inhibition efficiency determined by different methods. Figure 13.13(a) for the Fe/H_2SO_4 system demonstrates the manner in which R_t and R_p were determined, while Figure 13.13(b) shows the EC and the fit parameters for the experimental data in Figure 13.13(a) (49). Lorenz and Mansfeld (50) pointed out later that the Fe/H_2SO_4 + propargylic alcohol system, which for a long time was considered a model system for ac impedance studies, was in fact quite complicated due to the fact that the inhibitor desorption potential was close to the corrosion potential E_{corr}, which makes the use of dc or ac techniques for the determination of corrosion rates and inhibition efficiency very difficult.

13.4.2 Corrosion Inhibition

A very successful application of EIS has been in the study of the mechanism(s) of corrosion inhibition. The changes of R_p due to decreases of the corrosion rate and the electrode capacitance due to adsorption of the inhibitor molecules can be used to determine parameters such as corrosion inhibition efficiency as well as degree of

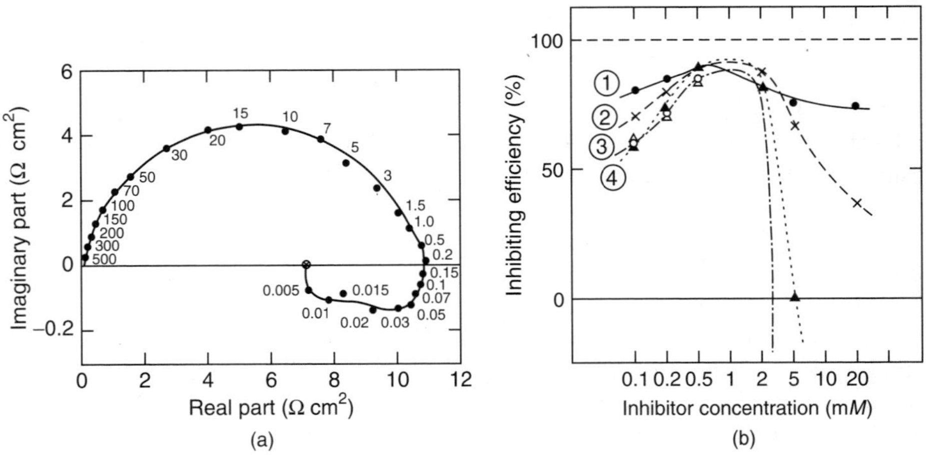

Figure 13.12 Complex plane plot for $Fe/0.5\,M\;H_2SO_4 + 2\,mM$ propargylic alcohol (a) and inhibition efficiency determined from weight loss (1), R_t (2), R_p (3), and C_{dl} measurements (4) (b). (From I. Epelboin, C. Gabrielli, and M. Keddam, H. Takeuochi *ASTM STP* 727, 150 [1981]. With permission.)

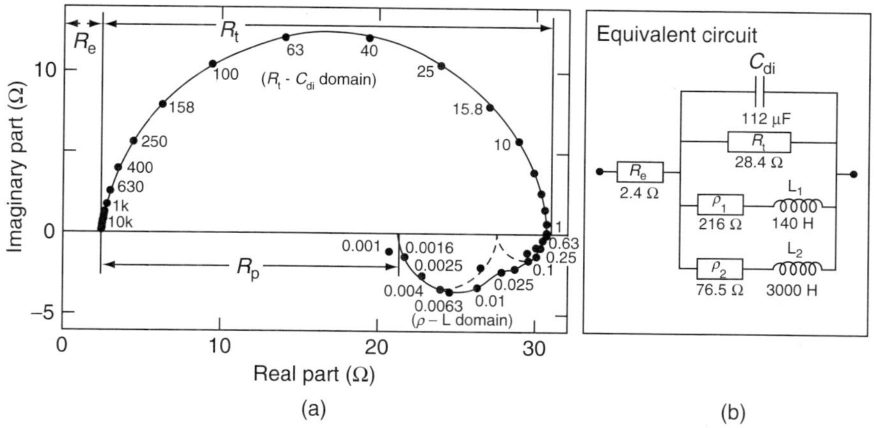

Figure 13.13 Complex plane plot for Fe/0.5 *M* H_2SO_4 (deaerated) (a) and EC with fit parameters (b). (From I. Epelboin, C. Gabrielli, and M. Keddam, H. Takenouti, *ASTM STP* 727, 150 [1981]. With permission.)

coverage with an inhibitor film. Changes of the corrosion mechanism can be detected by characteristic changes in the impedance spectra, for example, changes from a simple OTC model (OCTM) to a model that is typically found for (porous) polymer coatings on metals in neutral, aerated media (51). This type of inhibition has been called "interphase inhibition" by Lorenz and Mansfeld (52), who distinguished "interphase inhibition" from the "interface inhibition" that is usually observed in acid media. Mansfeld et al. (51) have presented an example for interphase inhibition for 4340 steel in 0.5 *M* Na_2SO_4 containing an inhibitor that was a mixture of phosphonic acid and fatty amine.

Lorenz and Mansfeld (50) investigated the system Fe/H_2SO_4 containing triphenylbenzylphosphonium chloride ($TPBP^+$) and other organic inhibitors. In addition to impedance spectra, data from small- and large-pulse polarization, R_p measurements, and solution analysis by atomic absorption were reported (11, 50). Figure 13.14 shows complex plane plots obtained at E_{corr} for an iron rotating disk electrode in 0.5 *M* H_2SO_4 in the absence and presence of $TPBP^+$. A depressed semicircle and an inductive loop were observed that were more pronounced in the presence of $TPBP^+$. Very good agreement between the results obtained with dc measurements (Tafel extrapolation and R_p from a linear sweep through E_{corr}), ac measurements (R_p according to Equation [13.11]), and solution analysis was found. The AA results were converted into R_p for $13\,\text{mV} < B < 17.4\,\text{mV}$ according to Equation (13.1). All polarization resistance values obtained by these different methods were lower than the intercept of the experimental data with the real axis of the complex plane plot that Epelboin et al. identified with R_t (49, 53). The difference between R_p and R_t increased with increasing inhibitor concentration. Lorenz and Mansfeld (50) concluded that in inhibitor-free solution and in the presence of $TPBP^+$ the corrosion rate was correlated with R_p and not with R_t as defined by Epelboin et al. (49, 53).

Nagiub and Mansfeld (54) studied the corrosion behavior of cartridge brass in 3% NaCl and in artificial seawater (AS) using EIS and electrochemical noise analysis (ENA). Benzotriazole (BTA), gluconic acid sodium salt (GASS), and polyphosphoric acid sodium salt (PP) were evaluated as corrosion inhibitors. While BTA provided excellent corrosion inhibition in both media, PP was not very effective. For GASS, maximum corrosion inhibition was obtained at a concentration of 0.01 *M*.

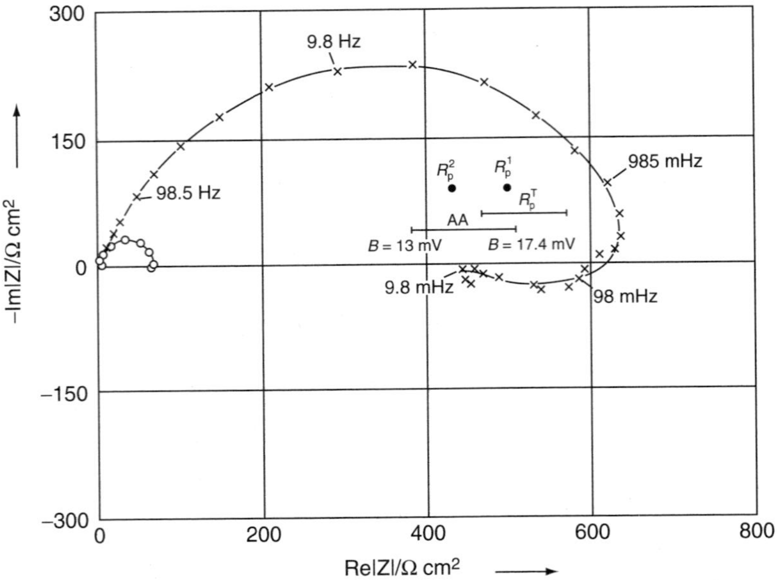

Figure 13.14 Complex plane plot for Fe/0.5 *M* H_2SO_4 + *x* m*M* $TPBP^+$, aerated, $x = 0$ (o) or 10 (*x*); R_p^T from linear sweep through E_{corr}, R_p^1 from galvanostatic polarization curves, R_p^2 from potentiostatic polarization curves, AA from atomic absorption data. (From W. J. Lorenz and F. Mansfeld, *Corros. Sci.* 21, 647 [1981]. With permission.)

In the Bode plots obtained for brass in 3% NaCl solution only one time constant was observed for the solution containing BTA with the minimum of the phase angle close to −90° and R_p close to $10^6\,\Omega$. For the solution without BTA a diffusion process was observed with the minimum of the phase angle of about −45° and R_p close to $10^3\,\Omega$. The impedance increased with time and the maximum R_p was obtained after 4 days, which indicates improvement of corrosion inhibition by BTA with time. A similar effect was observed for brass in AS solution (Figure 13.15). R_p was about $10^6\,\Omega$ for AS with BTA and about $10^3\,\Omega$ for AS without BTA. This result demonstrates the powerful effect of BTA as a corrosion inhibitor for brass in NaCl and AS.

Figure 13.16 and Figure 13.17 show experimental and fitted impedance data for brass exposed to AS + 0.01 *M* BTA for 2 days and to 3% NaCl for 4 days, respectively. Excellent agreement between the fitted and experimental data was obtained. For NaCl and AS solutions in the presence of BTA only one time constant was observed and the spectra were fitted using the OTCM shown in Figure 13.2. For brass in AS and NaCl a diffusion-controlled process was observed and the impedance spectra were fitted using the OFLD model (Figure 13.9b).

13.4.3 Corrosion Protection by Polymer Coatings

One of the most successful applications of EIS has been in the evaluation of protective polymer coatings on metals and their degradation during exposure to corrosive environments (3,11,55–57). The earliest applications of EIS over a wide range of frequencies in the study of polymer were discussed by Menges and Schneider (58, 59) and Kendig and Leidheiser, who presented their data in terms of a complex dielectric constant (60, 61). Bonora and coworkers (62–65) as well as de Wit and

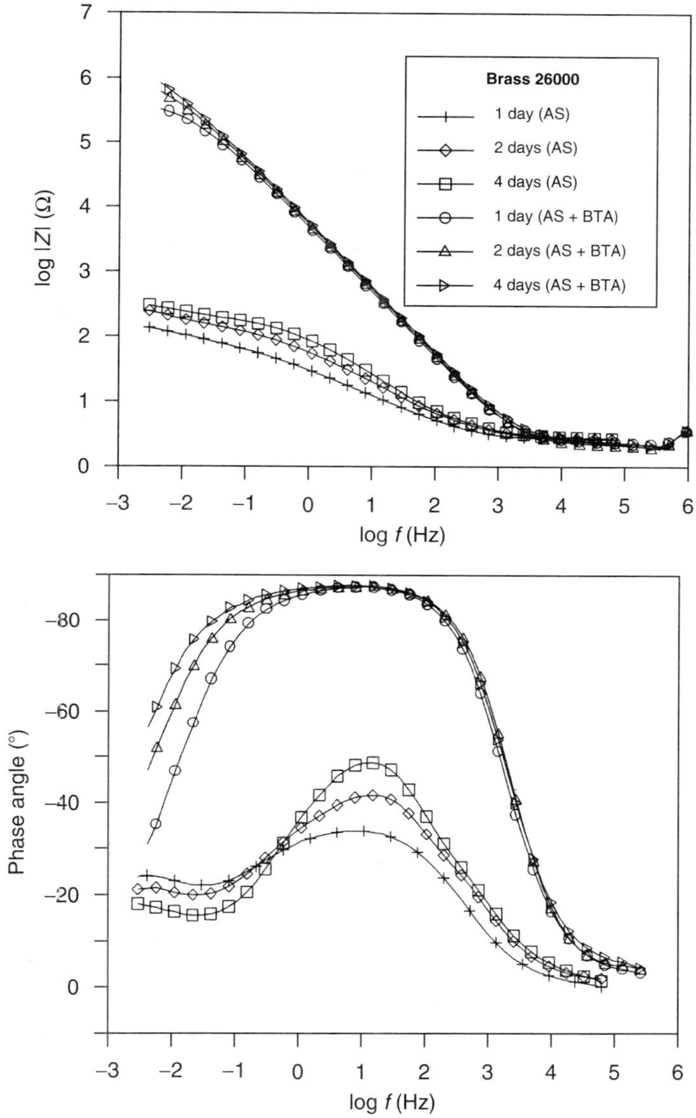

Figure 13.15 Bode plots for cartridge brass exposed to AS with and without 0.01 *M* BTA. (From A. Nagiub and F. Mansfeld, *Corros. Sci.* 43, 2147 [2001]. With permission.)

coworkers (66–68) have used EIS extensively to evaluate water uptake, coating degradation, the effects of mechanical deformation, etc. Walter (69), Rammelt and Reinhard (70) and Mansfeld (55) have provided reviews of the use of EIS in the study of polymer coated metals. Scully (71) and Kendig and Scully (72) have correlated short-term EIS studies with long-term coating performance. Murray (73) has given an extensive review citing a large number of references published between 1989 and 1994. The "coating model" shown in Figure 13.18(a) (73) has been used by many investigators. In this EC, C_c is the coating capacitance, R_{po} is the pore resistance, while C_{dl} and R_p are the capacitance and polarization resistance of the delaminated area, respectively, at which corrosion occurs. Figure 13.18(b) shows a Bode plot for an intact coating (curve 1) and a coating that has suffered some degradation (curve 2).

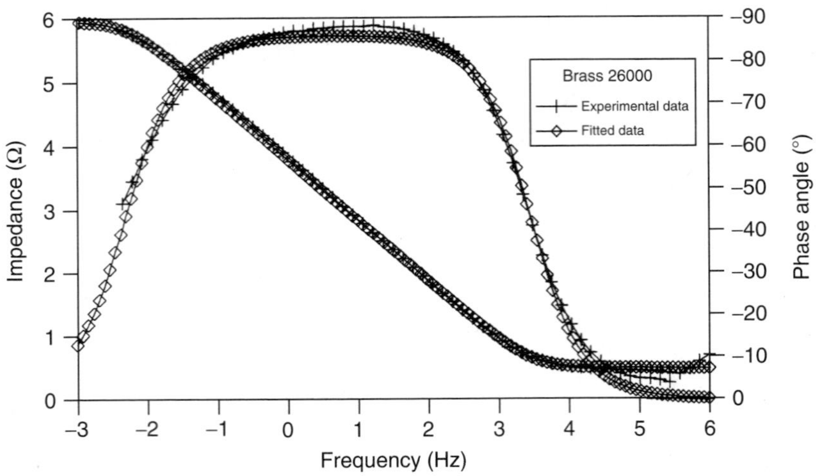

Figure 13.16 Comparison of experimental and fit data for brass exposed to AS + 0.01 *M* BTA for 2 days. (From A. Nagiub and F. Mansfeld, *Corros. Sci.* 43, 2147 [2001]. With permission.)

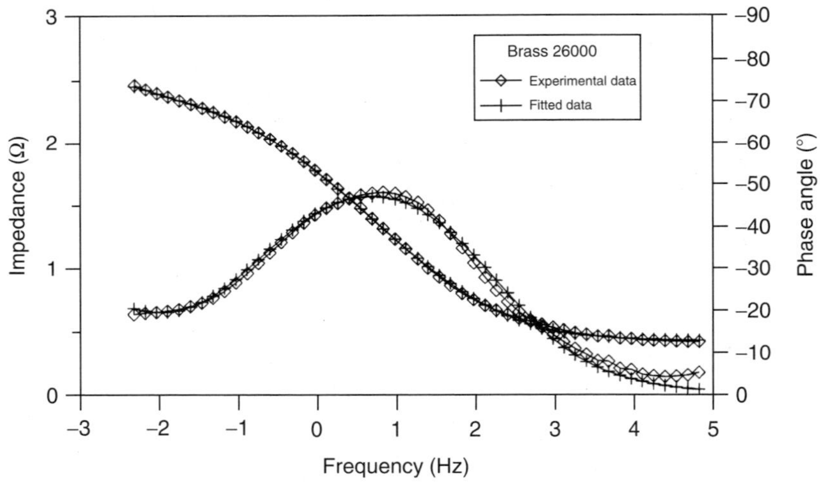

Figure 13.17 Comparison of experimental and fit data for brass exposed to 3% NaCl for 4 days. (From A. Nagiub and F. Mansfeld, *Corros. Sci.* 43, 2147 [2001]. With permission.)

Mansfeld and Lee (74, 75) have used theoretical impedance spectra such as those shown in Figure 13.18(b) to train artificial neural networks (ANN) to distinguish between good, intermediate, and poor conditions of in-service coating systems. The results of an evaluation of polymer coating properties with EIS, ENA, and ANN have been presented (75).

Water uptake of the coating in the initial stages of exposure can be followed by observing the changes of $C_c = \varepsilon\varepsilon_0 A/d$, where ε is the dielectric constant of the coating, $\varepsilon_0 = 8.84 \times 10^{-14}$ F/cm and d its thickness, and A is the exposed area (55, 76). The rate at which R_{po} decreases with increasing exposure time has been used by Mansfeld et al. (77, 78) to rank the efficiency of different surface pretreatment procedures for steel in providing corrosion protection. The change of C_{dl} with time can be used to estimate the delaminated or corroding area A_{corr} and to

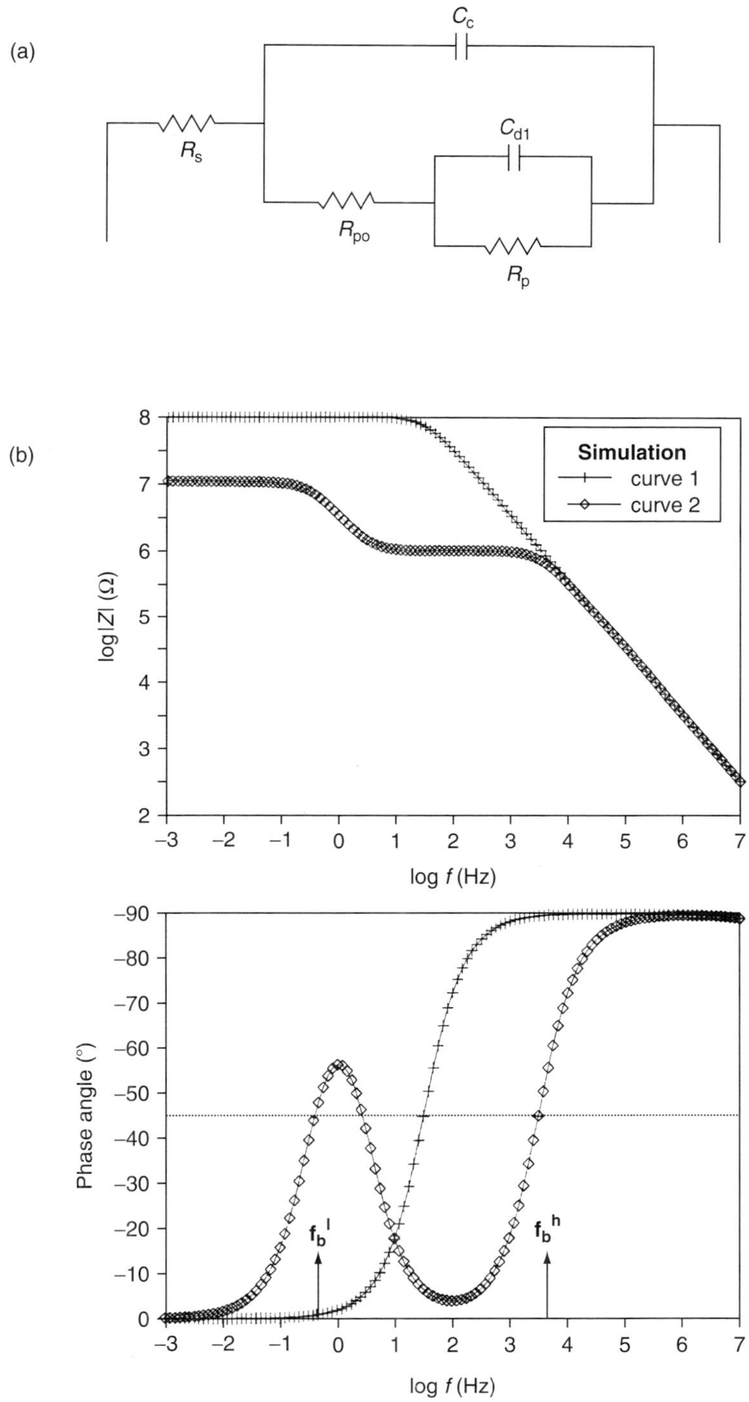

Figure 13.18 EC for the impedance of polymer coated metals (a) and Bode plot for the intact coating (curve 1) and damaged coating (curve 2) (b); curve 1: $R_s = 5\,\Omega$, $R_p = 1.10^8\,\Omega$, $C = 5.10^{-11}$ F; curve 2: $R_s = 5\,\Omega$, $R_p = 1.10^7\,\Omega$, $C_c = 5.10^{-11}$ F, $R_{po} = 1.10^6\,\Omega$, $C_{dl} = 5.10^{-8}$ F, $A = 1\,\text{cm}^2$.

calculate its specific polarization resistance R^0_{po} (ohm cm^2). Assuming $C^0_{dl} = 20\,\mu F/cm^2$, A_{corr} can be calculated as $A_{corr} = C_{dl}/C^0_{dl}$. The delamination ratio or damage function $\Delta = A_{corr}/A$, where A is the total exposed area of the polymer coated metal, has been used by Mansfeld et al. (21) for lifetime prediction of practical coating systems. For the example in Figure 13.18(b) $A_{corr} = 2.5 \times 10^{-3}\,cm^2$, $\Delta = 0.0025$, and $R^0_p = R_p A_{corr} = 2.5 \times 10^3$ ohm cm^2, which is similar to the polarization resistance commonly observed for bare steel corroding in neutral, aerated solutions.

The software module COATFIT (46–48) that has been developed for the analysis of EIS data that agree with the EC in Figure 13.18 has the advantage that fairly accurate starting values of the different fit parameters are not needed. Tsai and Mansfeld (79, 80) have discussed several parameters, including the breakpoint frequency f_b, that are useful indicators of coating delamination and corrosion at the coating/metal interface and can be determined directly from the experimental impedance spectra without the need for a quantitative analysis of the spectrum, as also discussed by Haruyama et al. (81) and Hack and Scully (82). Tsai and Mansfeld (80) have discussed the use of f_b as the basis of a method for field testing of protective coatings. In Figure 13.18(b) $f^h_b = (2\pi R_{po} C_c)^{-1}$ is the high-frequency breakpoint frequency that, for the parameters used to calculate the spectrum in curve 2, is determined as 3184 Hz. The low-frequency breakpoint frequency $f^l_b = (2\pi R_{po} C_{dl})^{-1} = k^l_b = 0.32$ Hz is independent of A_{corr}. The delaminated or corroding area A_{corr} and the damage function Δ can be determined from f^h_b, which is defined as (after some simplifications)

$$f^h_b = (2\pi R_{po} C_c)^{-1} = A_{corr}(2\pi R^0_{po} C_c)^{-1} = \Delta(2\pi R^0_{po} C^0_c)^{-1} = k^1_b \Delta \qquad (13.21)$$

where C^0_c is the specific coating capacitance (F/cm^2), which can be calculated for known values of ε and d, and R^0_{po} is the specific pore resistance. Figure 13.19 gives a comparison of the values of the damage function $\Delta = A_{corr}/A$ calculated according to Equation (13.21) and the results of visual observation according to ASTM D 610 for a Zn-rich primer/epoxy polyamide/latex coating system on hot-rolled steel (CR6) that had been exposed to natural seawater (NS) at Port Hueneme, CA (PH), and to

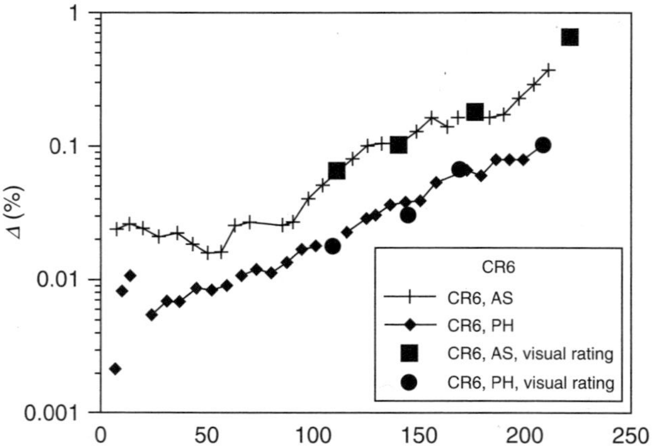

Figure 13.19 Time dependence of damage function Δ calculated from the experimental values of the breakpoint frequency f^h_b and by visual observation for a polymer coated steel exposed at Port Hueneme, (PH) (From F. Mansfeld, L. T. Han, C. C. Lee, and G. Zhang, *Electrochim. Acta* 43, 2933 [1998]. With permission.)

artificial seawater (AS) (21). The parameter k_b^1 in Equation (13.21) was determined by using the results of visual observation after 120 days to normalize the EIS data. Δ was found to increase continuously as coating damage increased. By fitting the experimental values to

$$\log \Delta = a + bt \tag{13.22}$$

or

$$\log \Delta = a(t - t_0) \tag{13.23}$$

where t_0 is the time for which $\Delta = 1\%$ (83), the possible lifetime of a coating system can be estimated.

While good agreement has been found between the experimental impedance spectra and the EC in Figure 13.18(a) for the thin ($d = 6$–$19\,\mu$m) polybutadiene coatings studied by Kendig et al. (78), better agreement of the EIS data with the OFLD model in Figure 13.9(b) has been found for thicker coating systems such as CR6 with a thickness of about 250 μm, as shown in Figure 13.19. Figure 13.20

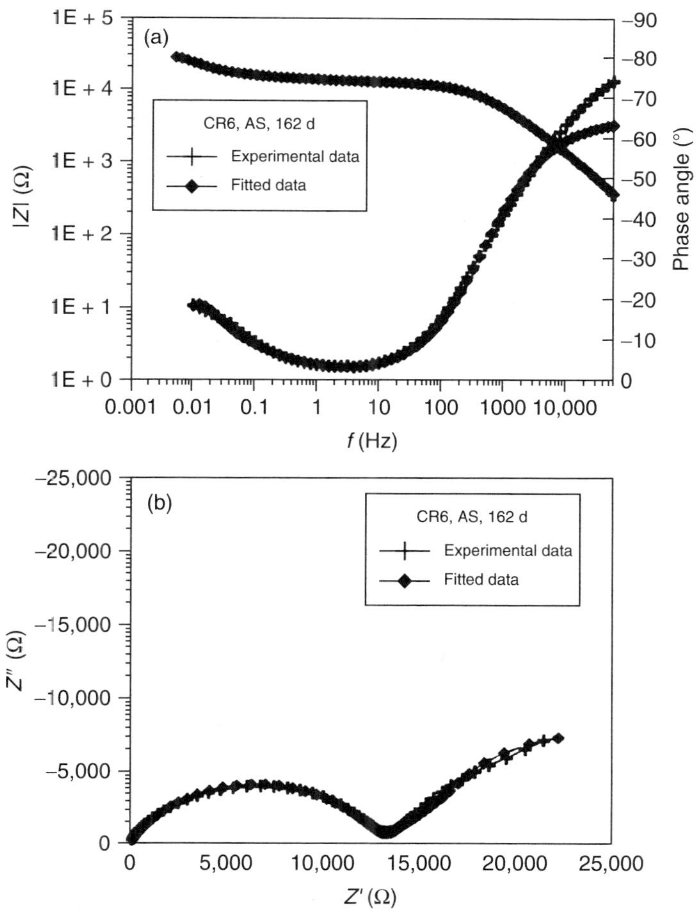

Figure 13.20 Comparison of experimental and fit data for coating system CR6 exposed to AS for 162 days. (From F. Mansfeld, L. T. Han, C. C. Lee, and G. Zhang, *Electrochim. Acta* 43, 2933 [1998]. With permission.)

gives a comparison of experimental and fit data for CR6 exposed to AS for 162 days. Very good agreement between the two sets of data was observed except for the deviation at the highest frequencies, which was considered to be due to instrumental artifacts.

13.4.4 Anodized Aluminum

EIS has been found to be a very useful tool for the evaluation of the properties of anodized Al alloys (84–87). The different processes that are used to seal the anodized samples have been evaluated by a number of researchers including a Spanish group (88–91). Mansfeld et al. (92) have proposed the EC in Figure 13.21(a) for anodized and hot water sealed Al samples. In this EC R_b is the resistance of the inner barrier layer and C_b is its capacitance. The pore resistance R_{po} has been replaced by a CPE $Z_{po} = K(j\omega)^n$, where K and n are fit parameters, to account for the statistical variations of the pore properties such as pore dimensions and pore solution conductivity (92). C_{po} is the capacitance of the outer porous oxide layer. C_b and C_{po} are actually determined as CPEs; however, it has been found that the deviations from ideal behavior were quite small, with n being close to -1 (92). The software module ANODAL (46–48) can be used to analyze impedance spectra for anodized, sealed Al alloys that agree with the EC in Figure 13.21(a). Figure 13.21(b) shows excellent

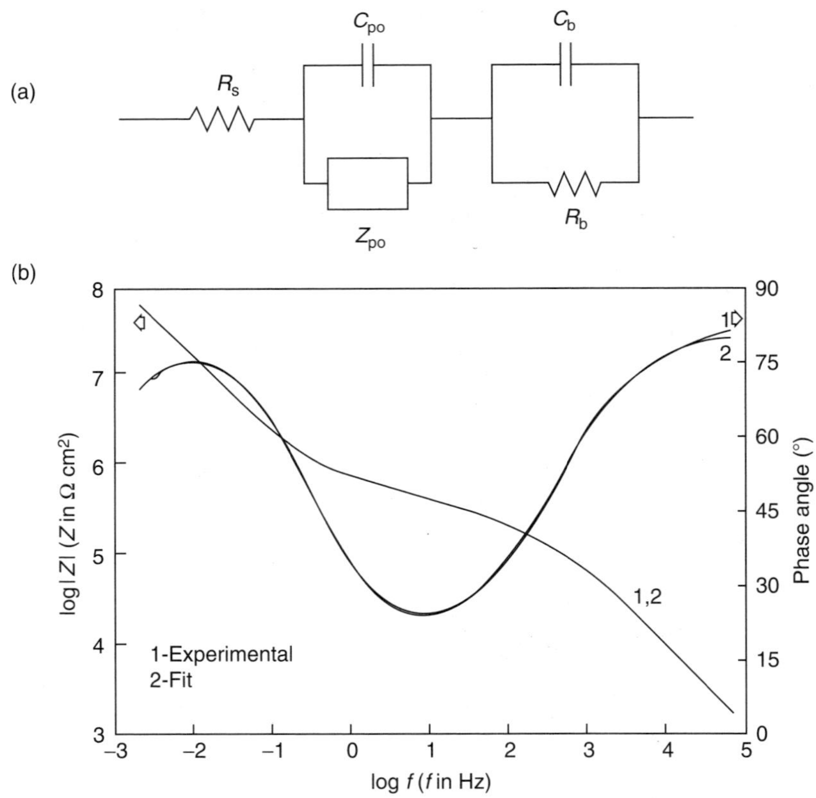

Figure 13.21 EC for anodized Al alloys (a) and results of the fit of the experimental data for Al 6061 (SAA + HWS) to the EC (b).

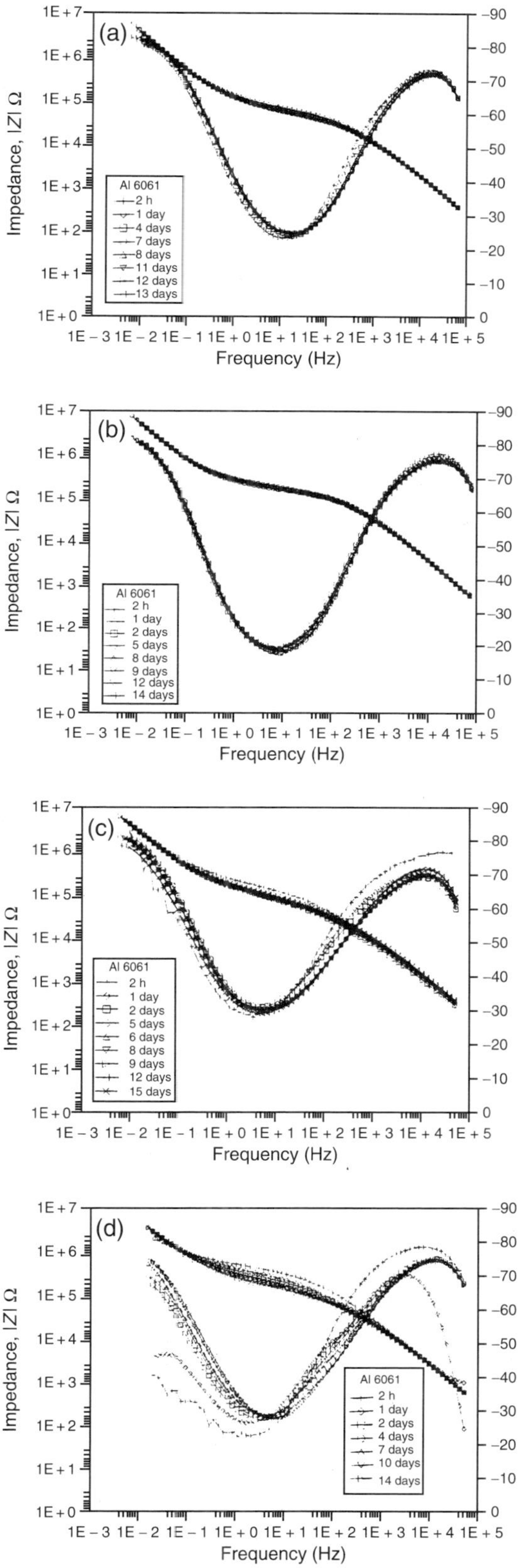

Figure 13.22 Impedance spectra for Al 6061 anodized in 15 wt% H_2SO_4 and sealed in boiling solutions of water (a), nickel acetate (b), cerium acetate (c), and yttrium acetate (d); exposure to 0.5 *N* NaCl. (From F. Mansfeld, G. Zhang, and C. Chen, *Plating and Surface Finish.* 84(12), 72 (1997). With permission.)

agreement between the experimental data for Al 6061 that was anodized in sulfuric acid (SAA) and hot water sealed (HWS). From the fitted values of C_b and C_{po} in Figure 13.21(b) the barrier layer thickness was calculated as 250 Å using $\varepsilon = 10$ and the porous oxide layer thickness as 20 μm for $\varepsilon = 55$. SEM observation of the cross-section of the anodized sample gave values between 20 and 21 μm.

Impedance spectra for Al 6061 that was anodized in 15 wt% H_2SO_4 and sealed in boiling solutions of water (a), nickel acetate (b), cerium acetate (c), and yttrium acetate (d) are shown in Figure 13.22. The very stable spectra obtained during exposure to 0.5 *N* NaCl for 13–15 days demonstrate the outstanding corrosion resistance of these SAA + HWS samples. Sealing in cerium acetate solutions produced surfaces with a corrosion resistance comparable to that determined for sealing in nickel acetate (92). Cerium and other rare earth metal salts have been proposed as alternative sealing solutions replacing the toxic chromate and nickel compounds used at present (93, 94).

For samples that were sealed in chromate solutions R_{po} is very low and the impedance spectra are dominated by the capacitance of the inner barrier layer (C_b). The spectra in Figure 13.23 for SAA Al 6061 that was sealed in a dilute chromate solution demonstrate the excellent corrosion resistance obtained in this process (92). No pits were observed after exposure to 0.5 *N* NaCl for 7 days.

13.4.5 Pitting of Aluminum Alloys

Mansfeld and coworkers have demonstrated that EIS is also very useful for detecting and monitoring of localized corrosion of Al alloys. The EC shown in Figure 13.24 is in agreement with the large number of EIS data that have been collected for Al alloys and Al-based metal matrix composites (95–107). It can be used to determine pit growth rates from impedance spectra collected at E_{corr} (48, 105). Impedance spectra

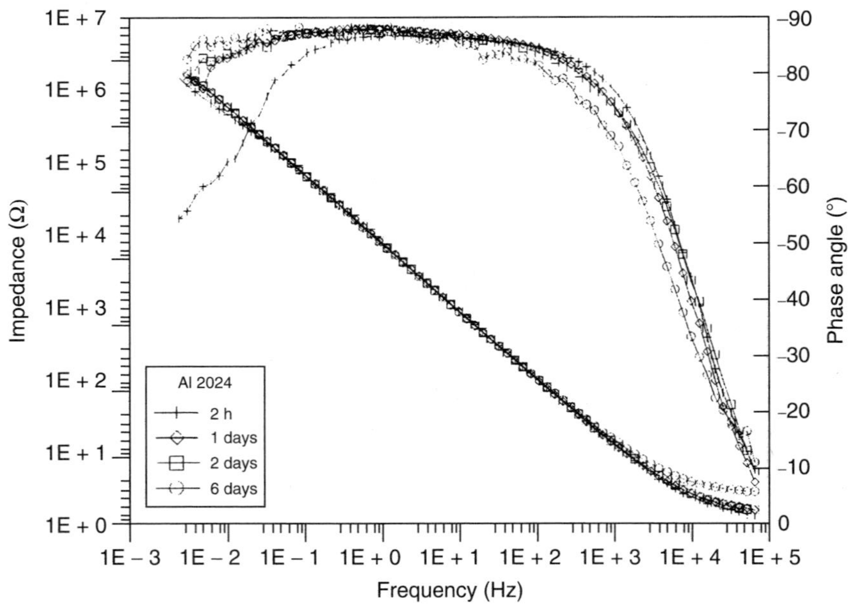

Figure 13.23 Impedance spectra for Al 6061 (SSA + HWS) during exposure to 0.5 *N* NaCl. (From F. Mansfeld, G. Zhang, and C. Chen, *Plating and Surface Finish*. 84(12), 72 (1997). With permission.)

that agree with the EC in Figure 13.24 can be analyzed wit the software module PITFIT (46–48).

In the model shown in Figure 13.24, R_p and C_p are the polarization resistance and capacitance, respectively, of the passive surface, while R_{pit} and C_{pit} are the corresponding values for the area where active pitting occurs. The frequency dependence of the impedance in the low-frequency range is described by the term $W = K(j\omega)^n/F$, where K and n are experimental parameters and F is the fraction of the surface area on which pitting occurs ($0 \leq F \leq 1$).

Theoretical spectra for $F = 0$ and $F = 0.005$ and for parameters in the EC that have been found to be representative of the pitting process for Al alloys are shown in Figure 13.25 (95–107). The decrease of the impedance in the capacitive region (i.e., the increase of the total capacitance), the change of the frequency dependence at very low frequencies, and the occurrence of a second maximum of the phase angle at these frequencies are characteristic for the pitting process for Al alloys.

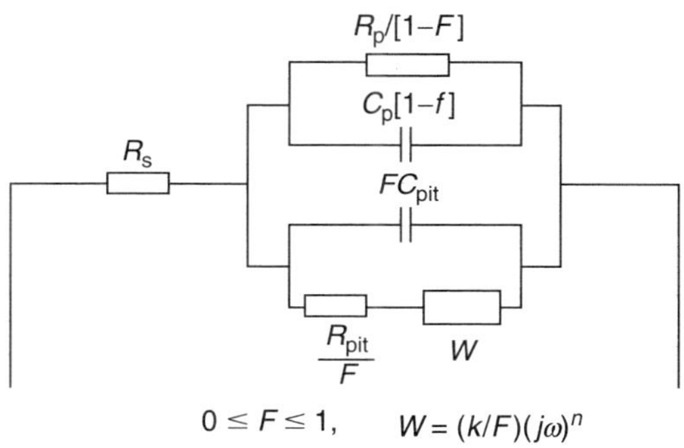

Figure 13.24 EC for the impedance of Al alloys undergoing active pitting.

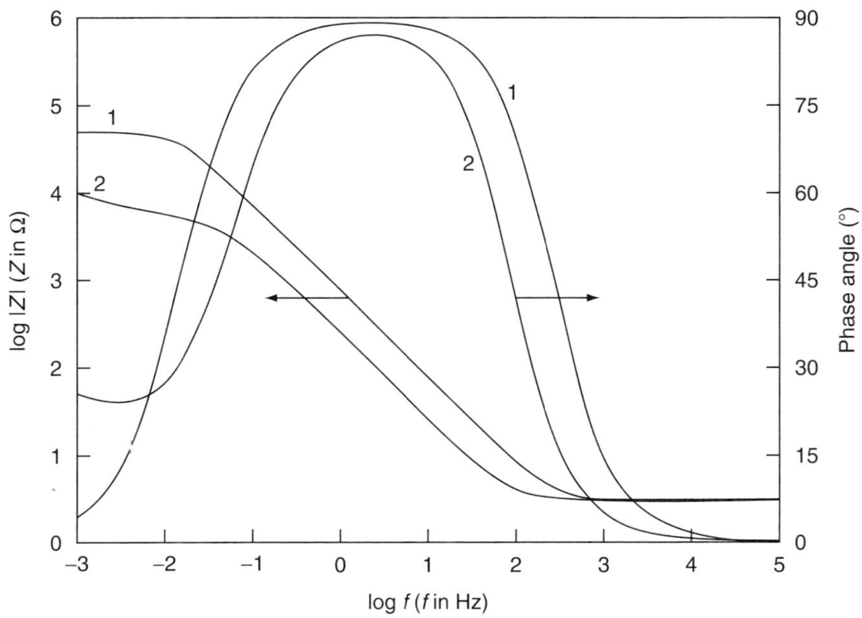

Figure 13.25 Simulated spectra for the EC in Figure 13.24 for $F = 0$ (curve 1) and $F = 0.005$ (curve 2); $R_s = 3\,\Omega$, $R_p = 5 \times 10^4\,\Omega$, $C_p = 2 \times 10^{-4}$ F, $R_{pit} = 25\,\Omega$, $C_{pit} = 0.08$ F, $K = 2.5\,\Omega$ (rad/sec)$^{-n}$, $n = 0.50$.

Figure 13.26 gives a comparison of experimental data (curve 1) and data obtained by a fit (curve 2) to the EC in Figure 13.24 for Al 6061 after exposure to 0.5 *N* NaCl for 24 h (48). The quality of the fit obtained by using PITFIT can be considered very good, taking into account the complexity of the impedance spectrum.

Analysis of the experimental impedance spectra with PITFIT results in the fit parameters R_s, $R_p/(1-F)$, $C_t = C_p(1-F) + FC_{pit}$, R_{pit}/F, K/F, and n. Since F is usually very small, $R_p/(1-F) = R_p$ and $C_t = C_p + FC_{pit}$. Figure 13.27 shows the time dependence of C_t for Al 7075-T6 in the as-received condition (curve 2) and after passivation in $CeCl_3$ (curve 1) (108). A very large increase of C_t was observed for the untreated Al 7075 in the first 2 days of exposure during which pits initiated and grew, while for the passivated sample C_t remained constant ($F = 0$) until pits were initiated after about 21 days, resulting in an increase of C_t.

In order to obtain further information concerning pit growth kinetics from the experimental values of R_{pit}/F, the value of F needs to be determined visually at the end of the experiment by microscopic observation. In addition, the value of FC_{pit} at the end of the test needs to be determined from the C_t data in Figure 13.27 as $FC_{pit} = C_t - C_p$ assuming that C_p did not change during the test. Based on a constant value of C^0_{pit} ($\mu F/cm^2$) it is then possible to calculate F as a function of time. R^0_{pit} ($\Omega\, cm^2$) is calculated by multiplying the fit parameter R_{pit}/F with $A_{pit} = 2FA$, which is the area of a hemispherical pit in a surface with a total area A.

The time dependence of the relative pit growth rate $1/R^0_{pit}$ has been expressed as

$$\log(R^0_{pit}) = a + b \log(t - t_0) \tag{13.24}$$

where t_0 is the time at which pitting was first detected with EIS. A plot according to Equation (13.24) for the data for the $CeCl_3$-treated Al 7075 in Figure 13.27 is shown

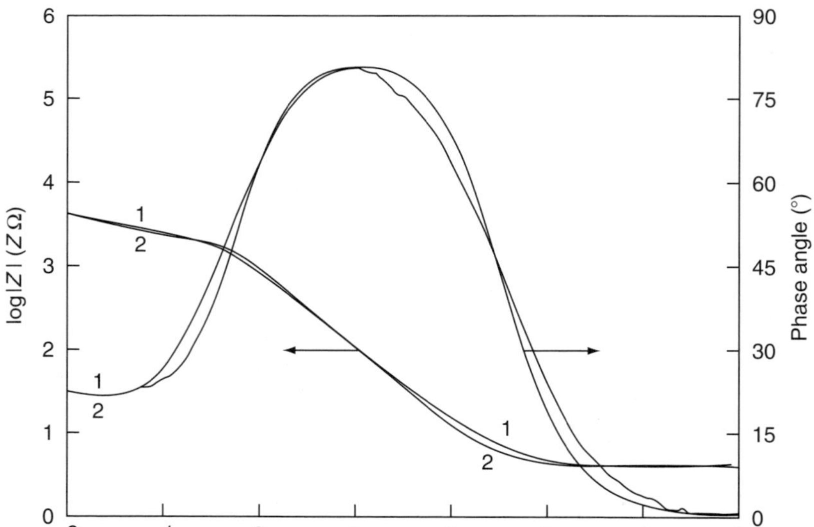

Figure 13.26 Bode plots for Al 6061-T6; curve 1 — after exposure to 0.5 *N* NaCl for 24 h; curve 2 — fitted spectra with $R_s = 4\,\Omega$, $R_p = 14{,}205\,\Omega$, $C_t = 175\,\mu F$, $R_{pit}/F = 2425\,\Omega$, $K/F = 850\,\Omega\,(rad/sec)^{-n}$, $n = -0.53$, $A = 20\,cm^2$. (From F. Mansfeld, Y. Wang, S. H. Lin, H. Xiao, and H. Shih, *ASTM STP* 1188, 297 [1993]. With permission.)

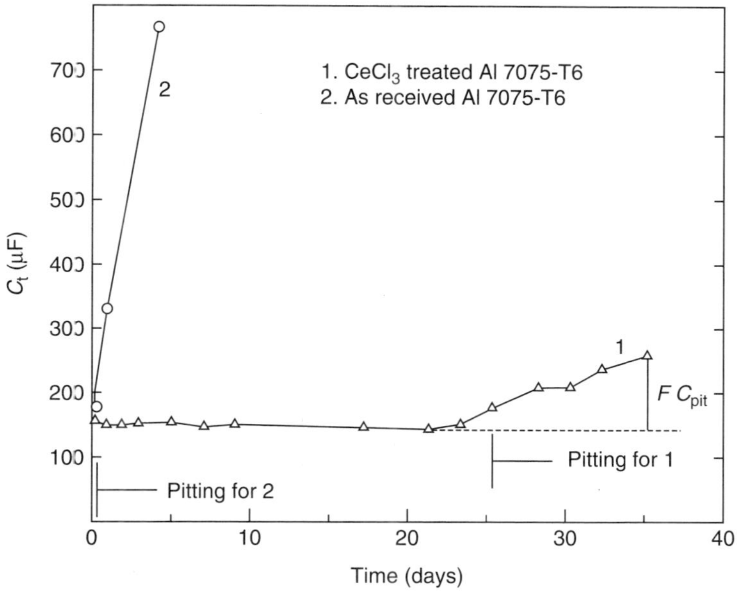

Figure 13.27 Time dependence of C_t for Al 7075–T6 (as-received vs. $CeCl_3$ treatment) during exposure to 0.5 *N* NaCl. (From F. Mansfeld, Y. Wang, S. H. Lin, H. Xiao, and H. Shih, *ASTM STP* 1188, 297 [1993]. With permission.)

in Figure 13.28. By curve fitting, $a = -3.00$, $b = -0.99$, and $t_0 = 21$ days were obtained (Figure 13.28). Since for most Al alloys studied by Mansfeld et al. (48) b was close to -1, it can be concluded that pit growth rates decreased linearly with time.

In monitoring pitting by EIS, it has to be noted that pits have to exceed a certain minimum size in order to produce the characteristic changes of the impedance spectra shown in Figure 13.25. Figure 13.29 shows theoretical impedance spectra as a

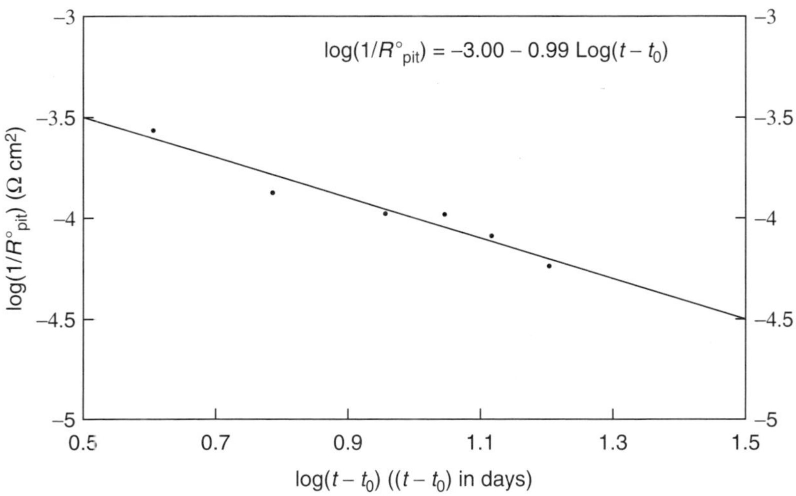

Figure 13.28 Plot of $\log(1/R^0_{pit})$ vs. $\log(t - t_0)$ for Al 7075–T6 passivated by immersion in $CeCl_3$). (From F. Mansfeld, Y. Wang, S. H. Lin, H. Xiao, and H. Shih, *ASTM STP* 1188, 297 [1993]. With permission.)

function of the radius r of a hemispherical pit between 0 and 0.5 cm, which corresponds to F values between 0 and 0.016 for $A = 1\,cm^2$. The characteristic changes of the spectra from those for a passive surface ($F = 0$) to those for a passive surface containing a growing pit occur for a pit radius larger than 0.015 cm ($F = 0.0016$). For a pit radius of 0.05 cm ($F = 0.016$) a significant increase of C_t and a minimum of the phase angle at about 0.2 Hz can be seen in Figure 13.29, curve 4.

One of the requirements for obtaining valid impedance data is stability of the system under investigation during the time of the measurement (11). Mansfeld and Fernandes (109) have pointed out that Al alloys exposed to NaCl are not stable in the early stages of immersion. Therefore, it was not possible to obtain valid impedance spectra during the first 24 h of exposure to NaCl.

13.4.6 Microbiologically Influenced Corrosion

There are several applications of EIS in studies of microbiologically influenced corrosion (MIC). Mansfeld and Little (100–112) as well as Schmitt (113) have provided reviews of the use of electrochemical techniques including EIS for evaluation and monitoring of MIC. Little and Mansfeld (118) have evaluated the corrosion behavior of stainless steel and copper alloys exposed to natural seawater.

The impact of microorganisms on the protective properties of 12 different polymer coating systems on hot-rolled steel has been studied by Mansfeld et al. (83). Samples were exposed to NS at PH and Key West, FL (KW), and to AS for up to 2 years. A special monitoring approach was developed for collection of electrochemical impedance and noise data from remote test sites (Figure 13.30) (20, 21, 83).

The impedance spectra for the alkyd coating system CR2 suggest that pronounced coating degradation occurred for both samples at PH and KW after about 4 and 5 months (Figure 13.31). This result is reflected in the sharp increase of f_b^h

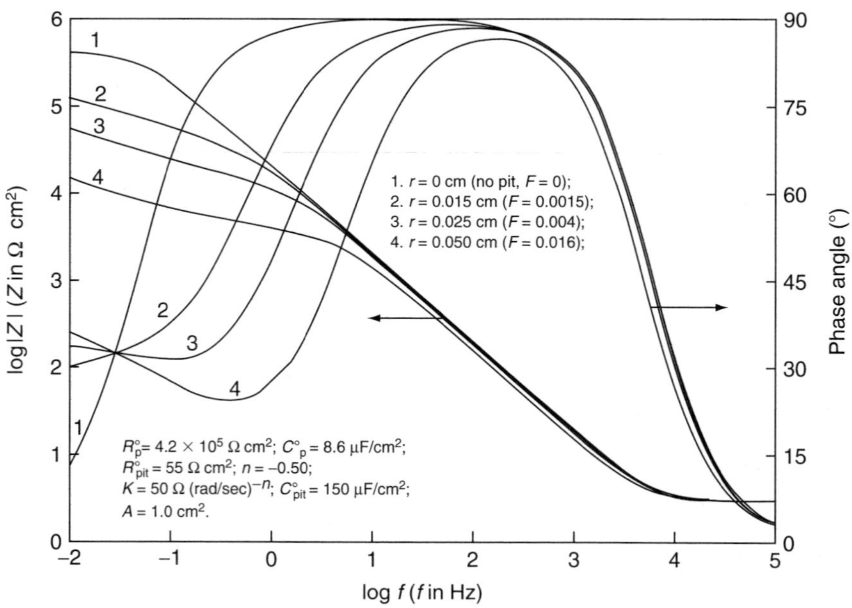

Figure 13.29 Theoretical Bode plots as a function of pit radius.

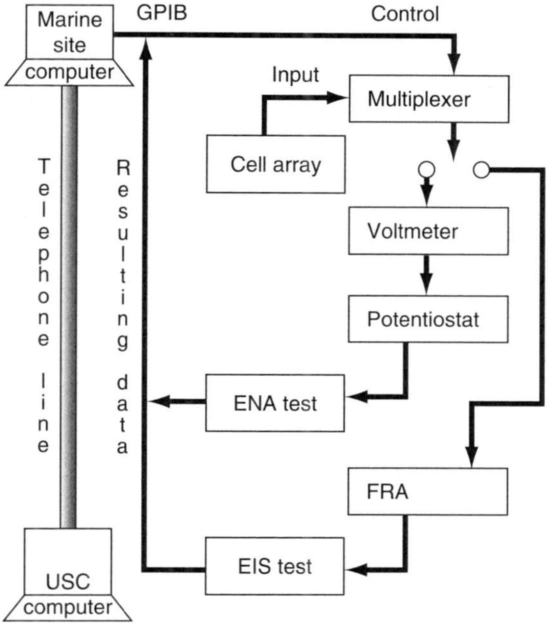

Figure 13.30 Experimental approach for the collection of electrochemical impedance and noise data from a remote test site. (From F. Mansfeld, H. Xiao, L. T. Han, and C. C. Lee, *Prog. Org. Coat.* 30, 89 [1997]. With permission.)

(Figure 13.32). The sample exposed in AS developed asymmetric behavior (115), therefore the impedance data were dominated by the sample with the more protective coating.

As discussed in Section 13.4.3, the impedance spectra for the thick coatings used by Mansfeld et al. (83) did not follow the coating model in Figure 13.18(a) but agreed well with the OFLD model in Figure 13.9(b) (Figure 13.20). A comparison of damage functions Δ obtained from experimental values of f_b^h and from visual observation has been given in Figure 13.19. Figure 13.33 shows plots of Δ vs. time according to Equation (13.23) for two polymer coated steel samples. The time t_0 to reach delamination of 1% of the total exposed area is given in Figure 13.33. Statistical analyses have been performed to determine the effects of coating composition on coating performance in general and attack by microorganisms in particular (83).

13.4.7 Corrosion Protection by Bacteria

From the time that MIC was discussed for the first time in the 1920s, it has been assumed that bacteria always accelerate corrosion rates by a number of different mechanisms (111). However, in 1984 Potekhina (116) reported that corrosion can be inhibited by aerobic bacteria. Research carried out by a team from the University of Connecticut, University of California at Irvine, and University of Southern California has shown that several types of bacteria can inhibit corrosion of different materials such as Al 2024, mild steel, and brass in corrosive solutions such as AS (117–121).

Figure 13.34(a) shows experimental impedance spectra obtained for Al 2024 during exposure to AS for 30 days (117–118). Only four of the spectra collected during this time are plotted in the Bode plots of Figure 13.34(a). The spectra suggest

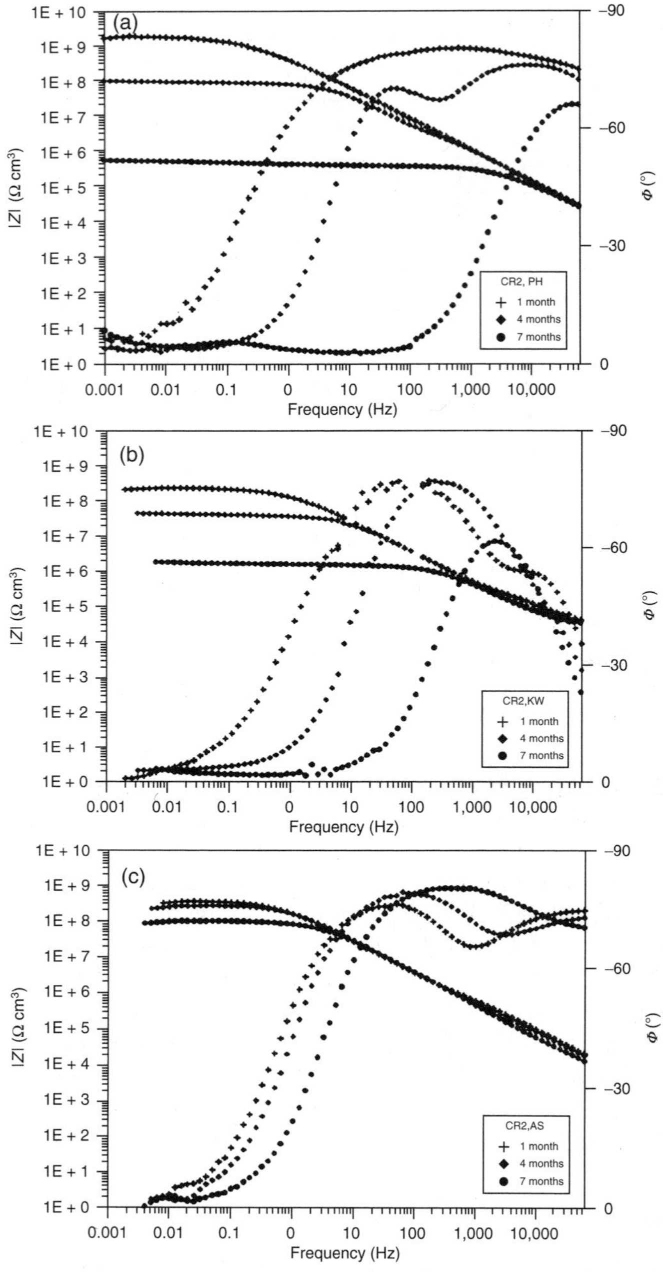

Figure 13.31 Impedance spectra for alkyd coating system on hot-rolled steel exposed to NS at PH (a), KW (b) and AS. (From F. Mansfeld, H. Xiao, L. T. Han, and C. C. Lee, *Prog. Org. Coat.* 30, 89 [1997]. With permission.)

that pitting occurred during the entire test period, as evidenced by the typical low-frequency minimum of the phase angle Φ, which is partially masked by the scatter of the data points below 0.01 Hz. Nevertheless, the spectra in Figure 13.34(a) are in agreement with the pitting model shown in Figure 13.24. Qualitatively, it can be observed that the polarization resistance of active pits R_{pit}, which is close to the

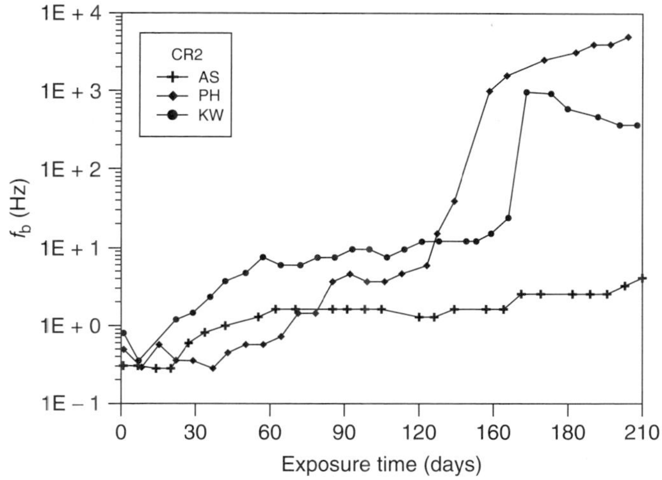

Figure 13.32 Time dependence of breakpoint frequency f_b^h for an alkyd coating system exposed to NS and AS. (From F. Mansfeld, H. Xiao, L. T. Han, and C. C. Lee, *Prog. Org. Coat.* 30, 89 [1997]. With permission.)

impedance value $|Z|$ at the frequency minimum at low frequencies, increased with increasing exposure time as the pit growth rate decreased.

In the presence of *Bacillus subtilis*, pitting also occurred in the first 2 days of exposure; however, after 3 days the spectra agreed with those for a passive surface, that is a simple one-time-constant model in R_p is in parallel with C (Figure 13.2). The fairly high values of R_p, which approach the megaohm range, suggest that pits formed in the initial stages of exposure have become passivated (Figure 13.34b). Very similar results were obtained in the presence of *B. subtilis* producing polyglutamate (Figure 13.34c) or polyaspartate (Figure 13.34d). The increased R_p values suggest that the inhibitors produced by the bacteria provided additional corrosion protection.

Figure 13.35 illustrates the time dependence of the relative corrosion rates expressed as $1/R_{pit}^0$ for the tests in the absence of bacteria and $1/R_{pit}^0$ for the tests in the presence of *B. subtilis*. These values have been obtained by normalizing the experimental R_p values with the total exposed area and the R_{pit} values with the time-dependent values of the pitted area A_{pit} determined by analysis of the impedance spectra as explained in Section 13.4.5. For the tests in the absence of bacteria, R_p could not be determined due to the lack of sufficient low-frequency data (Figure 13.34a). Figure 13.35 clearly demonstrates the inhibition of pitting corrosion in the presence of *B. subtilis* and the increased corrosion resistance in the early stages of exposure due to inhibitors produced by the bacteria. The lowest corrosion rates were observed for the biofilm producing polyaspartate (test 44).

For the tests in the presence of *B. subtilis* (tests 42 to 44) the values of A_{pit} determined by visual observation at the end of the exposure period were much less than those determined in the absence of bacteria for which the pitted area fraction F was 1.04%. The final F values for tests 42, 43, and 44 were 0.07%, 0.16%, and 0.06%, respectively.

The inhibition of pitting in the presence of bacteria could be due to exclusion of oxygen from the metal surface, which would reduce the rate of the cathodic reduction, resulting in a decrease of E_{corr} below the pitting potential E_{pit}. However, the

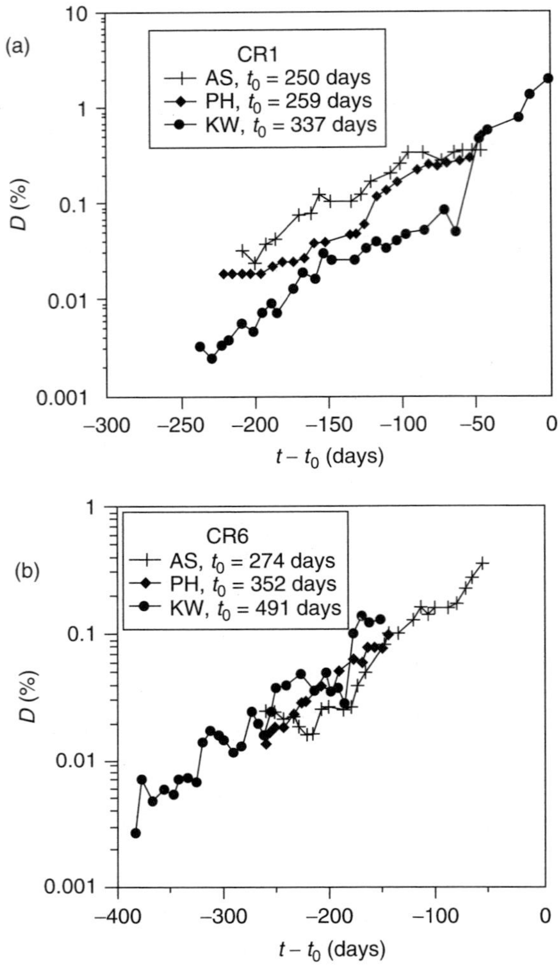

Figure 13.33 Plots of log Δ vs. $t - t_0$ for an alkyd coating system (CR1) (a) and an epoxy polyamide/latex coating system (CR 6) (b) on hot-rolled steel exposed to NS and AS. (From F. Mansfeld and C. C. Lee, *Corros. Sci.* 41, 439 [1999]. With permission.)

experimental values of E_{corr} were lowest in the absence of bacteria, while a certain degree of ennoblement was observed in the presence of bacteria (Figure 13.36) (117–121). Since the highest values of E_{corr} were observed in the presence of inhibitors, it can be concluded that the observed corrosion inhibition is due to a passivation effect that occurs in the presence of a biofilm. This beneficial effect is apparent even when the biofilm contains bacteria that were not engineered to produce inhibitors. Indeed, the observation that pitting occurred in all cases in the first 2 days of exposure suggests that formation of a stable biofilm is needed to stop the growth of active pits.

Nagiub and Mansfeld (122) have evaluated microbiologically influenced corrosion inhibition of Al 2024, cartridge brass, and mild steel exposed to AS caused by *Shewanella algae* and *S. ana*, which are classified under the group known as iron-reducing bacteria. Both strains are able to grow in seawater under aerobic and anaerobic conditions.

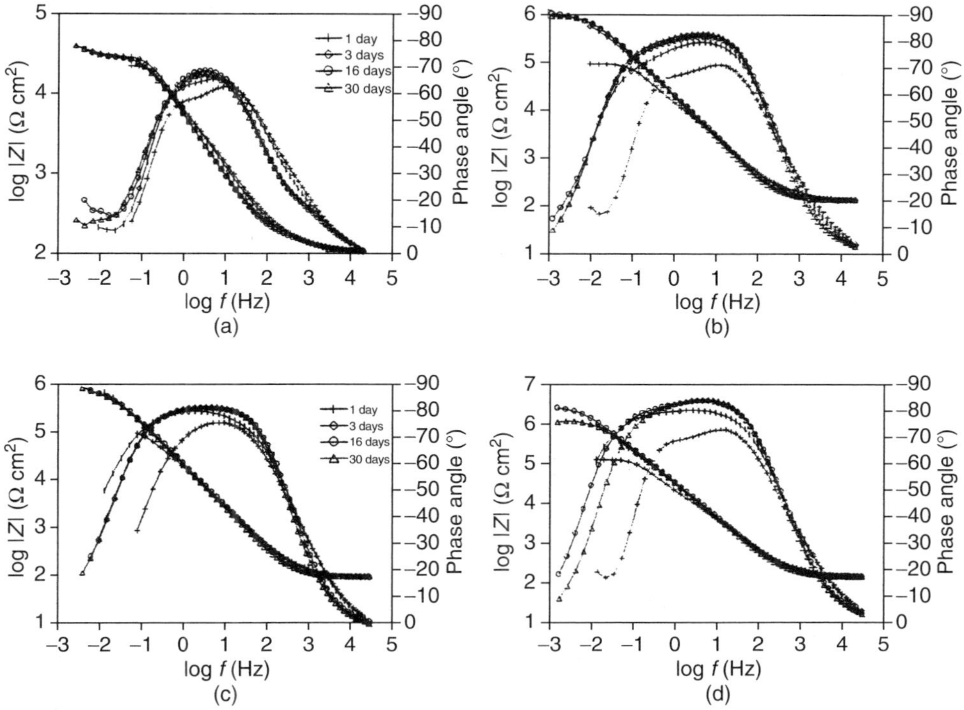

Figure 13.34 Bode plots for Al 2024 exposed to AS for different time periods: (a) sterile solution (test 45), (b) AS containing *Bacillus subtilis* WB600 (test 42), (c) AS containing *B. subtilis* WB600/pBE92 producing polyglutamate (test 43), and (d) AS containing *B. subtilis* WB600/pBE92 producing polyaspartate (test 44). With permission.)

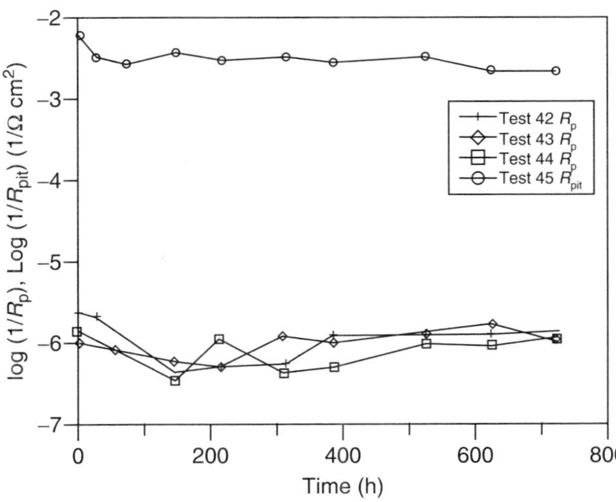

Figure 13.35 Time dependence of R_{pit}^{0} and R_{p}^{0} in sterile AS (test 45) and in AS containing *B. subtilis* WB600/pBE92 (tests 42 to 44). (From J. E. G. Gonzalez, F. J. H. Santana, and J. C. Mirza-Rosa, *Corros. Sci.* 40, 2141 [1998]. With permission.)

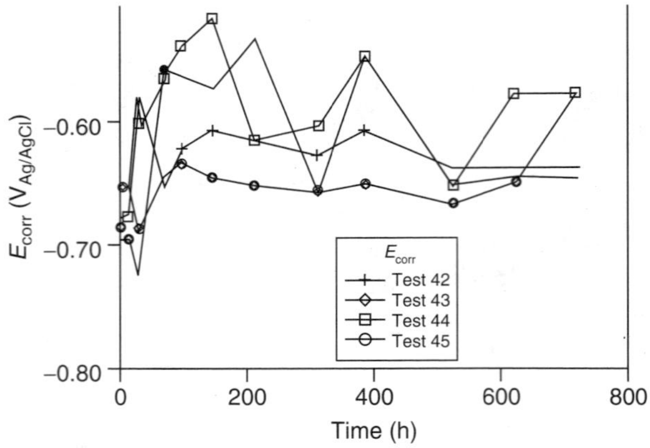

Figure 13.36 Time dependence of E_{corr} in sterile AS (test 45) and in the presence of *B. subtilis* WB600/pBE92 (test 42 to 44). (From D. Ōrnek, A. Jayaraman, T. K. Wood, Z. Sun, C. H. Hsu, and F. Mansfeld, *Corros. Sci.* 43, 2121 (2001). With permission.)

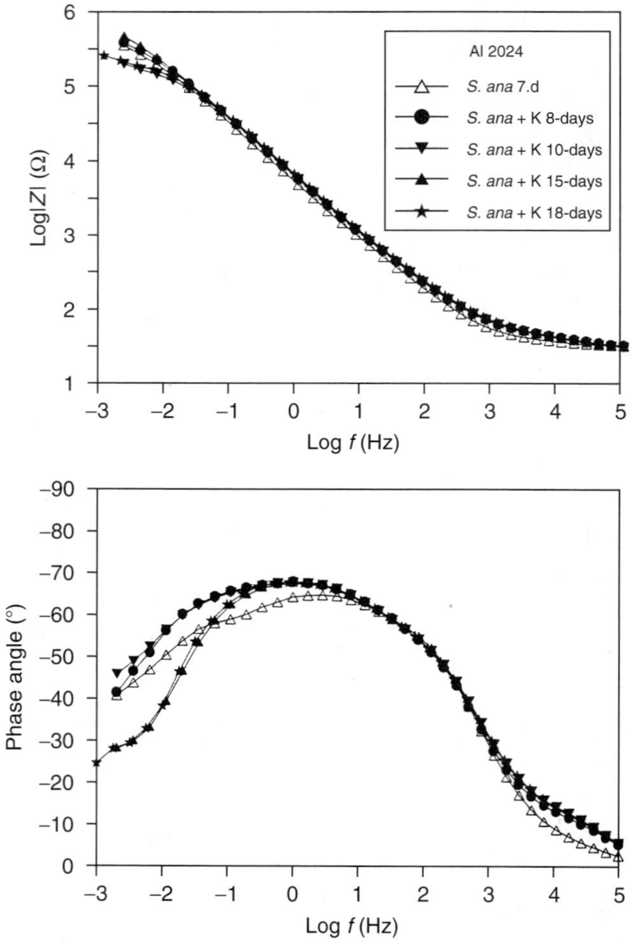

Figure 13.37 Bode plots for Al 2024 exposed to AS containing *S. ana*; addition of kanamycin after 7 days. (From A. Nagiub and F. Mansfeld, *Electrochim. Acta* 47, 2319 [2002]. With permission.)

S. algae or *S. ana* were added to AS containing growth medium and EIS data were collected for 1 week. The very similar impedance spectra (OTCM) obtained for Al 2024 demonstrated that both strains provided excellent corrosion protection. After that time 200 μg/ml kanamycin were added to the solution containing *S. ana* to kill the bacteria and EIS data were obtained for another week or longer. After addition of kanamycin significant changes in the spectra indicating that pitting had been initiated were observed only after total exposure for 15 days (Figure 13.37). Figure 13.38 shows the time dependence of E_{corr}, R^0_{pit}, and R^0_p for Al 2024 in AS and AS containing *S. algae* or *S. ana*. Also shown are results obtained after addition of kanamycin. In the presence of the bacterial strains, E_{corr} was more negative than in the abiotic solution (Figure 13.38). After addition of kanamycin an immediate increase of E_{corr} was observed. Due to passivation in the presence of the bacteria, R^0_p was much larger than R^0_{pit} (Figure 13.38). About 1 week after kanamycin had been added, R^0_{pit} had reached similar values as those determined in abiotic AS (Figure 13.39).

For mild steel, similar significant increases in the corrosion resistance were observed in the presence of both strains. A sharp increase in corrosion rates was observed about 1 week after addition of kanamycin (Figure 13.39). Contrary to the

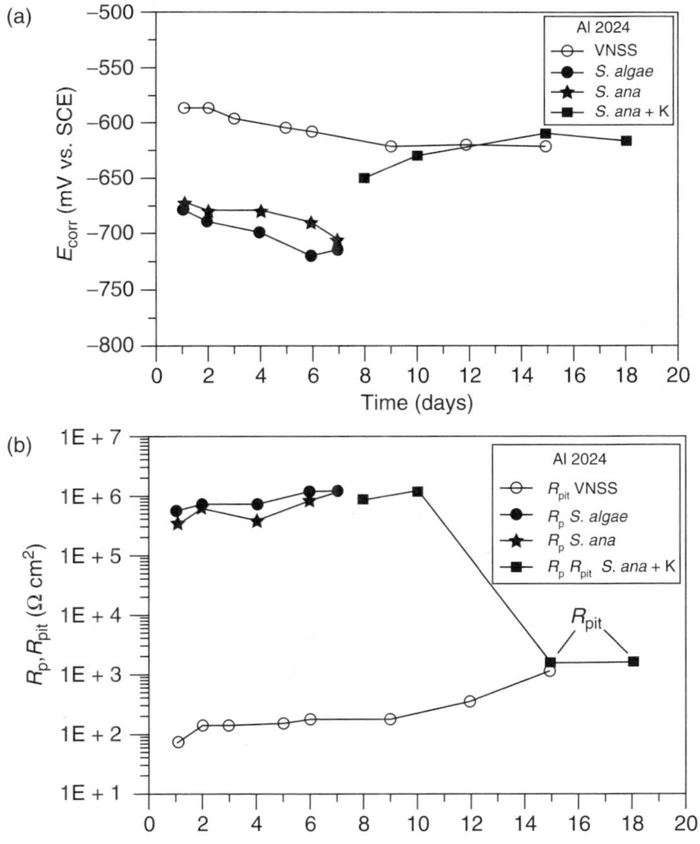

Figure 13.38 Time dependence of E_{corr} (a) and R_p and R_{pit} (b) for Al 2024 exposed to AS containing *S. algae* or *S. ana*. (From A. Nagiub and F. Mansfeld, *Electrochim. Acta* 47, 2319 [2002]. With permission.)

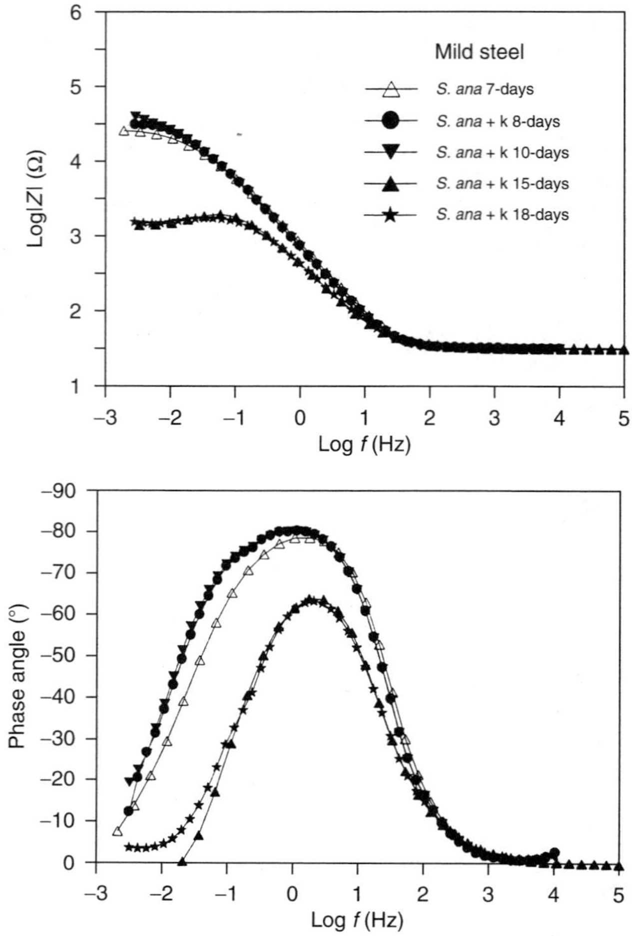

Figure 13.39 Bode plots for brass exposed to AS containing *S. ana*; addition of kanamycin after 7 days. (From A. Nagiub and F. Mansfeld, *Electrochim. Acta* 47, 2319 [2002]. With permission.)

results obtained for Al 2024 (Figure 13.38), E_{corr} was more positive in the solutions containing the bacteria (Figure 13.40). No significant changes in E_{corr} were observed after the addition of kanamycin to the solution containing *S. ana* despite the sharp drop of R_p after 15 days (Figure 13.40).

The two bacterial strains also increased the corrosion resistance of brass in AS, although the impedance spectra were somewhat different in the presence of *S. algae* and *S. ana* (122). A dramatic change in the spectra indicating increased corrosion activity was observed after exposure for 15 days (Figure 13.41). As shown in Figure 13.42 *S. ana* produced a larger decrease of E_{corr} and increase of R_p than *S. algae*. After addition of kanamycin, E_{corr} increased significantly, while R_p reached similar values as those found in abiotic AS after 15 days (Figure 13.42).

In the presence of *S. algae* or *S. ana* significant increases in the corrosion resistance were obtained for all three materials. E_{corr} decreased in the presence of bacteria, suggesting that the observed absence of pitting for Al 2024 and the increase

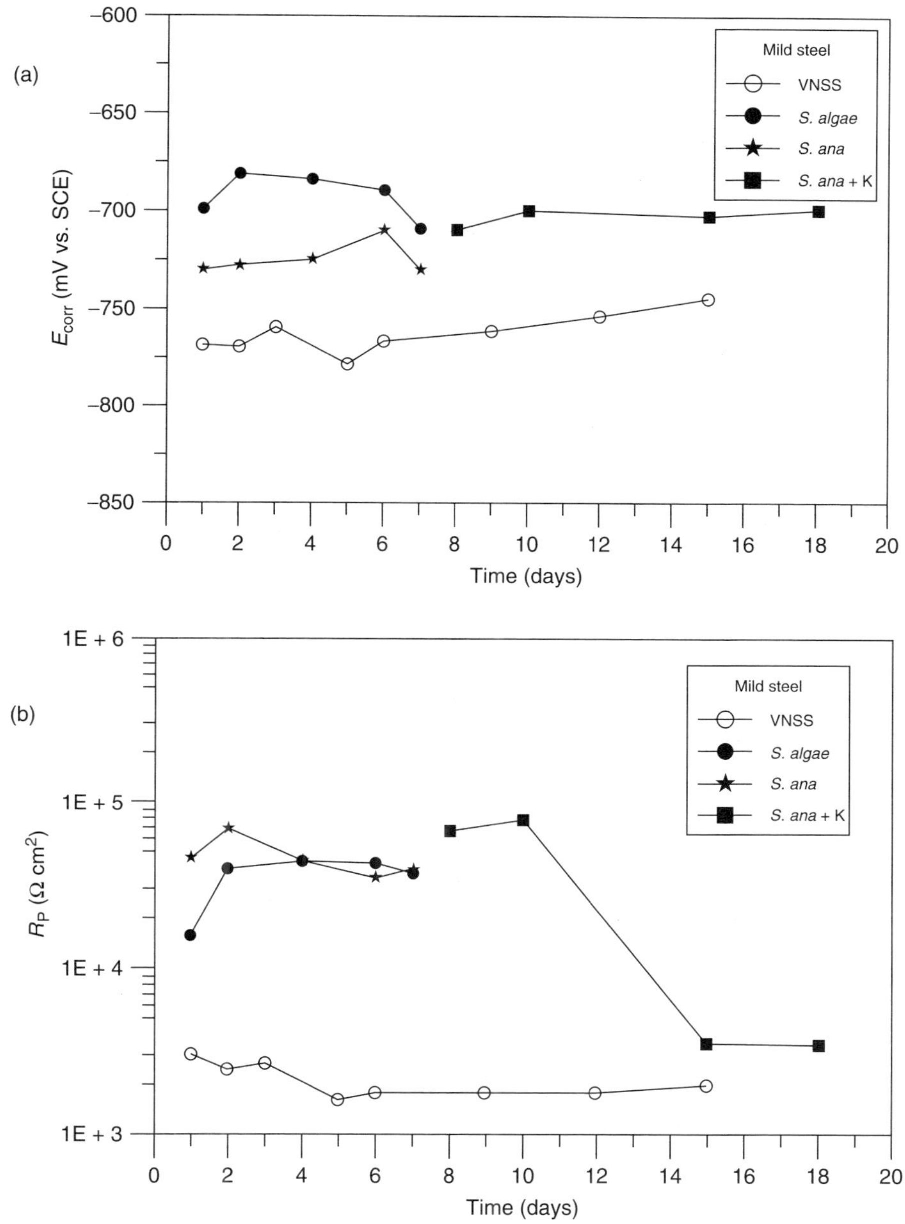

Figure 13.40 Time dependence of E_{corr} (a) and R_p (b) for brass exposed to AS containing *S. algae* or *S. ana*. (From A. Nagiub and F. Mansfeld, *Electrochim. Acta* 47, 2319 [2002]. With permission.)

in the corrosion resistance of brass were due to a reduction of the oxygen concentration at the electrode surface. Potentiodynamic polarization curves obtained in VNSS (Vätäänen nine salts solution) containing *S. algae* after exposure for 7 days showed indeed a significant reduction of the cathodic currents (122). For mild steel an increase in the corrosion resistance was accompanied by an increase of E_{corr}. In

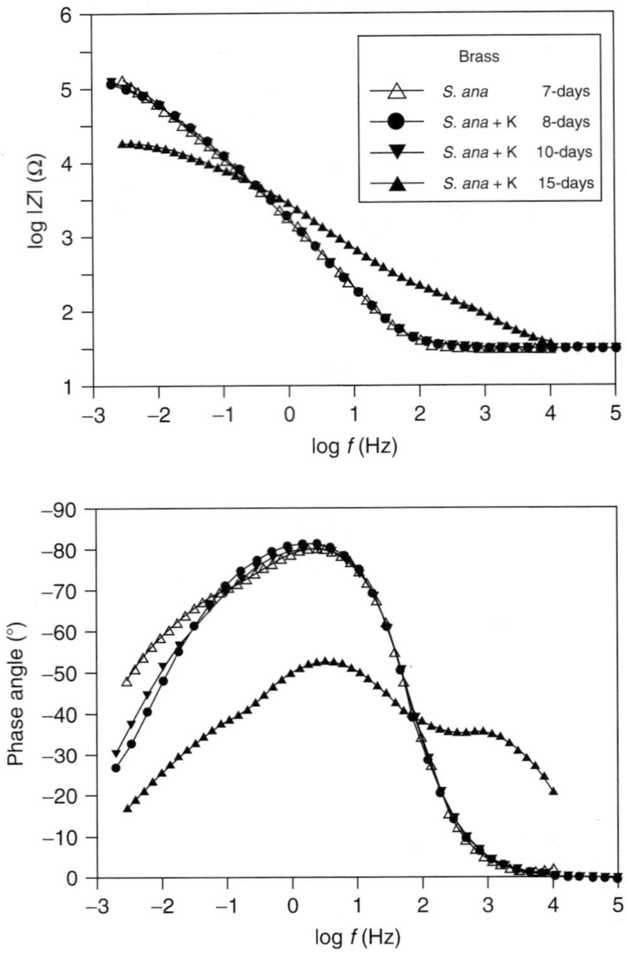

Figure 13.41 Bode plots for mild steel exposed to AS containing *S. ana*; addition of kanamycin after 7 days. (From A. Nagiub and F. Mansfeld, *Electrochim. Acta* 47, 2319 [2002]. With permission.)

this case anodic polarization curves demonstrated a small decrease of the anodic current in VNSS containing *S. algae* with little changes in the cathodic polarization curve. The observed decrease of corrosion rates accompanied by an increase of E_{corr} could be due to production of an inhibiting species or due to microbial respiration involving reduction of Fe^{3+} to Fe^{2+} accompanied by a reduction of the oxygen concentration at the electrode surface (123).

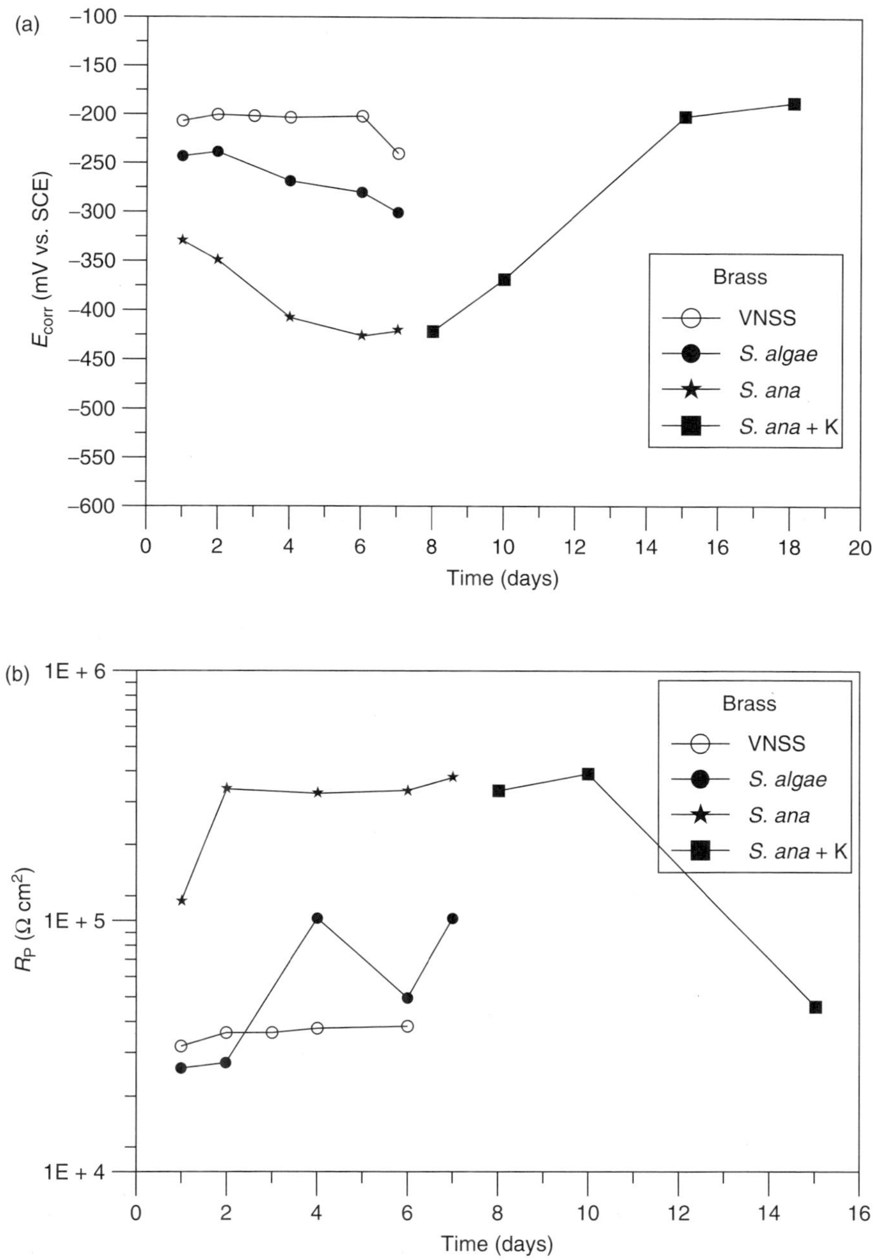

Figure 13.42 Time dependence of E_{corr} (a) and R_p (b) for mild steel exposed to AS containing *S. algae* or *S. ana*. (From A. Nagiub and F. Mansfeld, *Electrochim. Acta* 47, 2319 [2002]. With permission.)

REFERENCES

1. F. Mansfeld, *Corrosion* 44, 558 (1988).
2. I. Epelboin, C. Gabrielli, and M. Keddam, "Non-steady state techniques," in *Comprehensive Treatise of Electrochemistry* (E. Yeager, J. O'M. Bockris, B. E. Conway, and S. Sarangapani, eds.), Vol. 9, p. 62, Plenum Press, New York, 1984.
3. F. Mansfeld and M. W. Kendig, *Werkstoffe Korros.* 36, 473 (1985).
4. J. R. Macdonald, ed., *Impedance Spectroscopy*, J. Wiley, 1987.
5. *Electrochim. Acta* 35, 10 (1990).
6. *Electrochim. Acta* 38, 14 (1993).
7. *Electrochim. Acta* 41, 7–8 (1996).
8. *Electrochim. Acta* 44, 24 (1999).
9. *Electrochim. Acta* 47, 13–14 (2002).
10. J. R. Scully, D. C. Silverman, and M. W. Kendig, eds., "Electrochemical impedance: analysis and interpretation," *ASTM STP* 1188, 1993.
11. F. Mansfeld and W. J. Lorenz, "Electrochemical impedance spectroscopy (EIS): application in corrosion science and technology," in *Techniques for Characterization of Electrodes and Electrochemical Systems* (R. Varma and J. R. Selman, eds.), J. Wiley, 581–647, 1991.
12. *Mater. Sci. Forum* 44–45 (1989).
13. *Mater. Sci. Forum* 111–112 (1992).
14. *Mater. Sci. Forum* 192–194 (1995).
15. *Mater. Sci. Forum* 289–292 (1998).
16. *Electrochim. Acta* 46 (24–25) (2001).
17. F. Mansfeld, F. Huet, and O. Mattos, eds., "New trends in electrochemical impedance spectroscopy (EIS) and electrochemical noise analysis (ENA)," Proceedings, Vol. 2000–2024, The Electrochemical Society, 2001.
18. B. J. Little P. A. Wagner, and F. Mansfeld, "Microbiologically influenced corrosion," in *Corrosion Testing Made Easy*, (B. C. Syrett, ed) Vol. 5, NACE, 1997.
19. R. Cottis and S. Turgoose, "Electrochemical impedance and noise," in *Corrosion Testing Made Easy*, Vol. 7, (B. C. Syrett, ed) NACE, 1999.
20. F. Mansfeld, H. Xiao, L. T. Han, and C. C. Lee, *Prog. Org. Coat.* 30, 89 (1997).
21. F. Mansfeld, L. T. Han, C. C. Lee, and G. Zhang, *Electrochim. Acta* 43, 2933 (1998).
22. C. C. Lee and F. Mansfeld, *Corros. Sci.* 40, 959 (1998).
23. F. Mansfeld, C. C. Lee, and G. Zhang, *Electrochim. Acta* 43, 435 (1998).
24. F. Mansfeld, S. Lin, Y. C. Chen, and H. Shih, *J. Electrochem. Soc.* 135, 906 (1988).
25. F. Mansfeld, *Corrosion* 37, 301 (1981).
26. F. Mansfeld, M. W. Kendig, and S. Tsai, *Corrosion* 38, 570 (1982).
27. K. Juettner, W. J. Lorenz, W. Paatsch, M. Kendig, and F. Mansfeld, *Werkstoffe Korros* 36, 120 (1985).
28. F. Mansfeld, "The polarization resistance technique for measuring corrosion currents," in *Advances in Corrosion Science and Technology* (M. G. Fontana and R. W. Staehle, eds.), Vol. 6, p. 163, Plenum Press, New York, 1976.
29. M. W. Kendig, E. M. Meyer, G. Lindberg, and F. Mansfeld, *Corrosion* 23, 1007 (1983).
30. F. Mansfeld, M. W. Kendig, and W. J. Lorenz, *J. Electrochem. Soc.* 132, 290 (1985).
31. F. Mansfeld and M. W. Kendig, *Werkstoffe Korros* 34, 397 (1983).
32. E. Schmidt, J. Hitzig, J. Titz, K. Juettner, and W. J. Lorenz, *Electrochim. Acta* 31, 1041 (1986).
33. M. W. Kendig and F. Mansfeld, *Corrosion* 39, 466 (1983).
34. F. Mansfeld and M. W. Kendig, Proceedings of the European Symposium on 6th Corrosion Inhibitors, Ferrara, Italy, 1985.
35. M. W. Kendig, A. T. Allen, and F. Mansfeld, *J. Electrochem. Soc.* 131, 935 (1984).
36. F. Mansfeld, M. W. Kendig, and S. Tsai, *Corros. Sci.* 22, 455 (1982).
37. D. G. John, P. C. Searson, and J. L. Dawson, *Br. Corros. J.* 16, 455 (1981).

38. B. A. Boukamp, "Equivalent Circuit (Equivcrt.Pas)", University of Twenet, May 1989.
39. A. H. Windle, *Polymer Permeability*, Elsevier, Amsterdam, 1986.
40. F. M. Geenen, J. H. W. deWit, and E. P. M. van Westing, *Prog. Org. Coat.* 18, 299 (1990).
41. G. J. Brug, A. L. G. Van Den Eeden, M. Sluyters-Rehbach, and J. H. Sluyters, *J. Electroanal. Chem.* 176, 275 (1984).
42. E. van Westing, "Determination of Coating Performance with Impedance Measurements," Ph. D. thesis, Technical University of Delft, The Netherlands, 1992.
43. S. F. Mertens, C. Xhoffer, B. C. De Cooman, and E. Temmerman, *Corrosion* 53, 381 (1997).
44. G. O. Ilevbare and J. R. Scully, *Corrosion* 57, 134 (2001).
45. C. H. Hsu and F. Mansfeld, *Corrosion* 57, 747 (2001).
46. F. Mansfeld, C. H. Tsai and H. Shih, *ASTM STP* 1154, 186 (1992).
47. F. Mansfeld, H. Shih, H. Greene, and C. H. Tsai, *ASTM STP* 1188, 37 (1993).
48. F. Mansfeld, Y. Wang, S. H.. Lin, H. Xiao, and H. Shih, *ASTM STP* 1188, 297 (1993).
49. F. Mansfeld, C. H. Tsai and H. Shih, *ASTM STP* 1154, 186 [1992]
50. W. J. Lorenz and F. Mansfeld, *Corros. Sci.* 21, 647 (1981).
51. F. Mansfeld, M. Kendig, and W. Lorenz, *J. Electrochem. Soc.* 132, 290 (1985).
52. W. J. Lorenz and F. Mansfeld, *Electrochim. Acta* 127, 467 (1986).
53. I. Epelboin, M. Keddam, and H. Takenouti, *J. Appl. Electrochem.* 2, 71 (1972).
54. A. Nagiub and F. Mansfeld, *Corros. Sci.* 43, 2147 (2001).
55. F. Mansfeld, *J. Appl. Electrochem.* 25, 187 (1995).
56. F. Mansfeld and M. Kendig *ASTM STP* 866, 122 (1985).
57. F. Mansfeld, S. L. Jeanjaquet, and M. W. Kendig, *Corros. Sci.* 26, 727 (1986).
58. G. Menges and W. Schneider, *Kunststofftechnik* 12, 265 (1973).
59. G. Menges and W. Schneider, *Kunststofftechnik* 12, 343 (1973).
60. M. Kendig and H. Leidheiser, *J. Electrochem. Soc.* 121, 275c (1974).
61. M. Kendig and H. Leidheiser, *J. Electrochem. Soc.* 123, 982 (1976).
62. L. Fedrizzi, F. Deflorian, G. Boni, P. L. Bonora, and E. Pasini, *Prog. Org. Coat.* 29, 89 (1996).
63. F. Deflorian, L. Ferizzi, and P. L. Bonora, *Corros. Sci.* 38, 1697 (1996).
64. F. Deflorian, L. Ferizzi, and P. L. Bonora, "Organic Coating for corrosion control," *ACS Symp. Ser.* 689, 92 (1998).
65. F. Deflorian, L. Ferizzi, S. Rossi, and P. L. Bonora, *Electrochemical Approach to Selected Corrosion and Corrosion Control Studies* (P. L. Bonora and F. Deflorian, eds.), Eur. Fed. Corros. Publ., No. 28, IOM Comm., p. 49, 2000.
66. D. H. van der Weijde, E. P. M. van Westing, and J. H. W. de Wit, *Corros. Sci.* 36, 643 (1994).
67. D. H. van der Weijde, J. M. Ferrari, and J. H. W. de Wit, *Corros. Sci.* 36, 979 (1994).
68. D. H. van der Weijde, E. P. M. van Westing, and J. H. W. de Wit, *Electrochim. Acta* 41, 1103 (1996).
69. G. W. Walter, *Corros. Sci.* 26, 910 (1986).
70. U. Rammelt and G. Reinhard, *Prog. Org. Coat.* 21, 205 (1992).
71. J. R. Scully, *J. Electrochem. Soc.* 136, 979 (1989).
72. M. W. Kendig and J. R. Scully, *Corrosion* 46, 22 (1990).
73. J. N. Murray, *Prog. Org. Coat. 31*, 375 (1997).
74. F. Mansfeld and C. C. Lee, *Corros. Sci.* 41, 439 (1999).
75. F. Mansfeld and C. C. Lee, "Evaluation of polymer coating properties with EIS, ENA and ANN," in *Electrochemical Approach to Selected Corrosion and Corrosion Control Studies* (P. L. Bonora and F. Deflorian, eds.), Eur. Fed. Corros. Publ., No. 28, IOM Comm., p. 100, 2000.
76. D. M. Brasher and A. H. Kingsbury, *J. Appl. Chem.* 4, 62 (1954).
77. F. Mansfeld, M. Kendig, and S. Tsai, *Corrosion* 38, 478 (1982).
78. M. Kendig, F. Mansfeld, and S. Tsai, *Corros. Sci.* 23, 317 (1983).

79. F. Mansfeld and C. H. Tsai, *Corrosion* 47, 958 (1991).
80. C. H. Tsai and F. Mansfeld, *Corrosion* 49, 726 (1993).
81. S. Haruyama, M. Asari, and T. Tsuru, *Corrosion Protection by Organic Coatings* (M. Kendig and H. Leidheiser, Jr., eds.), The Electrochemical Society, Proceedings Vol. 87–2, p. 197, 1987.
82. H. P. Hack and J. R. Scully, *J. Electrochem. Soc.* 138, 33 (1991).
83. F. Mansfeld, C. C. Lee, L. T. Han, G. Zhang, B. J. Little, P. Wagner, R. Ray, and J. Jones-Meehan, "The Impact of Microbiologically Influenced Corrosion on Protective Polymer Coatings," Final Report to the Office of Naval Research, Contract No. N00014-94-1-0026, August 1988.
84. J. Hitzig, K. Juettner, W. J. Lorenz, and W. Paatsch, *Corros. Sci.* 42, 945 (1984).
85. F. Mansfeld and M. W. Kendig, *Corrosion* 41, 490 (1985).
86. J. Hitzig, K. Juettner, W. J. Lorenz, and W. Paatsch, *J. Electrochem. Soc.* 133, 887 (1986).
87. F. Mansfeld and M. W. Kendig, *J. Electrochem. Soc.* 135, 828 (1988).
88. R. Lizarbe, J. A. Gonzales, W. Lopez, and E. Otero, *Aluminium* 68, 140 (1992).
89. R. Lizarbe, W. Lopez, E. Otero, and J. A. Gonzales, *Rev. Metall. Madrid* 26, 359 (1990).
90. J. A. Gonzales, V. Lopez, E. Otero, A. Bautista, R. Lizarbe, C. Barba, and J. L. Baldonedo, *Corros. Sci.* 39, 1109 (1997).
91. J. A. Gonzales, V. Lopez, E. Otero, and A. Bautista, *J. Electrochem. Soc.* 147, 984 (2000).
92. F. Mansfeld, G. Zhang, and C. Chen, *Plat. Surf. Finish.* 84(12), 72 (1997).
93. F. Mansfeld, C. Chen, C. B. Breslin, and D. Dull, *J. Electrochem. Soc.* 145, 2792 (1998).
94. F. Mansfeld, "Surface modification of aluminum alloys: non-toxic processes for improved corrosion resistance," Proceedings of the Research Topical Symposium, *Corrosion in Advanced Materials and Systems*, p. 81, NACE, 1998.
95. F. Mansfeld, S. Lin, S. Kim, and H. Shih, *Corros. Sci.* 27, 997 (1987).
96. F. Mansfeld and H. Shih, *J. Electrochem. Soc.* 135, 1171 (1988).
97. F. Mansfeld, S. Lin, S. Kim, and H. Shih, *Werkstoffe Korros.* 39, 487 (1988).
98. F. Mansfeld, S. Lin, S. Kim, and H. Shih, *Corrosion* 45, 615 (1989).
99. F. Mansfeld, S. Lin, S. Kim, and H. Shih, *Mater. Sci. Forum* 44 & 45, 83 (1989).
100. H. Shih and F. Mansfeld, *Corrosion* 45, 610 (1989).
101. F. Mansfeld, S. Lin, S. Kim, and H. Shih, *Electrochim. Acta* 34, 1123 (1989).
102. F. Mansfeld, S. Lin, S. Kim, and H. Shih, *J. Electrochem. Soc.* 137, 78 (1990).
103. F. Mansfeld, *Electrochim. Acta* 35, 1533 (1990).
104. H. Shih and F. Mansfeld, *ASTM STP* 1134, 189 (1992).
105. F. Mansfeld, H. Xiao, and Y. Wang, *Werkstoffe Korros.* 46, 3 (1995).
106. F. Mansfeld and Y. Wang, *Br. Corros. J.* 29, 194 (1994).
107. C. Chen and F. Mansfeld, *Corros. Sci.* 39, 1075 (1997).
108. F. Mansfeld, Y. Wang and H. Shih, *J. Electrochem. Soc.* 138, L74 (1991).
109. F. Mansfeld and J. C. S. Fernandes, *Corros. Sci.* 34, 2105 (1993).
110. F. Mansfeld and B. Little, "A critical review of the application of electrochemical techniques to the study of MIC," Proceedings of the International Conference on Microbiologically Influenced Corrosion, Knoxville, TN, October 1990, p. 5.
111. B. Little, P. Wagner, and F. Mansfeld, *Int. Mat. Rev.* 36(6), 1 (1991).
112. F. Mansfeld and B. Little, *Corros. Sci. 32*, 247 (1991).
113. G. Schmitt, *Werkstoffe Korros.* 48, 586 (1997).
114. B. J. Little and F. Mansfeld, *Werkstoffe Korros.* 42, 331 (1991).
115. F. Mansfeld, C. Chen, C. C. Lee, and H. Xiao, *Corros. Sci.*, 38, 497 (1996).
116. J. S. Potekhina, *J. Prot. Met.* 3, 469 (1984).
117. D. Ōrnek, A. Jayaraman, T. K. Wood, Z. Sun, C. H. Hsu, and F. Mansfeld, *Corros. Sci.* 43, 2121 (2001).
118. F. Mansfeld, C. H. Hsu, D. Ōrnek, T. K. Wood, and B. C. Syrett, *J. Electrochem. Soc.* 149, B130 (2002).

119. D. Ōrnek, T. K. Wood, C. H. Hsu, and F. Mansfeld, *Corros. Sci.* 44, 2291 (2002).
120. F. Mansfeld, C. H. Hsu, Z. Sun, D. Ōrnek, and T. K. Wood, *Corrosion* 58, 187 (2002).
121. D. Ōrnek, T. K. Wood, Z. Sun, C. H. Hsu, and F. Mansfeld, *Corrosion* 58, 761 (2002).
122. A. Nagiub and F. Mansfeld, *Electrochim. Acta* 47, 2319 (2002).
123. M. Dubiel, C. C. Chen, C. H. Hsu, F. Mansfeld, and D. K. Newman, *Appl. Environ. Microbiol.* 68, 1440 (2002).

14
Electrochemical Noise Technique

Francois Huet
Centre National de la Recherche Scientifique, Laboratoire Interfaces et Systèmes Electrochimiques, Université Pierre et Marie Curie, Paris, France

Contents

14.1 INTRODUCTION

Many semimacroscopic phenomena related to corrosion, either localized or uniform, are stochastic by nature. This often complicates the use of the other *in situ* electrical techniques, such as polarization curves, electrochemical impedance spectroscopy (EIS), etc., which are described in different chapters of this book. Moreover, perturbation signals applied in some deterministic techniques may lead to an alteration of the process studied. A simple "listening" to the galvanic system through the analysis of the spontaneous fluctuations of the electrical quantities, current and potential, known as electrochemical noise (EN), appears as the method of choice for its study, with a number of attractive features for corrosion research and engineering.

The problems connected with EN measurements (ENM) can be divided into those concerning the detection of the noise signal above all undesirable influences, such as fluctuations induced by external sources and the intrinsic noise of the instrumentation, and the problems relative to the analysis of the EN measured. Examination of the literature or of the results obtained by different groups in a round-robin experiment on ENM, showing significant deviations among the participants (1), indicates evident measurement problems, often caused by nonstationary signals, instrumental noise, or aliasing in the analog-to-digital (A/D) conversion. In order to achieve a still broader acceptance of the EN technique, it is necessary to address and solve these problems. For this reason a large part of this chapter is devoted to a thorough description of the technique, with particular emphasis on measurement difficulties.

Interest in studying fluctuations in electrochemical systems can be traced back to the pioneering works of Tyagai, Barker, Rangarajan, Epelboin, and coworkers, and others. A large number of articles have been published dealing with the measurement and modeling of EN due to elementary processes (charge transfer, diffusion, etc.) for systems at equilibrium or out of equilibrium. These studies, which have been reviewed by Seralathan and Rangarajan (2), are not considered in this chapter because the noise generated by these processes is much lower than that generated by most corrosion processes, particularly at low frequencies, so that these studies have almost no application in this domain. From the early 1980s, most investigations using EN have concerned the corrosion field, with an increasing number of on-site applications, as shown in the First International Symposium on "ENM for Corrosion Applications" held in Montreal in 1994 (3), and in the regular technical symposia "Plant and Field Applications of Electrochemical Noise" at the annual meetings of the National Association of Corrosion Engineers (NACE) since 1998. Theoretical approaches for EN analysis (ENA), as those presented at the International Symposium "New Trends in EIS and ENA" of The Electrochemical Society held in Phoenix in 2000 (4), are also in constant progress.

The author is aware that many papers on EN in corrosion have not been mentioned in this chapter. The aim was not to review the literature, since this has been done in several papers (5–10). Only works illustrating the main aspects of ENM and data interpretation have been discussed here, with the purpose of providing the reader with a firm basis for understanding what information can or cannot be obtained from his experimental EN data. Particular attention has been given to the relatively new type of ENM in cells with two working electrodes (WEs) that has been introduced to simultaneously measure potential and current fluctuations. For this reason, it was not possible to consider the large amount of work on the mechanisms of localized corrosion, principally conducted on a single WE under potential control, aiming at understanding the role of the metallurgical defects and inclusions in the passive film or its capacitive behavior. For that, the reader is referred to the review papers mentioned above.

In this chapter, a brief historical review of the development of the EN technique is presented, followed by some generalities on the measurement of EN and on the main sources of noise in corrosion. After a detailed description of different measurement techniques, the state-of-the-art methods of data interpretation used to monitor the corrosion rate and distinguish localized from uniform corrosion are discussed. A mathematical background on random signals and digital signal processing, which provides the minimal knowledge for understanding the concepts specific to the EN technique, is presented in Appendix A.

14.2 HISTORICAL REVIEW OF THE DEVELOPMENT OF THE MEASUREMENT TECHNIQUE

Very few papers reported ENM in corrosion systems before the 1970s. Among them, the work of Iverson, a microbiologist interested in biocorrosion, published in 1968 is often referred to as the first EN investigation in corrosion (11). He constructed a circuit with a corroding electrode and a platinum electrode connected through a capacitor and a resistor. The fluctuations of the voltage across the resistor were recorded with a high-impedance voltmeter and a chart recorder. Moreover, using a low-input impedance audio amplifier with the two electrodes connected directly across the inputs, Iverson could hear with a loudspeaker the deep groans, louder or fainter, depending on the rate of corrosion, varied by changing the aggressivity of the electrolyte. Pioneering work by Hagyard and Prior in 1961 on the observation of the open-circuit potential of small aluminum electrodes in KCl solutions had previously shown the existence of voltage transients on oscillograph traces, which were considered to be due to the formation and disappearance of very small individual anodic areas on the electrodes (12). Also, Defranoux in 1963 had reported on "chronopotentiostatic" measurements performed on stainless steels (SS) in artificial seawater with a fast-recording galvanometer with a light spot. Current transients lasting about 1 sec were detected at a controlled anodic potential, and it was proposed that they were related to the process of pit initiation (13).

In the late 1970s, instrumentation was improved with the appearance of operational amplifiers (OAs) of increasingly better quality (high open-loop gain, wide-frequency bandwidth, low noise, etc.), enabling the design of low-noise amplifiers and regulation devices (potentiostat, galvanostat) (14, 15). This point was critical because of the very low amplitude of the EN generated by most corrosion processes in the high-frequency range investigated at that time (roughly between 1 Hz and 10 kHz). The characterization of the statistical properties of EN was also extended. While the mean and standard deviation (SD) of random signals were currently measured using analog instrumentation, such as dc and root-mean-square (RMS) voltmeters, the correlation function in the time domain and the power spectral density (PSD), or its square root, named amplitude spectral density (ASD), in the frequency domain began to be measured. Correlation functions were measured in the mid-1970s with numerical correlators for characterizing the noise generated on an iron electrode in sulfuric acid at the rest potential (16), or that generated by anodic dissolution or diffusion processes (17). A cross-correlation technique was used with two parallel and independent measurement channels to eliminate the spurious noises generated or picked up by the measurement channels in the correlation function (16). By taking the Fourier transform (FT) of the result, the PSD was then derived. Specific instrumentation based on a numerical correlator has even been developed for calculating third-order correlation functions (18). However, it has been recognized that artifacts could be encountered in the measurement of correlation functions. They were mainly due to the electronic noise of the regulation device, to the frequency components at 50 Hz (60 Hz in the U.S.) and harmonics due to pickup from the power supply, and also to the drift of the dc component of the signal. Frequency spectra have been measured by a direct method, first with analog instrumentation, by Okamoto et al. (19) and Epelboin et al. (20): the average of the square of the signal inside a narrow spectral window defined by a band-pass filter was measured with an RMS voltmeter and divided by the width of the window. By moving the window along the frequency axis, the PSD could be obtained in a wide frequency range, typically 1 Hz to 30 kHz. The cross-spectrum of the outputs X and Y of two independent measurement channels was also calculated by taking the time-average of the product $XY/10$ measured by an analog multiplier (20).

In practice, moving the spectral window of the filters was found to be too time consuming, therefore this direct method of PSD calculation was abandoned as soon as digital data acquisition systems (DAS) were commercially available in the late 1970s and early 1980s. PSD measurements were then carried out with Fourier spectrum analyzers comprising internal A/D converters (14, 21–26), or with digital voltmeters (DVMs) controlled by a desktop computer via an IEEE-488 interface bus (27–30). In the latter case, the FT was performed by the computer or the digital data were sent to the memory of a spectrum analyzer to take advantage of the short calculation time of the fast Fourier transform (FFT). With spectrum analyzers, ENM could be carried out in a wider-frequency bandwidth owing to the high speed of the A/D converters. Typically, PSDs or ASDs were measured from 0.1 Hz up to 2 kHz (21, 23), or even up to 30 kHz (24, 31), the lower limit being fixed by the cutoff frequency of the ac amplifiers or high-pass (HP) filters used to ensure the stationarity of the measured signals. Such measurements implied successive data acquisitions in overlapping frequency ranges; the quality of the overlap of the spectra was a good indication of the validity of the measurements.

Among the corrosion systems investigated, those involving generalized corrosion often revealed EN of very low amplitude at frequencies above 1 Hz. In order to reduce the influence of the noise generated by the amplifiers, some workers measured the numerical cross-spectrum (thereafter indicated as XPS) of the outputs of two identical and independent channels (24, 31). However, this approach could not eliminate the noise of the regulation device, which was sometimes preponderant. For example, Bertocci (21) showed that the current fluctuations generated by dissolution of copper in acidified copper sulfate had such a low amplitude that the measured noise was the deterministic response of the electrode to the noise voltage of the potentiostat. By taking the ratio of the electrode potential ASD to the current ASD he could measure the modulus of the impedance of the electrode. From the XPS and the coherence function between the electrode potential and the cell current, it was also possible to identify which part of the cell current was the response to the noise of the potentiostat and which part was caused by random fluctuations of the electrode parameters generated by corrosion processes (23).

Both techniques for measuring frequency spectra using spectrum analyzers and DVMs have been employed for a long time until the appearance in the 1990s of personal computers (PCs) containing a data acquisition card or controlling a DVM, and specific equipment for monitoring EN in the field. Unfortunately, for several years most of the equipments based on PC, even those commercially available, did not include the analog antialiasing (AA) filters necessary for proper A/D conversion (32), so that the quality of the EN data reported in corrosion was sometimes poorer than that of the results obtained 10 or 15 years earlier with spectrum analyzers containing filters. The problem was ignored for a while because most corrosion monitoring investigations were carried out at the open-circuit potential, where only slow processes are involved, and were limited to a single low-frequency range (often 1 mHz to 1 Hz), with the argument that the amplitude of EN at higher frequencies approaches the instrumentation noise level, also with problematical electromagnetic interference at power supply frequencies (9, 28, 33).

Historically, EN in corrosion has been studied first under potentiostatic or galvanostatic control. In the former case the potential of the electrode under test is fixed versus a reference electrode (RE), and the fluctuations of the current flowing through the electrode (electrochemical current noise [ECN]) are recorded. Many research works involved measurements in the passivity domain for studying the breakdown and repair of passive films. An alternative approach for ECN measurements has been developed by the Manchester group for investigations at the corrosion potential, by coupling two freely corroding electrodes through a zero-resistance ammeter (ZRA) (34). This allows the fluctuations of the coupling current to be measured without any RE, which is of great interest in field experiments. The second advantage is that the mean current flowing between identical electrodes remains close to zero, while for a single electrode potentiostatically controlled versus an RE, any change in the open-circuit potential of the electrode tends to polarize it out of the corrosion potential. In the galvanostatic mode, the current is fixed, and the fluctuations of the electrode potential (electrochemical potential noise [EPN]) are recorded against an RE. For investigations at the corrosion potential, no galvanostat is used in order to reduce the instrumentation noise, and the EPN is measured directly between the freely corroding electrode and the RE (35).

While in laboratory studies EPN or ECN is commonly measured using galvanostatic or potentiostatic techniques on a single corroding electrode, it has been shown by the Manchester group that such measurements do not permit obtaining of the corrosion rate. In 1986 this group proposed a new technique using two identical WEs of the same material, same size, and same surface preparation, connected through a ZRA, so as to have both WEs at the corrosion potential (34). The ECN is measured by the ZRA, while the EPN is measured with respect to a third electrode, either a true RE or an electrode identical to the WEs. Most recent applications of corrosion monitoring in the field and the laboratory are based on this technique.

14.3 BACKGROUND ON EN

Fluctuations of the electrical quantities, such as cell current and electrode potential, in electrochemical systems are referred to as EN. Basically, two types of EN are usually distinguished, depending on how the system under study is controlled. For systems polarized at constant current I_0 the potential V of the WE measured with respect to an RE is, at time t,

$$V(t) = V_{WE} - V_{RE} = V_0 + \Delta V(t) \qquad (14.1)$$

where V_0 is the dc component and ΔV the voltage noise, or EPN, representing the random spontaneous fluctuations of the electrode potential around V_0. For investigations at the corrosion potential ($I_0 = 0$) a galvanostat is not used in order to eliminate the influence of its instrumental noise on the voltage noise measured. In the same way, the current noise, or ECN, ΔI, is defined as the fluctuating part of the current $I(t)$ flowing through a WE polarized potentiostatically at potential V_0:

$$I(t) = I_0 + \Delta I(t) \tag{14.2}$$

where I_0 is the dc component. For cell configurations with two identical WEs connected through a ZRA, the current noise represents the fluctuations of the current flowing through the solution between the two electrodes and measured by the ZRA. In that case, the two electrodes are maintained at the corrosion potential as long as their open-circuit potentials are identical, so that the dc current I_0 is 0.

The analysis of the stochastic processes ΔV and ΔI may be performed in the time domain by investigating the shape, size, and occurrence rate of the time transients that are observed in certain corrosion situations, or by measuring the SDs or higher moments of the potential and current fluctuations, mainly for corrosion-monitoring applications. The analysis may also be carried out in the frequency domain by calculating the PSD Ψ_V and Ψ_I of the potential and current fluctuations. Appendix A presents a mathematical background on random signals giving the definition of the statistical quantities involved in ENA, such as the mean average, SD, RMS, correlation function, PSD, etc., and how these quantities are calculated from digitized signals.

The choice of the control device to fix the electrochemical system at the desired polarization point (I_0, V_0) depends on the shape of the stationary current–voltage curve. In some cases, as in the passivity domain, the electrode cannot be controlled galvanostatically, therefore the use of a potentiostat cannot be avoided. When both regulations can be employed, the question of the instrumentation noise of the regulation has to be considered in the choice. Some authors have emphasized the relatively large magnitude of the noise of the potentiostat, but it should be realized that any regulation device using active components such as OAs (potentiostat, galvanostat, ZRA) is actually noisy and provides a contribution in the measured EN, as shown below. Only resistive galvanostats based on batteries and resistors provide negligible contributions, but their use is limited to low polarization currents I_0 (36). Another consideration for the choice of the regulation concerns the simplicity of data interpretation. When measured successively potentiostatically and galvanostatically, ΔI and ΔV are obviously related since they are generated by the same corrosion processes. It will be shown further (see Equation [14.18]) that the FTs $\Delta I(f)$ and $\Delta V(f)$ of $\Delta I(t)$ and $\Delta V(t)$ obey the following relationship:

$$\Delta V(f) = -Z(f)\Delta I(f) \tag{14.3}$$

where $Z(f)$ is the impedance of the WE at frequency f, which gives for the PSDs

$$\Psi_V(f) = |Z(f)|^2\, \Psi_I(f) \tag{14.4}$$

According to Equation (14.3), the shape of the current and potential transients in the time records may differ because of the influence of the interfacial impedance. Hence, depending on the system investigated, it may be easier to interpret and model (shape of the transients, shape of the PSDs, etc.) ECN data than EPN data, or vice versa.

A new cell configuration has been introduced since 1986 for simultaneously measuring current and potential fluctuations (34). It is based on the use of two identical WEs connected through a ZRA that forces the two WEs to be at the same potential, which is the corrosion potential. While the fluctuations of the current crossing the solution are measured with the ZRA, the fluctuations of the potential of the WEs are measured against a third electrode, either a true RE or an electrode identical to the WEs. The definite advantage of such a cell configuration, widely used in corrosion monitoring applications in the field, is to have both current and potential fluctuations measured. However, this is to the detriment of the precision in the control of the polarization point of the WEs. Indeed, in practice the WEs do not always behave identically, so that a difference in their open-circuit potential may appear. In that case, even if they are actually at the same potential, this is not the corrosion potential, and one electrode acts as an anode, the other as a cathode, with a net mean current flowing in solution from the anode to the cathode. Moreover, the relationship between the current and potential fluctuations is not straightforward, in particular the ratio of their FTs does not give the electrochemical impedance of the WEs as in Equation (14.3). This point will be discussed extensively in the section on EN data interpretation.

Noise analysis has been developed for almost a century in many scientific fields (electronics, physics, chemistry, etc.) (37). Very common noise-generating processes, such as thermal noise, shot noise, and flicker noise, have been identified and can be useful for ENA.

14.3.1 Thermal Noise

Thermal noise (also called Johnson noise since it was first discovered by Johnson in 1926 (38)) results from thermal vibrations of electrons and charge carriers. According to the Nyquist theorem (later generalized in the fluctuation–dissipation theorem for any dissipative system), a conductor (metallic, electrolytic, etc.) at thermodynamic equilibrium generates a voltage noise across its terminals whose amplitude is related to the real part of its impedance (39):

$$\Psi_V(f) = 4k_B T \Re e(Z(f)) \tag{14.5}$$

where k_B is Boltzmann's constant ($k_B = 1.38 \times 10^{-23}$ J/K), T is the absolute temperature, and $\Re e$ denotes the real part of a complex number. For example, the PSD of the voltage noise generated by a resistor R at equilibrium ($I_0 = 0$) is

$$\Psi_V(f) = 4k_B TR \tag{14.6}$$

The PSD is frequency-independent, so the voltage noise is white (see the definition of white noise in Appendix A). To model the thermal noise of a more complex circuit, a noise voltage generator of amplitude given by Equation (14.6) is associated in series with any resistive element (e.g., see Figure 14.B.1). The other possibility is to associate a noise current generator in parallel to any resistive element, whose amplitude is derived from Equations (14.4) and (14.6):

$$\Psi_I(f) = 4k_B T \Re e\left(\frac{1}{Z(f)}\right) = 4k_B T \frac{1}{R} \tag{14.7}$$

The amplitude of the thermal voltage noise is low, except for high resistance values. For a resistance of 10 MΩ at 20°C, the PSD is $\Psi_V = 1.6 \times 10^{-13}$ V^2/Hz, which corresponds to an SD of $\sqrt{\Psi_V b} = 0.4$ μV (Equation (14.A.14)) in the frequency range usually investigated in corrosion monitoring applications (1 mHz – b = 1 Hz).

Thermal noise is the absolute minimum of the noise that can be expected. It is often at the origin of the noise of the amplifiers and controlling devices at frequencies above 1 or 10 Hz. Most practical corroding systems, even at the corrosion potential, are not at thermodynamic equilibrium, so that they produce EN well in excess of thermal noise. This was shown, for example, for iron dissolution in acidic medium at.the corrosion potential where hydrogen bubble evolution is the source of EN (20). To the author's knowledge, the only electrochemical systems for which EN was shown to be purely thermal concern redox reactions on inert electrodes at zero current (20, 40, 41).

Thermal noise is of great interest for testing and calibrating the entire channel of EN measurement, from the preamplifier to the software used for PSD calculation. The experiment consists in measuring the noise generated by an electronic circuit with passive components at equilibrium ($I_0 = 0$) and comparing its PSD with that of the thermal noise, $4k_BT\,\Re e(Z)$, calculated from the values of the passive components. Good agreement between the two curves validates the noise measurement, as in Figure 14.1, in which the low- and high-frequency PSD plateaus correspond to the thermal noises of the resistances R and r, respectively, and the PSD decrease is due to the shunting effect of the capacitance.

14.3.2 Shot Noise

Shot noise, first observed by Schottky in vacuum tubes (42), is due to the quantized nature of the charge carriers that move in very short times causing current spikes. If the charge carriers transporting an electric charge q move independently of each other (Poisson process), the average current is proportional to the mean number λ of charge emission per unit time

$$\langle I \rangle = \lambda q \tag{14.8}$$

and the PSD of the current fluctuations is given by

$$\Psi_I(f) = 2q\langle I \rangle = 2\lambda q^2 \tag{14.9}$$

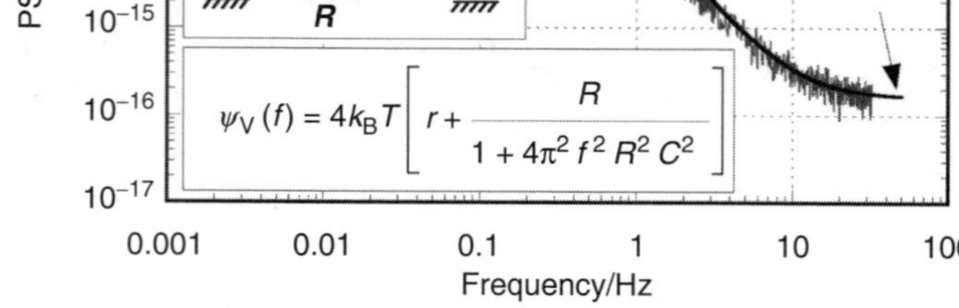

Figure 14.1 Calibration of the EN measurement channel. (A) PSD of the thermal noise generated by the R–C circuit given in the inset ($R = 100\,\mathrm{M\Omega}$, $r = 10\,\mathrm{k\Omega}$, $C = 15\,\mathrm{nF}$) and measured successively at two sampling frequencies of 10 and 100 Hz; ensemble averaging over 20 time records. (B) Theoretical PSD whose expression is given in the inset.

Ψ_I is frequency-independent because of the short times of charge transport, so shot noise in vacuum tubes is white. In electronic circuits shot noise also occurs; its amplitude is low because the charge q is that of an electron. For example, a current of 1 mA produces a shot noise of amplitude $\Psi_I = 3.2 \times 10^{-22}$ A^2/Hz, corresponding to an SD of $\sqrt{\Psi_I b} = 18$ pA in the range (1 mHz – $b = 1$ Hz). However, in localized corrosion, current bursts involving much larger amounts of charge often take place at the electrode, for example, during pit initiation, and then generate shot noise of high amplitude. In this case, the charge exchange is not instantaneous, so the shot noise is not white, the PSD dependence on frequency being affected by the shape of the current–time transients. In the simplest case of independent current bursts, Equations (14.8) and (14.9) can be generalized using the theory of Poisson point processes. For a current transient defined by $i(t,U)$ for $t > 0$ and 0 for $t < 0$, where U denotes a random parameter of the transient (e.g., amplitude, lifetime, time constant, etc.), the average current is (43)

$$\langle I \rangle >= \lambda \left\langle \int_{-\infty}^{+\infty} i(t,U)\mathrm{d}t \right\rangle_U = \lambda \langle q(U) \rangle_U \qquad (14.10)$$

where λ is the mean number of current bursts per unit time, q is the electric charge involved in the transient and $\langle\ \rangle_U$ indicates the average value over the values of U. The PSD of the current fluctuations obeys the following equation:

$$\Psi_I(f) = 2\lambda \left\langle |I(f,U)|^2 \right\rangle_U \qquad (14.11)$$

where $I(f,U)$ is the FT of the transient $i(t,U)$. As an example, the expressions of the average current and PSD of a sequence of exponentially decaying transients of random amplitude A, $i(t,A) = A\mathrm{e}^{-t/\tau}$, as sometimes encountered in pitting of a potential-controlled electrode, are

$$\langle I \rangle = \lambda \langle A \rangle \tau = \lambda \langle q \rangle \qquad (14.12)$$

$$\Psi_I(f) = 2\lambda \frac{\langle A^2 \rangle \tau^2}{1 + 4\pi^2 f^2 \tau^2} = 2\lambda \frac{\langle q^2 \rangle}{1 + 4\pi^2 f^2 \tau^2} \qquad (14.13)$$

These equations are formally similar to Equations (14.8) and (14.9) at zero frequency only. The PSD is white at low frequency and then decreases with increasing frequency. The shape of the PSD and how it is affected by the shape of the current transient will be discussed further in the section on EN data interpretation.

14.3.3 Flicker Noise

Flicker noise, also called $1/f$ noise, refers to noise phenomena with a spectrum in $1/f^{\alpha}$, with α usually, but not necessarily, close to unity. It has been first observed in vacuum tubes by Johnson in 1925 (44) and interpreted by Schottky in 1926 (45). Later, flicker noise was found to occur in most electronic devices and in a wide variety of systems, including electrochemical systems. Despite its nearly universal occurrence, it is unlikely that a universal physical mechanism is at the origin of flicker noise. Many sources of $1/f$ noise have been proposed, as shown by Van der Ziel (46) for electronic devices, such as fluctuations in the number of charge carriers, in their mobility, in temperature, etc., but in many cases, in particular in electrochemistry, the true reason for the presence of $1/f$ noise is rather unclear, even if it is obviously associated to a current flow in the system. In EN studies, flicker noise, which occurs in the instrumentation and limits its sensitivity at low frequency, appears to be a major limitation in the measurement of EN of low amplitude. It should also be noted that $1/f$ noise may naturally exist in EN spectra because of the frequency dependence of the interfacial impedance. For example, a white noise current generated by a diffusion-controlled system gives a PSD of potential noise in $1/f$, according to Equation (14.4), in the frequency range in which the modulus of the electrode impedance decreases in $1/\sqrt{f}$.

14.3.4 Sources of EN in Corrosion

Many sources of EN have been identified in corrosion systems, with some of them appearing in all systems while others are specific to the system under investigation:

1. *Fluctuations of the concentration of the reacting species at the metal–electrolyte interface*. These fluctuations may be first induced by elementary fluctuations in the flux of molecules or ions. The PSD of the resulting ECN can be deduced from the reaction mechanism, involving chemical and electrochemical reactions, by considering that the flux fluctuations can be modeled by white Poisson noise (47). This has been done for the dissolution of iron in sulfuric acidic medium buffered at pH 5, which involves one dissolution and one passivating intermediate species (25), and for a diffusion-limited redox reaction (41). In both cases, after HP filtering the signal below 0.1 Hz to eliminate the drift, the PSD level of the current noise above 1 Hz was found to be low; however, it was higher than the theoretical value. This discrepancy was caused by other EN sources, such as morphological changes in the dissolving surface in the former case, or fluctuations in electrolyte velocity due to natural convection in the latter. Fluctuations in concentration may cause EPN or ECN of much larger amplitude, as for metal dissolution at low pH where hydrogen bubbles evolve. The concentration fluctuations of dissolved hydrogen in the solution close to the electrode induced by the growth and departure of bubbles are known to be a major source of EN (48). Similar effects arise in corrosion systems under forced convection where fresh electrolyte is constantly transported to the electrode surface.
2. *Fluctuations in surface morphology*. Metals are generally polycrystalline materials, consisting of grains of different form, size, and orientation. Various chemical elements are concentrated on the grain boundaries. In addition, cracks, inclusions, voids, etc., are present inside metals. These microstructural features cause metals to be susceptible to random variations in physical properties and, consequently, in their corrosion behavior. Random local dissolution processes are, therefore, at the origin of surface roughness. The discontinuous dissolution of clusters of metal of different sizes, which produces bursts of charge involving thousands of electrons (yet not enough to be detected as individual transients), is likely to be the main source of the low-amplitude EN measured on electrodes corroding uniformly in quiescent electrolytes without hydrogen bubble evolution. Local surface morphology evolution is also probably responsible for the $1/f$ noise measured during silver electrodeposition (31).

3. *Fluctuations in electrode activity*. Breakdown and repair of passive films are widely known processes for generating current transients of large amplitude, due to the nucleation, growth, and death of metastable pits. Two types of current transients have been reported in the literature (6). The first one, defined by a slow rise followed by a sudden decay, seems to be characteristic of pitting corrosion of iron and iron-base alloys. This shape has been interpreted as a sudden decrease in the corrosion rate in the pit due to a rapid change between different steady states of these metals. The second transient type, defined by a sudden rise followed by a slow decay, seems to be characteristic of pitting corrosion of aluminum and to non-chemical initiation of the localized corrosion, such as abrasion, scratching, laser illumination. This transient is related to a sudden failure of the protective layer, due to a drastic change in the anodic reaction resistance on a small area of the electrode, followed by a repassivation process.
4. *Fluctuations in electrode active surface*. Various phenomena may lead to changes in the electrode active surface. For example, during stress corrosion cracking (SCC), the propagation of a crack provokes the formation of bare metal at the crack tip. Generation of bare surface in sudden bursts results in current or potential transients whose shape and sign depend on the anodic and cathodic reactions occurring on the bare surface in the crack tip solution. In contrast, hydrogen bubbles growing on an electrode in acidic medium screen the electrode surface, yielding locally nonuniform current distribution around the attached bubbles, which can be considered in first approximation as a decrease in active surface area. The growth and detachment of bubbles give rise to fluctuations in the size of the active surface, which lead to fluctuations in the ohmic component (changes in solution resistance) and in the kinetic component (changes in the charge-transfer and polarization resistances) of the EN signal. In addition to the fluctuations in dissolved hydrogen concentration mentioned above, bubble evolution also modulates hydrogen penetration into the metal, as shown by permeation experiments on a thin iron membrane in which all sudden increases in hydrogen permeation current were perfectly correlated to bubble detachments on the entrance side of the membrane (49). The various sources of EN induced by bubble evolution have been detailed in Ref. (50).
5. *Fluctuations in electrolyte velocity*. For corrosion processes controlled by mass transport, fluctuations in electrolyte velocity yield fluctuations in the concentration of reacting species at the surface, and therefore give rise to EN. In a model system, like the oxidation of potassium ferri-ferrocyanide at the diffusion limiting plateau on a platinum electrode, natural convection in a quiescent electrolyte has been shown to generate current fluctuations of very low amplitude, which could be reduced to an extremely low level by adding a gel layer on the electrode, thus "freezing" the electrolyte (41). On the contrary, EN of much larger amplitude was measured when stirring the electrolyte, or when using a rotating disk electrode so as to work in laminar or turbulent flows (36). This effect is likely to occur in pipelines in which turbulence is enhanced by the presence of oil/gas/brine mixtures. This effect has also, for example, been observed during corrosion of carbon steel in HCl at pH 3, where EN was drastically increased by electrolyte movement generated by stirring or by a submerged jet (51).

6. *Fluctuations due to external perturbations.* Localized corrosion in abrasion/erosion systems is provoked by continuous impingement of solid particles or gas bubbles that locally break down the passive film or remove the corrosion product film. Under potentiostatic control this gives rise to current transients very similar to those observed in pitting corrosion of aluminum triggered by chloride ions with sudden rise and slow decay. Mechanical damage produced by scratching or laser illumination also induces localized corrosion and, consequently, high levels of EN. On the contrary, impingement of oil droplets on a mild steel electrode in a brine solution was found to decrease both corrosion rate and EN amplitude because of the progressive formation of an "oily phase" on the metallic surface. When increasing the oil content in the brine solution, even if the corrosion rate estimated from electrochemical impedance measurements remained constant, the EN level increased with the progressive appearance of large moving oil macrostructures that temporarily screened the metallic surface (52). More generally, any modification in the operating conditions, either controlled as when corrosion inhibitors are added to the solution or uncontrolled as in pipelines with varying hydrodynamic flows of oil/gas/brine mixtures, is the source of variations in the EN signal.

14.3.5 Motivations of EN Investigations in Various Corrosion Domains

Most early investigations in corrosion using EN concern localized corrosion, essentially pitting corrosion, including both stable pitting and metastable pitting during the prepitting stage. The electrode was polarized potentiostatically in the passive region and current transients were clearly related to the initiation, growth, and death of pits. These early works have been reviewed in Refs. (5, 6), and the experimental techniques used in this field (statistical counting, ensemble averages, spectral analysis), together with a mathematical background on stochastic processes relevant to localized corrosion, have been summarized in Ref. (43). Fundamental approaches have been aimed at characterizing the mechanisms and the rate of pit generation from the evaluation of the specimens' pitting potential, incubation time, and survival probability by statistical treatment of the number and distribution in time of the current transients. Also, the shape, amplitude, and time constant of the transients have been employed to deduce the kinetics of the repassivation process, which is regarded as determining for the corrosion resistance of passive metals against localized attacks.

Under the impulse of the Manchester group in the 1980s, a number of corrosion investigations using EN have been devoted to the determination of the corrosion rate at the corrosion potential in view of field applications. Attempts to quantitatively relate corrosion rate either to fluctuations of the corrosion potential or to fluctuations of the current flowing between two identical electrodes often gave disappointing results. In contrast, measuring the fluctuations of both quantities, potential and current, and taking the ratio of their SD, so as to define the noise resistance R_n (34), was found to be much more successful. Indeed, R_n was often close to the polarization resistance R_p of the electrodes undergoing corrosion, so that the corrosion rate could be estimated by inserting R_n in the Stern–Geary equation in place of R_p. A second interest of EN in corrosion monitoring applications in the field lies in its great potential to be a diagnostic tool to distinguish, possibly in real time, between uniform corrosion and localized corrosion, and perhaps also between various kinds of localized corrosion, such as pitting corrosion, crevice corrosion, SCC, intergranular corrosion, etc., on the basis of shape and amplitude of the current fluctuations. It must be recognized that the classical deterministic techniques (linear polarization, electrochemical impedance, etc.), which average the signals in time, cannot provide such information. This explains why EN has been employed in many corrosion systems, in addition to pitting, as for example:[1]

Stress corrosion cracking: EN is induced by generation of bare metal surface at the tip of propagating cracks and by evolution of hydrogen bubbles nucleated in the tip because of electrolyte acidification. A review of ENM during SCC may be found in Ref. (54) (see also recent references in Ref. (55)). In case of hydrogen embrittlement by cathodic charging of a stressed steel electrode, EN due mostly to hydrogen bubble evolution was found to increase with the enhancement of hydrogen penetration into the metal related to modifications in the damages induced by stress and hydrogen embrittlement (56).

Crevice corrosion: EN is a promising technique for detecting the onset of crevice corrosion, which is typically revealed by a large drop in potential due to the creation of active surface in the crevice. It also gives information on the evolution of the corrosion processes: sharp decays of potential followed by slower rise are often detected before the crevice attack, indicating metastable pitting, while small and slow fluctuations reveal a more uniform corrosion process after the crevice attack (35, 57–59).

Uniform corrosion: In the absence of hydrogen evolution, the EN time records are close to the background noise often observed in electronics, without specific features. EN is due to bursts of current of very low amplitude related to the dissolution process that gives changes in the surface morphology. When hydrogen bubbles evolve, transients appear in the time records at each bubble detachment. However, at high evolution rates, the transients overlap, giving a background noise again. In the absence or in the presence of hydrogen evolution, EN has been shown to provide good estimates of the corrosion rate through the measurement of the noise resistance, especially for high corrosion rates (60–63).

Corrosion of coated metals: Many investigations on the breakdown/degradation processes on coated metal substrates have been carried out with EN methods, in particular by the Mansfeld and Bierwagen groups. For good coatings, no corrosion process takes place on the passive surface so that the EN level is extremely low and sensitive current amplifiers are required. Coating degradation is then monitored from the noise level (64–68).

Abrasion–erosion corrosion: EN has been used for characterizing the kinetics of the breakdown and healing processes of passive films continuously abraded by impinging solid particles from the analysis of EN transients (26, 69). Comparative study of ECN and acoustic emission signals sim-

[1]Considering the large number of papers on EN, the list of references given below is far from being exhaustive; the reader is referred to the reviews on EN applications in Refs. (5–10,53) for other references.

ultaneously measured in slurry erosion–corrosion experiments is also of great interest for investigating the synergy between corrosion damage and abrasive wear (70).

Corrosion in multiphase flows: Recent studies have shown that EN was a practical technique for monitoring corrosion in multiphase flows, which is strongly related to the flow pattern. High turbulence, as in the oil/brine/gas mixing zone in slug flows, may damage and remove the inhibitor film locally, thus enhancing corrosion there. The EN technique is also able to detect changes in flow regime, with significant differences between full pipe flow and slug flow, when using data acquisition rates of the order of 100 Hz or more, which is higher than that used in common field applications (71–73).

The EN technique has also been applied more or less successfully in other domains, for example, to study microbiologically induced corrosion (74, 75), intergranular corrosion in lead-acid battery electrodes (76), corrosion in reinforced concrete (77, 78), or to monitor continuously the formation and deterioration of corrosion inhibitor films, even when rapid changes in the operating conditions occur, which cannot be achieved by impedance measurements that can only be used in relatively stable electrochemical systems (79, 80).

14.4 MEASUREMENT TECHNIQUES

Because of the low level of the EN signal, great care must be taken in devising the whole measurement, from the electrochemical cell and the controlling device to the DAS, as discussed in detail later. Much valuable practical information on this topic may be found in the book on impedance and noise measurements of the NACE series *Corrosion Testing Made Easy* by Cottis and Turgoose (81).

14.4.1 Electrochemical Cell

Cell configurations with one or two WE are employed. Compared to other investigation techniques, nothing specific can be noted in the electrode preparation for ENM, except that the technique is much more sensitive to undesired effects, such as crevice corrosion between the sample and the insulator in which it is mounted (a thin layer of cataphoretic paint may be deposited on the lateral surface of the sample before mounting in the insulator to avoid this) or slow variation of the surface area during metal electrodeposition or dissolution that yields drifts in the potential and current signals and limits data analysis at low frequencies.

A first consideration for choosing the size of the WE comes from theoretical expectations on the EN level. For uniform corrosion, and more generally for processes involving a large number of corrosion events of low amplitude on the whole electrode surface area, the PSD of the current noise is proportional to the surface area S (SD proportional to $\sqrt{S}$), while the PSD of the potential noise is proportional to $1/S$ (SD proportional to $1/\sqrt{S}$) (8). Thus, large electrodes are preferable for measuring ECN and small electrodes are preferable for measuring EPN. However, this conclusion is not necessarily valid for localized corrosion such as pitting for which material and environment factors play an important role in the occurrence and number of pits on the WE. The EN level is then system-dependent and, therefore, it may be incorrect to normalize ECN data by dividing the amplitude of the current by the surface area. In pitting corrosion, another factor to be considered in the choice of the electrode size is the surface area available for the cathodic reaction, which must be large enough to provide the current necessary for stable pitting to occur (81). Other practical considerations concerning the probe design for monitoring various types and rates of corrosion in laboratory and field applications may be found in Ref. (9) with examples of typical probes. Finally, when monitoring corrosion of large structures with small probes, the occurrence of pits is much less probable on the probe than on the structure. To get an early warning of pitting corrosion from the probe, different approaches have been suggested, such as employing less corrosion-resistant materials for the probes, applying anodic polarization on the probes, or using probe surface finishes more susceptible to pitting corrosion (81).

True REs, such as saturated calomel electrodes (SCEs), saturated sulfate electrodes (SSEs), or silver/silver chloride electrodes, or a third electrode identical to the WEs in cell configurations with two WEs, are used for measuring or controlling the electrode potential. In the former case, the RE is usually considered as noiseless, but its noise may contribute to the measured EPN in systems generating EN of very low amplitude, as during uniform corrosion of copper in hydrochloric acid (82). Also, an RE is noisier when the electrolyte in contact with the reference element is polluted, even if the mean potential given by the RE is correct; therefore, the electrolyte inside the RE must be changed regularly, for example, once a day. The voltage noise of an RE may be determined with three identical REs of the same type by measuring the cross-spectrum of the voltage differences $V_{RE1} - V_{RE2}$ and $V_{RE1} - V_{RE3}$, which gives the PSD of the voltage noise of RE 1 (25).

14.4.2 Electrochemical Interfaces

ENMs can be performed with a single WE, under potentiostatic or galvanostatic control, or with two WEs connected through a ZRA. Figure 14.2(a) and (b) shows how to measure the ECN and EPN generated by a single WE. Under potentiostatic control, the current $(I_0 + \Delta I)$ is measured across the resistor R (Figure 14.2a), or

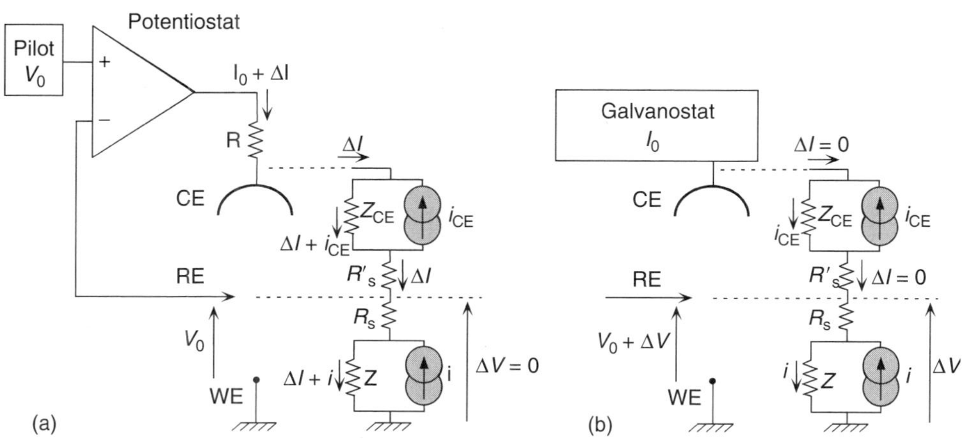

Figure 14.2 Electrical scheme and equivalent circuit of the measurement methods of ECN and EPN of a single WE. (a) Potentiostatic method. (b) Galvanostatic method.

using a current-to-voltage converter connected to the WE. For EPN measurements (Figure 14.2b), ΔV is measured between the RE and the WE. In case of measurement at the corrosion potential ($I_0 = 0$), the galvanostat is not needed. Both figures also exhibit the equivalent circuit between the counter electrode (CE) and the WE, in which the fluctuations due to anodic and cathodic processes occurring on the CE and the WE are modeled by current noise sources i and i_{CE} in parallel to the respective impedances Z and Z_{CE} (which include double-layer capacitances), in a manner similar to the thermal noise modeling explained above. The RE is here considered to be noiseless and the thermal noise of the solution resistance R_s is neglected. Ohm's law between the RE and the WE gives for Figure 14.2b

$$\Delta V(f) = Z(f)i(f) \tag{14.14}$$

In this equation, all terms are complex and frequency-dependent, $\Delta V(f)$ and $i(f)$ are the FTs of the real and time-dependent quantities $\Delta V(t)$ and $i(t)$. In the same way, ΔV in Figure 14.2(a), which is equal to 0 since the electrode potential is controlled, is

$$\Delta V(f) = R_s \Delta I(f) + Z(f)[\Delta I(f) + i(f)] = 0 \tag{14.15}$$

which gives

$$\Delta I(f) = -\frac{Z(f)i(f)}{R_s + Z(f)} = -i(f) \tag{14.16}$$

when the solution resistance is negligible, or, in the time domain

$$\Delta I(t) = -i(t) \tag{14.17}$$

Equations (14.14) and (14.16) show that both EPN ΔV and ECN ΔI do not depend on the current fluctuations i_{CE} generated by anodic or cathodic processes on the CE and, second, that they obey Equation (14.3), which can be rewritten as

$$\Delta V_g(f) = -Z(f)\Delta I_p(f) \tag{14.18}$$

where the indexes g and p emphasize that ΔV_g and ΔI_p are not measured in the same experiment (as in cell configurations with two WEs), but in two different experiments, under galvanostatic and potentiostatic control, respectively. This derivation also shows that the current noise flowing in the WE is $\Delta I + i = 0$ under potential control and i under current control. The current noise i generated by corrosion processes on the WE crosses the solution toward the CE in the former case, while it flows back toward cathodic areas on the WE in the latter.

The experimental arrangement used in cells with two identical WEs for measuring both EPN and ECN is shown in Figure 14.3. The potentials of the metallic phase of electrodes 1 and 2 are equal because of the ZRA, and in high-conductivity solutions, the potential is constant in the electrolyte between the WEs, so that both WEs work at the same electrochemical potential. If no net current flows from one electrode to the other, both WEs are actually at the corrosion potential. The analysis of the simultaneously measured ΔV and ΔI noises and their relationship as a function of the impedance of the WEs is more involved than for a single-WE cell; it will be given below in the section on data interpretation. Asymmetry in the WEs is often unavoidable because corroding electrodes, particularly if corrosion is localized, tend to develop differently with exposure time even if the initial conditions were the same. Moreover, cases may be envisaged where identical electrodes cannot be used, as in SCC investigations where only one electrode is under stress, or in crevice corrosion

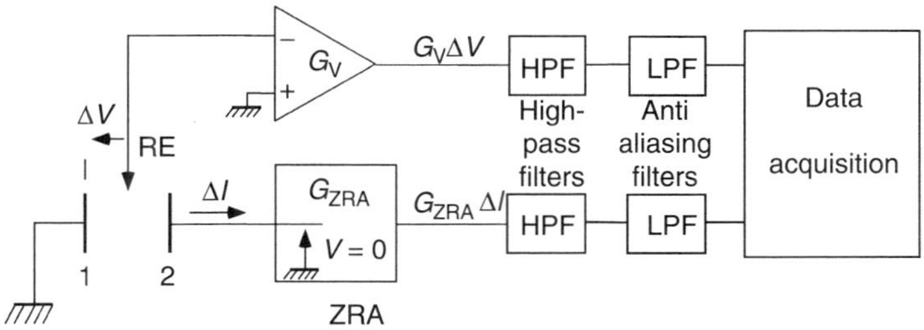

Figure 14.3 Experimental arrangement used for simultaneous measurements of EPN and ECN at the corrosion potential (G_V and G_{ZRA}: gains of the VA and ZRA, LPF: low-pass filter).

studies where crevice attack occurs on one electrode only (9). Instead of trying to prevent asymmetry, some authors have purposely employed it, in order to limit corrosion to one electrode, but still derive the noise resistance by measuring both current and voltage signals. For instance, exploiting asymmetry in the surface areas of the WEs has been suggested by Pistorius to measure the anodic dissolution rate on the smaller electrode with reduced interference from processes on the larger electrode (83). In another case a Pt microcathode was utilized as one of the electrodes (84) and in still another, asymmetry was produced by applying a bias potential between the two electrodes (85). These techniques will also be analyzed below in more detail.

14.4.3 Signal Conditioning

Because of the low amplitude of the EN signal and the difficulty in distinguishing it from the instrumental noise, it is important to use low-noise, highly sensitive instrumentation (amplifiers and ZRA), especially at frequencies above 1 Hz or at all frequencies for coated specimens where the current noise may be in the nanoampere range. The instrumentation must also have a high-input impedance for ENM on passive or coated electrodes. When measuring potential noise with a true RE, a dc component V_0 of 1 V (and sometimes more) may be present in the signal, which limits the gain of the voltage amplifier (VA). Whenever possible, it is advantageous to use amplifiers with three successive stages: a gain of 10 in the first one allows the influence of the noise generated by the subsequent stages to be minimized, the dc component is then eliminated by a low-noise dc-offset circuit, so that an amplifier of large gain can be used in the third stage. In some applications the electrode potential is measured against an electrode identical to the WE and, as a consequence, the dc component of the signal is close to zero. The EPN measured is then the sum of the voltage noises generated by the two electrodes and, in the case of independent voltage noises of the same amplitude, the SD of the EPN measured is $\sqrt{2}$ times that of the noise generated by the WE (see Equation (14.A.33) in Appendix A). However, this assumes that both electrodes are identically noisy, which is reasonable for generalized corrosion but rather questionable for localized corrosion in which corrosion events may occur on one electrode and not on the other. ECN measurements can also be performed with a VA (to which the above considerations apply) by measuring the voltage across a resistor (see Figure 14.2a), or with a ZRA. In that case, a low-noise potentiostat can be used, with the RE and CE outputs connected together and the dc voltage control set to 0 mV.

Stationarity is a required property of the EN signal, for second-order statistics, such as SD or PSD, to be valid. Therefore, drifting signals, that is, signals for which the dc baseline appears to steadily change, for example, because of progressive deterioration of the electrode undergoing corrosion, have to be conditioned before any statistical calculation. This can be done on the analog signal by inserting an analog HP filter in the measurement channel before data acquisition, as shown in Figure 14.3, to remove the components of the signal below a certain frequency, called cutoff frequency. Often analog filtering is present in instrumentation, such as amplifiers, which is "ac coupled," but the cutoff frequency is in general too high to be useful in EN work. Passive filters can be built inexpensively, but much better results can be achieved with active filters that can have higher down slopes below the cutoff frequency. The greatest drawback of analog filtering in real time is that, having a low cutoff frequency, they are subject to large oscillations when the signal shows sudden large transients. The oscillations then often drive amplifiers to saturation, and recovery times, being some multiples of the inverse of the cutoff frequency, may be annoyingly long. Digital HP filtering methods have also been proposed to detrend the signal after acquisition. These methods will be discussed below.

Most DAS are based on the digitization of analog signals, using either an A/D converter or a DVM. According to the Nyquist theorem, when sampling at frequency f_s, the maximum frequency that can be correctly analyzed is $f_{max} = f_s/2$. Therefore, if the analog signal contains components at frequencies higher than $f_s/2$, these components will appear in the digital time record at frequencies lower than $f_s/2$. To avoid this phenomenon, called aliasing, all signal components at frequencies above $f_s/2$ have to be eliminated before the A/D conversion by using a low-pass analog filter, called AA filter, with a cutoff frequency fixed at a value slightly lower than f_{max}, typically $0.7f_{max}$ to $0.8f_{max}$. The necessity of using AA filters is illustrated in Figure 14.4 for the measurement of white noise (frequency-independent PSD). White noise signals are typically encountered in localized corrosion, especially when the measurements are carried out only at low frequencies, as is usually done in corrosion-monitoring applications at the corrosion potential. Figure 14.4(a) shows the PSD of a white noise of 500 mV peak-to-peak (mV_{pp}) amplitude in the frequency range (0 to 1 kHz) delivered by a signal generator. Above 1 kHz, the signal was strongly attenuated by a low-pass filter inside the signal generator. The PSD measurements were carried out with a spectrum analyzer with (curves A, B, C, D) and without (curves E, F, G) AA filters, successively at four different sampling frequencies. It can be seen that the PSDs A, B, C, and D coincided perfectly when filters were used, whereas without filtering, the PSDs E, F, and G of the same signal had unacceptably different levels in overlapping frequency ranges. In that case, the signal power at high frequencies which should have been eliminated by an AA filter was aliased at lower frequencies. From these experiments, it can be surmised that all frequency-independent spectra in the literature measured without AA filtering have a wrong amplitude due to the presence of aliasing. Aliasing not only affects the PSD of the random signal, but also its digital time record. As an example, Figure 14.4(b) presents two time records of the white noise acquired with and without AA filter at sampling frequency $f_s = 128$ Hz. Their corresponding PSDs are curves D and G, respectively, in Figure 14.4(a). The SD of the filtered signal was 22.4 mV (curve A), while, without filtering, an incorrect value of 110.2 mV was found (curve B). It must be emphasized that, since aliasing cannot be eliminated from the time record after sampling, it is not possible, from curve B in Figure 14.4(b), to derive the correct value of the signal amplitude or its SD, a parameter that has to be estimated when monitoring corrosion from ENM. The aliasing problem was solved a

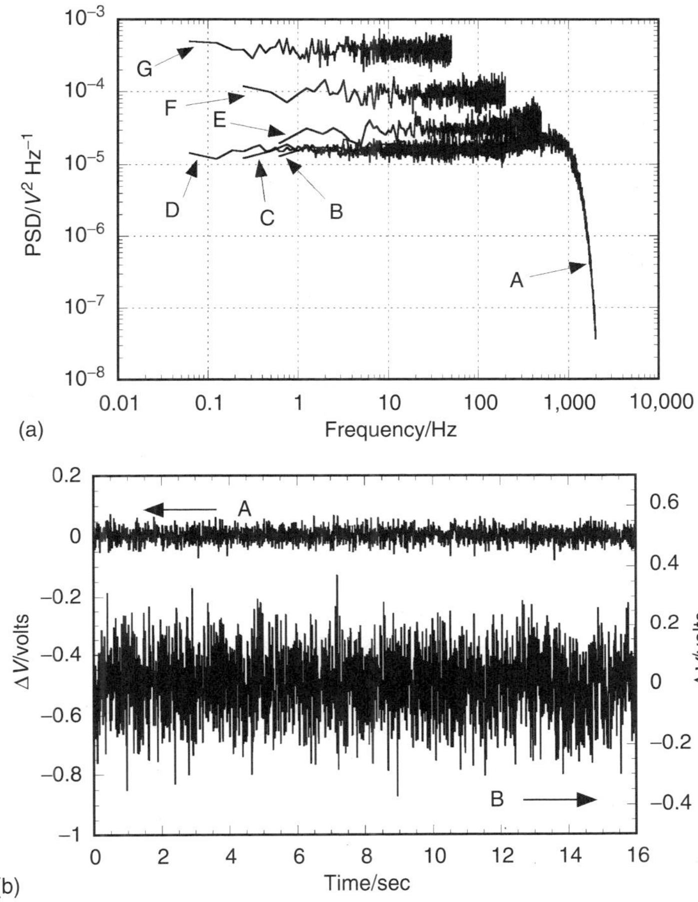

Figure 14.4 Influence of aliasing. (a) PSD of a white noise of 500-mV_{pp} amplitude in the frequency range (0 to 1 kHz), measured at sampling frequencies $f_s = 5120$ Hz (A), 1280 Hz (B, E), 512 Hz (C, F), and 128 Hz (D, G) with (A, B, C, D) or without (E, F, G) AA filtering. Ensemble averaging over 20 time records. (b) Time records sampled at frequency $f_s = 128$ Hz with (A) and without (B) AA filtering and corresponding to PSD curves D and G, respectively, in (a). (From IN Bastos, F Huet, RP Nogueira, P Rousseau. *J. Electrochem. Soc.* 147:671–677, 2000. With permission.)

long time ago in commercial spectrum analyzers but is currently ignored in the rapidly spreading noise measurements using data acquisition cards in PCs or DVMs. If most DVMs allow elimination of the power supply interference to increase the measurement accuracy (to the detriment of the sampling rate), the averaging procedure involved cannot prevent aliasing for all types of EN (32). It is therefore strongly recommended to those who use DVMs or acquire data with acquisition cards in PCs to add an AA filtering system, based on analog filters or a combination of analog and digital filters, as explained for example in Ref. (32).

14.4.4 Data Acquisition System

Even if some of the simpler EN variables (mean, SD, RMS value, PSD) may be measured using analog instrumentation (81, p. 61), most DAS are based on DVMs and A/D converters included in spectrum analyzers or in PCs, so that all calculations of parameters are performed with digitized data. Different types of A/D converters and interfaces for their connection to the DAS for data storage are presented in Ref. (81, p. 63). The resolution of the A/D conversion is given in the decimal number system (e.g., a three-digit DVM reading from −9.99 to 9.99 V has a resolution of 10 mV) or in the binary number system (e.g., a 12-bit A/D converter reading from −10 to 10 V has a resolution of $10\,\mathrm{V}/2^{11} = 4.88\,\mathrm{mV}$). Low resolution in ENM may be easily identified on a voltage or current time record by the presence of quantized levels, which indicates that the analog signal has not been sufficiently amplified before the A/D conversion. This may happen when measuring drifting signals with low-resolution DAS because the gain of the amplifier has to be low to avoid saturation. It is strongly recommended to use DAS with high resolution such as A/D converters working on 16 bits or more to minimize the quantization noise. For a 16-bit converter, PSDs varying over about nine decades in amplitude, quite rare in corrosion studies, can be correctly measured since the logarithm of the square of 2^{16} is higher than 9 (9.63).

14.4.5 Choice of Parameters and Measured Quantities

Since most corrosion processes are expected to be slow and generate EN signals of low amplitude at higher frequencies, investigations at the corrosion potential, especially in field applications, are conducted with DAS sampling at $f_s = 2\,\mathrm{Hz}$, giving a time interval of $\Delta t = 0.5$ sec, between two successive data points and an upper limit of frequency analyzed of $f_{\mathrm{max}} = 1\,\mathrm{Hz}$. However, in some cases, as for corrosion in multiphase flows, sampling rates of the order of 100 Hz or more are required to detect rapid variations in the flow regime and to study their influence on the corrosion processes at the WE. Another case concerns pitting corrosion at a controlled anodic potential, which was the object of a large number of investigations with the EN technique. The electric charge involved in the repair of the passive film after breakdown is provided by the double-layer capacitance of the WE, which is then quickly recharged by the potentiostat. Data acquisition rates of a few kilohertz, possible since the late 1970s with spectrum analyzers, are necessary to follow film repassivation.

The lowest frequency analyzed f_{min} is the inverse of the duration T of the time record $f_{\mathrm{min}} = 1/T$. It then depends on the number N of data points in the time record, which is usually taken as 1024, 2048, or 4096, a power of 2 to subsequently apply the FFT algorithm. For example, a value of $N = 2048$ with a sampling rate f_s of 2 Hz gives the commonly used frequency range (1 mHz to 1 Hz).

For a random signal $x(t)$, current or potential, various statistical quantities may be calculated from the digitized data x_n at time $n\Delta t$. Some basic quantities have been defined in Appendix A, some others are given below; their application to corrosion studies will be discussed in the section on data interpretation. The statistical quantities independent of the position n of x_n in the sequence of acquisitions are considered first. The mean $\langle x \rangle$, SD σ_x (or variance σ_x^2), RMS value x_{RMS}, and coefficient of variance CoV are derived from the first- and second-order moments of x, while the skewness (or skew) and kurtosis are derived from the third- and fourth-order moments. The quantities $\langle x \rangle$, σ_x, and x_{RMS}, are given by Equations (14.A.17), (14.A.19), and (14.A.8), respectively, and the values of CoV, skewness, and kurtosis are calculated from the following equations:

$$\mathrm{CoV} = \frac{\sigma_x}{\langle x \rangle} \tag{14.19}$$

$$\mathrm{Skewness} = \frac{1}{N-1}\sum_{n=0}^{N-1}\left(\frac{x_n - \langle x \rangle}{\sigma_x}\right)^3 \tag{14.20}$$

$$\mathrm{Kurtosis} = \left[\frac{1}{N-1}\sum_{n=0}^{N-1}\left(\frac{x_n - \langle x \rangle}{\sigma_x}\right)^4\right] - 3 \tag{14.21}$$

where the term -3 in Equation (14.21) makes the kurtosis equal to zero for a gaussian distribution. Skewness and kurtosis are two dimensionless factors that describe the distribution of the x_n values about the mean. The skewness characterizes the degree of asymmetry, a positive (respectively negative) value corresponds to an asymmetric tail extending out toward larger (respectively smaller) x. The kurtosis measures the "peakedness" (positive kurtosis) or flatness (negative kurtosis) of a distribution relative to the gaussian distribution (kurtosis $= 0$). The standard errors of the estimates of the skewness and kurtosis of a gaussian distribution calculated from $N = 2048$ data points are $\sqrt{15/N} = 0.085$ and $\sqrt{96/N} = 0.21$, respectively, and it is good practice to consider the values of skewness and kurtosis as significant only when they are several times as large as these values of standard error (86, p. 612).

For EN investigations in cells with two WEs, two other quantities, the noise resistance R_n and the localization index LI, have been introduced by the Manchester group (33, 34). They are calculated from the first- and second-order moments of the voltage and current fluctuations, as follows:

$$R_\mathrm{n} = \frac{\sigma_\mathrm{V}}{\sigma_\mathrm{I}} \tag{14.22}$$

$$\mathrm{LI} = \frac{\sigma_\mathrm{I}}{I_\mathrm{RMS}} = \frac{\sigma_\mathrm{I}}{\sqrt{\sigma_\mathrm{I}^2 + \langle I \rangle^2}} \tag{14.23}$$

All these statistical parameters may be calculated on acquired time records by subsequent off-line processing, but they can also be derived on-line by a recursive procedure consisting in calculating any parameter at time $n\Delta t$ from its value at time $(n-1)\Delta t$ and the new input data x_n. For example, the mean and SD are given by

$$\langle x \rangle_n = \frac{n-1}{n}\langle x \rangle_{n-1} + \frac{1}{n}x_n \tag{14.24}$$

$$\left(\sigma_x^2\right)_n = \frac{n-2}{n-1}\left(\sigma_x^2\right)_{n-1} + \frac{1}{n}\left(x_n - \langle x \rangle_{n-1}\right)^2 \tag{14.25}$$

For ensemble averaging the PSD in the periodogram method (Equation (14.A.21)), it is of great interest to use Equation (14.24) to have a PSD continuously updated without the necessity of storing the PSDs previously calculated.

Methods taking into account the position of x_n in the time record are essentially based on the analysis of transients observable, if any, in the potential and current time records (shape, amplitude, number), and on the calculation of second-order statistical functions (correlation function in the time domain and PSD in the frequency domain). Higher-order statistical functions, such as the third-order correlation function (18) or bispectrum (10), have been proposed, but they will not be discussed here since they have not yet found application in corrosion studies, possibly because of the great difficulty in extracting relevant information. Several methods have been proposed to calculate PSDs, but only those working on digital data, like the maximum entropy method (MEM) and the periodogram method based on FFT, are commonly used in the corrosion field (see Appendix A for definitions and calculation methods). It has been claimed that the MEM has several advantages over the FFT: it is faster than the FFT because it does not require ensemble averaging, so that acquiring a single time record is sufficient; it gives smoother spectra than the FFT; and third, it allows computation of the spectrum at frequencies lower than $f_{min} = 1/T$, which is the lowest frequency calculated by the FFT. As mentioned in Appendix A, a disadvantage of the MEM is the uncertainty introduced by the arbitrary choice of the order M, which is the number of coefficients used in the computation. For featureless spectra, such as those usually measured below 1 Hz in corrosion studies, the choice of M may not be critical. Figure 14.5 compares the PSDs measured on white noise by the two techniques: the PSDs are designated as FFT(K), to indicate the number K of averages taken in the FFT method, or MEM(M), to indicate the order used in the computation. The computation times of the MEM and FFT algorithms may be neglected in corrosion applications compared to the extended acquisition times required for corrosion monitoring, so that measuring an FFT(K) spectrum is K times longer than measuring an MEM(M) spectrum. The MEM computed with few coefficients produces a very smooth spectrum, preferable to that obtained by FFT, even after averaging. For this reason low values of M are commonly used in corrosion applications. However, the smoothness, or in other words, the poor frequency resolution obtained for low-order M allows characterization of white noise or featureless spectra, but not spectra having a more

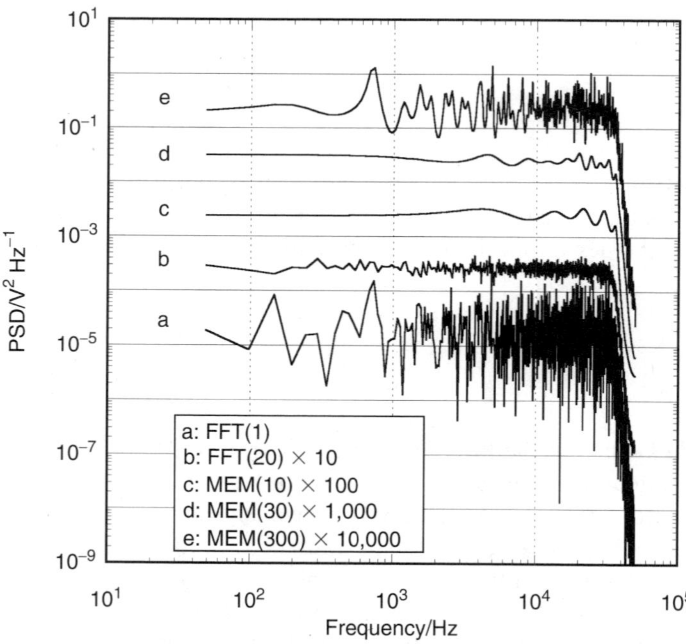

Figure 14.5 Comparison of MEM and FFT on white noise. PSD calculated with FFT(1) (a), FFT(20) (b), MEM(10) (c), MEM(30) (d), and MEM(300) (e). For clearer reading, the curves b, c, d, and e were multiplied by 10, 10^2, 10^3, and 10^4, respectively. (From U Bertocci, J Frydman, C Gabrielli, F Huet, M Keddam. *J. Electrochem. Soc.* 145:2780–2786, 1998. With permission.)

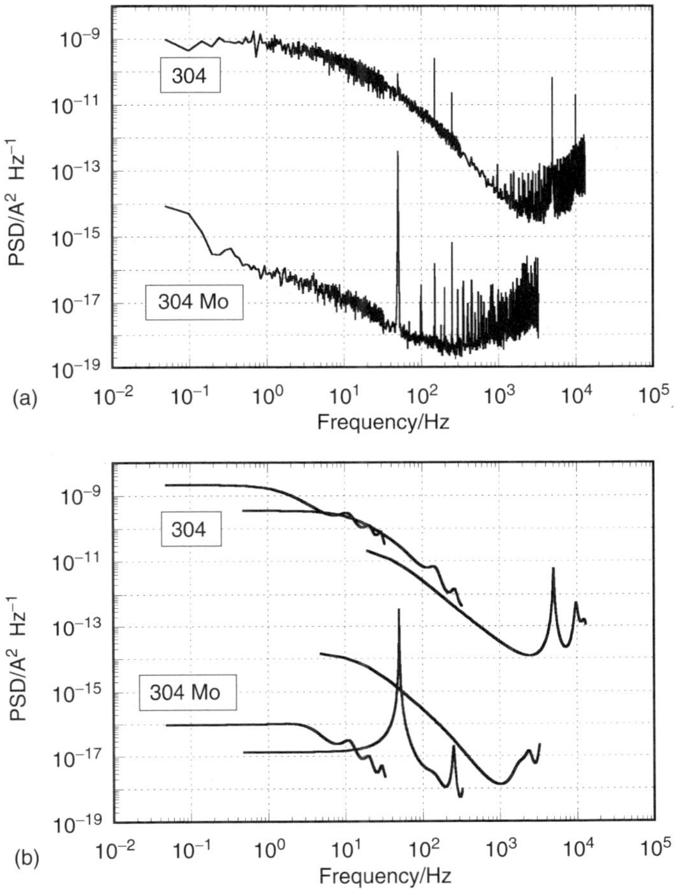

Figure 14.6 Comparison of MEM and FFT for deconvolving interfering signals. Corrosion of 304 SS with and without Mo implanted in 0.5 *M* H_2SO_4 + 0.05 *M* NaCl at potential 0.3 V versus an SSE. PSD of current fluctuations measured on three overlapping bandwidths with FFT(10) for the lowest bandwidth, FFT(20) otherwise (a) and MEM(10) (b). (From U Bertocci, J Frydman, C Gabrielli, F Huet, M Keddam. *J. Electrochem. Soc.* 145:2780–2786, 1998. With permission.)

complicated structure. As an example, Figure 14.6 shows the PSDs calculated by MEM and FFT of EN signals contaminated by the effect of ac power at 50 Hz and harmonics. The measurement of the current fluctuations of corroding 304 SS electrodes, with and without molybdenum implantation, under potential control in sulfuric acid with added chloride was performed on three overlapping bandwidths (87). The FFT spectra show a large number of peaks due to pickup from the power supply. This is particularly evident for the electrode with implanted Mo, because of its much lower EN level. Owing to Hann windowing (see Appendix A), which reduces the width of the peaks induced by the power supply, the shape of the PSD is discernible, and the interfering peaks could, if desired, be easily removed. For the PSDs obtained by MEM (10), the situation is much more confused because of the inability of this technique to reduce the width of the spectral lines. If the level of the EN signal is fairly large (304 SS electrode), in spite of the already mentioned poor tendency to overlap frequency ranges typical of a low-order *M*, it could be argued that the lack of frequency resolution of this method produces cleaner curves (although it is not known how the energy of the power supply peaks is distributed in the spectrum). However, in the case of 304 Mo SS electrode, the characteristics of the low-order MEM (poor overlapping and limited frequency resolution) combine to produce PSD curves of poor quality. Another result shown for both electrodes in Figure 14.6(b) is that the level of the low-frequency plateau measured at the intermediate sampling frequency is far below that obtained at the lowest sampling frequency. This means that the extension of the PSD computation to arbitrary frequencies below $f_{min} = 1/T$ produces misleading results, generating a low-frequency plateau, whether it exists or not. In the case of systems giving poorly stationary signals, misleading results can be produced with both methods, as shown below, and Hann windowing becomes absolutely necessary. However, the erroneous spectra obtained by the FFT method can be recognized more easily than for the MEM (see below), so that the FFT method is to be preferred if stationarity is not certain. In that case, if one wishes to improve the clarity of PSDs calculated with FFT(1), subsequent off-line smoothing may be applied (88).

Finally, for ENM in cells with two WEs, a quantity derived from the ratio of the PSDs of the EPN and ECN has been introduced by Bertocci et al. (89). Its terminology has varied with time; it is now commonly called "noise impedance" and is defined as

$$Z_n(f) = \sqrt{\frac{\Psi_V(f)}{\Psi_I(f)}} \tag{14.26}$$

When the PSDs are measured from only one potential and one current time record, which is often the case in practice, Z_n is strictly identical to the spectral noise response $R_{sn}(f)$ defined earlier by Xiao and Mansfeld (90) as the modulus of the ratio of the FTs of the potential and current fluctuations.

14.4.6 Measurement Problems

External sources of noise, such as mechanical vibrations and, above all, electromagnetic interference, have to be minimized, particularly if the EN level is low. Electromagnetic interference is often picked up from the ac power supply, but it may also come from the switching on and off of electric equipment in the experiment or in the environment. For example, to control the temperature of the electrolyte, it is strongly advised to use a thermostat switching on and off when the value of the ac power supply voltage is 0 V. In the laboratory, interference reduction may be obtained by using a Faraday cage and battery-powered instrumentation, but sometimes surrounding the electrochemical cell with an aluminum foil connected to the ground is sufficient. In contrast, interference reduction is a difficult task in some cases and practical solutions such as those suggested in Ref. (81, pp. 66–67) should be considered.

The effect of the instrumental noise in ENM can be described by means of a model of interaction of the instruments with the electrochemical system under study. As a simple example, the influence of the noise of a VA on the measurement of an EPN ΔV generated by a corroding system of impedance Z_s can be examined with the measurement scheme depicted in Figure 14.7. The noise of the VA of gain G_V is schematized by a voltage noise source e_n and a current noise source i_n. Appendix B gives a detailed analysis of these noise sources for the simplest design of a high-

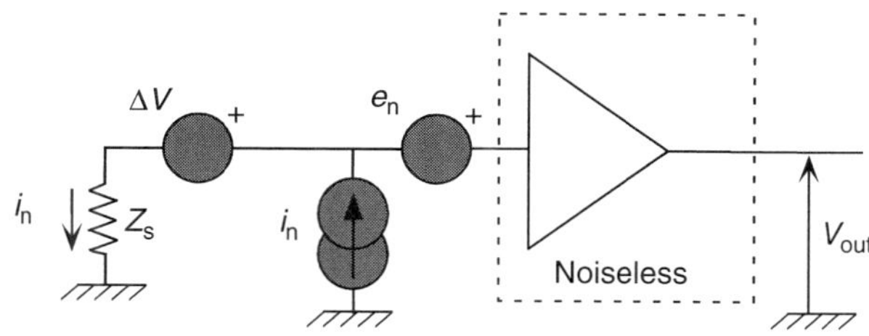

Figure 14.7 Modeling of the instrumental noise of the VA of gain G_V with the noise sources e_n and i_n.

impedance VA with a single OA and two resistors. For commercial amplifiers, the characteristics of e_n and i_n (amplitude, PSD) are given in technical specifications. The contribution of the other parts of the instrumentation, such as subsequent dc-offset circuit, amplifiers, filters, and A/D converters, can be neglected if there is sufficient amplification in the VA. The voltage output V_{out} is given by

$$V_{out} = G_V(\Delta V + e_n + Z_s i_n) \tag{14.27}$$

and the voltage measured ΔV_{meas} is obtained by dividing V_{out} by the gain of the VA

$$\Delta V_{meas} = \frac{V_{out}}{G_V} = \Delta V + e_n + Z_s i_n \tag{14.28}$$

The three noise terms are independent so that the PSD of ΔV_{meas} is

$$\Psi_{V_{meas}} = \Psi_V + \Psi_{e_n} + |Z_s|^2 \Psi_{i_n} \tag{14.29}$$

The contribution of the instrumental noise of the VA is then $\Psi_{e_n} + |Z_s|^2 \Psi_{i_n}$, which reduces to Ψ_{e_n} for low-impedance corroding systems, indicating that VAs with good characteristics in voltage noise have to be used. For high-impedance systems, as for passive or coated electrodes, VAs with current noise of low amplitude must be employed. A similar derivation of the contribution of the instrumental noise of the current amplifier on ECN measurements may be found in the appendix of Ref. (82). This analysis has been criticized by Cottis (10) on the basis that the noise at the inverting input of the OA has been ignored. However, there is no requirement for adding an extra noise source at this input, so that the author still considers that it is preferable to use an instrumentation amplifier to detect the voltage across the measuring resistor, rather than a conventional current amplifier based on a single OA.

The instrumental noise of the controlling device, essentially potentiostat and ZRA in corrosion studies, has to be considered too. A detailed analysis of the noise generated by a potentiostat is given in Ref. (25). Schematically, the voltage noise V_{reg} generated by the potentiostat around the mean controlled voltage V_0 produces a current noise $I_{reg} = V_{reg}/Z$ that is added to the ECN ΔI flowing through the impedance Z of the WE. If the noise sources of the different parts of the potentiostat (OA, current-measuring resistor, current amplifier) give an overall contribution i_n to the measured current I_{meas}, the PSD of I_{meas} is then

$$\Psi_{I_{meas}} = \Psi_I + \frac{\Psi_{V_{reg}}}{|Z|^2} + \Psi_{i_n} \tag{14.30}$$

$\Psi_{V_{reg}}$, which can be measured between the RE and the WE, and Ψ_{i_n} are usually constant above 10 or 100 Hz. With increasing frequency, $|Z|$ usually decreases, so $\Psi_{V_{reg}}/|Z|^2$ increases. This explains the increase in PSD of ECN often observed at high frequency as in Figure 14.6.

For ENM carried out in cells with two WEs using a VA and a ZRA, a full analysis of the instrumental noise and its effect on the measurement of EPN and ECN, and on the noise resistance derived from them, may be found in Ref. (82). The results on pure resistors and a very low-noise galvanic cell (copper in hydrochloric acid) indicate how to test the EN measurement system to understand the contributions of its various parts. The noise of the RE was also considered. It has been shown how to remove its influence by taking the cross-spectrum of the two RE signals (82, 91).

As already mentioned, removing fluctuations at frequencies well below $f_{min} = 1/T$, often referred to as "dc drift," before calculating SDs and PSDs is a necessary operation. This is illustrated in Figure 14.8, in which the PSD of the voltage fluctuations of a copper electrode in hydrochloric acid at the corrosion potential

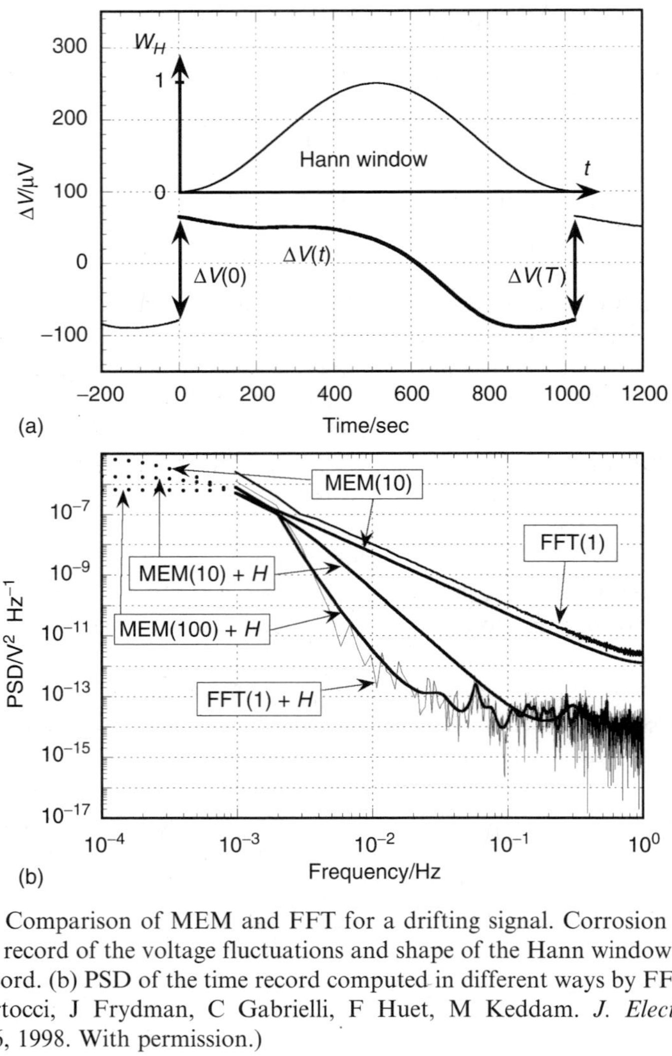

Figure 14.8 Comparison of MEM and FFT for a drifting signal. Corrosion of Cu in 1 *M* HCl. (a) Time record of the voltage fluctuations and shape of the Hann window to be applied to the time record. (b) PSD of the time record computed in different ways by FFT and MEM. (From U Bertocci, J Frydman, C Gabrielli, F Huet, M Keddam. *J. Electrochem. Soc.* 145:2780–2786, 1998. With permission.)

(82) is calculated with the FFT and MEM techniques. Other examples may be found in Ref. (92). If no detrending procedures are applied, both MEM(10) and FFT(1) PSDs are incorrect but the advantage of the FFT method is that the error is revealed by the uncharacteristic smoothness of the FFT(1) PSD in spite of no averaging. The drift of the signal, here defined as the difference $\Delta V(T)$ in amplitude between the first and the last point of the time record in Figure 14.8(a), is processed by the algorithm as a large, low-frequency signal. This explains the unexpected high level of the PSDs. Since the FFT assumes that the time record repeats itself identically from $-\infty$ to $+\infty$, the resulting PSD is that of a signal that looks roughly like a sawtooth. In the case of a linear drift with slope $a = \Delta V(T)/T$, it has been shown that the computed PSD, $\Psi_{x,\text{comp}}$ is equal to the sum of the PSD of a stationary process Ψ_x and of a term with $1/f^2$ slope, according to the expression (25, 93)

$$\Psi_{x,\text{comp}}(f) = \Psi_x(f) + \frac{a^2 T}{2\pi^2 f^2} \tag{14.31}$$

In the same way, the computed SD is also affected by the drift, as shown by the equation derived in Appendix C:

$$\sigma_{x,\text{comp}} = \sqrt{\sigma_x^2 + \frac{a^2 T^2}{12}} \tag{14.32}$$

The second term in Equations (14.31) and (14.32), which represents the effect of the drift, contains the slope a and the duration T of the time record. For the PSD, this component, which gives a straight line of slope -2 in the logarithmic plot, is deterministic and not stochastic, so that the error is zero. For this reason, even if the drift in Figure 14.8(a) is not linear, the FFT(1) spectrum in Figure 14.8(b) is very smooth. Another indication that the signal is not stationary is obtained by measuring the PSD at two different sampling rates. Since Equation (14.31) contains the total time T, which is different in the two cases, the two PSDs do not join in overlapping frequency ranges (see Ref. (92, figure 2)), so that one can only speak of pseudo-PSDs.

A way to eliminate the effect of discontinuities introduced by the amplitude difference between the first and the last data points is to multiply the time record by a "window," that is, a function that tapers to 0 at the beginning and at the end, and is equal to 1 in the middle. This operation reduces the leakage of the low frequencies to the higher frequencies in the calculated PSD. Several windows have been developed for signal analysis, but one of the most commonly used is the Hann window (see Appendix A and Figure 14.8a). Such an operation is commonly referred to as "hanning." Another important reason for hanning is that if the discontinuities are not properly eliminated, they will cause a "tail" of slope -2 in the PSD, which will increase the signal level at the highest frequencies, and mask completely steeper slopes, which could be an important indication of the shape of the EN fluctuations. PSDs computed using Hann windowing are indicated as FFT(K) + H and MEM(M) + H in Figure 14.8(b). The MEM(10) PSD in these conditions is very close to that of FFT(1), and therefore also incorrect; no improvement is obtained by increasing M from 10 to 100. Hanning eliminates the effect of the drift on the PSD: the spectrum FFT(1) + H is a good estimate of the PSD. However, the lack of a drop in amplitude due to the AA filter between 0.67 and 1 Hz indicates that the PSD level of 10^{-14} V^2 Hz^{-1} was below the measurement limit of the 16-bit A/D converter because of the rapid PSD decrease of more than seven decades below 0.01 Hz. With the second method, MEM(10) + H has insufficient frequency resolution, and only MEM(100) + H matches well the FFT spectrum, but the principal advantage of the MEM, that of producing smooth spectra, is lost.

It is sometimes argued that the drift is an informative part of the signal, actually caused by corrosion processes, and that, therefore, it should not be eliminated before calculating noise statistical quantities. Even if it is true that drift indicates the existence of processes at frequencies below those sampled, and that in many cases they may not be instrumental artifacts, this argument is not valid: the drift distorts the results. Removing dc trends can be achieved by analog HP filtering, as mentioned above, or by digital HP filtering. Various trend removal procedures have been proposed and compared in the literature, such as moving average, linear regression, digital HP filtering, and polynomial fitting (79, 92, 94, 95). As an example, a time record (Figure 14.9a, curve A) of the voltage fluctuations of an Al–Zn–Mg–Cu alloy in sodium chloride at the corrosion potential has been taken to compare in the time (Figure

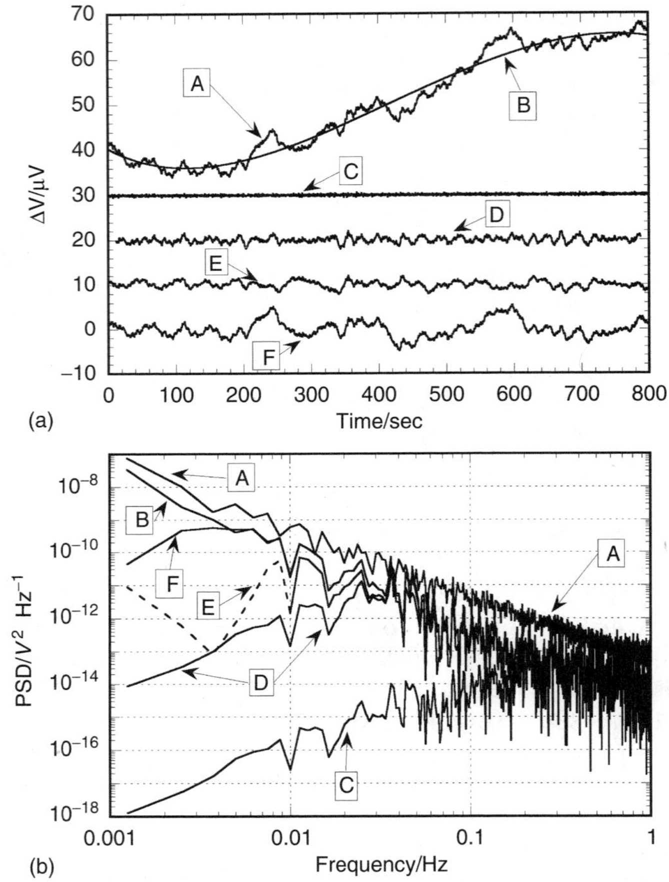

Figure 14.9 Comparison of trend removal procedures. Corrosion of an Al–Zn–Mg–Cu type 7050 alloy at corrosion potential in 0.5 *M* NaCl. (a) Time record sampled at 10 Hz (A), 5-order polynomial fitting curve (B), after 3-order MAR (C), after 30-order MAR (D), after analog HP filter at 0.01-Hz cutoff frequency (E), and after 5-order polynomial detrending (F). Curves A, B, C, D, and E have been shifted for clarity by +50, +50, +30, +20, and +10 μV, respectively. (b) PSD of the raw signal (A), after Hann windowing (B), after 3-order MAR and Hann windowing (C), after 30-order MAR and Hann windowing (D), after 0.01-Hz HP filter + Hann windowing (E), and after 5-order polynomial detrending + Hann windowing (F). Dotted line: PSD at frequencies below cutoff frequency. (From U Bertocci, F Huet, RP Nogueira, P Rousseau. *Corrosion* 58:337–347, 2002. With permission.)

14.9a) and frequency (Figure 14.9b) domains the following methods: 3-order moving average removal (MAR), 30-order MAR, analog HP filtering at a cutoff frequency of 0.01 Hz, and 5-order polynomial fitting (92). The drastic suppression of the low frequency by MAR, but also by the analog filter set at about 10 times the minimum frequency of 1/800 Hz, modifies the record so much that the detrended data resemble very little the original record. The PSDs computed after detrending (and Hanning, except for the raw data) show even more clearly how much of EN fluctuation is lost in the various operations. Although the curves are not smoothed by ensemble averaging, it is fairly clear that, except for the raw data, which give a PSD distorted by the drift, all detrended curves give the same PSD at high frequency, but the frequency range in which they agree depends on the detrending technique. It can also be seen that only the polynomially detrended data give reliable results down to a frequency of 3.75 mHz, (i.e., $3/T$). Thus, good results were obtained with these digital filtering techniques, except for the MAR technique, which is not appropriate for drift removal. Indeed, since estimating nonuniform drift on long times is impossible, a low order has to be used, which drastically reduces the analyzed frequency range, strongly affecting the PSD and giving too low SDs for most EN data. Similar conclusions were obtained by Mansfeld et al. (95), who also obtained PSDs with positive slope, as in Figure 14.9(b) (curves C and D). On the contrary, polynomial detrending appears to be quite suitable and trouble free. If the polynomial is of order 5, the method can successfully suppress any low-frequency component; the only loss is an attenuation of the frequencies $1/T$ and $2/T$. When the drift is fairly uniform, linear detrending, which may be considered as a polynomial of order 1, can also give satisfactory results. However, while it is evident that without detrending the value of the SD is dominated by the drift, less obvious is which detrending method gives the best results. Polynomial detrending seems to be the most adequate procedure as long as the order of the polynomial is kept low to avoid removing significant parts of the EN. Finally, correcting the effects of drift on SD and PSD also ensures good quality in the derived parameters, such as noise resistance, noise impedance, and localization index, as shown recently by Mansfeld et al. (95).

14.5 DATA INTERPRETATION

Various methods have been proposed for the interpretation of EN data to provide relevant information on corrosion processes. Direct methods to analyze the signal in the time and frequency domains will be described first. Then, indirect methods consisting of measuring the noise resistance and noise impedance to estimate corrosion rates will be examined. These methods concern cell configurations with two WEs for measuring both EPN and ECN simultaneously. Identical WEs are usually employed, but asymmetric cells have also been proposed. Finally, the various methods suggested to distinguish between generalized corrosion and localized corrosion are discussed.

14.5.1 Direct Methods

Data analysis may be performed first in the time domain by investigating the shape, size, and occurrence rate of the time transients that are often observed for localized corrosion, such as pitting, or for fluctuations caused by discrete events on the electrode surface, such as impingement of abrasive particles or the evolution of gas bubbles. It can also be performed by measuring the moments of the potential and current fluctuations (SD, skewness, kurtosis), mainly for corrosion-monitoring applications. However, the latter approach is extremely limited for data interpretation. The SD gives the amplitude of the fluctuations and is used to derive the noise resistance (see below), but it provides no mechanistic information on the corrosion process. Skewness and kurtosis have also been tentatively used for the detection of the corrosion type, but they suffer from several drawbacks. In the absence of current and voltage transients, their values are likely to be close to 0, as for signals with a gaussian distribution. Any deviation from 0 indicates the existence of transients, if it is not the result of signal drift, but considering the large estimation error mentioned above, it is quite probable that these parameters cannot identify any mechanism better than a direct analysis of the transients on the time records.

Data analysis may also be carried out in the frequency domain by calculating the PSD of the fluctuations. The interpretation of the PSD is often a challenge, particularly when no transients can be observed in the noise time records. When transients are observable, the interpretation is sometimes, but not always, easier because different transient shapes may give the same PSD. Indeed, the PSD is derived from the modulus of the FT of the time record and information on the phase angle is then lost. Illustration is given in Figure 14.10 that compares two practical corrosion cases in which the PSD can be derived analytically. One is pitting corrosion with exponentially decaying transients, the PSD of which is given in Equation (14.13). The

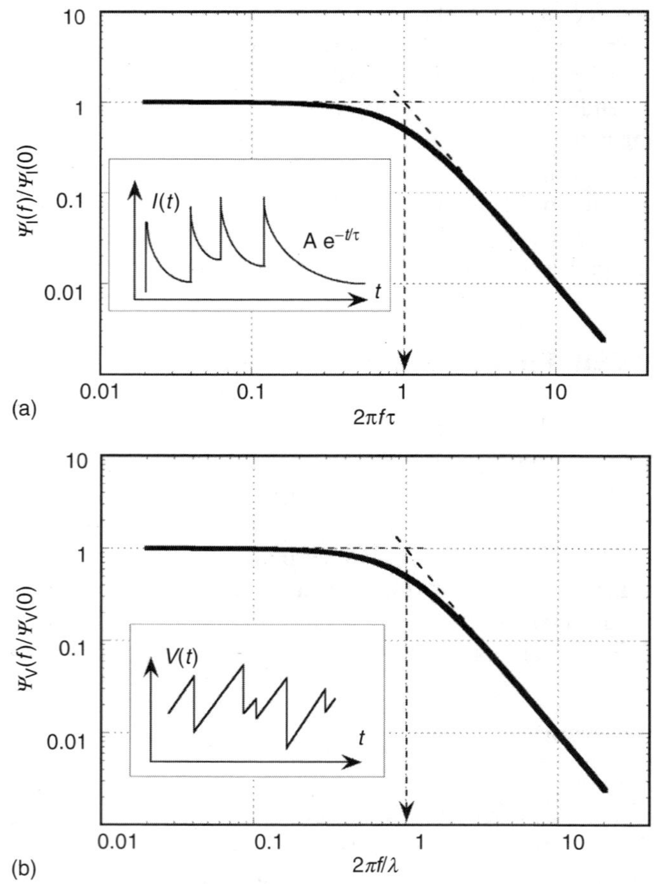

Figure 14.10 Comparison of the PSDs of the signals shown in the insets which are typical of (a) pitting corrosion and (b) uniform corrosion with hydrogen bubble evolution.

other concerns uniform corrosion with hydrogen bubble evolution in which each bubble departure induces a sudden potential jump, as observed, for example, on an iron electrode at the free corrosion potential in acidic medium. The PSD is given by the theory of the renewal processes (96)

$$\Psi_v(f) = \frac{2\lambda\langle A^2\rangle}{\lambda^2 + 4\pi^2 f^2} \tag{14.33}$$

where λ is the average rate of evolving bubbles and $\langle A^2\rangle$ the averaged square of the potential jump amplitudes. Both normalized PSDs in Figure 14.10 have strictly identical shapes, with a low-frequency plateau followed by a decrease in f^2. The cutoff frequency f_c at the intercept of the sloping part of the spectrum at high frequency (in the log f scale) with the zero-frequency limit of the PSD provides the time constant $\tau = 1/2\pi f_c$ of the repassivation process in the pitting case and the bubble evolution rate $\lambda = 2\pi f_c$ in the other case. Such information may be derived directly from the time record, but, as shown in Figure 14.11 in the pitting case, for

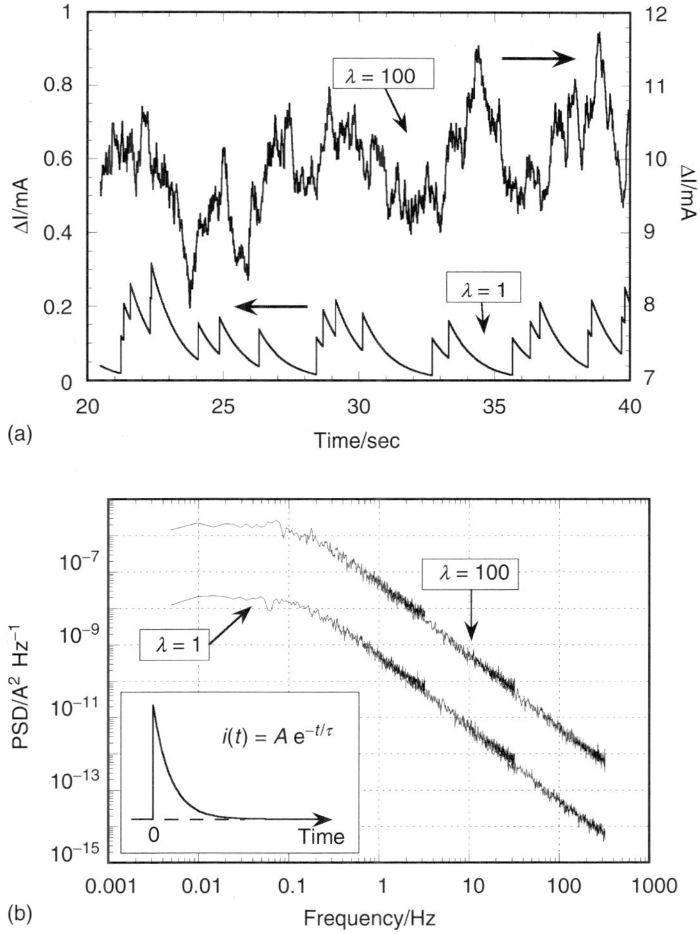

Figure 14.11 Influence of the number of transients on the time record (a) and PSD (b). Simulations of randomly occurring exponentially decaying transients defined in the inset for two values of the transients average rate: $\lambda = 1$ and 100 sec^{-1} ($A = 10^{-4}$ A, $\tau = 1$ sec). (From U Bertocci, F Huet, B Jaoul, P Rousseau. *Corrosion* 56:675–683, 2000. With permission.)

high pitting or nucleation rate λ, the transients overlap, so that it may be impossible to detect their exact shape, and exponentially decaying transients could even be mistaken for saw-tooth shapes. In that case, λ may be recovered from the amplitude of the spectrum since the shape of the PSD plot does not depend on λ, according to Equation (14.13). Figure 14.11 has been obtained by simulations consisting of superposing current transients occurring at random times T_j and of amplitude $i(t - T_j)$ at time $t - T_j$:

$$I(t) = \sum_j i(t - T_j, U_j) \tag{14.34}$$

to generate time records, the PSD of which is subsequently calculated (97, 98). The factor U_j is included so as to consider cases in which some characteristic of the transient (amplitude, decay constant, lifetime, etc.) is a randomly distributed quantity. For transients occurring at independent times T_j (Poisson process) and of simple shape, the PSD may be calculated analytically with Equation (14.11). In Figure 14.10 and Figure 14.11, the slope -2 of the PSD decrease at high frequencies is due to the sudden birth or the sudden death of the transients. Simulations and analytical calculations in Ref. (98) have shown that one of the most important features that influence the shape of the PSD is the existence of singular points in the analytical expression of the transient $i(t)$. The slope of the PSD depends on whether the discontinuities first appear in the function $i(t)$ or in one of its derivatives. If n is the lowest-order derivative that is discontinuous, then the PSD decreases in $1/f^{2(n+1)}$. The function $i(t)$ itself is its derivative of order 0. Therefore, transients like those in Figure 14.10 and Figure 14.11, which have a discontinuity in $i(t)$, generate a PSD with a slope of -2 whatever the shape of the decaying part. This explains the great number of experimental PSDs with a $1/f^2$ decrease, observed in localized corrosion or in general corrosion when hydrogen bubbles evolve. In Figure 14.12, higher slopes are exhibited in the PSDs corresponding to the transients indicated in the insets. The PSDs have been calculated from simulations and from their theoretical expression given in Ref. (98). Good agreement can be noticed, which validates the simulation procedure that can be carried out even for complex shapes of the transients. In Figure 14.12(a), the discontinuity at the beginning of the transient occurs first in the nth derivative and, as a consequence, the PSD actually decreases in $1/f^{2(n+1)}$, as shown for three values of n. PSDs with a steep decrease such as $1/f^6$ have been experimentally obtained by Oltra et al. in erosion–corrosion studies when measuring the acoustic emission response to the impingement of abrading particles on steels. The acoustic response was integrated by an RMS device, so that each transient triggered by a particle impact started very smoothly with a continuous first derivative (70). For a triangular transient, in which the discontinuity appears only in the first derivative, the slope is -4 at high frequency (Figure 14.12b). For an asymmetric triangle transient (curves B and C) an intermediate decrease in $1/f^2$ appears at intermediate frequencies because the transient contains two characteristic times (rise time and decay time). Indeed, the less symmetrical the triangle is, the longer is the decrease in $1/f^2$, because the steeper side of the triangle looks more and more like a discontinuity in the function. At higher frequencies, the PSD is sensitive to short time details in the transient so that the discontinuity at time $t = 0$ no longer appears in the function, but appears in its first derivative, leading to a $1/f^4$ decrease.

The shape of the PSD depends not only on the shape of the transients, but also on their statistical distribution in time when they do not occur at independent times,

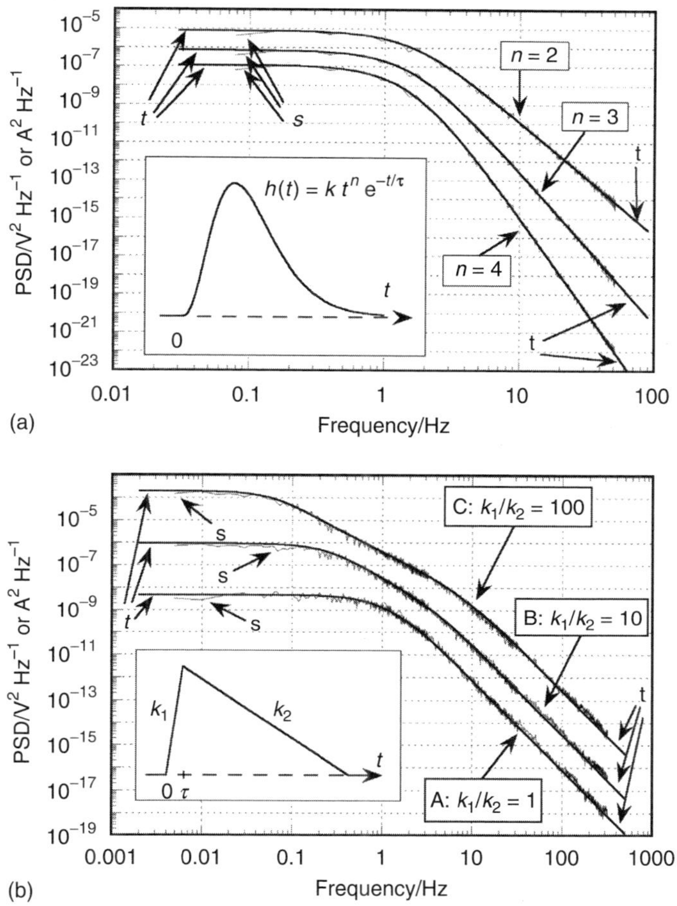

Figure 14.12 Influence of the shape of randomly occurring transients on the PSD (s: simulated, t: theoretical). (a) $k = 1$ V sec^{-n} or A sec^{-n}, $\tau = 0.1$ sec, $\lambda = 1$ sec^{-1}. (b) $k_2 = 10^{-3}$ V sec^{-1} or A s^{-1}, $\lambda = 1$ sec^{-1}, $k_1/k_2 = 1$ (curve A), 10 (curve B), 100 (curve C). τ is exponentially distributed with mean and SD: 0.1 sec (curve A), 0.05 sec (curve B), 0.02 sec (curve C). (From U Bertocci, F Huet, B Jaoul, P Rousseau. *Corrosion* 56:675–683, 2000. With permission.)

and on the distribution of their characteristic parameters (lifetime, decay constant, amplitude, etc.) (98). However, the treatment can account only for slopes described by even exponents while there are experimental instances in which the slopes do not have even or integer exponents, suggesting that this treatment is not general enough to cover all the possible phenomena that give rise to or influence EN, like, for example, mass transport by diffusion or natural convection. The limitation may also come from the assumption that the transients are independent of one another, which is often unrealistic, for instance, for passive film breakdown. The interested reader should consult Refs. (99, 100–102) for analysis of correlated transients.

It is sometimes argued that the slope of the PSD is related to the type of corrosion. For example, according to Eden, a slope of -2 in the current PSD measured below 1 Hz should correspond to uniform corrosion and a slope of 0 (white noise) to localized corrosion (33). However, even if this may be true for specific corroding systems, the result cannot be generalized since, for example, a PSD with slope 0 below 1 Hz may also correspond to the low-frequency plateau of the PSD of a system corroding uniformly with hydrogen bubble evolution (Figure 14.10b). Values of the PSD slope obtained in the literature and in Bagley's thesis are reported in Figure 14.13 as a function of the type of corrosion, showing clearly that the corrosion mechanism cannot be reliably identified only from the value of the PSD slope (103).

14.5.2 Indirect Methods

The interpretation of the noise resistance R_n and noise impedance Z_n measured in cells with two current-measuring electrodes depends on the measurement scheme, which may employ identical or nonidentical WEs for measuring the ECN, and a true RE ("noiseless" RE) or a third electrode that also undergoes corrosion (noisy RE) for measuring the EPN. The theoretical analysis has been presented in great detail in

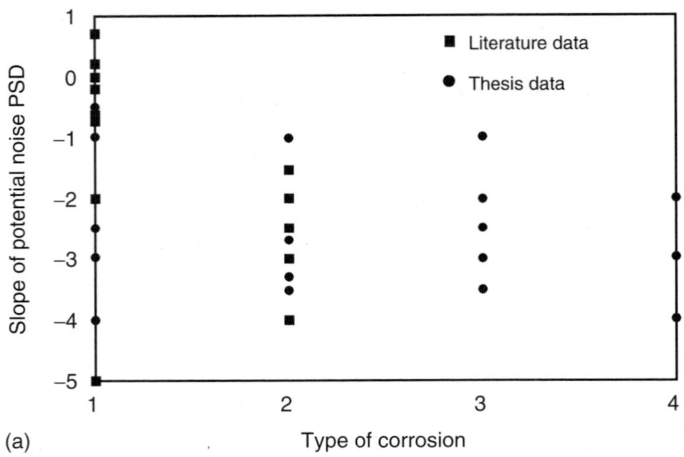

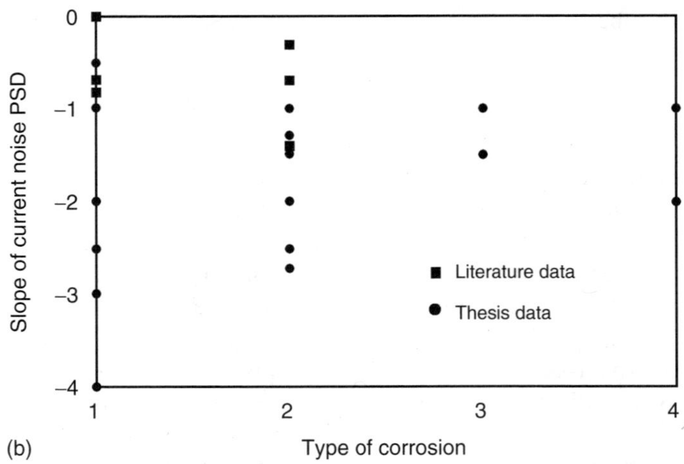

Figure 14.13 Slope of the PSD for various types of corrosion. (a) Potential noise. (b) Current noise. 1 = uniform corrosion, 2 = pitting corrosion, 3 = crevice corrosion, and 4 = combination of pitting and crevice corrosion. (From G Bagley. Ph.D. dissertation, University of Manchester, Manchester, U.K., 1998. With permission.)

a series of papers on the measurement of the noise resistance (82, 89, 104–108); it is summarized briefly below. The phenomena caused by corrosion of the WEs, such as bubble formation and detachment, metastable or stable pitting or crack advance, that give rise to voltage and current fluctuations, can be modeled by current noise sources, i_1 and i_2, in parallel with the impedances Z_1 and Z_2 (including double-layer capacitances) of the WEs. As in thermal noise modeling, voltage noise sources e_1 and e_2 in series with Z_1 and Z_2 can also be used; both methods are Thevenin equivalents, but it is often considered that the corrosion processes induce current fluctuations that lead to potential fluctuations, so only current sources are used in this presentation. The equivalent circuit representing the corrosion cell is given in Figure 14.14(a) for identical or nonidentical WEs and a noiseless RE. The potentials of the metallic phase of the WEs are equal because of the ZRA and are set arbitrarily to zero in Figure 14.14(a). The current through the RE is zero because the connection is to either an amplifier or a voltmeter with a high-input impedance. For simplification, the thermal noise of the solution resistance R_s and that of the noiseless RE, which usually are low, are disregarded. As mentioned above, all noise sources due to instrumentation, RE, and solution resistance have been taken into account in Ref. (82) to assess their contribution to the EN and to the derived parameters R_n and Z_n. The position of the RE with respect to the WEs does not affect the measurement of

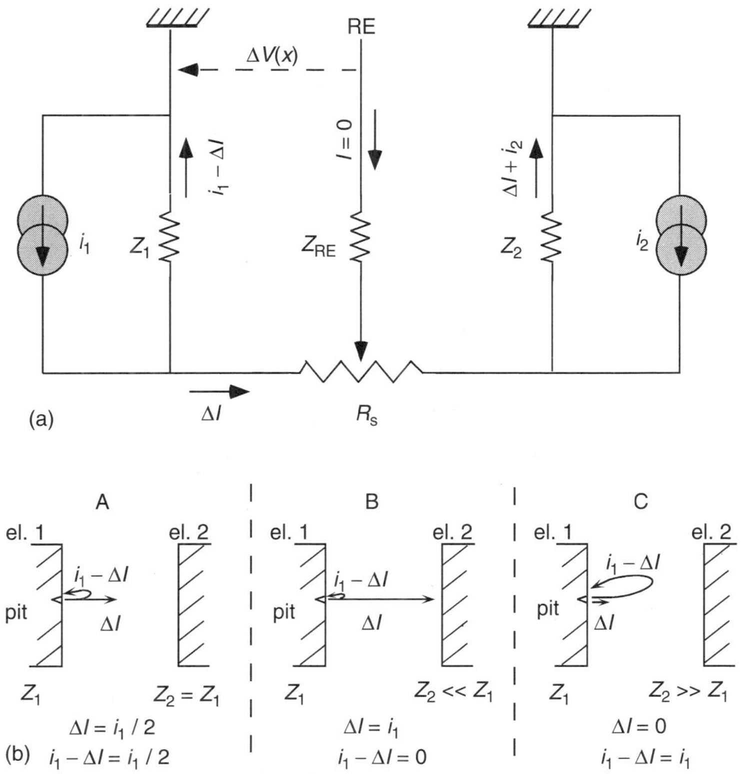

Figure 14.14 Modeling of EN sources. (a) Equivalent circuit for a cell with two WEs and a noiseless RE showing the current noise sources i_1 and i_2. (b) Partition of the current i_1 produced by a pit on WE 1 for symmetric WEs (A), and asymmetric WEs $Z_2 \ll Z_1$ (B), $Z_2 \gg Z_1$ (C). The solution resistance is neglected. (From A Bautista, F Huet. *J. Electrochem. Soc.* 146:1730–1736, 1999. With permission.)

ΔV as long as R_s is negligible compared to Z_1 and Z_2 (89), which is considered here. The noise sources i_1 and i_2 are not directly measurable. Only the fluctuations ΔI of the current flowing between the two electrodes and ΔV of the coupled electrode potential are accessible to measure. The purpose of the equivalent circuit is to describe the dynamic behavior of a system around its mean; hence, it applies for all two-WE cell configurations, that is, for symmetric or asymmetric electrodes, with or without the superimposition of a bias voltage between the WEs. Ohm's law applied to the equivalent circuit gives the following equations in the frequency domain

$$\Delta V(f) = -Z_1(i_1(f) - \Delta I(f)) = -Z_2(\Delta I(f) + i_2(f)) \quad (14.35)$$

All terms are complex and frequency-dependent. $\Delta I(f)$, $\Delta V(f)$, $i_1(f)$, and $i_2(f)$ are the FTs of the real and time-dependent quantities $\Delta I(t)$, $\Delta V(t)$, $i_1(t)$, and $i_2(t)$. Then, from Equation (14.35)

$$\Delta I(f) = \frac{Z_1(f)}{Z_1(f) + Z_2(f)} i_1(f) - \frac{Z_2(f)}{Z_1(f) + Z_2(f)} i_2(f) \quad (14.36)$$

$$\Delta V(f) = \frac{Z_1(f) Z_2(f)}{Z_1(f) + Z_2(f)} [i_1(f) + i_2(f)] \quad (14.37)$$

The current sources i_1 and i_2 have zero mean values and, therefore, so do the current and voltage signals. In the special case where the impedances Z_1, Z_2 are reduced to the polarization resistances R_{p1}, R_{p2} in the frequency range measured, Equations (14.36) and (14.37) may be replaced by equations with time-varying quantities

$$\Delta I(t) = \frac{R_{p1}}{R_{p1} + R_{p2}} i_1(t) - \frac{R_{p2}}{R_{p1} + R_{p2}} i_2(t) \quad (14.38)$$

$$\Delta V(t) = \frac{R_{p1} R_{p2}}{R_{p1} + R_{p2}} [i_1(t) + i_2(t)] \quad (14.39)$$

Equation (14.36) may also be written in the time domain for symmetric electrodes ($Z_1 = Z_2$)

$$\Delta I(t) = (i_1(t) - i_2(t))/2 \quad (14.40)$$

This relation may be illustrated as follows. If a corrosion event such as a pit, producing a current transient $i_1(t)$, occurs on electrode 1, while nothing happens on electrode 2 ($i_2(t) = 0$), a current $\Delta I(t) = i_1(t)/2$ flows toward electrode 2, while the same quantity $i_1(t)/2$ flows toward the cathodic area of electrode 1 around the pit. In Figure 14.14(a) this last current is modeled by the current $i_1 - \Delta I = i_1/2$ crossing the impedance Z_1 of electrode 1. On electrode 2, the current ΔI crosses the impedance Z_2 corresponding to a cathodic reaction on the whole surface, such as dissolved oxygen reduction or hydrogen evolution. The partition of i_1 in equal quantities $i_1/2$ on each WE is schematized in Figure 14.14(b). For asymmetric electrodes, i_1 divides as shown also in Figure 14.14(b): according to Equation (14.36), almost all the current i_1 flows toward electrode 2 when $Z_2 \ll Z_1$, whereas i_1 mainly flows back to electrode 1 when $Z_2 \gg Z_1$. It must be kept in mind that for asymmetric electrodes, Equation (14.36) contains impedance terms that are often complex and frequency-dependent in the frequency range measured, so that the shape of the current transient $i_1(t)$ is not identical to that of the measured current transient $\Delta I(t)$.

To experimentally check that in the case of electrodes having the same impedance the current generated by a localized corrosion event divides approximately equally onto each electrode, measurements have been carried out on a cell constituted of two circular 316 L SS electrodes immersed in a NaCl solution. In one of them (electrode 1) a hole was drilled to place in it a thin Fe wire insulated from the SS. The three electrodes were kept at the same potential by three ZRAs, so that it was possible to measure simultaneously the current on each electrode (107). The impedance of the composite electrode, SS + Fe wire, was found to be approximately equal to that of the original SS electrode. Because of the anodic character of the Fe with respect to the SS, corrosion soon occurred on it and due to the special circuit, it was possible to verify that the current transients generated in it flowed in equal quantities on the two SS electrodes, the slight dissymmetry caused by the solution resistance being negligible. The results are presented in Figure 14.15(a), where it can be seen that the current I_w on the Fe wire was composed of anodic transients, double in

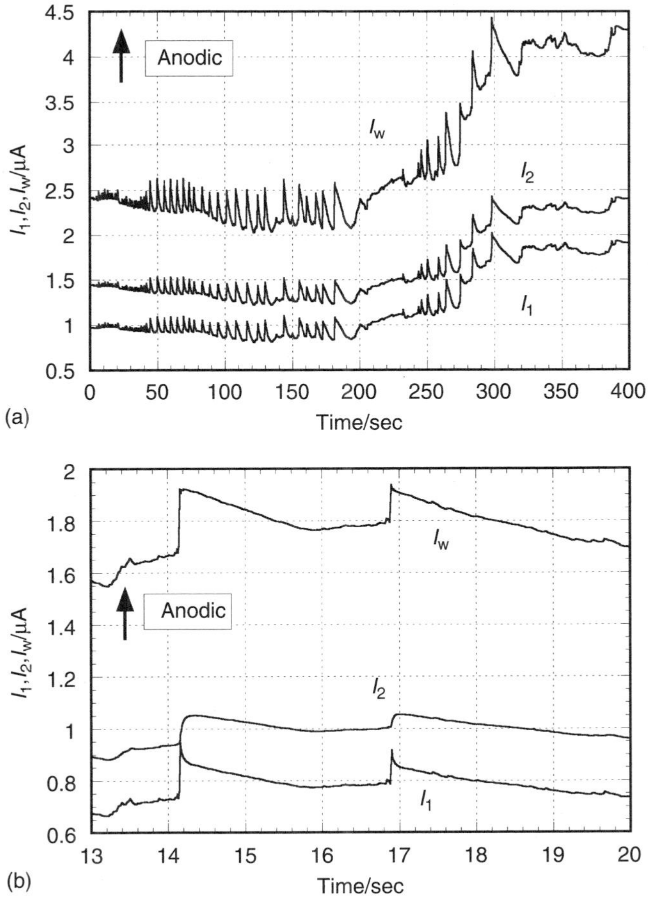

Figure 14.15 Partition of the current fluctuations between the electrodes. I_w is generated by corrosion of an Fe wire (diameter 250 µm) inserted in one of the two 316 L SS electrodes (diameter 1.5 cm) in NaCl at the corrosion potential; I_1 and I_2 are the corresponding currents crossing the two SS electrodes. Zeros of scales are arbitrary. (a) 0.06 *M* NaCl solution. (b) 0.006 *M* NaCl solution. (From A Aballe, F Huet. *J. Electrochem. Soc.* 149:B89–B96, 2002. With permission.)

amplitude compared with the current transients, I_1 and I_2, flowing to the two SS electrodes. Figure 14.15(b) presents the I_1, I_2, and I_w time records for a NaCl solution of higher resistivity. The I_w record exhibits two transients due to the wire dissolution that occurred suddenly and decayed slowly. The corresponding I_1 transients also started abruptly, but I_2 increased more slowly, because the double layer of electrode 1 was being charged by the fast variations of I_w since the solution resistance was high. Therefore, the shape of the transients recorded by the single ZRA used in conventional noise resistance measurements depends on the solution resistance and is not necessarily identical to the shape of the current transients produced by corrosion processes on the electrodes, as shown here by the different shapes of the I_2 and I_w transients. Consequently, when the solution resistance is not negligible, mechanistic information cannot be obtained through direct inspection of the measured current transients. In that case, the influence of the solution resistance is not correctly taken into account by the resistance R_s in the equivalent circuit of Figure 14.14(a). This is because R_s in this figure represents the solution resistance between the WEs while the solution resistance involved in the ENM concerns the electrolyte between electrode 2 and the iron wire simulating a pit on electrode 1. In practice, it is difficult to estimate this solution resistance because it depends on the size and position of the pit on electrode 1. A new equivalent circuit, based on that proposed by Danielson (109), was employed in Ref. (107) to give an approximate explanation of the experimental results.

Since the noise sources i_1 and i_2 of the two electrodes are uncorrelated, the PSDs of the measurable current and voltage noise, $\Psi_I(f)$ and $\Psi_V(f)$, which are proportional to $\Delta I \Delta I^*$ and $\Delta V \Delta V^*$ (see Appendix A), are directly given by Equations (14.36) and (14.37):

$$\Psi_I(f) = \left| \frac{1}{Z_1(f) + Z_2(f)} \right|^2 [|Z_1(f)|^2 \Psi_{i_1}(f) + |Z_2(f)|^2 \Psi_{i_2}(f)] \qquad (14.41)$$

$$\Psi_V(f) = \left| \frac{Z_1(f) Z_2(f)}{Z_1(f) + Z_2(f)} \right|^2 [\Psi_{i_1}(f) + \Psi_{i_2}(f)] \qquad (14.42)$$

The noise impedance Z_n, defined in Equation (14.26), is then

$$Z_n(f) = \sqrt{\frac{\Psi_V(f)}{\Psi_I(f)}} = |Z_1(f) Z_2(f)| \sqrt{\frac{\Psi_{i_1}(f) + \Psi_{i_2}(f)}{|Z_1(f)|^2 \Psi_{i_1}(f) + |Z_2(f)|^2 \Psi_{i_2}(f)}} \qquad (14.43)$$

As Equation (14.43) shows, Z_n depends on four parameters: the noise level of each electrode, expressed by their PSD Ψ_{i_1}, Ψ_{i_2}, and the impedance of each electrode, Z_1 and Z_2. For two identical electrodes having the same impedance ($Z_1 = Z_2 = Z$), Equation (14.43) is simplified. The noise impedance is equal to the modulus of the electrode impedance:

$$Z_n(f) = |Z(f)| \qquad (14.44)$$

It should be pointed out that Equation (14.44) is valid whatever the origin of the noises (localized or uniform corrosion, bubble evolution), and the shape of the impedance plot, even if the noise levels of the two electrodes, represented by Ψ_{i_1} and Ψ_{i_2}, are different. Therefore, the knowledge of Z_n is equivalent to the knowledge of the modulus of the impedance for the purpose of obtaining the corrosion rate. In that case, ENMs are equivalent to impedance measurements, in which the external signal perturbation has been replaced by the internal noise sources due to corrosion processes. It should be emphasized that only the modulus of the impedance is obtained, not the phase angle, which cannot be directly obtained from ENM. Indeed, Equations (14.36) and (14.37) with $Z_1 = Z_2$ show that there is no direct relation between $\Delta V(f)$ and $\Delta I(f)$ through the impedance $Z(f)$ as is the case for cells with a single WE (Equation (14.3)).

Often, "noiseless" REs cannot withstand harsh environment (high pressure, high temperature, etc.), particularly in field work, and a third electrode, made of the same material as the other two, is used. In this case, it is necessary to take into account the noise generated by the third electrode, which is modeled by a current noise source i_3. It has been shown that if the impedance of the three identical electrodes working at the corrosion potential is the same (denoted Z), the noise impedance Z_n is, for a negligible contribution of the solution resistance (89),

$$Z_n(f) = \sqrt{\frac{\Psi_V(f)}{\Psi_I(f)}} = |Z(f)|\sqrt{1 + \frac{4\Psi_{i_3}(f)}{\Psi_{i_1}(f) + \Psi_{i_2}(f)}} \tag{14.45}$$

In that case, Z_n is not equal to the modulus of the impedance of the current-measuring electrodes, irrespective of the noise level of the electrodes. Here, the hypothesis of identical noises is necessary for Z_n to be proportional to $|Z|$. Then

$$Z_n(f) = \sqrt{3}|Z(f)| \tag{14.46}$$

The factor $\sqrt{3}$ comes from the fact that the SD of the voltage measured with a noisy RE is $\sqrt{3}$ times the SD of the voltage measured with a noiseless RE (89, 91). This factor is not $\sqrt{2}$, as claimed by some authors on the basis that the SD of the corrosion potential of a single WE measured against an electrode identical to the WE is $\sqrt{2}$ times that measured against a noiseless RE (see the section on signal conditioning). The validity of the assumptions underlying Equation (14.46) can be considered doubtful, because it implies that the noise i_3 of the voltage-measuring electrode, crossed by a zero current, has the same amplitude as that of the noises i_1 and i_2 of two WEs connected through the ZRA, which might cause a nonzero mean current to flow between them. This assumption, moreover, cannot be experimentally checked since i_1 and i_2 cannot be separately measured under practical conditions of corrosion monitoring. Therefore, it might be questionable to relate Z_n to the corrosion rate of the material under study, since the estimation of $|Z|$ may be in error. In practice, if corrosion is uniform, the noises of the different electrodes have about the same amplitude, but when the processes are localized or discontinuous, the noise level may vary considerably from one electrode to the other, so that the applicability of Equation (14.46) is doubtful. For corrosion monitoring applications, it has been shown that the noise of the RE could be eliminated by using two REs, instead of one, and taking the cross-spectrum of the two voltage signals (91).

Examples of Z_n measurements with noiseless and noisy REs are given in Figure 14.16 and Figure 14.17. Figure 14.16 concerns uniform corrosion of iron in sodium sulfate at pH 3 where hydrogen bubbles evolve. Two REs were used, an SSE and an iron electrode (104). The time records (Figure 14.16a) show fairly sharp transients, consisting of a rise in potential and of current transients either positive or negative, depending on which electrode experienced the transient. It can be seen that all the voltage transients measured with the SSE, which is practically noiseless, coincide in time with the current transients. Some of the voltage transients measured with the Fe RE, as expected, have no corresponding current transient. These transients always

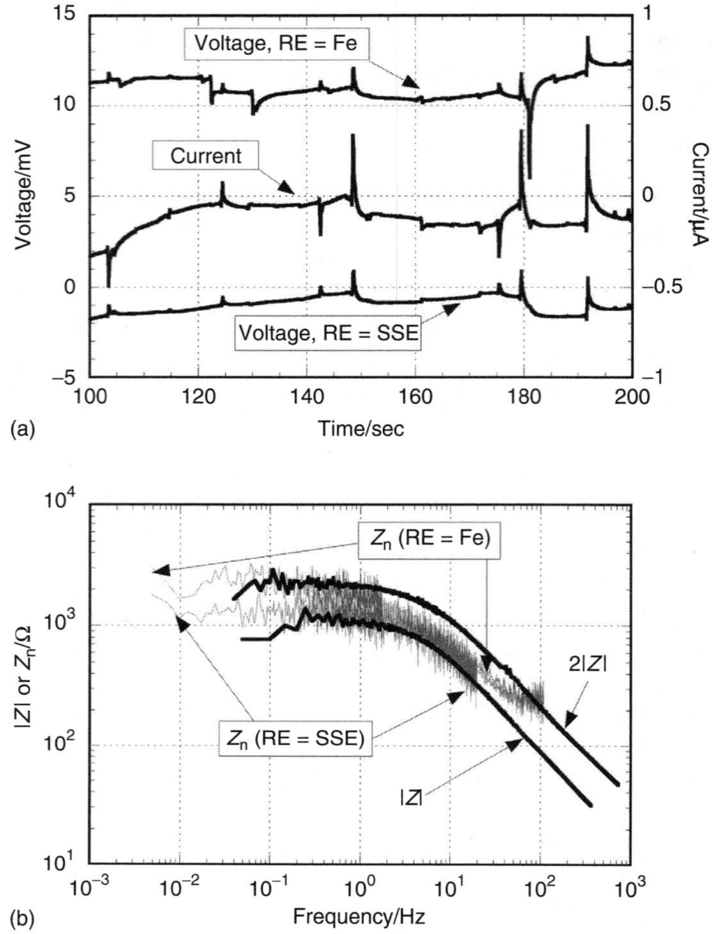

Figure 14.16 Uniform corrosion of Fe in 1 M Na_2SO_4 at pH 3 with hydrogen bubble evolution at the corrosion potential. (a) Time records of the voltage and current fluctuations (identical WEs, surface area 0.2 cm^2). Voltage traces taken simultaneously with two REs, Fe and SSE. $f_s = 10$ Hz. Zeros of scales are arbitrary. (b) Comparison of Z_n (RE = SSE) with $|Z|$, and Z_n (RE = Fe) with $2|Z|$. (From U Bertocci, C Gabrielli, F Huet, M Keddam, P Rousseau. *J. Electrochem. Soc.* 144:37–43, 1997. With permission.)

occur in a direction opposite to that of the correlated transients, which indicates that they are caused by a process identical to that occurring on the WEs, that is, hydrogen bubbles detaching from the RE. Figure 14.16(b) shows the Z_n curves derived from the PSDs, $|Z|$ measured with a single WE in a three-electrode cell using white noise as excitation signal (see Appendix A), and $2|Z|$ measured between the two WEs without RE. Although the fluctuations were only indirectly related to the corrosion process, the value of Z_n measured with the SSE coincides with the impedance modulus of a single electrode, at least at frequencies lower than 20 Hz, in agreement with Equation (14.44). Above 20 Hz, the discrepancy was due to the instrumentation noise. For the measurements with the Fe RE, the value of Z_n is close to $\sqrt{3}|Z|$ at frequencies below 1 Hz, in agreement with Equation (14.46), the difference between a factor of 2 and one of $\sqrt{3}$ being beyond the precision of the noise measurements. However, a net discrepancy can be detected above 1 Hz, which indicates that the three electrodes

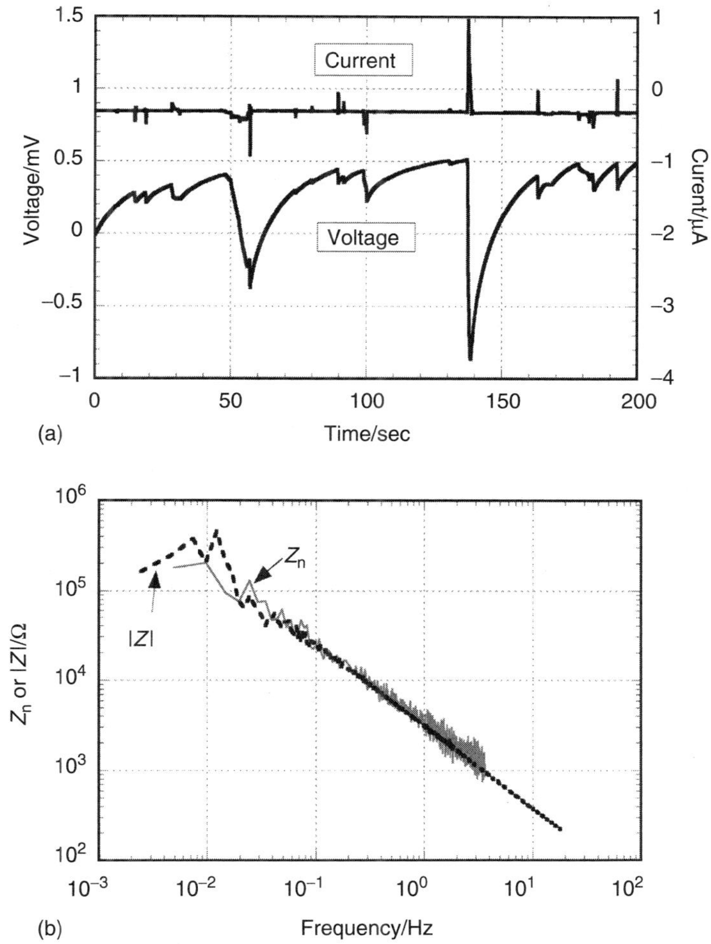

Figure 14.17 Pitting corrosion of Al in 1 *M* KCl. (a) Time records of the voltage and current fluctuations (identical WEs, surface area 5 cm^2). RE=SCE. $f_s = 10$ Hz. Zeros of scales are arbitrary. (b) Comparison of Z_n and $|Z|$. (From U Bertocci, C Gabrielli, F Huet, M Keddam, P Rousseau. *J. Electrochem. Soc.* 144:37–43, 1997. With permission.)

were not identically noisy. Figure 14.17 shows a typical example in which the modulus of the impedance of the WEs does not show any low-frequency plateau in the frequency range measured. It concerns pitting corrosion of aluminum in KCl solution (104). The time records show negative voltage transients decaying exponentially and faster current transients, also positive or negative depending on which WE the corresponding pit occurred. In this case, a noiseless RE was used and the Z_n and $|Z|$ curves coincide perfectly. Similar results have been found for coated electrodes by Mansfeld et al. (67, 110) and Bautista and Huet (105), even when the EN was extremely low and dominated by the instrumentation noise.

As mentioned before, another widely used quantity in the analysis of EN is the noise resistance R_n, which is attractive because it is calculated in the time domain and requires simpler instrumentation. However, its relationship with R_p, the polarization resistance that corrosion workers commonly employ to calculate corrosion rates, is not so obvious. R_n is calculated as the ratio of the SDs of the voltage and current fluctuations (Equation (14.22)), but, through the relationship between SD and PSD (Equation (14.A.14)), it can be expressed in the form

$$R_n = \frac{\sigma_V}{\sigma_I} = \sqrt{\frac{\int_{f_{min}}^{f_{max}} \Psi_V(f)\mathrm{d}f}{\int_{f_{min}}^{f_{max}} \Psi_I(f)\mathrm{d}f}} \tag{14.47}$$

where (f_{min}, f_{max}) is the frequency range analyzed. Hence, from Equations (14.26) and (14.44), R_n is given by

$$R_n = \sqrt{\frac{\int_{f_{min}}^{f_{max}} \Psi_I(f)|Z(f)|^2\mathrm{d}f}{\int_{f_{min}}^{f_{max}} \Psi_I(f)\mathrm{d}f}} \tag{14.48}$$

for identical electrodes at the same potential. If Z and Ψ_I are arbitrary functions of the frequency, Equation (14.48) shows that R_n is not related to the electrode impedance in a simple way, so as to be used to obtain the corrosion rate. From Equation (14.48) it is possible to understand when $R_n \approx R_p$ and when this is not true. Roughly speaking, if the Bode plot of $|Z|$ reaches the low-frequency asymptote R_p between f_{min} and f_{max}, as in Figure 14.16(b), R_n is very close to R_p. On the contrary, when R_p is very large and is not attained between f_{min} and f_{max}, as for passive electrodes or coated electrodes, R_n is substantially lower than R_p. In a more elaborate form, the outcome depends on the interplay between f_{min}, the slope of the current PSD Ψ_I, and the frequency at which $|Z|$ reaches the dc limit R_p. This can be discussed by means of Figure 14.18 for a corroding system, the impedance of which can be modeled at low frequencies by a capacitance C and a resistance R_p in parallel. The quantity $f_c = 1/2\pi R_p C$ is the critical frequency beyond which $|Z|$ decreases significantly. From the figure, one can estimate the frequency at which R_n will start to depart significantly from R_p, given the slope of the PSD, which was assumed to be constant between f_{min} and f_{max}, and the duration of the time record, which determines f_{min}. For example,

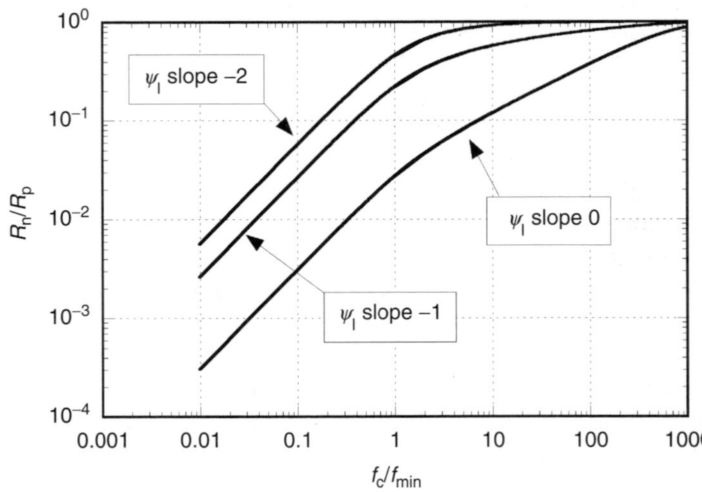

Figure 14.18 Relationship between R_n and R_p for a model system with impedance $R_p//C$. Ratio R_n/R_p as a function of the critical frequency $f_c = 1/2\pi R_p C$. The slope of Ψ_I in log scale is taken constant between f_{min} and f_{max} (f_{max}/f_{min}= 1024). (From A Aballe, A Bautista, U Bertocci, F Huet. *Corrosion* 57:35–42, 2001. With permission.)

when the slope of Ψ_I is -2 or higher, time records for which $f_c/f_{min} > 10$, that is, with a duration higher than $100R_pC$, may be sufficient in many cases to estimate R_p from R_n. However, if R_p is very high, it is often impossible in practice to acquire the signals during a sufficiently long time, so as to satisfy the condition $f_c/f_{min} > 10$, and, as a consequence, R_n may be much lower than R_p.

Similar considerations may be put forward for asymmetrical systems, but it is difficult to give general results since R_n depends on too many parameters. As a corollary of the fact that Z_n has a value between the larger and the smaller of the two electrode impedance moduli at any frequency — a direct consequence of Equation (14.43) — it can be shown that if $|Z_1|$ and $|Z_2|$ reach the polarization resistances R_{p1} and R_{p2} in the frequency range (f_{min}, f_{max}), the value of R_n is also bounded by the values of R_{p1} and R_{p2} (105).

The first example of asymmetrical systems is given when there is a large difference in size between the two electrodes. If the electrodes are otherwise identical, their impedances are inversely proportional to the surface area: $Z_2 = Z_1/k$ if k is the surface area ratio. For corrosion processes involving a large number of corrosion events of low amplitude, the PSD of the current noise sources is proportional to the surface area of the electrode (8), which gives $\Psi_{i_2} = k\Psi_{i_1}$. For a large electrode 2 and a small electrode 1 ($k \gg 1$), the current i_2, which is much larger than i_1, flows back to electrode 2 (see Figure 14.14b, case C), so that the ZRA measures the current generated on the less noisy electrode (electrode 1): $\Delta I = i_1$. This has been experimentally proven by Bertocci et al. (104) and Pistorius (83), who suggested the use of electrodes of very different sizes to facilitate the mechanistic interpretation of the noise transients. In contrast, the potential fluctuations ΔV are given by the noisier electrode (electrode 2), since ΔV is proportional to $i_1 + i_2$, according to Equation (14.37). This is why the noise impedance depends on the impedance of both WEs. From Equation (14.43), it is easily found that $Z_n = \sqrt{|Z_1 Z_2|}$, the geometric mean of the moduli of the electrode impedances, a relation that has been experimentally checked in Ref. (104). In the same way, R_n is the geometric mean of the polarization resistances R_{p1} and R_{p2} if Z_1 and Z_2 reach R_{p1} and R_{p2} in the frequency range analyzed.

Asymmetrical corroding electrodes often develop spontaneously when random corrosion events on one of the electrodes decrease its impedance, so that, by positive feedback, this electrode corrodes more rapidly than the other. This is certainly the case for coated electrodes. The effect on the EN measured can be illustrated from the experiments carried out by Mansfeld et al. (111) and Bierwagen et al. (112) consisting of drilling holes through coated electrodes (105), which gave different results. Without holes, the coated electrodes with very high impedance generate low current noises i_1 and i_2, so that the measured current noise ΔI is low. Both WEs have identical impedance Z so that Z_n measures $|Z(f)|$. A hole drilled through the coating of electrode 1 induces a much higher current noise i_1, indicating enhanced corrosion in the hole. If the impedance Z_1 has drastically decreased, i_1 flows toward the cathodic area of electrode 1 (Figure 14.14b, case C), so that the current ΔI only measures i_2 (Equation (14.36)) and remains low, as observed in Ref. (111). In this case, it can be shown that Z_n decreases and takes a value between $|Z_1|$ and $|Z_2|$, which can be much lower than $|Z_2|$ (see Ref. (105, figure 5)), despite the fact that the WEs are in series. On the contrary, if Z_1 has not decreased significantly, ΔI measures the part of i_1 that flows in solution (Equation (14.36)), and therefore increases (112). In this case, no change can be detected in Z_n that remains close to $|Z_2|$ (105). When holes are drilled on both electrodes, the high values of i_1 and i_2 yield a high measured noise ΔI, now given by Equation (14.40) since the electrodes again have impedances of the same order of magnitude. Thus, to detect corrosion, even when occurring only on a single electrode, it is necessary to monitor both ΔI and Z_n.

Asymmetry has also been employed purposely to have only one corroding electrode, while still having two WEs, one cathode and one anode, to measure both current and voltage fluctuations. Two methods have been proposed, as mentioned above: one consists in using a platinum microcathode as one of the WEs (84), in the other a bias voltage is applied between identical WEs (85). Both cases have been analyzed on the basis of Equation (14.43) (106). For example, if hydrogen bubbles are evolving on the cathode while the anode undergoes generalized corrosion, the noise of the cathode is orders of magnitude larger than that of the anode, so that Z_n becomes equal to the impedance modulus of the anode, $|Z_a|$. Under these conditions, while the time records appear to show only the cathodic processes, the impedance measured is that of the anode, using the noise of the cathode as input signal. An opposite case is provided by a cell where the anode is undergoing pitting, while the cathodic reaction is the reduction of oxygen dissolved in the solution. Since the anodic noise is preponderant, Equation (14.43) shows that Z_n is equal to the impedance modulus of the cathode $|Z_c|$. The anodic noise is the internal signal source utilized for the measurement of the impedance of the cathode. In intermediate cases the results of the measurement of Z_n are difficult to interpret (106). For the noise resistance R_n, analysis shows that if the impedances Z_a, Z_c are equal to the polarization resistances R_{pa}, R_{pc} in the frequency range measured, R_n is equal to the polarization resistance of the anode, R_{pa}, for a noisy cathode, and to the polarization resistance of the cathode, R_{pc}, for a noisy anode (106).

An informative example is given by aluminum coupled to a Pt microcathode in 1 M KCl. Because of the very small surface area of the cathode, coupling has a negligible influence on the potential of the anode, which is actually at its corrosion potential and where pitting generates large transients of the order of 1 mV (see Ref. (106, figure 4)). Figure 14.19(a) shows the PSDs Ψ_V and Ψ_I, the latter exhibiting an unexpected slope of -3 for a Poisson process that usually gives even numbers -2, -4 or even higher, as explained above. The PSD Ψ_{i_a} of the current noise source on the anode is also given. Indeed, for asymmetric cells, it is possible to infer the PSD of a single electrode from Ψ_I and Ψ_V when the impedances of the electrodes are known. In the present case, since the impedance Z_a of the anode was much lower than the impedance Z_c of the cathode, Equation (14.41) gives

$$\Psi_{i_a}(f) = \left|\frac{Z_c(f)}{Z_a(f)}\right|^2 \Psi_I(f) \qquad (14.49)$$

A slope of -2 is observed for Ψ_{i_a} in Figure 14.19(a), as it is often the case in pitting corrosion. The slope of -3 for Ψ_I can then be explained by the dependence in $1/f$ of $|Z_a/Z_c|^2$. It should be emphasized that Ψ_{i_a} differs from Ψ_I also in amplitude. Both the shape and the amplitude of the pitting transients on the anode were affected by the asymmetry in the impedances of the electrodes and did not correspond to the shape and amplitude of the transients measured by the ZRA. To test the characteristics of this kind of asymmetrical cell, the impedance of the two electrodes was measured separately, and compared with Z_n, as shown in Figure 14.19(b). As could be expected from Equation (14.43), the noisy anode works as the signal source, so that Z_n is equal to the impedance of the Pt microcathode. While examination of the time records and possibly of certain features of the individual PSDs might lead to useful information

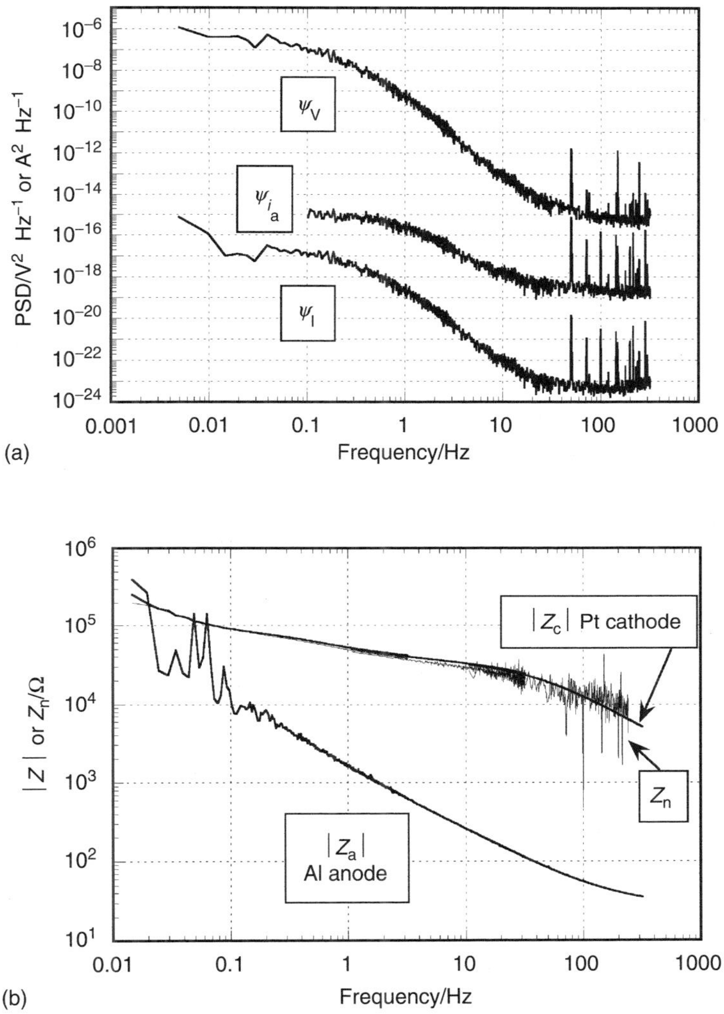

Figure 14.19 Noise impedance in asymmetric systems. Pitting corrosion of an Al disk (diameter 2.5 cm) at the corrosion potential connected to a platinum disk (diameter 1 mm) in 1 *M* KCl. $V_{\text{coupl}} = -0.813$ V/SCE. (a) PSDs Ψ_V, Ψ_I, and Ψ_{i_a} of the current noise source on the anode derived from Equation (14.49). (b) Impedance moduli of the platinum disk, the Al disk, and noise impedance. (From A Bautista, U Bertocci, F Huet. *J. Electrochem. Soc.* 148:B412–B418, 2001. With permission.)

on the pitting process at the anode, employing a microelectrode does not allow the impedance of the corroding electrode to be obtained.

14.5.3 Methods to Detect Localized Corrosion

Several methods have been proposed to distinguish localized and generalized corrosion, particularly for corrosion-monitoring applications. Most methods are based on the detection of current transients that often appear in the current fluctuations when localized processes occur. The monitoring of parameters, such as the slope of the PSD, the skewness, and the kurtosis, have already been discussed above. The coefficient of variance of the current fluctuations, $\text{CoV} = \sigma_I/\langle I \rangle$ (Equation (14.19)), was also evaluated. For general attack, CoV was found to be low because the mean current is usually high compared with the current fluctuations. On the contrary, for localized attack, CoV increased to 1 or greater because of the much higher value of σ_I (33). However, since this coefficient approaches infinity when $\langle I \rangle$ tends to 0, it was replaced by the localization index, LI $= \sigma_I / I_{RMS}$, which takes values ranging between 0 and 1, according to Equation (14.23). Values approaching 1 ($\sigma \gg \langle I \rangle$) are assumed to indicate localized corrosion, while very small values of about 10^{-3} are found for generalized corrosion ($\sigma \ll \langle I \rangle$). However, the inherent drawback of this method is that the mean current is theoretically equal to zero in cells with two identical electrodes, which gives a theoretical LI value of 1 according to Equation (14.A.8), whatever the type of corrosion. A low value of LI may be obtained only when the two WEs corrode differently, which induces a relatively high (at least higher than σ_I) net current that flows between the WEs. Even if this occurs frequently in generalized corrosion, LI must be used with great care, since it appears to be more a measure of the asymmetry in the assumed identical WEs, as suggested by Mansfeld and Sun (113).

Considering that dissolution occurs through sudden bursts of current, Cottis (10) has proposed to apply a shot noise model to derive the mean charge $\langle q \rangle$ involved in a current transient and the average rate λ (denoted by f_n by Cottis) of transients. Of course, a high rate associated with a small charge is assumed to correspond to generalized corrosion, while a low rate associated with a large charge corresponds to localized corrosion. In the first approach, q is supposed to be constant for all current bursts, so that λ and q may be estimated from the value of the corrosion rate $I_{corr} = \lambda q$ and that of the low-frequency plateau of the current PSD $\Psi_I(0) = 2\lambda q^2$ (Equations (14.8) and (14.9)). The method has great potentialities, but further developments are required. Indeed, the corrosion rate is usually estimated from the noise resistance, which is not systematically equal to R_p, as for coated or passive electrodes, and, above all, the value of $\Psi_I(0)$ must be known whereas the PSD does not exhibit any plateau at low frequencies in many corrosion systems. Moreover, even in the simpler case of cells with a single WE, situations are encountered in which the mean current is not directly related to the level of the current PSD (high current with low PSD, or the inverse), when, for instance, the mean current reflects the dissolution of large pits, while the noise reflects the activity of secondary pits in the large ones (24).

14.6 RECENT DEVELOPMENTS IN ENM

New data processing methods, such as wavelets (114, 115), third-order moments (10, 18), or those used in the theory of chaos (Lyapunov exponent) (116) and fractals (Hurst exponent) (117), have appeared in the EN literature, but will not be discussed here. Although of theoretical interest, their output is complicated, often difficult to interpret, and still without practical applications. The opinion of the author is that they have not proved to give new information on corrosion processes that cannot be obtained by the more conventional ENM techniques. Attempts have also been made to automate the interpretation of the huge amount of EN data provided by real-time monitoring by using neural networks, but training the network remains a great difficulty.

A promising approach for fundamental investigations consists in measuring the fluctuations of potential or current simultaneously with those of other signals that give complementary information on the electrochemical processes occurring at the WE. Various additional quantities have been measured together with EN. For example, the hydrogen permeation current in a Devanathan cell can be measured to correlate the fluctuations of electrode potential to that of hydrogen penetration into the metallic membrane (49). For the study of the kinetics of breakdown and repair of the passive film in abrasion–erosion by impinging solid particles, it is important to control the average rate of abrading particles. For this, Oltra et al. have recorded the current fluctuations of a phototransistor measuring the absorption of a light beam across the nozzle of the jet cell. The good correlation with the ECN signal revealed that the abrading process was a doubly stochastic process, that is, the average rate of abrading particles was fluctuating with time (26). Coupling of acoustic emission and ENM in such erosion–corrosion studies is also interesting for investigating the synergy between mechanical wear and corrosion rate. Typical acoustic signals integrated by an RMS device, so that each transient corresponds to a particle impact, are compared with ECN in Ref. (70). ENM can also be coupled with the rapidly spreading video image acquisition techniques, as performed by Legat and Leban, who employed a computer visualization technique combined with optical microscopy to correlate the EN signals with the actual corrosion state of the electrodes, revealed by digitized images acquired throughout the corrosion test (118). Another coupling technique, which has been developed by the Paris group, has the potentiality to be used in specific field applications, as in multiphase flows. It concerns the simultaneous measurement of potential (or current) and electrolyte resistance fluctuations in cells with one (119) or two WEs (120), allowing the ohmic and faradaic components of EPN or ECN to be separated. The elementary events generating the EN, such as bubble detachment, contact between WE, and flowing solid particles or oil droplets, can be identified more easily, particularly in two-phase systems (52, 121). Two examples of typical time records, measured during generalized corrosion of mild steel in an oil/brine mixture in a single-WE cell (52) and during corrosion of an aluminum alloy in the EXCO test solution in a two-WE cell (120), are presented in Figure 14.20. In the first case, it can be seen that the fluctuations of electrolyte resistance are much more sensitive to the oil/brine composition than the EPN, and in the second, that the steep jumps of electrolyte resistance marked the detachment of hydrogen bubbles, whose detection was not possible with the ECN and EPN signals alone.

Very recently, to avoid asymmetry between two supposedly identical electrodes and work on a single corroding electrode, but still measure both current and potential fluctuations, switching techniques have appeared (122, 123). EPN is measured at open circuit against an RE, then the WE potential is fixed at the last known value of the corrosion potential and ECN is measured. The switching is repeated periodically, measuring alternatively EPN and ECN, under galvanostatic control at zero current and potentiostatic control, respectively. At low switching rate, a full EPN time record is acquired, followed by a full ECN time record, but the simultaneity of the EPN and ECN measurements is lost, while at high switching rate, a single potential-data point followed by a single current-data point is sampled. These techniques are not well established, in particular the influence of the switching on the electrode behavior is not fully understood. In a recent presentation, Cottis et al. (124) concluded that either the EPN or the ECN measurement is invalid with these methods so that they do not provide any more information than the measurement of either EPN or ECN only.

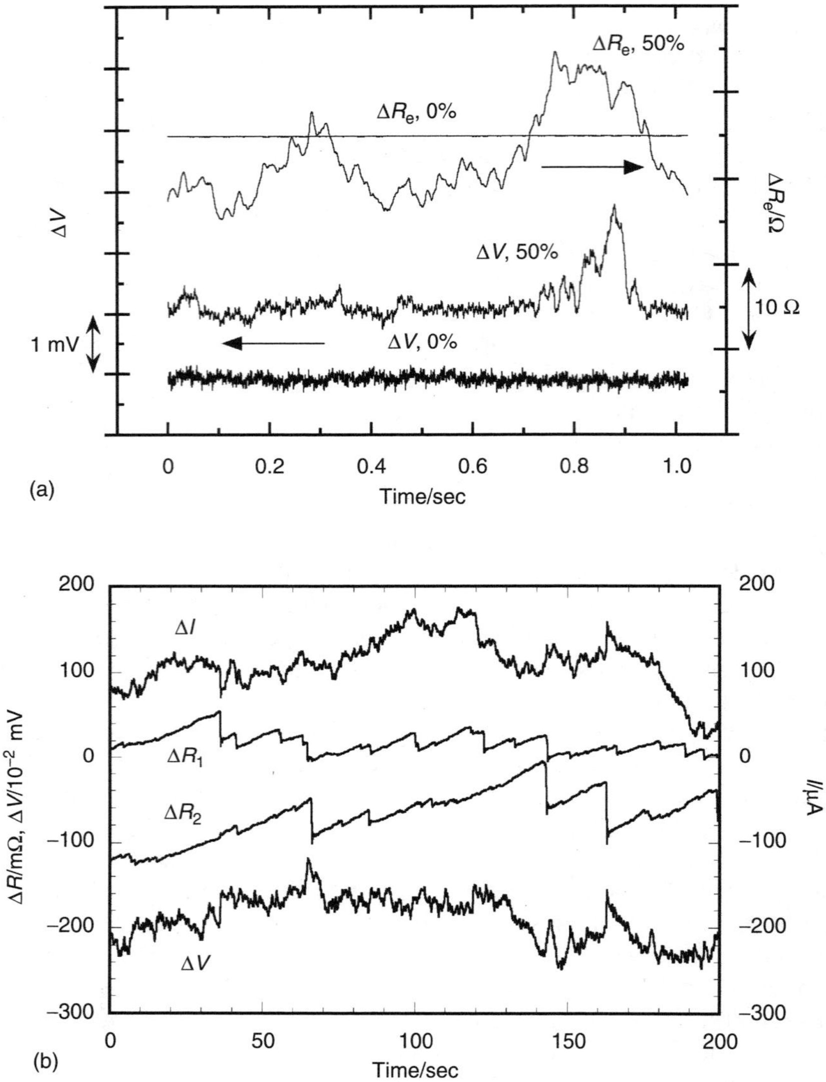

Figure 14.20 Simultaneous measurements of EN and electrolyte resistance fluctuations. (a) Corrosion of mild steel at the corrosion potential in 3 wt% NaCl with paraffin oil at a volumetric ratio of 0% and 50%. Time record of the fluctuations of ΔV and electrolyte resistance ΔR_e, $f_s = 2000\,\text{Hz}$. Arbitrary time scale origin. (From F Huet, RP Nogueira. CORROSION '03, Houston, NACE, 2003, Paper 416, pp 1–18. With permission. (b) Corrosion of UNS A92024 T351 aluminum alloy after 45 min of immersion in the EXCO test solution at 70°C. Time records of the fluctuations of ΔV, ΔI, and electrolyte resistance measured for each electrode ΔR_1 and ΔR_2. $f_s = 10\,\text{Hz}$. Zeros of scales are arbitrary. (From F Huet, C Kuntz, H Takenouti. CORROSION '98, Houston, NACE, 1998, Paper 378, pp 1–13. With permission.)

14.7 CONCLUSIONS

The recording and analysis of EN, begun a few decades ago, only recently has been adopted as a method for assessing and monitoring corrosion phenomena. The pioneering work of Eden et al. (34) has been instrumental in introducing the idea of a corrosion cell with two WEs, where both current and voltage fluctuations can be measured simultaneously. This has led to a broader acceptance of the EN technique for on-site applications in most corrosion domains, as shown in the regular symposia on EN at the annual NACE meetings. The question remained of putting the interpretation of the data on a firm basis. This is now possible, so that the working of both symmetrical and asymmetrical cells can be analyzed with confidence.

The analysis summarized in this chapter allows assessment of the usefulness of the two main quantities, noise resistance R_n and noise impedance Z_n, derived from this kind of noise measurement. While for identical electrodes at the same potential Z_n corresponds to the modulus of the electrode impedance in the frequency range studied, R_n is simply a number, which may or may not be equal to the polarization resistance of the electrodes under investigation, even for symmetric systems. The measurement of the impedance by means of an external broadband excitation is certainly not new, but in the EN case the source is internal, which simplifies considerably the necessary instrumentation, and makes its application attractive for on-line corrosion monitoring.

For asymmetric cells, the interpretation of Z_n is more involved, since Z_n depends on four quantities: the noise levels and the impedances of the two WEs. Usually, but not always, Z_n gives the impedance modulus of the less noisy of the two electrodes, and R_n its polarization resistance. Intuitively, the functioning of such a corrosion cell is that the noise generated by one of the electrodes acts as a source signal for the measurement of the impedance of the other electrode. Hence, corrosion rate estimation from Z_n or R_n must be restricted to systems where the noise delivered by the cathode predominates, otherwise the Z_n or R_n measured can be that of the cathode, and not that of the corroding electrode.

Although pitting corrosion is commonly detected by the appearance of transients in the ECN time records, much work is still needed to relate EN data to corrosion mechanisms, for example, to identify the various types of localized corrosion. In particular, the interpretation of the PSDs remains a controversial topic, especially when no transients are observable on the voltage and current time records. This requires first that the measurement errors discussed in this chapter be eliminated, to improve the quality of the PSDs over those often found in the literature.

It is generally accepted that EN can be used to monitor uniform corrosion and provides valuable information on the localized nature of the corrosion processes, so that EN is increasingly used as a complementary technique to traditional on-line monitoring methods, with the great advantage of detecting in real time corrosion excursions that can be related to changes in the operating conditions. Also, even if the corrosion rate calculated from R_n is not accurate, measuring the trend of R_n is usually enough for monitoring purposes.

ACKNOWLEDGMENTS

The author would like to thank his colleagues of the laboratory "Interfaces et Systèmes Electrochimiques" (UPR 15, CNRS) and Ugo Bertocci of the National Institute of Standards and Technology, Gaithersburg, MD, for many helpful discussions over the years.

APPENDIX A

Background on Random Signals and Digital Signal Processing

Generalities on Random Signals

For the sake of clarity to electrochemists working with the EN technique, it seems useful to give a precise definition of the various terms used in random signal processing (125). The time variations of a stationary random signal $x(t)$ are characterized by the moments of the order n, $\langle x^n(t)\rangle$, of x. The average value can be calculated by a time integral. For example, the mean value of x (first-order moment) is

$$\langle x\rangle = \lim_{T\to+\infty} \frac{1}{T}\int_0^T x(t)\mathrm{d}t \tag{14.A.1}$$

A random signal $x(t)$ can then be written as

$$x(t) = \langle x\rangle + \Delta x(t) \tag{14.A.2}$$

where $\Delta x(t)$ represents the fluctuation of x around $\langle x\rangle$. Obviously, the mean value of $\Delta x(t)$ is 0.

The (auto)correlation function $A_x(\tau)$ of a stationary and ergodic signal is defined as

$$A_x(\tau) = \langle x(t)x(t+\tau)\rangle = \lim_{T\to+\infty} \frac{1}{T}\int_0^T x(t)x(t+\tau)\mathrm{d}t \tag{14.A.3}$$

whereas the (auto)correlation function $C_x(\tau)$ of $\Delta x(t)$ (also called covariance function of x) is given by

$$C_x(\tau) = A_{\Delta x}(\tau) = \langle \Delta x(t)\Delta x(t+\tau)\rangle = \lim_{T\to+\infty} \frac{1}{T}\int_0^T \Delta x(t)\Delta x(t+\tau)\mathrm{d}t \tag{14.A.4}$$

Note that $C_x(\tau)$ is independent of the mean value $\langle x\rangle$, while $A_x(\tau)$ depends on $\langle x\rangle$ as (from Equations (14.A.2) to (14.A.4))

$$A_x(\tau) = C_x(\tau) + \langle x\rangle^2 \tag{14.A.5}$$

Both correlation functions do not depend on time t for a stationary signal, they depend on the time delay τ only. They give a quantitative indication of how much the signal values at two different times t and $t+\tau$ are correlated. Obviously, they are decreasing functions of τ with a maximum value at $\tau = 0$. However, when x has important periodic components (e.g., at 50 or 60 Hz because of the power supply), they also have relative maxima at values of τ multiple of the period (20 or 16.6 msec in the example). The faster $C_x(\tau)$ decreases to 0, the faster the time evolution of the fluctuations, that is, the higher the contribution of the high-frequency components in the signal. At the limit, $C_x(\tau)$ is a Dirac function (equal to 0 except for $\tau = 0$) so that two data points, even very close, are uncorrelated. In this case, the signal is totally random, it is called "white noise" (see an example in Figure 14.4b).

The variance of x, which is the square of the SD σ_x of x, is the second-order moment of Δx:

$$\sigma_x^2 = \langle(\Delta x(t))^2\rangle = \langle(x(t)-\langle x\rangle)^2\rangle = \langle x^2\rangle - \langle x\rangle^2 \tag{14.A.6}$$

and can be calculated from the correlation function $C_x(\tau)$ at time $\tau = 0$ (Equation (14.A.4))

$$\sigma_x^2 = C_x(0) = \lim_{T \to +\infty} \frac{1}{T} \int_0^T (\Delta x(t))^2 \mathrm{d}t \tag{14.A.7}$$

The variance appears as the average value of the square of the fluctuations Δx of x (mean power of Δx). Its unit is the square of the unit of x. The SD measures the amplitude of the fluctuations Δx. It can be related to the peak-to-peak amplitude of Δx: for a gaussian signal (most EN signals that do not exhibit transients are gaussian) the probability that Δx ranges between $-2\sigma_x^2$ and $+2\sigma_x^2$ is 95.4%. Another quantity of interest, very close to the SD, is the RMS value, x_{RMS}, of x, defined as the square root of the average value of the square of $x(t)$. Its expression is similar to that of the SD (Equation (14.A.6)), except that it contains the mean value of $\langle x \rangle$. According to Equation (14.A.6), the SD and RMS values of x are related as follows:

$$x_{\mathrm{RMS}} = \sqrt{\langle x^2(t) \rangle} = \sqrt{\sigma^2 + \langle x \rangle^2} \tag{14.A.8}$$

and are identical for signals with a zero mean value.

The contribution of the power of the component at frequency f of the signal is given by the PSD (or spectrum) $S_x(f)$ of x, which is the FT of the correlation function $A_x(\tau)$ (Wiener–Khinchin theorem):

$$S_x(f) = \int_{-\infty}^{+\infty} A_x(\tau) \mathrm{e}^{-2\mathrm{j}\pi f \tau} \mathrm{d}\tau \tag{14.A.9}$$

where $\mathrm{j} = \sqrt{-1}$. The PSD unit is volt square per hertz ($\mathrm{V^2/Hz}$) for a voltage signal. Giving PSDs in decibel (dB) units, as is sometimes done, should be avoided since they are inherently relative units. It is necessary to plot the PSD against frequency in a log–log scale in order to see the whole spectrum clearly. Amplitude spectral density, defined as the square root of the PSD (and expressed in $\mathrm{V}/\sqrt{\mathrm{Hz}}$ for a voltage signal), is sometimes used, but it has no specific advantage compared to the PSD.

$S_x(f)$ is a two-sided spectrum, that is, it is defined for negative and positive frequencies f. Equation (14.A.5) allows the PSD of the signal x and that of its fluctuations Δx to be related:

$$S_x(f) = S_{\Delta x}(f) + \langle x \rangle^2 \delta(f) \tag{14.A.10}$$

where

$$S_{\Delta x}(f) = \int_{-\infty}^{+\infty} C_x(\tau) \mathrm{e}^{-2\mathrm{j}\pi f \tau} \mathrm{d}\tau \tag{14.A.11}$$

and $\delta(f)$ is the Dirac function (its integral for f varying from $-\infty$ to $+\infty$ is equal to 1 even if $\delta(f)$ is equal to 0 for $f \neq 0$). This equation shows that the PSD depends on the mean value $\langle x \rangle$ only through the Dirac component at frequency 0 ($S_x(0)$ is proportional to $\langle x \rangle^2$). However, $S_x(f)$ is not a continuous function at frequency 0 and the low-frequency limit of the PSD $S_x^*(0) = S_{\Delta x}(0)$ is independent of $\langle x \rangle$, depending only on the fluctuating part of x. The important consequence of the analysis of EN data is that the first step will always consist in calculating the mean value of the data and for investigating the second- and higher-order statistic properties of the fluctuations of the data, this mean value will be removed from the raw data beforehand.

The correspondence of the properties of a random signal in the time and frequency domains can be illustrated first for white noise. Its correlation function $C_x(\tau)$ is a Dirac function, so its PSD is independent of the frequency f according to Equation (14.A.11) (the FT of $\delta(\tau)$ is equal to 1). Hence, the white noise, whose time records look totally random since all data points are uncorrelated, has low- and high-frequency components of the same amplitude. In contrast, typical EN in corrosion show slow variations of high amplitude superimposed to fast fluctuations of low amplitude, which corresponds in the frequency domain to high values of PSD at low frequency and low values at high frequency.

For real signals $x(t)$ the two-sided PSD $S_x(f)$ is an even function of f, therefore it is common in physics to introduce the one-sided spectrum $\Psi_x(f)$ defined for positive frequencies only as

$$\Psi_x(f) = 2S_{\Delta x}(f) \quad \text{for } f > 0 \tag{14.A.12}$$

This is the definition of the PSD used in this chapter.

The variance σ_x^2 can be calculated directly from the PSD, since the correlation function $C_x(\tau)$ is the inverse FT of $S_{\Delta x}(f)$ (Equation (14.A.11)), that is,

$$C_x(\tau) = \int_{-\infty}^{+\infty} S_{\Delta x}(f)\mathrm{e}^{+2\mathrm{j}\pi f\tau}\mathrm{d}f \tag{14.A.13}$$

It can be noticed that the f-convention used in the expression of the FT is easier than the ω-convention, as there are fewer factors 2π to remember in the FT and inverse FT. Then, with Equations (14.A.7) and (14.A.12)

$$\sigma_x^2 = C_x(0) = \int_{-\infty}^{+\infty} S_{\Delta x}(f)\mathrm{d}f = \int_{0}^{+\infty} \Psi_x(f)\mathrm{d}f \tag{14.A.14}$$

In practice, the signals are measured over a finite interval time T. A definition of the PSD, equivalent to Equation (14.A.9), is, for the one-sided spectrum

$$\Psi_x(f) = \lim_{T\to+\infty} \frac{2}{T}\langle|\Delta X(f,T)|^2\rangle \quad \text{for } f > 0 \tag{14.A.15}$$

where $\Delta X(f,T)$ is the finite FT of the fluctuations Δx of x

$$\Delta X(f,T) = \int_{0}^{T} \Delta x(t)\mathrm{e}^{-2\mathrm{j}\pi ft}\mathrm{d}t \tag{14.A.16}$$

Digital Signal Processing

The formulae given so far are valid for continuous quantities. However, with the introduction of digital instrumentation, it is now common practice to carry out measurements by sampling the signal at constant intervals Δt, taking N of these samples, so that the acquisition time of a time record is $T = N\Delta t$. The experimental parameters that characterize a measurement, then, are N, the number of points in a time record (N is often a power of 2: 1024, 2048, etc.), and Δt, the sampling interval, expressed as time/point. It is also common to give it as a sampling frequency $f_s = 1/\Delta t$ (in Hz). More precisely, its dimension is points/second.

If x_n denotes the value of x sampled at time $n\Delta t$, the digital expressions of the mean value of x, of the SD, and of the correlation function $C_x(m)$ at time $m\Delta t$ are, from Equations (14.A.1), (14.A.4), and (14.A.6), respectively,

$$\langle x\rangle = \frac{1}{N}\sum_{n=0}^{N-1} x_n \tag{14.A.17}$$

$$C_x(m) = \frac{1}{N}\sum_{n=0}^{N-m-1} (x_n - \langle x\rangle)(x_{n+m} - \langle x\rangle) \tag{14.A.18}$$

and

$$\sigma_x = \sqrt{\frac{1}{N-1}\sum_{n=0}^{N-1}(x_n - \langle x\rangle)^2} \tag{14.A.19}$$

In Equation (14.A.19) a factor $N - 1$ is preferred to a factor N to get an unbiased estimator of the SD. However, since the value of N is usually large, this has almost no consequence on the value of the SD.

The two-sided PSD is defined as the FT of the correlation function $C_x(\tau)$ (Equation (14.A.11)). The most computationally efficient method for obtaining it, however, is based on the periodogram method by means of the Cooley–Tukey FFT algorithm (125). According to the Nyquist theorem the maximum frequency analyzed in the digital time record is $f_{\max} = f_s/2$ (also called Nyquist frequency). It is worth reminding that an analog low-pass AA filter with a cutoff frequency lower than $f_s/2$ has to be used before the A/D conversion in order to eliminate all frequencies above $f_s/2$ that cause spectrum aliasing. The low limit of the frequency range analyzed, $f_{\min}$, is inversely proportional to the measurement time T of the time record: $f_{\min} = 1/T$. $1/T$ is also the value of the frequency resolution, Δf, so that the frequencies analyzed may be expressed as $m\Delta f$, with the integer m varying from 1 to $N/2$. It is reminded that the value of the PSD at frequency 0 ($m = 0$) depends on the mean value of x only. The digital expression of the FT of the fluctuations of x is

$$\Delta X(m\Delta f) = \Delta t \sum_{n=0}^{N-1}(x_n - \langle x\rangle)\mathrm{e}^{-2j\pi mn/N} \tag{14.A.20}$$

and the PSD is estimated with the periodogram method (Equation (14.A.15)) by using K time records of length T

$$\Psi_x(m\Delta f) = \frac{2}{T}\frac{1}{K}\sum_{j=1}^{K}|\Delta X_j(m\Delta f)|^2 \tag{14.A.21}$$

where $\Delta X_j\,(m\Delta f)$ is the FT of the fluctuations of the jth time record. The method is easily implemented on a computer, requires little time for computation, and is applicable to a large class of random signals. If $K = 1$, the standard error ε is 100% (125, p. 76). Therefore, in order to increase the accuracy, K spectra are ensemble averaged. In this case, the error becomes

$$\varepsilon = \frac{1}{\sqrt{K}} \tag{14.A.22}$$

but the disadvantage is that the acquisition time is multiplied by K, for the same value of the minimum frequency analyzed.

The SD can also be computed from the measured PSD. The digital expression of Equation (14.A.14) is

$$\sigma_x = \sqrt{\Delta f \sum_{m=1}^{N/2}\Psi_x(m\Delta f)} \tag{14.A.23}$$

This equation clearly shows that the SD depends on the analyzed frequency bandwidth, which may not be evident in the expression of the SD involving the raw data

(Equation (14.A.19)). Hence, the values of SD should always be reported together with the frequency range analyzed.

The computation of the FT assumes that the time record repeats itself exactly from time $-\infty$ to $+\infty$, so that the amplitude difference between the first and the last point becomes a sudden discontinuity between the last point and the next repeat of the record. A linear increase in $x(t)$ from 0 to T, for example, if repeated, looks like a sawtooth. A discontinuity in the time function generates a PSD with a slope of -2 (98), so that if steps are not taken to eliminate the jump, the log–log plot of the PSD will always decrease with a -2 slope. A widespread and effective method to eliminate the unwanted consequences of the discontinuity is to multiply the time record with an appropriately shaped "window" before computing the FFT. The operation is commonly called "windowing." Different windows have been proposed, each having somewhat different effects on the PSD. One of the most common windows is the Hann window, described by the following expression:

$$w_n = 0.5\left(1 - \cos\left[\frac{2\pi n}{N}\right]\right) \quad \text{for } n = 0, 1, 2, \ldots, N-1 \tag{14.A.24}$$

which is applied to the time record after subtracting its mean value. All PSD values are then multiplied by $8/3 = 2.667$, to compensate for the attenuation caused by the window (125, p. 75). Since the window tapers to zero at the two ends of the time record, the discontinuity between beginning and end disappears (see Figure 14.8a). The application of this window, usually referred to as "hanning," widens the frequency range where the PSD is correctly measured. It has also the advantage of narrowing the peaks at 50 or 60 Hz and harmonics due to the power supply or those generated by rotating electrodes. It should be pointed out that discontinuities between the beginning and the end of the time record can occur even when the signal is stationary. For this reason it is strongly recommended that always, as a last signal processing operation before computing the FFT, hanning be applied.

The frequency range analyzed ($f_{\min}$, $f_{\max}$) is limited to three frequency decades for $N = 2048$ data points acquired per time record. This may be sufficient for many investigations at the corrosion potential where the corrosion processes are slow. In this case, the frequency range studied is usually (1 mHz to 1 Hz). However, some corrosion processes may be faster, as those under hydrodynamic control like in pipelines or under potentiostatic control in the passive domain. In order to cover a wider frequency bandwidth, the PSDs are successively measured in overlapping frequency ranges by using time records acquired at different sampling frequencies. The connection between the PSDs must be good, with the same values where the frequencies overlap. The quality of the overlapping is a good indication of the stationarity of the signal.

An alternative method for estimating the PSD, known as the MEM, has been developed, first employed in geophysics (126, 127), to obviate the perceived drawback of the periodogram method of requiring much longer acquisition times. In analogy with the FT, two methods for estimating the PSD are possible: a direct method in which the estimation is performed on the time signals, and an indirect one, based on the correlation function (127). They give similar results, so that only the second method (Levinson algorithm), which is computationally faster, is presented.

The MEM is an estimation of the PSD suggested by information theory. It is assumed that the sampled data, x_n, can be described by an autoregressive model:

$$y_n = x_n - a_1 y_{n-1} - a_2 y_{n-2} - \cdots - a_M y_{n-M} \tag{14.A.25}$$

where y_n is the estimated value of x_n, whose average value is supposed here to be 0 for simplification, and M is the order of the model. The estimation of the PSD $\Psi_x(f)$ is carried out through its relation to the correlation function $C_x(m)$ calculated for the N values of m $(0 \leq m \leq N-1)$. It is further assumed that $\Psi_x(f)$ is independent of any hypotheses on the unknown values of $C_x(m)$ for $m \geq N$. This last point is equivalent to imposing maximum entropy, that is minimum information, to the PSD with respect to the values $C_x(m)$, $m \geq N$, hence the name of the method. It should be noted that, on the contrary, the FFT method assumes that $C_x(m) = 0$ for $m \geq N$. Also, while the FFT technique gives a spectrum in the range $(f_{\min}, f_{\max})$, the PSD can be calculated by the MEM at any frequency $f \leq f_{\max}$. The algorithm to calculate the PSD, which can be implemented on a computer, is based on the formula (127)

$$\Psi_x(f) = \frac{2\sigma^2 \Delta t}{\left| 1 + \sum_{k=1}^{M} a_k e^{-2j\pi f k \Delta t} \right|^2} \tag{14.A.26}$$

where σ^2, $a_1, \ldots, a_M$ are $M+1$ coefficients, which are calculated from a $(M+1) \times (M+1)$ matrix equation, whose terms are obtained from the correlation function

$$\begin{bmatrix} C_x(0) & C_x(1) & . & C_x(M) \\ C_x(1) & C_x(0) & . & C_x(M-1) \\ C_x(2) & C_x(1) & . & C_x(M-2) \\ . & . & . & . \\ C_x(M) & C_x(M-1) & . & C_x(0) \end{bmatrix} \begin{bmatrix} 1 \\ a_1 \\ a_2 \\ . \\ a_M \end{bmatrix} = \begin{bmatrix} \sigma^2 \\ 0 \\ 0 \\ . \\ 0 \end{bmatrix} \tag{14.A.27}$$

As Equation (14.A.27) shows, to estimate the PSD by the MEM, it is necessary to decide what is the order M of the autoregressive random process. This is unknown, so that, in contrast with the FFT method, the PSD by MEM is not uniquely determined by the time series, and its computation can be performed for various values of M. Several techniques allow determination of M by minimizing an "error" term which often exhibits local minima in the frequency range investigated (94). The MEM and FFT methods have been compared for the same EN data to assess their relative advantages (88); the main conclusions are given in the text.

Signal Processing for Two Random Signals

When two random signals $x(t)$ and $y(t)$ are considered, for example, representing the outputs of two measurement channels, specific tools allow selection of the signals that are correlated in x and y and rejection of the uncorrelated part. It is the cross-correlation function $\langle x(t)y(t+\tau)\rangle$ in the time domain and its FT, the XPS, in the frequency domain. The expression of the XPS is very similar to that of the PSD: while the PSD of x is proportional to $|\Delta X(f)|^2 = \Delta X(f)\Delta X^*(f)$ (Equations (14.A.15) and (14.A.21)), where * denotes the complex conjugate, the XPS of x and y is proportional to $\Delta Y(f)\Delta X^*(f)$, which is a complex number. In practice, the XPS at frequency $m\Delta f$ is estimated with the periodogram method:

$$\Psi_{xy}(m\Delta f) = \frac{2}{T}\frac{1}{K}\sum_{j=1}^{K} \Delta Y_j(m\Delta f)\Delta X_j^*(m\Delta f) \tag{14.A.28}$$

Another situation concerns the measurement of the frequency response function $H(f)$ of a linear system. For example, if the random signal $x(t)$ applied to the input of the system is a current and the output signal $y(t)$ is the voltage response, $H(f)$ is the impedance of the system. In the frequency domain, a simple relationship exists between the FT of the fluctuations $\Delta x(t)$ and $\Delta y(t)$

$$\Delta Y(f) = H(f)\Delta X(f) \tag{14.A.29}$$

By multiplying both sides of this equation by $\Delta X^*(f)$, the expression of $H(f)$ is directly obtained from the ratio of the input–output XPS to the PSD of the input signal

$$H(f) = \frac{\Psi_{xy}(f)}{\Psi_x(f)} \tag{14.A.30}$$

Since $\Psi_{xy}(f)$ is a complex quantity, this equation gives the complex value of $H(f)$ from which the modulus and phase of the frequency response function can be determined. In most spectrum analyzers $H(f)$ is calculated from Equation (14.A.30) rather than from the ratio of the FTs (Equation (14.A.29)) because this algorithm allows removal of a part of the influence of the external noise added by the instrumentation to the output signal $Y(t)$ of the linear system (125). In practice, a white noise input signal is applied and the XPS and PSDs of the input and output signals are simultaneously estimated with the periodogram method. The coherence function is then derived as

$$\Gamma(f) = \frac{|\Psi_{xy}(f)|^2}{\Psi_x(f)\Psi_y(f)} \tag{14.A.31}$$

$\Gamma(f)$ has no dimension, it ranges from 1, when the output signal is the deterministic response to the input signal, to 0 when there is no relationship between the input and output signals. A value close to 1 of the coherence function is then a clear indication that $H(f)$ has been correctly estimated from Equation (14.A.30). Moreover, Γ is also used to estimate the measurement relative error $\varepsilon_{|H|}(f)$ of the modulus of $H(f)$ (128):

$$\varepsilon_{|H|}(f) \leq \sqrt{\frac{1}{2K}\left(\frac{1}{\Gamma(f)} - 1\right)} \tag{14.A.32}$$

where K is the number of elementary spectra averaged in the periodogram method.

A last result concerns the sum z of two uncorrelated random signals x and y. The variance of z is the sum of the variances of the two signals, so that the SD of z is

$$\sigma_z = \sqrt{\sigma_x^2 + \sigma_y^2} \tag{14.A.33}$$

In the same way, the PSDs add

$$\Psi_z(f) = \Psi_x(f) + \Psi_y(f) \tag{14.A.34}$$

APPENDIX B

Noise Analysis of a Conventional VA

The scheme of the amplifier is given in Figure 14.B.1 with the noise source of each component. The noise of the OA is described as usual by a voltage noise source e_{na} and two current noise sources i_{n1} and i_{n2} positioned at the input terminals; the three noise sources are independent of each other and the current sources have the same amplitude (see, e.g., Ref. (129, figure 9)). The thermal noise generated by the resistors R_1 and R_2 are modeled by the voltage noise sources e_{n1} and e_{n2}. The two input terminals of the OA have the same electric potential, so Ohm's law gives

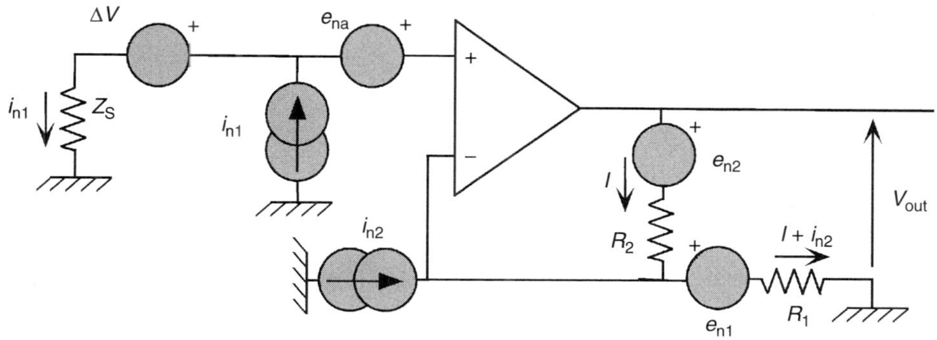

Figure 14.B.1 Detailed circuit of a conventional VA showing all instrumental noise sources.

$$\Delta V + Z_s i_{n1} + e_{na} = e_{n1} + R_1(I + i_{n2}) \quad (14.B.1)$$

where I is the current flowing in the resistor R_2. On the other hand, Ohm's law also gives the output voltage V_{out}

$$V_{out} = e_{n1} + e_{n2} + R_2 I + R_1(I + i_{n2}) \quad (14.B.2)$$

After simplification, these two equations lead to

$$V_{out} = \frac{R_1 + R_2}{R_1}\left[\Delta V + Z_s i_{n1} + e_{na} - \frac{R_2}{R_1 + R_2}(e_{n1} + R_1 i_{n2}) + \frac{R_1}{R_1 + R_2} e_{n2}\right] \quad (14.B.3)$$

Comparison with Equation (14.27) gives the gain of the VA and the noise sources e_n and i_n modeling the instrumental noise of the VA:

$$G_V = \frac{R_2}{R_1} + 1 \quad (14.B.4)$$

$$i_n = i_{n1} \quad (14.B.5)$$

$$e_n = e_{na} - \frac{R_2}{R_1 + R_2}(e_{n1} + R_1 i_{n2}) + \frac{R_1}{R_1 + R_2} e_{n2} \quad (14.B.6)$$

For a high gain G_V ($R_2 \gg R_1$), the PSD of e_n is, with Equation (14.6),

$$\Psi_{e_n} = \Psi_{e_{na}} + 4k_B T R_1 + R_1^2 \Psi_{i_{n2}} \quad (14.B.7)$$

The second and third terms can be neglected when low values of resistance R_1 are used (100 Ω or 1 kΩ). The voltage (respectively current) noise of the VA is then equal to the voltage (respectively current) noise generated by the OA.

APPENDIX C

SD of a Linearly Drifting Signal

The analytical expression of the PSD $\Psi_{x,comp}$ of a stationary random signal $x(t)$, to which a linear drift with slope a is superimposed, is given by Equation (14.31) when measured over a fixed period of time T. On the other hand, the SD $\sigma_{x,comp}$ can be derived from $\Psi_{x,comp}$ with Equation (14.A.23):

$$\sigma_{x,\text{comp}} = \sqrt{\Delta f \sum_{m=1}^{N/2} \Psi_{\text{x,comp}}(m\Delta f)} \tag{14.C.1}$$

where Δf is the frequency resolution $\Delta f = 1/T$. So, with Equation (14.31)

$$\sigma^2_{x,\text{comp}} = \Delta f \sum_{m=1}^{N/2} \Psi_x(m\Delta f) + \frac{a^2 T}{2\pi^2 \Delta f} \sum_{m=1}^{N/2} \frac{1}{m^2} \tag{14.C.2}$$

The sum of the last series can be calculated by using the residues method for a large value of N

$$\sum_{m=1}^{N/2} \frac{1}{m^2} = \sum_{m=1}^{+\infty} \frac{1}{m^2} = \frac{\pi^2}{6} \tag{14.C.3}$$

so that $\sigma_{x,\text{comp}}$ is

$$\sigma_{x,\text{comp}} = \sqrt{\Delta f \sum_{m=1}^{N/2} \Psi_x(m\Delta f) + \frac{a^2 T}{12\Delta f}} = \sqrt{\sigma_x^2 + \frac{a^2 T^2}{12}} \tag{14.C.4}$$

and Equation (14.32) is obtained.

REFERENCES

1. J Goellner, A Burkert, A Heyn, J Hickling. Using EN to detect corrosion: evaluation of a round-robin experiment. *Corrosion* 55:476–492, 1999.
2. M Seralathan, SK Rangarajan. Fluctuation phenomena in electrochemistry. Part I. The formalism. *J. Electroanal. Chem.* 208:13–28, 1986.
3. JR Kearns, JR Scully, PR Roberge, DL Reichert, JL Dawson, eds. *EN Measurement for Corrosion Applications*. West Conshohocken: American Society for Testing and Materials, STP 1277, 1996.
4. F Mansfeld, F Huet, OR Mattos, eds. *New Trends in EIS and ENA*. Pennington, NJ: The Electrochemical Society, Inc., Proceedings Volume 2000–24, 2000.
5. K Tachibana, G Okamoto. An experimental application of noise analysis to electrochemistry and corrosion. *Rev. Coat. Corros.* 3:229–267, 1981.
6. C Gabrielli, F Huet, M Keddam, R Oltra. A review of the probabilistic aspects of localized corrosion. *Corrosion* 46:266–278, 1990.
7. JL Dawson. EN measurement: the definitive in-situ technique for corrosion applications? In: JR Kearns, JR Scully, PR Roberge, DL Reichert, JL Dawson, eds. *EN Measurement for Corrosion Applications*. West Conshohocken: American Society for Testing and Materials, STP 1277, 1996, pp 3–35.
8. U Bertocci, F Huet. Noise analysis applied to electrochemical systems. *Corrosion* 51:131–144, 1995.
9. DA Eden. EN — The first two octaves. CORROSION '98, Houston, NACE, 1998, Paper 386, pp 1–31.
10. RA Cottis. Interpretation of EN data. *Corrosion* 57:265–285, 2001.
11. WP Iverson. Transient voltage changes produced in corroding metals and alloys. *J. Electrochem. Soc.* 115:617–618, 1968.
12. T Hagyard, MJ Prior. Potential of aluminium in aqueous chloride solutions, part 2. *Trans. Faraday Soc.* 57:2295–2298, 1961.

13. JM Defranoux. Méthodes électrochimiques d'étude du comportement des aciers inoxydables en milieu chlorure. *Corros. Sci.* 3:75–86, 1963.
14. RW Shideler, U Bertocci. A low-noise potentiostat for the study of small amplitude signals in electrochemistry. *J. Res. Natl. Bur. Stand.* 85:211–217, 1980.
15. I Epelboin, C Gabrielli, M Keddam, L Raillon. A study of potentiostat noise. *J. Electroanal. Chem.* 93:155–161, 1978.
16. G Blanc, C Gabrielli, M Keddam. Measurement of the EN by a cross correlation method. *Electrochim. Acta* 20:687–689, 1975.
17. G Blanc, I Epelboin, C Gabrielli, M Keddam. EN generated by anodic dissolution or diffusion processes. *J. Electroanal. Chem.* 75:97–124, 1977.
18. C Gabrielli, M Keddam, L Raillon. Random signals: third-order correlation measurement. *J. Phys. E: Sci. Instrum.* 12:632–636, 1979.
19. G Okamoto, T Sugita, S Nishiyama, K Tachibana. The analysis of passive current noise of stainless steels under potentiostatic conditions. *Boshoku Gijutsu* 23:439–451, 1974.
20. I Epelboin, C Gabrielli, M Keddam, L Raillon. Measurement of the power spectral density of EN: direct two-channel method. *J. Electroanal. Chem.* 105:389–395, 1979.
21. U Bertocci. Applications of a low noise potentiostat in electrochemical measurements. *J. Electrochem. Soc.* 127:1931–1934, 1980.
22. U Bertocci, J Kruger. Studies of passive film breakdown by detection and analysis of EN. *Surf. Sci.* 101:608–618, 1980.
23. U Bertocci. Separation between deterministic response and random fluctuations by means of the cross-spectrum in the study of EN. *J. Electrochem. Soc.* 128:520–523, 1981.
24. C Gabrielli, F Huet, M Keddam, R Oltra, C Pallotta. Stochastic aspects of mechanical and chemical breakdown of passivity. In: M Froment, ed. *Passivity of Metals and Semiconductors*. Amsterdam: Elsevier Science Publishers, 1983, pp 293–298.
25. C Gabrielli, F Huet, M Keddam. Investigation of electrochemical processes by an ENA. Theoretical and experimental aspects in potentiostatic regime. *Electrochim. Acta* 31:1025–1039, 1986.
26. R Oltra, C Gabrielli, F Huet, M Keddam. Electrochemical investigation of locally depassivated iron. A comparison of various techniques. *Electrochim. Acta* 31:1501–1511, 1986.
27. K Hladky, JL Dawson. The measurement of corrosion using 1/f noise. *Corros. Sci.* 22:231–237, 1982.
28. BS Skerry, DA Eden. Electrochemical testing to assess corrosion protective coatings. *Prog. Org. Coat.* 15:269–285, 1987.
29. JC Uruchurtu, JL Dawson. Noise analysis of pure aluminum under different pitting conditions. *Corrosion* 43:19–26, 1987.
30. PC Searson, JL Dawson. Analysis of EN generated by corroding electrodes under open-circuit conditions. *J. Electrochem. Soc.* 135:1908–1915, 1988.
31. C Cachet, C Gabrielli, F Huet, M Keddam, R Wiart. Growth mechanism for silver electrodeposition — a kinetic analysis by impedance and noise measurements. *Electrochim. Acta* 28:899–908, 1983.
32. IN Bastos, F Huet, RP Nogueira, P Rousseau. Influence of aliasing in time and frequency ENM. *J. Electrochem. Soc.* 147:671–677, 2000.
33. DA Eden. Comments on "ENA of iron exposed to NaCl solutions of different corrosivity" by F Mansfeld, H Xiao. *J. Electrochem. Soc.* 141:1402–1403, 1994.
34. DA Eden, K Hladky, DG John, JL Dawson. EN — simultaneous monitoring of potential and current noise signals from corroding electrodes. CORROSION '86, Houston, NACE, 1986, Paper 274, pp 1–9.
35. K Hladky, JL Dawson. The measurement of localized corrosion using EN. *Corros. Sci.* 21:317–322, 1981.

36. F Huet. Etude théorique et expérimentale du comportement stochastique d'une interface métal-électrolyte par l'analyse du bruit électrochimique en régime stationnaire. Ph.D. dissertation, Université Pierre et Marie Curie, Paris, France, 1984.
37. NG Van Kampen. *Stochastic Processes in Physics and Chemistry*. Amsterdam: North Holland, 1981.
38. JB Johnson. Thermal agitation of electricity in conductors. *Nature* 119:50–51, 1927.
39. H Nyquist. Thermal agitation of electric charge in conductors. *Phys. Rev.* 32:110–113, 1928.
40. C Gabrielli, F Huet, M Keddam, O Haas. *Ac* impedance and EN of strongly adsorbed electroactive species. Application to a redox polymer modified electrode. *Electrochim. Acta* 33:1371–1381, 1988.
41. C Gabrielli, F Huet, M Keddam. Fluctuations in electrochemical systems: II. Application to a diffusion limited redox process. *J. Chem. Phys.* 99:7240–7252, 1993.
42. W Schottky. Uber spontane stromschwankungen in verschiedenen elektrizitätsleitern. *Ann. Phys.* 57:541–567, 1918.
43. C Gabrielli, F Huet, M Keddam, R Oltra. Localized corrosion as a stochastic process: a review. In: HS Isaacs, U Bertocci, J Kruger, S Smialowska, eds. *Advances in Localized Corrosion*. Houston: NACE, 1990, pp 93–108.
44. JB Johnson. The Schottky effect in low frequency circuits. *Phys. Rev.* 26:71–85, 1925.
45. W Schottky. Small-shot effect and flicker effect. *Phys. Rev.* 28:74–103, 1926.
46. A van der Ziel. Flicker noise in electronic devices. *Adv. Electron. Electron Phys.* 49:225–297, 1979.
47. C Gabrielli, F Huet, M Keddam. Fluctuations in electrochemical systems: I. General theory on diffusion limited electrochemical reactions. *J. Chem. Phys.* 99:7232–7239, 1993.
48. J Dukovic, CW Tobias. The influence of attached bubbles on potential drop and current distribution at gas-evolving electrodes. *J. Electrochem. Soc.* 134:331–343, 1987.
49. Z Amrani, F Huet, M Jérôme, P Manolatos, F Wenger. Fluctuations of permeation rate through an iron membrane induced by hydrogen bubbles. *J. Electrochem. Soc.* 141: 2059–2061, 1994.
50. A Benzaïd, F Huet, M Jérôme, F Wenger, C Gabrielli, J Galland. ENA of cathodically polarised AISI 4140 steel. II. Identification of potential fluctuation sources for unstressed electrodes. *Electrochim. Acta* 47:4325–4332, 2002.
51. A Legat. Influence of electrolyte movement on measured EN. *Corrosion* 56:1086–1092, 2000.
52. F Huet, RP Nogueira. Comparative analysis of potential, current, and electrolyte resistance fluctuations in two-phase oil/water mixtures. CORROSION '03, Houston, NACE, 2003, Paper 416, pp 1–18.
53. B Baroux. Further insights on the pitting corrosion of stainless steels. In: P Marcus, ed. *Corrosion Mechanisms in Theory and Practice*. New York: Marcel Dekker, 2002, pp 311–347.
54. RE Ricker, U Bertocci. A review of EN and its application to the study of stress corrosion cracking. In: RH Jones, DR Baer, eds. *New Techniques for Characterizing Corrosion and Stress Corrosion*. Warrendale, PA: TMS-Miner Metals and Materials Society, 1996, pp 235–245.
55. A Benzaïd, F Huet, M Jérôme, F Wenger, C Gabrielli, J Galland. ENA of cathodically polarised AISI 4140 steel. I. Characterisation of hydrogen evolution on vertical unstressed electrodes. *Electrochim. Acta* 47:4315–4323, 2002.
56. A Benzaïd, F Huet, M Jérôme, F Wenger, C Gabrielli, J Galland. ENA of cathodically polarised AISI 4140 steel. III. Influence of hydrogen absorption for stressed electrodes. *Electrochim. Acta* 47:4333–4338, 2002.
57. AMP Simões, MGS Ferreira. Crevice corrosion studies on stainless steel using ENM. *Br. Corros. J.* 22:21–25, 1987.

58. MZ Yang, M Wilmott, JL Luo. Analysis of the EN for localized corrosion of type A516-70 carbon steel. *Corrosion* 54:869–876, 1998.
59. JA Wharton, BG Mellor, RJK Wood, CJE Smith. Crevice corrosion studies using ENM and a scanning electrode technique. *J. Electrochem. Soc.* 147:3294–3301, 2000.
60. JB Lumsden, MW Kendig, S Jeanjaquet. EN for carbon steel in sodium chloride solutions — effect of chloride and oxygen activity. CORROSION '92, Houston, NACE, 1992, Paper 224, pp 1–8.
61. F Mansfeld, H Xiao. ENA of iron exposed to NaCl solutions of different corrosivity. *J. Electrochem. Soc.* 140:2205–2209, 1993.
62. A Petek, V Doleček, V Vlachy. Stochastic analysis of current fluctuations during general corrosion of stainless steel in sulfuric acid. *Corrosion* 53:928–934, 1997.
63. V Brusamarello, A Lago, CV Franco. Analysis of different methods to calculate EN resistance using a three-electrode cell. *Corrosion* 56:273–282, 2000.
64. BS Skerry, DA Eden. Characterisation of coatings performance using ENA. *Prog. Org. Coat.* 19:379–396, 1991.
65. DJ Mills, GP Bierwagen, B Skerry, D Tallman. Investigation of anticorrosive coatings by the EN method. *Mater. Performance* 34-5:33–38, 1995.
66. H Xiao, LT Han, CC Lee, F Mansfeld. Collection of electrochemical impedance and noise data for polymer-coated steel from remote test sites. *Corrosion* 53:412–422, 1997.
67. F Mansfeld, CC Lee. The frequency dependence of the noise resistance for polymer-coated metals. *J. Electrochem. Soc.* 144:2068–2071, 1997.
68. GP Bierwagen, DE Tallman, S Touzain, A Smith, R Twite, V Balbyshev, Y Pae. EN methods applied to the study of organic coating and pretreatments. CORROSION '98, Houston, NACE, 1998, Paper 380, pp 1–13.
69. RJK Wood, JA Wharton, AJ Speyer, KS Tan. Investigation of erosion–corrosion processes using ENM. *Tribol. Int.* 35:631–641, 2002.
70. R Oltra, B Chapey, F Huet, L Renaud. Coupling of acoustic emission and EN measurement techniques in slurry erosion–corrosion studies. In: JR Kearns, JR Scully, PR Roberge, DL Reichert, JL Dawson, eds. *EN Measurement for Corrosion Applications*. West Conshohocken: American Society for Testing and Materials, STP 1277, 1996, pp 361–374.
71. Y Chen, M Gopal, WP Jepson. Comparison of ECN and EIS measurement for corrosion monitoring under multiphase flow conditions. CORROSION '97, Houston, NACE, 1997, Paper 276, pp 1–21.
72. JM Malo, J Uruchurtu, O Corona. Corrosion detection of mild steel in a two phase hydrocarbon-electrolyte system under flow conditions using EN. CORROSION '98, Houston, NACE, 1998, Paper 381, pp 1–13.
73. T Hong, Y Chen, YH Sun, WP Jepson. Monitoring corrosion in multiphase pipelines. *Mater. Corros.* 52:590–597, 2001.
74. SC Dexter, DJ Duquette, OW Siebert, HA Videla. Use and limitations of electrochemical techniques for investigating microbiological corrosion. *Corrosion* 47:308–318, 1991.
75. G Schmitt. Sophisticated electrochemical methods for MIC investigation and monitoring. *Mater. Corros.* 48:586–601, 1997.
76. EM Lehockey, AM Brennenstuhl, G Palumbo, P Lin. EN for evaluating susceptibility of lead-acid battery electrodes to intergranular corrosion. *Br. Corros. J.* 33:29–36, 1998.
77. RG Hardon, P Lambert, CL Page. Relationship between EN and corrosion rate of steel in salt contaminated concrete. *Br. Corros. J.* 23:225–228, 1988.
78. U Bertocci. A comparison of EN and impedance spectroscopy for the detection of corrosion in reinforced concrete. In: JR Kearns, JR Scully, PR Roberge, DL Reichert, JL Dawson, eds. *EN Measurement for Corrosion Applications*. West Conshohocken: American Society for Testing and Materials, STP 1277, 1996, pp 39–58.
79. YJ Tan, S Bailey, B Kinsella. The monitoring of the formation and destruction of corrosion inhibitor films using ENA. *Corros. Sci.* 38:1681–1695, 1996.

80. YJ Tan, B Kinsella, S Bailey. Monitoring batch treatment inhibitor performance using ENA. *Br. Corros. J.* 32:212–216, 1997.
81. RA Cottis, S Turgoose. *Corrosion Testing Made Easy: Electrochemical Impedance and Noise*. Houston: NACE, 1999.
82. U Bertocci, F Huet. Noise resistance applied to corrosion measurements: III. Influence of the instrumental noise on the measurements. *J. Electrochem. Soc.* 144:2786–2793, 1997.
83. PC Pistorius. Design aspects of ENM for uncoated metals: electrode size and sampling rate. *Corrosion* 53:273–283, 1997.
84. JF Chen, WF Bogaerts. Electrochemical emission spectroscopy for monitoring uniform and localized corrosion. *Corrosion* 52:753–759, 1996.
85. ML Benish, J Sikora, B Shaw, E Sikora, M Yaffe, A Krebs, G Martinchek. A new EN technique for monitoring the localized corrosion of 304 stainless steel in chloride-containing solutions. CORROSION '98, Houston, NACE, 1998, Paper 370, pp 1–14.
86. WH Press, SA Teukolsky, WT Vetterling, BP Flannery. *Numerical Recipes in C*. Cambridge, U.K: Cambridge University Press, 1994.
87. L Beaunier, J Frydman, C Gabrielli, F Huet, M Keddam. Comparison of spectral analysis with fast Fourier transform and maximum entropy method. Application to the role of molybdenum implantation on localized corrosion of type 304 stainless steel. In: JR Kearns, JR Scully, PR Roberge, DL Reichert, JL Dawson, eds. *EN Measurement for Corrosion Applications*. West Conshohocken: American Society for Testing and Materials, STP 1277, 1996, pp 114–128.
88. U Bertocci, J Frydman, C Gabrielli, F Huet, M Keddam. Analysis of EN by power spectral density applied to corrosion studies: MEM or FFT? *J. Electrochem. Soc.* 145:2780–2786, 1998.
89. U Bertocci, C Gabrielli, F Huet, M Keddam. Noise resistance applied to corrosion measurements: I. Theoretical analysis. *J. Electrochem. Soc.* 144:31–37, 1997.
90. H Xiao, F Mansfeld. Evaluation of coating degradation with electrochemical impedance spectroscopy and ENA. *J. Electrochem. Soc.* 141:2332–2337, 1994.
91. U Bertocci, F Huet, RP Nogueira. Use of multiple reference electrodes in ENM. *Corrosion*. 59:629–634, 2003.
92. U Bertocci, F Huet, RP Nogueira, P Rousseau. Drift removal procedures in the analysis of EN. *Corrosion* 58:337–347, 2002.
93. F Huet, U Bertocci, C Gabrielli, M Keddam. Noise measurement in corrosion. Proceedings of the Research Topical Symposia, CORROSION '97, Houston, NACE, 1997, pp 11–30.
94. T Schauer, H Greisiger, L Dulog. Details on MEM analysis of EN data and correlation with impedance measurements for organic coatings on metals. *Electrochim. Acta* 43:2423–2433, 1998.
95. F Mansfeld, Z Sun, CH Hsu, A Nagiub. Concerning trend removal in ENM. *Corros. Sci.* 43:341–352, 2001.
96. C Gabrielli, F Huet, M Keddam, A Sahar. Investigation of water electrolysis by spectral analysis. I. Influence of the current density. *J. Appl. Electrochem.* 19:683–696, 1989.
97. M Hashimoto, S Miyajima, T Murata. A spectrum analysis of potential fluctuation during passive film breakdown and repair on iron. *Corros. Sci.* 33:917–925, 1992.
98. U Bertocci, F Huet, B Jaoul, P Rousseau. Frequency analysis of transients in EN: mathematical relationships and computer simulations. *Corrosion* 56:675–683, 2000.
99. U Bertocci, M Koike, S Leigh, F Qiu, G Yang. A statistical analysis of the fluctuations of the passive current. *J. Electrochem. Soc.* 133:1782–1786, 1986.
100. U Bertocci, S Leigh, AC Van Orden, G Yang. Statistics of pit initiation: analysis of current transients during passive film breakdown. *Mater. Res. Soc. Symp. Proc.* 84:251–259, 1987.
101. B Wu, JR Scully, JL Hudson, AS Mikhailov. Cooperative stochastic behavior in localized corrosion. I. Model. *J. Electrochem. Soc.* 144:1614–1620, 1997.

102. TT Lunt, ST Pride, JR Scully, JL Hudson, AS Mikhailov. Cooperative stochastic behavior in localized corrosion. II. Experiments. *J. Electrochem. Soc.* 144:1620–1629, 1997.
103. G Bagley. The Measurement and the Analysis of EN. Ph.D. dissertation, University of Manchester, Manchester, U.K., 1998.
104. U Bertocci, C Gabrielli, F Huet, M Keddam, P Rousseau. Noise resistance applied to corrosion measurements: II. Experimental tests. *J. Electrochem. Soc.* 144:37–43, 1997.
105. A Bautista, F Huet. Noise resistance applied to corrosion measurements: IV. Asymmetric coated electrodes. *J. Electrochem. Soc.* 146:1730–1736, 1999.
106. A Bautista, U Bertocci, F Huet. Noise resistance applied to corrosion measurements: V. Influence of electrode asymmetry. *J. Electrochem. Soc.* 148:B412–B418, 2001.
107. A Aballe, F Huet. Noise resistance applied to corrosion measurements: VI. Partition of the current fluctuations between the electrodes. *J. Electrochem. Soc.* 149:B89–B96, 2002.
108. A Aballe, A Bautista, U Bertocci, F Huet. Measurement of the noise resistance for corrosion applications. *Corrosion* 57:35–42, 2001.
109. M Danielson. Modeling of certain electrode parameters on the EN response. *Corrosion* 53:770–777, 1997.
110. F Mansfeld, CC Lee, G Zhang. Comparison of electrochemical impedance and noise data in the frequency domain. *Electrochim. Acta* 43:435–438, 1998.
111. F Mansfeld, C Chen, CC Lee, H Xiao. The effect of asymmetric electrodes on the analysis of electrochemical impedance and noise data. *Corros. Sci.* 38:497–513, 1996.
112. GP Bierwagen, DE Tallman, J Zlotnick, CS Jeffcoate. Defects and heterogeneities in corrosion protective organic coatings films and their effects on film performance. In: GP Bierwagen, ed. *Organic Coatings for Corrosion Control.* Washington, DC: American Chemical Society, ACS Symposium Series 689, 1998, pp 123–136.
113. F Mansfeld, Z Sun. Localization index obtained from ENA. *Corrosion* 55:915–918, 1999.
114. JA Wharton, RJK Wood, BG Mellor. Wavelet analysis of ENM during corrosion of austenitic and superduplex stainless steels in chloride media. *Corros. Sci.* 45:97–122, 2003.
115. A Aballe, M Bethencourt, FJ Botana, M Marcos, JM Sánchez-Amaya. Use of wavelets to study EN transients. *Electrochim. Acta* 46:2353–2361, 2001.
116. A Legat, V Doleček. Chaotic analysis of EN measured on stainless steel. *J. Electrochem. Soc.* 142:1851–1858, 1995.
117. Á Horváth, R Schiller. Rescaled range analysis of the corrosion potential noise. *Corros. Sci.* 45:597–609, 2003.
118. A Legat, M Leban. Study of single fluctuations in EN signals. In: F Mansfeld, F Huet, OR Mattos, eds. *New Trends in EIS and ENA.* Pennington, NJ: The Electrochemical Society, Inc., Proceedings Volume 2000–24, 2000, pp 221–229.
119. C Gabrielli, F Huet, M Keddam. Real time measurement of electrolyte resistance fluctuations. *J. Electrochem. Soc.* 138:L82–L84, 1991.
120. F Huet, C Kuntz, H Takenouti. Measurement of electrolyte resistance fluctuations in corrosion applications. CORROSION '98, Houston, NACE, 1998, Paper 378, pp 1–13.
121. C Gabrielli, F Huet. Fluctuation analysis in electrochemical engineering processes with two-phase flows. *J. Appl. Electrochem.* 24:593–601, 1994.
122. V Jovancicevic. Controlled potentiostatic measurements of EN. CORROSION '00, Houston, NACE, 2000, Paper 419, pp 1–10.
123. B Röseler, CA Schiller. Strom-Potential-korrelierte Rauschmessung (CorrElNoise) — Ein neues Verfahren zur elektrochemischen Rauschanalyse. *Mater. Corros.* 52:413–417, 2001.
124. RA Cottis, M Kumaraguru, M Marti, HA Al-Mazeedi. Techniques for the measurement and interpretation of EN. Research in Progress Symposium, CORROSION '02, Houston, NACE, 2002, Extended Abstracts, pp 17–20.

125. JS Bendat, AG Piersol. *Engineering Applications of Correlation and Spectral Analysis*. New York: John Wiley & Sons, 1980.
126. JP Burg. Maximum entropy spectral analysis. Proceedings of the 37th Meeting of the Society of Exploration Geophysicists, Oklahoma City, 1967.
127. SM Kay, SL Marple Jr. Spectrum analysis — a modern perspective. *Proc. IEEE* 69:1380–1419, 1981.
128. C Gabrielli, F Huet, M Keddam. Comparison of sine wave and white noise analysis for electrochemical impedance measurements. *J. Electroanal. Chem*. 335:33–53, 1992.
129. Noise Analysis in Operational Amplifier Circuits, Application Report SLVA043A, Texas Instrument, 1999, http://focus.ti.com/docs/analog/analoghomepage.jhtml, accessed November 14, 2002.

15

Scanning Electrode Techniques for Investigating Near-Surface Solution Current Densities

R. Scott Lillard

Material Science and Technology Division, Los Alamos National Laboratory, Los Alamos, New Mexico

Contents

15.1 INTRODUCTION

Investigators have used a wide variety of methods to characterize local electrochemical processes such as those that might develop at intermetallics, second phase particles, grain boundaries, and other surface heterogeneities. As far back as the 1940s, Dix and Brown (1–3) used a masking technique to measure the potential of second-phase constituents and grain boundaries in Al–Cu (4% Cu) and Al–Mg alloys. Using a Bakelite varnish as a mask, samples were produced that exposed only the precipitate, grain boundary area, or grain centers to solution. While there have been many techniques over the years that allow analysis of the anode and cathode separately and have provided valuable insight into local electrochemical processes (the Dix and Brown mechanism for SCC in Al alloys survives even today), a method for measuring the local potential and current distributions for a freely corroding surface containing both anodic and cathodic processes is highly desirable. Thornhill and Evans (4, 5) developed one such method for measuring the potential distribution in solution above an iron electrode. The apparatus is shown in Figure 15.1. It consisted of two identical reference electrodes in solution (silver chloride for salt solutions and mercury sulfate for nonchloride solutions). One of these electrodes remained at a fixed position some distance from the sample surface while the other had a fine-tipped Luggin probe (tubulus) that was mounted on a series of stages that allowed it to be moved near to and across the sample surface. The reference electrodes were connected to a potentiometer and equipotential lines in solution were located by zeroing the current flow as measured by a galvanometer. The power of the Evans and Thornhill technique was demonstrated by Agar and Evans (6), who mapped the potential distribution across a partially immersed zinc sample. It was shown that cathodic reactions were isolated to the area on the electrode in the meniscus near the surface owing to the higher concentration of oxygen in this region while anodic reactions were well removed from the waterline (Figure 15.2). A method for calculating the local current distribution from the equipotential lines was also presented. This technique was the basis for the modern scanning reference electrode.

Today, a myriad of scanning electrode techniques have been developed to probe the electrochemical interface including: scanning electrochemical microscopy (7), scanning Kelvin probe microscopy (8), scanning confocal microscope (9), the atomic force microscope (10), and scanning tunneling microscopy (11). Many of these techniques have been described in other chapters of this book. In this chapter, we will review only those techniques that map the local current and potential

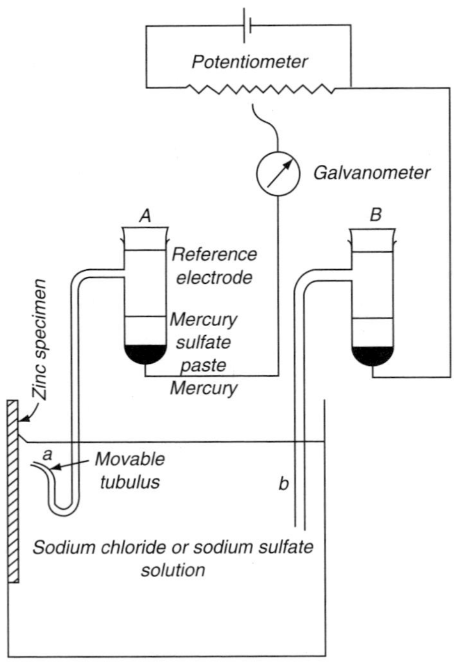

Figure 15.1 Apparatus used by U.R. Evans et al. to measure to the potential distribution across a zinc electrode. (From R. S. Thornhill, U. R. Evans, *Journal of the Chemical Society*, 614, 1938. IoM Communications, London. With permission.)

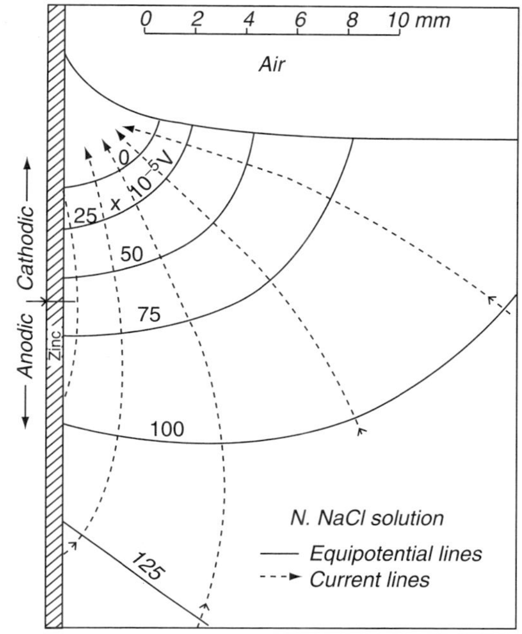

Figure 15.2 Equicurrent and equipotenial lines for a zinc electrode in sodium sulfate solution near the liquid/solution interface. Calculated from the potential distribution measured using the apparatus in Figure 15.1. (From R. S. Thornhill, U. R. Evans, *Journal of the Chemical Society*, 614, 1938. IoM Communications, London. With permission.)

distributions on a freely corroding surface. This includes the scanning reference electrode technique (SRET), the scanning vibrating electrode technique (SVET), and local electrochemical impedance spectroscopy (LEIS). Before we begin the discussion, it is necessary to develop a theoretical foundation for current and potential distributions for electrodes in solution. Here, we have chosen to use the disk electrode geometry for our analysis as it best represents experimental data. For probe calibration procedures, the derivation of a point source geometry is useful. The solution for a point and application to probe calibration is readily available in the literature (12).

15.2 CURRENT AND POTENTIAL DISTRIBUTIONS FOR DISK ELECTRODES

The La'Place equation for the potential gradient above a disk electrode in an insulating plane is given in cylindrical coordinates by the relationship:

$$\nabla^2\phi = \frac{\partial^2\phi}{\partial z^2} + \frac{1}{r}\frac{\partial}{\partial r}\left(r\frac{\partial\phi}{\partial r}\right) = 0 \qquad (15.1)$$

where ϕ is the potential, r is the radius of the electrode, and z is the vertical distance above the plane Nanis and Kesselman (13) used the Hankel transform of the La'Place equation to solve the primary potential distribution above an equipotential disk in an insulating plane:

$$\phi(r,z) = \frac{2\phi_0}{\pi} \sin^{-1}\left\{\frac{2}{[(r/a-1)^2+(z/a)^2]^{1/2}+[(r/a+1)^2+(z/a)^2]^{1/2}}\right\} \tag{15.2}$$

where a is the radial distance from the center of the disk. Primary current distribution neglects overpotential at the electrode/electrolyte interfaces and assumes the potential in solution immediately adjacent to the electrode surface is an equipotential plane. While this is rarely the case for corroding systems where the current distribution is governed by the polarizability of the electrode, the so-called secondary current distribution, primary current distribution is a limiting, worse case, scenario that provides mathematical simplicity.

The magnitude of potential distribution in Equation (15.2) as a function of radial distance from the center of a disk electrode is presented in Figure 15.3 for various vertical distances from the surface. For distances greater than a critical combination of z and r, the potential distribution becomes featureless; that is, it is not possible to detect local processes. Equipotential lines derived from Figure 15.3 are presented in Figure 15.4. These equipotential surfaces intersect the insulator at 90° as there is no current flow to the insulator. Current flow is always perpendicular to these equipotential surfaces in the primary current distribution model. The magnitude of the current distribution is given by (13):

$$\frac{J}{J_{av}} = \frac{1}{2}\frac{1}{\sqrt{1-(r/a)^2}} \tag{15.3}$$

and is presented in Figure 15.5. The asymptotic behavior for an electrode/insulator interface of 180° results from the assumption that the potential in solution immediately adjacent to the electrode surface is parallel to the electrode. As a result, very near the surface a zero impedance boundary exists in solution between the electrode and insulator; therefore, the current approaches infinity. Although the impedance becomes infinitely small at this interface it does so along an infinitesimally small line and, therefore, the current in the system remains finite. In comparison, for systems

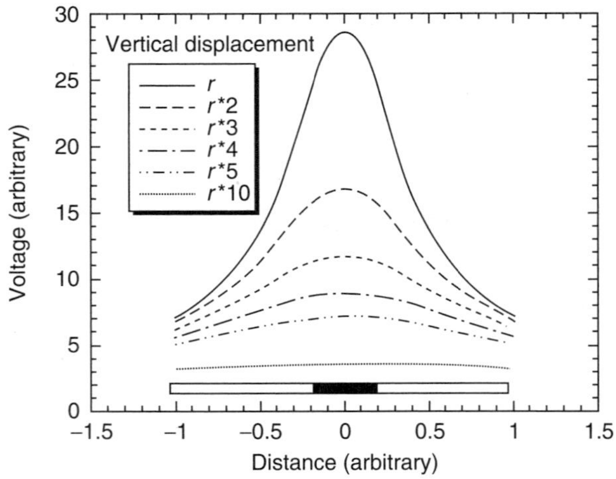

Figure 15.3 Magnitude of the potential distribution (primary current distribution) in Equation (15.2) as a function of radial distance from the center of a disk electrode for various vertical distances from the surface. The position of the electrode is represented in solid black. Thus, the electrode insulator interfaces are located at 0.2 and −0.2 units of distance.

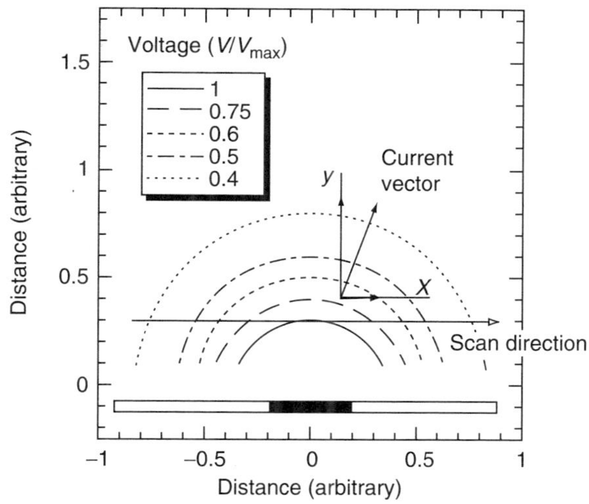

Figure 15.4 Equipotential lines derived from Figure 15.3. By definition, equipotential surfaces intersect the insulator at 90° as there is no current flow to the insulator. The position of the electrode is represented in solid black.

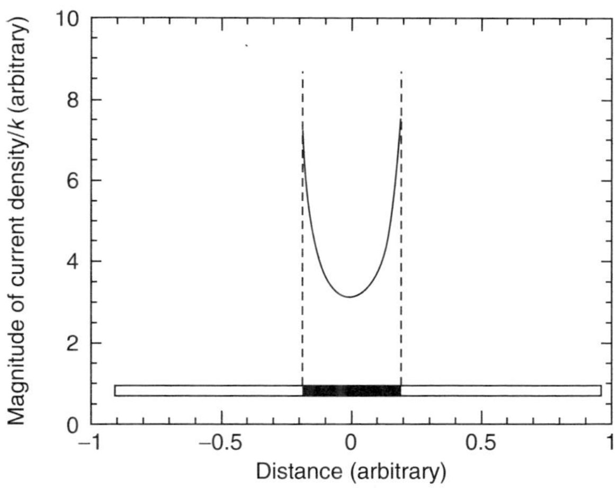

Figure 15.5 Magnitude of the current distribution (primary current distribution) for a disk electrode in an insulating plane. From Equation (15.3). The position of the electrode is represented in solid black. The asymptotic behavior at the electrode/insulator interface results from the assumption that the potential in solution immediately adjacent to the electrode surface is parallel to the electrode.

where the electrode/insulator interface is 90°, there is by definition no interface and the boundary condition is satisfied. In practice, removing 180° interfaces between electrodes and insulators, by masking, for example, minimizes the risk of crevicing by providing a more uniform current distribution at this interface.

The primary current distribution in Equations (15.2) and (15.3) yields the magnitude of the potential and current distributions. However, in practice, when mapping the actual distribution one must consider that current is a vector and has scalar components in the x, y, and z directions as seen in Figure 15.4. In this example, the current distribution is symmetric about the y-axis. When designing a current mapping technique it is possible to measure each of the scalar components individually and reconstruct the current vector. Although one might anticipate that the symmetries of these maps will be simplistic, in practice they are somewhat complex. For example, partial differentiation of Equation (15.2) with respect to z and r yields the normal and horizontal components of the current:

$$\frac{\partial\phi}{\partial z}, \frac{\partial\phi}{\partial r} = \frac{2\phi_0}{\pi} \frac{1}{\sqrt{1-u^2}} \frac{\partial u}{\partial z}, \frac{\partial u}{\partial r} \tag{15.4}$$

where

$$u = \frac{2}{[(r/a-1)^2+(z/a)^2]^{1/2}+[(r/a+1)^2+(z/a)^2]^{1/2}} \tag{15.5}$$

Plots of the normal and horizontal components of the current are presented in Figure 15.6 and Figure 15.7 for z close to the surface ($z/a < 0.6$). As anticipated for the 180° electrode insulator interface, the normal component of the current distribution is the local minimum near the center of the electrode while the current maxim are located at the interface between the electrode and the insulator. In comparison, one might mistake the horizontal component of the current as having both anodic and cathodic parts. This phenomenon can be understood by considering the signs of the current vectors and scalars in relation to the scanning direction in relation to the scanning relation (Figure 15.4). The current vector will always be positive for anodic processes and negative for cathodic processes (U.S. sign convention for current flow). By definition, at the center of the electrode the current is always normal to the surface as the radial distance a is equal to x ($y = 0$; converting to Cartesian coordinates $a^2 = x^2 + y^2$). Therefore, extending away from the y origin in opposite directions, the current will have opposite sign. For example, for a point source located at $y = 0$, at y equal to $0.5a$ and $-0.5a$ the current will be equal in magnitude and opposite in sign (assumes constant x, Equation (15.2)).

In summary, we have described a set of expressions for current flow in solution derived from primary current distribution. With these expressions, it was possible to describe the fundamental relationships for current distribution in solution. These

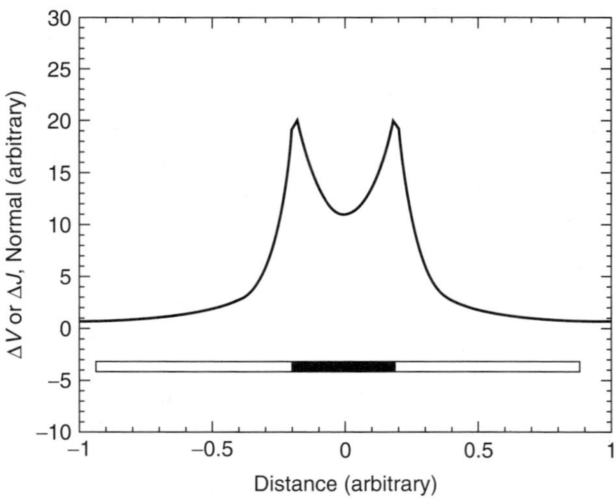

Figure 15.6 Normal component of the current distribution for a disk electrode in an insulating plane (primary current distribution). The position of the electrode is represented in solid black.

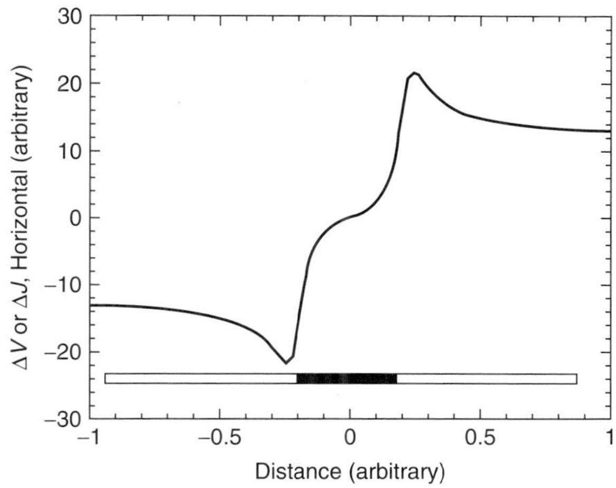

Figure 15.7 Horizontal component of the current distribution for a disk electrode in an insulating plane (primary current distribution). The position of the electrode is represented in solid black.

relationships have helped to uncover special conditions at the center of the electrode and the electrode/insulator interface. Further, these relationships have provided information on the shape dependence of the current distribution with probe geometry (normal vs. horizontal). Later in this chapter, it will be demonstrated that these relationships may also be used to derive the local current density at specific sites on the electrode surface from the solution current density as measured by SRET, SVET, and LEIS.

15.3 SCANNING REFERENCE ELECTRODE TECHNIQUE

15.3.1 Theory

It is important to stress that the SRET measures the potential distribution in solution owing to the flow of current (from localized corrosion, galvanic corrosion, or an externally applied potential). It does not measure the potential with respect to the sample surface (as was the case in the Dix and Brown study). In this manner, the SRET allows the investigator to quantify local current densities at the sample surface. Over the years there have been two separate methods for making local reference electrode measurements. The first method is similar to that of Evans (previously described); a single electrode (either a noble metal wire or a finely tipped salt bridge/Luggin probe) is scanned very near the sample surface. This electrode measures the potential as a function of position relative to a stationary electrode several centimeters from the sample surface. A schematic of the apparatus is shown in Figure 15.8. Potential data are collected by a computer-controlled differential electrometer that measures the potential difference between the scanning and stationary electrodes. In the second probe configuration two platinized noble metal wires are used. The wires are fixed in an epoxy or glass mount at a known separation distance (Figure 15.9). As this bielectrode probe is scanned across the sample surface, the potential difference between the two wires is recorded. In this method the stationary reference electrode is optional if potential control of the sample is desired.

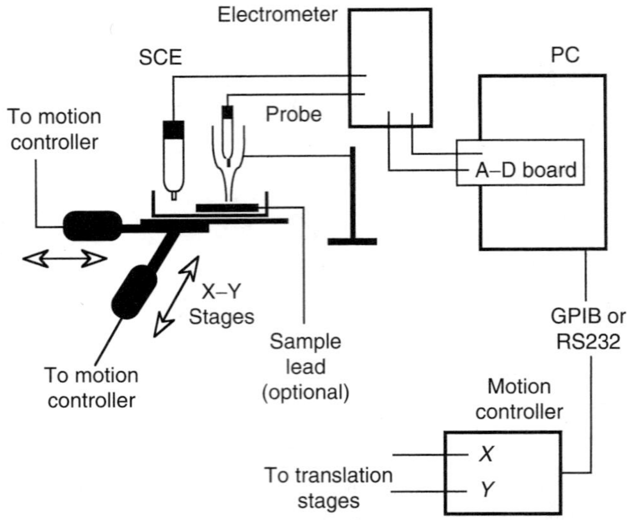

Figure 15.8 Diagram of apparatus commonly used in making scanning reference electrode measurements. This configuration shows the stationary reference electrode method for measuring potential. More commonly, a bi-electrode probe is used (Figure 15.9).

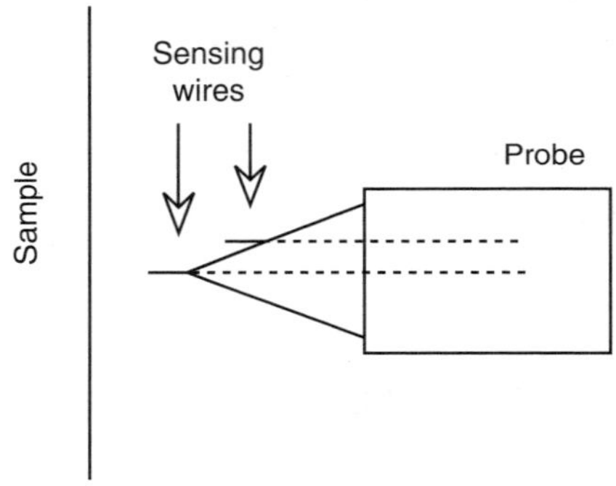

Figure 15.9 Bi-electrode probe used by some investigators in SRET to measure the local potential distribution near the solution/electrode interface. The potential difference between the sensing wires can be used in conjunction with Equation (15.6) to calculate the local solution current density.

In both single electrode and bi-electrode SRETs sensitivity to local current distributions will depend on several factors. One of these factors is the probe sample separation distance. For maximum sensitivity, the tip should be as close to the sample surface as possible. This concept was demonstrated in the current distribution discussion in Section 15.1 (Figure 15.6 and Figure 15.7). This can be seen from a practical standpoint in work of Trethewey et al. (14), where probe voltage as a function of sample separation distance was measured for a 10^{-4} cm^2 wire point source at a constant current of 2 mA (Figure 15.10) (14). However, one must also recognize that the tip has the potential to interfere with local current distributions if it is too close to the surface (i.e., shielding). The general rule-of-thumb is that a probe should be no closer to the surface than the outer diameter of its measuring tip. Thus, one would like to make the tip diameter as small as possible so that the tip/sample separation distance may be minimized. This can be accomplished relatively easily today with either tips used for tunneling electron microscopy or glass-encapsulated

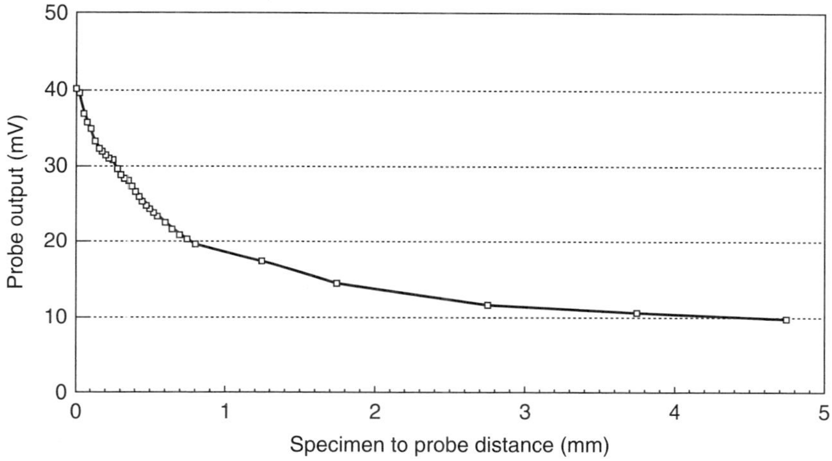

Figure 15.10 Dependence of SRET probe voltage on separation distance from electrode surface. For a gold bi-electrode probe. (From K. R. Trethewey, D. A. Sargeant, D. J. Marsh, A. A. Tamimi, *Corrosion Science*, 35, 127, 1993. Pergamon/Elsevier, Oxford. With permission.)

microprobes used by the biomedical industry. However, one is cautioned, as decreasing the tip size increases the $1/f$ noise associated with the measurement. To overcome this problem, investigators commonly coat the tip of the probe with platinum black to increase the surface area exposed to solution (polarize cathodically in 2% chloroplatinic acid at 100 mA/cm^2) (15). Spectral noise can be further reduced by buffering the signal with a suitable preamplification circuit (discussed later in this chapter).

While it is easy to visualize a small probe rapidly scanning back and forth in the x and y directions across the surface of a stationary sample, in practice this is seldom the way SRET data are collected. It is most common to place the probe on a stationary, vibration-free table and the sample/cell on an X/Y translation stage. This configuration minimizes vibration of the probe tip associated with the motors. Independent of whether the single tip or bi-electrode technique is used, scanning may be accomplished with a manual translation stage as Evans did, with stepper motors, or with servomotors. However, there is a trade-off between rapid probe movement provided by servomotors and the artifacts that may result. By synchronizing the voltage data collection speed with the servomotor sweep rate the location of the electrode at the time of the voltage measurement can be easily determined. Alternately, one may use stepper motors and move the probe across the sample surface a known distance, measure the new voltage, and then move to the next location; however, this method does not maximize data collection rates. Stepper motors also allow the investigator to minimize noise in the data, which is often the most important factor in scanning probe measurements. As will be shown later for LEIS, stepper motor control is mandatory and the data collection rate is much slower. Some commercial instruments employ a cylindrical sample that rotates. In this configuration, the sample to be mapped in SRET is a rod/cylinder that is mounted on a motor. The sample is then immersed in solution normal to the solution surface. The voltage-sensing probe (typically a bi-electrode probe) is mounted to a separated motor that provides motion control in the x-direction for adjusting the sample/probe separation distance and translation in the z-direction for potential mapping. A photo of such a system is shown in Figure 15.11. By setting the probe tip at a fiduciary mark on the sample before beginning the scan, for a

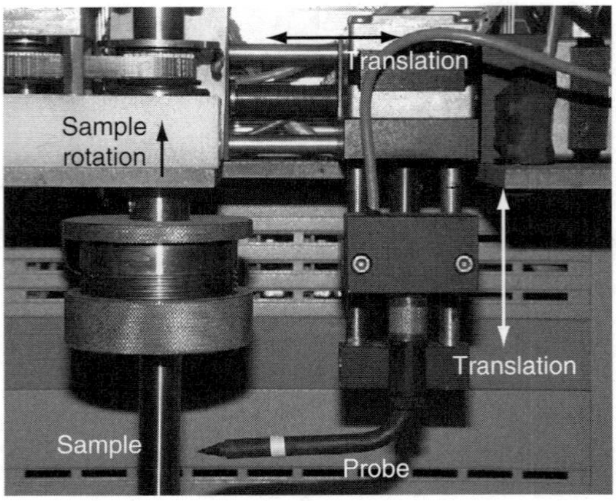

Figure 15.11 Photograph of a commercially available SRET apparatus that employs a cylindrical sample that is rotated in front of a bi-electrode probe that translates vertically across the sample. The rotation of the sample and translation of the probe allow the entire sample surface to be mapped.

known sample diameter and rotation rate, the tip position with respect sample surface can be calculated. The entire sample surface is mapped by stepping the tip along the z-direction, that is, along the length of the sample.

Collecting potential data from the tip in one of the above manners will provide a map of the potential distribution in solution over the sample surface for a fixed probe/sample separation distance. The data are often plotted in the form of contour maps or three dimension maps (x-position, y-position, z-voltage magnitude). The resulting voltage data can be analyzed in the "as collected" format to draw qualitative conclusion about the corrosion process, for example, locating anodic and cathodic sites, defects in a coating, or areas of relative activity. In fact, this is the most common way the data are used in the literature. Ideally, one would like to know somewhat more quantitatively what is occurring on the sample surface, such as the local current densities. In the single-tip SRET configuration the raw potential data can be converted to current density by using a calibration wire (16). This may be a noble metal wire placed in the same epoxy mount to the side of the sample, however, with a separated electrical lead. If this is not convenient, a separate calibration wire may be made. In both cases, the investigator translates the stage such that the SRET is scanned across this calibration wire, which is driven at constant current. By knowing the separation distance between the tip and the surface (by using a telescope, for example) and measuring the potential distribution in solution for the calibration wire the raw SRET potential data can be converted to a current density. For example, by measuring the peak potential above the calibration wire as a function of applied current density a calibration curve can be generated for several separation distances. Multiplication of the peak potentials in the scanning experiment by the slope from the appropriate calibration curve (i.e., at the same separation distance) results in approximation of the local current density. For the bi-electrode probe obtaining the local current density (i_{local}, A/cm^2) at any point in solution may be obtained from the relationship:

$$i_{\text{local}} = \frac{\Delta V_{\text{probe}}\sigma}{l} \qquad (15.6)$$

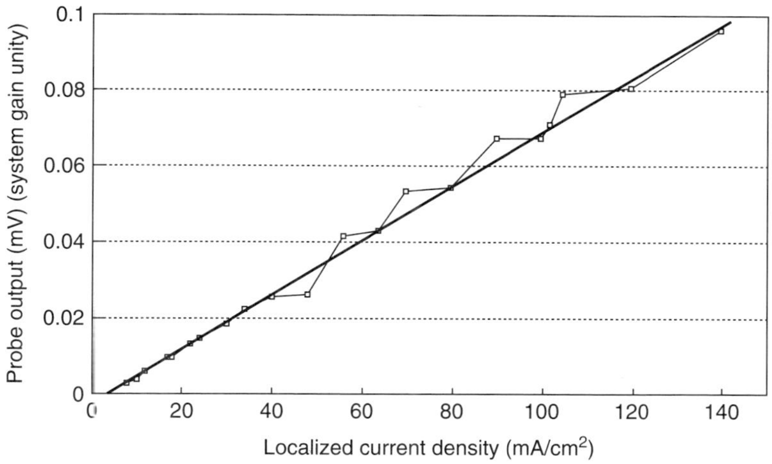

Figure 15.12 Current density calibration curve for gold bi-electrode probe (SRET). Measured for a gold wire being driven galvanostatically at 2.0 mA. (From K. R. Trethewey, D. A. Sargeant, D. J. Marsh, A. A. Tamimi, *Corrosion Science*, 35, 127, 1993. Pergamon/Elsevier, Oxford. With permission.)

where ΔV_{probe} is the potential difference between the bi-electrode probes in volts, σ is solution conductivity in $(\Omega\,cm)^{-1}$, and l is the separation distance between the bi-electrode probes in centimeters. An example of such a bi-electrode calibration is shown in Figure 15.12 (14).

It is worth commenting at this point on instrumentation and possible sources of error. If one is controlling the sample with a potentiostat or galvanostat during SRET experiments, it is important that either the electrometer or the potentiostat be electrically isolated from ground. If it is not, ground loops will develop resulting in erroneous data. These loops generally occur when the impedance path for current from the counter electrode of the potentiostat through the probe tip back to ground approaches the impedance of the current path through the working electrode of the potentiostat. For example, suppose one is using a potentiometer to measure probe voltage that has an input impedance of 10^6 Ω and a measuring resistor in their potentiostat of 10^5 Ω. If this potentiometer and the potentiostat operate on the same 120-V, 60-Hz power (e.g., the wall outlet in laboratories), then current "sees" a voltage divider. As a result, 10% of the current that the potentiostat/galvanostat is delivering does not flow across the measuring resistor, rather through the probe back to ground. This has two devastating consequences: (1) the true current at the sample surface is higher than what the potentiostat records; and (2) the scanning electrode becomes polarized due to an IR drop resulting in a mapping error. We expand on this problem further in Section 15.6 and propose methods for avoiding these problems.

Finally, there have been several other methods used to measure local solution current densities that we have not discussed. Most notably is a commercially available instrument that uses a bi-electrode stationary probe (17). One might consider this system a hybrid SRET as it does not measure potential differences at all, rather, it "bleeds-off" a small portion of the local solution current to a high impedance shunt circuit. At the time of publication of this article there was relatively little peer-reviewed literature available on this technique. Thus, neither the advantages nor limitations of this technique are known.

15.3.2 Application

The SRET has been used to investigate a wide variety of corrosion systems including pitting corrosion (14, 18–22), galvanic corrosion including dissimilar metals and welds (16, 23–25), corrosion inhibitors (26–28), and coating failure (29, 30).

The pitting corrosion of SS 304 in ferric chloride solution has been investigated with the SRET by Isaacs and Vyas (16, 20). In 0.4 *M* $FeCl_3$, pH 0.9, at the open circuit potential (OCP) active pitting corrosion of the specimen was observed with SRET before damage to the surface was visible to the naked eye. Initially, the exposed surface contained as many as 14 active pitting sites. Over the course of a 2-h exposure, the number of active sites was shown to be reduced to one; however, the peak current associated with this site was greater than any single site previously observed on the surface (Figure 15.13). Multiple experiments showed that the number of active sites on the surface was dependent on surface pretreatment: abraded, electropolished, oxidized at elevated temperature, or cathodic prepolarization. SRET showed that individual pits could easily be repassivated by exposure to a solution jet

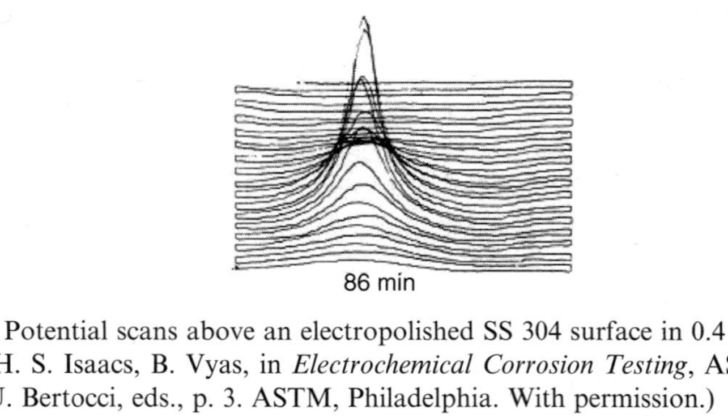

Figure 15.13 Potential scans above an electropolished SS 304 surface in 0.4 *M* $FeCl_3$ using SRET. (From H. S. Isaacs, B. Vyas, in *Electrochemical Corrosion Testing*, ASTM STP 727, F. Mansfield, U. Bertocci, eds., p. 3. ASTM, Philadelphia. With permission.)

that washed the pitting solution. Examination of samples after exposure with optical metallography found that the pits contained a dark green pitting solution. In addition, a thin oxide cover was observed to cover the pits. The data were interpreted in terms of an oxide cover model. In this model, the development of a critical pitting solution is necessary to maintain active pitting. The surface pretreatment was proposed to influence the properties of the oxide cover, some oxides rupturing more easily than others resulting in mass transport between the pit and bulk solutions and dilution of the critical pitting solution. In turn, this results in repassivation of the pits. In a separate study, Trethewey et al. (14) examined the pitting corrosion of SS 304 in seawater. Pit current densities were measured as a function of exposure time and applied potential. Trethewey demonstrated that local pit current densities easily exceed 250 mA/cm^2 while the surface averaged current density (as measured by the potentiostat) was only 0.85 mA/cm^2.

Voruganti et al. (25) used the SRET to study the behavior of marine steel weldments in ice break conditions. In this study, various steels and weld filler materials were compared in synthetic seawater at 2°C. To help verify the SRET data, samples were also exposed to long-term immersion testing and anodic dissolution. These samples were then examined for material loss and, thus, preferential corrosion. In this investigation, two distinctively different behaviors were observed in the SRET. In one case, peaks in the SRET output were observed in the heat affected zones (HAZ), indicating that the HAZ was anodic with respect to the base plates and weld filler material (Figure 15.14a). In these figures, two reference points are indicated: FL, the fusion line between the filler material and base plate, and CL, the weld centerline. The SRET data corresponded directly with material loss after immersion and anodic testing (Figure 15.14b). For a different combination of base plate composition, heat input, and weld filler composition, a single peak in the SRET output

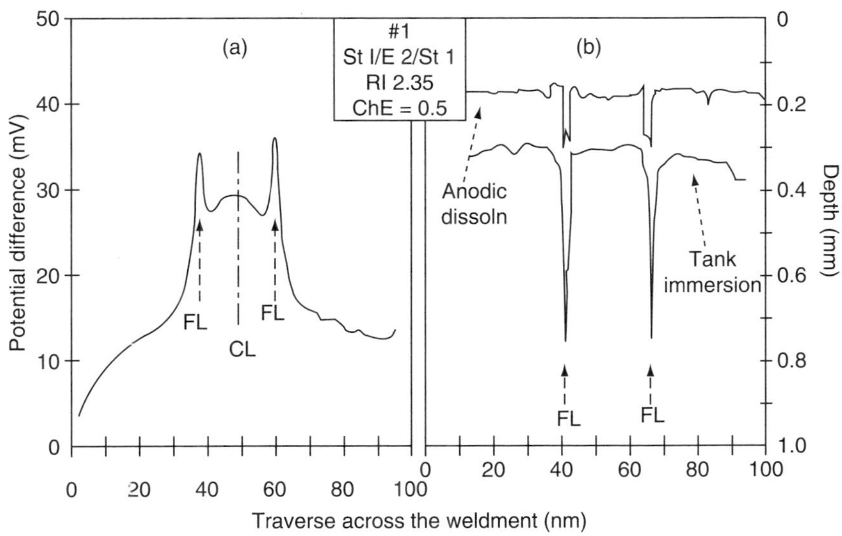

Figure 15.14 (a) SRET output as a function of scan distance across a steel weldment immersed in NaCl solution at the OCP. (b) Surface profiles after anodic dissolution in same solution. FL represents the fusion line in each figure and, therefore, the heat affected zone. Base plate was 32-mm thick steel consisting of 0.19 wt% C, 1.44 Mn, 0.33 Si, 0.01 P, 0.01 S, 0.26 Cu, 0.09 Ni, 0.04 Al, 0.1 Cr, and 0.07 V. Weld filler was 0.16 Si, 0.47 Cu, and 0.51 Ni. (From V. S. Voruganti, H. B. Luft, D. Degeer, S. A. Bradford, *Corrosion*, 47, 343, 1991. NACE, Houston. With permission.)

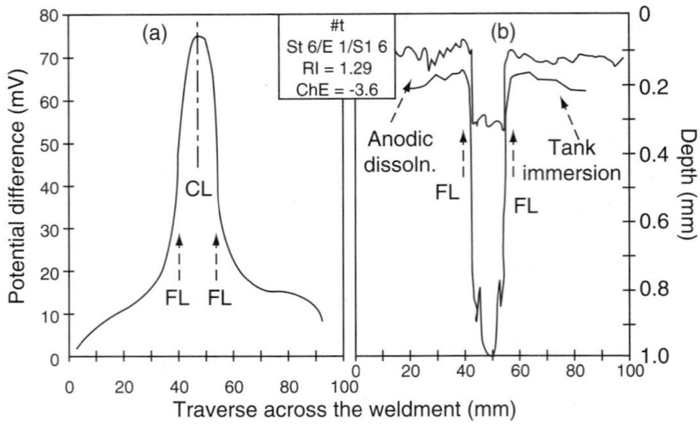

Figure 15.15 (a) SRET output as a function of scan distance across a steel weldment immersed in NaCl solution at the OCP. (b) Surface profiles after anodic dissolution in same solution. FL represents the fusion line in each figure. Base plate was 19-mm thick steel consisting of 0.23 wt% C, 0.71 Mn, 0.04 Si, 0.003 P, 0.013 S, 0.02 Cu, 0.0 Ni, 0.04 Al, 0.0 Cr, and 0.0 V. Weld filler was 0.58 Si, 0.04 Cu, and 0.03 Ni. (From V. S. Voruganti, H. B. Luft, D. Degeer, S. A. Bradford, *Corrosion*, 47, 343, 1991. NACE, Houston. With permission.)

was observed over the weld filler material indicating the filler material was anodic to the HAZ and base plate material (Figure 15.15a). The SRET data corresponded directly with material loss after immersion and anodic testing (Figure 15.15b). These data were rationalized in terms of heat input (hardenability) and chemical composition of the filler material relative to the base material. To quantify these, two parameters indexes, RI and ChE, were established. High RI values indicate greater microstructural changes in the base plate due to heat input and increased susceptibility to corrosion. Positive values for ChE indicate noble filler material while negative ChE values indicate active filler material. The weldment in Figure 15.14 had a relatively high RI value indicate and positive ChE value indicate, indicating an active HAZ and noble filler material. Thus, one would predict preferential corrosion in the HAZ as observed in the SRET data. In comparison, the weldment in Figure 15.15 had a relatively low RI value and negative ChE value, indicating active filler material as observed in the SRET data.

15.4 SCANNING VIBRATING ELECTRODE TECHNIQUE

15.4.1 Theory

The objective of the SVET is the same as the SRET, the measurement of the current distribution in solution above a freely corroding electrode. SVET has one great advantage over SRET, enhanced noise rejection allowing much smaller currents to be detected. However, it was not introduced by electrochemists, rather, by neurologists as a method for measuring nerve impulses at an axon (31). The theory is quite simple; a small probe is vibrated at a specified frequency above the sample surface. Typically, the probe is vibrated vertically with respect to the surface though we will consider other cases. In the presence of a potential gradient, such as that created by pitting corrosion, an AC voltage of the same frequency as the vibration stroke is detected by the probe, whose magnitude of voltage is proportional to the local solution current density.

In SVET, the probe may be vibrated either vertically or horizontally with respect to the sample surface to measure individual scalar components of the current or in both directions simultaneously to measure the two-dimensional current vector. Vibration is achieved with a bimorph piezoelectric transducer (bimorph PZT; two bonded piezoelectric ceramics) attached to, or sandwiched across, a cantilever beam. By applying an AC voltage to the bimorph piezoelectric, the beam can be made to deflect in the direction of interest at a known frequency. The concept is shown in Figure 15.16. A small insulated platinum wire, which acts as the probe, is sandwiched between the bimorph piezoelectric as suggested by Isaacs (12). By applying a voltage across the PZT the probe will deflect in the vertical direction (i.e., normal to the electrode surface in Figure 15.16). Deflection of the probe in the horizontal direction can be accomplished by changing the orientation of the probe or bimorph piezoelectric. The schematic in Figure 15.17 will achieve deflection of the probe in both the vertical and horizontal directions simultaneously (32). The system, referred to as the inverted "T" linkage, consists of two assemblies. In the first assembly, two PZTs mounted vertically to a rigid pier deflect the probe in the horizontal direction. To deflect the probe in the vertical direction, in the second assembly a PZT is attached orthogonally to the vertical ones via flexible silicone hinges. To drive the vertical vibration, a separate AC signal is used. This configuration provides nearly orthogonal motion of the probe in the vertical and horizontal directions.

A schematic of what a bi-directional SVET set-up might look like including the necessary equipment is presented in Figure 15.18. From a scanning perspective, the set-up is similar to the SRET; the PZT/probe assembly is held stationary while X/Y translation of the sample/cell is accomplished with stepper motor or with servomotor stages. The probe height is typically adjusted manually with a fine vernier. Lock-in amplifiers (LIA) are commonly used to analyze the magnitude of the AC voltage generated by the probe oscillating in a potential gradient.[1] Oscillators are used to provide the 100 to –1000 Hz, 2 to –10 V AC voltage necessary to drive the PZTs. This

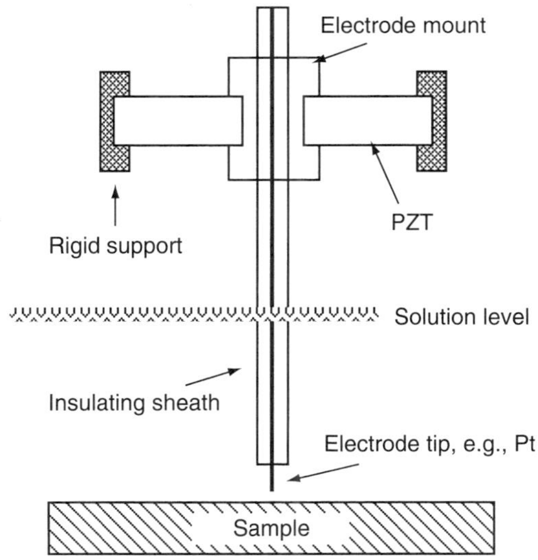

Figure 15.16 A conceptual diagram of a probe/PZT assembly for SVET that may be used to measure the normal component of the current density in solution above an electrode.

[1]Note that the quadrature is not of significance in this application.

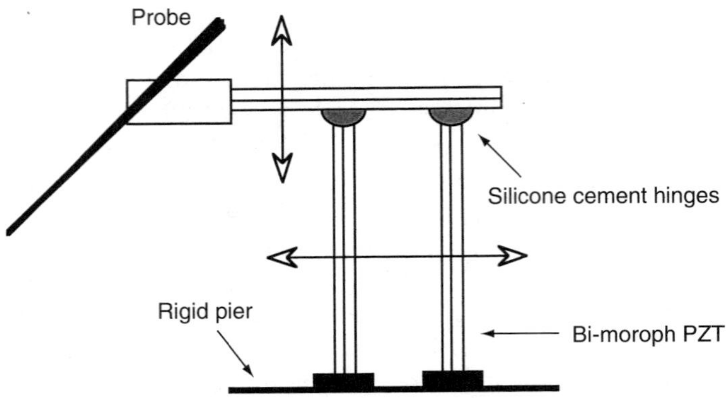

Figure 15.17 A conceptual diagram of a probe/PZT assembly for SVET that may be used to measure both the normal and horizontal components of the current density in solution above an electrode.

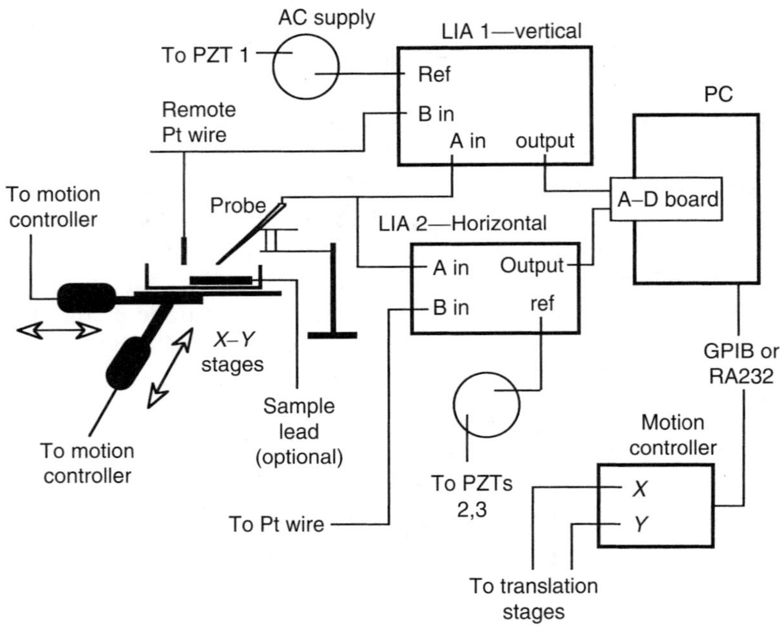

Figure 15.18 Diagram of apparatus commonly used in making scanning vibrating electrode technique to measure both the normal and horizontal components of the current distribution simultaneously.

voltage is also used as the LIA reference signal. The output voltage of the vertical and horizontal LIAs (the scalar components of the current density) is acquired with a PC and analog/digital board. Each scalar component is converted to solution current density using Equation (15.6), where l is the maximum deflection of the PZT (i.e., the magnitude of the probe stroke). These scalar components may be analyzed separately; however, it is most common to see the current density vector presented.

Ultimately, one wishes to know the current density at the sample surface; however, SVET (and SRET) only directly provide solution current density information. Given that the solution current density for equipotential planes has been described mathematically (Equations (15.2) to (15.5)), one would correctly conclude that given the distribution of potential or current in solution for a known distance above an electrode it is possible to deconvolute these data to determine the magnitude of the source on the sample surface. Isaacs (33) has described a baseline stripping procedure for deconvoluting surface current density from measured solution current density. In this procedure, the current is assumed to emanate from a point source:

$$\phi(x, y) = \frac{I}{2\pi\kappa}\left[\frac{1}{(x^2 + y^2 + h^2)^{1/2}} - \frac{1}{(x^2 + y^2 + (h + d)^2)^{1/2}}\right] \tag{15.7}$$

where I is the surface current (source), k is the solution conductivity, x and y are the Cartesian coordinates for location, h is the probe sample separation distance, and d is the magnitude of the probe stroke. The deconvolution is carried out by first determining the baseline current over the insulating mount. The location of the maxim (or minimum) current density with respect to the baseline is then determined. This location is defined by the scan step size with area $x_{\text{step}}y_{\text{step}}$. This location is then assigned a current value that is 25% of the maximum. The current distribution for this value (Equation (15.7)) is then subtracted from the data. This procedure is repeated until the baseline is 0.005 times the initial value. The currents assigned to each location are then summed and divided by the area to determine the surface current density.

Zou et al. (34) have proposed a fast Fourier transform FFT method for deconvoluting SVET data to determine surface current density. Though we often think of FFT in terms of the time (s) and frequency domain (1/s, cycles/s), it is perfectly acceptable to work in units of distance (m and 1/m or wavelength) as applicable here. In this method, both the equipotential disk and point source potential distributions (Equations (15.2) and (15.7)) were examined. The potential distribution in solution above the electrode is defined as the convolution of two functions

$$\phi(x, y) = i(x', y') * g(x, y) \tag{15.8}$$

where ϕ is the surface current density, x' and y' denote surface positions, and $g(x, y)$ describes the solution component and is defined for a point source (Equation (15.7)) or equipotential disk (Equation (15.2)) as

$$g(x, y) = \phi(x, y)/I \tag{15.9}$$

As the data collected in SVET are a discrete array at regularly spaced intervals, the discrete Fourier transform may be used to deconvolute the data:

$$\Delta\phi(x, y) = \sum_{n=-\infty}^{\infty}\sum_{m=-\infty}^{\infty} i(n\Delta_{x'}, m\Delta_{y'})g(x - n\Delta_{x'}, y - m\Delta_{y'}) \tag{15.10}$$

In this manner, Zou et al. demonstrated that the FFT and baseline stripping methods produced comparable results. The FFT had the advantage of being computationally less burdensome and provided a separate means for analyzing the data (i.e., wavelength). However, the accuracy of both methods was shown to be sensitive to noise in the experimental data.

15.4.2 Application

The SVET has been used extensively to investigate localized corrosion (35, 36), galvanic corrosion (33, 37, 38), coatings (39–46), inhibitors (47, 48), interfaces and grain boundaries (49–55), second-phase particles (56), stress corrosion cracking (57, 58), and microbial corrosion (59, 60).

Ogle et al. (61) used SVET in conjunction with a scanning pH probe to characterize cut-edge corrosion on galvanized steel. Factory painted steel sheet is trimmed before being placed on a roll exposing a bare cutedge to the environment. In these experiments, the sample consisted of commercial electrogalvanized low carbon steel commonly used in the automotive industry. Prior to the experiment, the SVET probe was calibrated with point current sources; in the case of the pH probe standard pH buffers were used. For cut edges, local anodic and cathodic current densities corresponded with that anticipated from mixed potential theory; the current densities over Fe regions were cathodic while current densities over the electrogalvanized Zn were anodic (Figure 15.19). Variations in solution pH above the sample were also mapped with an Ag/AgCl micro-pH probe. These data indicate that the Fe regions were predominately alkaline (pH 10) in comparison to the Zn region (pH 7.6). The alkaline pH over the iron portion of the electrode and observed precipitation of Zn in this region was consistent with cathodic reduction of water. Current density maps at the OCP after a period of immersion revealed the lack of cathodic activity at the Fe surface where hydrated Zn oxide films had precipitated. This fact was further confirmed through local polarization curves generated with SVET. Thus, it was concluded that Zn corrosion products act as cathodic inhibitors further protecting the electrogalvanized steel.

Isaacs (41) has used SVET to characterize defects in ion vapor-deposited aluminum on steel. In that work, steel specimens that had been coated with ion vapor deposited aluminum were studied in a borate solution (pH 8). The SVET was used to first locate defects in the coating indicated by areas of high anodic or cathodic activity. To accomplish this, the surfaces were mapped at potentials where hydrogen or oxygen evolution readily occurs on steel and not aluminum. These potential regions were determined by generating potentiodynamic polarization curve for bare

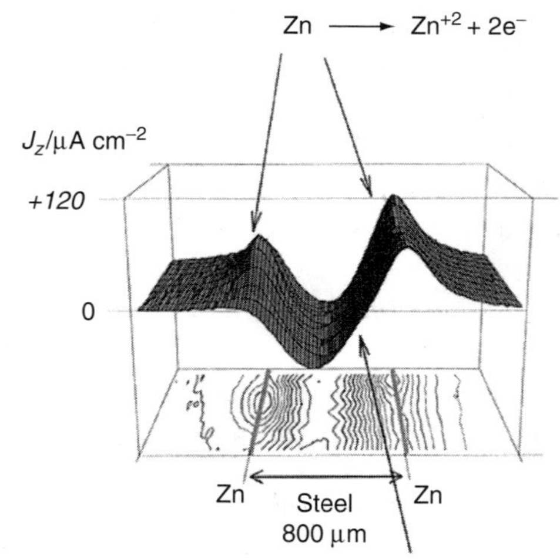

Figure 15.19 Normal component of the solution current density (SVET) above a galvanized steel sheet held on end such that the Zn/steel/Zn interface at the edge was exposed to NaCl solution. (From K. Ogle, V. Bauda, L. Garrigues, X. Phillippe, *Journal of the Electrochemical Society*, 147, 3654, 2000. Electrochemical Society, Pennington. With permission.)

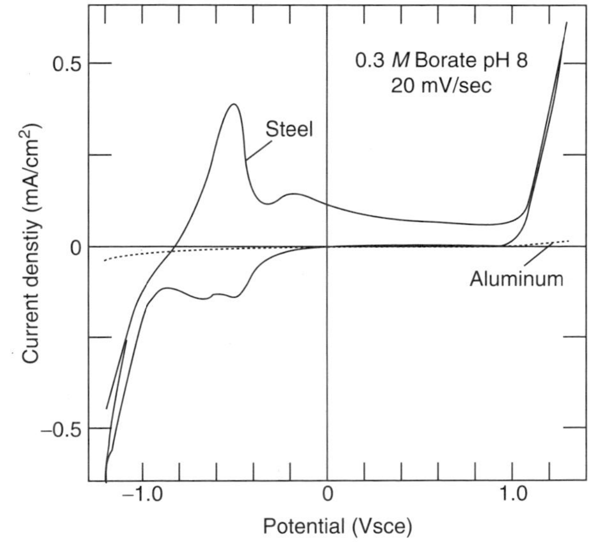

Figure 15.20 Potentiodynamic polarization curves for steel and aluminum in pH 8 borate buffer solution. (From H. S. Isaacs, *Corrosion*, 43, 594, 1987. NACE, Houston. With permission.)

steel and bare aluminum in borate solution (Figure 15.20). Once a defect was located, a local potentiodynamic polarization curve for the defect was made by ramping the potential of the electrode and measuring the current from the defect locally with the probe (Figure 15.21). As the local potentiodynamic polarization curve for the defect site did not resemble either the polarization curve for ion vapor deposited aluminum or steel in this solution, it was concluded that the defects observed by SVET were not pinholes in the coating. Subsequent local potentiodynamic polarization curves generated in the same manner with SVET for intentional defects in the coating confirmed this. Energy depressive spectroscopy of the areas of anodic activity located by

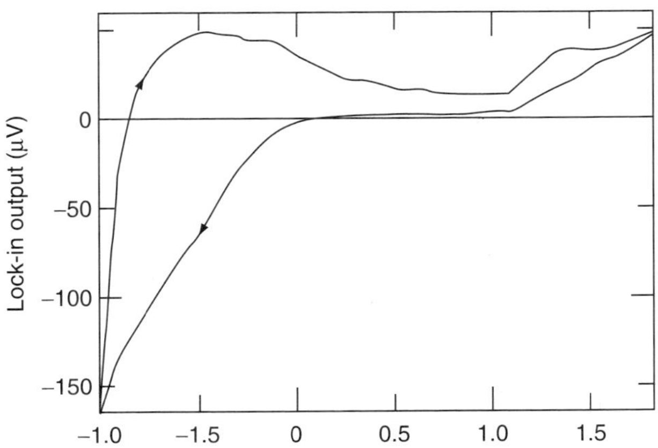

Figure 15.21 Psuedo-potentiodynamic polarization curve generated with SVET for a defect in steel coated with ion vapor deposited aluminum. Solution was pH 8 borate buffer. (From H. S. Isaacs, *Corrosion*, 43, 594, 1987. NACE, Houston. With permission.)

SVET found that these sites were, in fact, inclusions in the coating primarily comprising of Si, Ca, and Ti.

The inhibition of corrosion on low carbon steel in the presence of cerium salts was investigated by Isaacs et al. (48). In that study both the normal and horizontal components of the solution current density over C1020 steel in 0.01 *M* sodium sulfate solution were mapped and the corresponding solution current vectors were reported. The probe calibration was confirmed by mapping the current distribution above a coextrude Al/Cu sample. The current density measured in an SVET line scan across a portion of the sample that contained both anodic and cathodic areas is shown in Figure 15.22. The addition of 0.005 *M* $CeCl_3$ solution resulted in a change in the current distribution at this location, such that the anodic currents were replaced by cathodic currents of much smaller magnitude. Changes in the Cl^- ion activity or oxygen transport as explanations for the observed changes in current distribution were ruled out given the order of the solution replacement method used (see Figure 15.22). Further, the suppression of the anodic current density was accompanied by an increase in the open circuit potential indicating that the cerous ion inhibits the anodic reaction contrary to prior claims that $CeCl_3$ was a cathodic inhibitor.

Kasper and Crowe (38) have used SVET to investigate an iron–copper galvanic couple. The couple was created by swaging copper around a 6-mm diameter iron

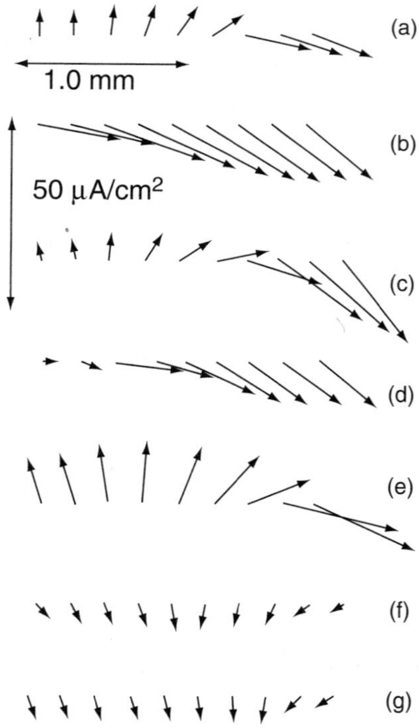

Figure 15.22 Current vectors from SVET line scans across the same location on a steel electrode showing change from anodic to cathodic current after addition of $CeCl_3$ via solution replacement technique: (a) 0.01 *M* Na_2SO_4, (b) fresh 0.01 *M* Na_2SO_4, (c) same as "b" after 10 min, (d) 0.025 *M* NaCl, (e) same as "d" after 10 min, (f) 0.005 *M* $CeCl_3$, (g) same as "f" 20 min later. (From H. S. Isaacs, A. J. Davenport, A. Shipley, *Journal of the Electrochemical Society*, 138, 390, 1991.The Electrochemical Society, Pennington. With permission.)

wire. The final electrode had an o.d. of 50 mm. Current density was mapped by a single-axis probe (normal component) and a dual-axis probe (normal and horizontal components). The data were collected at the OCP in acidified chloride solution. The current densities mapped by SVET were compared to predictions using a finite element model of the current distribution. The model relied on potentiodynamic polarization data for copper and iron in the test solution. From these data potential-dependent polarization resistances were determined using a local tangent slope approach. These polarization resistances were introduced into the COSMIC/NAS-TRAN code using CWEDGE elements to simulate geometry and geometric symmetries. Finally, the model assumed Dirichlet and Neumann boundary conditions. To examine the influence of probe/sample separation distance on the measured current distribution separation distances of 1150 and 650 μm were examined. For both separation distances, the total current density as mapped by the SVET was in fairly good agreement with the model values for any given position. However, at a separation distance of 650 μm, the measured current density normal to the electrode surface at the copper/iron interface ($r = 3.2$ mm) was distributed more broadly than predicted by the model (Figure 15.23). In addition, differences between the magnitude of the measured and modeled normal current densities at any given location were observed. The best agreement between the finite element model and SVET data was obtained for the horizontal component of the current density at a separation distance of 650 microns (Figure 15.24). It was concluded that the finite element technique was a useful tool for predicting the current distributions in galvanic couples.

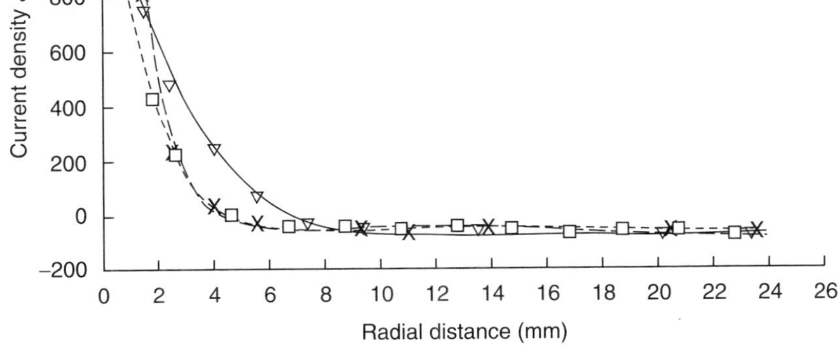

Figure 15.23 Normal component of the solution current density (SVET) above a copper electrode immersed in NaCl. Both the measured (□) values and values calculated from finite element modeling (×, ▽) are shown. (From R. G. Kasper, C. R. Crowe, in *Galvanic Corrosion*, ASTM STP 978, H. P. Hack, ed. With permission.)

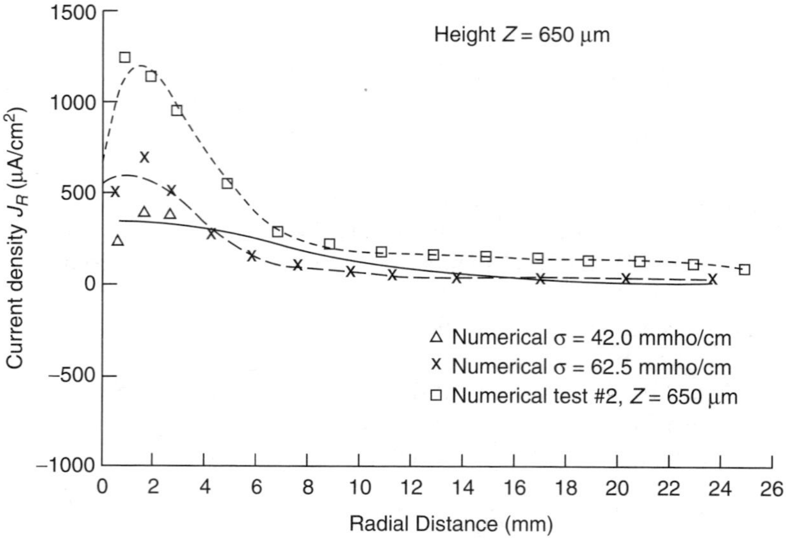

Figure 15.24 Horizontal component of the solution current density (SVET) above a copper electrode immersed in NaCl. Both the measured (□) values and values calculated from finite element modeling (×, ▽) are shown. (From R. G. Kasper, C. R. Crowe, in *Galvanic Corrosion*, ASTM STP 978, H. P. Hack, ed. With permission.)

15.5 LOCAL ELECTROCHEMICAL IMPEDANCE SPECTROSCOPY

15.5.1 Theory

Electrochemical impedance spectroscopy (EIS) is a powerful nondestructive technique that has been used extensively to evaluate electrochemical systems. A thorough explanation of EIS has been presented elsewhere in this book. To summarize, in the conventional approach a small (mV) sinusoidal voltage perturbation is applied between the sample and a distant reference electrode. EIS data are generated by measuring the transfer function of the reference voltage to the current response of the electrode as a function of applied frequency. Because the current response from the entire electrode surface is used in this technique, a surface-averaged measurement results. In the case of a nonuniform electrode (owing to defects or pitting corrosion, for example), difficulties may arise when analyzing EIS data if the time constants for the individual process occurring at the surface are not well separated in the frequency domain or the current from one of the processes is particularly low. Thus, to investigate local processes due to second-phase particles, grain boundaries, or other surface heterogeneities, a method for generating local impedance data is desirable.

Early attempts to generate local impedance data used a microelectrode that contained reference and counter electrodes. This probe was used to polarize the sample locally as a function of frequency (62–65). By scanning this microelectrode across the sample surface at a distance of approximately 30 μm, with the same techniques used to generate SRET and SVET data, the authors attempted to generate EIS data for discrete areas on the sample surface. Although the authors were successful in determining qualitative differences for coating defects and other heterogeneities, no quantitative EIS data were produced by these methods. This was due, in part, to the inability to confine the current to the area of interest. As one might anticipate, the input impedance of the microelectrodes was relatively high and current tended to "leak" to low impedance areas on the sample surface. The high impedance of the electrodes also limited the sensitivity of the technique.

In this chapter, we will discuss two separate quantitative methods for generating LEIS data. In the first method Lillard et al. (66) used a traditional three-electrode EIS set-up to generate LEIS data (66). In this technique, a sinusoidal voltage perturbation is applied between the sample and a distant reference electrode. The current required to apply this potential is driven between the sample and a traditional (remote) counter electrode (e.g., a graphite rod or platinum mesh electrode). A bi-electrode probe is used to map the current density over the sample surface as a function of applied frequency (Figure 15.25). Scanning of the probe is accomplished by fixing the probe at a stationary point and translating the electrode/cell below the probe with *X*/*Y* stages as discussed for SRET and SVET. For any given frequency, the solution current density is obtained from Ohm's law and the AC potential difference between the probe electrodes:

$$i(\omega)_{\text{local}} = \frac{\Delta V(\omega)_{\text{probe}} \cdot \sigma}{l} \tag{15.11}$$

where $i(\omega)_{\text{local}}$ is the local AC solution current density at the probe tip, $\Delta V(\omega)$ is the AC voltage difference between the probe electrodes, and l is the separation distance between the probe electrodes at the tip. It has been demonstrated that the AC profiles are similar to those for equipotential disks (67). From the local solution current

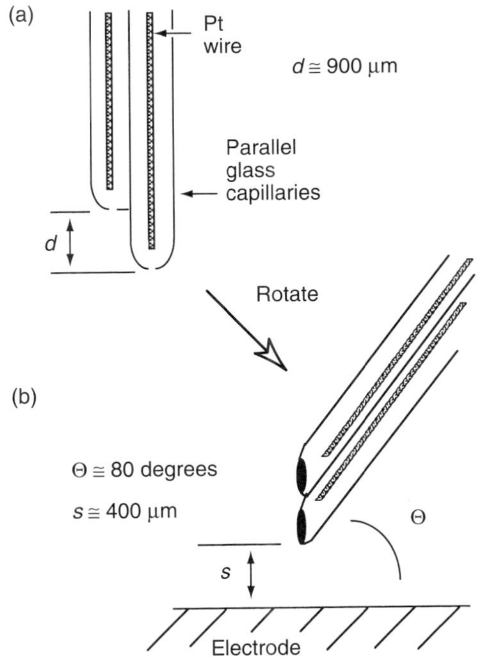

Figure 15.25 Bi-electrode probe used to make LEIS measurements. Probe consisted of two glass capillaries that were drawn down to a final OD of approximately 40 mm. Pt wires for making AC potential measurements were coated with Pt-black. (a) Configuration of as drawn capillaries, and (b) configuration for measuring the vertical component of the solution current density.

density the area normalized magnitude of the impedance, $|Z(\omega)|_{local}$, at the surface can be determined from the relationship:

$$|Z(\omega)|_{local} = \frac{V(\omega)_{applied}}{\Delta V(\omega)_{probe}} \cdot \frac{l}{\sigma} \tag{15.12}$$

where $V(\omega)_{applied}$ is the magnitude of the applied AC voltage perturbation between the reference electrode and the surface. There are several assumptions implicit in Equation (15.12) that are important to note. The first assumption is that the AC potential is uniform across the electrode surface and equal to the applied voltage $V(\omega)_{applied}$. This assumption seems reasonable given that the surface is being controlled by a potentiostat. Further, given that the current will minimize the resistance back to the counter electrode by using as much of the solution volume available to it, the IR drop very near the surface (i.e., at the probe location) may be much larger per unit length than at larger distances such as at the reference electrode. Therefore, most of the IR drop due to the solution resistance likely occurs very near the surface.

Equation (15.12) also assumes that the current density measured by the probe in solution at some distance away from the electrode is equal to the current density at the surface directly below the probe. We know this to be incorrect, for example for primary current distribution, Equation (15.2), there will be a finite current density over the insulator portion of a conductor/insulator electrode, Figure 15.4. The magnitude of the resulting error in Equation (15.12) will increase for nonconducting solutions and at greater separation distances. Conversely, error in Equation (15.12) will be minimized in those systems where there is a small probe/surface separation distance and a conducting solution. To further minimize the error introduced by Equation (15.12), the investigator may also employ a deconvolution technique such as those described for SVET in Section 15.4.1. A baseline stripping procedure and an FFT method for accomplishing this in the AC domain have been described by Zou et al. (34). In terms of the Fourier transform the deconvolution expression for ac currents is given by

$$\Delta\phi(\omega, \nu) = I(\omega, \nu)G(\omega, \nu) \tag{15.13}$$

where $\Delta\phi$ is the potential difference measured by the probe and the capital letters represent the Fourier transform of Equation (15.8).

In addition to the bi-electrode probe technique for making LEIS measurements, a vibrating probe, such as that used in SVET, can be used to take LEIS data. A conceptual diagram for what such a set-up might look like is presented in Figure 15.26. In this diagram, the FRA generates an AC voltage perturbation that is applied between the working and reference electrodes by the potentiostat. This signal is the reference voltage in the LEIS measurement and, therefore, the output of the potentiostat voltage monitor is sent to the second (reference) input channel of the FRA (V2). The local AC current distribution is detected by the vibrating probe; however, the probe signal now has two AC components, one related to the vibration frequency as before (ω_{probe}) and the second related to the AC perturbation being generated by the FRA ($\omega_{applied}$). To measure the local AC solution current density, the signal from the probe is first passed through an LIA referenced to ω_{probe}. The output voltage of the LIA (magnitude) will have a time response that is related to the local impedance of the interface at $\omega_{applied}$. The magnitude and phase of $\omega_{applied}$ (local) may be quantified by sending the output of the LIA to the first (comparison) channel of the FRA (V1). The local impedance is calculated from the transfer function of V2/V1 and Equation (15.12). To obtain the true surface impedance deconvolution of the

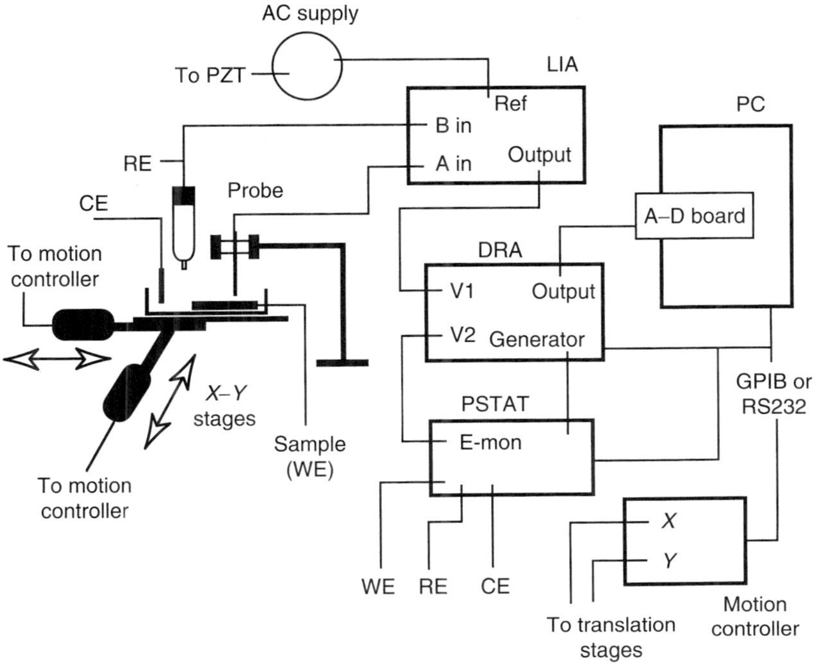

Figure 15.26 Conceptual diagram of apparatus that could be used to make LEIS-SVET measurements.

data is necessary. The advantages of using a vibrating probe over a bi-electrode probe are, at this point, uncertain and have not been demonstrated. There are, however, disadvantages of using a vibrating probe to collect LEIS data. For example, probe vibration frequency limits the upper EIS frequency that can be measured.

Bayet et al. (68, 69) have used a vibrating probe technique to collect LEIS data galvanostatically. In the Bayet et al. method the local AC potential is used in Equation (15.12) as opposed to the average electrode potential. This local potential was obtained from the probe tip. This was accomplished in galvanostatic mode by differentiating the potential between the working electrode and probe. Using the local potential has the advantage of decreasing the error owing to IR drop in solution. For example, a large fraction of the total current from an electrode owes to a low impedance site such as a defect in a coating or an active corrosion pit. There are also other advantages such as changes in local concentration overpotentials (such as those that may develop at the mouth of the pit); however, one would hope that the probe was removed sufficiently from the diffusion boundary layer as not to perturb these gradients. A thorough comparison of the bi-electrode and vibrating electrode techniques has been performed by Bayet and Garrigues et al. (70).

15.5.2 Application

The bi-electrode probe method for generating LEIS data has been used to investigate second-phase particles (71), coating defects (72–75), and pitting corrosion (76). Wittmann et al. (73) examined artificial defects to simulate those that occur during the production of polyvinyl acetate/polyvinyl chloride coatings. These defects included both chemical and physical anomalies such as pinholes, underfilm corrosion,

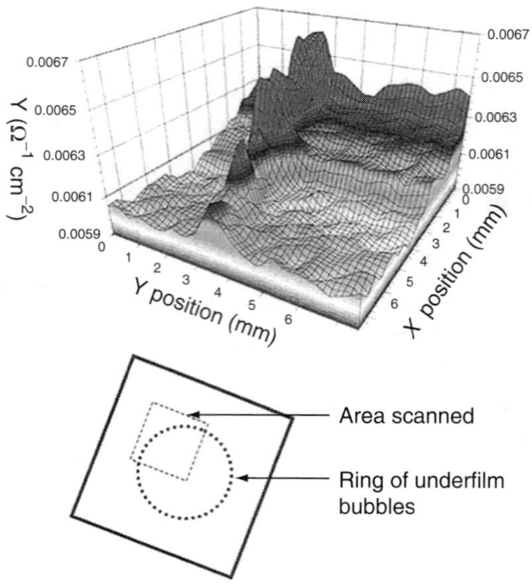

Figure 15.27 Admittance map of vinyl coating on steel in NaCl with LEIS (15 mV, 1000 Hz). Scan area and location of air bubbles attached to steel under coating are also shown. (From M. W. Wittman, R. B. Leggat, S. R. Taylor, *Journal of the Electrochemical Society*, 146, 4071, 1999. The Electrochemical Society, Pennington. With permission.)

and small underfilm air bubbles to simulate coating delamination. Admittance data ($1/Z$) from LEIS maps easily located underfilm bubbles attached to the metal substrate (seen in Figure 15.27 as peaks; minima in Z). In comparison, for bubbles that were created in the polymer film (free from the surface), the admittance values above the defects were much closer to nondefect polymer making them more difficult to resolve. These findings were consistent with long-term testing, which found that surface bubbles are less of a corrosion "liability" than in film (detached) bubbles. Wittmann et al. were also able to resolve in LEIS maps both machine oil spots under the coating and regions of high vinyl or epoxy concentrations. Each of these defects simulate relevant to the manufacturing industry. In a separate study of polymer coatings, Lillard et al. (72) used LEIS to monitor the impedance of urea — formaldehyde modified epoxy coatings at weld seams in aerosol containers. A typical LEIS map of the weld area and an intentional defect are shown in Figure 15.28. From individual LEIS maps generated over a large frequency range, local Bode magnitude and phase plots as a function of immersion time were constructed. These Bode magnitude plots were used to monitor the behavior of a specific site as a function of immersion time. A summary plot of the low-frequency impedance is shown in Figure 15.29. As seen in this figure, the polymer coating immediately over the weld seam degrades quickly. Because the welding of these cans destroys the coating in the immediate vicinity of the weld, the area must be recoated. It was proposed that the curing process for this area is less efficient (a flame cure is used).

LEIS has also been used to investigate pitting corrosion. Annergren et al. (76) used the bi-electrode LEIS technique to study the initiation and propagation of corrosion pits in an Fe–Cr alloy (approximately 13 wt% Cr) in NaCl. Propagation was examined by using a potential pulse technique whereby the electrode was pulsed with a potentiostatic square wave for a period of time during which traditional EIS

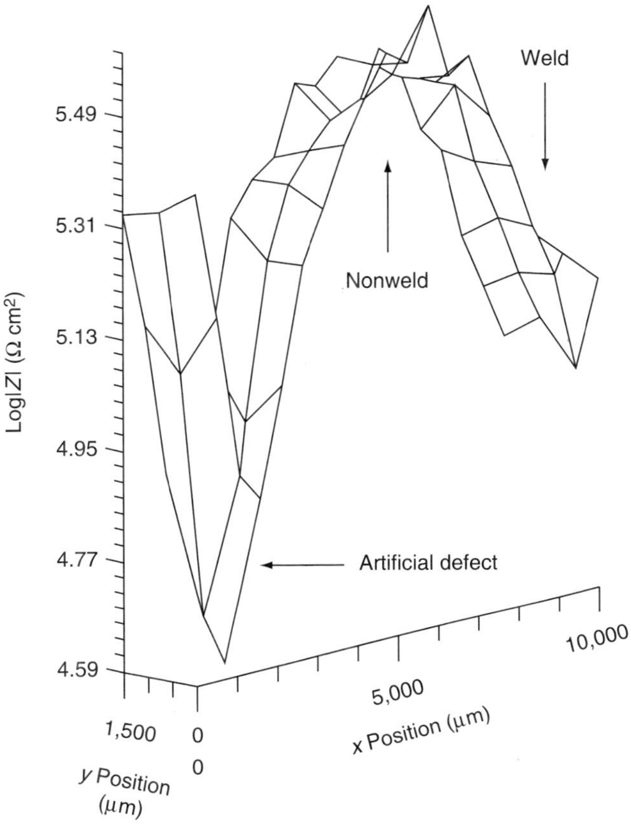

Figure 15.28 LEIS map of steel coated with urea–formaldehyde modified epoxy in NaCl buffer pH 7. Locations of an artificial defect, coating, and coating over a weld are shown. Data were collected at 2.4 Hz. (From R. S. Lillard, J. Kruger, W. S. Tait, P. J. Moran, *Corrosion*, 51, 251, 1995. NACE, Houston. With permission.)

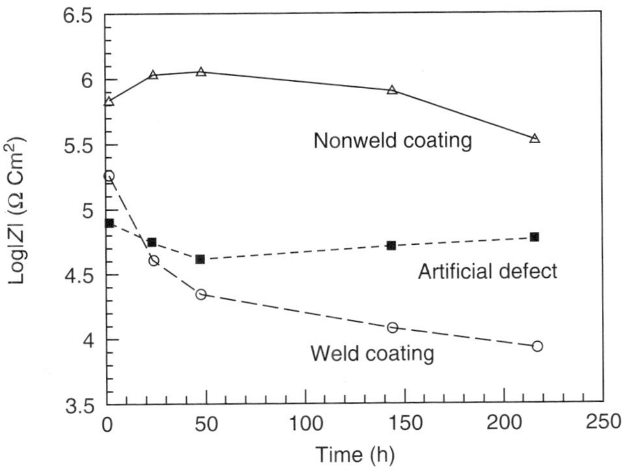

Figure 15.29 Bode magnitude data from LEIS of steel coated with urea–formaldehyde modified epoxy in NaCl buffer pH 7. Low-frequency impedance (1 Hz) as a function of immersion time. (From R. S. Lillard, J. Kruger, W. S. Tait, P. J. Moran, *Corrosion*, 51, 251, 1995. NACE, Houston. With permission.)

data and LEIS data were collected. The DC current density that resulted from the pulse was also recorded. For propagating pits, clear differences between the traditional EIS and LEIS Nyquist diagrams were observed. An example is shown in Figure 15.30a and b. The traditional Nyquist plot (Figure 15.30a) was characterized by two capacitive loops whereas the LEIS Nyquist plot generated concurrently was characterized by a high-frequency capacitive loop, an intermediate negative imped-

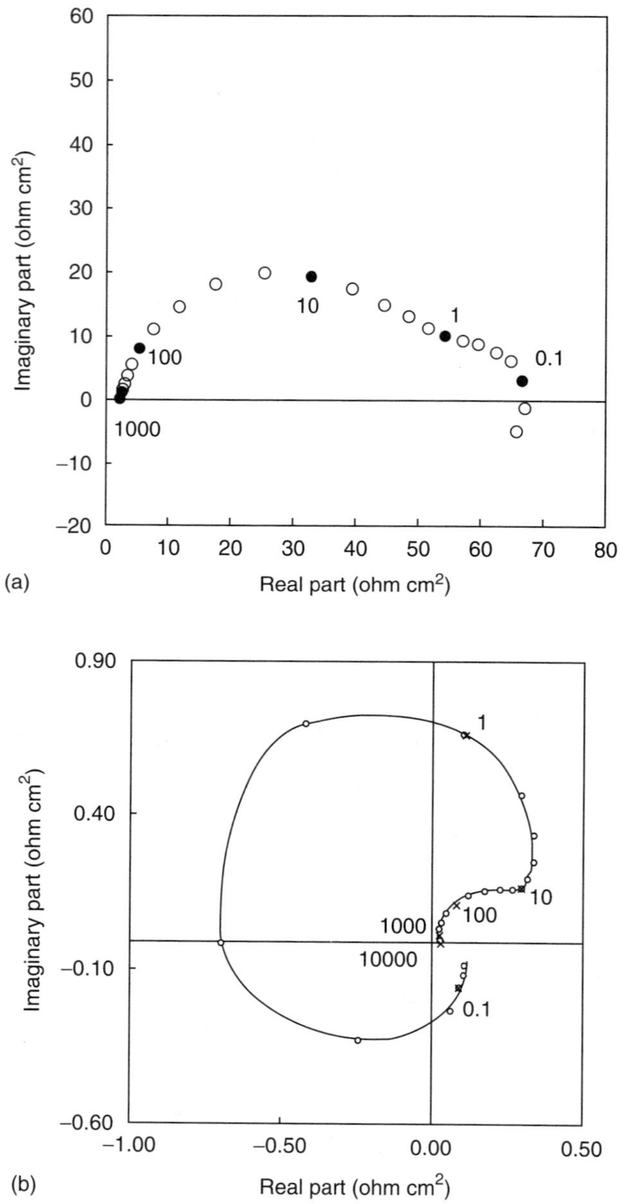

Figure 15.30 Nyquist plots for an Fe–Cr–Mo alloy in NaCl: (a) conventional three-electrode EIS data; (b) data generated with LEIS over a small corrosion pit in the sample. Data sets were generated simultaneously on the same sample. (From I. Annergren, F. Zou, D. Thierry, *Electrochemica Acta*, 44, 4383, 1999. Pergamon/Elsevier, Oxford. With permission.)

ance loop, and a low-frequency inductive loop (Figure 15.30b). The LEIS Nyquist data were consistent with salt film formation observed during anodic dissolution of Fe–Cr pencil electrodes in H_2SO_4. It is not clear from the Annegren work if a single equivalent circuit could be used to model both the traditional EIS and LEIS data; however, these LEIS data reinforce the fact that traditional EIS methods are surface-averaged techniques and, the sum of each of the processes may produce a result that is misleading. That is, in many cases with out a priori knowledge of each of the reactions taking place on the surface it is difficult at best to know what parameters should be considered when analyzing surface-averaged EIS data.

15.6 CONCLUDING REMARKS

This chapter has described several methods for obtaining information from local solution current densities above an electrochemically active surface. From these techniques it is possible to obtain both qualitative and quantitative information about local current densities at electrode surfaces (SRET and SVET) and local impedances (LEIS). However, it is important to point out several pitfalls that investigators often fall into (including this one!) without due diligence and quite a bit of experience. The first group of mistakes concerns what the techniques are actually measuring; that is, potential differences in solution that result from local current distributions. The techniques do not directly measure the potential or the current density at the sample surface. Quantitative local current densities (or impedances) at the sample surface may only be obtained from a deconvolution of "raw" SRET or SVET data. The second pitfall concerns probe geometry. Many authors report observing local anodes and cathodes on surfaces when, in fact, they are more likely measuring a current vector other than the normal component of the solution current density. If the probe tips are not directly above one another (90° with respect to the surface) or the vibrating reed is misaligned, the measurement will include a portion of the horizontal component of the solution current density. The apparent result is two neighboring anodic and cathodic peaks. The current associated with these neighboring peaks will pass through zero and have positive and negative portions resembling Figure 15.7. The anodic and cathodic peaks may be of equal magnitude (45° misalignment) or a large anodic peak and small cathodic peak (small deviation from 90°). This mistake can easily be avoided by placing a small calibration wire directly in the epoxy mount (made of a noble metal and placed away from the electrode being investigated). Under galvanostatic control, the investigator can calibrate for sensitivity and probe alignment.

Finally, while there are many excellent systems that are commercially available, the best systems are often the ones built by the investigators themselves as they will become intimately acquainted with their systems capabilities. The down side of building a system is that one must become familiar to some degree with the art of electronics, especially input impedances of DVMs, FRAs, or spectrum analyzers as ground loops may result. One way that ground loops may present themselves is by an apparent change in local solution current density with a change in input impedance, for example, when changing the measuring resistor on a potentiostat. They can be more precisely diagnosed with a calibration wire driven at a known current density. Consider the simplified diagram in Figure 15.31. This diagram represents the case where an investigator has unwittingly placed a probe in solution and connected it directly to the sensing instrument (an FRA or DVM). The input impedance of this

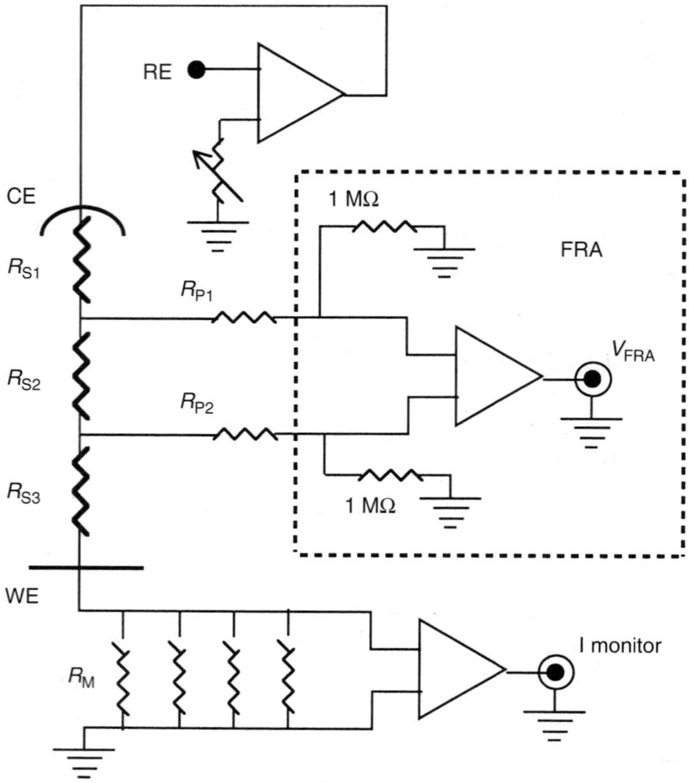

Figure 15.31 Simplified diagram of an improperly isolated scanning probe experiment.

sensing device is 1 MΩ (the impedance back to ground) as shown in the figure. The potential of the sample being examined (WE) is controlled by a potentiostat and corresponding reference (RE) and counter electrodes (CE). The probe wires are represented by R_{p1} and R_{p2} (small values if platinized tips are used), the solution resistances are represented by R_{S1}, R_{S2}, and R_{S3}, while R_m represents the measuring resistors within the potentiostat. The probe is measuring the potential difference across a small volume of solution very near the surface represented by R_{S2}. To control the potential between the WE and RE (either DC or AC), the potentiostat passes current between the CE and WE. In Figure 15.31, the current does not necessarily have to pass back through the current monitor of the potentiostat to reach ground. It has a "choice," either pass through the potentiostat or the FRA/DVM. Its choice will depend on the impedance in each loop (i.e., the input impedances of each device) as each of these loops is in parallel. The fraction of current in each loop is inversely proportional to the sum of the impedance in that loop (Ohm's law for resistors in parallel). For small R_m (1 Ω), the error associated with a ground loop through the FRA/DVM will be small. For large R_m (100 kΩ), the error will be significant especially if the impedance of the sample is large (such as a passive electrode or coated sample). Similarly, the error may be large for a high-impedance working electrode such as a polymer-coated substrate. The ground loop can be avoided by effectively increasing the input impedance of the FRA/DVM. This is accomplished by placing an instrumentation amplifier (such as those manufactured by Analog devices or Burr-Brown) between the probe and the sensing device.

Instrumentation amplifiers not only have the advantage of having a high input impedance (10^{12} or $10^{14}\,\Omega$ sec) but also allow the investigator to adjust the gain of the signal from the probe tips. Thus, small probe signals may be effectively amplified.

ACKNOWLEDGMENTS

Work on this chapter was performed by the University of California under the auspices of the United States Department of Energy contract W7405-ENG36. The author would like to thank Alan Shipley of Applicable Electronics for his helpful comments on the manuscript and Hugh Isaacs (BNL) and Pat Moran (USNA) for guiding my early research in this field.

REFERENCES

1. E. H. Dix, *Transactions of AIME*, **137**, 11 (1940).
2. E. H. Dix, *Transaction of the American Society for Metals*, **42**, 1057 (1950).
3. E. H. Dix, R. H. Brown, in *Metals Handbook*, 7th edition (1948), L. Taylor, ed., p. 228, American Society for Metals, Cleveland (1954).
4. R. S. Thornhill, U. R. Evans, *Journal of the Chemical Society*, 614 (1938).
5. R. S. Thornhill, U. R. Evans, *Journal of the Chemical Society*, 2109 (1938).
6. J. N. Agar, U. R. Evans, *The Journal of the Iron and Steel Institute*, **141**, 219 (1940).
7. C. H. Paik, H. S. White, R. C. Alkire, *Journal of the Electrochemical Society*, **147**, 4120 (2000).
8. M. Stratmann et al., *Progress in Organic Coatings*, **27**, 261 (1996).
9. M. A. Alodan, W. H. Smyrl, *Journal of the Electrochemical Society*, **145**, 1571 (1998).
10. J. R. Waldrop, M. W. Kendig, *Journal of the Electrochemical Society*, **145**, L11 (1998).
11. N. Casillas, S. R. Snyder, W. H. Smyrl, H. S. White, *Journal of Physical Chemistry*, **95**, 7002 (1991).
12. H. S. Isaacs, *Journal of the Electrochemical Society*, **138**, 722 (1991).
13. L. Nanis, W. Kesselman, *Journal of the Electrochemical Society*, **118**, 454 (1971).
14. K. R. Trethewey, D. A. Sargeant, D. J. Marsh, A. A. Tamimi, *Corrosion Science*, **35**, 127 (1993).
15. E. Gileadi, E. Kirowa-Eisner, J. Penciner, *Interfacial Electrochemistry*, p. 216, Addison-Wesley, Reading, MA (1975).
16. H. S. Isaacs, B. Vyas, in *Electrochemical Corrosion Testing*, ASTM STP 727, F. Mansfeld, U. Bertocci, eds., p. 3, ASTM, Baltimore (1981).
17. S. J. Badger, S. B. Lyon, S. Turgoose, *Journal of the Eletrochemical Society*, **145**, 4074 (1998).
18. S. J. Bates, S. R. Gosden, D. A. Sargeant, *Materials Science and Technology*, **5**, 356 (1989).
19. N. Cui, L. J. Qiao, J. L. Luo, S. Chiovelli, *British Corrosion Journal*, **35**, 210 (2000).
20. H. S. Isaacs, in *Localized Corrosion*, R. W. Staehle, B. F. Brown, J. Kruger, A. Argrawal, eds., p. 158, NACE, Houston (1974).
21. H. N. McMurray, *Ironmaking and Steel Making*, **23**, 254 (1996).
22. D. A. Sargeant, J. G. C. Haines, S. Bates, *Materials Science and Technology*, **5**, 487 (1989).
23. R. Akid, D. J. Mills, *Corrosion Science*, **43**, 1203 (2001).
24. H. N. McMurray, S. Macgill, B. D. Jeffs, *Ironmaking and Steel Making*, **23**, 183 (1996).
25. V. S. Voruganti, H. B. Luft, D. Degeer, S. A. Bradford, *Corrosion*, **47**, 343 (1991).
26. S. M. Powell, H. N. McMurray, D. A. Worsley, *Corrosion*, **55**, 1040 (1999).

27. S. Powell, *Surface Engineering*, **16**, 169 (2000).
28. I. M. Zin, R. L. Howard, S. J. Badger, J. D. Scantlebury, S. B. Lyon, *Progress in Organic Coatings*, **33**, 203 (1998).
29. P. J. Kinlen, V. Menon, Y. W. Ding, *Journal of the Electrochemical Society*, **146**, 3690 (1999).
30. F. J. Maile, T. Schauer, C. D. Eisenbach, *Progress in Organic Coatings*, **38**, 117 (2000).
31. L. F. Jaffe, R. Nuccitelli, *Journal of Cell Biology*, **63**, 614 (1974).
32. C. Scheffey, *Review of Scientific Instruments*, **59**, 787 (1988).
33. H. S. Isaacs, *Corrosion Science*, **28**, 547 (1988).
34. F. Zou, H. S. Isaacs, D. Thierry, *Corrosion Science*, **42**, 1149 (2000).
35. S. Fujimoto, T. Shibata, *Denki Kagaku*, **64**, 967 (1996).
36. H. S. Isaacs, Y. Ishikawa, *Journal of the Electrochemical Society*, **132**, 1288 (1985).
37. C. R. Crowe, R. G. Kasper, *Journal of the Electrochemical Society*, **133**, 879 (1986).
38. R. G. Kasper, C. R. Crowe, in *Galvanic Corrosion*, ASTM STP 978, H. P. Hack, ed., p. 118, ASTM, Philadelphia (1988).
39. V. J. Gelling, D. E. Tallman, G. P. Bierwagen, *Journal of the Electrochemical Society*, **147** (2000).
40. V. J. Gelling, D. E. Tallman, G. P. Bierwagen, *Journal of the Electrochemical Society*, **147**, 3667 (2000).
41. H. S. Isaacs, *Corrosion*, **43**, 594 (1987).
42. M. Khobaib, A. Rensi, T. Matakis, M. S. Donley, *Progress in Organic Coatings*, **41**, 266 (2001).
43. K. Masuhara, H. Fukumoto, H. Mizuki, K. Koshiishi, K. Mori, *Journal of the Iron and Steel Institute of Japan*, **81**, 727 (1995).
44. H. Mizuki, *Journal of the Iron and Steel Institute of Japan*, **73**, S406 (1987).
45. I. Sekine, *Progress in Organic Coatings*, **31**, 73 (1997).
46. I. Sekine, T. Suzuki, M. Yuasa, K. Handa, L. Silao, *Progress in Organic Coatings*, **31**, 185 (1997).
47. H. Fukumoto, H. Mizuki, K. Masuhara, *Journal of the Iron and Steel Institute of Japan*, **77**, 1034 (1991).
48. H. S. Isaacs, A. J. Davenport, A. Shipley, *Journal of the Electrochemical Society*, **138**, 390 (1991).
49. S. Bohm, M. Challis, T. Heatley, D. A. Worsley, *Transaction of the Institute of Metal Finishing*, **79**, 16 (2001).
50. M. Ishikawa, S. Yoshitakes, M. Morita, Y. Matsuda, *Journal of the Electrochemical Society*, **141**, L159 (1994).
51. H. N. Mcmurray, S. M. Powell, D. A. Worsley, *British Corrosion Journal*, **36**, 42 (2001).
52. D. A. Worsley, A. Belghazi, S. M. Powell, *Ironmaking & Steelmaking*, **26**, 387 (1999).
53. D. A. Worsley, A. Williams, J. S. G. Ling, *Corrosion Science*, **43**, 2335 (2001).
54. M. Yamashita et al., *Materials Science Forum*, **294**, 739 (1999).
55. F. Zou, C. Barreau, R. Hellouin, D. Quantin, D. Thierry, *Materials Science Forum*, **289**, 83 (1998).
56. X. Jiang, T. Yoshimura, Y. Ishikawa, *Journal of the Electrochemical Society*, 1001 (1992).
57. H. S. Isaacs, *Journal of the Electrochemical Society*, **135**, 2180 (1988).
58. H. Uchida, M. Yamashita, S. Inoue, K. Koterazawa, *Materials Science & Engineering A*, **A319–321**, 496 (2001).
59. M. J. Franklin, D. C. White, H. S. Isaacs, *Corrosion Science*, **32**, 945 (1991).
60. G. Schmitt, *Werkstoffe und Korrosion*, **48**, 586 (1997).
61. K. Ogle, V. Baudu, L. Garrigues, X. Phillippe, *Journal of the Electrochemical Society*, **147**, 3654 (2000).
62. H. S. Isaacs, M. W. Kendig, *Corrosion*, **36**, 269 (1980).
63. H. S. Isaacs, in *Novel NDE Methods for Materials*, B. B. Rath, ed., p. 63, The Metallurgical Society of AIME, New York (1982).
64. J. V. Standish, H. Leidheiser, *Corrosion*, **36**, 390 (1980).

65. M. C. Hughes, J. M. Parks, in *Corrosion Control by Organic Coatings*, H. Leidheiser Jr., ed., p. 45, NACE, Houston (1981).
66. R. S. Lillard, P. J. Moran, H. S. Isaacs, *Journal or the Electrochemical Society*, **139**, 1007 (1992).
67. F. Zou, D. Thierry, H. S. Isaacs, *Journal of the Electrochemical Society*, **144**, 1957 (1997).
68. E. Bayet, F. Huet, M. Keddam, K. Ogle, H. Takenouti, *Journal of the Electrochemical Society*, **144**, L87 (1997).
69. E. Bayet, F. Huet, M. Keddam, K. Ogle, H. Takenouti, *Electrochimica Acta*, **44**, 4117 (1999).
70. E. Bayet et al., in *Local In Situ Methods for Investigating Electrochemical Interfaces*, PV 99–28, S. R. Taylor, A. C. Hillier, M. Seo, eds., The Electrochemical Society, Pennington (1999).
71. J. Kruger, R. S. Lillard, C. C. Streinz, M. J. Moran, *Materials Science & Engineering A*, **198**, 11 (1995).
72. R. S. Lillard, J. Kruger, W. S. Tait, P. J. Moran, *Corrosion*, **51**, 251 (1995).
73. M. W. Wittmann, R. B. Leggat, S. R. Taylor, *Journal of the Electrochemical Society*, **146**, 4071 (1999).
74. A. M. Mierisch, J. Yuan, R. G. Kelly, S. R. Taylor, *Journal of the Electrochemical Society*, **146**, 4449 (1999).
75. S. R. Taylor, *Progress in Organic Coatings*, **43**, 141 (2001).
76. I. Annergren, F. Zou, D. Thierry, *Electrochimica Acta*, **44**, 4383 (1999).

FURTHER READING

SRET

R. Akid, D. J. Mills, "A comparison between conventional macroscopic and novel microscopic scanning electrochemical methods to evaluate galvanic corrosion," *Corrosion Science*, **43**, 1203 (2001).

P. J. Kinlen, Y. Ding, D. C. Silverman, "Corrosion protection of mild steel using sulfonic and phosphonic acid-doped polyanilines," *Corrosion*, **58**, 490 (2002).

H. N. McMurray, "Localized corrosion behavior in aluminum–zinc alloy coatings investigated using the scanning reference electrode technique," *Corrosion*, **57**, 313 (2001).

SVET

D. Battocchi, J. He, G. P. Bierwagen, D. E. Tallman, "Emulation and study of the corrosion behavior of Al alloy 2024-T3 using a wire beam electrode (WBE) in conjunction with scanning vibrating electrode technique (SVET)," *Corrosion Science*, **47**, 1165 (2005).

J. Elvins, J. A. Spittle, D. A. Worsley, "Relationship between microstructure and corrosion resistance in Zn–Al alloy coated galvanised steels" *Corrosion Engineering, Science and Technolog*, **38**, 197 (2003).

H. N. McMurray, J. R. Searle, B. P. Wilson, D. A. Worsley, "Investigating changes in corrosion mechanism induced by laser welding galvanised steel specimens using scanning vibrating electrode technique," *British Corrosion Journal*, **37**, 225 (2002).

J. H. Sullivan, D. A. Worsley, "Zinc runoff from galvanised steel materials exposed in industrial/marine environment," *British Corrosion Journal*, **37**, 282 (2002).

B. Vullemin, X. Philippe, R. Oltra, V. Vignal, L. Coudreuse, L. C. Dufour, E. Finot, "SVET, AFM and AES study of pitting corrosion initiated on MnS inclusions by microinjection," *Corrosion Science*, **45**, 1143 (2003).

D. A. Worsley, H. N. McMurray, J. H. Sullivan, I. P. Williams, "Quantitative assessment of localized corrosion occurring on galvanized steel samples using the scanning vibrating electrode technique," *Corrosion*, **60**, 437 (2004).

LEIS

G. Baril, C. Blanc, M. Keddam, N. Pebere, "Local electrochemical impedance spectroscopy applied to the corrosion behavior of an AZ91 magnesium alloy," *Journal of the Electrochemical Society*, **150**, B488 (2003).

A. M. Mierisch, S. R. Taylor, V. Celli, "Understanding the degradation of organic coatings through local electrochemical impedance methods II. Modeling and experimental results of normal field variations above disk electrodes," *Journal of the Electrochemical Society*, **150**, B309 (2003).

L. V. S. Philippe, G. W. Walter, S. B. Lyon, "Investigating localized degradation of organic coatings. Comparison of electrochemical impedance spectroscopy with local electrochemical impedance spectroscopy," *Journal of the Electrochemical Society*, **150**, B111 (2003).

16
Application of Scanning Kelvin Probe in Corrosion Science

M. Rohwerder and M. Stratmann
Department of Interface Chemistry and Surface Engineering, Max–Planck–Institut für Eisenforschung, Düsseldorf, Germany

P. Leblanc and G.S. Frankel
Fontana Corrosion Center, The Ohio State University, Columbus, Ohio

Contents

16.1 INTRODUCTION

The scanning Kelvin probe (SKP) is a technique introduced into corrosion science only recently by the group of Stratmann (1–10). The technique is unique insofar as it allows a noncontact measurement of electrode potentials, and therefore, also corrosion potentials. It is used instead of electrochemical reference electrodes and can measure electrode potentials through insulating dielectric media, such as air or polymeric films. It is mainly used where standard electrochemical techniques, which require a finite electrical resistance between working and reference electrodes, will fail.

The Kelvin probe is a rather old technique that has been used for decades by surface physicists to measure work functions. The scanning version of the classical capacitor setup was first introduced by Parker and Warren (11), who studied lateral variations in work functions on gold and graphite with a resolution of several millimeters. Since then, the SKP technique has been steadily improved, and a resolution of several tens of micrometers is easily achieved (12, 13). Mäckel et al. (14, 15) prepared probe tips with a flattened surface of just a few micrometers in diameter and controlled the distance between tip and sample surface to about 50 nm, resulting in a spatial resolution of 5 μm. An impressive lateral resolution of 100 nm was reported by Nabhan et al. (16), who also used active distance control. A high-resolution technique that is much easier to use is the scanning Kelvin probe force microscope (SKPFM), which is based on atomic force microscopy. The SKPFM technique was introduced into corrosion science in the late 1990s by Frankel et al. (17–20) (see Section 16.5).

The SKP technique is very susceptible to artifacts, especially at high resolution. Much less problematic is the classical parallel plate capacitor setup, where the lateral dimensions of the capacitor are several orders of magnitude larger than the distance between the sample and the probe. This classical setup is, in fact, one of the oldest techniques for measuring the work functions of materials, first introduced by Kohlrausch (see Ref. (21)). If the capacitance between the probe plate and the sample surface is known, the work function difference between the probe (reference material) and the sample can be calculated, although the measurement of the charge is not easily performed. This problem can be overcome by measuring the discharge current when the distance between the probe and the sample is varied, first introduced by Lord Kelvin. By moving the capacitor manually and measuring the current flow, the work function difference can be calculated (22). Zisman (23) developed the technique further to the vibrating capacitor technique, in which the probe plate vibrates periodically, thus causing a steady AC current to flow. The vibrating capacitor technique is the approach commonly used now.

The problems related to high-resolution SKP and SKPFM will be discussed in Section 16.5.

In the following, a short overview over advantages and disadvantages of SKP will be given. Then the theoretical basis is provided for a deeper understanding of the information obtainable by SKP. Finally, different examples for application are presented to demonstrate the power of the SKP for corrosion science and related fields of research.

16.1.1 Advantages and Disadvantages of SKP

As will be shown, the physical relation between the quantity measured by the Kelvin probe — the "Volta potential" — and the electrode potential at the buried interface is straightforward, but includes other terms such as dipole or membrane potentials. This may complicate the analysis of measured data but will also provide additional valuable information.

The Kelvin probe has the following advantages:

1. Electrochemical properties of surfaces that are covered by extremely thin electrolyte layers can be studied. This is not possible with conventional reference electrodes, which disturb the composition and thickness of the electrolyte film at the point of measurement. This is of importance for studies on atmospheric corrosion, fuel cells, or other electrocatalytically important electrodes, and corrosion of electronic materials.
2. Local corrosion beneath insulating films, such as organic or inorganic coatings, can be studied.
3. The local structure of surface films determined by oriented dipoles can be determined.
4. Localized corrosion phenomena on surfaces covered by highly resistive films or electrolytes can be monitored. This property may be used to detect defects such as inclusions in metallic matrices or to analyze corrosion in the presence of such defects.

Disadvantages of the Kelvin probe are as follows:

1. The Kelvin probe shows no improvement with respect to conventional reference electrodes if the surface is covered by low resistive films or thick electrolyte layers. Under those circumstances, potential coupling between different surface sites will result and the Kelvin probe will detect only mean electrode potentials as any other reference electrode.
2. The Kelvin probe will mainly detect electrode potentials and not the rate of electrochemical or corrosion reactions. The Kelvin probe may, however, be used as part of a more complex electronic circuit to measure polarization curves on surfaces covered by very thin electrolyte layers (see below).
3. The interpretation of data generated by the Kelvin probe is not always straightforward, as other potential differences may interfere with the electrode potential.

16.2 THEORY OF SKP

16.2.1 Theory of the Kelvin Probe

In principle, the Kelvin probe consists of a metallic reference electrode, which is separated from the sample by a dielectric medium and connected to the sample by a metallic wire (Figure 16.1). If the work function of the sample is given as W_{sample} and the work function of the reference electrode as W_{ref}, then

$$W_{\text{Sample}} = -\tilde{\mu}_{\text{e}}^{\text{Sample}} = -\left[\mu_{\text{e}}^{\text{Sample}} - F\chi^{\text{Sample}}\right] \tag{16.1}$$

where $\mu_{\text{e}}^{\text{Sample}}$, $\tilde{\mu}_{\text{e}}^{\text{Sample}}$ are chemical and electrochemical potentials of an electron in the sample, F is the Faraday constant, and χ^{Sample} is the surface potential of the sample.

$$W_{\text{Ref}} = -\tilde{\mu}_{\text{e}}^{\text{Ref}} = -\left[\mu_{\text{e}}^{\text{Ref}} - F\chi^{\text{Ref}}\right] \tag{16.2}$$

where $\mu_{\text{e}}^{\text{Ref}}$, $\tilde{\mu}_{\text{e}}^{\text{Ref}}$ are chemical and electrochemical potentials of the electron in the reference, F is the Faraday constant, and χ^{Ref} is the surface potential of the reference probe.

As after connection of the two metals the electrochemical potential of the electrons within both phases will be identical, and a charging of one sample with respect to the other (Volta potential difference) will be observed:

$$\tilde{\mu}_{\text{e}}^{\text{Sample}} = -W_{\text{Sample}} - F\Psi^{\text{Sample}} = \tilde{\mu}_{\text{e}}^{\text{Ref}} = -W_{\text{Ref}} - F\Psi^{\text{Ref}} \tag{16.3}$$

where Ψ^{Sample} is the Volta potential of the sample, Ψ^{Ref} is the Volta potential of the reference probe, and $\tilde{\mu}_{\text{e}}^{\text{Ref}}$ is the electrochemical potential of the electron in the reference probe; and

$$\Delta\Psi_{\text{Sample}}^{\text{Ref}} = \Psi^{\text{Ref}} - \Psi^{\text{Sample}} = -\frac{1}{F}(W_{\text{Ref}} - W_{\text{Sample}}) \tag{16.4}$$

Therefore, for a given and constant work function of the reference metal, the work function of the sample will be determined by a measurement of the Volta potential difference between the reference and the sample, $\Delta\,\Psi_{\text{Sample}}^{\text{Ref}}$. It can be measured by the vibrating condenser Kelvin probe method. Sample and reference electrodes form a condenser and the reference electrode is forced to vibrate about a mean distance from the sample $\bar{d}$ by an amplitude Δd. A simplified expression for the capacity C of an ideal parallel plate capacitor is given by

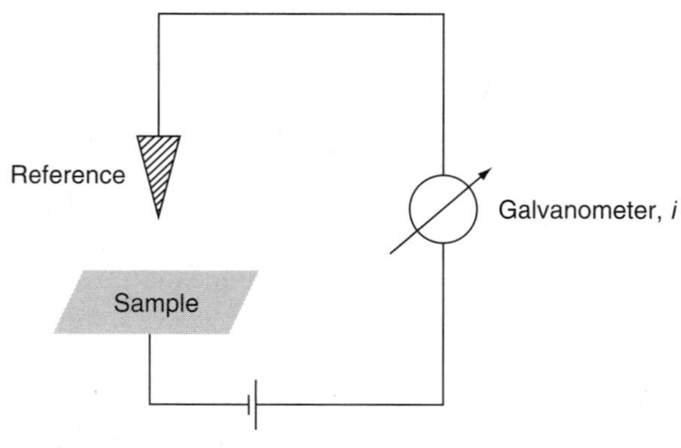

Figure 16.1 Schematic setup of the scanning Kelvin probe.

$$C = \varepsilon\varepsilon_0 \frac{A}{\bar{d} + \Delta d \sin(\omega t)} \tag{16.5}$$

where ε is the dielectric constant of the medium, ε_0 is the permittivity of free space, ω is the frequency of vibration for the reference metal, A is the surface area of the reference plate; and the induced AC current in the external circuit is

$$\begin{aligned} i_{AG} &= \Delta\Psi^{Ref}_{Sample} \frac{dC}{dt} \\ &= \Delta\Psi^{Ref}_{Sample} (\varepsilon \cdot \varepsilon_0 \cdot A \cdot \Delta d \cdot \omega) \frac{\cos(\omega t)}{(\bar{d} + \Delta d \sin(\omega t))^2} \\ &\approx \Delta\Psi^{Ref}_{Sample} (\varepsilon \cdot \varepsilon_0 \cdot A \cdot \Delta d \cdot \omega) \frac{\cos(\omega t)}{(\bar{d})^2} \quad \text{for } \bar{d} \gg \Delta d \end{aligned} \tag{16.6}$$

If an external voltage U is switched into the external circuit (see Figure 16.1), then

$$i = (\Delta\Psi - U) \frac{dC}{dt} \tag{16.7}$$

where $i = 0$ for $\Delta\,\Psi = U$.

In the conventional nulling technique, the voltage U is changed until the current i vanishes, and for this condition $\Delta\,\Psi^{Ref}_{Sample} = U$ is measured. The accuracy of the measurement is determined by the sensitivity S:

$$S = \frac{di_{rms}}{dU} \approx \frac{\varepsilon \cdot \varepsilon_0 \cdot A \cdot \Delta d \cdot \omega}{(\bar{d})^2} \tag{16.8}$$

Therefore, for small areas of the reference plate, A, sensitivity problems will arise. As mentioned earlier, this has been overcome recently by improvement of the experimental setup.

16.2.2 Relation Between the Volta Potential and the Electrode Potential

The theoretical relation between the measured Volta potential and the corrosion potential has been discussed for different situations in detail (24). Here only some of the results will be summarized.

16.2.2.1 Electrolyte-Covered Metal Surfaces

The following potential differences have to be considered (see Figure 16.2):

$\Delta\Phi^{Me}_{El}$: Galvani potential difference Φ between the metal and the electrolyte; the Galvani or inner potential of each phase being defined as the sum of the Volta potential Ψ (resulting from a charge of the phase) and the surface potential χ.

χ_{El}: The surface potential of the electrolyte. This surface potential is not directly measurable but according to theoretical considerations is small and positive due to a preferential orientation of the oxygen side of water dipoles toward the gas phase. Typical estimations of the surface potential are in the order of +0.05 to +0.1 V.

$\Delta\Psi^{Ref}_{El}$ is the Volta potential difference and is a measurable quantity.

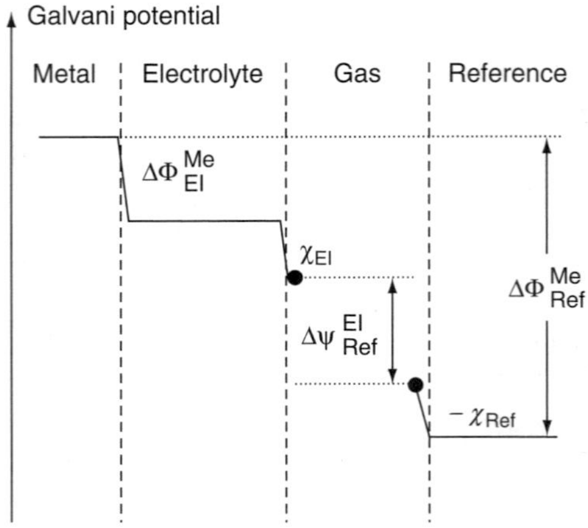

Figure 16.2 Potential distribution over the system metal/electrolyte/gas/reference probe.

χ_{Ref}: The surface potential of the reference probe itself. The surface potential will depend strongly on the adsorbed water layer on top of the reference metal and will be constant only under identical experimental conditions.

It can be shown that the following equation holds (24):

$$E_{Corr} = \varepsilon_{1/2} - \varepsilon_{1/2}^{Standard} = \left\{ \frac{W_{Ref}}{F} - \chi_{El} - \varepsilon_{1/2}^{Standard} \right\} + \Delta\Psi_{El}^{Ref} \tag{16.9}$$

if one takes into account the definition of the absolute half cell potential according to Trasatti ($\varepsilon_{1/2} = \{\Delta\Phi_{El}^{Me} - (1/F)\,\mu_e^{Me}\}$) and the definition of the work function ($W_{Ref} = -\,(\mu_e^{Ref} - F\chi_{Ref})$), $\varepsilon_{1/2}^{Standard}$ being the absolute half cell potential of the chosen electrochemical reference standard (such as calomel, Ag/AgCl). Equation (16.9) shows the linear relation between the Volta potential difference and the corrosion potential as long as the term in brackets is regarded as constant. This is true for a given electrolyte, as then the dipole potential of the liquid phase will be constant, and for a reference electrode with constant work function.

16.2.2.2 Oxide-Covered Metal Surface

The Galvani potential difference between the metal and the electrolyte is composed of three terms:

$$\Delta\Phi_{El}^{Me} = \Delta\Phi_{ox}^{Me} + \Delta\Phi_{ox} + \Delta\Phi_{El}^{ox} \tag{16.10}$$

taking into account the potential drops at the two interfaces and the potential drop across the oxide phase itself. However, as long as only the relation between the corrosion potential and the Volta potential difference is concerned, all potential drops will contribute to the electrode potential and the Volta potential difference in the same way and therefore Equation (16.9) still holds. Neither the electrode potential nor the Volta potential difference can distinguish between an oxide-covered surface and an oxide-free surface.

16.2.2.3 Metals in Humid Air

In not overly humid air, the metal surface will be covered only by a few monolayers of water as long as no hygroscopic impurities are adsorbed. Under those circumstances, regular electrochemistry will not be valid anymore, as electron or ion transfer reactions will result in charge distribution at the interface, which cannot be compensated by ions of a supporting electrolyte. Therefore, the reactions change the local electrode potential (contrary to standard electrochemistry), which must result in a retardation of the reaction rate.

For nonreactive materials (ion transfer reactions are negligible), an electron transfer reaction to adsorbed oxygen molecules results in charged oxygen ions, and therefore, in a dipole layer that retards further electron transfer. Reactive substrates allow additional ion transfer reactions in compensation of the charge distribution, which results immediately in oxide formation. Therefore, for reactive materials the "electrode potential" that is measured by the Kelvin probe will reflect the redox properties of the oxide scale. In particular, the "electrode potential" will be shifted to a value at which the rate of any electrochemical reaction, such as the reduction of oxygen, is nearly zero. Therefore, this potential is an indirect measure of the electron transfer properties of the formed passive layer. The more noble a metal is, the higher is the potential where its oxide reaches passivity.

Indeed, it has been stated by Frankel et al. that a relation exists between the Volta potential of a metal as measured in humid air and its "practical nobility" (see Section 16.5.2.2).

The Volta potential difference between two metals in humid air is defined as

$$\Delta\Psi_{\mathrm{Me}}^{\mathrm{Ref}} = \left\{\frac{1}{F}\mu_{\mathrm{Me}} - \chi_{\mathrm{Me}}\right\} - \left\{\frac{1}{F}\mu_{\mathrm{Ref}} - \chi_{\mathrm{Ref}}\right\}$$

and therefore changes in the Volta potential difference mainly reflect changes in the dipole structure of the adsorbed water layer, which is the basis of many classical Kelvin probe studies under ultrahigh vacuum conditions.

If, however, the measurement of $\Delta\Psi_{\mathrm{Me}}^{\mathrm{Ref}}$ is performed in the presence of oxygen and the work function of the metal is small, then an additional transfer of electrons from the metal to oxygen is possible. This results in a drastic change of the surface potential, at least partly due to charged oxygen monolayers, and an increase of the work function. For less noble metals, this will result in the transport of ions within the high electric field and the formation of a dense and thin oxide layer. Under these circumstances, the measured Volta potential reflects the electronic properties of the surface oxide, which in the case of iron oxides is directly linked to the chemical surface composition (24):

$$\begin{aligned}\Delta\Psi_{\mathrm{ox}}^{\mathrm{Ref}} &= -\frac{\Delta\mu^{0}_{\mathrm{Fe^{2+}/Fe^{3+}}}}{F} - \frac{W_{\mathrm{Red}}}{F} + \chi_{\mathrm{ox}} + \frac{RT}{F}\ln\frac{[\mathrm{Fe}^{3+}]''}{[\mathrm{Fe}^{2+}]''}\\ &\approx \mathrm{const.} + \frac{RT}{F}\ln\left(\frac{[\mathrm{Fe}^{3+}]''}{[\mathrm{Fe}^{2+}]''}\right)\end{aligned} \qquad (16.11)$$

In a simple understanding of Equation (16.11), the oxide surface on iron will be oxidized in air until the oxidation level and therefore the Volta potential is high enough to prevent any further oxygen reduction.

16.2.2.4 Polymer-Coated Metals

In general, most polymer coatings used for corrosion protection are considered insulators. However, experimentally it has been shown that, in particular, in a humid atmosphere, the Kelvin probe will measure the potential just outside the polymer surface. Therefore, some electronic conductivity between the polymer surface and the inner interface must exist for highly swollen polymers, at least for the currents involved in the Kelvin probe measurement (10^{-8} A/cm^2). If the polymer phase shows no conductivity at all, then the Kelvin probe should measure through the polymer, that is, the polymer acts just as a dielectric. However, a potential measurement with the Kelvin probe is difficult on insulating materials due to surface charging.

The correlation of the Volta potential measured on the surface of a polymer coating to the electrode potentials at the polymer/substrate interface is a complex issue. For semiconducting organic coatings, the apparent work function measured with the Kelvin probe in some cases depends strongly on the underlying metal (25), that is, the metal work function seemed to determine the work function of the coating. This is due to Fermi-level pinning, which prevents full alignment of the Fermi levels of coating and substrate and thus leaves an energy gap Δ between the two Fermi levels. Thus, the apparent work function of the semiconducting coating measured with the Kelvin probe, $W_{\text{sem app}}$, is different from the true work function of the semiconducting coating, W_{sem}, by the amount Δ, or $W_{\text{sem app}} = W_{\text{sem}} - \Delta$. Because the energy gap Δ depends strongly on the metal work function, the metal influences the work function measured with the Kelvin probe.

In the following, we will focus solely on the case in which the Fermi levels of the polymer and metal (oxide) at the interface are in alignment and the polymer is basically an insulator (only sufficient conductivity for currents in the range of 10^{-8} A/cm^2 is required).

In a first approximation, the case of a polymeric coating is analogous to the case of an electrolytic film. Hence, only the dipole potential of water is replaced by the dipole potential of the polymer χ_{Pol}:

$$E_{\text{Corr}} = \left\{ \frac{W_{\text{Ref}}}{F} - \chi_{\text{Pol}} - \varepsilon_{1/2}^{\text{standard}} \right\} + \Delta\Psi_{\text{Pol}}^{\text{Ref}}$$

The Volta potential difference is determined by the "corrosion potential" of the buried interface. Therefore, the corrosion properties of buried interfaces can be measured by this technique. However, the physical meaning of this electrode potential is not as obvious as for other cases, because it cannot be interpreted by conventional electrochemical kinetics. In the absence of any faradaic current, the electrode potential could be determined by dipole orientation of segments of the polymer chain. If, however, faradaic reactions like the oxygen reduction occur at the inner interface, the interface will be polarized until the rate of the oxygen reduction is negligible. This is true for polymer-coated gold surfaces.

In the case of delamination, an additional electrolyte layer is placed between the metallic substrate and the polymeric coating (see Figure 16.3). In this situation, which is of considerable technological interest, an additional potential drop at the electrolyte/polymer interface — the Donnan potential $\Delta\phi_{\text{D}}$ — has to be regarded:

$$E_{\text{corr}} = \underbrace{\Delta\phi_{\text{D}}} + \underbrace{\left[\frac{W^{\text{Ref}}}{F} - \chi^{\text{Pol}} + \varepsilon_{1/2}^{\text{standard}} \right]} + \Delta\Psi_{\text{Pol}}^{\text{Ref}}$$

Therefore, the Volta potential difference $\Delta\Psi_{\text{Pol}}^{\text{Ref}}$ only measures the corrosion potential at the inner metal/electrolyte interface buried below the polymeric coating if the Donnan potential is known or negligible.

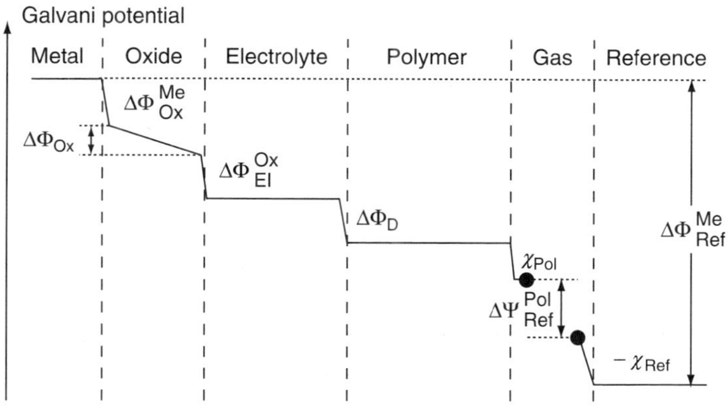

Figure 16.3 Potential profile over the system metal/oxide/electrolyte/polymer/humid air/reference probe.

In order to check the importance of the Donnan potential, calibration experiments have been performed by measuring simultaneously the electrode potential E_{corr} and the Volta potential difference $\Delta\Psi_{\text{Pol}}^{\text{Ref}}$ as a function of the composition of the electrolyte (8). Then, taking into account the theory of Donnan potentials, the following equation was derived:

$$E_{\text{corr}} - \Delta\Psi_{\text{Pol}}^{\text{Ref}} = \underbrace{\Delta\phi_{\text{D}}}_{=f(c_{\text{electrolyte}})} + \underbrace{\left[\frac{W^{\text{Ref}}}{F} - \chi^{\text{Pol}} + \varepsilon_{1/2}^{\text{standard}}\right]}_{\neq f(c_{\text{electrolyte}})}$$

$$\approx \frac{zRT}{F}\ln x + \left[\frac{W^{\text{Ref}}}{F} - \chi^{\text{Pol}} + \varepsilon_{1/2}^{\text{standard}}\right] \qquad (16.12)$$

with χ being the ratio of the fixed charge density within the polymer and the concentration of the (1:1) electrolyte.

Both potentials on the left-hand side are measurable quantities and can be determined as a function of the concentration of ions within the electrolyte. Since the Donnan potential is the only parameter on the right-hand side of Equation (16.12), that depends on the concentration of ions within the electrolyte, a plot of the difference ($E_{\text{corr}} - \Delta\Psi_{\text{Pol}}^{\text{Ref}}$) versus $c_{\text{electrolyte}}$ will show the significance of the two terms on the right-hand side of Equation (16.12) and therefore demonstrate if the Donnan potential significantly affects the measured electrode potential.

16.3 EXPERIMENTAL CONSIDERATIONS

16.3.1 Standard Scanning Kelvin Probe

A typical setup for measuring the local Volta potential is shown in Figure 16.4. Basically, the Kelvin probe is fixed to a computer-controlled *xyz*-stage and the sample is fixed in its local position. The probe, consisting of a sharp needle (e.g., corrosion-resistant Cr–Ni alloy), is forced to vibrate by a magnetically driven loudspeaker system. The probe is set to ground. The sample is connected to a current voltage converter with a very high amplification factor by a very short cable and is separated from ground by the input resistance of the operational amplifier (typically 10^8 ohm). The acoustic and electric shielding of the preamplifier is very important for

Figure 16.4 View into the SKP chamber: needle-tip, sample, and $Cu/CuSO_4$ reference.

a good signal/noise ratio. The output of the preamplifier is connected to the input of a lock-in amplifier and is analyzed with respect to a reference frequency that is identical to the vibration frequency of the probe. The output of the lock-in amplifier is then proportional to the current with the same frequency as the probe and a preselected phase shift. This output voltage is integrated and switched between ground and the sample; therefore, changing the voltage difference between the sample and the probe until the AC signal vanishes and the lock-in output falls to zero. The voltage between the sample and the probe necessary to reduce the AC signal to zero is regarded as the Volta potential difference $\Delta\Psi_{\text{sample}}^{\text{Ref}}$. During measurement, the chamber of the Kelvin probe is kept at very high relative humidity (typically >95%) in order to keep the electrochemical reactions running. This requires a corrosion-resistant construction for the chamber, the probe, the sample holder, etc. (e.g., 316 Ti steel) and may give rise to experimental problems. The resistance between the sample and ground has to be kept at values $>10^8$ ohm, which is not an easy task in very humid atmospheres. Materials that charge up during operation (e.g., teflon-like polymers) must be avoided in the construction of the SKP, as any surface charge will significantly perturb the signal.

During the operation of the Kelvin probe, several objectives have to be taken into consideration.

16.3.1.1 Sample Preparation

If only a metallic substrate will be analyzed, little precaution is necessary. However, if the measured Volta potential difference will be interpreted as a corrosion potential, then the corrosion conditions of the sample under investigation must be well defined. Studies of samples covered by bulk electrolytes with the Kelvin probe are usually not performed because standard electrochemical techniques are possible. Typical applications are atmospheric corrosion of samples covered by thin electrolyte layers, galvanic coupling between dissimilar metals (e.g., to check the cathodic protection of cut edges by the nondisturbed surface of galvanized steel), and measurement of the potential distribution in the vicinity of welding zones.

In the case of polymer-coated materials, the polymer surface must not be charged. This frequently requires measurements at humidity close to 100% and long-term exposure to such conditions prior to any useful measurement. Charging of the polymer surface usually shows up as huge potential fluctuations during potential mapping. Possible long-lasting charging effects may also be checked by depositing a large charge onto the polymer surface (e.g., by charged Teflon) and then following the charge decay by the Kelvin probe itself. In case of a slow charge reduction, the measured Volta potentials have to be taken with care.

The Kelvin probe has been used to follow the delamination of organic coatings in the presence of defects. This will be successful only if the polymer surface is not galvanically coupled to the defect, for example, by surface contaminations. In particular, the surface resistance ($R_{\|}$ in Figure 16.5) has to be high in comparison to the resistance of the polymer itself perpendicular to the surface ($R_{\perp}$) even at high humidity. This requires extremely clean polymer surfaces and impedes any useful measurement after salt spray testing of the polymer-coated substrate. Usually, these measurements are performed at a relative humidity close to 100%, the defect itself being covered by a bulk electrolyte of known composition. Obviously, the electrolyte has to be strictly limited to the defect and may not wet the polymer surface. This can be achieved by suitable sample layouts such as the layout shown in Figure 16.5.

16.3.1.2 Calibration

The Volta potential difference $\Delta\Psi$ has been shown to be related to the electrode or corrosion potential in a linear manner. However, a calibration constant is needed to calculate the electrode potential from a Volta potential measurement. For aqueous electrolytes, the calibration constant is easily obtained by measuring the Volta potential of a metal electrode that is exposed to an electrolyte containing the metal cation in a defined concentration. The electrode potential of the metal/metal cation system is known from the Nernst equation and the calibration constant is obtained

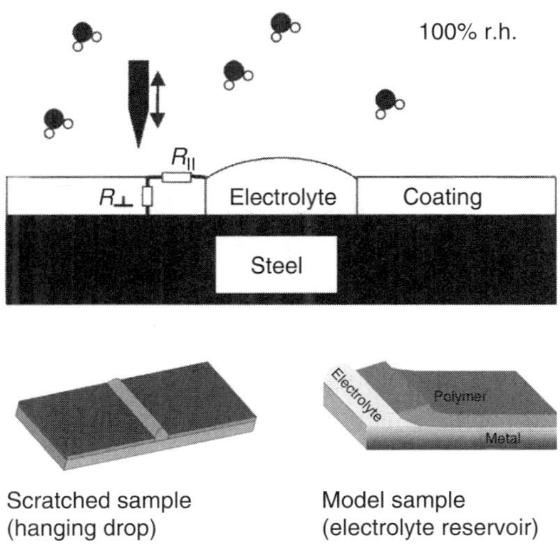

Figure 16.5 Defect preparation for measuring polymer-coated materials with a scanning Kelvin probe: the defect is either prepared as a scratch where the electrolyte is applied as a "hanging drop" (bottom left), or by pulling one edge of the coating up and thus forming a small pocket where the electrolyte is contained in a kind of reservoir (bottom right).

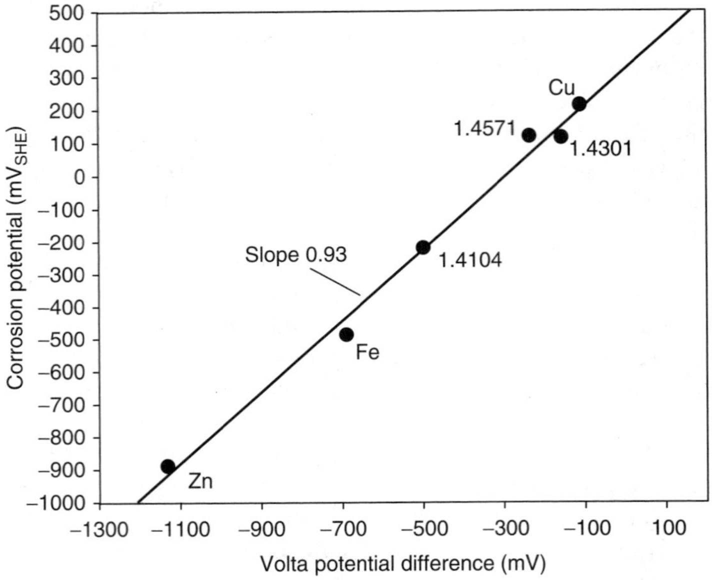

Figure 16.6 Calibration of the Kelvin probe with metal/metal cation systems of known electrode potential. (From A. Leng, H. Streckel and M. Stratmann, *Corros. Sci.*, 41, 547, 1999. With permission.)

by plotting the known electrode potential versus the Volta potential for different metal/metal cation systems (see Figure 16.6). However, it should be kept in mind that the calibration constant contains the work function of the reference metal and depends strongly on the actual surface condition of the reference material. Therefore, the calibration procedure has to be performed frequently and typically with only one simple metal/metal cation system like $Cu/CuSO_4$.

For polymer-coated materials, similar experiments have been performed with a thin electrolyte layer between the metallic sample and the polymer film (8). Again a linear relation has been observed between the electrode potential and the Volta potential, which allowed determination of a calibration constant.

16.3.1.3 Elimination of Parasitic Signals

The Kelvin probe is prone to parasitic signals that can originate either from electromagnetic coupling caused by the AC current driving the loudspeaker system or from acoustic coupling that also originates from the vibrator and leads to vibration of small wires within the preamplifier.

The first parasitic signal can be eliminated easily by careful shielding of the driver and by minimizing the power necessary to vibrate the needle. This is usually achieved by a driver construction that has a strong and sharp resonance frequency on the order of kilohertz. At resonance, the needle oscillates with maximum amplitude at minimum power; for off resonance conditions no vibration is observed. Checking the output of the lock-in amplifier under off resonance conditions allows assessment of the parasitic signal that originates from lack of electrical shielding.

Parasitic signals that originate from acoustic coupling are far more difficult to observe and to eliminate. These signals appear under resonance conditions and therefore are hard to distinguish from the real signal of the Kelvin probe. Usually, this signal shows up even if the tip of the needle is moved far away from the sample surface. More precisely, this signal, which is independent of the distance between sample and reference needle, is quantified by Equation (16.7). The output of the lock-in amplifier is measured as a function of the external voltage U, resulting in linear plot of i(lock-in) versus U. The measurement is performed for varying distances d resulting in a linear plot for each distance. The intercept of all linear plots will be on the abscissa in the absence of any parasitic signal and will be off the abscissa in the presence of parasitic signals.

16.3.1.4 Local Resolution

The principal considerations for the optimization of the resolution achievable with the standard SKP setup are as follows. The local resolution of the SKP will strongly depend on the shape of the tip and the mean distance between needle and sample surface. In order to avoid stray capacities from the tip sides, the needle should be formed like a small cylinder with a well-defined tip area and not like the tip of a scanning tunneling microscope. The vibration amplitude should be small and the tip–sample distance as small as possible.

The output of the lock-in amplifier will deviate strongly from a simple sine wave if the minimum tip–sample distance is small in comparison to the mean distance, that is, if $\Delta d \approx \bar{d}$. A Fourier analysis of the lock-in signal shows that, with decreasing distance, the $2f$ component of the signal will be considerably larger than the $1f$ component. If, therefore, for small distances the $2f$ component is used instead of the $1f$ component as input of the integrator, improved local resolution is observed and parasitic signals are eliminated as they usually have only a predominant $1f$ component.

Theoretical considerations by McMurray and Williams (26) show that, even for a hypothetically indefinitely small point probe, the resolution Δ_{L50} would be limited to about 90% of the tip–sample distance. Here, Δ_{L50} is the lateral distance for which 25 to 75% of the total Volta potential difference between two bordering surface features is measured by SKP. For a measurement of the full potential difference, much larger lateral distances are required, depending on the tip–sample distance and the real tip size (see Figure 16.30). For a 125-μm plane-ended cylindrical tip at a mean probe height of about 100 μm, the measurement of the full potential difference requires about 1 mm! It should be noted that the response might be sharper when the SKP is operated in the second harmonic mode and that these estimations are certainly on the pessimistic side. However, the point is that the measurement of the full potential contrast requires a certain feature size.

16.3.1.5 Tip Preparation

For the application of SKP for corrosion and delamination studies, the Volta potential of the SKP reference electrode should be as stable as possible. Ni/Cr alloy wires (500 μm) have been found to be suitable. In order to prepare tips for higher resolution, the wire is etched, then embedded in a resin, and subsequently ground to a planar and smooth finish. Then the resin is removed. Reference tips with a diameter of 100 μm are easily prepared. In contrast to a tip that is just etched, tips with a flat bottom form, in first approximation, a plate capacitor with the surface, thus providing a well-defined performance.

16.3.2 Measurement of Polarization Curves with a Kelvin Probe as Reference Electrode

If a metal electrode is polarized by a current i, then its potential will shift according to the Butler–Volmer equation and this potential shift may be measured by a Kelvin Probe:

$$\varepsilon_{1/2}(i) = \left\{ \Delta\Phi_{\mathrm{El}}^{\mathrm{Me}}(i) - \frac{1}{F}\mu_{\mathrm{e}}^{\mathrm{Me}} \right\} = \left\{ \frac{W_{\mathrm{Ref}}}{F} - \chi_{\mathrm{El}} - \varepsilon_{1/2}^{\mathrm{standard}} \right\} + \Delta\Psi_{\mathrm{El}}^{\mathrm{Ref}}(i) \qquad (16.13)$$

The polarization, which is defined by the electrode potential in the presence of a current and the open circuit potential (OCP), is directly given by the change of the Volta potential as measured by the Kelvin probe:

$$\eta = \varepsilon_{1/2}(i) - \varepsilon_{1/2}(i=0) = \Delta\Phi_{\mathrm{El}}^{\mathrm{Me}}(i) - \Delta\Phi_{\mathrm{El}}^{\mathrm{Me}}(i=0) \qquad (16.14)$$

Therefore, it is possible to measure and control the polarization of an electrode by passing a current between a counter electrode and the electrode under investigation, and using the Kelvin probe as a reference electrode in a circuit operating similar to a potentiostat (Figure 16.7). This setup is particularly useful if the sample is covered by a thin or an ultrathin electrolyte layer; so the measurement would not be possible by conventional potentiostatic circuits. Obviously, ohmic drops have to be minimized. In the setup shown in Figure 16.7, this is achieved by a thin metallic plate embedded into an epoxy matrix, which also contains two parallel counter electrodes in shallow electrolytic reservoirs. Only the edge of the metallic plate is exposed to the atmosphere; the electrolytic reservoirs are connected via a thin electrolyte film of variable thickness. The thickness of the electrolyte on top of the sample is controlled by the volume of the electrolyte in the counter electrode compartments.

The Kelvin probe is situated directly above the sample and controls the electrode potential via an electronic circuit shown schematically in Figure 16.7. In this setup, the voltage between the sample and Kelvin probe is fixed to a predefined value

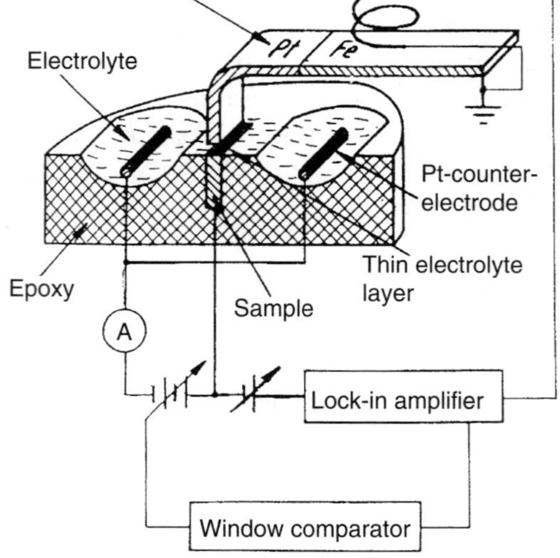

Figure 16.7 Experimental setup for measuring current density potential curves with a Kelvin probe as reference (3–5).

and the output of the lock-in amplifier is used to change the voltage between the counter electrode and the working electrode such that the alternating current between working and reference electrodes vanishes. Therefore, in this setup the external voltage is not adjusted to the Volta potential difference but rather the Volta potential difference is adjusted by a polarization to the external voltage.

The setup allows the measurement of well-behaved polarization curves for electrolyte thicknesses down to 2 μm. The setup has been successfully applied to analyze the kinetics of the oxygen reduction reaction, the kinetics of the metal dissolution reaction, and the kinetics of corrosion inhibition in the presence of very thin electrolyte layers (3–5).

The technique is only limited by the ohmic resistance of submicrometer thick electrolyte layers and by crevices that can disturb the currents for reactive metal electrodes.

16.4 APPLICATIONS

16.4.1 Atmospheric Corrosion

16.4.1.1 Free Corrosion Conditions

Stratmann and coworkers have developed a number of new experimental techniques to investigate the electrochemical reaction mechanism during the atmospheric corrosion in particular of iron and steel under wet/dry conditions. The rate of metal dissolution has been measured by a magnetic balance technique, the rate of oxygen reduction by a gas volumetric technique, and the corrosion potential by the Kelvin probe (3–5). Figure 16.8 shows the rates of metal dissolution and oxygen reduction during a typical wet/dry transition (3). The cyclic corrosion conditions are obvious and allow clear differentiation of the stages of wetting, wet surface, and drying. The highest rates are observed during drying, but further studies have shown that these

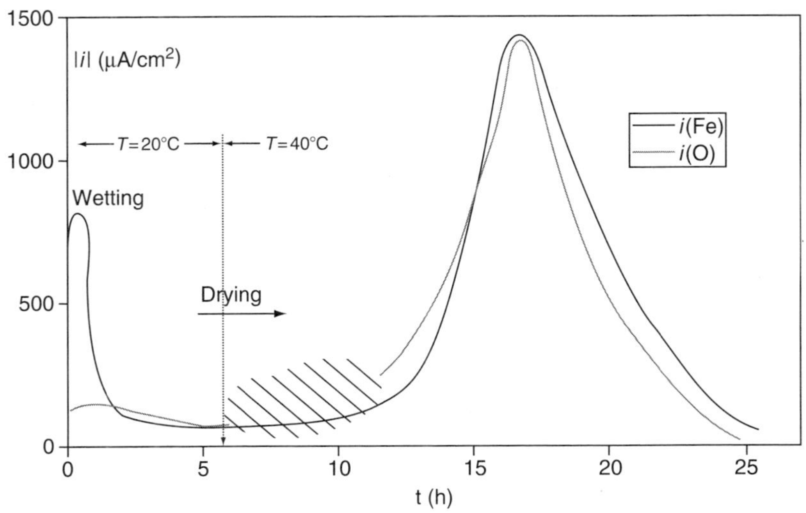

Figure 16.8 Deduced rates for metal dissolution (full line) and oxygen reduction (gray line). The shaded area indicates a time interval of uncertain values for *i*(O), which was deduced from gas-volumetric measurements and is subject to higher error rates when the temperature is increased to 40°C.

rates are directly linked to solid-state reactions, which occur during the first stage of wetting (3). For most of the time, the anodic iron dissolution is compensated by cathodic oxygen reduction. However, for the first peak after wetting, the metal dissolution is significantly larger than the oxygen reduction, the difference being due to the reduction of iron oxide.

The Kelvin probe has been primarily used to investigate the drying stage of the atmospheric corrosion. For these investigations, the Kelvin probe is placed in a small and closed volume and the partial pressure of oxygen is measured with respect to a reference chamber of constant pressure. During a wet/dry transition, the corrosion rate is measured by $\partial p_{O_2}/\partial t$ and the corrosion potential by the Kelvin probe.

Figure 16.9 shows a potential transient measured during a dry/wet/dry cycle of a pure iron surface contaminated with SO_2. In principle, these potential changes correspond to active/passive transitions and it is quite obvious from Figure 16.9 that the electrode potential measured in the wet stage of corrosion differs significantly from that of a surface covered only by some monolayers of water. As corrosion rates can be measured simultaneously with the corrosion potentials, "polarization curves" can be constructed during a wet/dry transition without touching the surface under investigation. Figure 16.10 shows such a plot for a pure iron substrate for repeated wet/dry transitions. It is obvious that active/passive transitions dominate the corrosion process. During the first stage of drying, the corrosion rate increases due to an increasing rate of oxygen reduction. During the second stage of drying, the rate of metal dissolution is reduced, which is accompanied by a further anodic shift of the electrode potential.

Most interesting is the fact that the active/passive characteristics change with increasing number of wet/dry cycles. In particular, for weather-resistant steel, activation is observed during the wet stage of the corrosion cycle and therefore the active/passive transition, which is associated with the highest corrosion rates, disappears. The lack of activation during wetting is best observed on model alloys (Fe–3.4% Cu) (Figure 16.11), but is also typical for technical alloys. This fact, which is seen only by the Kelvin probe (a scanning instrument was not used for these studies), provides a new approach for the design of new and superior corrosion resistant alloys.

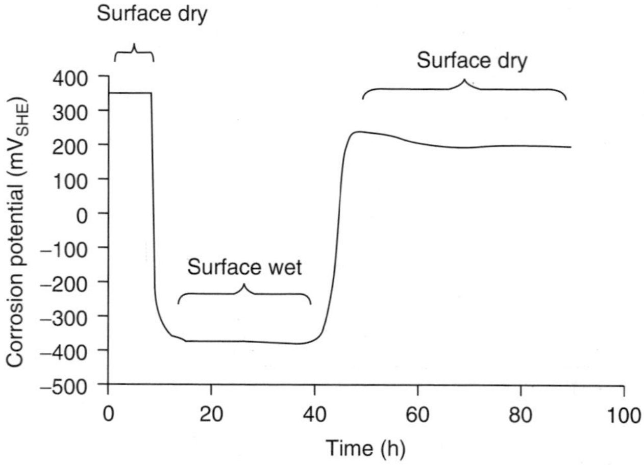

Figure 16.9 Corrosion potential of iron as measured with a Kelvin probe during a dry/wet/dry transition. (From Stratmann, M. (1994), *Die Korrosion von Metalloberflächen unter dünnen Elektrolytfilmen*. Düsseldorf: VDI. With permission.)

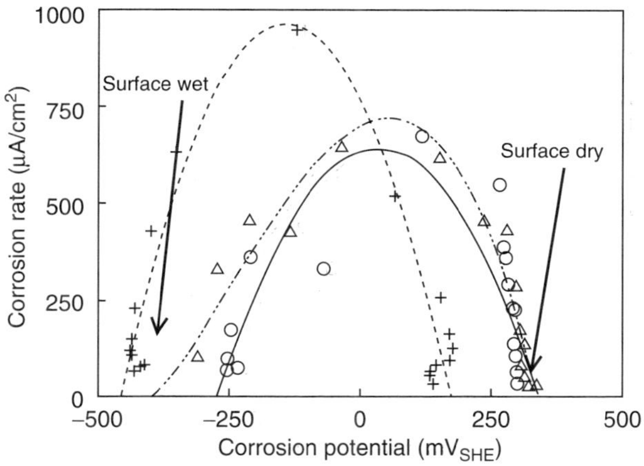

Figure 16.10 Corrosion current as a function of the corrosion potential during the first (+), fourth (△), and seventh (○) dry/wet cycle of an iron surface contaminated with 1 g/cm^2 SO_2. The corrosion rate was measured by $\partial p_{O_2}/\partial t$ (from which the corrosion current is calculated) and the corrosion potential was measured by a Kelvin probe. (From M. Stratmann and H. Streckel, *Corros. Sci.*, 30, 697, 1990. With permission.)

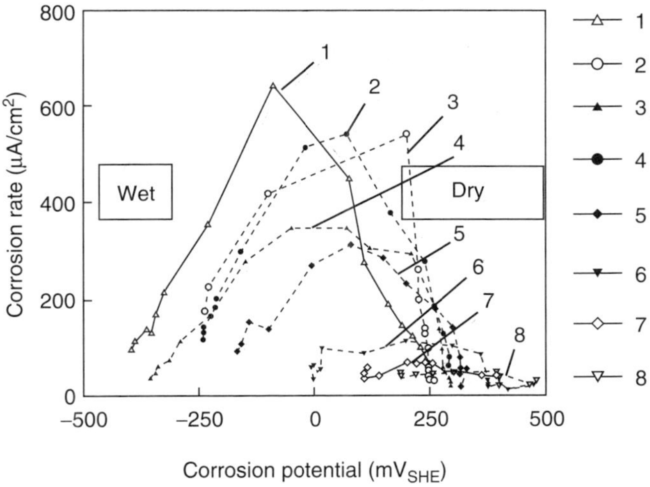

Figure 16.11 Corrosion current as a function of the corrosion potential for several successive wet/dry cycles. For Fe–3.4% Cu with 1 g/cm^2 SO_2. Measured as described in caption to Figure 16.10. (From M. Stratmann and H. Streckel, *Corros. Sci.*, 30, 697, 1990. With permission.)

16.4.1.2 Polarization Curves

Using the Kelvin probe potentiostat described above, well-behaved polarization curves can be measured even for reactive surfaces covered by thin electrolyte layers.

As an example, Figure 16.12 shows polarization curves for oxygen reduction on Pt for constant oxygen activity but varying electrolyte thickness (5). The plot differs for the Tafel regimes of oxygen reduction and water decomposition and the diffusion-limited plateau of the oxygen reduction at medium electrode potentials. The diffusion-limited current density depends on the activity as expected, but on the electrolyte thickness in an unexpected manner (Figure 16.13) (5). For large thicknesses,

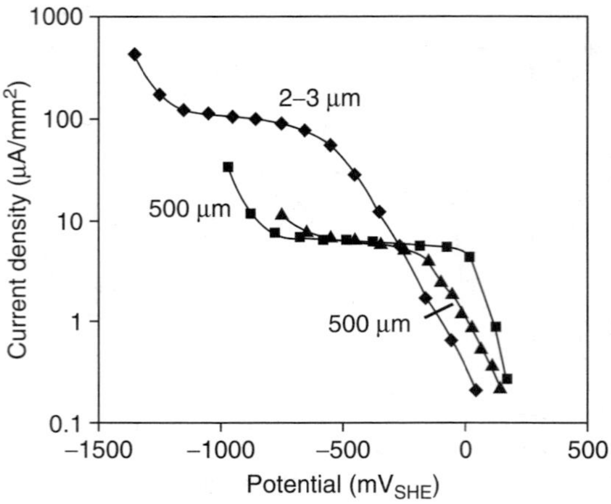

Figure 16.12 Kinetics of the oxygen reduction on platinum as a function of the thickness of the electrolyte (1 *M* Na_2SO_4); $p_{O_2} = 1$ bar. (From M. Stratmann, H. Streckel, K. T. Kim and S. Crockett, *Corros. Sci.*, 30, 715, 1990. With permission.)

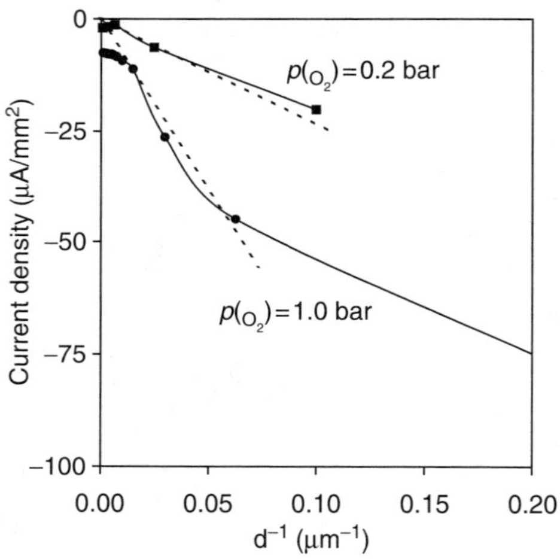

Figure 16.13 Dependence of the diffusion-limited current for the oxygen reduction on platinum (electrode potential: $-0.65\ V_{SHE}$) on the reciprocal thickness of the electrolyte (1 *M* Na_2SO_4). (From M. Stratmann, H. Streckel, K. T. Kim and S. Crockett, *Corros. Sci.*, 30, 715, 1990. With permission.)

the rate of oxygen reduction is independent of electrolyte thickness as turbulences within the thin film define the diffusion length. For electrolyte thicknesses between 100 and 10 μm, the rate is inversely proportional to the thickness (Fick's law) and for even thinner films the rate is again independent of the electrolyte thickness as now the rate of oxygen uptake into the electrolyte is rate determining (5).

Similar polarization curves of oxygen reduction on iron electrodes can be measured, but most interesting are such curves in combination with anodic polarization plots of the metal dissolution reaction (Figure 16.14). Increasing rates of oxygen reduction and decreasing rates of metal dissolution define the corrosion scenario. Corrosion potentials and corrosion rates calculated from these polarization plots demonstrate an excellent coincidence with corrosion rates and potentials as measured under open circuit conditions (Figure 16.15).

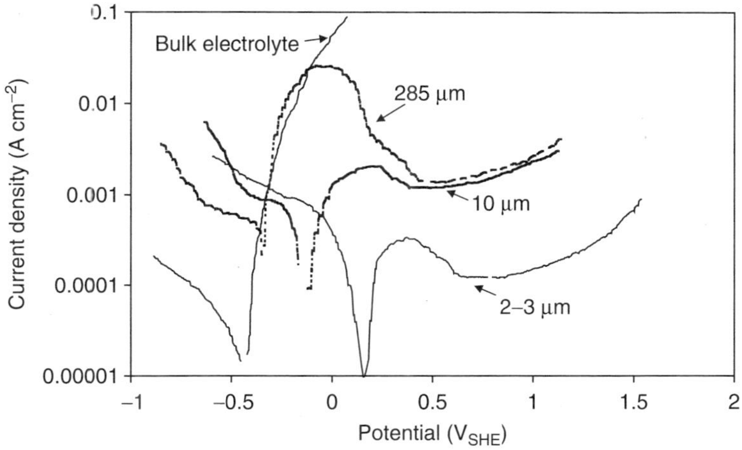

Figure 16.14 Cathodic and anodic polarization curves for iron as measured with a Kelvin probe potentiostat for thin electrolyte layers. (From M. Stratmann, K. T. Kim and H. Streckel, *Zeitschrift fur Metallkunde*, 81, 715, 1990. With permission.)

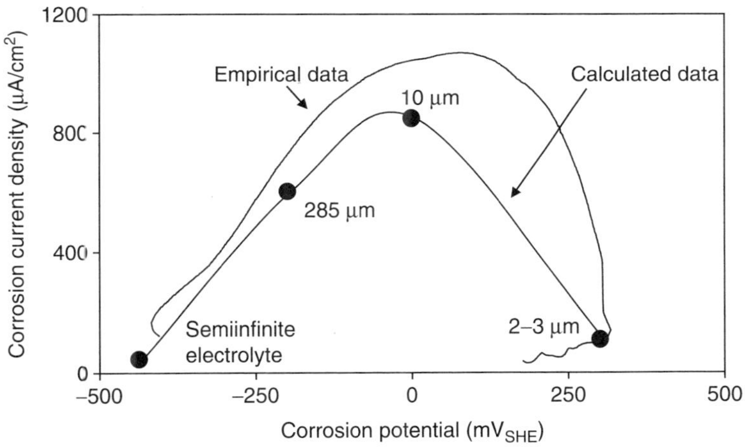

Figure 16.15 Corrosion currents as extrapolated from Figure 16.14 as a function of the corrosion potential in comparison to directly measured values during a wet/dry transition.

The Kelvin probe potentiostat is, therefore, an excellent tool that allows for the collection of kinetic data for atmospheric corrosion conditions and interpretation of the OCPs measured by the conventional Kelvin probe.

16.4.1.3 Galvanic Coupling

The SKP has also been used to map the corrosion potential of samples undergoing atmospheric corrosion (1). In particular, the cathodic protection of steel by zinc (galvanized onto the steel) has been analyzed as a function of the electrolyte thickness and composition (1). A typical result is shown in Figure 16.16 for pure water as electrolyte. The region of the steel surface that is galvanically protected can easily be seen; for electrolyte thicknesses below 60 μm, the cathodic protection breaks down completely due to the increasing rate of oxygen reduction on the iron surface and the increasing electrolytic resistance within the thin electrolyte film.

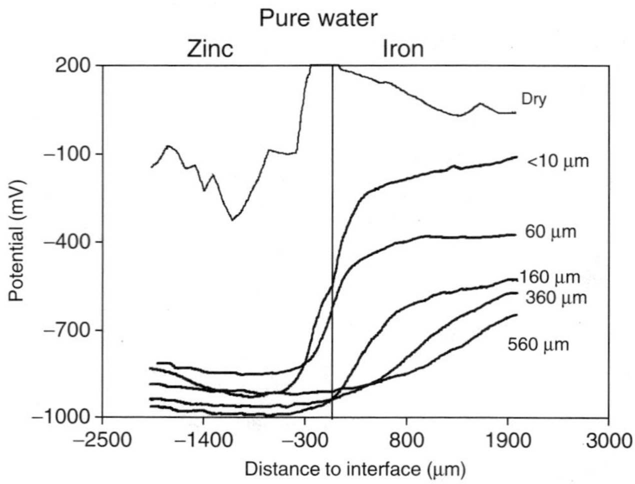

Figure 16.16 Potential profile across the phase boundary iron/zinc as a function of the electrolyte thickness.

These studies are extremely important for the development of metallic coatings, but even more important for the development and improvement of nontoxic pigments in organic coatings that reduce the rate of metal dissolution and oxygen reduction on the bare metal surface (such as cut edges) and therefore allow cathodic protection even for thin and nonconducting electrolyte layers.

16.4.1.4 Corrosion Inhibition

A last example of a successful application of the Kelvin probe for atmospheric corrosion studies is the investigation of vapor-phase inhibitors that are used to protect metallic parts during shipping, storing, etc. Figure 16.17 shows current–voltage curves measured with the Kelvin probe potentiostat for an iron surface

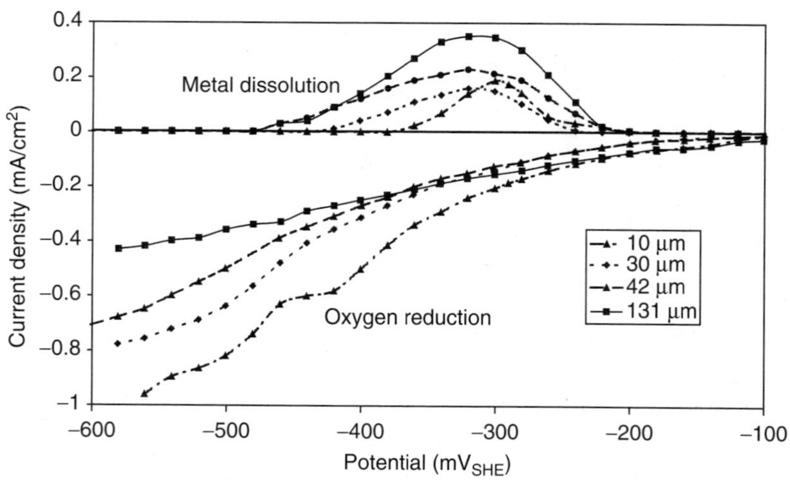

Figure 16.17 Partial polarization curves for iron as measured with a Kelvin probe potentiostat in the presence of the inhibitor ammonium benzoate for electrolyte layers (0.01 *M* Na_2SO_4 with 0.1 *M* ammonium benzoate) of different thicknesses. (From A. Leng and M. Stratmann, *Corros. Sci.*, 34, 1657, 1993. With permission.)

covered with an electrolyte containing ammonium benzoate as inhibitor (6). The upper part shows the anodic metal dissolution and the lower part the cathodic oxygen reduction. With decreasing thickness, the rate of oxygen reduction increases as expected, but also the rate of metal dissolution decreases due to the formation of dense benzoate inhibiting surface layers. Both effects result in an active-/passive-type behavior.

If the sum of both rates is analyzed for a 131-μm thick and a 10-mm thin electrolyte film (Figure 16.18), it is obvious that stable passivity is observed only for the case of thin electrolytes as the cathodic currents are larger than the critical current density a necessary for passivation. Therefore, the inhibitor works properly only for

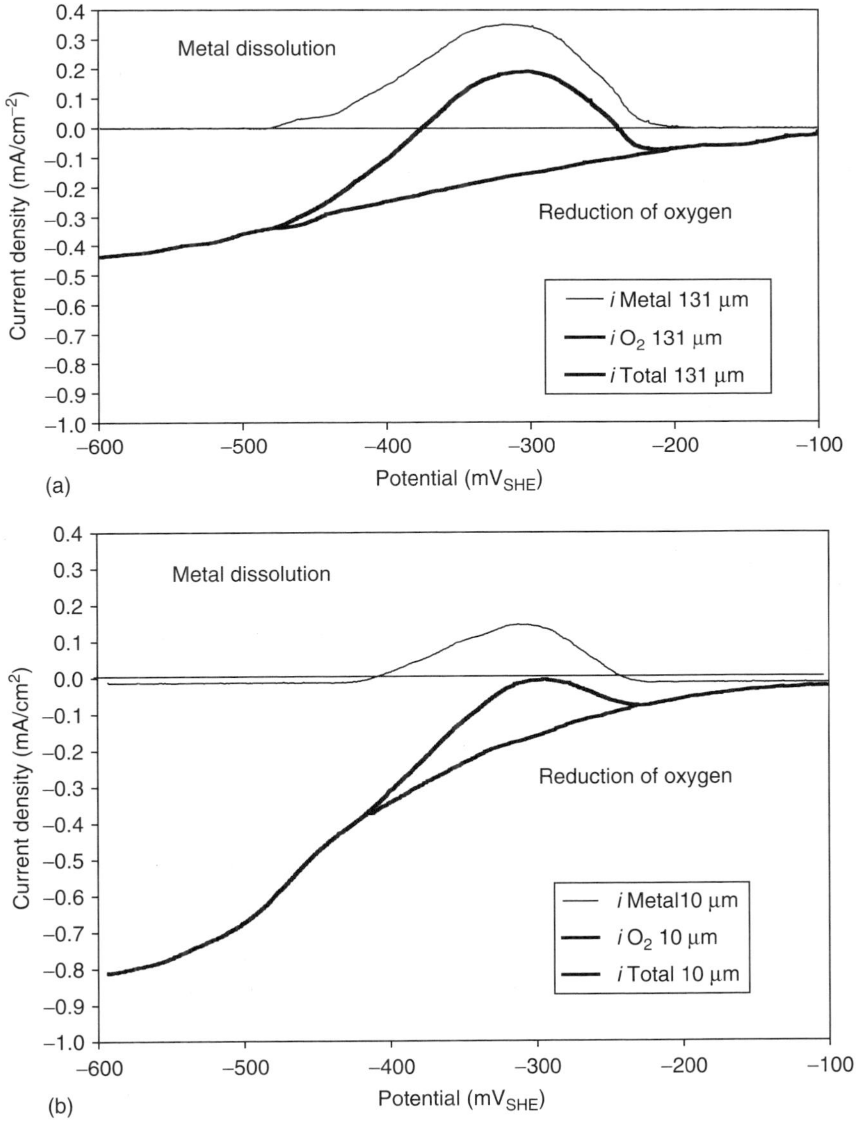

Figure 16.18 Total current (gray line) and anodic (i.e., the metal dissolution [black]) and cathodic (i.e., the oxygen reduction [dark gray]) net currents for (a) 131 μm thick and (b) 10 μm electrolyte film (0.01 *M* Na_2SO_4, 0.1 *M* ammonium benzoate).

those conditions. This is characteristic for a vapor-phase inhibitor, which has to work in the presence of thin and even adsorbed electrolyte layers.

16.4.2 Delamination

16.4.2.1 Corrosion Protection by Organic Coatings

The corrosion rate of reactive metals can be reduced significantly by organic coatings, that is, by paint coatings or even by a modification of the metal surface by only monolayers of organic molecules. Examples are the use of inhibitors or the corrosion protection by lacquers and other organic coatings (27–29), which are used, for example, to protect cars against atmospheric corrosion, pipelines against corrosion in humid soil, and ships against corrosion in sea water. It has long been believed that the corrosion protection is due to the barrier properties of the coating, which impedes the penetration of water and oxygen (30, 31) to the metal/polymer interface. However, many coatings are highly permeable to water and oxygen, and hence it is not the barrier effect on the diffusion process that gives rise to the corrosion stability, but the specific electrochemical properties of the metal/polymer interface; in particular, the formation of an extended diffuse double layer (32, 33). In the presence of defects (pores, pinholes, etc.), which may penetrate through the coating, the diffusion barrier is lowered, and the delamination rate of the polymer at the defect is determined by the formation of galvanic elements (34–38). In many cases, the local anode of this element is the defect and the local cathode, at which predominantly oxygen is reduced, is given by the delamination frontier. This process is called *cathodic delamination*. In other cases — in particular for Al-based alloys — the galvanic element is just opposite and the delamination frontier is characterized by the local anode. This is called *anodic delamination*, often occurring as *Filiform corrosion*. In the first case, the stability of the metal/polymer interface is determined by inhibition of the oxygen reduction at this interface, because during the electrochemical reduction of oxygen very aggressive species are liberated ($OH^{\cdot}$, OH^{-}), which will immediately attack and destroy chemical bonds within the polymer (39–42). The corrosion inhibition of the coating, therefore, depends more on the composition, structure, and chemical bonds of the metal/polymer interface than on the thickness of the coating.

SKP is an ideal tool for the *in situ* monitoring of the degradation process at the buried interface. In the following, model experiments are discussed that elucidate the mechanisms of pure cathodic and pure anodic delaminations. However, it should be mentioned that pure anodic or cathodic delaminations do not occur in real life. In fact, in most technical systems the pure cases are the exception and, of course, delamination is usually very slow, which makes *in situ* SKP studies very tedious. This is why industry tests, for example, outdoor exposure, usually are applied on the fully developed coatings (including extensive pretreatments of the surface, such as phosphate conversion coating and chromate rinse), whereas the SKP has been applied up to now only to the most simple coatings without further surface treatment and under constant exposure conditions. The industry tests usually include cyclic climatic conditions, causing also cyclic active–passive–active transitions. Obviously, a huge gap exists between the laboratory experiments and real systems. The first experiments with SKP under technical conditions and on fully developed systems have been carried out only recently. Rohwerder et al. (43) studied the delamination of technical samples under cyclic humidity conditions. These first experiments with a specially developed new Kelvin probe give only the first hints on the complex delamination mechanism, but they already demonstrate the importance also of anodic reactions on the stability of the interface. Further investigations are necessary.

Williams and McMurray (44) used the SKP technique to study the influence of chromate (CrO_4^{2-}) on the kinetics and mechanism of the delamination processes affecting polyvinyl butyral (PVB) coatings on hot dip galvanized steel. Soluble chromates in the electrolyte in the prepared defect reduced delamination rates by less than 25%, because CrO_4^{2-} anions were excluded from the underfilm electrolyte layer by the delamination cell electric field (see Section 16.4.2.2). In contrast, dispersions of particulate $SrCrO_4$ in the PVB coatings allowed CrO_4^{2-} diffusion directly into the underfilm electrolyte layer and profoundly inhibited delamination. It was proposed that replacement of underfilm O_2 reduction by a self-limiting CrO_4^{2-} reduction process is the most significant factor in decreasing delamination rates.

The effect of cerium cations in the electrolyte on the defect or in dispersions of silica and bentonite-based cerium cations in the PVB coating showed similar behavior. With the help of SKP, Williams et al. (45) proved that the outward migration of hydroxide ions in the electric field alongside the interface stopped inward migration of cerium cations from the defect, while the cerium cations from the dispersed silica and bentonite in the PVB coating reached the active sites and effectively inhibited coating delamination. Williams et al. (46) also successfully applied SKP for the study of filiform corrosion.

In the following, more fundamental work on the delamination of model coatings from galvanized steel and on the filiform corrosion on Al performed with SKP is summarized.

16.4.2.2 Cathodic Delamination

Cathodic delamination always starts at defects within the coating. At an isolated metal/electrolyte interface, metal is oxidized (anodic reaction) and oxygen is reduced (cathodic reaction) at comparatively high rates. The electrode potential is the value where the sum of anodic and cathodic currents equals zero. In this case, the diffusion rate of oxygen through the electrolyte layer limits the corrosion rate. At an isolated intact metal/polymer interface, the kinetics for electrochemical reactions are very low, metal oxidation being completely inhibited and no electrons being available for oxygen reduction. This results in a positively charged interface, where the electrical potential does not allow any further oxygen reduction. If these two interfaces are brought into contact, for example, via a defect in a coating, electrons flow through the metal from the defect to the intact interface, where they are available for oxygen reduction, provided there is ionic conduction along the interface to complete the circuit. Consequently, the potential at the polymer/metal interface in the defect shifts significantly more negative. During this process, the diffusion of cations and anions along the interface turns into a nearly exclusive transport of cations from the defect to the delamination site in order to compensate the flow of electrons (see Figure 16.19). Anion transport is inhibited by the electric field along the interface (8–10, 47).

Intermediate products of the oxygen reduction are radicals such as $OH^{\bullet}$. Radicals are aggressively reactive molecules that are likely the main cause for the degradation of the polymer/metal bond, that is, the adhesion between polymer and metal gradually diminishes in their presence (10). When this happens, the interface turns into a delamination site. The major product of the oxygen reduction is OH^{-},

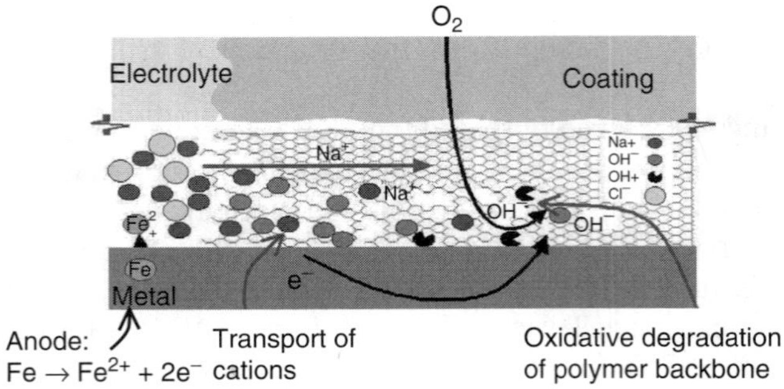

Figure 16.19 Schematic presentation of the mechanisms of cathodic delamination. The flow of electrons through the metal to the delaminating interface is compensated by the migration of cations in the electric field alongside the interface. Side products of the oxygen reduction are radicals such as $OH^{\bullet}$ that destroy the chemical bonds at the interface.

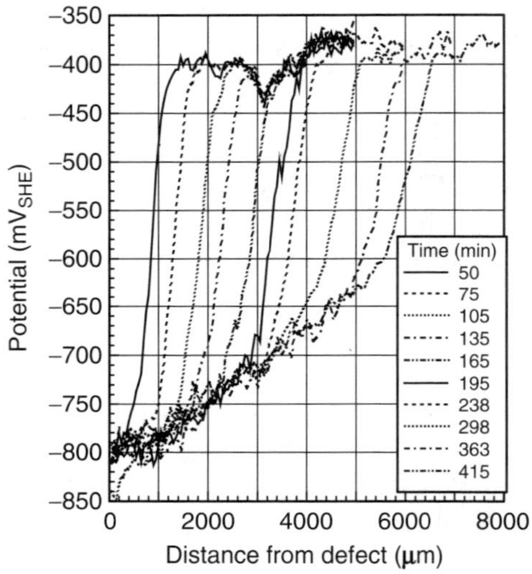

Figure 16.20 Potential profiles measured during delamination from a galvanized steel surface (defect only down to zinc).

which significantly increases the pH at the interface. The high pH at the interface contributes to the polymer degradation by saponification. If iron is the substrate material, it is passive at such high pH values and at the electrode potentials typical at the delaminating interface. As a result, iron dissolution does not occur, even though the bond between polymer and metal is gradually destroyed. The only anodic reactions occur at the defect site itself (8–10).

This is different if the substrate is zinc or galvanized steel. For galvanized steel, two situations have to be distinguished: (a) the defect penetrates only to the zinc surface so that zinc is present at both the metal/polymer interface and inside the defect; and (b) the defect penetrates to the steel substrate, which sets up an additional galvanic element between zinc at the interface and iron at the defect, the interface being the sacrificial anode necessary to cathodically protect the iron defect (48–50).

The first case is similar to the delamination of organic coatings from steel (48). A typical potential profile measured during delamination is shown in Figure 16.20. The interpretation of these potential profiles is similar to those of a polymer-coated steel substrate. At the intact interface, the electrode potential is rather high. The situation described above changes completely if ions are present at the substrate/polymer interface. Then a finite ionic conductivity is established, resulting in a galvanic coupling between the intact interface and the defect site. This makes oxygen reduction at the intact interface possible, the electrons for this cathodic reaction being provided by an enhanced anodic metal dissolution at the defect. This change of reactivity results in a negative shift of the electrode potential at the previously intact polymer interface, as measured by the SKP (see Figure 16.20).

An alkaline electrolyte forms at the intact substrate/polymer interface as a consequence of the preferential oxygen reduction, whereas the electrolyte at the defect becomes acidic. As mentioned above, the adhesion of the polymer coating on the metal is destroyed by the side products of the oxygen reduction. Due to the galvanic current along the delaminating metal/polymer interface, an ohmic potential drop is observed between the local anode (defect) and the front of the delaminated zone (cathode). Obviously, the potential drop increases with increasing distance to the defect. It is not yet understood why this ohmic drop shows up as a linear increase of the potential (Figure 16.20). Since the galvanic current of the delaminated interface near the defect should be higher than that further away from the defect, a nonlinear increase should be expected. This remains as one of the open questions to be solved, as well as an explanation for the observed tendency of cathodic delamination to proceed in distinctive steps (51).

As zinc is amphoteric, subsequent reactions will start in the alkaline electrolyte at the interface. These reactions differ significantly from the case of iron, which is rather stable under alkaline conditions. This becomes obvious when the potential profiles along the interface are measured after a sudden drop of the oxygen activity in the atmosphere. Figure 16.21 shows such profiles measured on a sample that had been delaminated in air to a length of approximately 7 mm (see Figure 16.20) (48). The

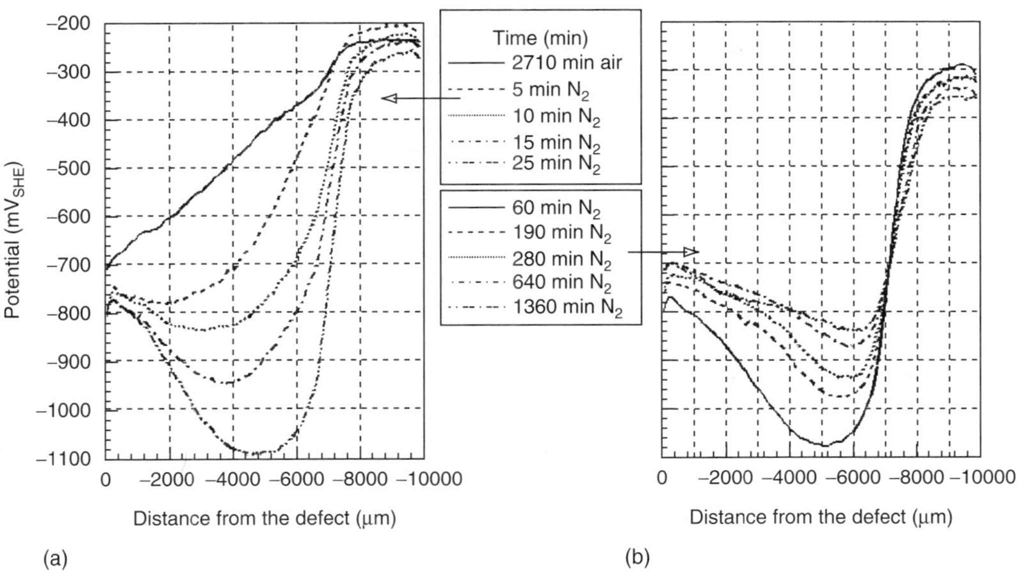

Figure 16.21 Potential profiles during delamination of a polymer from galvanized steel (a) before and (b) after a sudden decrease of the oxygen activity in the atmosphere. (From W. Fürbeth and M. Stratmann, *Corros. Sci.*, 43, 207, 2001. With permission.)

potential at the delamination frontier drops quickly and then slowly relaxes back to a value similar to that of the defect. In the absence of oxygen and therefore in the absence of any faradaic current, the Kelvin probe measures the local equilibrium potential. Therefore, the huge potential shifts shown in Figure 16.21 reflect a local change of the chemical environment at the interface, which changes with time and distance to the defect. Immediately after removing oxygen from the atmosphere, the chemistry at the interface is still given by the galvanic element, which had been in operation just before the exchange. The decrease in potential at the delamination front to a value more negative than the potential at the defect is easily explained by the alkaline pH, which stabilizes a very negative equilibrium potential. The observed potential of −1.1 V SHE (see Figure 16.21a) would reflect a pH of 11 according to the potential–pH diagram of the zinc/water system (52). If now the atmosphere is kept oxygen-free for a longer time, the pH gradient will lead to a diffusion of hydroxyl ions from the delaminated area to the defect. This results in a decrease of the pH in the delaminated area toward near-neutral values, so that after longer times the equilibrium potential of the delaminated zone and the defect will be rather equal (see Figure 16.21b).

If the defect penetrates to the steel substrate, the potential profiles show two plateaus (Figure 16.22) (49). The first one corresponds to the anodic dissolution of zinc and the second one to the cathodic delamination front, which precedes the zone of anodic dissolution. Cathodic delamination takes place even if iron is exposed to the electrolyte in the defect. As soon as the interface is destroyed by cathodic delamination, the zinc starts to cathodically protect the defect and then a local anode is formed within the delaminated zone close to the defect. The original defect site is then decoupled from the cathodic delamination site, because the corroding zinc in the already delaminated area causes a lower potential than the corroding steel in the original defect. Hence, cation migration from the original defect to the cathodic delamination site is no longer possible. The corroding zinc then fuels the cathodic delamination frontier. This can be seen clearly when the anodic and cathodic delamination frontiers are plotted versus time. The cathodic delamination front slows down when it is far away from the anodic site, and it speeds up again when the anodic site catches up (Figure 16.23) (49).

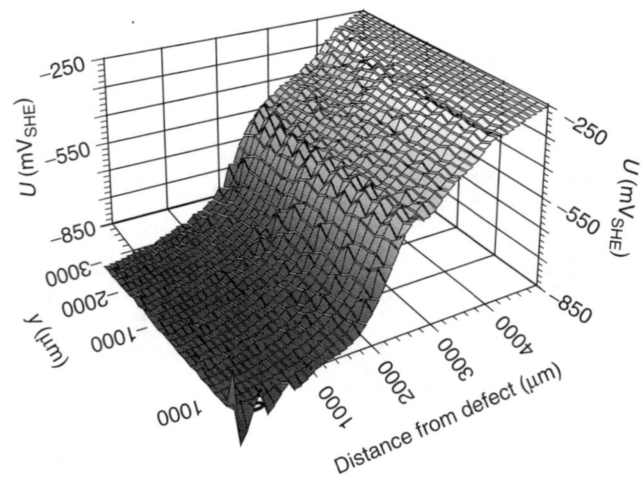

Figure 16.22 Potential map during delamination for galvanized steel, the defect being prepared down to the steel substrate. (From W. Fürbeth and M. Stratmann, *Corros. Sci.*, 43, 229, 2001. With permission.)

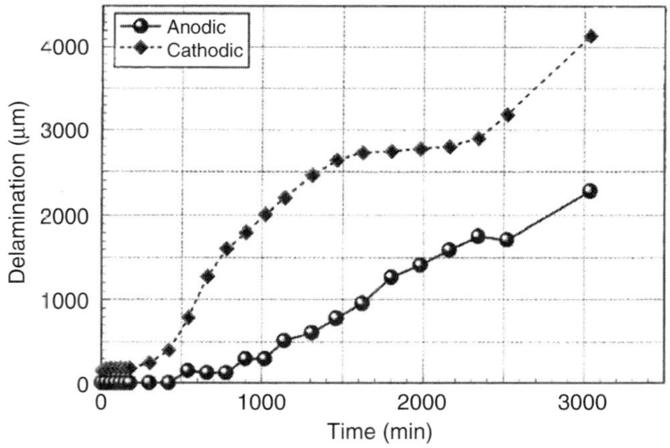

Figure 16.23 Delamination kinetics for the cathodic and the anodic delamination fronts for a defect down to steel. (From W. Fürbeth and M. Stratmann, *Corros. Sci.*, 43, 229, 2001. With permission.)

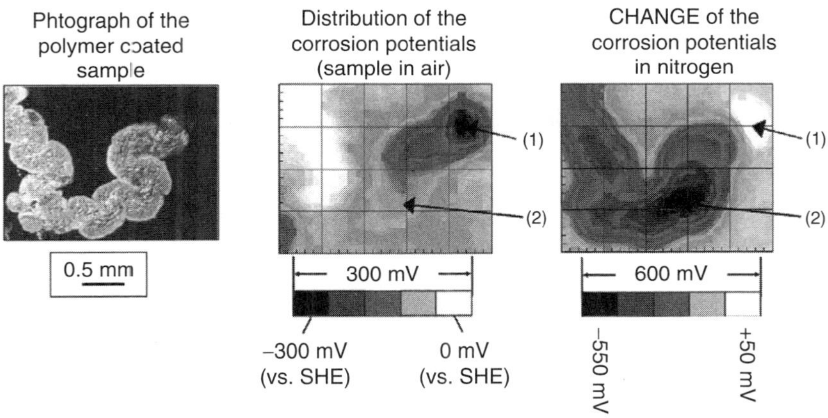

Figure 16.24 Filiform corrosion on polymer-coated aluminum alloy. Left: photograph; middle and right: corrosion potential distributions in air (middle) and map of potential change after switching to nitrogen atmosphere (right) as measured with the SKP at constant RH (90%). The local anode (1) is clearly visible as active area (negative potential) in the potential map taken in air; the local cathode (2) is not visible from this map, but clearly shows up by a large shift toward more negative potentials when the atmosphere is switched to nitrogen (right). The tail of inactive corrosion products is marked by positive potentials in air.

16.4.2.3 Filiform Corrosion

Homogeneous delamination is not observed on Al substrates as described above for steel or galvanized steel substrates. Instead, delamination takes place locally, mostly as "filiform corrosion." The electrochemical mechanism of filiform corrosion is described by a differential aeration cell between the front of the filament's "active head" (low oxygen concentration) and the back (open to air through the cracked/porous tail of dry corrosion products). The head is the local anode and the tail the local cathode, which is counter to the cathodic delamination described earlier. Several publications on SKP studies of filiform corrosion convincingly proved this concept (46, 53, 54). In Figure 16.24, an optical photograph of the filiform head is compared to the potential distribution in air (center) and to a map of the potential change when switching to nitrogen (right) (53). In humid air, the potential distribution indicates an area of more negative corrosion potentials at the front of the filament, followed by a more positive area immediately behind it. This kind of potential distribution is expected for a galvanic corrosion cell with a local anode (metal dissolution; more negative corrosion potential) and a local cathode (oxygen reduction; more positive potential). While the local anode (1) is clearly visible as active area (negative potential) in the potential map taken in air, the local cathode (2) is not visible, but clearly shows up by a large shift toward more negative potentials when the atmosphere is switched to nitrogen (image on the right-hand side in Figure 16.24). While a potential shift toward more negative values is always expected for cathodes when the atmosphere is switched to oxygen-free atmosphere, the magnitude of the shift (nearly 600 mV) indicates that this shift may be due to the high pH caused by oxygen reduction at the local cathode. Under high pH conditions, the Al electrode will have a rather negative electrode potential.

This shows that SKP is an ideal tool for the investigation of filiform corrosion. Especially the initial stages of filiform corrosion and an explanation of the seemingly erratic filamant growth are of interest for future research. For the case of PVB-coated Al alloy, Williams et al. (46) observed with the help of SKP that, in the initial phase, areas between 0.1 and 0.25 mm^2 from the vicinity of the prepared defect show a decrease in potential (see Figure 16.25). These regions of decreased potential are consistent with a depassivation of the aluminum surface. They are the initial phases of filament heads (46).

Concerning the course of the filament, it is important to consider that filiform corrosion is observed on Al alloy substrates, but usually not on pure Al. Al alloys often contain inclusions (e.g., intermetallic iron- or copper-rich phases), and it is generally accepted that the oxygen reduction takes place at the intermetallic inclusions. These features might also determine the direction of the seemingly erratic filiform growth. SKP could help to solve this open question. However, since these surface features are in the micrometer or even submicrometer range, the resolution of standard SKP is not sufficient. SKPFM might be the right tool.

16.5 SCANNING KELVIN PROBE FORCE MICROSCOPE

16.5.1 Introduction

The development of techniques linked to the atomic force microscope (AFM) has enabled the evaluation of physical and chemical properties of submicrometer structures. The advantages of AFM include the ability to provide quantitative information regarding the topography of a surface and to perform those measurements *in situ*. However, perhaps the real power of AFM is that it forms the basis for associated techniques that provide enhanced capabilities. One such technique is SKPFM, which was first introduced by Nonnemacher et al. (56, 57) for microelectronic applications. SKPFM has only recently been applied to corrosion studies (17–20, 55).

SKPFM is an AFM-based technique with submicrometer resolution, and it is a robust technique that is relatively quick and simple to perform. It simultaneously provides topographic and potential maps of a region on a sample surface. This section will summarize findings from a number of recent publications. The nature of the potential measured by SKPFM will be discussed and it will be shown that the

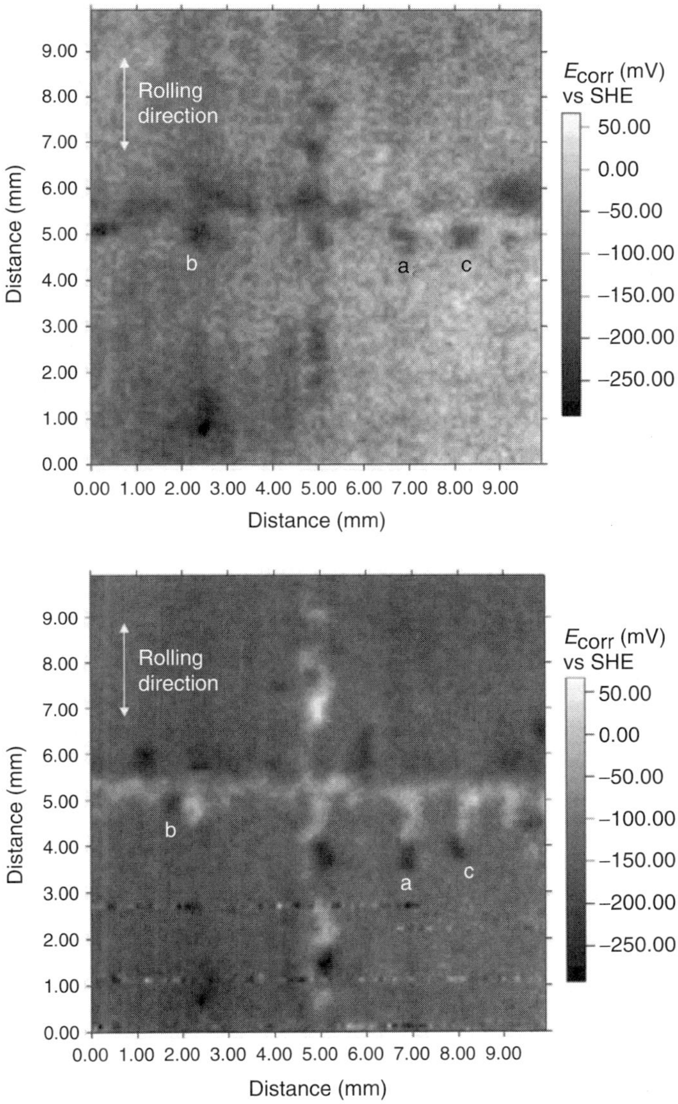

Figure 16.25 Evolution of the potential distribution on PVB-coated AA2024-T3 aluminum alloy at 50°C and 95% RH following transient exposure to HCl vapor (left image after 10 h, right image after 35 h). Areas of decreased potential become filiform heads (a)–c). Areas in the wake of the advancing filament heads passivate significantly with time. (From G. Williams, H. N. McMurray, D. Hayman and P. C. Morgan, *PHYSCHEMCOMM* 6, 1, 2001. With permission.)

overall nature justifies the description of it as the Volta potential difference and the use of the name Kelvin probe.

The principle and details of the SKPFM measurement are described in detail in Refs. (17, 18, 58–60). In short, it involves scanning the surface in tapping mode to determine the topography on a line-by-line basis. The metal-coated or doped silicon cantilever is then lifted a fixed distance from the surface, typically 100 nm, and the tip is rescanned across the surface at this fixed height in "lift mode." On the rescan, the tapping piezo is turned off, but an AC voltage is applied to the tip, which stimulates oscillations of the cantilever in the presence of an electric field. The magnitude of the oscillations at the stimulating frequency ω, monitored by the AFM detection scheme, is nulled on a point-by-point basis during the lift mode rescan by adding to the tip a DC voltage that balances the field. The total electric force on the tip is equal to (61):

$$F_e = 1/2\left\{\left(V_{DC} - \Delta\Psi^{Ref}_{Sample}\right)^2 + 1/2V^2_{AC}\right\}C' + C'\left(V_{dc} - \Delta\Psi^{Ref}_{Sample}\right)V_{AC}\sin(\omega t) + 1/4C'V^2_{AC}\cos(2\omega t)$$

The force modulation at frequency ω becomes zero for $V_{DC} = \Delta\Psi^{Ref}_{Sample}$; however, a DC component and a component at 2ω remain.

This approach of potential distribution measurement is not possible in an aqueous solution, because the large voltages applied to the tip cause faradaic reactions in solution. In the following, all potential mappings were performed in air. The tips are only pseudoreferences since their potential may vary with changes in their surface oxide.

Furthermore, SKPFM apparently does not measure the full potential contrast between two surface features, but only a fraction of it (61, 62). The main reason for this is the effect of stray capacitance, derived from the entire cantilever area, and not simply the tip, on the Kelvin signal. The passage of the cantilever over substrate areas far from the tip, with inhomogeneous topographic and compositional features, can significantly influence the Kelvin signal and result in incorrect results. Hochwitz et al. (61) report deviations by a factor of two or more.

Additionally, the SKPFM generates a rather large AC voltage modulation. The adjustment of the external DC bias between the tip and the sample in order to nullify the ω component of the electric force exerted on the tip, that is, to compensate the Volta potential difference between the tip and the sample, does not lead to a field-free sample/air interface. On the contrary, high electric AC fields are active, in the order of 5 V/10 nm. These extremely high fields might have an effect on the physical conditions of the surface, especially for semiconducting samples.

A final problem with SKPFM is the limitation of the field of view. AFM piezoelectric scanners typically are limited to 100 μm in range or smaller. Relevant electrochemical processes sometimes take place on a larger scale more appropriate for a standard SKP instrument.

Despite these concerns, and a still incomplete understanding of the origin of the signal generated from the SKPFM, measurements made in air can provide important information regarding the corrosion behavior in solution.

16.5.2 SKPFM for Corrosion Studies

16.5.2.1 Motivation

SKPFM is particularly useful for studying corrosion and delamination phenomena on a microscopic or even submicroscopic scale. SKPFM generates a map of the potential distribution across a sample with a resolution of at least 100 nm. This allows the study of alloy samples with heterogeneous microstructure, such as AA2024-T3. As will be shown, the OCP of various pure metals in solution seems to be linearly related to the Volta potential value measured in air immediately after exposure. Assuming a correlation between the measured potential distribution across a surface and relative practical nobility, this approach is useful for understanding the interaction between different regions of a heterogeneous sample, such as the different types of intermetallic particles and the matrix in AA2024-T3. The shape, position, compositional inhomogeneities, and local nobility of these particles can be measured.

In general, pitting corrosion is usually driven by microstructural heterogeneities (63). For instance, in Al alloys, alloying elements added for increased strength are often segregated to and enriched in intermetallic particles (64, 65). Such particles can be large constituent phases on the order of tens of micrometers in size, or precipitated hardening particles nanometers in size. Localized corrosion typically initiates at the larger particles (micrometer size or larger), but the role of the particle in the localized corrosion process depends on the particle type. In AA2024-T3, there exist two primary types of large intermetallic particles: AlCuFeMn particles and S-phase Al_2CuMg particles (64, 65). The FeMn-containing particles have a range of composition and are often themselves heterogeneous. These particles are typically considered to be cathodic to the matrix (66). The S-phase particles are more homogeneous, and are thought to be active relative to the matrix owing to the high Mg concentration. It has been suggested that Mg and Al can dealloy from S-phase particles, leaving a porous Cu-rich residue that might break apart and redistribute Cu across the sample surface, providing a large active cathode (64).

Clearly, it is of interest to be able to understand the exact role of these intermetallic particles in the localized corrosion process on Al alloys. Owing to their small size, techniques with high spatial resolution are required to do so. A number of standard techniques with submicrometer resolution exist, such as scanning electron microscopy (SEM), energy dispersive x-ray analysis (EDS), Auger electron spectroscopy (AES), and transmission electron microscopy (TEM). The advantage of SKPFM is that it can be applied *in situ* and gives simultaneous information about topography and Volta potential.

16.5.2.2 Calibration

The tips are only pseudoreferences since their potential may vary with changes in their surface oxide. In order to avoid errors associated with variations in the tips or instabilities in the instrument electronics, the potential measurements have been calibrated by comparison to the potential measured on a pure Ni surface after immersion in double ionized (DI) H_2O because Ni has a stable potential (17). Consecutive measurements in air on a stable reference sample using different tips coated with the same metal showed potential difference of less than 50 mV, giving an indication of the reproducibility of this method.

Experiments on Pure Metals. In order to calibrate and to determine the usefulness of SKPFM, measurements were made on a number of pure metal samples. The samples were exposed to DI water or aqueous solutions at open circuit for 30 min, removed, rinsed, and dried prior to measurement of the Volta potential in air. The values were compared to the OCP values measured on those samples prior to removal from deionized water, 0.5 M NaCl or 0.1 M Na_2SO_4 solution, Figure 16.26 (17, 55). In this figure, Volta potentials are reported versus the potential measured on a pure Ni sample, which served as a calibration standard. The potential measured by SKPFM is seen to correlate directly with the OCP measured in solution for metals with a wide range of practical nobility. This signal is thus a measure of practical nobility, with the advantage that it can be mapped over a surface with high spatial

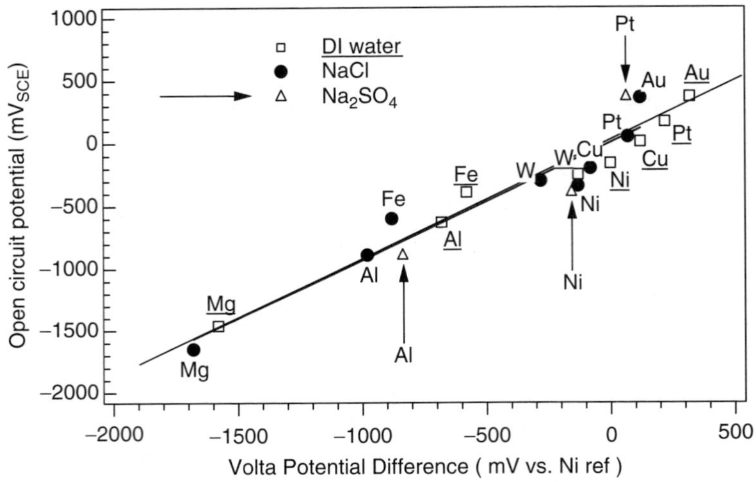

Figure 16.26 Comparison of the potential measured in air by scanning Kelvin probe force microscopy with open circuit measured in solution: in DI water (squares and underlined elemental symbols), in 0.5 *M* NaCl solution (filled dots), and in 0.1 *M* Na_2SO_4 (triangles and indicated with an arrow). (From V. Guillaumin, P. Schmutz and G.S. Frankel, *J. Electrochem Soc.*, 148, B163, 2001. With permission.)

resolution. This correlation between measured Volta potential and OCP also holds for a single metal exposed to different solutions in which it exhibits different OCPs. For samples exposed to the chloride and sulfate solutions, both the OCP in solution and the Volta potential difference in air after solution exposure were shifted in the active direction by around 150 mV relative to the values measured in and following DI water exposure. This suggests that adsorption of charged species at the electrode surface in chloride or in sulfate changed the dipole structure in the double layer and influenced the measured Volta potential measured in air by the same value.

To distinguish the various components of the potential, Ni and Pt samples were held at an applied cathodic potential in sulfate solution and then withdrawn from solution under potential control (55). For Ni, the measured Volta potential exhibited a slow decay in air associated with discharge of the oxide and a more permanent component associated with adsorbed species. This more permanent component was similar in magnitude (~150 mV) to the difference in Volta potential observed after OCP immersion in the sulfate solution relative to the value after immersion in DI water. The decay was much faster for a Pt electrode, because it does not carry an oxide layer. However, the same permanent offset was observed. These studies validate the use of SKPFM, which must be applied *ex situ*, for relating the potential distribution across metal alloy surfaces to their corrosion behavior, particularly for oxide-covered metals, such as are often of interest in corrosion studies. As will be shown below, the corrosion behavior of alloys can be related to the Volta potential distribution.

According to theoretical considerations, the measured Volta potential should vary with tip–sample separation distances below 100 nm owing to the influence of image or dipole charges at the electrode surface (67). Ni shows almost no distance dependence over the full range of separation distances (55). The potential on Ni is also independent of the time in air. This suggests the nickel oxide surface is very stable, and is the reason why Ni was chosen as a reference for the Volta potential difference measurements. Other materials, for example, Al, show distinct dependencies on distance. The reasons for the different behavior are not fully understood yet.

However, the fact that the potential measured by SKPFM for distances greater than 100 nm from the surface is constant corresponds to the conceptual definition of the Volta potential difference for a metal/solution interface. So the overall nature of the potential measured by SKPFM justifies the description of it as the Volta potential difference.

16.5.2.3 Spatial Resolution of SKPFM on Heterogeneous Surfaces

The advantage of the SKPFM over standard scanning Kelvin probes (2, 3) is the improved spatial resolution owing to the small size of the probe and the separation of the probe from the sample surface. Figure 16.27 shows topographic and potential maps for a region of an as-polished sample of AA2024-T3 that was prepared without exposure to water to minimize corrosion during the preparation (17). Considerable debris resulting from the nonaqueous polishing and rinsing of the sample is visible in the topographic image on the left. Also visible are scratches and some holes that might have resulted from corrosion or particle pullout. Finally, there are some raised regions associated with intermetallic particles that are harder than the alloy matrix and polish slower. On the right is the Volta potential map of the same region. The intermetallic particles are clearly evident with high contrast. Figure 16.28 shows an SEM image of the same area (17). The bright spots in the SEM image were found by EDS analysis to be intermetallic particles. They match up perfectly with the high potential features in the potential map of Figure 16.27. The large, blocky particles are the AlCuFeMn type and the three round particles, which are barely visible in the SEM image, are S-phase particles. The secondary electron intensity in the SEM image is a convolution of topographic and chemical (z-number) effects. In contrast, the topographic and potential maps produced by the SKPFM separate these effects.

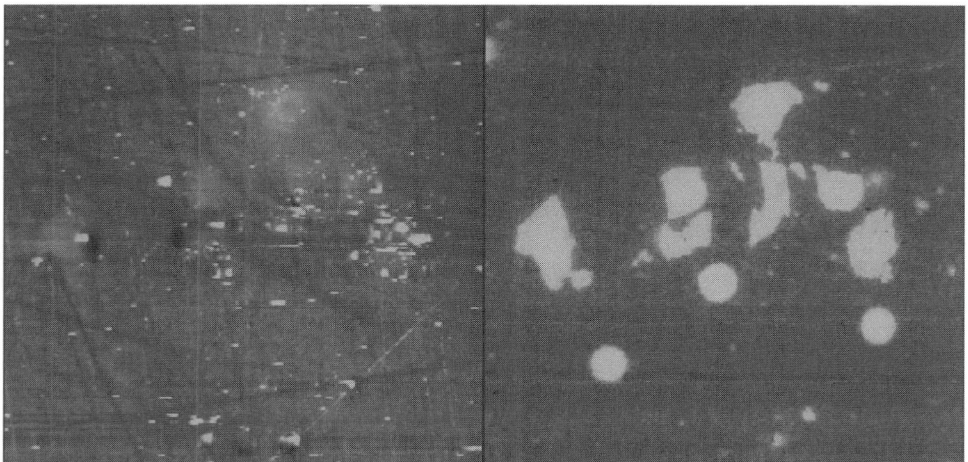

Figure 16.27 AFM image of intermetallic particles in AA2024-T3. Both images are $80 \times 80\,\mu m$. Left, topographical map with 100 nm z-scale; right, Volta potential map with 1.5 V z-scale. (From P. Schmutz and G. S. Frankel, *J. Electrochem. Soc.*, 145, 2285, 1998. With permission.)

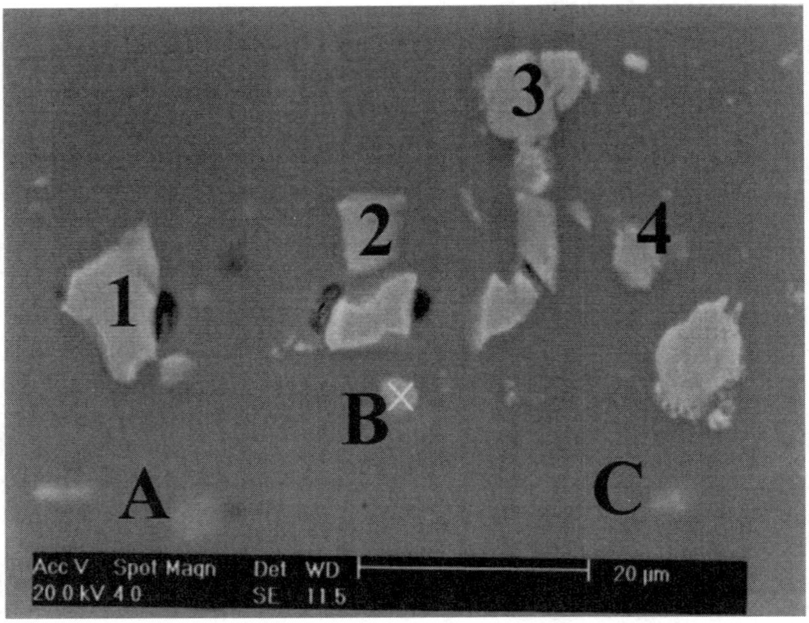

Figure 16.28 SEM image of the same region of the same sample as shown in the figure. (From P. Schmutz and G.S. Frankel, *J. Electrochem. Soc.*, 145, 2285, 1998. With permission.)

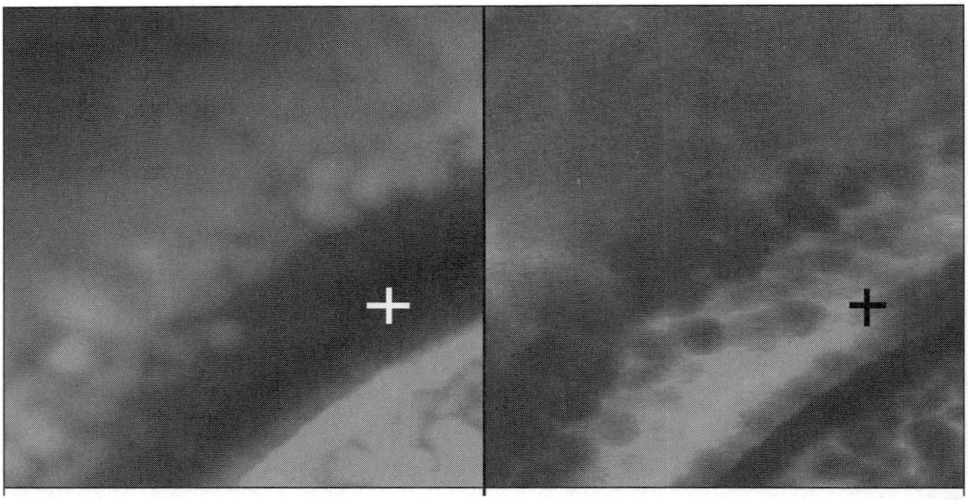

Figure 16.29 Topographic (left, *z*-scale 600 nm) and Volta potential (right, *z*-scale 0.5 V) maps of a region of AA2024-T3 sample exposed to 0.5 *M* NaCl for 10 min. The area of both maps is 1 × 1 µm. An AlCuFeMn particle is located at the bottom right of the images. (From P. Schmutz and G. S. Frankel, *J. Electrochem. Soc.*, 145, 2298, 1998. With permission.)

The limits of the spatial resolution capability of the SKPFM have not been fully investigated. Figure 16.29 shows an example of a small scan taken at the edge of an AlCuFeMn particle after exposure of the sample to 0.5 *M* NaCl for 10 min (18). Trenching of the matrix around the cathodic particle was observed as well as some attack within the particle. The region shown in Figure 16.29 is the trench along the side of the particle, which is in the bottom right part of the figure. The trench is seen to be associated with a high potential, likely as a result of Cu enrichment. Potential and topographic features smaller than 100 nm can be seen in the images. It is likely that smaller features could be resolved by this technique if a scanner with improved resolution were used.

But, as already mentioned, even for very sharp tips the lateral resolution Δ_{L50} for measuring just 50% of the potential contrast is limited by physical constrictions to 90% of the tip–sample distance (26). For the measurement of the total potential contrast the resolution is even much worse (see Figure 16.30). For SKPFM the unfavorable shape of the tip (an ideal tip would be cylindrical with a flat bottom, so that tip and sample form a plate capacitor; AFM tips are pyramidal) and the influence of the cantilever are additional detrimental factors.

16.5.2.4 Behavior of Particles in AA2024-T3

The high Volta potential exhibited by the S-phase particles in the as-polished condition relative to the matrix (Figure 16.27) is opposite of what would be expected given

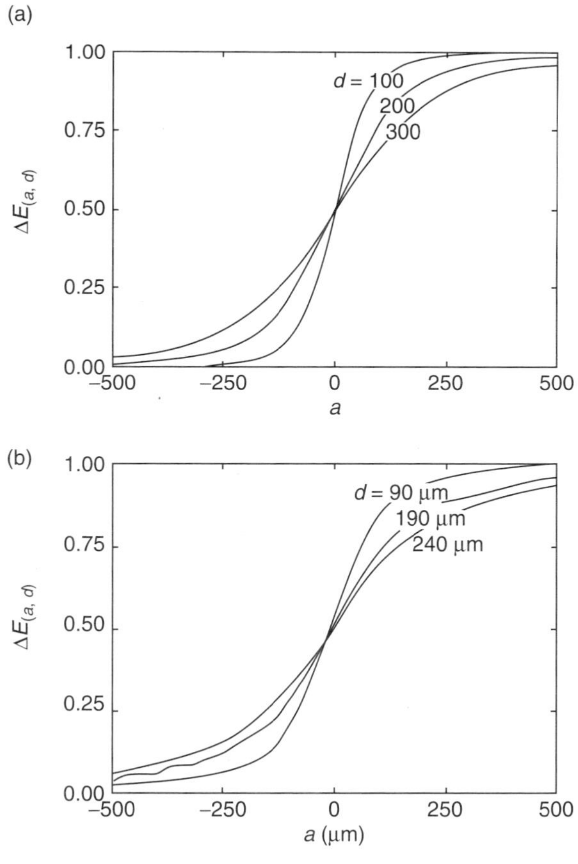

Figure 16.30 (a) Lateral response for a point probe scanned over a step in specimen Volta potential. The normalized change in Kelvin potential $\Delta E(a, d)$ is plotted as a function of the dimensionless lateral distance *a* from the step, for various values of dimensionless probe height *d*. (b) Plots of experimental $\Delta E(a, d)$ vs. *a* for various values of mean probe height *d*, obtained using a 125-μm plane-ended cylindrical probe. (From H. N. McMurray and G. Williams, *J. Appl. Phys.*, 91(3), 1673, 2002. With permission.)

the relatively low potential reported for grains of S-phase in a bulk fabricated analog sample (66) and the correlation presented above relating Volta potential measured by the SKPFM to corrosion potential. Auger depth profiles of the matrix, Al_2CuMg and AlCuFeMn particles, collected simultaneously on a sample of AA2024-T3 directly after polishing in a nonaqueous polishing slurry, indicate that the noble potentials observed on the intermetallic particles after polishing using SKPFM are not caused by surface enrichment of copper as might be expected. The surface of the intermetallic particles was covered with Al–Mg oxide (or hydroxide); Cu was depleted on the surface of both kinds of particles relative to the concentration beneath the surface. The source of the Volta potential signal is apparently more complicated than simple compositional variations.

The oxide film formed on the reactive S-phase particles during nonaqueous polishing provides some protectiveness to the surface. It has been shown that S-phase particles do not dissolve immediately upon exposure to 0.1 or 0.5 *M* NaCl (18, 20). With time of exposure in chloride solution, the high potential contrast exhibited by the as-polished sample decreases as the potentials of the matrix and particles merge to a single value. This is shown in Figure 16.31 for various regions of samples exposed to 0.5 *M* NaCl (20). Regions that are active pits go to a high potential similar in value to pure Cu exposed to the same solution. It was shown that S-phase particles not associated with a pit in the as-polished condition are eventually attacked when their potential reaches a value similar to that of the matrix. At this point, the protectiveness of the oxide film on the particle is reduced and localized attack at the particle can commence.

16.5.2.5 Composition–Volta Potential Correlation

Figure 16.26 shows that the Volta potential measured by SKPFM varies for different metals, and Figure 16.27 shows that the Volta potential varies across the surface of a heterogeneous alloy. However, the Volta potential measured by SKPFM is not a measure of the composition of the alloy or surface oxide, and there is not necessarily

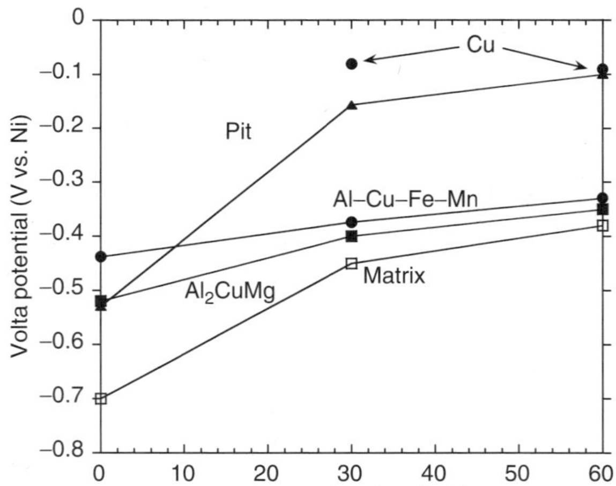

Figure 16.31 Volta potential change of pits, uncorroded matrix, and uncorroded intermetallic particles in AA2024-T3 following immersion in 0.5 *M* NaCl for different periods of time. Volta potential of Cu given for comparison. (From P. Leblanc and G. S. Frankel, *J. Electrochem. Soc.*, 149(6), B239, 2002. With permission.)

a relationship between the two. For instance, a difference in Volta potential was observed for different grains of sputter-etched Mg (99.95%), even though the composition of the surface oxide should be essentially identical from grain to grain (55). Another example involves a sample of AA2024-T3 that was treated in 0.5 *M* NaCl + 1×10^{-4} *M* $Na_2Cr_2O_7$ (55). The oxide formed on Al_2CuMg S-phase particles was found to contain a higher concentration of Cr than that on the matrix phase. However, despite this difference in composition, the Volta potential on the particles was found to be essentially identical to that of the matrix.

A final example also involves the S-phase particles in AA2024-T3, which were shown to have a high potential relative to the matrix in the as-polished condition in Figure 16.27. Partial removal of the oxide from an as-polished surface by sputter etching (removal of only a few nm) results in the reversal in contrast for some, but not all, of the S-phase particles at the surface (20, 55). In other words, after sputter etching and subsequent exposure to air, a select number of these particles have a Volta potential lower than that of the matrix. The oxide composition did not change as a result of the light sputtering, based on the Auger electron spectroscopic analysis of the sample surface prior to air exposure (the sputtering was performed in a scanning Auger microscope). Subsequent exposure of the sputter-etched sample to air and then 0.5 *M* NaCl for 30 min resulted in dissolution of some, but not all, of the particles that had exhibited a low potential after sputter etching (20). None of the particles that exhibited an unchanged Volta potential corroded during solution exposure.

These examples, along with the observations of AA2024-T3 corrosion described above, indicate that the Volta potential measured by SKPFM relates strongly to corrosion behavior, though not always to surface composition. Certain investigators in the field of corrosion have placed considerable emphasis on surface analysis and the surface composition. However, the Volta potential distribution seems to be a more important property when it comes to predicting and understanding the corrosion behavior.

Most likely differences in nanoscopic surface roughness (also of great impact on Volta potential) and differences in the electronic structure of the oxide (which is influenced by even the slightest changes of the defect structure not detectable by XPS or AES) are the reasons for the observed differences in the Volta potential of analytically (seemingly) equivalent oxides. These properties will have a direct effect on corrosion behavior, which correlates well with the observations.

16.5.3 SKPFM and Delamination

16.5.3.1 Introduction

While the basic mechanisms of anodic and cathodic delamination, that is, the electrochemically driven de-adhesion of polymer coatings from the metal substrate, are already well understood (see Section 16.4.2) (8–10, 27, 28, 47–50, 68), only little is known about the microscopic and submicroscopic aspects of delamination. The main difficulty for a detailed investigation of delamination with suitably high resolution is that de-adhesion is a process occurring at a buried interface. On the macroscopic scale, SKP (resolution of about a few tens of micrometers) has been the most important method to elucidate the fundamental processes of cathodic and anodic delamination, which in combination with other methods such as photoelectron spectroscopy (XPS) and measurement of the adhesion strength as a function of distance from the defect, helped to develop detailed models (8–10, 47–50). The development of the SKPFM opened application of the Kelvin probe technique to the submicroscopic scale. In this section, the first results of these submicroscopic investigations are presented and discussed in view of earlier results.

Delamination of organic coatings from metal surfaces can occur in a number of different ways, for example, as pure cathodic delamination, as anodic delamination, as filiform corrosion, or a mixture of these. In fact, in most technical systems the pure cases are the exception and, of course, delamination is usually very slow. Hence, either very long times are required for an investigation of the delamination kinetics or alternative methods are needed to provide extremely high lateral resolution. As will be shown in this section, in fast delaminating systems the length scales of the delamination features may range between several hundreds of micrometers and several millimeters, while in systems that show slow delamination the reaction zones can be confined to submicroscopic distances. This underlines the importance of investigation methods with submicroscopic resolution. Another example is the filiform corrosion on aluminum alloys; while the aluminum matrix is protected by the passive alumina scale, oxygen reduction can occur on the intermetallic particles embedded in the aluminum matrix. Even though it is certain that this is the driving force for the filiform corrosion, the exact role of the intermetallic particles (IP) on the kinetics of filiform corrosion and the path of the filaments remains unknown. A standard Kelvin probe does not allow a high enough resolution necessary to show the submicroscopic or only few micrometers-large IPs.

In both cases, SKPFM can play an important role in deriving new information on delamination processes on the submicroscopic scale. However, as for all SKP techniques, the resolution for SKPFM is strongly dependent on the distance between the tip and the surface. Since the surface of interest here is the interface between polymer and metal surface, a good resolution requires the preparation of special model samples that are characterized by ultrathin polymer coatings and specially prepared defects that show a very sharp borderline to the intact coating.

As discussed earlier, SKPFM was successfully used for *ex situ* corrosion studies on metal and alloy surfaces not covered by organic coatings (17, 18, 55). The application to delamination problems, however, gives rise to new questions on the interpretation of SKPFM images. Detailed studies are still needed to elucidate the imaging mechanisms of SKPFM under these conditions.

One important criterion for high-resolution studies is the preparation of suitable model samples with ultrathin polymer coatings. High resolution will only be possible if the coatings have a thickness on the order of a few tens of nanometers. Preliminary tests indicate that the nature of the polymer as well as the thickness of the polymer coating have a significant effect on the resolution. Figure 16.32 shows SKPFM images of copper markers on gold covered by polyurethane or a plasma polymer of different thickness. The resolution of the 70-nm plasma polymer coating is the best. At 210 nm the resolution is much worse than for the 680 nm polyurethane coating. This is the subject of current research. However, since the preparation of polyurethane coatings with thickness in the 100-nm range is very difficult and 20-nm plasma polymer coatings are no problem the first experiments on cathodic delamination were carried out on model systems with ultrathin plasma polymers (51).

For the experiments on cathodic delamination, n-doped silicon tapping mode tips were used. The measurements were carried out in a custom-made glass cell purged with humid air to control the relative humidity of the environment

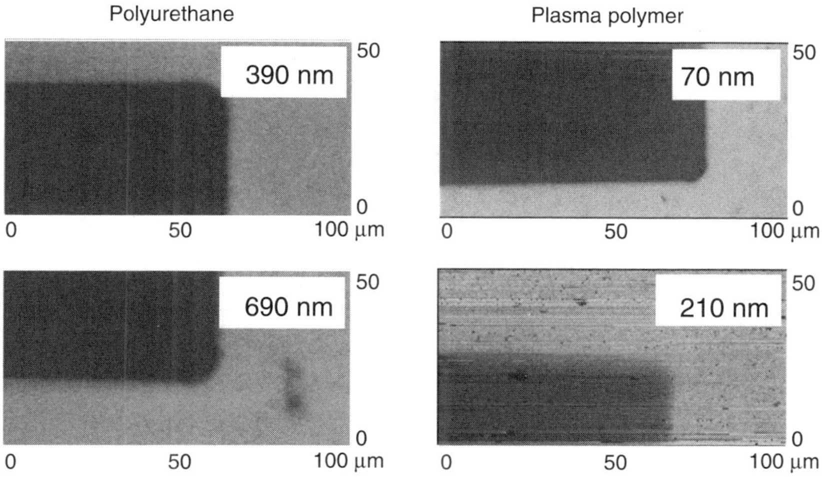

Figure 16.32 Kelvin-mode SKPFM images of evaporated nickel (low Volta potential) on gold, coated with polyurethane, or a plasma polymer film of different thickness. Especially for the case of a 210-nm plasma polymer the nickel corner is significantly blurred, while for the 680-nm polyurethane coating it is still sharp.

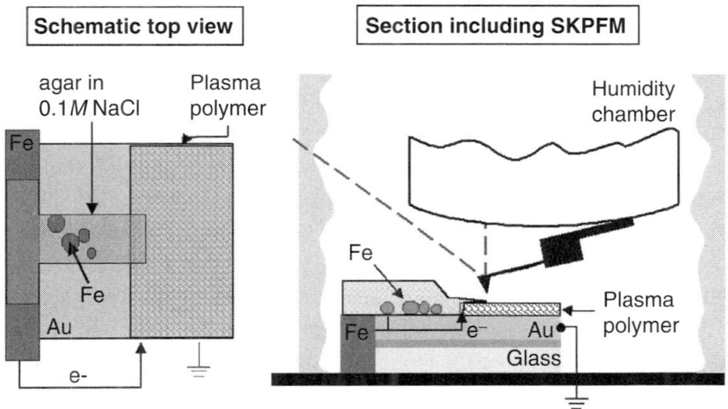

Figure 16.33 Schematic sample setup for model cathodic delamination investigations by *in situ* SKPFM. The uncoated part of the gold surface (defect) is polarized by the corroding iron particles.

(see Figure 16.33). Flame-annealed gold was covered by an ultrathin plasma polymer film made from hexamethyldisilane and electrochemically coupled to iron to set the electrode potential of the defect gold to $-0.4\,V_{SHE}$. The electrolyte was 0.1 *M* NaCl in agar applied onto the surface. The gold surface, which allows electron transfer but no ion transfer reactions, behaves similar to a passive iron electrode below the organic coating and served as the local cathode during delamination while the iron acted as the local anode.

This special sample setup successfully initiated cathodic delamination, which proceeds similar to the cathodic delamination on iron. Figure 16.34 compares the AFM topography (a) with an SKPFM potential image (b) (51). The images show the delamination front at approximately 5000 μm from the defect. The delamination area is seen in the AFM map as elevated topography because the ultrathin polymer film is lifted up by undermining by the electrolyte. The delamination front as deduced

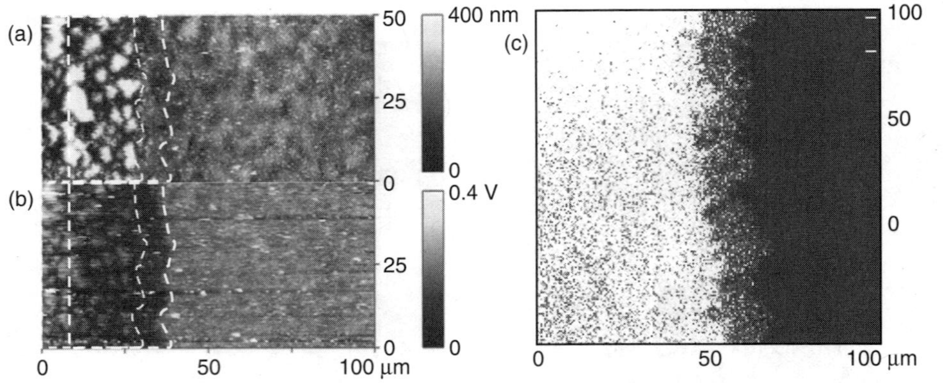

Figure 16.34 (a) Topography of a plasma-polymer-coated gold surface during delamination. (b) Correlated potential map. (c) TOF-SIMS: correlated sodium distribution. The scans were obtained about 5000 μm from the defect (i.e., from the border between uncoated and coated gold (see Figure 16.33), i.e., the delamination had already advanced by this distance. (From M. Rohwerder, E. Hornung and M. Stratmann, *Electrochim. Acta*, 48, 1235, 2003. With permission.)

from the height profile is marked by a white dashed line. In the Volta potential map the delamination is also clearly visible as a region of low potential, similar to the conventional SKP results. However, the front of the negative electrode potential exceeds the delamination front by a constant distance of approximately 10 μm, indicating that an electrochemical delaminating zone advances the already mechanically delaminated area by these 10 μm.

The magnified cross-section of the potential map for a similar sample is shown in Figure 16.35. A potential transient is clearly visible over a distance of 7 μm (Figure 16.35b), which is comparable to the difference between the de-adhesion front and the sharp potential decline. Obviously, this potential transient marks the reaction zone that leads to de-adhesion. The size of this reaction zone is much smaller than the resolution of standard SKP (Figure 16.35a). However, the potential drop detected by SKPFM is quite small (225 mV) in comparison to the potential drop as measured by standard SKP on the same sample (500 mV). This is most likely due to the above-mentioned influence of the cantilever and other stray capacitance effects.

The dimension of the reaction zone was further confirmed by time-of-flight secondary ion mass spectrometry (TOF-SIMS) measurements (see Figure 16.34c). In contrast to the continuous potential profile measured by SKPFM, the chemistry between the reaction zone and the delaminated zone according to TOF-SIMS changes abruptly for reasons not yet understood (51). The chemistry and electrochemistry going on within this reaction zone is the subject of ongoing studies.

The importance of submicroscopic studies for a deeper understanding of cathodic and anodic delamination has been discussed in more detail in Ref. (51), where further results of cathodic delamination studies obtained with SKPFM on polymer-covered metal substrates are presented. A number of open questions on electrochemical-driven delamination remain to be answered (51), such as why the delamination seems to proceed in distinctive steps, what the real extent of the active delamination zone might be, or, in the case of filiform corrosion, what determines the course of the filament growth. SKPFM might play an important role in answering these questions.

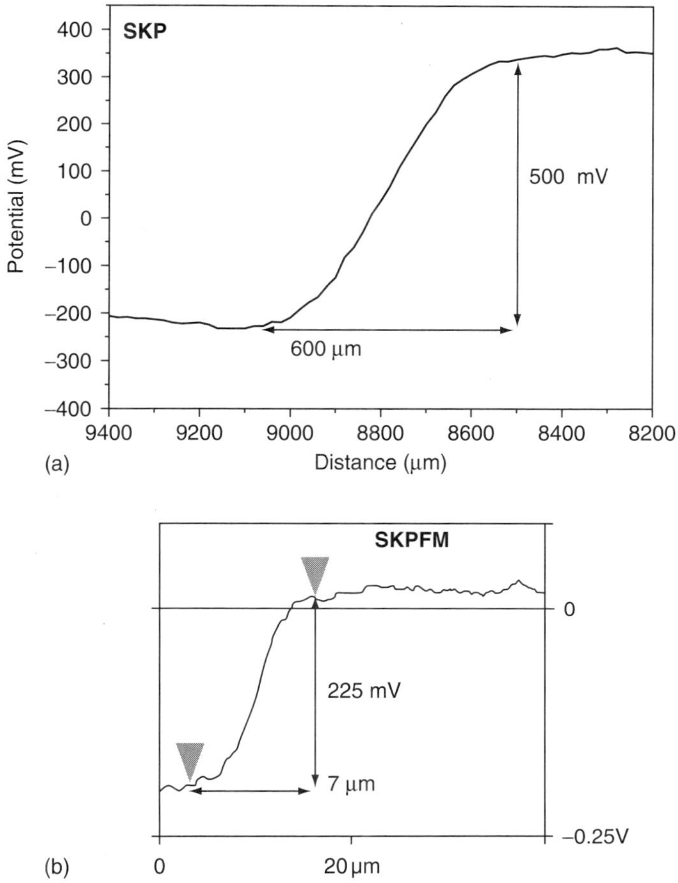

Figure 16.35 Potential scan at the delamination front of a plasma-polymer-coated Au surface: (a) scanning Kelvin probe; (b) scanning Kelvin probe force microscopy. (From M. Rohwerder, E. Hornung and M. Stratmann, *Electrochim. Acta*, 48, 1235, 2003. With permission.)

16.6 SUMMARY

This chapter demonstrated the potential of Kelvin probe techniques for corrosion science. SKP and SKPFM allow the measurement of the corrosion potential not only at the surface of materials but also at buried interfaces, for example, beneath a thin or even ultrathin electrolyte film or a polymer coating. Generally, a calibrated SKP allows the measurement of exact potentials (8, 24), while SKPFM usually only gives relative potentials (potential differences of features in one scan window), since the tip potential is prone to changes. Only if considerable care is exercised is it possible to calibrate the SKPFM (55).

SKP was shown to be useful for fundamental studies on model samples and to give new insight even to very fundamental corrosion phenomena, such corrosion under electrolyte droplets, where it provides direct information about distribution and activity of local cathodes and anodes and how these are affected by substrate composition and inhibitors in the electrolyte (see, e.g., Ref. (69)). But while SKP can also routinely be applied to technical samples (e.g., even for measurements on welded steel parts (70), SKPFM usually requires the preparation of special model samples. The reason for this is that SKPFM only makes sense if one is interested in high lateral resolution, since AFM piezoelectric scanners typically are limited to 100 μm in range or smaller. Relevant electrochemical processes usually take place on a larger scale more appropriate for a standard SKP instrument. However, if submicroscopic aspects of delamination processes are to be investigated, ultrathin model coatings must be used, since otherwise the resolution for the lateral potential distribution would not be sufficient (51).

In summary, SKP has already been successfully applied to a considerable number of corrosion problems, especially in atmospheric corrosion and coating delamination. It has already gained a good standing as a valuable method for *in situ* investigations. However, it is not easy to use, that is, it requires a lot of experience, and is comparatively expensive. That is why it is still not widely used.

The SKPFM, on the other hand, is available as an upgrade for a number of commercial AFM systems, which are quite widely distributed. That is why an increasing number of researchers are starting now to use SKPFM, also for the investigation of corrosion problems. However, it should be kept in mind that valuable information can only be obtained from well-conceived model experiments carried out on well-designed model samples.

REFERENCES

1. M. Stratmann, K.T. Kim, H. Streckel, *Z. Metallkd.* **81** (1990) 715.
2. S. Yee, M. Stratmann, and R.A. Oriani, *J. Electrochem. Soc.* **138** (1991) 55.
3. M. Stratmann and H. Streckel, *Corros. Sci.* **30** (1990) 681.
4. M. Stratmann and H. Streckel, *Corros. Sci.* **30** (1990) 697.
5. M. Stratmann, H. Streckel, K.T. Kim, and S. Crockett, *Corros. Sci.* **30** (1990) 715.
6. A. Leng and M. Stratmann, *Corros. Sci.* **34** (1993) 1657.
7. M. Stratmann, *Bull. Electrochem.* **8** (1992) 30.
8. A. Leng, H. Streckel, and M. Stratmann, *Corros. Sci.* **41** (1999) 547.
9. A. Leng, H. Streckel, and M. Stratmann, *Corros. Sci.* **41** (1999) 579.
10. A. Leng, H. Streckel, K. Hofmann, and M. Stratmann, *Corros. Sci.* **41** (1999) 599.
11. J.H. Parker and R.W. Warren, *Rev. Sci. Instrum.* **33** (1962) 948.
12. H. Baumgärtner and H.D. Liess, *Rev. Sci. Instrum.* **59** (1988) 802.
13. A. Broniatowski, W. Nabhan, B. Equer, and G. de Rosny, *Mater. Sci. Forum* **207–209** (1996) 149.
14. R. Mäckel, H. Baumgärtner, and J. Ren, *Rev. Sci. Instrum.* **64** (1993) 694.
15. H.-D. Liess, R. Mäckel, and J. Ren, *Surf. Interface Anal.* **25** (1997) 855.
16. W. Nabhan, B. Equer, A. Broniatowski, and G. de Rosny, *Rev. Sci. Instrum.* **68** (1997) 3108.
17. P. Schmutz and G.S. Frankel, *J. Electrochem. Soc.* **145** (1998) 2285.
18. P. Schmutz and G.S. Frankel, *J. Electrochem. Soc.* **145** (1998) 2298.
19. P. Schmutz and G.S. Frankel, *J. Electrochem. Soc.* **146** (1999) 4461.
20. P. Leblanc and G.S. Frankel, *J. Electrochem Soc.* **149**(6) (2002) B239.
21. I.F. Patai and M.A. Pomerantz, *J. Franklin Inst.* **252** (1951) 239.
22. W. Thomson, *Philos. Mag.* **46** (1898) 82.
23. W.A. Zisman, *Rev. Sci. Instrum.* **3** (1932) 367.
24. G. Grundmeier, K.-M. Jüttner, and M. Stratmann, in M. Schütze (ed.) *Corrosion and Environmental Degradation*, Vol. I, Wiley-VCH, Weinheim, 2000, p. 285.
25. Y. Harima, K. Yamashita, H. Ishii, and K. Seki, *Thin Solid Films* (2000) 237.
26. H.N. McMurray and G. Williams, *J. Appl. Phys.* **91**(3) (2002) 1673.

27. H. Leidheiser, Jr. (ed.), *Corrosion Control by Organic Coatings*, NACE, Houston, 1981.
28. R.A. Dickie and F.L. Floyd (eds.), *Polymeric Materials for Corrosion Control*, American Chemical Society, Washington, 1986.
29. H. Leidheiser, Jr., in F. Mansfeld (ed.), *Corrosion Mechanisms*, Marcel Dekker, New York, 1987.
30. A.J. Kinloch, in K.W. Allen (ed.), *Adhesion 3*, Applied Science, London, 1979.
31. R. Feser and M. Stratmann, *Werkstoffe Korros.* **42** (1991) 187.
32. R. Feser, PhD thesis, University of Clausthal-Zellerfeld, 1990.
33. M. Stratmann, R. Feser, and A. Leng, *Electrochim. Acta* **39** (1994) 1207.
34. H. Leidheiser, Jr., W. Wang, and L. Igetoft, *Prog. Org. Coat.* **11** (1983) 19.
35. H. Leidheiser, Jr., and W. Wang, in H. Leidheiser, Jr. (ed.), *Corrosion Control by Organic Coatings*, NACE, Houston, 1981, p. 70.
36. H. Leidheiser, Jr., in R.A. Dickie and F.L. Floyd (eds.), *Polymeric Materials for Corrosion Control*, American Chemical Society, Washington, 1986, p.124.
37. E.L. Koehler, *Proc. 4th Int. Congr. on Metallic Corrosion*, 1972, p. 736.
38. E.L. Koehler, in H. Leidheiser, Jr. (ed.), *Corrosion Control by Organic Coatings*, NACE, Houston, 1981, p. 87.
39. J.J. Ritter and J. Kruger, in H. Leidheiser, Jr. (ed.), *Corrosion Control by Organic Coatings*, NACE, Houston, 1981, p. 28.
40. H. Leidheiser, Jr., L. Igetoft, W. Wang, and K. Weber, *Proc. 7th Int. Conf. on Organic Coatings*, Athens, Greece, 1981.
41. J.S. Hammond, J.W. Holubka, and R.A. Dickie, *J. Coat. Technol.* **51** (1979) 45.
42. J.E. Castle and J.F. Watts, in H. Leidheiser, Jr., (ed.), *Corrosion Control by Organic Coatings*, NACE, Houston, 1981, p. 78.
43. M. Rohwerder et al., *GALVATECH* (2001), 585.
44. G. Williams and H.N. McMurray, *J. Electrochem. Soc.* **148**(10) (2001) B377.
45. G. Williams, H.N. McMurray, and D.A. Worsley, *J. Electrochem. Soc.* **149**(4) (2002) B154.
46. G. Williams, H.N. McMurray, D. Hayman, and P.C. Morgan, *PHYSCHEMCOMM* **6** (2001) 1.
47. M. Rohwerder and M. Stratmann, *MRS Bull.* **24** (1999) 43.
48. W. Fürbeth and M. Stratmann, *Corros. Sci.* **43** (2001) 207.
49. W. Fürbeth and M. Stratmann, *Corros. Sci.* **43** (2001) 229.
50. W. Fürbeth and M. Stratmann, *Corros. Sci.* **43** (2001) 243.
51. M. Rohwerder, E. Hornung, and M. Stratmann, *Electrochim. Acta* **48** (2003) 1235.
52. M. Pourbaix, *Atlas of Electrochemical Equilibria in Aqueous Solutions*, 2nd edition, NACE, Houston, 1974.
53. W. Schmidt and M. Stratmann, *Corros. Sci.* **40** (1998) 1441.
54. J.H.W. de Wit, *Electrochim. Acta* **46** (2001) 3641.
55. V. Guillaumin, P. Schmutz, and G.S. Frankel, *J. Electrochem. Soc.* **148** (2001) B163.
56. M. Nonnenmacher, M.P. O'Boyle, and H.K. Wickramasinghe, *Appl. Phys. Lett.* **58** (1991) 2921.
57. M. Nonnenmacher, O. Wolter, J. Greschner, and R. Kassing, *J. Vac. Sci. Technol.* **B9** (1991) 1358.
58. H. Jacobs, P. Leuchtmann, O. Homan, and A. Stemmer, *J. Appl. Phys.* **84**(3) (1998) 1168.
59. H. Jacobs, H. Knapp, and A. Stemmer, *Rev. Sci. Instrum.* **70**(3) (1999) 1756.
60. F. Robin, H. Jacobs, O. Homan, A. Stemmer, and W. Bachtold, *Appl. Phys. Lett.* **76**(20) (2000) 2907.
61. T. Hochwitz, A.K. Henning, C. Levey, C. Daghlian, and J. Slinkman, *J. Vac. Sci. Technol.* **B14**(1) (1996) 457.
62. A.K. Henning, T. Hochwitz, J. Slinkman, J. Never, S. Hoffmann, P. Kaszuba, and C. Daghlian, *J. Appl. Phys.* **77**(5) (1995) 1888.
63. Z. Szklarska-Smialowska, *Pitting Corrosion of Metals,* NACE, Houston, 1986.

64. R.G. Buchheit, R.P. Grant, P.F. Hlava, B. McKenzie, and G.L. Zender, *J. Electrochem. Soc.* **144** (1997) 2621.
65. V. Guillaumin and G. Mankowski, *Corr. Sci.* **41** (1999) 421.
66. R.G. Buchheit, *J. Electrochem. Soc.* **142** (1995) 3994.
67. J.O.M. Bockris and A.K.N. Reddy, *Modern Electrochemistry*, Plenum Press, New York, 1970.
68. D.M. Brewis, D. Briggs, *Industrial Adhesion Problems*, Orbital Press, Oxford, 1985.
69. C. Chen, C.B. Breslin, F. Mansfeld, *Mater. Corros.* **49** (1998) 569.
70. L.T. Han and F. Mansfeld, *Corros. Sci.* **39** (1997) 199.

17

The Microcell Technique

Thomas Suter and H. Böhni
Swiss Federal Institute of Technology, Zürich, Switzerland

Contents

17.1 INTRODUCTION

Electrochemical corrosion tests on samples of a few square centimeters provide information only on the behavior of a large material surface. They are not very useful for testing local corrosion processes occurring on passive materials. Therefore, a new microelectrochemical technique (the microcell technique) for corrosion studies in the micrometer range was developed (Figure 17.1).

The technique is based on a new approach. The use of a glass microcapillary that serves as an electrochemical cell permits reducing the diameter of the exposed surface to 1 μm. This makes it possible to test the corrosion resistance of single heterogeneities in structure. A reference and counter electrode are connected to the capillary, allowing electrochemical control of the investigated surface.

To prevent the electrolyte from leaking, the grounded tip of the microcapillary is coated with silicon rubber. The deformability and the hydrophobic behavior of the silicon seal are a major advantage of the measuring technique developed. The great deformability permits the testing of samples with rough surfaces as they occur in

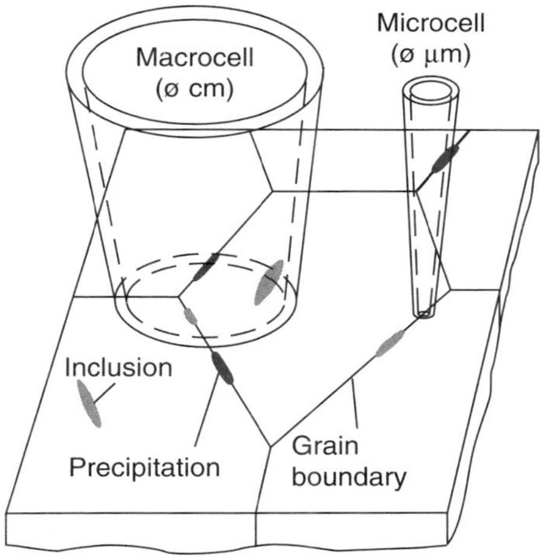

Figure 17.1 Macro- and microelectrochemical measurements on heterogenous substrates.

practice and due to the hydrophobic behavior of the silicon, crevice corrosion does not occur even in cases of large surface roughness.

Due to the decrease in sample surface the background noise can be greatly reduced: with a high current resolution of about 10 fA it is possible to detect processes that would otherwise be drowned in the current noise. On the basis of the measured currents, local processes in the micrometer and nanometer ranges can be detected.

In the case of 18/8 CrNi stainless steels and the aluminum alloy Al 2024-T3, localized corrosion is mainly caused by inclusions. The microcell technique has been applied to this type of weak points. Their influence has been studied in solutions with and without chlorides under different conditions and at different scales. The microcell technique was also applied to study the influence of the corrosion resistance of the single-phase austenite and ferrite of duplex stainless steels on the entire alloy.

Apart from inhomogeneities weld joints are a major target of corrosion. The microcell technique has been used successfully to test welds of austenitic stainless steels. Local measurements of different areas of welds have indicated that different zones suffer corrosion attack depending on the type of weld. The microcell technique proves to be easy to perform and apply. Unlike immersion tests, which may take several days, it yields a reliable judgment of the quality of a weld in only a few minutes.

When testing for localized corrosion, large-scale measurements often prove to be inadequate. To understand the mechanisms that cause pitting, combined micro- and macroinvestigations are usually necessary to provide the relevant information.

17.2 Overview of Micro- and Nanoelectrochemical Techniques for Corrosion Studies

The durability of technical passive alloys (e.g., stainless steels or aluminum alloys) is often limited by chemical or structural heterogeneities. Inclusions play a key role in being potential initiation sites of pitting and crevice corrosion (1–12). Many electrochemical methods to study localized corrosion are based on large-scale experiments with exposed areas in the square millimeter to square centimeter range. To understand the mechanisms of pit initiation and propagation, it is useful to also study the corrosion processes at the micro- and nanometer ranges. Such investigations often provide a better insight into the initiation mechanism of localized corrosion. The electrochemical techniques for localized corrosion tests in microscopic and nanoscopic dimensions can be divided into two major groups:

1. *Scanning techniques*: Immersed samples (immersed area is in the square millimeter to square centimeter range) are scanned using microelectrodes or ultramicroelectrodes. These methods allow us to obtain information about the local distribution of one or several parameters during corrosion experiments. The experiments might be performed under open-circuit conditions or under potential or current control. Depending on the particular technique, a lateral resolution down to a few nanometers is possible. However, many scanning methods do not allow measuring of local corrosion currents. The potentiostat controls the current flow of the whole immersed surface area.
2. *Small-area techniques*: By decreasing the size of the exposed area it is possible to localize the electrochemical process. This can be achieved by thin embedded wires, photoresist techniques, a droplet cell, or small glass capillaries touching only small areas of the specimen surface. These techniques allow, in contrast to the scanning methods, polarization of the microscopic surface areas. All common electrochemical techniques, for example, potentiostatic or potentiodynamic measurements, galvanostatic measurements, and cyclic voltammetry, can be applied. Local corrosion currents can be evaluated directly. With a high-resolution potentiostat, corrosion processes on a molecular level can be recorded, even if the small-area techniques do not show such a high lateral resolution as the scanning methods.

17.2.1 Scanning Techniques at the Micrometer Range

There are a number of scanning techniques that can be used to study localized corrosion at the micrometer range. The principle of some of these techniques is shown in Figure 17.2. Local techniques are often ranked by their lateral resolution. The authors believe that the lateral resolution is an important feature of a technique,

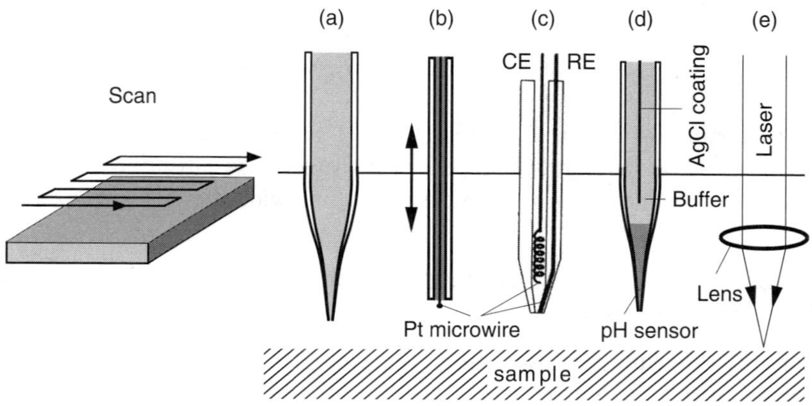

Figure 17.2 Scanning techniques used to study localized corrosion at the micrometer range: (a) SRET, scanning reference electrode technique (*in situ*); (b) SVET, scanning vibrating electrode technique (*in situ*); (c) LEIS, local electrochemical impedance spectroscopy (*in situ*); (d) ion selective microelectrode (*in situ*); and (e) photoelectrochemical imaging (*in situ*).

but not the most important one. Nevertheless, a rough estimate of the lateral resolution will be given for each technique.

17.2.1.1 SRET, SVET, and LEIS

In 1938 Evans and Thornhill developed the first local technique, the scanning reference electrode technique (SRET) (Figure 17.2a–c) (13–15). Two reference electrodes, one stationary and the other fixed on a translation stage, were used to measure local potential differences on a metal or alloy surface. Evans and Thornhill were able to identify scratches on a steel surface as anodic areas (14). The technique was also applied to evaluate the distribution of anodic and cathodic areas on a zinc plate immersed in 1 *mM* NaCl (15). The technique was continuously improved in the first half of the 20th century. Budd and Booth published in 1965 an excellent review dealing with early local corrosion methods, including the SRET (16). In the following years the size of the reference electrode was decreased more and more, down to about 1 μm (17). Isaacs et al. have probably been the most active and successful researchers who developed and used microelectrochemical scanning techniques to study corrosion processes. He applied the SRET to investigate pitting, intergranular corrosion, and the corrosion behavior of welds on stainless steels (18, 19).

Isaacs was the first to use the scanning vibrating electrode technique (SVET) (Figure 17.2b) to study stress corrosion cracking, the quality of coatings, and galvanic corrosion on soldered copper (20–23). The SVET is based on a vibrating microreference electrode. Potential differences are measured with a lock-in amplifier. The potential differences can be converted into current densities if the specific electrolyte resistance is known.

Isaacs et al. also developed the local electrochemical impedance spectroscopy (LEIS) (Figure 17.2c). The LEIS technique allows measuring of the local AC impedance by scanning the surface with a microcapillary that contains both a microreference and a micro-counterelectrode (24, 25).

The SRET, the SVET, and the LEIS yield maps of local potentials, current densities, and impedance data. The lateral resolution of these techniques is >50 μm. The resolution is mainly determined by the electrolyte conductivity and the distance between the scanning microelectrode and the sample surface. A high lateral resolution demands a short distance between the scanning microelectrode and the sample surface and a low electrolyte conductivity. Therefore, most scanning technique measurements have been performed in diluted electrolytes.

17.2.1.2 Ion Selective Microelectrodes

Additional information concerning localized corrosion can be gained with potentiometric and ion selective microelectrodes (Figure 17.2d). They are applied to obtain one-dimensional H^+ and Cl^- profiles in corrosion pits on metals (26, 27) or to record two-dimensional H^+ profiles on metal or alloy surfaces during stable pitting (28, 29). Measurements with potentiometric and ion selective microelectrodes indicate a lateral resolution of about 50 μm.

17.2.1.3 Photoelectrochemical Imaging

Photoelectrochemcial imaging (Figure 17.2e) involves the mapping of the photocurrent or photovoltage at an electrode–electrolyte interface as a function of the position of an illuminating spot of a focused laser beam (30). The electrode may be a passive metal or a semiconductor. The technique produces position-sensitive information by locally perturbing the surface. Photoelectrochemcial imaging is often performed in a "step-scan, lock-in" mode, in which either the focused laser beam or the sample is moved from one imaging point to the next. The illumination is intensity modulated and the signal is measured by a phase-sensitive detector (lock-in amplifier). The image acquisition time is limited by the need to rest at each image point for the time required for a sufficient number of illumination cycles to give satisfactory detection of the signal and noise rejection.

Butler used the technique to study passive oxide layers on polycrystalline titanium and high-resistance polycrystalline GaP photoelectrodes (31). He proposed that localized heating from a focused laser beam could be used to image the current distribution on metal electrodes. The oxide film growth of polycrystalline titanium was also investigated by the group of Smyrl (32) and by the group of Schultze (33, 34). Kozlowski and Smyrl used Auger depth profiling to determine the local film thickness of the anodic oxide grown on a Ti substrate (32). The oxide thickness was studied as a function of substrate crystallography and final growth voltage. The results were related to local photocurrent measurements. The nonuniform photoresponse was attributed to variations in the defect density of the oxide. Schultze and Thietke measured local photocurrent–potential curves on single grains (33) and compared the photoelectrochemical images of single grains with the results obtained by of ellipsometry, coulometry, and capacitance measurements (34). It was found that the photocurrent images could be correlated with the surface reactivity of the Ti surfaces.

Shukla and Stimming demonstrated the application of the photoelectrochemical imaging technique with examples of images of corroded surfaces, corroding surfaces, and a noncorroding alloy (35). The various factors which influence the photocurrent signal were discussed. Williams' group studied the passivation of iron, both of a pure metal and of an impure material in which grain boundary segregation was suspected (36). The photocurrent images showed significant heterogeneity in NaOH electrolyte, but much less heterogeneity in Na_2SO_4 electrolyte. It was speculated that the passive layer formed in Na_2SO_4 had a duplex morphology, with a uniform overlayer dominating the photoresponse, whereas in NaOH the film comprised a single barrier layer nucleated directly onto the metal.

Buzza and Alkire used a focused laser beam to initiate single corrosion pits as well as geometrically ordered arrays of pits on pure aluminum immersed in 1 *M* NaCl at pH 11 (37). The current was measured during potentiostatic growth and the shape of each pit was recorded. It was found that there was a critical applied potential below which pits repassivated, and that the critical potential decreased with pit size. The results suggested the hypothesis that a necessary condition for repassivation was the dissolution of a surface film. For arrays of pits grown under galvanostatic conditions, it was observed that the smallest pits repassivated. In additional work, Verhoff and Alkire developed several mathematical models, which included both migration and diffusion as transport modes (38, 39). The models were used to interpret the experimental results of the laser initiated pits experiments and to predict concentration profiles that resulted from dissolution based on the assumption of a hemispherical corrosion pit geometry.

The lateral resolution of the photoelectrochemical imaging technique is in the range of a few micrometers and single corrosion pits on aluminum with an initial diameter of about 5 μm can be initiated with a focused laser beam.

17.2.1.4 Kelvin Probe Technique

The Kelvin probe technique allows measuring of the Volta potential difference (Kelvin probe potential, V_{KP}) of samples covered with only a thin layer of electrolyte using a probe that does not touch the electrolyte. The Volta potential is the potential generated by free charges on the metal surface. A NiCr rod with a tip diameter of a few hundred micrometers can be used for measuring the corrosion potential E_{corr}. The sinusoidal change of distance between the probe tip and the corroding surface produces an alternating current due to the electric field across the gaseous gap between tip and specimen. When this electric field is exactly opposed by an applied potential V_{KP}, so that the alternating current becomes zero, the electrochemical open-circuit potential, E_{corr}, is equal to V_{KP} plus a constant and can be evaluated. The constant can be determined by calibrating the tip over a saturated solution of $CuSO_4$ on copper. The Volta potential difference varies linearly with the corrosion potential of the sample/solution interface. Stratmann et al. have demonstrated the possibilities of the Kelvin probe technique in corrosion studies of iron and aluminum (40–42). They were also able to detect local corrosion attack under organic coatings (43). The lateral resolution of the Kelvin probe technique is in the range of a few 100 μm.

17.2.2 Scanning Techniques at the Nanometer Range

After the development of the scanning tunneling microscope (STM) and atomic force microscopy (AFM) instrumentation by Binnig and Rohrer (44, 45), a new level of structural insight into surfaces in solution became possible. These techniques, known collectively as scanning probe microscopy (SPM), have been widely used to track the evolution of the surface morphology during corrosion processes. Some recent review articles (46, 47) and the proceeding volume "localized in-situ methods for investigating electrochemical interfaces" (48) provide a good overview of different electrochemical applications, including corrosion. There are a number of SPM techniques used to study localized corrosion in the nanometer range. The principle of some of these techniques is shown in Figure 17.3.

17.2.2.1 Scanning Tunneling Microscope

First, *in situ* STM corrosion measurements were performed by Fan and Bard (49) who recorded the surface changes of stainless steel immersed in aqueous chloride media under open-circuit conditions. The results showed that defects or edges are potential initiation points for corrosion and that extended corrosion leads to a moundlike structure. Similar findings were presented by Miyasaka and Ogawa (50), who investigated stainless steel in 10% and 3% NaCl. Their results showed that the STM technique is a promising tool for studies of the initial stage of pitting. Marcus and coworkers used the STM for *ex situ* (in air) studies on single-crystal surfaces obtaining images with atomic resolution. They tracked the oxide film formation on Ni(1 1 1) by anodic polarization at different potentials in 0.05 *M* H_2SO_4, and the passive film formation on Fe-22Cr(11) in 0.05 *M* H_2SO_4 (51, 52). Furthermore, they studied the atomic structure of metastable pits formed on a Ni(1 1 1) single-crystal surface (53). *Ex situ* and *in situ* STM measurements on well-defined systems (e.g., single crystal) reveal topographic details on the atomic level, whereas measurements on technical alloys have a resolution of a few nanometers (Figure 17.3a).

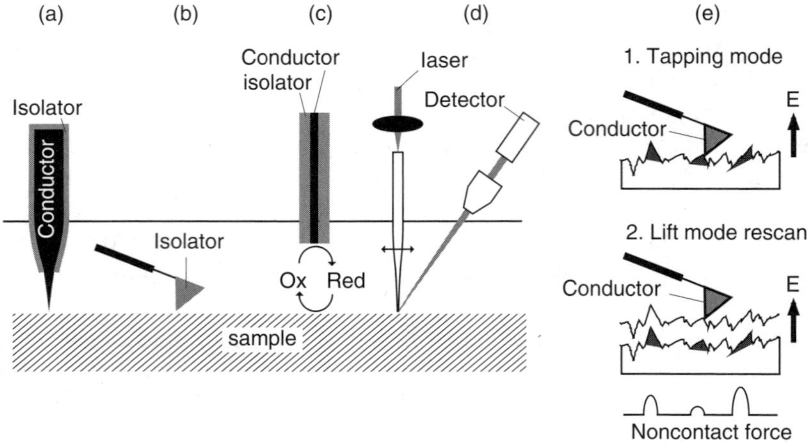

Figure 17.3 Scanning techniques used to study localized corrosion at the nanometer range: (a) STM, scanning tunneling microscopy (*in situ*); (b) AFM, atomic force microscopy (*in situ*); (c) SECM. scanning electrochemical microscopy (*in situ*); (d) SNOM. scanning near-field optical microscopy (*in situ*); and (e) SKPFM, scanning kelvin probe force microscopy (*ex situ*).

17.2.2.2 Atomic Force Microscopy

The first *in situ* SPM corrosion measurements were STM studies, but the *in situ* AFM technique is nowadays more frequently used. The main advantages of the AFM technique are the easier handling and the possibility to investigate nonconducting surfaces. The passive films of metals or alloys are semiconductors (e.g., stainless steel) or isolators (e.g., aluminum). STM studies on passive metals or alloys are possible (54–56), but the STM technique demands a more complex sample preparation and the STM images obtained are more difficult to interpret. Many researchers performed *in situ* AFM measurements to clarify the role of inclusions or second-phase particles on stainless steel and aluminum alloys during the pit initiation phase (57–59). Although stainless steels and Al alloys contain different kinds of inclusions, it was generally observed that they act as preferential starting points for localized corrosion. The lateral resolution of the AFM is similar to that of the STM (Figure 17.3b).

17.2.2.3 Scanning Electrochemical Microscopy

More than morphology information can be gained with scanning electrochemical microscopy (SECM) (Figure 17.3c). The SECM can be used in different modes of operation (47). Corrosion studies are usually performed with a four-electrode configuration. A micro- or ultra-microelectrode tip serves as a second working electrode besides the sample. The microelectrode tip might be a micrometer-sized disk or a fiber embedded in an insulator. The potential of the tip is controlled versus the reference electrode by adjusting the current flux between the tip and the counter electrode. SECM measurements are performed in solutions containing redox species. When a positive potential is applied to the microelectrode, the redox species are oxidized at the tip at a rate controlled by the diffusion of the redox species. If the tip is far from the sample surface, the current becomes steady state. When the tip is close to the sample surface, the oxidized species formed diffuse to the substrate, where they might be reduced again. This process produces an additional flux of species to the tip and hence an increase in tip current. This phenomenon is called "positive feedback."

The feedback mode was used to identify precursor sites for pitting (60–62). Casillas et al. immersed titanium foils in potassium bromide solution (60, 61) and showed that the oxidation of bromide occurred rapidly at a few microscopic surface sites on the TiO_2 film. The reactions on the sample surface and at the tip were

$$2\ Br^- \longrightarrow Br_2 + 2e^- \qquad \text{reaction on the sample surface} \tag{17.1}$$

$$Br_2 + 2e^- \longrightarrow 2\ Br^- \qquad \text{reaction at the tip} \tag{17.2}$$

After determining the position of the active sites, pitting was nucleated by increasing the electrode potential to more positive values. It was observed that pitting occurred preferentially at surface sites of high electrochemical activity.

The generation collection mode is another SECM mode used in corrosion studies. One working electrode (sample) generates some species which are collected at the second working electrode (microelectrode). The SECM tip can be used to identify corrosion products or to obtain a concentration map as the tip is scanned across the surface (63, 64). Paik et al. investigated the dissolution of MnS inclusions in stainless steels and Ni 200 during initiation of pitting corrosion (64). Since the sulfur species produced could not be detected directly by the SECM tip due to slow electron transfer kinetics, the iodide/triodide couple ($3I^- \Longleftrightarrow I_3^- + 2e^-$) was used as redox mediator. It was possible to measure the dissolution rate of an inclusion as its size, composition, and pit chemistry evolve with time. The reactions at the inclusion (depending on the sulfur species produced) and at the tip were

$$I_3^- + HS^- \longrightarrow 3I^- + H^+ + S \qquad \text{reaction at the inclusion} \tag{17.3a}$$

or

$$I_3^- + 2\ S_2O_3^{2-} \longrightarrow 3I^- + 2\ S_4O_6^{2-} \qquad \text{reaction at the inclusion} \tag{17.3b}$$

$$3I^- \longrightarrow I_3^- + 2e^- \qquad \text{reaction at the tip} \tag{17.4}$$

Still and Wipf used the SECM tip as the source of electrogenerated, aggressive chlorides. Single pits could be initiated on stainless steel, aluminum, and passivated iron samples (63). The chlorides were produced at the tip through the reaction

$$CCl_3COOH + H_2O + 2e^- \longrightarrow CHCl_2COOH + OH^- + Cl^- \tag{17.5}$$

The lateral resolution of SECM is mainly determined by the size of the electrode tip, the electrolyte conductivity, and the distance between the scanning microelectrode and the sample surface. A high lateral resolution demands a small tip, a short distance between the scanning microelectrode and the sample surface, and a low electrolyte conductivity. Although almost all SECM experiments employed large electrode tips (e.g., 10 to 25 μm diameter), electrodes tips with a diameter as small as 8 nm have been fabricated and can potentially be used as SECM tips (47). A lateral resolution of a few nanometers might be possible with such small tips. However, the lateral resolution of the SECM technique used in corrosion studies is in the range between 10 and 100 μm.

17.2.2.4 Scanning Near-Field Optical Microscopy

In contrast to other scanning probe techniques scanning near-field optical microscopy (SNOM) (Figure 17.3d) allows the concurrent acquisition of topographic and optical information with resolution beyond the diffraction limit of light (65). This is achieved by shining light through an optical fiber with a subwavelength aperture on the sample surface and collecting the light reflected from the surface. The aperture to surface separation distance has to be within 10 to 20 nm. This distance control is possible with a shear force feed back system. By scanning the aperture over the surface the topography and the optical information of the surface are collected. The use of light allows for the detection and identification of surface properties and changes that are not accessible with topography alone (66). Conventional optical techniques like reflection and fluorescence microscopy have been implemented successfully.

For the application of the SNOM in electrochemical applications the tuning fork based shear force feedback was implemented (67, 68). In recent work it was demonstrated with SNOM that fluorescent dyes mark anodic dissolution sites on Al 2024 by forming a precipitate (69, 70). Additionally, it could be shown that the selective dissolution of aluminum around cathodic sites results in a change in the color of the matrix that can be explained with an increased copper content on the surface (69, 71). The additional implementation of spectroscopic techniques like Raman will allow extension of the number of applications in electrochemistry. A drawback of this technique, however, is the complex and very time-consuming handling. The lateral resolution of the SNOM technique used in corrosion studies is in the range of a few 100 nm.

17.2.2.5 Scanning Kelvin Probe Force Microscopy

The *ex situ* scanning Kelvin probe force microscopy (SKPFM) technique (Figure 17.3c) allows concurrent acquisition of topographic and Volta potential information. Schmutz and Frankel developed and used this technique in corrosion research (72). The SKPFM can be performed with a commercial AFM using an electrically conducting (metal coated) AFM tip. The surface topography and potential distribution are measured simultaneously on a line-by-line basis. The topographic profile is first determined along a line. The tip is then lifted off the surface by 50 nm, and the line is rescanned with the tip maintained at the exact distance at every point along the line by turning off the feedback loop of the AFM and applying the recorded signal from the topographical profile to the scanning piezoelectric device. An AC voltage is applied to the AFM tip which excites cantilever oscillation in the presence of an external electric field. At each point of the rescan, a DC voltage is applied to the AC voltage until oscillation is nulled. Since this technique is a nulling method and applies a voltage to the tip the output signal from the instrument is inverted. It was shown that the inverted DC signal corresponds to the Volta potential. The SKPFM is operated in air, but a few monolayers of absorbed water allow measuring of a potential that corresponds well with the practical nobility of the surface in solution. The technique was applied to determine the sites of localized corrosion at intermetallic particles on Al 2024-T3 samples after exposure to chloride solutions. Experiments without and with inhibitors were performed (12, 73). The results indicated that Al–Cu–Mg particles, which show a noble Volta potential in air because of an altered surface film, are actively dissolved in chloride solution after a short incubation time. The lateral resolution of the SKPFM technique is $< 1\ \mu m$.

17.2.3 Small-Area Measurements

Small-area measurements in corrosion research are achieved by thin embedded wires, photoresist techniques, a droplet cell, or small glass capillaries touching only small areas of the specimen surface. The principle of these techniques is shown in Figure 17.4.

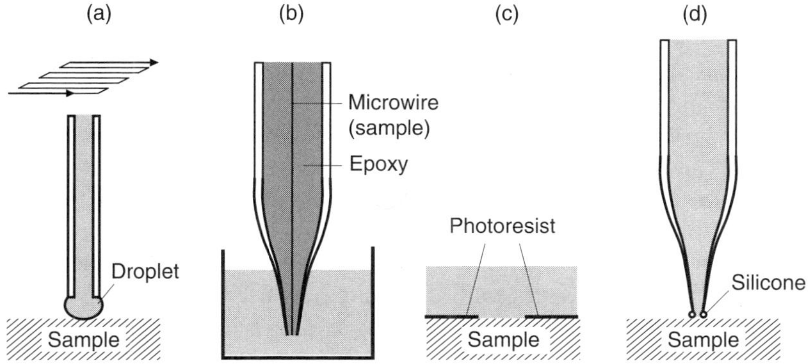

Figure 17.4 Small-area techniques used to study localized corrosion: (a) scanning droplet cell; (b) embedded thin wire technique; (c) photoresist technique; and (d) microcapillary cell.

17.2.3.1 Scanning Droplet Technique

Dix first developed in 1940 a small-area technique, the "scanning droplet technique" (Figure 17.4a) (74). An electrolyte droplet at the end of a glass capillary was used to scan the specimen surface. The size of the droplet was adjusted by controlling the flow through the capillary. The technique has been used to determine corrosion potentials without errors due to polarization effects. Dix and coworkers investigated sensitized $AlCu_4$ alloys in 0.1 *M* potassium chloride. The diameter of the droplet was about 1 mm. The results indicated that the grain boundaries were 90 mV less noble than the centers of the grains. Filipov and, later, Devijagina and Tomashov (mentioned in Ref. (16)) used this technique to study the susceptibility of stainless steel welds to intergranular corrosion. To avoid concentration changes in the electrolyte droplet by evaporation, the technique was operated in a tapping mode and a new droplet was used for each potential reading. The scanning droplet technique was recently improved by Lohrengel et al. (75–77). The improved technique, named the "scanning droplet cell," enables not only potential measurements, but also surface modification. According to the conventional three-electrode arrangement, most common potentiostatic and galvanostatic measuring techniques are possible. The scanning droplet cell was used for the surface analysis of die cast alloys, solder pads, and the monitoring of demetallization (78).

17.2.3.2 Embedded Wire Technique

The use of embedded wires is another possibility to perform small-area measurements (Figure 17.4b). Nickel and stainless steel wires in the millimeter and submillimeter range were used in earlier work to study pit propagation (79, 80) and the formation of a salt film (81, 82) in artificial pits. Thin wires with diameters down to 10 μm were used in more recent works to explore pit initiation processes. Figure 17.4(b) shows a possible configuration of a thin embedded wire. The wire is glued into the drawn end of a glass capillary and the tip of the capillary is grounded or polished. Riley et al. used a similar configuration to measure current fluctuations in the picoampere range on 10-μm stainless steel wires in a solution with chlorides. The observed shape of the current peaks was characterized by an immediate increase followed by a slow decrease (83). The current fluctuations were proposed to be pit nucleation events. Burstein and Mattin determined critical factors for the onset of pitting on 50-μm stainless steel wires (84) and Laycook et al. combined their results measured on 10- and 50-μm stainless steel wires with the findings of other authors to propose a "general pitting theory" (85).

Akiyama and Frankel used not thin wires, but thin embedded foils as artificial crevice electrodes (86). They studied dissolution kinetics of pits and crevices in aluminum and the effect of dichromate ions on the dissolution kinetics. The artificial crevice electrodes were potentiostatically polarized over a range of potentials in 0.1 *M* NaCl solution.

The advantages of the thin wire (foil) technique are the easy handling and well-defined exposed areas. However, only few materials are commercially available as thin wires. Furthermore, wires exhibit often special metallurgical structures.

17.2.3.3 Lacquers or Photoresist Technique

Covering the residual surface with lacquers or photoresist is a further possibility to decrease the size of the exposed area (Figure 17.4c). In 1950 Smith and Pingel presented the "micro solution-potential measuring technique." The polished surface of the metal specimen was covered with a transparent plastic film. The film was punctured with the aid of a specially adapted hardness tester. Smith and Pingel used the technique to study the corrosion behavior of heat treated Al–Cu alloys immersed in a sodium chloride solution. They compared the corrosion potentials measured across grain boundaries with the ones taken in the center of single grains. Budd and Booth improved the technique to reduce the possibility of crevice corrosion underneath the lacquer (16). Beck and Chan applied the technique to track pit propagation on stainless steel in sodium chloride (87). Lajain measured polarization curves on steel welds in 0.5 *M* KCl solution (88, 89). His results correlated well with the findings of macroscopic experiments.

Brandel was one of the first to use photoresist for sample preparation (90, 91). She was able to expose only 2-μm small squares or rectangles and to measure reproducible polarization curves on hard soldering joints. Due to the limited current resolution of the potentiostat, however, it was not possible to investigate only one stripped feature and therefore it was necessary to expose simultaneously several rectangles of a sample. Alkire and Wong applied the photoresist technique to study propagation of single pits on an 18/8 CrNi steel in 1 *M* H_2SO_4 + 0.1 *M* NaCl at +600 mV(SCE) (92). The technique was also applied to investigate the dissolution of inclusions on stainless steels and aluminum alloys (5).

The advantage of the photoresist techniques is the possibility to expose small and well-defined features with different and complicated shapes. A drawback is the often poor adherence of the lacquer or photoresist on the substrate, leading to crevice effects during the experiments. Furthermore, the technique demands a time-consuming sample preparation.

Nowadays, scanning techniques are quite common in corrosion research. The high lateral resolution and the availability of commercial apparatus led to a large number of SPM investigations in recent years. However, small-area techniques are also promising tools to study initiation processes of localized corrosion. They allow direct correlation of the measured corrosion currents of a local spot with the corresponding corrosion attack observed by SEM or AFM after the experiment, whereas a direct correlation is not always possible using scanning experiments. Since a larger area is exposed to the electrolyte, several activation and repassivation processes might occur simultaneously at different spots on the sample surface. Therefore, it might be difficult to correlate the measured corrosion current of the whole immersed surface area with the topographic information. In this chapter a small-area technique (Figure 17.4d) is presented, that uses glass microcapillaries as electrochemical microcell (93–96). The technique, developed at the Swiss Federal Institute of Technology, Zürich, allows a wide range of different microelectrochemical measurements on local areas down to about 1 μm^2. The possibilities and the limitations of this technique will be demonstrated. Furthermore, different microcell studies of the last few years are summarized. The results of microcell studies on different systems will be presented:

- Studies of pit initiation at single inclusions on AISI 304 stainless steel (2, 93–97)
- Studies of pit initiation at single inclusions on Al 2024-T3 (98, 99)
- Studies of the corrosion resistance of super duplex stainless steels (SDSS) (100, 101)
- Microelectrochemical testing of stainless steel welds (96, 102)
- Microcell studies in microelectronics (103).

17.3 THE MICROCELL TECHNIQUE

17.3.1 Setup

The setup for microelectrochemical investigations based on the microcapillary is shown schematically in Figure 17.5(a). The entire setup is mounted on a microscope, allowing for precise positioning of the capillary. The fully assembled microcell is shown in Figure 17.5(b).

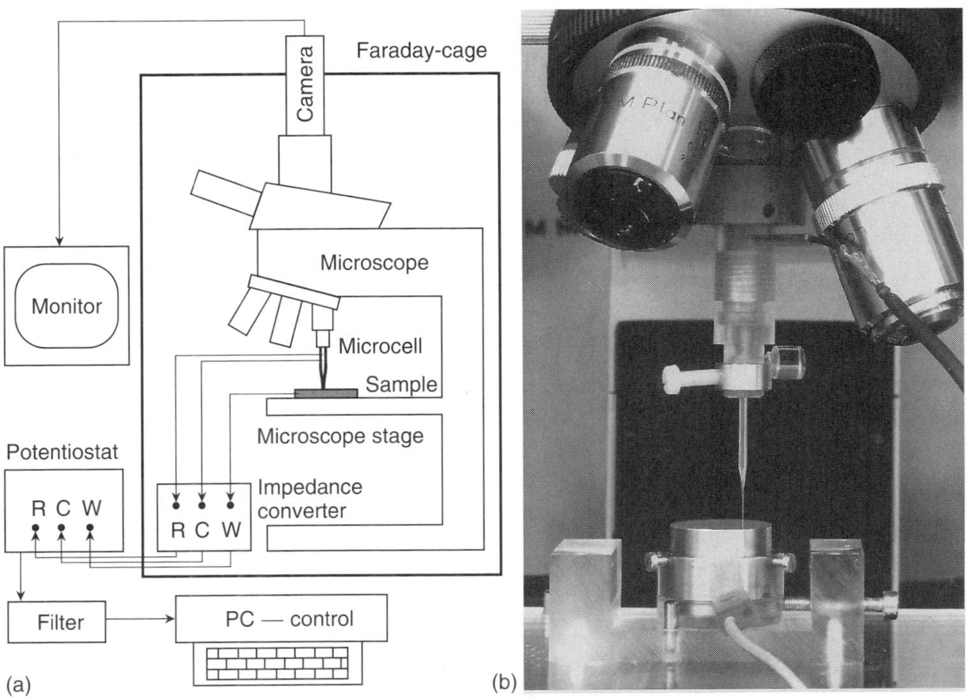

Figure 17.5 Microelectrochemical setup: a microcapillary is used as electrochemical cell. (a) Schematic drawing of the microcell technique. (b) Picture of the assembled microcell.

The microcell basically consists of a pulled microcapillary filled with electrolyte. The tip diameter of the capillary can be varied from about 1 to 1000 μm depending on the experimental requirements. Between the front end of the microelectrode and the surface of interest a layer of silicone rubber is applied as sealant. The microcell is fixed at the revolving nosepiece replacing an objective, and the specimen is mounted on the microscope stage. This arrangement enables the search for a site with different magnifications before switching to the microcapillary. In this way, simple, precise, and fast positioning of the microcell is possible. A reference and counterelectrode are connected to the capillary allowing electrochemical control of the investigated surface.

17.3.2 Microcapillary and Seal Preparation

Thin microcapillaries are obtained by heating borosilicate glass tubes and pulling them when they reach the glass temperature. For sealing, the glass capillaries are dipped in a one-component silicone lacquer. A stream of ethanol is pressed through the micropipette in order to flush out the silicone inside the capillary without destroying the fine tip. Very thin layers of silicone are applied to the capillary tip by repeating this process many times. The hardening process proceeds at a slow time scale, which allows the silicone surface to energetically minimize under the effects of surface tension, resulting in smooth and dense coatings (Figure 17.6a). This process allows production of capillaries with tip diameters below 1 μm.

The quality of the silicone sealant is a most critical factor, as it determines to a high degree the quality of an electrochemical experiment. Two properties of the silicone sealant are very important:

Figure 17.6 Microcapillary (100 μm) with silicone sealant. (a) Surface quality of the silicone sealant. (b) Deformability of the silicone sealant.

1. The hydrophobic property of the silicone prevents the electrolyte from penetrating under the seal. Therefore, crevice corrosion under the silicon seal is avoided.
2. The high deformability of the silicone sealant enables measurements on rough and even extremely curved surfaces. Figure 17.6(b) shows the deformability of the silicone sealant when the microcapillary is pressed on a surface.

17.3.3 Resolution of the Potentiostat

The resolution of the potentiostat is the second critical factor of the microcell technique. Microelectrochemical corrosion measurements are mainly performed to study localized corrosion on passive systems (stainless steels, Al alloys, Cu alloys, Ni alloys, etc.). The polarization resistance of these materials is in the range of 10^5 to 10^6 $\Omega\ cm^2$. Hence, the sample resistance of corrosion measurements with an exposed area of 1 μm^2 is in the range of 10^{13} to 10^{14} Ω. Low-level (DC voltage and DC current) measurements at such high source resistances require measuring instruments with a very high input resistance and a very low input current. The theoretical limit of sensitivity in any measurement is determined by the ideal noise (thermal noise) generated by the resistances present in the circuit (104). The best choice for low-level measurements are usually electrometers which are available with an input resistance up to 10^{16} Ω and an input current down to 10^{-17} A. Corrosion measurements, however, demand an instrument (potentiostat) that is able to polarize the sample and to measure the corrosion current. Therefore, voltage noise and current noise are critical factors. The response time of an electrometer is in the range of a few seconds, whereas the response time of a potentiostat is in the range of a few 100 μsec. A modified low-noise battery operated potentiostat (Jaissle 1002T-NC-3) has been used for the microelectrochemical corrosion experiments . The input resistance is better than 10^{15} Ω and the input current about 10^{-14} A (20°C). In combination with a good shielding, corrosion currents as low as 10 fA (10^{-14} A) can be detected.

The evaluation of the critical factors for microcell measurements showed that the quality of the silicone sealant determines the quality of a microelectrochemical corrosion experiment. The resolution of the potentiostat, however, determines the smallest useful diameter of a microcapillary. Two requirements fix the size of the smallest useful tip diameter. First, for corrosion measurements the input resistance of the potentiostat should be about 100 times higher than the sample resistance. Second, to obtain polarization curves with a reasonable amount of detail the current sensitivity of the potentiostat should be at least 10 times lower than the passive current. These parameters result in a minimum useful tip diameter of a microcapillary between 1 and 5 μm.

It may be noted that these findings are only valid for passive materials with a high polarization resistance. Electrochemical experiments on systems with a lower sample resistance and higher currents (e.g., copper deposition, corrosion on active or noble metals or alloys) allow the use of glass capillaries with a finer aperture.

17.3.3.1 Testing the Resolution of the Potentiostat

While checking the current detection limit of a potentiostat one has to consider not just the noise of the potentiostat but also the thermal noise (Johnson noise) of the test resistors. It is

$$U(\text{RMS}) = \sqrt{4kTR\Delta\nu} \tag{17.6}$$

$$U(\text{RMS}) \sim \sqrt{R} \tag{17.7}$$

$$i(\text{RMS}) \sim 1/\sqrt{R} \tag{17.8}$$

where U is the potential, i is the current, k is the Boltzman constant, T is the temperature, $\Delta\nu$ is the frequency band width, and R is the resistance.

With increasing resistance the voltage noise increases by the square root of the resistance, while the current noise decreases by the same factor. Therefore, especially manufactured 10-GΩ (10^{10}) resistors were used for testing the resolution of the potentiostat. Figure 17.7 shows the shape of the nominal voltage curve and of the curve of the measured current. The potential increase of both ramps is 0.2 mV/sec. The resulting current curve proves that the detection limit is better than 30 fA (current drop at 81 sec.) The noise of the current curve corresponds well with the expected noise of the 10-GΩ resistors. This means that the noise of the potentiostat is not higher than 20 fA. In order to perform more exact test measurements, resistors in the Terraohm (10^{12}) range would be necessary.

17.3.4 Handling the Microcapillary

17.3.4.1 Searching an Area of Interest

The microcell apparatus enables searching of an area of interest with different magnifications before approaching the microcapillary. Pulled microcapillaries are not straight; they are always slightly curved. Therefore, the tip of the microcapillary

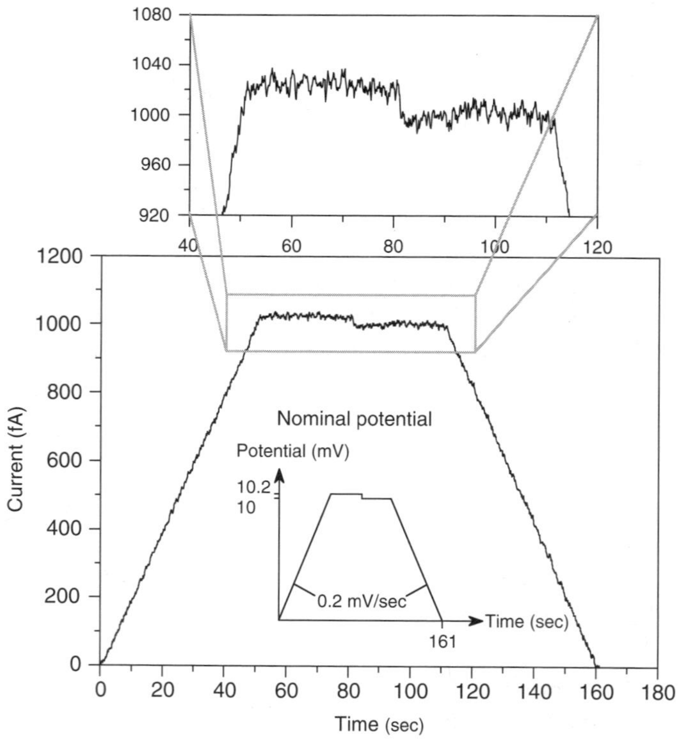

Figure 17.7 Current resolution of the potentiostat; smaller figure, nominal potential; larger figure, measured current.

is not in exact line with the axis of the objective. However, with the help of a video camera and a monitor, an exact positioning (on 1 μm) of the microcapillary is possible (Figure 17.8).

Steps:

1. Searching an area of interest, for example, an inclusion
2. Putting the microcapillary on a site far away from the inclusion; applying a high corrosion current for a short time
3. Marking the corroded site on the monitor
4. Moving the sample; the inclusion has to be inside the marked spot on the monitor
5. Positioning the microcapillary on the inclusion.

A modified procedure allows also investigation of the inhomogeneities, which are not visible using an optical microscope, for example, grain boundaries. The sample is marked with a grid of microhardness dots. The sites between the microhardness dots are investigated. After several measurements the sample is etched. The

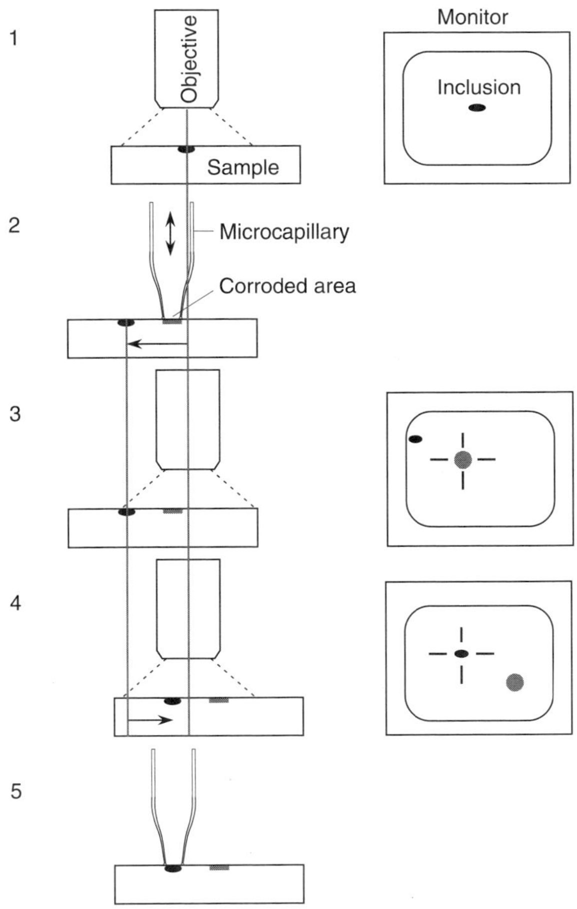

Figure 17.8 Exact positioning of a microcapillary on an inclusion.

etch patterns indicate which areas were investigated. This procedure is of course quite time consuming.

17.3.4.2 Putting the Microcapillary on the Sample

When putting the microcapillary on the sample the silicone seal should always show the same deformation. This guarantees a reproducible exposed area. Therefore, and to avoid destroying the fine capillary tip, a tilted microscope is used to put the microcapillary on the sample (Figure 17.9)

17.3.5 Measurements with Modified Microcell

Modification of the microcells allows the evaluation of various parameters during a corrosion experiment or performing of corrosion measurements under different conditions (Figure 17.10).

pH measurements: A 25-μm tungsten wire insulated with 5 μm polyimide can be used to measure the pH at variable positions over the sample surface during corrosion experiments (26). The wire, with a freshly cut tip, is inserted into the microcapillary and positioned with a fine threaded micrometer screw with a resolution of ±2 μm. Prior to an experiment the tungsten wire is calibrated in different buffer solutions to check the reproducibility. Only the results of measurements with a reproducibility better than 0.5 pH units are usually accepted. The experimentally determined response time of the tungsten wire is in the order of 1 sec, which is about 10 times longer than the time of the diffusion of the hydrogen ions to the probe.

Measurements with temperature control: A Peltier element is used to control the temperature of the sample during a corrosion experiment. The Peltier element allows varying of the temperature between 0°C and 100°C. A special temperature control unit with a linear control characteristic was developed in order to avoid interference with low current measurements.

Measurements with applied mechanical stress: A special sample holder is used for experiments with applied stress. For constant deflection tests, thin samples are used as specimens. Stress is applied to the specimen by holding it down at the border and bending it over a centered screw. The amplitude of the applied stress is calculated.

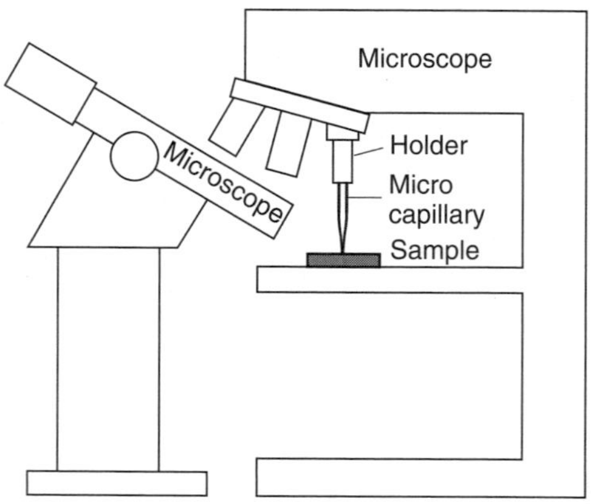

Figure 17.9 A tilted microscope is used to put the microcapillary on the sample.

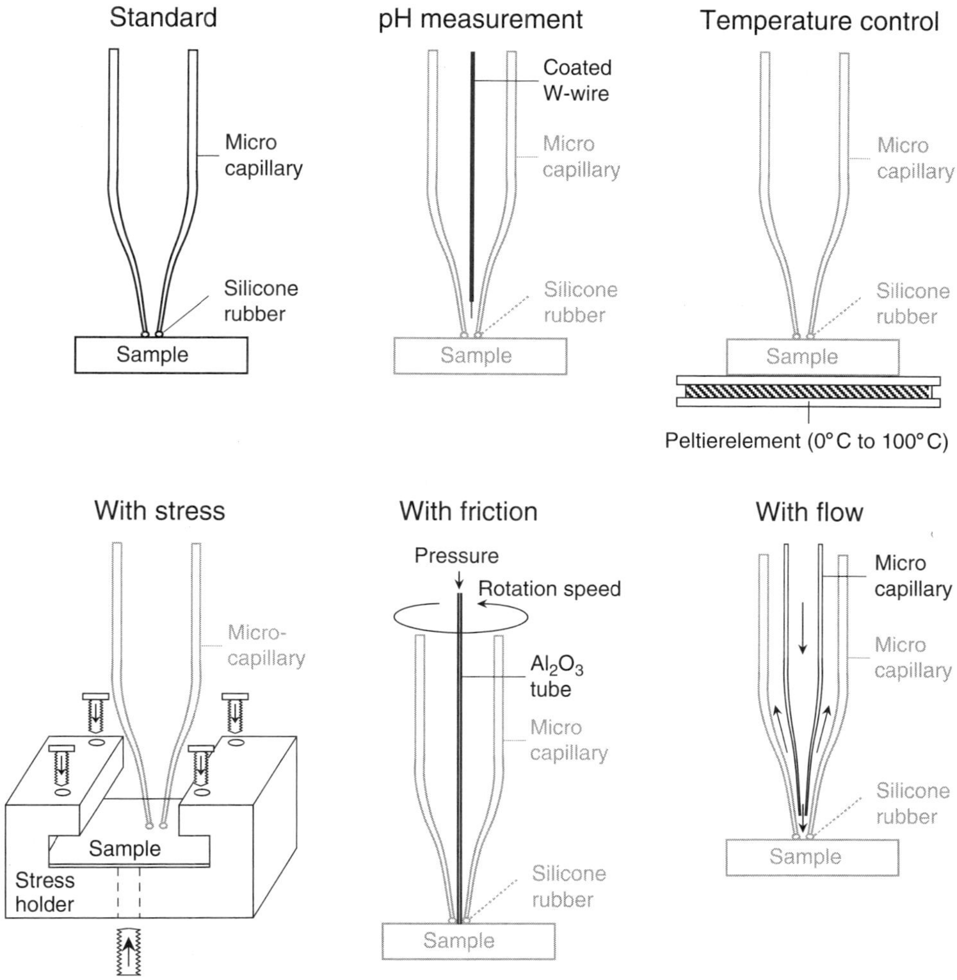

Figure 17.10 Modified microcells. They allow evaluation of additional parameters during a corrosion experiment or performing of corrosion measurements under different conditions.

The calculations are verified with the help of the minimum deformation which led to indications of plastic deformation after relaxation of the applied stress.

Measurements with friction: A rotating aluminum oxide tube inside the microcapillary allows corrosion measurements with friction. This modification allows testing of the behavior of a metal probe in an electrolyte under rubbing conditions on the micrometer scale. Rotational speed, load, and applied electrochemical potential or current can be varied so that it is possible to analyze the corrosion rate during rubbing at various controlled conditions. In order to quantify the tribocorrosion, profilometer measurements of the treated zone were performed. An accuracy of 10% was found. With this microtribometer the testing object could be directly placed on the sample holder without any particular pretreatment. Experiments on samples with complex geometry such as gears, piston rings, or implants can easily be performed (27).

Measurements with flow: An additional microcapillary inside the microcell enables testing of the vertical fluid flow. The bottleneck shape of the microcapillary allows application of very high flow rates. The inner capillary is positioned with a fine threaded micrometer screw with a resolution of $\pm 2\,\mu m$.

17.4 MICROCELL STUDIES OF PIT INITIATION AT SINGLE INCLUSIONS ON A 304 STAINLESS STEEL

Corrosion measurements at different scales provide different information. Large-scale experiments show the general corrosion behavior of a material. Microelectrochemical experiments using capillaries with a tip diameter in the range of $100\,\mu m$ allow investigation of sites that contain a single weak point, which might initiate pitting. Inclusions, especially MnS inclusions, are the main weak points of stainless steels. Measurements with a 100-μm microcapillary enable the study of pit initiation at MnS inclusions of different composition, structure, and size. Electrochemical measurements using capillaries with tip diameters in the range of 1 to $3\,\mu m$ allow investigation of different zones of a single inclusion. Hence, the corrosion behavior of the weakest zone of an inclusion can be determined. Experiments with a larger exposed area do not allow such investigations. The pitting potential of large-scale experiments is usually fixed by the corrosion behavior of the most active inclusion, whereas the pitting potential measured in an experiment on a single inclusion is usually determined by the weakest zone of the inclusion, as was the case for the Al alloy described above. A more detailed description of the stainless steel studies is given elsewhere (2, 93–97).

17.4.1 Experimental

Microelectrochemical investigations were carried out on three different 18-Cr, 10-Ni stainless steel (SS) specimens having different sulfur contents. The low-sulfur 304 SS plate (0.003% S) contained few, small MnS inclusions. The medium-sulfur 304 SS rod (0.017%) contained numerous MnS inclusions with an average diameter of about $3\,\mu m$. The high-sulfur 304 steel plate with a sulfur content of 0.026% contained large shallow MnS inclusions which, owing to the rolling process, were distended to a shape that was typically $30\,\mu m$ by $10\,\mu m$ by $3\,\mu m$. Prior to all electrochemical experiments, the samples were mechanically polished with wet SiC abrasive paper down to 4000 grit, rinsed with distilled water, and ultrasonically cleaned in ethanol for 5 min. All solutions were prepared from reagent grade chemicals and distilled water. After positioning the microcell, a cathodic potential of $-500\,mV$ (SCE) was applied for 2 min, after which potentiodynamic polarization curves were measured at a scan rate of 1 mV/sec. A modified high-resolution potentiostat with a current detection limit of 10 fA (Jaissle 1002T-NC-3) was used. The positioning of the microcell over different single inclusions allowed several measurements on the same specimen. Before and after the microelectrochemical measurements, the investigated sites were examined by scanning electron microscopy (SEM), energy dispersive x-ray spectroscopy (EDX), and AFM.

17.4.2 Influence of the Size of the Exposed Area on Pit Initiation

Figure 17.11 shows how the pitting potential might change if the exposed area is decreased. Local polarization curves were measured at random sites on the low-sulfur 304 plate in 1 *M* NaCl. The diameter of the microcapillaries varied from 50 to

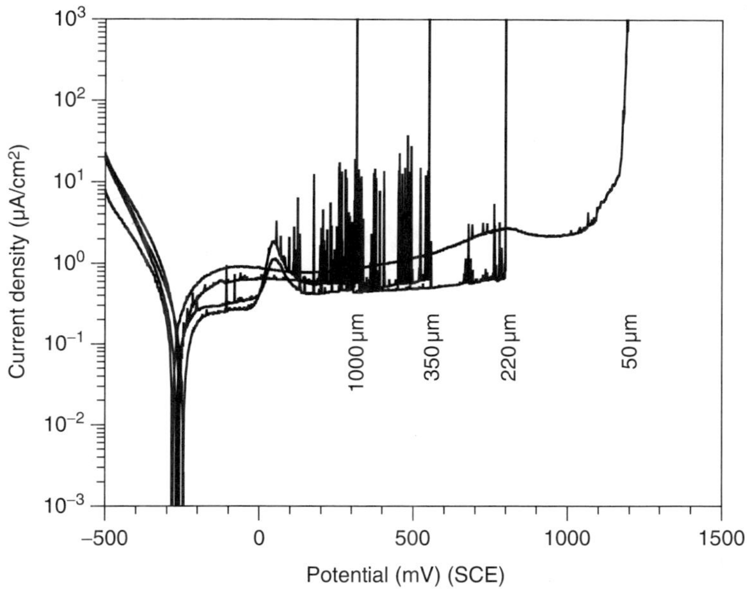

Figure 17.11 Pitting potentials of an SS 304 plate (0.003% S) as a function of the diameter of the microcell. The polarization curves were measured at random sites in 1 *M* NaCl (From H. Böhni, T. Suter, and A. Schreyer, *Electrochim. Acta*, **40**, 1361 [1995]. With permission.)

1000 μm. Common large-scale measurements show a pitting potential of about +300 mV (SCE). Measurements performed with 1000 μm microcapillaries show similar pitting potentials. Diminishing the exposed surface to an area of 50 μm in diameter led to an increase of the pitting potential to about +1200 mV (SCE).

Microelectrochemical experiments revealed that the pitting potential is an area-dependent value. Additional measurements showed that the increase of the pitting potential is caused by a decrease of the number of weak points when the exposed area is diminished.

17.4.3 Studies of Pit Initiation at Single MnS Inclusions

Figure 17.12(a) shows a polarization curve measured on the medium-sulfur 304 rod. A capillary with a tip diameter of about 100 μm was used to select a single MnS inclusion. The electrolyte was 1 *M* NaCl. The curve reveals onset of pitting at +300 mV (SCE). Large-scale measurements showed stable pitting at a similar potential. Microelectrochemical measurements performed on sites without MnS inclusion did not indicate pitting at potentials below the onset of oxygen evolution. Figure 17.12(b) shows the SEM picture of the inclusion after the corrosion experiment. It indicates that the inclusion/bulk interface is mainly attacked. Additional measurements and measurements from other authors support this finding (56, 105–107).

17.4.4 Studies of Pit Initiation at Different Zones of a Single MnS Inclusion

Figure 17.13(a) shows polarization curves measured on the high-sulfur 304 plate. A capillary with a tip diameter of about 2.5 μm was chosen to study the corrosion

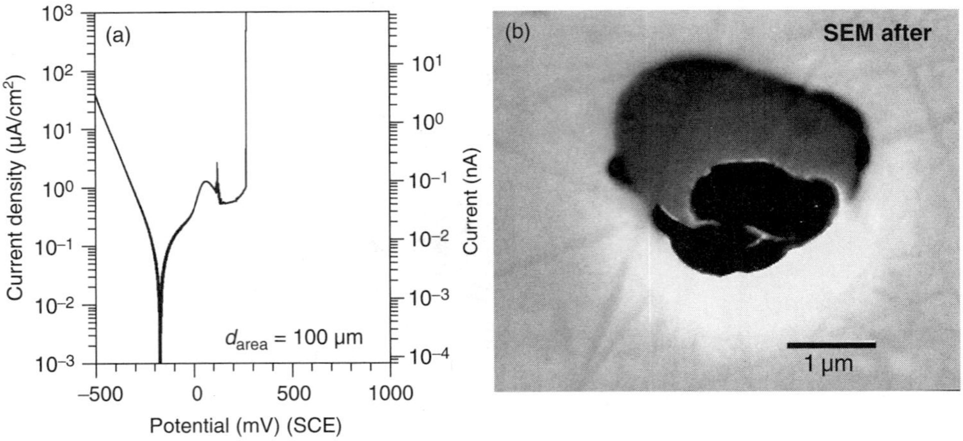

Figure 17.12 Pitting behavior of a single MnS inclusion on an SS 304 rod (0.017% S): (a) polarization curve measured in 1 *M* NaCl, and (b) SEM picture of the attacked MnS inclusion taken after the corrosion experiment. (From T. Suter and H. Böhni, in *Critical Factors in Localized Corrosion II*, P. M. Natishan, R. G. Kelly, G. S. Frankel, and R. C. Newman, Editors, PV 95–15, p. 127, The Electrochemical Society Proceedings Series, Pennington, NJ (1996). With permission.)

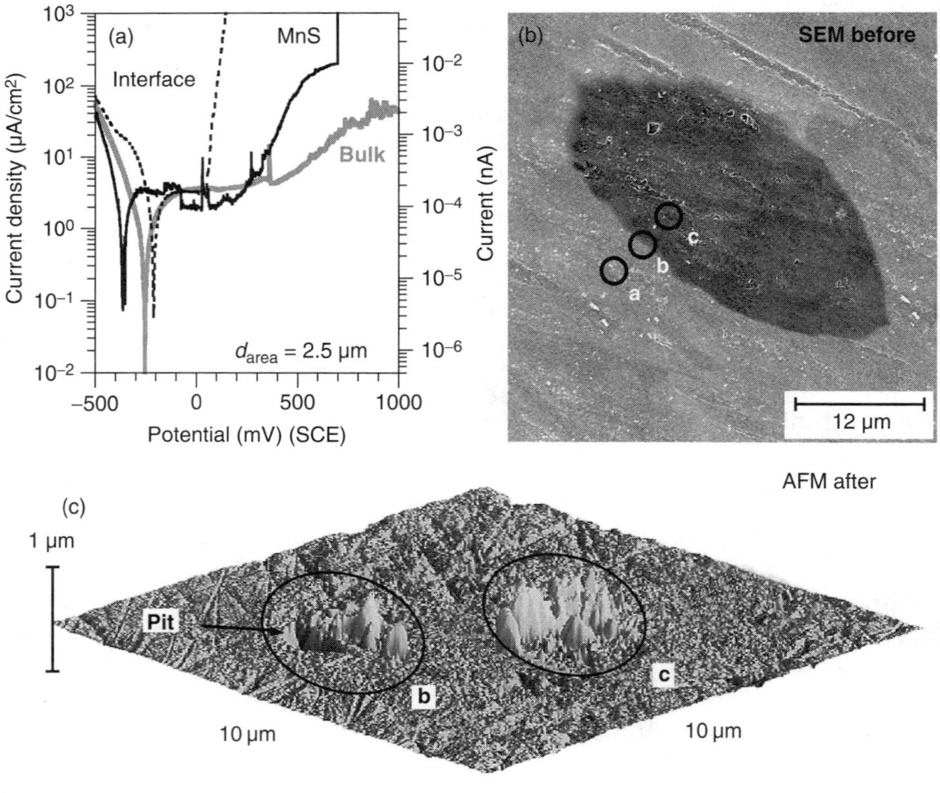

Figure 17.13 Corrosion behavior of different zones of a single MnS inclusion on an SS 304 plate (0.026% S) in 1 *M* NaCl. (a) Polarization curves measured on different spots of the MnS inclusion using a microcapillary with a tip diameter of about 2.5 μm. The investigated zones were the adjacent bulk matrix, the interface, and the center of MnS inclusion. (b) SEM picture of a MnS inclusion taken before the corrosion measurements. The black circles show spots that were investigated. (c) AFM picture of the spots b and c taken after the corrosion measurements. (From T. Suter and H. Böhni, *Electrochim. Acta*, **47**, 191 [2001]. With permission.)

behavior of different zones of a single MnS inclusion in 1 *M* NaCl. The investigated spots are indicated with white circles on the SEM picture taken before the corrosion experiments (Figure 17.13b).

The electrochemical experiments were discontinued at a current limit of 10 nA (corresponding to a current density of about 200 mA/cm^2). After a measurement, the capillary was positioned on the next spot. After all three corrosion measurements, the investigated inclusion was marked, rinsed with ethanol, and examined by SEM, EDX, and AFM. The polarization curves of the spots a to c provide information on the pitting behavior of different zones of the inclusion. The measurements performed on the adjacent bulk matrix indicate no pitting at potentials below +1000 mV (SCE). The current fluctuations in the passive range were probably caused by activation and repassivation processes. The measurements performed on the inclusion/bulk interface show that pit initiation occurred in the potential range between +100 and +300 mV (SCE). Curves measured on the interface showed numerous current fluctuations caused by pit nucleation processes. The pitting potentials of the polarization curves measured on the MnS inclusion are shifted to values about 200 mV more positive compared to the interface.

The fact that the center part of an MnS inclusion shows active behavior only at higher potentials supports the hypothesis that MnS inclusions are covered by an oxide film. The oxide film shows most defects at the interface inclusion/bulk due to the change of the crystallographic structure. This might explain why different overpotentials are necessary to activate the interface and the center part of an inclusion. The AFM picture shows the spots b and c after the corrosion experiments (Figure 17.13c). The measurement on the interface (spot b) indicates that a pit or microcrevice was formed on the adjacent bulk matrix, but not on the MnS inclusion. On the contrary, "nm hills" were formed at the border of the inclusion. On spot c (center of inclusion) "nm hills" are visible, but no pits which raises the question: "Which process led to the formation of these nm hills?" EDX measurements indicated a high sulfur, but a low manganese content for the "hills." It might be that the precipitation of dissolved sulfur species formed the hills. Another possibility is a preferential dissolution of Mn. Since dissolved Mn species are more voluminous, the S-rich remnant was probably pushed out forming a crust on the surface.

17.4.5 Measurements with Modified Microcell on Stainless Steel

17.4.5.1 pH Measurements

A coated 25-μm tungsten wire was used to determine the pH before and after pit initiation at a MnS inclusion in 1 *M* NaCl. A capillary with a tip diameter of about 100 μm was used to select a single MnS inclusion on the high-sulfur SS 304 plate. The pH was measured 20 μm over the sample surface (Figure 17.14). Prior to each polarization experiment a cathodic potential of −500 mV (SCE) was applied for 2 min. The pretreatment led to a pH of about 9.5. During the experiment the pH decreased slowly back to the value of the bulk electrolyte. A value of pH 5 was measured during the dissolution of the MnS inclusion and a pH value around 2 after the onset of stable pitting. Diffusion calculations were performed to estimate the pH inside the pit nucleation site. The results showed that a measured pH of 2 correlated with a pit pH of 1.5 and a measured pH of 5 correlated with a pit pH of 4.5 (97).

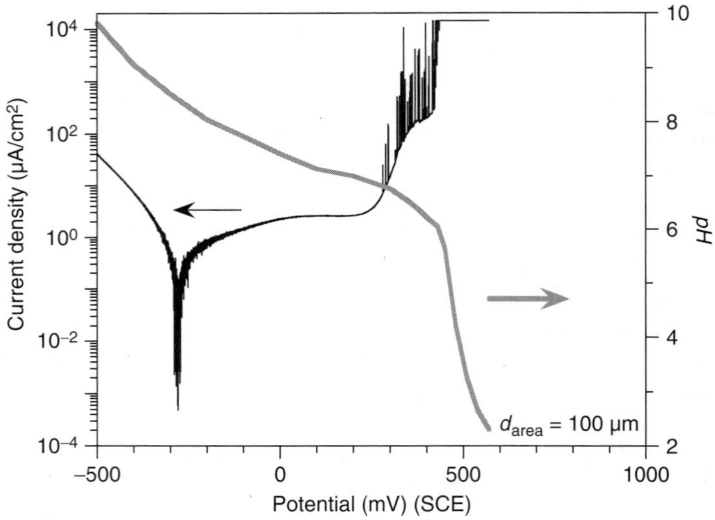

Figure 17.14 Polarization curve and pH values measured on a single MnS inclusion of an SS 304 plate (0.026% S) in 1 *M* NaCl. The pH was measured 20 μm over the inclusion. (From T. Suter, E. G. Webb, H. Böhni, and R. C. Alkire, *J. Electrochem. Soc.*, **148**, 186 [2001]. With permission.)

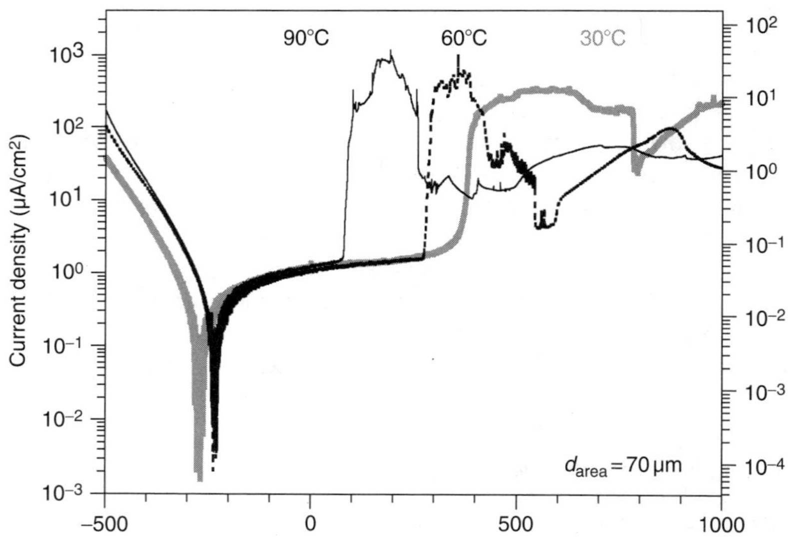

Figure 17.15 Polarization curves measured on single MnS inclusions of an SS 304 rod (0.017% S) in 1 *M* Na_2SO_4 at different temperatures. The temperature of the SS sample was 30°C, 60°C, and 90°C.

17.4.5.2 Measurements with Temperature Control

Figure 17.15 shows polarization curves measured in 1 *M* Na_2SO_4 on the medium-sulfur 304 plate (108, 109) at different temperatures. A capillary with a tip diameter of about 70 μm was chosen to study the dissolution behavior of single MnS inclusions in 1 *M* Na_2SO_4 at elevated temperature. In sodium sulfate, MnS inclusions are active, but the 304 SS does not pit (1).

The curves reveal that the dissolution behavior is strongly influenced by the temperature. Higher temperatures shift the onset of the MnS dissolution to more negative potentials and accelerate the MnS dissolution. However, the amount of dissolved MnS, indicated by the current charge, is independent of temperature.

17.4.5.3 Measurements with Applied Mechanical Stress

Microelectrochemical corrosion measurements were performed at single large MnS inclusions in order to evaluate the influence of applied mechanical stress on the pitting behavior of stainless steels at room temperature. The rolling process of the high-sulfur SS 304 plate leads to stretched inclusions. Depending on their orientation, different corrosion behavior is observed. Experiments on unstressed specimens with single shallow MnS inclusions showed metastable pitting in 1 *M* NaCl, whereas stable pitting was initiated at deep inclusions. A more detailed description of the stainless steel studies with applied mechanical stress is given in a previous paper (97).

Figure 17.16(a) shows the polarization curves for sites with single large and shallow MnS inclusions measured in 1 *M* NaCl without stress and with an applied stress equal to 80% of the yield strength. The curve for an inclusion without applied stress indicates MnS dissolution between +300 and +550 mV (SCE). Large current peaks were observed 100 mV after the onset of MnS dissolution. The curve indicates metastable corrosion, but not stable pitting. SEM pictures (Figure 17.16b) showed several small corrosion sites at the inclusion/bulk interface. The MnS inclusion did not dissolve completely in 1 *M* NaCl; it dissolved just partly. The interface

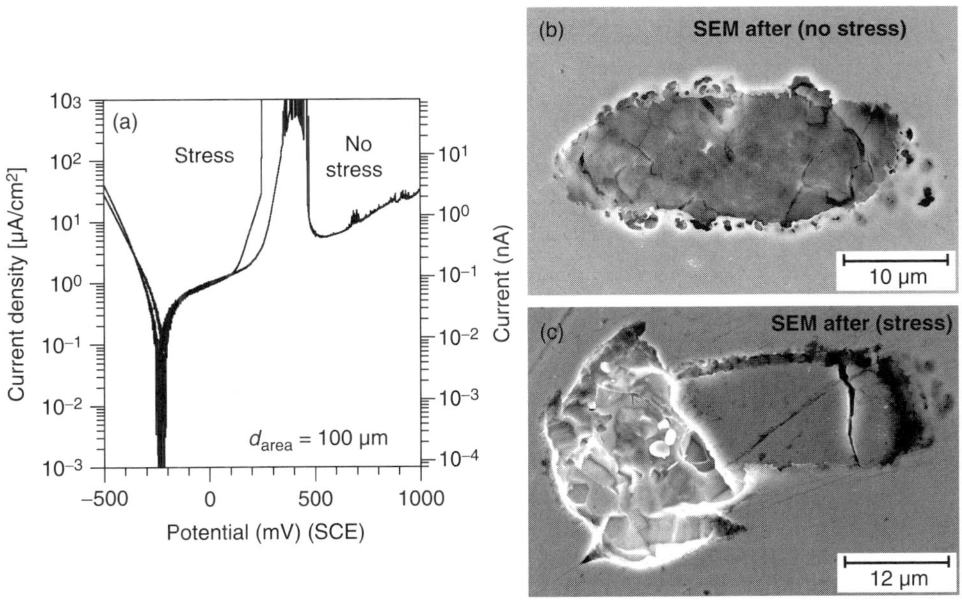

Figure 17.16 Dissolution and pitting behavior of single shallow MnS inclusion in 1 *M* NaCl without and with applied stress. (a) Polarization curves measured on single shallow MnS inclusions of an SS 304 plate (0.026% S) in 1 *M* NaCl. One experiment was performed with applied stress and the other without. (b) SEM image of the shallow MnS inclusion after the electrochemical experiment without applied stress. (c) SEM image of the shallow MnS inclusion after the electrochemical experiment with applied stress. ((From T. Suter, E. G. Webb, H. Böhni, and R. C. Alkire, *J. Electrochem. Soc.*, **148**, 186 [2001]. With permission.)

was mainly attacked, but the center of the inclusion dissolved just slightly. Measurements on single round MnS inclusions with a diameter >1 μm showed stable pitting in 1 *M* NaCl (Figure 17.12) (5, 107). The investigated shallow inclusions, however, showed only metastable pitting. The shallow geometry of a MnS inclusion decreased the probability of the onset of stable pitting (110).

Metastable corrosion events occurred along the inclusion/bulk interface. Attack underneath the thin edges of the inclusion led to chipped borders and hairline cracks. The cracks were 100 to 300 nm wide. The initiation points of the cracks were regions of corrosion attack at the interface. After a short period of active steel dissolution, the pressure of the products formed was high enough to chip the thin edges (<0.5 μm) of the shallow MnS inclusions, and an electrolyte exchange occurred. The formation of an aggressive electrolyte for stable pitting was not possible.

With applied stress (Figure 17.16a), MnS dissolution started at +100 mV (SCE), which is a more negative potential than that measured without stress. The curve indicates the onset of stable pitting at a potential that is 150 mV more positive. No current peaks of metastable events were observed before the onset of stable pitting. After the initiation of stable pitting, the corrosion currents increased immediately to values over 10 μA.

Figure 17.16(c) shows the SEM picture taken after the corrosion experiments with applied stress. The inclusion showed a crack running through the entire particle and a pit at one end, otherwise the MnS remnant dissolved only slightly (note that a diagonal polishing scratch on the surface was still visible after the corrosion experiment). If we assume uniform dissolution of the MnS inclusions before the onset of stable pitting, then the integrated current charge of about 100 nC would correspond to the dissolution of a 20-nm-thick MnS layer. Under applied stress, cracks were formed within all investigated shallow and large MnS inclusions. SEM and AFM investigations showed that active pitting started at the bottom of the cracks formed. Due to mechanical stress, microcrevices are formed at or within the inclusions, where the concentration of aggressive species such as chloride or hydrogen ions reaches a critical value for stable pit propagation. Our studies at large shallow MnS inclusions showed that mechanical stress accelerates the onset of pitting and in this way the formation of potential nucleation sites for stress corrosion cracking.

17.5 MICROCELL STUDIES OF PIT INITIATION AT SINGLE INCLUSIONS ON A 2024-T3 AL ALLOY

The heterogenous microstructure of Al 2024-T3 alloy has been optimized for desirable mechanical properties, but such heterogeneities render the alloy more susceptible to localized corrosion. Numerous constituent particles (>300,000/cm^2) can be divided into two major types based on composition: (a) AlCuFeMn inclusions, which are more noble than the Al matrix and act as cathodes, and (b) AlCuMg inclusions, which are less noble than the Al matrix and act as anodes (9–12). Corrosion, however, leads to a preferential dissolution of Mg and Al. The Cu rich remnants act as cathodes (111–113). The AlCuMg inclusions are generally considered to be S-phase particles.

The electrochemical methods to study corrosion attacks at inclusions are often based on large-scale experiments with exposed areas in the square millimeter to square centimeter range. Recent work by different groups (28, 114–123) showed that local investigations at the micro- and nanometer range provide a better insight into the initiation mechanism of pitting on Al alloys. The microcell technique has been applied to track the initiation processes at single inclusions on Al 2024-T3 alloy. A more detailed description of the Al 2024-T3 studies is given in previous papers (98, 99).

17.5.1 Experimental

Electrochemical investigations were carried out on Al 2024-T3 disks with a diameter of 12 mm and a thickness of 4 mm. The nominal element concentrations (wt%) for the Al 2024-T3 are 3.8% to 4.9% Cu, 1.2% to 1.8% Mg, 0.3% to 0.9% Mn, 0.5% Fe and Si, 0.15% Zn and Ti, 0.1% Cr. Prior to all electrochemical experiments, the samples were mechanically ground with successively finer SiC abrasive paper lubricated with water and finally polished with an aqueous 5-μm alumina slurry. The samples were rinsed with distilled water and ultrasonically cleaned in ethanol for 15 min. SEM pictures of polished surfaces showed some microcracks between inclusions and matrix after polishing and some partly dissolved inclusions (119). Only regions with unattacked inclusions and with no microcracks between inclusion and matrix were selected for study with the microcapillary. After positioning the microcell, a potential of −1000 mV (SCE) was applied for 1 min. Potentiodynamic polarization curves were then measured at a scan rate of 1 mV/sec. The electrolytes were 0.01 *M* and 1 *M* NaCl solutions, aerated and unbuffered (pH = 5.8 to 6.2). All solutions were prepared from reagent grade chemicals and distilled water.

Since this work focused on pit initiation and not propagation, the electrochemical experiments were usually discontinued at a current limit of 10 nA. After a microelectrochemical measurement, the investigated site was marked, rinsed with ethanol, and analyzed by SEM and EDX (EDX, Hitachi S-4700 and Zeiss DSM 960).

Figure 17.17 shows a typical surface where capillaries of different sizes were used to investigate different regions as indicated by the circles. A 20-μm capillary was used to test the corrosion behavior of single inclusions. A 20-μm region contained a single inclusion and the adjacent matrix. In that way it was possible to study the corrosion behavior of the interface inclusion/matrix which might be the weakest zone of an inclusion. A grid of reference marks on each specimen allowed easy location of selected areas and an exact positioning of the microcapillary over a chosen particle. A 50-μm capillary was used to evaluate the possible interaction between four or five inclusions and a 100-μm capillary for measurement at random sites. Areas with a diameter of 100-μm typically contained 20 to 30 inclusions.

17.5.2 Influence of the Size of the Exposed Area on Pit Initiation

Potentiodynamic polarization curves were measured to evaluate the pitting potentials of pure Al 99.999% and Al 2024-T3 in NaCl by varying the size of the exposed areas and the chloride concentrations. Figure 17.18 compares the curves of large-scale (1-cm diameter) and small-area (0.01-cm diameter) measurements of pure Al 99.999% and of Al 2024-T3 in 0.01 *M* NaCl. It may be recognized that the exposed areas of corrosion tests with a 100-μm microcapillary were smaller by four orders of magnitude. For the small-area measurements a random site on the specimen was selected with a microcapillary. When the diameter of the exposed area was decreased (from 1 cm to 100 μm) the pitting potentials of both pure Al and alloy materials shifted to higher values. On pure Al it was not possible to detect inclusions with

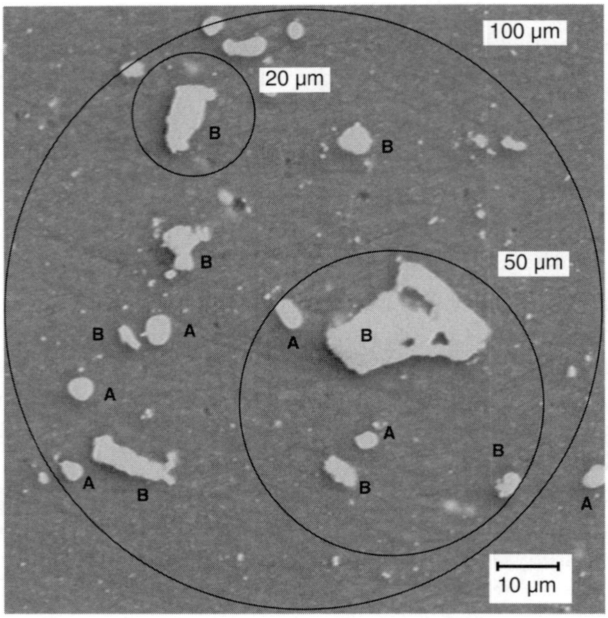

Figure 17.17 Microstructure of Al 2024-T3: A = AlCuMg inclusions, B = AlCuFeMn inclusions (From T. Suter and R. C. Alkire, *J. Electrochem. Soc.*, **148**, 36 [2001]. With permission.)

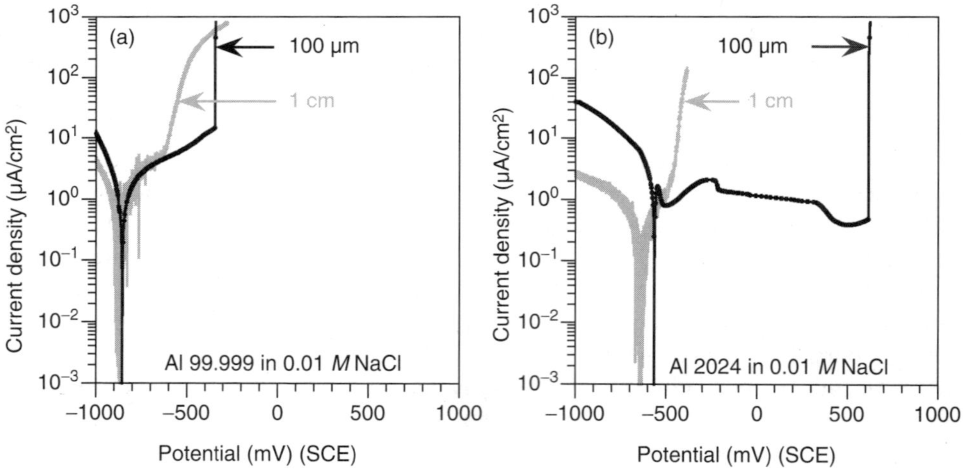

Figure 17.18 Polarization curves of Al 99.999% (a) and Al 2024-T3 (b) in 0.01 *M* NaCl. The size of the exposed area was varied. (From T. Suter and R. C. Alkire, *J. Electrochem. Soc.*, **148**, 36 [2001]. With permission.)

SEM/EDX investigations. Hence, it is speculated that the weak points were grain boundaries, dislocations, or inclusions smaller than 50 nm, undetectable by SEM. On Al 2024-T3, inclusions were the main weak points. Although areas of 100-μm diameter still included 20 to 30 inclusions, the pitting potential of 5 out of 10 measurements exceeded 0 mV (SCE). Several measurements in 0.01 *M* NaCl showed that pitting started at only a few AlCuMg inclusions (or the adjacent matrix) within 200 mV from the pitting potential of large-scale measurements.

Inclusions, even inclusions of apparently the same composition, show a wide range of corrosion behavior. Some inclusions were active at low potential, whereas others were inactive even at very high potentials. In general, the pitting potential is fixed by the most active inclusions. Large areas (1 cm^2 with several hundred thousand inclusions) include all kinds of inclusions, whereas the inclusions of smaller areas (0.0001 cm^2 with 20 to 30 inclusions) show less variation in their corrosion behavior. Since smaller areas usually do not include the most active inclusions, the pitting potential is increased, that is, only few measurements show a pitting potential close to that of large scale measurements.

It may also be seen in Figure 17.18 that, unlike pure Al, the cathodic currents of Al 2024-T3 increased and the corrosion potentials shifted to higher values when polarization was performed on smaller areas (100 μm). During large-scale experiments pit nucleation occurred at several inclusions at very negative potentials. Most of these events did not lead to stable pitting, but their anodic nucleation currents lowered the cathodic currents. Few events, however, led to stable pitting and fixed the corrosion potentials close to the pitting potential at values around −600 mV (SCE). The inclusions of the smaller areas were often inactive up to high potentials. The reduction in the number of active inclusions led not only to higher pitting potentials, but also to higher corrosion potentials and higher cathodic currents as well.

The reproducibility of the pitting potential for several large-area experiments in 0.01 and 1 *M* NaCl was about 20 mV. For small-area measurements, the pitting potentials were shifted to higher values and furthermore occurred in a much wider potential range (1200 mV for 0.01 *M* NaCl and 400 mV for 1 *M* NaCl). This range was larger in the less aggressive 0.01 *M* NaCl solution and the alloy Al 2024-T3 showed a larger range than pure Al. In less aggressive 0.01 *M* NaCl, active inclusions induced pitting occasionally, but in some cases there was no pitting at all. For 1 *M* NaCl, it was more likely for pitting to occur.

17.5.3 Pit Initiation at Areas with Four to Five Inclusions

A 50-μm microcapillary was used to determine the corrosion behavior of three types of selected areas: (a) areas with only AlCuMg inclusions, (b) areas with only AlCuFeMn inclusions, and (c) areas with both types of inclusions. Figure 17.19(a) shows an investigated area (c) with two AlCuMg inclusions (A1 and A2) and two AlCuFeMn inclusions (B1 and B2). The electrolyte was 1 *M* NaCl. Pitting was observed to initiate at inclusion A1. Such areas typically contained four inclusions (>1 μm) on average. It may be seen in Figure 17.19(b) that the pitting potentials of the areas with only AlCuMg inclusions and with both types of inclusions were usually 200 to 300 mV lower than the pitting potentials of areas with only AlCuFeMn particles.

The onset of pitting of an area with both types of particles always occurred at an AlCuMg inclusion. This behavior indicated that the less corrosion resistant sites were AlCuMg inclusions, and that before such an inclusion was completely dissolved, pitting started at the adjacent matrix. The pitting potentials of areas with only AlCuMg inclusions were in the same potential range as areas containing both types of inclusions. It is suggested that measurements with an external cathode (counter electrode) do not show pitting accelerated by a possible interaction between AlCuMg and AlCuFeMn inclusions. Under open-circuit conditions, however, an interaction between AlCuMg and AlCuFeMn inclusions cannot be excluded.

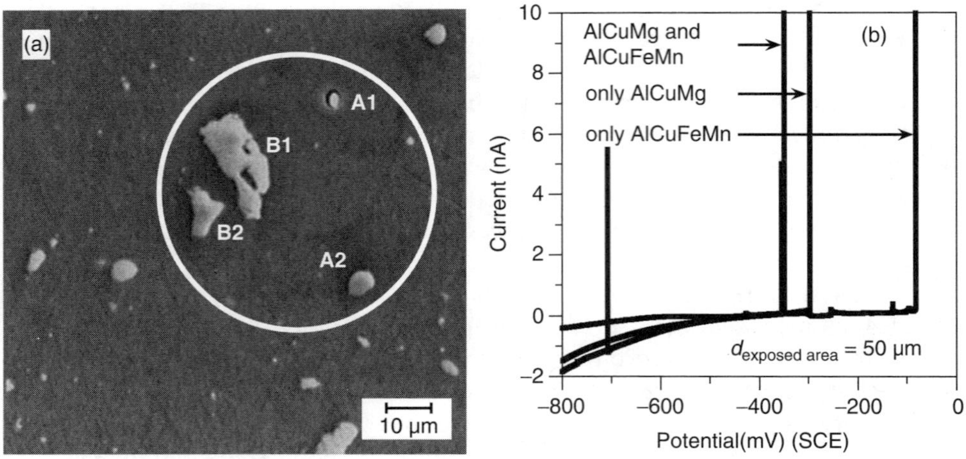

Figure 17.19 (a) Area with two AlCuMg (A1, A2) and two AlCuFeMn (B1, B2) inclusions. Pitting was initiated at particle A1. (b) Polarization curves at sites with four to five inclusions, Al 2024-T3 in 1 *M* NaCl. (From T. Suter and R. C. Alkire, *J. Electrochem. Soc.*, **148**, 36 [2001]. With permission.)

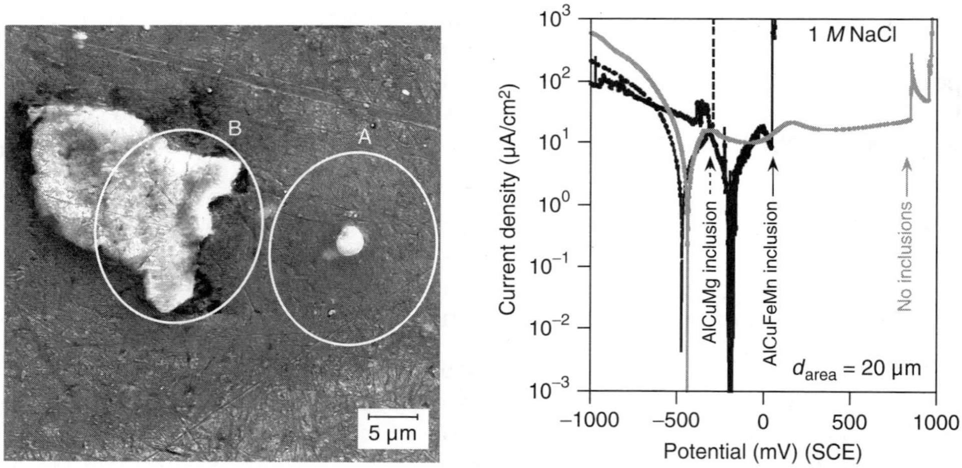

(a) (b)

Figure 17.20 Area of Al 2024-T3 with an AlCuMg (A) and an AlCuFeMn (B) inclusion: (a) The SEM picture shows the area after the microelectrochemical measurements. (b) Polarization curves of the AlCuMg (dotted line) and the AlCuFeMn (single line) inclusion and of a site without inclusions (gray line); Al 2024-T3 in 1 *M* NaCl. (From T. Suter and R. C. Alkire, *J. Electrochem. Soc.*, **148**, 36 [2001]. With permission.)

AlCuFeMn inclusions might act as additional cathodes and therefore accelerate the onset of pitting at AlCuMg inclusions.

17.5.4 Pit Initiation at Areas with Single Inclusions

To study pit initiation at single inclusions, a 20-μm capillary was used and the site was subsequently analyzed with SEM and EDX. Figure 17.20(a) shows SEM images of an area with an AlCuMg inclusion (region A) and an AlCuFeMn inclusion (region

B) after corrosion experiments in an electrolyte of 1 *M* NaCl. The corresponding polarization curves (Figure 17.20b) show that pit initiation occurred about 350 mV earlier on the region containing the AlCuMg inclusion.

For 15 experiments performed on each type of inclusion, the pitting potentials of measurements on single AlCuMg inclusions were found to vary from −700 to −150 mV (SCE) and the pitting potentials of measurements on single AlCuFeMn inclusions from −200 to +950 mV (SCE). Figure 17.20b also shows a polarization curve for an area without inclusions ($d_{inclusion} < 0.5\ \mu m$) for which it was typical that no pitting below +600 mV (SCE) was observed. Three out of six measurements showed pitting between +600 and +1000 mV (SCE), whereas the other three measurements showed no pitting at all. The measurements at areas without inclusions indicated that copper probably improves the resistance against pitting if it is homogeneously dissolved in the matrix (9).

Figure 17.21 shows EDX measurements on both an AlCuMg and an AlCuFeMn inclusion after the microelectrochemical tests. The edges and a center were investigated.

The smallest volume that can be analyzed with EDX is about 1 μm^3. Therefore, also the bulk contributed to the EDX measurements, showing their qualitative character. The ratios of the Cu and Mg peaks of the AlCuMg inclusion indicate often a preferential dissolution of Mg at the edge.

The edges of most AlCuMg inclusions showed a preferential dissolution of Mg, whereas the composition of the center changed hardly at all. In combination with SEM images, the EDX measurements indicated that AlCuMg inclusions started to dissolve at the edges. The ratios between Cu, Fe, and Mn peaks of the AlCuFeMn inclusion demonstrated preferential dissolution of Fe and Mn at the edge. The edges

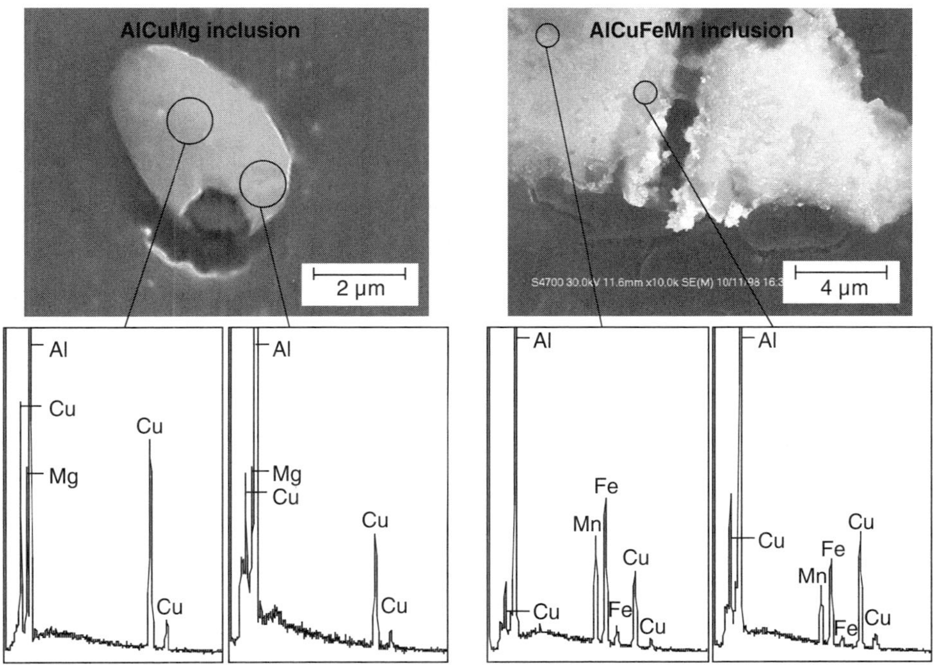

Figure 17.21 EDX measurements on the center and the edge of an AlCuMg and an AlCuFeMn inclusion after the microelectrochemical measurements. (From T. Suter and R. C. Alkire, *J. Electrochem. Soc.*, **148**, 36 [2001]. With permission.)

of five out of seven AlCuFeMn inclusions showed a preferential dissolution of Fe and Mn, while the composition of the center did not change. In addition, all EDX measurements indicated preferential dissolution of Al at AlCuMg and AlCuFeMn particles. Schmutz and Frankel showed that AlCuFeMn particles are heterogenous (12). Hence, it is not possible to ascribe directly local differences in composition to differences in local dissolution. Therefore, the composition of different zones of the inclusions was determined before and after the microelectrochemical measurement. The EDX results indicated that the edges of most inclusions showed a higher copper content after the electrochemical experiment.

It has been reported that adsorbed copper around inclusions was often observed in large-scale corrosion tests (111, 113). In this work the EDX results of several microelectrochemical experiments (maximum current = 10 nA) indicated the presence of copper in only a few cases. Microelectrochemical measurements stopped at a higher current (1 μA maximum, pit propagation) showed copper around single inclusions. It is possible that the adsorbed copper of large-scale experiments under open-circuit potential conditions (111, 113) originated from a few inclusions which were active at low potentials. Based on several SEM investigations and the EDX measurements the following sequences of events are suggested for the onset of pitting (maximum current = 10 nA) of the AlCuMg inclusions and the AlCuFeMn inclusions.

17.5.4.1 Onset of Pitting (Maximum Current = 10 nA) at AlCuMg Inclusions

Magnesium and aluminum dissolve preferentially at the edges leading to a microcrevice at the interface. If pit initiation occurs at lower potentials, pitting starts at the adjacent matrix of the AlCuMg inclusions. If pit initiation occurs at higher potentials, the inclusions dissolve more and pitting starts underneath the dealloyed Cu-rich remnant. Sometimes the formation of microcracks in the adjacent matrix was observed. The microcracks acted possibly as preferential starting points for pitting.

17.5.4.2 Onset of Pitting (Maximum Current = 10 nA) at AlCuFeMn Inclusions

The first step of pit initiation at lower potential occurs at the interface between the Al matrix and AlCuFeMn inclusion. Aluminum dissolves, accompanied by preferential dissolution of Fe and Mn at the edge of the AlCuFeMn inclusion. The cathodic reactions are considered to be both hydrogen and oxygen reduction. Hydrogen evolution likely forms microcracks in the bulk material, which provides initial sites for pitting (Figure 17.21). Pitting at AlCuFeMn inclusions started in about 20% of the experiments at a rather high potential. If pitting occurred at higher potentials, the interface between matrix and inclusion was the starting point for pitting. The preferential dissolution of Fe and Mn on the AlCuFeMn inclusion was even stronger. The inclusions looked quite porous after the experiment.

Figure 17.22 provides a summary of the pitting potentials of several microelectrochemical measurements in 1 *M* NaCl. The potential range for the onset of pitting was similar for areas with a single AlCuMg inclusion and with both types of inclusions. The potential of areas with only a single AlCuFeMn inclusion was shifted to higher values and its range was much wider. Areas without inclusions exhibited pitting at very high potential values.

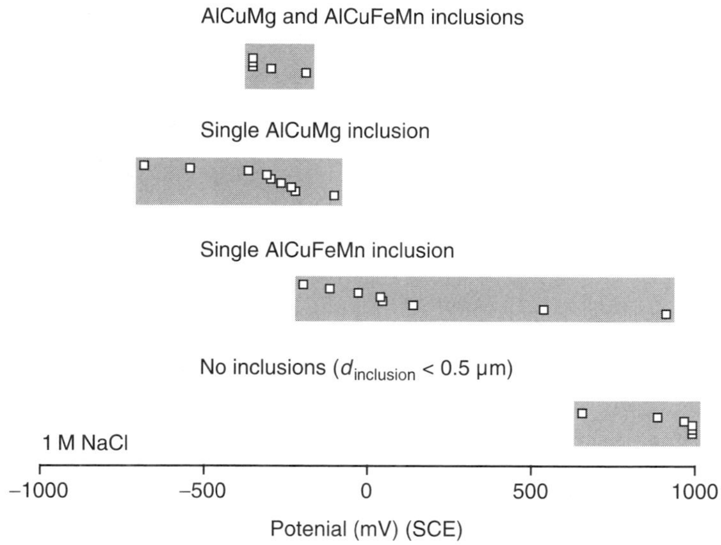

Figure 17.22 Pitting potentials of areas with AlCuMg and AlCuFeMn inclusions (50-μm area), with only a single AlCuMg inclusion (20-μm area), with only a single AlCuFeMn inclusion (20-μm area), and without inclusions (20-μm area). (From T. Suter and R. C. Alkire, *J. Electrochem. Soc.*, **148**, 36 [2001]. With permission.)

17.5.5 Studies of Pit Initiation at Different Zones of a Single Inclusion

In a recent study with a 3-μm microcapillary different spots of the matrix near an AlCuMg inclusion were investigated, as shown in Figure 17.23.

The polarization curves were taken at different distances from the matrix/inclusion interface. The results clearly show that the pitting potential increases approximately linearly with increasing distance from the interface. A pitting potential of over 1000 mV_{SCE} is attained at a distance of about 40 μm. This is completely different from stainless steels where already at a distance of 5 μm the bulk properties are measured and pitting was not observed any more. Further, local surface analyses have to be performed on Al 2024-T3 to determine whether these findings can be explained, for instance, by a copper depleted zone around the inclusion, since dissolved copper strongly affects the pitting behavior of aluminum.

17.6 MICROCELL STUDIES OF THE CORROSION RESISTANCE OF SDSS

The microstructure of duplex stainless steels (DSS) consists of two phases, austenite and ferrite. Both phases are present in separate domains and in approximately equal volume fractions. SDSS are duplex stainless steels for which the pitting resistance equivalent number (PREN) (= wt% Cr + 3.3 wt% Mo + 20 wt% N) exceeds 40 (124–126). In SDSS, the corrosion properties of both ferrite and austenite depend strongly on the actual chemical composition. The main alloying elements (chromium, molybdenum, nickel, and nitrogen) are not equally distributed in ferrite and austenite; the austenite is enriched in nickel and nitrogen, while the ferrite is enriched in chromium and molybdenum. The partitioning of these elements influences the

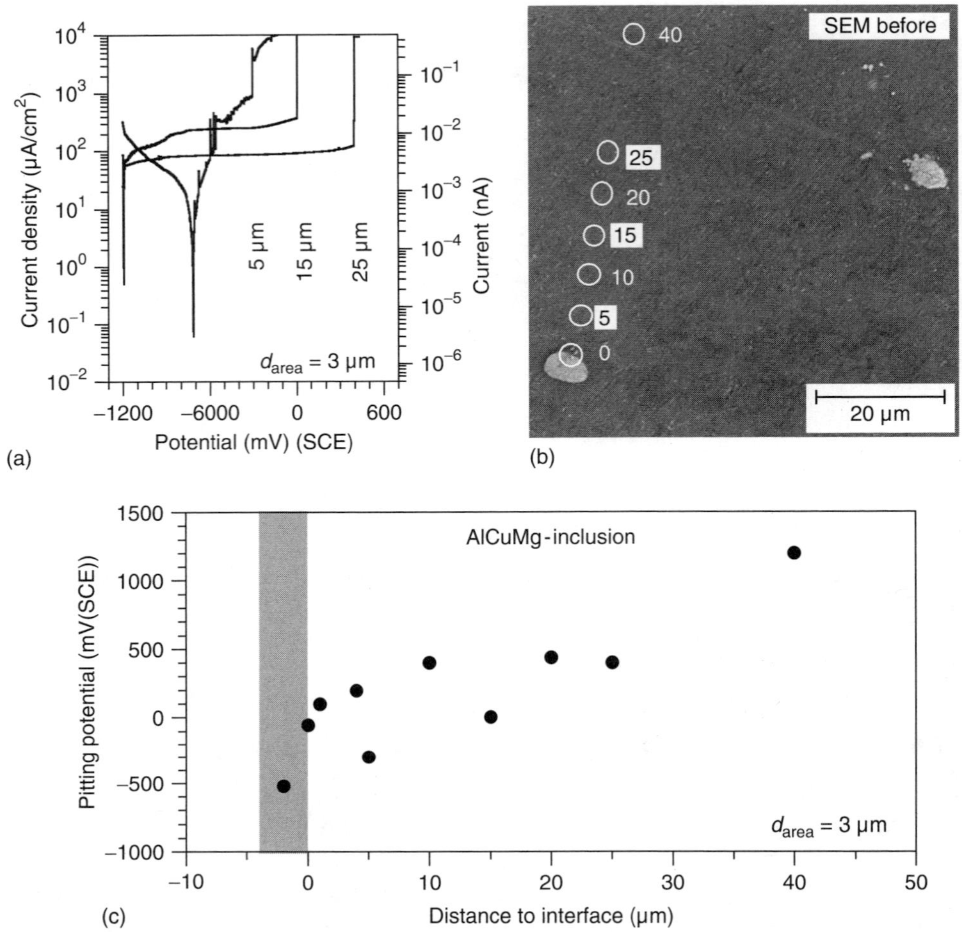

Figure 17.23 Corrosion behavior of different spots of the matrix near an AlCuMg inclusion on an Al-2024-T3 plate in 1 *M* NaCl. (a) Polarization curves measured at different distances from the matrix/inclusion interface using a microcapillary with a tip diameter of about 3 μm. (b) SEM picture of the AlCuMg inclusion taken before the corrosion measurements. The white circles show spots that were investigated. (c) Pitting potentials of investigated spots at different distances from the interface. The gray area indicates the position of the inclusion.

corrosion resistance of both the single phases and the entire alloy. DSS and in particular SDSS offer a very favorable combination of mechanical properties, especially high strength and high corrosion resistance. Newly developed SDSS exhibit extraordinary resistance to chloride induced corrosion.

These properties open a large field of applications in marine and petrochemical industries. A more detailed description of the SDSS studies is given in previous papers (100, 101).

17.6.1 Experimental

Three highly alloyed SDSS with slight variations in chromium, molybdenum, nickel, and manganese were used. The chemical compositions of the alloys are given in Table 17.1 and a typical structure of such an SDSS is shown in Figure 17.24(a). The

Table 17.1 Chemical Composition of the Investigated SDSS Alloys

SDSS	C (wt%)	Cr (wt%)	Ni (wt%)	Mn (wt%)	Mo (wt%)	N (wt%)
Alloy 1	0.016	25.1	7.1	3.8	3.7	0.38
Alloy 2	0.011	24.7	7.1	0.18	3.7	0.26
Alloy 3	0.013	23.5	7.1	4	4.4	0.41

Note: In the subsequent diagrams the numbers refer to these alloys.

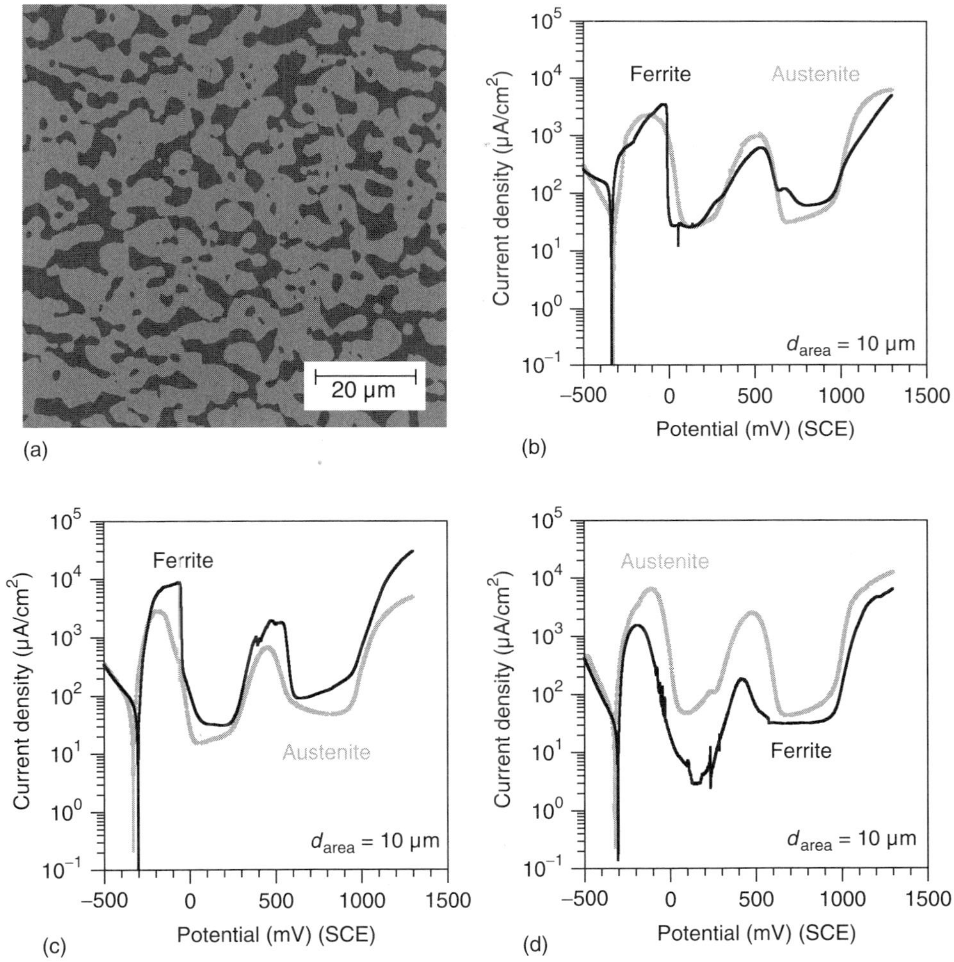

Figure 17.24 Corrosion behavior of the single phases of super duplex stainless steels. (a) Typical structure of an SDSS with the two phases ferrite and austenite. (b) Polarization curves of austenite (PREN = 44.2) and ferrite (PREN = 44.8) of alloy 1 (solution annealed at 1080°C) in 3 *M* HCl. Both phases show a similar corrosion resistance. (c) Polarization curves of austenite (PREN = 44.7) and ferrite (PREN = 42.3) of alloy 2 (solution annealed at 1050°C) in 3 *M* HCl. The austenite phase shows a higher corrosion resistance. (d) Polarization curves of austenite (PREN = 45.8) and ferrite (PREN = 47) of alloy 3 (solution annealed at 1080°C) in 3 *M* HCl. The ferrite phase shows a higher corrosion resistance. (From R. A. Perren, T. Suter, P. J. Uggowitzer, L. Weber, K. Magdowski, H. Böhni, and M. O. Speidel, *Corros. Sci.*, **43**, 707 [2001]. With permission.)

nitrogen content of these alloys was in the range of 0.26 to 0.41 wt%. Nitrogen is strongly enriched in the austenite (127) and thus improves the corrosion resistance by increasing the PREN of the austenite (128). Moreover, it has been reported that nitrogen also reduces the tendency for sigma phase precipitates by reducing the chromium enrichment in the ferrite (129, 130). Therefore, a high nitrogen content would allow a further increase of chromium and molybdenum, but the low solubility of nitrogen in the ferrite might cause precipitation of chromium nitrides.

The alloys were produced in a vacuum induction furnace (10-kg melt). After a homogenization treatment at 1230°C for 10 h (in air), the alloys were forged between 1200°C and 1150°C into bars of the dimension 15 mm by 65 mm. The final step was a heat treatment (solution annealing) performed at different temperatures. The alloys were chosen for microelectrochemical investigation because of the difference in the PREN of the individual phases. The aim was to study the corrosion behavior as a function of the PREN of the individual phases and to study the influence of precipitates on the corrosion behavior.

To measure the corrosion behavior of the individual phases, microcapillaries with a tip diameter of 10 μm were used. The very high resistance against pitting of SDSS in combination with the small size of the exposed single phases (no or only very few weak points) made microelectrochemical measurements of pitting potentials in SDSS in dilute chloride solution at room temperature impossible. Therefore, 3 *M* HCl was chosen as electrolyte, and the criteria for the corrosion resistance were the critical current density, the passivation potential, and the passive current density. Measurements with 1 *M* HCl usually revealed three zero current intersections and complicated the interpretation of the experiments. A stronger acid (3 *M* HCl) enlarged the active area of the anodic part reaction and provided only one zero current intersection. All microelectrochemical measurements revealed to some extent a second peak at a potential of about +500 mV (SCE), independent of the tested specimen. This second peak was probably caused by the oxidation of iron ions in the solution from the +II to the +III oxidation state. Because of the vertically mounted microcapillary the dissolved ions were not able to escape and caused the second peak in the polarization curves.

17.6.2 The Corrosion Behavior of Single Phases in SDSS without Precipitates

It has been suggested that the corrosion resistance of SDSS is determined by the corrosion resistance of the weaker phase (127, 131). Hence, an optimized alloy should have equal PRENs for both phases, ferrite and austenite. Different PRENs of the single phases were observed for two reasons. On the one hand, the variation of the chemical composition, in particular of the nitrogen content, leads to alloys with different PRENs of the single phases. On the other hand, the partitioning of the main alloying elements as a function of the solution annealing temperature leads to different PREN values.

Polarization curves of macroelectrochemical experiments looked as if the specimens consisted only of one phase and did not reveal the distinct contributions from the austenite and the ferrite phase in the active range. In this context the small average phase size of the examined SDSS of 20 μm (or less) must be mentioned.

The microelectrochemical measurements, however, showed that the PREN is quite reliable to assess the corrosion resistance of the individual phases of an SDSS. Figure 17.24(b) shows the polarization curves of an alloy with similar PREN of both phases ($PREN_{austenite}$ = 44.2; $PREN_{ferrite}$ = 44.8). The current densities and the passivation behavior of ferrite and austenite were similar. Figure 17.24(c) shows the polarization curve of an alloy with a lower PREN of the ferrite and a higher PREN of the austenite ($PREN_{austenite}$ = 44.7; $PREN_{ferrite}$ = 42.3). This is reflected in the differences in the passivation potential, the critical current densities as well as the passive current densities. The situation becomes reversed in Figure 17.24(d). Since the ferrite has a higher PREN it exhibits in this case a better corrosion resistance, while the austenite is slightly weaker with respect to corrosion ($PREN_{austenite}$ = 45.8; $PREN_{ferrite}$ = 47).

17.6.3 The Corrosion Behavior of Single Phases in SDSS with Precipitates

In addition to ferrite and austenite, a large variety of undesired secondary phases may be formed in SDSS during isothermal aging or quenching. The undesirable formation of secondary phases reduces the overall corrosion resistance and makes the alloy more susceptible to localized corrosion attack. Moreover, such precipitates may also affect the mechanical properties, mainly the toughness. The chemical composition and the heat treatment temperature have the most important influence on the precipitation behavior of SDSS since they determine the phase volume fraction of ferrite and austenite and the partitioning of the main alloying elements, chromium, molybdenum, nickel, and nitrogen (127). SDSS alloy 3 was chosen to investigate the influence of three important precipitates (sigma phase, chromium nitride, and secondary austenite) on the corrosion resistance. Different heat treatments (solution annealing) led to the formation of the precipitates. The corrosion behavior of the precipitation free state of the SDSS alloy 3 is shown in Figure 17.24(d). The three investigated precipitates were formed in the ferrite phase and reduced therefore the corrosion resistance of this phase.

Sigma phase: Sigma phase (Figure 17.25a) is by far the most important secondary phase because of its relatively large volume fraction and its detrimental influence on toughness and corrosion resistance. Especially in SDSS, the high contents of chromium and molybdenum, the main promoting elements for sigma phase formation, facilitate the precipitation of the sigma phase (132). In practice, when the annealing times are generally too short to approach thermodynamic equilibrium, the sigma phase can be formed. Figure 17.25(b) illustrates the polarization curves of alloy 3, solution annealed at 900°C for 30 min, where almost the whole ferrite is decomposed into sigma phase and eutectoid secondary austenite. The measurements were performed in the area of the primary austenite and in the former ferrite region, that is, in the sigma phase and in the secondary austenite. The polarization curve of the sigma phase and the secondary austenite does not show passivation and high current densities, while the primary austenite passivates readily and reveals much lower current densities.

Chromium nitrides: Rapid cooling (water quenching) from high solution temperatures causes supersaturation of nitrogen in the ferrite and leads to the formation of chromium nitride (133). Chromium nitrides of this type nucleate at single dislocation lines. They are accumulated mainly in the center of ferritic regions and are rarely found close to the ferrite–austenite boundary (Figure 17.26a). During quenching, nitrogen close to the phase boundary may escape into the adjacent austenite phase, which provides a much higher solubility for nitrogen. The frequently observed accumulation and alignment of chromium nitrides at ferrite subgrain boundaries

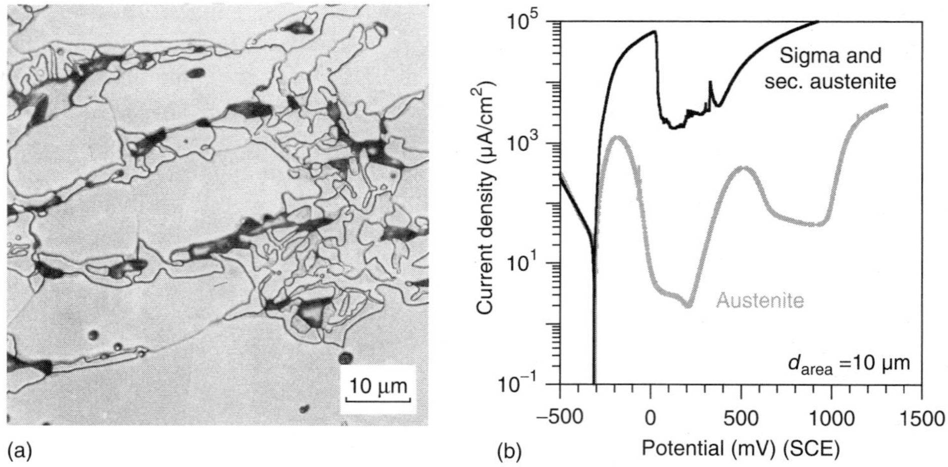

Figure 17.25 Influence of undesired sigma phase on the corrosion behavior of SDSS: (a) decomposition of ferrite (dark) into sigma phase and eutectoid secondary austenite (alloy 3, solution annealed at 950°C) and (b) polarization curves of sigma phase and eutectoid secondary austenite of alloy 3 (solution annealed at 950°C) in 3 *M* HCl. (From R. A. Perren, T. Suter, C. Solenthaler, G. Gullo, P. J. Uggowitzer, H. Böhni, and M. O. Speidel, *Corros. Sci.*, **43**, 727 [2001]. With permission.)

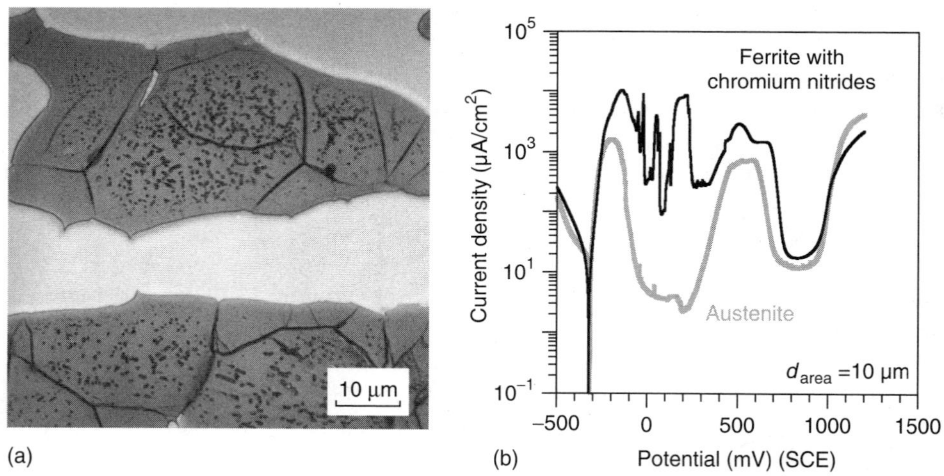

Figure 17.26 Influence of undesired chromium nitrides on the corrosion behavior of SDSS: (a) chromium nitrides (black needles) in the ferrite matrix (alloy 3, solution annealed at 1250°C), and (b) polarization curves of ferrite with chromium nitrides of alloy 3 (solution annealed at 1250°C) in 3 *M* HCl. (From R. A. Perren, T. Suter, C. Solenthaler, G. Gullo, P. J. Uggowitzer, H. Böhni, and M. O. Speidel, *Corros. Sci.*, **43**, 727 [2001]. With permission.)

correspond to the high dislocation density of those boundaries and can be explained through the diffusion impediment of nitrogen at these dislocations during quenching. Figure 17.26(b) shows that chromium nitrides formed in the ferrite phase during quenching change the shape of the polarization curve of the ferrite drastically. All microelectrochemical experiments revealed large transients in the active–passive transition range and the potential range between −100 and +400 mV (SCE). These transients corresponded to the amount of the chromium nitrides in the ferrite.

Secondary austenite: Secondary austenite in the shape of Widmannstätten needles in the ferrite region (Figure 17.27a) is formed by heating slowly through a temperature range around 1000°C to annealing temperatures below 1150°C (134). The temperature range depends on the alloy composition. Secondary austenite of the Widmannstätten type, precipitates in very thin needles (2 to 5 μm), was measured with the surrounding ferrite phase. Figure 17.27(b) exhibits the polarization curve of the austenite phase and ferrite-secondary austenite phase of alloy 3, solution annealed at 1050°C. A contribution from both phases was observed for the curve of the ferrite-secondary austenite phase as indicated by two active peaks. The first peak of the polarization curve belongs to the surrounding ferrite matrix and the second peak to the secondary austenite (Widmannstätten type).

17.6.4 The Corrosion Behavior of Phase Boundaries

The potentiodynamic polarization experiments with the microelectrochemical method at phase boundaries, that is, in the region where austenite and ferrite meet, revealed two different phenomena.

The first phenomenon is the superposition of the two polarization curves of the single phases (in the active range) to a new polarization curve of the phase boundary region, which contains in the active range a significant fraction of each phase. Higher fractions of one single phase in the measured area reduce the relative contribution of the other phase to the active range. In Figure 17.28(a), the current density curves of alloy 3, solution annealed at 1250°C, are displayed. The polarization curves of the single-phase austenite and ferrite with chromium nitride were already mentioned in the previous section (Figure 17.26b). The solid line in Figure 17.28(a) shows the

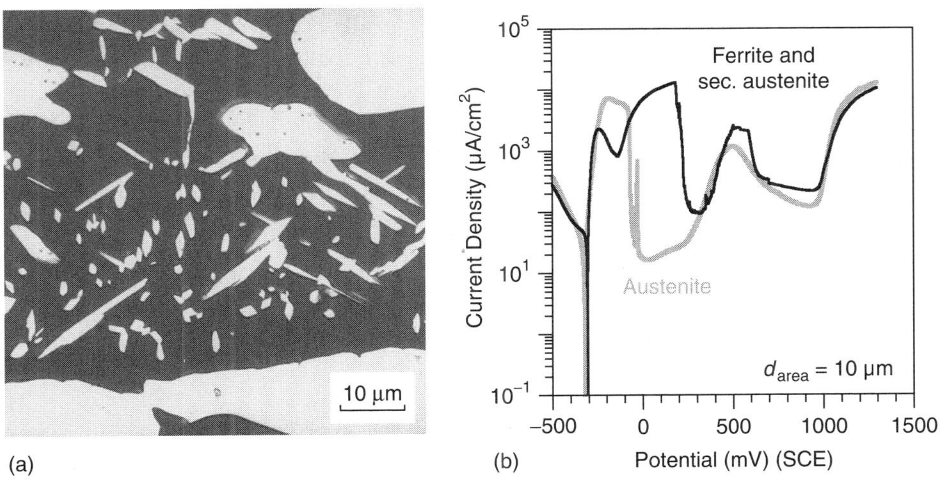

Figure 17.27 Influence of undesired secondary austenite on the corrosion behavior of SDSS: (a) secondary austenite (Widmannstätten needles) in the ferrite matrix (alloy 3, solution annealed at 1040°C) and (b) polarization curves of ferrite and secondary austenite of alloy 3 (solution annealed at 1040°C) in 3 *M* HCl. (From R. A. Perren, T. Suter, C. Solenthaler, G. Gullo, P. J. Uggowitzer, H. Böhni, and M. O. Speidel, *Corros. Sci.*, **43**, 727 [2001]. With permission.)

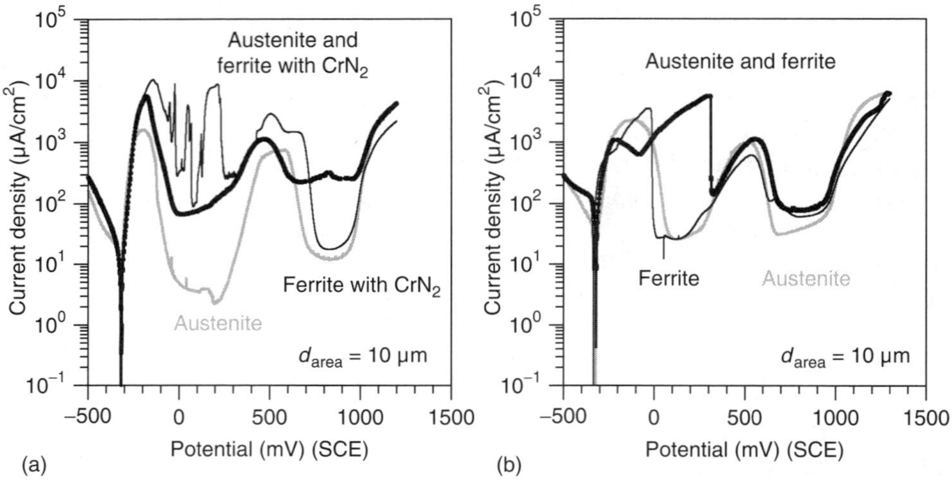

Figure 17.28 Corrosion behavior of the phase boundaries of SDSS: (a) polarization curves of austenite, ferrite, and both phases (austenite and ferrite with chromium nitrides) of alloy 3 (solution annealed at 1250°C) in 3 *M* HCl and (b) polarization curves of austenite, ferrite, and both phases (austenite and ferrite) of alloy 1 (solution annealed at 1080°C) in 3 *M* HCl. (From R. A. Perren, T. Suter, P. J. Uggowitzer, L. Webber, R. Magdowiski, H. Böhni, and M. O. Speidel, *Corros. Sci.*, **43**, 707 [2001]. With permission.)

polarization curves measured in the phase boundary region austenite and ferrite with chromium nitride.

The second phenomenon is the separation of the active ranges of the single phases in the polarization curve of the phase boundary. This separation results in a shift of the passivation potential of the weaker phase (with respect to the PREN) toward higher potentials, while the passivation potential of the stronger phase is shifted to lower potentials. In Figure 17.28(b) the polarization curves of both individual phases and of the phase boundary of alloy 1 are displayed. The active peak of the austenite is shifted to more negative potentials, whereas the active peak of the ferrite is shifted to more positive potentials. Here, a local corrosion element protects the stronger phase (austenite), while the weaker phase (ferrite) is now even more susceptible to corrosion. The overall corrosion resistance of such a local element is much worse compared to the simple superposition in the first case. A separation of the contribution of both phases was usually observed when the difference of the PREN of the single phases was small.

17.7 MICROELECTROCHEMICAL TESTING OF STAINLESS STEEL WELDS

Weld joints are often a major target of corrosion. The welding process often results in decreased material properties in the heat affected zone and the weld itself. The microcell technique has been successfully used to test welds of stainless steels. Local measurements of different areas of welds have indicated that different zones suffer corrosion attack depending on the type of weld. The measuring technique proved easy to handle and apply. Unlike immersion tests, which may take several days, it yields a reliable judgment of the quality of a weld in only a few minutes. In many applications, it is required that the corrosion resistance of the entire construction is tested. The microcell technique allows also a nondestructive testing of welds. The stainless steel weld tests are described in more detail elsewhere (96).

Figure 17.29 and Figure 17.30 show two examples of microelectrochemical testing of welds. The corrosion resistance of laser welds was evaluated to optimize a new laser welding process to join 304 SS sheets. Figure 17.29(a) shows SEM pictures of a laser weld. The production pieces with the laser weld were tested as received in 1 *M* NaCl solution. In Figure 17.29(b) an example of polarization curves measured on different spots of the laser welding is shown. It is clear that the corrosion resistance of the matrix and the zone adjacent to the weld trench is significantly higher than that of the weld trench itself. The bottom of the trench showed a low corrosion resistance. It was found that the holes at the bottom of the trench (Figure 17.29a) significantly decreased the corrosion resistance. A tow high laser intensity caused these holes. This meant that the process parameters had to be optimized.

In another case, Ciba-Geigy AG desired to cut the costs of the metal inert gas MIG welding process for 316L stainless steels (about 3% Mo). It was intended to replace the expensive, highly alloyed (6% Mo) welding-rod material with a cheaper, lower alloyed (less Mo) one. To verify the necessary Mo content of the welding-rod material, several MIG welds with different Mo contents were investigated.

The corrosion resistance of a tested weld zone was compared with the evaluated Mo content of the corresponding zone. Figure 17.30(a) shows the investigated spots on an MIG welding of a 316L stainless steel. Figure 17.30(b) indicates the

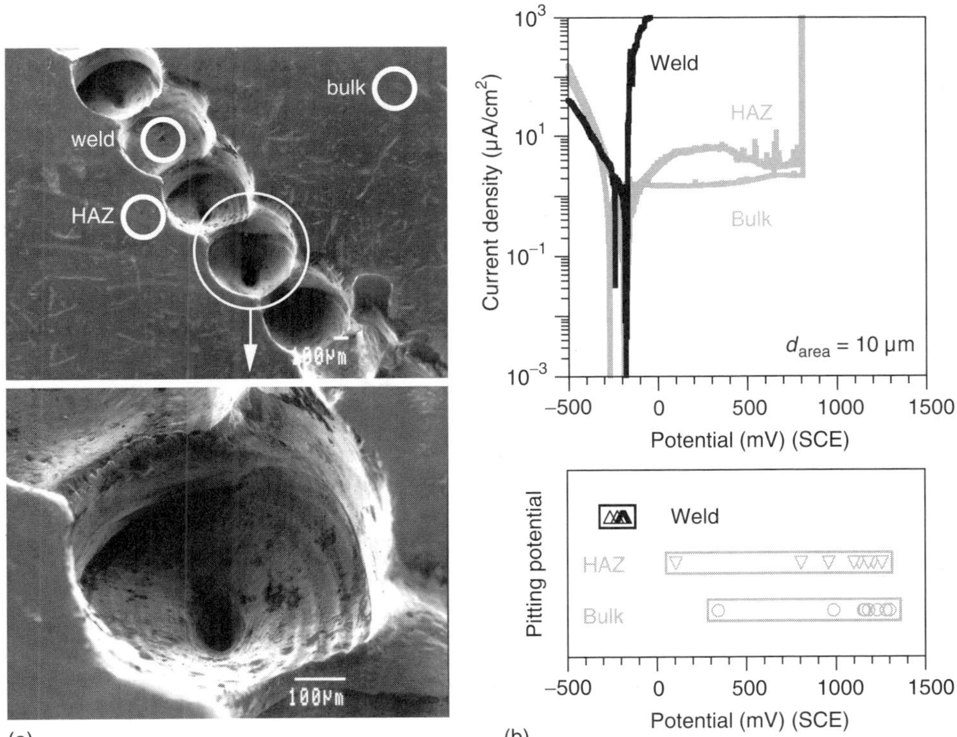

Figure 17.29 Microelectrochemical testing of a laser weld on a 304 SS: (a) SEM pictures of the laser weld and (b) polarization curves and pitting potentials of different zones of the laser weld measured in 1 *M* NaCl.

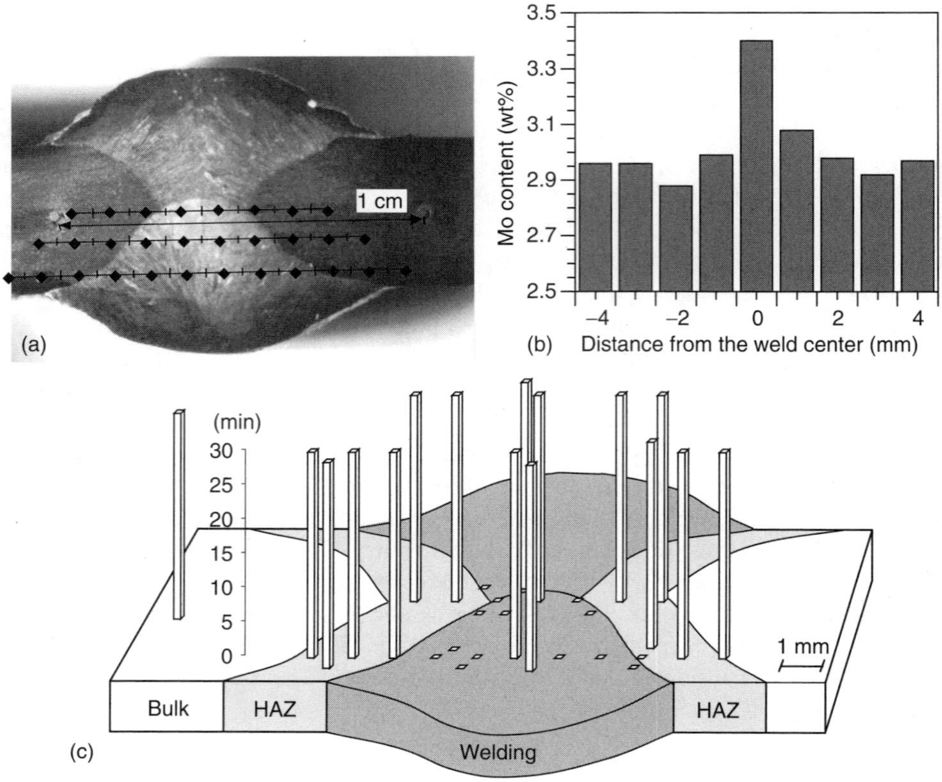

Figure 17.30 Microelectrochemical testing of an MIG weld on a 316L SS. (a) Investigated spots (black diamonds). (b) Mo content of the MIG welding. (c) Time to failure (the maximum measuring time was 30 min). The MIG weld was tested at +700 mV (SCE) in 1 *M* NaCl.

Mo content of the weld. The center of the weld contains about 3.4% Mo, whereas the edges of the weld contain only around 2.9% Mo. Potentiostatic measurements at +700 mV (SCE) were performed with a 1-mm microcapillary in 1 *M* NaCl in order to verify the corrosion resistance of the weld. In Figure 17.30(c), the time to failure (onset of stable pitting) is shown. The maximum measuring time was limited to 30 min. Since the center of the weld (3.5% Mo) did not show stable pitting, it can be concluded that a Mo content of 3.5 % is sufficient to prevent localized corrosion for the given conditions.

17.8 MICROCELL STUDIES IN MICROELECTRONICS

There are various experimental techniques that provide information on the physical and chemical properties of microengineered parts. However, the determination of electrochemical properties with high resolution still remains a problem. Local surface conditions like the catalytic activity, the insulating properties of cover layers, and the behavior of metal combinations on a micrometer scale are important quality factors. There are only few techniques available that perform similar experiments with the same high resolution. Most other techniques, like electrochemical AFM or SECM, require exposure and polarization of the entire device in the electrolyte.

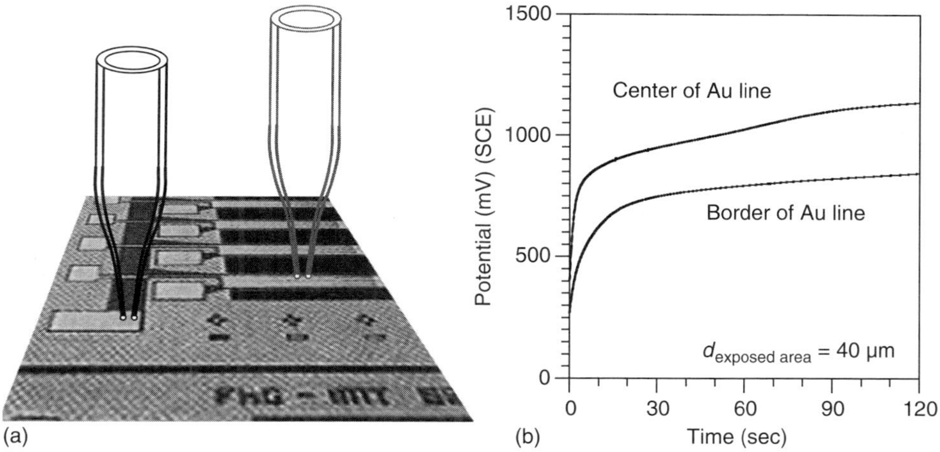

Figure 17.31 Application of microcell technique in microelectronics: (a) microelectrochemical tests on an electronic device, and (b) potential time curve during galvanostatic measurements at the center and at the edge of a gold line; $i = 10\ \mu A/cm^2$; the test solution was borate buffer, pH 8.4. (From T. Suter and H. Böhni, in *Electrochim. Acta*, **47**, 191 [2001]. With permission.)

The use of photolithographic methods requires highly flat surfaces and involves a significant amount of work for masking the areas. The microcell technique allows a combination of high throughput with highly confined resolution and electrochemical polarization. An example of the application of the microcell technique in microelectronics is shown in Figure 17.31. By measuring the catalytic activity with microelectrochemistry it could be demonstrated that the edges of the Au lines had lower overpotentials than the center, indicating a higher amount of impurities.

17.9 SUMMARY

- The micro- and nanoelectrochemical techniques which are used to investigate initiation and propagation processes of localized corrosion can be divided into two major groups:
  1. Scanning techniques (immersed samples are scanned using microelectrodes or ultra-microelectrodes)
  2. Small-area techniques (only a microscopic area of the sample is exposed to the electrolyte). Scanning techniques show often a better lateral resolution, whereas the small-area techniques allow polarization of microscopic surface areas. Therefore, a direct correlation between the measured corrosion currents of a local spot and the corresponding corrosion attack is possible.

- Microelectrochemical techniques using microcapillaries as microcells are an elegant method to study local phenomena on metal surfaces. Due to the enhanced current resolution down to picoamperes and femtoamperes local processes in the nanometer range can easily be measured. Incorporation of microsensors into the microcapillary cell generates additional chemical parameters such as pH values simultaneously with electrochemical measurements. This provides important analytical information especially, with respect to nonstationary processes.

- The results of localized corrosion experiments on stainless steels in 1 *M* NaCl clearly showed that in moderately concentrated solutions, the presence of MnS inclusions is of primary importance for pit initiation, whereas the size and geometry as well as the amount of chlorides and the temperature are critical for the transition from metastable to stable pit growth. Pit initiation studies at MnS inclusions on 304 stainless steels were performed at different scales. Electrochemical experiments using a capillary with a tip diameter in the range of 100 μm allowed evaluation of the corrosion behavior of sites with one single inclusion. Electrochemical measurements using a capillary with a tip diameter in the range of 1 μm allowed investigation of different spots of a single inclusion. Hence, the corrosion behavior of the weakest zone could be determined.
- The onset of pitting at single inclusions on the aluminum alloy 2024-T3 has been studied in 1 *M* NaCl. The results showed that pitting started at only a few AlCuMg inclusions within 200 mV of the pitting potential of large-scale measurements. Pit initiation was found to occur at the adjacent matrix of AlCuMg inclusions. Compared to the AlCuMg inclusions the onset of pitting at AlCuFeMn inclusions occurred at potentials 200 to 600 mV more positive. Microcracks in the bulk material, presumably formed by hydrogen evolution, were found to be starting points for pitting. The onset of pitting at areas without inclusions occurred at potentials higher than +500 mV (SCE). If copper was homogenously dissolved in the matrix the resistance against pitting was strongly improved.
- The individual corrosion behavior of both single phases in SDSS was determined. The results showed a good correlation with the empirical PREN of the corresponding single phase. The microelectrochemical experiments in 3 *M* HCl have revealed two different kinds of interactions between the ferrite and the austenite phases at the phase boundary, a superposition and a separation of the two polarization curves of the single phases. Inappropriate heat treatment of SDSS may cause precipitation of the undesired phases, which drastically decrease the corrosion resistance and which may also reduce the toughness. The most important precipitates, namely, sigma phase, chromium nitrides, and secondary austenite, decreased the corrosion resistance of the ferrite phase of the SDSS.
- The microcell technique is a powerful tool to test the corrosion resistance of welds and to verify the electrochemical behavior of electronic devices.

ACKNOWLEDGMENTS

The Al studies and some SS studies (pH measurements and measurements with applied stress) were performed in collaboration with the laboratory of Prof. Richard Alkire at the University of Illinois, USA. The SDSS studies were performed in collaboration with the Institute of Metallurgy of the Swiss Federal Institute of Technology Zürich, Switzerland.

REFERENCES

1. Z. Szlarska-Smialowska, *Pitting Corrosion of Metals*, NACE, Houston, TX (1986).
2. H. Böhni, T. Suter, and A. Schreyer, *Electrochim. Acta*, **40**, 1361 (1995).
3. J. Stewart and D. E. Williams, *Corros. Sci.*, **33**, 457 (1992).
4. E. Lott and R. C. Alkire, *J. Electrochem. Soc.*, **136**, 973 (1989).
5. R. Ke and R. Alkire, *J. Electrochem. Soc.*, **139**, 1573 (1992).
6. Z. Szklarska-Smialowska and E. Lunarska, *Werkstoffe Korros.*, **32**, 478 (1981).
7. G. S. Eklund, *J. Electrochem. Soc.*, **121**, 467 (1974).
8. H. Keller and H. J. Grabke, *Werkstoffe Korros.*, **32**, 540 (1981).
9. I. L. Mueller and J. R. Galvele, *Corros. Sci.*, **17**, 179 (1977).
10. K. Urushino and K. Sugimoto, *Corros. Sci.*, **19**, 225 (1979).
11. G. S. Chen, M. Gao and R. P. Wei, *Corrosion*, **52**, 8 (1996).
12. P. Schmutz and G. S. Frankel, *J. Electrochem. Soc.*, **145**, 2295 (1998).
13. U. R. Evans, Nature, **142**, 160 (1938).
14. R. S. Thornhill and U. R. Evans, *J. Chem. Soc.*, 614 (1938).
15. R. S. Thornhill and U. R. Evan, *J. Chem. Soc.*, 2109 (1938).
16. M. K. Budd and F. F. Booth, *Metalloberfläche* **19**, 197 (1965).
17. P. Doig and J. W. Edington, *Br. Corros. J.*, **2,** 88 (1974).
18. H. S. Isaacs and G. Kissel, *J. Electrochem. Soc.*, **119**, 1628 (1972).
19. H. S. Isaacs and B. Vyas, in *Electrochemical Corrosion Testing*, F. Mansfeld, U. Bertocci, Editors, p. 3, ASTM, Philadelphia, PA (1981).
20. H. S. Isaacs, *Corrosion*, **43**, 594 (1987).
21. H. S. Isaacs, *Corros. Sci.*, **28**, 547 (1988).
22. H. S. Isaacs, *J. Electrochem. Soc.*, **135**, 2180 (1988).
23. H. S. Isaacs, *Corros. Sci.*, **29**, 313 (1989).
24. H. S. Isaacs and M. W. Kendig, *Corrosion*, **36**, 269 (1980).
25. R. S. Lillard, P. J. Moran, and H. S. Isaacs, *J. Electrochem. Soc.*, **139**, 1007 (1992).
26. J. L. Luo, Y. C. Lu, and M. B. Ives, *J. Electroanal. Chem.*, **326**, 51 (1992).
27. F. Assi, Ph.D. Thesis No. 13715, ETH Zürich (2000).
28. J. O. Park, C. H. Paik, and R. C Alkire, *J. Electrochem. Soc.*, **143**, L174 (1996).
29. J. O. Park, C. H. Paik, Y. H. Huang, and R. C Alkire, *J. Electrochem. Soc.*, **146**, 517 (1999).
30. D. E. Williams, A. R. J. Kucernak, and R. Peat, *Electrochim. Acta*, **38**, 57 (1993).
31. M. A. Butler, *J. Electrochem. Soc.*, **131**, 2185 (1984).
32. M. Kozlowski and W. H. Smyrl, *Electrochim. Acta*, **34**, 1763 (1989).
33. J. W. Schultze and J. Thietke, *Electrochim. Acta*, **34**, 1769 (1989).
34. S. Kudelka and J. W. Schultze, *Electrochim. Acta*, **42**, 2817 (1997).
35. D. Shukla and U. Stimming, *Werkstoffe Korros.*, **40**, 43 (1989).
36. D. E. Williams, A. R. J. Kucernak, and R. Peat, *Electrochim. Acta*, **38**, 71 (1993).
37. D. W. Buzza and R. C. Alkire, *J. Electrochem. Soc.*, **142**, 1104 (1995).
38. M. Verhoff and R. C. Alkire, *J. Electrochem. Soc.*, **147**, 1359 (2000).
39. M. Verhoff and R. C. Alkire, *J. Electrochem. Soc.*, **147**, 1366 (2000).
40. M. Stratmann and H. Steckel, *Ber. Bunsenges. Phys. Chem. I.*, **92**, 1244 (1988).
41. M. Stratmann, K.T. Kim, and H. Steckel, *Z. Metallkd.*, **81**, 715 (1990).
42. R. E. Lobing, D. J. Siconolfi, J. Maisano, G. Grundmeier, H. Steckel, R. P. Frankenthal, M. Stratmann, and J. D. Sinclair, *J. Electrochem. Soc.*, **143**, 1175 (1996).
43. S. Yee, R. A. Oriani, and M. Stratmann, *J. Electrochem. Soc.*, **138**, 55 (1991).
44. G. Binnig and H. Rohrer, *Hel. Phys. Acta*, **55**, 726 (1982).
45. G. Binnig, H. Rohrer, C. Gerber, and E. Weibel, *Phys. Rev. Lett.*, **49**, 57 (1988).
46. A. A. Gewirth and B. K. Niece, *Chem. Rev.*, **97**, 1129 (1997).
47. M. V. Mirkin, *Mikrochim. Acta*, **130**, 127 (1999).

48. S. R. Taylor, A. C. Hillier, and M. Seo, Editors, Localized in-situ methods for investigating electrochemical interfaces, PV 99–28, The Electrochemical Society Proceedings Series, Pennington, NJ (1999).
49. F. F. Fan and A. J. Bard, *J. Electrochem. Soc.*, **136**, 166 (1989).
50. A. Miyasaka and H. Ogawa, *Corros. Sci.* **31**, 99 (1990).
51. V. Maurice, H. Talah, and P. Marcus, *Surf. Sci.*, **304**, 98 (1994).
52. V. Maurice, W. P. Yang, and P. Marcus, *J. Electrochem. Soc.*, **143**, 1182 (1996).
53. V. Maurice, L. H. Klein, and P. Marcus, *Electrochem. Solid State Lett.*, **4**, B1 (2001).
54. S. Matsuoka, H. Masuda, Y. Ikeda, K. Akaika, and Y. Ochi, *JSME Int. J.*, **35**, 456 (1992).
55. R. C. Bhardwaj, A. Gonzalez-Martin, and J. O. Bockris, *J. Electrochem. Soc.*, **138**, 1901 (1991).
56. M. P. Ryan, R. C. Newman, and G. E. Thompson, *J. Electrochem. Soc.*, **141**, L164 (1994).
57. M. A. Baker, *Surf. Interface Anal.*, **20**, 535 (1993).
58. R. M. Rynders, C. H. Paik, R. Ke, und R. C. Alkire, *J. Electrochem. Soc.*, **141**, 1439 (1994).
59. A. Schreyer, T. Suter, L. Eng, and H. Böhni, in *Electrochemical Nanotechnology*, W. J. Lorenz and D. Pliet, Editors, p. 199, Wiley-VCH, Wheinheim, Germany (1997).
60. N. Casillas, S. Charlebois, W. H. Smyrl, and H. S. White, *J. Electrochem. Soc.*, **141**, 636 (1994).
61. P. James, N. Casillas, and W. H. Smyrl, *J. Electrochem. Soc.*, **143**, 3853 (1996).
62. Y. Zhu and D. E. Williams, *J. Electrochem. Soc.*, **144**, L43 (1997).
63. J. W. Still and D. O. Wipf, *J. Electrochem. Soc.*, **144**, 2657 (1997).
64. C. H. Paik, S. B. Basame, H. S. White, and R. C. Alkire, in *Localized In-Situ Methods for Investigating Electrochemical Interfaces*, S. R. Taylor, A. C. Hillier, and M. Seo, Editors, PV 99-28, p. 122, The Electrochemical Society Proceedings Series, Pennington, NJ (1999).
65. E. Betzig, J. K. Trautman, T. D. Harris, J. S. Weiner, and R. L. Kostelak, *Science*, **251**, 1468 (1991).
66. J. Kerimo, M. Büchler, and W. H. Smyrl, in *Optical Metrology*, G. A. Al-Jumaily, Editor, PV CR72, p. 232, SPIE Optical Engineering Press, Bellingham, WA (1999).
67. P. I. James, L. F. Garfias-Mesias, P. J. Moyer, and W. H. Smyrl, *J. Electrochem. Soc.*, **145**, L 64 (1998).
68. L. F. Garfias-Mesias and W. H. Smyrl, *J. Electrochem. Soc.*, **146**, 2495 (1999).
69. M. Büchler, J. Kerimo, F. Guillaume, and W. H. Smyrl, in *Localized In-Situ Methods For Investigating Electrochemical Interfaces*, PV 99–28, p. 55, The Electrochemical Society Proceedings Series, Pennington, NJ (1999).
70. F. Guillaume, L. F. Garfias-Mefias, M. Büchler, and W. H. Smyrl, in *Critical Factors in Localized Corrosion III*, R. G. Kelly, G. S. Frankel, P. M. Natishan, and R. C. Newman, Editors, PV 98–17, p. 155, The Electrochemical Society Proceedings Series, Pennington, NJ (1998).
71. M. Büchler, J. Kerimo, F. Guillaume, and W. H. Smyrl, *J. Electrochem. Soc.*, **147**, 3691 (2000).
72. P. Schmutz and G. S. Frankel, *J. Electrochem. Soc.*, **145**, 2285 (1998).
73. P. Schmutz and G. S. Frankel, *J. Electrochem. Soc.*, **146**, 4461 (1999).
74. E. H. Dix, *Trans. AIMME*, **137**, 11 (1940).
75. M. M. Lohrengel, *Electrochim. Acta*, **42**, 3265 (1997).
76. J. W. Schultze and M. M. Lohrengel, *Galvanotechnik*, **51**, 735 (1997).
77. J. W. Schultze and V. Tsakova, *Electrochim. Acta*, **44**, 3605 (1999).
78. A. W. Hassel and M. M. Lohrengel, *Electrochim. Acta*, **42**, 3327 (1997).
79. R. C. Newman and H. S. Isaacs, *J. Electrochem. Soc.*, **130**, 1621 (1983).
80. R. C. Newman, *Corros. Sci.*, **25**, 341 (1985).
81. P. Heimgartner and H. Böhni, *Corrosion*, **41**, 715 (1985).

82. F. Hunkeler, A. Krolikowski, and H. Böhni, *Electrochim. Acta*, **32**, 615 (1987).
83. A. M. Riley, D. B. Wells, and D. E. Williams, *Corros. Sci.*, **32**, 1307 (1991).
84. G. T. Burstein und S. P. Mattin, in *Critical Factors in Localized Corrosion II*, P. M. Natishan, R. G. Kelly, G. S. Frankel, and R. C. Newman, Editors, PV 95–15, p.1, The Electrochemical Society Proceedings Series, Pennington, NJ (1996).
85. N. J. Laycock, M. H Moayed, and R. C. Newman, in *Critical Factors in Localized Corrosion II*, P. M. Natishan, R. G. Kelly, G. S. Frankel, and R. C. Newman, Editors, PV 95–15, p. 68, The Electrochemical Society Proceedings Series, Pennington, NJ (1996).
86. E. Akiyama and G. S. Frankel, *J. Electrochem. Soc.*, **146**, 4095 (1999).
87. T. R. Beck und S. G. Chan, *Corrosion.*, **37**, 665 (1981).
88. H. Lajain, *Z. Werkstofftech.*, **2**, 19 (1971).
89. H. Lajain, *Werkstoffe Korros.*, **32**, 537 (1981).
90. W. Brandl, *Beitrag zum Messen elektrochemischer Kennwerte im Mikrobereich: Korrosion an thermischen Spritzschichten und Loetverbindungen*, 1st Edition, Vol. 15, DVS-Verlag, Duesseldorf (1987).
91. W. Brandl, *Electrochim. Acta*, **37**, 2263 (1992).
92. R. C. Alkire and K. P. Wong, *Corros. Sci.*, **28**, 411(1988).
93. T. Suter, T. Peter, and H. Böhni, in *Electrochemical Methods in Corrosion Research V*, M. G. S. Ferreira and A. M. P. Simoes, Editors, Materials Science Forum, Proceedings Vol. 192–194, p. 25, Trans Tech Publications, Uetikon, Switzerland (1995).
94. T. Suter and H. Böhni, 12th Int. Corr. Cong., Proceedings Volume 3B, p. 1367, NACE, Houston, Texas (1993).
95. T. Suter and H. Böhni, *Electrochim. Acta*, **42**, 3275 (1997).
96. T. Suter, Ph. D. Thesis No. 11962, ETH Zürich (1997).
97. T. Suter, E. G. Webb, H. Böhni, and R. C. Alkire, *J. Electrochem. Soc.*, **148**, 186 (2001).
98. T. Suter and R. C. Alkire, in *Critical Factors in Localized Corrosion III*, T. G. Kelly, G. S. Frankel, P. M. Natishan, and R. C. Newman, Editors, PV 98–17, p. 118, The Electrochemical Society Proceedings Series, Pennington, NJ (1999).
99. T. Suter and R. C. Alkire, *J. Electrochem. Soc.*, **148**, 36 (2001).
100. R. A. Perren, T. Suter, P. J. Uggowitzer, L. Weber, R. Magdowski, H. Böhni, and M. O. Speidel, *Corros. Sci.*, **43**, 707 (2001).
101. R. A. Perren, T. Suter, C. Solenthaler, G. Gullo, P. J. Uggowitzer, H. Böhni, and M. O. Speidel, *Corros. Sci.*, **43**, 727 (2001).
102. T. Suter and H. Böhni, NACE Research Topical Symposium, Corrosion 2001, J. S. Frankel and J. R. Scully, Editors, p. 191, NACE International Publication (2001).
103. T. Suter and H. Böhni, *Electrochim. Acta*, **47**, 191 (2001).
104. *Low Level Measurements, Precision DC Current, Voltage and Resistance Measuremnts*, 5th Edition, Keithly, Company Publication (1998).
105. G. Wranglén, *Corros. Sci.* **14**, 331 (1974).
106. M. B. Ives and S. C. Srivastava, in *Advances in Localized Corrosion*, H. S. Isaacs, U. Bertocci, J. Kruger, S. Smialowska, Editors, Vol. 9, p. 295, NACE, Houston, TX (1990).
107. T. Suter and H. Böhni, in *Critical Factors in Localized Corrosion II*, P. M. Natishan, R. G. Kelly, G. S. Frankel, and R. C. Newman, Editors, PV 95–15, p. 127, The Electrochemical Society Proceedings Series, Pennington NJ (1996).
108. S. Matsch, T. Suter, and H. Böhni, Proceedings in Electrochemical Methods in Corrosion Research VI, P. L. Bonora and L. Deflorin Editors, Materials Science Forum, Proceedings Vol. 289–292, p. 1127, Tech Publications, Uetikon, Switzerland (1998).
109. S. Matsch and H. Böhni, Elektrokhimiya, *Russ. J. Electrochem.*, **36**, 1122 (2000).
110. V. Scotto, G. Ventura, and E. Traverso, *Corros. Sci.*, **19**, 237 (1979).
111. T. J. Warner, M. P. Schmidt, F. Sommer, and D. Bellot, *Z. Metallkd.*, **86**, 494 (1995).
112. V. Guillaumin and G. Mankowski, *Corros. Sci.*, **41**, 421 (1999).
113. C. M. Liao and R. P. Wei, *Electrochim. Acta*, **45**, 881 (1999).

114. R. G. Buchheit, R. P. Grant, P. F. Hlava, B. Mckenzie, and G. L. Zender, *J. Electrochem. Soc.*, **144**, 2621 (1997).
115. W. Schmidt and M. Stratmann, *Corros. Sci.*, **40**, 1441 (1998).
116. C. M. Liao, J. M. Olive, M. Gao, and R. P. Wei, *Corrosion*, **54**, 451 (1998).
117. R. P. Wei, C. M. Liao, and M. Gao, *Metall. Mater. Trans. A Phys. Metall. Mater. Sci.*, **29**, 1153 (1998).
118. M. Gao, C. R. Feng, and R. P. Wei, *Metall. Mater. Trans. A Phys. Metall. Mater. Sci.*, **29**, 1145 (1998).
119. M. A. Alodan and W. H. Smyrl, *J. Electrochem. Soc.*, **145**, 1571 (1998).
120. K. Kowal, J. Deluccia, J. Y. Josefowicz, C. Laird, and C. Farrington, *J. Electrochem. Soc.*, **143**, 2471 (1996).
121. A. M. Mierisch, J. Yuan, R. G. Kelly, and S. R. Taylor, *J. Electrochem. Soc.*, **146**, 4449 (1999).
122. A. J. Aldykiewicz, A. J. Isaacs, and A. J. Davenport, *J. Electrochem. Soc.*, **142**, 3342 (1999).
123. A. Kollics, A. E. Thomas, and A. Wieckowski, *Langmuir*, **11**, 4605 (1995).
124. J.-O. Nilsson, *Mater. Sci. Technol.* **8**, 685 (1992).
125. J. Charles, Proceedings of the Duplex Stainless Steels Beaune Conference, Beaune, Vol. 1, p. 3 (1991).
126. J. Charles, Proc. Conf. Appl. Stainless Steels, Vol. 2, p. 587 (1992).
127. L. Weber, P. J. Uggowitzer, *Mater. Sci. Eng.*, **A 242**, 222 (1998).
128. H. Tsuge, Y. Tarutani, and T. Kodo, *Corrosion*, **44**, 305 (1988).
129. G. G. Kolchin, B. S. Ermakov, R. I. Grechin, and Y. V. Gladnev, *Steel USSR*, **17**, 235 (1987).
130. L.-A. Norström, S. Pettersson, and S. Nordin, *Z. Werkstofftech.*, **12**, 229 (1981).
131. H. Vannevik, J.-O. Nilsson, J. Frodigh, and P. Kangas, *ISIJ Int.*, **36**, 807 (1996).
132. Y. Maehara, Y. Ohmori, J. Murayama, N. Fujino, and T. Kunitake, *Metall. Sci.*, **17**, 541 (1983).
133. J. H. Kullmann, Ph.D. Thesis, Institut für Werkstoffkunde und Materialprüfung, TU Vienna (1990).
134. B. Josefsson, J.-O. Nilsson, and A. Wilson, Proceedings of the Duplex Stainless Steels Beaune Conference, Beaune, Vol. 1, p. 67 (1991).

18

Photoelectrochemical Techniques in Corrosion Studies

Francesco Di Quarto, Monica Santamaria, and C. Sunseri
Dipartimento di Ingegneria Chimica dei Processi e dei Materiali,
Università di Palermo, Palermo, Italy

Contents

18.1 INTRODUCTION

The discovery of photoeffects at the metal–electrolyte interface can be traced to the studies of Becquerel in the first half of the 19th century (1). Although earlier studies on photoelectrochemical behavior of oxidized metals were reviewed at the beginning of the 1940s (2), there is no doubt that the beginning of photoelectrochemical science can be marked with the work of Brattain and Garrett (3) that appeared in the mid-1950s. The state of the art of these initial studies can be found in the review by Gerischer (4) and in the classical book by Myamlin and Pleskov (5). It may be of interest to mention that in the appendix of this book a paper (6) summarized the most immediate problems in electrochemistry of semiconductors as: "The important problem remains of extending the theoretical concepts and experimental investigation methods to polycrystalline semiconductors, in particular, oxide films on so-called barrier metals and to oxide and sulfide electrodes and also to use the concepts of the electrochemistry of semiconductors for investigating the phenomena of the passivation of metals." This last task was initially undertaken by Oshe and Rozenfel'd (7), who proposed to characterize the nature of passive films on metals by using a photopotential method initially proposed by Williams (8) for bulk semiconductors. Some applications of such a method in passivity studies can be found in Refs. (9, 10) and references therein. The inadequacy of the method of Oshe and Rozenfeld in characterizing a complex metal/oxide/electrolyte interface was initially shown by Hackermann et al. (11). This fact and the onset of more refined theories of photocurrent generation at illuminated semiconductor/electrolyte interface in the mid-1970s made obsolete the characterization of a passive metal/electrolyte interface by the method of Oshe and Rozenfeld.

As a result of an intensive research effort on the photoelectrochemical behavior of semiconductor electrodes, initially aimed to check their use as photoelectrodes in electrochemical solar cells, photoelectrochemical techniques started to be exploited as an analytical tool for *in situ* characterization of semiconductor/electrolyte interfaces (12–15). By considering that in many cases the investigated photoelectrodes were oxides, it was evident to electrochemists that passive films and corrosion layers having semiconducting or insulating behavior could be studied by using such techniques (16–19). Photoelectrochemical techniques, currently referred to as photocurrent spectroscopy (PCS), have gained importance in passivity studies (20–23) for investigating physicochemical properties of photoactive electrodes. This is due to the fact that PCS is a nondestructive technique based on the analysis of the electrochemical response (photocurrent) of the passive electrode/electrolyte interface under irradiation with photons of suitable energy and intensity. The choice of analyzing the photocurrent under potentiostatic control is preferred because of the prominent role of the electrode potential in the establishment of electrochemical equilibrium involving different metal oxidation states and reactivity of the passive films (24).

The physicochemical characterization of passive film and corrosion layers is important for the comprehension of the corrosion behavior of metals and alloys. PCS is able to provide information on solid-state properties (optical band gap and conductivity type) and the energetics (flat band potential, internal photoemission thresholds) of a metal/passive film/electrolyte interface. When compared to other optical techniques, PCS offers the further advantage that the photocurrent response of the passive film is directly related to the amount of absorbed photons. This means that the technique is not demanding in terms of surface finishing and allows the monitoring of long-lasting corrosion processes, where changes of surface reflectivity are expected owing to possible roughening of metal surfaces covered by corrosion products. As for the risk of electrode modifications under illumination, it can be minimized by improving the sensitivity of the signal detection by using a lock-in amplifier, coupled to a mechanical light chopper, which allows to study very thin films (1 to -2 nm thick) also under relatively low-intensity photon irradiation.

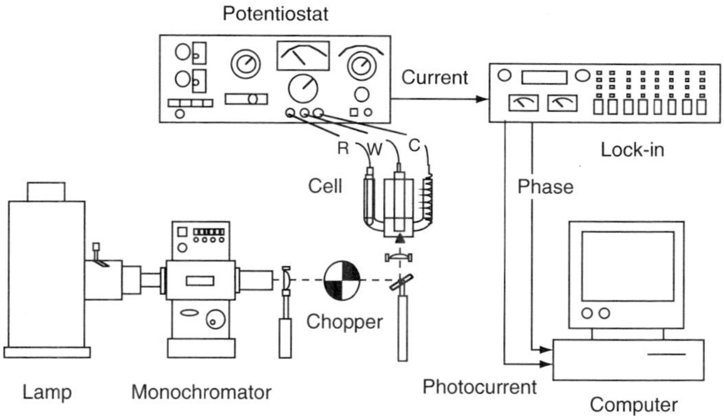

Figure 18.1 Schematic experimental setup currently employed for PCS study. (From *Handbook of Thin Film Materials*, Chapter 8, Vol. 2, p. 373, 2002. Academic Press. With permission.)

In Figure 18.1, we report the scheme of a typical experimental setup currently employed for PCS study. The use of a lamp-monochromator system allows getting a photon flux at variable wavelength in a rather wide range of photon energies (5.5 to 1.5 eV). A laser source at constant wavelength and variable energy density is preferred to the lamp-monochromator system for imaging studies, where high spatial resolution is the main requisite (described below; see also Ref. (25)). In these experiments, the photoelectrochemical cell is mounted on an X–Y table driven by stepper motors controlled by a computer and allowing different step sizes (1 to $-10\,\mu$m).

The use of a chopper together with a lock-in amplifier is necessary to extract weak photocurrent signals in presence of large dark current. When the photocurrent is measured by means of the lock-in technique (see Figure 18.1), there is a periodically chopped photon flux, that is, of an illumination of the junction, which is interrupted several times per second. In this case, the lock-in measures a signal whose intensity depends on the ratio between the chopping angular frequency, ω_c, and the time constant of the electrical equivalent circuit of the junction, τ, including also the electrolyte resistance ($\tau = R_t C_t$, with R_t and C_t representing the total resistance and capacitance of the junction). The effect of the chopping frequency on the recorded photocurrent at the passive film/electrolyte junction has been reported previously (23). More details on the lock-in technique can be found in Ref. (26).

It has been shown that, depending on the investigated systems, the PCS technique can provide information on (23):

1. The energetics at the metal/passive-film/electrolyte interfaces (flat band potential determination, conduction and valence band edges location, internal photoemission thresholds).
2. The electronic structure and indirectly (through the optical band gap values) chemical composition of passive films *in situ* and under controlled potential in long-lasting experiments.
3. The mechanisms of generation (geminate recombination effects) and transport of photocarriers in amorphous materials.

4. The kinetics of growth under illumination of photoconducting films and interference effects in absorbing materials at constant growth rate.

The first two points are relevant for corrosion studies if we consider that both the ion transfer (ITR) and electron transfer reactions (ETR) are controlled by the electronic properties of the passive film and the energetics at the metal/film and film/electrolyte interfaces (27–34). As for the last two points, they provide important information on the solid-state properties of passive films as well as on the interaction between ionic and electronic transport processes during film formation (23). PCS also presents limitations owing to the following aspects:

1. The technique is able to scrutinize only photoactive corrosion layers.
2. The investigation of surface layers having optical band gap lower than 1 eV or larger than 5.5 eV requires a special setup or they are experimentally not accessible in aqueous solutions.
3. Structural information and chemical composition of the layers are not accessible directly and complementary investigation based on other techniques are required.

As for the first two limitations they are rather apparent than real. With the exception of noble metals (Ir, Ru, etc.), which are covered by conducting oxides only at high electrode potentials, most of metals are thermodynamically unstable by immersion in aqueous solution, and they become covered by oxide or hydroxide films having often insulating or semiconducting properties. Moreover, with the exception of very few oxides grown on metals of lower electronegativity, the most common base metal oxides have band gap values largely lying within the optical window experimentally accessible by PCS (23).

The third limitation is the principal one. It is within the scope of this work to show that PCS can provide indirectly structural and compositional information of passive films if we assume an interpretative model of their photoelectrochemical behavior, which accounts also for the complex electronic structure of amorphous materials (35–39). Complementary information accessible by other *in situ* and *ex situ* techniques can help in this task. The aim of this chapter is limited to:

1. Provide a general interpretative framework of the photoelectrochemical behavior of passive metal electrodes by discussing some features that are related to the extreme thinness and amorphous nature of the passivating layers.
2. Highlight a more recent quantitative use of PCS for characterizing the chemical nature of passive films and corrosion layers.

A very short theoretical background on the photoelectrochemistry of crystalline semiconductors will be provided for readers not acquainted with the subject, in order to show the differences between the photoelectrochemical behavior of passive films and bulk crystalline semiconductors. A more extensive and detailed introduction to the principles of electrochemistry and photoelectrochemistry of semiconductors can be found in classical books and workshop discussions published on the subject (4, 5, 40–45). As for some topics on the photoelectrochemical behavior of thin films not dealt in this chapter, the interested reader can refer to Ref. (23). Theoretical interpretations of the experimental results will be presented on the basis of simple models developed initially for passive films grown on valve metals (Al, Ta, Zr, Nb, Ti, W) and their alloys. In order to show the usefulness of PCS in the study of complex systems also, we will discuss some results pertaining to passive films grown on metals and alloys particularly interesting for corrosion scientists (Fe, Cr, Ni, stainless steels). Moreover, some very recent results of a quantitative use of PCS for the chemical characterization of passive films on metallic alloys and conversion coatings will be presented.

18.2 SEMICONDUCTOR/ELECTROLYTE JUNCTION

The structure of a semiconductor/electrolyte (SC/El) interface differs from the more familiar metal/electrolyte (M/El) interface, owing to the large difference in the concentration of mobile carriers in the two materials. In the case of a metal/solution interface, the potential drop occurs mainly across the electrolytic double layer so that a change in the electrode potential, U_E, with respect to a reference electrode, affects the Fermi level of the metal with respect to the electronic levels in solution by changing directly the photoemission threshold, E_{th}, of the metal/solution interface according to the relationship (42):

$$E_{th} = E_{th,\,0} + eU_E \tag{18.1}$$

where $E_{th,0}$ is the value of the photoemission threshold at an arbitrary zero electrode potential value. In the case of a SC/solution interface, the potential drop occurs mainly across the space-charge region of the SC electrode. The potential drop across the electrolytic double layer is almost constant in the presence of moderate doping level and in the absence of a large density of charged surface states.

These assumptions will be used for a simple introduction to the behavior of the SC/El junction in dark and under illumination. In these conditions, the onset photocurrent energy threshold of the SC/El interface coincides with the band gap of the SC, E_g, which is independent of the electrode potential. In the following, we present a simplified picture of an ideal SC/El interface in dark and under illumination, which allows for a preliminary approach to the more complex behavior of the passive film/electrolyte interface.

18.2.1 The Structure of the SC/El Interface at Equilibrium

According to Figure 18.2, the total potential drop across the SC/El interface can be written as

$$\Delta\Phi_{SC/El} = \{\phi_{SC}(\infty) - \phi_{SC}(0)\} + \{\phi_{SC}(0) - \phi_{SC}(-\infty)\} \tag{18.2}$$

where the first term in the right-hand side represents the Galvani potential drop across the SC, $\Delta\Phi_{SC} = \{\phi_{SC}(\infty) - \phi_{SC}(0)\}$, and the second term represents the Galvani potential across the electrolytic double layer $\Delta\Phi_H$. The region of SC necessary for screening such a potential drop defines the space-charge region, X_{SC}. The width of this region changes with the potential drop according to the following equation:

$$X_{SC} = X_{SC}^0(|\Delta\Phi_{SC}| - kT/e)^{1/2} \tag{18.3}$$

where $X_{SC}^0 = (2\varepsilon\varepsilon_0/eN_{D,A})^{1/2}$ represents the space-charge width into the SC electrode at 1 V of band bending and its value depends on the concentration of mobile carriers

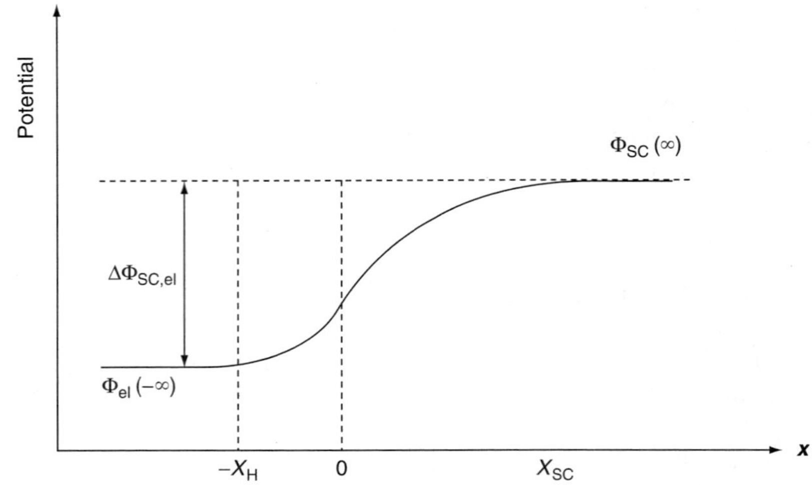

Figure 18.2 Galvani potential profile across the semiconductor/electrolyte (SC/El) interface. (From *Handbook of Thin Film Materials*, Chapter 8, Vol. 2, p. 373, 2002. Academic Press. With permission.)

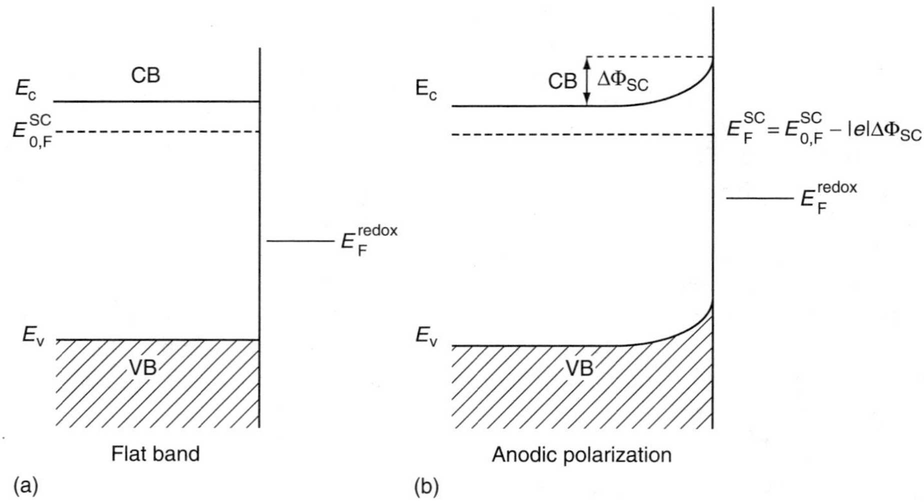

Figure 18.3 Electron energy levels in the SC/El junction (a) at the flat band potential and (b) under anodic polarization for an n-type semiconductor. E^{SC}_F and E^{red}_F are the Fermi levels of the semiconductor and the redox couple in solution, respectively.

present in the SC. The number of mobile carriers in a SC is a function of the doping level (N_D for n-type or N_A for p-type SC) of the material. The other symbols have the usual meaning. In Figure 18.3 and Figure 18.4, we report the schematic diagram of an n-type and a p-type SC/electrolyte junction in the energy–distance coordinates under slightly depleted and flat band conditions ($X_{sc} = 0$). An anodic (n-type, $\Delta\Phi_{SC} > 0$) or cathodic (p-type, $\Delta\Phi_{SC} < 0$) polarization corresponds to the conditions depicted in Figure 18.3b and Figure 18.4b, whilst $\Delta\Phi_{SC} = 0$ (no potential drop within the SC) corresponds to the special flat band condition, $U_E = U_{fb}$. In terms of electrode potential, U_E, we can write:

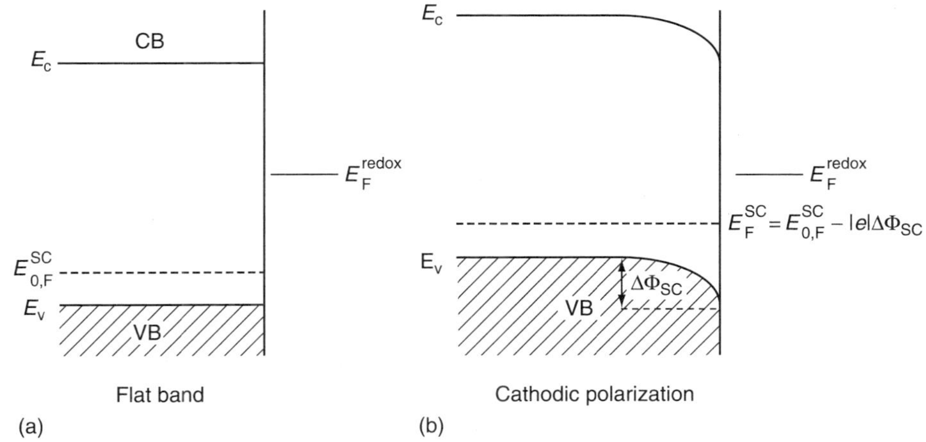

Figure 18.4 Electron energy levels in the SC/El junction (a) at the flat band potential and (b) under cathodic polarization for a p-type semiconductor. E_F^{SC} and E_F^{red} are the Fermi levels of the semiconductor and the redox couple in solution, respectively.

$$\Delta\Phi_{SC} = U_E - U_{fb} \tag{18.4}$$

The determination of the flat band potential is a preliminary task in finding the energetics of any SC/El interface. Once the U_{fb} value is known, with respect to a reference electrode, it is possible to locate the Fermi level of the SC in the electrochemical scale, E_F^0 (El), by means of the relationship:

$$E_F^0(El) = -eU_{fb} \tag{18.5}$$

The location of the remaining energy levels of the junction is found by using the relationships for n- and p-type semiconductors:

$$E_C = E_F^0 + kT \ln\left(\frac{N_C}{N_d}\right) \quad \text{for n-type SC} \tag{18.6a}$$

$$E_V = E_F^0 - kT \ln\left(\frac{N_V}{N_a}\right) \quad \text{for p-type SC} \tag{18.6b}$$

$$E_g = E_C - E_V \tag{18.6c}$$

where N_C and N_V are the effective density of states at the bottom of the SC conduction band and at the top of the SC valence band, N_d and N_a are the donor and acceptor concentrations in the SC, E_C and E_V are the conduction and valence band edges, respectively, and E_g is the band gap of the SC.

A rather popular method, also in corrosion studies, to get the flat band potential of a SC/El junction is based on the use of Mott–Schottky (M–S) analysis of differential capacitance measurements of the SC/El junctions. Although the validity of M–S theory has been tested rigorously for several crystalline SC/El interfaces (46–53), in the presence of amorphous semiconducting material, the simple M–S analysis can give misleading information. A detailed discussion of the inadequacy of traditional M–S analysis for interpreting the capacitance data of the amorphous SC/El interface can be found in Ref. (23). A different way to obtain the U_{fb} value of an ideal SC/El junction relies on the determination of the onset photocurrent potential under special conditions as shown in Refs. (54–58). For this we present a model of illuminated SC/El junction, initially proposed by Butler (54), for deriving the flat band potential.

18.2.2 SC/El Junctions Under Illumination: The Gärtner–Butler Model

The modeling of photocurrent vs. potential curves at fixed irradiating wavelength (photocharacteristics) for a crystalline SC/El junction has been carried out by several authors (54–58), starting from the seminal paper of Gärtner (59) on the behavior of an illuminated solid-state Schottky barrier.

In Figure 18.5, the absorption process of incident light in the bulk of an SC is sketched: J_0 (in cm^{-2} sec^{-1}) is the photon flux entering the SC (corrected for the reflections losses at the SC/El interface), which is absorbed following the Lambert–Beer law. The number of electron–hole pairs generated per second and unit volume at any distance from the SC surface, $g(x)$, is given by

$$g(x) = J_0 \alpha e^{-\alpha x} \tag{18.7}$$

where α (in cm^{-1}), the light absorption coefficient of the semiconductor, is a function of the impinging wavelength. It is assumed that each absorbed photon, having energy

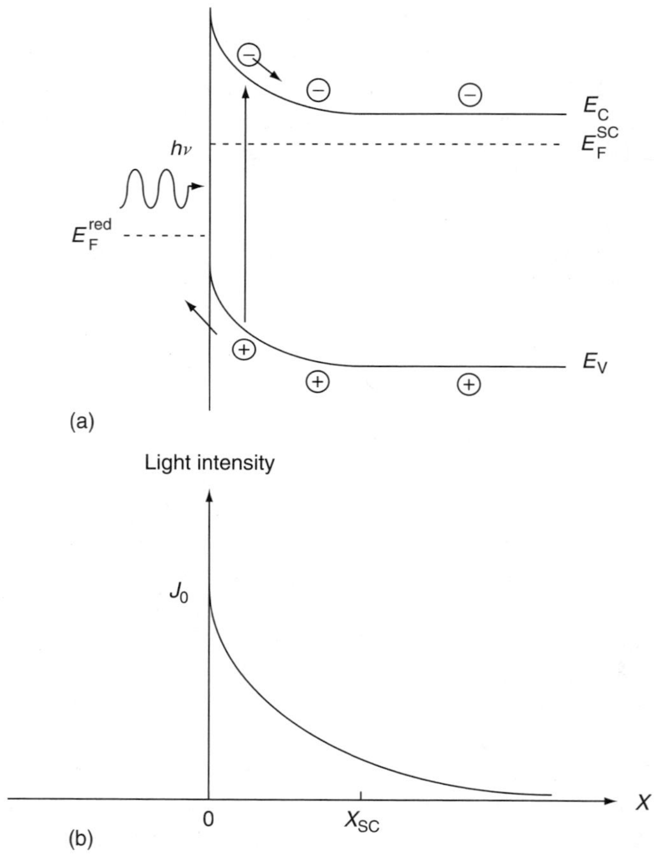

Figure 18.5 Schematic representation of a crystalline n-type SC/El interface under illumination, showing the electron–hole pair generation (a) and the change of light intensity due to absorption within the semiconductor (b). (From *Handbook of Thin Film Materials*, Chapter 8, Vol. 2, p. 373, 2002. Academic Press. With permission.)

$h\nu > E_g$, forms a free electron–hole couple. In the Gärtner–Butler model, the total photocurrent collected in the external circuit is calculated as the sum of two terms: a migration term, I_{drift}, and a diffusion term, I_{diff}. The first term takes into account the contribution of the minority carriers generated into the space-charge region; the second term accounts for the minority carriers entering the edge of the space-charge region from the bulk field free region ($x > X_{SC}$) of SC. No light reflection at the rear interface is assumed, so that the light entering is absorbed within the SC. Moreover, a rather strong assumption in the model is the absence of minority carriers' recombination within the space-charge region of the SC as well as at the surface of the SC. Gärtner showed that the following expression of photocurrent could be derived under such simplifying assumptions:

$$I_{ph} = eJ_0\left[\frac{1 - \exp(-\alpha X_{SC})}{1 + \alpha L_h}\right] + ep_0\frac{D_h}{L_h} \tag{18.8}$$

where D_h and L_h are the hole diffusion coefficient and the diffusion length, respectively, and p_0 the hole equilibrium concentration in a n-type SC. The same equation holds for p-type SCs, with D_n and L_n instead of D_h and L_h, and n_0 (electrons equilibrium concentration) instead of p_0. For a wide band gap SC (e.g., n-type WO_3), the equilibrium concentration of minority carriers, p_0, into the bulk is very small; thus, Butler derived the equation for the photocurrent in a crystalline SC/El junction (54) by using Equations (18.3) and (18.4) as

$$I_{ph} = eJ_0\left[1 - \frac{\exp\left(-\alpha X_{sc}^0\sqrt{U_E - U_{fb} - (kT/e)}\right)}{1 + \alpha L_h}\right] \tag{18.9}$$

By assuming $\alpha X_{SC} \ll 1$ and $\alpha L_h \ll 1$, Equation (18.9) can be written as

$$I_{ph} = eJ_0\alpha X_{SC}^0\left(U_e - U_{fb} - \frac{kT}{e}\right)^{1/2} \tag{18.10}$$

Equation (18.10) represents a quadratic dependence of the photocurrent on the electrode potential, which can be used for finding the flat band potential of the junction. In fact, by neglecting the term kT/e, a plot of $(I_{ph})^2$ vs. U_E should intercept the voltage axis at the flat band potential, U_{fb}, regardless of the employed λ as long as the $\alpha X_{SC} \ll 1$ condition is obeyed. In Figure 18.6, we report such a plot for a crystalline WO_3 anodic film in contact with an acidic solution showing that Equation (18.10) is obeyed in the high electrode potential region allowing the determination of a zero photocurrent potential value in quite good agreement with the values reported in literature for WO_3. In the low electrode potential region, the square root dependence has been attributed to the possible kinetic control due to the recombination of photogenerated carriers (discussed later).

A second important aspect embodied in Equation (18.10) is the direct proportionality between the measured photocurrent and the light absorption coefficient. In the vicinity of the optical absorption threshold of the SC (42) the relationship between the absorption coefficient and optical band gap of the material, E_g, can be written as

$$\alpha = A\,\frac{(h\nu - E_g^{opt})^{n/2}}{h\nu} \tag{18.11}$$

and it is possible to derive the following expression:

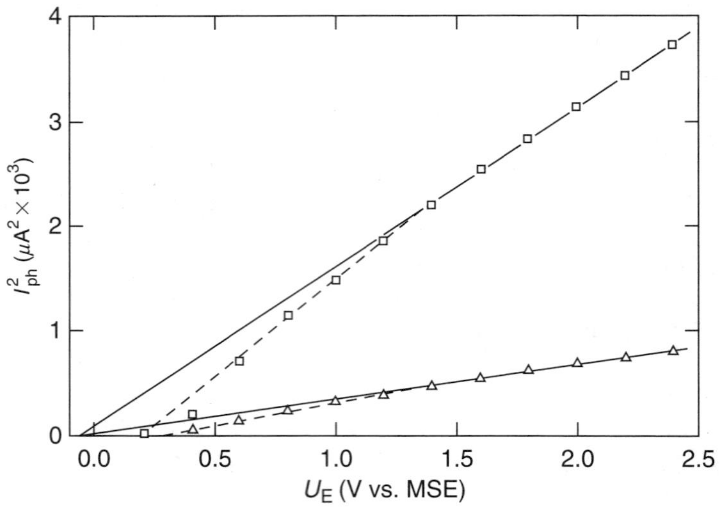

Figure 18.6 $(I_{ph})^2$ vs. U_E plots recorded for a WO_3 film grown at 8 mA/cm^2 in 0.1 *N* H_3PO_4 electrolyte up to 100 V and crystallized for 3 h at 350°C under argon atmosphere. Electrode surface: 0.053 cm^2. Irradiating wavelengths: □ $\lambda = 300$ nm; Δ $\lambda = 380$ nm. (From *Solar Energy Materials*, 11: 419, 1985, reprinted from Elsevier Science. With permission.)

$$Qh\nu = A(h\nu - E_g^{opt})^{n/2}(X_{SC}^0\sqrt{|U_e - U_{fb}|}) \tag{18.12}$$

$Q = (I_{ph})/eJ_0$ represents the photocurrent collection efficiency and E_g^{opt} the optical threshold for the onset of photocurrent at the illuminated electrode. Equation (18.12) shows that, at constant electrode potential, it is possible to get the optical band gap of the material from the dependence of the photocurrent on the wavelength (referred as the photocurrent spectrum of the junction) of incident light at constant photon flux. In fact, by plotting $(Qh\nu)^{2/n}$ vs. $h\nu$ at constant electrode potential ($U_E - U_{fb} =$ const.), we get a characteristic photon energy $h\nu_0 = E_g^{opt}$. For an ideal SC/El junction, E_g^{opt} coincides with the minimum distance in energy between the filled states of VB and empty states of CB (band gap, E_g) and n can assume different values depending on the nature of the optical transitions between states of the VB and states of the CB.

Optical transitions at energies near the band gap of a crystalline material may be direct or indirect. In the first case, no intervention of other particles is required, apart from the incident photon and the electron of the VB; in the second case, the optical transition is assisted by the intervention of lattice vibrations. By assuming a parabolic electronic density of states distribution, DOS ($N(E) \propto E^{1/2}$) near the band edges, in the case of direct transitions n assumes values equal to 1 or 3, depending on whether the optical transitions are allowed or forbidden in the quantum mechanical sense (42). In the case of indirect optical transitions, the value of n in Equation (18.11) is equal to 4. It will be shown in the case of amorphous materials that the measured optical band gap does not necessarily coincide with the band gap of crystalline material but it can be still related to the material composition and morphology.

Apart from the initial assumption of Gärtner of an ideal Schottky barrier, we have to mention that several hypotheses underlie to the use of Equations (18.8) to (18.12) for interpreting photoelectrochemical data. Other authors have introduced the possible existence of kinetic control at the SC surface or within the space-charge region (55–58) showing that:

1. In the presence of strong surface recombination effects the onset photocurrent may occur at much higher band bending than that foreseen by the Gärtner–Butler equation (55, 56, 58).
2. A square root dependence of the photocurrent on the electrode potential is still compatible with the existence of some mechanism of recombination (first-order kinetics) within the space-charge region (57), so that the determination of the flat band potential from the square of the photocurrent vs. electrode potential plot must be taken with some caution.

These aspects must be carefully considered when the photocharacteristics are used for deriving the flat band potential of SC/El junctions, especially in the case of corrosion layers that are far from the ideal behavior of crystalline semiconductors previously assumed. Finally, we have to remark that all previous equations pertain to the steady-state values of the DC photocurrent. The equations derived for steady state remain valid also for chopped conditions provided that the lock-in measured signal remains proportional to the steady-state chopped value.

The transient behavior of the photocurrent as a function of time and chopping frequency has been investigated by several researchers (60–63), providing new techniques like intensity modulated photocurrent spectroscopy, which can be employed in the study of recombination processes and charge transfer kinetics at the SC/El junction.

18.3 THE PASSIVE FILM/ELECTROLYTE INTERFACE

On going from crystalline thick SC electrodes to very thin insulating or semiconducting corrosion films on metals, new experimental features are observed that require the extension of previous interpretative models and the introduction of new theoretical concepts in order to account for novel results not observed for crystalline bulk materials. The studies of electronic properties of passive films were initially performed by using the same interpretative models valid for crystalline SC/El junctions. Unfortunately, the application of such models (Mott–Schottky theory of space-charge differential capacitance and Gärtner–Butler model for the photocurrent curves) to experimental data pertaining to amorphous or strongly disordered passive films can give misleading information (23). In the following, we discuss the main differences in the electronic and optical properties of amorphous and crystalline materials (64–67) and how they can affect the photoelectrochemical behavior and band gap values of passive films.

18.3.1 Electronic Properties of Disordered Passive Films

In Figure 18.7(a) and (b), we report two models of DOS usually employed for interpreting the optical and electronic behavior of insulating or semiconducting amorphous materials. We stress that amorphous materials usually retain the same short-range order with respect to their crystalline counterparts and that the main differences come out from the absence of the long-range order typical of crystalline phases. It is now generally accepted that the band structure model retains its validity also in the absence of long-range lattice periodicity. This means that the long-range

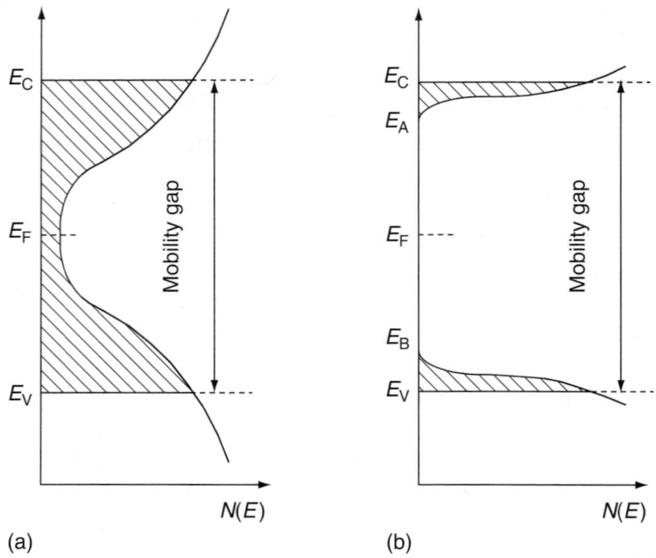

Figure 18.7 Density of electronic states as a function of energy according to different amorphous semiconductor models: (a) Cohen–Fritzsche–Ovishinsky model; (b) Mott and Davis model. Hatched areas represent the localized gap states, E_C and E_V are the conduction and valence band mobility edges, respectively. (From *Ber. Bunsenges. Phys. Chem.*, 90: 549, 1986, reprinted from VCH Verlagsgesellschaft mbH. With permission.)

disorder perturbs but does not annihilate the band structure: its main effect is the presence of a finite density of states within the so-called "mobility gap" ($E_C - E_V$) of the amorphous semiconductor (a-SC) or insulator. The DOS distribution of Figure 18.7(a), initially proposed by Cohen–Fritzsche–Ovishinsky (CFO model) (64), takes into account the presence of defect states within the semiconductor, which originate a continuous distribution of electronic states within the mobility gap (like in a-Si:H). The DOS distribution of Figure 18.7(b), due to Mott and Davis (67), can be attributed to an ideal amorphous material in which only a long-range lattice disorder is taken into account. Other models have been suggested for explaining the behavior of different classes of amorphous materials, but they involve only minor modifications to those of Figure 18.7, when the existence of specific defects in the investigated material is considered. From Figure 18.7 it is evident that the main differences in the DOS distribution of amorphous materials with respect to the crystalline ones can be attributed to the following features:

1. Existence of a finite DOS within the mobility gap, defined by two sharp mobility edges, E_C and E_V, in the conduction band (CB) and in the valence band (VB), respectively.
2. The free electron-like DOS, $N(E) \propto E^{1/2}$, valid for energy levels close to the CB and VB edges, is no more valid below E_C or above E_V. In these energy regions, the presence of a tail of states, decreasing exponentially (66) or linearly (67), has been suggested for explaining the optical properties of different amorphous materials.
3. Different mechanisms of charge carrier transport are invoked in extended (above E_C or below E_V) or localized (within the mobility gap) electronic states. A free carrier-like mechanism of transport is involved in the first

case, whilst a transport by hopping (thermally activated) is assumed in localized states.

The electronic structure outlined above affects the generation and transport processes of photocarriers in amorphous material and provides the reason for the photoelectrochemical behavior of the a-SC/El junction. The main differences in the photocurrent response of disordered thin films with respect to the case of bulk crystalline semiconductors arise from the following facts:

1. The optical band gap of an amorphous material may or may not coincide with that of the crystalline counterpart, depending on the presence of different types of defects that can modify the DOS distribution.
2. At variance with crystalline materials, the generation process of free carriers by the absorption of photons having energy equal or higher than the optical band gap of the film may depend on the electric field, owing to the presence of initial (geminate) recombination effects.
3. The small thickness of passive film makes possible the optical excitation at the inner metal/film interface. This allows injecting photocarriers from the underlying metal into the VB or CB of a thick film (internal photoemission), or directly into the electrolyte (external photoemission) in the case of very thin films (1 to -2 nm thickness).
4. The presence of reflecting metal/film and film/electrolyte interfaces makes possible the onset of multiple reflections, even for photons having energy higher than the optical absorption threshold. This fact provides interference effects in the photocurrent vs. film thickness curves. This last aspect will be not treated here for brevity; but interested readers can refer to Ref. (23). In the following, we derive an equation for the photocurrent in amorphous SC and insulators by taking into account the influence of the amorphous nature on the electronic properties of materials.

18.3.2 Amorphous Film/Electrolyte Junction Under Illumination

Due to the low mobility of carriers in amorphous materials, it is reasonable to assume that a negligible contribution to the measured photocurrent arises from the field-free region of the semiconductor. In this case, an expression for the migration term in the space-charge region of an amorphous semiconducting passive film can be derived in a way quite similar to that followed by Gärtner but by introducing also a recombination probability for the photocarriers generated in the space-charge region of the a-SC. It has been shown that under steady-state conditions the following general expression can be written for the photocurrent in an a-SC as a function of the space-charge region width and electric field (23):

$$I_{\mathrm{ph}} = e\Phi_0 \frac{\alpha L_{\mathrm{d}}}{1+\alpha L_{\mathrm{d}}}\left[1-\exp\left(-X_{SC}\frac{1+\alpha L_{\mathrm{d}}}{L_{\mathrm{d}}}\right)\right]\eta_{\mathrm{g}}(r_0,\ F_{\mathrm{av}}) \tag{18.13}$$

where $\eta_{\mathrm{g}}(r_0,\ F_{\mathrm{av}})$ is the efficiency of free carriers generation in the presence of geminate recombination effects. L_{d} is the drift length of the photocarriers ensemble in the average field approximation:

$$L_{\mathrm{d}} = \mu\tau F_{\mathrm{av}} \tag{18.14}$$

where μ and τ are the drift mobility and the lifetime of the photocarriers, respectively, and F_{av} is the mean electric field in an a-SC (68). An expression of X_{sc} and F_{av} in an a-SC under the hypothesis of constant DOS distribution has been calculated in the low band bending approximation ($e\Delta\Phi_{SC} < E_g/2$) (68) as well as in the high band bending approximation ($e\Delta\Phi_{SC} > E_g/2$) (69). Equation (18.13) does not take into account the possible reflections at the rear interface so that Φ_0 represents the photon flux entering at the SC/El interface.

We like to stress two aspects in the expression of the photocurrent in an a-SC/El junction. The first one is that from Equation (18.13) a direct proportionality follows between the photocurrent and the absorption light coefficient for $\alpha L_d \gg 1$ (no recombination) and $\alpha X_{SC} \ll 1$ (slightly absorbed light), as previously derived for crystalline materials (see Equation 18.10). On the other hand, for $\alpha L_d \ll 1$ a direct proportionality between I_{ph} and α is still assured by the fractional term $\alpha L_d/(1 + \alpha L_d)$. According to these considerations, we can still assume for amorphous SCs a direct proportionality between the photocurrent yield, $Q = I_{ph}/e\Phi_0$, and the light absorption coefficient, α, in the vicinity of the absorption edge under constant electrode potential. Like for crystalline materials, this allows the replacement of α with the photocurrent yield in deriving the optical band gap of amorphous semiconducting films from the photocurrent spectra (see Figure 18.8).

A second aspect we like to stress is that Equation (18.13) contains as a particular case the expression of the photocurrent for an amorphous insulator/electrolyte junction after substitution of the film length, d_f, to the space-charge region X_{SC}. Accordingly, we can write the following relationship for the insulating film/electrolyte junction:

$$I_{ph} = e\Phi_0\eta_g(r_0, F_{av})\frac{\alpha L_d}{1 + \alpha L_d}\left[1 - \exp\left(-d_f\frac{1 + \alpha L_d}{L_d}\right)\right] \tag{18.15}$$

The usual expression for the mean electric field in insulator $F_{av} = (U_E - U_{fb})/d_f$ holds in the absence of trapping phenomena (23, 70). In both cases, the variation of the

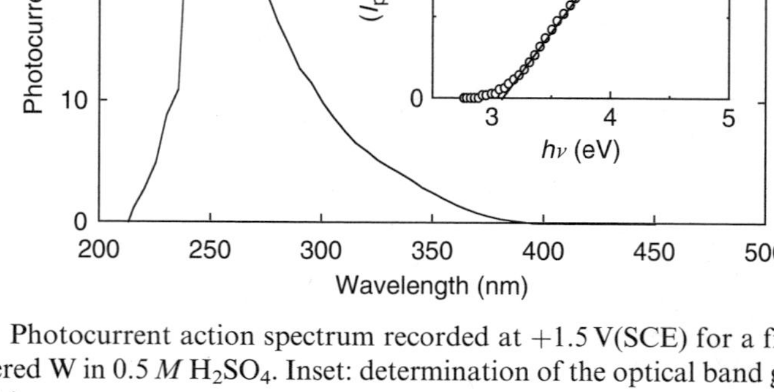

Figure 18.8 Photocurrent action spectrum recorded at +1.5 V(SCE) for a film grown up to 50 V on sputtered W in 0.5 *M* H_2SO_4. Inset: determination of the optical band gap by assuming indirect transitions.

efficiency of generation with the electric field and photons energy, trough r_0, can account for changes in E_g^{opt} values measured at different electrode potentials as well as for the dependence from the incident photon energy of the photocurrent vs. potential curves. This last aspect has been deeply discussed both for semiconducting (a-WO_3 and a-TiO_2) and insulating (a-Ta_2O_5) anodic films in previous works (23, 68, 71). In Figure 18.9 and Figure 18.10, we report the fitting curves of experimental data for films grown on titanium and tantalum metals. The details of fitting procedure can be found in Refs. (23, 68). Here, we have to mention that the flat band potential can now be derived as a fitting parameter of the photocharacteristics. As for the efficiency of free carriers generation, $\eta_g(r_0, F_{av})$, an expression previously suggested by Pai and Enck for amorphous materials was used (72). This last expression shows that very low efficiency of free carrier generation is expected at low electrical fields and thermalization lengths r_0, that is, at photon energy near the mobility gap (68). This finding could affect in some extent the measured optical band gap value of amorphous anodic films. In the presence of surface layers having regions of different composition it is still possible to generalize the previous expression for the photocurrent, thus allowing to obtain information on the flat band potential values of different phases present on the electrode surface (73).

We like to mention that the previous considerations and final equations hold also for any low-mobility material, including microcrystalline and crystalline materials, for which a small diffusion length is expected.

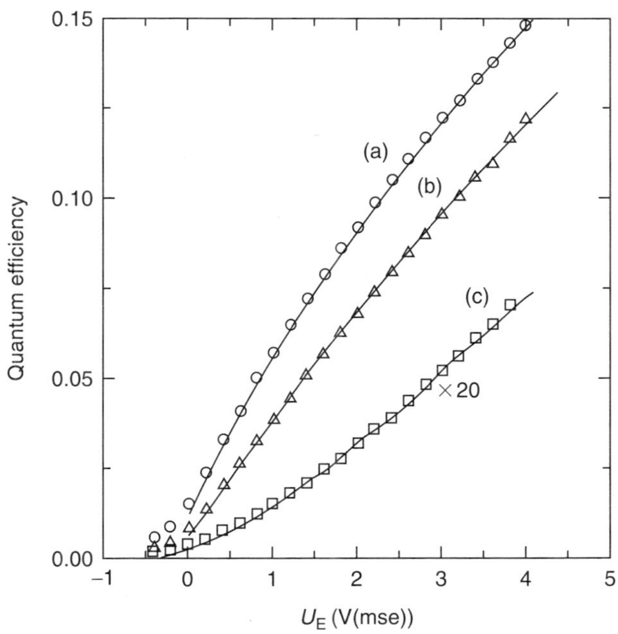

Figure 18.9 Fitting of the experimental photocharacteristics for the a-TiO_2 film grown on Ti at 2 mV/sec in 0.5 M H_2SO_4 according to Equation (18.13). Symbols are the experimental values. Lines are the theoretical curves calculated using the parameters $x_0 = 40$ Å, $\varepsilon = 15$, $\mu\tau = 2\ 10^{-13}$ cm^2/V. (a) $\lambda = 250$ nm, $\alpha = 8 \times 10^5$ cm^{-1}, $r_0 = 25$ Å; (b) $\lambda = 270$ nm, $\alpha = 6 \times 10^5$ cm^{-1}, $r_0 = 18$ Å; (c) $\lambda = 370$ nm, $\alpha = 1.5 \times 10^4$ cm^{-1}, $r_0 = 10$ Å. (From *Electrochimica Acta*, Vol. 38, No. 1, pp. 29–35 (1993), reprinted from Elsevier Science. With permission.)

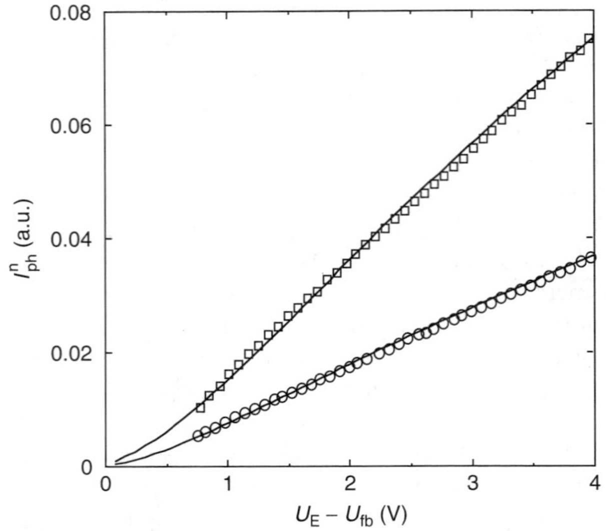

Figure 18.10 Fitting of the experimental photocharacteristics for the Ta_2O_5 film grown at $8\,mA/cm^2$ in 0.1 *M* NaOH solution up to 10 V(MSE) ($d_{ox} = 20\,nm$) according to Equation (18.15). Symbols are the experimental values. Lines are the theoretical curves calculated using the parameters: $\mu\tau = 10^{-13}\,cm^2/V$, $\varepsilon = 11$. ○ $\lambda = 230\,nm$, $\alpha = 3.5 \times 10^5\,cm^{-1}$, $r_0 = 20\,Å$; □ $\lambda = 300\,nm$, $\alpha = 1.5 \times 10^4\,cm^{-1}$, $r_0 = 15\,Å$. (From *Handbook of Thin Film Materials*, Chapter 8, Vol. 2, p. 373, 2002. Academic Press. With permission.)

18.3.3 Optical Gap in Amorphous Materials

It was previously mentioned that optical transitions at energies near the band gap of a crystalline material may be direct (allowed or forbidden) or indirect. In the case of amorphous materials, owing to the relaxation of the *k*-conservation selection rule, "no intervention of phonons is invoked to conserve momentum and all energy required is provided by the incident photons" (67). By assuming again a parabolic DOS distribution in the vicinity of the mobility edges of both the conduction and valence bands (above E_C and below E_V, with reference to Figure 18.7), it has been shown that for amorphous materials the following relationship holds (66):

$$\alpha h\nu = \text{const}(h\nu - E_g^{m})^2 \qquad (18.16)$$

where $E_g^{m} = E_C - E_V$ is now the mobility gap of the a-SC (see Figure 18.7). The exponent 2 is reminiscent of the indirect optical transitions in crystalline material but now photons interact with the solid as a whole: this type of transition in amorphous materials is termed nondirect. Because some tailing of states is theoretically foreseen for an a-SC by any proposed model of DOS, E_g^{m} represents an extrapolated rather than a real zero in the density of states. On the other hand, in the presence of a DOS distribution varying linearly with energy in the ranges $E_C - E_A$ and $E_B - E_V$ of Figure 18.7(b), it is possible to get the following relationship for the absorption coefficient (67) of amorphous material:

$$\alpha h\nu = \text{const}(h\nu - E_g^{opt})^2 \qquad (18.17)$$

where E_g^{opt} now represents the difference of energy ($E_A - E_V$) or ($E_C - E_B$) in Figure 18.7(b), whichever is smaller, whilst the constant assumes values close to

10^5 eV^{-1} cm^{-1}. The range of energy in which Equation (18.17) should be valid is in the order of 0.4 eV or less (67). In order to distinguish between these two different models of optical transitions, both giving a similar dependence of absorption coefficient on the photon energy, we will refer to the first one as Tauc's approximation and to the second one as the Mott–Davis approximation. From the first one, we derive an estimation of the mobility gap and from the second one the optical gap of amorphous materials. If $(\alpha h\nu)^{0.5}$ vs. $h\nu$ plots display a linear region larger than 0.4 eV it seems more correct to interpret the data on the basis of Tauc's model of optical transitions. The coexistence of both types of transitions has been reported for thin anodic films grown on niobium (74), with the presence of a mobility gap in the order of 3.5 eV in the high photon energy range extending around 1 eV and an optical gap (in Mott's sense) of about 3.05 eV.

In the case of anodic films on valve metals, an exponential decrease in the photocurrent yield (Urbach tail) as a function of photon energy is frequently observed at photon energies lower than the mobility gap. A possible origin of such dependence can be attributed to a variation of the light absorption coefficient according to the following law:

$$\alpha = \alpha_0 \exp\left(-\gamma \frac{E_0 - h\nu}{kT}\right) \tag{18.18}$$

with γ and α_0 constant. This relationship, which has been found to hold also for crystalline materials, has been rationalized in the case of a-SCs by assuming an exponential distribution of localized states in the band edge tails (75). In these cases, E_0 marks the energy where ln α vs. $h\nu$ (Urbach plot) ceases to be linear. This value frequently coincides with the mobility gap value determined according to Equation (18.13). A typical example is reported in Figure 18.11 for an a-WO_3 film: the value of E_g^m, equal to about 3.10 eV, is in good agreement with the value of E_0 ($\cong$3.15 eV) derived from the Urbach plot. Other explanations have been suggested for this behavior in the case of crystalline materials (67).

In agreement with a general statement reported in Ref. (67) it has been suggested (23) that, in the absence of appreciable differences in short-range order of amorphous and crystalline counterparts, the mobility gap of amorphous anodic films should be equal or larger than the band gap of the crystalline counterpart. Such a

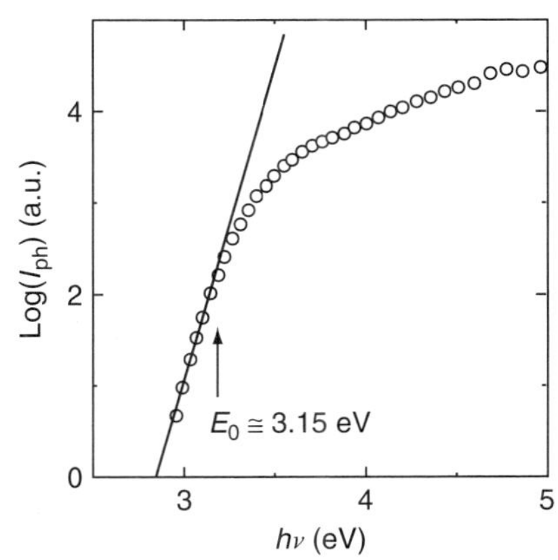

Figure 18.11 Urbach tail relative to the anodic film of Figure 18.8.

Table 18.1 Measured Optical Gap, E_g^m, for Passive Films on Pure Metals Compared with the Band Gap of the Crystalline Counterpart, E_g^{cryst}. ΔE_{am} is the Difference Between E_g^m and E_g^{cryst}

Phase	E_g^m (eV)	E_g^{cryst} (eV)	ΔE_{am} (eV)
ZrO_2	4.70–4.80[a]	4.50[a]	0.20–0.30
Ta_2O_5	3.95–4.05[a]	3.85[a]	0.10–0.20
Nb_2O_5	3.30–3.40[a]	3.15[a]	0.15–0.25
TiO_2	3.20–3.35[a]	3.05 (rutile)[a] 3.20 (anatase)[a]	0.15–0.20
WO_3	2.95–3.15[a]	2.75[a]	0.20–0.40
MoO_3	2.95–3.10[a]	2.90[b]	0.05–0.20
Cr_2O_3	3.30–3.55[a]	3.30[a]	0.0–0.25
NiO	3.43[a]	3.45–3.55[a]	0
Cu_2O	1.86[a]	1.86[a]	0
Fe_2O_3	1.90–1.95[a]	1.90[a]	0–0.05
$Fe_{0.25}Ti_{0.75}O_{1.875}$	2.95[c]	2.80[*]	0.15
$Fe_{0.1}Ti_{0.9}O_{1.95}$	3.15[c]	3.00[*]	0.15

[*] Estimated from the corresponding crystalline phases according to Equations (18.22) and (18.23).
[a] Data from Ref. (35).
[b] Data from Ref. (38).
[c] Data from Ref. (76).

difference in optical band gap value can be assumed as a measure of the influence of lattice disorder on optical gap of the films.

In Table 18.1, we report the mobility gap and band gap values of some passive films grown on valve metals. The difference $\Delta E_{am} = (E_g^m - E_g^{cryst})$, in the range of 0.1 to 0.35 eV, is in agreement with the expected extension of the localized states regions near the band edges due to the lattice disorder (67). Values of E_0 (see Equation (18.18)) nearly coincident with the mobility gap, E_g^m, have been frequently derived for passive films on valve metals. It seems quite reasonable to suggest, for such a class of amorphous materials, a band model similar to that shown in Figure 18.7(b) with an exponentially decreasing DOS in the mobility gap of the films at energies lying below E_C and above E_V. A mobility gap of passive film lower than the band gap of the crystalline counterpart must be interpreted as an indication that differences are present in the short-range order of the two phases. A different short-range order can imply the formation of a defective structure, with a high density of localized states within the mobility gap as well as changes in the density of the passive film, which is known to affect also the value of the optical gap in amorphous materials (64–67). The experimental findings on passive films and corrosion layers suggest that large differences in optical gap values, between amorphous and crystalline counterparts, should be traced out to a different chemical environment around the metallic cation or to the presence of large amount of defects within the passive films, providing electronic states within the mobility gap. A remarkable case is reported in Ref. (77), where the incorporation of organic species into anodic films, grown on electropolished Al samples in tartrate-containing solution, provided a DOS distribution within the band gap of a-Al_2O_3 and thus allowing the onset of anodic photocurrent at photon energies ($h\nu \cong 3.0$ eV) well below the band gap of Al_2O_3 ($E_g > 6.30$ eV).

18.3.4 Photoemission Phenomena at the Metal/Passive Film Interface

In this section, we discuss the role of the inner metal/film interface in the generation processes of photocarriers for thin and thick passive films. In the presence of thin passive films, it is possible that under illumination a large fraction of photons impinging the film/solution interface arrive at the metal/film interface by exciting metal electrons to higher energy levels and leaving vacant states below the Fermi level of the metal. The fate of the excited states into the metal depends on the occurrence of different physical deactivation processes at this interface. Apart from the thermal deactivation by scattering of excited electrons with the lattice vibrations, photoemission phenomena of excited photocarriers can be observed. In the case of very thin passive films ($d_{ox} < 2\,\text{nm}$) external (into the electrolytic solution) photoemission processes become possible by tunneling of excited electrons or holes at the metal surface throughout the film. A hole photoemission process has been suggested in the case of a gold electrode covered with a very thin oxide (78). The photoemission of electrons directly from the metal to the ground state of liquid water has been observed more frequently through very thin oxide films covering metals (79–81). When such an external photoemission process occurs, in the absence of diffuse double layer effects or adsorbed large molecules, it has been shown that emission photocurrent from the metal to acceptor species in solution depends both on photon energy and electrode potential according to the so-called "5/2 power law" (42):

$$\text{I}_{\text{ph}} \propto (h\nu - h\nu_0 - eU_E)^{5/2} \tag{18.19}$$

In Equation (18.19), U_{E} is the electrode potential measured with respect to a reference electrode and $h\nu_0$ is the photoelectric threshold at $U_{\text{E}} = 0$. At a fixed wavelength, we can derive the photoelectric threshold $h\nu_0 = h\nu - eU^*(\text{ref})$, where $U^*(\text{ref})$ is the intercept potential measured with respect to the reference electrode employed, by extrapolating to $I_{\text{ph}} = 0$ the plot of $(I_{\text{ph}})^{0.4}$ vs. U_{E} (see Figure 18.12). From this value, it is possible to derive the energetic level of a hydrated electron into the electrolyte

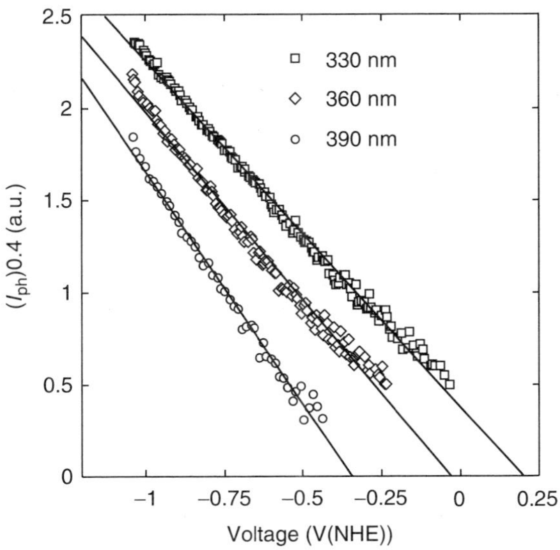

Figure 18.12 $(I_{\text{ph}})^{0.4}$ vs. electrode potential relative to electropolished Al recorded by irradiating the electrode at different wavelengths in $0.5\,M$ K_2SO_4.

(the so-called "conduction band" of liquid water) with respect to the vacuum level, using the relationship:

$$E_c(H_2O) = h\nu_0 + eU_{ref}(vac) \quad (18.20)$$

where $U_{ref}(vac)$ is the potential of the reference electrode with respect to the vacuum scale ($U_{NHE}(vac) = -4.5\,eV$). On the other hand, at constant electrode potential, U_E, the photoemission threshold can be obtained from the energy spectrum of the photoemission current yield (42), by plotting $Q^{0.4}$ vs. $h\nu$ (see Figure 18.13). In this case, the threshold energy value should shift 1 eV/V by changing the electrode potential (see the inset of Figure 18.13). Following this procedure, the location of the "conduction band" of liquid water, at -1.0 ± 0.1 eV with respect to the vacuum level, has been performed for Al, Zr, and Ni (79–81) covered by very thin initial films and in agreement with the analogous determination reported in Ref. (78). The previous relationships help to discriminate between cathodic photocurrents due to photoemission processes and those originating from a band-to-band excitation in p-type semiconducting thin films. In fact, in this last case, an optical band gap value independent (or slightly depending trough η_g) of the electrode potential is expected.

In the case of thicker films ($d_{ox} > 5$ nm), where the external photoemission processes are forbidden, the possibility of an internal photoemission process due to the injection of photoexcited electrons (or holes) from the metal into the CB (or VB) of the passive film must be considered. In such a case the internal photocurrent emission varies with the photon energy according to the so-called Fowler photoemission law (42):

$$Q = \text{const}(h\nu - E_{th})^2 \quad (18.21)$$

where E_{th} is the internal photoemission threshold energy, which can be obtained from a plot of $Q^{0.5}$ vs. the photon energy at constant photon flux. This threshold is a measure of the distance in energy between the Fermi level of the metal and the edge

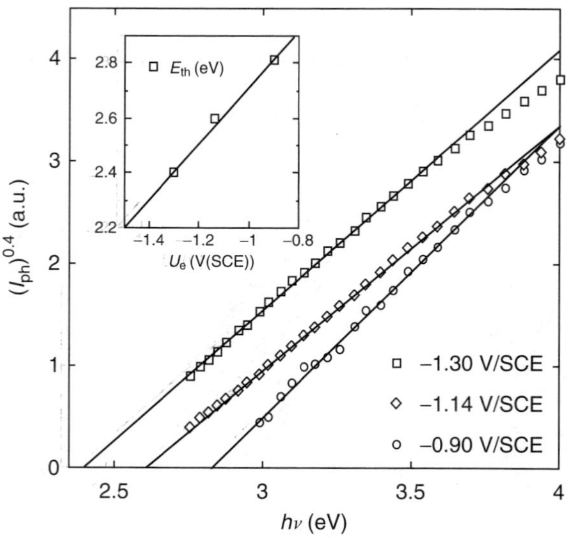

Figure 18.13 $Q^{0.4}$ vs. photon energy plots derived from the long wavelength region of the photocurrent spectra recorded at different cathodic polarizations for electropolished Al immersed in 0.5 *M* K_2SO_4. Inset: energy threshold for the external photoemission as a function of the electrode potential.

of film CB (electron photoemission) or VB (hole photoemission). The occurrence of electron or hole internal photoemission in the case of insulating films is established by the direction of the electric field, and in turn by the electrode potential value with respect to the photocurrent sign inversion potential. In the absence of trapping effects, the inversion photocurrent potential can be used to determine the flat band potential of insulating passive films. In the case of insulating anodic films on valve – metals, internal electron photoemission processes are usually observed under cathodic polarization and under illumination with photons having energy lower than the optical band gap of the film (79, 81). In Figure 18.14, we report the determination of the internal photoemission threshold for an anodic oxide film grown on Hf. In the case of semiconducting films, no evidence of internal photoemission process is expected, owing to the absence of any electric field at the metal/film interface, as long as the space-charge region of the SC is less than the film thickness. The knowledge of the internal photoemission thresholds allows locating the energy level of the conduction band of the oxide films with respect to the Fermi level of the underlying metal, once the work function of the metal is known. In Ref. (23), the internal thresholds for cathodic photoemission of a series of insulating oxide films grown on different valve metals have been reported.

18.4 BAND GAP AND OXIDE FILM COMPOSITION

18.4.1 Crystalline Binary Oxides

In spite of the relevant information on the structure of the passive film/electrolyte junctions obtainable by PCS, a more widespread use of this technique in corrosion studies has been hampered by the lack of a viable theory that relates the measured E_g values to the passive film composition. In general terms, such a task is a very challenging one, also for advanced theoretical methods based on quantum mechanical calculations. It is within the aims of this chapter to show that it is possible to relate the band gap values of numerous oxides to their composition by means of a semiempirical approach. In a previous work (35), it was shown that such a task could be accomplished by proposing the following general correlation between the band gap of crystalline oxides and the difference of electronegativity of their constituents:

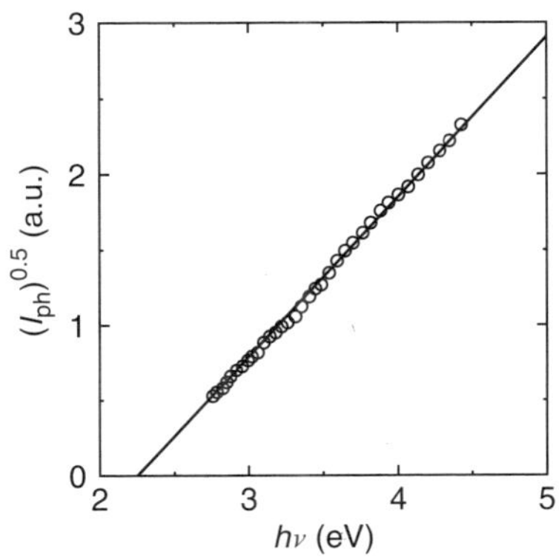

Figure 18.14 Fowler plots ($Q^{0.5}$ vs. $h\nu$) relative to an anodic film grown on Hf in 0.1 *M* ammonium borate electrolyte up to +10 V at 1 mA cm^{-2}, polarized at $U_E = -1$ V/SCE.

$$E_g = 2[E_I(\chi_M - \chi_O)^2 + \Xi] \qquad (18.22)$$

where, according to Phillips (82), E_I is the extra-ionic energy unit orbitally dependent, assumed "to vary with hybridization configuration, that is, with different atomic coordination in different crystal structures." χ_M and χ_O are the electronegativities of metal and oxygen, respectively. By plotting the band gap values of numerous oxides as a function of $(\chi_M - \chi_O)^2$ it comes out that the proposed correlation was able to provide:

1. Two clearly separated interpolating lines having slopes of about 2.17 and 1.35 eV and intercept values equal to 2.71 and 1.49 for sp-metal and d-metal oxides, respectively, apart from a few exceptions (see below).
2. A better fitting of the experimental data with respect to previous proposed correlation as evidenced by the higher correlation coefficient values.

From the d-metal correlation it follows that metallic oxides ($E_g \cong 0$) are expected to form on metals having Pauling's electronegativity value around 2.45, in agreement with the common experience that noble metal oxides at higher oxidation states (RuO_2, IrO_2 notably) usually display metallic conductivity. From a practical point of view, two more aspects are interesting for corrosion studies:

1. In this correlation, NiO stays neatly on the sp-metal oxides interpolating line, whilst Cr_2O_3, FeO, Fe_2O_3, Cu_2O, and CuO are well interpolated as d-metal oxides.
2. Three nontransitional-metal oxides (PbO, In_2O_3, Tl_2O_3) are better interpolated like d-metal oxides. According to this an intriguing d-/sp-metal oxides dividing line along the diagonal Zn, In, Pb/Ga, Sn, Bi appears, with some of sp-metals (In, Tl, Pb) of higher atomic number showing a d-like behavior in terms of optical band gap vs. $(\chi_M - \chi_O)^2$ correlation. Like all semiempirical approaches, the correlation cannot account for such a different behavior.

As for the electronegativity values in all calculations, Pauling's scale of electronegativity (83), integrated with the Gordy–Thomas values (84), has been used with the exception of Tl(III) for which the value given by Allred (85) has been preferred. The electronegativity value of different elements, calculated by using the experimental band gap values of the corresponding oxides and according to the best fitting straight lines, differ from those reported in Refs. (83–85) for a quantity of about 0.05, which is more or less the uncertainty given by the authors. According to this procedure, in Table 18.2, the estimated electronegativity for a group of lanthanides metals, obtained by using the d-metal correlation and the band gap values reported in the same table, is reported. The electronegativity values for such a group of metals stay within the limits 1.1 to 1.3 usually reported for f-block elements and in fair agreement with those reported by Allred (85) for the same elements at different oxidation states. Although the proposed correlation seems to work nicely also for f-block elements, some uncertainty still remains as for the parameters of the sp-metal oxides' correlation owing to the limited numbers of oxide band gap values used to derive it as well as to the difficulty to get reliable optical band gap values for s,p-metals having very low electronegativity parameters.

Table 18.2 Experimental Band Gap Values and Electronegativity Parameter Estimated by Using the d-Metal Correlation for a Group of Lanthanides Metals

Phase	E_g (eV)	Reference	χ_M
Sm_2O_3	5.0	86	1.31
Dy_2O_3	5.0	86	1.31
Yb_2O_3	5.2	86	1.27
Gd_2O_3	5.3	86	1.26
La_2O_3	5.4	87	1.24
CeO_2	5.5	88	1.22

18.4.2 Ternary Crystalline Oxides

The most interesting aspect embodied in the proposed correlation is the possibility to use such relationships for predicting the band gap of mixed oxides, thus opening a new route to the quantitative characterization of corrosion layers on metallic alloys. In Ref. (35), it was suggested that Equation (18.22) could be extended to ternary oxides, $A_aB_bO_o$, containing only d,d-metal or sp,sp-metal oxides, by substituting to the metal electronegativity, χ_M, the arithmetic mean for the cationic group, χ_c, defined as

$$\chi_c = \frac{a\chi_A + b\chi_B}{a + b} \tag{18.23}$$

where a and b are the stoichiometric coefficients of the cations in the ternary oxide, and χ_A and χ_B their electronegativities. In Table 18.3, the experimental band gap values for a number of ternary d,d-metal oxides covering a quite a large range of band gap values (1.90 to 4.0 eV) and difference of metal electronegativity ($\Delta\chi \cong 0.7$) are reported. A very good agreement is observed between the experimental data obtained by PCS and the theoretical data. Unfortunately, we are not aware of a similar large range of band gap values for sp-metal mixed oxides. The few available

Table 18.3 Experimental Optical Band Gap Values for d-Metal and d–d-Metal Alloy Oxides and Comparison Between Metal Electronegativity Estimated According to Equations (18.22) and (18.23), χ_{exp}, and Pauling Electronegativity

Phase	E_g^{exp} (eV)	Reference	χ_{exp}	$\chi_{Pauling}$
Y_2O_3	5.50	89	1.22	1.20
Sc_2O_3	5.40	89	1.24	1.30
$CuYO_2$	3.50	90	1.55	1.55
$MnTiO_3$	3.10	35	1.65	1.60
$La_2Ti_2O_7$	4.00	35	1.48	1.445
$FeTiO_3$	2.85	35	1.71	1.725
$Y_3Fe_5O_{12}$	3.00	35	1.68	1.64
$CuScO_2$	3.30	91	1.62	1.60
$Fe_{18}Ti_2O_{31}$	2.09	39	1.87	1.88
$Fe_8Ti_2O_{16}$	2.17	39	1.85	1.85
$Fe_6Ti_4O_{17}$	2.35	39	1.81	1.80
$Fe_{10}Ti_{10}O_{35}$	2.50	39	1.78	1.775
$Fe_{18}Ti_{22}O_{71}$	2.60	39	1.76	1.76

Table 18.4 Experimental Optical Band Gap Values for sp–d Metal Alloy Oxides and Comparison Between Metal Electronegativity Estimated According to Equations (18.22) and (18.23), χ_{exp}, and Pauling Electronegativity

Phase	E_g^{exp} (eV)	Reference	χ_{exp}	$\chi_{Pauling}$
$Bi_{0.7}Y_{0.3}O_{1.5}$	3.0	92	1.68	1.69
$SrZrO_3$	5.40	35	1.24	1.20
$MgTiO_3$	3.70	35	1.54	1.425
La_2NiO_4	4.0	93	1.48	1.43
$Mg_{0.19}Zn_{0.81}O$	3.76	94	1.53	1.52
$Mg_{0.27}Zn_{0.73}O$	3.92	94	1.50	1.49
$Mg_{0.36}Zn_{0.67}O$	4.19	94	1.45	1.46

data seem to confirm the validity of sp-metal correlation also for ternary oxides with differences in the experimental and calculated E_g values in the order of 10%.

As for ternary sp,d-metal oxides, it was found that the d-metal correlation was able to fit quantitatively their band gap values provided that the difference in the electronegativity between the metallic cations is less than 0.5 (see Ref. (35) and Table 18.4). Due to the limited number of systems investigated the limits of applicability of such a correlation to the sp,d-metal mixed oxides remain unsolved. Very recent results both for crystalline bulk Mg–Zn (94) oxides and amorphous anodic oxide on Al–W (95) alloys seem to suggest that when the atomic fraction of the d-metal in the ternary oxides reaches values lower than 20% the experimental E_g data are better interpolated by the sp-metal oxide correlation. These aspects need further investigations aimed to better define the limits of validity of the proposed semiempirical correlation.

18.4.3 Amorphous Oxide Films

The previous correlation was extended to amorphous anodic films and corrosion layers by taking into account the influence of the amorphous nature on the optical band gap of passive films as previously discussed. Numerous investigations on anodic oxide films of valve metals have shown that amorphous oxides usually displayed optical band gap values larger than the crystalline counterpart in the absence of other specific defects. According to these results, it has been proposed to take into account the influence of the amorphous nature of passive films on their optical band gap values by modifying the correlation, used for crystalline oxides, as follows (38):

$$E_{gf} - \Delta E_{am} = A(\chi_c - \chi_O)^2 + B \tag{18.24}$$

where E_{gf} is the optical band gap of a passive film and ΔE_{am} represents the difference between E_{gf} and the optical band gap of the crystalline counterpart in electronvolts. The choice of A, B, and ΔE_{am} depends on the nature (sp or d) of the metal cations as well as on their relative atomic fraction in the case of mixed oxides. As for the influence of the lattice disorder on E_g, a value of $\Delta E_{am} < 0.35$ to 0.40 eV seems able to account for the experimental data (see Table 18.1). ΔE_{am} values in the order of 0.35 to 0.40 eV are typical of truly amorphous oxides (MoO_3, WO_3, Al_2O_3,), whilst lower values are expected for anodic film oxides having a tendency to grow in microcrystalline forms (ZrO_2, TiO_2, Ta_2O_5). In Figure 18.15, a fitting is reported,

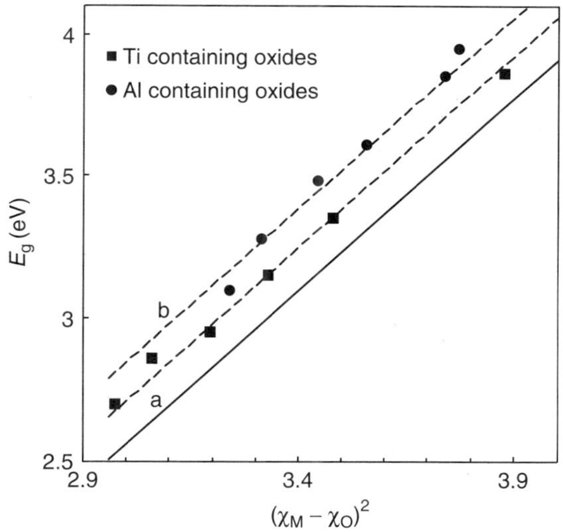

Figure 18.15 Experimental band gap values (see Table 18.5) relative to amorphous ternary oxides as a function of $(\chi_M - \chi_O)^2$. ———: Theoretical prediction for crystalline d-metal oxide according to Equation (18.22); -------: theoretical prediction for amorphous d-metal oxide according to Equation (18.24) with $\Delta E_{am} = 0.15$ eV (a) and 0.29 eV (b).

according to Equation (18.24), of the band gap values as a function of the compositional parameter χ_c obtained according to Equation (18.23) for two different amorphous ternary oxide systems containing both sp,d-metal and d,d-metal cations obtained by anodizing metallic alloys (Al–W magnetron sputtered alloys) or by the chemical route (see Table 18.5). In both cases, a ΔE_{am} value in agreement with the theoretical expectation is derived from the fitting of experimental data as well as a larger value of ΔE_{am} for anodic films made by oxides having a tendency to grow in truly amorphous state. Further evidence in favor of the proposed correlation for anodic films grown on Al–Ti, Al–Ta, Mo–Ta, Al–Nb, Fe–Zr metallic alloys can be found in Refs. (38, 39, 97). We have to mention that apart from the nature of the oxide

Table 18.5 Experimental Band Gap Values of Amorphous Mixed Oxides as a Function of their Composition

Oxide composition (considering metals only)	$(\chi_M - \chi_O)^2$	E_g (eV)	Reference
Pb–15 at.%Ti	2.97	2.7	96
Pb–30 at.%Ti	3.06	2.86	96
Fe–75 at.%Ti	3.19	2.95	96
Fe–90 at.%Ti	3.33	3.15	96
Al–23 at.%Ti	3.88	3.86	39
Al–34 at.%Ta	4.00	4.15	39
Al–29 at.%W	3.77	3.95	95
Al–33 at.%W	3.74	3.85	95
Al–57 at.%W	3.56	3.61	95
Al–72 at.%W	3.44	3.48	95
Al–90 at.%W	3.31	3.28	95

other experimental parameters can affect the lattice disorder degree of passive films and then the ΔE_{am} value in anodic films (32, 39).

18.4.4 Correlation for Hydroxides and Oxyhydroxide Films

By analogy with the procedure followed for mixed oxides, in Refs. (35, 36) it was suggested that for the OH group an average electronegativity value, χ_{an}, could be defined by using the arithmetic mean between the values pertaining to oxygen and hydrogen. Accordingly, a value of $\chi_{an} = 2.85$ was assumed for the OH group and in agreement with values reported in literature (98). On the basis of experimental data on a number of hydroxides, the following correlation (36) has been proposed:

$$E_g = 1.21(\chi_M - \chi_{OH})^2 + 0.90 \tag{18.25}$$

for hydroxides grown on sp-metal, and

$$E_g = 0.65(\chi_M - \chi_{OH})^2 + 1.38 \tag{18.26}$$

for hydroxides grown on d-metals. It is noteworthy that once again different relationships hold for sp-metal and d-metal hydroxides, as previously reported for metal oxides. Although this last correlation suffers some limitations owing to the rather limited number of investigated systems, it is able to rationalize some results reported in the literature. In fact, from Equation (18.25) an optical band gap around 2.25 eV is derived for $Ni(OH)_2$ by assuming $\chi_{Ni} = 1.80$ according to the Pauling scale, and under the hypothesis that also for Ni-hydroxides the s,p-metal correlation is valid for NiO. The optical gap estimated according to Equation (18.25) is in good agreement with the value estimated from photocurrent spectra reported in the literature for passive films on Ni ($E_g = 2.20$ eV) (18) and for $Ni(OH)_2$ electrochemically deposited and soaked in KOH solution (99). These findings confirm once more that in the proposed correlation Ni(II) oxide and hydroxide conform to the sp-metal behavior.

A further support to the proposed correlation of hydroxides comes from PCS data obtained from conversion coating grown on Mg metal in a stannate bath, where the formation of a mixed $MgSn(OH)_6$ layer has been reported (100), as well as from PCS characterization of a $La(OH)_3$ anodic film grown in NaOH (101, 102). The band gap values, $E_g = 3.0 \pm 0.1$ eV for $MgSn(OH)_6$ and 3.15 ± 0.1 eV for $La(OH)_3$, measured, respectively, for the two layers agree reasonably well to the expected values according to the proposed correlation (Equations (18.25) and (18.26)).

The last point we would like to discuss is the possibility to relate the band gap values of passive films to their hydroxylation degree. On a purely heuristic basis, a connection formula between the band gap of oxides and the band gap of the corresponding oxyhydroxides, $MO_{(y-m)}OH_{(2m)}$ has been suggested (33, 35). According to this suggestion, it has been proposed that the band gap of anhydrous and hydroxylated oxide phases can be related by the relationship:

$$E_g^{hyd} = \frac{E_g^{anh}}{1 + k_m x_{OH}} \tag{18.27}$$

where k_m is a constant that can be calculated for each system once the band gap values of the anhydrous oxide and hydroxide are known. x_{OH} is the fraction of the hydroxylic group in the oxyhydroxide phase defined as

$$x_{OH} = \frac{2m}{y + m} \tag{18.28}$$

More recently, some authors (103), by assuming as generally valid the proposed correlation between optical band gap and difference of electronegativity of cation and average anion group electronegativity (35), proposed for an oxyhydroxide phase the following relationship for the dependence of the optical band gap from the hydroxylation degree $\alpha = m/y$:

$$E_g(\alpha) = A\left[\frac{(\chi_O - \chi_M) + (\chi_H - \chi_M)\alpha}{1+\alpha}\right]^2 + B \tag{18.29}$$

where A and B can be calculated once the values of $\alpha = 0$ and $\alpha = 1$ are substituted in the previous relation at the corresponding E_g value of oxide and hydroxide. It is encouraging that for the investigated system ($SnO_{2-m}(OH)_{2m}$) very small differences in the estimated band gap values are obtained by using indifferently Equation (18.28) or (18.29) (104).

18.4.5 PCS Analysis of Passive Films and Corrosion Layers on Base Metals and Alloys

One of the most challenging tasks for any experimental technique is to be able to provide useful information on the electronic properties and chemical composition of passive films on base metals (like Cu, Fe, Cr, Ni) and their alloys (stainless steels (SS), Fe–Cr alloys, etc.). The previous correlation has been used to relate quantitatively PCS data to passive films composition formed on some of these metals (32, 33, 35, 37, 80, 105).

In more complex systems, further information on the oxidation state of each metallic cation and their atomic fraction would be necessary in order to use quantitatively the previous correlation. Moreover, in some cases the expected optical band gap values of oxides and hydroxides of the same metal, at different oxidation states, are very close or almost coincident so that it becomes difficult to get quantitative information from PCS measurements alone.

In order to highlight this point, in Table 18.6(a) we report the band gap values of crystalline oxides and hydroxides of a group of base metals (Cu, Cr, Fe, Ni) experimentally measured by the PCS technique or derived from the literature. In Table 18.6(a), we also report the Pauling electronegativity parameter for the different

Table 18.6a Band Gap Values of Crystalline Oxides and Hydroxides of a Group of Base Metals. Comparison between Pauling Electronegativity and PSC Estimated Electronegativity

Phase	E_g (eV)	Reference	$\chi_{Pauling}$	χ_{PCS}
Cr_2O_3	3.30	106	1.60	1.62
$Cr(OH)_3$	2.43	105	1.60	1.62
CrO_3	2.0	106		1.89
ZnO	3.34	107	1.60	1.60
Cu_2O	1.86	108	1.90	1.90
CuO	1.40	106	2.00	2.04
FeO	2.40	106	1.80	1.80
Fe_2O_3	1.90	106	1.90	1.91
NiO	3.80 (direct)	106	1.80	1.80
	3.58 (indirect)	109		1.77
$Ni(OH)_2$	2.31	23	1.80	1.77

Table 18.6b Band Gap Values of Some Base Metal Hydroxides Estimated According to Equation (18.28)

Phase	χ_{PCS}	E_g (eV)
$Zn(OH)_2$	1.60	2.39
CuOH	1.90	1.97
$Cu(OH)_2$	2.04	1.80
$Fe(OH)_2$	1.80	2.10
$Fe(OH)_3$	1.91	1.95

oxidation states. Apart the electronegativity value of Cr(VI) for which we are not aware of other reliable values in the literature, we like to stress that the electronegativity data obtained by Equation (18.22) are in very good agreement with those reported in the literature (83–85) so that we suggest to use these last values as preferred PCS electronegativity parameters for the characterization of passive films on pure base metals and on their alloys. Accordingly, we report in Table 18.6(b) the estimated band gap values of the corresponding hydroxides as derived from the proposed correlation and by using the PCS preferred χ_M values.

It can be inferred from the data in Table 18.6(a) and (b) that oxides and hydroxides of d-metal having electronegativity around 1.90 display very close band gap values, thus making it very difficult to distinguish by PCS alone which type of passive films is formed. This is the case of copper and iron passive films for which at least three different phases at different hydroxylation degrees exist with very close optical band gap values. In such a case further information gathered by other (possibly *in situ*) analytical techniques are necessary to rationalize PCS data of passive films.

Other authors (110–112) have proposed a different route for the identification of passive film. According to these authors, a comparison of PCS data (as well as of the general semiconducting behavior) between passive films anodically grown and sputtered oxides could help to identify the nature of passive films. This approach may be useful provided that any difference in the defect structure and hydroxylation degree for oxide phases grown in such different ways is negligible.

The data of Table 18.6(a) and (b) and the previous considerations could help to rationalize some of the experimental findings on PCS analysis concerning passive films on iron, for which band gap values ranging between 1.90 and 2.10 eV are reported in the literature (Ref. (112) and references therein). A band gap value of 1.90 eV has been estimated in Ref. (112) for passive films formed on iron in borate buffer solution (pH 8.4) at high potential (U_E = 0.6 to 0.8 V/SCE). This value matches well the one reported in the same work for evaporated Fe_2O_3 and the one reported in Table 18.6(a) for crystalline Fe_2O_3, thus allowing identification of the passive film composition as Fe(III) oxide. This conclusion is in agreement with recent XANES data suggesting the formation of microcrystalline iron oxide structure (LAMM phase, Ref. (113)) on passive iron anodized under similar conditions. The higher E_g measured at lower polarizing voltage ($U_E < 0.4$ V/SCE) can be ascribed to the presence of crystallographic disorder, as in amorphous passive films, or to the partial reduction of Fe(III) to Fe(II) with a subsequent variation of film composition. According to Equations (18.22) and (18.23), a band gap of 2.0 eV is expected for a 10% Fe(II) containing oxide in agreement with the value reported in Ref. (112) for oxidized magnetite, while $E_g = 2.1$ eV can be estimated for Fe_3O_4 in contrast with the value reported in the same work. The E_g variation with U_E, reported in Ref. (112), is predicted by the above-mentioned equations as a function of the Fe(II) content into the film, and it should agree with the description of passive film on iron, described as a Fe(II) deficient magnetite of formula $Fe(II)_{1-x}Fe(III)_2O_{4-x}$, with Fe(II) content in the range of 10 to 25% depending on the passivation potential (112).

By considering that a band gap value of 1.95 eV, almost coincident with that of Fe_2O_3, is estimated in Table 18.6(b) for $Fe(OH)_3$, we can conclude that $E_g = 1.90$ to 1.95 eV is expected for iron passive films having composition equal to FeOOH, and thus PCS is not sensitive to variation of the hydration degree of Fe passive films due to the small dependence of the band gap values on the OH content. These conclusions account for the following experimental findings:

1. The band gap of passive films on iron grown in strongly alkaline solution (33, 114) is not appreciably different from the E_g values of films grown in neutral or acidic solution, where less hydrated phases are expected to form.
2. Passive iron electrodes covered with iron hydroxide (115) show an energy threshold for the photocurrent onset coincident with that measured in the absence of the iron hydroxide outer layer.

The use of PCS in the identification of passive films on iron–chromium–nickel alloys and stainless steel appears even more difficult, despite the appreciable differences in the band gap values of the corresponding pure crystalline oxides. The possible formation of a passive film with an unknown hydration degree and with different metals at different oxidation state makes the PCS analysis very complex. Different photoelectrochemical data on such systems are reported in the literature and the interested reader can refer to Refs. (10, 73, 114, 116–125) and references therein.

Owing to space limitations, we will discuss briefly and qualitatively some selected PCS data pertaining to such systems, which can be compared to some extent to more recent quantitative analysis based on *in situ* and *ex situ* techniques (XANES, XPS, EQCM) (126–128).

The formation of $Cr(OH)_3$ on passive Cr at low potentials ($U_E < -0.6$V/MSE) and in a wide range of pH ($0 < pH < 9$) has been suggested on the basis of the measured optical band gap value ($E_g = 2.45 \pm 0.1$ eV; see Refs. (105, 129)). Such a hypothesis is in agreement with XPS and XANES data reported in Refs. (126–128) for films grown on Cr and on Fe–Cr alloys in acidic solutions as well as in slightly alkaline solutions (130).

As for PCS data of passive films grown on Fe–Cr alloys (Cr content <30 at.%) in borax buffer solution ($8.4 < pH < 9.2$) band gap values ranging from 1.90 to 2.20 eV with increasing Cr content have been reported in Refs. (119, 123). In a more recent work (131), by using steady-state PCS data, band gap around 3.0 eV was reported for passive film on Fe–Cr alloys (Cr content <20 at.%) at variance with the previous ones. Such a difference in E_g values could be attributable, in principle, to a loss of sensitivity in the last method, which could not have detected the lower photocurrent values usually measured for photon energies near the band gap of passive films. This last hypothesis could also explain higher E_g values measured, for passive films grown on experimental ferritic SS (116) in chloride-containing solution, at low anodic potentials ($E_g = 2.80$ eV). On the other hand, E_g values around 2.60 eV were measured, at higher electrode potentials ($U_E = 0.6$ V/SCE), for passive films formed on analogous ferritic SS containing small amount (<4 wt.%) of Mo and Ni (73).

Due to the large range of E_g values present in the literature for passive films on Fe–Cr alloys, the loss of sensitivity of PCS for photon energies near the band gap of passive films does not seem a satisfactory explanation. The data on passive films composition obtained by *in situ* and *ex situ* analytical techniques reported in Refs. (126–128) for Fe–Cr passive films grown at pH values <4.5 can be useful to justify these results.

XANES data on various Fe–Cr alloys (126) strongly suggest that at pH 4.5 (acetate buffer solution) films formed at low potentials ($U_e = -0.3$ V/MSE) are enriched, with respect to the base alloy, of Cr(III) usually present in the films as hydrated species, whilst at higher potentials ($U_E = 0.2$ V/MSE) the passive films display both a minor Cr enrichment as well as a decrease in hydroxide content. These findings suggest that passive films on Fe–Cr alloys are oxides containing both the metals with a Fe/Cr ratio and a hydration degree dependent on experimental conditions. The photoelectrochemical results given above agree with this hypothesis. According to Equations (18.22), (18.23), and (18.27), band gap around 2.1 ± 0.1 eV strongly points toward the presence on the surface of Cr–Fe alloys of passive films constituted of mixed iron–chromium hydroxides richer in iron, with larger E_g values attributable to passive film at higher Cr content. Even larger E_g values could be attributed to anhydrous mixed oxide passive films. Moreover, extending the previous considerations to passive films on ferritic SS, the decrease of their optical band gap with increasing electrode potential can be explained by an increase of the iron content.

As for passive films on austenitic SS (Fe–18Cr–8Ni) and Ni–Cr alloys (alloy 600 series), formed at high anodic potential ($U_E = 0.8$ V/SCE) in borate buffer solution at pH 9.2, a rather constant optical band gap ($E_g = 2.35 \pm 0.1$ eV) has been reported in Refs. (123–125), the highest values being for passive films on SS with higher Ni content or on Ni–Cr alloys. The E_g value suggests that also in this case a chromium hydroxide rich phase is the main component of the passive films. At lower potentials ($U_E < 0.1$ V/SCE) higher band gap values ($E_g = 2.75 \pm 0.1$ eV) have been found (120, 122). However, passive films on austenitic SS with higher molybdenum content (>4 wt.%) usually displayed a more limited variation in the E_g values as a function of electrode potentials (116, 120). This finding was particularly evident in the case of the superaustenitic commercial alloy 254-SMO (20Cr–18Ni–6Mo), which displayed an almost constant E_g value (2.45 ± 0.1 eV) in a rather large range of electrode potential ($-0.1 < U_E < 0.9$ V/SCE) at pH 6.5 and in the presence of chloride (116). The beneficial action of Mo in improving the pitting resistance of SS is well known (132) and, according to PCS data, it could be ascribed to its ability to keep on the surface of such an alloy a passive film having an almost constant composition (very close to $Cr(OH)_3$, according to the E_g values) at different potentials. This interpretative hypothesis is in qualitative agreement with the results of STM and XPS analysis reported in Ref. (127) showing that the passive film grown on Fe–18Cr–13Ni in 0.5 *M* H_2SO_4 at 0.5 V/SHE keeps a stable thicker outer layer containing almost chromium hydroxide also at longer polarization times. It seems reasonable to conclude that in the case of 254-SMO alloy the external layer of the passive film formed at pH 6.5 has a composition close to $Cr(OH)_3$, as reported for the outer layer of the Fe–Cr–Ni alloy discussed in Ref. (127).

Higher E_g values (>2.40 eV) are expected for passive films on austenitic SS at higher potentials in the presence of a dehydration process and for the formation of anhydrous Cr-mixed oxides (127, 133).

The possibility to find by PCS the presence of bilayer structures or passive films with a variable hydration degree as a function of distance from the external surface has been discussed in Refs. (134, 135). The possibility of PCS to indicate the presence of different phases on the surface of Cu–Zr alloys has been discussed in Ref. (35).

18.5 PHOTOELECTROCHEMICAL IMAGING

In the previous sections, we discussed PCS data relative to average photoelectrochemical behavior of large surface areas. The possibility to get a locally resolved photoelectrochemical imaging of passive metals was initially proposed by Butler who performed a photoelectrochemical mapping of a passive Ti electrode by moving a laser spot of suitable wavelength on the surface of passive Ti at constant electrode potential (25, 136).

Since this first contribution different research groups (137–152) exploited the possibility to use a local photocurrent probe to characterize, with micrometer and submicrometer resolutions, different areas of passive electrodes. At constant electrode potential it comes out that the intensity of photocurrent signal is a complex function of different variables (see Equations (18.13) and (18.15)) like film thickness, electric field, photon energy, light intensity, and kinetics of charge transfer at the passive film–electrolyte interface. The interpretation of photoelectrochemical imaging data is, mainly, within the theoretical framework previously exposed. Some more technical issues in local photoelectrochemical techniques and imaging theory, not discussed in previous sections, can be found in Refs. (141, 143–145, 147).

To our knowledge, the most investigated system in imaging studies is the Ti/passive film/electrolyte system although few studies on passive chromium (151), iron (145), and stainless steels (118, 149, 150) are reported in the literature. All these studies have shown the role of surface defects and of particle inclusions in modifying the photoelectrochemical behavior of passive electrodes.

The presence of photoactive inclusions in SS (149–150) and photoelectrochemical active sites on passive titanium metal and their possible role in determining the pitting behavior of passive film has been addressed by combining photoelectrochemical imaging and other local electrochemical techniques with resolution in the order of 0.1 μm (152).

Further improvements in PCS technique allowing to scrutinize passive electrodes at this scale of resolution by incident light of continuous wavelength would permit the onset of a micro-area-photocurrent spectroscopy technique. Such an improvement could help to better characterize in terms of electronic properties and chemical composition very fine local changes of composition so that the nature of inclusions and their role in favoring local corrosion attack of metals and alloys can be unraveled (153, 154).

18.6 CONCLUSIONS

In this work, we presented a relatively simple approach to the fundamentals of photoelectrochemical techniques currently used for characterizing passive films and corrosion layers on metal and alloys. Some of the interpretative models suggested for a rationalization of experimental PCS data are still under scrutiny; thus, they must be used with some caution. The results obtained up to now strongly support the proposed models although further investigations are necessary which could modify or validate them for a limited number of systems. Nevertheless, it is in our opinion that the PCS technique remains one of the most versatile techniques at our disposal for obtaining information on the solid-state properties of passive layers in corrosion studies.

ACKNOWLEDGMENTS

The financial support of Palermo University and Becromal S.p.A. (Milano) during the years is gratefully acknowledged. The authors thank Prof. S. Piazza for his collaboration through the years and R. Masi and F. Criminisi for performing part of very recent experimental work.

REFERENCES

1. E. Becquerel and C.R. Hebd, *Séan. Acad. Sci.* 9: 56, 1839.
2. W. Copeland, O.D. Black, and A.B. Garrett, *Chem. Rev.* 31: 177, 1942.
3. W.H. Brattain and C.G.B. Garrett, *Bell Syst. Tech. J.* 34: 129, 1955; *Phys. Rev.* 99: 177, 1955.
4. H. Gerischer, *Advances in Electrochemistry and Electrochemical Engineering*, Vol. 1, p. 139, edited by P. Delahay, Interscience Publishers, New York, 1961.
5. V.A. Myamlin and Yu.V. Pleskov, *Electrochemistry of Semiconductors*, Plenum Press, New York, 1967.
6. V.A. Tyagai and Yu.V. Pleskov, *Elektrokhimiya* 1: 1167, 1965.
7. E.K. Oshe and I.L. Rozenfel'd, *Elektrokhimiya* 4: 1200, 1968.
8. R. Williams, *J. Chem. Phys.* 32: 1505, 1960.
9. W. Paatsch, *J. Phys.* 38: C5-151, 1977.
10. E. Angelini, M. Maja, and P. Spinelli, *J. Phys.* 38: C5-261, 1977.
11. S.M. Wilhelm, Y. Tanizawa, C-Y Liu, and N. Hackermann, *Corros. Sci.* 22: 791, 1982.
12. A. Fujishima and K. Honda, *Nature* 238: 37, 1972.
13. Faraday Discussion No.70, The Royal Society of Chemistry, London, 1980.
14. A.J. Nozik (ed.) *Photoeffects at Semiconductor–Electrolyte Interfaces*, ACS Symposium Series 146, American Chemical Society, Washington, DC, 1981.
15. M. Gratzel, *Nature* 414: 338, 2001.
16. F. Di Quarto, G. Russo, C. Sunseri, and A. Di Paola, *J. Chem. Soc. Faraday Trans. I* 78: 3433, 1982.
17. M. Froment (ed.) *Passivation of Metals and Semiconductors*, in *Proceedings of the 5th International Symposium on Passivity*, Elsevier, Oxford, 1983.
18. U. Stimming, *Electrochim. Acta* 3: 415, 1986.
19. L. Peter, *Comprehensive Chemical Kinetics*, Vol. 29, p. 382, edited by R.G. Compton, Elsevier Science, Oxford, 1989.
20. N. Sato and K. Hashimoto (eds.) *Passivation of Metals and Semiconductors*, in *Proceedings of the 6th International Symposium on Passivity*, Pergamon Press, Oxford, 1990.
21. K.E. Heusler (ed.) *Passivation of Metals and Semiconductors*, in *Proceedings of the 7th International Symposium on Passivity*, Trans Tech. Publications Ltd, Zurich, 1995.
22. M.B. Ives, J.L. Luo, and J.R. Rodda (eds.) *Passivation of Metals and Semiconductors*, in *Proceedings of the 8th International Symposium on Passivity*, The Electrochem Soc. Inc., PV 99-42, Pennington, NJ, 2001.
23. F. Di Quarto, C. Sunseri, S. Piazza, and M. Santamaria, in *Handbook of Thin Films*, edited by H.S. Nalwa, Vol. 2, p. 373, Academic Press, New York, 2002.

24. M. Pourbaix, *Atlas of Electrochemical Equilibria in Aqueous Solutions*, Pergamon Press, Oxford, 1966.
25. M.A. Butler, *J. Electrochem. Soc.* 131: 2185, 1984.
26. M.L. Meade, *Lock-in Amplifiers: Principles and Applications*, Peter Peregrinus Ltd, London, 1983.
27. W. Schmickler and J.W. Schultze, *Modern Aspects of Electrochemistry*, Vol. 17, p. 357, edited by J.O'M Bockris, B.E. Conway, and R.E. White, Plenum Publishing Corporation, New York, 1986.
28. H. Gerischer, *Electrochim. Acta* 35: 1677, 1990.
29. H. Gerischer, *Corros. Sci.* 29: 257, 1989.
30. H. Gerischer, *Corros. Sci.* 31: 81, 1990.
31. N. Sato, *Passivation of Metals and Semiconductors*, in *Proceedings of the 6th International Symposium on Passivity*, p. 1, edited by N. Sato and K. Hashimoto, Pergamon Press, Oxford, 1990.
32. F. Di Quarto, S. Piazza, and C. Sunseri, *Current Topics in Electrochemistry*, Vol. 3, p. 357, edited by J.C. Alexander, Council of Scientific Research Integration, Trivandrum, 1994.
33. F. Di Quarto, S. Piazza, and C. Sunseri, *Mater. Sci. Forum* 192–194: 633, 1995.
34. J.W. Schultze and M.M. Lohrengel, *Electrochim. Acta* 45: 2499, 2000.
35. F. Di Quarto, C. Sunseri, S. Piazza, and M.C. Romano, *J. Phys. Chem. B* 101: 2519, 1997.
36. F. Di Quarto, M.C. Romano, M. Santamaria, S. Piazza, and C. Sunseri, *Russ. J. Electrochem.* 36: 1203, 2000.
37. L. Anicai, S. Piazza, M. Santamaria, C. Sunseri, and F. Di Quarto, *ATB Metallurgie* XXXX, 3–4: 175, 2000.
38. M. Santamaria, D. Huerta, S. Piazza, C. Sunseri, and F. Di Quarto, *J. Electrochem. Soc.* 147: 1366, 2000.
39. F. Di Quarto, M. Santamaria, S. Piazza, and C. Sunseri, *Symposium AA Structure–Property Relationships of Oxide Surfaces and Interfaces*, edited by C.B. Carter, X. Pan, K. Sickafus, H.L. Tuller, FT. Wood, MRS Proceedings, Vol. 654, AA4.8.1.
40. H. Gerischer, *Physical Chemistry. An Advanced Treatise*, Vol. IXA, p. 463, edited by H. Eyring, D. Henderson, and W. Jost, Academic Press, New York, 1970.
41. S.R. Morrison, *Electrochemistry at Semiconductor and Oxidized Metal Electrodes*, Plenum Press, New York, 1980.
42. Yu.Ya. Gurevich, Yu.V. Pleskov, and Z.A. Rotenberg, *Photoelectrochemistry*, Consultants Bureau, New York, 1980; Yu.V. Pleskov and Yu.Ya. Gurevich, *Semiconductor Photoelectrochemistry*, Consultants Bureau, New York, 1986.
43. A. Hamnett, *Comprehensive Chemical Kinetics*, Vol. 27, p. 61, edited by R.G. Compton, Elsevier Science, Oxford, 1987.
44. J.O'M. Bockris and S.U.M. Khan, *Surface Electrochemistry*, Plenum Press, New York, 1993.
45. N. Sato, *Electrochemistry at Metals and Semiconductor Electrodes*, Elsevier Science, Amsterdam, 1999.
46. J.F. Dewald, *Bell Syst. Tech. J.* 39: 615, 1960.
47. J.F. Dewald, *J. Phys. Chem. Solids* 14: 155, 1960.
48. W.H. Laflère, R.L. Van Meirhaeghe, F. Cardon, and W.P. Gomes, *Surf. Sci.* 59: 401, 1976.
49. K. Tomkiewicz, *J. Electrochem. Soc.* 126: 1505, 1979.
50. W.H. Laflère, R.L. Van Meirhaeghe, F. Cardon, and W.P. Gomes, *J. Appl. Phys. D* 13: 2135, 1980.
51. G. Cooper, J.A. Turner, and A.J. Nozik, *J. Electrochem. Soc.* 129: 1973, 1982.
52. H.O. Finklea, *J. Electrochem. Soc.* 129: 2003, 1982.
53. H. Gerischer and R. McIntyre, *J. Chem. Phys.* 83: 1363, 1985.
54. M.A. Butler, *J. Appl. Phys.* 48: 1914, 1977.

55. R.H. Wilson, *J. Appl. Phys.* 48: 4292, 1977.
56. H. Reiss, *J. Electrochem. Soc.* 125: 937, 1978.
57. W.J. Albery, P.N. Bartlett, A. Hamnett, and M.P. Dare-Edwards, *J. Electrochem. Soc.* 128: 1492, 1981.
58. F. El Guibaly and K. Colbow, *J. Appl. Phys.* 53: 1737, 1982.
59. W.W. Gärtner, *Phys. Rev.* 116: 84, 1959.
60. W.J. Albery and P.N. Bartlett, *J. Electrochem. Soc.* 129: 2254, 1982.
61. J. Li and L.M. Peter, *J. Electroanal. Chem.* 193: 27, 1985.
62. J. Li and L.M. Peter, *J. Electroanal. Chem.* 199: 1, 1986.
63. P.C. Searson, D.D. Macdonald, and L.M. Peter, *J. Electrochem. Soc.* 139: 2538, 1992.
64. M.H. Cohen, H. Fritzsche, and S.R. Ovishinsky, *Phys. Rev. Lett.* 22: 1065, 1969.
65. D. Adler, *Amorphous Semiconductors*, CRS Press, Cleveland, 1971.
66. J. Tauc, *Amorphous and Liquid Semiconductors*, Plenum Press, London, 1974.
67. N.F. Mott and E.A. Davis, *Electronic Processes in Non-crystalline Materials*, 2nd Ed., Clarendon Press, Oxford, 1979.
68. F. Di Quarto, S. Piazza, and C. Sunseri, *Electrochim. Acta* 38: 29, 1993.
69. C. Da Fonseca, M.G. Ferreira, and M. Da Cunha Belo, *Electrochim. Acta* 39: 2197, 1994.
70. F. Di Quarto, S. Piazza, R. D'Agostino, and C. Sunseri, *J. Electroanal. Chem.* 228: 119, 1987.
71. F. Di Quarto, S. Piazza, and C. Sunseri, *J. Chem. Soc. Faraday Trans. 1* 85: 3309, 1989.
72. D.M. Pai and R.C. Enck, *Phys. Rev. B* 11: 5163, 1975.
73. C. Sunseri, S. Piazza, A. Di Paola, and F. Di Quarto, *J. Electrochem. Soc.* 134: 2410, 1987.
74. S. Piazza, C. Sunseri, and F. Di Quarto, *J. Electroanal. Chem.* 293: 69, 1990.
75. G.D. Cody, *Semiconductors and Semimetals*, Vol. 21, part B, p. 11, edited by J.I. Pankove, Academic Press, London, 1984.
76. H. Kim, N. Hara, and K. Sugimoto, *J. Electrochem. Soc.* 146: 955, 1999.
77. F. Di Quarto, S. Piazza, A. Splendore, and C. Sunseri, *Oxide Films on Metals and Alloys*, p. 311, edited by B.R. McDougall, R.S. Alwitt, T.A. Ramanarayanan, PV 92-22, The Electrochemical Society Inc., 1992.
78. T. Watanabe and H. Gerischer, *J. Electroanal. Chem.* 122: 73, 1981.
79. F. Di Quarto, G. Tuccio, A. Di Paola, S. Piazza, and C. Sunseri, *Oxide Films on Metals and Alloys*, p. 25, edited by K. Hebert and G.E. Thompson, PV 94-25, The Electrochemical Society Inc., 1994.
80. C. Sunseri, S. Piazza, and F. Di Quarto, *Mater. Sci. Forum*, 185–188: 435, 1995.
81. F. Di Quarto, S. Piazza, C. Sunseri, M. Yang, and S.M. Cai, *Electrochim. Acta* 41: 2511, 1996.
82. J.C. Phillips, *Bonds and Bands in Semiconductors*, Academic Press, New York, 1973.
83. L. Pauling, *The Nature of Chemical Bond*, Chapter 3, Cornell University Press, Ithaca, NY, 1960.
84. W. Gordy and W.J.O. Thomas, *J. Phys. Chem.* 24: 439, 1956.
85. L. Allred, *J. Inorg. Nucl. Chem.* 17: 215, 1961.
86. D.G. Schlom and J.H. Haeni, *MRS Bull.* 27: 198, 2002.
87. V. Emeline, G.V. Kataeva, V.K. Ryabchuk, and N. Serpone, *J. Phys. Chem. B* 103: 9190, 1999.
88. J.B. Torrance, P. Lacorre, C. Asavaroengchai, and R.M. Metzger, *Physica C* 182: 351, 1991.
89. V. Emeline, G.N. Kuzmin, D. Purevdorj, V.K. Ryabchuk, N. Serpone, *J. Phys. Chem. B* 104: 2989, 2000.
90. M.K. Jayaraj, A.D. Draeseke, J. Tate, and A.W. Sleight, *Thin Solid Films* 397: 244, 2001.
91. N. Duan, A.W. Sleight, M.K. Jayaraj, and J. Tate, *Appl. Phys. Lett.* 77: 1325, 2000.
92. H. Mizoguchi, H. Kawazoe, H. Hosono, and S. Fujitsu, *Solid State Commun.* 104: 705, 1997.

93. H. Eisaki, S. Uchida, T. Mizokawa, H. Namatame, A. Fujimori, J. Van Elp, P. Kuiper, G.A. Sawatzky, S. Hosoya, and H. Katayama-Yoshida, *Phys. Rev. B* 45: 12513, 1992.
94. W. Teng, J.F. Muth, U. Ozgur, M.J. Bergman, H.O. Everitt, A.K. Sharma, C. Jin, and J. Narayan, *Appl. Phys. Lett.* 76: 979, 2000.
95. M. Santamaria, F. Di Quarto, G.E. Thompson, and P. Skeldon, *ATB Metallurgie* XXXX, 3–4: 431, 2000–2001.
96. M. Vithal, P. Nachimutu, T. Banu, and Jagannathan, *J. Appl. Phys.* 81: 7922, 1997.
97. A.R. Newmark and U. Stimming, *Langmuir* 3: 905, 1987.
98. W. Smith, *J. Chem. Educ.* 67: 559, 1990.
99. M.K. Carpenter and D.A. Corrigan, *J. Electrochem. Soc.* 136: 1022, 1989.
100. M.A. Gonzalez-Nunez, P. Skeldon, G.E. Thompson, and H. Karimzadeh, *Corrosion* 55: 1136, 1999.
101. R. Masi, Chem. Eng. Degree Thesis, Univeristà di Palermo, November 2001.
102. F. Criminisi, Chem. Eng. Degree Thesis, Università di Palermo, April 2002.
103. A. Moina and G.O. Ybarra, *J. Electroanal. Chem.* 504: 175, 2001.
104. F. Di Quarto, S. Piazza, M. Santamaria, and C. Sunseri, *UPB Sci. Bull. Ser. B* 63: 45, 2001.
105. F. Di Quarto, S. Piazza, and C. Sunseri, *Corros. Sci.* 31: 721, 1990.
106. P.A. Cox, *Transition Metal Oxides*, Claredon Press, Oxford, 1992.
107. H. Kang, Y.R. Park, and K.J. Kim, *Solid State Commun.* 115: 127, 2000.
108. F. Di Quarto, S. Piazza, and C. Sunseri, *Electrochim. Acta* 30: 315, 1985.
109. M.P. Dare-Edwards, J.B. Goodenough, A. Hamnett, and N.D. Nicolson, *J. Chem. Soc. Faraday Trans.*, 277: 643, 1981.
110. S Virtanen, P Schmuki, H. Bohni, P. Vuoristo, and T. Mantyla, *J. Electrochem. Soc.* 142: 3067, 1995.
111. P. Schmuki, M. Buchler, S Virtanen, H Bohni, R. Muller, and L.J. Gauchler, *J. Electrochem. Soc.* 142: 3336, 1995.
112. M. Buchler, P. Schmuki, H Bohni, T. Stenberg, and T. Mantyla, *J. Electrochem. Soc.* 145: 378, 1998.
113. A.J. Davenport, L.J. Oblonsky, M.P. Ryan, and M.F. Toney, *J. Electrochem. Soc.* 147: 2162, 2000.
114. G. Dagan, W-M. Shen, and M. Tomkiewicz, *J. Electrochem. Soc.* 139: 1855, 1992.
115. M. Buchler, P. Schmuki, and H. Bohni, *J. Electrochem. Soc.* 145: 609, 1998.
116. A. Di Paola, F. Di Quarto, and C. Sunseri, *Corros. Sci.* 26: 935, 1986.
117. T.D. Burleigh and R.M. Latanision, *J. Electrochem. Soc.* 134: 137, 1987.
118. A. Di Paola, D. Shukla, and U. Stimming, *Electrochim. Acta* 36: 345, 1991.
119. M.J. Kloppers, F. Bellucci, and R.M. Latanision, *Corrosion* 48: 229, 1992.
120. P. Schmuki and H. Bohni, *J. Electrochem. Soc.* 139: 1908, 1992.
121. P. Schmuki, H. Bohni, and F. Mansfeld, *J. Electrochem. Soc.* 140: L119, 1993.
122. A.M.P. Simoes, M.G.S. Ferreira, B. Rondot, and M. Da Cunha Belo, *J. Electrochem. Soc.* 137: 82, 1990.
123. N.E. Hakiki, M. Da Cunha Belo, A.M.P. Simoes, M.G.S. Ferreira, *J. Electrochem. Soc.* 145: 3821, 1998.
124. N.E. Hakiki, M. Da Cunha Belo, and M.G.S. Ferreira, *Electrochim. Acta* 44: 2473, 1999.
125. M.F. Montemor, M.G.S. Ferreira, N.E. Hakiki, and M. Da Cunha Belo, *Corros. Sci.* 42: 1635, 2000.
126. L.J. Oblonsky, M.P. Ryan, and H.S. Isaacs; *J. Electrochem. Soc.* 145: 1922, 1998.
127. P. Marcus and V. Maurice, *Interfacial Electrochemistry*, pp. 541–558, edited by A. Wieckowski, Marcel Dekker, New York, 1999.
128. D. Hamm, C-O.A. Olsson, and D. Landolt, *Corros. Sci.* 44: 1009, 2002.
129. C. Sunseri, S. Piazza, and F. Di Quarto, *J. Electrochem. Soc.* 137: 2411, 1990.
130. M. Kerkar, J. Robinson, and A.J. Forty, *Faraday Discuss.* 89: 31, 1990.
131. E-A. Cho, H-S. Kwon, and D.D. Macdonald, *Electrochim. Acta* 47: 1661, 2002.
132. M. Kaneko and H.S. Isaacs, *Corros. Sci.* 44: 1825, 2002.

133. L. Wegrelius, F. Falkenberg, and I. Olefjord, *J. Electrochem. Soc.* 146: 1397, 1999.
134. S. Piazza, C. Sunseri, and F. Di Quarto, *Corrosion* 58: 436, 2002.
135. F. Di Quarto, M. Santamaria, N. Mallandrino, V. Laget, R. Buchheit, and K. Shimizu, PV 2002–13, *Corrosion Science: A Retrospective and Current Status, In Honour of Robert P. Frankenthal*, edited by G. Frankel, J.R. Scully, H.S. Isaacs, and J.D. Sinclair, Electrochemical Society.
136. M.A. Butler, *J. Electrochem. Soc.* 130: 2358, 1983.
137. D. Shukla, T. Wines, and U. Stimming, *J. Electrochem. Soc.* 134: 2086, 1987.
138. P.S. Tyler, M.R. Kozlowski, W.H. Smyrl, and R. Atanasoski, *J. Electroanal. Chem.* 237: 295, 1987.
139. M. Kozlowski, W.H. Smyrl, Lj. Atanasoska, and R. Atanasoski, *Electrochim. Acta* 34: 1765, 1989.
140. J.W. Schultze and J. Thietke, *Electrochim. Acta* 34: 1769, 1989.
141. J.W. Shultze, K. Bade, and A. Michaelis, *Ber. Bunsenges. Phys. Chem.* 95: 1349, 1991.
142. M.R. Kozlowski, P.S. Tyler, W.H. Smyrl, and R.T. Atanasoski, *J. Electrochem. Soc.* 136: 442, 1989.
143. D.E. William, A.R.J. Kucernak, and R. Peat, *Faraday Discuss.* 94: 149, 1992.
144. D.E. William, A.R.J. Kucernak, and R. Peat, *Electrochim. Acta* 38: 57, 1993.
145. A.R.J. Kucernak, R. Peat, and D.E. William, *Electrochim. Acta* 38: 71, 1993.
146. N. Casillas, P. James, and W.H. Smyrl, *J. Electrochem. Soc.* 142: L16, 1995.
147. S. Preusser, U. Stimming, and S. Tokonuga, *J. Electrochem. Soc.* 142: 102, 1995.
148. S. Kudelka and J.W. Schultze, *Electrochim. Acta* 42: 2817, 1997.
149. D.E. Williams, T.F. Mohiuddin, and Y.Y. Zhu, *J. Electrochem. Soc.* 145: 2664, 1998.
150. P. Schmuki and H. Bohni, *J. Electrochem. Soc.* 141: 362, 1994.
151. P.C. Searson and R.M. Latanision, *Electrochim. Acta* 35: 445, 1990.
152. L.F. Garfias and W.H. Smyrl, *J. Electrochem. Soc.* 146: 2495, 1999.
153. M.P. Ryan, D.E. Williams, R.J. Chater, B.M. Hutton, and D.S. McPhail, *Nature* 415: 770, 2002.
154. Z. Szklarska-Smialowska, *Corros. Sci.* 44: 1143, 2002.

19
Electrochemical Quartz Crystal Microbalance

C.-O. A. Olsson and D. Landolt
Materials Department, Swiss Federal Institute of Technology, Lausanne, Switzerland

Contents

19.1 INTRODUCTION

The quartz crystal microbalance (QCM) is a mass sensor with a resolution of less than an atomic monolayer. The use of quartz crystals for this purpose was suggested by Sauerbrey in 1958 (1). Soon, they were in wide spread use as rate monitors in vacuum deposition systems. Initially, the heavy damping impeded their use as sensors in liquid systems. Operation in liquids was obtained in the early 1980s (2, 3); since then, the number of applications in solution has increased rapidly. The electrochemical quartz crystal microbalance (EQCM) uses one of the thin film electrodes of a QCM as an anode or a cathode in an electrochemical cell. This arrangement permits the study of different electrochemical phenomena *in situ*. One of the first applications with this arrangement was electrodeposition of silver (2). In the field of corrosion, the (E)QCM has found a wide range of applications; for example, atmospheric corrosion, general corrosion in solution, corrosion inhibition, and passive film growth.

The QCM can be used to monitor mass changes in the nanogram range with a millisecond time resolution. At present, a commercial oscillator circuit can be obtained for a few thousand dollars. Data evaluation can start with a straightforward comparison of mass changes with other experimental parameters and proceed to extensive modeling of the interaction of several viscous overlayers with the vibrating surface.

When combined with techniques giving chemical information on the surface properties, the QCM becomes a key method for *in situ* measurements of desorption, adsorption, and film growth. The intention of this chapter is to provide a short introduction to QCM theory and to describe some practical examples of its application to corrosion science. More extensive reviews are available, for example, the reviews by Buttry and Ward (4, 5). Alternative designations of the QCM technique are thickness shear mode (TSM) resonators or surface acoustic wave (SAW) devices.

19.2 THEORY

19.2.1 The Sauerbrey Equation

For a shear wave at the base oscillatory mode, the quartz thickness equals half the wavelength. The eigenfrequency for this mode is given by

$$f_0 = \frac{v_{tr}}{2d} = \frac{\sqrt{\mu_q/\rho_q}}{2d} \tag{19.1}$$

where v_{tr} is the sound velocity in AT-cut quartz in the transverse direction, given by $v_{tr} = \sqrt{\mu_q/\rho_q}$, where μ_q is the shear modulus and ρ_q is the density; d is the thickness of the quartz crystal. For small changes in mass, the frequency shift is linearly dependent on the mass, since

$$\Delta f_0 = -\frac{v_{tr}}{2d^2}\Delta d = -\frac{2f_0^2}{v_{tr}}\Delta d = -\frac{2f_0^2}{\sqrt{\mu_q\rho_q}}\Delta m \tag{19.2}$$

where the relationship $\Delta d = \Delta m/\rho_q$ was used to obtain the final form. Δm is the mass change per area (g/cm^2). This equation is commonly referred to as the Sauerbrey equation (19.1). It is valid for small mass changes, for which the difference in acoustic impedance, $z = \sqrt{\rho\mu}$, between the quartz and the deposit can be disregarded. Cernosek et al. (6) have shown that these conditions are fulfilled for most practical sensor applications. The mass sensitivity is high: for a 10 MHz quartz, a resolution of 1 Hz corresponds to about 4.5 ng/cm^2, which is equivalent to 4% of a monolayer of metallic iron. Driving the quartz at an overtone can give a large increase in sensitivity (7).

The oscillating quartz crystal can be described with an electrical analog, cf. Figure 19.1, which shows an *LRC* circuit, sometimes referred to as a Butterworth–van Dyke (BVD) circuit. C_0 represents stray capacitance in the cable as well as between the sample holder and solution. At the resonance frequency, the inductive (L) and capacitive (C_1) contributions cancel and the effective admittance becomes $1/R$. To model viscous loading, more complex models can be applied. Successful implementations of transmission line models have been made by, for

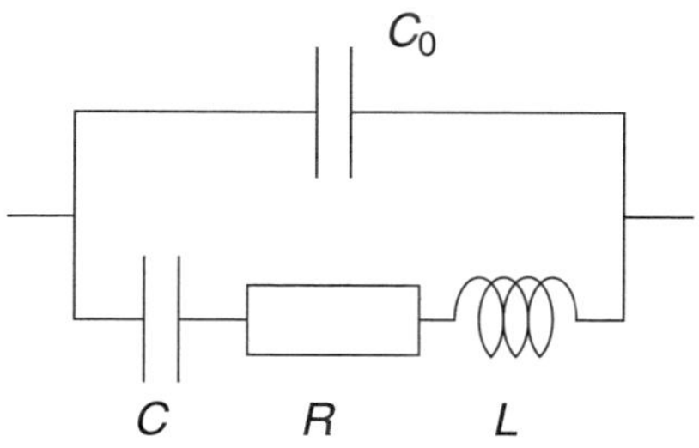

Figure 19.1 Equivalent circuit.

example, Martin et al. (8). For most practical applications, however, a simple model with a loading consisting of lumped elements has been shown to be sufficient (6).

19.2.2 Viscous Loading

When an oscillating quartz is in contact with a viscous medium, the coupling will induce a heavily damped shear wave that will influence the resonance frequency of the entire system. This effect has been studied analytically by Stockbridge (9) and others (10–12). Assuming no-slip conditions, the viscous loading frequency shift can be calculated from

$$\Delta f = -\frac{f_0^{3/2}}{\sqrt{\pi \rho_q \mu_q}} \sqrt{\rho_l \eta_l} \qquad (19.3)$$

where ρ_l is the density and η_l the absolute viscosity of the solution. Analytical solutions for more complex systems can also been found, for example for bilayer structures with different viscosities (13).

In addition to the frequency shift, immersion of a quartz crystal into solution also affects the shape of the admittance peak. This is illustrated in Figure 19.2, where the total admittance has been measured for different glycerol–water mixtures using a network analyzer (8, 14). For heavier loadings, the resonance peak decreases and becomes less well defined. The change in admittance with increased viscous loading can also be estimated by measuring the energy dissipation of the quartz crystal. Some commercial oscillator circuits provide access to this parameter as the total system admittance. A third possibility is to disconnect the driving circuit from the quartz at regular intervals and monitor the decay of the amplitude in the system. This procedure has the advantage of providing a direct measure of the dissipation factor D. A comparison of these three methods has been made by Geelhood et al. (15).

Recording the dissipation factor also gives access to the friction between the fluid and the quartz through the ratio $-\Delta f/\Delta D$ (10). The dissipation is linearly related to the factor $\sqrt{\rho_L \eta_L}$ (16, 17). This can be used to estimate the part of the frequency shift that is due to viscous loading, cf. Equation (19.3).

The QCM can also be sensitive to changes in viscous properties in the double layer. As the viscosity of the electrolyte within the double layer differs from its bulk value, a change in thickness also affects the frequency shift. This was studied by Kern and Landolt (18) on gold and iron electrodes. When

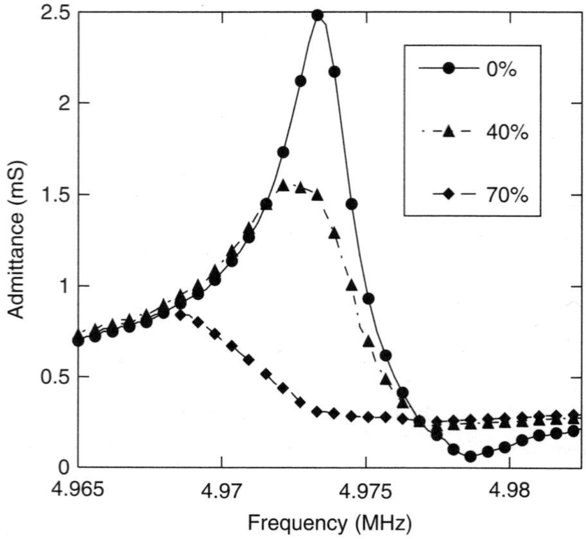

Figure 19.2 The effect of viscous loading on a quartz crystal resonator. The curves were obtained for a glycerol–water mixture of 0, 40, and 70%. (After H. L. Bandey, S. J. Martin, R. W. Cernosek, A. R. Hillman, *Anal. Chem.*, **71**, 2205 (1999). With permission.)

going from dilute to more concentrated perchlorate solutions, they showed that the frequency shift followed closely with the theory of a shrinking double layer.

19.2.3 Roughness Response

Another important parameter that needs consideration is the surface roughness (4, 19–21). A formalism for dealing with rough quartz surfaces has been introduced by Urbakh and Daikhin (20). They suggested that the effect of surface roughness could be estimated by studying the response of the quartz to solutions with different viscoelastic behavior. For slightly rough surfaces, a typical roughness height h is much smaller than the decay length δ of the shear wave in the solution. With their roughness model, Urbakh and Daikhin (21) showed that whereas slightly rough surfaces give a response that is highly dependent on viscosity, that is, the coupling between the quartz surface and the solution, strongly rough surfaces with pits or grooves give a frequency shift which is proportional to the volume of the inhomogeneities of the surface. All roughness studied was considerably smaller than the wavelength for shear mode oscillations, that is, 0.1 to 1 mm (20).

19.2.4 Other Parameters that Influence the Frequency Change

Other parameters that influence the frequency response are stress, temperature, and pressure (4). To ascertain that the frequency shift obtained from the microbalance corresponds to a mass change that can be evaluated using Equation (19.1), all these parameters need to be controlled rigorously.

19.3 EXPERIMENTAL ASPECTS

19.3.1 Quartz Crystals

Quartz (SiO_2) is a highly anisotropic material. There are a number of different cuts that give a wide variety of elastic properties. The one most frequently used for QCM applications is the AT cut, which was the first (A) temperature (T) compensated cut to be used. Other cuts are: LC (linear cut) which can be used as a temperature sensor and BT, which is another temperature-compensated cut, but with a stress response factor opposite to that of the AT cut. Another useful cut is the stress-compensated (SC) cut, which is insensitive to stress in the deposited material.

Figure 19.3 shows a schematic of a keyhole design quartz crystal. The quartz is normally operated in shear mode at 5 or 10 MHz; the frequency is determined by the thickness of the quartz. From a sensitivity aspect, a high base frequency is desirable, since the mass sensitivity is proportional to f_0^2. For an AT-cut 10 MHz quartz, the thickness is about 170 μm. The quartz diameter is typically 1.25 cm.

19.3.2 Samples and Sample Holders

In the EQCM, the working electrode of the electrochemical part of the system is combined with one of the oscillating electrodes of the quartz. The electrodes are usually deposited on the quartz using physical vapor deposition (PVD) or electrodeposition. It is important that the electrode adheres well to the surface. It is equally important that the electrode is free from defects, as they affect the coupling between the quartz and the liquid. For a long time, this limited the use of the QCM to laboratory materials. Recently, it has been shown that it is possible to perform QCM measurements also on thin foil of commercial polycrystalline materials. Herranen and Carlsson (22) studied titanium, whereas Ehahoun et al. (23) studied corrosion of a 304 type stainless steel. The thin foil samples were produced using techniques developed for transmission electron microscopy. They were then glued onto the quartz, which was operated at 2 MHz to allow for stable oscillation.

A simple way of fixing the quartz crystal is to mount it between two O-rings. This type of fixture has been described by, for example, Cavic and Thompson (24),

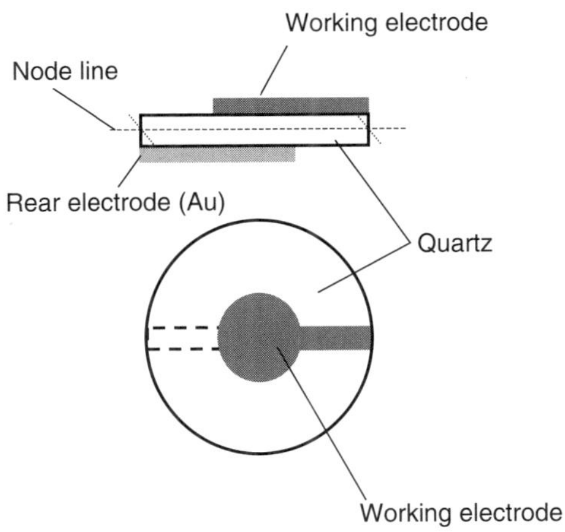

Figure 19.3 Schematic of a keyhole design quartz crystal.

who used it in a flow cell. However, apart from inducing stress in the substrate, the risk for crevice corrosion at the wet side of the quartz/O-ring interface is increased. Another method is to use an epoxy resin (e.g., Araldite) or a silicon-based polymer (e.g., Wacker Elastosil) to fix the crystal. This gives a stress-free joint to the sample holder, but might be more difficult to remove for subsequent surface analysis.

To perform experiments under controlled mass transport conditions the quartz crystal can be mounted on a rotating disk electrode shaft (25). The rotating disk design provides controlled hydrodynamic conditions (26). Alternatively, an impinging jet design can be used to obtain controlled hydrodynamic conditions (27, 28).

19.3.3 EQCM Setup and Calibration

An EQCM oscillator circuit must permit the application of a potential or a current to the working electrode. This requires electrical insulation between the primary and quartz driving side of the oscillator circuit. With a conventional oscillator circuit, the potential of the solution side of the quartz is not defined.

Figure 19.4(a) shows a block diagram of two different methods of constructing a QCM measuring circuit. Originally, frequency was measured by taking the difference between the quartz in the measuring solution and a reference quartz. The resulting frequency to be measured will thus stay in the kilohertz range. This difference is then converted to a voltage and recorded. This setup does not necessitate a separate frequency counter and gives a time resolution sufficient for most applications. A detailed description of this type of oscillator circuit can be found in Ref. (3). The strong development in electronics since the introduction of the microbalance has made it possible to measure the frequency directly. For this arrangement, the time resolution of the QCM setup is closely related to the internal clock frequency of the counter — the higher the clock frequency, the better the time resolution. Today, frequency counters with internal clock frequencies above 1 GHz are standard. This is more than sufficient to obtain a millisecond resolution for a 10 MHz quartz crystal. The measurable frequency range for this approach will be determined by the oscillator circuit.

The frequency counter approach is further detailed in Figure 19.4(b). The high frequency is fed to the quartz through a coaxial cable, where the working electrode side is connected to the screen. It is also simultaneously connected to the working electrode output of the potentiostat. With this configuration, the working electrode can be held at signal ground. The high frequency is led through the center lead and is connected to the back side of the crystal. The oscillator in this setup is the commercial Maxtek PLO 10i. It can drive the quartz crystal in the frequency range of 5 to 10 MHz, while compensating for stray capacitance. It gives two outputs: a high-frequency signal (thick line) for which the period is measured with a frequency counter, and a voltage from which the crystal resistance can be calculated. The potentiostat is controlled from the computer. Data are then logged with an IEEE 488 interface (gray). For increased acquisition rates, frequency data can be stored temporarily in the internal memory of the frequency counter.

For EQCM applications, it is possible to perform a calibration by comparing the charge transferred during a deposition with the mass uptake. It can be made as an electrodeposition of copper from a 0.5 M $CuSO_4$ + 0.5 M H_2SO_4 + 1.1 M C_2H_5OH solution, which gives a current efficiency close to 100% (29). The deposited mass can then be calculated based on the charge, following Faraday's law. By dividing this result by the measured frequency shift, a calibrated value for the linear coefficient in

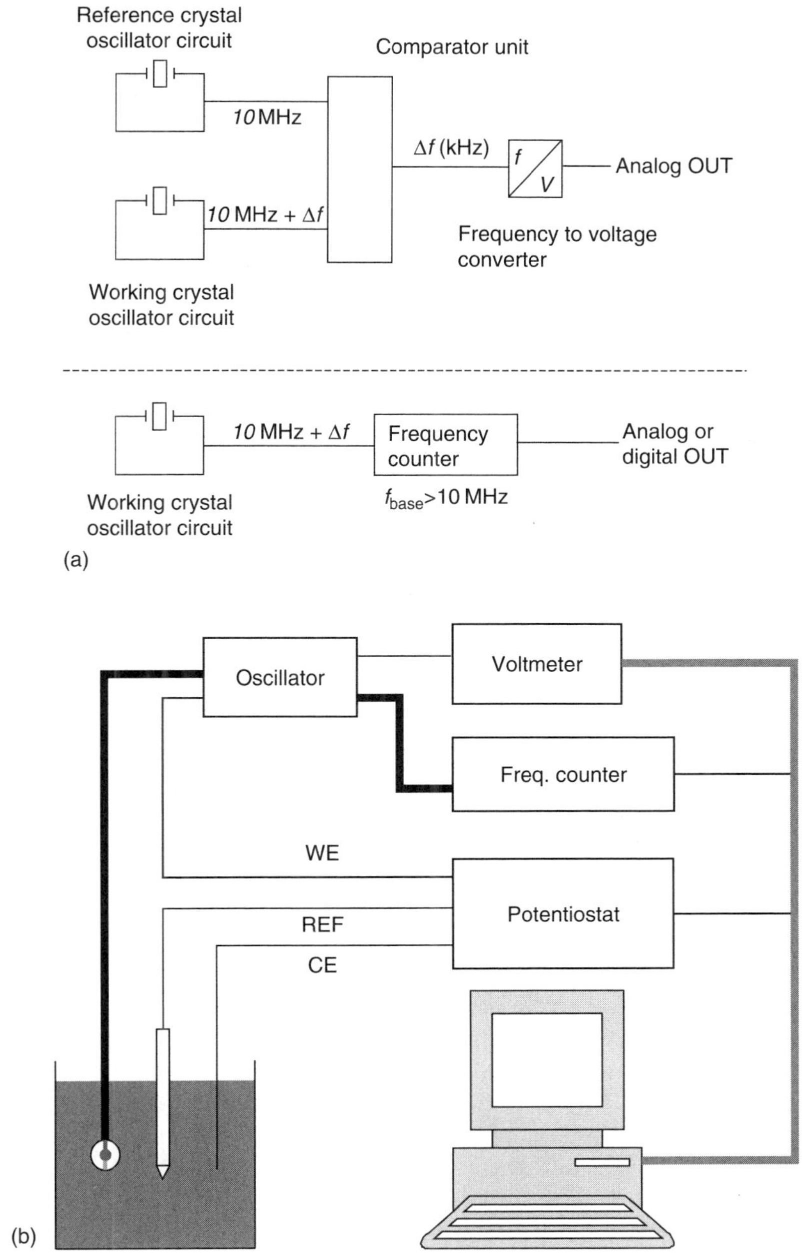

Figure 19.4 (a) Block diagram illustrating two frequency measurement techniques. (b) Schematic of a computer-controlled EQCM setup.

the Sauerbrey equation (19.1) is obtained. This procedure compensates for differences in acoustic impedance between the deposit and the base quartz crystal, as well as for effects related to the viscous loading.

The sensitivity of the quartz to a mass or dissipation change is not equally distributed over the entire oscillating area; the sensitivity is higher in the center and follows a Gaussian shape. Rodahl and Kasemo (16) illustrated this effect by placing droplets at different positions on the quartz surface. Another method to estimate the differential sensitivity coefficient is to use a polymer mask with a hole at a specific location. By depositing silver dots at different distances from the center, Gabrielli et al. (30) measured the differential sensitivity of the quartz. Knowledge of the differential mass sensitivity is useful when studying localized phenomena such as pitting (31, 32).

19.4 APPLICATION OF THE EQCM TO THE STUDY OF CORROSION REACTIONS

The (E)QCM has been applied to the *in situ* study of a variety of corrosion reactions by measuring the mass changes resulting from dissolution or film growth. For many such applications, the (E)QCM should be combined with methods yielding additional information such as surface analysis, for example, x-ray photoelectron spectroscopy (XPS), infrared (IR) spectroscopy, or electrochemical impedance spectroscopy. In the following, the application of the (E)QCM to the study of atmospheric corrosion, uniform corrosion in solution, reaction of inhibitors, and the growth of passive oxide films will be illustrated with examples from the literature.

19.4.1 Atmospheric Corrosion

Exposure of a metal to humid atmosphere leads to the formation of corrosion products, which usually stay on the surface. The corresponding mass increase of the sample can be monitored with a QCM. The method was used by different groups for the investigation of the atmospheric corrosion of copper under laboratory conditions (33–36). Aastrup and Leygraf (34) describe an experimental setup consisting of a chamber with continuous gas flow. It is equipped with a QCM and an FTIR spectrometer. The QCM yields *in situ* information of reaction kinetics, while infrared reflection absorption spectroscopy (IRAS) yields information on surface chemistry. Investigating the formation of cuprous oxide films on copper, the authors could detect films down to 1 nm with IRAS and 0.2 nm with the QCM. In later publications, they investigated the role of relative humidity (33), and the effect of sulfur dioxide, ozone, and nitrogen dioxide on the interfacial reactions of copper (36). An atomic force microscope (AFM) added to the chamber permitted the study of topographical changes. Figure 19.5 illustrates results obtained during exposure of a copper thin film to a synthetic corrosive atmosphere (36). The total mass change was higher when SO_2 was added to air of 80% humidity. The most dramatic effect, however, results from adding ozone. The amounts of sulfate, water, and cuprous oxide were estimated from IRAS to chemically identify the reaction products for different exposure conditions. A similar approach was used by Grundmeier et al. (37) for the study of the formation of ultra thin polysiloxane films on oxidized iron during glow discharge. The resulting coatings were then tested for their corrosion protection properties.

A practical application of the QCM for corrosion monitoring in a telephone switching office has been described by Schubert et al. (38). They placed QCM electrodes of different metals such as copper, silver, nickel, and gold in air ducts and measured the dynamic response over several months. For their study, they compared the EQCM data with surface compositions obtained with Auger Electron Spectroscopy (AES). The results showed a distinctly higher corrosion rate in the return air tract than in the supply air tract.

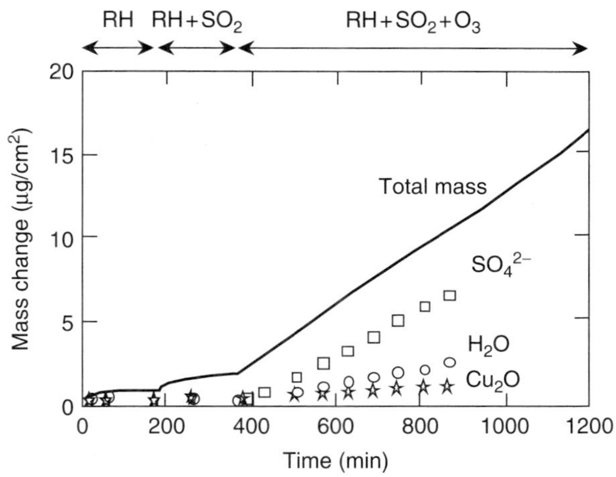

Figure 19.5 Atmospheric corrosion of copper. Total mass gain measured by QCM and the mass gain of Cu_2O, H_2O, and $CuSO_4$ deduced from IRAS. After 180 min 200 ppb of SO_2 were added and after another 180 min 200 ppb SO_2 + 200 ppb of O_3 were added. (After T. Aastrup, M. Wadsak, C. Leygraf, M. Schreiner, *J. Electrochem. Soc.*, **147**, 2543 (2000). With permission.)

19.4.2 Uniform Corrosion in Electrolytes

The QCM permits the measurement of the rate of corrosion under open-circuit conditions. Provided only dissolved products are formed, a linear relationship between QCM frequency change and mass loss due to uniform corrosion is observed. This was verified by Chandler et al. (39) for corrosion of cobalt in sulfuric acid in the presence of oxygen and for copper corrosion in aerated hydrochloric acid.

More detailed information on corrosion reactions is obtained by varying the potential. Seo et al. (40, 41) used the EQCM to study the uniform corrosion of iron in deaerated neutral borate and phosphate electrolytes. The quartz crystal electrode was placed in the wall of a flow cell, allowing recirculation of the electrolyte. At open-circuit, a linear relationship between frequency change and time was observed, indicating that only dissolved products were formed. The rate of corrosion was higher in phosphate than in borate solutions, and addition of chloride was found to increase the measured corrosion rate. According to the mixed potential theory, the total current density i_{tot} measured on a corroding metal is the sum of the anodic and cathodic partial current densities i_a and i_c, respectively

$$i_{tot} = i_a + i_c \tag{19.4}$$

The EQCM permits the measurement of i_a as a function of potential, since according to Faraday's law, the anodic partial current density is proportional to the rate of mass loss. The difference between the measured total current density and the anodic partial current density corresponds to the cathodic partial current density, cf. Equation (19.4). Figure 19.6 shows the partial current–voltage curves for iron dissolution in a phosphate solution of pH 6.48 determined in this manner (40). The results nicely illustrate the mixed potential theory of corrosion, which states that the measured current results from anodic and cathodic partial reactions and at the corrosion potential their rate is equal.

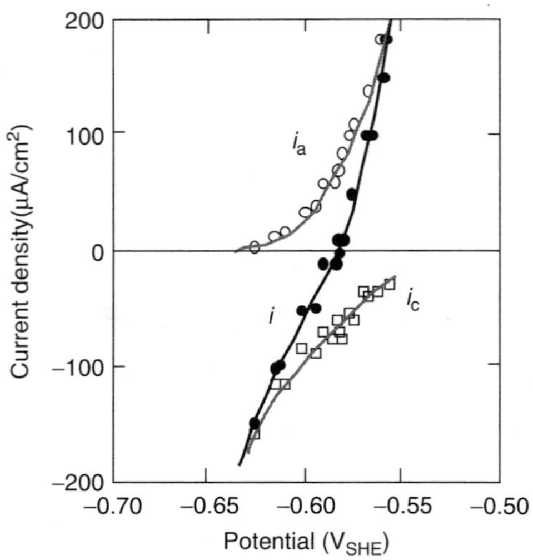

Figure 19.6 Application of the EQCM to the determination of partial current densities of iron in a phosphate electrolyte of pH 4. (After M. Seo, K. Yoshida, K. Noda, *Mater. Sci. Forum*, **192–194**, 753 (1995). With permission.)

Several groups applied the EQCM together with other electrochemical methods such as impedance (28, 42) or cyclic voltammetry (43). Some used the EQCM in combination with *in situ* detection of reaction products, similar to the rotating ring-disk electrode technique (27, 28, 44). The EQCM electrode was placed in the wall of an electrochemical cell and surrounded by a ring (28) or a split ring (27, 44) electrode separated from the disk by a narrow gap. The ring-disk electrodes were fabricated by sputter deposition onto the quartz crystal through a mask. An immersed electrolyte jet impinging perpendicularly onto the disk provided reproducible mass transport conditions. The EQCM setup was used to study copper dissolution in chloride media by simultaneously monitoring the mass change and the amount of dissolved monovalent reaction products. Gabrielli et al. (28) described an impinging jet cell with stationary ring-disk EQCM electrodes and used it in conjunction with impedance for the study of copper dissolution in NaCl and $NaHCO_3$ solution. The experimental arrangement permitted the identification of surface reaction intermediates. Several authors applied the EQCM to the study of alloy corrosion (42, 45–47). Weil et al. (42) performed EQCM and impedance measurements during anodic dissolution of a Au–Cu alloy in acid sodium sulfate solution (42). The frequency shift due to selective copper dissolution was smaller than that calculated from Faraday's law. The deviation was attributed to water trapped in the porous surface structure developing during dissolution. Leinartas et al. (45) studied the corrosion rate of Au–Pd–In alloys in physiological solutions and found a mass increase at open circuit indicating an accumulation of corrosion products. The same authors also observed a mass increase of sputter deposited Fe–Cr–Ni–Ta alloys exposed to a neutral chloride solution (46).

19.4.3 Evaluation of Corrosion Inhibitors

The (E)QCM is well suited for the evaluation of corrosion inhibitors by measuring the decrease in corrosion rate due to the addition of an inhibitor. However, difficulties may arise when the corrosion products are only partly dissolved. Typical results

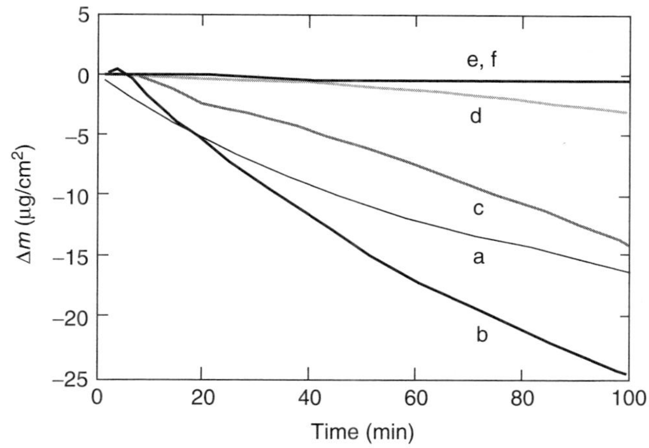

Figure 19.7 Mass loss of a nickel electrode at open circuit in 1 *M* $HClO_4$ in the presence of different corrosion inhibitors: (a) no inhibitor, (b) acridine, (c) benzyl quinolinium chloride, (d) dodecyl quinolinium chloride, (e) tributylbenzyl ammonium iodide, (f) potassium iodide. (After F. Zucchi, M. Fonsati, G. Trabanelli, *J. Appl. Electrochem.*, **29**, 347 (1999). With permission.)

of a study with nickel in $HClO_4$ in the presence of different inhibitors are shown in Figure 19.7 (48). The measured mass change is the difference between film formation and metal dissolution, but for the experimental conditions shown, the latter was dominant as evidenced by the observed mass decrease. Among the inhibitors tested, tributylbenzyl ammonium iodide (TNBI) appears to be the most efficient, whereas acridine (A) stimulated dissolution. Corrosion inhibition of copper has been studied widely using the EQCM (49–53). Telegdi et al. (52) compared the effectiveness of aromatic sulfoxides and derivatives of benzo-hydroxamic acid and azole. The same authors also studied biofilm formation on iron. Bacterial adhesion led to a measurable mass increase, the extent of which depended on the presence or not of biocides. Hydrophilic and ionic biocides were found more effective in reducing bacterial adhesion than hydrophobic biocides.

In the presence of corrosion inhibitors that form surface films the interpretation of mass changes in terms of corrosion rate is not straightforward and additional methods are needed to interpret results, such as cyclic voltammetry (50), or impedance (51, 53). Frignani et al. (51), who studied film formation on copper in the presence of benzotriazole with impedance and EQCM, found reasonable agreement between the two methods, but for this they had to assume an unreasonably low dielectric constant for the films. Garcia et al. (54) studied the effect of a phosphonate inhibitor and of flow rate on scale formation on a gold EQCM electrode exposed to mineralized water. The working electrode was placed in the wall of a flow channel cell and scale formation was stimulated by electrochemical polarization. The authors showed that adding the inhibitor or increasing the flow rate reduced the positive mass change associated with scale formation. Calcium carbonate scale formation on gold was studied by Gabrielli et al. (55, 56) using a jet flow cell to control hydrodynamics. *In situ* measured mass increases were found in good agreement with those obtained from titration experiments in spite of the fact that calicum carbonate scales were not smooth.

The high mass sensitivity of the EQCM permits its application to the study of adsorption from solution. Kern and Landolt (26, 57, 58), using this technique, studied the adsorption behavior of organic model inhibitors on iron and gold in deaerated neutral perchlorate solution. They measured the mass change resulting from adding increasing amounts of inhibitor to a perchlorate solution using a rotating EQCM to assure reproducible mixing. Adding inhibitor to the solution resulted in a positive frequency shift indicating a mass decrease of the electrode. The results were interpreted by postulating that the adsorbing inhibitor displaces water from the region near the electrode surface. Indeed, the EQCM response includes a region of solution of about 17 nm in these electrolytes. Replacing the oriented water layer at the surface by organic molecules corresponds to an effective decrease in density that can explain the observed frequency increase of the EQCM. A similar conclusion was reached by Gileadi and coworkers (59–61), who studied the adsorption of pyridine on gold and silver from aqueous electrolytes and butanol solutions. As described in the theoretical part of this chapter, the response of a quartz crystal in a solution depends not only the mass change, but also of the solution viscosity, density, and on diffuse double layer effects (18). In addition, surface roughness and slippage at the adsorbate–solution interface (61) need also be considered. For the conditions of the experiments described, these effects were duly taken into consideration. Commercial inhibitor packages for iron often contain an acid and a base. Using different combinations of two acids (benzoic acid [BA] and ω-benzoyl alkanoic acid [BAA]) and two bases (*N*-ethyl morpholine [EM] and NaOH), the EQCM provided information on the individual adsorption behavior of the acid or base components (26). Comparing EQCM and impedance data to different adsorption models, it was found that the Langmuir–Freundlich isotherm gave the best fit, cf. Figure 19.8 (26). In a subsequent study, XPS was used to obtain independent *ex situ* information on the adsorption behavior of typical model compounds (57). The XPS data were found to be in qualitative agreement with EQCM results, in spite of the difficulties inherent to *ex situ* analysis under ultra-high vacuum (UHV).

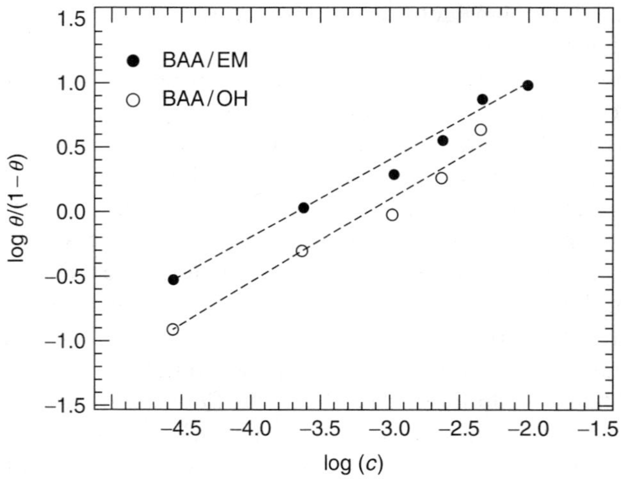

Figure 19.8 Adsorption isotherms of corrosion inhibitors deduced from EQCM data for iron in 0.1 *M* $NaClO_4$. BAA/EM: ω-benzoyl alkanoic acid + *N*-ethyl-morpholine base, BAA/OH: ω-benzoyl alkanoic acid + NaOH. (After P. Kern, D. Landolt, *Electrochim. Acta*, **47**, 589 (2001). With permission.)

19.4.4 Growth and Dissolution of Passive Oxide Films

Passive films on metals and alloys grow by incorporation of oxygen anions from the solution (Figure 19.9). This leads to a mass increase of the electrode, which can be monitored with the EQCM. For a well-known film stoichiometry, a comparison of the measured mass change with the charge passed allows for separation of film formation and metal dissolution reactions.

Let us assume that an anodic potential step is applied to a metal M in its passive state, leading to formation of an oxide MO_ν, where ν is a stoichiometric coefficient. The mass of metal oxidized at the metal/oxide interface, Δm_I (g/cm^2), is proportional to the charge density q (C/cm^2), as given by

$$\Delta m_\mathrm{I} = \frac{qM_\mathrm{M}}{nF} \tag{19.5}$$

Here the subscript I indicates the metal/oxide interface, M_M is the atomic mass of the metal $n = 2$ is the charge number, and F is the Faraday constant. Only a fraction R_g (growth fraction) of the oxidized metal stays in the film, while the rest dissolves at the film/solution interface II (Figure 19.9). The growth fraction is defined as

$$R_\mathrm{g} = \frac{\Delta m_\mathrm{I,M} + \Delta m_\mathrm{II,M}}{\Delta m_\mathrm{I,M}} = 1 + \frac{\Delta m_\mathrm{II,M}}{\Delta m_\mathrm{I,M}} \tag{19.6}$$

In this equation, $\Delta m_\mathrm{II,M}$ is negative, since all quantities leading to film growth are considered positive. This definition corresponds to the EQCM measurement: the mass going into the film leads to a mass increase and is thus positive, dissolution is negative. The measured mass change, Δm, includes the mass increase due to incorporation of anions into the growing film, $\Delta m_\mathrm{II,anion}$, and the loss of metal ions due to dissolution at the film/solution interface, $\Delta m_\mathrm{II,M}$

$$\Delta m = \Delta m_\mathrm{II,anion} + \Delta m_\mathrm{II,M} \tag{19.7}$$

The film stoichiometry provides a relationship between the quantities $\Delta m_\mathrm{II,anion}$, $\Delta m_\mathrm{I,M}$, and $\Delta m_\mathrm{I,M}$:

$$\frac{M_\mathrm{M}}{\nu} \Delta m_\mathrm{II,anion} = M_\mathrm{O}(\Delta m_\mathrm{I,M} + \Delta m_\mathrm{II,M}) \tag{19.8}$$

From these relations, one can express the film growth fraction as

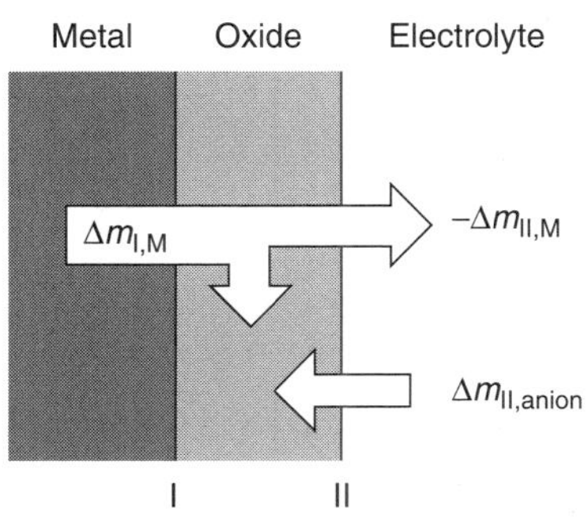

Figure 19.9 Schematic of partial reactions involved in anodic oxide growth and dissolution of a passive film.

$$R_g = \frac{M_M + nF(\Delta m/q)}{M_M + \nu M_O} \tag{19.9}$$

The measured mass/charge ratio, $\Delta m/q$, thus yields the film growth fraction directly. In a similar way, one can define a differential or instantaneous growth fraction r_g, which is used to characterize the film growth in real time during an anodization experiment:

$$r_g = \frac{M_M + nF[(\partial m/\partial t)/i]}{M_M + \nu M_O} \tag{19.10}$$

The mass change of the oxide film, Δm_{film} (g/cm^2), follows from the measured mass change and the charge:

$$\Delta m_{film} = \Delta m_{II,anion} + \Delta m_{I,M} + \Delta m_{II,M} = \Delta m + \Delta m_{I,M} = \Delta m + q\frac{M_M}{nF} \tag{19.11}$$

From the film mass change, it is possible to estimate the thickness change as $\Delta d_{film} = \Delta m_{film}/\rho_{film}$, where ρ_{film} is the film density (g/cm^3).

Figure 19.10 shows results obtained by applying an anodic potential change to passive chromium in a sulfuric acid solution (62). The measured quantities, current and mass change, are plotted as a function of time. Also shown are the calculated

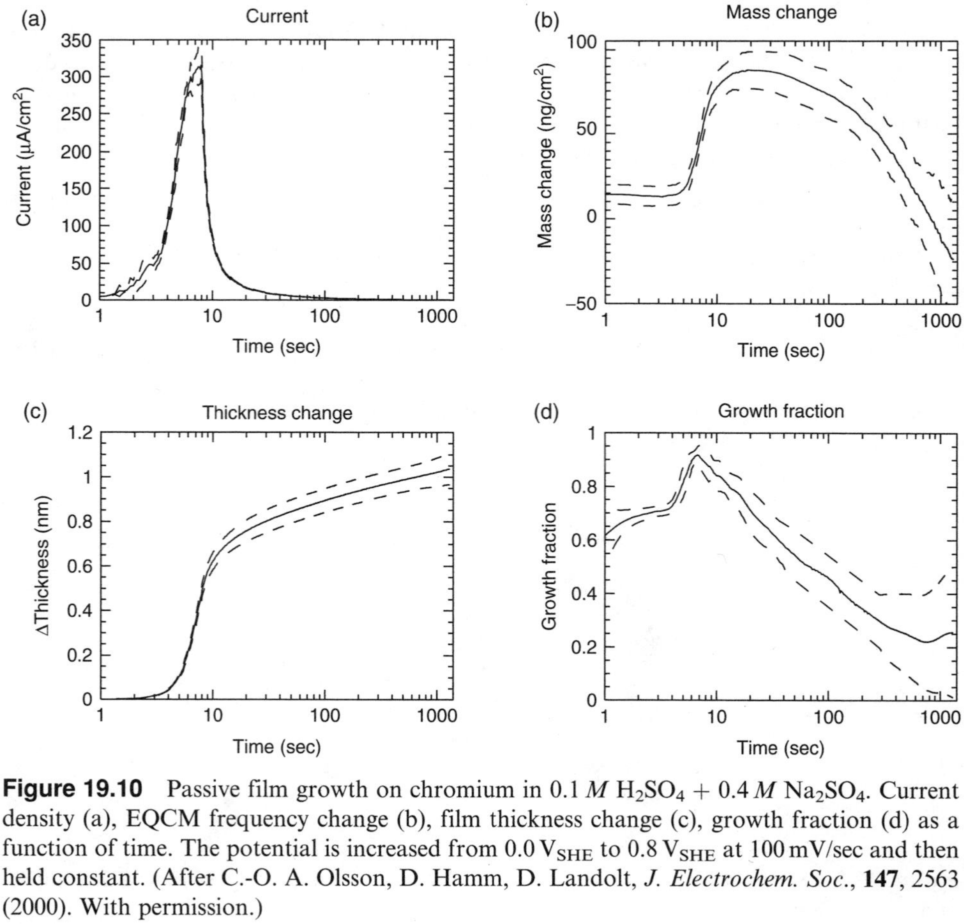

Figure 19.10 Passive film growth on chromium in 0.1 *M* H_2SO_4 + 0.4 *M* Na_2SO_4. Current density (a), EQCM frequency change (b), film thickness change (c), growth fraction (d) as a function of time. The potential is increased from 0.0 V_{SHE} to 0.8 V_{SHE} at 100 mV/sec and then held constant. (After C.-O. A. Olsson, D. Hamm, D. Landolt, *J. Electrochem. Soc.*, **147**, 2563 (2000). With permission.)

film thickness change and the growth fraction assuming formation of Cr_2O_3. XPS measurements showed that the thickness of the covering hydroxide was potential independent. The growth fraction goes through a maximum close to unity and then decreases to a low value as film growth slows down and the relative influence of dissolution increases. The film growth rate was compared to growth models (63), assuming growth rate control by high field conduction (HFM) or by interfacial reaction kinetics (IFM). Interestingly, the two different models yielded similar film growth curves for variations in sweep rates and potentials, indicating that they are not very sensitive to the different mechanisms. It can be noted that a growth limiting reaction at the interface is not explicitly dependent on the absolute film thickness, whereas this would be the case for high field rate limiting. For the HFM, the driving force for growth is obtained by dividing the applied potential by the film thickness. By applying two consecutive steps with an equilibration period in-between, a distinction between the two models was made possible; for chromium, the IFM provided a better description of growth behavior (63).

In a similar way, the EQCM can be used to study film growth on passive alloys. Figure 19.11 illustrates the mass change measured during a potential sweep covering the active, passive, and transpassive region of a Fe–25Cr alloy in an acid sulfate electrolyte (64). The slope of the mass curve is a measure of the dissolution rate. It is highest in the active and transpassive potential regions, but a small negative slope is observed also in the passive region indicating an overall mass loss. This behavior can be understood by considering that passive films on Fe–Cr alloys in acid solutions are strongly enriched in chromium (65). The data of Figure 19.11 suggest that the enrichment results from selective dissolution of iron. To further study this behavior, experiments similar to those described above for chromium were performed with a Fe–25Cr alloy (66). For the iron–chromium system, the differences in chemistry between the alloying elements are sufficiently small to allow data evaluation in the same way as for a pure metal using average mass and valence. A comparison of the data of Figure 19.12 with those of Figure 19.10 shows that contrary to pure chromium, the alloy exhibits a mass loss when the potential increases. The growth

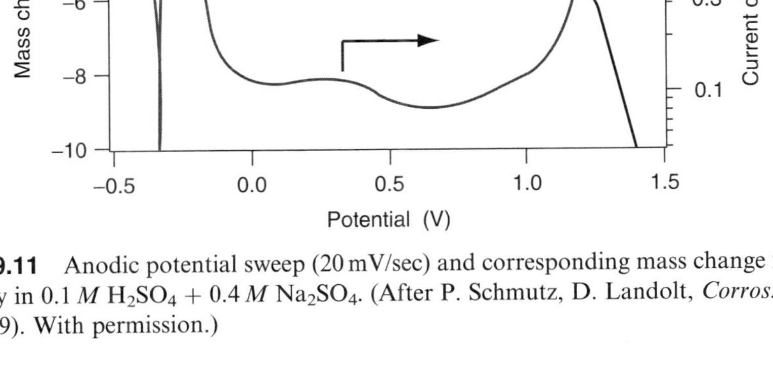

Figure 19.11 Anodic potential sweep (20 mV/sec) and corresponding mass change for a Fe–25Cr alloy in 0.1 *M* H_2SO_4 + 0.4 *M* Na_2SO_4. (After P. Schmutz, D. Landolt, *Corros. Sci.*, **41**, 2143 (1999). With permission.)

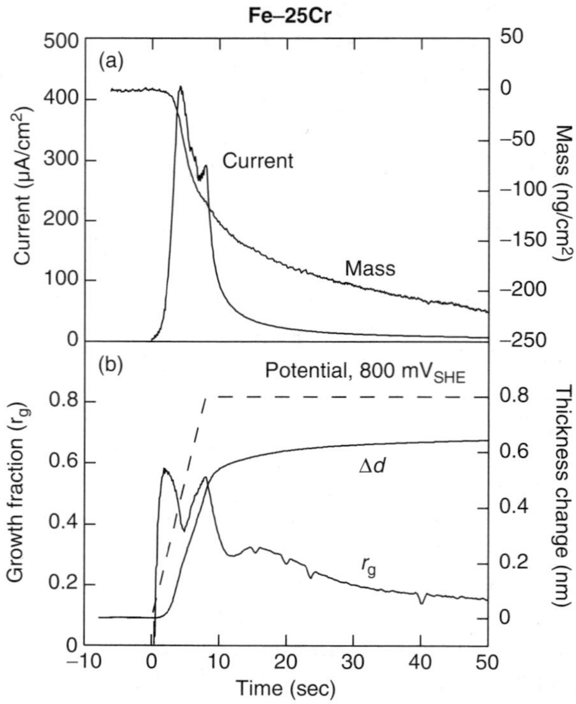

Figure 19.12 Current and mass response (a) and film thickness change and growth fraction (b) of Fe–25Cr subjected to a potential step in the passive region. (After D. Hamm, C.-O. A. Olsson, D. Landolt, *Corros. Sci.*, **44**, 1009 (2002). With permission.)

fraction is smaller than for pure chromium but still positive, indicating that in spite of an overall mass loss, the passive film grows in thickness. A semiquantitative estimate, originally proposed by Schmutz and Landolt (65), suggests that under the experimental conditions of Figure 19.11, more than 90% of iron oxidized at the metal/film interface goes into solution, while only little chromium dissolves leading to chromium enrichment in the passive film. Chemical analysis of the solution using plasma emission and surface analysis by XPS confirmed this behavior (67). It may be mentioned that in alkaline solutions, Fe–Cr alloys behave differently since iron oxide is less soluble. Following an anodic potential step in the passive region, EQCM yields a mass increase of the electrode and XPS shows that the passive film becomes enriched in iron oxide rather than chromium oxide (65).

It is well known that the corrosion resistance of stainless steels in acidic chloride media can be greatly improved by adding certain alloying elements, notably Mo. Recently, the EQCM in conjunction with XPS analysis was employed for comparing the effect of Mo, W, and Cr on the growth and dissolution behavior of passive films formed on sputter deposited 304 type stainless steel exposed to sulfuric acid (68). Alloys were produced as thin nanocrystalline films with 18Cr–9Ni (AISI 304) as the base composition. To investigate the effect of different alloying elements 20 at% of Cr, Mo, and W were added to the matrix. It was found that all alloys exhibited a mass loss when the potential was swept within the passive region in the anodic direction. The mass loss was most pronounced for the alloy containing molybdenum and least for the alloy with added chromium. In Figure 19.13 the mass response is shown for an anodic sweep rate of 10 mV/sec followed by a potential arrest. Corres-

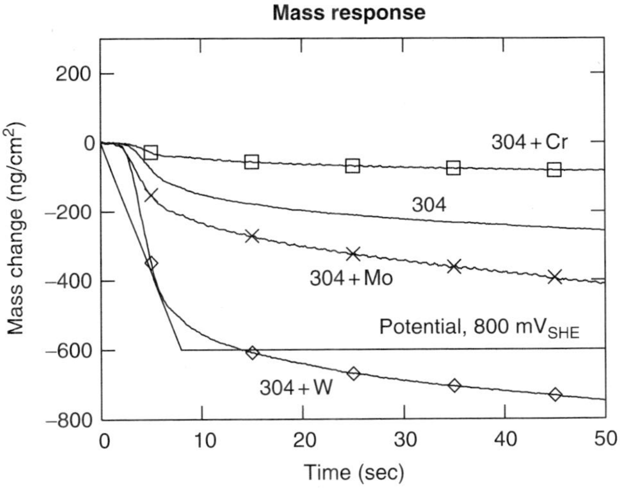

Figure 19.13 Mass change and film growth resulting from a potential increase in the passive potential region of sputter deposited 304 type stainless steel with addition of Cr, Mo, or W. (After C.-O. A. Olsson, D. Landolt, *J. Electrochem. Soc.*, **148**, B438 (2001). With permission.)

pondingly, the increase in passive film thickness evaluated from mass and charge measurements was most pronounced for the Mo-alloy and least for 304Cr. In accordance with this observation, the growth fraction was highest for 304Mo and lowest for 304Cr. These results suggest that addition of Mo or W increased the ionic conductivity of the passive film, leading to a higher anodic current density and to an enhancement of film growth and selective dissolution rates. To establish whether this behavior is related to the beneficial effect of Mo and W for pitting corrosion resistance would require similar experiments in chloride media.

19.5 CONCLUDING REMARKS

The (E)QCM is a versatile method to estimate real-time mass changes *in situ*. These mass changes reflect different reactions, such as adsorption, dissolution, or film growth. To obtain more complete information on the processes occurring on the surface, the (E)QCM can be combined with different analytical techniques, for example, surface spectroscopy, IR absorption, or electrochemical impedance spectroscopy. Its high sensitivity permits the study of mass changes corresponding to fractions of a monolayer. However, a quartz sensor is also sensitive to the coupling between its surface and the surrounding environment. For a correct data interpretation, it is important that the effects of viscous loading, double layer structure, surface roughness, and temperature are taken into consideration.

In the field of corrosion, the EQCM has been successfully applied to a wide variety of problems. One example is the study of uniform dissolution in identifying partial currents by comparing the mass loss with the total current. It has also been used to calculate adsorption isotherms for organic inhibitors. Another application is the study of passive films on stainless steels, where it has been shown that a combination with XPS gives the possibility to study film growth kinetics for alloys with several alloy elements showing widely different chemical behavior.

REFERENCES

1. G. Sauerbrey, *Z. Phys.*, **155**, 206 (1959).
2. T. Nomura, M. Iijima, *Anal. Chim. Acta*, **131**, 97 (1981).
3. S. Bruckenstein, M. Shay, *Electrochim. Acta*, **30**, 1295 (1985).
4. D. A. Buttry, M. D. Ward, *Chem. Rev.*, **92**, 1355 (1992).
5. D. A. Buttry, in *Electrochemical Interfaces: Modern Techniques for In-Situ Interface Characterisation*, H. D. Abruna, Ed. (VCH, New York, 1998), p. 529.
6. R. W. Cernosek, S. J. Martin, A. R. Hillman, H. L. Bandey in IEEE International Frequency Control Symposium, Orlando, FL (IEEE, Piscataway, NJ, 1997), p. 96.
7. K. Matsuura, Y. Ebara, Y. Okahata, *Thin Solid Films*, **273**, 61 (1996).
8. S. J. Martin, V. E. Granstaff, G. C. Frye, *Anal. Chem.*, **63**, 2272 (1991).
9. C. D. Stockbridge, in *Vacuum Microbalance Techniques* (Plenum, New York, 1966), vol. 5, p. 147.
10. M. Rodahl, B. Kasemo, *Sens. Actuators A — Phys.*, **54**, 448 (1996).
11. K. K. Kanazawa, J. G. Gordon, *Anal. Chem.*, **57**, 1770 (1985).
12. S. J. Martin, G. C. Frye, K. O. Wessendorf, *Sens. Actuators A — Phys.*, **44**, 209 (1994).
13. M. V. Voinova, M. Rodahl, M. Jonson, B. Kasemo, *Phys. Scr.*, **59**, 391 (1999).
14. H. L. Bandey, S. J. Martin, R. W. Cernosek, A. R. Hillman, *Anal. Chem.*, **71**, 2205 (1999).
15. S. J. Geelhood, C. W. Frank, K. K. Kanazawa, *J. Electrochem. Soc.*, **149**, H33 (2002).
16. M. Rodahl, B. Kasemo, *Sens. Actuators B — Chem.*, **37**, 111 (1996).
17. H. Muramatsu, E. Tamiya, I. Karube, *Anal. Chem.*, **60**, 2142 (1988).
18. P. Kern, D. Landolt, *J. Electroanal. Chem.*, **500**, 170 (2001).
19. S. J. Martin, G. C. Frye, A. J. Ricco, *Anal. Chem.*, **65**, 2910 (1993).
20. M. Urbakh, L. Daikhin, *Phys. Rev. B*, **49**, 4866 (1994).
21. M. Urbakh, L. Daikhin, *Langmuir*, **10**, 2836 (1994).
22. M. Herranen, J.-O. Carlsson, *Corros. Sci.*, **43**, 365 (2001).
23. H. Ehahoun, C. Gabrielli, M. Keddam, H. Perrot, Y. Cètre, L. Diguet, *J. Electrochem. Soc.*, **148**, B333 (2001).
24. B. A. Cavic, M. Thompson, *Anal. Chim. Acta*, **469**, 101 (2002).
25. P. Kern, D. Landolt, *J. Electrochem. Soc.*, **147**, 318 (2000).
26. P. Kern, D. Landolt, *Electrochim. Acta*, **47**, 589 (2001).
27. M. Itagaki, J. Kadowaki, K. Watanabe, *Electrochemistry*, **68**, 684 (2000).
28. C. Gabrielli, M. Keddam, F. Minouflet-Laurent, H. Perrot, *Electrochem. Solid State Lett.*, **3**, 418 (2000).
29. K. K. Kanazawa, O. R. Melroy, *IBM J. Res. Dev.*, **37**, 157 (1993).
30. C. Gabrielli, M. Keddam, R. Torresi, *J. Electrochem. Soc.*, **138**, 2657 (1991).
31. R. Oltra, I. O. Efimov, *J. Electrochem. Soc.*, **141**, 1838 (1994).
32. I. Efimov, M. Itagaki, M. Keddam, R. Oltra, H. Takenouti, B. Vuillemin, *Mater. Sci. Forum*, **185–188**, 937 (1995).
33. M. Wadsak, M. Schreiner, T. Aastrup, C. Leygraf, *Surf. Sci.*, **454–456**, 246 (2000).
34. T. Aastrup, C. Leygraf, *J. Electrochem. Soc.*, **144**, 2986 (1997).
35. J. Itoh, T. Sasaki, M. Seo, T. Ishikawa, *Corros. Sci.*, **39**, 193 (1997).
36. T. Aastrup, M. Wadsak, C. Leygraf, M. Schreiner, *J. Electrochem. Soc.*, **147**, 2543 (2000).
37. G. Grundmeier, E. Matheisen, M. Stratmann, *J. Adhes. Sci. Technol.*, **10**, 573 (1996).
38. R. Schubert, B. Meagher, C. C. Chang, *J. Electrochem. Soc.*, **142**, 3157 (1995).
39. C. Chandler, J. B. Ju, R. Atanasoski, W. H. Smyrl, *Corrosion*, **47**, 179 (1991).
40. M. Seo, K. Yoshida, K. Noda, *Mater. Sci. Forum*, **192–194**, 753 (1995).
41. M. Seo, K. Yoshida, K. Noda, *Mater. Sci. Eng. A*, **198**, 197 (1995).
42. U. Pittermann, R. Reining, K. G. Weil, *J. Electrochem. Soc.*, **141**, 3416 (1994).
43. M. Alodan, W. H. Smyrl, *Electrochim. Acta*, **44**, 299 (1998).
44. M. Itagaki, J. Kadowaki, K. Watanabe, *Anal. Sci.*, **16**, 1049 (2000).

45. K. Leinartas, P. Miecinskas, A. Sudavicius, D. Jelinskiene, R. Juskenas, V. Lisauskas, B. Vengalis, E. Juzeliunas, *J. Appl. Electrochem.*, **31**, 1079 (2001).
46. K. Leinartas, M. Samuleviciene, A. Bagdonas, A. Sudavicius, V. Lisauskas, E. Juzeliunas, *Electrochem. Commun.*, **3**, 494 (2001).
47. K. Wang, H. W. Pickering, K. G. Weil, *Electrochim. Acta*, **46**, 3835 (2001).
48. F. Zucchi, M. Fonsati, G. Trabanelli, *J. Appl. Electrochem.*, **29**, 347 (1999).
49. M. Fonsati, F. Zucchi, G. Trabanelli, *Electrochim. Acta*, **44**, 311 (1998).
50. D. Jope, J. Sell, H. W. Pickering, K. G. Weil, *J. Electrochem. Soc.*, **142**, 2170 (1995).
51. A. Frignani, M. Fonsati, C. Monticelli, G. Brunoro, *Corros. Sci.*, **41**, 1217 (1999).
52. J. Telegdi, A. Shaban, E. Kalman, *Electrochim. Acta*, **45**, 3639 (2000).
53. M. Hepel, E. Cateforis, *Electrochim. Acta*, **46**, 3801 (2001).
54. C. Garcia, G. Courbin F. Ropital, C. Fiaud, *Electrochim. Acta*, **46**, 973 (2001).
55. C. Gabrielli, M. Keddam, A. Khalil, G. Maurin, H. Perrot, R. Rosset, M. Zidoune, *J. Electrochem. Soc.*, **145**, 2386 (1998).
56. C. Gabrielli, G. Maurin, G. Poindessous, R. Rosset, *J. Crystal Growth*, **200**, 236 (1999).
57. P. Kern, D. Landolt, *Corros. Sci.*, **44**, 1809 (2002).
58. P. Kern, D. Landolt, *J. Electrochem. Soc.*, **148**, B228 (2001).
59. V. Tsionsky, L. Daikhin, G. Zilberman, E. Gileadi, *Faraday Discuss.*, **107**, 337 (1997).
60. G. Zilberman, V. Tsionsky, E. Gileadi, *Electrochim. Acta*, **45**, 3473 (2000).
61. L. Daikhin, E. Gileadi, V. Tsionsky, M. Urbakh, G. Zilberman, *Electrochim. Acta*, **45**, 3615 (2000).
62. C.-O. A. Olsson, D. Hamm, D. Landolt, *J. Electrochem. Soc.*, **147**, 2563 (2000).
63. C.-O. A. Olsson, D. Hamm, D. Landolt, *J. Electrochem. Soc.*, **147**, 4093 (2000).
64. P. Schmutz, D. Landolt, *Corros. Sci.*, **41**, 2143 (1999).
65. P. Schmutz, D. Landolt, *Electrochim. Acta*, **45**, 899 (1999).
66. D. Hamm, C.-O. A. Olsson, D. Landolt, *Corros. Sci.*, **44**, 1009 (2002).
67. D. Hamm, K. Ogle, C.-O. A. Olsson, S. Weber, D. Landolt, *Corros. Sci.*, **44**, 1443 (2002).
68. C.-O. A. Olsson, D. Landolt, *J. Electrochem. Soc.*, **148**, B438 (2001).

Index